SUPERACID CHEMISTRY

SUPERACID CHEMISTRY

SECOND EDITION

George A. Olah
G. K. Surya Prakash
Árpád Molnár
Jean Sommer

WILEY
A John Wiley & Sons, Inc., Publication

Published by John Wiley & Sons, Inc., Hoboken, New Jersey
Published simultaneously in Canada

Library of Congress Cataloging-in-Publication Data:

Superacid chemistry / by George A. Olah . . . [et al.]. – 2nd ed.
p. cm.
Rev. ed. of: Superacids / George A. Olah, G.K. Surya Prakash, Jean Sommer. c1985.
Includes index.
ISBN 978-0-471-59668-4 (cloth)
1. Superacids. I. Olah, George A. (George Andrew), 1927- II. Olah, George A. (George Andrew), 1927- Superacids.
QD477.O53 2009
546′.24–dc22

2008026776

Printed in the United States of America
10 9 8 7 6 5 4 3 2 1

In memory of Katherine Bogdanovich Loker, Benefactor and Friend

CONTENTS

PREFACE TO THE SECOND EXTENDED AND UPDATED EDITION

More than 20 years passed since the publication of our book on Superacids. The book became out of print and much progress since was made in the field, which is gaining increasing interest and significance. Hence, it seems warranted to provide the interested reader with a comprehensively updated review and discussion of the field with literature coverage until early 2008. The title has been changed to "Superacid Chemistry" to reflect enormous progress in the field. Some aspects of superelectrophilic activation are also discussed (for more elaborate coverage, readers are referred to G. A. Olah and D. A. Klump, "Superelectrophiles and Their Chemistry" Wiley-Interscience, 2008). Our friend and colleague, Árpád Molnár joined us as a coauthor and made an outstanding contribution to the revised new edition of our book, which we hope will be of interest and use to the chemical community. Our publisher is thanked for arranging the new revised edition.

George A. Olah
Los Angeles, California

G. K. Surya Prakash
Los Angeles, California

Árpád Molnár
Szeged, Hungary

Jean Sommer
Strasbourg, France

October 2008

PREFACE TO THE FIRST EDITION

The chemistry of superacids, that is, of acid systems stronger than conventional strong mineral Brønsted acids such as sulfuric acid or Lewis acids like aluminum trichloride, has developed in the last two decades into a field of growing interest and importance. It was J. B. Conant who in 1927 gave the name "superacids" to acids that were capable of protonating certain weak bases such as carbonyl compounds and called attention to acid systems stronger than conventional mineral acids. The realization that Friedel–Crafts reactions are, in general, acid catalyzed with conjugate Lewis–Brønsted acid systems frequently acting as the de facto catalysts extended the scope of acid catalyzed reactions. Friedel–Crafts acid systems, however, are usually only 10^3 to 10^6 times stronger than 100% sulfuric acid. The development in the early 1960s of Magic Acid, fluoroantimonic acid, and related conjugate superacids, 10^7 to 10^{10} times stronger than sulfuric acid, added a new dimension to and revival of interest in superacids and their chemistry. The initial impetus was given by the discovery that stable, long-lived, electron-deficient cations, such as carbocations, acidic oxonium ions, halonium ions, and halogen cations, can be obtained in these highly acidic systems. Subsequent work opened up new vistas of chemistry and a fascinating, broad field of chemistry is developing at superacidities. Because acidity is a term related to a reference base, superacidity allows extension of acid-catalyzed reactions to very weak bases and thus extends, for example, hydrocarbon chemistry to saturated systems including methane.

Some years ago in two review articles (*Science 206*, 13, 1979; *La Recherche 10*, 624, 1979), we briefly reviewed some of the emerging novel aspects of superacids. However, we soon realized that the field was growing so fast that to be able to provide a more detailed survey for the interested chemist a more comprehensive review was required. Hence, we welcomed the suggestion of our publisher and Dr. Theodore P. Hoffman, chemistry editor of Wiley-Interscience, that we write a monograph on superacids.

We are unable to thank all of our friends and colleagues who directly or indirectly contributed to the development of the chemistry of superacids. The main credit goes to all researchers in the field whose work created and continues to enrich this fascinating area of chemistry. Professor R. J. Gillespie's pioneering work on the inorganic chemistry of superacids was of immense value and inspiration to the development of the whole field. Our specific thanks are due to Drs. David Meidar and Khosrow Laali, who helped with the review of solid superacid systems and their reactions. Professor E. M. Arnett is thanked for reading part of our manuscript and for his thoughtful comments.

Finally we would like to thank Mrs. R. Choy, who tirelessly and always cheerfully typed the manuscript.

George A. Olah
Los Angeles, California

G. K. Surya Prakash
Los Angeles, California

Jean Sommer
Strasbourg, France

CHAPTER 1

General Aspects

1.1. DEFINING ACIDITY

1.1.1. Acids and Bases

The concept of acidity was born in ancient times to describe the physiological property such as taste of food or beverage (in Latin: *acidus*, sour; *acetum*, vinegar). Later during the development of experimental chemistry, it was soon realized that mineral acids such as sulfuric, nitric, and hydrochloric acids played a key role in chemical transformations. Our present understanding of acid-induced or -catalyzed reactions covers an extremely broad field ranging from large-scale industrial processes in hydrocarbon chemistry to enzyme-controlled reactions in the living cell.

The chemical species that plays a unique and privileged role in acidity is the hydrogen nucleus, that is, the proton: H^+. Since its 1*s* orbital is empty, the proton is not prone to electronic repulsion and by itself has a powerful polarizing effect. Due to its very strong electron affinity, it cannot be found as a free "naked" species in the condensed state but is always associated with one or more molecules of the acid itself or of the solvent. Free protons exist only in the gas phase (such as in mass spectrometric studies). Regardless, as a shorthand notation, one generally depicts the proton in solution chemistry as "H^+." Due to its very small size (10^5 times smaller than any other cation) and the fact that only the 1*s* orbital is used in bonding by hydrogen, proton transfer is a very facile chemical reaction and does not necessitate important reorganization of the electronic valence shells. Understanding the nature of the proton is important while generalizing quantitative relationships in acidity measurements.[1,2]

The first clear definition of acidity can be attributed to Arrhenius, who between 1880 and 1890 elaborated the theory of ionic dissociation in water to explain the variation in strength of different acids.[3] Based on electrolytic experiments such as conductance measurements, he defined acids as substances that dissociate in water and yield the hydrogen ion whereas bases dissociate to yield hydroxide ions. In 1923, J. N. Brønsted generalized this concept to other solvents.[4] He defined an acid as a species that can donate a proton and defined a base as a species that can accept it. This

Superacid Chemistry, Second Edition, George A. Olah, G. K. Surya Prakash, Árpád Molnár, and Jean Sommer

definition is generally known as the Brønsted–Lowry concept. The dissociation of an acid HA in a solvent S can be written as an acid–base equilibrium [Eq. (1.1)].

$$HA + S \rightleftharpoons A^- + SH^+ \tag{1.1}$$

The ionization of the acid HA in solvent S leads to a new acid HS^+ and a base A^-. Equation (1.1) has a very wide scope and can be very well applied to neutral and positively and negatively charged acid systems. The acid–base pair that differs only by a proton is referred to as the conjugate acid–base pair. Thus, H_2O is the conjugate base of the acid H_3O^+. An obvious consequence of the concept is that the extent to which an acid ionizes depends on the basicity of the solvent in which the ionization takes place. This shows the difficulty in establishing an absolute acidity scale. Acidity scales are energy scales, and thus they are arbitrary with respect to both the reference point and the magnitude of units chosen.

Fortunately, many of the common solvents by themselves are capable of acting as acids and bases. These amphoteric or amphiprotic solvents undergo self-ionization [e.g., Eqs. (1.2) and (1.3)], which can be formulated in a general way as in Eq. (1.4).

$$2\,H_2O \rightleftharpoons H_3O^+ + OH^- \tag{1.2}$$

$$2\,HF \rightleftharpoons H_2F^+ + F^- \tag{1.3}$$

$$2\,HA \rightleftharpoons H_2A^+ + A^- \tag{1.4}$$

This equilibrium is characterized by the autoprotolysis constant K_{ap}, which under the usual high dilution conditions can be written as in Eq. (1.5).

$$K_{ap} = [H_2A^+][A^-] \tag{1.5}$$

Indeed the extent of dissociation of the solvent is very small (in HF, $K_{ap} \approx 10{-}11$; in H_2O, $K_{ap} = 10{-}14$). The pK_{ap} value that gives the acidity range will be discussed later.

G.N. Lewis extended and generalized the acid–base concept to nonprotonic systems.[5,6] He defined an acid as a substance that can accept electrons and defined a base as a substance that can donate electrons. Lewis acids are electron-deficient molecules or ions such as BF_3 or carbocations, whereas Lewis bases are molecules that contain readily available nonbonded electron pairs (as in ethers, amines, etc.) [Eq. (1.6)].

$$BF_3 + {:}O(CH_3)_2 \rightleftharpoons BF_3{:}O(CH_3)_2 \tag{1.6}$$

Of course, in a generalized way, the proton H^+ is also a Lewis acid and the Brønsted acids and bases also fall into the Lewis categories.

Considering the general equation (1.4) for the auto-ionization of solvent HA, one can define an acid as any substance that will increase $[H_2A^+]$ and define a base as any substance that will increase $[A^-]$ and thus decrease $[H_2A^+]$. This definition, which includes both Lewis' and Brønsted's concepts, is used in practice while measuring the acidity of a solution by pH.

A number of strategies have been developed for acidity measurements of both aqueous and nonaqueous solutions. We will briefly review the most important ones and discuss their use in establishing acidity scales.

1.1.2. The pH Scale

The concentration of the acid itself is of little significance other than analytical, with the exception of strong acids in dilute aqueous solutions. The concentration of H^+ itself is not satisfactory either, because it is solvated diversely and the ability of transferring a proton to another base depends on the nature of the medium. The real physical quantity describing the acidity of a medium is the activity of the proton a_{H^+}. The experimental determination of the activity of the proton requires the measurement of the potential of a hydrogen electrode or a glass electrode in equilibrium with the solution to be tested. The equation is of the following type [Eq. (1.7)], wherein C is a constant.

$$E = C - \frac{RT}{F}\log_{10}(a_{H^+}) \tag{1.7}$$

It was Sørensen's idea[7] to use this relationship, which can be considered as a basis to the modern definition of the pH scale of acidity for aqueous solutions. The pH of a dilute solution of acid is related to the concentration of the solvated proton from Eq. (1.8). Depending on the dilution, the proton can be further solvated by two or more solvent molecules.

$$\text{pH} = -\log[\text{HS}^+] \tag{1.8}$$

When the acid solution is highly diluted in water, the pH measurement is convenient, but it becomes critical when the acid concentration increases and, even more so, if nonaqueous media are employed. Since a reference cell is used with a liquid junction, the potential at the liquid junction also has to be known. The hydrogen ion activity cannot be measured independently, and for this reason the equality of Eq. (1.9) cannot be definitely established for any solution.

$$\text{pH} = -\log_{10}(a_{H^+}) \tag{1.9}$$

Under the best experimental conditions, the National Bureau of Standard has set up a series of standard solutions of pH from which the pH of any other aqueous solution can be extrapolated as long as the ionic strength of the solution is not higher than 0.1 M. For more concentrated solutions, the pH scale will no longer have any real significance. In extending the limit to 1 M solutions, it is apparent that the available range of acidity is directly related to the autoprotolysis constant [Eq. (1.5)], because the minimum value of pH in a solution is zero and the maximum value is $pK_{ap} = p(H_2A^+) + p(A^-)$. Thus, the range of pH (ΔpH) is pK_{ap} (for water, 14 pH units). These limiting conditions are rather unfortunate because many chemical transformations are achieved beyond this range and under much less ideal conditions.

1.1.3. Acidity Functions

Considering the limited applicability of the pH scale, a quantitative scale is needed to express the acidity of more concentrated or nonaqueous solutions.

A knowledge of the acidity parameter should permit one to estimate the degree of transformation of a given base (to its protonated form) in its conjugate acid. This should allow one to relate these data to the rate of acid-catalyzed reactions. Hammett and Deyrup[8] in 1932 were the first to suggest a method for measuring the degree of protonation of weakly basic indicators in acid solution. The proton transfer equilibrium in the acid solution between an electro-neutral weak base B and the solvated proton can be written as in Eq. (1.10).

$$\mathrm{B} + \mathrm{H_2A^+} \rightleftharpoons \mathrm{BH^+} + \mathrm{AH} \qquad (1.10)$$

Bearing in mind that the proton is solvated (AH_2^+) and that AH is the solvent, the equilibrium can be written as in Eq. (1.11).

$$\mathrm{B} + \mathrm{H^+} \rightleftharpoons \mathrm{BH^+} \qquad (1.11)$$

The corresponding thermodynamic equilibrium constant is $K_{\mathrm{BH^+}}$, which is expressed as in Eq. (1.12), in which a is the activity, C the concentration, and f the activity coefficient.

$$K_{\mathrm{BH^+}} = \frac{a_{\mathrm{H^+}} \cdot a_{\mathrm{B}}}{a_{\mathrm{BH^+}}} = \frac{a_{\mathrm{H^+}} \cdot C_B}{C_{\mathrm{BH^+}}} \cdot \frac{f_{\mathrm{H}}}{f_{\mathrm{BH^+}}} \qquad (1.12)$$

From this equation, Eq. (1.13) follows.

$$\frac{C_{\mathrm{BH^+}}}{C_{\mathrm{B}}} = \frac{1}{K_{\mathrm{BH^+}}} \cdot a_{\mathrm{H^+}} \cdot \frac{f_{\mathrm{B}}}{f_{\mathrm{BH^+}}} \qquad (1.13)$$

Because the first ratio represents the degree of protonation, Hammett and Deyrup[8,9] defined the acidity function H_0 by Eq. (1.14).

$$H_0 = -\log a_{\mathrm{H^+}} \cdot \frac{f_{\mathrm{B}}}{f_{\mathrm{BH^+}}} = -\log K_{\mathrm{BH^+}} + \log \frac{C_{\mathrm{B}}}{C_{\mathrm{BH^+}}} \qquad (1.14)$$

Equation (1.14) can be written for further discussion in the more usual form of Eq. (1.15).

$$H_0 = \mathrm{p}K_{\mathrm{BH^+}} - \log \frac{[\mathrm{BH^+}]}{[\mathrm{B}]} \qquad (1.15)$$

From Eq. (1.14) it is clear that in dilute aqueous solution, as the activity coefficients tend to unity, the Hammett acidity function becomes identical with pH. On the other hand, by making the fundamental assumption that the ratio $f_{\mathrm{B}}/f_{\mathrm{BH^+}}$ is the same for different bases in a given solution, Hammett postulated that the H_0 function was unique for a particular series of solutions of changing acidity. The first application was made for the H_2SO_4–H_2O system using a series of primary anilines as indicators. By starting with the strongest base B_1, the $pK_{B_1H^+}$ was measured in dilute aqueous solution. The pK of the next weaker base B_2 was then determined by measuring the ionization ratio of the two indicators in the same acid solution using the relation of Eq. (1.16).

$$pK_{B_1H^+} - pK_{B_2H^+} = \log\frac{[B_1H^+]}{[B_1]} - \log\frac{[B_2H^+]}{[B_2]} \tag{1.16}$$

The ionization ratio was measured by UV–visible spectroscopy. With the help of successively weaker primary aromatic amine indicators, the strongest base being *para*-nitroaniline ($pK = 1.40$) and the weakest trinitroaniline ($pK = -9$), Hammett explored the whole H_2O–H_2SO_4 range up to 100% sulfuric acid and the perchloric acid–water solution up to 60% of acid. Similar acidity functions such as H_-, H_+, H_{2+} were proposed related to acid–base equilibria in which the indicator is negatively, positively, or even dipositively charged. The validity of all of these functions is based on the simple assumption that the activity coefficient ratio is independent of the nature of the indicator at any given solvent composition. In this case the log $[BH^+]/[B]$ plots against H_0 should be linear with a slope of -1.00 for all neutral bases. This is not the case for groups of indicators with different structures, and especially for different basic sites, which often show significant deviations. For this reason, it is well recognized now that the above assumption does not have a general validity. The measurement of a Hammett acidity function should be limited to those indicators for which log $[BH^+]/[B]$ plotted against H_0 gives a straight line with a negative unit slope. These indicators are called Hammett bases.

Equilibria other than proton transfer have also been used to determine acidity functions. One of these is based on the ionization of alcohols (mainly arylmethyl alcohols) in acid solution following the equilibrium in Eq. (1.17).

$$ROH + H^+ \rightleftharpoons R^+ + H_2O \tag{1.17}$$

The corresponding acidity function described as H_R is then written in Eq. (1.18).

$$H_R = pK_{R^+} - \log\frac{[R^+]}{[ROH]} \tag{1.18}$$

This H_R function, also called J_0 function, has also been used to measure the acidity of the sulfuric acid–water and perchloric acid–water systems. It shows a large deviation from the H_0 scale in the highly concentrated solutions as shown in Figure 1.1.

However, all these and other acidity functions are based on Hammett's principle and can be expressed by Eq. (1.19), in which B and A are the basic and the conjugate acidic form of the indicator, respectively. They become identical with the pH scale in highly dilute acid solutions. The relative and absolute validity of the different acidity functions have been the subject of much controversy and the subject has been extensively reviewed.[1,10–14]

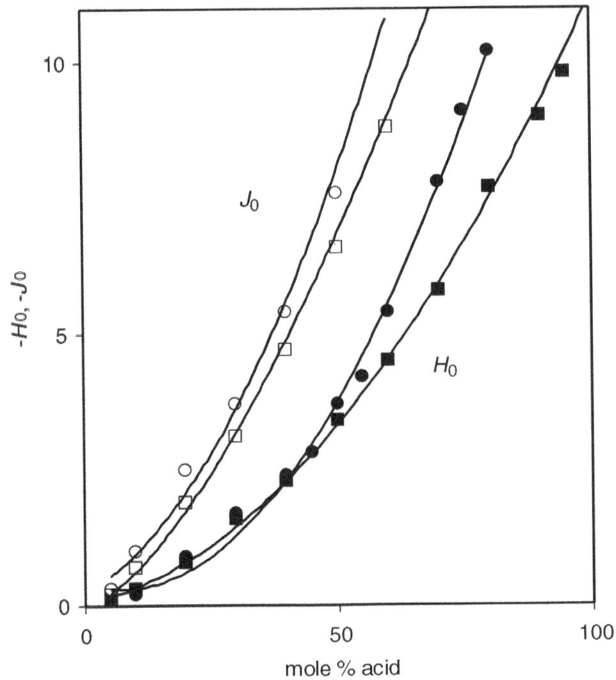

Figure 1.1. H_0 and J_0 functions for H_2SO_4–H_2O and $HClO_4$–H_2O systems. $HClO_4$: ○, ref. 11; ●, ref. 13; H_2SO_4: □, ref. 12; ■, ref. 14.

$$H_x = pK_A - \log\frac{A}{B} \tag{1.19}$$

Whatever may be the limitations of the concept first proposed by Hammett and Deyrup in 1932 [8] until now, no other widely used alternative has appeared to better assess quantitatively the acidity of concentrated and nonaqueous strongly acidic solutions.[9] The experimental methods that have been used to determine acidity functions are discussed in Section 1.4.

1.2. DEFINITION OF SUPERACIDS

It was in a paper (including its title) published in 1927 by Hall and Conant[15] in the *Journal of the American Chemical Society* that the name "superacid" appeared for the first time in the chemical literature. In a study of the hydrogen ion activity in a nonaqueous acid solution, these authors noticed that sulfuric acid and perchloric acid in glacial acetic acid were able to form salts with a variety of weak bases such as ketones and other carbonyl compounds. These weak bases did not form salts with the aqueous solutions of the same acids. The authors ascribed this high acidity to the ionization of these acids in glacial acetic acid, increasing the concentration of $CH_3COOH_2^+$, a species less solvated than H_3O^+ in the aqueous acids. They proposed to call these solutions "superacid solutions." Their proposal was, however, not further followed up or used until the 1960s, when Olah's studies of obtaining stable solutions of highly electron-deficient ions, particularly carbocations, focused interest on very high-acidity nonaqueous systems.[16,17] Subsequently, Gillespie proposed an arbitrary but since widely accepted definition of superacids,[18,19] defining them as any acid system that is stronger than 100% sulfuric acid, that is, $H_0 \leq -12$. Fluorosulfuric acid and trifluoromethanesulfonic acid are examples of Brønsted acids that exceed the acidity of sulfuric acid with H_0 values of about -15.1 and -14.1, respectively.

To reach acidities beyond this limit, one has to start with an already strong acid ($H_0 \approx -10$) and add to it a stronger acid that increases the ionization. This can be achieved either by adding a strong Brønsted acid (HB) capable of ionizing in the medium [Eq. (1.20)] or by adding a strong Lewis acid (L) that, by forming a conjugate acid, will shift the autoprotonation equilibrium by forming a more delocalized counterion of the strong acid [Eq. (1.21)].

$$\mathrm{HA} + \mathrm{HB} \rightleftharpoons \mathrm{H_2A^+} + \mathrm{B^-} \tag{1.20}$$

$$2\,\mathrm{HA} + \mathrm{L} \rightleftharpoons \mathrm{H_2A^+} + \mathrm{LA^-} \tag{1.21}$$

In both cases, a remarkable acidity increase is observed from the H_0 value of the neat HA as shown in Figure 1.2 for HSO_3F.

It is this large acidity jump, generally more than 5 H_0 units, that raises a strong acid solution into the superacid region. Therefore, it is clear that the proposed reference of $H_0 = -12$ for the lower limit of superacidity is only arbitrary. It could as well be $H_0 = -15.1$ with HF or HSO_3F as solvent.

Gillespie's definition of superacids relates to Brønsted acid systems. Because Lewis acids also cover a wide range of acidities extending beyond the strength of conventionally used systems, Olah suggested the use of anhydrous aluminum chloride as the arbitrary reference and we categorize Lewis superacids as those stronger than aluminum chloride[17] (see, however, subsequent discussion on the difficulties of measuring the strength of a Lewis acid).

It should be also noted that in biological chemistry, following a suggestion by Westheimer,[20] it is customary to call catalysis by metal ions bound to enzyme systems as "superacid catalysis." Because the role of a metal ion is analogous to a proton, this arbitrary suggestion reflects enhanced activity and is in line with previously discussed Brønsted and Lewis superacids.

1.2.1. Range of Acidities

Despite the fact that superacids are stronger than 100% sulfuric acid, there may be as much or more difference in acidity between various superacid systems than between neat sulfuric acid and its 0.1 *M* solution in water.

Acidity levels as high as $H_0 = -27$ have been estimated on the basis of exchange rate measurements by NMR for an HSO_3F–SbF_5 mixture containing 90 mol% SbF_5.[21] In fact, the HF–SbF_5 is considered one of the strongest superacid system based on

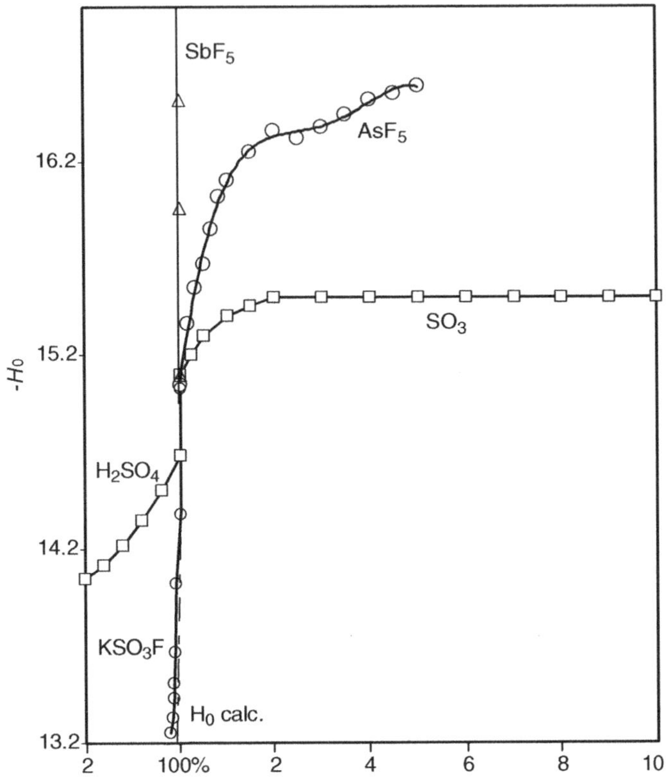

Figure 1.2. Acidity increase near the H_0 value of neat HSO_3F.[19]

various measurements. Meanwhile, however, Sommer and coworkers found that the weakest basic indicator of the *para*-methoxybenzhydryl cation family (4,4'-dimethoxy; $pK_{BH^+} \sim -23$) could not be diprotonated even in the strongest HF–SbF_5 acid.[22] For this reason it appears that one should not expect acidity levels higher than approximately $H_0 = -24$ in the usual superacid systems (Figure 1.3). Predictions of stronger acidities are all based on indirect estimations rather than direct acid–base equilibria measurements. It is important to recognize that the naked proton "H^+" is not present in the condensed phase because even compounds as weakly basic as methane or even rare gases bind the proton.[23,24]

A quantitative determination of the strength of Lewis acids to establish similar scales (H_0) as discussed in the case of protic (Brønsted-type) superacids would be most useful. However, to establish such a scale is extremely difficult. Whereas the Brønsted acid–base interaction invariably involves a proton transfer reaction that allows meaningful comparison, in the Lewis acid–base interaction, involving for example Lewis acids with widely different electronic and steric donating substituents, there is no such common denominator.[25,26] Hence despite various attempts, the term "strength of Lewis acid" has no well-defined meaning.

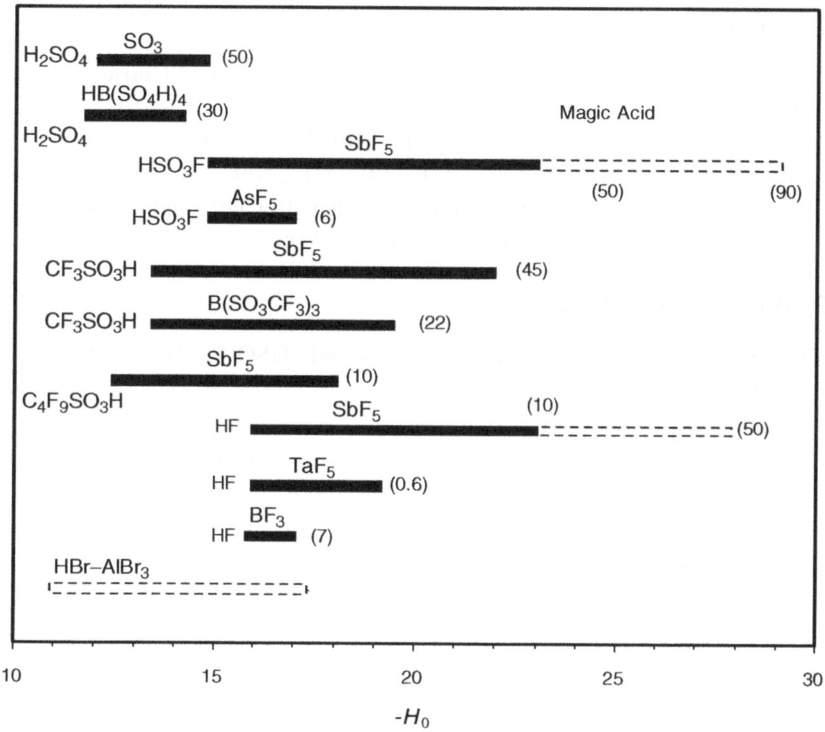

Figure 1.3. Acidity ranges for the most common superacids. The solid and open bars are measured using indicators; the broken bar is estimated by kinetic measurements; numbers in parentheses indicate mol% Lewis acid.

Regardless, it is important to keep in mind that superacidity encompasses both Brønsted and Lewis acid systems and their conjugate acids. The qualitative picture of Lewis acid strengths will be discussed in Section 1.4.7.

The acidity strength of solid acids is still not well known and is difficult to measure. Claims of superacidity in solids are numerous and will be discussed later in Chapter 2. Among the reviews related to acidity characterization of solids, those of Corma,[27] Farneth and Gorte,[28] and Fripiat and Dumesic[29] are quite significantly representative.[30]

1.3. TYPES OF SUPERACIDS

As discussed, superacids, similar to conventional acid systems, include both Brønsted and Lewis acids and their conjugate systems. Protic (Brønsted-type) superacids include strong parent acids and the mixtures thereof, whose acidity can be further enhanced by various combinations with Lewis acids (conjugate acids). The following are the most frequently used superacids.

1.3.1. Primary Superacids

1. Brønsted superacids such as perchloric acid ($HClO_4$), halosulfuric acids (HSO_3Cl, HSO_3F), perfluoroalkanesulfonic acids (CF_3SO_3H, R_FSO_3H), hydrogen fluoride, and carborane superacids [$H(CB_{11}HR_5X_6)$].
2. Lewis superacids, such as SbF_5, AsF_5, PF_5, TaF_5, NbF_5, BF_3, tris(pentafluorophenyl) borane, boron tris(trifluoromethanesulfonate), and aprotic organic superacids developed by Vol'pin and co-workers.

1.3.2. Binary Superacids

1. Binary Brønsted superacids such as $HF–HSO_3F$, $HF–CF_3SO_3F$, and $HB(HSO_4)_4$.
2. Conjugate Brønsted–Lewis superacids:
   a. Combination of oxygenated Brønsted acids (H_2SO_4, HSO_3F, CF_3SO_3H, R_FSO_3H) with Lewis acids (SO_3, SbF_5, AsF_5, TaF_5, and NbF_5);
   b. Hydrogen fluoride in combination with fluorinated Lewis acids such as SbF_5, PF_5, TaF_5, NbF_5, and BF_3;
   c. Conjugate Friedel–Crafts acids such as $HBr–AlBr_3$ and $HCl–AlCl_3$.

1.3.3. Ternary Superacids

Examples are $HSO_3F–HF–SbF_5$, $HSO_3F–HF–CF_3SO_3H$, and $HSO_3F–SbF_5–SO_3$.

1.3.4. Solid Superacids

The acid–base character of solids was studied very early by Tanabe's group[31,32] and was first described in a landmark volume.[33]

Solid superacids can be further divided into various groups depending on the nature of the acid sites. The acidity may be a property of the solid as part of its chemical structure (possessing Lewis or Brønsted sites; the acidity of the latter can be further enhanced by complexing with Lewis acids). Solid superacids can also be obtained by deposition on or intercalation of strong acids into an otherwise inert or low-acidity support.

1. Zeolitic acids.
2. Polymeric resin sulfonic acids including sulfonic acid resins complexed with Lewis acids and perfluorinated polymer resin acids (Nafion–H and Nafion–silica nanocomposites).
3. Enhanced acidity solids including Brønsted and Lewis acid-modified metal oxides and mixed oxides, as well as metal salts complexed with Lewis acids.
4. Immobilized superacids and graphite-intercalated superacids.

As with previous classifications, these are also arbitrary and are suggested for practical discussion of an otherwise rather complex field. The superacid character of solids is discussed later in subsequent subchapters, and individual superacid systems are discussed in Chapter 2.

1.4. EXPERIMENTAL TECHNIQUES FOR ACIDITY MEASUREMENTS (PROTIC ACIDS)

From Eq. (1.14) it is apparent that the main experimental difficulty in determining acidities is the estimation of the ratio between the free base and its protonated ionic form of a series of indicators, their so-called ionization ratios.

1.4.1. Spectrophotometric Method

In the early work of Hammett and Deyrup[8] the measurement of the ionization ratio was based on the color change of the indicator. The solutions containing the indicator were compared at 25°C in a colorimeter with a standard reference. This reference was water, when the indicator was colorless in its acid form, and 96% sulfuric acid (or 70% perchloric acid), when the indicator was colorless in the basic form.

For example, when the indicator was colored in water the authors define a stoichiometric color intensity relative to water $I_w = C_w/C_a$, where C_a and C_w are the stoichiometric concentrations of indicator in solution A and in water. On the other hand, the specific color intensity of the colored form relative to water is defined as $S_w = [B]_w/[B]_a$, where $[B]_w$ is the concentration of the colored base in water and $[B]_a$ is concentration in solution A. Because the indicator exists only in its basic form in water, $[B]_w = C_w$; and in solution A, $C_a = [B]_a + [BH^+]_a$. The ionization ratio is given by Eq. (1.22).

$$\frac{[BH^+]}{[B]} = \frac{S_w - I_w}{I_w} \tag{1.22}$$

Despite seven decades of technical and scientific progress, the original Hammett method has not become obsolete. The colorimeter has been replaced by modern spectrophotometers that can be operated at selected wavelengths extending the spectra beyond visible into the ultraviolet region of the electromagnetic spectrum. The experimental variable, which is wavelength-dependent, is the optical density D. D is related to the concentration by the Beer–Lambert law [Eq. (1.23)].

$$D_{i,\lambda} = \varepsilon_{i,\lambda} C_i \cdot l \tag{1.23}$$

C_i is the concentration of the absorbing species, l is the length of the cell, and ε_i is the molar absorptivity (or extinction coefficient). If at a given wavelength λ, ε_{BH^+}, ε_B, and ε_λ are the extinction coefficients, respectively, of acid form of the indicator, its basic form, and of the unknown solution, the ionization ratio is given by Eq. (1.24).

$$\frac{[BH^+]}{[B]} = \frac{\varepsilon_B - \varepsilon_\lambda}{\varepsilon_\lambda - \varepsilon_{BH^+}} \tag{1.24}$$

For a greater precision in this determination, λ should be chosen so as to have the maximum difference between ε_{BH^+} and ε_B. For this reason, the areas between the absorption line and the baseline of both acidic and basic forms of the indicator should be measured and compared.

Whereas the precision of the method is generally excellent, a number of drawbacks may appear with some indicators. First, Eq. (1.24) is true only with the assumption that ε_{BH^+} is solvent-independent (it is clear that ε_B and ε_{BH^+} cannot be measured separately in the same solution). The medium effect on the absorption spectrum (mainly the wavelength shift of λ_{max}) can be easily taken into account in the measurements. However, large changes in the absorption spectrum during the increase in ionization are difficult to correct. Another difficulty that might appear is the structural change of the indicator, during or after protonation. The change in temperature, however, has been shown in the H_2SO_4–H_2O system to have little effect on the H_0 value,[34] but the pK_{BH^+} and the ionization ratios are more sensitive.

The pK_{BH^+} value is easy to determine, when the ionization ratio can be measured in dilute aqueous solution [Eq. (1.25)].

$$pK_{BH^+} = \text{limit}_{(HA\to 0)} \log \frac{[BH^+]}{[B]} - \log[H_3O^+] \tag{1.25}$$

It is to be noted that when the acid solution is very dilute, the presence of the indicator modifies the acidity: $[H_3O^+] = [HA] - [B]$, and thus the concentration of the indicator has to be taken into account.

As is apparent from Eq. (1.14), an indicator is only useful over an acidity range where its ionization ratio can be measured experimentally with sufficient precision. For spectrophotometric method, this means approximately 2 log units per indicator. Accordingly, the direct determination of the pK_{BH^+} value of an indicator in concentrated solution is not possible. It is actually achieved by the method developed by Hammett in his early work using a series of indicators of overlapping range.[8] Taking into account the overlapping of each indicator with the preceding and the following one, each of which is useful for 2 log units, it appears that several indicators are necessary (approximately as many indicators as the number of desired log units). This is illustrated in Figure 1.4.

Paul and Long[35] have tabulated pK_{BH^+} values for indicators, which were used to establish Hammett acidity functions for aqueous acids between the years 1932 and 1957. The data were summarized as a set of "best values" of pK_{BH^+} for the bases. Since then, subsequent work seems to suggest that some of these values are incorrect. This is particularly the case for some of the weaker bases whose quoted pK_{BH^+} were based on a stepwise extrapolation of results of some indicators that have since been proven to be unsatisfactory based on the strict definition of H_0. These data, as well as those for weaker bases that have been studied since, covering the whole acidity range from dilute acid to the superacid media are collected in Table 1.1.

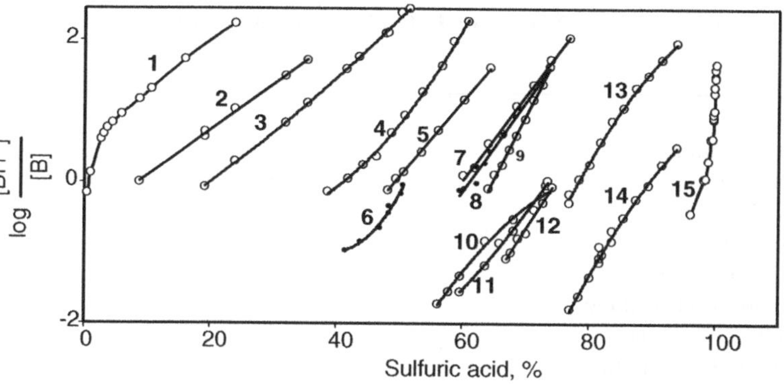

Figure 1.4. The ionization ratio as measured for a series of indicators in the 0–100% H_2SO_4–H_2O system.[8]

Up to a H_0 value of -10, all indicators are primary amines and are therefore suitable for the measurement of the Hammett H_0 function. For stronger acids, new indicators such as nitro compounds have to be used. Although the acidity function scale based upon nitro compounds as indicators may not be a satisfactory extension of the aniline indicator scale, Gillespie and Peel[18] have shown that the most basic nitro compound indicator, *para*-nitrotoluene overlaps in a satisfactory manner with the weakest indicator in the aniline series, 2,4,6-trinitroaniline. Thus, the acidity measurements using the nitro compounds may be considered to give the best semiquantitative picture of the acidity of the various superacid systems.

UV spectroscopy of adsorbed Hammett bases has also been used to estimate the acidity of solids such as zeolites.[38]

1.4.2. Nuclear Magnetic Resonance Methods

NMR spectroscopy, which was developed in the late 1950s as a most powerful tool for structural analysis of organic compounds, has also proven to be useful for acidity determinations. The measurement of the ionization ratio has been achieved by a variety of methods demonstrating the versatility of this technique. If we consider the general acid–base equilibrium Eq. (1.26) obtained when the indicator B is dissolved in the strong acid HA, then k_p, and k_d, respectively, are the rates of protonation and deprotonation.

$$\mathrm{HA} + \mathrm{B} \underset{k_d}{\overset{k_p}{\rightleftharpoons}} \mathrm{BH^+} + \mathrm{A^-} \qquad (1.26)$$

The thermodynamic equilibrium constant K is related to these rates according to Eq. (1.27).

$$K = \frac{[\mathrm{BH^+}][\mathrm{A^-}]}{[\mathrm{HA}][\mathrm{B}]} = \frac{k_p}{k_d} \qquad (1.27)$$

Table 1.1. Selected pK_{BH^+} Values for Extended Hammett Bases

Base	pK_{BH^+}	References
3-Nitroaniline	2.50	35
2,4-Dichloroaniline	2.00	35
4-Nitroaniline	0.99	35
2-Nitroaniline	−0.29	35
4-Chloro-2-nitroaniline	−1.03	35
5-Chloro-2-nitroaniline	−1.54	14
2,5-Dichloro-4-nitroaniline	−1.82	36
2-Chloro-6-nitroaniline	−2.46	36
2,6-Dichloro-4-nitroaniline	−3.24	36
2,4-Dichloro-6-nitroaniline	−3.29	35
2,6-Dinitro-4-methylaniline	−4.28	35
2,4-Dinitroaniline	−4.48	35
2,6-Dinitroaniline	−5.48	36
4-Chloro-2,6-dinitroaniline	−6.17	36
6-Bromo-2,4-dinitroaniline	−6.71	35
3-Methyl-2,4,6-trinitroaniline	−8.37	36
3-Bromo-2,4,6-trinitroaniline	−9.62	36
3-Chloro-2,4,6-trinitroaniline	−9.71	36
2,4,6-Trinitroaniline	−10.10	18,19
para-Nitrotoluene	−11.35	18,19
meta-Nitrotoluene	−11.99	18,19
Nitrobenzene	−12.14	18,19
para-Nitrofluorobenzene	−12.44	18,19
para-Nitrochlorobenzene	−12.70	18,19
meta-Nitrochlorobenzene	−13.16	18,19
2,4-Dinitrotoluene	−13.76	18,19
2,4-Dinitrofluorobenzene	−14.52	18,19
2,4,6-Trinitrotoluene	−15.60	18,19
1,3,5-Trinitrobenzene	−16.04	18,19
(2,4-Dinitrofluorobenzene)H^+	−17.57	18,19
(2,4,6-Trinitrotoluene)H^+	−18.36	37

In NMR spectroscopy, when a species (for example, here $[BH^+]$) is participating in an equilibrium, its spectrum is very dependent on its mean lifetime (τ).[39,40] The inverse of the mean lifetime is a first-order rate constant, called the rate of exchange ($k = 1/\tau$), which can be obtained from the line-shape analysis of the NMR bands if $1\ s^{-1} \leq k$ $10^3\ s^{-1}$. Three cases can thus be envisaged:

1. "Slow Exchange" Conditions: $k \leq 10^{-2}\ s^{-1}$. The species can be observed as if no exchange were taking place.
2. Measurable Exchange Conditions: $1\ s^{-1} < k < 10^3\ s^{-1}$. The rate of exchange can be calculated from the line-shape analysis of the NMR bands of the exchanging species.

3. "Fast Exchange" Conditions: $k > 10^4\ s^{-1}$. The observed NMR bands appear as the weighted average of the species participating in the equilibrium.

Depending on these conditions, various NMR methods have been proposed and used to calculate the ionization ratio of weak bases in a superacid medium.

1.4.2.1. Chemical Shift Measurements. Under "slow-exchange" conditions, the ionization ratio cannot be measured. In fact, one of the major advantages of the superacidic media is the ease with which weak bases can be fully protonated and directly observed by NMR. Because it is known that the protonation rates are practically diffusion-controlled ($\sim 10^9$ liter mol^{-1} s^{-1}), under these conditions ($\leq 10^{-2}\ s^{-1}$) the indicator is "totally" in the acidic form described by the NMR spectrum and no variable is available to measure the ionization ratio.

Under "fast-exchange" conditions, however, the NMR spectrum presents a weighed average of the bands of the exchanging species, and with the sensitivity limits (~5–95%) the ionization ratio can be measured taking the chemical shift as a variable. The calculation is simply based on the observed chemical shift of the average line (δ_{obs}), provided that the chemical shift of the base indicator (δ_B) and of its acid form (δ_{BH^+}) are known [Eq. (1.28)]. This is generally obtained by increasing or decreasing the acidity of the medium.

$$\delta_{obs} = \frac{\delta_{BH^+}[BH^+] + \delta_B[B]}{[BH^+] + [B]} \qquad (1.28)$$

By plotting the chemical shift variation against the acidity, one observes a typical acid–base titration curve (Figure 1.5) and the pK_{BH^+} of the indicator can be determined this way. This NMR method, which was first proposed by Grunwald et al.,[41] has been applied by Levy et al.[42] using various ketones and α-haloketones for the determination of ketone basicity and evaluation of medium acidity.

Compared with spectrophotometry, the NMR method has a number of advantages: (i) The procedure is very rapid, and it can be used by observing the variation of chemical shifts of diverse nuclei such as ^1H, ^{13}C, ^{19}F, and ^{17}O. (ii) It is insensitive to colored impurities and slight decomposition of the indicator. (iii) In principle, it can be used over the whole range of known acidity. The medium effect, which may be important in ^1H NMR, becomes negligible in the case of ^{13}C NMR spectroscopy. The method can be used with a wide variety of weak bases having a lone-pair containing heteroatoms as well as simple aromatic hydrocarbons.

However, because the measurement of the ionization ratio requires the presence of a minimum of 5% of one of the forms of the indicator, it necessitates the availability of a family of structurally similar compounds of varying basicity to cover a large domain of acidity. This condition has been met by Sommer and co-workers[22,43] using the *para*-methoxybenzhydryl cations as useful indicators for the strongest superacids.

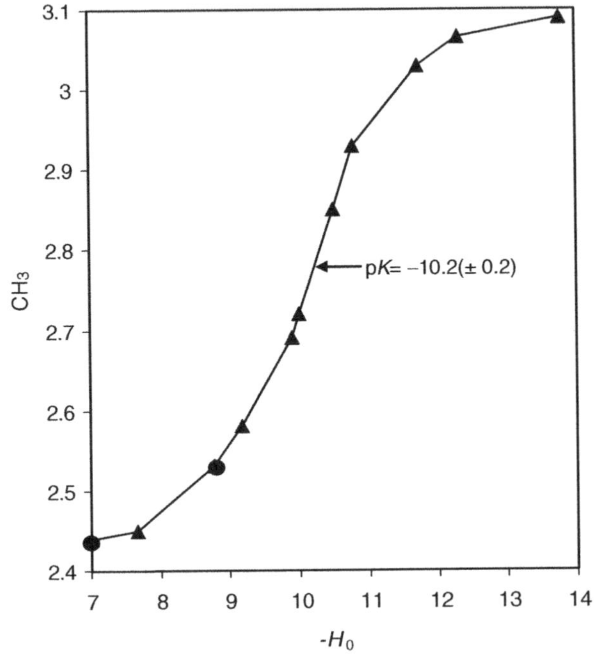

Figure 1.5. Acidity-dependent ^{1}H NMR chemical shift variations: protonation curve for acetaldehyde.[42] ● $CF_3COOH–H_2SO_4$, ▲ $CF_3COOH–CF_3SO_3H$.

The basicity of these indicators can be controlled by suitable substitution of the phenyl rings effecting ionization of the corresponding benzydrols [Eq. (1.29)]. The protonation equilibria [Eq. (1.30)] is measurable by ^{1}H or ^{13}C NMR spectroscopy.

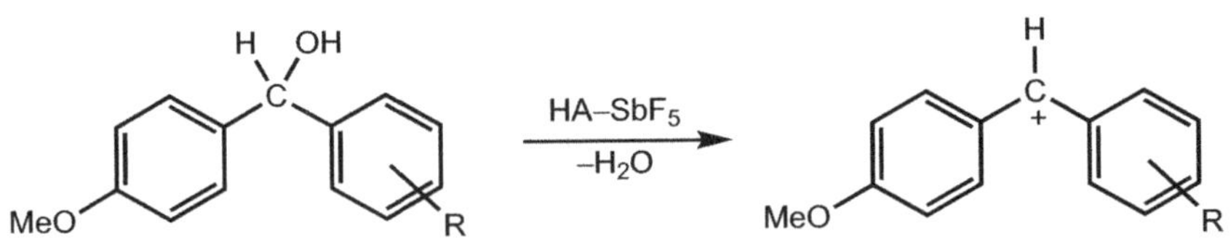

(1.29)

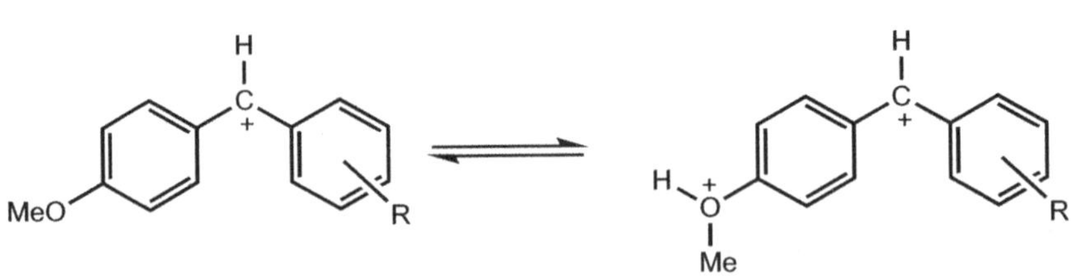

R = **1**: 4'-MeO, **2**: 4'-Me, **3**: 4'-H, **4**: 4'-Br,
5: 4'-CF_3, **6**: 4'-$MeOH^+$

(1.30)

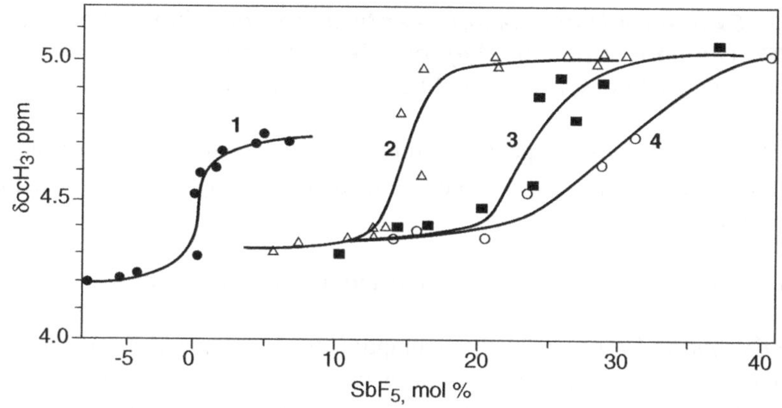

Figure 1.6. Protonation curves of indicators **1–4** in the $HSO_3F–SbF_5$ system.[22]

The protonation curves of indicators **1**–**4** in the $HSO_3F–SbF_5$ system (Figure 1.6) show how the decreasing basicity of the indicator necessitates an increasing amount of SbF_5 for half-protonation. The decreasing slope of the inflection point shows also that the increase in acidity by SbF_5 addition becomes smaller as more SbF_5 is added to the Brønsted acid (as a consequence of oligomeric anion formation).

The same indicators have been used to compare the relative acidity of the three most used superacids. As shown on Figure 1.7, the half-protonation of indicator **3** necessitates 40, 25, and 8 mol% SbF_5, respectively, in triflic acid, fluorosulfuric acid, and hydrogen fluoride systems. These results show the supremacy of the $HF–SbF_5$ system in which small concentrations of SbF_5 induce a dramatic increase in acidity (see also Section 2.2.2.7).

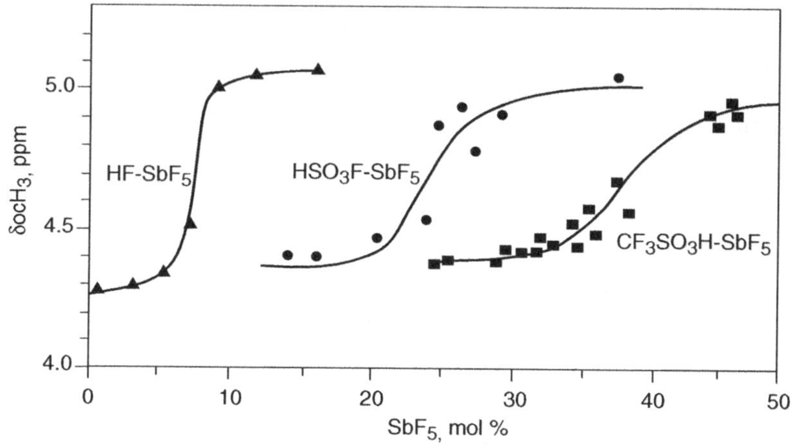

Figure 1.7. Protonation curves of indicator **3** in the superacid systems (from left to right) $HF–SbF_5$, $HSO_3F–SbF_5$, and $CF_3SO_3H–SbF_5$.[22]

1.4.2.2. Exchange Rate Measurements Based on Line-Shape Analysis (DNMR: Dynamic Nuclear Magnetic Resonance). Under the measurable exchange rate conditions, two possibilities have been considered:

1. The change in line shape can be directly related to the proton exchange.
2. The change in line shape is due to a separate exchange process related to the proton exchange.

Both methods have been exploited to determine ionization ratios.

Direct Exchange Rates. With the assumption that k_p is a constant over the range of measured acidity and k_{BH^+} of a series of overlapping bases remains measurable [Eq. (1.31)] (each base covering approximately 3 log units for a given concentration), Gold et al.[21] explored the acidity of the HSO_3F–SbF_5 system containing up to 90 mol% SbF_5. The highest acidity was estimated at $H_0 = -23$ approximately.[43]

$$H_0 = pK_{BH^+} - \log \frac{k_p \cdot [HA]}{k_{BH^+}} \tag{1.31}$$

Indirect Exchange Rates. In this case, the line shape is indirectly related to the acid-base equilibrium. Besides measuring intermolecular processes like the proton exchange rates, DNMR often has been used to measure intramolecular processes like conformational changes that occur on the same time scale. When the activation energy of such a process is very different in the acidic and basic forms for an indicator, DNMR can be used to measure the ionization ratio.

Due to a partial π-character, aromatic carbonyl compounds have an activation energy barrier for rotation around the phenyl–carbonyl bond, the value of which is substantially increased upon protonation.[44] In *para*-anisaldehyde a second protonation of the methoxy group will drastically decrease their barrier. The temperature-dependent NMR spectrum will reflect both exchange processes, intra- and intermolecular, as shown in Scheme 1.1.

A careful analysis of the NMR line shape provides the BH_2^{2+}/BH^+ ratio. Because of a large difference in activation energies between the rotation barriers in the mono- and diprotonated aldehyde, the observed rates are very sensitive to the concentration of BH^+. Thus ionization ratios of the order of 10^{-4} could be measured and approximately 4 log units of H_0 could be covered with the same indicator.[45]

By combining this method with the previously discussed chemical shift method, which is sensitive in the 0.05–20 range for the ionization ratio, the acidity could be measured over more than 5 H_0 units with the same indicator. Figure 1.8 shows the complementarity of both methods.

These are only approximate estimations, but are in reasonable agreement with more recent H_0 determination by other methods.[43]

Despite the evident advantages of the NMR methods, two points must be considered concerning the results of the acidity measurements. First, the concentration of the

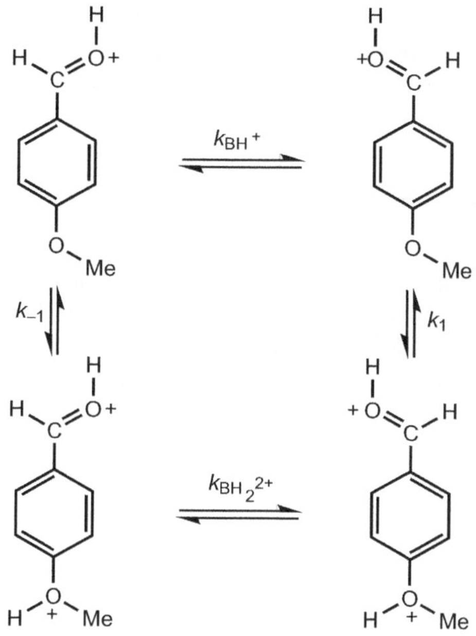

Scheme 1.1.

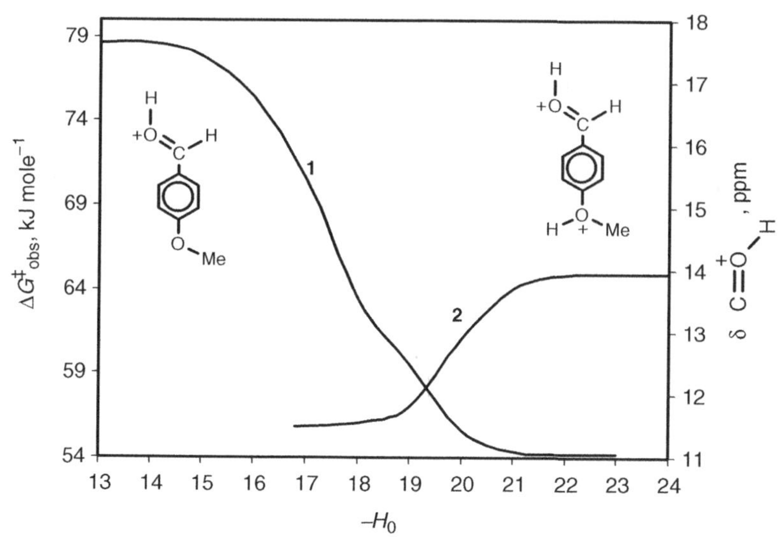

Figure 1.8. Complementarity of NMR methods in determining the variation of the acidity with the ionization of the indicator.[46] Curve **1:** line-shape analysis; curve **2:** chemical shift measurement.

indicator cannot be neglected as in the UV method, especially when the BH_2^{2+} is in low concentration. Second, aldehydes and ketones that have been generally used in the NMR methods are not true Hammett bases and the acidity that is derived should be considered only in a relative sense.

1.4.3. Electrochemical Methods

Electrochemistry provides a number of techniques for acidity measurements. The hydrogen electrode is the most reliable method in nonreducible solvents. It has been shown, however, that its reliability is limited to relatively weak acid solutions. A more general method was proposed by Strehlow and Wendt[47] in the early 1960s. They suggested a method to measure the potential variation of a pH-dependent system with respect to a reference system whose potential was solvent-independent. The measurement was made with a cell using $Pt/H_2/H_2O–H_2SO_4$, ferrocene–ferricinium (1:1)/Pt, containing sulfuric acid solution up to 100%. Strehlow defined an acidity function $R_0(H)$ [Eq. (1.32)], in which E_x and E_1 are the electromotive forces of the cell at proton activities x and unity, respectively.

$$R_0(H) = \frac{F}{2.303RT} E_x - E_1 \qquad (1.32)$$

Like all the other acidity functions, $R_0(H)$ equals pH in dilute aqueous solution. In strong acids, this function should be a logarithmic measure of the proton activity as long as the normal potential of the redox system, ferrocene–ferricinium, is constant. This was, however, not the case in very strong acid solutions because ferrocene underwent protonation. Other electrochemical pH indicators have been proposed, such as quinine–hydroquinone or semiquinone–hydroquinone, the basicity of which can be modified by substitution on the aromatic ring. These electrochemical indicators have been used with success by Trémillon and co-workers[48] for acidity measurements in anhydrous HF and HF containing superacids.

In principle, the $R_0(H)$ function is of limited interest for kinetic applications because the indicators are chemically very different from the organic substrates generally used. On the other hand, as the measurements are based on pH determination, the length of the acidity scale is limited by the pK value of the solvents. However, very interesting electrochemical acidity studies have been performed in HF by Trémillon and co-workers, such as the acidity measurement in anhydrous HF solvent and the determination of the relative strength of various Lewis acids in the same solvent. By studying the variation of the potential of alkane redox couples as a function of acidity, the authors provide a rational explanation of hydrocarbon behavior in the superacid media.[48]

1.4.4. Chemical Kinetics

The idea that the acidity function may be useful in determining the rates of acid-catalyzed reactions was the main reason for development of the method first proposed by Hammett and Deyrup.[8] The parallelism between reaction rate and H_0 was noticed by Hammett in the early phase of his studies.[49] Especially when the protonation of the substrate parallels the protonation of the Hammett bases, the observed rate constant can be plotted versus $-H_0$ with unit slope. The validity of this principle for a large number of acid-catalyzed reactions and its limitation due to deviations in the protonation behavior has been reviewed extensively.[50] This method has also been applied to obtain a qualitative classification of the relative acidity of various superacid solutions.

Brouwer and van Doorn[51] used this approach in the early 1970s. In the NMR study of the interconversion rates of alkyl tetrahydrofuryl ions **7** and **8** (Scheme 1.2), proceeding via dicarbenium ion intermediates, they measured the overall rate of rearrangement in various superacid combinations of HF, HSO_3F, and SbF_5.

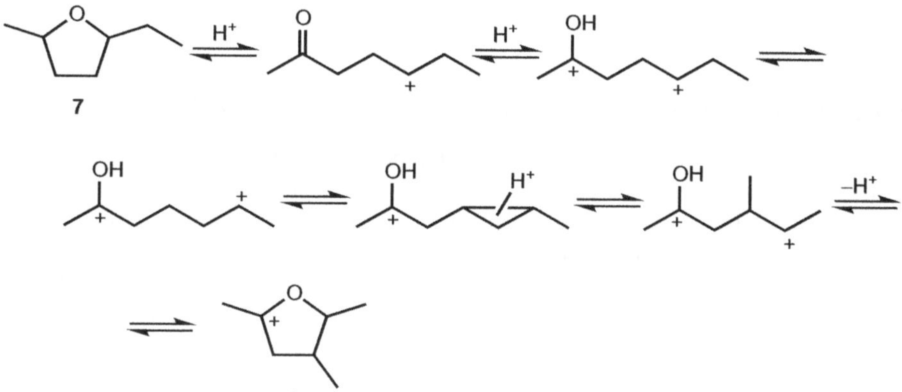

Scheme 1.2.

By making the assumption that the rates were only proportional to the concentration of the dication and taking into account the temperature dependence of the rate, they could estimate the relative acidity of these systems. By repeating these experiments with closely related reactions and varying the acid composition, they were able to estimate the relative acidity of 1:1 HF: SbF_5, 9:1 HF–SbF_5, 1:1 HSO_3F–SbF_5, and 5:1 HSO_3F–SbF_5 as 500:1:10^{-1}:10^{-5}. These estimations have since been shown to be very approximate in comparison with the results obtained by other techniques. Moreover, rates can be affected by other factors than acidity such as solvation effects and temperature. Until now, however, no H_0 value has been measured for the 1:1 HF–SbF_5 medium.

The Friedel–Crafts acid systems HCl–$AlCl_3$ and HBr–$AlBr_3$ are widely used superacids of great practical significance, and various techniques have been used to rank Lewis acids such as $AlCl_3$ and $AlBr_3$ in strong Brønsted acids such as HSO_3F, CF_3SO_3H, HF, HBr, and HCl.[52,53] However, despite opposite claims, their acidity is lower than those of the fluoroacids discussed.

1.4.5. Heats of Protonation of Weak Bases

Arnett[54] had shown that several problems still exist in the currently available methods dealing with the behavior of weak bases in solution. For example, pK_A values of a wide variety of carbonyl compounds given in the literature vary over an unacceptably wide range. The variations are due not only to the activity coefficient problems, but also to practical difficulties such as the effect of media on position of the UV absorption peaks.[55] The previously discussed NMR method seems to alleviate these problems. An alternative method was proposed by Arnett and co-workers measuring the heats of protonation of number of weak bases in HSO_3F medium for which pK_A values are known from other methods. They found a good correlation of these heats of protonation with recorded pK_A values. The heat of protonation method has the advantage over the acidity function procedure that all measurements are made in the same solvent. These studies were applied to systems such as Magic Acid® (HSO_3F–SbF_5) but not extended to HF-based superacids.

1.4.6. Theoretical Calculations and Superacidity in the Gas Phase

The knowledge of acidities or basicities independent from solvation effects is of general interest to chemists because it gives important information on the solvent effects. It also allows the study of the intrinsic ability of the chemical structure of the acid or base to stabilize the anion/cations involved in the acid–base reaction and a quantitative structure–property relationship. In the last two decades, with the development of techniques of high-pressure mass spectrometry,[56] flowing afterglow,[57] and ion cyclotron resonance,[58] the study of ion–molecule reactions became possible in the gas phase. These techniques operating under very different conditions of pressure and time domain gave good agreement for the relative basicity measurements via proton transfer equilibria determination.

Extension of these studies in the superacid field has been reported,[59] which proposes a quantitative intrinsic superacidity scale for sulfonic acids based on measurement of the proton transfer equilibrium between the superacid and its conjugate base. The free energies (and enthalpies) of deprotonation (kcal mol^{-1}) have been estimated (within $\pm$2.5 kcal mol^{-1}) as follows: HPO_3 303 (311); H_2SO_4 302 (309); HSO_3F 300 (307); CF_3SO_3H 299 (306). These results show that the acidity order of the Brønsted superacids measured in the gas phase mirrors the acidity order found in solution. However, the method will be difficult to apply for measuring the large acidity domain of these acids when combined as usual with strong Lewis acids such as SbF_5.

Koppel et al.[60] have established a series of overlapping values of relative gas-phase acidities of a large number of very strong CH, OH, and SH Brønsted acids by using the pulsed FT ion cyclotron resonance (ICR) equilibrium constant method. The new intrinsic acidity scale covers a wide range from $(CF_3)_2NH$ (ΔG_{acid} = 324.3 kcal mol^{-1}) to $(C_4F_9SO_2)_2NH$ (ΔG_{acid} = 284.1 kcal mol^{-1}) and is anchored to the thermodynamic ΔG_{acid} value (318.1 kcal mol^{-1}) of HBr. In several cases, the gas-phase acidity of compounds which make up the scale exceeds the acidity of such traditionally strong mineral acids as HCl, HBr, HI, or H_2SO_4 by more than 30 powers of 10. Based on these results, the acid $(C_4F_9SO_2)_2NH$ may be the strongest measured gas-phase superacid.

Subsequently, the development of both theoretical DFT methods and more sophisticated *ab initio* high-level MP2-type calculations has also spurred investigations in the superacid field.

Later, in another series of papers, Koppel et al.[61] used a theoretical approach at the G2 or G2(MP2) level/and also with the DFT method (B3LYP//6-311+G** level) to calculate the intrinsic acidities and gas-phase deprotonation enthalpies for 39 neutral strong or superstrong Brønsted acids, Brønsted–Lewis conjugate superacids, and even acidic cluster of zeolites. Comparison of the calculated gas-phase acidity with the H_0 values of the neat acid showed a fairly linear correlation. Similar DFT studies were carried out in calculating the Bronsted acidity of polycyanated hydrocarbons in the gas phase.[62]

ΔG values of deprotonation are as low as 249–250 kcal mol^{-1} for HSO_3F or HF associated with SbF_5 or SO_3. The strongest superacid was found to be dodecafluorocarborane acid $CB_{11}F_{12}H$ with a ΔG of 209 kcal mol^{-1} even suggesting that the dodeca(trifluoromethyl)carborane acid $CB_{11}(CF_3)_{12}H$ would be below the 200 kcal mol^{-1} level!

Mota and co-workers have investigated the nature of superacid electrophilic species in HF–SbF_5 by density functional theory[63] and measured the ability of the system to protonate light alkanes (methane, ethane, propane, and isobutane).[64]

More recently, Gutowski and Dixon[65] have recalculated [G3(MP2) theory] the intrinsic gas-phase acidities of a series of 21 Brønsted acids. The computed results are in excellent agreement with experimental gas-phase acidities in the range 342–302 kcal mol^{-1} to within $<$ 1 kcal/mol for 14 out of 15 acids. However, acids with experimental acidities lower than 302 kcal mol^{-1} were found to have large deviations compared to the G3(MP2) results.

1.4.7. Estimating the Strength of Lewis Acids

A quantitative method to determine the strength of Lewis acids and to establish similar scales as discussed in the case of Brønsted acids would be very useful. However, establishing such a scale is extremely difficult and challenging. Whereas the Brønsted acid–base interaction always involves proton transfer, which allows a meaningful quantitative comparison, no such common relationship exists in the Lewis acid–base interaction. The result is that the definition of strength has no real meaning with Lewis acids.

The "strength" or "coordinating power" of different Lewis acids can vary widely against different Lewis bases. Thus, for example, in the case of boron trihalides, boron trifluoride coordinates best with fluorides, but not with chlorides, bromides, or iodides. In coordination with Lewis bases such as amines and phosphines, BF_3 shows preference to the former (as determined by equilibrium constant measurements).[66] The same set of bases behaves differently with the Ag^+ ion. The Ag^+ ion complexes phosphines much more strongly than amines. In the case of halides (F^-, Cl^-, Br^-, and I^-), fluoride is the most effective base in protic acid solution. However, the order is reversed in the case of Ag^+; iodide forms the most stable complex, whereas fluoride forms the least stable one.

The Lewis acidity with respect to strong bases such as NH_3 is greater for BCl_3 than for BF_3. In contrast, toward weak bases such as CO, BF_3 is a stronger acid than BCl_3.[67]

Despite the apparent difficulties, a number of qualitative relationships were developed to categorize Lewis acids.

Pearson proposed a qualitative scheme in which a Lewis acid and base is characterized by two parameters, one of which is referred to as strength and the other is called softness. Thus, the equilibrium constant for a simple acid–base reaction would be a function of four parameters, two for each partner. Subsequently, Pearson introduced the *hard* and *soft* acids and bases (HSAB) principle[68,69] to rationalize behavior and reactivity in a qualitative way. Hard acids correspond roughly in their behavior to class *a* acids as defined by Schwarzenbach[70] and Chatt.[66] They are characterized by small acceptor atoms that have outer electrons not easily excited and that bear considerable positive charge. Soft acids, which correspond to class *b* acids, have acceptor atoms with lower positive charge, large size, and easily excited outer electrons. Hard and soft bases are defined accordingly. Pearson's HSAB principle states that hard acids prefer to bind to hard bases while soft acids prefer to bind to soft bases. The principle has proven useful in rationalizing and classifying a large number of chemical reactions involving acid–base interactions in a qualitative manner,[71] but it gives no basis for a quantitative treatment.

There are many attempts made in the literature[72,73] to rate qualitatively the activity of Lewis acid catalysts in Friedel–Crafts-type reactions. However, such ratings largely depend on the nature of the reaction for which the Lewis acid catalyst is employed.

The classification of Lewis superacids as those stronger than anhydrous aluminum trichloride is only arbitrary. Just as in the case of Gillespie's classification of Brønsted superacids,[18,19] it is important to recognize that acids stronger than conventional Lewis acid halides exist with increasingly unique properties. Again the obvious difficulty is that reported sequences of Lewis acid strengths were established against widely varying bases. Still in applications such as ionizing alkyl halides to their corresponding carbocations, in heterocations systems, catalytic activity, and so on, Lewis acid halides such as SbF_5, AsF_5, TaF_5, and NbF_5, clearly show exceptional ability far exceeding those of $AlCl_3$, BF_3, and other conventional Lewis acid halides. Moreover, these super Lewis acid halides also show remarkable coordinating ability to Brønsted acids such as HF, HSO_3F, and CF_3SO_3H, resulting in vastly enhanced acidity of the resulting conjugate acids.

The determination of the strength of the Lewis acids MF_n, has been carried out in various solvents using the conventional methods. Numerous techniques have been applied: conductivity measurements,[74–79] cryoscopy,[80–83] aromatic hydrocarbon extraction,[53,84] solubility measurements,[85–87] kinetic parameters determinations,[52,88,89] electroanalytical techniques (hydrogen electrode),[90–93] quinones systems as pH indicators,[94–97] or other electrochemical systems,[98,99] IR,[100,101] and acidity function (H_0) determinations with UV–visible spectroscopy,[8,9,14,19,102–105] or with NMR spectroscopy.[20–22,44–46,106–108] Gas-phase measurements are also available.[109–111] Comparison of the results obtained by different methods shows large discrepancies (Table 1.2).

Table 1.2. Relative Strength of MF_n-Type Lewis Acids

Relative Strength	Solvent	Method	References
$BF_3 > TaF_5 > NbF_5 > TiF_4 > PF_5 > SbF_5 > WF_6 \gg SiF_4 \sim CrF_3$	HF	Xylene extraction by *n*-heptane	83
$SbF_5 > AsF_5$, $BF_3 > BiF_5 > TaF_5 > NbF_5 \gg SbF_3 \sim AlF_3 \sim CrF_3$	HF	Solubility and salt formation	85
SbF_5, $PF_5 > BF_3$	HF	Solubility	84
$SbF_5 > TaF_5 \sim NbF_5$	HF	Conductivity and H_0 determination	74
$SbF_5 > AsF_5 \gg PF_5$	HF	Conductivity and cryoscopy	79
$TaF_5 > SbF_5 > BF_3 > TiF_4 > HfF_4$	HF	Reactions/rates parameter selectivity	52
$OsF_5 > ReF_5 > TaF_5 > MoF_5 > NbF_5 \gg MoF_4$	HF	Conductivity and Raman spectra	81
$SbF_5 > TaF_5 > BF_3 > SO_3$	HF	Potentiometry (quinones)	95
$AsF_5 > TaF_5 > BF_3 > NbF_5 > PF_5$	HF	Potentiometry (hydrogen electrodes)	89
$SbF_5 > BF_3$	HF	Infrared spectra	99
$SbF_5 > AsF_5 > SO_3$	HSO_3F	H_0 determination	19
$SbF_5 > BiF_5 > AsF_5 \sim TiF_4 > NbF_5 \sim PF_5$	HSO_3F	Conductivity	75
$SbF_5 > AsF_5 > BF_3 > PF_5$	HSO_3F	Infrared spectra	100
$AsF_5 > BF_3 > PF_5 > SF_4$, SF_5	CH_2Cl_2	^{19}F NMR	105
$BF_3 > TaF_5$	Toluene	^{19}F NMR	106
$SbF_5 > AsF_5 > TaF_5 > NbF_5 > BF_3$	Toluene	Conductivity and cryoscopy of SeF_4–MF_n	80
$SbF_5 > AsF_5 > BF_3 > PF_5$	Gas phase	Complexing with F_3NO	108
$AsF_5 > PF_5 > BF_3 > SiF_4 > AsF_3$	Gas phase	Reaction rates: $SF_6^- + MF_n \rightarrow MF_{n+1}^- + SF_5^{\cdot}$	109
$BF_3 > SiF_4 > PF_5 > PF_3$	Gas phase	Affinity measurements for F^-	110

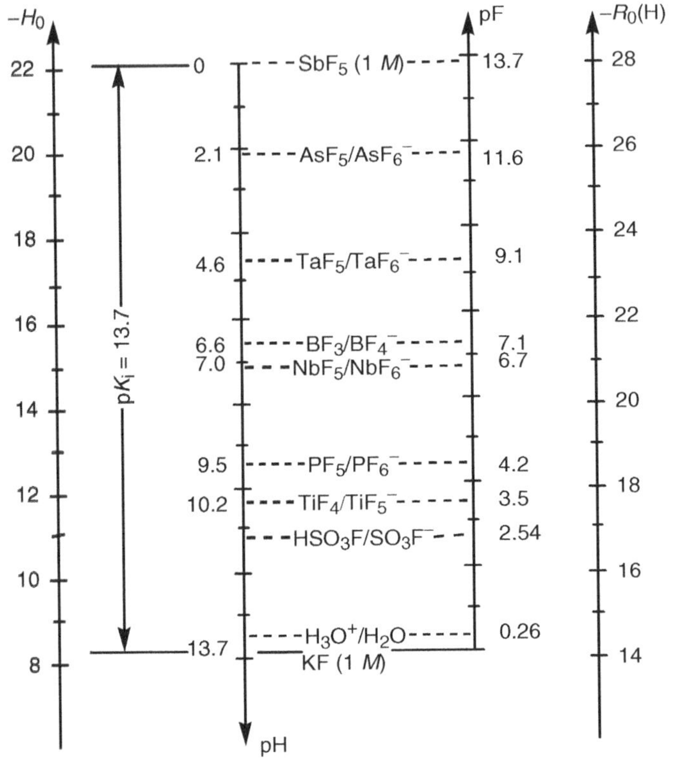

Figure 1.9. Relative strength of some strong Lewis acids as measured in HF on the pH scale by electrochemical titration.[48]

The acidity scale in anhydrous hydrogen fluoride has been the subject of electrochemical investigations by Trémillon and coworkers[48] and is presented in Figure 1.9. The figure also indicates the acidity constants of various Lewis acids allowed to buffer the medium to a pH value as calculated by Eq. (1.33), or in dilute solution by Eq. (1.34).

$$pH = pK_A - \log \frac{^{a}MF_n}{^{a}MF_{n+1}} \qquad (pK_A = -\log K_A) \tag{1.33}$$

$$pH = pK_A - \log \frac{[MF_n]}{[MF_{n+1}]} \tag{1.34}$$

In hydrogen fluoride, the Lewis acid strength is in the following decreasing order: $SbF_5 > AsF_5 > TaF_5 > BF_3 > NbF_5$.

As in all areas, the theoretical tools developed in the last decade was also used to address this question. A theoretical approach with the semiempirical MNDO method

Table 1.3. Abbreviated pF^- Scale[26]

Compound	pF^-	Compound	pF^-
SbF_5	12.03	ClF_5	7.47
AlF_3	11.50	BrF_3	7.35
$AlFCl_2$	11.50	SiF_4	7.35
AlF_2Cl	11.47	SeF_4	7.12
$AlCl_3$	11.46	SOF_4	6.60
$TeOF_4$	10.79	$XeOF_4$	6.37
InF_3	10.75	TeF_6	6.15
GaF_3	10.70	POF_3	5.86
AsF_5	10.59	XeF_4	5.71
SnF_4	9.82	SF_4	5.67
cis-IO_2F_3	9.66	COF_2	4.99
PF_5	9.49	PF_3	4.49
$SeOF_4$	8.69	HF	3.68
TeF_4	8.34	NO_2F	1.92
BF_3	8.31	NOF	1.74
GeF_4	8.30		

combined with ^1H NMR chemical shift measurements has also been used to compare the relative acidity of 18 Lewis acids complexing crotonaldehyde.[73] However, this scale does not completely agree with the pF scale determined in HF solution by electrochemical titration (Figure 1.9).

More recently, a quantitative scale for Lewis acidity based on fluoride ion affinities was calculated using *ab initio* calculations at the MP2/B2 level of theory.[26] Due to its high basicity and small size, the fluoride ion reacts essentially with all Lewis acids; thus the fluoride affinity (or reaction enthalpy) may be considered as a good measure for the strength of a Lewis acid. An abbreviated pF^- scale is given in Table 1.3. This scale was used recently by Christe and Dixon[112] for estimating the stability of salts of complex fluoro anions and cations. The pF value represents the fluoride affinity in kcal mol^{-1} divided by 10.

1.4.8. Experimental Techniques Applied to Solid Acids

Since solid acid catalysts are used extensively in chemical industry, particularly in the petroleum field, a reliable method for measuring the acidity of solids would be extremely useful. The main difficulty to start with is that the activity coefficients for solid species are unknown and thus no thermodynamic acidity function can be properly defined. On the other hand, because the solid by definition is heterogeneous, acidic and basic sites can coexist with variable strength. The surface area available for colorimetric determinations may have widely different acidic properties from the bulk material; this is especially true for well-structured solids like zeolites. It is also not possible to establish a true acid–base equilibrium.

Moreover, the accessibility of sites causes discrepancies in the different methods and to measure the acidity of solids. Because several reviews on this subject have been published in recent years,[27–29] we will just illustrate this problem with sulfated zirconia (SZ), probably the most studied single catalyst in the last 20 years.

SZ was claimed to be a solid superacid by Hino and Arata[113] in 1980 on the basis of its ability to isomerize *n*-butane at low temperature. Since then, various authors using all experimental techniques available tried to verify the superacidity character. Whereas the color change of Hammett indicators suggested a value of -14 to -16 on the H_0 scale,[114] the use of these indicators is considered invalid for surface acidity measurements. The color change of the indicators used (in the pK_a range of -10 to -16) is always from colorless to yellow; this makes the visual appreciation very subjective especially on catalysts, which are generally not colorless. Moreover, the change in color may be due to sites able to transfer electrons to the aromatic ring of the indicators, and also Brønsted and Lewis sites may both contribute.

On the basis of less subjective UV spectroscopy using the same indicators, Hall and co-workers[38] concluded that neither SZ nor zeolites (such as HY, HZSM-5, and H-MOR) were superacids. However, on the basis of ^1H NMR spectroscopy and Raman spectroscopy, Knözinger and co-workers[115] suggested that superacidic protons were present on SZ; but here again, chemical shifts that depend on various factors should not be directly related to acidity. Early EPR studies by Vedrine and co-workers[116] have shown the formation of charge transfer complexes with benzene followed by the formation of radical cations. Since benzene has a high ionization potential, their observation was interpreted as being very probably due to strong Lewis acidity.

On the basis of its catalytic activity in isobutane conversion, this catalyst was described as zirconia-supported oleum,[117] but this hypothesis implies that the reaction mechanism is known, which is not the case. On the same basis, Fraenkel[118] suggested also that SZ was a very strong solid superacid.

In contrast, another ^1H NMR study suggested that the acidity was lower than that of zeolite HZSM-5.[119] Using a combination of solid-state NMR and theoretical methods, the same authors concluded on the nonsuperacidic character for SZ.[120] Similarly, CO adsorption experiments monitored by microcalorimetry and FT–IR concluded to a lower Brønsted acidity in comparison with H-zeolites[121] and showed that acidity is comparable to that of sulfuric acid.[122] On the basis of FT–IR analysis of adsorbed CO and acetonitrile, Sachtler and co-workers[123] arrived at the same conclusion and suggested that the exceptional activity of SZ can be attributed to its ability to stabilize transition states on the surface. According to a recent diffuse reflectance IR study, SZ does not exhibit higher acid strength than zeolites.[124]

Results of temperature-programmed desorption (TPD) of ammonia[125] or argon[126] were attributed to superacid sites. TPD of very weak bases such as substituted benzenes has been used successfully to compare the superacid character of a series of sulfate-treated mixed oxides.[127] However, the validity of TPD measurements and the results of IR study of adsorbed pyridine for acidity determinations have been questioned by various authors[128] as nonspecific to the acid site, considering that different probes may probe different sites providing only qualitative information and measure only an overall acidity. In fact the question whether the alleged superacidity is related to Brønsted or Lewis acid sites is still debated.

More recently, the reactivity of SZ has been assigned to its oxidizing ability,[129–131] which should not be surprising because it has often been considered as SO_3 adsorbed on zirconium oxide. However, that sulfated zirconia is not only an oxidant but also a strong protic acid has been demonstrated by Sommer, Walspurger, and co-workers[132] on the basis H/D exchange experiments with neopentane.

Concerning the acidity of zeolites, Koltunov et al.[133] have shown in a series of papers that reactions involving superelectrophiles could be achieved with excellent yields.

It appears that despite the lack of reliability of acidity determination of solid acids by spectroscopic means and in the absence of knowledge of the nature of the initial step in alkane activation by solid acids the qualification of superacid solids has been and continues to be used, despite the absence of a clear definition of solid superacidity.

Considering the impressive amount of literature on sulfated zirconia and solid superacids,[125,134–139] it will be difficult to impose a definition a posteriori. On the other hand, due to the large difference in acidity and in structure between various liquid superacids, there is no unique chemistry of hydrocarbons in liquid superacids. For this reason it is not possible to suggest a unequivocal definition of solid superacidity at the present stage. Nevertheless, it seems clear from all the data presently available that at high temperatures the chemical reactivity of the proton bound to the surface shows a close resemblance to the one observed at low temperature in liquid superacidic media as will be seen in Chapter 5.

REFERENCES

1. C. H. Rochester, *Acidity Functions*, Academic Press, New York, 1970.
2. F. Coussemant, M. Hellin and B. Torck, *Les Fonctions d'Acidité et Leurs Utilisations en Catalyse Acido-Basique*, Gordon and Breach, New York, 1969.
3. S. Arrhenius, *Z. Phys. Chem.* **1**, 631 (1887).
4. J. N. Brønsted, *Recl. Trav. Chim. Pays-Bas* **42**, 718 (1923).
5. G. N. Lewis, *Valency and Structure of Atoms and Molecules*, American Chemical Society Monographs, the Chemical Catalog Co., New York, 1923.
6. W. B. Jensen, *The Lewis Acid–Base Concepts*, Wiley, New York, 1980.
7. S. P. L. Sørensen, *Biochem. Z.* **24**, 131, 201 (1909).
8. L. P. Hammett and A. J. Deyrup, *J. Am. Chem. Soc.* **54**, 2721 (1932).

9. For discussion on other acidity functions, see T. H. Lowry, and K. S. Richardson, *Mechanism and Theory in Organic Chemistry*, Harper and Row, New York, 1981, pp. 237–247.
10. D. S. Noyce and M. J. Jorgenson, *J. Am. Chem. Soc.* **84**, 4312 (1962).
11. N. C. Deno, H. E. Berkheimer, W. L. Evans, and H. J. Peterson, *J. Am. Chem. Soc.* **81**, 2344 (1959).
12. N. C. Deno, J. J. Jaruzelski, and A. Schriesheim, *J. Am. Chem. Soc.* **77**, 3044 (1955).
13. K. Yates and H. Wai, *J. Am. Chem. Soc.* **86**, 5408 (1964).
14. R. S. Ryabova, I. M. Medvetskaya, and M. I. Vinnik, *Russ. J. Phys. Chem.* **40**, 182 (1966).
15. N. F. Hall and J. B. Conant, *J. Am. Chem. Soc.* **49**, 3047 (1927).
16. For a comprehensive early review, see G. A. Olah, *Angew. Chem. Int. Ed. Engl.* **12**, 173, (1973) references cited therein.
17. For reviews on superacids and see, G. A. Olah, G. K. S. Prakash, and J. Sommer, *Science* **206**, 13 (1979), G. A. Olah and J. Sommer *La Recherche* **10**, 624 (1979).
18. R. J. Gillespie and T. E. Peel, *Adv. Phys. Org. Chem.* **9**, 1 (1971).
19. R. J. Gillespie and T. E. Peel, *J. Am. Chem. Soc.* **95**, 5173 (1973).
20. F. H. Westheimer, *Spec. Publ. Chem. Soc.* **8**, 1 (1957).
21. V. Gold, K. Laali, K. P. Morris, and L. Z. Zdunek, *Chem. Commun*, 769 (1981).
22. D. Touiti, R. Jost, and J. Sommer, *J. Chem. Soc., Perkin* **2**, 1793 (1986).
23. F. Cacace, *Pure Appl. Chem.* **69**, 227 (1997).
24. P. Ahlberg, A. Karlsson, A. Goeppert, S. O. N. Lill, P. Dinér, and J. Sommer, *Chem. Eur. J.* **7**, 1936 (2001).
25. D. Fărcaşiu, P. Lukinskas, A. Ghenciu, and R. Martin, *J. Mol. Catal. A: Chem.* **137**, 213 (1999).
26. K. O. Christe, D. A. Dixon, D. McLemore, W. W. Wilson, J. A. Sheehy, and J. A. Boatz, *J. Fluorine Chem.* **101**, 151 (2000).
27. A. Corma, *Chem. Rev.* **95**, 559 (1995).
28. W. E. Farneth and R. J. Gorte, *Chem. Rev.* **95**, 615 (1995).
29. *Acidity in Aluminas, Amorphous and Crystalline Silico-Aluminas*, J. J. Fripiat and J. A. Dumesic, eds., Topics Catal. **4**, 1–171 (1997).
30. See also: V. Semmer, P. Batamack, C. Dorémieux-Morin, and J. Fraissard, *Topics Catal.* **6**, 119 (1998).
31. K. Tanabe, *Solid Acids and Bases, Their Catalytic Properties*, Kodansha, Tokyo, Academic Press, New York, 1970.
32. K. Tanabe, M. Misono, Y. Ono, and H. Hattori, *New Solid Acids and Bases, Their Catalytic Properties*, Studies in Surface Science and Catalysis, Vol. 51, B. Delmon and J. T. Yates,eds. Kodansha, Tokyo, Elsevier, Amsterdam, 1989.
33. K. Tanabe, in *Catalysis: Science and Technology*, Vol. 2, J. R. Anderson and M. Boudart, eds., Springer, New York, 1981, Chapter 5, p. 231.
34. C. D. Johnson, A. R. Katritzky, and S. A. Shapiro, *J. Am. Chem. Soc.* **91**, 6654 (1969).
35. M. A. Paul and F. A. Long, *Chem. Rev.* **57**, 1 (1957).

36. M. J. Jorgenson and D. R. Hartter, *J. Am. Chem. Soc.* **85**, 878 (1963).
37. D. Brunel, A. Germain, and A. Commeyras, *Nouv. J. Chim.* **2**, 275 (1978).
38. B. Umansky, J. Engelhardt, and W. K. Hall, *J. Catal.* **127**, 128 (1991).
39. H. Kessler, *Angew. Chem.* **82**, 237 (1970).
40. H. Günther, *NMR Spectroscopy*, 2nd ed. Wiley, Chichester, 1992.
41. E. Grunwald, A. Loewenstein, and S. Meiboom, *J. Chem. Phys.* **27**, 630 (1957).
42. G. C. Levy, J. D. Cargioli, and W. Racela, *J. Am. Chem. Soc.* **92**, 6238 (1970).
43. R. Jost, and J. Sommer, *Rev. Chem. Intermed.* **9**, 171 (1988).
44. T. Drakenberg, J. Sommer, and R. Jost, *J. Chem. Soc., Perkin* **2**, 363, (1980).
45. J. Sommer, P. Rimmelin, and T. Drakenberg, *J. Am. Chem. Soc.* **98**, 2671 (1976).
46. J. Sommer, P. Canivet, S. Schwartz, and P. Rimmelin, *Nouv. J. Chim.* **5**, 45 (1981).
47. H. Strehlow, and H. Wendt, *Z. Phys. Chem. (Frankfurt am Main)* **30**, 141 (1960).
48. P.-L. Fabre, J. Devynck, and B. Trémillon, *Chem. Rev.* **82**, 591 (1982) and references cited therein.
49. L. P. Hammett, *Chem. Rev.* **16**, 67 (1935).
50. M. Liler, *Reaction Mechanisms in Sulfuric Acid and other Strong Acid Solutions*, Academic Press, New York, 1971.
51. D. M. Brouwer and J. A. van Doorn, *Recl. Trav. Chim. Pays-Bas* **91**, 895 (1972).
52. G. M. Kramer, *J. Org. Chem.* **40**, 298, 302 (1975).
53. D. Fărcaşiu, S. L. Fisk, M. T. Melchior, and K. D. Rose, *J. Org. Chem.* **47**, 453 (1982).
54. E. M. Arnett, *Prog. Phys. Org. Chem.* **1**, 223 (1963).
55. J. T. Edwards and S. C. Wong, *J. Am. Chem. Soc.* **99**, 4229 (1977).
56. P. Kebarlé, *Ann. Rev. Phys. Chem.* **28**, 445 (1977).
57. H. I. Schiff and D. K. Bohme, *Int. J. Mass Spectrom. Ion Phys.* **16**, 167 (1975).
58. J. L. Beauchamp, *Ann. Rev. Phys. Chem.* **22**, 517 (1971).
59. A. A. Viggiano, M. J. Henchman, F. Dale, C. A. Deakyne, and J. F. Paulson, *J. Am. Chem. Soc.* **114**, 4299 (1992).
60. I. A. Koppel, R. W. Taft, F. Anvia, S.-Z. Zhu, L.-Q. Hu, K.-S. Sung, D. D. DesMarteau, L.M. Yagupolskii, Y. L. Yagupolskii, N. V. Ignat'ev, N. V. Kondratenko, A. Yu. Volkonskii, V. M. Vlasov, R. Notario, and P.-C. Maria, *J. Am. Chem. Soc.* **116**, 3047 (1994).
61. I. A. Koppel, P. Burk, I. Koppel, I. Leito, T. Sonoda, and M. Mishima, *J. Am. Chem. Soc.* **122**, 5114 (2000); I. A. Koppel, P. Burk, I. Koppel, and I. Leito, *J. Am. Chem. Soc.* **124**, 5594 (2002).
62. R. Vianello and Z. B. Maksić, *New J. Chem.* **32**, 413 (2008) and references cited therein.
63. P. M. Esteves, A. Ramírez-Solís, and C. J. A. Mota, *J. Am. Chem. Soc.* **124**, 2672 (2002).
64. P. M. Esteves, A. Ramírez-Solís and C. J. A. Mota, *J. Phys. Chem. B* **105**, 4331 (2001).
65. K. E. Gutowski and D. A. Dixon, *J. Phys. Chem. A* **110**, 12044 (2006).
66. S. Ahrland, J. Chatt, and N. R. Davies, *Quart. Rev.* **12**, 265 (1958).
67. B. D. Rowsell, R. J. Gillespie, and G. L. Heard, *Inorg. Chem.* **38**, 4659 (1999).

68. R. G. Pearson, *J. Am. Chem. Soc.* **85**, 3533 (1963); *Science* **151**, 172 (1966).
69. T.-L. Ho, *Hard and Soft Acids and Bases Principle in Organic Chemistry*, Academic Press, New York, 1977.
70. G. Schwarzenbach and M. Schellenberg, *Helv. Chim. Acta* **48**, 28 (1965).
71. T.-L. Ho, *Chem. Rev.* **75**, 1 (1975).
72. G. A. Olah, *Friedel–Crafts Chemistry*, Wiley, New York, 1973, Chapter 5.
73. P. Laszlo and M. Teston, *J. Am. Chem. Soc.* **112**, 8750 (1990).
74. M. Kilpatrick and F. E. Luborsky, *J. Am. Chem. Soc.* **76**, 5865 (1954).
75. H. H. Hyman, L. A. Quarterman, M. Kilpatrick, and J. J. Katz, *J. Phys. Chem.* **65**, 123 (1961).
76. R. J. Gillespie, K. Ouchi, and G. P. Pez, *Inorg. Chem.* **8**, 63 (1969).
77. R. J. Gillespie and K. C. Moss, *J. Chem. Soc. A*, 1170 (1966).
78. M. Azeem, M. Brownstein, and R. J. Gillespie, *Can. J. Chem.* **47**, 4159 (1969).
79. M. Kilpatrick and T. J. Lewis, *J. Am. Chem. Soc.* **78**, 5186 (1956).
80. P. A. W. Dean, R. J. Gillespie, R. Hulme, and D. A. Humphreys, *J. Chem. Soc. A* 341 (1971).
81. R. J. Gillespie and A. Whitla, *Can. J. Chem.* **48**, 657 (1970).
82. R. T. Paine and L. A. Quarterman, *J. Inorg. Nucl. Chem., H. H. Hyman Mem. Suppl.* **85** (1976).
83. A. A. Woolf, *J. Chem. Soc.*, 279 (1955).
84. D. A. McCaulay, W. S. Higley, and A. P. Lien, *J. Am. Chem. Soc.* **78**, 3009 (1956).
85. S. Kongpricha and A. F. Clifford, *J. Inorg. Nucl. Chem.* **20**, 147 (1961).
86. A. F. Clifford, H. C. Beachell, and W. M. Jack, *J. Inorg. Nucl. Chem.* **5**, 57 (1957).
87. A. F. Clifford and E. Zamora, *Trans. Faraday Soc.* **57**, 1963 (1961).
88. G. A. Olah, J. Shen, and R. H. Schlosberg, *J. Am. Chem. Soc.* **92**, 3831 (1970); **95**, 4957, (1973).
89. D. M. Brouwer and J. A. van Doorn, *Recl. Trav. Chim. Pays-Bas* **89**, 553 (1970).
90. R. Gut and K. Gautschi, *J. Inorg. Nucl. Chem., H. H. Hyman Mem. Suppl.*, 95 (1976).
91. J. Devynck, A. Ben Hadid, and P.-L. Fabre, *J. Inorg. Nucl. Chem.* **41**, 1159 (1979).
92. J. Herlem and A. Thiebault, *J. Electroanal. Chem.* **84**, 99 (1977).
93. G. I. Kaurova, L. M. Grubina, and Ts. A. Adzhemyan, *Elektrokhimiya* **3**, 1222 (1967).
94. A. Ben Hadid, Doctoral thesis, University of Paris V1, 1980.
95. J. Devynck, A. Ben Hadid, P. L. Fabre, and B. Trémillon, *Anal. Chim. Acta* **100**, 343 (1978).
96. J. P. Masson, J. Devynck, and B. Trémillon, *J. Electroanal. Chem.* **64**, 175 193 (1975).
97. A. Ben Hadid, P. Rimmelin, J. Sommer, and J. Devynck, *J. Chem. Soc., Perkin Trans.* 2, 269 (1982).
98. J. Devynck, P. L. Fabre, B. Trémillon, and A. Ben Hadid, *J. Electroanal. Chem.* **91**, 93 (1978).
99. R. Gut and J. Rueede, *J. Coord. Chem.* **8**, 47 (1978).

100. M. Couzi, J. C. Cornut, and P. V. Huong, *J. Chem. Phys.* **56**, 426 (1972).
101. F. O. Sladsky, P. A. Bulliner, and N. Bartlett, *J. Chem. Soc. A.* 2179 (1969).
102. L. P. Hammett and A. J. Deyrup, *J. Am. Chem. Soc.* **55**, 1900 (1933).
103. M. I. Vinnik, *Russ. Chem. Rev.* **35**, 1922 (1966).
104. H. H. Hyman, M. Kilpatrick, and J. J. Katz, *J. Am. Chem. Soc.* **79**, 3668 (1957).
105. H. H. Hyman and R. A. Garber, *J. Am. Chem. Soc.* **81**, 1847 (1959).
106. S. Brownstein, *Can. J. Chem.* **47**, 605 (1969).
107. S. Brownstein, *J. Inorg. Nucl. Chem.* **35**, 3567 (1973).
108. J. Sommer, S. Schwartz, P. Rimmelin, and P. Canivet, *J. Am. Chem. Soc.* **100**, 2576 (1978).
109. W. B. Fox, C. A. Wamser, R. Eibeck, D. K. Huggins, J. S. MacKenzie, and R. Juurik, *Inorg. Chem.* **8**, 1247 (1969).
110. J. C. Haartz and D. H. McDaniel, *J. Am. Chem. Soc.* **95**, 8562 (1973).
111. T. C. Rhyne and J. G. Dillard, *Inorg. Chem.* **10**, 730 (1971).
112. K. O. Christe and D. A. Dixon, unpublished results.
113. M. Hino and K. Arata, *J. Chem. Soc., Chem. Commun.*, 851 (1980).
114. K. Arata, *Adv. Catal.* **97**, 165 (1990).
115. T. Riemer, D. Spielbauer, M. Hunger, G. A. H. Mekhemer, and H. Knözinger, *J. Chem. Soc., Chem. Commun.*, 1181 (1994).
116. F. R. Chen, G. Coudurier, J.-F. Joly, and J. Vedrine, *J. Catal.* **143**, 616 (1993).
117. F. Babou, G. Coudurier, and J. C. Vedrine, *J. Chim. Phys. Phys. Chim. Biol.* **921**, 1457 (1995).
118. D. Fraenkel, *Chem. Lett.* 917 (1999).
119. J. Zhang, J. B. Nicholas, and J. F. Haw, *Angew. Chem. Int. Ed.* **39**, 3302 (2000).
120. J. F. Haw, J. Zhang, K. Shimizu, T. N. Venkatraman, D.-P. Luigi, W. Song, D. H. Barich, and J. B. Nicholas, *J. Am. Chem. Soc.* **122**, 12561 (2000).
121. C. Morterra, G. Cerrato, V. Bolis, S. Di Ciero, and M. Signoretto, *J. Chem. Soc., Faraday Trans.* **93**, 1179 (1997).
122. F. Babou, G. Coudurier, and J. C. Vedrine, *J. Catal.* **152**, 341 (1995).
123. V. Adeeva, J. W. de Haan, J. Jänchen, G. D. Lei, V. Schünemann, L. J. M. van de Ven, W. M. H. Sachtler, and R. A. van Santen, *J. Catal.* **151**, 364 (1995).
124. B. S. Klose, F. C. Jentoft, R. Schlögl, I. R. Subbotina, and V. B. Kazansky, *Langmuir* **21**, 10564 (2005).
125. G. D. Yadav and *J. J.* Nair, *Microporous Mesoporous Mater* **33**, 1 (1999).
126. H. Matsuhashi and K. Arata, *J. Chem. Soc., Chem. Commun.*, 387 (2000).
127. C.-H. Lin and C.-Y. Hsu, *J. Chem. Soc., Chem. Commun.*, 1479 (1992).
128. W.-H. Chen, H.-H. Ko, A. Sakthivel, S.-J. Huang, S.-H. Liu, A.-Y. Lo, T.-C. Tsai, and S.-B. Liu, *Catal. Today* **116**, 111 (2006).
129. D. Fărcaşiu, A. Ghenciu, and J. Q. Li, *J. Catal.* **158**, 116 (1996).
130. A. Ghenciu and D. Fărcaşiu, *J. Chem. Soc., Chem. Commun.* 169 (1996).
131. C. R. Vera, J. C. Yori, and J. M. Parera, *Appl. Catal., A* **167**, 75 (1998).
132. S. Walspurger, Y. Sun, A. S. S. Sido, and J. Sommer, *J. Phys. Chem. B* **110**, 18368 (2006).

133. K. Yu. Koltunov, S. Walspurger, and J. Sommer, *Chem. Commun.*, 1754 (2004); *Tetrahedron Lett.* **46**, 8391 (2005).

134. X. Song and A. Sayari, *Catal. Rev.–Sci. Eng.* **38**, 329 (1996).

135. K. Arata, *Appl. Catal., A* **146**, 3 (1996).

136. T.-K. Cheung and B. C. Gates, *Topics Catal* **6**, 41 (1998).

137. V. Adeeva, H.-Y. Liu, B.-Q. Xu, and W. M. H. Sachtler, *Topics Catal* **6**, 61 (1998).

138. D. Fărcaşiu and J. Q. Li, *Appl. Catal., A* **175**, 1 (1998).

139. Z. Hong, K. B. Fogash, and J. A. Dumesic, *Catal. Today* **51**, 269 (1999).

CHAPTER 2

Superacid Systems

As discussed in Chapter 1, superacids encompass both Brønsted and Lewis types and their conjugate combinations. In this chapter, we will review the physical and chemical properties of the most significant superacids. The development of high-field multinuclear NMR instrumentation as well as new tools facilitating theoretical approaches notably increased the level of our knowledge concerning the structure of superacids, particularly those associated with HF or HSO_3F.

2.1. PRIMARY SUPERACIDS

2.1.1. Brønsted Superacids

Using Gillespie's arbitrary definition, Brønsted superacids are those whose acidity exceeds that of 100% sulfuric acid ($H_0 = -12$). The physical properties of the most commonly used Brønsted superacids are summarized in Table 2.1.

2.1.1.1. Perchloric Acid. Historically, it was Conant's study[1] of the protonating ability of weak bases (such as aldehydes and ketones) by perchloric acid that first called attention to the "superacid" behavior of certain acid systems.

Commercially, perchloric acid is manufactured either by reaction of alkali perchlorates with hydrochloric acid[2] or by direct electrolytic oxidation of 0.5 *N* hydrochloric acid.[3] Another commercially attractive method is by the direct electrolysis of chlorine gas (Cl_2) dissolved in cold dilute perchloric acid.[4] Perchloric acid is commercially available in a concentration of 70% (by weight) in water, although 90% perchloric acid also had limited availability (due to its explosive hazard, it is no longer provided in this strength); 70–72% $HClO_4$, an azeotrope of 28.4% H_2O, 71.6% $HClO_4$, boiling at 203°C is safe for usual applications; however, because it is a strong oxidizing agent, it must be handled with care. Anhydrous acid (100% $HClO_4$) is prepared by vacuum distillation of the concentrated acid solution with a dehydrating agent such as $Mg(ClO_4)_2$. It is stable only at low temperatures for a few days, decomposing to give $HClO_4{\cdot}H_2O$ (84.6% acid) and ClO_2.

Superacid Chemistry, Second Edition, George A. Olah, G. K. Surya Prakash, Árpád Molnár, and Jean Sommer

Table 2.1. Physical Properties of Brønsted Superacids

	$HClO_4$	HSO_3Cl	HSO_3F	CF_3SO_3H	HF
Melting point, °C	−112	−81	−89	−34	−83
Boiling point, °C	110 (explosive)	151–152 (decomposes)	162.7	162	20
Density (25°C), g cm^{-3}	1.767[a]	1.753	1.726	1.698	0.698
Viscosity (25°C), cp	—	3.0[b]	1.56	2.87	0.256
Dielectric constant	—	60 ± 10	120		84
Specific conductance (25°C), ohm^{-1} cm^{-1}	—	$0.2–0.3 \times 10^{-3}$	1.1×10^{-4}	2.0×10^{-4}	1×10^{-6}
$-H_0$ (neat)	~13.0	13.8	15.1	14.1	15.1

[a] At 20°C.
[b] At 15°C.

Perchloric acid is extremely hygroscopic and is a very powerful oxidizer. Contact of organic materials with anhydrous or concentrated perchloric acid can lead to violent explosions. For this reason, the application of perchloric acid has serious limitations. The acid strength, although not reported, can be estimated to be around $H_0 = -13$ for the anhydrous acid.

Formation of various perchlorate salts such as $NO_2^+ClO_4^-$, $CH_3CO^+ClO_4^-$, and $R^+ClO_4^-$, via ionization of their appropriate neutral precursors in perchloric acid can also lead to serious explosions. The probable reason is not necessarily the thermal instability of the ionic perchlorates but instead their equilibria with highly unstable and explosive covalent perchlorates [Eqs. (2.1)–(2.3)].

$$NO_2^+\ ClO_4^- \rightleftharpoons NO_2OClO_3 \quad (2.1)$$

$$CH_3CO^+\ ClO_4^- \rightleftharpoons CH_3COOClO_3 \quad (2.2)$$

$$R^+\ ClO_4^- \rightleftharpoons ROClO_3 \quad (2.3)$$

Actually, no specific advantage exists in using perchlorate salts, when comparable safe conjugate fluoride salts such as BF_4^-, SbF_6^-, and $Sb_2F_{11}^-$ are available. The use of perchlorate salts always necessitates extreme care and precautions.

The chemistry of perchloric acid along with its applications has been well reviewed.[5] The main use of perchloric acid has been the use of its salts (such as $NH_4^+ClO_4^-$) as powerful oxidants in pyrotechniques and rocket fuels.

2.1.1.2. Chlorosulfuric Acid. Chlorosulfuric acid, the monochloride of sulfuric acid, is a strong acid containing a relatively weak sulfur–chlorine bond. It can be prepared by the direct combination of sulfur trioxide and dry hydrogen chloride gas.[6] The reaction is very exothermic and reversible, making it difficult to obtain chlorosulfuric acid free of SO_3 and HCl. Upon distillation, even in good vacuum, some dissociation is unavoidable. The acid is a powerful sulfating and sulfonating agent as well as a strong dehydrating agent and a specialized chlorinating agent. Because of the above properties, chlorosulfuric acid is rarely used for its protonating superacid properties.

Gillespie and co-workers[7,8] have measured systematically the acid strength of the H_2SO_4–$ClSO_3H$ system using aromatic nitro compounds as indicators. They found an H_0 value of -13.8 for 100% $ClSO_3H$.

An extensive review about the physical and chemical properties of chlorosulfuric acid and its use as a reagent is available.[9]

2.1.1.3. *Fluorosulfuric Acid.*

Fluorosulfuric acid, HSO_3F, is a mobile colorless liquid that fumes in moist air and has a sharp odor. It may be regarded as a mixed anhydride of sulfuric acid and hydrogen fluoride. It has been known since 1892[10] and is prepared commercially from SO_3 and HF in a stream of HSO_3F. It is readily purified by distillation, although the last traces of SO_3 are difficult to remove. When water is excluded, it may be handled and stored in glass containers, but for safety reasons the container should always be cooled before opening because the gas pressure may have developed from hydrolysis [Eq. (2.4)]. Fluorosulfuric acid generally also contains hydrogen fluoride as an impurity, but according to Gillespie the hydrogen fluoride can be removed by repeated distillation under anhydrous conditions. The equilibrium in Eq. (2.5) always produces traces of SO_3 and HF in stored HSO_3F samples. When kept in glass for a long time, SiF_4 and H_2SiF_6 are also formed (secondary reactions due to HF).

$$HSO_3F + H_2O \rightleftharpoons H_2SO_4 + HF \qquad (2.4)$$

$$HSO_3F \rightleftharpoons SO_3 + HF \qquad (2.5)$$

Fluorosulfuric acid is employed as a catalyst and chemical reagent in various chemical processes including alkylation, acylation, polymerization, sulfonation, isomerization, and production of organic fluorosulfates.[11] It is insoluble in carbon disulfide, carbon tetrachloride, chloroform, and tetrachloroethane, but is soluble in nitrobenzene, diethyl ether, acetic acid, and ethyl acetate and it dissolves most organic compounds that are potential proton acceptors. The IR, Raman, and NMR spectra of fluorosulfuric acid have been reported.[12,13] The ^{19}F chemical shift is 44.9 ppm downfield from CCl_3F.[14] The acid can be dehydrated to give $S_2O_5F_2$.[15] Electrolysis of fluorosulfuric acid gives $S_2O_6F_2$ or $SO_2F_2 + F_2O$, depending on conditions employed.

HSO_3F has a wide liquid range (mp = −89.0°C, bp = +162.7°C), making it advantageous as a superacid solvent for the protonation of a wide variety of weak bases.

Fluorosulfuric acid ionizes in H_2SO_4 according to Eq. (2.6). The H_0 acidity function has been measured by Gillespie and Peel[8] for the H_2SO_4–HSO_3F system using fluoro- and nitroaromatic bases. They found a strong increase in acidity in the vicinity of 100%

HSO_3F which can be attributed to the self-ionization of the acid [Eq. (2.7)].

$$HSO_3F + H_2SO_4 \rightleftharpoons H_3SO_4^+ + SO_3F^- \quad (2.6)$$

$$2\,HSO_3F \rightleftharpoons H_2SO_3F^+ + SO_3F^- \quad (2.7)$$

The neat acid is ascribed an H_0 value of -15.1. This ranks it as the strongest known simple Brønsted acid. Trifluoromethanesulfonic acid (HSO_3CF_3), in which fluorine is replaced by a CF_3 group, has a slightly lower acid strength (-14.1). HSO_3F, the most widely used superacidic solvent system, has a low viscosity and good thermal stability and wide liquid range (~250°C from mp to bp). The acidity of HSO_3F can be increased further by the addition of Lewis acid fluorides (*vide infra*).

2.1.1.4. Perfluoroalkanesulfonic Acids. Perfluoroalkanesulfonic acids were first reported in 1956[16] and subsequently have been prepared by electrochemical fluorination (ECF) of the corresponding alkanesulfonyl halides and subsequent hydrolysis[17–19] [Eq. (2.8)]. The boiling points, density, and H_0 values of these acids are compared in Table 2.2.

$$RSO_2F \xrightarrow[\text{ECF}]{\text{HF}} R_FSO_2F \xrightarrow[\text{aq. KOH}]{} R_FSO_3K \xrightarrow{H_2SO_4} R_FSO_3H \quad (2.8)$$

Trifluoromethanesulfonic Acid. Trifluoromethanesulfonic acid (CF_3SO_3H, triflic acid), the first member in the perfluoroalkanesulfonic acid series, has been studied extensively, and excellent reviews describing its physical and chemical properties have been published.[18,19] Besides its preparation by electrochemical fluorination of methanesulfonyl halides,[20] triflic acid may also be prepared from

Table 2.2. Characteristics of Perfluoroalkanesulfonic Acids

Compound	bp, °C (760 mmHg)	Density (25°C)	H_0 (22°C)
CF_3SO_3H	161	1.70	-14.1[a]
$C_2F_5SO_3H$	170	1.75	-14.0[a]
$C_4F_9SO_3H$	198	1.82	-13.2[a]
$C_5F_{11}SO_3H$	212	[b]	
$C_6F_{13}SO_3H$	222	[b]	-12.3[a,c]
$C_8F_{17}SO_3H$	249		
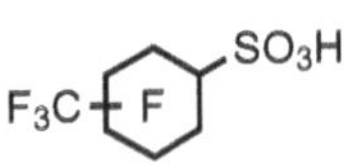	241		
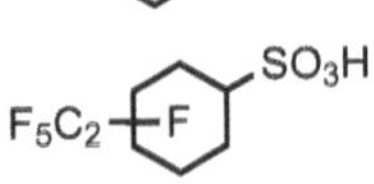	257		

[a]Values from ref. 17.
[b]Solid at 25°C.
[c]Recalculated from the value measured at 35°C.

trifluoromethanesulfenyl chloride[21] [Eq. (2.9)]. A simplified preparation starts with readily available CS_2 according to Eq. (2.10).[22]

$$CF_3SSCF_3 \xrightarrow{Cl_2} CF_3SCl \xrightarrow[H_2O]{Cl_2} CF_3SO_2Cl \xrightarrow[\text{aq. KOH}]{} CF_3SO_3H \quad (2.9)$$

$$CS_2 \xrightarrow{Cl_2} CCl_3SCl \longrightarrow CF_3SCl \longrightarrow CF_3SO_3H \quad (2.10)$$

In 2000, Rhodia began the production of triflic acid by a new process, which includes sulfination of potassium trifluoroacetate, oxidation of the resulting potassium triflinate, followed by acidification and purification[23] [Eq. (2.11)].

$$CF_3COOK \xrightarrow{SO_2} CF_3SO_2K \xrightarrow{\text{oxidation}} CF_3SO_3K \xrightarrow{H^+} CF_3SO_3H \quad (2.11)$$

CF_3SO_3H is a stable, hygroscopic liquid that fumes in moist air and readily forms the stable monohydrate (hydronium triflate), which is a solid at room temperature (mp 34°C, bp 96°C/1 mmHg). The ^{19}F NMR chemical shift for the neat acid is 78.5 ppm upfield from CCl_3F.[24] Conductivity measurements[25] in glacial acetic acid have shown triflic acid to be one of the strongest simple protic acids known, similar to HSO_3F and $HClO_4$. The acidity of the neat acid as measured by UV spectroscopy with a Hammett indicator shows indeed an H_0 value of -14.1.[17] It is miscible with water in all proportions and soluble in many polar organic compounds such as dimethyl formamide, dimethyl sulfoxide, and acetonitrile. It is generally a very good solvent for organic compounds that are capable of acting as proton acceptors in the medium. The exceptional leaving group properties of the triflate anion ($CF_3SO_3^-$) makes triflate esters excellent alkylating agents. The acid and its conjugate base do not provide a source of fluoride ion, even in the presence of strong nucleophiles. Furthermore, because it lacks the sulfonating properties of oleums and HSO_3F, it has gained a wide range of application as a catalyst in organic synthesis— for example, in Friedel–Crafts chemistry, protection group chemistry, the synthesis of heterocycles, carbohydrate chemistry, polymerization, and organometallic chemistry (see Chapter 5).

Higher Perfluoroalkanesulfonic Acids. Higher homologous perfluoroalkanesulfonic acids are hygroscopic, oily liquids or waxy solids. They are prepared by the distillation of their salts from H_2SO_4, giving stable hydrates that are difficult to dehydrate. The acids show the same polar solvent solubilities of trifluoromethanesulfonic acid but are quite insoluble in benzene, heptane, carbon tetrachloride, and perfluorinated liquids. Many of the perfluoroalkanesulfonic acids have been prepared by the electrochemical fluorination reaction [20] (or conversion of the corresponding perfluoroalkane iodides to their sulfonyl halides). α,ω-Perfluoroalkanedisulfonic acids have been prepared by aqueous alkali permanganate oxidation of the compounds $RSO_2(CF_2CF_2)_nSO_2R$.[26] $C_8F_{17}SO_3H$ and higher perfluoroalkanesulfonic acids are surface-active agents and form the basis for a number of commercial fluorochemical surfactants.[27] Because of toxicity problems perfluorooctanesulfonic acid is phased out, however, from Teflon manufacturing process.

Extreme care should be taken while handling the perfluoroalkanesulfonic acids. Studies suggest that extreme irritation and permanent eye damage could occur following eye contact, even if the eyes are flushed immediately with water. Acute inhalation toxicity studies (in albino rats) indicate that high vapor or mist concentrations can cause significant respiratory irritation. All contacts of the acids and their esters with the skin should be avoided. Usual procedures for treatment of strong acid burns should be applied.

Contact of the perfluoroalkanesulfonic acids with cork, rubber, cellulose, and plasticized materials will cause discoloration and deterioration of these materials. Samples are best stored in glass ampoules or glass bottles with Kel-F or Teflon plastic screw-cap linings.

Acidity measurements on perfluoroalkanesulfonic acids have been reported by Commeyras and co-workers[17] (Table 2.2). All the acids show a strong UV absorption band around 283 nm. Up to C_4 chain length, the acids are liquids at room temperature, and the H_0 measurements using Hammett bases were carried out at 22°C. Perfluorohexanesulfonic acid $C_6F_{13}SO_3H$ melts at 33°C and the H_0 value has been corrected for the temperature difference. Perfluorooctanesulfonic acid is a waxy solid melting at 90°C, and its acidity has not been measured. Surprisingly, there is relatively little decrease in acidity with increase in the perfluoroalkyl chain length, and the first CF_2 group adjacent to the sulfonic acid moiety is most responsible for the acid strength. Thus, even higher solid homologous members are capable of acting as superacidic catalysts. As will be seen in Section 2.2.2.6, the acidity of these acids can be further enhanced by complexation with Lewis acids such as SbF_5 and TaF_5.

Harmer et al.[28,29] have reported the synthesis of 1,1,2,2-tetrafluoroethanesulfonic acid, $CF_2HCF_2SO_3H$, and 1,1,2,3,3,3-hexafluoropropanesulfonic acid, CF_3CFHCF_2-SO_3H, by the addition of sulfite to fluorinated double bonds [Eq. (2.12)]. The product acids have lower volatility, which makes their handling easier, and are expected to have similar acidity values to those of their fully fluorinated counterparts. In fact, catalytic conversions of a range of transformations were found to be comparable to triflic acid (see Chapter 5).

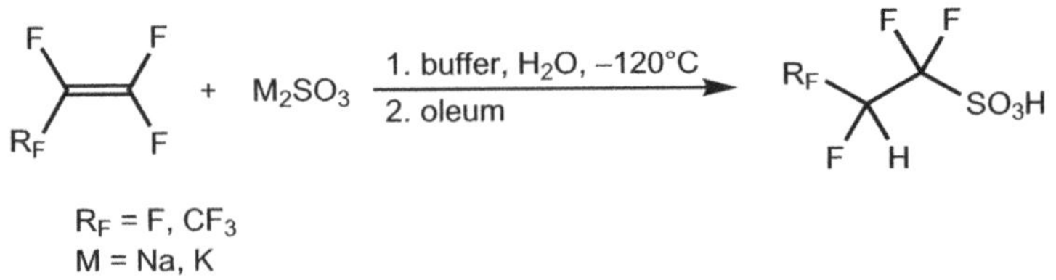

(2.12)

2.1.1.5. *Hydrogen Fluoride.* Anhydrous hydrogen fluoride is generally prepared by action of concentrated sulfuric acid on calcium fluoride "Fluorspar" (>98% CaF_2). The estimated world production is about 1 million tons mostly to prepare chlorofluorocarbons and synthetic cyalite necessary for aluminum production. It is also used as catalyst for alkylation in the petrochemistry and for uranium processing UF_4/UF_6.

Extreme caution should be used in handling anhydrous HF. It can cause severe burns that may not be noticed immediately but will be very painful later; HF dehydrates the skin, and F^- *removes* Ca^{2+} *from tissues and delays healing. Immediate thorough water washing of any exposed skin should be followed by application of calcium gluconate gel or benzalkonium chloride (trade name: Zephiran Chloride), and medical attention is essential.*

Due to its highly corrosive nature, HF should be handled in plastic, Teflon, PTFE, or Monel apparatus. Due to strong hydrogen bonding, it forms a two-dimensional polymer in the liquid phase. It is also a useful solvent both for organic and inorganic compounds. Until the more recent work by Gillespie and Liang[30] the H_0 value for the neat anhydrous acid (−15.1) was generally underestimated (−11) due to the very high sensitivity of the autoprotonation equilibrium to trace impurities of water [Eq. (2.13)]. Its association with SbF_5 generates the strongest superacid system (*vide infra*).

$$2\,HF \rightleftharpoons H_2F^+ + F^- \qquad K = 10^{-10} \tag{2.13}$$

An easier-to-handle, convenient liquid alternative of anhydrous hydrogen fluoride is pyridinium poly(hydrogen fluoride) (PPHF), also known as Olah's reagent.[31] The PPHF reagent (30 wt% pyridine–70 wt% HF), which is stable up to 50°C, is in equilibrium with a small amount of free hydrogen fluoride and, consequently, is the liquid equivalent of HF and serves as a general-purpose fluorinating agent. A similar solid polymeric reagent, poly-4-vinylpyridinium poly(hydrogen fluoride) (PVPHF), has also been developed.[32]

Stable dialkyl ether poly(hydrogen fluoride) complexes (R_2O–$[HF]_n$, R = Me, Et, *n*-Pr) have recently been developed by Prakash, Olah, and colleagues.[33] DFT calculations suggest a cyclic poly(hydrogen fluoride) bridged structure. Dimethyl ether–5 HF (DMEPHF) was shown to be a convenient and effective fluorinating agent (see Section 5.10.1).

2.1.1.6. Carborane Superacids $H(CB_{11}HR_5X_6)$. Recently, new carbon superacids, icosahedral carboranes $H(CB_{11}HR_5X_6)$ (where X = chlorine, bromine or iodine; R = H, Me, Cl), have been described by Reed et al.[34], whose conjugate base, the carborane anion $(CB_{11}HR_5X_6)^-$, is quite inert due to low nucleophilicity.

As a nonoxidizing, low-nucleophilicity system, reaction of carborane acid $H(CB_{11}H_6X_6)$ (Figure 2.1) with C_{60} gives HC_{60}^+ as a stable ion in solution.[35] Arenium ions are also remarkably stable in this medium.[36] Previously investigated only at subambient temperatures in highly superacidic media, protonated benzene is readily isolated as a crystalline salt, thermally stable to >150°C. Solid-state ^{13}C NMR spectra are similar to those observed earlier in solution, indicating that lattice

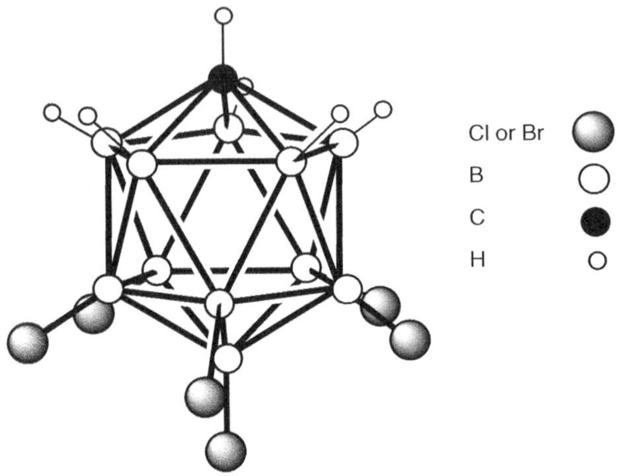

Figure 2.1. The structure of carborane superacid $H(CB_{11}H_6X_6)$.

interactions are comparable to solution solvation effects.[35,36] A new and cheap synthesis of the $[closo\text{-}CB_{11}H_{12}]^-$ anion, which serves as a starting material for the preparation of many extremely weakly nucleophilic anions, has been reported by Michl and co-workers.[37]

2.1.2. Lewis Superacids

In Chapter 1, we arbitrarily defined Lewis superacids as those that are stronger than anhydrous aluminum chloride in their reactivity, the most commonly used Friedel–Crafts catalyst. Of course, Lewis acidity is only a relative term concerning specific bases and involved counterions (association, steric hindrance, etc.). The physical properties of some of the Lewis superacids are given in Table 2.3.

2.1.2.1. Antimony Pentafluoride. Antimony pentafluoride (SbF_5) is a highly associated, colorless, very viscous liquid at room temperature. It is hygroscopic and fumes in moist air. Its viscosity at 20°C is 460 cP, which is close to that of glycerol. The pure liquid can be handled and distilled in glass if moisture is excluded. Commercial antimony pentafluoride is shipped in steel cylinders or in perfluoroethylene bottles for laboratory quantities.

Table 2.3. Physical Properties of Some Lewis Superacids

	SbF_5	AsF_5	TaF_5	NbF_5
Melting point, °C	7.0	−79.8	97	72–73
Boiling point, °C	142.7[a]	−52.8	229	236
Specific gravity (15°C), g cm^{-3}	3.145	2.33[b]	3.9	2.7

[a]Anomalous high value because of association of molecules.
[b]At the bp.

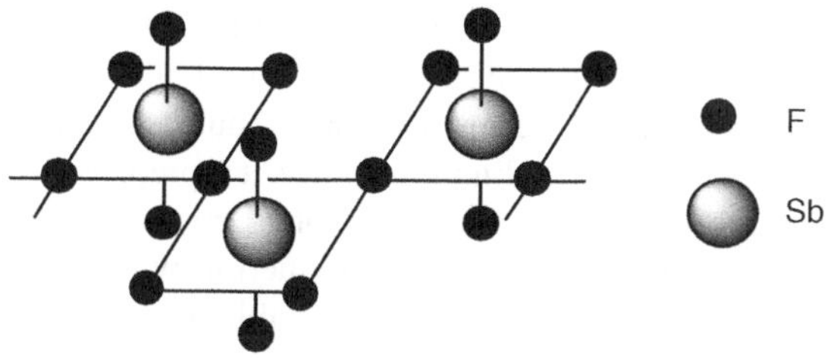

Figure 2.2. The polymeric structure of liquid SbF_5.

The polymeric structure of the liquid SbF_5 has been established by ^{19}F NMR spectroscopy[38] and is shown to have the frameworks depicted in Figure 2.2. A *cis*-fluorine bridged structure is found in which each antimony atom is surrounded by six fluorine atoms in an octahedral arrangement.

Ab initio molecular dynamics were used to investigate the structure of liquid SbF_5. The results confirm the high tendency of SbF_5 to oligomerize and impart a highly ionic character to the Sb–F bond. It confirms also the *cis*-bridged chain polymer as the most stable structure.[39]

Antimony pentafluoride is also a powerful oxidizing and an effective fluorinating agent. It is even able to oxidize small alkanes below room temperature. It readily forms stable intercalation compounds with graphite (*vide infra*), and it spontaneously inflames phosphorus and sodium. It coordinates or associates with water to form $SbF_5 \cdot 2\ H_2O$, an unusually stable solid hydrate (probably a hydronium salt, $H_3O^+SbF_5OH^-$) that reacts violently with excess water to form a clear solution. Slow hydrolysis can be achieved in the presence of dilute NaOH and forms $Sb(OH)_6^-$. Sulfur dioxide and nitrogen dioxide form 1:1 adducts, SbF_5–SO_2 and SbF_5–NO_2, respectively,[40,41] as do practically all nonbonded electron pair donor compounds. The exceptional ability of SbF_5 to complex and subsequently ionize nonbonded electron pair donors (such as halides, alcohols, ethers, sulfides, amines, etc.) to carbocations, recognized first by Olah (Chapter 3), has made it one of the most widely used Lewis halides in the study of cationic intermediates and catalytic reactions.

Vapor density measurements suggest a molecular association corresponding to $(SbF_5)_3$ at 150°C and $(SbF_5)_2$ at 250°C. On cooling, SbF_5 gives a nonionic solid composed of trigonal bipyramidal molecules. Antimony pentafluoride is prepared by the direct fluorination of antimony metal or antimony trifluoride (SbF_3). It can also be prepared by the reaction of $SbCl_5$ with anhydrous HF, but the exchange of the fifth chloride is difficult and the product is generally SbF_4Cl.[42] As shown by conductometric, cryoscopic, and related acidity measurements, it appears that antimony pentafluoride is by far the strongest Lewis acid known in the condensed phase. Thus, it is preferentially used in preparing stable ions and conjugate superacids (*vide infra*). Antimony pentafluoride is also a strong oxidizing agent, allowing, for example, preparation of arene dications. At the same time, its easy reducibility to antimony trifluoride represents a limitation in many applications, although it can be refluorinated.

2.1.2.2. *Arsenic Pentafluoride.* Arsenic pentafluoride (AsF_5) is a colorless gas at room temperature, condensing to a yellow liquid at −53°C. Vapor density measurements indicate some degree of association, but it is a monomeric covalent compound with a high degree of coordinating ability. It is prepared by reacting fluorine with arsenic metal or arsenic trifluoride. As a strong Lewis acid fluoride, it is used in the preparation of ionic complexes and in conjunction with Brønsted acids forms conjugate superacids. It also forms with graphite stable intercalation compounds that show electrical conductivity comparable to that of copper.[43] Great care should be exercised in handling any arsenic compound because of potential high toxicity.

2.1.2.3. *Phosphorus Pentafluoride.* Phosphorus pentafluoride, which may be prepared by fluorinating phosphorus pentachloride with AsF_3 or CaF_2, is a colorless gas (bp −93°C) that reacts with Lewis bases to form a six-coordinated anion [Eq. (2.14)]. However in HF, PF_5 is a non-electrolyte[44] and hexafluorophosphoric acid HPF_6 is available as a 60% solution in water (which cannot be considered a superacid). In the pF^- scale established by Christe[45] to measure the strength of Lewis acids on the basis of fluoride affinity (see Section 1.4.7) it is rated 9.49, lower than AsF_5(10.59) but higher than BF_3(8.31).

$$PF_5 + F^- \longrightarrow PF_6^- \tag{2.14}$$

2.1.2.4. *Tantalum and Niobium Pentafluoride.* The close similarity of the atomic and ionic radii of niobium and tantalum are reflected by similar properties of tantalum and niobium pentafluorides. They are thermally stable white solids that may be prepared either by the direct fluorination of the corresponding metals or by reacting the metal pentachlorides with HF. Surprisingly, even reacting metals with HF gives the corresponding pentafluorides. They both are strong Lewis acids complexing a wide variety of donors such as ethers, sulfides, amines, halides, and so on. They both coordinate with fluoride ion to form anions of the type $(MF_6)^-$. TaF_5 is a somewhat stronger acid than NbF_5 as shown by acidity measurements in HF. The solubility of TaF_5 and NbF_5 in HF and HSO_3F is much more limited than that of SbF_5 or other Lewis acid fluorides, restricting their use to some extent. At the same time, their high redox potentials and more limited volatility make them catalysts of choice in certain hydrocarbon conversions, particularly in combination with solid catalysts. Their chemistry has been extensively reviewed.[46,47]

2.1.2.5. *Boron Trifluoride.* BF_3 is a pungent colorless gas (bp −99.9°C) that can be prepared by reacting B_2O_3 or borates with CaF_2 and concentrated H_2SO_4.[48] A modern two-stage process gives a much better yield [Eq. (2.15)]. It is a strong Lewis acid that forms adducts with most compounds having an available electron lone pair. Unlike other boron halides, BF_3 does not undergo hydrolysis with water under usual conditions to boric acid and HF. Moreover, because it is a volatile gas, it can be easily recovered for reuse.

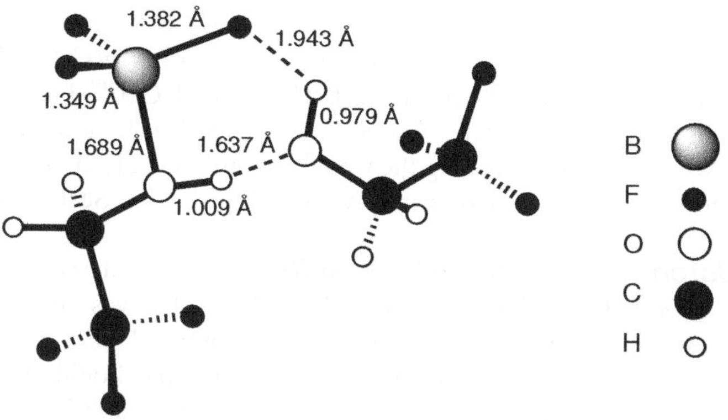

Figure 2.3. Geometrical parameters by DFT calculations of the BF_3–$2CF_3CH_2OH$ complex.

$$Na_2B_4O_7 + 12\,HF \xrightarrow[-6\,H_2O]{} [Na_2O(BF_3)_4] \xrightarrow[-2\,NaHSO_4,\ -H_2O]{2\,H_2SO_4} 4\,BF_3 \qquad (2.15)$$

It is commonly available as diethyl etherate $(C_2H_5)_2O{\cdot}BF_3$ and is widely used as catalyst in Friedel–Crafts and other acid-catalyzed reactions in organic chemistry.[49]

Prakash, Olah, and co-workers have explored the use of boron trifluoride monohydrate (BF_3–H_2O) and the complex BF_3–$2CF_3CH_2OH$ in acid-catalyzed transformations. Both BF_3–H_2O and BF_3–$2CF_3CH_2OH$ are strong acids and very efficient superacidic catalysts comparable in acid strength to at least that of 100% sulfuric acid.[50–52] The structure of the BF_3–$2CF_3CH_2OH$ adduct has been studied using multinuclear NMR spectroscopy and DFT calculations (Figure 2.3). The activity of both complexes was successfully tested on various acid-catalyzed organic reactions (see Chapter 5).

2.1.2.6. *Tris(pentafluorophenyl) Borane.* $B(C_6F_5)_3$ was first prepared by Stone, Massey, and Park in 1963 by treatment of pentafluorophenyllithium with BCl_3 at low temperature[53,54] [Eq. (2.16)]. Since the lithium reagent may eliminate LiF explosively, the use of the Grignard reagent, pentafluorophenylmagnesium bromide, is a safer alternative. It is thermally robust, resistant toward oxidation by molecular oxygen, and water-tolerant. It is best known for its role as activator in Ziegler–Natta chemistry. However, its special properties as a strong Lewis acid are increasingly used in organic and organometallic chemistry.[55,56] Its Lewis acidity, determined by the NMR method developed by Childs,[57] is described as between that of BF_3 and BCl_3.[58]

$$C_6F_5Br + n\text{-BuLi} \xrightarrow[\substack{\text{pentane}\\ -\text{butane}}]{} C_6F_5Li \xrightarrow[\substack{-78^\circ\text{C to RT}\\ -\text{LiCl}}]{BCl_3} B(C_6F_5)_3 \qquad (2.16)$$

Another related boron Lewis superacid, $B(CF_3)_3$, is worth mentioning, even though it has not been isolated.[59] However, the corresponding anion $B(CF_3)_4^-$ is known and found in simple salts (Ag, Li, K, etc.). $B(CF_3)_4^-$ has been shown to be weakly coordinating as determined by examining the stability of $[Ag(CO)_x]^+[B(CF_3)_4]^-$ ($x = 1-4$) complexes generated from $Ag[B(CF_3)_4]$ and CO.[60] Furthermore, the complex $B(CF_3)_3CO$ in hydrogen fluoride is considered to be a strong conjugate Brønsted–Lewis superacid furnishing the weakly coordinating $[(CF_3)_3BF]^-$ anion in the generation of homoleptic metal carbonyl complexes[59] (see Section 4.5.1).

2.1.2.7. Boron Tris(trifluoromethanesulfonate). Boron tris(trifluoromethanesulfonate) [boron tris(triflate), $B(OSO_2CF_3)_3$, $B(OTf)_3$] was first prepared by Olah and co-workers from boron chloride and bromide[61] [Eq. (2.17)]. Boron tris(triflate) is a volatile, colorless, extremely hygroscopic liquid (mp 43–45°C, bp 68–73°C/0.5 torr), which distills in vacuum without decomposition. It is readily soluble in such solvents as CH_2Cl_2, Freon-113, SO_2, and SO_2ClF. Data for acidity measurement are not available; but on the basis of its catalytic activity in Friedel–Crafts reactions (see Section 5.2.3) and the ability to generate stable, long-lived carbocations in SO_2ClF solution at low temperature, $B(OSO_2CF_3)_3$ is a super Lewis acid.

$$BX_3 + 3\,CF_3SO_3H \longrightarrow B(OSO_2CF_3)_3 + 3\,HX \qquad X = Br, Cl \tag{2.17}$$

2.1.2.8. Aprotic Organic Superacids (Vol'pin's Systems). In the late 1980s, Vol'pin and his group[62] noticed the high activity of complexes of acyl halides with aluminum halides toward low-temperature transformation of alkanes and cycloalkanes. They proposed to call these media Aprotic Organic Superacids (AOS).[63] The activity necessitated 2 moles of aluminum halide per mole of acyl halide. Multinuclear NMR studies showed that the 1:1 complexes give exclusively donor–acceptor complexes in solution, whereas the 2:1 complexes form equilibrium mixtures of acylium salts [Eq. (2.18)]. Later on, they also included systems such as $CHBr_3$–$2AlBr_3$ and CCl_4–$2AlBr_3$, which were very active at room temperature for alkane isomerization, cracking, and various functionalizations[62,64] (see Chapter 5).

$$RCOX + 2\,AlX_3 \longrightarrow RCOX \rightarrow Al_2X_6 \underset{}{\overset{CH_2X_2}{\rightleftharpoons}} RCO^+\,Al_2X_7^- \qquad R = \text{alkyl, aryl};\ X = Br, Cl \tag{2.18}$$

The key of alkane transformation was assigned to the formation of CX_3^+-type cations that are electrophilic enough (probably due to a second complexation of AlX_3), to abstract a hydride anion from linear and cycloalkanes. When these cations are generated in superacidic media, a protosolvation induces a superelectrophilic character, which was supported by Olah on the basis of high-level *ab initio* calculations.[65] The generation of these cations was also used by various teams[66,67] to initiate selective low temperature alkane activation.

2.2. BINARY SUPERACIDS

2.2.1. Binary Brønsted Superacids

2.2.1.1. Hydrogen Fluoride–Fluorosulfuric Acid. Fluorosulfuric acid containing up to 5% of HF acting as a protic coacid was found to be very efficient for isomerization of *n*-butane to isobutane at room temperature, whereas pure HSO_3F did not show this activity. Moreover, at higher HF concentration the activity diminishes substantially.[68]

2.2.1.2. Hydrogen Fluoride–Trifluoromethanesulfonic Acid. The acidity of this binary Brønsted acid system has not been measured, but the superacidic properties are mentioned in numerous patents concerning fluorination, olefin alkylation, and hydrocarbon conversion.

2.2.1.3. Tetra(Hydrogen Sulfato)Boric Acid–Sulfuric Acid. $HB(HSO_4)_4$ prepared by treating boric acid, $B(OH)_3$, with sulfuric acid ionizes in sulfuric acid as shown by acidity measurements[7] [Eq. (2.19)].

$$HB(HSO_4)_4 + H_2SO_4 \rightleftharpoons H_3SO_4^+ + B(HSO_4)_4^- \qquad (2.19)$$

The increase in acidity is, however, limited to $H_0 = -13.6$ as a result of insoluble complexes that precipitate when the concentration of the boric acid approaches 30 mol%. Figure 2.4 shows the composition-related acidity increase for the system in comparison with oleum.

2.2.2. Conjugate Brønsted–Lewis Superacids

2.2.2.1. Oleums–Polysulfuric Acids. SO_3-containing sulfuric acid (oleum) has been long considered as very strong mineral acid and one of the earliest superacid systems to be recognized. The concentration of SO_3, in sulfuric acid can be determined by weight or by electrical conductivity measurement.[69] The vapor pressure of oleum rises rapidly with the increase in concentration of SO_3 and increase in temperature[70] as shown in Figure 2.5.

Lewis and Bigeleisen,[71] who first determined the Hammett acidity function values for oleums, used a method to derive them from the vapor pressure measurements. Brand et al.,[72] however, subsequently by the use of nitro compound indicators showed that the H_0 values are not directly related to the vapor pressure. The most accurate H_0 values for oleums so far have been published by Gillespie et al.[7] (Table 2.4).

The increase in acidity on addition of SO_3 to sulfuric acid is substantial, and an H_0 value of -14.5 is reached with 50 mol% SO_3. The main component up to this SO_3 concentration is pyrosulfuric (or disulfuric) acid $H_2S_2O_7$. Upon heating or in the

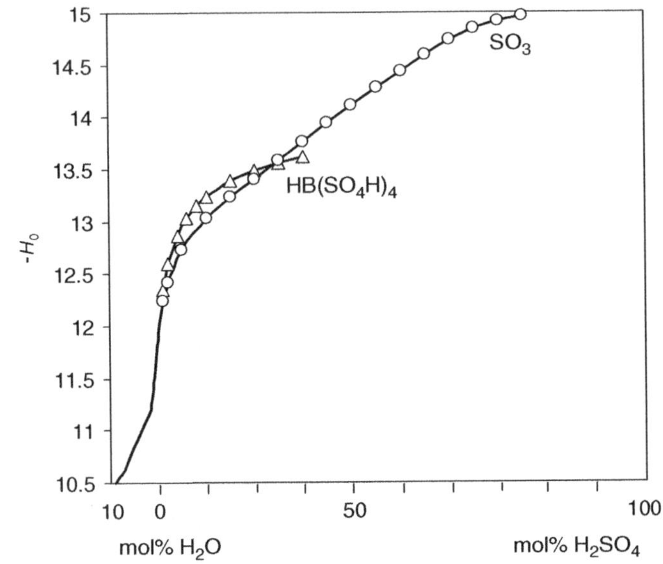

Figure 2.4. H_0 acidity function values for H_2SO_4–$HB(SO_4H)_4$ and H_2SO_4–SO_3.[7]

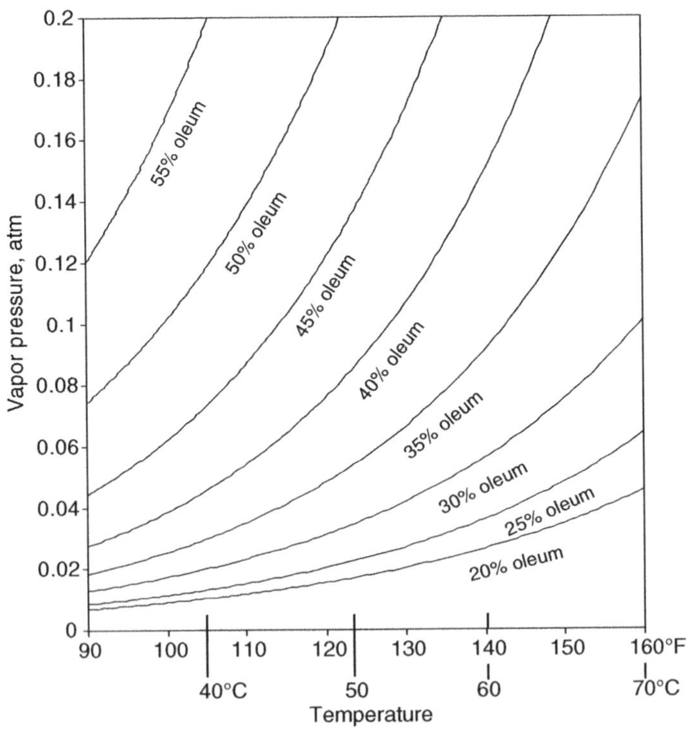

Figure 2.5. Vapor pressure of oleum.

Table 2.4. H_0 Values for the H_2SO_4–SO_3 System

Mol% SO_3	H_0	Mol% SO_3	H_0	Mol% SO_3	H_0
1.00	−12.24	25.00	−13.58	55.00	−14.59
2.00	−12.42	30.00	−13.76	60.00	−14.74
5.00	−12.73	35.00	−13.94	65.00	−14.84
10.00	−13.03	40.00	−14.11	70.00	−14.92
15.00	−13.23	45.00	−14.28	75.00	−14.96
20.00	−13.41	50.00	−14.44		

presence of water, it decomposes and behaves like a mixture of sulfuric acid and sulfur trioxide. In sulfuric acid, it ionizes as a stronger acid [Eq. (2.20)].

$$H_2S_2O_7 + H_2SO_4 \rightleftharpoons H_3SO_4^+ + HS_2O_7^- \quad K = 1.4\ 10^{-2} \tag{2.20}$$

At higher SO_3 concentration, a series of higher polysulfuric acids such as $H_2S_3O_{10}$, $H_2S_4O_{13}$, and so on, are formed and a corresponding increase in acidity occurs. However, as can be seen from Table 2.4, the acidity increase is very small after reaching 50 mol% of SO_3 and no data are available beyond 75%.

Despite its high acidity, oleum has found little application as a superacid catalyst, mainly because of its strong oxidizing power. Also, its high melting point and viscosity have considerably hampered its use for spectroscopic study of ionic intermediates and in synthesis, except as an oxidizing or sulfonating agent.

2.2.2.2. *Fluorosulfuric Acid–Antimony Pentafluoride ("Magic Acid").*

Of all the superacids, a mixture of fluorosulfuric acid and antimony pentafluoride, named by Olah as "Magic Acid," is probably the most thoroughly investigated concerning measurements of acidity and also the most widely used medium for the spectroscopic observation of stable carbocations (see Chapter 3). The fluorosulfuric acid–antimony pentafluoride system was developed in the early 1960s by Olah for the study of stable carbocations and was studied by Gillespie for the generation of electron-deficient inorganic cations. The name Magic Acid originated in Olah's laboratory at Case Western Reserve University in the winter of 1966. The HSO_3F–SbF_5 mixture was extensively used in his group to generate stable carbocations. J. Lukas, a German postdoctoral fellow, put a small piece of Christmas candle left over from a lab party into the acid system and found that it dissolved readily. He then ran an ^1H NMR spectrum of the solution. To everybody's amazement, he obtained a sharp spectrum of the *tert*-butyl cation. The long-chain paraffin, of which the candle is made, had obviously undergone extensive cleavage and isomerization to the more stable tertiary ion. It impressed Lukas and others in the laboratory so much that they started to nickname the acid system Magic Acid. The name stuck and soon others started to use it too. It is now a registered trade name and has found its way into the chemical literature.

The acidity of the Magic Acid system as a function of SbF_5 content has been measured successively by Gillespie,[8] Gold,[73] Sommer,[74] and their co-workers. The increase in acidity is very sharp at low SbF_5 concentration and was estimated to

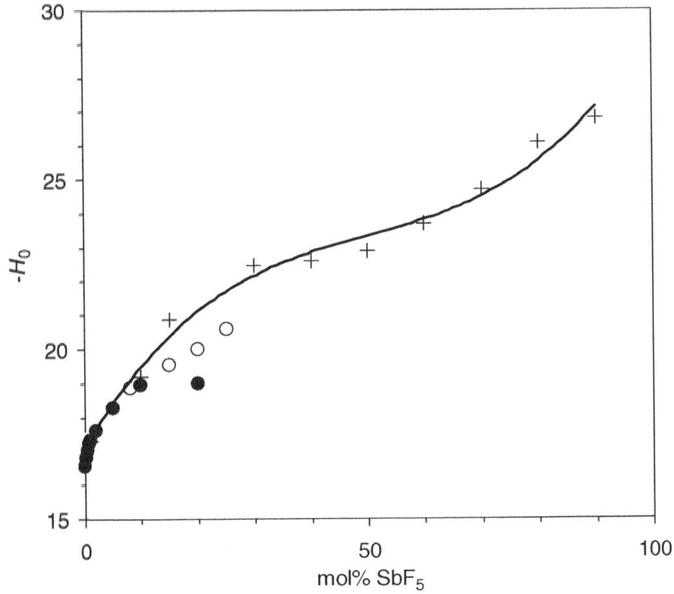

Figure 2.6. Dependence of H_0 acidity function values for the HSO_3F–SbF_5 system upon SbF_5 addition (up to 90 mol%) from ref. 8 (●), ref. 73 (+), and ref. 74 (○).

continue up to the value of −27 for the 90 mol% SbF_5 solution as shown in Figure 2.6.[8,73–75] This estimation was, however, questioned on the basis of the contrasting results obtained by various authors using different experimental techniques.[76] It seems that a value of H_0 −22 to −23 is probably a fair estimation of the upper acidity limit of this most classic superacid system (Figure 2.7).

The initial ionization of HSO_3F–SbF_5 is shown in Eq. (2.21). At higher concentrations of SbF_5, larger polyantimony fluorosulfate ions are formed [Eq. (2.22)].

$$2\,HSO_3F + SbF_5 \rightleftharpoons H_2SO_3F^+ + SbF_5(SO_3F)^- \quad (2.21)$$

$$SbF_5 + SbF_5(SO_3F)^- \rightleftharpoons Sb_2F_{10}(SO_3F)^- \quad (2.22)$$

Due to these equilibria, which have been discussed by several authors,[77,78] the composition of the HSO_3F–SbF_5 system is quite complex and depends on the SbF_5 content. The first ^{19}F NMR study by Commeyras and co-workers[14] has shown such dependence and is presented in Figure 2.8 for the major structural components, which are depicted in Table 2.5.

Free SbF_5 is not observed as it complexes the Brønsted acid and leads to its initial ionization [Eqs. (2.23) and (2.24)].

$$HSO_3F + SbF_5 \rightleftharpoons HFSO_3SbF_5 \quad (2.23)$$

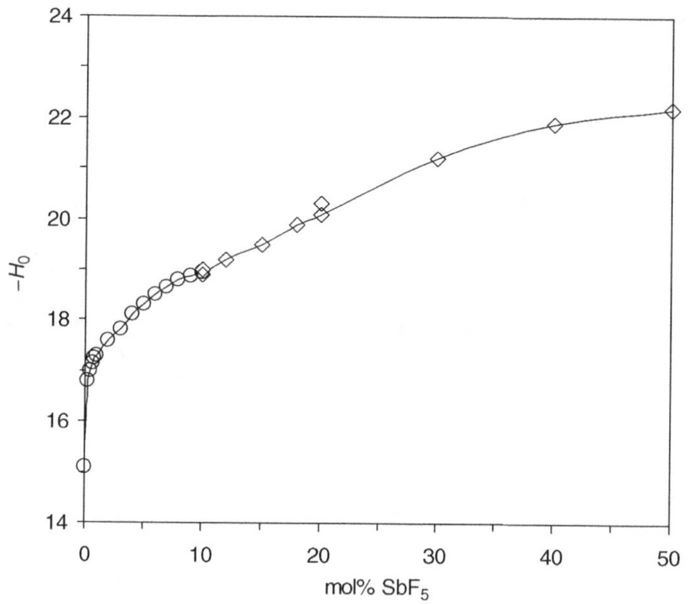

Figure 2.7. Dependence of H_0 acidity function values for the HSO_3F–SbF_5 system [ref. 8 (○) and ref. 76 (◇)].

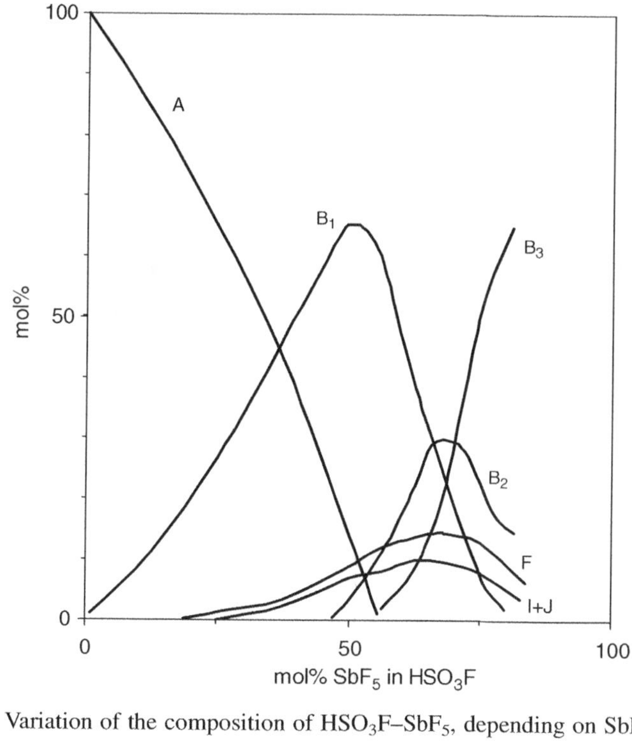

Figure 2.8. Variation of the composition of HSO_3F–SbF_5, depending on SbF_5 content.[14]

Table 2.5. Main Components of the Magic Acid System

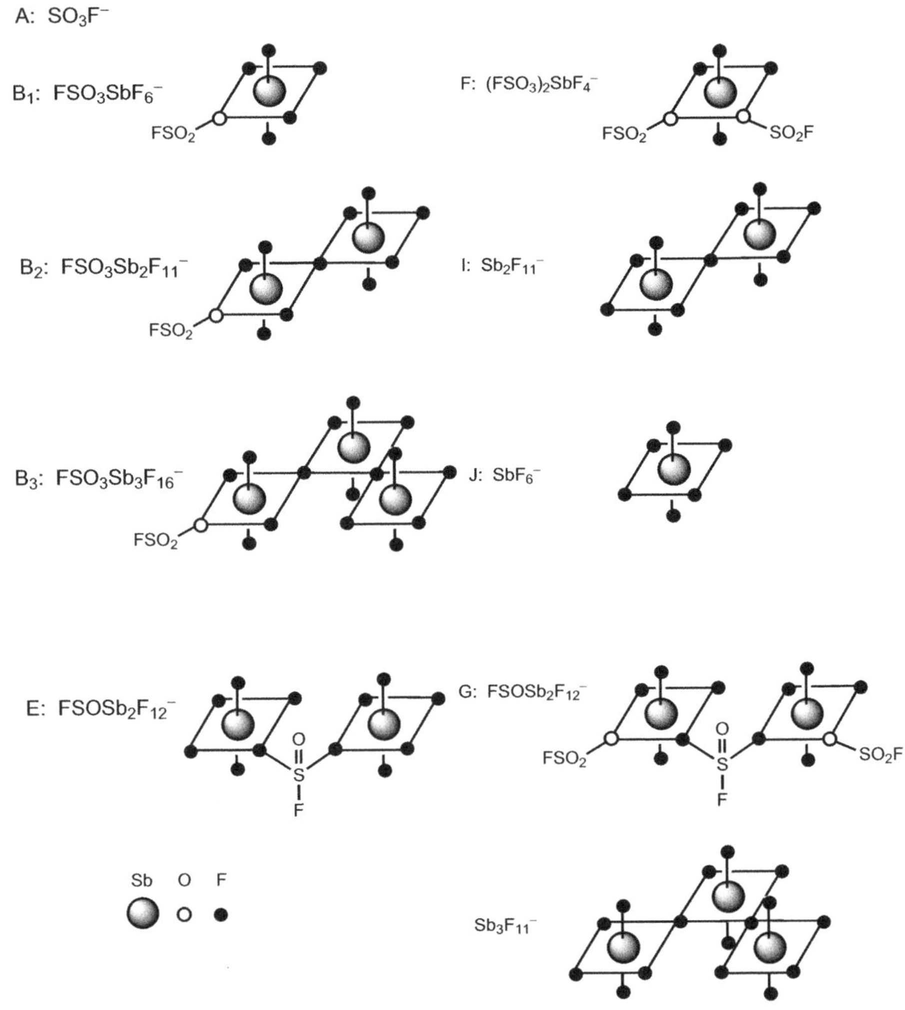

$$HSO_3F \; + \; HSO_3F{-}SbF_5 \rightleftharpoons H_2SO_3F^+ \; + \; FSO_3SbF_5^- \qquad (2.24)$$

An interesting point in this system is that HSO_3F is completely ionized at 55% SbF_5 indicating that the SbF_5 binds preferentially to the acid instead of increasing the size of the anions.

Aubke and co-workers[79–81] reinvestigated the "Magic Acid" system by modern 500 MHz ^1H and 471 MHz ^{19}F NMR methods varying the molar fractions of SbF_5 in HSO_3F from 0.01 to 0.49 in the absence of diluents such as SO_2 or SO_2ClF. Using NMR tubes fitted with Teflon lining, the formation of HF and H_2O via interaction with glass was avoided. In this concentration range, after keeping the samples for several months, only six anionic species were found besides SO_3F^-: B_1, F, J, I, E, G (Table 2.5). As suggested earlier, the main species at the 1:1 concentration is B_1.

The major reason for the wide application of this superacid system compared with others (besides its very high acidity) is probably the large temperature range in which it can be used. In the liquid state, NMR spectra have been recorded from temperatures as low as −160°C (acid diluted with SO_2F_2 and SO_2ClF)[82] and up to +80°C (neat acid in sealed NMR glass tube).[73] Glass is attacked by the acid very slowly, when moisture is excluded. The Magic Acid system can also be an oxidizing agent that results in reduction to antimony trifluoride and sulfur dioxide. On occasion, this represents a limitation.

2.2.2.3. *Fluorosulfuric Acid–Sulfur Trioxide.* Freezing point and conductivity measurements[12] show that SO_3 behaves as a nonelectrolyte in HSO_3F. Acidity measurements show a small increase in acidity that is attributed to the formation of fluoropolysulfuric acids HS_2O_6F and HS_3O_9F up to $HS_7O_{21}F$.[12] Evidence for the existence of these acids has been obtained by ^{19}F NMR measurements in SO_2ClF solutions at −100°C.[77] The acidity of these solutions reaches a maximum of −15.52 on the H_0 scale for 4 mol% SO_3 and does not increase any further[8] (Figure 2.9).

2.2.2.4. *HSO_3F–$MF_n(SO_3F)_{5-n}$; n = 3, 4, M = Nb, Ta.* Superacids based on the conjugation of HSO_3F with niobium and tantalum fluorosulfates acting as Lewis acids were also studied by Aubke and co-workers.[80] In contrast with NbF_5 and TaF_5, these fluorosulfate Lewis acids are soluble in HSO_3F over a large concentration range. Results obtained by UV spectroscopy of dissolved Hammett bases seem to indicate

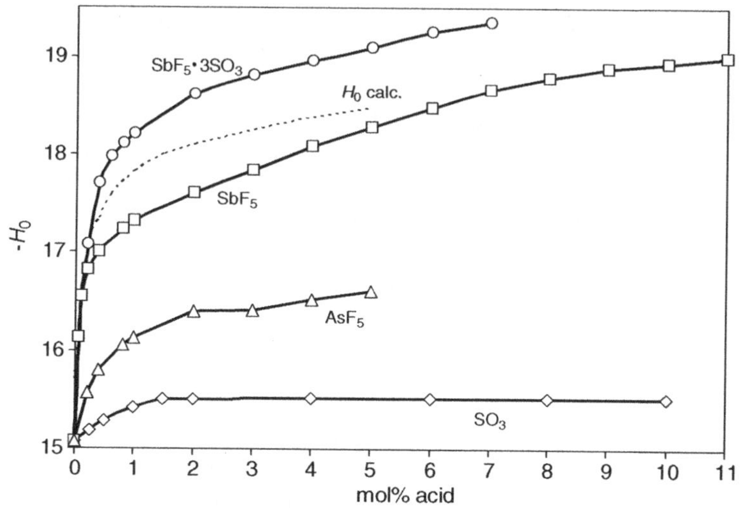

Figure 2.9. H_0 acidity function values of the HSO_3F solvent system.[8]

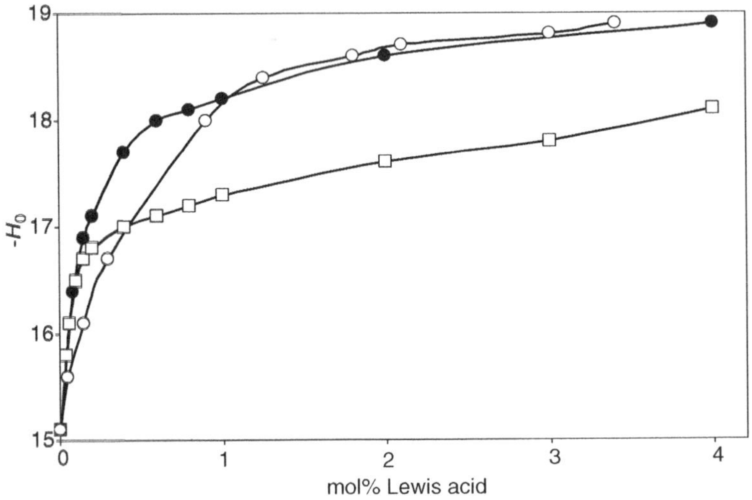

Figure 2.10. Hammett acidity of $Ta(SO_3F)_5$ (○), SbF_5 (□), and $SbF_2(SO_3F)_3$ (●) in HSO_3F at ambient temperature.[80]

that these systems are even more acidic than Magic Acids at low concentration (Figure 2.10).

2.2.2.5. Fluorosulfuric Acid–Arsenic Pentafluoride. AsF_5 ionizes in HSO_3F.[83] The $AsF_5SO_3F^-$ anion has the octahedral structure as shown for the antimony analog (B_1, Table 2.5). Values for the H_0 acidity function up to 4 mol% AsF_5 show a larger increase as compared with SO_3 (Figure 2.9)[8] but smaller when compared with SbF_5.[80,81]

2.2.2.6. Perfluoroalkanesulfonic Acid-Based Systems. ***$C_nF_{n+1}SO_3H$–SbF_5.*** CF_3SO_3H–SbF_5 ($n = 1$) was introduced by Olah as an effective superacid catalyst for isomerizations and alkylations. The composition and acidity of systems where $n = 1, 2, 4$ have been thoroughly studied by Commeyras and co-workers.[14] The ^{19}F NMR spectra are very similar for all of these systems and closely resemble those obtained with fluorosulfuric acid, as described in Section 2.2.2.6. The change in composition of the triflic acid–antimony pentafluoride system depending on the SbF_5 content has been studied. For the 1:1 composition, the main counteranion is $[CF_3SO_3SbF_5]^-$ and for the 1:2 composition $[CF_3SO_3(Sb_2F_{11})]^-$ is predominant, with the structures proposed for the anions shown in Figure 2.10. With increasing SbF_5 concentration, the anionic species grow larger and anions containing up to 5 SbF_5 units have been found (Figure 2.11). Under no circumstances was free SbF_5 detected.

It has not been possible to measure the acidity of the CF_3SO_3H–SbF_5 system by spectrophotometry because of its very strong absorption in the UV–visible spectrum. In comparison with the strength of the related perfluoroalkanesulfonic acids, its acidity should be very close or slightly higher (by 1 H_0 unit) than the acidity of the perfluoroethane- and perfluorobutanesulfonic acid–SbF_5 mixtures, which was

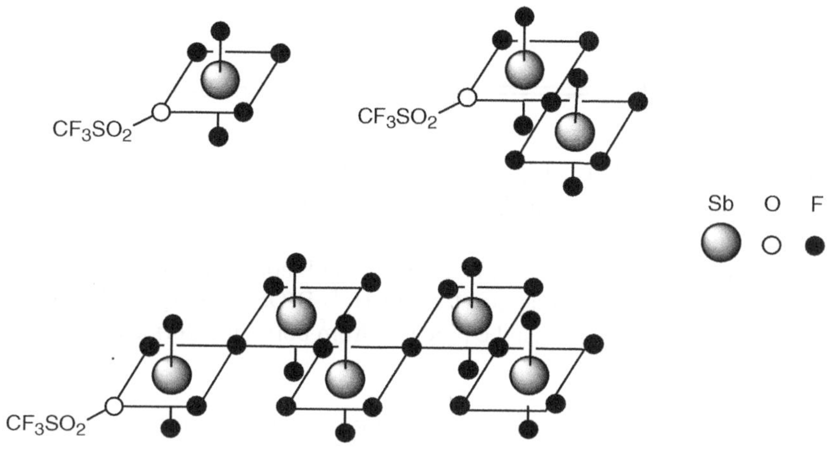

Figure 2.11. Components of the CF_3SO_3H–SbF_5 acid system.

measured by Commeyras and co-workers[14] (Figure 2.12). Because these authors were using the same spectroscopic technique as previously used by Gillespie, the measurements were limited to $H_0 = -18.5$ due to lack of suitable weak indicator bases. However, on the basis of ^{19}F NMR structural studies, a moderate increase in acidity is expected beyond the 50 mol% SbF_5 concentration corresponding to the autoprotolysis

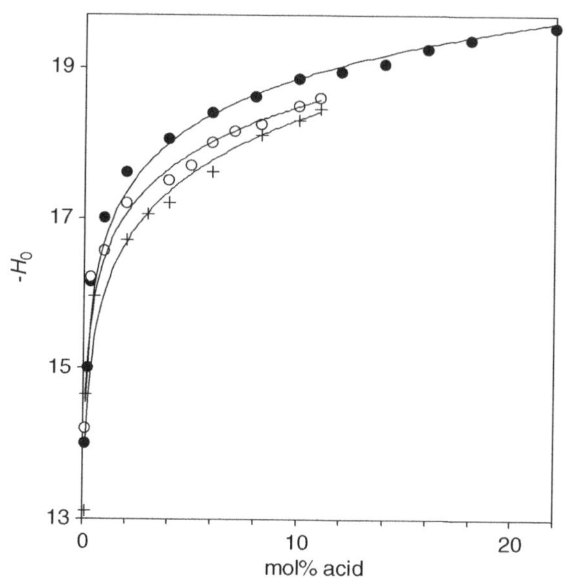

Figure 2.12. H_0 acidity function values for various perfluoroalkanesulfonic acid-based systems. (○) CF_3SO_3H–SbF_5 (ref. 14); (+) $C_4F_9SO_3H$–SbF_5 (ref. 14); (●) $CF_3SO_3H(CF_3SO_3)_3B$ (ref. 84).

equilibrium [Eq. (2.25)].

$$2\ HR_FSO_3SbF_5 \rightleftharpoons R_FSO_3(SbF_5)^- + H_2R_FSO_3SbF_5^+ \quad (2.25)$$

CF_3SO_3H–$B(OSO_2CF_3)_3$. The acidity of triflic acid can also be substantially increased by addition of boron triflate $B(OSO_2CF_3)_3$ as indicated by Engelbrecht and Tschager[84] (Figure 2.12). The increase in acidity is explained by the ionization equilibrium [Eq. (2.26)]. Consequently, the triflic acid–boron tris(triflate) (triflatoboric acid) system may also be formulated as $CF_3SO_3H_2^+$–$B(OSO_2CF_3)_4^-$ or $(CF_3SO_3H)_2$–$B(OSO_2CF_3)_3$. The acid may also be prepared from boron trichloride and triflic acid [Eq. (2.27)].

$$B(OSO_2CF_3)_3 + 2\ CF_3SO_3H \rightleftharpoons CF_3SO_3H_2^+ + B(OSO_2CF_3)_4^- \quad (2.26)$$

$$4\ CF_3SO_3H + BCl_3 \xrightarrow[-3\ HCl]{} CF_3SO_3H\text{–}B(OSO_2CF_3)_3 \quad (2.27)$$

Acidity measurements were again limited for the lack of suitable indicator base and even 1,3,5-trinitrobenzene, the weakest base used, was fully protonated ($H_0 \approx -18.5$) in the 22 mol% solution of boron triflate. Extrapolation of this system to 40% $B(CF_3SO_3)_3$ in CF_3SO_3H would lead to an H_0 value of -20 in the Hammett scale. Consequently, the acidity is comparable to that of the ternary system HSO_3F–SbF_5–SO_3.

2.2.2.7. *Hydrogen Fluoride–Antimony Pentafluoride (Fluoroantimonic Acid).* The HF–SbF_5 (fluoroantimonic acid) system is considered the strongest liquid superacid and also the one that has the widest acidity range. Due to the excellent solvent properties of hydrogen fluoride, HF–SbF_5 is used advantageously for a variety of catalytic and synthetic applications[85] (see Chapter 5). Anhydrous hydrogen fluoride is an excellent solvent for organic compounds with a wide liquid range. Antimony pentafluoride ionizes anhydrous HF and the proton is solvated by the medium according to [Eq. (2.28)]

$$2\ SbF_5 + 2\ HF \rightleftharpoons H_2F^+ + SbF_6^-\ (Sb_2F_{11}^-) \quad (2.28)$$

Due to the small autoprotolysis constant ($K_{ap} \sim 10$–12), the H_0 value of neat anhydrous HF was difficult to measure. Trace amount of impurities increased the H_0 value by several orders of magnitude, and for this reason the value of $H_0 = -11$ was generally found in the literature. In 1987, Gillespie and Liang[30] used a set of nitroaromatic indicators and found that the H_0 value of anhydrous HF was -15.1, in good agreement with the prediction of Devynck[86] based on the electrochemical method (see Section 1.4.3). Figure 2.13 shows the dramatic change in acidity of anhydrous HF when 1 mol base or 1 mol acid is added.

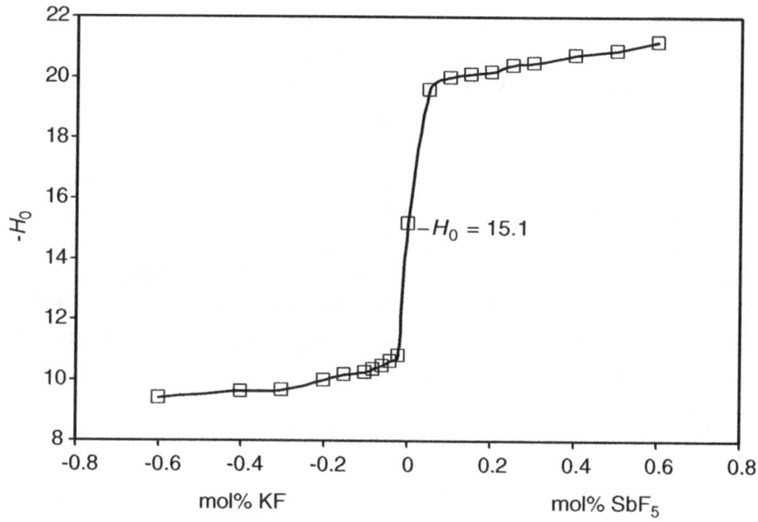

Figure 2.13. Comparison of calculated (full line) and experimental (squares) $-H_0$ values for the KF–HF–SbF system.[30]

Considering the fact that the H_0 values of neat anhydrous HF and pure HSO_3F are the same (~ -15.1), it is interesting to notice the much higher sensitivity of HF to SbF_5 addition: A $1\,M$ solution of HF–SbF_5 is about 104 times stronger acid than a $1\,M$ solution of SbF_5 in HSO_3F. In order to reach an H_0 value of -21, 25 times more SbF_5 has to be added to HSO_3F than to HF (Figure 2.14).

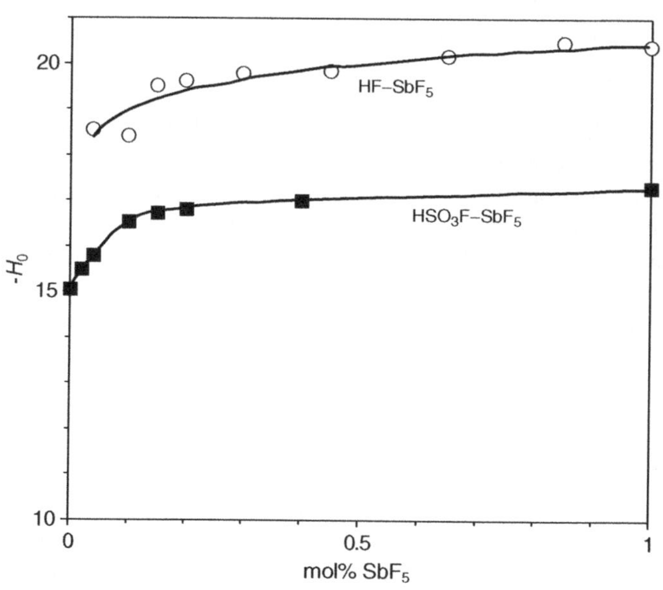

Figure 2.14. Comparison of H_0 acidity function values for HF–SbF_5 (ref. 74) and HSO_3F–SbF_5 (ref. 8).

Nevertheless, the acidity measurements based on the use of the benzhydryl cation indicator family[87] show that the upper acidity limit of the HF–SbF_5 system is reached with 10 mol% of SbF_5. The weakest indicator available is protonated 4,4′-dimethoxy-benzhydryl cation ($pK_{BH^+} \sim -23$) and could not be further protonated, even in the most concentrated HF–SbF_5 solutions.

Since the first quantitative study by Kilpatrick and Lewis,[88–90] the ionic composition of the strongest superacid system, HF–SbF_5, has been investigated by various analytical techniques. The results obtained from conductometric,[83] cryoscopy,[91] and vapor-phase measurements,[92] as well as infrared[93,94] and ^{19}F NMR spectroscopy,[87] all agree that SbF_5 is fully ionized in dilute HF solutions, first yielding the SbF_6^- anion. With increasing concentration of SbF_5, increasing amounts of polymeric SbF_5 and increasing amounts of polymeric anions ($Sb_2F_{11}^-$, $Sb_3F_{16}^-$, etc.) are formed.

On the basis of IR studies, Bonnet and Masherpa[95] concluded that HF-solvated $H_3F_2^+$ was the predominant species in the range 0–20 mol% SbF_5 followed by $H_3F_2^+$ in the 20–40 mol% range, replaced progressively by H_2F^+ above 40 mol% SbF_5. ^{19}F NMR studies show, however, that the proton exchange between the different cations $H_{n+1}F_n^+$ is very fast and it appears that the proton is always solvated to a maximum.

The HF–SbF_5 system has also been reinvestigated by high-field NMR spectroscopy[87] as well as by the theoretical methods such as *ab initio* molecular dynamics[96] and DFT theory.[97]

The anionic composition of the HF–SbF_5 system is less complex than the Magic Acid system because it comprises only $F^-(SbF_5)_n$ adducts with $1 < n < 4$. It is interesting to note that, in contrast with sulfonic acid based systems, SbF_5 prefers complexing fluoroantimonate ions instead of ionizing HF over the whole concentration range. In both HSO_3F- and CF_3SO_3H-based superacids, complete ionization of the acids was observed when the concentration of SbF_5 reached 50 mol%. In contrast, in the HF–SbF_5 system the ^{19}F NMR signal of the unionized HF can be observed up to 80 mol% SbF_5.

On the other hand, above 20 mol% SbF_5, a small but increasing amount of unionized SbF_5 can be observed, which may rationalize the change in the mechanism of alkane activation from the protolytic to the oxidative pathway, when the concentration of SbF_5 increases over 20 mol% (see Section 5.1.1).

The anionic composition of the HF-SbF_5 system as a function of SbF_5 concentration is shown in Figure 2.15.

The structure of the anions has already been suggested by Dean and Gillespie[98] in 1969 on the basis of 94.1 MHz ^{19}F NMR spectral analysis and the structure of $Sb_2F_{11}^-$ has been confirmed by X-ray structural analysis[99] (Figure 2.16).

The cationic species—that is, solvation of proton by HF—has also been the subject of various theoretical approaches in association with the development of the superacid field.

Kim and Klein[96,100] have investigated SbF_5 in liquid HF by *ab initio* molecular dynamics simulation. They observed the formation of SbF_6^- ion and a very fast diffusion of the proton along the hydrogen-bonded HF chains. The cationic species is the protonated HF chain, in good agreement with experimental results.[101]

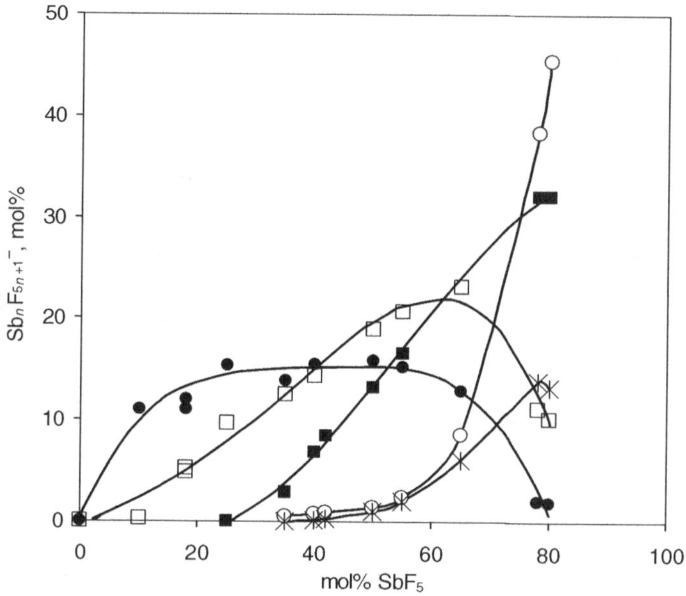

Figure 2.15. The anionic composition of the HF–SbF_5 system. (●) SbF_6^-; (□) $Sb_2F_{11}^-$; (■) $Sb_3F_{16}^-$; (*) $Sb_4F_{21}^-$; (○) SbF_5.[87]

Esteves et al.[97] have carried out a high-level [B3LYP/6-31++G**+RECP(Sb)] density functional study of the HF–SbF_5 system (Figure 2.17).

2.2.2.8. Hydrogen Fluoride–Phosphorus Pentafluoride. In the early 1960s, Clifford and Kongpricha[102] measured the thermodynamic solubility constant of PF_6^- in HF and concluded that in contrast with BF_3, PF_5 was a strong acid in HF. H_0 measurements in sulfolane solution showed HPF_6 to be a stronger acid than $HClO_4$

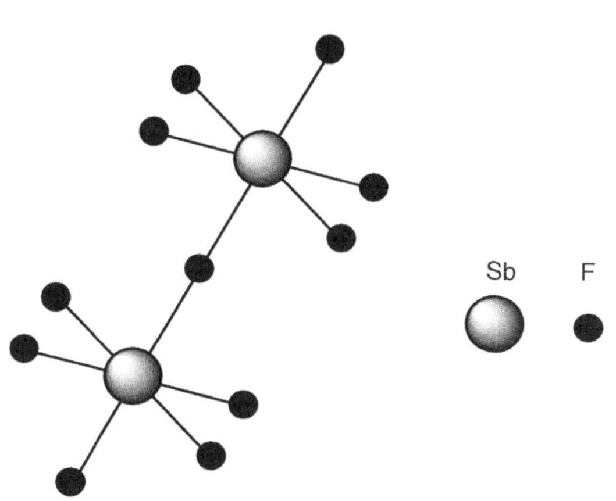

Figure 2.16. Structure of the $Sb_2F_{11}^-$ anion.

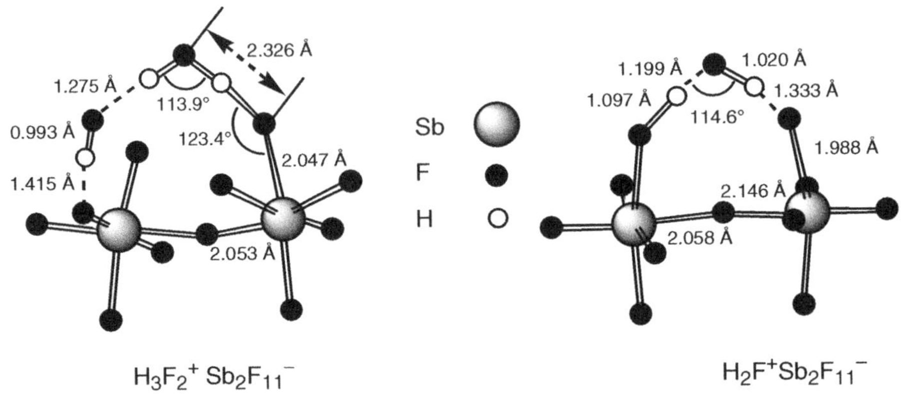

Figure 2.17. Geometrical parameters by DFT calculations of the $H_3F_2^+Sb_2F_{11}^-$ and $H_2F^+Sb_2F_{11}^-$ complex.

and HSO_3F.[103] However, the more recent theoretical calculations based on *ab initio* methods at the G-31 G** level seem to indicate[44] that the adduct HF–PF_5 is only weakly bound (note the long HF—F bond distance, Figure 2.18) and the PF_5 moiety is only weakly distorted and not prone to exchange its fluorine atom with HF.

2.2.2.9. *Hydrogen Fluoride–Tantalum Pentafluoride.* HF–TaF_5 is a catalyst for various hydrocarbon conversions of practical importance. In contrast to antimony pentafluoride, tantalum pentafluoride is stable in a reducing environment. The HF–TaF_5 superacid system has attracted attention mainly through the studies concerning alkane–alkene alkylation[104] and aromatic protonation.[105] Generally, heterogeneous mixtures such as the 10:1 and 30 : 1 HF–TaF_5 have been used because of the low solubility of TaF_5 in HF (0.9% at 19°C and 0.6% at 0°C). For this reason, acidity measurements have been limited to very dilute solutions, and an H_0 value of −18.85 has been found for the 0.6% solution. Both electrochemical studies[106,107] and aromatic protonation studies[108] indicate that the HF–TaF_5 system is a weaker superacid compared to HF–SbF_5.

2.2.2.10. *Hydrogen Fluoride–Boron Trifluoride (Tetrafluoroboric Acid).* Boron trifluoride ionizes in anhydrous HF [Eq. (2.29)]. The stoichiometric compound

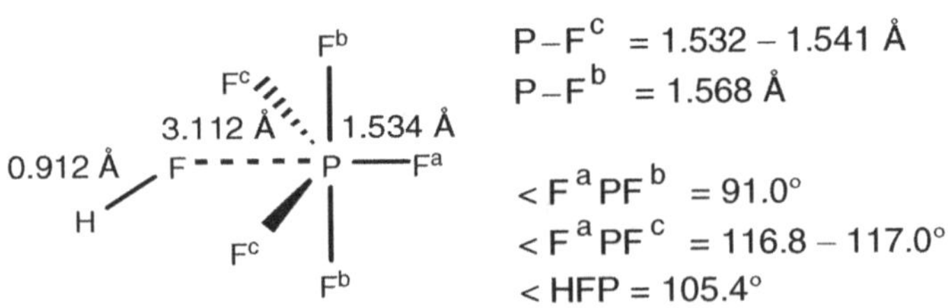

Figure 2.18. Calculated geometrical parameters of HF–PF_5.

exists only in excess of HF or in the presence of suitable proton acceptors. The HF–BF_3 (fluoroboric acid)-catalyzed reactions cover many of the Friedel–Crafts-type reactions.[109] One of the main advantages of this system is the high stability of HF and BF_3. Both are gases at room temperature and are easily recovered from the reaction mixtures.

$$BF_3 + 2\,HF \rightleftharpoons BF_4^- + H_2F^+ \qquad (2.29)$$

The large number of patents in this field demonstrates the industrial interest in this superacid system (such as isomerization of xylenes and carbonylation of toluene). The HF–BF_3 system in the presence of hydrogen has also been found to be an effective catalyst for ionic hydrodepolymerization of coal to liquid hydrocarbons.[110,111] The basicity of many aromatic hydrocarbons has been measured in the HF–BF_3 system by NMR[112] and vapor pressure determinations.[113] Acidity measurements of the system have been limited to electrochemical determinations, and a 7 mol% BF_3 solution was found to have an acidity of $H_0 = -16.6$. This indicates that BF_3 is a much weaker Lewis acid as compared with either SbF_5 or TaF_5. Nevertheless, the HF–BF_3 system is strong enough to protonate many weak bases and is an efficient and widely used catalyst.[114]

An *ab initio* study of $BF_3 + (HF)_n$ clusters[115] has shown that up to $n = 3$, only weakly bonded van de Waals associations are found but with $n = 4$–7, cyclic clusters were formed in which BF_3 is hydrogen bonded to HF with 3 fluorine atoms. Microwave studies[116] and IR investigations[117] show that the intermolecular BF bond is nevertheless somewhat shorter than expected on the basis of pure van de Waals interactions.

2.2.2.11. Conjugate Friedel–Crafts Acids (HX–AlX_3, etc.). The most widely used Friedel–Crafts catalyst systems are HCl–$AlCl_3$ and HBr–$AlBr_3$. These systems are indeed superacids by the present definition. However, experiments directed toward preparation from aluminum halides and hydrogen halides of the composition $HAlX_4$ were unsuccessful in providing evidence that such conjugate acids are not formed in the absence of proton acceptor bases.

The hydrogen chloride–aluminum chloride system has been investigated by Brown and Pearsall[118] under a variety of conditions, including temperatures as low as −120°C by vapor pressure measurements. No evidence was found for a combination of the two components.

The catalytic activity of hydrogen bromide itself has been recognized for alkylation, acylation, and polymerization reactions. The experimental difficulty associated with its narrow liquid range is probably the reason why there is no acidity measurements available for neat HBr. H_0 measurements on a 1 *M* solution in anhydrous sulfolane[103] have shown the following sequence of decreasing acidity: $HBF_4 > HClO_4 > HSO_3F > HBr > H_2SO_4 > HCl$. This sequence is in accord with conductometric measurements performed in glacial acetic acid by Gramstad.[25] Gold and co-workers[119,120] have shown that HBr in CF_2Br_2 is capable of protonating a variety of ketones and alcohols at sufficiently low temperatures (−90°C and below) at which the conjugate acids can be observed by NMR under slow-exchange conditions.

This indicates that the acidity of HBr is between -10 and -13 on the H_0 scale. The acidity can be increased further by addition of a Lewis acid such as TaF_5 or $AlBr_3$.[121,122] The relative acidity of the Lewis acids in HBr has been found to be in the order $AlBr_3 > GaBr_3 > TaF_5 > BBr_3 > BF_3$, a sequence deduced from selectivity parameter measurements in 2 *M* solution,[121] assuming an ideal behavior of the individual components. An advantage of HBr as a Brønsted acid is its nonoxidizing nature, but at the same time the bromide ion is a strong nucleophile and neat HBr is a very poor solvent for most of the organic substrates.

Aluminum bromide is sparingly soluble in HBr (1.77 g mol^{-1} at $-80°C$), but its solubility increases substantially in the presence of aromatic or aliphatic hydrocarbons. In this case, two phases separate—the upper layer is pure HBr and the lower layer consists of a sludge, for which the following composition has been found[108]: $(R^+Al_2Br_7^-)$–$AlBr_3$–HBr in the ratio 1 : 0.8 : 0.7. The commonly encountered sludges or "red oil" in Friedel–Crafts hydrocarbon conversion processes can thus be considered as a solution of carbocations in superacidic HBr–$AlBr_3$ and HCl–$AlCl_3$ systems. These systems are generally very complex, containing organic oligomers and alkylates. In the ethylation of benzene with the HCl–$AlCl_3$ system, the sludge was shown by NMR spectroscopy to contain the heptaethylbenzenium ion. Protonation studies based on ^{13}C NMR chemical-shift measurements have shown that the HBr–$AlBr_3$ system is capable of protonating benzene at 0°C. Comparing these results with those obtained using HF–TaF_5, Fărcaşiu et al.[108] claimed that HBr–$AlBr_3$ was an acid of comparable strength to HF–SbF_5, a ranking also proposed by Kramer based on his selectivity parameter.[121] It is, however, highly unlikely that HBr–$AlBr_3$ has strength comparable to that of HF–SbF_5 because it is incapable of protonating many weak bases or ionizing precursors to long-lived carbocations (such as alkyl cations). An extensive discussion of Friedel–Crafts superacids is available and will not be repeated here.[109]

2.3. TERNARY SUPERACIDS

2.3.1. HSO_3F–HF–SbF_5

When Magic Acid is prepared from fluorosulfuric acid not carefully distilled (which always contains about 2–5% of HF), upon addition of SbF_5 the ternary superacid system HSO_3F–HF–SbF_5 is formed.[123,124] Because HF is a weaker Brønsted acid, it ionizes fluorosulfuric acid, which, upon addition of SbF_5, results in a high-acidity superacid system at low SbF_5 concentrations. ^{19}F NMR studies on the system have indicated the presence of SbF_6^- and $Sb_2F_{11}^-$ anions, although these can result from the disproportionation of $SbF_5(FSO_3)^-$ and $Sb_2F_{11}(FSO_3)^-$ anions.

The ternary HSO_3F–HF–SbF_5 system was recognized as a highly efficient superacid catalyst in its own right by McCaulay in octane upgrading of light naphtha streams exhibiting improved selectivity and lifetime.[125]

2.3.2. HSO_3F–HF–CF_3SO_3H

As mentioned in Section 2.3.1, fluorosulfuric acid used in common laboratory practice always contains HF. A mixture of HSO_3F and CF_3SO_3H is, in fact, the ternary superacid HSO_3F–HF–CF_3SO_3H.

2.3.3. CF_3SO_3H–HF–Lewis Acid

Olah has applied liquid ternary catalysts comprising triflic acid, HF, and a Lewis acid (BF_3, PF_5, AsF_5, SbF_5, TaF_5, NbF_5) to upgrade natural gas liquids containing saturated straight-chain hydrocarbons to highly branched hydrocarbons (gasoline upgrading).[126]

2.3.4. HSO_3F–SbF_5–SO_3

When sulfur trioxide is added to a solution of SbF_5 in HSO_3F, there is a marked increase in conductivity that continues until approximately 3 moles of SO_3 have been added per mole of SbF_5.[127] This increase in conductivity has been attributed to an increase in $H_2SO_3F^+$ concentration arising from the formation of a much stronger acid than Magic Acid. Acidity measurements have confirmed the increase in acidity with SO_3–SbF_5 in the HSO_3F system (Figure 2.9). This has been attributed to the presence of a series of acids of the type $H[SbF_5(SO_3F)]$ (see Table 2.5, anion F), $H[SbF_3(SO_3F)_3]$, $H[SbF_2(SO_3F)_4]$ of increasing acidity. Thus, of all the fluorosulfuric acid-based superacid systems, sulfur trioxide-containing acid mixtures are, however, difficult to handle and cause extensive oxidative side reactions when contacted with organic compounds.

2.4. SOLID SUPERACIDS

Considering the exceptional activity of liquid superacids and their wide application in hydrocarbon chemistry, it is not surprising that work was also extended to solid superacids. The search for solid superacids has become an active area since the early 1970s, as reflected primarily by the existence of extensive patent literature.

Solid acid catalysts such as mixed oxides (chalcides) have been used extensively for many years in the petroleum industry and organic synthesis. Their main advantage compared with liquid acid catalysts is the ease of separation from the reaction mixture, which allows continuous operation, as well as regeneration and reutilization of the catalyst. Furthermore, the heterogeneous solid catalysts can lead to high selectivity or specific activity. Due to the heterogeneity of solid superacids, accurate acidity measurements are difficult to carry out and to interpret. Up until now, the most useful way to estimate the acidity of a solid catalyst is to test its catalytic activity in well-known acid-catalyzed reactions.

Solid acidic oxide catalysts generally do not show intrinsic acidity comparable with liquid superacids, and therefore generally high temperatures are required to achieve catalytic activity.

Various solid acids were qualified in the literature as superacids on the basis of very different arguments. The most studied "solid superacid," sulfated zirconia, and related sulfated oxides were considered as superacids because of their ability to convert *n*-butane into isobutane at low temperatures.

In the late 1990s, a series of reviews have been devoted to sulfated zirconia and its potential application in industry.[128–134]

Zeolites such as HZSM-5 were considered as superacids on the basis of the initial product distribution in accord with C–H and C–C bond protolysis when isoalkanes were reacted at 500°C (the Haag and Dessau mechanism).[135] The reactivity was assigned to superacidic sites in the zeolite framework.[136] The superacid character of other solid acids was claimed on the basis of Hammett indicator color change[137,138] or on the basis of UV spectrophotometric measurements.[139,140] In 2000, a special issue of Microporous and Mesoporous Materials[141] was devoted to the superacid-type hydrocarbon chemistry taking place on solid acids as suggested by the late Werner Haag.

As already pointed out earlier (see Section 1.4.8), a clear definition of solid superacidity is needed. On the other hand, for catalysts to be able to activate alkanes at low temperatures, such as sulfated zirconias and heteropoly acids, the redox properties should not be neglected in the activation step.[142,143]

Nevertheless, a large number of solid acids may be considered as solid superacids on the basis of their ability to convert saturated hydrocarbons at moderate to low temperatures.

Moreover, recently the application of zeolites in organic synthesis demonstrated their ability to achieve reactions necessitating usually superelectrophilic intermediates as in superacid media.[144] These results cannot be rationalized on the basis of the acidic character alone and needs the understanding of confinement effects[145] in which case the zeolite cage behaves like a nanosized reactor favoring the contact between reactant and specific acidic sites. The electrostatic effects may also be important.

The following subchapters cover various solid superacids, including perfluorinated sulfonic acid resins (Nafion resins). Furthermore, in the past, various attempts have been made to obtain solid superacids by either (a) enhancing the intrinsic acidity of a solid acid by treatment with a suitable co-acid or (b) physically or chemically binding a liquid superacid to an otherwise inert surface. We will briefly review some of these attempts because most of these catalysts rapidly lose activity and need to be regenerated.

2.4.1. Zeolitic Acids

Zeolites are a well-defined class of crystalline, naturally occurring or synthetic aluminosilicates. They have a three-dimensional structure arising from a framework of $[SiO_4]^{4-}$ and $[AlO_4]^{5-}$ tetrahedra linked by their corners of shared oxygens. The frameworks are generally very open and contain channels and cavities in which cations and water molecules are located. The alkali cations have a high degree of mobility and can be easily exchanged, and the water molecule can be lost and regained the origin of their chemical properties.

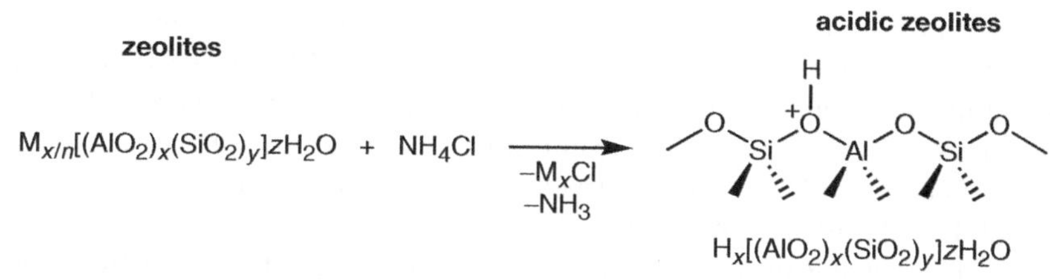

Figure 2.19. Synthesis and structure of acidic zeolites.

Whereas the synthesis of zeolite occurs in nature and in the laboratory under strongly basic conditions (pH 9–11), they are widely used as catalysts in hydrocarbon chemistry under their acidic form. In order to obtain acidic zeolites, the alkali cations (K, Li, Na, Ca, etc.) are first exchanged by $NH_4^+Cl^-$ followed by heating which, after release of ammonia, leaves the proton loosely attached to the framework on the Si–O–Al bridging group (Figure 2.19).

Quantifying precisely the acidity of zeolites or other acidic solids is a goal, which up to now has not been satisfactorily achieved. The main problem is the lack of an acceptable scale of solid acidity comparable to pK_a scale for aqueous solutions or proton affinities for gas-phase reactions. For this reason all available physical and theoretical methods of investigation have been applied over the years on this subject and a large number of papers have been published. Several reviews are available.

Whereas the number of Brønsted acid sites can be easily determined,[146–149] their acidity may vary depending on their position in the framework and also by interaction with Lewis acid sites. For these reasons and despite the various techniques tested, there are no general and reliable methods to measure the acidity of solid acids. The synthesis of active acidic zeolites used in industry still relies on a very empirical base, the most important character being their catalytic activity.

2.4.2. Polymeric Resin Sulfonic Acids

2.4.2.1. Lewis Acid-Complexed Sulfonic Acid Resins. In most of the acidic ion exchangers, the active sites are the sulfonic acid groups that are attached to a solid backbone such as sulfonated coal, sulfonated phenol–formaldehyde resins, or sulfonated styrene–divinylbenzene cross-linked polymers. The latter sulfonated resins are the only type acidic enough to gain importance as a catalyst for acid-catalyzed electrophilic reactions. The most widely used are the Dowex 50 (Dow Chemicals) and Amberlite IR-120 (Rohm and Haas)-type resins, which are made by sulfonation of cross-linked polystyrene–divinylbenzene beads. Dowex 50 is comparable in acid strength to HCl[150] and therefore is considered as a solid acid of moderate strength.

The acidity of these resins, however, increases significantly by treating them with Lewis acid halides. Gates and co-workers[151–153] have prepared a superacid catalyst from $AlCl_3$ and beads of macroporous, sulfonated polystyrene–divinylbenzene. The catalyst was prepared by exposing the macroporous beads of the polymeric Brønsted acid at 110°C to a stream of nitrogen containing sublimed $AlCl_3$. HCl was liberated and the resulting polymer had an S:Al:Cl ratio of 2:1:2. Electron microprobe X-ray analysis has shown that Al and Cl were uniformly dispersed throughout each polymer bead. By analogy with the known structure of liquid superacids, structure **1** was suggested [Eq. (2.30)]. This $AlCl_3$-complexed polystyrenesulfonic acid resin catalyst was found to be capable of isomerizing and cracking *n*-hexane at 85°C in a flow reactor with an overall initial conversion of 80%.

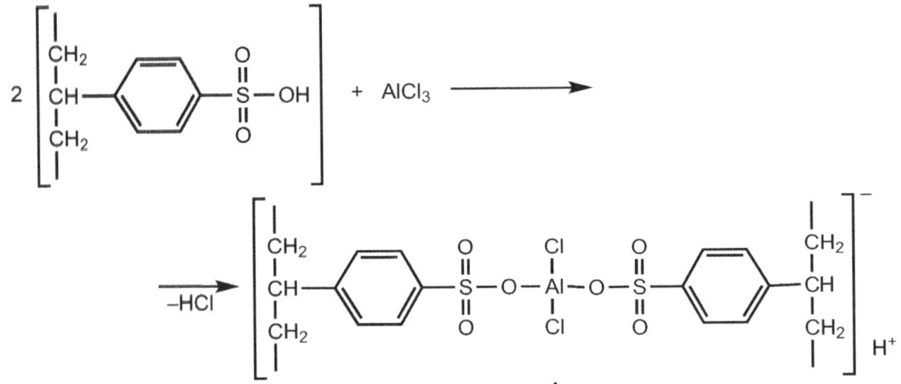

1

(2.30)

2.4.2.2. *Perfluorinated Polymer Resin Acids.* A convenient solid perfluorinated sulfonic acid can be readily generated from DuPont's commercially available potassium salts of Nafion brand ion membrane resins (a copolymer of perfluorinated epoxide and vinylsulfonic acid, Figure 2.20). Nafion-H-catalyzed reactions gained substantial interest and are reviewed in Chapter 5. Similarly, C_8 to C_{18} perfluorinated sulfonic acids, such as perfluorodecanesulfonic acid [$CF_3(CF_2)_9SO_3H$, PDSA], are very strong acids and are useful catalysts for many organic syntheses.[154]

Nafion membrane materials were originally developed for electrochemical applications.[155] The acid strength ($-H_0$) is estimated to be between 11 and 13[156]; that is, the acidity of Nafion is comparable to that of concentrated sulfuric acid. In these systems, the sulfonic acid group is attached to a CF_2 or CF group in a perfluorinated backbone.

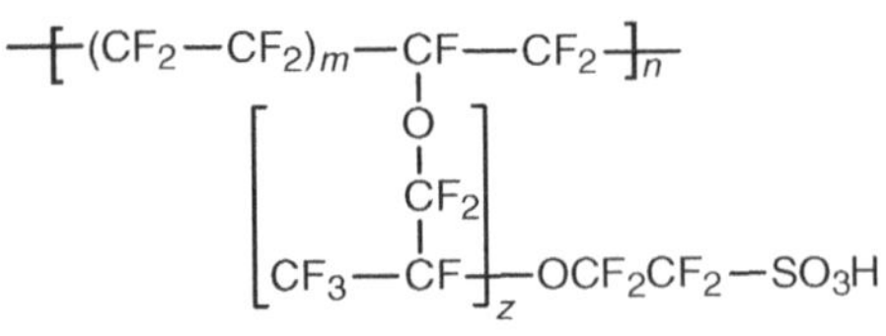

Figure 2.20. The structure of Nafion resin.

This structure provides high acidity sites, and at the same time the perfluorinated polymer itself is highly inert and resists acid cleavage. The maximum operating temperature in continuous operation is 175°C; however, Nafion can be used at even higher temperature in aqueous systems and proton-donating organic solvents.[157] Nafion is a noncorrosive solid marketed as the potassium salt in the form of granules (beads).

Other perfluorinated ionomer membranes, chemically very similar to Nafion, are also available commercially. Aciplex, manufactured by the Asahi Chemical Company, is very similar to Nafion, except that it has perfluoropropanesulfonic acid side chains. Flemion (Asahi Glass Company), in contrast, possesses perfluorobutanoic acid functions.

Two significant drawbacks of Nafion-H in catalytic applications are its very low surface area ($0.02\,m^2\,g^{-1}$) and the hindered accessibility of the active sites (sulfonic acid groups) located inside the pockets of the polymeric backbone. Consequently, the specific activity of Nafion, namely, the number of reacting molecules per unit weight, is extremely low.

To overcome these difficulties, Harmer and co-workers[158–160] have used the *in-situ* sol–gel technique to develop Nafion–silica nanocomposite solid acid catalysts. During the process, nanometer-size silica particles are formed, which are aggregated to produce a porous silica network entrapping the Nafion particles. The nanocomposites have large surface areas (typically $150–500\,m^2\,g^{-1}$)—that is, approximately four orders of magnitude larger than that of the starting Nafion polymer resin with Nafion particles having a typical particle size in the range 2–20 nm.[160,161] Synthesis conditions, particularly the choice of the silica source, allow us to tailor the microstructure, surface area, and acidity of the nanocomposites. Furthermore, Nafion loading can also be varied easily up to about 40 wt%. Another method, repeated impregnation of spinodal porous silica with aqueous methanol solution of Aciplex resin, has also been used to prepare Aciplex–SiO_2 samples.[162] Furthermore, Nafion was also embedded into ordered mesoporous MCM-41 silica material.[163,164]

The high acidity of the unique sulfonic acid function of Nafion can also be utilized by tethering perfluoroalkanesulfonic acid groups to the surface of various silicas. Two approaches have been described to prepare such hybrid organic–inorganic materials. The grafted materials **3** have been prepared by reacting preformed silica materials (MCM-41, SBA-15) with sultone **2** [Eq. (2.31)].[165] Co-condensation of tetraethoxysilane with **4** in the presence of dodecylamine template, in turn, furnished the HMS-based material **5** [Eq. (2.32)].[166,167]

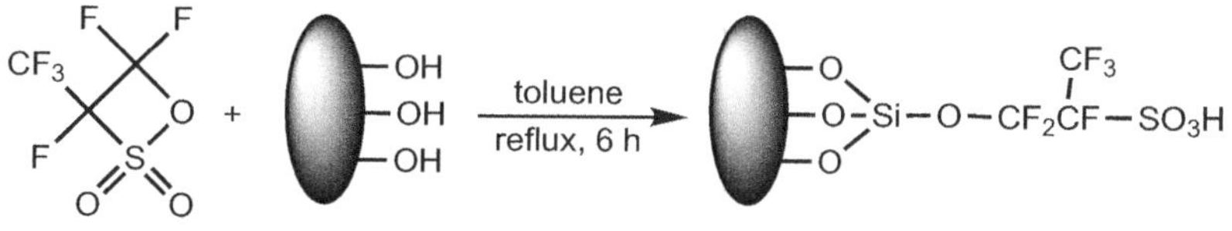

(2.31)

2 SBA-15 or MCM-41 **3**

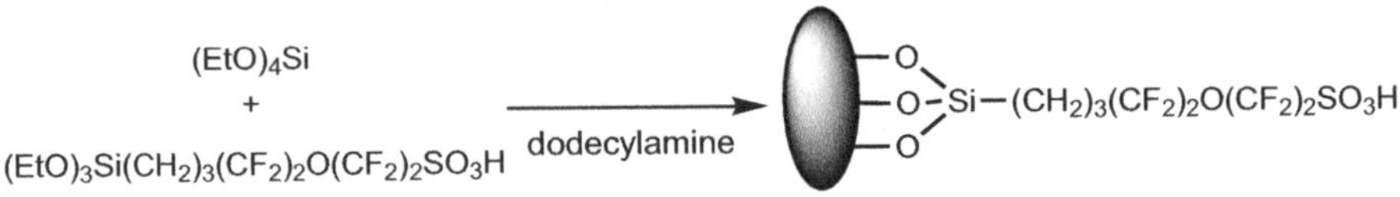

4 5

(2.32)

The new Nafion-nanocomposite catalysts are produced by DuPont and marketed as Nafion SAC materials with Nafion loading between 10% and 20%. Additional information for perfluorinated sulfonic acid resin nanocomposites including characterization by a variety of physical and chemical methods can be found in a recent review paper.[168]

2.4.3. Enhanced Acidity Solids

The acidic sites of solid acids may be of either the Brønsted (proton donor, often OH group) or Lewis type (electron acceptor). Both types have been identified by IR studies of solid surfaces using the pyridine adsorption method. The absorption band at 1460 cm^{-1} is assigned to pyridine coordinated with the Lewis acid site, and another absorption at 1540 cm^{-1} is attributed to the pyridinium ion resulting from the protonation of pyridine by the Brønsted acid sites. Various solids displaying acidic properties, whose acidities can be enhanced to the superacidity range, are listed in Table 2.6.

The natural clay minerals are hydrous aluminum silicates with iron or magnesium replacing aluminum wholly or in part, and with alkali or alkaline earth metals present as essential constituents in some others. Their acidic properties and natural abundance have favored their use as catalysts for cracking of heavy petroleum fractions. With the exception of zeolites and some specially treated mixed oxides for which superacid properties have been claimed, the acidity as measured by the color changes of absorbed Hammett bases is generally far below the superacidity range. They are inactive for alkane isomerization and cracking below 100 °C and need co-acids to reach superacidity.

2.4.3.1. Brønsted Acid-Modified Metal Oxides: TiO_2–H_2SO_4; ZrO_2–H_2SO_4. By exposing freshly prepared $Ti(OH)_4$ to 1 *N* sulfuric acid followed by calcination in air at 500°C for 3 h, Hino and Arata[137] have obtained a solid catalyst active for isomerization of butane in low yields at room temperature. The acid strength as indicated by the color change of a Hammett indicator (2,4-dinitrobenzene) was found to be as high as $H_0 = -14.5$.

Subsequently, the same authors[138] described the preparation of a solid superacid catalyst with acid strength of $H_0 = -16$ with a sulfuric acid-treated zirconium oxide. They exposed $Zr(OH)_4$ to 1 *N* sulfuric acid and calcined it in air at approximately 600°C. The obtained catalyst was able to isomerize (and crack) butane at room temperature. The acidity was examined by the color change method using Hammett indicators added to a powdered sample placed in sulfuryl chloride. The

Table 2.6. Solid Acids

1. Natural clay minerals:
   kaolinite, bentonite, attapulgite, montmorillonite, clarit, Fuller's earth, zeolites, and synthetic clays or zeolites
2. Metal oxides or sulfides:
   ZnO, CdO, Al_2O_3, CeO_2, ThO_2, ZrO_2, SnO_2, PbO, Aa_2O_3, Bi_2O_3, Sb_2O_5, V_2O_5, Cr_2O_3, MoO_3, WO_3, CdS, ZnS
3. Metal salts:
   $MgSO_4$, $CaSO_4$, $SrSO_4$, $BaSO_4$, $CuSO_4$, $ZnSO_4$, $CdSO_4$, $Al_2(SO_4)_3$, $FeSO_4$, $Fe_2(SO_4)_3$, $CoSO_4$, $NiSO_4$, $Cr_2(SO_4)_3$, $KHSO_4$, $(NH)_2(SO_4)_3$, $Zn(NiO_3)_2$, $Ca(NiO_3)_2$, K_2SO_4, Bi $(NiO_3)_3$, $Fe(NiO_3)_3$, $CaCO_3$, BPO_4, $AlPO_4$, $CrPO_4$, $FePO_4$, $Cu_3(PO_4)_2$, $Zn_3(PO_4)_2$, $Mg_3(PO_4)_2$, $Ti_3(PO_4)_2$, $Zr_3(PO_4)_2$, $Ni_3(PO_4)_2$, AgCl, CuCl, $CaCl_2$, $AlCl_3$, $TiCl_3$, $SnCl_2$, CaF_2, BaF_2, $AgClO_4$, $Mg(ClO_4)_2$
4. Mixed oxides:
   $SiO_2-Al_2O_3$, SiO_2-TiO_2, SiO_2-SnO_2, SiO_2-ZrO_2, SiO_2-BeO, SiO_2-MgO, SiO_2-CaO, SiO_2-SrO, SiO_2-ZnO, $SiO_2-Ga_2O_3$, $SiO_2-Y_2O_3$, $SiO_2-La_2O_3$, SiO_2-MOO_3, SiO_2-WO_3, $SiO_2-V_2O_5$, SiO_2-ThO_2, Al_2O_3-MgO, Al_2O_3-ZnO, Al_2O_3-CdO, $Al_2O_3-B_2O_3$, $Al_2O_3-ThO_2$, $Al_2O_3-TiO_2$, $Al_2O_3-ZrO_2$, $Al_2O_3-V_2O_5$, $Al_2O_3-MoO_3$, $Al_2O_3-WO_3$, $Al_2O_3-Cr_2O_3$, $Al_2O_3-Mn_2O_3$, $Al_2O_3-Fe_2O_3$, $Al_2O_3-Co_3O_4$, Al_2O_3-NiO, Al_2O_3-NiO, TiO_2-CuO, TiO_2-MgO, TiO_2-ZnO, TiO_2-CdO, TiO_2-ZrO_2, TiO_2-SnO_2, $TiO_2-Bi_2O_3$, $TiO_2-Sb_2O_3$, $TiO_2-V_2O_5$, $TiO_2-Cr_2O_3$, TiO_2-MnO_3, $TiO_2-Fe_2O_3$, $TiO_2-Co_3O_4$, TiO_2-NiO_2, ZrO_2-CdO, $ZnO-MgO$, $ZnO-FE_2O_3$, $MoO_3-CoO_3-Al_2O_3$, $MoO_3-NiO-Al_2O_3$, TiO_3-SiO_2-MgO, $MoO_3-Al_2O_3-MgO$
5. Cation exchange resins, polymeric perfluorinated resinsulfonic acids

existence of both Brønsted and Lewis sites was shown by the IR spectra of absorbed pyridine.

The X-ray photoelectron and IR spectra showed that the catalyst possessed bidentate sulfate ion coordinated to the metal. The specific surface areas were much larger than those of the zirconium oxides, which had not undergone the sulfate treatment. The interesting feature of these catalysts is the high temperature at which they are prepared, which means that they maintain their acidity at temperatures as high as 500°C and should thus be easy to regenerate and reuse.

Considering the high activity obtained after treatment of these oxides with H_2SO_4, it is rather surprising that a large variety of oxides and mixed oxides show almost no increase in activity after treatment with much stronger HSO_3F.[169] $SiO_2-Al_2O_3$ treated with Magic Acid was, however, found to be moderately active at room temperature in cracking butane, but its activity was much less than the Lewis acid-modified oxides (*vide infra*).

2.4.3.2. *Lewis Acid-Modified Metal Oxides and Mixed Oxides.* In 1976, Tanabe and Hattori[170] reported the preparation of solid superacids such as $SbF_5-TiO_2-SiO_2$, SbF_5-TiO_2, and $SbF_5-SiO_2-Al_2O_3$, whose acid strengths were in the range of -13.16 to -14.52 on the H_0 scale. In a subsequent work, the same authors reported a thorough study of the preparation and activity measurement of a large number of oxides and mixed oxides, treated with a variety of superacids. The SbF_5-treated oxides

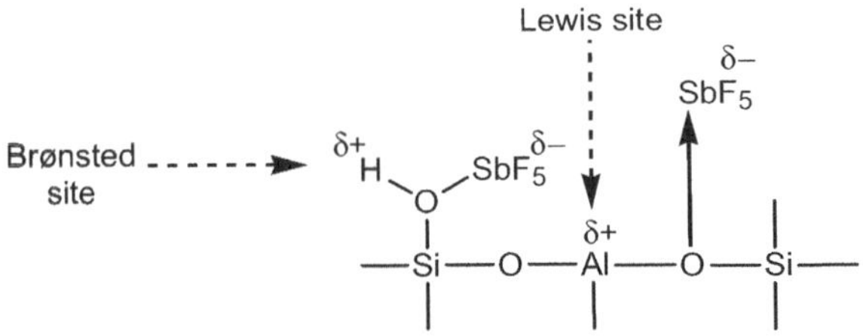

Figure 2.21. The structure of SbF_5-treated SiO_2–Al_2O_3.

such as TiO_2 and SiO_2, as well as mixed oxides like TiO_2–ZrO_2, TiO_2–SiO_2, SiO_2–Al_2O_3, were found to be most effective in the isomerization and cracking reactions of butane[170] and pentanes.[171] The catalysts were prepared by exposing the powdered metal oxides to SbF_5 vapor at room temperature followed by degassing. The adsorption–desorption cycle was repeated a number of times, and finally the catalyst was subjected to high vacuum to remove last traces of free SbF_5 at a given temperature. The amount of SbF_5 remaining on the catalyst was measured by the weight increase of the catalyst.

The highest activity for butane cracking at room temperature was obtained with the TiO_2 - and SiO_2-containing systems. For pentane and 2-methylpentane isomerizations at 0°C, the most active catalyst was SbF_5–TiO_2–ZrO_2 with a selectivity close to 100% for skeletal isomerization.

The high activity of these mixed oxides was ascribed to oxygen coordination of SbF_5 enhancing the acidity of both Brønsted and Lewis acid sites (Figure 2.21).

The activity and stability of aluminum chloride-treated alumina and silica–alumina as alkane isomerization catalysts has been investigated by Oelderik and Platteeuw.[172] The degree of hydration of the carrier and the effects of hydrogen chloride and hydrogen pressure on the carrier have been studied. The authors concluded[172] that $AlCl_3$ reacts with the hydrated carrier to give a surface-bound compound that is converted to an acidic site by HCl absorption. However, prolonged treatment of the carrier with excess $AlCl_3$ results in HCl gas evolution, until no acidic site is left on the catalyst. The activity of the catalyst system for hexane isomerization, however, declines exponentially with time.

Chlorinated alumina is still one of the most useful industrial catalyst for light alkane isomerization with generally a small amount of platinum being added in order to prevent coking.[173–177]

2.4.3.3. Lewis Acid-Complexed Metal Salts. Mixtures of aluminum chloride and metal chloride are known to be active for the isomerization of paraffins at room temperature.[178] Ono and co-workers[179–183] have shown that the mixtures of aluminum halides with metal sulfates are much more selective for similar reactions at room temperature.

$AlCl_3$–Metal Sulfates. Room temperature isomerization of pentane has been carried out with a series of mixtures containing aluminum chloride with sulfates of metals such as Ti, Fe(III), Ni, Cu, Mn, Fe(II), Al, Zr, Co, Mg, Ca, Ba, Pd, with conversions in excess of 10% after 3 h.[180] The most effective catalyst was an equimolar mixture of $AlCl_3$ and $Ti_2(SO_4)_3$ with a conversion of 46% and a selectivity to isopentane of 84%. The catalysts were prepared by kneading a mixture of aluminum chloride and a dehydrated metal salt in a porcelain mortar in a dry nitrogen atmosphere. Even better catalyst performance was found for $AlBr_3$–$Ti_2(SO_4)_3$ (87% conversion, >99% selectivity).[181]

The $AlCl_3$–$CuSO_4$ mixture has been more thoroughly investigated in the 5–23°C temperature range.[182] The acidity was estimated to be approximately $H_0 = -14$ and the activity was found to be proportional to the amount of $CuSO_4$ and also to the specific surface area of the $CuSO_4$ used. It was claimed that the addition of water had no effect on the catalytic activity, which seems to indicate that the active species are essentially different from those in the $AlCl_3$–H_2O system.

$AlCl_3$–Metal Chlorides. The catalytic activity of metal chloride–aluminum chloride mixtures for pentane isomerization has been studied by Ono.[179] The highest conversion (31% after 3 h at room temperature) was found when $AlCl_3$ was combined with $TiCl_3$ (the latter was prepared by reducing $TiCl_4$ with Al and supposed to have a composition of $TiCl_3$–1/3$AlCl_3$), $MnCl_2$ (12%), and $CuCl_2$ (11%), but no such catalyst reached the activity of metal sulfate–aluminum halide mixtures.

2.4.4. Immobilized Superacids (Bound to Inert Supports)

Ways have been found to immobilize and/or to bind superacidic catalysts to an otherwise inert solid support. Several types are described in this section.

2.4.4.1. Superacids Immobilized on Solid Supports. The considerable success of Magic Acid and related superacids in solution chemistry and interest to extend the scope and utility of acid-catalyzed reactions, particularly hydrocarbon transformations, logically led to the attempts to adopt this chemistry to solid systems allowing heterogeneous catalytic processes.

The acidity of perfluorinated sulfonic acids can be increased further by complexation with Lewis acid fluorides, such as SbF_5, TaF_5, and NbF_5.[183] They have been found to be effective catalysts for *n*-hexane, *n*-heptane isomerization, alkylation of benzene, and transalkylation of alkylbenzenes (see Chapter 5).

Attempts have recently been made to prepare solid acids by loading triflic acid into various inert oxides including silica,[184] titania,[185,186] and zirconia.[187,188] Silica functionalized with anchored aminopropyl groups was also used to immobilize triflic acid.[189] These new catalysts have been tested in a variety of organic transformations, such as alkane–alkene alkylation, Friedel–Crafts acylation, alkene dimerization, and acetalization. Silica nanoboxes prepared by dealumination of Na–X- and Ca–A-type zeolites were also loaded with triflic acid up to 32 wt%.[190] The materials were thoroughly characterized but have not been tested as catalysts.

Harmer et al.[29] have applied the sol–gel technique to prepare 1,1,2,2-tetrafluoroethanesulfonic acid supported on silica, which proved to be an excellent catalyst for several processes such as alkylation and acylation of aromatics, isomerization, oligomerization, and Fries rearrangement. This material has activity similar to that of triflic acid but is much easier to handle.

Immobilization of fluoroboric acid (HBF_4) on silica was also successfully performed,[191,192] and the resulting material has shown good characteristics in catalytic transformations.

2.4.4.2. Graphite-Intercalated Superacids. The 1970s witnessed a renewal of interest and a virtual explosion in research in the preparation, structure, and properties of graphite intercalation compounds. A number of reviews[193,194] have been devoted to this subject, showing that intercalation compounds can be useful materials as catalysts, reagents, electronic conductors, battery components, and lubricating agents. This research activity has continued ever since.[195–198]

Graphite possesses a layered structure that is highly anisotropic. It consists of sheets of sp^2 carbon atoms in hexagonal arrays with a C–C bond distance of 1.42 Å consistent with a one-third double bond and two-thirds single bond character (Figure 2.22). The distance between the layers is 3.35 Å and is in accord with the fact that the graphite sheets are held together only by weak van der Waals forces. The atomic layers are not directly superimposable but alternate in the pattern A–B–A–B– as shown in Figure 2.22 for the most stable form, called hexagonal graphite. Two special types of synthetic graphite are considered of great interest: highly oriented pyrolytic graphite (HOPG) and graphite fibers. HOPG is a highly crystalline material in which all the carbon layers are nearly parallel in the whole sample, which may be of sizeable dimension. The graphite fibers are another form of synthetic graphite that show unusual mechanical properties in the composite material (used extensively in aviation industry).

The intercalation process has been a much disputed topic, but there is now sufficient experimental proof that the intercalation step is generally a redox process in which ions

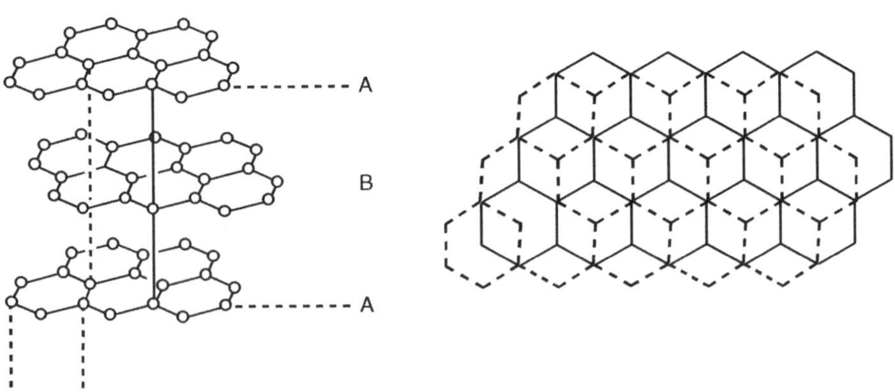

Figure 2.22. Graphite structure.

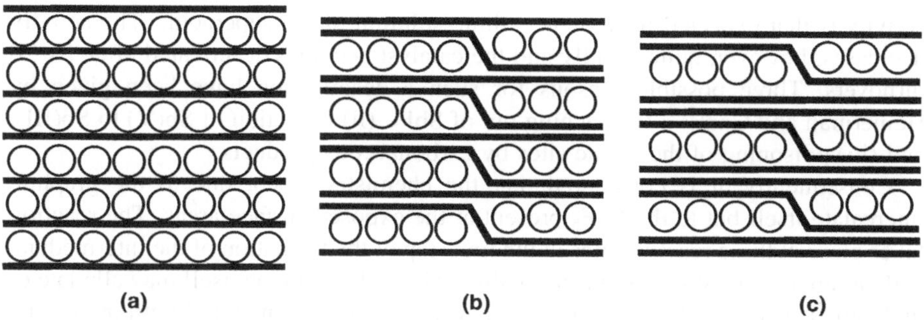

Figure 2.23. Staging in graphite intercalates: (**a**) first stage; (**b**) second stage; (**c**) third stage.

and neutral species enter the interlamellar space without disrupting the layered topology of graphite. X-ray studies show that the compounds at equilibrium favor the existence of a sequence of filled and empty interlamellar voids. This forms the basis for the concept of staging. In the first-stage (stage 1) compound, all the interlamellar regions are filled by the intercalant molecules (Figure 2.23). In the second- and third-stage compounds, each second or third interlamellar region, respectively, is filled by the intercalant and so on (Figure 2.23).

AsF_5 and SbF_5 Graphite Intercalates. Graphite reacts with AsF_5 (at 3 atm, 25°C) and SbF_5 to form the first-stage compounds C_8–AsF_5 and $C_{6.5}$–SbF_5, respectively. At higher temperatures, higher-stage compounds can be prepared. For the AsF_5 compound, a composition shown in Eq. (2.33) has been proposed.[199]

$$8\,C + AsF_5 \longrightarrow (C_8^{0.33+})(AsF_6^-)_{0.33}(AsF_5)_{0.5}(AsF_3)_{0.17} \qquad (2.33)$$

During the insertion process, the metal is partially reduced from M(V) to the M(III) oxidation state and at the same time the neutral species are inserted. The same assumptions hold good for the SbF_5 inserted graphite for which ^{19}F NMR,[201,202] ^{121}Sb Mossbauer,[202,203] and X-ray studies[204] have been conducted. Similar to the reactions of SbF_5 in liquid phase, an additional equilibrium is believed to take place in SbF_5–graphite intercalates [Eq. (2.34)], which is of significance in catalytic applications.

$$C_n^+\,MF_6^- + MF_5 \longrightarrow C_n^+\,(M_2F_{11})^- \qquad (2.34)$$

The catalytic properties of SbF_5–graphite have been investigated.[205–208] The chemistry is basically the same as that of SbF_5 itself with two major differences: (i) As a solid, it can be more easily separated from the reaction mixture, and (ii) the activity is that expected from a very dilute solution of the superacid. The activity decreases, however, quite rapidly and the catalyst becomes deactivated after 10–20 turnovers. Three possible reasons for such a deactivation have been given by Heinerman and Gaaf[209]: (i) leaching out of SbF_5, (ii) reduction of Sb(V) to Sb(III), and (iii) poisoning of the acidic sites by carbonaceous products.

Consequently, it is considered that the chemistry is not due to the graphite intercalate itself but to the SbF_5 present at the exposed surface areas. The reaction takes place at these reactive sites in the first step of the formation of the intermediate carbenium ion and SbF_5 is reduced to SbF_3. The carbenium ion itself may alkylate or undergo cleavage; and higher-molecular-weight polymeric material with increasing C:H ratio builds up on the catalytic site, thereby deactivating the catalyst slowly.[209] X-ray diffraction studies and elemental analysis show that large quantities of SbF_5 exist in the deactivated catalyst[208] and that the catalyst is not easily exfoliated (the lamellar structure is conserved and large regions of the first-stage compound can still be detected).

Miscellaneous Superacid Intercalates. The intercalation of $AlCl_3$ and $AlBr_3$ in graphite in the presence of chlorine or bromine has been long known.[210,211] The intercalates have been found to be milder catalysts than the pure Lewis acids.[212] However, it was shown that the Lewis acids are readily leached out from the intercalates during Friedel–Crafts-type reactions, and it is not clear whether the intercalates themselves or more probably the surface-exposed Lewis acids are the de facto catalysts. Other superacids such as NbF_5, TaF_5, HF–SbF_5, and $AlBr_3$–Br_2 have also been used as graphite intercalates for various superacid-catalyzed reactions such as reduction of alkyl halides,[213] alkylation of aromatics,[214] and isomerization and cracking of C_5, C_6 alkanes.[215]

2.4.4.3. SbF_5-Fluorinated Graphite, SbF_5-Fluorinated Al_2O_3. It has been shown that improved catalytic activity and stability is obtained with SbF_5-treated fluorinated graphites instead of graphite.[209] Fluorinated graphites are not uniform in structure and contain many imperfections and nonfluorinated areas. It has also been demonstrated that SbF_5 is not intercalated into higher fluorinated graphite with a F:C ratio of 1.1; similar catalytic activity can be obtained by reacting SbF_5 with fluorinated γ-alumina.[209] In less fluorinated samples it seems that SbF_5 is intercalated only in the nonfluorinated regions. The catalytic activity, however, is due to SbF_5 bound to the surface-exposed areas. The isomerization of pentane and hexane has been studied as a model reaction to assess the potential of SbF_5–graphite and SbF_5–fluorographite. A kinetic measurement of the deactivation process showed that nonfluorinated graphite–SbF_5 deactivated much more rapidly than fluorographite–SbF_5. It was suggested that in SbF_5–fluorographite, because of strong interaction with the carbon-bound fluorine and SbF_5, the catalyst is strongly immobilized. Better catalytic activity could be obtained by using fluorographite containing some graphite (with optimum at $CF_{0.8}$), and it has been assumed that the initially intercalated SbF_5 leaches out of the layer during the isomerization and binds to the surface of fluorinated graphite, thereby creating a new active site. The stability of the catalyst has been found comparable to SbF_5 bound to fluorinated γ-alumina. The characteristics of the hydroisomerization of pentane over SbF_5 on fluorinated graphite or fluorinated alumina, such as disproportionation reactions, influence of hydrogen pressure, stability, and deactivation mechanism are similar to those found with the homogeneous HF–SbF_5 liquid acid system.[209]

REFERENCES

1. N. F. Hall and J. B. Conant, *J. Am. Chem. Soc.* **49**, 3047 (1927).
2. J. C. Pernert, U. S. Patent 2,392,861 (1946).
3. H. M. Goodwin and E. C. Walker, III, *Chem. Metall. Eng.* **25**, 1093 (1921). *Trans. Am. Electrochem. Soc.* **40**, 157 (1921).
4. W. Mueller and P. Joenck, *Chem. Eng. Tech.* **35**, 78 (1963).
5. G. N. Dorofeenko, S. V. Krivuch, V. I. Dulenko, and Yu. A. Zhadanov, *Russ. Chem. Rev.* **34**, 88 (1965).
6. A. W. Williamson, *Proc. Royal. Soc. (London)* **7**, 11 (1854).
7. R. J. Gillespie, T. E. Peel, and E. A. Robinson, *J. Am. Chem. Soc.* **93**, 5083 (1971).
8. R. J. Gillespie and T. E. Peel, *J. Am. Chem. Soc.* **95**, 5173 (1973).
9. R. J. Cremlyn, *Chlorosulfonic Acid A Versatile Reagent*, Royal Society of Chemistry, Cambridge, (2002).
10. T. E. Thorpe and W. Kirman, *J. Chem. Soc., Trans.* **61**, 921 (1892).
11. G. A. Olah, G. K. S. Prakash, Q. Wang, and X.-Y. Li, in *Encyclopedia of Reagents for Organic Synthesis*, Vol. 4, L.A. Paquett, ed., Wiley, Chichester, 1995, p. 2567.
12. R. J. Gillespie and E. A. Robinson, *Can. J. Chem.* **40**, 675 (1967).
13. R. Savoie and P. A. Giguère, *Can. J. Chem.* **42**, 277 (1964).
14. D. Brunel, A. Germain, and A. Commeyras, *Nouv. J. Chim.* **2**, 275 (1978).
15. R. C. Paul, K. K. Paul, and K. C. Malhotra, *J. Inorg. Nucl. Chem.* **31**, 2614 (1969).
16. T. Gramstad and R. N. Haszeldine, *J. Chem. Soc.* 173 (1956).
17. J. Grondin, R. Sagnes, and A. Commeyras, *Bull. Soc. Chim. Fr.* 1779 (1976).
18. R. D. Howells and J. D. McCown, *Chem. Rev.* **77**, 69 (1977).
19. P. J. Stang and M. R. White, *Aldrichimica Acta* **16**, 15 (1983).
20. T. J. Brice and P. W. Trott, U. S. Patent 2,732,398 (1956).
21. R. N. Haszeldine and J. M. Kidd, *J. Chem. Soc.* 2901 (1955).
22. G. A. Olah and G. K. S. Prakash, unpublished results.
23. I. Jouve (Rhodia Organics), personal communication.
24. C. H. Dungan and J. R. van Wazer, Compilation of Reported ^{19}F NMR Chemical Shifts, 1951 to mid-1967, Wiley-Interscience, New York, 1970.
25. T. Gramstad, *Tidsskr. Kjemi. Bergves. Metall.* **19**, 62 (1959), *Chem. Abstr.* **54**, 66368 (1960).
26. R. B. Ward, U. S. Patent 3,346,606 (1967).

27. R. A. Guenthner and M. L. Vietor, *Ind. Eng. Chem. Prod. Res. Div.* **1**, 165 (1962).
28. M. A. Harmer, C. P. Junk, Z. Schnepp, U. S. Patent application (2005).
29. M. A. Harmer, C. Junk, V. Rostovtsev, L. G. Carcani, J. Vickery, and Z. Schnepp, *Green Chem.* **9**, 30 (2007).
30. R. J. Gillespie and J. Liang, *J. Am. Chem. Soc.* **110**, 6053 (1988).
31. G. A. Olah, J. T. Welch, Y. D. Vankar, M. Nojima, I. Kerekes, and J. A. Olah, *J. Org. Chem.* **44**, 3872 (1979).
32. G. A. Olah and X.-Y. Li, *Synlett.* 267 (1990).
33. I. Bucsi, B. Török, A. I. Marco, G. Rasul, G. K. S. Prakash, and G. A. Olah, *J. Am. Chem. Soc.* **124**, 7728 (2002).
34. C. A. Reed, *Acc. Chem. Res.* **31**, 133 (1998). *Chem. Commun.* 1669 (2005).
35. C. A. Reed, K.-C. Kim, R. D. Bolskar, and L. J. Mueller, *Science* **289**, 101 (2000).
36. C. A. Reed, K.-C. Kim, E. S. Stoyanov, D. Stasko, F. S. Tham, L. J. Mueller, and P. D. W. Boyd, *J. Am. Chem. Soc.* **125**, 1796 (2003).
37. A. Franken, B. T. King, J. Rudolph, P. Rao, B. C. Noll, and J. Michl, *Coll. Czech. Chem. Commun.* **66**, 1238 (2001).
38. C. J. Hoffman, B. E. Holder, and W. L. Jolly, *J. Phys. Chem.* **62**, 364 (1958).
39. S. Raugei and M. L. Klein, *J. Chem. Phys.* **116**, 7087 (2002).
40. E. E. Aynsley, R. D. Peacock, and P. L. Robinson, *Chem. Ind.* 1117 (1951).
41. R. D. Peacock and I. L. Wilson, *J. Chem. Soc. A*, 2030 (1969).
42. O. Ruff and W. Plato, *Ber. Dtsch. Chem. Ges.* **37**, 673 (1904).
43. E. R. Falardeau, G. M. T. Foley, C. Zeller, and F. L. Vogel, *J. Chem. Soc., Chem. Commun.* 389 (1977).
44. A. F. Janzen, X. Ou, and M. G. Sowa, *J. Fluorine Chem.* **83**, 27 (1997).
45. K. O. Christe, D. A. Dixon, D. McLemore, W. W. Wilson, J. A. Sheehy, and J. A. Boatz, *J. Fluorine Chem.* **101**, 151 (2000).
46. F. Fairbrother, *The Chemistry of Niobium and Tantalum*, Elsevier, London, 1967.
47. D. Brown, in *Comprehensive Inorganic Chemistry*, Vol 3, J. C. Bailar, Jr., H. J. Emeléus, R. Nyholm, and A.F. Trotman-Dickenson, eds., Pergamon Press, Compendium Publishers, Elmsford, New York, 1973, Chapter 35, p. 553.
48. N. N. Greenwood and B. S. Thomas, in *Comprehensive Inorganic Chemistry*, Vol. 1, J. C. Bailar, Jr., H. J. Emeléus, R. Nyholm, and A. F. Trotman-Dickenson, eds., Pergamon Press, Oxford, 1973, Chapter 7, p. 957
49. G. A. Olah and Á. Molnár, *Hydrocarbon Chemistry*, 2nd ed., Wiley-Interscience, Hoboken, NJ, 2003.
50. C. H. Rochester, *Acidity Functions*, Academic Press, London, New York, 1970, p. 52.
51. D. Fărcaşiu and G. Ghenciu, *J. Catal.* **134**, 126 (1992).
52. G. K. S. Prakash, T. Mathew, E. R. Marinez, P. M. Esteves, G. Rasul, and G. A. Olah, *J. Org. Chem.* **71**, 3952 (2006).
53. A. G. Massey, A. J. Park, and F. G. A. Stone, *Proc. Chem. Soc.* 212 (1963).
54. A. G. Massey and A. J. Park, *J. Organomet. Chem.* **2**, 245 (1964), *Organomet. Synth.* **3**, 461 (1986).

55. G. Erker, *Dalton Trans.* 1883 (2005).
56. W. E. Piers and T. Chivers, *Chem. Soc. Rev.* **26**, 345 (1997).
57. R. F. Childs, D. L. Mulholland, and A. Nixon, *Can. J. Chem.* **60**, 801, 809 (1982).
58. S. Döring, G. Erker, R. Fröhlich, O. Meyer, and K. Bergander, *Organometallics.* **17**, 2183 (1998).
59. M. Finze, E. Bernhardt, A. Terheiden, M. Berkei, H. Willner, D. Christen, H. Oberhammer, and F. Aubke, *J. Am. Chem. Soc.* **124**, 15388 (2002).
60. E. Bernhardt, G. Henkel, H. Willner, G. Pawelke, and H. Bürger, *Chem. Eur. J.* **8**, 11 (2002).
61. G. A. Olah, K. Laali, and O. Farooq, *J. Org. Chem.* **49**, 4591 (1984).
62. M. E. Vol'pin, S. Akhrem, and A. V. Orlinkov, *New J. Chem.* **13**, 771 (1989).
63. I. Akhrem, A. Orlinkov, and M. Vol'pin, *J. Chem. Soc., Chem. Commun.* 671 (1993).
64. I. S. Akhrem and A. V. Orlinkov, *Russ. Chem. Bull. (Engl. Transl.)* **47**, 740 (1998); *Chem. Rev.*, **107**, 2037 (2007).
65. G. A. Olah, G. Rasul, A. K. Yudin, A. Burrichter, G. K. S. Prakash, A. L. Chystiakov, I. V. Stankevich, I. S. Akhrem, N. P. Gambaryan, and M. E. Vol'pin, *J. Am. Chem. Soc.* **118**, 1446 (1996).
66. H. Vančik, K. Percač, and D. E. Sunko, *J. Am. Chem. Soc.* **112**, 7418 (1990).
67. J.-C. Culmann, M. Simon, and J. Sommer, *J. Chem. Soc., Chem. Commun.* 1098 (1990).
68. G. A. Olah, O. Farooq, A. Husain, N. Ding, N. J. Trivedi, and J. A. Olah, *Catal. Lett.* **10**, 239 (1991).
69. R. J. Gillespie and S. Wasif, *J. Chem. Soc.* 215 (1953).
70. F. D. Miles, H. Niblock, and G. L. Wilson, *Trans. Farad. Soc.* **35**, 345 (1940).
71. G. N. Lewis and J. Bigeleisen, *J. Am. Chem. Soc.* **65**, 1144 (1943).
72. J. C. D. Brand, W. C. Horning, and M. B. Thornley, *J. Chem. Soc.* 1374 (1952).
73. V. Gold, K. Laali, K. P. Morris, and L. Z. Zdunek, *J. Chem. Soc., Chem. Commun.* 769 (1981).
74. J. Sommer, P. Canivet, S. Schwartz, and P. Rimmelin, *Nouv. J. Chim.* **5**, 45 (1981).
75. V. Gold, K. Laali, K. P. Morris, and L. Z. Zdunek, *J. Chem. Soc., Perkin Trans 2*, 865 (1985).
76. D. Touiti, R. Jost, and J. Sommer, *J. Chem. Soc., Perkin Trans 2*, 1793 (1986).
77. P. A. W. Dean and R. J. Gillespie, *J. Am. Chem. Soc.* **92**, 2362 (1970).
78. A. Commeyras and G. A. Olah, *J. Am. Chem. Soc.* **91**, 2929 (1969).
79. D. Zhang, M. Heubes, G. Hägele, and F. Aubke, *Can. J. Chem.* **77**, 1869 (1999).
80. W. V. Cicha, D. Zhang, and F. Aubke, *J. Fluorine Chem.* **89**, 117 (1998).
81. J. Kühn-Velten, M. Bodenbinder, R. Bröchler, G. Hägele, and F. Aubke, *Can. J. Chem.* **80**, 1265 (2002).
82. G. A. Olah, G. K. S. Prakash, M. Arvanaghi, and F. A. L. Anet, *J. Am. Chem. Soc.* **104**, 7105 (1982).
83. R. J. Gillespie and T. E. Peel, *Adv. Phys. Org. Chem.* **9**, 1 (1971).
84. A. Engelbrecht and E. Tschager, *Z. Anorg. Allg. Chem.* **433**, 19 (1977).
85. G. A. Olah, G. K. S. Prakash, Q. Wang, and X.-Y. Li, in *Encyclopedia of Reagents for Organic Synthesis*, Vol. 4, L.A. Paquett, ed., Wiley, Chichester, 1995, p. 2715.

86. J. Devynck, P. L. Fabre, B. Trémillon, and A. Ben Hadid, *J. Electroanal. Chem.* **91**, 93 (1978).
87. J.-C. Culmann, M. Fauconet, R. Jost, and J. Sommer, *New J. Chem.* **23**, 863 (1999).
88. M. Kilpatrick and T. J. Lewis, *J. Am. Chem. Soc.* **78**, 5186 (1956).
89. H. H. Hyman, T. J. Lane, and T. A. O'Donnell, in *Abstracts of the 145th National Meeting of the American Chemical Society*, American Chemical Society, Washington D. C., 1963, p. 63T.
90. H. H. Hyman, L. A. Quarterman, M. Kilpatrick, and J. J. Katz, *J. Phys. Chem.* **65**, 123 (1961).
91. P. A. W. Dean, R. J. Gillespie, R. Hulme, and D. A. Humphreys, *J. Chem. Soc. A*, 341 (1971).
92. K. D. Abney, G. R. Ball, and P. G. Eller, *J. Fluorine Chem.* **51**, 165 (1991).
93. M. Couzi, J.-C. Cornut, and P. V. Huong, *J. Chem. Phys.* **56**, 426 (1972).
94. B. Bonnet, C. Belin, J. Potier, and G. Masherpa, *C.R. Séances Acad. Sci.* **C281**, 1011 (1975).
95. B. Bonnet and G. Mascherpa, *Inorg. Chem.* **19**, 785 (1980).
96. D. Kim and M. L. Klein, *J. Phys. Chem. B.* **104**, 10074 (2000).
97. P. M. Esteves, A. Ramírez-Solís, and C. J. A. Mota, *J. Am. Chem. Soc.* **124**, 2672 (2002).
98. P. A. W. Dean and R. J. Gillespie, *J. Am. Chem. Soc.* **91**, 7264 (1969).
99. D. Mootz and K. Bartmann, *Angew. Chem. Int. Ed. Engl.* **27**, 391 (1988). *Z. Naturforsch.* **46B**, 1659 (1991).
100. D. Kim and M. L. Klein, *J. Am. Chem. Soc.* **121**, 11251 (1999).
101. P. Ahlberg, A. Karlsson, A. Goeppert, S. O. N. Lill, P. Dinér, and J. Sommer, *Chem. Eur. J.* **7**, 1936 (2001).
102. A. F. Clifford and S. Kongpricha, *J. Inorg. Nucl. Chem.* **20**, 147 (1961).
103. R. W. Alder, G. R. Chalkley, and M. C. Whiting, *Chem. Commun.*, 405 (1966).
104. J. Sommer, M. Muller, and K. Laali, *New J. Chem.* **6**, 3 (1982).
105. D. Fărcaşiu, *Acc. Chem. Res.* **15**, 46 (1982) and references cited therein.
106. R. Gut and K. Gautschi, *J. Inorg. Nucl. Chem., H.H. Hyman Mem. Supp.* 95 (1976).
107. J. Devynck, A. Ben Hadid, and P.-L. Fabre, *J. Inorg. Nucl. Chem.* **41**, 1159 (1979).
108. D. Fărcaşiu, S. L. Fisk, M. T. Melchior, and K. D. Rose, *J. Org. Chem.* **47**, 453 (1982).
109. For reviews see G.A. Olah, *Friedel–Crafts and Related Reactions*, Vols 1–4, Wiley, New York, 1963–1965, and G. A. Olah, *Friedel-Crafts Chemistry*, Wiley, New York, 1973.
110. G. A. Olah, M. R. Bruce, E. H. Edelson, and A. Husain, *Fuel* **63**, 1130 (1984).
111. K. Shimizu, I. Saito, H. Kawashima, and S. Sasaki, *Energy Fuels* **13**, 197 (1999).
112. C. MacLean and E. L. Mackor, *Disc. Farad. Soc.* **34**, 165 (1962).
113. E. L. Mackor, A. Hofstra, and J. H. van der Waals, *Trans. Farad. Soc.* **54**, 186 (1958).
114. P. M. Brouwer, E. L. Mackor, and C. MacLean, *Recl. Trav. Chim. Pays-Bas.* **84**, 1564 (1965).
115. D. Kim and M. L. Klein, *Chem. Phys. Lett.* **308**, 235 (1999).

116. J. A. Phillips, M. Canagaratna, H. Goodfriend, A. Grushow, J. Almlöf, and K. R. Leopold, *J. Am. Chem. Soc.* **117**, 12549 (1995).
117. K. Nauta, R. E. Miller, G. T. Fraser, and W. J. Lafferty, *Chem. Phys. Lett.* **322**, 401 (2000).
118. H. C. Brown and H. W. Pearsall, *J. Am. Chem. Soc.* **73**, 4681 (1951).
119. J. Emsley, V. Gold, M. J. B. Jais, and L. Z. Zdunek, *J. Chem. Soc. Perkin Trans 2*, 881 (1982).
120. J. Emsley, V. Gold, and M. J. B. Jais, *J. Chem. Soc., Chem. Commun.*, 961 (1979).
121. G. M. Kramer, *J. Org. Chem.* **40**, 298, 302 (1975).
122. G. M. Kramer and G. B. McVicker, *Acc. Chem. Res.* **19**, 78 (1986).
123. G. A. Olah, O. Farooq, A. Husain, N. Ding, N. J. Trivedi, and J. A. Olah, *Catal. Lett.* **10**, 239 (1991).
124. G. A. Olah, G. K. S. Prakash, Q. Wang, and X.-Y. Li, in *Encyclopedia of Reagents for Organic Synthesis*, **Vol. 4**, L.A. Paquett ed., Wiley, Chichester, 1995, p. 2569.
125. D. A. McCaulay, U. S. Patent 4,144,282 (1979).
126. G. A. Olah, U. S. Patent 4,472,268 (1984).
127. R. C. Thompson, J. Barr, R. J. Gillespie, J. B. Milne, and R. A. Rothenbury, *Inorg. Chem.* **4**, 1641 (1965).
128. X. Song and A. Sayari, *Catal. Rev.–Sci. Eng.* **38**, 329 (1996).
129. K. Arata, *Appl. Catal., A.* **146**, 3 (1996).
130. T.-K. Cheung, and B. C. Gates, *Topics Catal.* **6**, 41 (1998).
131. V. Adeeva, H.-Y. Liu, B.-Q. Xu, and W. M. H. Sachtler, *Topics Catal.* **6**, 61 (1998).
132. D. Fărcaşiu and J. Q. Li, *Appl. Catal., A.* **175**, 1 (1998).
133. Z. Hong, K. B. Fogash, and J. A. Dumesic, *Catal. Today* **51**, 269 (1999).
134. G. D. Yadav and J. J. Nair, *Microporous Mesoporous Mater.* **33**, 1 (1999).
135. W. O. Haag and R. H. Dessau, *Proceedings, 8th International Congress Catalysis*, Vol. II, West-Berlin, 1984, DECHEMA, Frankfurt am Main, 1984, p. 305
136. C. Mirodatos and D. Barthomeuf, *J. Chem. Soc., Chem. Commun.* 39 (1981).
137. M. Hino and K. Arata, *J. Chem. Soc., Chem. Commun.* 1148 (1979).
138. M. Hino and K. Arata, *J. Chem. Soc., Chem. Commun.* 851 (1980).
139. B. S. Umansky and W. K. Hall, *J. Catal.* **124**, 97 (1990).
140. B. Umansky, J. Engelhardt, and W. K. Hall, *J. Catal.* **127**, 128 (1991).
141. Microporous Mesoporous Mater. 35–36, pp. 1–641 (2000).
142. M. Misono and T. Okuhara, *Chemtech.* **23**(11), 23 (1993).
143. N. Mizuno and M. Misono, in *Catalytic Activation and Functionalization of Light Alkanes, Advances and Challenges*, NATO ASI series 44, E. G. Derouane, J. Haber, F. Lemos, F. R. Ribero, and M. Guisnet, eds., Kluwer, Amsterdam, 1998, p. 311.
144. K. Yu. Koltunov, S. Walspurger, and J. Sommer, *Chem. Comm.* 1754 (2004).
145. S. Walspurger, Y. Sun, A. S. S. Sido, and J. Sommer, *J. Phys. Chem. B.* **110**, 18368 (2006).
146. W. E. Farneth and R. J. Gorte, *Chem. Rev.* **95**, 615 (1995).
147. A. Corma, *Chem. Rev.* **95**, 559 (1995).
148. J. A. Martens and P. A. Jacobs, in *Handbook of Heterogeneous Catalysis*, Vol. 3, VCH-Wiley, Weinheim, 1997, Chapter 5.4.2, p. 1137.

149. Acidity in Aluminas, Amorphous and Crystalline Silico-Aluminas, J. J. Fripiat and J. A. Dumesic, eds., *Topics Catal.* 4, 1–171 (1997).
150. G. A. Olah, *Friedel–Crafts Chemistry*, Wiley, New York, 1973, p. 356.
151. V. L. Magnotta, B. C. Gates, and G. C. A. Schuit, *J. Chem. Soc., Chem. Commun.*, 342 (1976).
152. V. L. Magnotta and B. C. Gates, *J. Catal.* **46**, 266 (1977).
153. G. A. Fuentes and B. C. Gates, *J. Catal.* **76**, 440 (1982).
154. G. A. Olah, P. S. Iyer, and G. K. S. Prakash, *Synthesis*, 513 (1986).
155. S. J. Sondheimer, N. J. Bunce, and C. A. Fyfe, *JMS—Rev. Macromol. Chem. Phys.* **C26**, 353 (1986).
156. G. A. Olah, G. K. S. Prakash, and J. Sommer, *Science* **206**, 13 (1979).
157. K. Arata, *Adv. Catal.* **37**, 165 (1990).
158. Q. Sun, M. A. Harmer, and W. E. Farneth, *Chem. Commun.*, 1201 (1996).
159. M. A. Harmer, W. E. Farneth, and Q. Sun, *J. Am. Chem. Soc.* **118**, 7708 (1996).
160. M. A. Harmer and Q. Sun, *Adv. Mater.* **10**, 1255 (1998).
161. H. Wang and B.-Q. Xu, *Appl. Catal. A.* **275**, 247 (2004).
162. K. Okuyama, *J. Mater. Chem.* **10**, 973 (2000).
163. M. Fujiwara, K. Kuroaka, T. Yazawa, Q. Xu, M. Tanak, and Y. Souma, *Chem. Commun.*, 1523 (2000).
164. M. Fujiwara, K. Shiokawa, and Y. Zhu, *J. Mol. Catal. A: Chem.*, **264**, 153 (2007).
165. M. Alvaro, A. Corma, D. Das, V. Fornés, and H. García, *Chem. Commun.*, **956** (2004). *J. Catal.* **231**, 48 (2005).
166. M. A. Harmer, Q. Sun, M. J. Michalczyk, and Z. Yang, *Chem. Commun.*, 1803 (1997).
167. D. J. Macquarrie, S. J. Tavener, and M. A. Harmer, *Chem. Commun.*, 2363 (2005).
168. Á. Molnár, *Curr. Org. Chem.* **12**, 159 (2008).
169. H. Hattori, O. Takahashi, M. Takagi, and K. Tanabe, *J. Catal.* **68**, 132 (1981).
170. K. Tanabe and H. Hattori, *Chem. Lett.*, 625 (1976).
171. O. Takahashi, T. Yamauchi, T. Sakuhara, H. Hattori, and K. Tanabe, *Bull. Chem. Soc. Jpn.* **53**, 1807 (1980).
172. J. M. Oelderik and J. C. Platteeuw, in *Proceedings, 3rd International Congress Catalysis*, Vol. 1, W. H. M. Sachtler, G. C. A. Schmit, and P. Zwietering, eds., North Holland, Amsterdam, 1965, pp. 736–746
173. A. A. Castro, O. A. Scelza, E. R. Benvenuto, G. T. Baronetti, and J. M. Papera, *J. Catal.* **69**, 222 (1981).
174. P.-M. Bernard and M. Primet, *J. Chem. Soc., Faraday Trans.* **86**, 567 (1990).
175. V. Adeeva and W. M. H. Sachtler, *Appl. Catal. A*, **163**, 237 (1997).
176. F. Schmidt, *Appl. Catal. A*, **221**, 15 (2001).
177. H. Weyda and E. Köhler, *Catal. Today* **81**, 51 (2003).
178. L. Schmerling and J. A. Vesely, U. S. Patents 3,846,503 and 3,846,504 (1974).
179. Y. Ono, T. Tanabe, and N. Kitajima, *J. Catal.* **56**, 47 (1979).
180. Y. Ono, T. Tanabe, and N. Kitajima, *Chem. Lett.* 625 (1978).

181. Y. Ono, S. Sakuma, T. Tanabe, and N. Kitajima, *Chem. Lett.* 1061 (1978).
182. Y. Ono, K. Yamaguchi, and N. Kitajima, *J. Catal.* **64**, 13 (1980).
183. G. A. Olah, U. S. Patents 4,547,474 (1985) and 4,613,723 (1986).
184. A. de Angelis C. Flego, P. Ingallina, L. Montanari, M. G. Clerici, C. Carati, and C. Perego, *Catal. Today* **65**, 363 (2001).
185. L. R. Pizzio, *Mater. Lett.* **60**, 3931 (2006).
186. D. O. Bennardi, G. P. Romanelli, J. C. Autino, and L. R. Pizzio, *Appl. Catal. A* **324**, 62 (2007).
187. M. Chidambaram, C. Venkatesan, P. R. Rajamohanan, and A. P. Singh, *Appl. Catal. A* **244**, 27 (2003).
188. M. Chidambaram, D. Curulla-Ferre, A. P. Singh, and B. G. Anderson, *J. Catal.* **220**, 442 (2003).
189. A. N. Pârvulescu, B. C. Gagea, G. Poncelet, and V. I. Pârvulescu, *Appl. Catal. A*, **301**, 133 (2006).
190. L. Lu, R. L. Van Mao N. Al-Yassir A. Muntasar, and N. T. Vu, *Catal. Lett.* **105**, 139 (2005).
191. A. K. Cahkraborti and R. Gulhale, *Tetrahedron Lett.* **44**, 3521 (2003).
192. B. P. Bandgar, A. V. Patil, and O. S. Chavan, *J. Mol. Catal. A: Chem.* **256**, 99 (2006).
193. N. Bartlett and B. W. McQuillan, Intercalation Chemistry, in *Material Science and Technology Series*, M.S. Whittingham and A. J. Jacobson, eds., Academic Press, New York, 1982.
194. W. C. Forsman, T. Dziemianowicz, K. Leong, and D. Carl, *Synth. Metals* **5**, 77 (1983).
195. M. S. Dresselhaus and G. Dresselhaus, *Mol. Cryst. Liq. Cryst.* **244**, 1 (1994). *Adv. Phys.* **51**, 1 (2002).
196. D. D. L. Chung, *J. Mater. Sci.* **37**, 1475 (2002).
197. M. Caragiu and S. Finberg, *J. Phys.: Condens. Matter* **17**, R995 (2005).
198. A. D. Pomogailo, *Inorg. Mater.* **41**, Suppl. 1, S47 (2005).
199. L. B. Ebert, *J. Mol. Catal.* **15**, 275 (1982).
200. L. Facchini, J. Bouat, H. Sfihi, A. P. Legrand, G. Furdin, J. Melin, and R. Vangelisti, *Synth. Metals* **5**, 11 (1982).
201. L. B. Ebert, R. A. Huggins, and J. I. Brauman, *J. Chem. Soc., Chem. Commun.*, 924 (1974).
202. J. G. Ballard and T. Birchall, *J. Chem. Soc., Dalton Trans.*, 1859 (1976).
203. J. M. Friedt, L. Soderholm, R. Poinsot, and R. Vangelisti, *Synth. Metals* **8**, 99 (1983).
204. P. Touzain, *Synth. Metals* **1**, 3 (1979).
205. F. Le Normand E. Fajula, F. Gault, and J. Sommer, *Nouv. J. Chim.* **6**, 411 (1982) and references cited therein.
206. K. Laali and J. Sommer, *Nouv. J. Chim.* **5**, 469 (1981).
207. K. Laali, M. Muller, and J. Sommer, *J. Chem. Soc., Chem. Commun.*, 1088 (1980).
208. R. Vangelisti, V. Polo, J. Bukala, and J. Sommer, *Mol. Cryst. Liquid Crys.* **244A**, 183 (1994).

209. J. J. L. Heinerman and J. Gaaf, *J. Mol. Catal.* **11**, 215 (1981).
210. W. Rüdorff and R. Zeller, *Z. Anorg. Allg. Chem.* **279**, 182 (1955).
211. T. Sasa, Y. Takahashi, and T. Mukaibo, *Bull. Chem. Soc. Jpn.* **45**, 2250 (1972).
212. J. -M. Lalancette, M.-J. Fournier-Breault, and R. Thiffault, *Can. J. Chem.* **52**, 589 (1974).
213. G. A. Olah and J. Kaspi, *J. Org. Chem.* **42**, 3046 (1977).
214. G. A. Olah, J. Kaspi, and J. Bukala, *J. Org. Chem.* **42**, 4187 (1977).
215. N. Yoneda, T. Fukuhara, T. Abe, and A. Suzuki, *Chem. Lett.*, 1485 (1981).

CHAPTER 3

Carbocations in Superacid Systems

This chapter begins with a short historic retrospect about the development of the carbocation concepts and covers the techniques used for their generation, observation, and characterization under superacidic long-lived conditions. This is followed by an extensive coverage of the multitude of trivalent (classical) and equilibrating (degenerate) and higher (five or six) coordinate (nonclassical) carbocations.

Since the inception of the superelectrophilic concept in the 1970s[1] and 1980s,[2] first formulated as protosolvation of cationic intermediates, superelectrophiles as highly reactive dicationic and tricationic intermediates have been successfully observed and characterized.[3–5] Consequently, selected examples of superelectrophiles are also covered in this chapter where appropriate, whereas various organic transformations, in which the involvement of superelectrophilic intermediates is invoked or superelectrophiles are observed, are treated in Chapter 5.

3.1. INTRODUCTION

3.1.1. Development of the Carbocation Concept: Early Kinetic and Stereochemical Studies

In the early 1900s, the pioneering work of Norris, Kehrmann, Baeyer, and others on triarylcarbenium salts generated wide interest in such species.[6–9] They were, however, long considered only of interest to explain the color of these dyes. One of the most daring and fruitful ideas born in organic chemistry was the suggestion that in the course of reactions, carbocations might be intermediates that start from nonionic reactants and lead to nonionic covalent products. It was H. Meerwein[10] who in 1922, while studying the kinetics of the rearrangement of camphene hydrochloride to isobornyl chloride, reported the important observation that the reaction rate increased in a general way with the dielectric constant of the solvent. Furthermore, he found that metallic chlorides such as $SbCl_5$, $SnCl_4$, $FeCl_3$, $AlCl_3$, and $SbCl_3$ (but not BCl_3 or $SiCl_4$), as well as dry HCl, which promote ionization of triphenylmethyl chloride by

Superacid Chemistry, Second Edition, George A. Olah, G. K. Surya Prakash, Árpád Molnár, and Jean Sommer

the formation of ionized complexes, considerably accelerated the rearrangement of camphene hydrochloride. Meerwein concluded that the conversion of camphene hydrochloride to isobornyl chloride actually does not proceed by way of migration of the chlorine atom but by a rearrangement of a cationic intermediate. Thus, the modern concept of carbocation intermediates was born.

Ingold, Hughes, and their collaborators in England, beginning in the late 1920s, carried out detailed kinetic investigations on what later became known as nucleophilic substitution at saturated carbon and polar elimination reactions.[11] The well-known work relating to S_N1 and later E1 reactions established the carbocation concept in these reactions. Whitmore,[12] in a series of papers that began in 1932, generalized Meerwein's rearrangement theory to many organic chemical reactions, although he cautiously avoided writing positive signs on any of the intermediates.

Kinetic and stereochemical evidence helped to establish carbocation intermediates in organic reactions. These species, however, were generally too short-lived and could not be directly observed by physical means.

3.1.2. Observation of Stable, Long-Lived Carbocations

The transient nature of carbocations arises from their extreme reactivity with nucleophiles. The use of low-nucleophilicity counterions, particularly tetrafluoroborates (BF_4^-), enabled Meerwein in the 1940s to prepare a series of oxonium and carboxonium ion salts, such as $R_3O^+BF_4^-$ and $HC(OR)_2{}^+BF_4^-$, respectively.[13] These Meerwein salts are effective alkylating agents, and they transfer alkyl cations in S_N2-type reactions. However, simple alkyl cation salts ($R^+BF_4^-$) were not obtained in Meerwein's studies. The first acetyl tetrafluoroborate—that is, acetylium tetrafluoroborate—was obtained by Seel[14] in 1943 by reacting acetyl fluoride with boron trifluoride at low temperature [Eq. (3.1)].

$$CH_3COF \quad + \quad BF_3 \longrightarrow CH_3CO^+BF_4^- \qquad (3.1)$$

In the early 1950s, Olah and co-workers[15] started a study of the intermediates of Friedel–Crafts reactions and inter alia carried out a systematic investigation of acyl fluoride–boron trifluoride complexes. They were able to observe a series of donor–acceptor complexes as well as stable acyl cations. Subsequently, the investigations were also extended to other Lewis acid halides. In the course of these studies, they increasingly became interested in alkyl halide–Lewis acid halide complexes. The study of alkyl fluoride–boron trifluoride complexes by electrical conductivity measurements indicated the formation of ionic complexes in the case of *tert*-butyl and isopropyl fluoride at low temperatures, whereas methyl and ethyl fluoride formed molecular coordination complexes. Subsequently, Olah and co-workers initiated a systematic study of more suitable acid and low-nucleophilicity solvent systems. This resulted in the discovery of (a) antimony pentafluoride as a suitable very strong Lewis acid, (b) related superacid systems such as Magic Acid[16]

(HSO_3F–SbF_5), and (c) low-nucleophilicity solvent systems such as SO_2, SO_2ClF, and SO_2F_2, which finally allowed obtaining alkyl cations as stable, long-lived species.[17] Subsequently, by the use of a variety of superacids, a wide range of practically all conceivable carbocations became readily available for structural and chemical studies.[18,19]

3.1.3. General Concept of Carbocations

Electrophilic reactions since Meerwein's pioneering studies are generally considered to proceed through cationic (i.e., carbocationic) intermediates. The general concept[20] of carbocations encompasses all cations of carbon-containing compounds, which sometimes were differentiated into two limiting classes: (i) trivalent ("classical") carbenium ions and (ii) five or higher coordinate ("nonclassical") carbonium ions. Whereas the differentiation of limiting trivalent carbenium and pentacoordinate carbonium ions serves a useful purpose to establish the significant differences between these ions, it is also clear that in most specific systems there exists a continuum of charge delocalization. In fact, in all carbocations (even in the parent CH_3^+), there is a continuum of degree of charge delocalization, and thus to think in limiting terms is rather meaningless. Participation by neighboring groups can be not only by *n*- and π-donors, as most generally recognized, but also by σ-ligands. There is in principle no difference between these electron donors. σ-Participation in properly oriented systems is not only possible, but is unavoidable. The only question is its degree and not whether it exists.[21]

It is well known that trivalent carbenium ions play an important role in electrophilic reactions of π- and *n*-donors systems. Similarly, pentacoordinate carbonium ions are the key to electrophilic reactions of σ-donor systems (single bonds). The ability of single bonds to act as σ-donors lies in their ability to form carbonium ions via delocalized two-electron, three-center (2*e*–3*c*) bond formation. Consequently, there seems to be in principle no difference between the electrophilic reactions of π- and σ-bonds except that the former react more easily even with weak electrophiles, whereas the latter necessitate more severe conditions.

On the basis of the study of carbocations by direct observation of long-lived species, it became increasingly apparent that the carbocation concept is much wider than previously realized and necessitated a general definition.[21] Therefore, such a definition was offered based on the realization that two distinct, limiting classes of carbocations exist (Figure 3.1).

1. Trivalent ("classical") carbenium ions contain an sp^2-hybridized electron-deficient carbon center that tends to be planar in the absence of constraining skeletal rigidity or steric interference. (It should be noted that *sp*-hybridized, linear acyl cations and vinyl cations also show substantial electron deficiency on carbon.) The carbenium carbon contains six valence electrons, and thus it is highly electron-deficient. The structure of trivalent carbocations can always be adequately described by using two-electron, two-center bonds (Lewis valence bond structures).

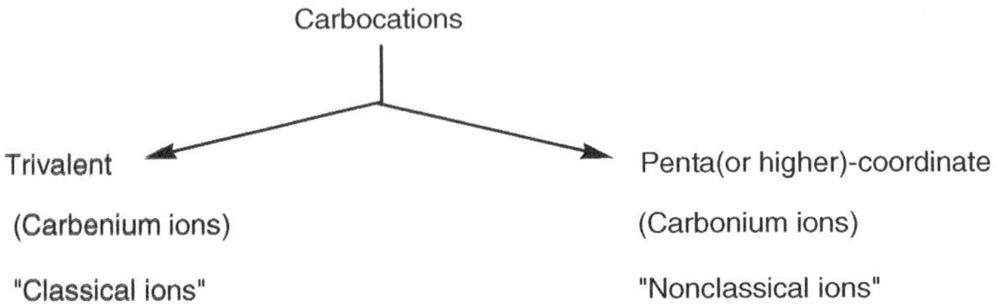

Figure 3.1. Classification of carbocations.

2. Penta(or higher)-coordinate ("nonclassical") carbonium ions, which contain five (or higher) coordinate carbon atoms, cannot be described by two-electron single bonds alone, but also necessitates the use of two-electron, three-center (or multicenter) bond(s). The carbocation center is always surrounded by eight electrons, although two (or more) of them are involved in multicenter bonds, and thus such ions are overall electron-deficient (due to electron-sharing of two binding electrons between three (or more) centers).

Lewis's concept that a chemical bond consists of a pair of electrons shared between two atoms became the foundation of structural chemistry, and chemists still tend to name compounds as anomalous when their structures cannot be depicted in terms of such bonds alone. Carbocations with too few electrons to allow a pair for each "bond" came to be referred to as "nonclassical," a label still used even though it is now recognized that, like many other substances, they adopt the delocalized structures appropriate for the number of electrons they contain.

Expansion of the carbon octet via $3d$-orbital participation does not seem possible; there can be only eight valence electrons in the outer shell of carbon.[22,23] Thus, the covalency of carbon cannot exceed four. Penta(or higher)-coordination implies a species with five (or more) ligands within reasonable bonding distance from the central atom.[24] The transition states long ago suggested for S_N2 reactions represent such cases, but involving 10 electrons around the carbon center. Charge–charge repulsions in the S_N2 transition state forces the entering and leaving substituents as far apart as possible leading a trigonal bipyramid with a long $4e$–$3c$ bond allowing little possibility of a stable intermediate. In contrast, S_E2 substitution reactions involve $2e$–$3c$ interactions but have been in the past mainly restricted to organometallic compounds (e.g., organomercurials).[25]

The direct observation of stable penta(or higher)-coordinate species with eight electrons around the carbon center in solutions was not reported until recent studies of long-lived "nonclassical" ions in superacid solvent systems.

Neighboring group interactions with the vacant p orbital of the carbenium ion center can contribute to ion stabilization via charge delocalization. Such phenomena can involve atoms with unshared electron pairs (n-donors), C—H and C—C hyperconjugation, bent σ-bonds (as in cyclopropylcarbenium ions), and π-electron systems (direct conjugative or allylic stabilization). Thus, trivalent carbenium ions can show varying degrees of delocalization without becoming pentacoordinate carbonium ions. The limiting classes defined do not exclude varying degrees of delocalization, but in fact imply a spectrum of carbocation structures.

In contrast to the rather well-defined trivalent ("classical") carbenium ions, "nonclassical ions"[26] have been more loosely defined. In recent years, a lively controversy centered on the classical–nonclassical ion problem.[27–37] The extensive use of "dotted lines" in writing carbonium ion structures has been (rightly) criticized by Brown,[31] who carried, however, the criticism to question the existence of any σ-delocalized (nonclassical) ion. For these ions, if they exist, he stated "...a new bonding concept not yet established in carbon structures is required."

Clear, unequivocal experimental evidence has by now been obtained for nonclassical ions such as the norbornyl cation.[38–40] The bonding concept required to define "nonclassical ions" is simply to consider them as penta(or higher)-coordinated carbonium ions involving at least one two-electron three-center (or multicenter) bond, of which CH_5^+ (the methonium ion–carbonium ion) is the parent, as CH_3^+ (the methenium ion, methyl cation, carbenium ion) is the parent for trivalent carbenium ions. An example of a hexacoordinate carbonium ion is the pyramidal dication of Hogeveen.[41]

Concerning the carbocation concept, it is regrettable that for a long time in the Anglo-Saxon literature the trivalent planar ions of the CH_3^+ type were named as carbonium ions. If the name considered analogous to the other -onium ions (ammonium, sulfonium, phosphonium ions, etc.), then it should relate to the higher-valency or coordination-state carbocations. The higher-bonding-state carbocations, however, clearly are not the trivalent but the penta(or higher)-coordinate cations. The German and French literatures indeed frequently used the "carbenium ion" naming for the trivalent cations. If one considers these latter ions as protonated carbenes, the naming is indeed correct.[42] It should be pointed out, however, that the "carbenium ion" naming depicts only trivalent ions and thus should not be a general name for all carbocations. IUPAC's Organic Chemistry Division has reviewed the nomenclature of physical organic chemistry[43] and recommends the use of the "carbocation" for naming all positive ions of carbon. The namings "carbenium" or "carbonium ion," similar to the "carbinol" naming of alcohols, is discouraged.

3.2. METHODS OF GENERATING CARBOCATIONS IN SUPERACIDS SYSTEMS

Common superacid systems that are generally employed in the preparation of carbocations are the Brønsted acids such as HSO_3F and HF and Lewis acids such as SbF_5. Also, Lewis and Brønsted acid combinations such as HSO_3F–SbF_5 (Magic Acid), HSO_3F–SbF_5 (4:1), and HF–SbF_5 (fluoroantimonic acid) have been used. Other superacid systems such as HF–BF_3, AsF_5, CF_3SO_3H, HF–PF_5, HCl–$AlCl_3$, HBr–$AlBr_3$, H_2SO_4, and $HClO_4$ have been successfully adopted. Sometimes a metathetic reaction involving a halide precursor and $AgBF_4$ or $AgSbF_6$ has also been successful. The most convenient non-nucleophilic solvent systems used in the preparation of carbocations are SO_2, SO_2ClF, and SO_2F_2. To be able to study ionic solutions at very low temperatures (ca. −160°C) by NMR spectroscopy, Freons like $CHCl_2F$ may be used as a co-solvent to decrease the viscosity of the solution.

The success of a carbocation preparation is superacid generally depends on the technique employed. For most of the stable systems, Olah's method[44] of mixing the precooled progenitor dissolved in appropriate solvent system along with the superacid using a simple vortex stirrer to mix the components is generally convenient. However, care should be taken to avoid moisture and local heating. Low temperatures between −78°C (using acetone–dry ice) and −120°C (using liquid N_2–ethanol slush), which suppress side reactions like dimerization and oligomerizations, are most commonly employed.

Methods for the generation of very reactive carbocations have been developed most notably by Ahlberg[45] and Saunders et al.[46] The former group describes an ion-generation apparatus that consists of a Schlenk tube-like vessel attached to an NMR tube in which the carbocation is prepared at low temperatures. This methodology has been further improved.[47,48]

The method of Saunders,[46] however, is more sophisticated. It uses deposition of the starting reagents from the gas phase (using high vacuum) on a surface cooled to liquid nitrogen temperature to produce stable solutions of carbocations. The details of the method have been published in detail.[49] Myhre and Yannoni[50] have successfully utilized the above method to generate carbocations in an SbF_5 matrix at very low temperatures for their solid-state ^{13}C NMR spectroscopic work.

Kelly and Brown[51] describe a method for the preparation of concentrated solutions of carbocations (~1 *M*) in SbF_5–SO_2ClF. This method, which employs a syringe technique, allows quantitative conversion of precursors soluble in SO_2ClF at −78°C into the corresponding carbocations.

3.3. METHODS AND TECHNIQUES IN THE STUDY OF CARBOCATIONS

3.3.1. Nuclear Magnetic Resonance Spectra in Solution

One of the most powerful tools in the study of carbocations is nuclear magnetic resonance (NMR) spectroscopy. This method yields direct information—through chemical shifts, coupling constants, and the temperature dependence of band shapes—about the structure and dynamics of carbocations.

Although the initial developments in the observation of stable carbocations in solution relied heavily upon ^1H NMR spectroscopy, ^{13}C NMR spectroscopy has proven to be the single most useful technique. ^{13}C NMR permits the direct observation of the cationic center, and the chemical shifts and coupling constants can be correlated to the cation geometry and hybridization. At the beginning of the ^{13}C NMR studies in the early 1960s, Olah and co-workers used the INDOR technique to obtain spectra of carbocations at natural ^{13}C abundance. In the past decades, the development of the Fourier transform NMR spectrometers has made available ^1H and ^{13}C NMR spectra of very dilute solution of cations. The signal-to-noise (S/N) ratio has been further improved by the introduction of high-field superconducting magnets. With such

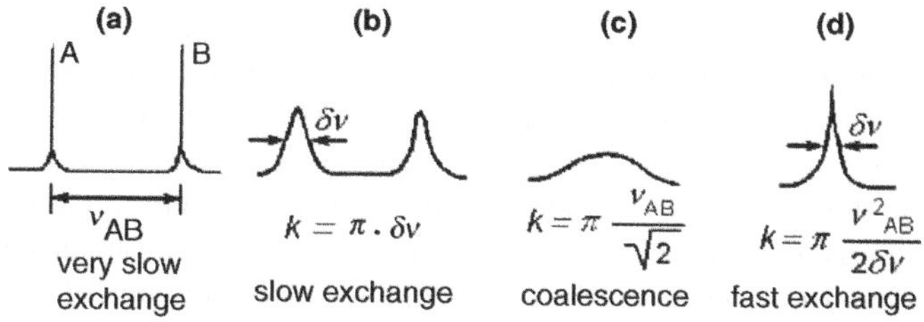

Figure 3.2. Results in the NMR spectrum of exchange of two equally populated sites at different rates. The rate constants (k) for the exchange at the conditions **b**, **c**, and **d** could be approximately estimated using the attached formulas and corrected values of δν.

high-sensitivity, high-resolution spectrometers, degenerate rearrangement reaction rate constants of the order $10^7\ s^{-1}$, which corresponds to a free energy of activation of 3.3 kcal mol^{-1} at −160°C, have been measured.

Degenerate rearrangements of carbocations, if they are fast enough, result in temperature-dependent NMR spectra. At slow exchange, the signals of the exchanging nuclei show up as separate absorptions. If the exchange rate is gradually increased by raising the temperature, the signals first broaden and, upon a further rise in the temperature, they coalesce. Still further increase of the exchange rate results in sharpening of the broad coalesced signals (Figure 3.2).

The formula in Figure 3.2*c* shows that the larger the shift difference (ν_{AB}) between the exchanging signals, the larger the rate constant needed to get coalescence at a specific temperature. Thus, since chemical shift differences in ^{13}C NMR are usually much larger (in Hz) than in ^1H NMR, ^{13}C NMR spectroscopy permits the quantitative study of much faster processes than can be investigated by ^1H NMR spectroscopy.

Very slow exchange could be detected and the rate measured by the transfer of spin saturation, a tool that could be useful in the elucidation of reaction mechanisms causing exchange. One of the signals participating in very slow exchange is saturated by an extra RF field while the rest of the spectrum is observed. If exchange of the spin-saturated nuclei takes place at a rate comparable to that of the nuclear-spin relaxation (T_1), then transfer of the spin saturation by the degenerate reaction will partially saturate the other exchanging nuclei. From the degree of transferred spin saturation and measured T_1, the rate of exchange could be evaluated. This technique was devised by Forsén and Hoffman[52,53] using ^1H NMR spectroscopy, and it has been utilized in the study of carbocations. The method has some limitations with ^1H NMR due to the small shift differences and large couplings between the protons. However, the method has recently been applied using ^{13}C NMR spectroscopy (proton decoupled) for study of complex carbocation rearrangements.[54]

3.3.2. ^{13}C NMR Chemical Shift Additivity

An empirical criterion based on additivity of ^{13}C NMR chemical shifts for distinguishing classical trivalent and higher coordinate carbocations has been developed by

Schleyer, Olah, and co-workers.[55] In this method, the sums of all ^{13}C chemical shifts of carbocations with their respective hydrocarbon precursors are compared. A trivalent carbocation has a sum of chemical shifts of at least 350 ppm higher than the sum for the corresponding neutral hydrocarbon. This difference can be rationalized by partly attributing it to the hybridization change to sp^2 and to the deshielding influence of an unit positive charge in the trivalent carbocation. Since higher coordinate carbocations (nonclassical ions) have penta- and hexa-coordinate centers, the sum of their chemical shifts relative to their neutral hydrocarbons is much smaller, often by less than 200 ppm.

3.3.3. Isotopic Perturbation Technique

The deuterium isotopic perturbation technique developed by Saunders and co-workers[56–60] is capable of providing a convenient means to differentiate the rapidly equilibrating or bridged nature of carbocations.

In 1971, Saunders and Vogel[56–58] discovered that by asymmetrically introducing deuterium into some reversible rearrangement processes of carbocations, large splittings were produced in their NMR spectra. Although the ions were interconverting extremely rapidly and thus gave averaged spectra, the isotope made the energies of the two interconverting species slightly different and thus the equilibrium constant between them was no longer 1. Each ion, therefore, spent a little more time on one side of the equilibrium barrier than on the other side. The weighted-average peaks of the two carbon atoms that were interchanged by the rearrangement process no longer coincided. Splitting values in the ^{13}C NMR spectrum of over 100 ppm were observed as a result of perturbation by deuterium.[61]

However, when deuterium was introduced into systems that are recognized as static, single-minimum, nonequilibrating species,[62] such as the cyclohexenyl cation, no large splittings were observed and, in contrast to the behavior of the equilibrating systems, there were no observable changes in the spectra with temperature. In fact, the isotope-induced changes in the spectra were not very different from changes that occur in any simple molecule on introducing deuterium, and they were roughly only 50 times smaller than effects produced in the equilibrating systems.

These observations led to the method of observing changes in NMR spectra produced by asymmetric introduction of isotopes (isotopic perturbation) as a means of distinguishing systems involving equilibrating species passing rapidly over a low-energy barrier from molecules with a single energy minimum, intermediate between the presumed equilibrating structures. Application of this method for individual carbocations will be discussed later. This method also allows accurate determination of equilibrium isotope effects.

3.3.4. Solid-State ^{13}C NMR

The use of magic-angle spinning (MAS) and cross-polarization (CP) techniques has enabled high-resolution ^{13}C NMR spectra to be obtained in the solid state. Yannoni, Myhre, and Fyfe have obtained solid-state NMR spectra of frozen carbocation solutions (such as alkyl cations, the norbornyl cation, etc.) at very low temperatures[50,63,64] (even at 5 K). At such low temperatures, rearrangements and most molecular motions can be frozen out. Olah's group has studied a series of stable carbocations, inlcuding the 1-adamantyl cation, acyl cations, and so on, by ^{13}C CPMAS at ambient temperature. In these studies solid-state effects seem to be unimportant on chemical shifts (as compared to solution data) but are significant on rates of exchange processes.[63,64]

3.3.5. X-ray Diffraction

X-ray diffractometry is the most powerful method to determine atomic coordinates of molecules in the solid state. X-ray crystal structure analysis was, however, rarely applied in the early years of development of persistent, long-lived alkyl carbocations and studies were only performed to investigate structures of carbocations of aryl derivatives and aromatic systems.[65] This is due to the low thermal stability of alkyl carbocations and to the difficulties in obtaining single crystals of carbocations suitable for analysis. Since then, however, methods and instrumentation have improved significantly and X-ray crystal structure analysis has become a powerful tool to solve structural problems of carbocations.[65,66]

3.3.6. Tool of Increasing Electron Demand

The nature of electronic effects in cationic reactions has been probed by application of the Gassman–Fentiman tool of increasing electron demand.[67] Aryl-substituted cationic centers can be made more electron demanding (i.e., electrophilic) by introduction of electron-withdrawing substituents into the aryl ring.

When a cationic center becomes sufficiently electrophilic, it may draw on electrons from neighboring π- and σ-bonds and thus delocalize positive charge density. The onset of participation of π- and σ-bonds can be detected as a departure from linearity in a Hammett-type plot as the electron-withdrawing ability of the aryl substituent increases. In stable ion studies, ^{13}C NMR chemical shifts are generally used as a structural probe reflecting the charge density at the cation center (in closely related homologous cations, other factors that may affect chemical shift may be assumed constant).

The tool of increasing electron demand has proved useful in detecting the onset of π- and σ-delocalizations under stable-ion conditions. It is, however, an extremely coarse technique since the aryl group can still delocalize charge into its π-system, even with strongly electron-withdrawing substituents. Only in cases where neighboring σ- and π-bonds can effectively compete with the aryl ring in stabilizing the cationic center are significant deviations from linearity observed in the Hammett-type plot.

3.3.7. Core Electron Spectroscopy

X-ray photoelectron spectroscopy measures the energy distribution of the core electrons emitted from a compound on irradiation with X rays.[68,69] The electron binding energy (E_b) is a function of the chemical environment of the atom. In particular, the atomic charge on carbon can be directly correlated to the carbon 1*s* electron binding energy.[70] The cationic center of a classical carbocation (e.g., *tert*-butyl cation) has a carbon 1*s* E_b approximately 4 eV *higher* than a neutral sp^3 carbon atom.[70] Electron deficiencies of different degrees in different carbocations give different carbon core electron 1*s* binding energies; the delocalization of charge from a cationic center lowers the carbon 1*s* binding energy.

Core electron spectroscopy for chemical analysis (ESCA) is perhaps the most definitive technique applied to the differentiation between nonclassical carbocations from equilibrating classical species. The time scale of the measured ionization process is of the order of 10^{-16} s so that definite species are characterized, regardless of (much slower) intra- and intermolecular exchange reactions—for example, hydride shifts, Wagner–Meerwein rearrangements, proton exchange, and so on.

3.3.8. Infrared and Raman Spectroscopy

Infrared and Raman spectra of stable carbocations have been obtained[71,72] and are in complete agreement with their electron-deficient structures. Infrared spectra of alkyl cations and their deuteriated analogs correspond to the spectra predicted by calculations based on molecular models and force constants. Thus, these spectra can be used in the identification of stable carbocations.

Laser Raman spectroscopy, particularly with helium–neon lasers, is another powerful tool in the study of carbocations. Because Raman spectra give valuable information on symmetry, these spectra help to establish, in detail, structures of the ions and their configurations.

3.3.9. Electronic Spectroscopy

The observation of stable alkyl cations in antimony pentafluoride solutions also opened up the possibility of investigating the electronic spectra of these solutions. It has been reported[73] that solutions of alkyl cations in $HSO_3F–SbF_5$ solution at −60°C showed no absorption maxima above 210 nm. In view of this observation, it was resolved that previous claims relating to a 290-nm absorption of alcohols and alkenes in sulfuric acid solutions were due to condensation products or cyclic allylic ions and not to simple alkyl cations.[74]

3.3.10. Low-Temperature Solution Calorimetric Studies

Arnett and co-workers,[75,76] in a series of investigations, have determined heats of ionization (ΔH_i) of secondary and tertiary chlorides and alcohols in $SbF_5–SO_2ClF$ and $HSO_3F–SbF_5–SO_2ClF$ solutions, respectively, at low temperatures. They have also measured heats of isomerizations of secondary to tertiary carbocations in the superacid media. These measured thermochemical data have been useful to determine the intrinsic thermodynamic stability of secondary and tertiary carbocations.

3.3.11. Quantum Mechanical Calculations

One of the main aims of quantum mechanical methods in chemistry is the calculation of energies of molecules as a function of their geometries. This requires the generation of potential energy hypersurfaces. If these surfaces can be calculated with sufficient accuracy, they may be employed to predict equilibrium geometries of molecules, relative energies of isomers, the rates of their interconversions, NMR chemical shifts, vibrational spectra, and other properties. Carbocations are ideally suited for calculations because relative energies of well-defined structural isomers are frequently not easily determined experimentally. It should, however, be kept in mind that theoretical calculations usually refer to isolated ion structures in the gas phase.

Over the years, several computational methods have been developed. The variational theory can be used either without using experimental data other than the fundamental constants (i.e., *ab initio* methods) or by using empirical data to reduce the needed amount of numerical work (i.e., semiempirical data methods). There are various levels of sophistication in both *ab initio* [HF(IGLO), DFT GIAO-MP2, GIAO-CCSD(T)] and semiempirical methods. In the *ab initio* methods, various kinds of basic sets can be employed, while in the semiempirical methods, different choices of empirical parameters and parametric functions exist. The reader is referred to reviews of the subject.[18,77]

Recent developments in computational chemistry have established the exact structure of carbocations by combining computational and experimental results.[78,79] Furthermore, accurate ^{1}H and ^{13}C NMR chemical shifts of carbocations and other organic molecules can be calculated with the application of recent coupled cluster methods, such as GIAO-CCSD(T).[80]

3.4. TRIVALENT CARBOCATIONS

3.4.1. Alkyl Cations

3.4.1.1. Early Unsuccessful Attempts. Until the early 1960s, simple alkyl cations were considered only as transient species.[15] Their existence has been inferred from the kinetic and stereochemical studies of reactions. No reliable physical measurements, other than electron impact measurements in the gas phase (mass spectrometry), were known. The formation of gaseous organic cations under electron bombardment of alkenes, haloalkanes, and other precursors has been widely investigated in mass spectrometric studies.[81] No direct observation of carbocations in solutions was achieved prior to the early 1960s.

The observation of alkyl cations such as the *tert*-butyl cation [trimethylcarbenium ion, $(CH_3)_3C^+$] **1** and the isopropyl cation [dimethylcarbenium ion, $(CH_3)_2CH^+$] **2** was a long-standing challenge. The existence of alkyl cations in systems containing alkyl halides and Lewis acids has been inferred from a variety of observations, such as vapor pressure depressions of CH_3Cl and C_2H_5Cl in the presence of $GaCl_3$,[82] conductivity of $AlCl_3$ in alkyl chlorides,[83] and conductivityof alkyl fluorides in BF_3,[84] as well as the effect of ethyl bromide on the dipole moment of $AlBr_3$.[85] However, in no case had well-defined, stable cation complexes been established even at very low temperatures.

Electronic spectra of alcohols and alkenes in strong proton acids (H_2SO_4) were obtained by Rosenbaum and Symons.[86] They observed, for a number of simple aliphatic alcohols and alkenes, absorption maxima around 290 nm and ascribed this absorption to the corresponding alkyl cations.

Finch and Symons,[87] on reinvestigation of the absorption of aliphatic alcohols and alkenes in sulfuric acid solution, showed that the condensation products formed with acetic acid (used as solvent for the precursor alcohols and alkenes) were responsible for the spectra and not the simple alkyl cations. Moreover, protonated mesityl oxide was identified as the absorbing species in the system of isobutylene, acetic acid, and sulfuric acid.

Deno[88] has carried out an extensive study of the fate of alkyl cations in undiluted H_2SO_4 and oleum. He obtained equal amounts of a saturated hydrocarbon mixture (C_4–C_{19}) insoluble in H_2SO_4 and a mixture of cyclopentenyl cations (C_9–C_{10}) in the H_2SO_4 layer. These cations exhibit strong UV absorption around 300 nm.

Therefore, it must be concluded that earlier attempts to prove the existence of stable, well-defined alkyl cations were unsuccessful in experiments using sulfuric acid solutions and inconclusive in the interaction of alkyl halides with Lewis acid halides. Proton elimination reactions or dialkyl halonium ion formation may have affected the early conductivity studies.

3.4.1.2. Preparation from Alkyl Fluorides in Antimony Pentafluoride Solution and Spectroscopic Studies. In 1962, Olah et al.[71,72,89–91] first directly observed alkyl cations in solution. They obtained the *tert*-butyl cation **1** (trimethylcarbenium ion) when *tert*-butyl fluoride was dissolved in excess antimony pentafluoride, which serves as both Lewis acid and solvent. Later, the counterion was found to be, under these conditions, primarily the dimeric $Sb_2F_{11}^-$ anion[92,93] [Eq. (3.2)] whereas in SbF_5–SO_2 or SbF_5–SO_2ClF solutions, SbF_5^- and $Sb_2F_{11}^-$ are both formed.

$$(CH_3)_3CF \quad + \quad (SbF_5)_2 \longrightarrow \underset{\mathbf{1}}{(CH_3)_3C^+ \; Sb_2F_{11}^-} \qquad (3.2)$$

The possibility of obtaining stable alkyl pentafluoroantimonate salts from alkyl fluorides (and subsequently other alkyl halides) in antimony pentafluoride solution (neat or diluted with sulfur dioxide, sulfuryl chloride fluoride, or sulfuryl fluoride) or in other superacids (for a definition, see Chapter 1) such as HSO_3F–SbF_5 (Magic Acid), HF–SbF_5 (fluoroantimonic acid), HF–TaF_5 (fluorotantalic acid), and the like, was evaluated in detail, extending studies to all isomeric C_3–C_8 alkyl halides, as well to a number of higher homologs.[94]

Propyl, butyl, and pentyl fluorides with antimony pentafluoride gave the isopropyl, *tert*-butyl, and *tert*-pentyl cations (as their fluoroantimonate salts) **2**, **1**, and **3**, respectively [Eqs. (3.3)–(3.5)].

$$C_3H_7F \;+\; (SbF_5)_2 \longrightarrow CH_3-\overset{+}{C}H-CH_3\ Sb_2F_{11}^- \quad \mathbf{2} \tag{3.3}$$

$$C_4H_9F \;+\; (SbF_5)_2 \longrightarrow CH_3-\overset{+}{C}(CH_3)-CH_3\ Sb_2F_{11}^- \quad \mathbf{1} \tag{3.4}$$

$$C_5H_{11}F \;+\; (SbF_5)_2 \longrightarrow CH_3-\overset{+}{C}(CH_3)-CH_2CH_3\ Sb_2F_{11}^- \quad \mathbf{3} \tag{3.5}$$

The secondary butyl and amyl cations can be observed only at very low temperatures, and they rearrange readily to the more stable tertiary ions. Generally, the most stable tertiary or secondary carbocations are observed from any of the isomeric alkyl fluorides in superacidic solvent systems.

The main feature of the proton NMR spectra of alkyl fluorides in antimony pentafluoride is the substantial deshielding of the protons in the carbocations as compared with the starting alkyl fluorides (Figure 3.3 and Table 3.1).

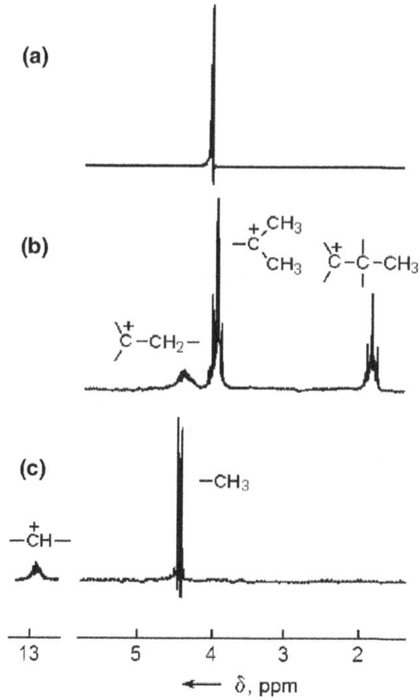

Figure 3.3. ^{1}H NMR spectra of **(a)** the *tert*-butyl cation 1 [trimethylcarbenium ion, $(CH_3)_3C^+$], **(b)** the *tert*-pentyl cation 3 [ethyldimethylcarbenium ion, $C_2H_5(CH_3)_2C^+$], and **(c)** the isopropyl cation 2 [dimethylcarbenium ion, $(CH_3)_2HC^+$] (60 MHz, SbF_5–SO_2ClF solution, −60°C).

Table 3.1. Characteristic ^1H NMR Parameters of Alkyl Cations in SbF_5–SO_2ClF Solution at −70°C

		δ^1H					
Cation		CH^+	$J_{^+CH}$	$J_{^+CCH}$	α-CH_2	α-CH_3	β-CH_2
$(CH_3)_2HC^+$	**2**	13	169	3.3		4.5	
$(CH_3)_3C^+$	**1**			3.6		4.15	
$C_2H_5(CH_3)_2C^+$	**3**				4.5	4.1	1.94
$(C_2H_5)_2CH_3C^+$	**4**				4.44	4.16	1.87

To prove that stable alkyl cations, and not exchanging donor–acceptor complexes, were obtained, Olah and co-workers also investigated as early as 1962 the ^{13}C NMR of the potentially electropositive carbenium carbon atom in alkyl cations.[94] The $^{13}C^+$ shift in the *tert*-butyl cation $(CH_3)_3C^+$ **1** in SbF_5–SO_2ClF solution at −20°C is at $\delta^{13}C$ 335.2 (all ^{13}C NMR shifts are from ^{13}C tetramethylsilane) with a long-range coupling to the methyl protons of 3.6 Hz.

The $^{13}C^+$ shift of the isopropyl cation **2** under identical conditions, is $\delta^{13}C$ 320.6 with a long-range coupling to the methyl protons of 3.3 Hz. The direct ^{13}C–H coupling is 169 Hz (indicating sp^2 hybridization of the carbenium carbon atom), whereas the long-range proton–proton coupling constant is 6.0 Hz (see Figure 3.2). Substitution of the methyl group in the *tert*-butyl cation by hydrogen causes an upfield shift of 10.4 ppm. Although the ^{13}C NMR shift of the carbocation center of the *tert*-butyl cation is more deshielded than that of the isopropyl cation (by about 10 ppm), this can be explained by the methyl substituent effect, which may sometimes amount up to 22 ppm. The *tert*-butyl cation thus is more delocalized and stable than the secondary isopropyl cation.

The ^{13}C NMR shift in the *tert*-pentyl cation $[C_2H_5(CH_3)_2C^+]$ **3** is at $\delta^{13}C$ 335.4, which is similar to the that of the *tert*-butyl cation. The shift difference is much smaller than the 17 ppm found in the case of the related alkanes, although the shift observed is in the same direction. The ^{13}C NMR chemical shifts and coupling constants J_{C-H} of C_3 to C_8 alkyl cations **1–13** are shown in Tables 3.2 and 3.3.[95]

It is difficult to interpret these large deshieldings in any way other than as a direct proof that (i) the hybridization state of the carbon atom at the carbenium ion center is sp^2 and (ii) at the same time, the sp^2 center carries a substantial positive charge.

Data in Tables 3.2 and 3.3 are characterized by substantial chemical shift deshieldings and coupling constants (J_{C-H}) that indicate sp^2 hybridization. Also subsequently, Myhre and Yannoni[50] have obtained ^{13}C NMR spectrum of *tert*-butyl cation **1** in the solid state, which agrees very well with the solution data.[95]

The temperature-dependent ^1H NMR spectrum of isopropyl cation **2** (prepared from isopropyl chloride in SbF_5–SO_2ClF solution) demonstrated[96] rapid interchange of two types of protons. Line-shape analysis showed the reaction to be intramolecular, with an activation energy barrier of $16.4 \pm 0.4\ kcal\ mol^{-1}$. Based on these observations, Saunders and Hagen[96] suggested that the rearrangement involves *n*-propyl

Table 3.2. ^{13}C **Chemical Shifts of the Static** C_3 **to** C_8 **Alkyl Cations**[95]

Cation		1	2	3	4	5	6	7	8
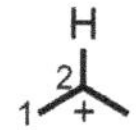	**2**	51.5 (*q*)	320.6 (*s*)						
	**1**	47.5 (*q*)	335.2 (*s*)						
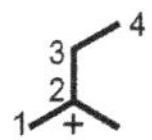	**3**	44.6 (*q*)	335.4 (*s*)	57.5 (*t*)	9.3 (*q*)				
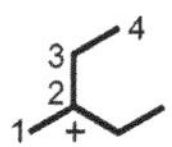	**4**	41.9 (*q*)	336.4	54.5 (*t*)	9.3 (*q*)				
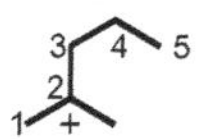	**5**	45.0 (*q*)	333.4	64.4 (*t*)	20.9 (*t*)	12.6 (*q*)			
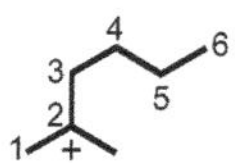	**6**	44.9 (*q*)	332.9 (*s*)	62.8 (*t*)	29.3 (*t*)	22.6 (*t*)	13.0 (*q*)		
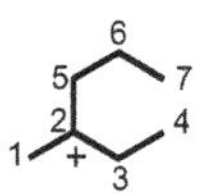	**7**	42.1 (*q*)	334.7 (*s*)	55.1 (*t*)	9.1 (*q*)	61.6 (*t*)	20.2 (*t*)	12.5 (*q*)	
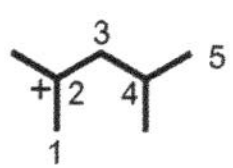	**8**	45.4 (*q*)	332.1 (*s*)	70.1 (*t*)	31.4 (*d*)	21.7 (*q*)			
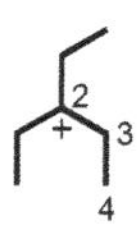	**9**		336.8 (*s*)	51.8 (*t*)	8.6 (*q*)				

(continued)

Table 3.2. *(Continued)*

Cation		1	2	3	4	5	6	7	8
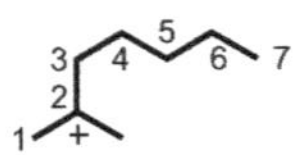	**10**	44.6 (*q*)	332.5 (*s*)	62.7 (*t*)	27.4 (*t*)	31.1 (*t*)	22.3 (*t*)	13.1 (*q*)	
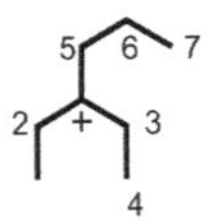	**11**		334.7 (*s*)	51.6 (*t*)	8.1 (*q*)	58.8 (*t*)	18.3 (*t*)	11.6 (*q*)	
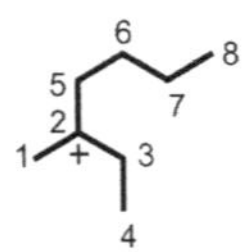	**12**	42.0 (*q*)	334.3 (*s*)	54.7 (*t*)	8.7 (*q*)	59.7 (*t*)	28.5 (*t*)	22.0 (*t*)	12.9 (*q*)
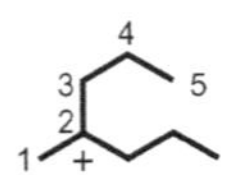	**13**	42.1 (*q*)	332.8 (*s*)	61.9 (*t*)	19.7 (*t*)	12.1 (*q*)			

Table 3.3. $J_{^{13}C-H}$ **Coupling Constants of the Static C_3 to C_8 Alkyl Cations**[95]

Cation		1	2	3	4	5	6	7	8
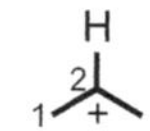	**2**	131.7	171.3						
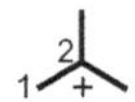	**1**	130.8							
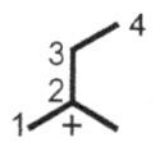	**3**	131.8		127.4	130.8				
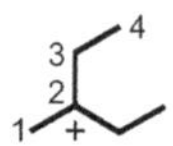	**4**	131.7		124.8	129.6				
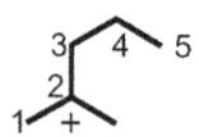	**5**	132.1		126.6	131.8	129.1			
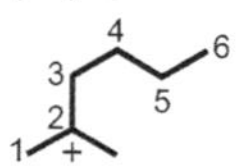	**6**	131.4		126.7	131.4	127.5	126.2		
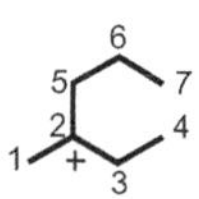	**7**	131.5		124.0	128.8	119.2	126.2	123.9	
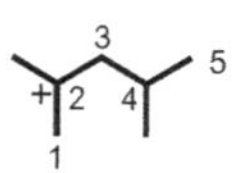	**8**	131.6		124.7	137.2	124.1			
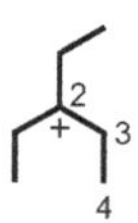	**9**			123.2	129.8				

(continued)

Table 3.3. *(Continued)*

Cation		1	2	3	4	5	6	7	8
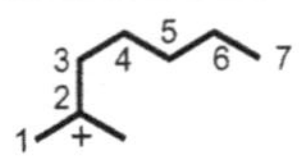	**10**	131.4		127.2	~131	~131	133.1	126.4	
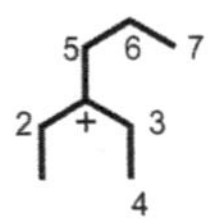	**11**			121.6	130.3	121.1	125.6	124.3	
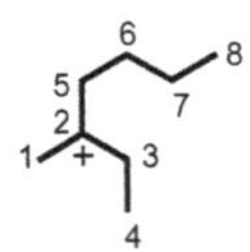	**12**	132.5		~122	129.2	122.4			127.0
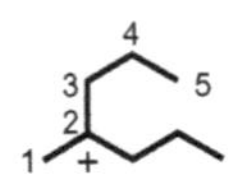	**13**	132.0		126.2	132.85	129.5			

All values are measured in hertz.

cation **14** as an intermediate [Eq. (3.6)] and that the activation energy provides an estimate of the energy difference between primary and secondary carbocations. Another mechanism involving protonated cyclopropane intermediates could not be excluded with the above results.

(I)

$$\underset{\mathbf{2}}{CH_3-\overset{+}{C}H-CH_3} \rightleftharpoons \underset{\mathbf{14}}{CH_3-CH_2-\overset{+}{C}H_2} \rightleftharpoons \underset{\mathbf{2}}{CH_3-\overset{+}{C}H-CH_3} \qquad (3.6)$$

Strong support for the involvement of protonated cyclopropane intermediates came from the work of Olah and White.[94] The isopropyl cation obtained from [2-^{13}C]2-chloropropane (50% ^{13}C) was studied by ^{1}H NMR. The ^{13}C label scrambled uniformly over 1 and 2 positions at $-60°$C within a few hours (half life $\approx$1 h).

The relative rates of hydrogen and carbon interchange have been measured by Saunders et al.[97] using a mixture of [1,1,1-$^2H_3$]- and [2-^{13}C]-labeled isopropyl cations at $-88°$C. The changes in the relative areas of different peaks as well as ^{13}C satellites were observed, and the time dependence of the concentrations of different labeled isomers were simulated using a computer [using mechanisms (**I**), (**II**), and (**III**) in Eqs. (3.6)–(3.8)] A combination of mechanisms (**I**) and (**II**) or of mechanisms (**I**) and (**III**) could match the measurements. The rate for (**I**) was found to 1.5 $\pm$ 0.5 times that of (**II**) or (**III**). Thus, proton interchange is only slightly faster than the carbon interchange. A later study by Saunders et al.[98] using the double-labeled 2-propyl-[2-^{2}H,2-^{13}C] cation confirmed these results. Quenching of the D_3-isopropyl ion by methylcyclopentane and NMR analysis of the D_3-propane product mixture gave preliminary results consistent with mechanisms (**I**) and (**II**) alone. Labeling experiments indicating the intermediacy of protonated cyclopropanes have also been performed by Lee and Woodcock[99] and by Karabatsos et al.[100]

(II)

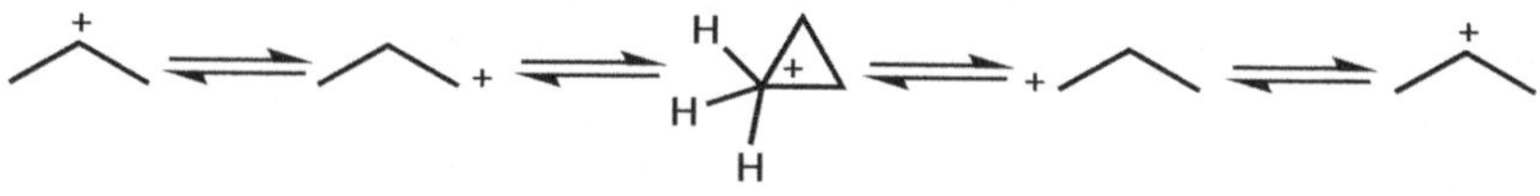

(3.7)

(III)

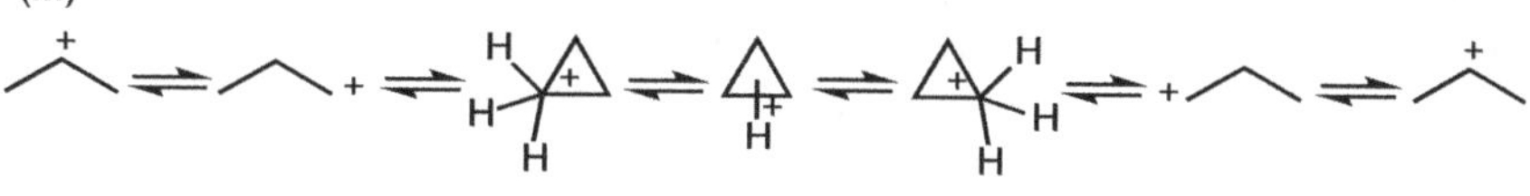

(3.8)

Ab initio theoretical calculations by Pople and co-workers[101] and at the semiempirical level by Dewar and co-workers[102,103] did not give consistent results. Subsequent *ab initio* calculations (SCF level including electron correlation) by Lischka and Köhler[104] are inconsistent with earlier *ab initio* work. Their calculations have confirmed the stability of the isopropyl cation **2** and the instability of face-protonated cyclopropane **15**. Edge-protonated cyclopropane **16** is found to be a saddle point on the potential energy surface of lower energy (by 5 kcal mol^{-1}) than the corner-protonated species **17**.

15 **16** **17**

Subsequent high-level *ab initio* theoretical studies by Dewar et al.[105] (MP4SQ/6-31G* level) and by Koch, Schleyer, and co-workers[106] (MP2/6-311G** level) showed that both the corner-protonated and edge-protonated cyclopropane cations have unsymmetrical structure. Furthermore, the isopropyl cation was shown to be chiral[106–108] with a twisted structure of C_2 symmetry (**18**) [MP4/6-311G**//MP2/6-311G**+ZPE]. Structure **18** is the global minimum lying 7.2 kcal mol^{-1} below the corner-protonated cyclopropane, which is also an energy minimum structure on the potential energy surface[107] and only about 2 kcal mol^{-1} more stable than the edge-protonated species.[109] Both *n*-propyl cations are saddle points.

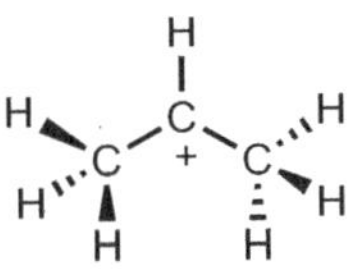

18

Fornarini, Matîre, and co-workers[110] have recently used Fourier transform ion cyclotron resonance (FT–ICR) mass spectrometry assaying the multiphoton dissociation behavior (IR–MPD) of the $C_3H_7^+$ ion. This study has confirmed the conclusions of the computational results discussed above. The IR spectra recorded in solution and in a solid matrix display close resemblance to the spectral characteristics found by the IR–MPD study. Theoretical studies also indicated that the virtually free methyl rotation allows the interconversion of the two enantiomers of the isopropyl cation.

Intermolecular secondary–secondary hydride transfer between **2** and propane in SbF_5–SO_2ClF solution has been observed by Hogeveen and Gaasbeek.[111] The reaction was rapid on the NMR time scale, and a single peak was obtained from the two types of methyl groups down to at least $-100°C$ ($\Delta G^{\ddagger} \leq 6$ kcal mol^{-1}).

Similar scrambling has been documented in the case of secondary butyl **19** cation (see later discussion), and *tert*-pentyl cation **3** and 1-methylcyclopent-1-yl cation.[97,112] Subsequently, it was reported that even *tert*-butyl cation **1** undergoes such scrambling.[113] The *tert*-butyl cation **1** (60% ^{13}C enriched at the cationic center) prepared from 2-choro-2-methylpropane in HSO_3F–SbF_5–SO_2ClF solution undergoes complete carbon scrambling in about 20 h at +70°C (Figure 3.4). The most likely mechanism one can invoke for the scrambling process is the rearrangement through primary isobutyl cation **20** via the delocalized protonated methylcyclopropane intermediate (or transition state) **21** to the secondary butyl cation **19** (Scheme 3.1). The results seem to indicate a lower limit of E_a ~30 kcal mol^{-1} for the

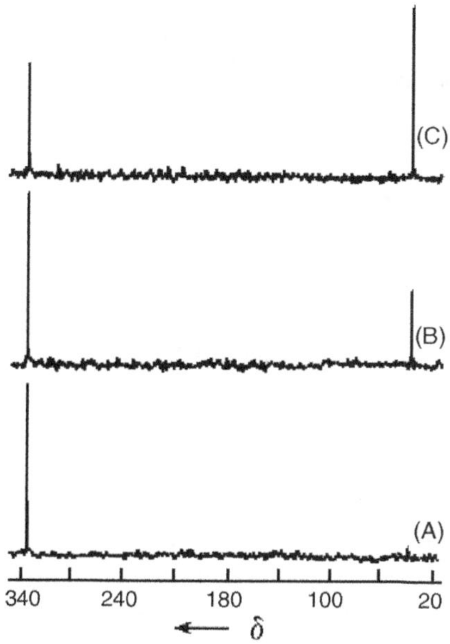

Figure 3.4. ^{13}C scrambling in *tert*-butyl cation (A) immediately after preparation, (B) after 2.5 h at +70°C, (C) after 9.5 h at +70°C.

scrambling process,[97,113] which could correspond to the energy between *tert*-butyl (**1**) and primary isobutyl (**20**) cations.

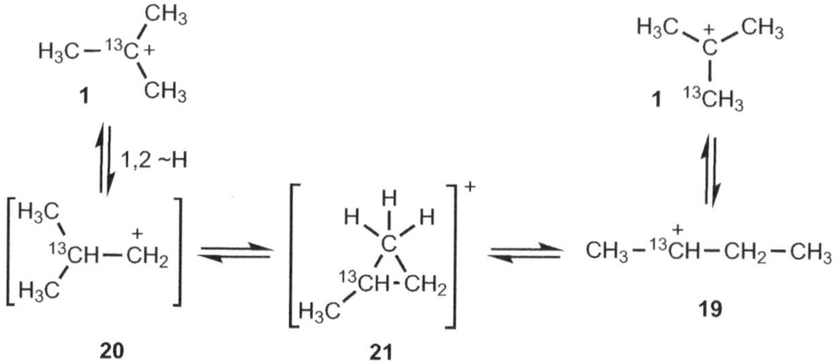

Scheme 3.1

Saunders and co-workers have recently reported results strongly supporting the protonated cyclopropane intermediate in the isotope scrambling process.[114] They found no isomerization of the 2-butyl-1,2-$^{13}C_2$ cation (**19a**) at −78°C, whereas the isotopomer 2-butyl-2,3-$^{13}C_2$ cation (**19b**) showed rapid formation of all other isotopomers except **19a** (Scheme 3.2). These results are consistent with the involvement of protonated cyclopropane with the interchange of either C(1) and C(2) or C(3) and C(4) with the breaking of only the C(2)–C(3) bond.

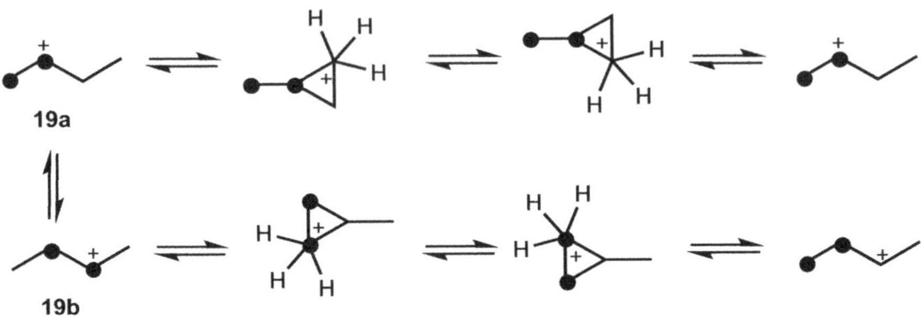

Scheme 3.2

Ab initio theoretical calculations for the 4-methylpent-2-yl cation by Fărcaşiu et al.[115] have shown both a distorted 1-protonated 1,1,2-trimethylcyclopropane and the open cation to be energy minima along the reaction coordinate at the B3LYP/6-31G** level. In contrast, only the protonated cyclopropane was found to be an energy minimum in the MP2/6-31G** optimization. Whereas the open cation was a transition structure at this level, a coupled cluster geometry optimization (CCSG/6-31G**) showed that the open ion is also a true energy minimum.

Infrared and Raman spectra of alkyl cations give valuable information of their structures (Table 3.4).[71–73] The Raman spectroscopic data provide strong evidence that *tert*-butyl cation in Magic Acid solution prefers a conformation leading to overall C_{3v} point group symmetry (Table 3.4 and Figure 3.5). Thus the $(CH_3)_3C^+$ ion exists in these solutions with a planar carbon skeleton and with one hydrogen atom of each CH_3 group above the plane. The other two hydrogen atoms are arranged symmetrically below the $^+C{-}C_3$ plane to the right and left of the C_3 axis. Raman spectra observed for the *tert*-pentyl cation, the pentamethylethyl cation, and the tetramethylethyl cation also show similar structures. The Raman spectroscopic studies thus provide, in addition to ^{13}C NMR data, direct evidence for the planar carbenium center of alkyl cations.

The gas-phase infrared spectrum of the *tert*-butyl cation generated from *tert*-butyl chloride by a high voltage discharged in a pulsed supersonic expansion has recently been reported by Duncan and co-workers.[116] The infrared photodissociation spectrum measured by means of argon tagging matches those predicted for the C_1 or C_s geometries and differs significantly from those calculated for isomeric $C_4H_9^+$ cations (methyl-bridged and proton-bridged 2-butyl cations). Despite the significantly differing conditions, the band positions of the gas-phase spectrum and those determined by Olah et al.[71] in superacid solution show remarkable similarity (Table 3.5). A significant difference is the resolution of the band at 1290 cm^{-1} into a triplet.

Evidence for planarity or near planarity of the sp^2 center of trivalent alkyl cations thus comes from the combined results of NMR (^1H and ^{13}C), IR, and Raman spectroscopy.[71–73,97]

In the electronic spectra (UV), alkyl cations in HSO_3F–SbF_5 at −60°C show no absorption maxima above 210 nm.[74]

X-ray photoelectron (ESCA) spectra of carbenium ions have also been obtained in frozen superacid solutions, generally in SbF_5–SO_2 solution or as isolated salts. Sulfur dioxide was subsequently removed by the usual freeze–thaw procedure. A thin layer of the viscous SbF_5 solution was deposited on the precooled sample holder, in a dry nitrogen

Table 3.4. Raman and IR Frequencies of the *tert*-Butyl Cation and [D_9]-*tert*-Butyl Cation and Their Correlation with Those of $(CH_3)_3B$ and $(CD_3)_3B$[71,72]

Species	ν_1, ν_{12}, ν_7, ν_{19}	ν_1, ν_{13}	ν_{21}	ν_{14}	ν_{15}	ν_{17}	ν_5	ν_{16}	ν_6	ν_9	ν_{10}	ν_{18}
$(CH_3)_3C^+$ (**1**)	2947	2850		1450		1295			667		347	306
$(CH_3)_3B$	2975	2875	1060	1440	1300	1150	906	866	675	973 (486?)	336[a]	320
$(CD_3)_3C^+$	2187	2090		1075		980			720		347	300
$(CD_3)_3B$	2230	2185		1033	1018	1205			620	870	(289)[b]	(276)[b]

[a]IR frequency.
[b]Calculated.

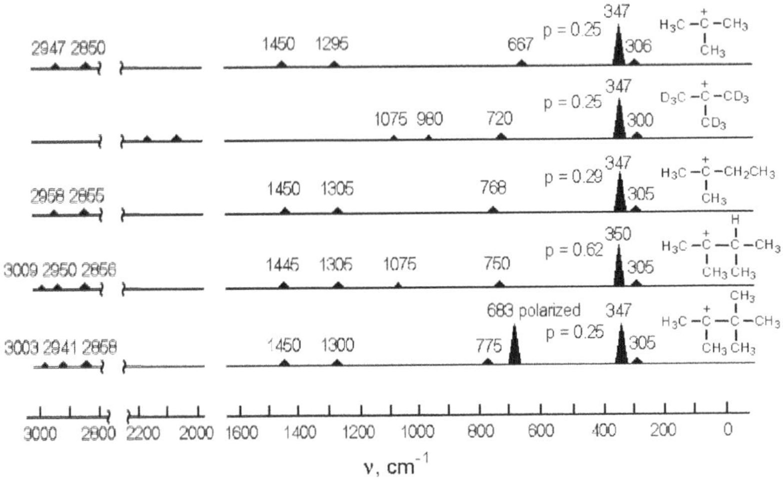

Figure 3.5. Schematic representation of Raman spectra of alkyl cations.

atmosphere. The spectra were recorded at liquid nitrogen temperature.[117] The photoelectron spectrum of *tert*-butyl cation **1** is shown in Figure 3.6. The lower traces in Figure 3.6 represent the results given by a curve resolver. The peak area ratio is close to 1 : 3.

The experimental carbon 1*s* binding energy difference (3.9 eV) between the carbenium ion center and the remaining three carbon atoms is in the limit of that predicted by *ab initio* calculation (4.45 eV). Comparable results were obtained for the *tert*-pentyl cation ($\Delta E_{b+\text{C-C}} = 4 \pm 0.2$ eV).

Attempts have been made to prepare sterically highly crowded tertiary cations (see later discussion). The possibly most hindered trialkylmethyl cation, the tris(1-adamantyl)methyl ion **22**, has been prepared from the corresponding chloride[118] [Eq. (3.9)]. Proton loss is not favored in this case because this would lead to the formation of a bridgehead alkene. Consequently, cation **22** could be observed as a persistent, long-lived cation.

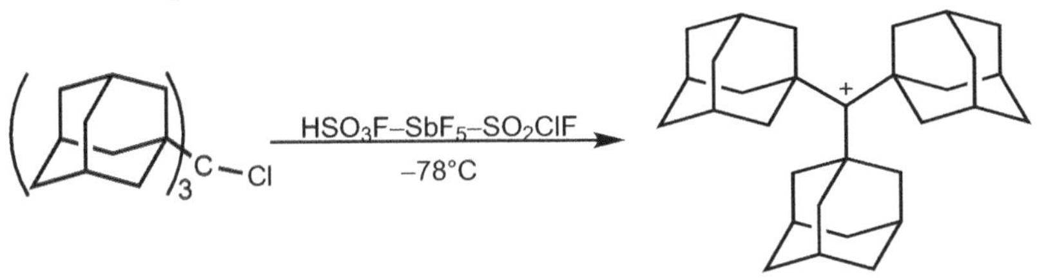

22

(3.9)

Table 3.5. IR Frequencies of the *tert*-Butyl Cation[a]

Condition	Frequencies (cm^{-1})							Reference
Gas phase	970	1098	1278	1302	1316	1465	2834	116
Theoretical	936	1062	1253	1283	1325	1445	2833	116
In superacid	962	1070		1290		1455	2830	71

[a]Only major absorption bands are given.

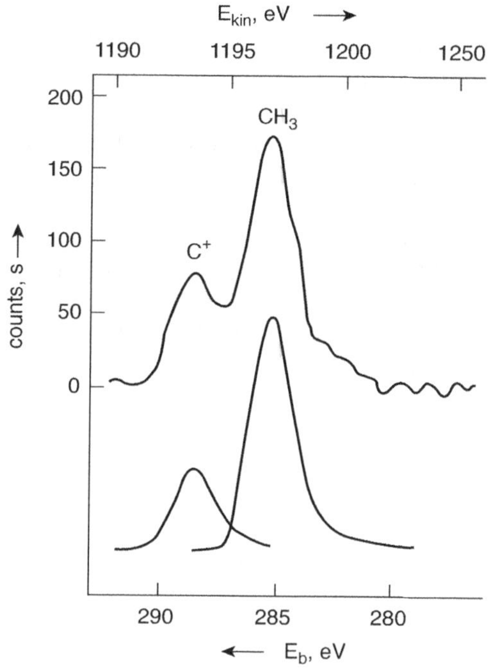

Figure 3.6. Carbon 1*s* photoelectron spectrum of *tert*-butyl cation **1**.

Yannoni, Myhre, and co-workers[119] applied nutation NMR spectroscopy to determine the geometry of the *tert*-butyl cation using double ^{13}C-labeled isotopomers at 77 K. The central carbon–methyl carbon bond distance was found to be 1.46–1.47 Å—that is, almost 0.1 Å shorter than the usual single-bond value. This bond length is in close agreement with the 1.457–1.459 Å found by Schleyer, Koch, and co-workers at the MP2(full)/6-31G** level.[120] In this study, three forms of the *tert*-butyl cation **1** (C_s, C_{3h}, and C_{3v}) were calculated and compared. The C_s form was found to be more stable than C_{3h} by only 0.1 kcal mol^{-1} [MP4sdtq/6-31G**//MP2(full)/6-31G**], but after correction for zero-point energy the order reverses. This indicates that the potential energy surface of **1** is extremely flat. In contrast, structure C_{3v} is 1.1 kcal mol^{-1} higher in energy than C_{3h}.

According to the X-ray crystal structure analysis of the *tert*-butyl cation ($Sb_2F_{11}^-$ salt) determined by Hollenstein and Laube,[121,122] the average bond distance is 1.442 Å, the C–C–C bond angle is 120°, and the carbon skeleton has approximate D_{3h} symmetry. Kato and Reed[123] have recently used methyl carboranes [$Me(CHB_{11}Me_5X_6)$, X = Cl, Br)] to generate tertiary carbocations, including the *tert*-butyl and *tert*-pentyl cations, and determined their X-ray crystal structure. The *tert*-butyl cation is planar within experimental error (sum of bond angles = 360°), and the C–C bond distances are 1.429, 1.438, and 1.459 Å. As observed in both studies, the shorter C–C bond distances reflect partial double-bond character from substantial C–H hyperconjugation. In the *tert*-pentyl cation **3**, a dihedral angle of 25.8° was detected between the carbocation plane and the C(2)–C(3)–C(4) plane. This shows the near absence of C–C hyperconjugation, which was, however, suggested to exist by calculation having the C(2)–C(3)–C(4) plane aligned (parallel) with the vacant *p* orbital.[124]

A recent computational study by Olah, Prakash, and Rasul (*ab initio* methods at the MP2/cc-pVTZ level) has also shown that the structure of the *tert*-pentyl cation is stabilized by both C–C and C–H hyperconjugations.[125] Both structures **23** and **24** have been found to be minima on the potential energy surface, and they are very close energetically (structure **23** is more stable by a mere 0.5 kcal mol^{-1}). Their interconversion has a kinetic barrier of only 1.5 kcal mol^{-1}. The calculated average ^{13}C NMR shift value of the cationic center ($\delta^{13}C$ 334.6, GIAO-MP2 method) matches extremely well with both the experimental value ($\delta^{13}C$ 335.4)[95] and that calculated by Schleyer *et al.* ($\delta^{13}C$ 335.7, IGLO method).[124]

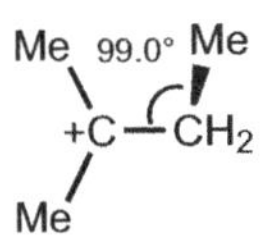

23

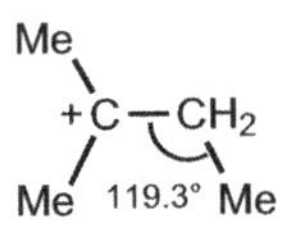

24

3.4.1.3. Preparation from Other Precursors. Alkyl cations can be formed not only from halide precursors (the earlier investigation of generation from alkyl fluorides was later extended to alkyl chlorides, bromides, and even iodides) but also from alkenes in superacids like $HF–SbF_5$ [Eq. (3.10)].

$$\mathrm{RCH{=}CH_2} \xrightarrow[\mathrm{HF{-}SbF_5}]{\mathrm{HSO_3F{-}SbF_5}} \mathrm{R\overset{+}{C}H{-}CH_3} \qquad (3.10)$$

Tertiary and reactive secondary alcohols in superacids like $HSO_3F–SbF_5$ (Magic Acid), HSO_3F, and $SbF_5–SO_2–SO_2ClF$ also ionize to the corresponding carbocations.[126] The generation of alkyl cations from alcohols indicates the great advantages of increasing acidity and of using acid systems with low freezing points. Deno showed that the use of sulfuric acid and oleum results in formation of cyclized allylic ions from simple aliphatic alcohols.[88] With the use of extremely strong acid, $HSO_3F–SbF_5$, tertiary and many secondary alcohols can be ionized to the corresponding alkyl cations [Eq. (3.11)].

$$\mathrm{(CH_3)_3COH} \xrightarrow{\mathrm{HSO_3F{-}SbF_5}} \underset{\mathbf{1}}{\mathrm{(CH_3)_3C^+}} \qquad (3.11)$$

Primary and less reactive secondary alcohols are protonated in $HSO_3F–SbF_5$ solution at low temperature (−60°C) and show very slow reaction rates[127] [Eqs. (3.12) and (3.13)].

$$CH_3CH_2CH_2OH \xrightarrow{HSO_3F\text{–}SbF_5} CH_3CH_2CH_2\overset{+}{O}H_2 \qquad (3.12)$$

$$CH_3\text{–}\underset{\displaystyle OH}{CH}\text{–}CH_3 \xrightarrow{HSO_3F\text{–}SbF_5} CH_3\text{–}\underset{\displaystyle \overset{}{O}H_2^{+}}{CH}\text{–}CH_3 \qquad (3.13)$$

Temperature-dependence studies of the NMR spectra of protonated alcohols from −60 to +60°C allow the kinetic of dehydration to be followed[128] [Eq. (3.14)].

$$CH_3CH_2CH_2CH_2OH \xrightarrow[SO_2]{HSO_3F\text{–}SbF_5} CH_3CH_2CH_2CH_2\overset{+}{O}H_2 \xrightarrow[-H_3O^+]{\Delta,\ H^+} [CH_3CH_2CH_2CH_2^+] \longrightarrow$$

$$\xrightarrow{\text{rearrangement}} \underset{\displaystyle \mathbf{1}}{(CH_3)_3C^+} \qquad (3.14)$$

Antimony pentafluoride itself (neat or in SO_2 or SO_2ClF solution) also ionizes alcohols to form alkyl carbocations [Eq. (3.15)].

$$R\text{–}OH \xrightarrow[SO_2 \text{ or } SO_2ClF]{SbF_5SO_2} R^+\ SbF_5OH^- \qquad (3.15)$$

To overcome difficulties and achieve ionization of primary (and less reactive secondary) alcohols at low temperatures, it was found, in some cases, that it is advantageous to convert them to the corresponding haloformates or halosulfites with carbonyl halides [Eq. (3.16)] or thionyl halides [Eq. (3.17)]. These, in turn, ionize readily in SbF_5–SO_2 solution and lose CO_2 or SO_2.[129]

$$\underset{X\,=\,Cl,\,F}{ROH} \xrightarrow{COX_2} ROCOX \xrightarrow[\substack{SO_2 \text{ or } SO_2ClF \\ -78°C}]{SbF_5SO_2} R^+\ SbF_5X^- + CO_2 \qquad (3.16)$$

$$\underset{X\,=\,Cl,\,F}{ROH} \xrightarrow{SOX_2} ROSOX \xrightarrow[\substack{SO_2 \text{ or } SO_2ClF \\ -78°C}]{SbF_5SO_2} R^+\ SbF_5X^- + SO_2 \qquad (3.17)$$

Sterically highly crowded tertiary cations have long eluded characterization. In an early hydrolysis study of tris-*tert*-butylmethyl *para*-nitrobenzoate, Bartlett and Stiles[130] reported the formation of a C_{13} alkene formed as a result of repeated successive alkyl migrations. Under superacid conditions, tris(*tert*-butyl)methanol gave a mixture of equilibrating cations formed by a cleavage mechanism.[131] The secondary bis(1-adamantyl)methyl cation **25a**, in turn, undergoes fast rearrangement to yield a set of equilibrating 4-(1-adamantyl)-3-homoadamantyl cations.[132] However, tertiary ions [methyl-, ethyl-, and phenyl-bis(1-adamantyl)methyl cations **25b–25d**], with the exception of the *tert*-butyl-bis(1-adamantyl)methyl cation (**25e**), could be prepared and characterized under superacidic conditions.[132]

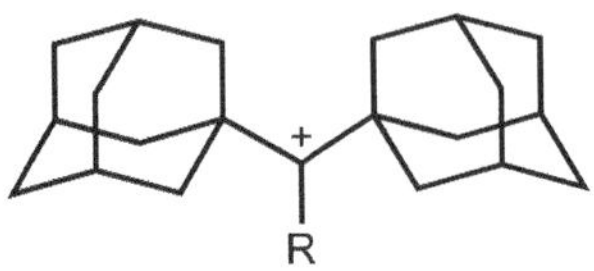

R = H **a**, Me **b**, Et **c**, Ph **d**, *tert*-Bu **e**

25

Aliphatic ethers are protonated in strong acids, at low temperatures; the exchange rates of the acidic proton are slow enough to permit direct observation by NMR spectroscopy.[133] Temperature-dependent NMR spectral studies allow us again to follow the kinetics of ether cleavage to form alkyl cations [Eq. (3.18)].

$$CH_3OCH_2CH_2CH_2CH_3 \xrightarrow[-60°C]{HSO_3F-SbF_5} CH_3\overset{+}{O}(H)CH_2CH_2CH_2CH_3 \xrightarrow{H^+} CH_3\overset{+}{O}H_2 + [CH_3CH_2CH_2CH_2^+] \xrightarrow{\text{rearrangement}} \underset{\mathbf{1}}{(CH_3)_3C^+} \quad (3.18)$$

Superacids such as $HSO_3F–SbF_5$ also act as very efficient hydrogen-abstracting agents, allowing the generation of carbocations from saturated hydrocarbons[134] [Eq. (3.19)].

$$CH_3-\underset{CH_3}{\underset{|}{CH}}-CH_3 \xrightarrow[SO_2ClF]{HSO_3F-SbF_5} \underset{\mathbf{1}}{(CH_3)_3C^+} \quad (3.19)$$

Alkyl cations can also be generated by decarbonylation of tertiary acylium ions, like the pivaloyl cation **26** [Eq. (3.20)].[91] This reaction corresponds to the reverse of Koch–Haaf acid synthesis, which is known to involve carbocation intermediates. Indeed the reaction of the *tert*-butyl cation with carbon monoxide gives the pivaloyl cation.[91,135]

$$(CH_3)_3CCOF + SbF_5 \longrightarrow \underset{\mathbf{26}}{[(CH_3)_3CCO]^+SbF_6^-} \xrightarrow{\Delta} \underset{\mathbf{1}}{(CH_3)_3C^+} + CO \quad (3.20)$$

Protonation and ionization of mercaptans (thiols) and thioethers (sulfides) have been similarly studied. Thiols and sulfides can also be used, similar to their oxygen analogs, as precursors for alkyl cations[136] [Eq. (3.21)]. Ionization with SbF_5-type superacids generally necessitates somewhat more forcing conditions (higher temperatures). Alkyl thiohaloformates also form alkyl cations via fragmentative ionization[137] [Eq. (3.22)].

$$\mathrm{RSH} \xrightarrow[\text{or } HSO_3F\text{–}SbF_5]{SbF_5 \text{ or } HF\text{–}SbF_5} \mathrm{R^+\,Y^-} \qquad Y = SbF_5SH,\ SbF_6,\ SbF_5FSO_3 \tag{3.21}$$

$$\underset{X = F,\ Cl}{\mathrm{RSCOX}} \xrightarrow{SbF_5\text{–}SO_2ClF} \mathrm{R^+\,SbF_5X^-} + \mathrm{COS} \tag{3.22}$$

Amines also can be used as precursors for the generation of alkyl cations. The classic method of deaminative formation of carbocations involves some type of diazotation reaction producing an equimolar amount of water [Eq. (3.23)].

$$\mathrm{RNH_2} \xrightarrow[-H_2O]{HNO_2} [\mathrm{RN_2^+}] \longrightarrow \mathrm{R^+} + \mathrm{N_2} \tag{3.23}$$

Newer methods overcome this difficulty. The corresponding sulfinylamine or isocyanate is first prepared and then reacted with stable nitrosonium salts to give the corresponding carbocations[138] [Eqs. (3.24) and (3.25)].

$$\mathrm{RNH_2} \xrightarrow{SOCl_2} \mathrm{RNSO} \xrightarrow{NO^+SbF_6^-} \mathrm{R^+SbF_6^-} + \mathrm{N_2} + \mathrm{SO_2} \tag{3.24}$$

$$\mathrm{RNH_2} \xrightarrow{COCl_2} \mathrm{RNCO} \xrightarrow{NO^+SbF_6^-} \mathrm{R^+SbF_6^-} + \mathrm{N_2} + \mathrm{CO_2} \tag{3.25}$$

Lambert et al.[139] generated the β-triethylsilyl-substituted carbocation **27** reacting solvated triethylsilylium ion with 1,1,-diphenylethylene [Eq. (3.26)]. β-Silyl-substituted carbocations are of high interest because of the β-effect, that is, the ability of a silicon atom to stabilize a nonadjacent positive charge of a carbocationic center via hyperconjugation. Results have been summarized in reviews.[140–142]

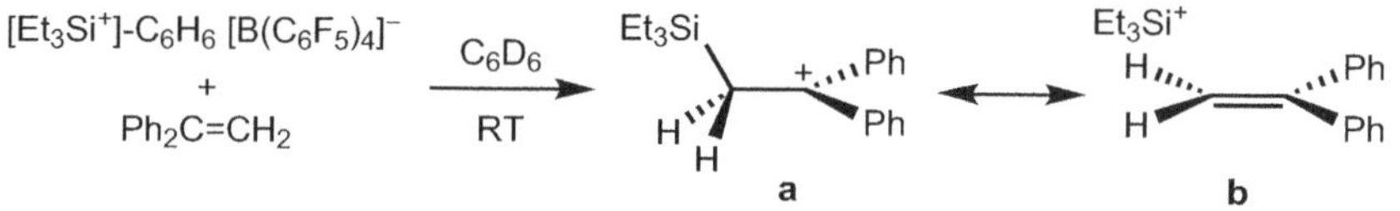

27

(3.26)

The resonance of the cationic carbon (δ^{13}C 225.4) is definitive for the cation structure **27a**. The single ^{29}Si shift (δ 46.2) is of higher frequency than neutral tetrahedral silicon, indicating that silicon bears some positive charge, that is, a small contribution from resonance structure **27b**. The $Bu_3GeCH_2C^+Ph_2$ analog, in turn, showed increased hyperconjugation. No evidence was found for an Si-bridging structure.

3.4.1.4. Observation in Different Superacids. Whereas antimony pentafluoride-containing superacids (such as $HF–SbF_5$, $HSO_3F–SbF_5$, $CF_3SO_3H–SbF_5$, etc.) are the preferred solvents for obtaining alkyl cations, other nonoxidizing superacids such as $HF–BF_3$ and $HF–TaF_5$ can also, on occasion, be used successfully. The stability of carbocations in these solvents is generally somewhat lower.

3.4.2. Cycloalkyl Cations

Tertiary cycloalkyl cations, such as the 1-methylcyclopent-1-yl cation **28**, show high stability in strong acid solutions. This ion can be obtained from a variety of precursors (Figure 3.7).[143,144] It is noteworthy to mention that not only cyclopentyl- but also cyclohexyl-type precursors give 1-methylcyclopent-1-yl cation **28**. This indicates that the cyclopentyl cation has higher stability, which causes isomerization of the secondary cyclohexyl cation to the tertiary methylcyclopentyl ion.

The methylcyclopentyl cation **28** undergoes both carbon and hydrogen scrambling.[97,112] An activation energy of $15.4 \pm 0.5\ \text{kcal mol}^{-1}$ was measured for a process that interchanges α and β hydrogens in **28**. Above 110°C, coalescence to a single peak was observed in the ^1H NMR spectrum of **28**. An activation energy barrier of $18.2 \pm 0.1\ \text{kcal mol}^{-1}$ was found for the methyl and ring hydrogen interchange. The

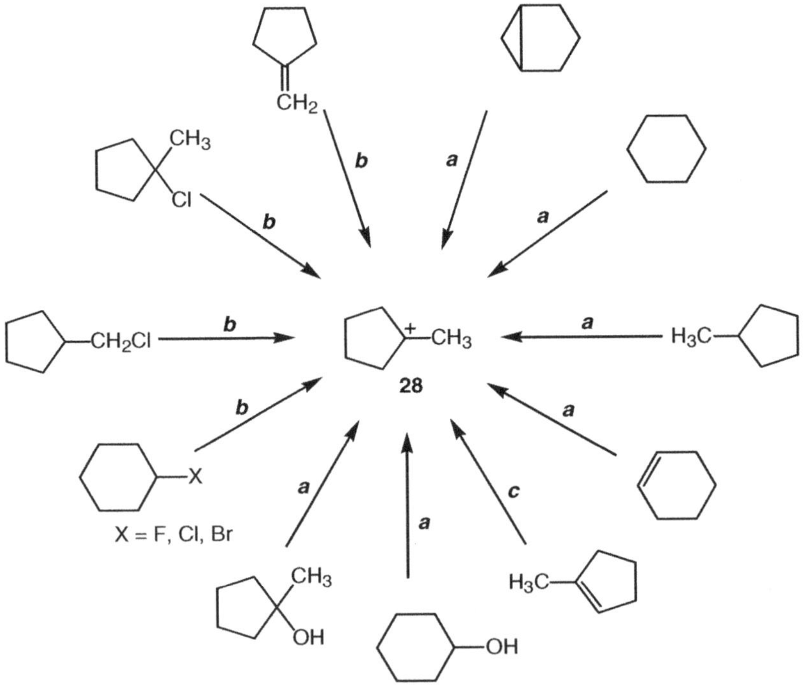

Figure 3.7. Preparation of 1-methylcyclopent-1-yl cation **28** from various precursors. Reaction conditions: (a) $HSO_3F–SbF_5$, (b) $SbF_5–SO_2$, (c) $HF–SbF_5–SO_2$.

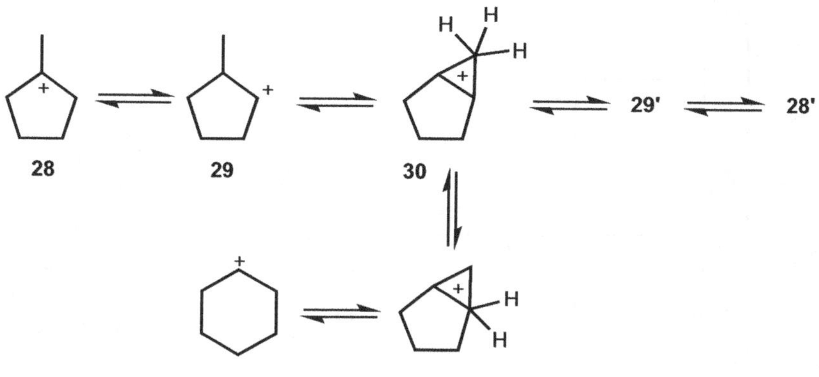

28 29 30

32 31

Scheme 3.3

protonated cyclopropane intermediate **30** required for the process (Scheme 3.3) might undergo an additional reversible ring opening to secondary cyclohexyl cation **32**. To investigate this possibility, Saunders and Rosenfeld[145] measured deuterium and carbon scrambling rates simultaneously using a mixture containing ^{13}C- and deuterium-labeled methyl groups in methylcyclopentyl cation **28**. The two observed rates were related in exactly the manner predicted if the protonated bicyclohexane **31** opens to **32** more rapidly than it returns to **30** via a corner-to-corner hydrogen shift.

A recent X-ray characterization of the cyclopentyl cation **33** generated with methyl carboranes has shown a perfect planar structure around the carbocation center [bond angles: $CH_3{-}C^+{-}CH_2 = 124.9$ and $125.2°$, $CH_2{-}C^+{-}CH_2 = 109.9°$; bond distances: $CH_3{-}C^+ = 1.46$ Å, $CH_2{-}C^+ = 1.45$ Å).[123]

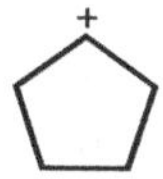

33

The cyclopentyl cation **33** shows in its proton NMR spectrum in $SbF_5{-}SO_2ClF$ solution, even at $-150°C$, only a single absorption line at δ^1H 4.75.[143,146,147] This observation indicates a completely degenerate ion with a very low barrier to the secondary–secondary hydrogen shift (see Section 3.5.2.1). Although stable secondary cyclohexyl cation **32** is unknown under long-lived stable ion conditions, a static secondary cyclohexyl cation **34** with an α-spiro-cyclopropyl group (spiro[2.5]oct-4-yl cation) has been prepared and studied by various routes[148,149] (Scheme 3.4). However, this ion derives its stability by substantially delocalizing its positive charge into the neighboring cyclopropyl group (i.e., it is also a cyclopropylcarbinyl cation). The bisected nature of the cyclopropyl group is indicated by a single ^{13}C NMR signal for the cyclopropyl methylene group. Ion **34** is stable up to $-10°C$, where it rearranges to the equilibrating bicyclo[3.3.0]oct-1-yl cation **35**.

Sorensen and co-workers[150] have prepared tertiary cycloalkyl cations of different ring sizes, $n=4$ (small ring), $n=5$–7 (common rings), $n=8$–11 (medium rings),

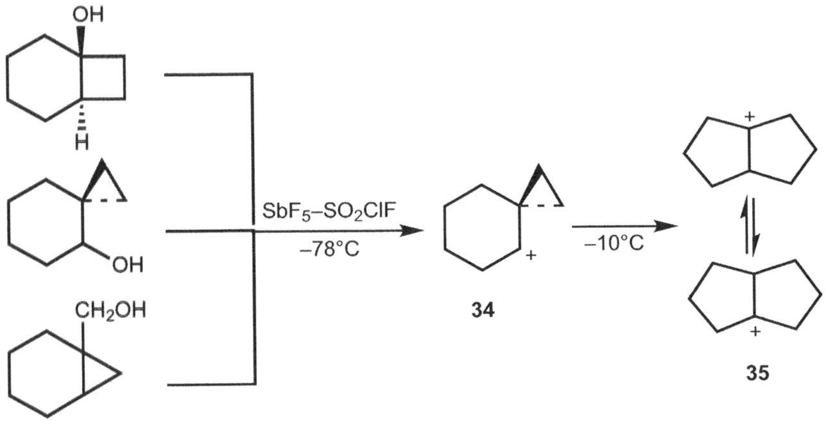

Scheme 3.4

$n = 12$–20 (large rings). These ions were in general found to undergo ring expansion or contraction reactions, often in multiple or repetitive steps as shown in the following sequence. At very low temperature, some of these cycloalkyl cations ($n = 8$–11) show μ-hydrido bridging (see Section 3.5.2.6).

4^+-Pr	⟶	5^+-Et + 6^+-Me
5^+-Et	⟶	6^+-Me
7^+-Pr	⟶	6^+-Bu
8^+-Et	⟶	6^+-Bu
9^+-Me	⟶	8^+-Et + 7^+-Pr + 6^+-Bu
10^+-Me	⟶	6^+-Pen
10^+-Et	⇌	11^+-Me
11^+-Me	⇌	10^+-Et ⇌ 6^+-Hex
10^+-Pr	⟶	12^+-Me + 6^+-Hept
11^+-Et	⟶	12^+-Me + 10^+-Pr
12^+-Pr	⟶	13^+-Et
13^+-Et	⟶	14^+-Me
14^+-Et	⟶	15^+-Me
12^+-R	⟶	ring expansion

Tertiary (2-propyl)cycloalkyl cations were prepared from the corresponding alcohols[151] using the technique developed by Saunders.[46] The ions undergo fast nondegenerate 1,2-hydride shifts, and the direction of equilibria investigated by ^{13}C NMR spectroscopy was found to depend on the ring size. For the cyclopentyl derivative, the cyclopentyl cation **36b** is more stable [Eq. (3.27)], whereas for the cyclohexyl derivative the cyclohexyl cation **37a** is the more stable structure [Eq. (3.28)]. Calculations [MP2/6-31G(d) level] found three minimum energy structures for the cyclohexyl derivative, including two conformers of **37b**: one with the empty p orbital occupying an equatorial-like position, the other with an axial-like position. In both structures, one of the methyl groups is oriented for C—C hyperconjugation with the empty *p*-orbital of C^+. Similar conformations called hyperconjomers were also calculated for the 1-methylcyclohexyl cation.[152]

36a **36b** (3.27)

37a **37b** (3.28)

A series of aryl-substituted cycloalkyl cations **38** and **39** and other aryl-substituted cyclic systems (**40**, **41**, **42**) have been studied in connection with the application of the tool of increasing electron demand.[67,153–156]

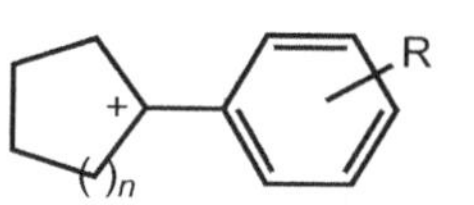

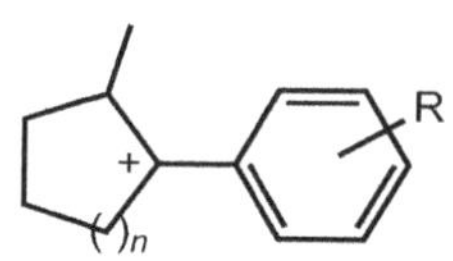

n = 0–2 X = H, 4-Me, 4-MeO, 4-Br, 4-Cl, 4-F *n* = 1, 2
3-CF_3, 4-CF_3, 3,5-$(CF_3)_2$, etc.

38 **39**

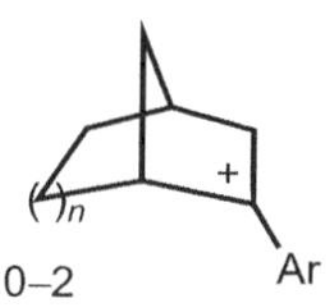

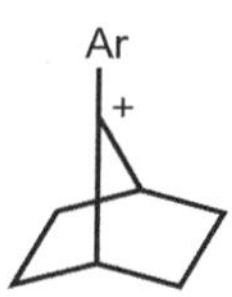

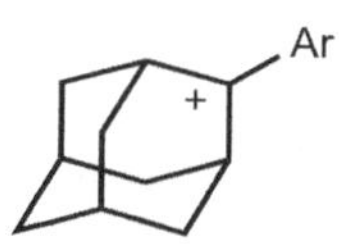

n = 0–2

40 **41** **42**

The X-ray crystal structure of the 2-phenyladamant-2-yl cation (**43**) determined by Laube and Hollenstein[157] shows remarkable features. Namely, the displacement of the C(2) bridge by 7.8°, the additional bending of the phenyl ring out of plane by 5.6°, and a slight pyramidalization of $\Delta = 0.049$ Å (pyramidalization Δ: distance of a tricoordinate atom from the plane through its neighbors). Furthermore, the C_α–C_β bond distance is shorter compared with the reference value (1.45 Å versus 1.51 Å), whereas the C_β–C_γ bond is elongated (1.58 Å versus 1.53 Å). These features are consistent with the C—C hyperconjugation between the cation center and the C_β–C_γ bond.

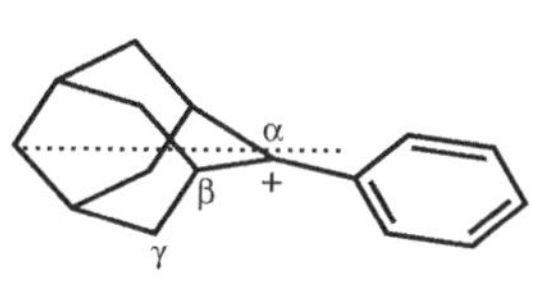

43

The direct observation of the cyclopropyl cation **44** has evaded all attempts, owing to its facile ring opening to the energetically more favorable allyl cation **45** [Eq. (3.29)].[158]

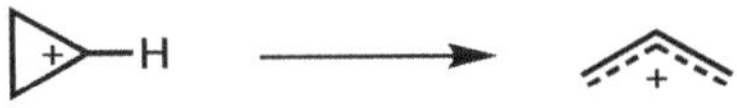

(3.29)

44 **45**

Many cyclopropyl cation precursors indeed readily rearrange to allyl cations under stable ion conditions.[159–162] However, a distinct cyclopropyl cation **46** showing a significant 2π aromatic nature has been prepared by the ionization of 11-methyl-11-bromotricyclo[4.4.1.0^{1,6}]undecane in SbF_5–SO_2ClF solution at −120°C[163] [Eq. (3.30)].

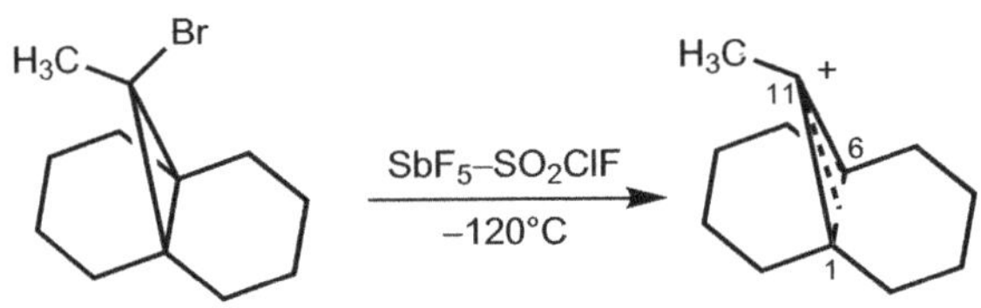

(3.30)

46

The direct observation of ion **46** is of particular interest in that it clearly does not involve a significantly opened cyclopropane ring, which could lead to the formation of an allylic cation. Thus, it must be considered as a bent cyclopropyl cation.[164–170] It is, however, clear that the C_1–C_6 σ-bond must to some degree interact with the empty *p* orbital at C_{11} and that "homoconjugation" between them becomes the important factor in stabilizing such a "bent" cyclopropyl species.

The parent secondary cyclobutyl cation **47** undergoes threefold degenerate rearrangement via σ-bond delocalization involving nonclassical bicyclobutonium ion-like system[171,172] (see Section 3.5.2.5). Similar behavior is also observed for the 1-methylcyclobutyl cation **48a**.[173–176] The 1-phenylcyclobutyl cation **48b**, on the other hand, is a trivalent tertiary carbocation.

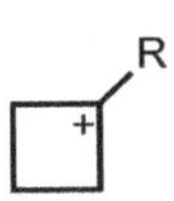

R = Me **a**
R = Ph **b**

47 **48**

3.4.3. Bridgehead Cations

Bredt's rule in its original form[177,178] excluded the possibility of carbocation formation at bridgehead positions of cycloalkanes. Indeed, bridgehead halides, such as apocamphyl chloride, proved extremely unreactive under hydrolysis conditions.[179] However, 1-bromoadamantane very readily gives the bridgehead carboxylic acid under the usual conditions of Koch–Haaf acid synthesis.[180] 1-Fluoromoadamantane is ionized in SbF_5 to give the stable bridgehead 1-adamantyl cation **49** [Eq. (3.31)].[181,182]

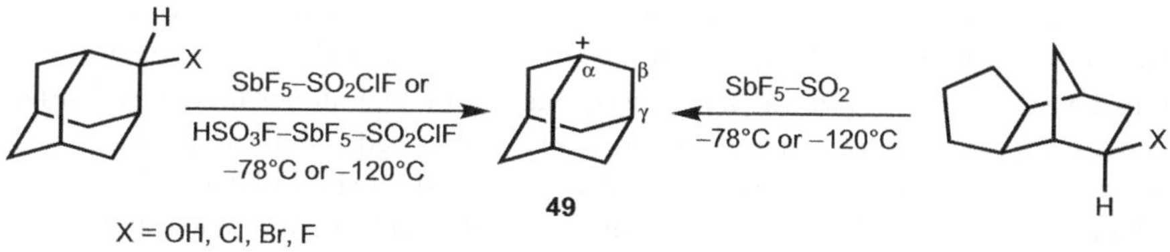

(3.31)

The proton NMR spectrum of the 1-adamantyl cation **49** in SbF_5 solution at 25°C consists of resonances at δ^1H 5.40, 4.52, and 2.67 with peak areas of 3:6:6 (Figure 3.8*a*). The ^{13}C NMR spectrum (Fig. 3.8*b*) shows the γ-carbons more deshielded than the β-carbon atoms, indicating strong C—C bond hyperconjugation with the empty *p* orbital. The bridgehead 1-adamantyl cation **49** can also be prepared

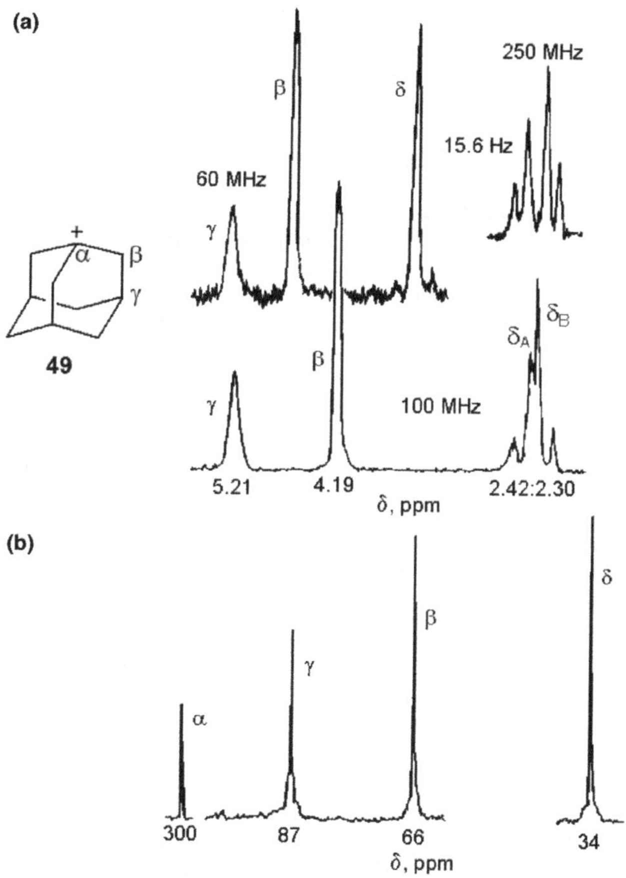

Figure 3.8. (**a**) ^{1}H NMR spectrum of the 1-adamantyl cation at 60 MHz, 100 MHz, and 250 MHz; (**b**) Fourier-transform ^{13}C NMR spectrum of the 1-adamantyl cation (in HSO_3F–SbF_5).

from 2-adamantyl as well as trimethylenenorbornyl precursors[182] [Eq. (3.31)]. A subsequent solid-state ^{13}C NMR spectral study of 1-adamantyl hexafluoroantimonate salt using magic-angle spinning and cross-polarization techniques indicates similarities between solid-state spectra and previously obtained solution spectra.[182] The 1-adamantyl cation **49** has been calculated to be more stable by 11 kcal mol^{-1} than the 2-adamantyl cation.[183] The IR spectrum of cation **49** was reported[184] and found to be in reasonably good agreement with the theoretical spectrum.[183]

Several methyl-substituted bridgehead adamantyl cations **50** have been prepared and characterized. The X-ray structure of the 3,5,7-trimethyladamant-1-yl cation has been determined by Laube.[185,186] The C_α–C_β bond distance (1.439 Å) is slightly shorter than, whereas the C_β–C_γ bond length (1.612 Å) is somewhat longer than, the corresponding reference value (1.502 Å and 1.538 Å, respectively). This supports the C–C hyperconjugation in the bridgehead adamantyl cation. The sp^2 carbocationic center has a pyramidal structure with a pyramidalization value $\Delta = 0.212$ Å. The unusual geometry of the cation center is also indicated by the C_β–C_α–$C_{\beta'}$ and C_α–C_β–C_γ bond angles (117.9° and 99.7°, respectively). The bridgehead homoadamantyl cation **51** has been obtained[187] from both adamantylcarbinyl and homoadamantyl precursors [Eq. (3.32)].

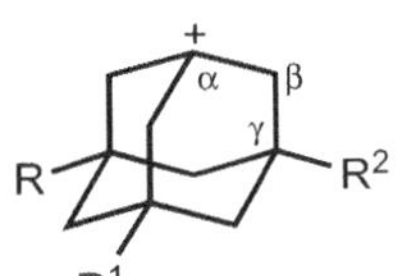

R = CH_3, R^1, R^2 = H
R, R^1 = CH_3, R^2 = H
R, R^1, R^2 = CH_3

50

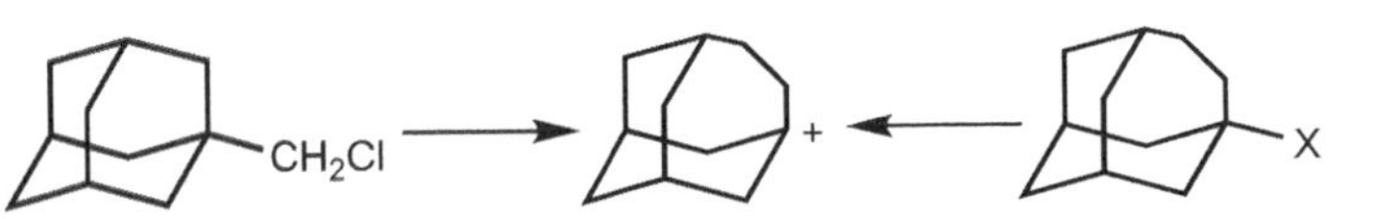

(3.32)

51 X = OH, Cl, Br

Bridgehead bicyclo[4.4.0]decyl, bicyclo[4.3.0]nonyl, and bicyclo[3.3.0]octyl cations are found to be rapid equilibrating ions (see Section 3.5.2.1).[188] The isomeric bridgehead congressane (diamantane) cations **52** and **53** have been prepared and observed.[182] The diamant-4-yl cation **52** rapidly rearranges to the diamant-1-yl cation **53** at −60°C, possibly through intermolecular hydride shifts. Bridgehead bicyclo[3.3.3]undec-1-yl cation **54** has also been observed by ^1H and ^{13}C NMR spectroscopy.[189]

52

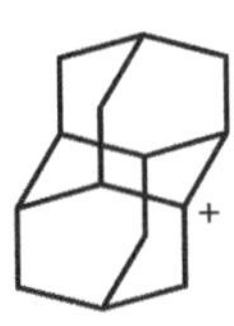

53

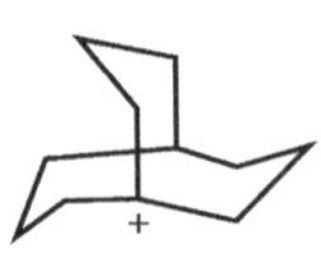

54

Olah, Prakash, and Stephenson[190] prepared isomeric triamantyl cations. Protolytic ionization of the parent triamantane yielded a mixture of the triamant-2-yl and triamant-3-yl cations **55** and **56**, respectively. The two cations, however, could be cleanly generated from 2- and 3-triamantanol, respectively, in SbF_5–SO_2ClF at −80°C. Ionization of 9-triamantanol, in turn, gave exclusively bridgehead cation **56**.

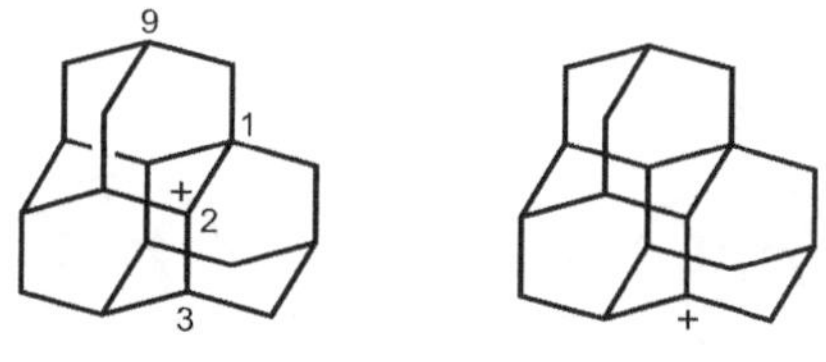

55 **56**

Fokin, Schreiner, and co-workers[191] have calculated the stability of tertiary diamondoidyl cations relative to the 1-adamantyl cation (B3LYP/6-31G* level). The relative stabilities were found to increase continuously with cage size: The penta- and hexamantyl cations are about 10 kcal mol^{-1} more stable than the 1-adamantyl cation. Furthermore, when isomeric cations are compared, the position of the cationic center also affects stabilities. Namely, the closer the cationic carbon to the geometrical center of the molecule, the higher the stability. For example, 4-diadamantyl cation is more stable than 1-diadamantyl cation (3 kcal mol^{-1}). Also illustrative is the comparison of the least and most stable tetramantyl cations: Cation **57** is less stable by 4.5 kcal mol^{-1} than cation **58**, and the distances of the cationic centers to the geometrical center are, respectively, 3.309 and 1.282 Å.

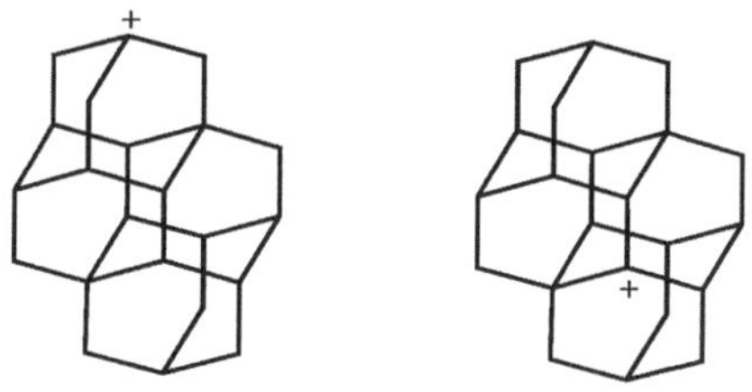

57 **58**

Bridgehead bicyclo[2.2.1]hept-1-yl cation (1-norbornyl cation) has not yet been directly observed; 1-chloronorbornane yields the stable 2-norbornyl cation in SbF_5–SO_2 solution.[192] Thus, ionization to the bridgehead carbocation must be followed by a fast shift of hydrogen from C(1) to C(2) (either intramolecular or intermolecular), the driving force for which is obviously the tendency to relieve strain in the carbocation.

Starting with dodecahedrane or substituted derivatives, Olah, Prakash, Paquette, and coworkers prepared dodecahedryl cation **59** under superacid conditions[193,194]

[Eq. (3.33)]. The ^{13}C NMR spectrum showed six absorptions [the most deshielded at $\delta^{13}C$ 363.9 (s)] clearly indicating that dodecahedryl cation **59** is a static ion. It showed no tendency to undergo degenerate hydrogen scrambling (no change in the ^1H NMR line shapes was observed up to 0°C). Upon standing in the superacid medium at −50°C, cation **59** slowly and irreversibly transformed into dodecahedrane-1,16-diyl dication (see Section 3.4.12).

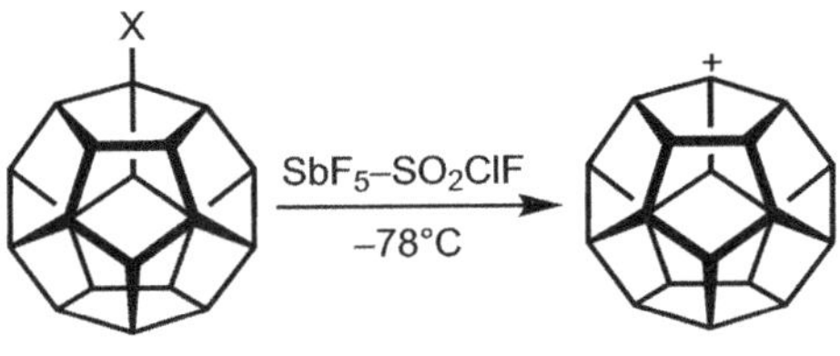

X = H, Cl, OH **59** (3.33)

3.4.4. Cyclopropylmethyl Cations

Solvolysis studies of Roberts and Mazur[195] and Hart and Sandri[196] showed both the unusual stability of cyclopropylmethyl cations and the ease with which such ions rearrange. Cyclopropyl groups have a strong stabilizing effect on neighboring carbocation center by delocalizing charge through bent σ-bond. The direct observation[197] of a variety of cyclopropylmethyl cations in cyclic, acyclic, and polycyclic systems by NMR spectroscopy provides one of the clearest examples of delocalization of positive charge into a saturated system.[198,199]

The first cyclopropylmethyl cation directly observed was the tricyclopropylmethyl cation **60** by Deno et al.[197] Its ^1H NMR spectrum in H_2SO_4 consists of a single sharp line at δ^1H 2.26.[200] In the 300-MHz ^1H NMR spectrum in SbF_5–SO_2ClF solution, however, the methine and methylene protons are well-resolved[201] (Figure 3.9).

Since then, a wide variety of cyclopropylmethyl cations have been prepared and studied by ^{13}C and ^1H NMR spectroscopy.[201–203] These studies have led to the conclusion that cyclopropylmethyl cations adopt bisected geometry and are static in nature with varying degrees of charge delocalization into the cyclopropane ring. Most interesting of these ions is the dimethylcyclopropylmethyl cation **61** (Figure 3.10). The methyl groups are nonequivalent and show a ^1H NMR shift difference of 0.54 ppm. The energy difference between bisected and eclipsed structures is estimated to be 13.7 kcal mol^{-1} (by temperature-dependent NMR studies)[204] and is quite close to the 12.3 kcal mol^{-1} energy obtained by molecular orbital calculations at the minimal basis set STO-3G.[174]

Previously discussed spiro[2.5]oct-4-yl cation **34**[148] is also a cyclopropylmethyl cation with substantial positive-charge delocalization into the cyclopropyl group. Other representative cyclopropylmethyl cations that have been prepared in the

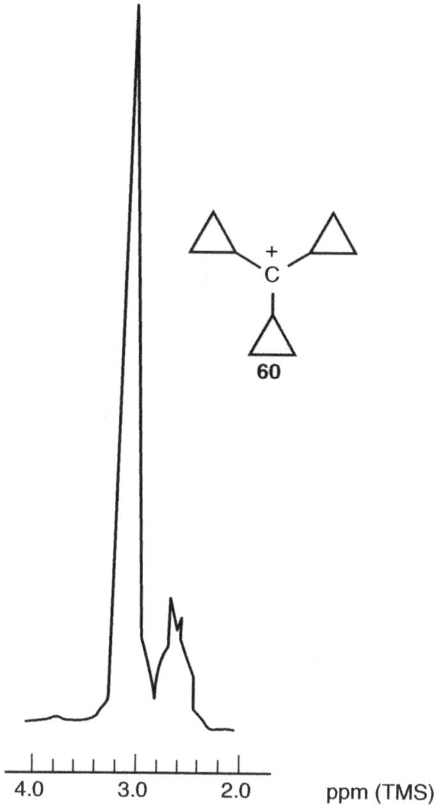

Figure 3.9. ^{1}H NMR spectrum (300 MHz) of the tricyclopropylmethyl cation in SbF_5–SO_2ClF at −60°C.

superacid media and characterized are cations **62**,[203] **63**,[173,174] **64**,[205] **65**,[205] **66**,[206] **67**,[207] and **68**.[208]

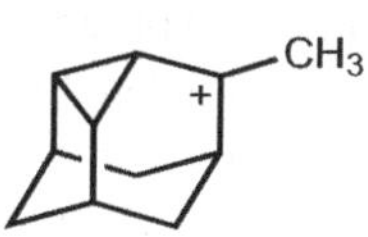

62

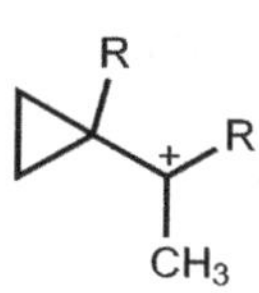

R = H, CH_3

63

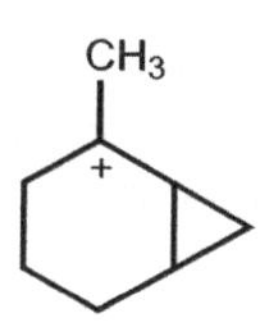

64

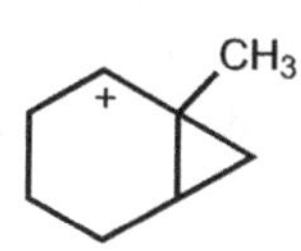

65

R = H, CH_3, Ph

66

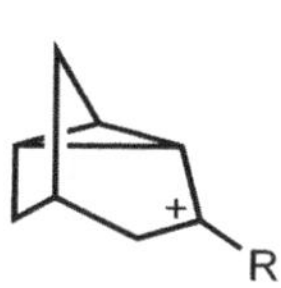

R = CH_3, Ph

67

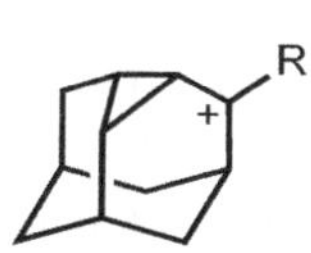

R = H, CH_3, Ph

68

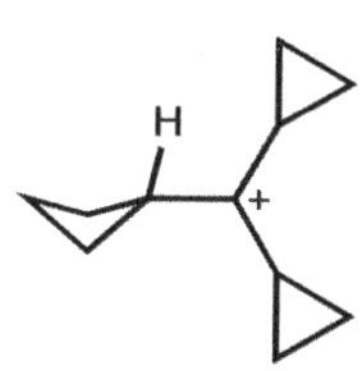

69

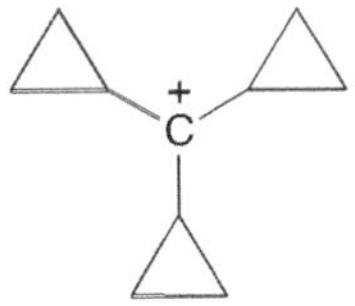

60

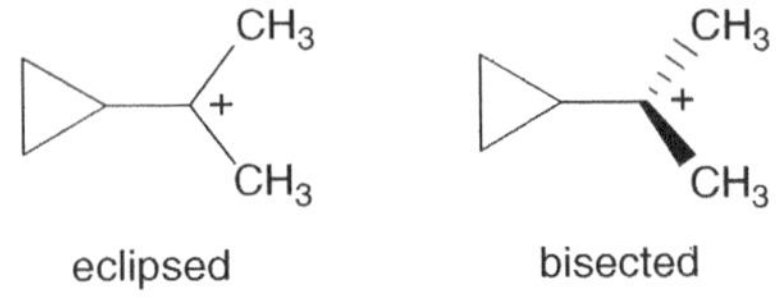

61

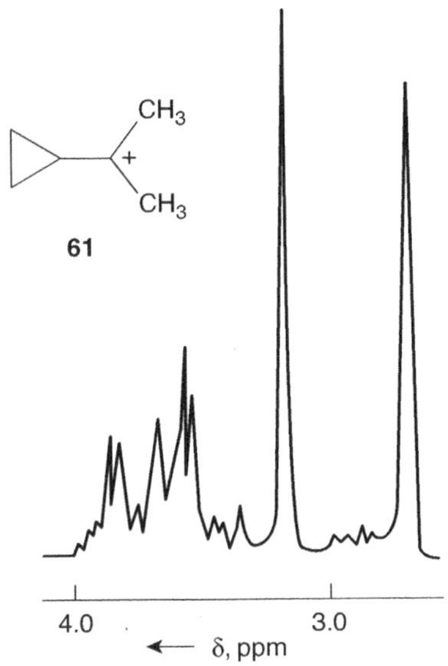

Figure 3.10. 100-MHz ^{1}H NMR spectrum of the dimethylcyclopropylmethyl cation.

Primary and secondary cyclobutylmethyl cations are nonclassical in nature and rearrange to thermodynamically more stable products (e.g., cyclopentyl cations). In search for a persistent cyclobutylmethyl cation, Prakash, Olah, and co-workers[209,210] were able to prepare the more stabilized tertiary cyclobutyldicyclopropylmethyl cation **69** by ionizing the corresponding alcohol at −90°C. Variable temperature ^{13}C NMR studies and theoretical calculations showed that the ion is classical and exists preferentially in a bisected conformation at −80°C with significant delocalization into the neighboring cyclopropyl rings (B3LYP/6-31G*).

Olah, Prakash, and co-workers[211] generated and characterized the 2-substituted spirocyclopropane-norbornane cations **70** [Eq. (3.34)]. Cation **70c** proved to be highly stable species, which is attributed to the superior stabilizing effect of the cyclopropyl groups. The methyl- and phenyl-substituted cations **70a** and **70b** rearranged to the corresponding allylic cations **71** even at slight warming [Eq. (3.34)]. ^{13}C NMR chemical shift values support the observed order of stability: As the cationic center is increasingly stabilized by methyl to phenyl to cyclopropyl, the quaternary carbon is progressively less deshielded (δ^{13}C 69.2, 61.2, and 58.8 for **70a**, **70b**, and **70c**, respectively). A wide range of studies have indicated[198,201,212–214] that a cyclopropyl group is equal or more effective than a phenyl group in stabilizing an adjacent carbocation center. Schleyer and co-workers[215] found (orbital deletion procedure at HF/6-311G** level) that the hyperconjugation energy in the cyclopropylmethyl cation is as large as the conjugation effect of the allyl cation. Consequently, cyclopropane is just as effective as an electron donor as a C=C double bond. The parent secondary cation (**70**, R = H) could not be observed; only the corresponding rearranged allylic cation was obtained.[149]

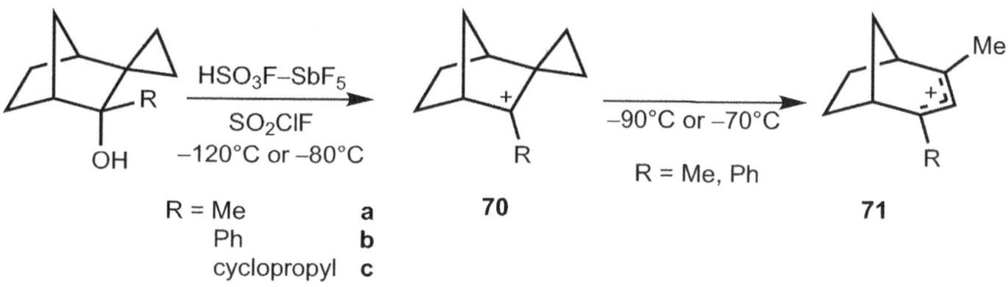

(3.34)

The cyclopropylmethyl cation has recently been generated in the gas phase from both homoallyl chloride and cyclopropylmethyl chloride and studied using collisional activated dissociation mass spectrometry.[216] Interestingly, cyclobutyl chloride, which yields the cyclopropylmethyl cation in the condensed phase, gives a substantial amount of the bicyclobutonium ion in the gas phase.

In contrast to "classical" tertiary and secondary cyclopropylmethyl cations (showing substantial charge delocalization into the cyclopropane ring but maintaining its identity), primary cyclopropylmethyl cations show completely σ-delocalized nonclassical carbonium ion character (see Section 3.5.2.5). Also, some of the secondary cyclopropylmethyl cations undergo rapid degenerate equilibrium (see later discussion).

3.4.5. Alkenyl Cations

Many alkenyl cations have now been directly observed particularly by Deno, Richey and co-workers,[88,217,218] Sorensen,[219] Olah and co-workers,[220–227] and Carpenter.[228] Deno[229] has reviewed the chemistry of these ions. Allylic cations particularly show great stability with generally insignificant 1,3-overlap, except in the case of cyclobutenyl cations[230] (vide infra). Representative observed alkenyl cations are **72**,[229] **73**,[221–223] **74**,[221,222] **75**,[221,223] **76**,[223] and **77**.[226]

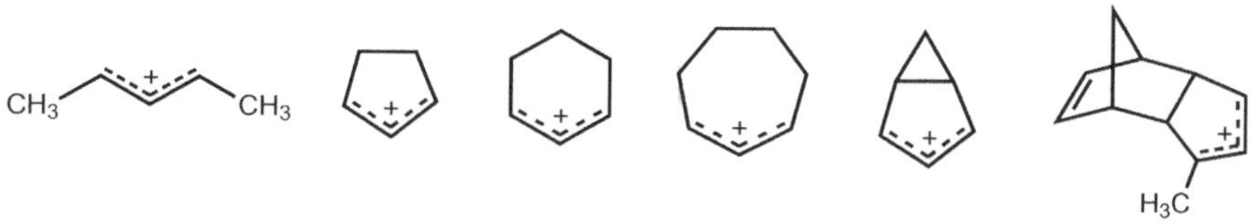

72 **73** **74** **75** **76** **77**

The formation of allyl cations from halocyclopropanes via ring opening of the unstable cyclopropyl cations also has been investigated[159,161,162] [Eq. (3.35)].

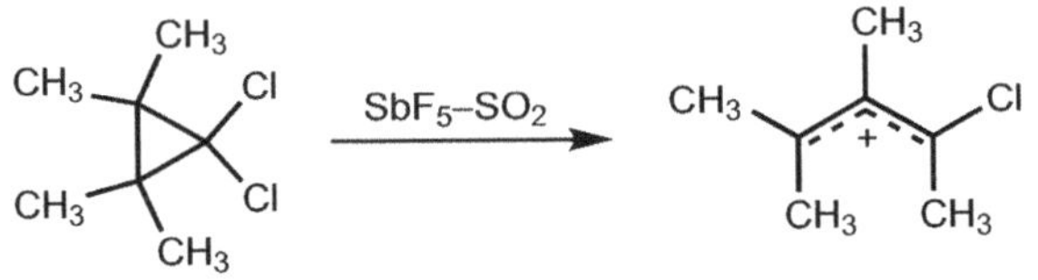

(3.35)

In fact, this approach has been used by Schleyer and coworkers to generate the parent allyl cation in superacid cryogenic matrix.[231] Codeposition of cyclopropyl bromide and SbF_5 at 110 K resulted in a clean IR spectrum with the prominent signal at 1578 cm^{-1}. This was assigned to the asymmetric C–C–C allyl stretching vibration and is in good agreement with calculated values of 1592 cm^{-1} (MP2/6-31G*) and 1558 cm^{-1} (CISD/6-31G**). A previous report for the allyl cation formed in zeolites and detected by ^{13}C CP-MAS NMR was questioned on the basis of the lower than expected ^{13}C NMR shift of $\delta^{13}C$ 218.[232] A recent study by Duncan and co-workers (infrared photodissociation spectroscopy by means of rare gas tagging)[233] has confirmed previous findings of the resonance-stabilized, charge-delocalized structure of C_{2v} symmetry.[164] The 2-propenyl cation has also been detected and was shown to have C_s symmetry and a nearly linear C–C–C backbone with hyperconjugative stabilization by the methyl group. Schleyer and co-workers[234] also detected the *trans*-1-methylallyl cation using the matrix isolation technique.

The 2-chloroallyl cation generated by matrix isolation has been studied by FT–IR spectroscopy and calculations [MP2(fc)/6-311G(d,p) level].[235] Structure **78** of C_s symmetry with bridging chlorine, proposed earlier,[162] was found to be less stable by 7.5 kcal mol^{-1} than cation **79** of C_{2v} symmetry and could not be found by spectroscopy. Furthermore, in contrast to 1- or 3-chloroallyl cations, the centrally positioned chlorine does not contribute to charge delocalization through back-donation as a consequence of the π-orbital noninteraction between the *n* electrons of Cl and the LUMO of the allyl cation.[236]

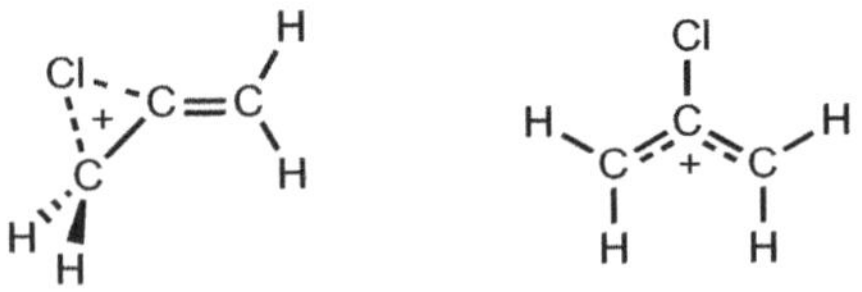

78 **79**

Allyl cations **80** and **81** have been generated and studied by NMR spectroscopy.[237] Although sterically crowded, cation **80** proved to be surprisingly stable up to 80°C. The rotational barrier estimated on the basis of the coalescence temperature of the ^{13}C NMR signals is 16.8 kcal mol^{-1}, in good agreement with the calculated value (MNDO, 16.5 kcal mol^{-1}). In contrast, the rotational barrier of cation **81** was found to be less than 5 kcal mol^{-1}. This was attributed to the charge stabilizing effect of the adjacent cyclopropyl groups, resulting in an unsymmetrically charge-delocalized species.

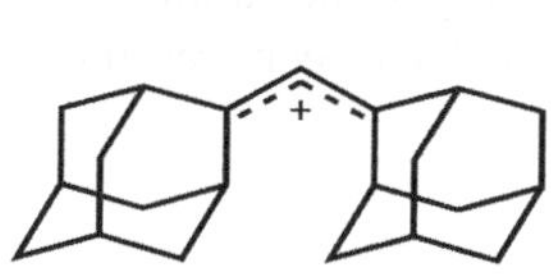

80

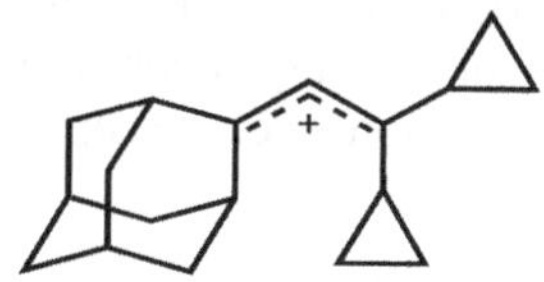

81

Protonation of allenes also leads to allyl cations, allowing one to obtain ions—for example, the dimethylallyl and tetramethylallyl cations **82** [Eq. (3.36)]—that are otherwise difficult to generate from allylic precursors.[238]

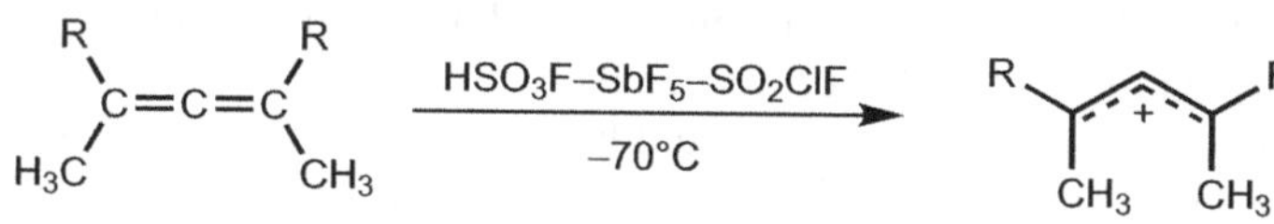

R = H, CH_3

82

(3.36)

3.4.6. Alkadienyl and Polyenylic Cations

Deno, Richey, and their co-workers[88] have observed a substantial number of alkadienyl cations. Sorensen[239] has observed divinyl and trivinyl cations **83** and **84**.

83

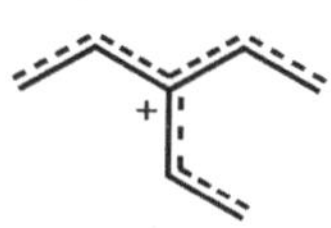

84

Alkadienyl cations such as the 1,3,5-trimethylheptadienyl cation **85** show great tendency to cyclize [Eq. (3.37)] and these reactions have been followed by NMR.[240] Several novel fulvenes have been protonated to their corresponding dienyl cations[241] (e.g., **86** and **87**).

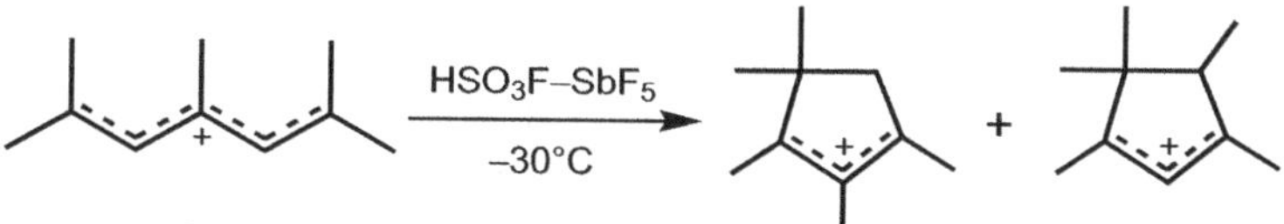

85

(3.37)

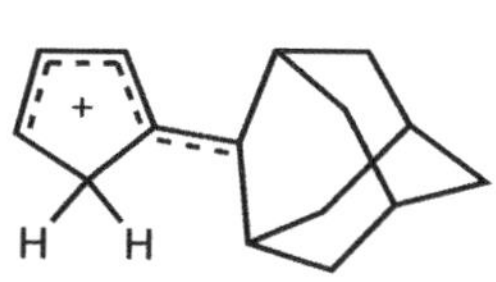

86

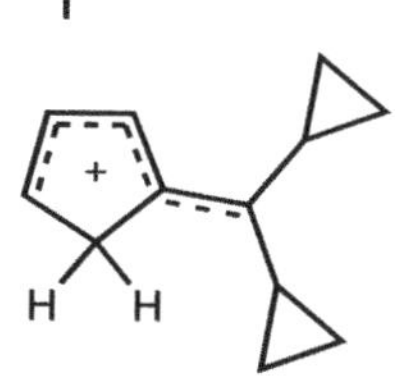

87

Sorensen and co-workers[242] have studied the stereochemistry of ring closure of arylallyl cations to bicyclic trienyl cations. Similar studies on 1-phenylallyl cations have been carried out by Olah et al.[225] Persistent polyenylic cations generated upon incorporation of cinnamyl alcohols in zeolitic matrix have been identified on the basis of IR and UV–visible spectroscopy.[243]

3.4.7. Arenium Ions

Cycloalkadienyl cations, particularly cyclohexadienyl cations (benzenium ions), the intermediate of electrophilic aromatic substitution, frequently show remarkable stability. Protonated arenes can be readily obtained from aromatic hydrocarbons[244–251] in superacids and studied by ^{1}H and ^{13}C NMR spectroscopy.[252,253] Olah et al.[252] have even prepared and studied the parent benzenium ion ($C_6H_7^+$) **88**. Representative ^{1}H NMR spectra of benzenium[253] and naphthalenium ions[254] **88** and **89** are shown in Figures 3.11 and 3.12, respectively.

Isomeric mono-, di-, tri-, tetra-, penta-, and hexaalkylbenzenium and halobenzenium ions have been observed.[251,252,255] Alkylation, nitration, halogenation, and so on, of hexamethylbenzene give the related ions. Doering, Saunders, and co-workers[256] have studied the dynamic NMR spectra of heptamethylbenzenenium ion **90** (R = Me), and Fyfe and co-workers[257] have obtained the solid-state ^{13}C NMR spectrum.[257]

Recent progress in experimental techniques allowed the isolation of various benzenium ions, including the parent ion **88** as stable salts, which permitted their characterization by X-ray crystallography. Kochi and co-workers[258] generated

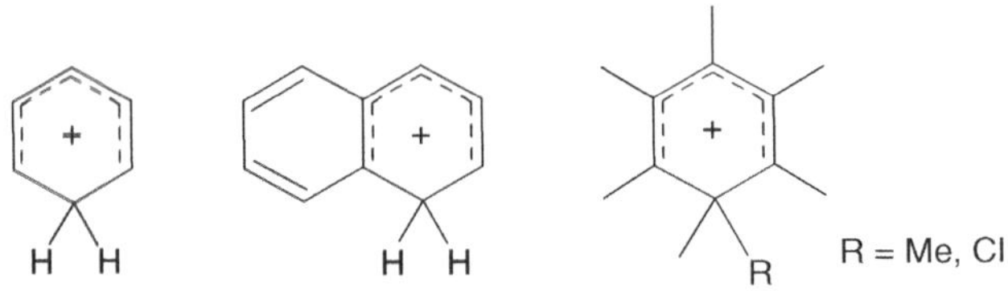

88 **89** **90**

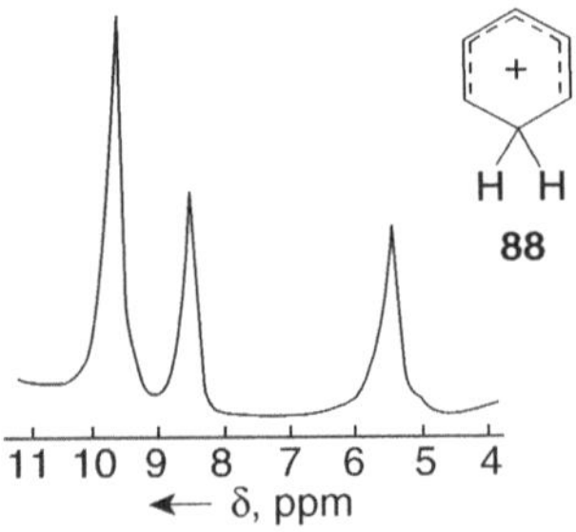

Figure 3.11. The 270-MHz ^{1}H NMR spectrum of the "static" benzenium ion in $HSO_3F–SbF_5–SO_2ClF–SO_2F_2$ solution at −140°C.

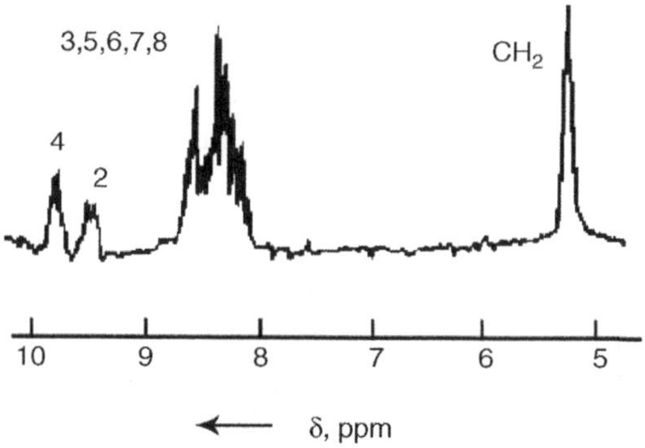

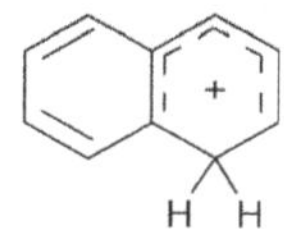

89

Figure 3.12. 100-MHz ^{1}H NMR spectrum of the naphthalenium ion **89** at −90°C.

the chlorohexamethylbenzenium cation (**90**, R = Cl) and isolated it as the hexachloroantimonate salt, whereas Reed et al.[259,260] transformed benzene, toluene, *meta*-xylene, mesitylene, and penta- and hexamethylbenzene to the corresponding benzenium ions with carborane superacids [$H(CB_{11}HR_5X_6)$, R = H, Me; X = Cl, Br] and applied NMR spectroscopy, IR spectroscopy, and X-ray crystallography for characterization. According to Mulliken population analysis, the positive charge is delocalized onto the H atoms with the greatest partial charges on the hydrogens at the sp^3 carbon. Crystallographic data show that the cyclohexadienyl resonance form is the best representation from the point of view of bond lengths as shown for *meta*-xylenium ion **91**. All bond angles are within 4° of the idealized value of 120°. The benzenium ion skeleton is essentially planar with a slight tilt of the protonated carbon the largest being 4.3° for the mesitylenium ion.

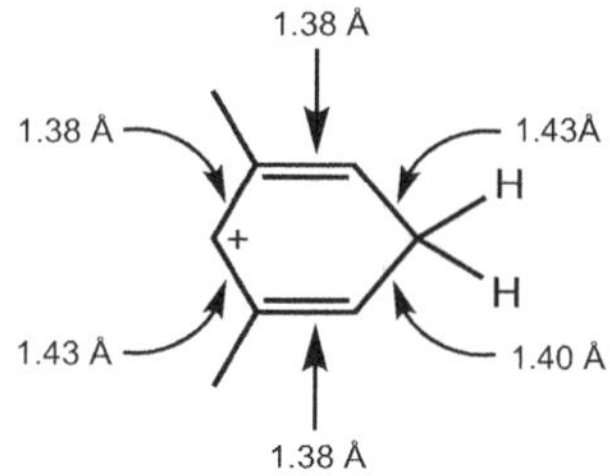

91

The Si-substituted [$Et_3Si(C_6H_4Me)]^+$ cation was originally erroneously identified as being a silyl cation (silicenium ion, **92b**)[261] on the basis of a long Si–C_{toluyl} bond distance of 2.18 Å. Calculated bond distances and bond angles (HP/6-31G* level)[262] are in close agreement with those found experimentally. Similarly, there is a particularly good agreement between the observed (δ^{29}Si 81.8) and calculated (δ^{29}Si 82.1) ^{29}Si NMR chemical shifts, in contrast with the calculated highly deshielded shift of 355.7 ppm of the still elusive Me_3Si^+ ion. These observations indicate that the actual species is the *para*-triethylsilyltoluenium ion **92a** with contribution from the resonance form **92b**. In a subsequent study, (IGLOII′/B3LYP/6-31G*)[263] chemical shift values of δ^{29}Si 79.7 and 79.8 for two isomeric forms of **92a** were calculated, whereas the chemical shift for the silicenium ion **92b** was computed to be δ^{29}Si 371.3.

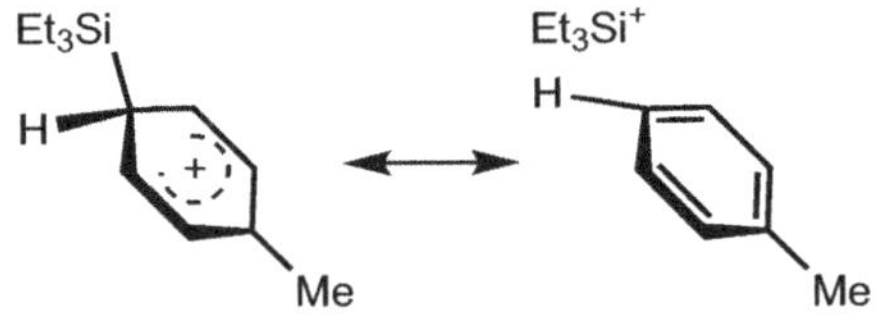

92

Novel bis-silylated arenium ions **93** have been generated and characterized by Müller and co-workers.[264] A single ^{29}Si NMR signal with shift values (δ^{29}Si 19.1–25.6) close to those found for bis-β-silyl-substituted vinyl cations (δ^{29}Si 22–24) indicates in each case the formation of a symmetric species. The strongly deshielded *ortho* and *para* carbons ($\Delta\delta^{13}C_{ortho} = 28.9-39.6$ and $\Delta\delta^{13}C_{para} = 21.6-35.1$) relative to the starting silanes lend further evidence to the cation structure. Furthermore, the strong high-field shifts of the *ipso* carbons ($\Delta\delta^{13}C_{ipso} = 36.5-49.4$) relative to the precursors are indicative of the rehybridization of C_{ipso} from sp^2 to sp^3. The calculated structures of cations clearly show that these are arenium ions, and NMR shift calculations [GIAO/B3LYP/6-311G(d,p)//B3LYP/6-31G(d)+ΔZPE level] corroborate the validity of the structures.

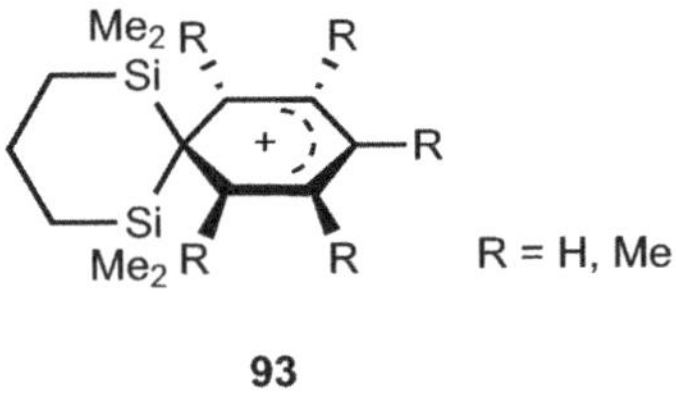

93

Lickiss and co-workers[265] have synthesized and characterized the bis-silylated cation **94**. Structural analysis of crystals isolated shows an almost symmetrical bridging of the phenyl ring [Si(1)–C_{ipso} and Si(2)–C_{ipso} distances = 2.104 and 2.021 Å, respectively). In contrast to **93** and other arenium ions that have alternating bond lengths attributed to the cyclohexadienyl resonance structure (long C_{ipso}–C_{ortho} and C_{ortho}–C_{meta}, and short C_{meta}–C_{para} bonds), the bond distances of cation **94** are rather similar (C_{ipso}–C_{ortho} = 1.39 and 1.40 Å, C_{ortho}–C_{meta} = 1.36 and 1.37 Å, C_{meta}–C_{para} = 1.35 and 1.37 Å). Furthermore, calculations have shown [B3LYP/6-31G(d)] that the positive charge is predominantly on the four-membered ring. The authors concluded that the 1,3-phenyl-bridging structure **94b** is a better description for the cation than the bis-silylated benzenium ion **94a**.

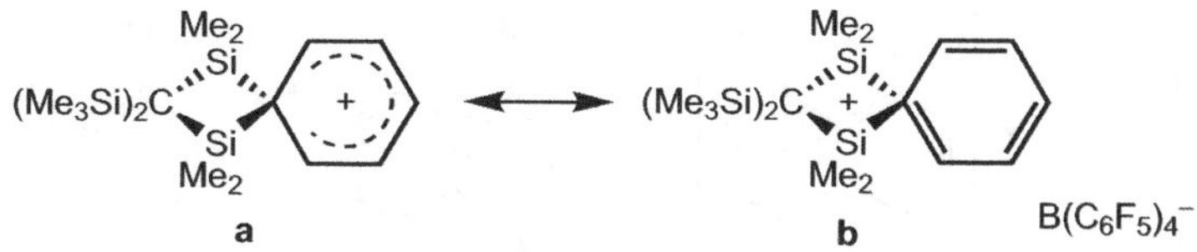

94

The novel arenium ion **95** was synthesized[266] by one-electron oxidation of the triphenylene-based starting compound to form a radical cation which abstracted a chlorine atom with a concomitant rearrangement to yield the hexachloroantimonate salt. The arenium ion character is apparent from the ^{13}C spectrum (three signals at $\delta^{13}C$ 212.9, 187.6, and 173.6) and from the bond distances, which are very close to those shown for ion **91**. Cation **95** can be stored at room temperature for months. This exceptional stability was attributed to the annelation to the two bicyclo[2.2.2]octane units and the spiroconjugation effect of the fluorenyl moiety.[267]

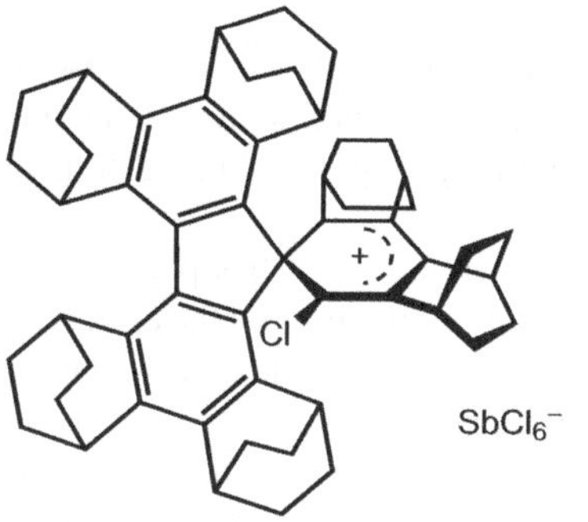

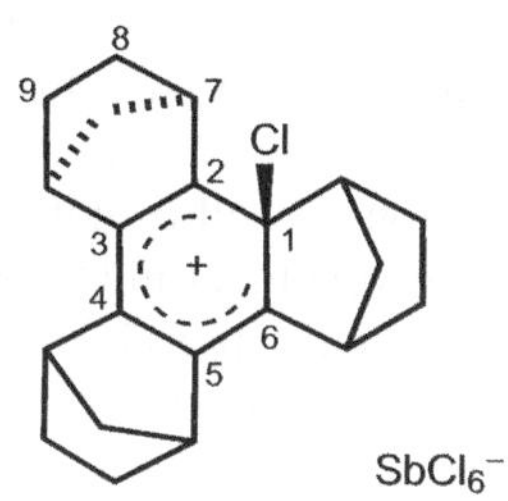

95 **96**

Kochi and co-workers[268] have obtained the related novel chloro-arenium cation **96**. Single-crystal X-ray crystallography revealed that the Cl–C bond is relatively long (1.86 Å), and the Cl atom is bonded to the aromatic carbon from the less hindered face to minimize steric interactions with the transannular C(8)–C(9) ethano bridge. The central ring has a distorted cyclohexadienyl structure with two short bonds [C(2)–C(3) = 1.380 Å, C(5)–C(6) = 1.353 Å] and an unusually long C(1)–C(6) bond (1.517 Å) when compared to the C(1)–C(2), C(3)–C(4), and C(4)–C(5) bonds [1.460, 1.409, and 1.413 Å, respectively]. Furthermore, the ring has a quasi-envelope conformation with C(1), C(2), C(3), and C(4) being coplanar and C(5) and C(6) lying below the plane. The cation **96** may be viewed as a positive chlorine stabilized (solvated) by an aromatic ligand as in a Wheland intermediate.

Lammertsma and Cerfontain[269] performed a detailed study on the ions generated from isomeric dimethylnaphthalenes. An interesting observation is the transformation upon increasing the temperature of cation **97** through an intermediate to cations **98** and **99**. The driving force is the relief of steric strain in the *peri* position [Eq. (3.38)]. For the same reason, polymethylnaphthalenes are protonated at the α-position, and similar rearrangements likewise lead to the product of least strain[270] [Eq. (3.39)].

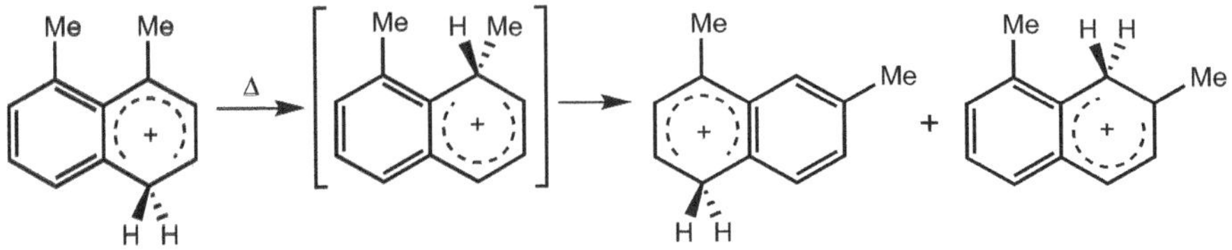

97 98 99

(3.38)

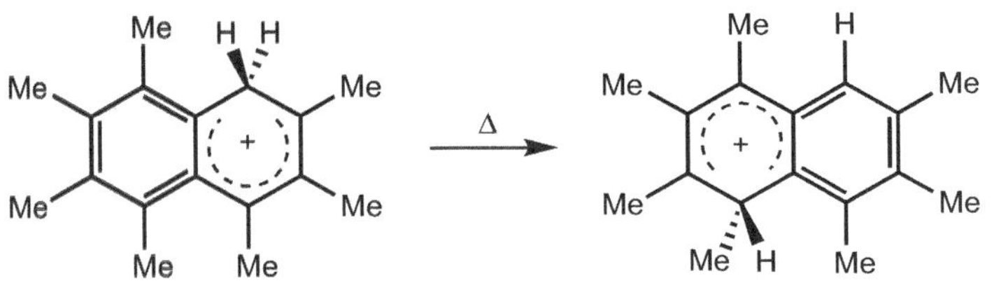

(3.39)

Anthracenium[253,255,271,272] and phenanthrenium ions[271–274] have also been well-studied. A detailed investigation of 9-alkylanthracenes showed that protonation always occurs at C(10).[275] An additional methyl group at C(1) or C(8) results in the C(9) protonation as well (relief of *peri* strain).

Of other polycyclic aromatic hydrocarbons, Laali[271,272] has performed extensive studies on pyrenium cations including detailed investigations on the effect of the nature and position of substituents on the site of protonation. As Hückel calculation predicts, α-protonation is most favored (**100** and **101**) irrespective of the position of substituents.[276] Generation of benz[*a*]anthracenium,[277] cyclopenta[*a*]phenanthrenium,[278] and benzo[*a*]pyrenium cations[279] in HSO_3F–SO_2ClF at low temperature has also been explored.

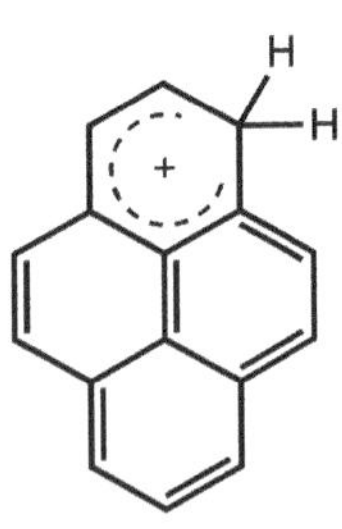

100

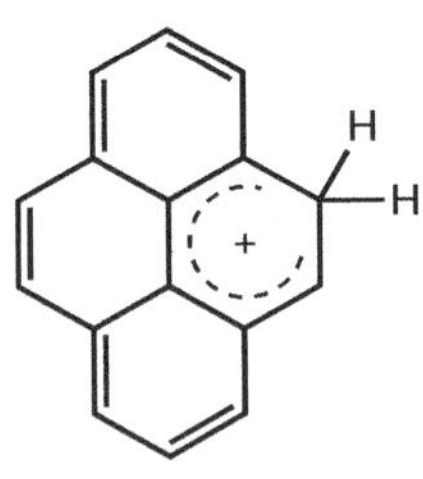

101

In a recent comprehensive study with respect to the substituent effects of benz[*a*] anthracene carbocations,[280] exclusive protonation at C(7) (**102**) and C(12) (**103**, bay region protonation) in the C ring was shown to occur. The relative stability of the resulting carbocations, however, strongly depends on the substitution pattern. Substrates methyl-substituted in the A ring give mixtures of the two cations. Protonation of 5-, 6-, and 7-methyl- and 7-ethyl-substituted compounds, in turn, yields exclusively the corresponding 12-protonated **103** derivatives. Interestingly, the protonation of the 12-methyl derivative favors C(7) (cation ratio = 6.7:1), but the amount of thermodynamically more stable *ipso*-protonated isomer increases over time and becomes the main product (1:16). 7,12-Disubstituted compounds exhibit similar behavior. Protonation in the bay region (formation of **103**) is also the main reaction with compounds with a single methyl group in the D ring.

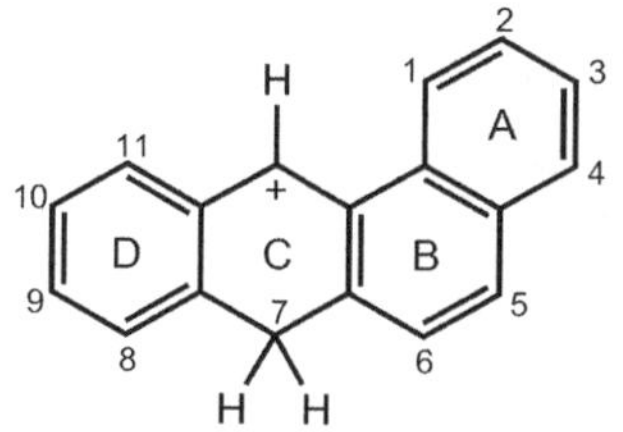

102

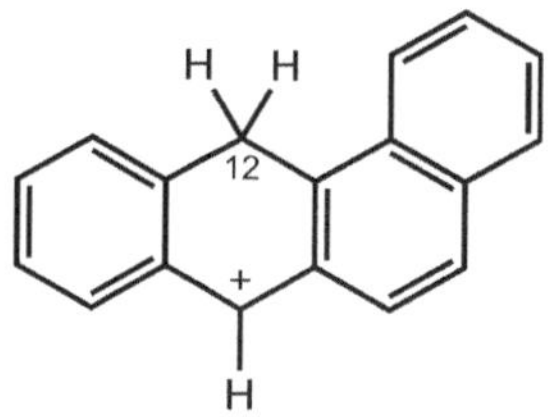

103

Protonation of substituted benzo[*e*]dihydropyrenes with HSO_3F or $HSO_3F–SbF_5$ (4:1) in SO_2ClF at −78C° results in the formation of dication **104a** and trication **104b**.[281,282] However, cations by two-electron oxidation are also formed (see Section 3.4.14).

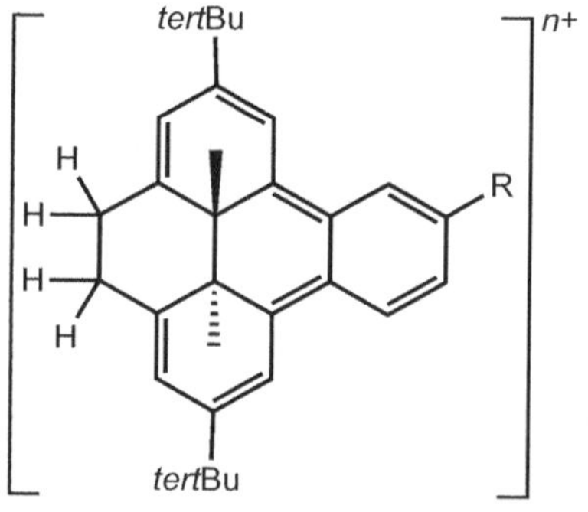

104 **a** R = H, *n* = 2
b R = CO_2EtH, *n* = 3

The persistent 4-methyl[6]helicenium cation **105** has been observed and characterized by Laali and Hauser.[283] Although the parent ion could be generated under mild conditions (TfOH–SO_2ClF), it decomposed rapidly.

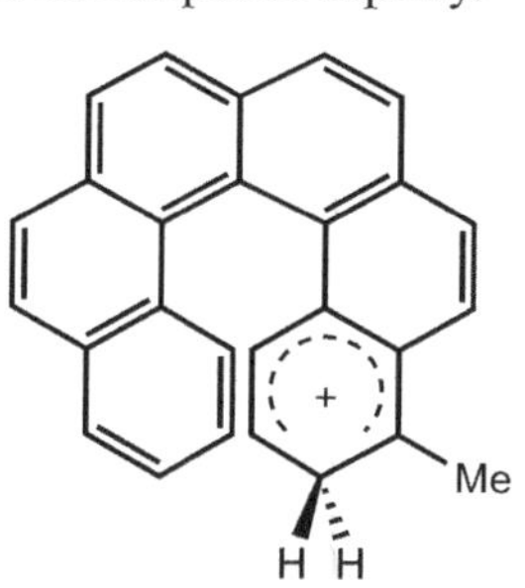

105

Warner and Winstein[284] have obtained the stable monocation **106** by protonating 1,6-methano[10]annulene in HSO_3F. Cram and Cram[285] have been successful in protonating [2.2]-*para*-cyclophane. The initial protonated species **107** rearranges to the *meta*-protonated species **108** [Eq. (3.40)].

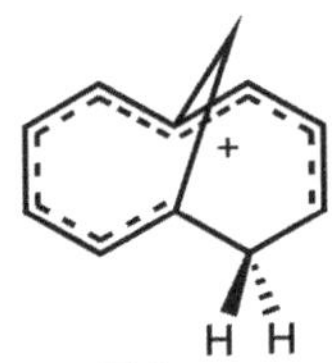

106

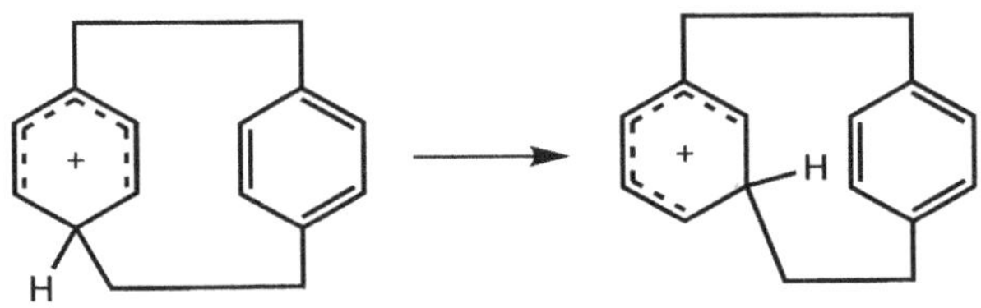

107 **108** (3.40)

Protonation of bismethano[14]annulene and its bridged analogs allowed to generate and characterize the corresponding **109** and **110** cations.[286] Protonation always occurs at C(7).

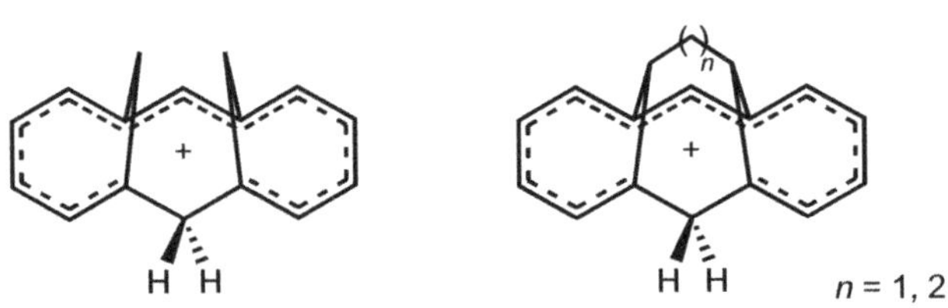

109 **110**

3.4.8. Ethylenearenium Ions

The classical–nonclassical ion controversy[27–37] initially also included the question of the so-called "ethylenephenonium" ions.

Cram's original studies[287] established, based on kinetic and stereochemical evidence, the bridged ion nature of β-phenylethyl cations in solvolytic systems. Spectroscopic studies (particularly ^{1}H and ^{13}C NMR)[288–291] of a series of stable long-lived ions proved the symmetrically bridged structure and at the same time showed that these ions do not contain a pentacoordinate carbocation center (thus are not "nonclassical ions"). They are spiro[2.5]octadienyl cations **111** (spirocyclopropylbenzenium ions)—in other words, cyclopropyl-annulated cations in which the cationic center belongs to a cyclohexadienyl cation (benzenium ion).

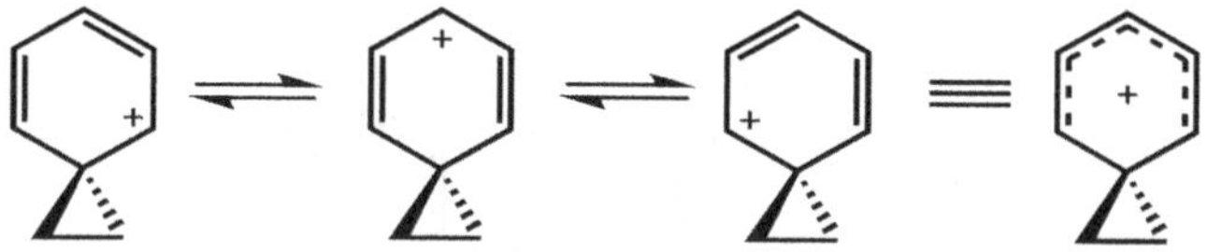

111

The nature of the spiro carbon atom is of particular importance in defining the carbocation nature of the ions. ^{13}C NMR spectroscopic studies have clearly established the aliphatic tetrahedral nature of this carbon, thus ruling out a "nonclassical" pentacoordinate carbocation. The positive charge is delocalized into both the dienylic framework and the spirocyclopropane moiety. As theoretical calculations demonstrated,[292] these charge delocalizations result in almost identical C–C bond lengths in the six-membered ring (1.384–1.419 Å) and long C_{ipso}–CH_2 bond distances in the cyclopropane ring (1.625 Å). Recent calculations (B3LYP/6-31G* method) clearly showed[293] that back-bonding from the phenyl cation moiety to the ethylene fragment determines the formation of the three-membered cycle, which renders the shielding of the *ipso* carbon similar to that for an sp^3 carbon.

The formation of the ethylenebenzenium ion **111** from β-phenylethyl precursors [Eq. (3.41)] can be depicted as cyclialkylation of the aromatic π-systems and not of the C_{ipso}–C_β bond, which would give the tetracoordinate ethylenephenonium ion. Recently, protonation of benzocyclobutene under superacidic conditions[294] has been found to be a new independent route to generate **111** [Eq. (3.41)]. Rearrangement of the β-phenylethyl to an α-phenylethyl (styryl) ion, on the other hand, takes place through a regular 1,2-hydrogen shift. Rearrangement and equilibria of ions formed from side-chain-substituted β-phenylethyl chlorides have also been explored.[291]

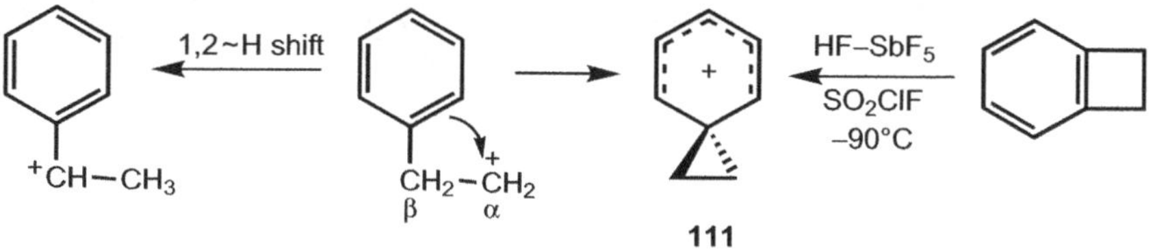

111

(3.41)

The ethylenebenzenium ion **111** was found to undergo isomerization to yield the α-phenylethyl (styryl) ion with an activation energy of $E_a = 13$ kcal mol^{-1}.[295] In a recent study[296] a value of 26.7 kcal mol^{-1} was computed [B3LYP/6-3111G(d,p) level] and the styryl ion was found to be more stable by 13.9 kcal mol^{-1}.

Olah and co-workers[297,298] have prepared and characterized a series of 4-substituted ethylenenaphthalenium ions **112**. Eberson and Winstein[299] have reported the ^{1}H NMR spectrum of ethyleneanthracenium ion **113** (R = H). A series of 9-substituted anthracenium ions have been studied by ^{13}C NMR spectroscopy.[300]

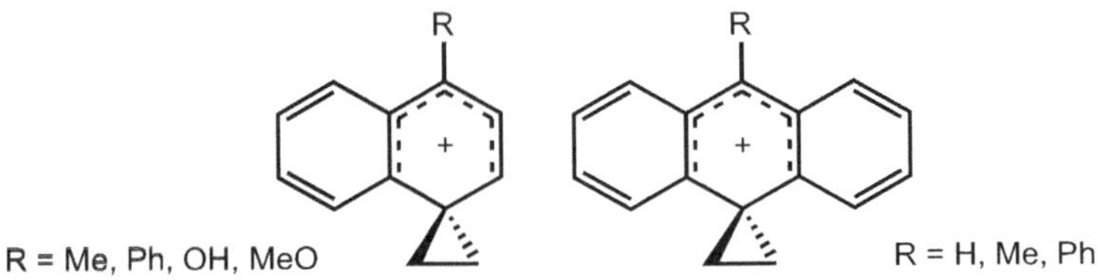

112 **113**

The benzonorbornenyl cation **114a** can be considered as the spiro cation **114b**.[301] The unsubstituted 9-benzonorbornenyl cation **115** (R = H) has not been observed but the NMR characterization of derivatives (R = Me, *para*-F-phenyl, OH) has been reported.[302] Recently, Laube has prepared the $Sb_2F_{11}^-$ salt of the 9-methyl derivative and reported X-ray crystal structure analysis.[303] A remarkable feature of cation **115** (R = Me) is the strong bending of the C(9) bridge toward the aromatic ring [C(9)–C(4a) and C(9)–C(8a) bond distances = 1.897 Å, bond angle between C(9) and the C(4a)–C(8a) bond = 95.7°]. Computed values [B3LYP/6-311+G(d,p)] show an even stronger bending (1.981 Å and 98.6°). These values, when compared to those of the parent benzonorbornene (2.35 Å and 126°), indicate a strong interaction. Furthermore, the aromatic ring shows bond alternations [C(5)–C(6) and C(7)–C(8) = 1.358 Å, C(6)–C(7) = 1.423 Å, C(4a)–C(5) and C(8)–C(8a) = 1.395 Å], and the ring is nonplanar with both C(4a) and C(8a) having a pyramidalization value of $\Delta = 0.077$ Å.

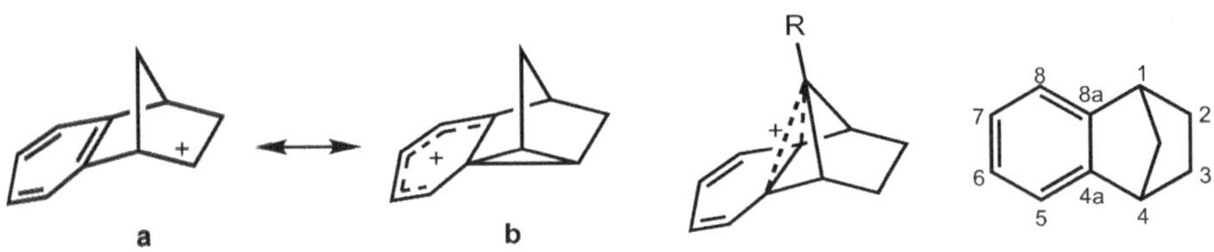

114 **115**

3.4.9. Propargyl and Allenylmethyl Cations (Mesomeric Vinyl Cations)

No *bona fide* unsubstituted vinyl cations have yet been experimentally observed under long-lived stable ion conditions.[304–308] However, experimental observations in the gas phase and calculations are available.[309] According to spectroscopic studies,[310] of the two structures of the vinyl cation (protonated acetylene) the planar, bridged nonclassical structure **116b**, with the two protons slightly tilted toward the bridging proton, is preferred over the classical (Y-shaped) ion **116a**. The energy difference between the two structures is about 5 kcal mol^{-1}.[311,312] Subsequent Coulomb-explosion imaging (CEI)[313] and simulated CEI studies[314] showed, however, that the structure is not planar, but the protons *trans*-bent with respect to the plane of the bridge moiety.

116

In contrast to vinyl cations, propargyl cations **117a** exist in mesomeric allenyl forms **117b**, which can serve as models for vinyl cations.

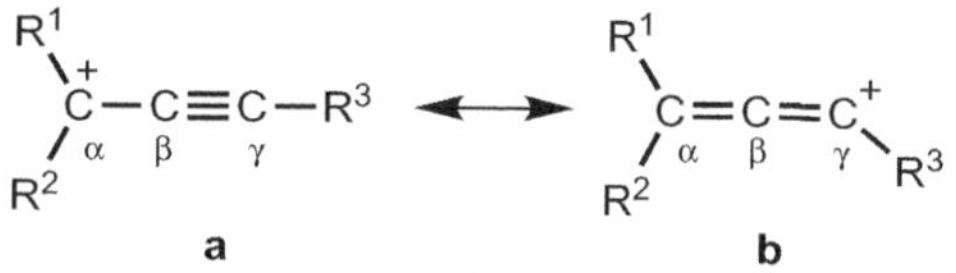

117

Extensive work has been carried out on these propargyl cations with a wide variety of substituents.[214,315–317] Olah et al.[316] found that in cation **118** the positive charge is extensively delocalized; that is, the contribution of the **117b** resonance form is significant. Komatsu et al.[318] showed that both the α and γ carbons become more shielded in ions bearing two or three phenylethynyl groups (cation **119**) due to the more extended charge dispersion.

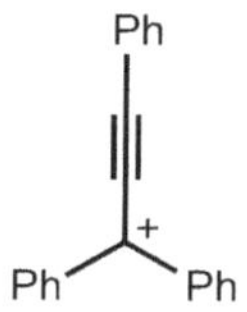

118

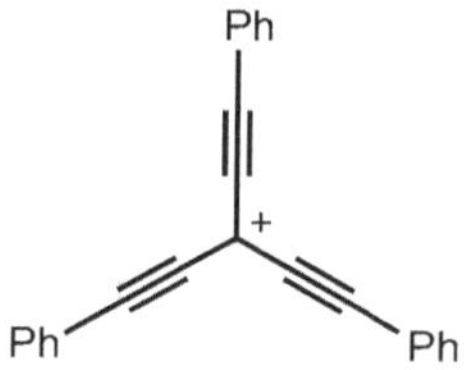

119

In the early 1980s, Siehl and co-workers[319,320] prepared two allenylmethyl cations **120** and **121** using the codeposition technique developed by Saunders et al.[49] The cations thus prepared also exhibit extensive mesomeric vinyl cation character.

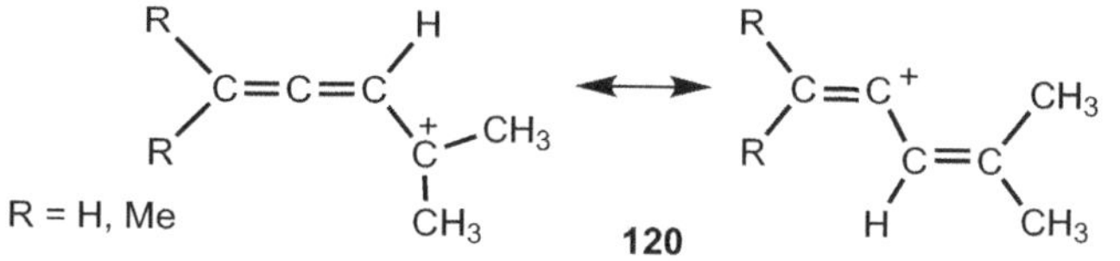

120

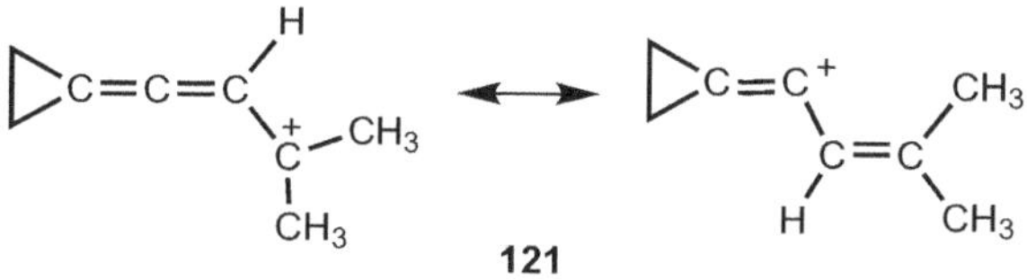

121

Figure 3.13. Orbital representation of vinyl cations.

In vinyl cations (Figure 3.13), as in trisubstituted carbenium ions, the positive charge is stabilized by electron donating substituents or by aryl or vinyl groups via π-conjugation. Further stabilization may be achieved by σ participation— that is, by hyperconjugation of α substituents, by complexation to a metal, or by the β-silyl effect.

Siehl has made extensive efforts to generate a range of stabilized vinyl cations.[321] Protonation of an alkyne bearing an aromatic α substituent and a substituent of sufficient steric bulk at the β position can yield stable vinyl cations (**122**).[322] Both the parent mesityl-vinyl cation **123**[322] and the α-(*para*-methoxyphenyl)-substituted cation **124**[323] were observed when the corresponding β-Si-substituted cations underwent cleavage at higher temperature (−100°C and −115°C, respectively). The cyclopropylcyclopropylidenemethyl cation **125** has been prepared[324] by the protonation of the corresponding allene [bis(cyclopropylidene)methane], and both NMR spectroscopic data and calculations (MP2/6-31G*) indicate charge delocalization due to hyperconjugation with the banana bond of the cyclopropylidene group.

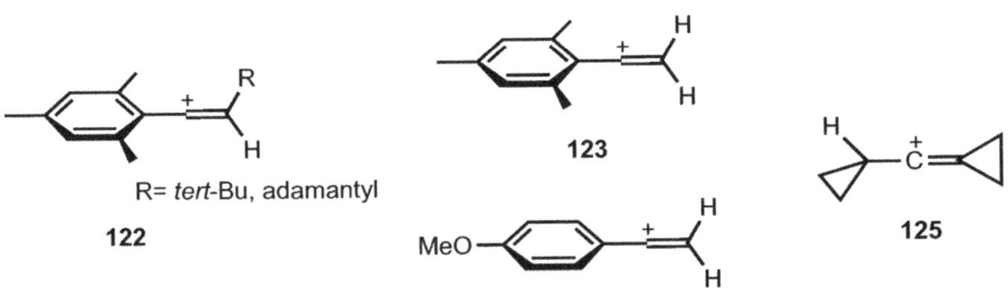

122 **123** **124** **125**

Recently, Siehl and co-workers[325,326] have made computational studies of the ^{13}C NMR shift of a series of α-vinyl-substituted vinyl cations (1,3-dienyl-2-cations, Figure 3.14). They found that inclusion of electron correlation effects are important to get reliable data. The small differences (Δ = 1–2 ppm) between calculated [CCSD(T)/tzp/dz] and observed NMR chemical shifts suggest that the geometry of the cations is not significantly affected by the medium.

Siehl[327] and, recently, Müller et al.[328] have generated and observed a variety of β-silylated vinyl cations. The first successful observation by ^{13}C NMR of a persistent β-silyl-substituted vinyl cation was reported by Siehl et al.[329] [Eq. (3.42)]. The

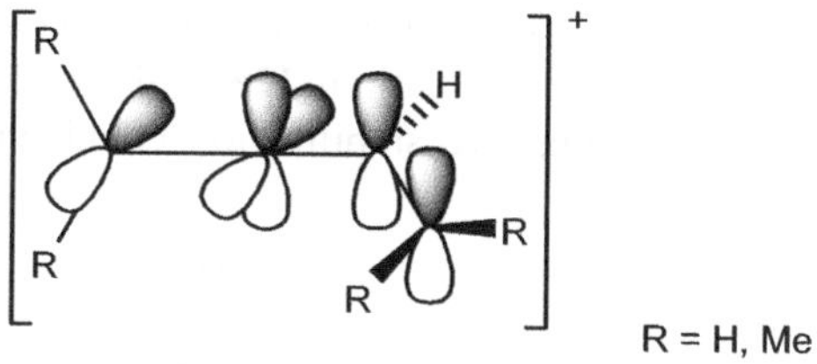

Figure 3.14. Orbital representation of 1,3-dienyl-2-cations.

resonance of the cationic carbon (δ^{13}C 208.7) shows strong shielding comparable to those of vinyl cations stabilized by hyperconjugation with cyclopropyl substituents. Similar shielding effect of the β-silyl substituent was found by calculations (IGLO). Due to the hyperconjugation effect, cation **126** prefers the conformation in which the empty 2*p* orbital of the cationic carbon is parallel to the C–Si bond allowing maximal overlapping. The rotation barrier in the parent model compound about the C^+–CH_2SiH_3 bond was calculated to be 14.5 kcal mol^{-1}.

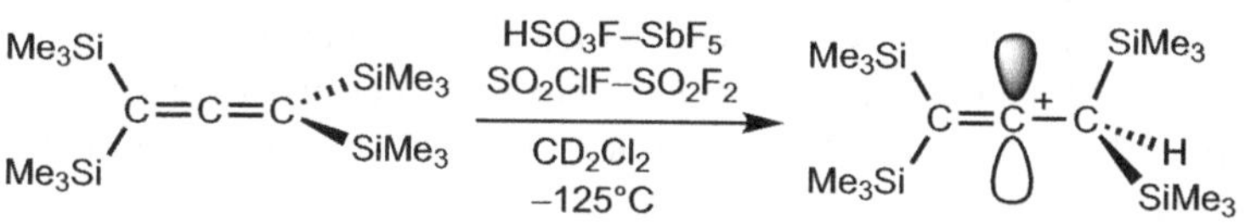

126

(3.42)

Subsequently, Siehl and Kaufmann[322] reported the synthesis and NMR spectroscopic characterization of a range of persistent α-mesityl- and β-silyl-substituted vinyl cations (**127**) by protonating the corresponding alkynes at −130°C (HSO_3F–SbF_5 in SO_2ClF–SO_2F_2). The ^{13}C NMR chemical shifts values of the C^+ carbons show that all silyl-substituted vinyl cations shielded more strongly (~30–33 ppm) than the silicon-free analogs, which is indicative of the magnitude of the β-silyl effect.

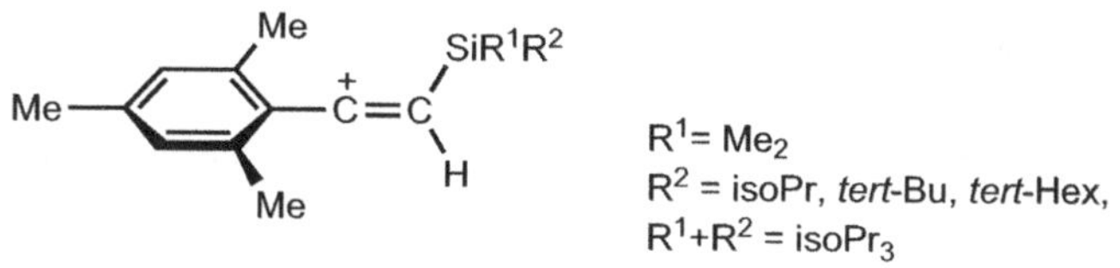

127

Unusually stable vinyl cations have been prepared by Müller, Siehl, and coworkers[330] by the intramolecular addition of transient silylium ions to alkynes at room temperature. Both the ^{13}C NMR chemical shifts for the vinyl carbons ($\delta^{13}C_\alpha$ = 185.8, $\delta^{13}C_\beta$ = 84.1) and the high-field shift of the *ipso* carbon (δ 113.7) found for **128** (R = Ph) are characteristic of aryl-substituted vinyl cations. The *ortho* and *para* carbon atoms are highly deshielded, which indicate that a significant amount of the positive charge is transferred to the phenyl ring. The phenyl-substituted cation was stable for weeks at room temperature, whereas the methyl-substituted analog decomposed in several days. The difference is due to the lack of charge delocalization in the case of the latter. In contrast to the high stability of cations **128**, cations **127** with only a single β-silyl substituent for hyperconjugative stabilization could only be generated in superacid media at −130°C.[322]

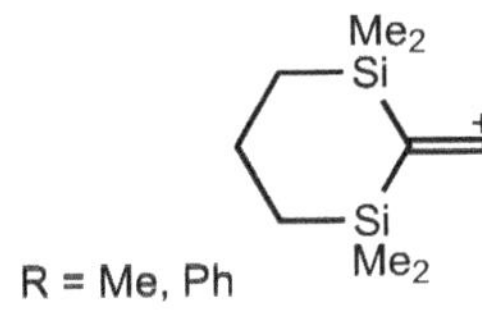

128

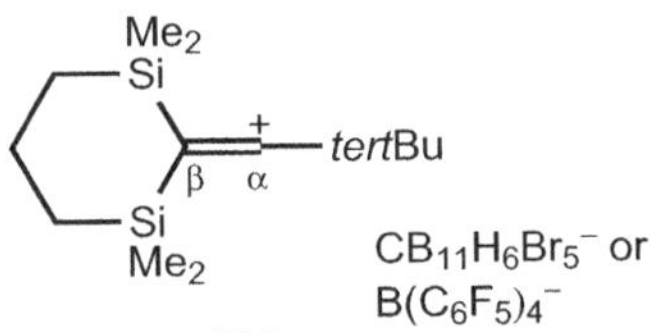

129

Recently, the X-ray structure of vinyl cation **129** ($CB_{11}H_6Br_5^-$ salt) has been determined.[331] The disilacyclohexane ring adopts a regular chair conformation with the silicon atoms and the C_α and C_β atoms nearly coplanar (Si–C_α–C_β–Si angle = 173.9°). The molecule is linear around the vinyl bond (C_β–C_α–C = 178.8°) with an unusually short C–C double bond (1.22 Å) indicating that the carbocation carbon is *sp* hybridized. Furthermore, the C_β–Si bonds (1.98 and 1.95 Å) are about 0.1 Å longer than a regular C_{sp2}–Si bond, which can be attributed to the β-silyl hyperconjugation. Molecular structures calculated by density functional theory [B3LYP/6-31G(d)] and *ab initio* [MP2/6-31G(d)] method closely match the experimental solid-state geometry.

Müller et al.[332] have characterized a number of α-aryl-, β,β′-disilyl-substituted vinyl cations (**130**). Both ^{13}C NMR data and calculations show the effect of both π-conjugation with the aromatic moiety and σ-delocalization of the β-Si–C bonds. π-Delocalization is manifested by the marked low-field shift of the *ortho* and *para* carbon atoms compared to those of the precursor alkynes, whereas σ-delocalization is shown by the deshielding of the ^{29}Si NMR resonance relative to the precursor silylalkyne ($\Delta\delta^{29}$Si 29.5–41.4). Computed structures [B3LYP/6-31G(d) level] are all very similar, that is, calculation is not able to show the subtle interplay between σ- and π-delocalization, which is evident from the NMR spectroscopic data.

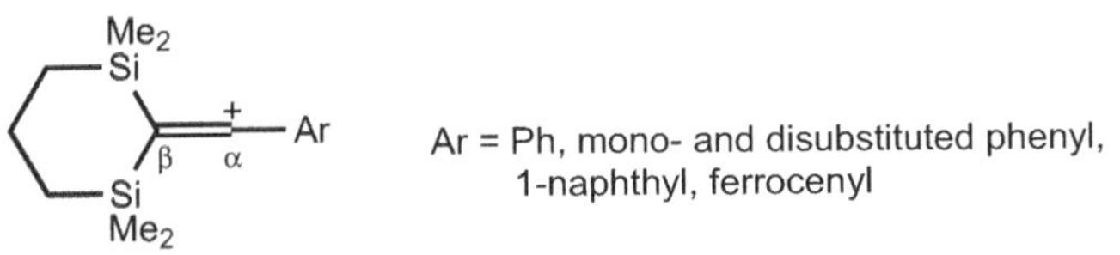

130

The 1-(*para*-methoxyphenyl)-2-(triisopropylsilyl)vinyl cation **131** has also been generated, characterized, and studied[323,333] to elucidate the importance of α–π aryl and β–σ hyperconjugative stabilization. A comparison was made with the *para*-methoxyphenylvinyl cation **124**, which has a higher demand for α-aryl π-stabilization. The *para* carbon of cation **131** is 7 ppm more shielded compared to that of cation **124**, indicating that the β-silyl stabilization effect is operative. The experimentally determined rotation barrier of the *para*-methoxy substituent for **131** (<8 kcal mol^{-1}) and **124** (9.0 kcal mol^{-1}) shows a less significant double-bond character of the MeO–C(4) bond in ion **131**, that is, the existence of β-silyl effect. These findings demonstrate that σ-bond hyperconjugation contributes to charge distribution even in highly π-stabilized α-(*para*-methoxyphenyl)vinyl carbocations.

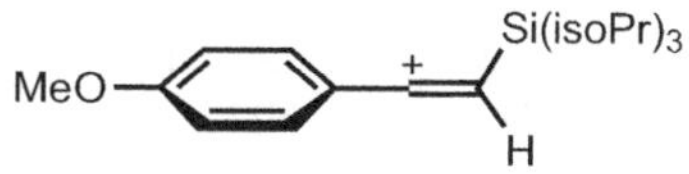

131

To study the possible stabilizing effect of β-silyl cations, Olah and co-workers[334] prepared the 2-[(1-trimethylsilyl)vinyl]-2-adamantyl cation **132** [Eq. (3.43)] as well as the parent silicon-free carbocation. In contrast to the above observations, NMR data [the (C1′), (C2), and (C2′) carbons are more deshielded in **132** than in the parent ion] showed that cation **132** is destabilized compared with the silicon-free analog. Furthermore, at −100°C the C(1) and C(3) carbons were found to be equivalent, whereas in the parent ion they were nonequivalent. This indicates a rapid rotation about the C(1)–C(3) bond in **132**, which can be rationalized by assuming the intermediacy of the β-silyl-stabilized cation **133**. The difference between cation **132** and those having β-silyl-stabilization discussed above may be the orthogonal arrangement of the β-C–Si bond and the *p*-orbital of the carbocation center.

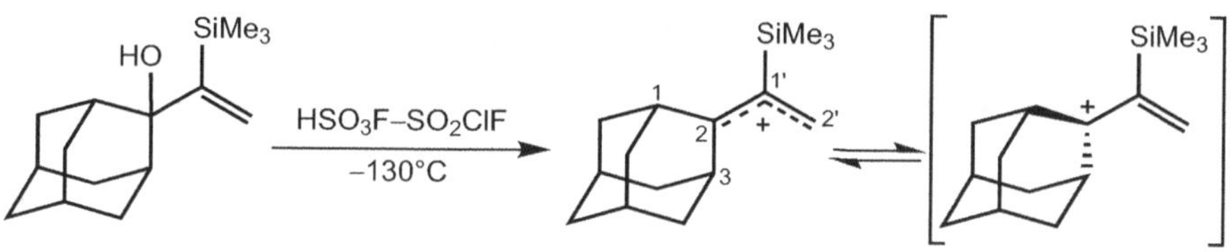

132 **133**

(3.43)

3.4.10. The Phenyl Cation

The phenyl cation (**134**) first postulated by Waters[335] is a highly reactive species of low stability and plays a fundamental role in organic chemistry—for example, in the chemistry of diazonium ions. According to gas-phase studies and calculations, its stability is between that of the ethyl cation and the vinyl cation.[336] Since it is an extremely electrophilic and short-lived species, it could not be isolated or observed directly in the condensed phase. For example, solvolytic and dediazoniation studies under superacidic conditions by Laali et al.[337,338] failed to find evidence of the intermediacy of the phenyl cation. Hyperconjugative stabilization via *ortho*-Me_3Si or *ortho*-*tert*-Bu groups, however, allowed Sonoda and co-workers[339] generate and trap the corresponding substituted phenyl cations.

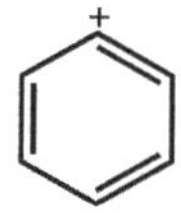

134

Theoretical studies have shown that the singlet minimum of the parent phenyl cation lies about 20–25 kcal mol^{-1} lower than the triplet minimum.[340] Recent *ab initio* calculations have shown that electron-donating substituents in the *para* position increase the relative stability of the triplet state.[340] Benzannelation has a similar effect. Protonation of the substituents, however, dramatically increases the preference for the singlet state, with $-SH_2^+$, $-SMeH^+$, and $-NH_3^+$ exhibiting the largest effect. The effect of the $-N_2^+$ group is similar to the onium cationic substituents. Dramatic stabilization of the singlet state was also observed in *ortho*-Me_3Si-phenyl cations generated by photolysis of Me_3Si-substituted chlorobenzenes.[341] Lodder and co-workers[342] performed photolysis of various phenyl cation precursors in the presence of anisole and correlated distinct product patterns with the singlet/triplet nature of the phenyl cations.

Winkler and Sander[343] have recently succeded in preparing the phenyl cation by co-deposition of iodobenzene or bromobenzene with a microwave-induced argon plasma in cryogenic matrix. Cation **134** was characterized by IR spectroscopy by an intense absorption at 3110 cm^{-1}. DFT studies and isotope labeling experiments show that this characteristic absorption is caused by a C–H stretching vibration involving the *ortho* hydrogens. The 4-methoxyphenyl cation has recently been generated in solution by photolysis of 4-chloroanisole and detected by flash photolysis.[344]

3.4.11. Arylmethyl and Alkylarylmethyl Cations

The first stable, long-lived carbocation observed was the triphenylmethyl (trityl) cation **135**.[6–9]

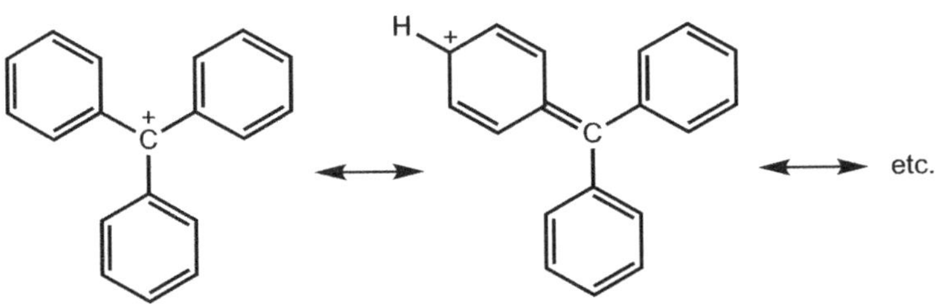

135

This ion is still the best-investigated carbocation; and its propeller-shaped structure, shown, for example, by its X-ray characterization (aromatic ring angles = 54°),[345] is well-recognized. Strong contribution from *para*- (and *ortho*-) quinonoidal resonance forms are responsible for much of the reactivity of the ion. Diphenylmethyl cations (benzhydryl cations) are considerably less stable than their tertiary analogs. Reindl et al.[346] used force field (MMP2) and *ab initio* (MP2/6-31G*) methods to show, however, that stabilization by the phenyl group is attenuated. Thus, the resonance stabilization value of cation **135** (average value of −41.6 kcal mol^{-1} for each phenyl ring) is much less than that of the benzhydryl ion **136** (−51.4 kcal mol^{-1}) and the benzyl cation (−76.4 kcal mol^{-1}). Although UV spectra in dilute sulfuric acid solutions have been obtained early,[347] only in the 1960s has the benzhydryl ion **136** been observed in higher concentrations in superacid solutions [HSO_3Cl,[348] SbF_5,[349] and $HSO_3F–SbF_5$[350]]. Recently, 4,4′-disubstituted benzhydryl cations have been generated by the oxidative C–H bond dissociation of the corresponding diarylmethanes.[351] The accumulation of the cations was confirmed by ^{13}C NMR spectroscopy: A signal for the 4,4′-difluorobenzhydryl cation was detected at $\delta^{13}C$ 192.6, which is close to the value ($\delta^{13}C$ 193.3) observed in superacid medium ($HSO_3F–SbF_5–SO_2ClF$).[352]

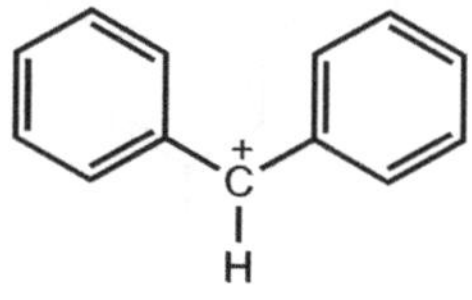

136

Mono- and dialkylarylmethyl cations can be obtained readily from the corresponding alcohols, alkenes, or halides in superacid solutions, such as $HSO_3F–SbF_5$,[350] HSO_3Cl and SbF_5,[348,349] and oleum.[88] Structures **137–139** are representative alkylarylmethyl cations.

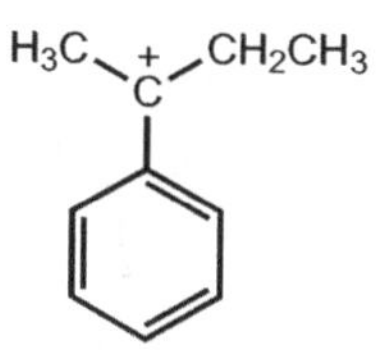

137

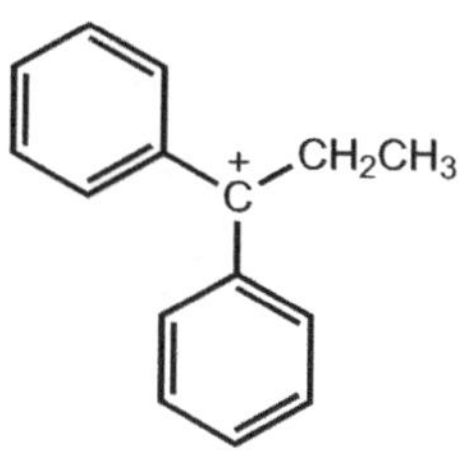

138

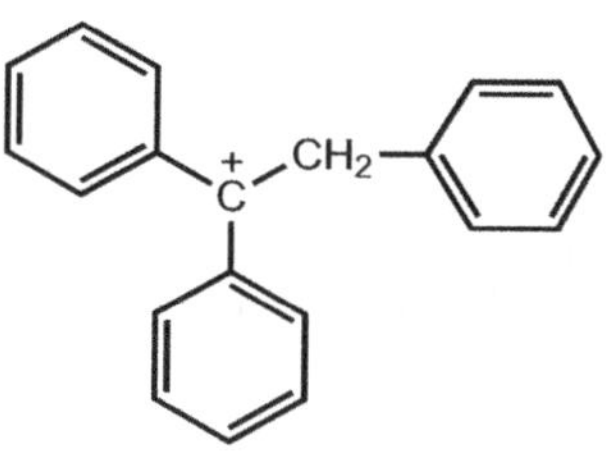

139

Because of the high stability of the tertiary ions, these are preferentially formed in the superacid systems from both tertiary and secondary, and even primary, precursors.[353] If, however, the tertiary carbocation is not benzylic, rearrangement to a

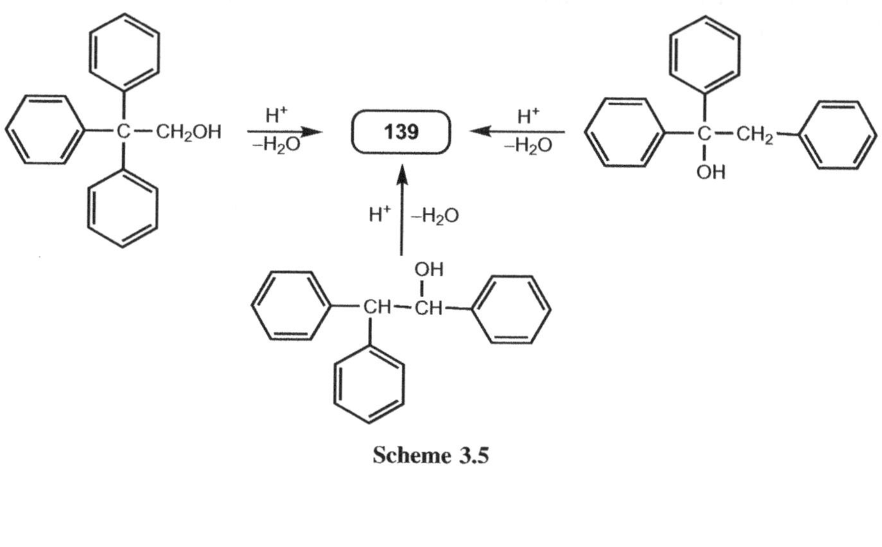

Scheme 3.5

secondary benzylic ion can be observed[350,353] (Scheme 3.5). With suitable substituent groups (which also prevent transalkylations), secondary styryl cations (**140**–**142**) were found as stable, long-lived ions.[291,354]

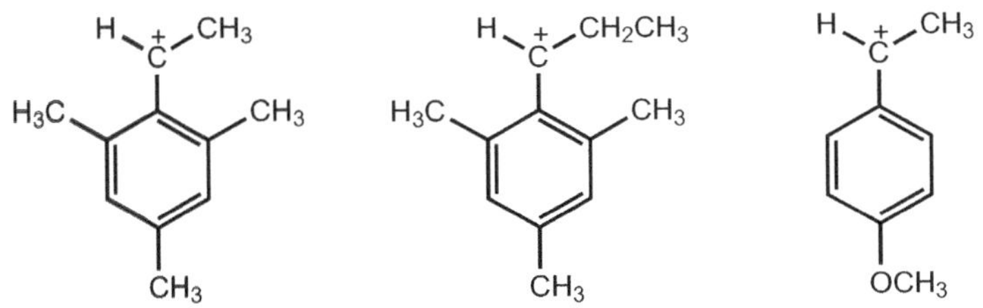

140 **141** **142**

Although the unsubstituted benzyl cation is still elusive, many substituted derivatives have been observed (cations **143**–**146**).[355,356]

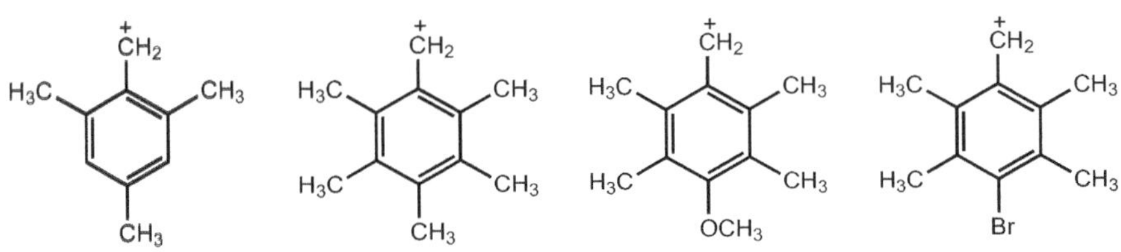

143 **144** **145** **146**

In a cation such as the (2,4-di-*tert*-butyl-6-methyl)benzyl cation **147**, a high rotational barrier around the sp^2-hybridized atom is observed. The methylene protons are found magnetically nonequivalent in the ^1H NMR spectrum.[356] Recent combined experimental and theoretical studies for the related cation **143** suggest[357] that structure **143b** is an important resonance contributor.

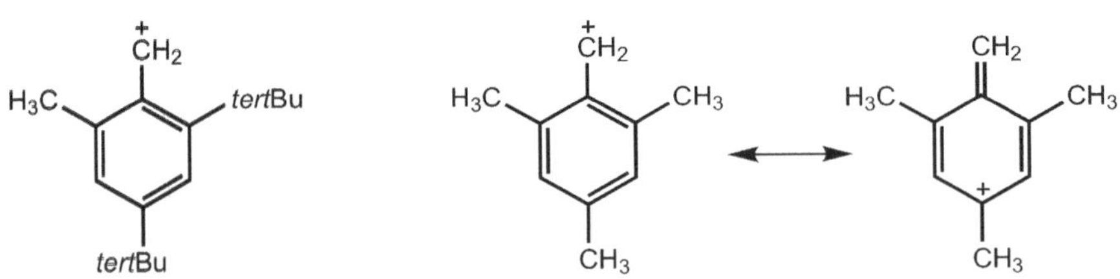

147 **a** **b**

143

No rearrangement of benzyl cations in acid solutions to tropylium ions has been found, although this rearrangement is observed in the gas phase (mass spectrometry),[358,359] and the tropylium ion was shown by *ab initio* studies to lie about 7 kcal mol^{-1} lower in energy than the benzyl cation **148a**.

Laube, Olah, and Bau have reported an X-ray crystallographic study of the cumyl cation (**148c**) hexafluoroantimonate salt.[360] Cation **148c** is nearly planar (phenyl ring twist angle = 8°) and has a short C^+–C_{ipso} bond (1.41 Å), and bond distances in the phenyl ring show strong benzylic delocalization. Furthermore, a slight shortening of the C^+–Me bonds (0.025 Å) indicates weak C–H hyperconjugation. The structure and stability of a range of benzylic cations, including **135**, **136**, and **148**, have also been calculated.[346] It was found that methyl substitution at the cation center of the benzyl cation **148a** increases stabilization of the carbocation carbon by C–H hyperconjugation. As a result, the C^+–C_{ipso} bond elongates (**148a** = 1.37 Å, **148b** = 1.39 Å, **148c** = 1.41 Å, MP2 level). These theoretical results, in agreement with the solid-state structure and other findings discussed above, suggest that the aromaticity of the benzene ring is substantially reduced. Therefore, the best description of the benzyl cation **148a** is structure **148′**.

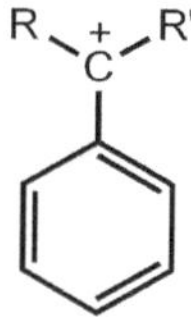

a R, R' = H
b R = H, R' = Me
c R, R' = Me

148

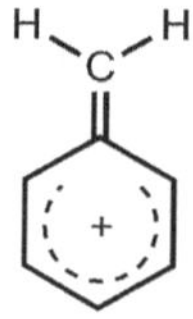

148'

Sterically crowded arylmethyl cations have been prepared by Olah, Prakash, and co-workers.[361,362] 1,1′-Diadamantylbenzyl cations **149** generated from the corresponding substituted methanols were studied to explore the extent of *p*–π conjugation, affected by twisting of the aromatic ring brought about by steric crowding. The Hammett plot of *para*-substituted cations **149** showed negligible slope. Furthermore, the ^{13}C NMR chemical shift of the 1,1′-diadamantylbenzyl cation (**149**, R = H, $\delta^{13}C$ 286. 5) was found to be much deshielded compared with the that of the dimethylbenzyl cation ($\delta^{13}C$ 255.0), and the *para*-carbon chemical shifts were practically unaltered with respect to those of the parent alcohols. All these data indicate the virtual absence of π resonance stabilization.

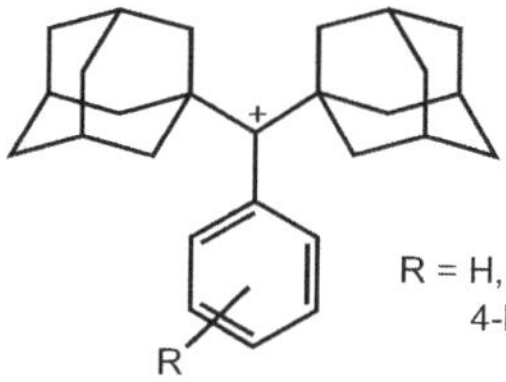

R = H, 2-Me, 3-Me, 3-*tert*-Bu
4-Me, 4-MeO, 4-F, 4-CF_3

149

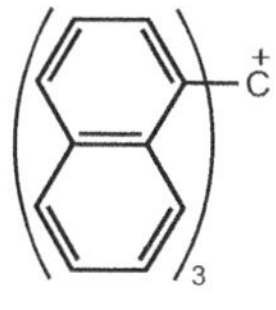

150

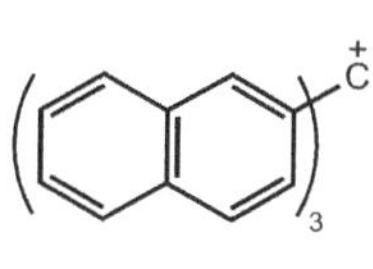

151

The highly crowded tris(1-naphthyl)methyl cation **150** and tris(2-naphthyl)methyl cation **151** were prepared and used to abstract hydride ion from cycloheptatriene to generate tropylium ion.[363] Hydride abstraction, however, could be performed only with the less crowded cation **150**.

Suzuki and co-workers[364,365] studied *peri*-disubstituted triarylmethane–triarylmethylium naphthalene derivatives to explore the possibility of the bridging C–H structure. In cation **152** (Ar′ = Ph), the H–C^+ and H–C distances were found to be 1.09 and 2.39 Å, respectively, by X-ray structure analysis. Such geometric features of the solid-state structure indicate negligible contribution from the delocalized *3c–2e* bonding arrangement **152a**. Consequently, these cations prefer the localized structure **152b**. PM3 calculations indicated that the delocalized structure **152a** is the transition state for the degenerate interconversion of **152b**.

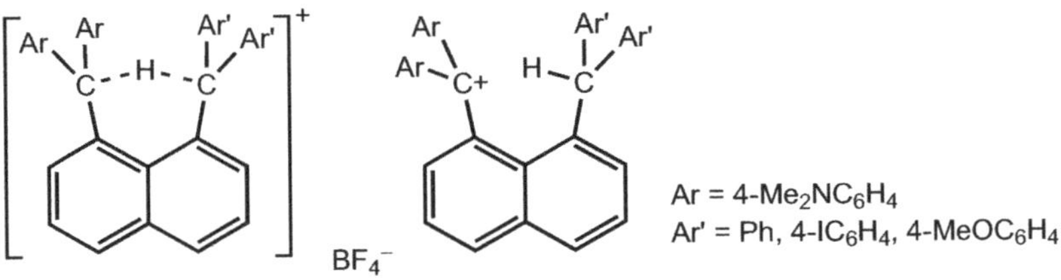

152

Laursen, Krebs, and co-workers synthesized a series of triangulenium cations with the general formula **153** with varied substitution patterns. These are highly stable cations with very high pK_{R+} values and used as textile and laser dyes and cellular stains for diagnostic purposes. According to X-ray characterization, the geometry for trioxatriangulenium cations is planar,[366,367] whereas the triazatriangulenium cations have a disc shape.[368]

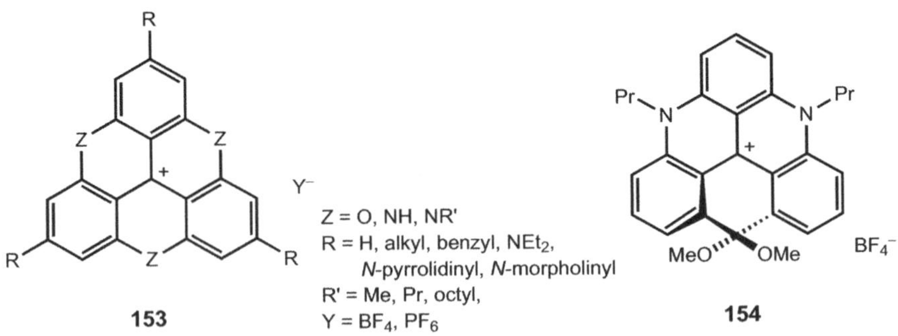

153 **154**

The same group has recently reported[369] the synthesis and characterization of the [4]heterohelicenium cation **154**. X-ray analysis confirmed that strong repulsion between the methoxy substituents prevents the system being planar. Consequently, cation **154** is chiral and indeed crystallizes as a racemate. The oxygen atoms are positioned exactly above one another, and this determines a pitch of 2.7 Å of the system ([6]helicene has a pitch of 3.2–3.3 Å). Cation **154** is configurationally stable, which allowed the resolution of the racemate. The barrier for racemization was determined to be $\Delta G^{\ddagger} = 41.3\,\text{kcal mol}^{-1}$ at 200°C ($t_{1/2} = 182.7$ h), which compares favorably with that of [6]helicene.

The naphthacenyl cation **155**[370] and the indanyl cation **156**[371] have been investigated by X-ray crystallography. The rather long C(3a)–C(7a) bond (1.456 Å) and short C(3a)–C(4) bond (1.346 Å) suggest charge delocalization into the aromatic ring.

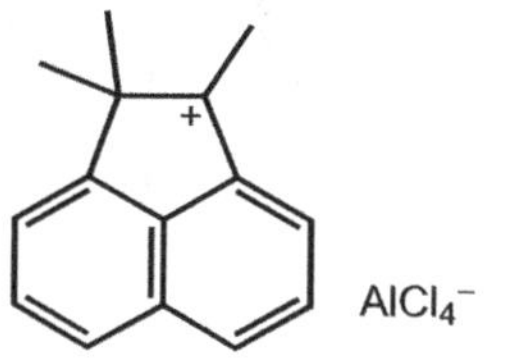

155

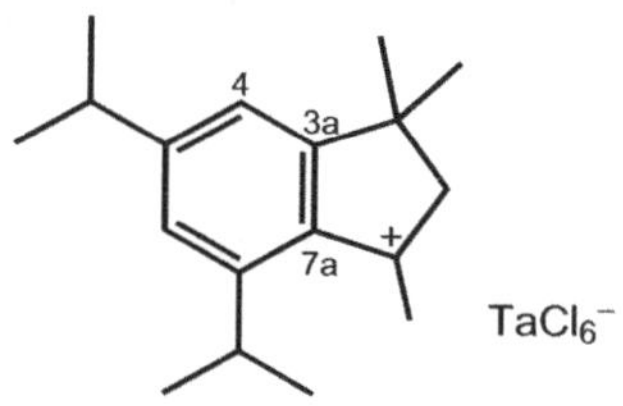

156

Laali et al.[372,373] have characterized carbocations generated from substituted polycyclic aromatic compounds. The related cation **157** is a true aryl-methyl-type ion, whereas cations **158** have arenium ion character because the strongly electron-withdrawing α-CF_3 group enhances charge delocalization into the pyrenyl and phenyl groups.

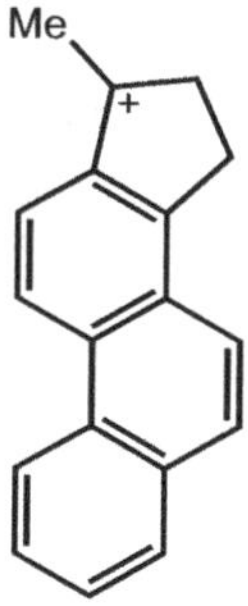

157

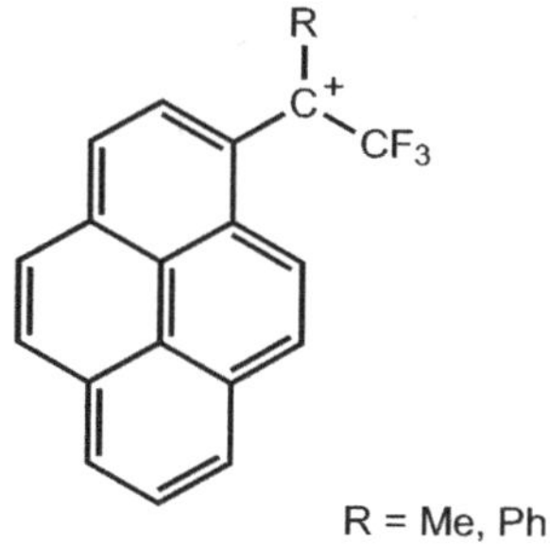

158

In a series of papers, Takekuma and co-workers[374–377] have reported the synthesis of a variety of 3-guaiazulenylmethyl cations **159** [Eq. (3.44)] with full spectroscopic characterization (UV–visible, IR, multinuclear NMR) and X-ray crystal structure analysis. The corresponding 2-furyl-, 2-thienyl-, and 2-pyrrolyl-substituted hexafluorophosphates have also been obtained.[377] The molecular structure of the salts **159** reveals that the plane of the 3-guaiazulenyl group is twisted from the plane of the phenyl ring (dihedral angles = 20.7–40.1°). Further data indicate that although the positive charge mainly localized on the methyl carbon, the charge is slightly transferred to the guaiazulenyl group and the phenyl group (resonance structures **160a–160c** for the 4-dimethylamino derivative).

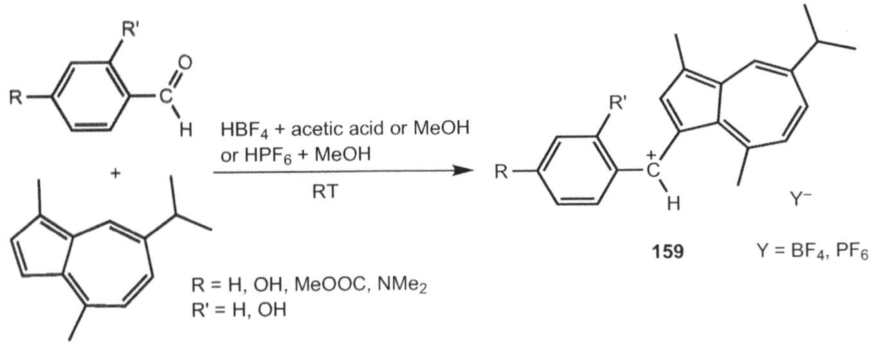

(3.44)

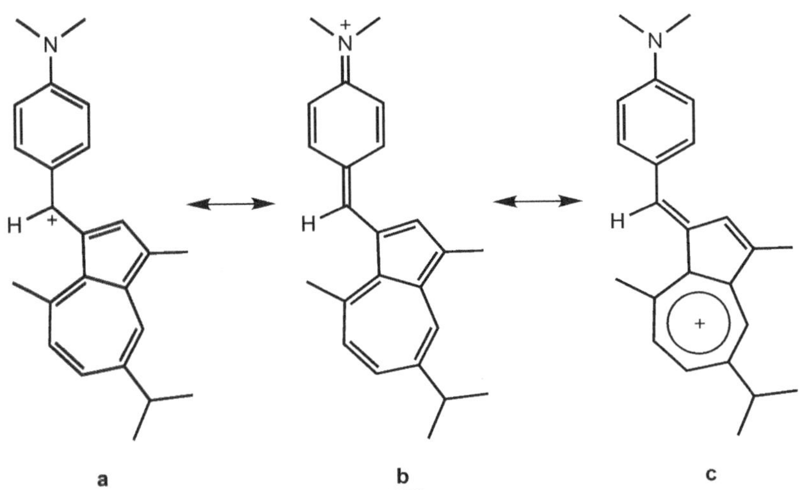

160

The S_N1-type displacement reactions of chiral benzylic alcohols with aromatics catalyzed by triflic acid or $HBF_4 \cdot OEt_2$ have been observed by Prakash, Bach, and co-workers[378] to exhibit high facial diastereoselectivity. This indicates the involvement of a carbocationic intermediate with a restricted conformation. Although the *tert*-butyl-substituted tertiary carbocation underwent β-elimination, they succeded in observing the ethyl-substituted cation **161** in SbF_5–SO_2ClF solution at −70°C by NMR. The ^{13}C NMR spectrum exhibits 12 carbon signals with the most deshielded detected at $\delta^{13}C$ 262.2 relative to acetone-d_6. The six carbon atoms of the aromatic rings are magnetically nonequivalent, which is due to the restricted rotation about the phenyl–C^+ bond. More importantly, NOE experiments showed spatial proximity of *ortho* hydrogens with the methyl group and the hydrogen of the stereogenic center, respectively. Such conformational restriction accounts for the observed facial discrimination.

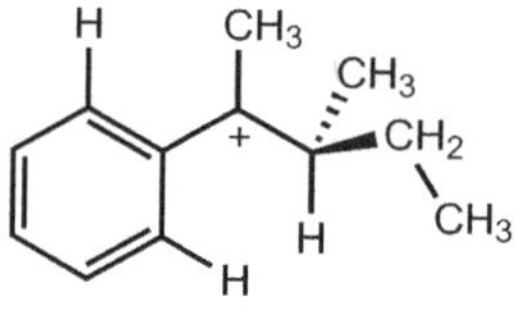

161

3.4.12. Carbodications and Polycations

Interest in carbocations has not been confined to monopositive carbon species (carbomonocations). The study of carbodications has more recently been of substantial interest and the topic has been recently reviewed.[379,380]

Early reports[381] that a carbodication had been observed from pentamethyltrichloromethyl-benzene turned out to be incorrect. The species obtained was the dichloropentamethylbenzyl cation **162**.[382–384] Ionization of the 2,6-bis(chloromethyl)mesitylene, in turn, did yield the remarkably stable, unique 2,6-dimethylmesitylene-2,6-diyl dication **163**.[357,385] Based on the NMR data, the diallylic cation **163b** appears to be the major resonance contributor to the structure, and this highly stabilized dienyl–allyl dication nature ensures the high stability. The structure is similar to that of the experimentally still elusive bisallylic benzene dication **164**, although polycyclic analogs were obtained (see Section 3.4.14). Calculations (MINDO/3) by Dewar and Holloway showed[386] that the benzene dication **164** favors a C_{2h} chair conformation, whereas Schleyer and co-workers[387] found (HF/3-21G) an elongated cyclic double-allyl D_{2h} geometry. The allyl units are uncoupled, thereby keeping apart the two pairs of π-electrons and minimizing mutual repulsions. The resulting unequal bond lengths force the ring out of planarity. A subsequent study by Krogh-Jespersen (HF/6-31G*//6-31G*)[388] favored again the C_{2h} chair conformation.

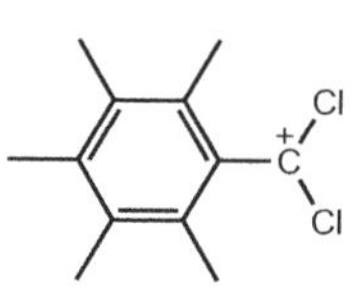

162

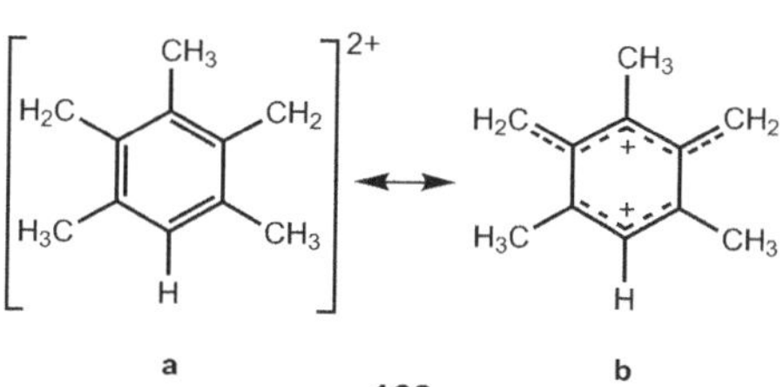

163

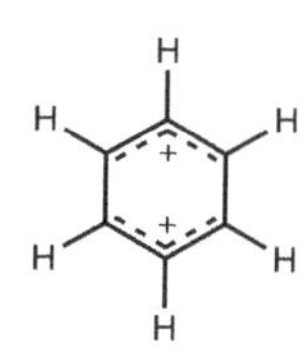

164

If two carbocation centers are separated by a phenyl ring, a variety of carbodications and carbotrications can be obtained (ions **165–167**).[389–391]

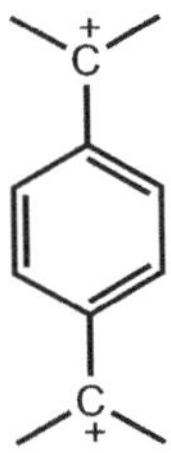

165

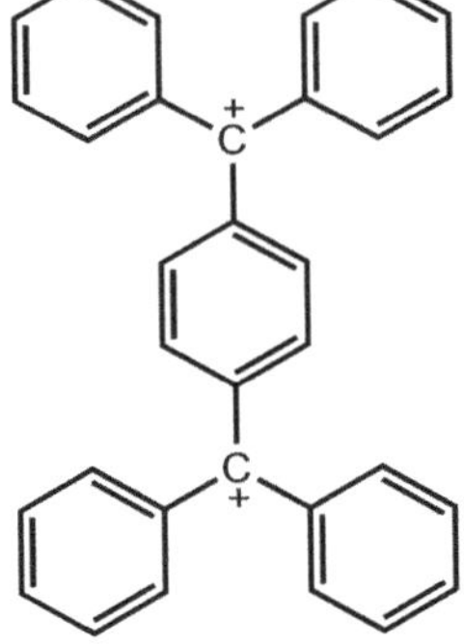

166

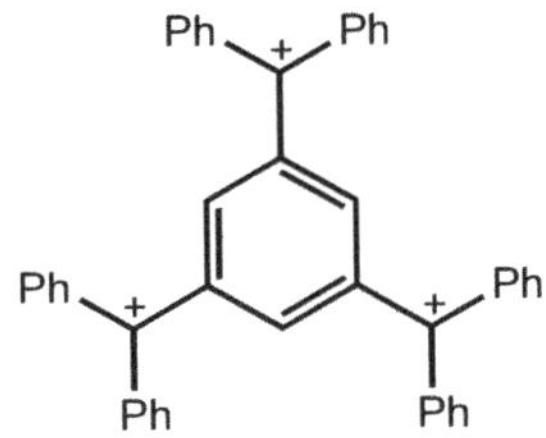

167

Separation of two carbocation centers by at least two methylene groups in open-chain carbodications renders them stable and observable.[387,392] Indeed, such stable carbodications **168** have been subjected to a comprehensive NMR spectroscopic study.[391]

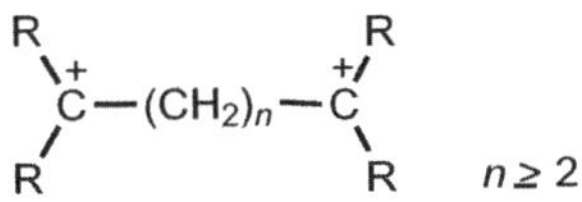

168

Carbodications have also been observed in more rigid systems such as the apical, apical congressane (dimantane) dication **169**[182] and the polycyclic bridgehead dication **170**.[182] The bicyclo[2.2.2]octane-1,4-diyl dication **171** was claimed to have been prepared in an earlier study.[393] In a subsequent reinvestigation,[394] however, dication **171** could not be detected. Instead, only the monocation monodonor–acceptor complex was obtained.

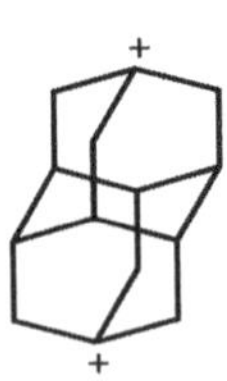

169

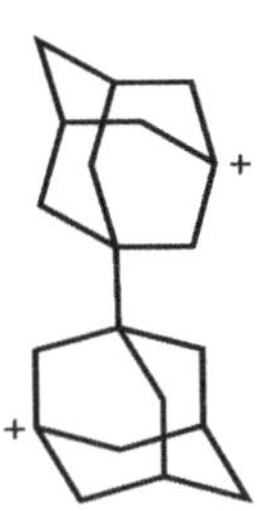

170

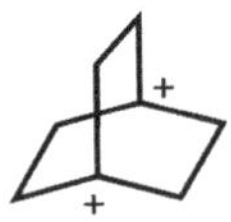

171

On the other hand, the bicyclo[3.3.3]undecane-1,5-diyl dication **172** (manxyl dication) was observed first by Olah et al.[189] (the ^{13}C NMR spectrum is shown in Figure 3.15) [Eq. (3.45)]. Until the generation of the 1,16-dodecahedryl dication

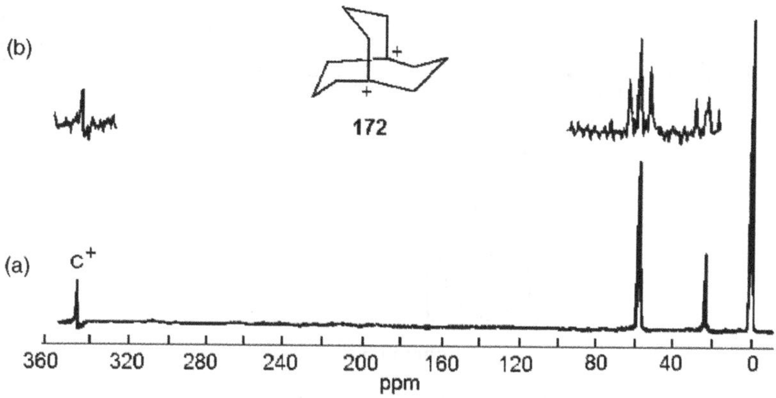

Figure 3.15. The 25-MHz ^{13}C NMR spectrum of the bicyclo[3.3.0]undecane-1,5-diyl dication **172** in SbF_5–SO_2ClF solution. (a) Proton decoupled, (b) proton coupled.

(see Section 3.4.3), dication **172** was the most deshielded species observed (δ^{13}C 346.2) (Figure 3.15). Subsequently, it was also investigated by Taeschler and Sorensen[395] and was found to rearrange to dication **173**, which was independently prepared form a mixture of two isomeric dienes.

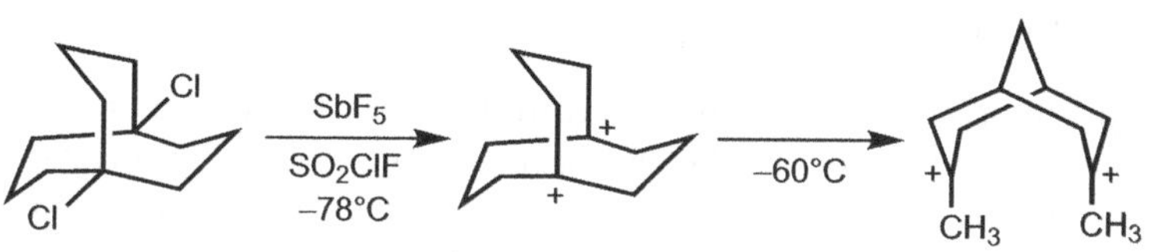

172 **173** (3.45)

Sorensen and co-workers[396] prepared and characterized dication **174** in an attempt to transform it into a μ-hydrido bridged monocation (see Section 3.5.2.6) (Scheme 3.6). The two lowest-field ^{1}H NMR signals (δ 4.76 and 4.60) were assigned to the bridgehead protons. ^{13}C NMR characterization showed a C_s symmetry and fluxional behavior due to a ring flip. The computed distance of the cationic centers (2.81 Å, close to the sum of the van der Waals radii) is practically identical to that found for the manxyl dication **172** (2.80 Å). Hydride transfer to dication **174** did not result in the formation of the desired bridged species; instead, the H–out monocation **175** was formed, which rearranged to cation **176**.

When the dodecahedryl cation **59** prepared by ionization of dodecahedryl derivatives or the parent hydrocarbon was left standing in the superacid medium for 6–7 h at −50°C, it slowly and irreversibly transformed into dodecahedrane-1,16-diyl dication **177** [Eq. (3.46)]. Dication **177** is of D_{3d} symmetry and characterized by three NMR absorption (δ^{13}C 379.2, 78.8, 59.8). The ^{13}C chemical shifts of the positively charged centers in cation **59** (δ^{13}C 363.9) and dication **177** (δ^{13}C 379.2) are the most deshielded ever observed. According to calculations (SCF-MO),[194] the dodecahedrane skeleton is incapable of accommodating a planar carbocation

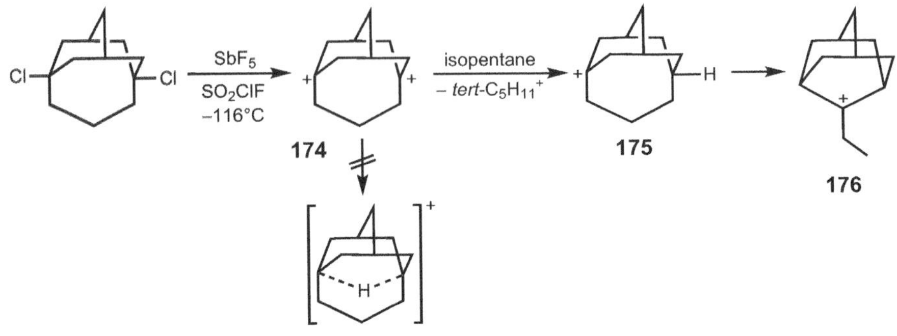

Scheme 3.6

geometry. Consequently, dodecahedryl cation **59** and dodecahedrane-1,16-diyl dication **177** can be considered true sp^3-hybridized carbocations.

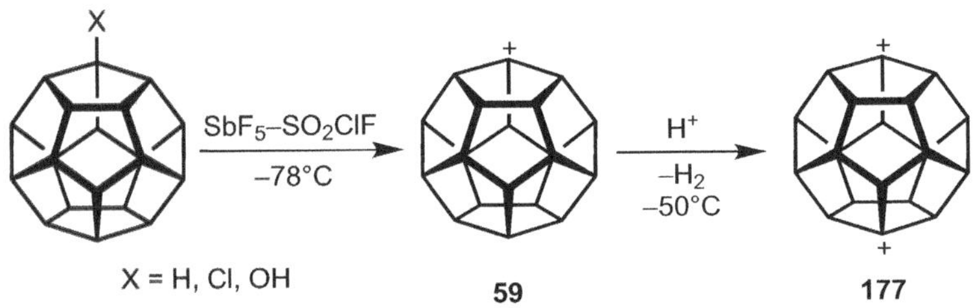

(3.46)

Attempts to observe circumambulatory rearrangement in the 2,6-*anti*-tricyclo [5.1.0.0^{3,5}]octane-2,6-diyl dication **178** have been unsuccessful.[397] The dication **178** would appear to rearrange instantaneously to the homotropylium ion **179** by proton elimination. However, substituted dications of type **178** (e.g., **180**) are quite stable; they are static, and a substantial part of the charge is delocalized into the cyclopropane rings.

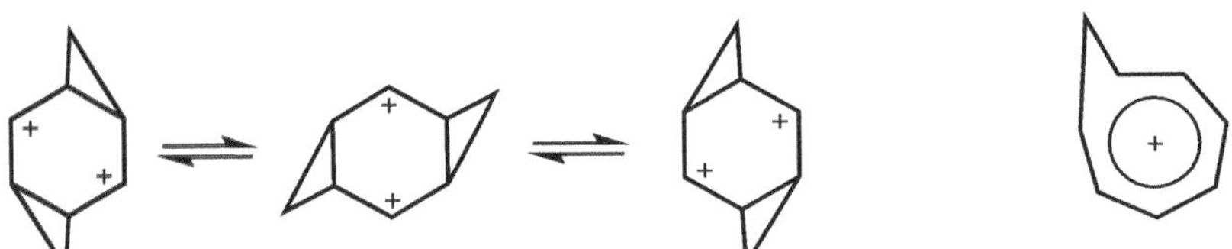

178 **179**

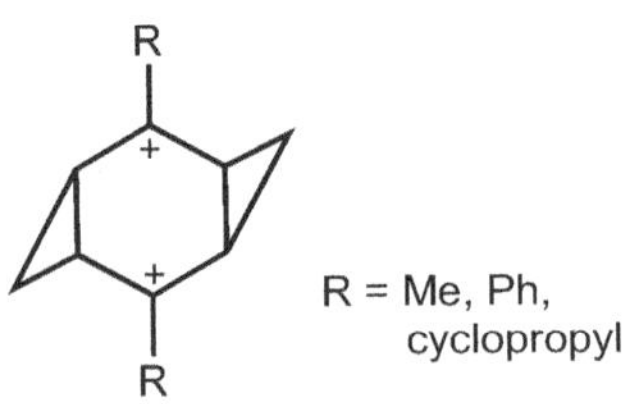

180

Several 2,6-disubstituted adamantane-2,6-diyl dications **181** have been prepared and characterized.[398] They are stable only when they contain charge delocalizing substituents such as cyclopropyl or phenyl groups. Recently, a DFT study (B3LYP/6-31G** level) of the adamantanediyl dications $C_{10}H_{14}^{2+}$ has been reported.[399] Interestingly, the adamantane-1,2-diyl dication **182** is not an energy minimum structure on the potential energy surface, whereas the adamantane-1,3-diyl dication with two bridgehead tertiary cationic centers (**183**) was found to be the most stable, being more stable by 14.6 kcal mol^{-1} than **182**. Furthermore, despite charge separation in structure **184**, this cation is less stable, although only by 0.4 kcal mol^{-1}, than structure **183**.

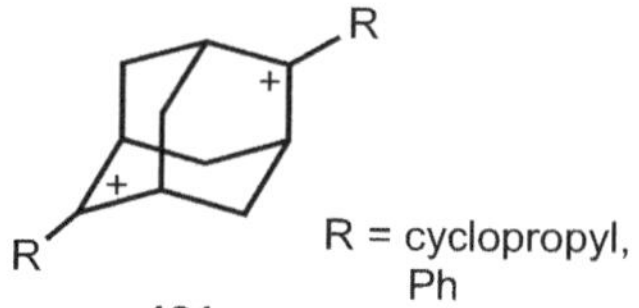

181

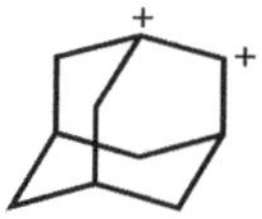
182

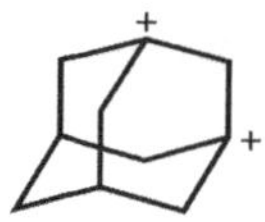
183

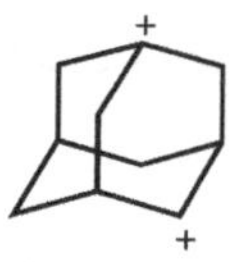
184

Additional di- and tetracations employing the adamantane skeleton were prepared and characterized by Olah, Prakash, and co-workers. The 1,5-distonic dications **185** are stabilized by both the α,α-disubstitution and bridgehead hyperconjugation.[400] Interestingly, however, the dication with cyclopropyl groups was unstable. In the exceptionally rare, tetrahedrally arrayed tetracation **186**[401] the bridgehead carbons and the formal cationic sites are shielded, whereas carbons of the phenyl rings are more deshielded relative to **185**. This indicates that in order to offset the additional charge–charge repulsion in **186** relative to **185**, increased delocalization of the positive charge into the phenyl groups is effected.

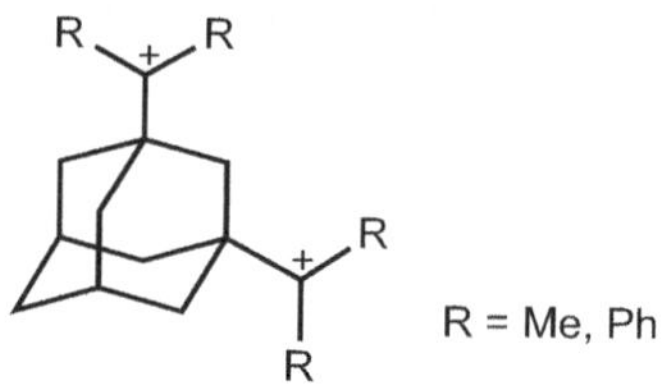

185

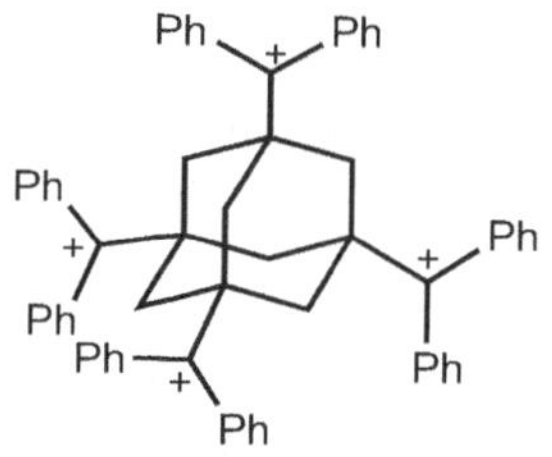

186

Ionization of the corresponding diols with HBF_4 or oxidation of 9,9,10,10-tetraaryldihydrophenanthrenes results in the formation of the 2,2′-bis(triarylmethylium) dications **187**.[402] These were characterized by UV spectroscopy and found to show tricolor electrochromic behavior in reversible reduction–oxidation cycles.

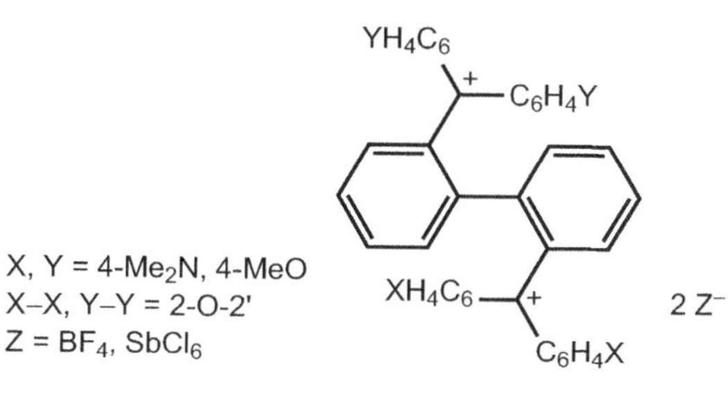

187

By ionizing the respective alkenediol, Olah and co-workers[403] were able to prepare the (hexaphenyltrimethylene)methane dication **188** [Eq. (3.47)].[403] The dication possesses C_3 symmetry and adopts a propeller-like structure, but no evidence for "Y-aromatic" stabilization was found. Furthermore, the phenyl groups themselves are twisted, which prevents the optimum overlap between the π-system and the vacant *p*-orbitals at the three C(2) carbon atoms. Calculated charge density distributions and the ^{13}C NMR data show that C(1) is not significantly deshielded for an sp^2 atom, indicating that the three C(2) carbon atoms and their phenyl substituents bear the charge of dication **188** similar to the trityl cation **135**.

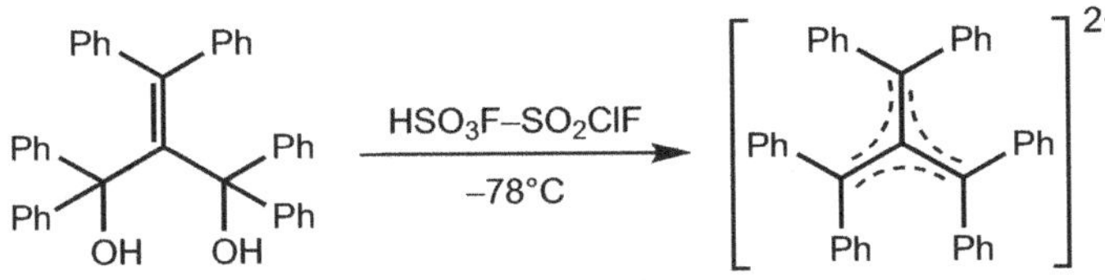

188

(3.47)

Tetracation **189** and hexacation **190** have been obtained by ionizing the corresponding tetrahydroxy and hexahydroxy precursors, respectively, with triflic acid or tetrafluoroboric acid.[404] These robust polytrityl cations are stable at room temperature for prolonged periods. Owing to their highly symmetric structure, the ^{1}H NMR spectra of the polycations are simple and similar to the parent trityl cation. Likewise, the ^{13}C NMR spectra reveal a single resonance for the cationic carbons at δ^{13}C 208.15 (**189**) and 208.22 (**190**) (δ^{13}C 209.73 for the parent trityl cation).

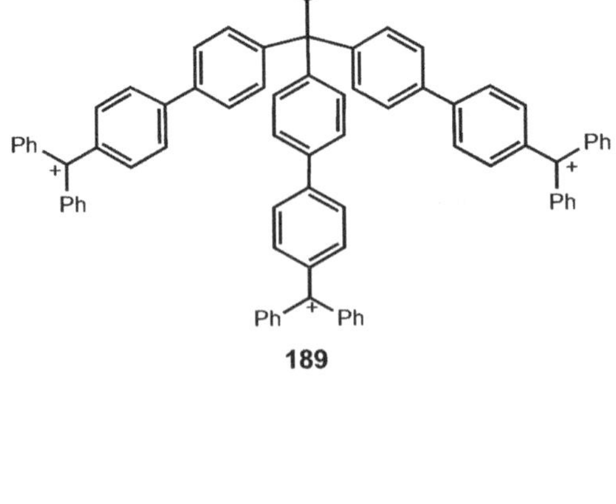

189

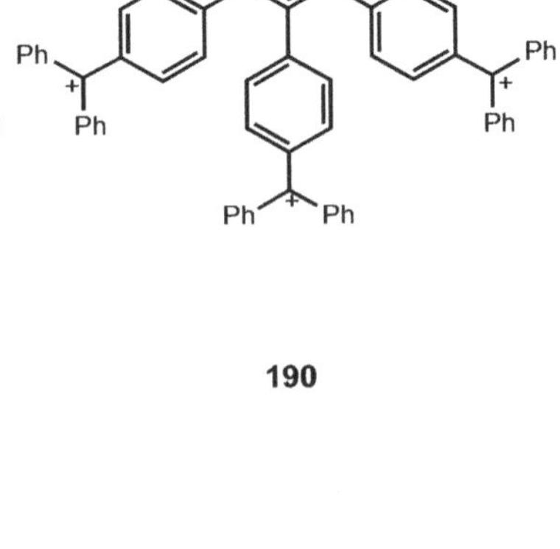

190

On the basis of previous observations on the high stabilizing ability of the cyclopropyl group, Olah et al.[391] made further successful attempts to prepare other dications with cyclopropyl groups. Dication **191** (R = Ph) shows significantly enhanced delocalization into the aromatic ring compared with the corresponding analogous monocation (diphenylcyclopropylmethyl cation) with partial stabilization by the cyclopropane ring. The ^{13}C NMR shifts of the carbocation center of dication **191** (R = cyclopropyl)[405] and **192**[227] show enhanced shielding as compared to that of the tricyclopropylmethyl cation **60** and 1,1-dicyclopropylethyl cation by 16 ppm and 12 ppm, respectively. This, again, indicates that these dications are significantly stabilized by charge delocalization involving the cyclopropane rings.

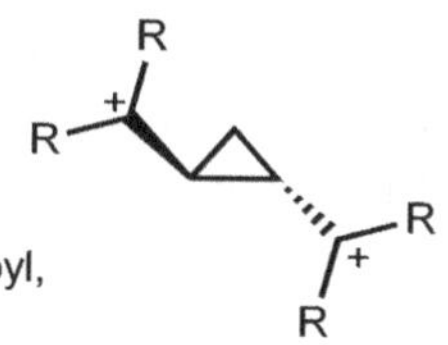

R = cyclopropyl,
Ph

191

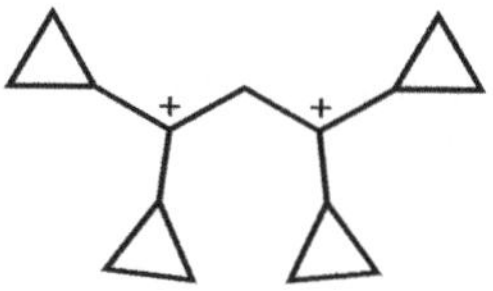

192

The effect of the introduction of two electron-deficient centers into the bicyclo [2.2.1]heptyl skeleton (norbornyl framework) has been explored.[406] A ^{13}C NMR spectroscopic study of several substituted 2,5-diaryl-2,5-norbornadiyl dications **193** reveals the regular dicarbenium ion nature of the system.

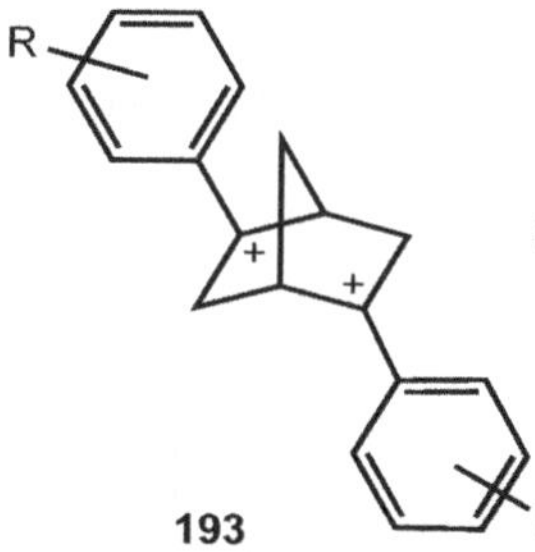

R = H, 4-Me, 4-MeO,
4-CF_3, 3,5-$(CF_3)_2$, etc.

193

The bicyclic bisallylic dication **194** has been obtained from a variety of precursors under strong acid conditions.[407–409]

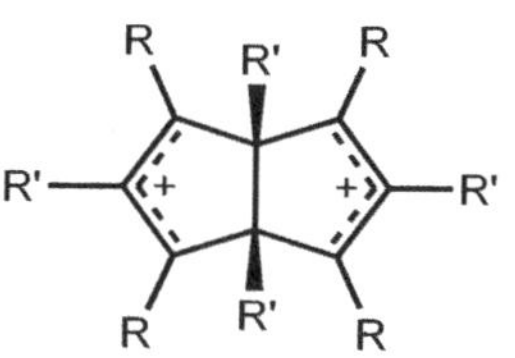

R= R' = H, Me
R= H, R' = Me

194

The *endo*-3,10-dimethyltricyclo[5.2.1.$0^{2,6}$]deca-4,8-dien-3,10-diyl dication **195** has been prepared[410] in $HSO_3F–SbF_5–SO_2ClF$ at −120°C. It possesses a novel bishomoaromatic/allylic dication structure. At higher temperatures, **195** rearranges stereospecifically to the symmetrical *cis-anti-cis*-3,10-dimethyltricyclo[5.3.0.0^{2,6}] deca-4,8-dien-3,10-diyl dication **196** [Eq. (3.48)].

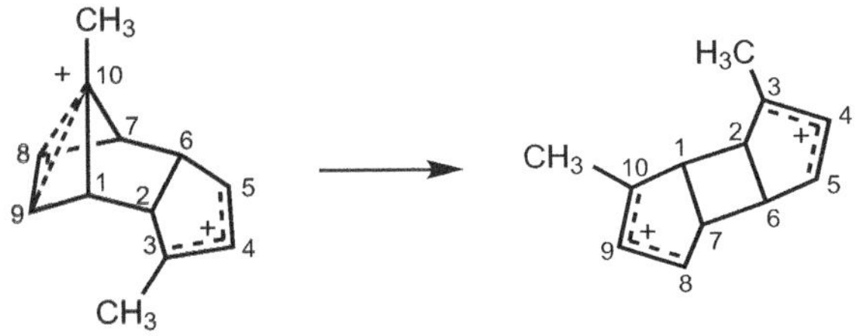

195 **196** (3.48)

Lammertsma and Cerfontain[411] have obtained the cyclopropyldicarbinyl dication **197** by diprotonating the 1,6-methano[10]annulene in much stronger Magic Acid at −60°C [Eq. (3.49)]. If the same protonation was carried out at −120°C, they were able to obtain the previously discussed monocation **106**.[284]

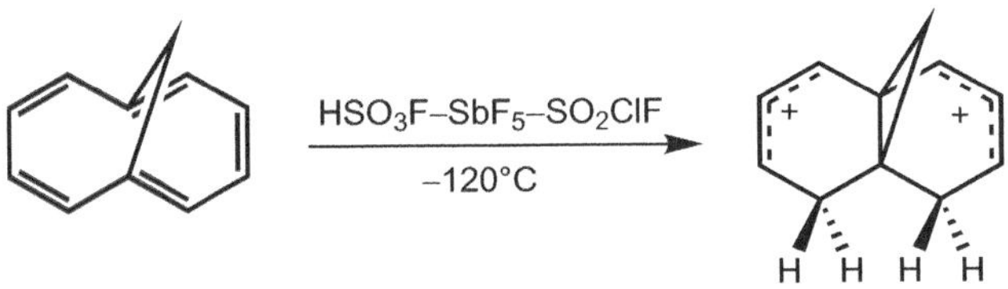

197 (3.49)

Lammertsma[412] has also obtained dication **198** by the protonation of hexahydropyrene in Magic Acid solution. This is the first example of a β,β′-diprotonated naphthalene derivative. Koptyug and co-workers[270,413] have studied α,α′-diprotonated hexa- and octamethylnaphthalenes **199** under superacidic conditions.

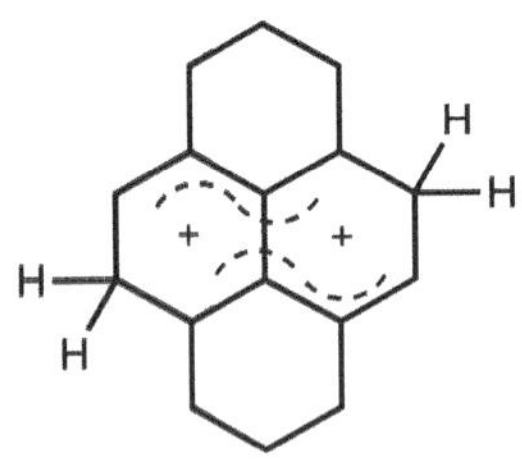

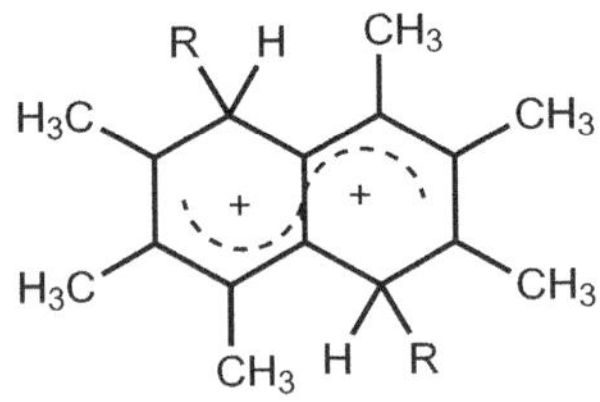

R = H, Me

198 **199**

Many aromatic stabilized dications have been prepared and characterized by NMR spectroscopy (see subsequent discussion).

Dicationic systems in which two diarylmethyl cations are linked by an 1,2-diphenylethyl,[414] biphenyl,[415] binaphthyl,[415] or naphthalene-1,8-diyl[416,417] backbone have attracted great interest recently because of their unique electrochromic properties. The dications **200** were prepared from the corresponding diols[416,417] [Eq. (3.50)] and could be reversibly transformed into the acenaphthylene system. The tight geometrical constraints induce distortion of the naphthalenediyl skeleton: the C^+–C(1)–C(8a) and C^+–C(8)–C(8a) angles (125.36° and 126.60°, respectively) are larger than the ideal 120° (data for dication **200**, R = 4-MeO). Each carbocation center adopts a trigonal–planar arrangement but the planes of the two centers form large dihedral angles with the plane of the naphthalene skeleton (59.7° and 59.6°). This indicates that there exists only a modest conjugation between the sp^2 cation centers and the naphthalene backbone. This unique molecular structure results in a rather short separation of the carbenium carbons (3.076 Å), which is the shortest distance reported for such systems.

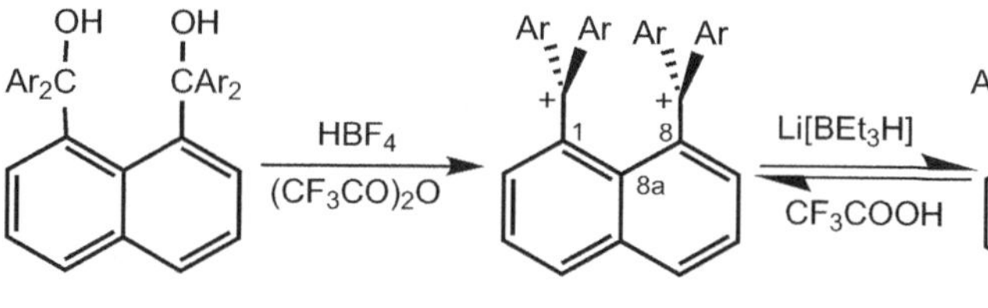

200

(3.50)

Oxidation of tetrasubstituted ethylenes, that is the formal removal of two electrons, yields tetrasubstituted ethylene 1,2-dications. Stable salts can be generated, however, only from tetraarylethylenes or the corresponding dihalides and diols since the aromatic rings in the resulting dication allow an efficient delocalization of the positive charge. The parent dication has been obtained by Olah et al.[391] [Eq. (3.51)] and the crystal structure of the tetraanisyl dichloroiodate salt was reported.[418]

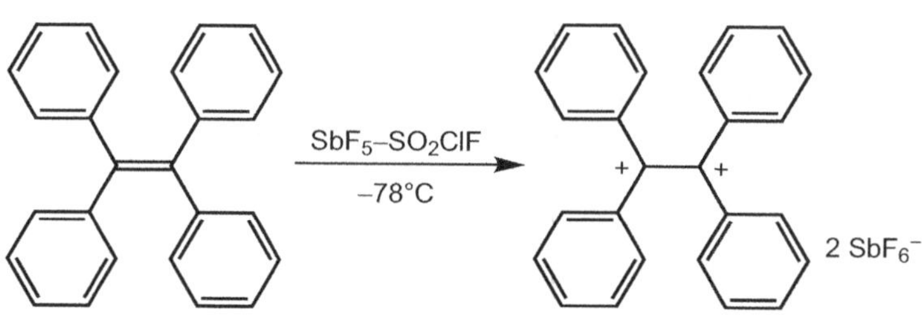

(3.51)

Kochi and co-workers[419] have prepared tetraanisyl dication **201** [R = MeO; $(SbCl_6^-)(Sb_2Cl_7^-)$] by chemical oxidation, whereas electrochemical or chemical oxidation has been applied by Mori and Inoue[420] to obtain chiral dications **201** (R = 2-Bu, 2-octyl; $SbCl_6^-$). The X-ray crystal structure of the tetraanisyl cation ($SbCl_6^-$ salt) shows significant increases upon oxidation in both the central C–C bond length and the dihedral angle about this bond from 1.359 Å and 3.8° of the neutral precursor to 1.503 Å and 61.6° of the dication. Calculated values for the AM1 optimized structure [B3LYP/6-31G(d) level] of the *R,R,R,R*-tetra-2-butyl derivative are 1.503 Å and 57.5°.[420] Furthermore, the geminal anisyl groups are inequivalent with one group being more coplanar than the other (19° versus 28°). Bond distances of the anisyl group also change with oxidation as a result of contribution from the quinoidal structure. Specifically, the C(1)–C(2) and C(3)–C(4) bonds become longer (1.391 versus 1.413 Å and 1.397 Å versus 1.434 Å) whereas the C_α–C(1) and C(2)–C(3) bonds become shorter (1.492 versus 1.411 Å and 1.387 versus 1.362 Å). Interestingly, however, one pair of the vicinal coplanar anisyl groups is more involved in charge delocalization than the other pair, allowing maximum delocalization of the positive charge (structure **202**).

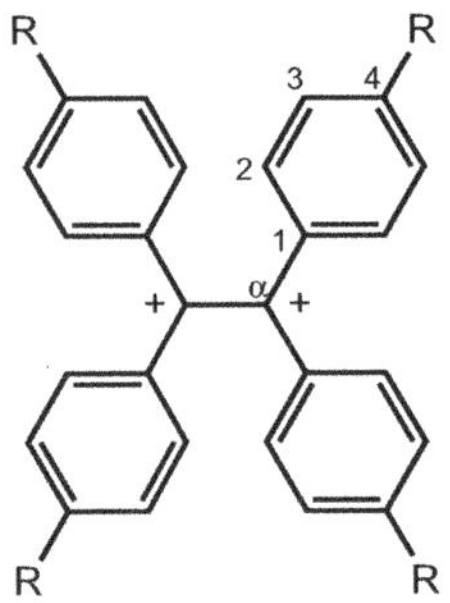

201

2 $SbCl_6^-$ or
$(SbCl_6^-)(Sb_2Cl_7^-)$

R = MeO, 2-Bu, 2-octyl

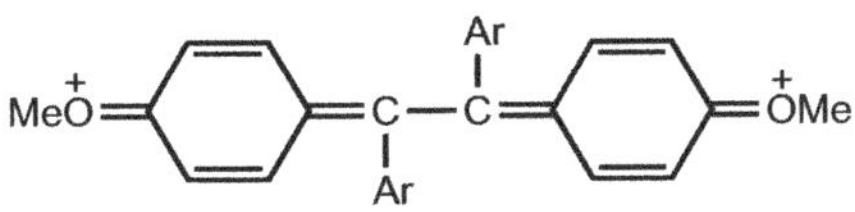

Ar = anisyl

202

In addition to the monocations **159** discussed in Section 3.4.11, Takekuma and coworkers[421–423] have isolated the bis(3-guaiazulenyl)-substituted dications **203** and reported ^1H and ^{13}C NMR spectral data. The molecular structure of the cation **203** (X = 1,2-phenylene, Y = PF_6)[422] shows that both 3-guaiazulenylmethyl substituents twisted to the same side from the benzene ring, and one of the 3-guaiazulenyl groups is planar, whereas the other is not. Furthermore, the benzene ring is slightly distorted as result of the large steric and electrostatic interactions of the substituents. NMR data of dications with heteroaromatic central unit indicate that additional resonance structures with the positive charge delocalized into the 3-guaiazulenyl and heteroaromatic rings also contribute to the structure.

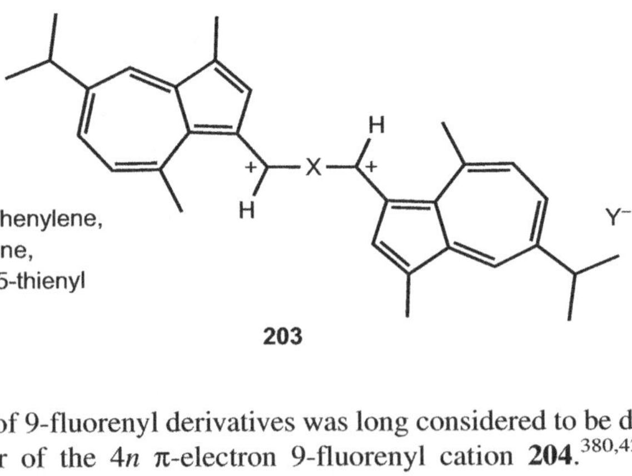

X = none, 1,2-phenylene,
1,4-phenylene,
2,5-furyl, 2,5-thienyl
Y = BF_4, PF_6

203

The low reactivity of 9-fluorenyl derivatives was long considered to be due to the antiaromatic character of the 4*n* π-electron 9-fluorenyl cation **204**.[380,424] Later, however, antiaromatic destabilization of the fluorenyl cation **204** was calculated to be small,[425] and its stability was shown to be close to that of the diphenylmethyl cation **136** (benzhydryl ion).[426] Fluorenyl cations are quite readily formed and studied in flash photolysis experiments but have not been observed under stable ion conditions. Mills et al.[427] have recently obtained a wide range of fluorenyl dications such as the tetrabenzo[5.5]fulvalene dication **205**. These were generated by oxidizing the corresponding alkenes in SbF_5–SO_2ClF solution and could be studied by low-temperature NMR spectroscopy.[428] The fluorenyl rings are orthogonal and σ–p conjugation occurs from the C–C bond of one ring to the empty p orbital of the other. Both experimental results (NMR, magnetic susceptibility exaltation, electrochemical oxidation potentials)[427–429] and calculations[430,431] indicate that such dicationic systems are indeed antiaromatic.[432] However, the antiaromaticity of dication **205** is only slightly enhanced relative to that of two fluorenyl cations.[433]

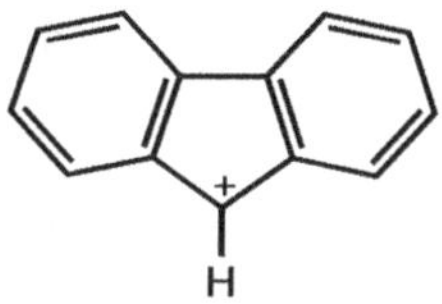

204

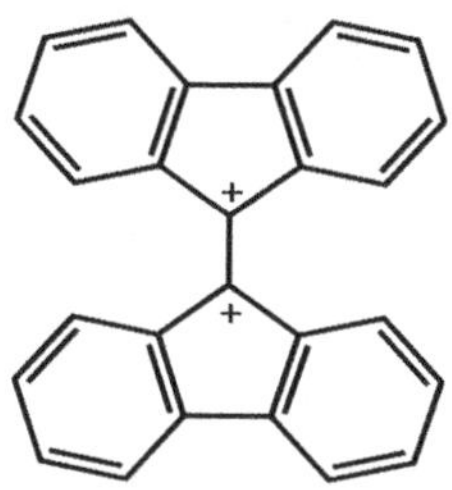

205

3.4.13. Aromatic Stabilized Cations and Dications

If a carbocation or a dication at the same time is also a Hückeloid $(4n+2)\pi$ aromatic system, resonance can result in substantial stabilization. The simplest 2π aromatic system is the Breslow's cyclopropenium ion **206**.[434–439] Recently, electronic and infrared spectra of the parent ion cyclo-$C_3H_3^+$ (**206**, R = H) in neon matrices[440] and the X-ray characterization of the tris(trimethylsilyl) derivative were reported.[441] The destabilizing effect of the silyl groups was found to be significantly smaller than in vinyl cations. The ion was computed to be more stable than the parent cyclopropenium ion by 31.4 kcal mol^{-1} [MP3(fc)/6-311G**//6-31G* + ZPVE level]. The alkynylcyclopropenylium ions **207** have been reported recently.[442]

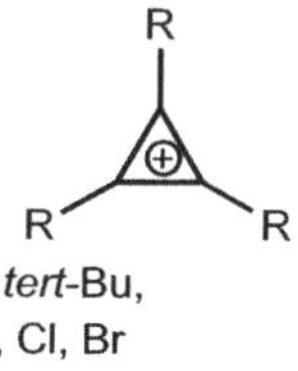

R = H, Me, Pr, *tert*-Bu, Ph, $SiMe_3$, Cl, Br

206

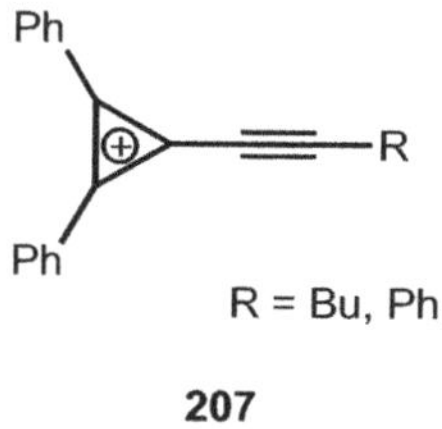

R = Bu, Ph, $SiMe_3$, Si(isoPr)$_3$

207

The benzo analogs of **206**, benzocyclopropenium ions **208**, are also known.[443–446] Although the 2π aromatic parent cyclobutadiene dication is still elusive, substituted analogs **209** have been prepared and characterized.[447–449] *Ab inito* calculations perdicted that, in contrast to expectations, 2π electron Hückel aromatic cyclobutadiene dications have puckered structure.[450,451] Evidence came from a comparison of experimental chemical shift data of tetramethylcyclobutadiene dication (**209′**) with those calculated by IGLO (double ξ basis set) for the puckered structure.[452] The calculated δ^{13}C shift values for ring carbons and methyl carbons (δ^{13}C 209 and δ^{13}C 18.7, respectively) are nearly identical with those determined experimentally for **209′** (δ^{13}C 209.7 and δ^{13}C 18.8, respectively).[447]

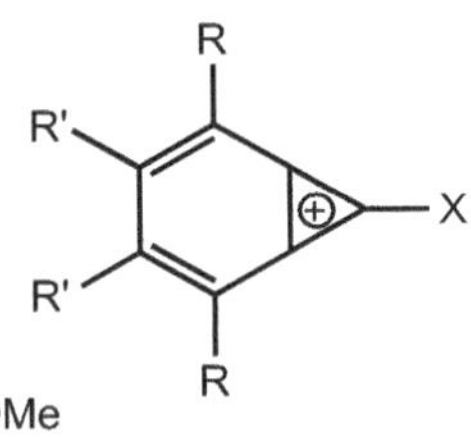

R, R' = H, Ph
X = Cl, F, COOMe

208

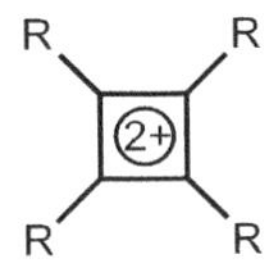

R = H, F, Me, Ph

209 **209'** R = Me

Dications with two cyclopropenium ion moieties are known. NMR characterization of cation **210** showed[453] significant upfield shifts of the protons of the phenyl rings and the carbons of the cyclopropenylium ring, whereas downfield shifts was noted of most of the naphthalene ring carbons when compared with cation **211**. These changes presumably reflect charge delocalization to the naphthalene π-system from the cyclopropenylium rings to reduce electrostatic repulsion between the cationic rings. This information gives strong evidence that the diphenylcyclopropenylium units in dication **210** have essentially face-to-face conformation. Furthermore, pK_{R+} data indicated that cation **210** is more destabilized than cations **211** and **212**,[454] wherein the two cationic units are spaced farther away.

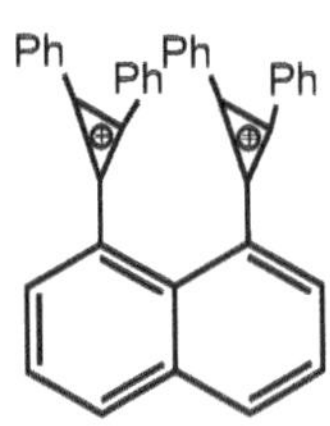

210

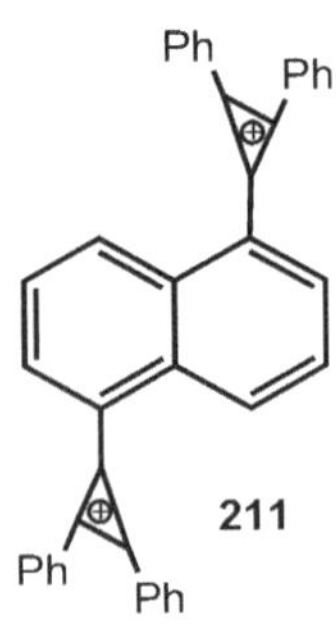

211

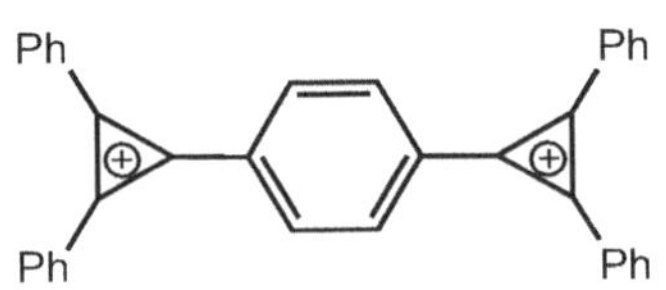

212

Among 6π aromatic systems, tropylium ion **213**[455,456] and numerous substituted derivatives have been prepared and characterized (**214**,[457] **215**,[458] **216**,[459] and **217**[459]). The aromatic stabilization energy of tropylium ion **213** was found to be

$-15.7\,\text{kcal mol}^{-1}$, only slightly lower than that of benzene ($-16.4\,\text{kcal mol}^{-1}$) and significantly higher, than that of the phenyl ring in benzyl cation **148a** ($-11.2\,\text{kcal mol}^{-1}$).[346]

Crystallographic characterization[460] of cation **215** revealed that the seven-membered ring adopts a shallow boat conformation whereas the phenyl substituents possess an average dihedral angle of 80° relative to the plane of the tropylium skeleton. Since the phenyl rings can adopt either clockwise or anticlockwise orientations, cation **215** is chiral.

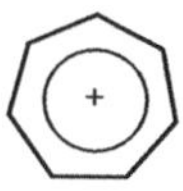

213

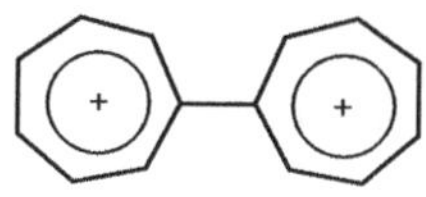

214

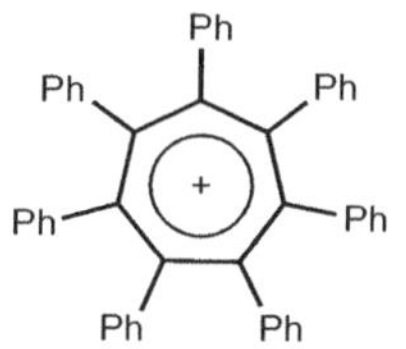

215

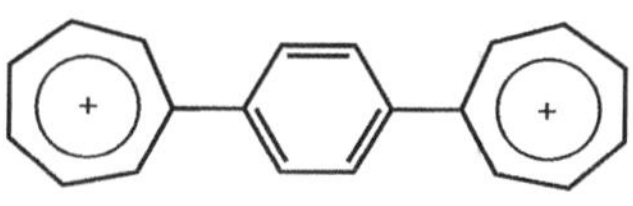

216

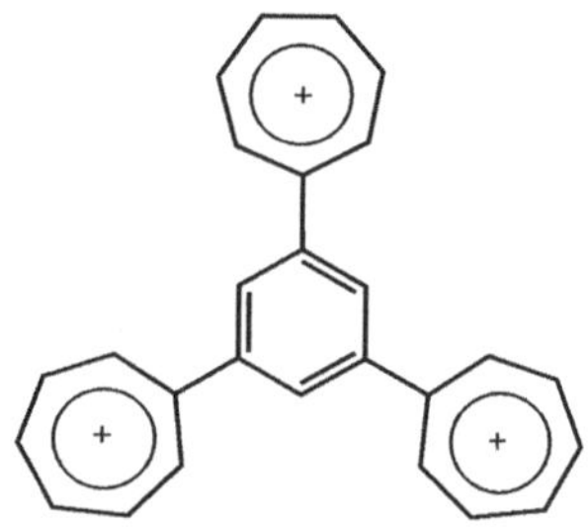

217

Both ^{1}H NMR and ^{13}C spectra of cation **218** show upfield shifts of the tropylium ring as compared with the mono-annelated cation, indicating decreased charge density on the cationic ring.[461] Similar cations with other fused rings have been reported recently.[267,462] In dication **219**, two **218** ion fragments are connected by a triple bond.[463] A characteristic feature of the ^{1}H NMR spectrum of the dication **219** compared to monocation **218** is that the signal of one of the bridgehead protons extending over the triple bond is deshielded by 0.37 ppm. The results of X-ray crystallography indicate that the tropylium rings are slightly bent into a boat conformation and the two rings are twisted by an angle of 44°.

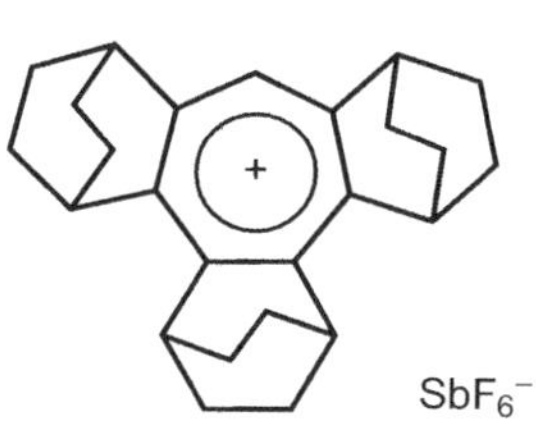

218

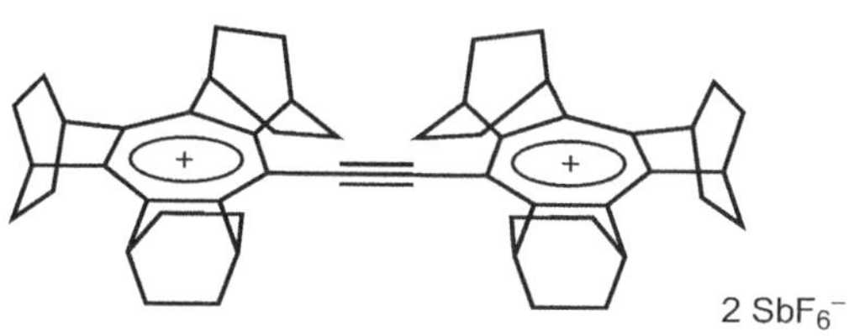

219

Alkylazuelenes are protonated at C(1) or C(3) position to form the stable azulenium cations **220**,[464] which can be viewed as vinyl-substituted tropylium ions [Eq. (3.52)].

The highly reactive bridged homoazulene could be easily protonated to yield the homotropylium ion **221** [Eq. (3.53)].[465]

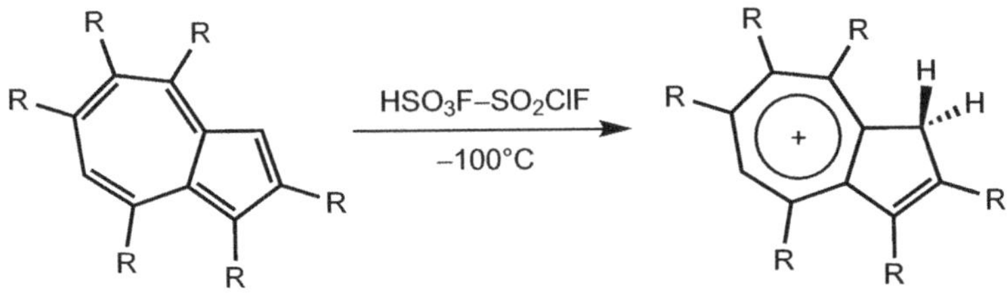

(3.52)

R = H, Me, isoPr, *tert*-Bu, Ph, CHO

220

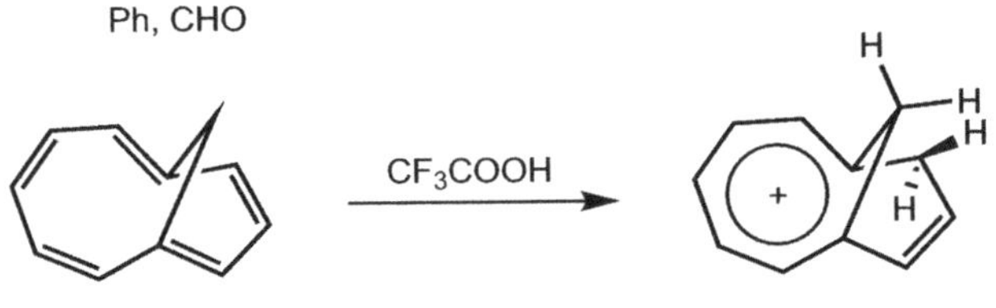

(3.53)

221

A variety of 1,1-disubstituted azulenes including 1,1-spiro compounds have been synthesized and then transformed by hydride abstraction to the corresponding highly stable tropylium cations **222**.[466] NMR characterization of the spiro-ethylene derivatives (**222**, $R-R=CH_2CH_2$)[467] and a comparison with the ethylene benzenium ion indicate less delocalization of the positive charge into the cyclopropane ring (higher field shifts of the cyclopropane carbon and methylene proton signals) and the least bond alternation in the seven-membered ring (equivalent coupling constants of the vicinal protons). These data show the significant difference of the bonding arrangement of the two systems and the total chemical shift value of 323 ppm for the unsubstituted spiro-ethylene derivative indicates a classical ion nature. The high thermodynamic stability of these ions is attributed to stabilization by the inductive and σ–π conjugation of the 1,1-substituents and the π–π conjugative effect of the double bond of the five-membered ring.

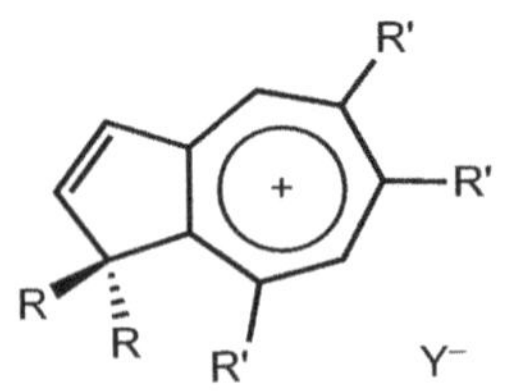

R = Me, Et, Pr
R–R = CH_2CH_2
R' = H, isoPr, *tert*-Bu
Y = BF_4, TsO

222

The synthesis and characterization of a series of azulenyl-substituted cations (**223**) showed[468,469] that these are extremely stable ions, which is attributed to the dipolar structure of the azulene rings.

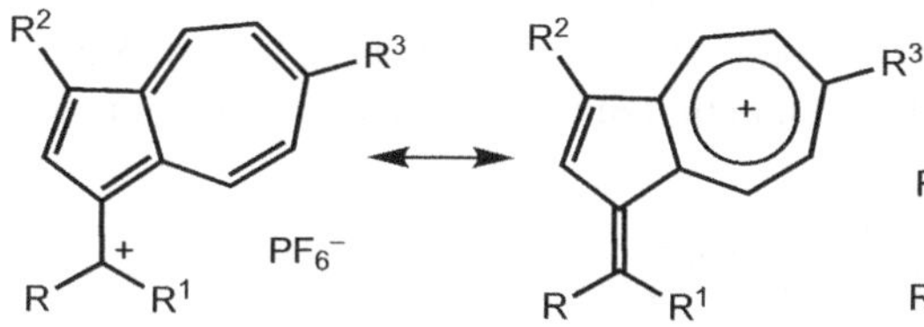

R, R[1] = substituted phenyl or azulenyl

R[2], R[3] = H, Me, *tert*-Bu, MeO, NMe_2, *N*-morpholinyl, COOMe

223

The β-triethylsilyl methyl tropylium ion **224** was prepared as shown in Eq. (3.54).[470] The ^{13}C and ^{29}Si NMR spectral features demonstrate that the β-Si substituent interacts strongly with the π-system of the tropylium ion by hyperconjugation. The ^{29}Si NMR spectrum has a single resonance at δ^{29}Si 17.3 significantly deshielded from the precursor (δ^{29}Si 6.6) indicative of some dispersal of the positive charge to the silicon. The X-ray structure showed that the C(8)–Si bond is orthogonal to the tropylium ring thereby maximizing hyperconjugation. As a result, the C(8)–Si bond is lengthened (1.929 Å) compared with the average Si–CH_2(Et) (1.867 Å) and the short C(1)–C(8) (1.479 Å) bonds. Since the C–C bond distances alternate significantly, contribution of the resonance form was suggested.

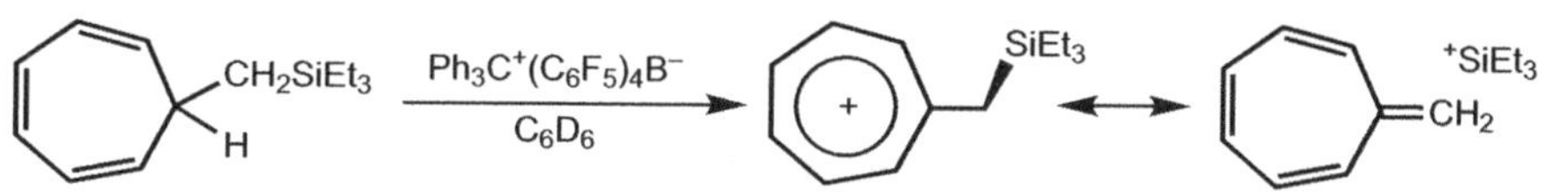

224

(3.54)

Cyclooctatetraene dications **225**[407–409] and benzocyclobutadiene dications **226**[471,472] are well studied. The parent cyclooctatetraene dication is still elusive, despite repeated attempts to prepare it under a variety of superacid conditions.

R = R' = H, Me
R = H, R' = Me
R = H, R' = Ph

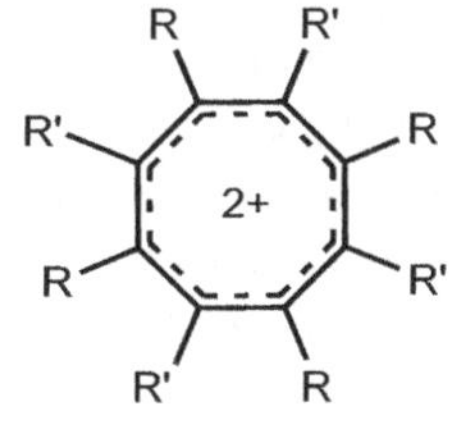

225

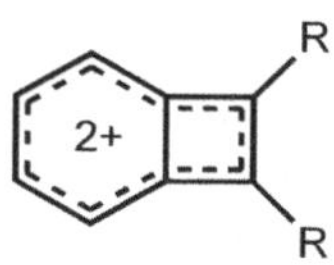

R = Me, OH

226

The 10π dibenzocyclobutadiene dication **227** was prepared[472] by the two-electron oxidation of biphenylenes in excess of SbF_5–SO_2ClF solutions at −10°C, and the octamethyl derivative **228** was also generated under similar conditions.[473]

R = H, Me

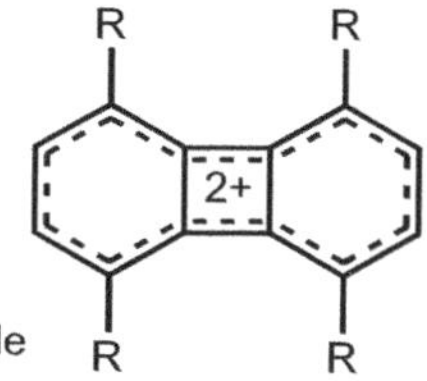

227

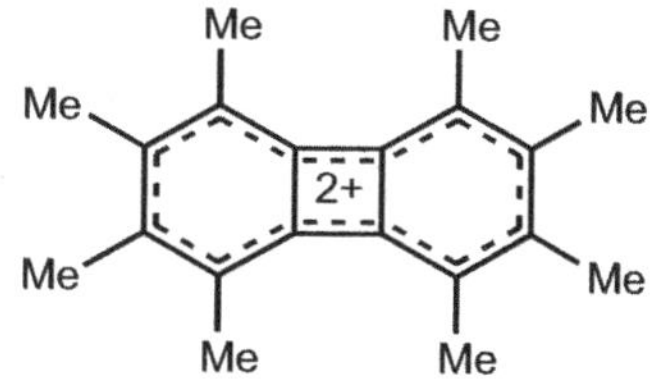

228

Since protonation of cyclooctatetraene is known to yield the homotropylium ion (see Section 3.5.3.1), Schröder and co-workers[474] reasoned that the homo[15]annulenyl cation **229** can be formed by the protonation of the [16]annulene (Scheme 3.7).

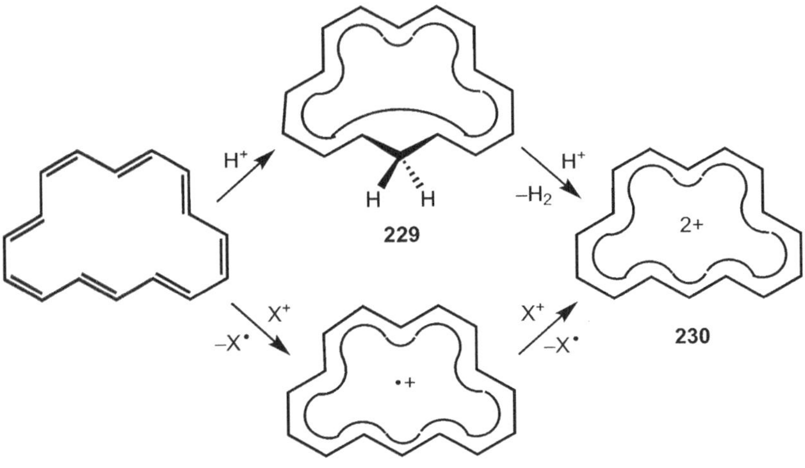

Scheme 3.7

Instead, the [16]annulenediyl dication **230** was obtained along with polymeric products in $HSO_3F–SO_2Cl–CD_2Cl_2$ media at −80°C. The ^1H and ^{13}C NMR data are consistent with the formation of the 14π dication **230**; however, attempts to trap the dication with $NaOAc–CH_3OH$ gave only polymeric products. Two possible mechanisms for the formation of dipositive ion **230** have been considered.[474] The first involves initial protonation to the homo[15]annulenyl cation **229** which further undergoes protolytic cleavage to the dication **230** (by loss of a molecule of hydrogen). The second route is associated with the stepwise oxidation of annulene by the proton or the conjugate acid of sulfur dioxide ($O{=}S{=}OH^+$), where the radical cation **231** is involved as an intermediate. Schröder and co-workers[474] seem to prefer the second mechanisms since no hydrogen gas has been detected in the reaction.

The aromatic nature of the presently discussed carbocations and dications have been further established by subjecting their NMR parameters to charge density chemical shift relationship[449,475–477] originally developed by Spiesecke and Schneider.[478,479] Furthermore, NICS (nucleus-independent chemical shift) developed by Schleyer et al.[480] offers a simple and efficient probe for aromaticity.

3.4.14. Polycyclic Arene Dications

The ease of oxidation of polycyclic aromatic hydrocarbons in the gas phase[481] as well as in solution is well-documented.[482–495] In strong acid solutions, monopositive radical ions and/or dipositive ions, also known as oxidation dications, have been reported.[271,494,495] Similar species have been observed in anodic oxidations of aromatic compounds in low nucleophilicity solvents.[482] Simple Hückel molecular orbital theory predicts that arenes whose highest occupied molecular orbitals (HOMOs) are at higher energy levels (smaller E_{HOMO} values) should be prone to two-electron oxidation to dipositive ions. On the other hand, the arenes with low-lying HOMOs should be more difficult to ionize.[495] Certain polycyclic arenes have also been protonated to the corresponding dipositive ions.

Although benzene does not undergo two-electron oxidation reactions upon treatment with SbF_5–SO_2ClF, but gives instead the benzenium ion **88** by protonation[253] (due to HF impurity in the system), naphthalene has been reported to give the corresponding radical monocation upon treatment with SbF_5–SO_2ClF at −78°C.[494,495] However, the presence of methyl substituents on the ring lowers the ionization potential to a point that stable carbodications can be formed. Thus, the tetramethyl- and octamethylnaphthalene dications **232** and **233** have been prepared[495] from the corresponding arenes.

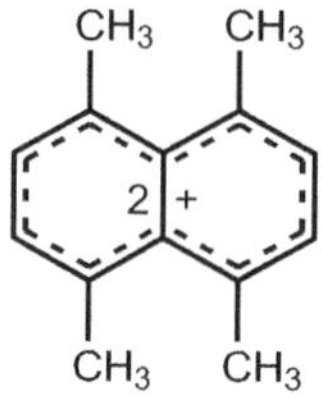

232

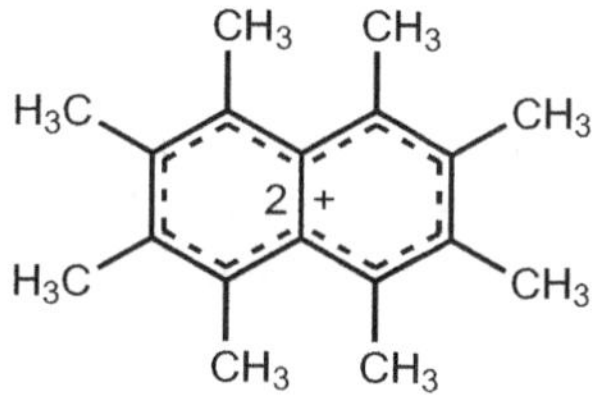

233

Anthracene and substituted anthracenes are readily oxidized in SbF_5–SO_2ClF to the corresponding carbodications **234**.[487,489,494,496] They have been a subject of a ^{13}C NMR spectroscopic study.[496] Also, a variety of carbodications of higher homologous polycyclic arenes have been generated in SbF_5–SO_2ClF and studied by ^{13}C NMR spectroscopy by Forsyth and Olah.[494]

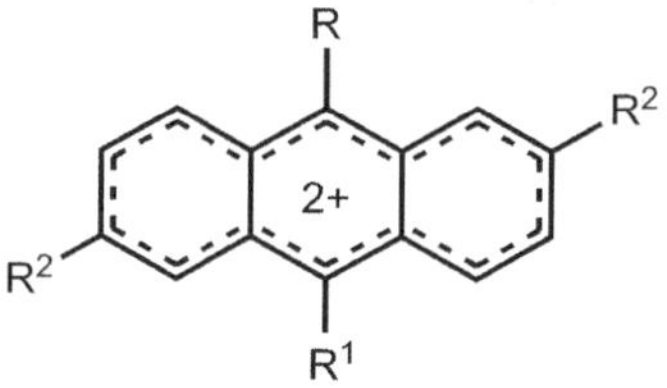

234

R, R^1, R^2 = H
R, R^2 = H, R^1 = CH_3
R, R^1 = CH_3, R^2 = H
R, R^1 = H, R^2 = CH_3
R, R^2 = H, R^1 = Br
R, R^2 = H, R^1 = Cl

The preparation of the 8C-6π aromatic dication **236** of the unknown pentalene **235** has proven to be unusually difficult. However, the dibenzoannulated derivatives have been prepared by Rabinowitz and co-workers.[497,498]

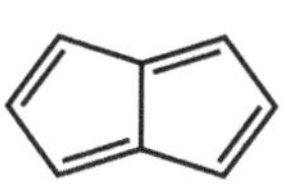

235

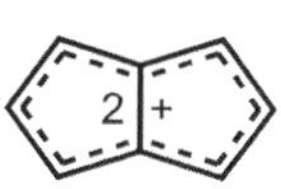

236

Upon treatment of dibenzo[*b*,*f*]pentalene or the 1,9-dimethyldibenzo[*b*,*f*]pentalene with SbF_5–SO_2ClF at −78°C, the two-electron oxidation product dibenzo[*b*,*f*] pentalene dications **237** were obtained[497,498] [Eq. (3.55)]. The observed ^1H and ^{13}C NMR spectral deshieldings of **237** as compared with those of their progenitors clearly establish their dicationic nature.

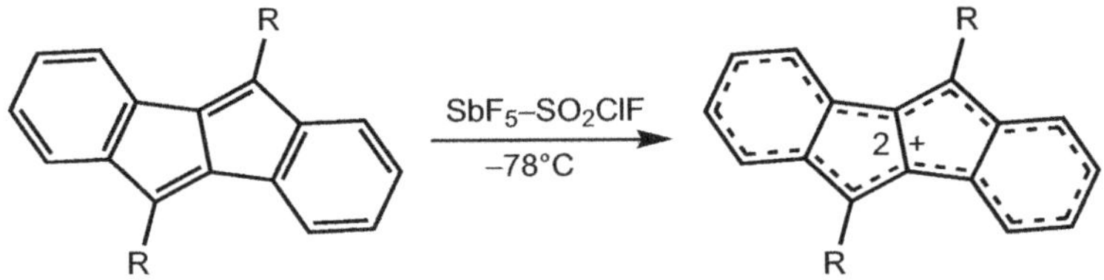

R = H, Me **237** (3.55)

Laali and coworkers have made experimental and computational studies on oxidation dications of a range of polycyclic aromatic compounds[271] including various benzo[*a*]pyrenes,[279,281,282] isomeric azulenophenalenes,[279] and benzo[*a*]anthracenes.[499] As mentioned, substituted benzo[*e*]dihydropyrenes gave diprotonated and triprotonated cations (see Section 3.4.7). Dication **104a** was shown, however to slowly transform to oxidation dication **238a**.[281] Ethanophenantrenium–carboxonium trication **238b**, formed by two-electron oxidation and ester protonation, in turn, could only be generated in the more oxidizing superacid HSO_3F–SbF_5 (1:1).[282]

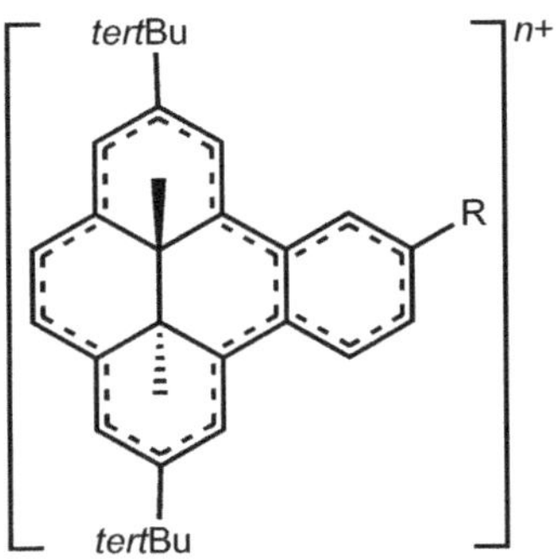

238 **a** R = H, *n* = 2
b R = CO_2EtH, *n* = 3

3.4.15. Fullerene Cations

In contrast to highly stable and prolific fullerene anionic species, fullerene cations are rare. The first fullerene cation was prepared in 1996 by Reed and co-workers[500] by single-electron oxidation of C_{76} to form radical cation $C_{76}^{\bullet +}$ isolated in solid form as the $CB_{11}H_6Br_6^-$ salt [Eq. (3.56)]. The cation was identified in solution by a characteristic visible–near-infrared absorption ($\lambda_{max} = 780$ nm), FT–IR and EPR spectroscopy. $C_{60}^{\bullet +}$ was generated in an analogous way later.[501] Reed et al.[501] also succeded in generating HC_{60}^+ by protonation of C_{60} with $H(CB_{11}H_6Cl_6)$ in *ortho*-dichlorobenzene [Eq. (3.57)] and isolating the salt in solid form. The ^{13}C NMR spectrum of HC_{60}^+ (**239**) shows two singlets assigned to the unique sp^3-hybridized carbon ($\delta^{13}C$ 56) and the cationic carbon ($\delta^{13}C^+$ 182), respectively. Additional features indicate a structure of C_s symmetry. Mueller et al. have applied the scalar coupling-driven uniform-sign cross-peak NMR method (UC2Qf COSY) to experimentally characterize HC_{60}^+.[502] Two cross-peaks at $\delta^{13}C$ 55.4 and $\delta^{13}C^+$ 182.0 are indicative of the direct bond between the protonated sp^3-hybridized site and the sp^2 cationic site. Additional cross-peaks ($\delta^{13}C$ 55.4–148.2 and 140.0–182.0) could be assigned to neighboring ^{13}C resonances.

$$C_{76} + Ar_3N^{\bullet +}\ CB_{11}H_6Cl_6^- \xrightarrow[\text{1,1,2,2-tetrachloroethane}]{\text{ortho-dichlorobenzene or}} C_{70}^{\bullet +}\ CB_{11}H_6Br_6^- + Ar_3N \qquad (3.56)$$

Ar = 2,4-dibromophenyl

$$C_{60} + HCB_{11}H_6Cl_6) \xrightarrow[\text{ortho-dichlorobenzene}]{} HC_{60}^+\ CB_{11}H_6Cl_6^- \qquad (3.57)$$

239

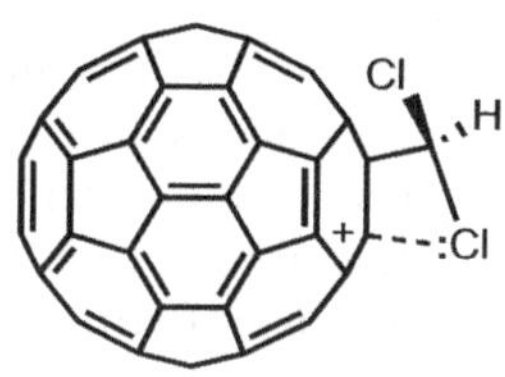

240

Subsequently, Kitagawa and Takeuchi[503,504] generated and characterized substituted fullerene cations by treating substituted fullerenols with triflic acid[503,505] [Eq. (3.58)]. The resulting cations, which are stable in triflic acid for weeks, give signals for the cationic centers at $\delta^{13}C^+$ 185.6 (R = Me), 180.3 (R = CH_2Cl), 175.6 (R = $CHCl_2$), 171.8 (R = CCl_3), and 174.9 (R = CCl_2CH_2Cl). These resonances are at higher fields when compared to those of aryl-substituted carbocations such as Ph_3C^+ ($\delta^{13}C^+$ 221.6) and 9-phenylfluorenyl ($\delta^{13}C^+$ 224.2), which is an indication of charge delocalization into the C_{60} cage. An additional factor is an interaction between the lone pair of chlorine and the cationic center (**240**), which is manifested by the high-field shift of the carbon resonance with increasing number of chlorine atoms attached to the α carbon.

$$RC_{60}OH \xrightarrow[-H_2O]{CF_3SO_3H} RC_{60}^+ \qquad (3.58)$$

R = Me, CH_2Cl, $CHCl_2$, CCl_2CH_2Cl

Birkett and co-workers[506] have applied a different approach using chloride abstraction with $AlCl_3$ from $Ar_5C_{60}Cl$ to generate pentaarylated fullerene cations

[Eq. (3.59)]. ^{1}H and ^{13}C NMR data are consistent with cation **242** with the cationic center observed at $\delta^{13}C^+$ 173.74 (Ar = Ph) and $\delta^{13}C^+$ 171.66 (Ar = 4-FC_6H_4). This is accounted for by the transformation of the initially formed antiaromatic cyclopentadienyl cation **241** through 1,2-aryl migration.

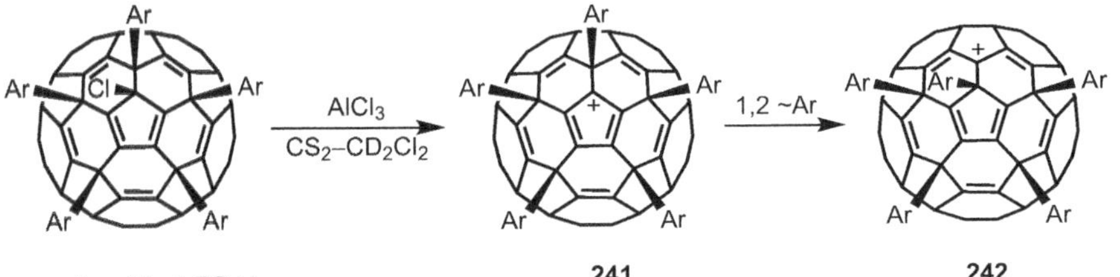

(3.59)

Another method developed by Kitagawa and co-workers[507] uses the more easily accessible fullerenes RC_{60}–H and dimer RC_{60}–C_{60}R [R = $CH_2P(O)(OEt)_2$] as precursors. Treatment of RC_{60}–H in H_2SO_4–CF_3SO_3H (1:4) results in the formation of a solution with resonances at δ^{13}C 53.66 and $\delta^{13}C^+$ 174.67 for the substituted sp^3 and the sp^2 cationic sites, respectively. Since the rate of cation formation depends not on acidity but on the oxidizing ability of the acids used (H_2SO_4, FSO_3H, CF_3SO_3H), cation formation was postulated to occur via two one-electron oxidation steps. The signal of sp^2 cationic site appeared as a doublet indicating the coordination of the substituent to the cationic center. Furthermore, some positive charge on phosphorus can be deduced from the ^{31}P NMR resonance at δ^{31}P 31.08, which is downfield by 7 ppm from that of the parent compound. DFT calculations found dicationic structure [$(EtO)_2(OH)P^+$–C_{60}^+] **243** to be the most stable. Further support comes from GIAO calculations, which reproduced the experimental chemical shift value quite accurately ($\delta^{13}C^+$ 174.60).

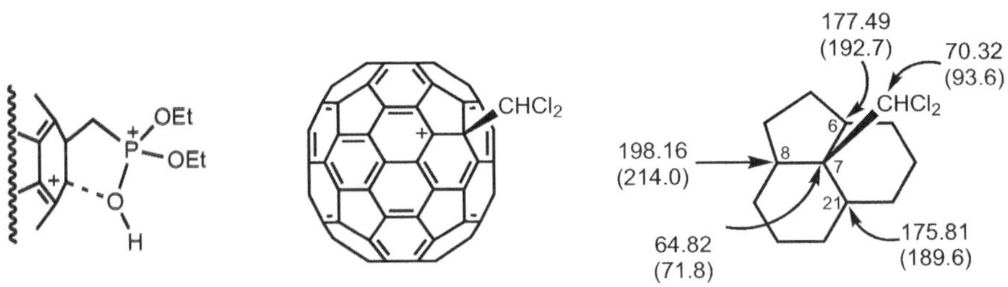

243 **244** **245**

Kitagawa et al.[508] used the corresponding fullerenol to generate cation $Cl_2CHC_{70}^+$ (**244**) in CF_3SO_3H. The characteristic ^{13}C NMR experimental resonances (in ppm) together with the GIAO calculated values (in parentheses) are shown in formula **245**. Sixty-two peaks in the ^{13}C NMR spectrum assigned to sp^2-hybridized carbon atoms indicate C_1 symmetry. This symmetry and energies for five isomeric cations calculated by the DFT method support the structure to be cation **244**.

3.4.16. Heteroatom-Stabilized Cations

In contrast to hydrocarbon cations, heteroatom-substituted carbocations are strongly stabilized by electron donation from the unshared electron pairs of the heteroatoms adjacent to the carbocation center (**246**).[17,509]

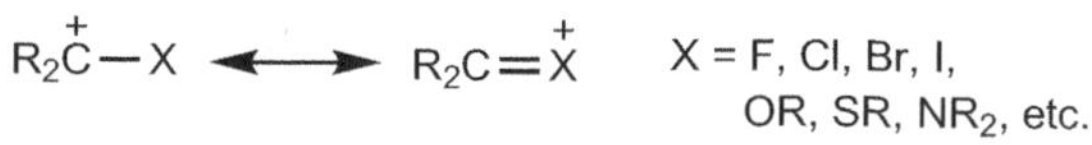

246

The stabilizing effect is enhanced when two, or even three, electron-donating heteroatoms coordinate with the electron-deficient carbon atom (Scheme 3.8).

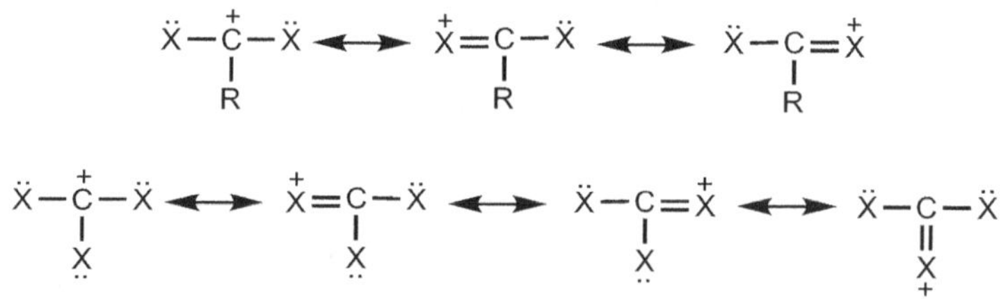

Scheme 3.8

Carbocations with α-heteroatom substituents such as trimethylsilyl and nitro groups that lack a stabilizing lone pair of electrons have also been prepared and studied.[375,510]

3.4.16.1. Halogen as Heteroatom. In 1966 Olah, Cupas, and Comisarow[511] reported the first α-fluoromethyl cation. Since then, a large variety of fluorine-substituted carbocations have been prepared. α-Fluorine has a particular ability to stabilize carbocations via back-donation of its unshared electron pairs into the vacant *p* orbital of the carbocationic carbon atom. ^{19}F NMR spectroscopy is a particularly efficient tool for the structural investigations of these ions.[512,513] The 2-fluoro-2-propyl cation **247** (NMR spectra, Figure 3.16) and 1-phenylfluoroethyl cation **248** are representative examples of the many reported similar ions.[514]

Christe et al.[515] have reported the first crystal structures of fluoro-substituted carbocations **247** and **249**. The C_2CF skeleton of cation **247** (AsF_6^- salt) is planar and the C^+–F bond is significantly shorter than the length of C_{sp^2}–F bonds (1.285 versus 1.333 Å). The average C–C bond distance is also considerably shorter than the average length of C_{sp^3}–C_{sp^2} bonds (1.432 versus 1.510 Å). These structural characteristics indicate substantial electron back-donation from fluorine to the carbenium center and a significant methyl hyperconjugation. Bond distances in cation **249**, isolated in the form of AsF_6^- and $As_2F_{11}^-$ salts, show varied changes. Whereas the C^+–C_{ipso} bond is short, similar to those found in the trityl cation and the related cation **250**,[516] the C^+–F bond shortens only slightly (1.31 versus 1.333 Å). The observed ^{19}F chemical shift (δ^{19}F 18)

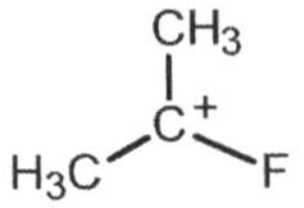

247

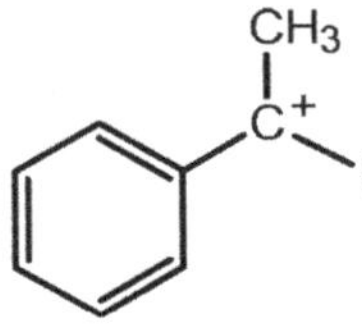

248

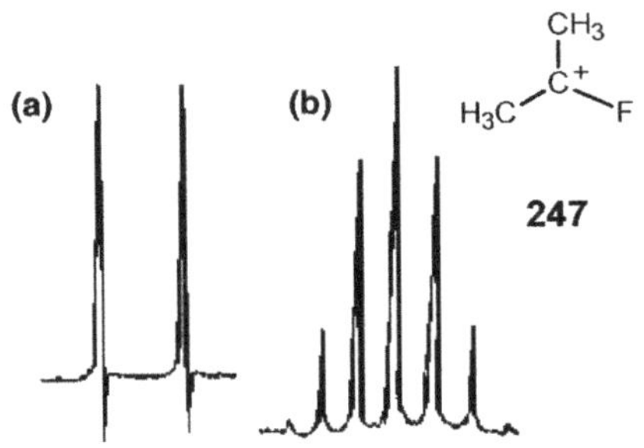

Figure 3.16. **(a)** ^{1}H NMR spectrum of the 2-fluoro-2-propyl cation **247** at 60 MHz, $J_{H-F} = 25.4$ Hz **(b)** ^{19}F NMR spectrum of the same ion at 56.5 MHz, $J_{H-F} = 25.4$ Hz.

is in good agreement with that reported for Ph_2CF^+ earlier (δ^{19}F 11.5).[511] A comparison of C^+–F bond length of cation **249** with C–X bond distances of cation **250**[516] and $CH_3CX{=}CH_2$ haloalkenes demonstrates that chlorine is a better electron back-donor than fluorine. In both cation **247** and cation **249**, the carbocation carbon centers form along their $2p_z$ axes two fluorine bridges with two different anions.

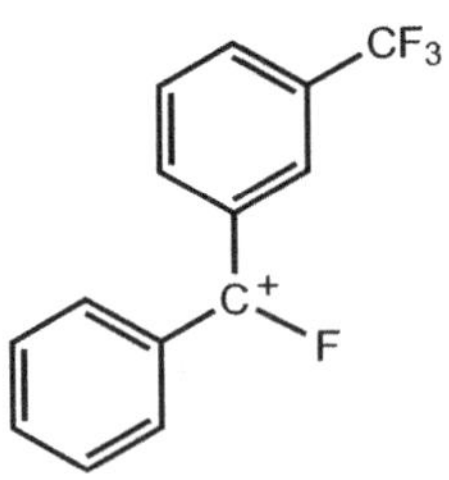

249

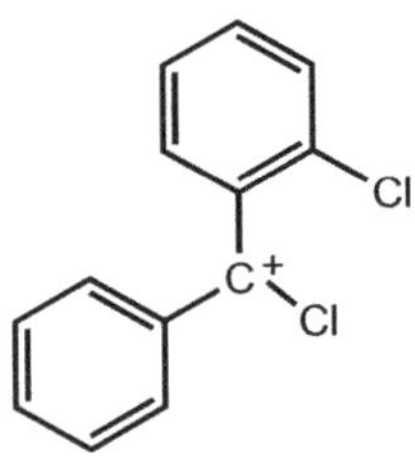

250

The X-ray characterization of the SbF_6^- salt of cation **250** by Laube *et al.*[516] corroborates the above findings. The short C^+–Cl distance (1.668 Å versus 1.734 Å for C_{sp^2}–Cl) indicates significant chlorine back-donation resulting in partial double bond character in agreement with experimental findings and theoretical calculations.

Trifluoromethyl-substituted[517] and perfluorophenyl-substituted carbocations[518,519] have also been prepared and studied. Because of the relatively large fluorine chemical shifts, anisotropy and ring current effects play a relatively much smaller role than they do in the case of proton shifts. Therefore, a better correlation of charge distribution with chemical shifts can be obtained. The trifluorocyclopropenium ion **251** has also been reported.[520] The ^{19}F NMR spectrum exhibits a single resonance at $\delta^{19}F$ −63.1, which is deshielded by 57.8 ppm relative to the neutral perfluorocyclopropene ($\delta^{19}F$ −120.9). A review has summarized the chemistry of fluorinated allyl cations including NMR characterization[521] and the chemistry of other perfluorinated carbocations has also been reviewed.[522]

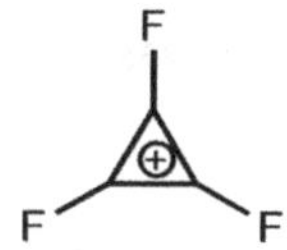

251

Prakash, Rasul, Olah, and co-workers[523] have performed *ab initio* IGLO/GIAO-MP2 studies of a series of fluorocarbocations. For the CH_3CHF^+ cation, not observed yet under stable ion conditions, the classical open carbenium ion has been found to be the global minimum on the potential energy surface (MP2/6-31G* level). The C–C bond of the methyldifluorocarbenium ion **252**, observed by Olah and Mo,[524] is longer than that in the CF_3CHF^+ cation (1.452 versus 1.433 Å). This is due to the back-donation of two fluorine atoms and the stabilization of the ion by resonance. The C_2 symmetry structure is the most stable conformer for the 2-fluoro-2-propyl cation **247**. The long H_2C–CH_3 bond of 1.578 Å of cation **253** was shown to be aligned parallel with the empty *p*-orbital of C^+ which allows maximum C–C hyperconjugation. The global minimum of cation **254** is a twisted structure of C_2 symmetry. The mono-fluorocyclopropenium ion **255** shows a longer C–C bond (1.385 Å) opposite to the fluorine-substituted carbon and a shorter C–C bond (1.367 Å) adjacent to the fluorine-substituted carbon. The short C–F bond (1.274 Å), which is shorter than those of ions **247**, **253**, and **254**, is indicative of a substantial interaction between fluorine and the cyclopropyl ring. The structural features of the difluorocyclopropenium ion and trifluorocyclopropenium ion **251** are very similar to those of cation **255**. The GIAO-MP2-calculated ^{19}F NMR chemical shifts of the fluorocarbenium ions studied show an overall good correlation with the experimental values.

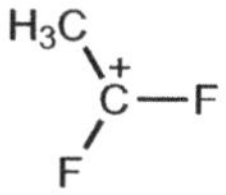

252

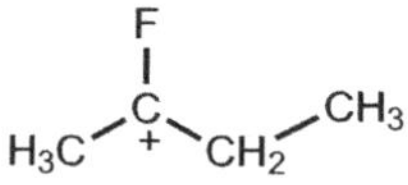

253

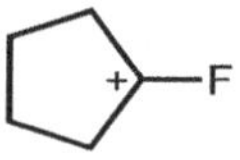

254

255

A series of chloromethyl cations were observed, including phenylchloromethyl cations[382–384,525,526] and perchlorotriphenylmethyl ion **256**.[527] West and Kwitowski[528] have characterized the perchloroallyl cation **257**. A series of chloro- as well as bromo- and iodomethyl cations have been observed **(258–260)** and the general stabilizing effect of halogen attached to carbocation center has been demonstrated.[529]

Olah, Halpern, and co-workers[530,531] were able to study these effects in detail using ^{13}C NMR spectroscopy.

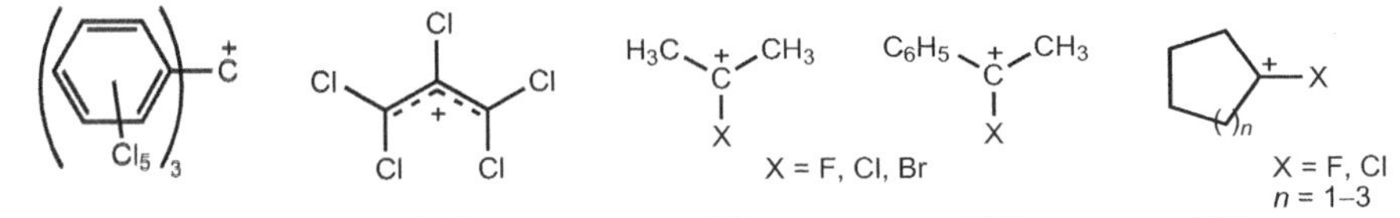

256 257 258 259 260

Olah et al.[532,533] studied trihalomethyl cations (CX_3^+, X = Cl, Br, I) under stable ion conditions. ^{13}C NMR chemical shift values correlate well with the decreasing order of back-donation (Cl > Br > I). Similar correlation was also found for dimethylhalocarbenium ions **258**. The CF_3^+ fluoro analog, however, could not be observed under any conditions. This can be attributed to a combination of unfavorable thermodynamics (generation of CF_3^+ from CF_4 is endothermic by about 20 kcal mol^{-1}) and the lack of stable alternate starting materials and a suitably strong Lewis acid.[534]

Further studies included *ab initio* calculations of the protonation of trihalomethyl cations [MP2 and MP4(SDTQ) levels with 6-31G* basis set].[535] The protonated trifluoromethyl dication CF_3H^{2+} lies 79.1 kcal mol^{-1} higher on the potential energy surface than the trifluoromethyl cation CF_3^+ and has a kinetic barrier of 17.4 kcal mol^{-1} for deprotonation. The charge repulsion between C^+ and the protonated F atom results in the lengthening of the corresponding C—F bond by 0.177 Å and the concomitant shortening of the other two C—F bonds by 0.032 and 0.036 Å. The protonated trichloromethyl dication CCl_3H^{2+} is only 4.0 kcal mol^{-1} less stable than the trichloromethyl cation CCl_3^+, but it has a significantly higher kinetic barrier for deprotonation (65.5 kcal mol^{-1}). The changes in the C–Cl bond lengths are similar to those for the CF_3H^{2+} dication. The structure and energetics of the protonated tribromomethyl dication CBr_3H^{2+} and protonated triiodomethyl dication CI_3H^{2+} are similar to the other two protonated trihalomethyl dications.

Since fluorine is the most electronegative element, it should inductively destabilize carbocations. The stability of fluoromethyl cations in the gas phase decreases in the order $CFH_2^+ > CF_2H^+ > CF_3^+ > CH_3^+$. The trend in solution, however, could be different, due to solvent effects, ion pairing, and so on. Indeed, fluorine has been shown to provide stabilization for carbocations. The existence of $CH_3CF_2^+$, in contrast to the elusive ethyl cation $CH_3CH_2^+$, is a clear evidence that replacement of H atoms by F atoms provides stabilization for carbocations.[524] Furthermore, it was found that in perfluorobenzyl cation $C_6F_5CF_2^+$ fluorine atoms in resonance positions (*ortho* and *para*) are more deshielded than those in *meta* positions.[536] This indicates carbocation stabilization by back-donation.

It has been shown in a recent theoretical study[537] (*ab initio* calculations at the MP2/VTZ+D+P level) that π-donor ability in all halogens increases with F < Cl < Br < I, which is the opposite reported by Olah et al.[532,533] on the basis of NMR chemical shifts. Recently,[515] this discrepancy has been clarified as being caused by the assumption that the σ effect from F to the heavier halogens are identical for neutral halocarbons (CH_3CHXCH_3, $CH_3CX{=}CH_2$) and carbenium ions. In fact, the carbenium carbon is highly electron-deficient and more electronegative than chlorine.[534,537] Consequently, in carbenium ions, chlorine becomes both a π donor and a σ donor. In contrast, the more electronegative fluorine is only a π donor and strongly withdraws electrons through the σ effect.

Schrobilgen and co-workers[538,539] have recently prepared the trichloromethyl and tribromomethyl cations by use of the noble-gas oxidant $XeOTeF_5^+Sb(OTeF_5)_6^-$ [Eq. (3.60)]. The ^{13}C NMR chemical shifts of cations **261-Cl** ($\delta^{13}C$ 237.1) and **261-Br** ($\delta^{13}C$ 209.7) are significantly deshielded relative to CCl_4 ($\delta^{13}C$ 96.4) and CBr_4 ($\delta^{13}C$ −29.7), which is consistent with the earlier studies of Olah.[532,533] The cations of D_{3h} symmetry were found to be planar in the solid state (sum of X−C−X bond angles = 360°) with no significant interactions between C^+ and the fluorine atoms of the anion. Cation **261-Cl** is disordered in the crystal and gives a slightly wider range of bond lengths and angles than that of cation **261-Br**. The C−X bond lengths are shorter than in CCl_4 (1.592–1.672 Å versus 1.751 Å) and CBr_4 (1.783–1.851 Å versus 1.91 Å). The geometries, natural charges, and bond orders of all four CX_3^+ (X = F, Cl, Br, I) cations have been calculated using HF and MP2 methods.[523,533,538]

$$CX_4 + XeOTeF_5^+\,Sb(OTeF_5)_6^- \xrightarrow[-78°C]{SO_2ClF} CX_3^+\,Sb(OTeF_5)_6^- \qquad (3.60)$$

X = Cl, Br → X = Cl **261-Cl**; X = Br **261-Br**

Recently, the triiodomethyl cation salt $CI_3^+Al[OC(CF_3)_3]_4^-$ has been obtained by the reaction of CI_4 with $AgAl[OC(CF_3)_3]_4$.[540] The ^{13}C NMR spectrum of the cation in CH_2Cl_2 solution shows a single resonance at $\delta^{13}C$ 97. The C−I bond length in the solid state is significantly shorter than that of CI_4 (2.013 versus 2.159 Å). The salt contains isolated ions with trigonal planar geometry (sum of I−C−I angles = 360.0°). The I···F contacts are all longer then the sum of the van der Waals radii, which indicates that the positive charge is delocalized onto the iodine atoms.

The mixed-ligand cations $CBr_n(OTeF_5)_{3-n}^+$ ($n = 1, 2$), which contain the highly electronegative fluoro analog $OTeF_5^-$ (pentafluorooxotellurate, teflate) anion, were also characterized by ^{13}C and ^{19}F NMR spectroscopy.[538,539] Evidence for two fluorine-containing trihalomethyl cations has also been obtained.[539] Oxidation by $XeOTeF_5^+Sb(OTeF_5)_6^-$ of $CFCl_3$ resulted in the formation of $CFCl_2^+$ characterized by multinuclear NMR spectroscopy. The ^{13}C and ^{19}F are significantly deshielded ($\delta^{13}C$ 214.3, $\delta^{19}F$ 168.6) as compared to those in the parent compound $CFCl_3$ ($\delta^{13}C$ 117.1, $\delta^{19}F$ −1.1). Particularly revealing is the large increase in $^1J_{(^{19}F-^{13}C)}$ coupling (429 versus 335 Hz) indicative of the increased s character of the sp^2-hybridized carbon center. The $CFCl_2^+Sb(OTeF_5)_6^-$ salt could be isolated and characterized by low-temperature Raman spectroscopy. Oxidation of CF_2Br_2 presumably leads to CF_2Br^+ as the initial product, which, however, undergoes rapid halogen exchange to yield $CFBr_2^+$. Characteristic NMR data obtained at −80°C are similar to those of ion $CFCl_2^+$.

3.4.16.2. Oxygen as Heteroatom. Ions **262** discussed in this section are called unsaturated oxonium ions or carboxonium ions.[541] This latter name reflects the hybrid nature of oxonium ion (**262a**)–carbenium ion (**262b**) resonance. Carboxonium ions are classified according to substituent R. When R is hydrogen, the ions are hydroxylated ions, or secondary or acidic carboxonium ions. Ions with R = alkyl are called tertiary or alkylcarboxonium (alkoxycarbenium) ions, or nonacidic carboxonium ions. A third group of carbenium ions with oxygen participation are the acylium ions (acyl cations).

R = H, alkyl

a **b**

262

Acidic Carboxonium Ions (Hydroxylated Cations). Acidic carboxonium ions are generally obtained by protonation of carbonyl compounds.[541,542] Aldehydes and ketones protonate on the carbonyl oxygen atom in superacid media at low temperatures, and the corresponding carboxonium ions can be directly observed[543–551] [Eqs. (3.61) and (3.62)].

$$\mathrm{RCHO} \xrightarrow{\mathrm{HSO_3F{-}SbF_5{-}SO_2}} \mathrm{RCH{=}\overset{+}{O}H} \qquad (3.61)$$

$$\mathrm{R_2CO} \xrightarrow{\mathrm{HSO_3F{-}SbF_5{-}SO_2}} \mathrm{R_2C{=}\overset{+}{O}H} \qquad (3.62)$$

Even protonated formaldehyde has been observed.[548] Protonated acetaldehyde shows two isomeric forms, with the proton on oxygen being *syn* or *anti* to the methine proton (**263**).[548]

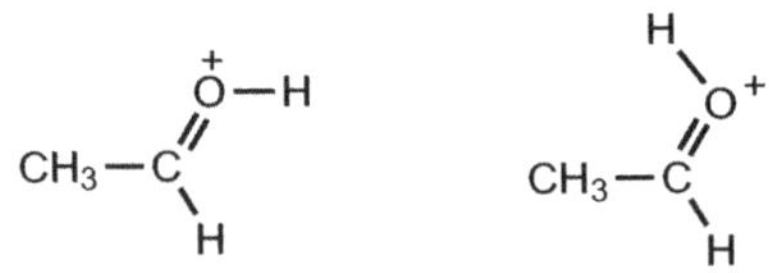

80% *syn* 20% *anti*

263

The hydroxymethyl cation forms of protonated ketones (**264**) and aldehydes (**265**) contribute to the resonance hybrid. Based on ^{13}C NMR studies,[94,548–551] the degree of contribution of the hydroxymethyl cation forms can be quite accurately estimated. Similar studies have been carried out using ^{17}O NMR spectroscopy.[552] Recent theoretical studies (MP2/6-31G* level)[553] for protonated acetone have supported the conclusion that the hydroxymethyl cation resonance form is only a minor contributor to the overall structure. Similar conclusion was arrived at by low-temperature IR spectroscopy in superacid solutions ($HF–SbF_5$, $HF–BF_3$)[554] and IR photodissociation spectroscopy of protonated acetone in the gas phase.[555,556] Olah and co-workers[557] also measured one-bond $^{13}C–^{13}C$ coupling constants in a series of protonated benzaldehydes and acetophenones.

264

265

Diprotonation is of simple carbonyl compounds has been studied theoretically (MP2/6-31G*//HF/6-31G* level).[558,559] Diprotonation is increasingly favored, going from formaldehyde to acetaldehyde to acetone, which is attributed to methyl hyperconjugation. The C–O bond length of the diprotonated species becomes increasingly longer (1.305, 1.335, and 1.359 Å), indicating increasing single-bond character.

The X-ray structure is of carboxonium ions **266**[560] and **267**[561] have been determined. The main features, a substantial lengthening of the C–O bonds and a shortening of the $C_{C=O}–C$ bonds, were observed upon protonation of the starting ketones.

R = Ph, cyclopropyl

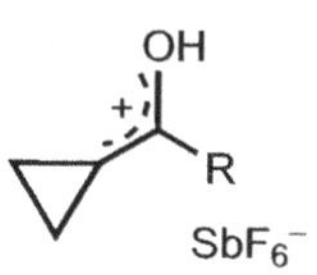

266

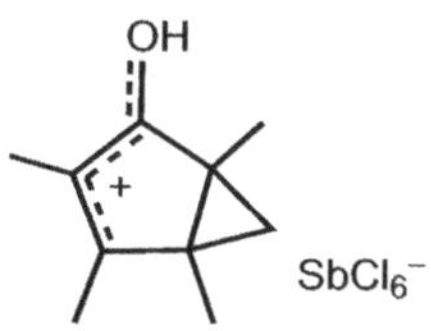

267

The generation of the stable cyclopropenylcarbinyl cation **268** under superacid conditions is an exceptional example[562] [Eq. (3.63)]. It appears that in this case the bulky *tert*-butyl group prevents the usually fast rearrangement of cyclopropenylcarbinyl cations to the corresponding homocyclopropenylium ions.

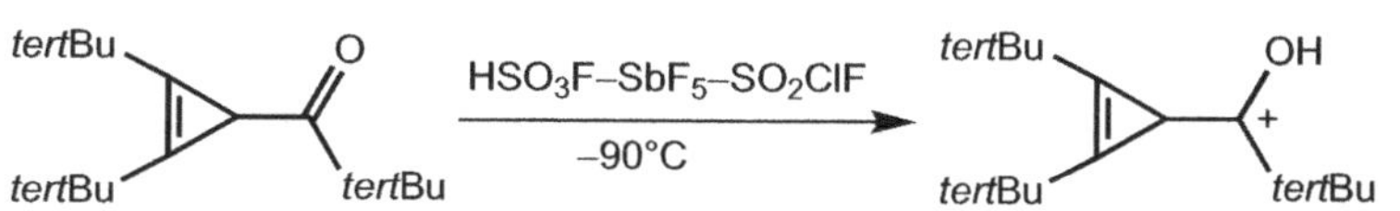

268

(3.63)

Protonation of homologous oxocarbon compounds (compounds in which all carbon atoms are bonded to carbonyl or enolic oxygens or their hydrated or deprotonated equivalents[563]) have been studied by Olah and co-workers.[564] They have found that protonated deltic acid (2,3-dihyroxy-2-cyclopropenone, **269**) and diprotonated squaric acid (1,2-dihydroxycyclobutenedione, **270**) prefer planar aromatic delocalized structures. In contrast, triprotonated croconic acid (4,5-dihydroxy-4-cyclopentene-1,2,3-trione, **271**) and tetraprotonated rhodizonic acid (5,6-dihydroxy-5-cyclohexene-1,2,3,4-tetraone, **272**) exhibit only carboxonium type structures.

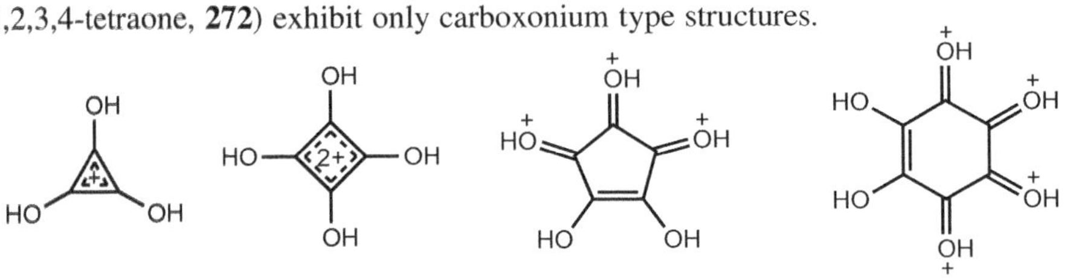

269 **270** **271** **272**

Olah, Prakash, and co-workers[565] have prepared the elusive mono-*O*-protonated deltic acid **269** by protolysis of di-*tert*-butoxy deltate in fivefold excess of Magic Acid [Eq. (3.64)]. The ^{1}H and ^{13}C NMR resonances of the solution at δ^1H 7.86 and δ^{13}C 128.7 were assigned to cation **269**. The C_{3h} symmetry structure with C–C and C–O bond lengths of 1.389 and 1.301 Å was calculated to be the global minimum. The computed ^{13}C NMR chemical shifts (IGLO/II/MP2/6-31G* level) of all four protonated oxocarbon cations **269–272** are in good agreement with the experimental values.

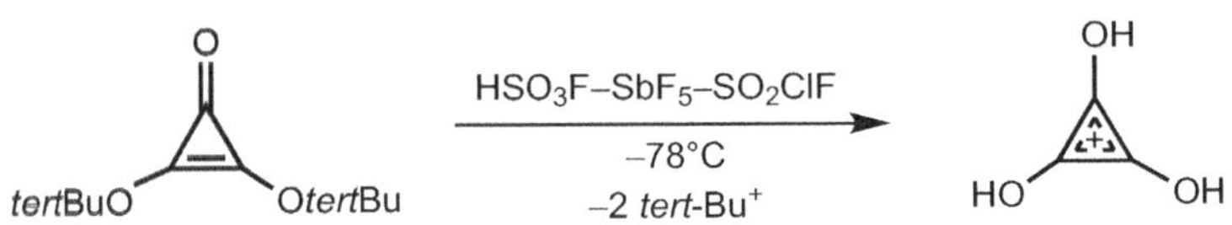

269 (3.64)

^{13}C NMR spectral investigations have been extended[566] to the study of heteroaromatic stabilized 6π 3-dioxolium and 10π benzo-3-dioxolium ions **273** and **274**.

273 **274**

Carboxylic acids are protonated in superacid media, such as HSO_3F–SbF_5, HF–SbF_5, or HF–BF_3.[541,542,567–570] The NMR spectrum of acetic acid in such media at low temperature shows two OH resonances, indicating that carbonyl protonation is favored and that hindered rotation about the resultant C–OH bond occurs[547,567] [**275**, Eq. (3.65)]. The predominant conformer observed is the *syn,anti*, although about 5% of the *syn,syn* isomer has also been seen. These isomers can be readily identified from the magnitudes of the proton vicinal coupling constants. No evidence for the *anti,anti* isomer has been found in either protonated carboxylic acids, esters, or their analogs.

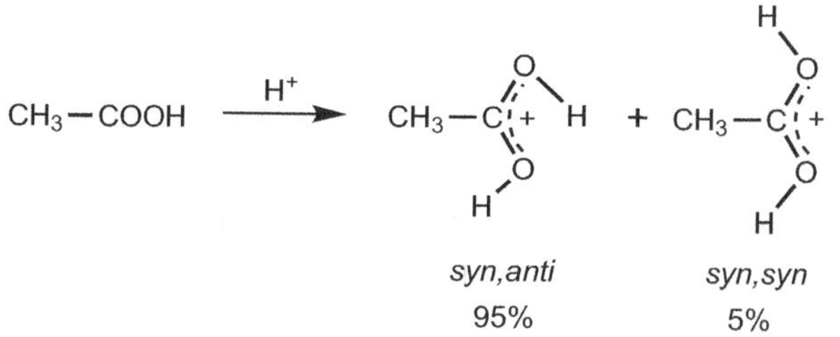

(3.65)

275

Esters behave in an analogous fashion, with carbonyl protonation being predominant [Eq. (3.66)]. Thus, protonated methyl formate **276** is present in $HSO_3F–SbF_5–SO_2$ solution as two isomers in a ratio of 90% to 10%.[571,572]

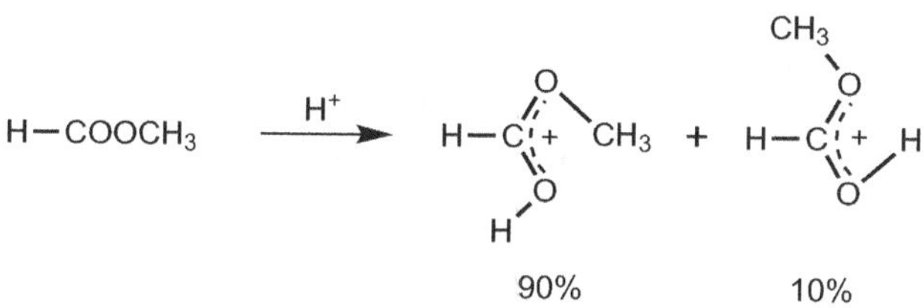

(3.66)

276

The X-ray crystal structure of protonated formic acid, acetic acid, and methyl formate has recently been determined by Minkwitz and co-workers. In agreement with NMR data discussed above, the nearly equal C—O bond lengths of protonated formic acid (1.239 and 1.255 Å),[573] protonated acetic acid **275** (1.251–1.291 Å for various salts),[574] and protonated methyl formate **276** (1.260 and 1.264 Å)[575] show efficient delocalization of the positive charge. Minkwitz et al.[576] have also studied protonation of oxalic acid in $HF–SbF_5$ and isolated the hexafluoroantimonate of the mono- and diprotonated acid at −75°C and −40°C, respectively. Structural characteristics are very similar to those of the other cations discussed.

By raising the temperature of solutions of protonated carboxylic acids and esters, unimolecular cleavage reactions are observed. These reactions can be considered within the framework of the two unimolecular reaction pathways for acid-catalyzed hydrolyses of esters, involving either alkyl–oxygen or acyl–oxygen cleavage. The advantage of studies of these reactions in superacid media, as compared with the solvolytic conditions, is that the cleavage step can be isolated and studied in detail because the cleavage products generally do not undergo any further reaction.

For example, in the case of protonated carboxylic acids in $HSO_3F–SbF_5–SO_2$ solution, a reaction analogous to the rate-determining step in the unimolecular cleavage of esters is observed leading to the acyl cation and oxonium (hydronium) ion [Eq. (3.67)].

$$RCO\overset{+}{O}H_2 \xrightarrow[-20°C]{H^+} RC\overset{+}{O} + H_3\overset{+}{O} \qquad (3.67)$$

Unimolecular cleavage in this case corresponds to the dehydration of the acid, but in the case of protonated esters the cleavage pathway depends on the nature of the alkoxy group [Eqs. (3.68)–(3.70)].

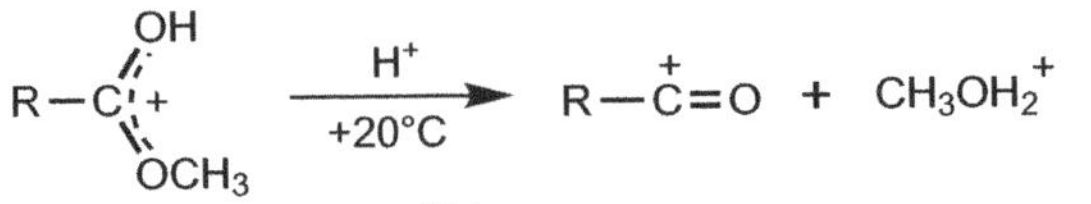
(3.68)

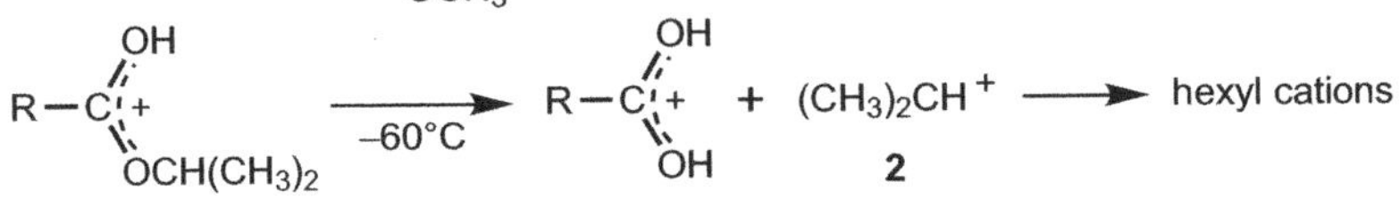
(3.69)

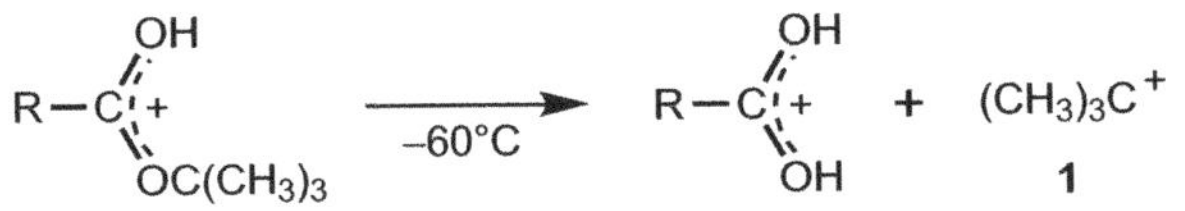
(3.70)

^{13}C NMR spectroscopic and theoretical studies (DFT, *ab initio*, and IGLO) of a series of protonated cycloalkylcarboxylic acids were performed by Prakash, Olah, and co-workers.[577] The study showed that the carboxylic carbons are deshielded to a limited degree from those of the corresponding carboxylic acids. This indicates that the cycloalkyl groups, including the cyclopropyl group, have little effect on the ^{13}C NMR chemical shift of the carbocationic carbons of carboxonium ions. Cycloalkylcarboxonium ions, consequently, are primarily stabilized by delocalization involving neighboring oxygen atoms.

Protonation and diprotonation of formic acid and acetic acid have also been studied theoretically (HF and MP2/6-312G* levels).[553,558] Both cations can be visualized as the donor–acceptor complex of H_2O and diprotonated carbon monoxide ($HCOH^{2+}$) and protonated acetyl cation (CH_3COH^{2+}), respectively. Both dications **277** and **278** are characterized by a long C–OH_2 bond and a shorter C–OH bond, suggesting that the latter retains substantial double-bond character.

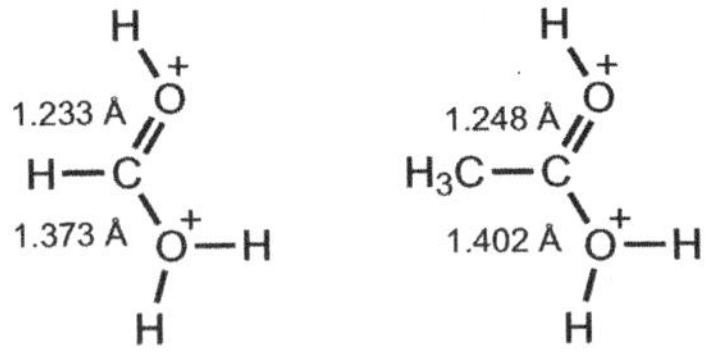

277 **278**

Ab initio molecular orbital theory [MP4(SDTQ)/6-31G*//MP2/6-31G* level] has been used to study the mono- and diprotonation of acetic acid and methyl acetate in superacidic media.[578] The large energy difference between the acyl-*O*-protonated and alkyl-*O*-protonated ester rules out any equilibrium between the two species. Diprotonation of the ester gives two types of gitonic dications. Of these, **279** is the most stable acyl-*O*-diprotonated species and **280** is the most stable alkyl–acyl diprotonated species, being more stable by 3.5 kcal mol^{-1} than the corresponding *E* rotamer. Acyl-*O*-diprotonation (dication **279**) results in lengthening of the acyl O–C bond and shortening of the alkyl O–C bond, which is opposite to what is observed in the case of dication **280**.

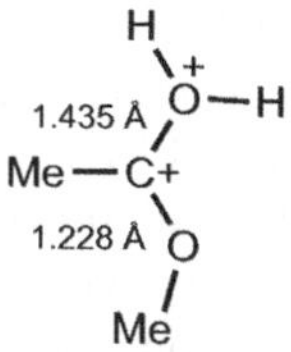

279

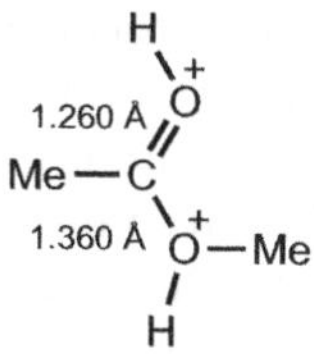

280

Cubylcarboxonium ions have been also studied by Prakash, Olah, and coworkers.[579,580] The parent cation **281** prepared under superacid conditions was stable at low temperature but decomposed to cubylacylium cation **282** as a result of further protonation and dehydration [Eq. (3.71)]. In addition to cation **281**, di- and tetracarboxonium ions and the corresponding protonated methyl esters were also observed as long-lived species stable under superacidic conditions. Experimental evidence and theoretical data indicated that the strained cubyl system effectively stabilizes the carbocationic centers through C–C bond hyperconjugation (**283**). On the basis of ^{13}C data, three conformers of protonated dimethyl cubane-1,4-dicarboxylate (**284**–**286**) could be identified.

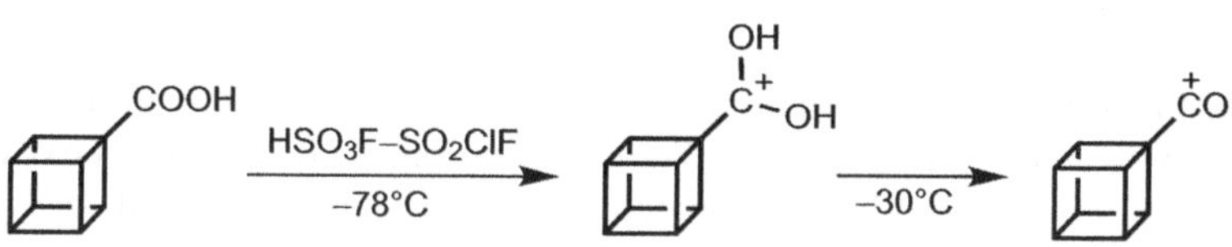

(3.71)

281 **282**

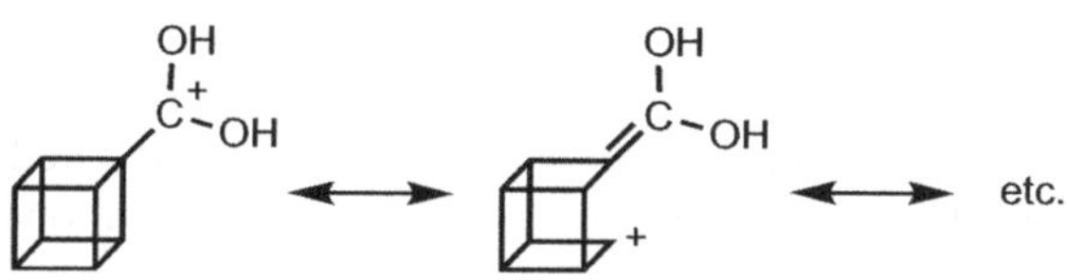

283

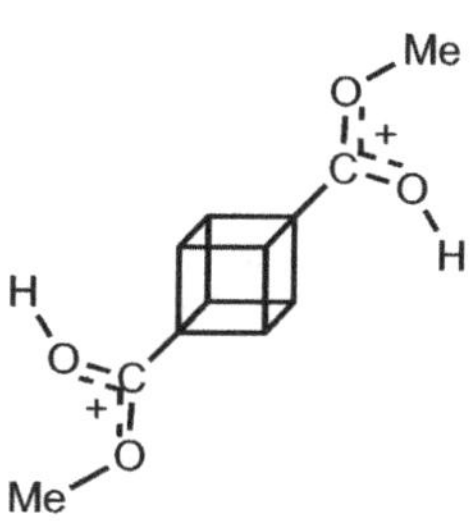

syn,anti syn,anti

284

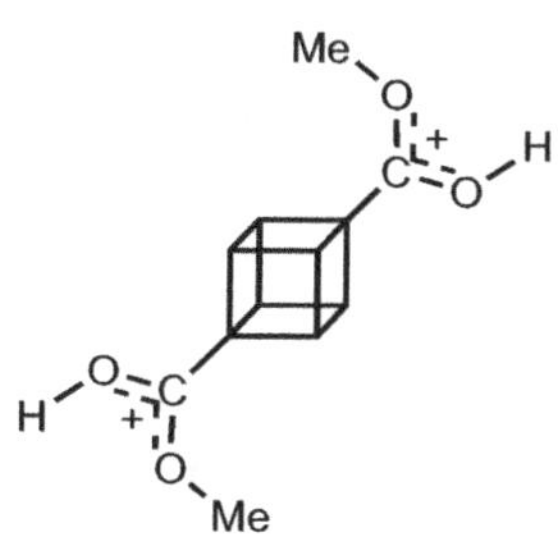

anti,syn anti,syn

285

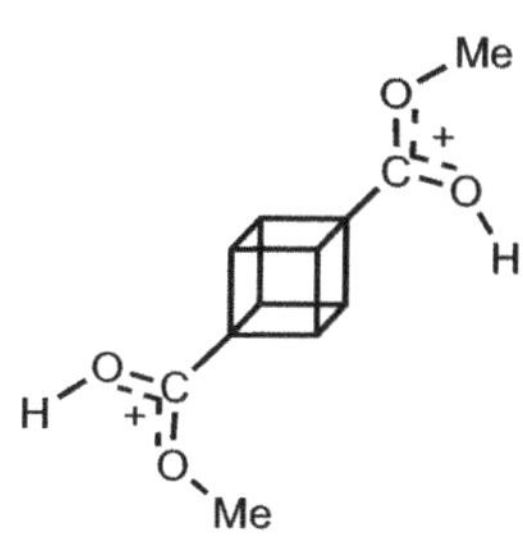

syn,anti anti,syn

286

Olah, Prakash, and co-workers[581] have studied the protonation of trifluoroacetone and methyl trifluoroacetate under superacidic conditions. Only one of the possible conformers of protonated trifluoroacetone **287** and protonated methyl trifluoroacetate **288** were generated. Protonation of hexafluoroacetone, in turn, was not successful. The deshielding at the carbonyl carbon (38 ppm) in **287** is smaller than that observed in protonated acetone (44 ppm). According to MO calculations [MP2(fu)/6-31G* level], the C–O bond in protonated trifluoroacetone is elongated, but to a smaller degree than in protonated acetone (3.6% versus 4.1%). The contribution of the carbenium ion structure **288b** is also greatly reduced as manifested by the higher deshielding of protonated methyl acetate (22 ppm) with respect to the parent methyl acetate as compared with the slight deshielding (14.5 ppm) for the fluorinated counterpart. These features are explained by the stronger contribution of the oxonium ion resonance form **287** and **288a** brought about by the strong inductive electron-withdrawing effect of the trifluoromethyl group.

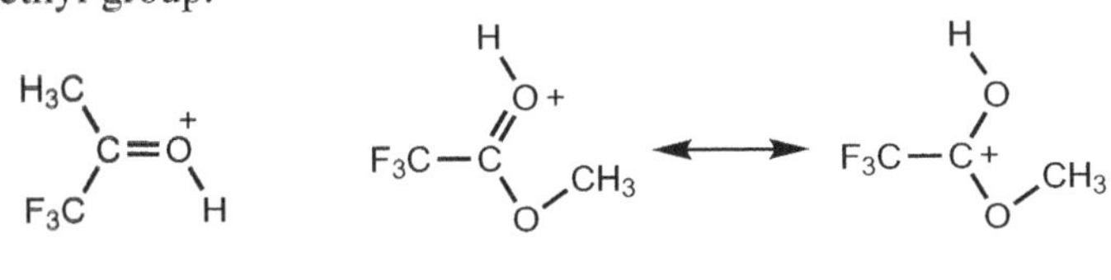

287 **a** **288** **b**

Protonated lactones may be prepared by direct protonation[582] or by protonating hyroxycarboxylic acids[583] [Eq. (3.72)]. The resulting protonated lactones exist as two conformers similar to other protonated carbonyl compounds, and they are usually stable in excess superacids.[582] A notable exception is α-angelicalactone, which undergoes acyl–oxygen cleavage to form protonated acetyl propionyl dication **289** (Scheme 3.9).[582] Similar conclusions have been arrived at in a recent study for $C_5H_6O_2$ isomeric unsaturated lactones using sulfuric acid, with the exception that the ring-opened product was transformed to diprotonated ketoacid **290** (Scheme 3.9).[584] Triflic acid, in turn, brings about polymerization under identical

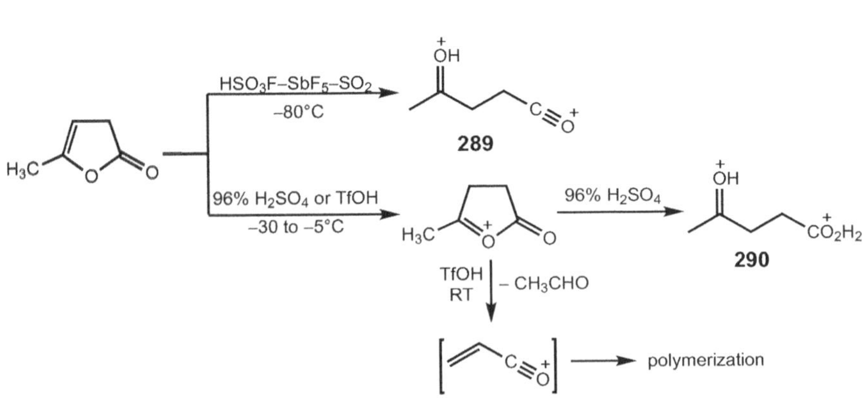

Scheme 3.9

conditions. Acyl–oxygen cleavage is also characteristic of protonated β-lactones to yield the corresponding acyl cations[585] [Eq. (3.73)].

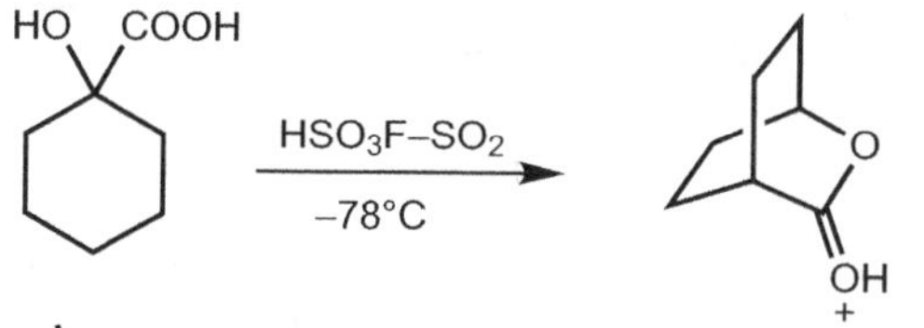

(3.72)

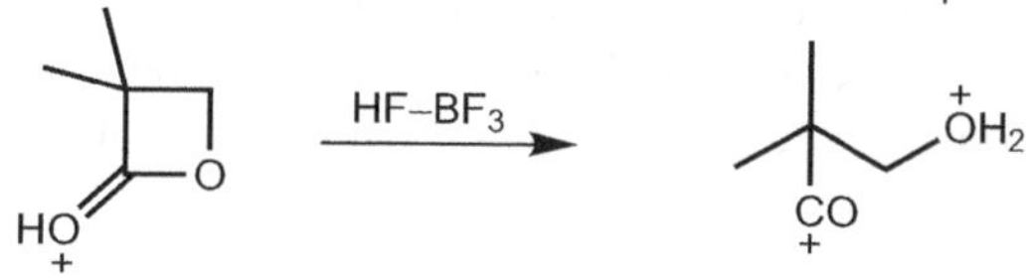

(3.73)

Carboxylic acid anhydrides cleave immediately when protonated even in $HSO_3F–SbF_5$ solution at low temperature (−80°C).[541,569] Monoprotonated anhydrides **291**, however can easily be generated reacting oxocarbenium ions (acyl cations) with carboxylic acids in SO_2 solution. Rapid intramolecular proton transfer between the carbonyl groups was observed by NMR.

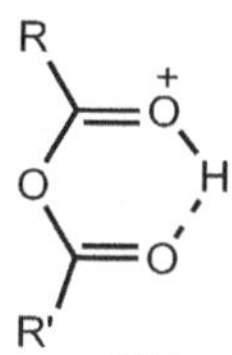

291

Dialkyl carbonates have been studied in $HSO_3F–SbF_5$ solution and have been shown to be protonated on the carbonyl group giving the dialkoxyhydroxymethyl cation **292**.[586] Di-*tert*-butyl carbonate cleaves immediately at −80°C with alkyl–oxygen fission, giving the *tert*-butyl cation **1** and the protonated carbonic acid **293** [Eq. (3.74)]. The structure of the latter has been established from the ^{13}C NMR spectrum of the central carbon atom, which shows a quartet, 4.5-Hz long-range C–H coupling constant, with the three equivalent hydroxyl protons.[586] Diisopropyl and diethyl carbonate cleave at a higher temperature, also via alkyl–oxygen cleavage, with initial formation of protonated alkyl hydrogen carbonates [Eq. (3.75)]. The alkyl hydrogen carbonates are also formed by protonation of their metal salts.

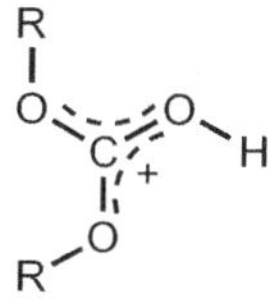

292

R = Me, Et, isoPr, *tert*-Bu, Ph
R-R = CH_2CH_2, CH=CH

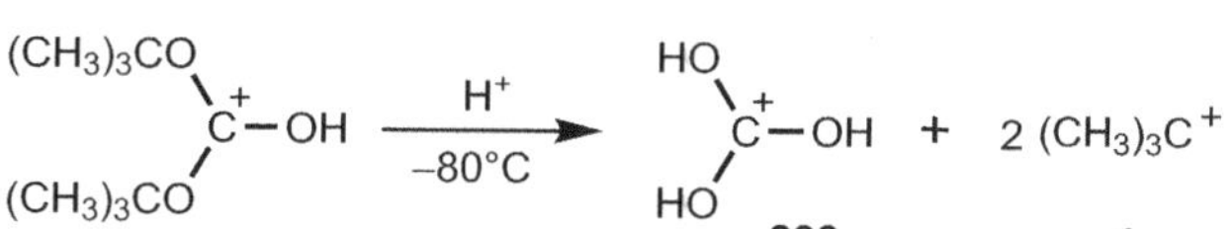

293 **1**

(3.74)

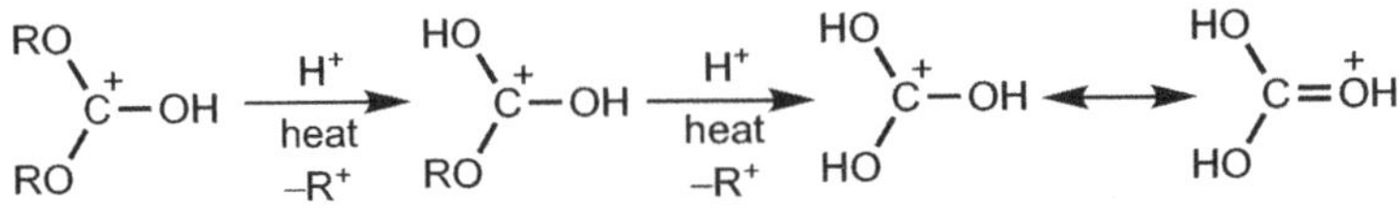

293 (3.75)

Carbon dioxide also reacts under superacid conditions[587] and protonated carbonic acid **293** can also be obtained[588] by dissolving inorganic carbonates and hydrogen carbonates in $HSO_3F–SbF_5$ at −80°C [Eq. (3.76)]. It is stable in solution to about −0°C, where it decomposes to the hydronium ion and carbon dioxide.

$$CO_3^{2-} \text{ or } HCO_3^- \xrightarrow[-80°C]{HSO_3F–SbF_5} \underset{\mathbf{293}}{H_3CO_3^+} \xrightarrow{\Delta} CO_2 + H_3O^+ \qquad (3.76)$$

Minkwitz and co-workers have recently prepared the salts of protonated carbonic acid **293** (trihydroxycarbenium hexafluorometalates)[589] and protonated dimethyl carbonate **294** (dimethoxyhydroxycarbenium hexafluorometalates)[590] [Eq. (3.77)]. The cation **293** of C_s symmetry in the solid state shows C−O bond lengths of 1.231 Å (AsF_6^- salt), which are comparable to those found in protonated formic acid (1.255 and 1.239 Å).[573] This allows the conclusion that the positive charge is delocalized over the whole cation. Likewise, the C−O bond lengths of the CO_3 unit (SbF_6^- salt) do not show significant differences (1.26, 1.274, and 1.300 Å). Furthermore, the two Me−O−C−O torsional angles reveal that the methyl groups do not lie in the CO_3 plane. In the ^1H NMR spectra a singlet for the protons of the hydroxyl groups of cation **293** is observed at δ^1H 11.55[586] and 11.7,[589] whereas a singlet for the hydroxyl group of cation **294** is detected at δ^1H 11.27–11.8.[586,590] It is worth mentioning here that, in addition to these two examples, the X-ray structure of the trioxygenated carbocation $C(OTeF_5)_3^+$ was also determined.[538] In the salt $[C(OTeF_5)_3][Sb(OTeF_5)_6]\cdot 3SO_2ClF$, the cation has C_1 crystallographic symmetry, which is very close to the optimized C_{3h} gas-phase geometry. Two of the three SO_2ClF molecules have short C---O contacts (2.690 and 2.738 Å).

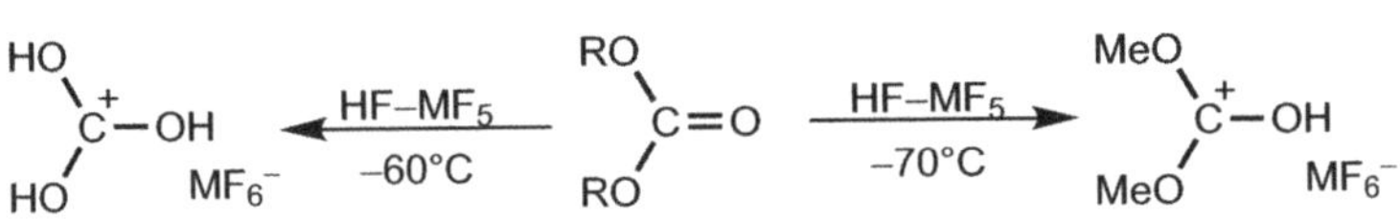

293 R = $SiMe_3$ M = As, Sb R = Me **294** (3.77)

NMR studies and *ab initio* and IGLO calculations of the mono- and diprotonation of carbonic acid have been reported by Olah, Prakash, Rasul, and co-workers.[588] The symmetrical C_{3h} **293′** structure was found to be the global minimum for the monoprotonated species. On the basis of ^1H and ^{13}C NMR data (single sharp peaks at δ^1H 12.05 and $\delta^{13}C$ 165.4), this structure was suggested by Olah and White[586] for the observed species in superacidic solution. The same species generated in the gas phase by electron ionization of diethyl carbonate has recently been found by single-photon IR photodissociation spectroscopy.[591] Structure **295** of C_{2v} symmetry is the lowest-energy structure of diprotonated carbonic acid, which can be considered as a donor–acceptor complex of H_2O and diprotonated CO_2. The C–OH bond length of dication **295** is between the normal C–O and C=O bond lengths indicative of the delocalization of the second positive charge into the two oxygen atoms of the C–OH moieties.

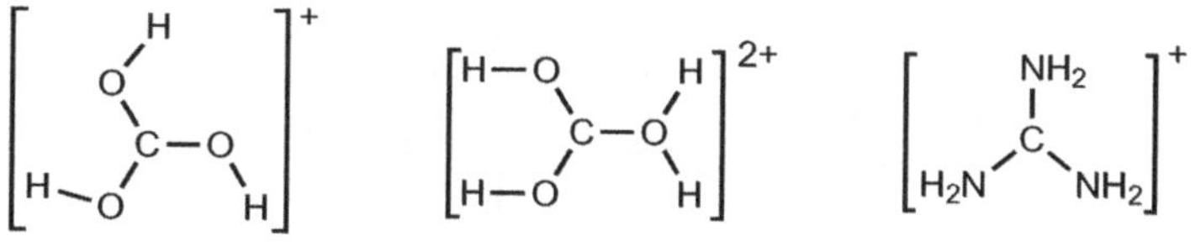

293' **295** **296**

It is worthwhile to point out the close similarity of protonated carbonic acid (trihydroxymethyl cation, **293**) with the guanidinium ion, its triaza-analog (**296**). The latter has been observed in superacidic solution by NMR. Both are highly resonance-stabilized through their onium forms. The observation of protonated carbonic acid as a stable chemical entity with substantial resonance stabilization also may have implications in understanding of some of the more fundamental biological carboxylation processes. Obviously, the in vitro observation in specific, highly acidic solvent systems cannot simply be extrapolated to different environments (biological systems). However, it is possible that on the active receptor sites of enzyme systems (for example, those of the carbonic anhydrase type), local hydrogen ion concentration may be very high, as compared with the overall "biological pH." In addition, on the receptor sites a very favorable geometric configuration may help to stabilize the active species, a factor that cannot be reproduced in model systems in vitro.

Alkylcarboxonium (Alkoxycarbenium) Cations. The ionic structure of pyrylium salts was clearly stated by Hantzsch[592] as early as 1922. In pyrylium salts **297**, there is a contribution from carbocation structures, a fact apparent in the behavior toward strong nucleophiles leading to phenols.

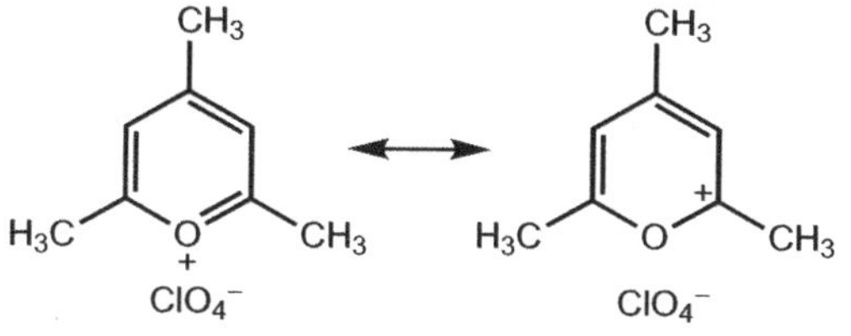

297

A recent interesting example is the chiral dication **298** obtained by treatment of the corresponding diol with HBF_4 (only one enantiomer is shown).[593] It has a pseudo-C_2

symmetry, and the two halves of the cation are twisted by 73° around the binaphthyl axis. The two cationic planes overlap in a face-to-face manner (the distance between the two C(9) carbons is 3.53 Å).

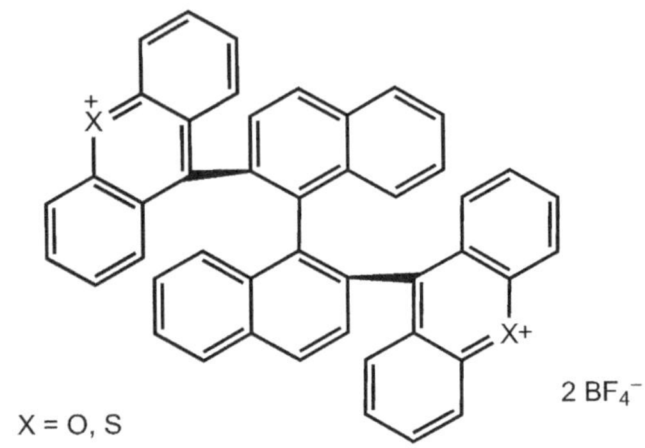

298

Resonance, similar to that in pyrylium salts, was shown[594,595] to exist between oxonium ion (**299a**) and carbenium ion (**299b**) forms in alkylated ketones, esters, and lactones that were obtained via alkylation with trimethyl- or triethyloxonium tetrafluoroborates[596] [Eq. (3.78)]. Ramsey and Taft[597] used ^{1}H NMR spectroscopy to investigate the nature of a series of secondary and tertiary carboxonium ions (**300**–**302**).

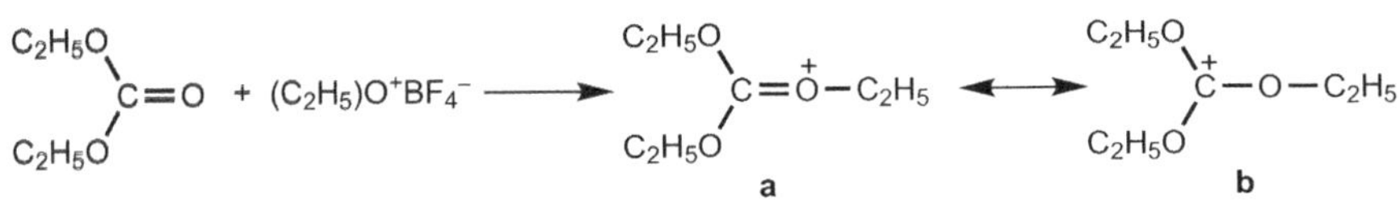

299

(3.78)

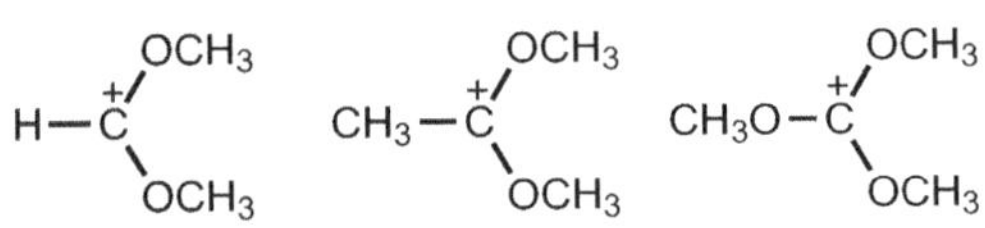

300 **301** **302**

As mentioned, the stabilizing effect of heteroatoms of carbenium ions increases with the number of electron-donating heteroatoms. It follows, therefore, that the stability of carboxonium ions increases in the order monoalkoxy < dialkoxy < trialkoxy. This has been demonstrated by Taft and co-workers,[598] who determined relative stabilization energies, and by Kresge, Larsen, and co-workers,[599] who measured enthalpies of formation. In a recent paper, however, the validity of this stability order has been questioned[600] on the basis of the rates of hydrolysis of tetraphenyl orthocarbonate and triphenyl orthoformate. The stabilizing effect of oxygen has been elucidated by Kiprof et al. through analysis of natural bond order and hydride exchange reactions.[601] Enthalpies of isodesmic hydride exchange reactions betweenx *tert*-butyl cation and methoxyalkanes (calculated at the G2 level of theory) demonstrated that the addition of methoxy substituents to a carbocation increases stability through π interaction with the oxygen. However, the incremental stabilization decreases dramatically (Scheme 3.10).

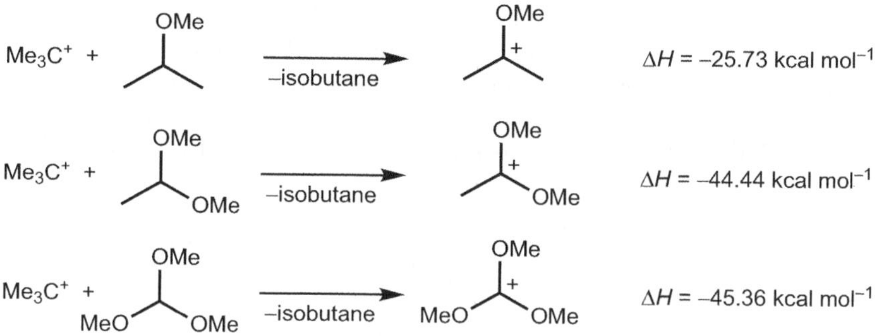

Scheme 3.10

The dimethylation of methyl acetate (methylation of the 1,1-dimethoxyethyl cation) was calculated to form two isomeric gitonic dications (**303**), which are thermodynamically more stable than the neutral ester.[578] They are expected to be involved in the observed methyl exchange of 1,1-dimethoxyethyl cation. Rotamer **303b** is more stable than the isomeric dication **303a**, by 10.0 kcal mol^{-1} (MP2/6-31G*//MP2/6-31G*+ZPE level).

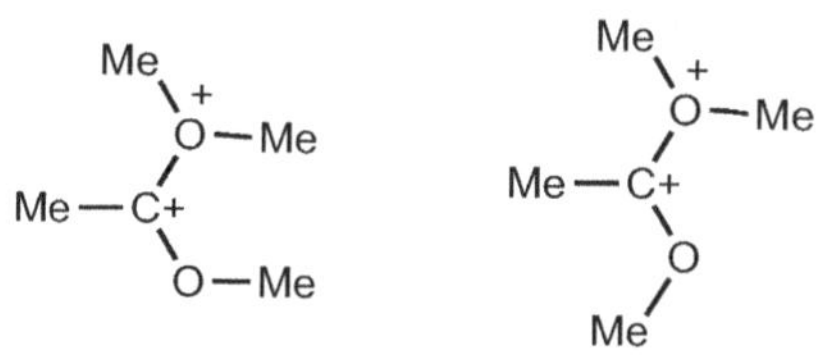

a b

303

Olah and co-workers[602,603] have obtained primary carboxonium ions such as methoxymethyl and phenoxymethyl cations and their halogenated derivatives (**304–307**).

$CH_3O\overset{+}{C}H_2 \longleftrightarrow CH_3\overset{+}{O}{=}CH_2$ **304**

$CH_3O\overset{+}{C}HCl \longleftrightarrow CH_3\overset{+}{O}{=}CHCl$ **305**

$ClCH_2O\overset{+}{C}H_2 \longleftrightarrow ClCH_2\overset{+}{O}{=}CH_2$ **306**

$C_6H_5O\overset{+}{C}H_2 \longleftrightarrow C_6H_5\overset{+}{O}{=}CH_2$ **307**

$CH_3O\overset{+}{C}HX \longleftrightarrow CH_3\overset{+}{O}{=}CHX$ **308** X = F, Cl

Olah and Bollinger[602] also carried out a ^{1}H and ^{13}C NMR spectroscopic investigation of a series of haloalkyl carboxonium ions **308**. Ionization of α,α-difluoromethyl and α,α-dichloromethyl methyl ether in SbF_5–SO_2 solution yields isomeric methoxyfluoro- and methoxychloro-carbenium ions **308** [Eq. (3.79)]. Of the rotamers the *trans* isomer is obtained in higher amounts (70 : 30 for F and 81 : 19 for Cl). The calculated structures of the isomeric methoxyfluorocarbenium ions indicate[523] that these ions are predominantly carboxonium ions rather than carbenium ions. Interestingly, at the MP4(SDTQ)//6-31G*//MP26-31G* level of calculation, the *cis* isomer of the methoxyfluorocarbenium ion was found to be more stable by 3.2 kcal mol^{-1}.

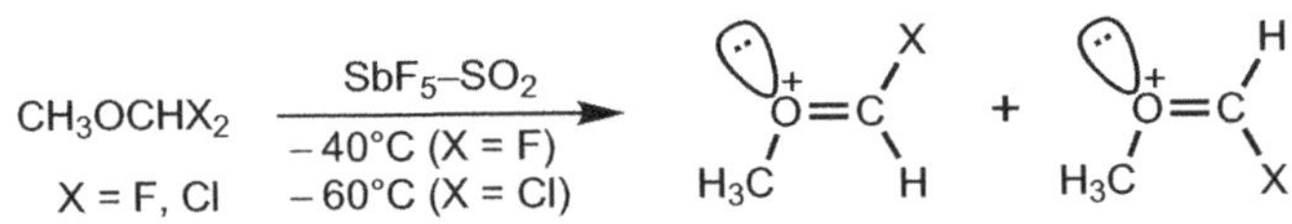

trans *cis*

308

(3.79)

Minkwitz et al.[604] have prepared the hexafluorometalates of cations **308**. The reaction of α,α-dichloromethyl methyl ether in HF–Lewis acid solution at −78°C leads to the formation of chloro cation **308-Cl**, whereas at −65°C fluoro derivative **308-F** is isolated as a result of chlorine–fluorine exchange [Eq. (3.80)]. Interestingly, the chlorine atom and the methyl group are *trans* in the hexafluoroantimonate salt of cation **308-Cl**, whereas the fluorine atom and the methyl group are *cis* in cation **308-F**. The arrangement of the C−O−C−H atoms is nearly planar with F/Cl−C−O−C torsion angles of 2.84° (**308-F**) and 179.0° (**308-Cl**). The C−O bond distances (1.224 and 1.479 Å for cation **308-F**, and 1.252 and 1.517 Å for cation **308-Cl**) reveal dominant oxonium ion character.

$$CH_3\overset{+}{O}{=}CHCl \xleftarrow[-78^\circ C]{HF\text{–}MF_5} CH_3OCHCl_2 \xrightarrow[-65^\circ C]{HF\text{–}MF_5} CH_3\overset{+}{O}{=}CHF \qquad (3.80)$$

308-Cl M = As, Sb **308-F**

Olah et al.[603] have observed the formation of cation **309** (protonated fluoromethanol) upon treatment of formaldehyde in HF–SbF_5 [Eq. (3.81)]. When Minkwitz et al.[605] attempted to isolate salts of the ion, however, the hydroxymethyl(methylidene) oxonium ion **310** was obtained [Eq. (3.81)]. Crystal structure analysis of the hexafluoroarsenate salt shows that cations and anions are connected by short H···F distances, forming a three-dimensional network. The bond lengths of the C–O=C fragment (1.226 and 1.470 Å) are longer than those in formaldehyde (1.208 Å) and dimethyl ether (1.410 Å). The C−O−C bond angle is 121.2°.

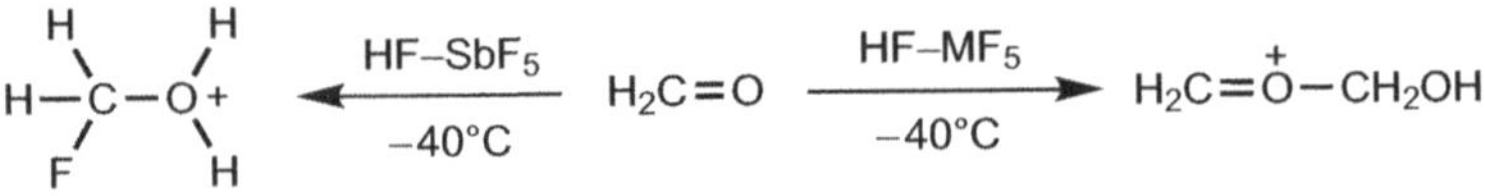

M = As, Sb

309 **310**

(3.81)

Alkylated carboxonium ions have also been prepared by direct electrophilic oxygenations of alkanes, alcohols, and so on, by ozone or hydrogen peroxide in superacidic media[606] [Eq. (3.82)].

$$(CH_3)_3CH \xrightarrow[HSO_3F\text{–}SbF_5]{H_2O_2 \text{ or } O_3} [(CH_3)_3CO^+] \longrightarrow (CH_3)_2C{=}\overset{+}{O}CH_3 \quad (3.82)$$

Smit *et al.*[607,608] have isolated a series of cyclic carboxonium salts (**311**, **312**) by acylation of cycloalkenes [Eqs. (3.83) and (3.84)].

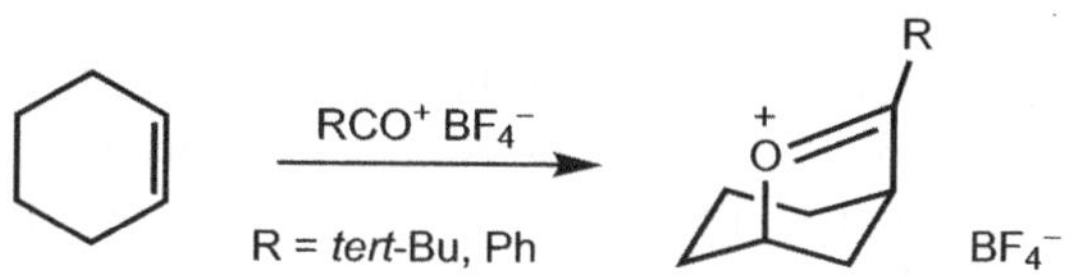

311 (3.83)

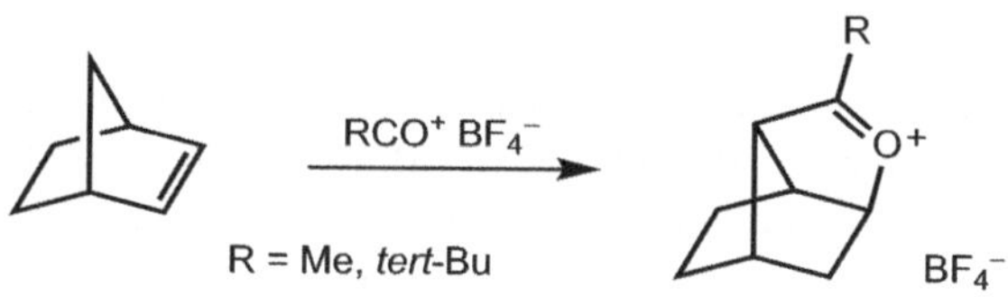

312 (3.84)

Four- and five-membered cyclic carboxonium ions have been prepared by intramolecular alkylations[541,596,609] using halogen- or hydroxy-substituted carbonyl compounds[610,611] [Eq. (3.85)], γ,δ-unsaturated carbonyl derivatives,[612,613] or substituted cyclopropanes[614,615] [Eq. (3.86)].[541,596]

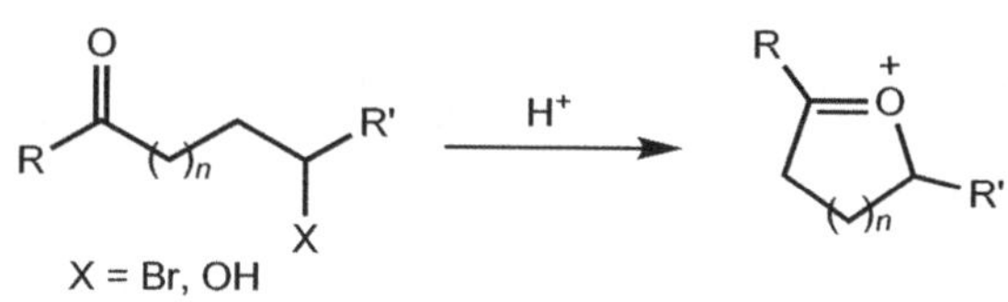

(3.85)

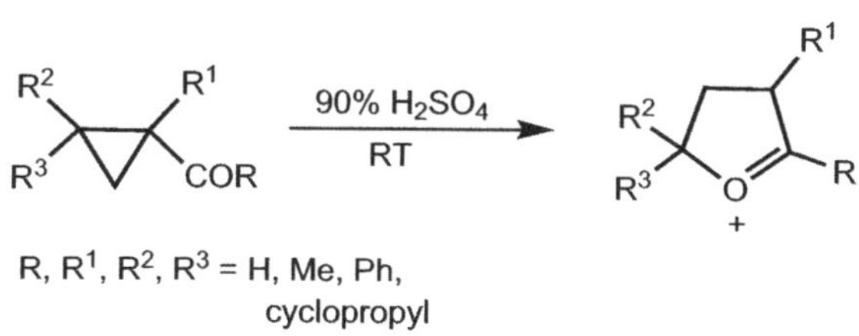

(3.86)

Cyclic cations such as 4-oxo-1,3-dioxolane-2-ylium ion **313** and 4-oxo-1,3-dioxane-2-ylium ion **314** can be prepared in intramolecular reaction of α-acetoxyacyl chlorides[616] or β-acetoxyacyl chlorides[617] with $SbCl_5$. Meerwein's method,[618] that is, the reaction of bromoethyl esters with $AgBF_4$, was applied to generate cations **315**[619,620] and dications **316**.[621,622]

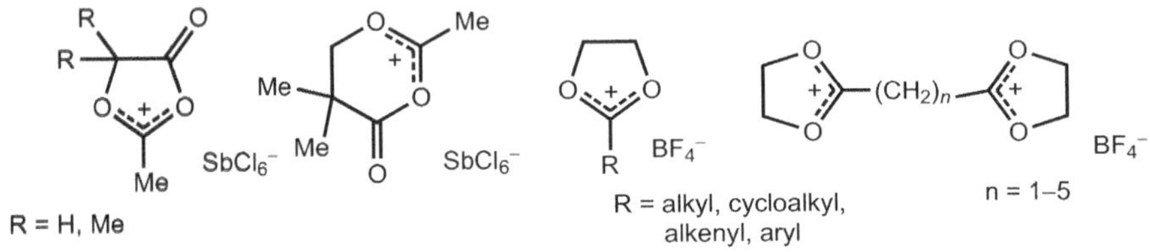

313 **314** **315** **316**

Another method, hydride abstraction of the corresponding 1,3-dioxolanes using trityl tetrafluoroborates, could not be applied to generate dications **317** and trication **318**. Instead, they were prepared by the ionization of the corresponding 2-methoxyethyl benzoates in HSO_3F–triflic acid.[623] Interestingly, all four dications gave very similar ^{13}C NMR chemical shifts for the carbocationic centers ($\delta^{13}C = 182.2-182.7$ for the dications, and 182.9 for trication **318**), which is indicative of extensive charge delocalization into the neighboring oxygens. The calculated stabilities of the *meta* and *para* isomers are very similar (B3LYP6-31G*//B3LYP6-31G*+ZPE level), whereas the *ortho* isomer is significantly less stable (15.0 kcal mol^{-1}), which is attributed to charge–charge repulsion and higher steric constraints.

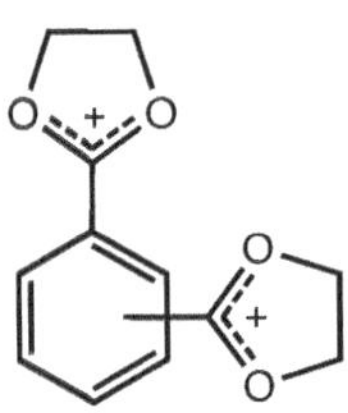

317

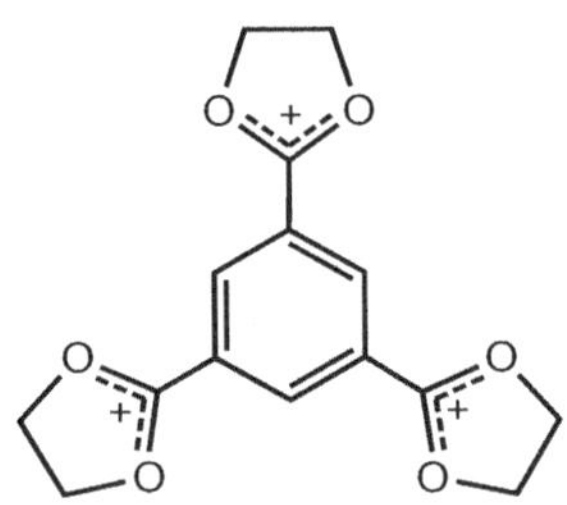

318

Alkylcarboxonium ions may be synthesized easily by the dehalogenation of α-haloethers. In Magic Acid, first the oxygen is protonated to oxonium ion **319**, which, upon increasing temperature, decomposes to yield alkylcarboxonium ion **320** [Eq. (3.87)].[624] In a similar manner, protonation of α-alkoxyacetic chlorides[625] or acids[626] also leads to carboxonium ions. The latter approach was used to generate cyclic alkylcarboxonium ion **321** [Eq. (3.88)].

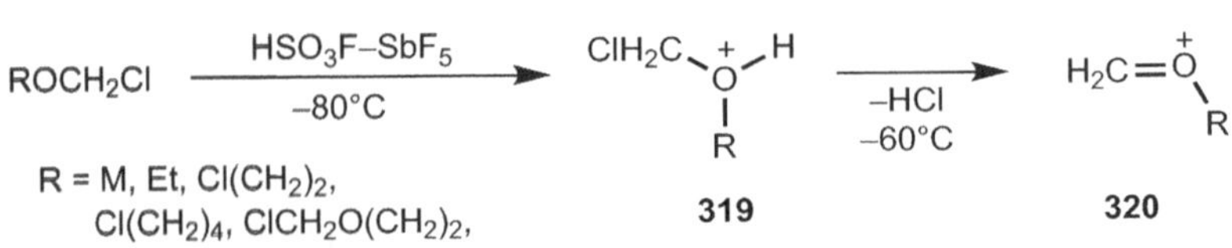

(3.87)

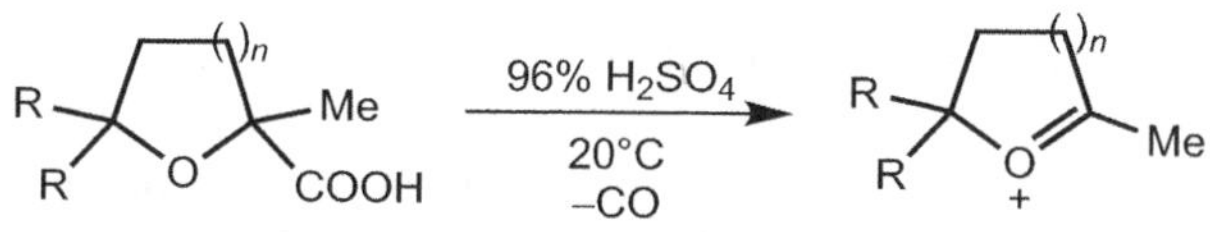

(3.88)

R = H, Me
n = 1, 2

321

Acetals and orthoesters can be conveniently transformed to mono- and dialkoxycarbenium ions by cleavage of an alkoxy anion[596,609] [Eq. (3.89)]. The method can also be applied to transform tetraalkoxymethanes to trialkoxycarbenium ions. This method was reported early by Meerwein to generate alkoxycarbenium ions.[627] Acetals can also be transformed to dialkoxycarbenium ions using hydride-abstracting agents such as trityl cations[608] [Eq. (3.90)] or trialkyloxonium ions.[609]

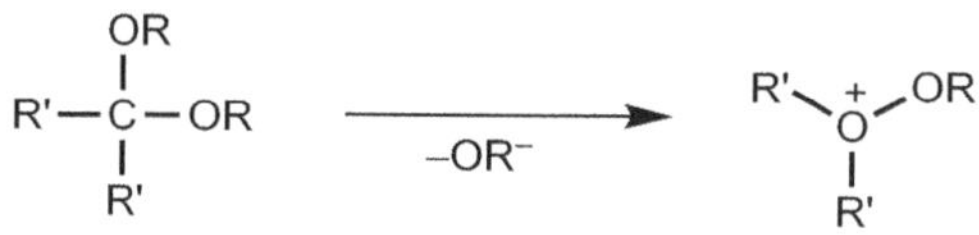

(3.89)

R' = alkyl, aryl, OR

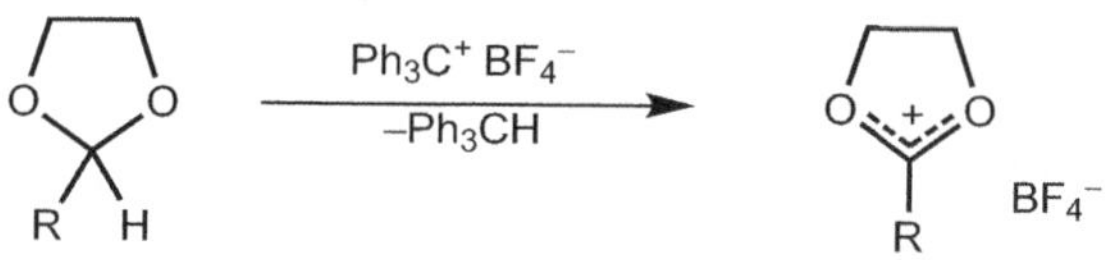

(3.90)

315

Results with respect to the synthesis of fluorinated alkylcarboxonium cations have recently been reported. Olah, Prakash, and co-workers[581] have methylated trifluoroacetone and methyl trifluoroacetate in $CH_3F–SbF_5–SO_2$ solution at −60°C and observed the methylated ions as long-lived species by NMR spectroscopy.[581] Conformers **322** and **323** were identified by calculation [MP2(fu)/6-31G* and MP2/6-31G*+ZPE levels, respectively]. The magnitude of the deshielding at the carbonyl carbons, again, indicates a stronger contribution of the corresponding oxonium ion resonance forms in both cases.

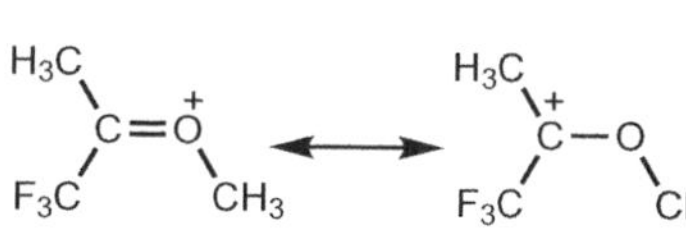

322

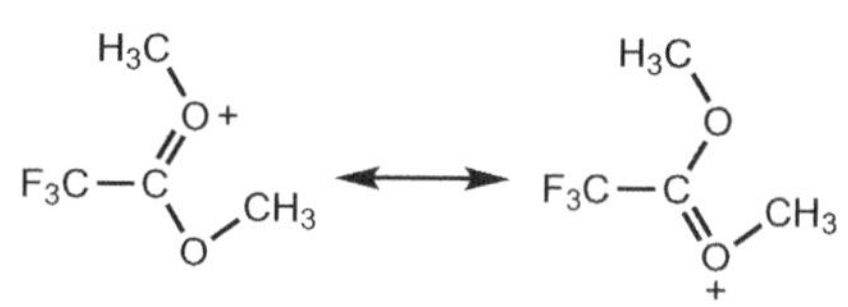

323

Stable fluorinated bis(alkoxy)methyl cations **324** and tris(alkoxy)methyl ion **325** were prepared reacting the corresponding difluoroformals and fluoroorthoester with excess SbF_5.[628,629]

R_F = $CClF_2CH_2$, CF_3CH_2, $(CF_3)_2CH$

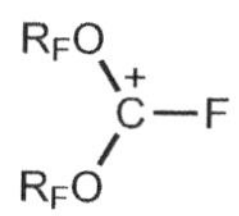

324

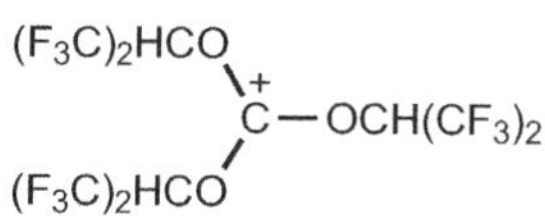

325

The X-ray structure of a number of alkoxycarbenium ions has been determined.[66] An interesting example is 2-methoxy-1,7,7,-trimethylbicyclo[2.2.1]hept-2-ylium tetrafluroborate **326**.[630] It is a substituted 2-norbornyl cation and, indeed, the C(2)–C(1)–C(6) bond angle (98.8°) and the C(1)–C(6) bond distance (1.603 Å) indicate σ-bond charge delocalization, that is, the contribution of the **326b** resonance form.

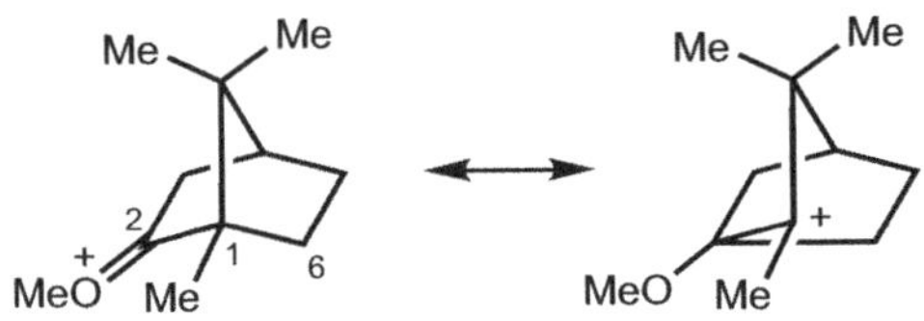

a **b**

326

Recently, a variety of *O*-silylated carboxonium ions have been prepared and characterized by NMR spectroscopy.[541] Kira et al.[631] used the Corey hydride transfer method, whereas Olah, Prakash, and co-workers applied triphenylmethyl tetrakis (pentafluorophenyl)borate to silylate esters,[632] ketones, enones, and carbonates[633] [Eq. (3.91)]. The ions thus produced are resonance hybrids of oxocarbenium (**327b**) and carboxonium (**327a**) ions with the latter as the major contributors. Calculated (DFT/IGLO) ^{29}Si NMR chemical shifts agree well with the experimental data.

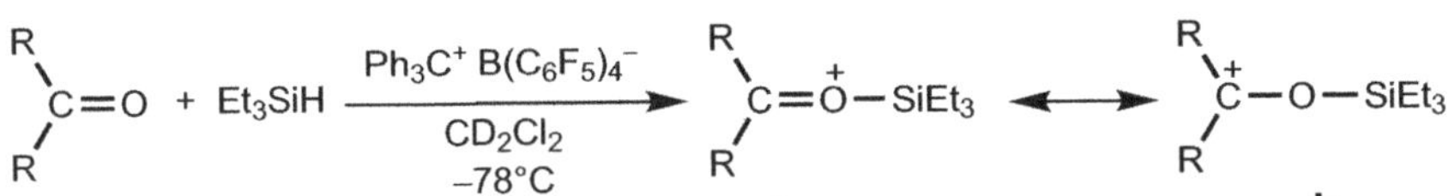

R = alkyl, aryl, MeO **a** **b**

327

(3.91)

Acylium Ions (Acyl Cations). In a recent study de Rege, Gladysz, and Horváth[634] have observed protonated carbon monoxide by IR and NMR spectroscopy at high CO pressure. Protonation occurs at the carbon to give formyl cation **328** [Eq. (3.92)]. A fast proton exchange observed with increasing temperature was attributed to the formation of isoformyl cation $^{+}$COH and diprotonated CO (protoformyl dication, COH_2^{+}).

$$\text{CO} \xrightarrow[\substack{25°\text{C}, \\ P_{\text{CO}} < 100\ \text{atm}}]{\text{HF–SbF}_5} \text{O}{=}\overset{+}{\text{C}}{-}\text{H} \qquad (3.92)$$

328

Protonation of carbon monoxide have been studied theoretically.[558,635–638] Both the formyl cation **328** and the isoformyl cation **329** are protonated in exothermic processes with cation **328** being more stable by 38 kcal mol^{-1}. The calculated energy barrier of rearrangement (36 kcal mol^{-1} at MP3/6-31G** level), however, suggests that cation **329**, once formed, is a relatively long-lived species. Furthermore, deprotonation of diprotonated CO has significant kinetic barriers (Scheme 3.11). Calculated optimized structures (HF/6-31G* level) and energy values (in hartree)[558] of CO, monoprotonated CO (HCO^{+}), and diprotonated CO (protoformyl dication, COH_2^{+}), all with $C_{\infty v}$ symmetry, are given in Figure 3.17. According to calculations, the

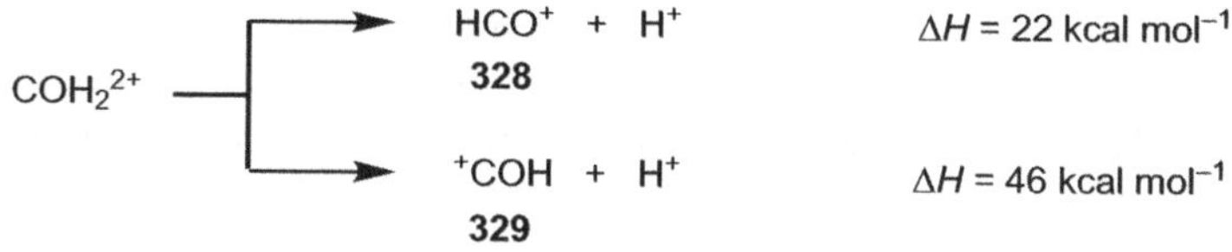

Scheme 3.11

protoformyl (hydroxymethine) dication is kinetically stable and has a very short C–O bond and a deprotonation barrier of >20 kcal mol^{-1}.

According to a recent *ab initio* molecular dynamic simulation study[639] formation of formyl cation **328** is optimally favored in the 1:1 HF–SbF_5 solution. No evidence, however, was found for the formation of the isoformyl cation **329** and diprotonated CO. The fast proton exchange observed earlier was suggested to occur between HCO^+ and $HF(HF)_n$ or SbF_6^-.

Christe, Prakash, Olah, and co-workers[534] have made a comparative experimental and theoretical study of the halocarbonyl cations (XCO^+). $ClCO^+$, $BrCO^+$, and ICO^+ were observed in the condensed state in SO_2ClF solution by ^{13}C NMR spectroscopy,[640] and FCO^+ was also reported to be generated in HSO_3F–SbF_5.[641] The new study has shown, however, that FCO^+ cannot be stabilized in the condensed phase with presently known Lewis acids. $ClCO^+$, in turn, could be generated with Sb_3F_{15} having sufficient acidity to form salt **330** [Eq. (3.93)]. Aubke and co-workers[642] have also reported the generation in SbF_5 medium and characterization by vibrational spectroscopy of the salts of $ClCO^+$ with oligomeric fluoroantimonate anions.

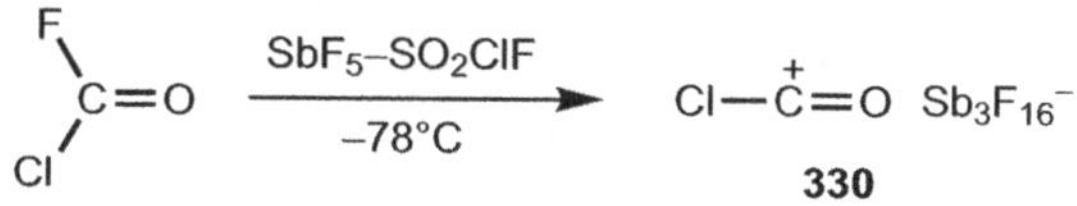

330 (3.93)

In 1943, Seel observed[14] the first stable acyl cation. Acetyl fluoride with boron trifluoride gave a complex (decomposition point 20°C) that was characterized as the acetyl tetrafluoroborate salt **331** [Eq. (3.94)].

$$CH_3COF + BF_3 \longrightarrow CH_3\overset{+}{C}{=}O \;\; BF_4^- \qquad (3.94)$$

331

The identification was based on analytical data and chemical behavior. Only in the 1950s were physical methods like infrared and NMR spectroscopy applied, making further characterizations of the complex possible. Since 1954, a series of other acyl and substituted acyl cations have been isolated and identified.[91,643–647]

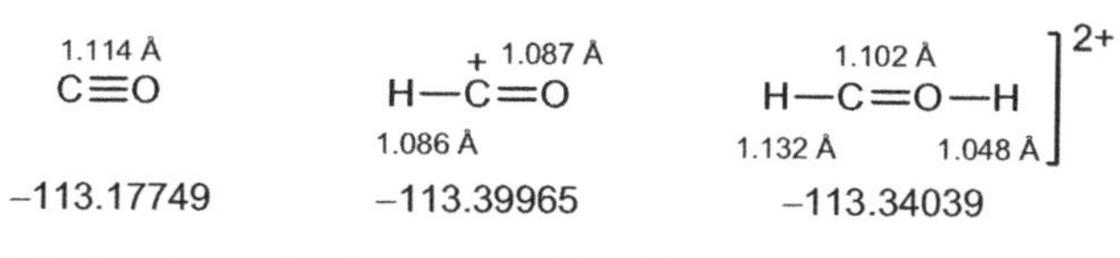

Figure 3.17. Calculated optimized structures (HF/6-31G* level) and energy values (in hartree) of CO, along with monoprotonated and diprotonated CO.

The hexafluoroantimonate and hexafluoroarsenate complexes were found to be particularly stable.[91,645,646] Deno et al.[648] investigated solutions of carboxylic acids in sulfuric acid and oleum. They observed protonation at lower acid concentrations and dehydration, giving acyl cations, at higher acidities [Eq. (3.95)].

$$\mathrm{RCOOH} \xrightarrow{H_2SO_4} \mathrm{RCO\overset{+}{O}H_2} \xrightarrow{\text{oleum}} \mathrm{R\overset{+}{C}O} + \mathrm{H_3\overset{+}{O}} \tag{3.95}$$

The investigation of acyl cations in subsequent work was substantially helped by NMR. Not only ^{1}H, but also ^{2}H, ^{13}C, ^{19}F, and ^{17}O resonance studies established the structure of these ions.[91,552,645,646,648–650] These investigations based on ^{13}C and proton resonance showed that acyl cations, such as the CH_3CO^+, acetyl ion **331**, are not simple oxonium ions (acylonium complexes), but are resonance hybrids of the oxonium ion, acyl cation, and the ketene-like nonbonded mesomeric forms. The X-ray crystallographic study of the $CH_3CO^+\ SbF_6^-$ salt[651] substantiated this suggestion and provided convincing evidence for the linear structure of the crystalline complex.

$$\mathrm{CH_3{-}C{\equiv}\overset{+}{O}} \longleftrightarrow \mathrm{CH_3{-}\overset{+}{C}{=}O} \longleftrightarrow \underset{\mathrm{H^+}}{\mathrm{CH_2{=}C{=}O}}$$

331

The experimental ^{13}C chemical shift of the carboxonium carbon at δ^{13}C 150.3 is well-reproduced by calculation (δ^{13}C 158.5 by the GIAO-MP2 method at the qz2p/dzp level).[553] With respect to ^{17}O NMR results, IGLO(II) and GIAO-MP2(qz2p/dzp) calculations give a shielding of 56 ppm (δ^{17}O 340.8) for the acetyl cation with respect to carbon monoxide (δ^{17}O 397.0), which is an excellent agreement with the experimentally observed shielding of 50 ppm (δ^{17}O 299.5) relative to carbon monoxide (δ^{17}O 350.0). This reveals that the C–O bonding character of the acetyl cation is less than that of a triple bond but significantly more than that of a C–O double bond.

Investigation of acyl cations has been extended to the study of cycloacylium ions **332–336**,[652] diacylium ions (dications) **337–339**,[569,653] and unsaturated acylium ions **340–342**.[654] Computational studies at various levels of theory have also been performed.[553,655–657]

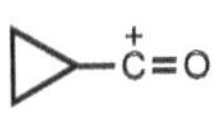

332

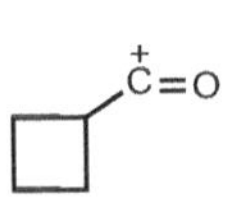

333

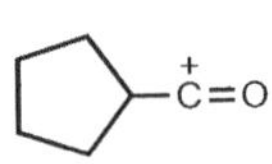

334

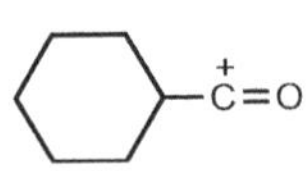

335

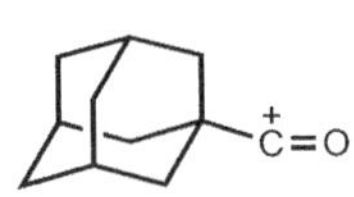

336

$\mathrm{O{=}\overset{+}{C}(CH_2)_3\overset{+}{C}{=}O}$

337

$\mathrm{O{=}\overset{+}{C}(CH_2)_4\overset{+}{C}{=}O}$

338

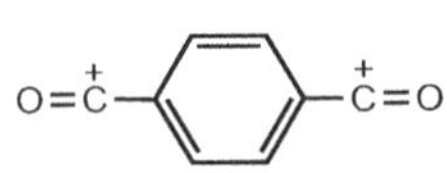

339

$\mathrm{CH_2{=}C(CH_3)\overset{+}{C}{=}O}$

340

$\mathrm{CH_2{=}CH\overset{+}{C}{=}O}$

341

$\mathrm{CH_3CH{=}C(CH_3)\overset{+}{C}{=}O}$

342

The structure of alkenoyl cations (unsaturated acylium ions) was studied by Olah et al.[658] by NMR spectroscopic methods. They found only a limited contribution from structure **343b** and a substantial contribution of the delocalized ketene-like structure **343c**, which is due to the ability of the π-electrons of the double bond to stabilize the positive charge. Substitution at the β-carbon increases further the importance of **343c** relative to **343a**. Diprotonation of propenoyl and isopentenoyl cations studied theoretically [*ab initio* GIAO–CCSD(T) method][659] has been shown to result in the formation of dication **344** (tertiary carbenium–acylium dication); that is, the positive charges are localized primarily on CO and the β-carbon.

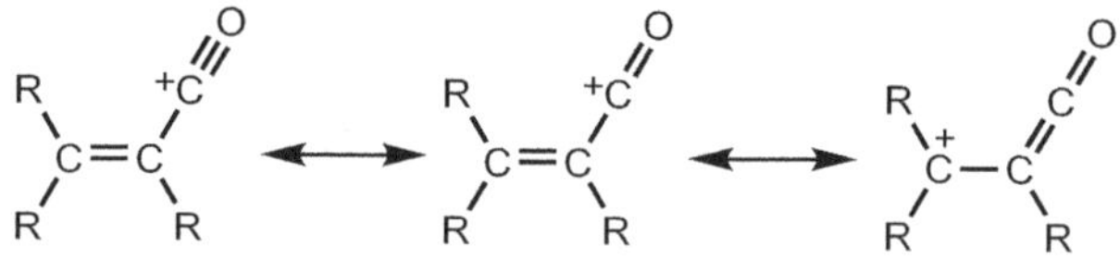

R = H, Me, Ph, cyclopropyl

a **b** **c**

343

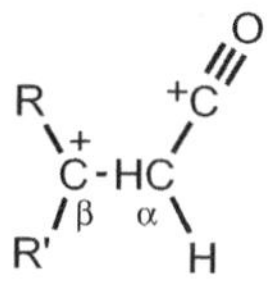

R = R' = H, Me

344

^{13}C NMR spectroscopic and theoretical studies (DFT, *ab initio*, IGLO) of a series of cycloacylium ions were performed by Prakash et al.[577] The study showed that the cycloalkyl groups have little effect on the shift of the carbocationic carbon. Furthermore, charge calculations showed that the delocalization into the cycloalkyl group is greater than in the protonated carboxylic acid (carboxonium ion), where two oxygen atoms participate in delocalization.

Cubylacylium cation **282** discussed above and diacylium ion **345** were generated from the corresponding acid chlorides in SbF_5 [Eq. (3.96)].[579] Calculated charge densities showed that the formal cationic carbon of these ions, when compared with the adamantylacylium ion, bears less positive charge; that is, the cubyl cage participates in delocalization more efficiently than does the adamantyl skeleton.

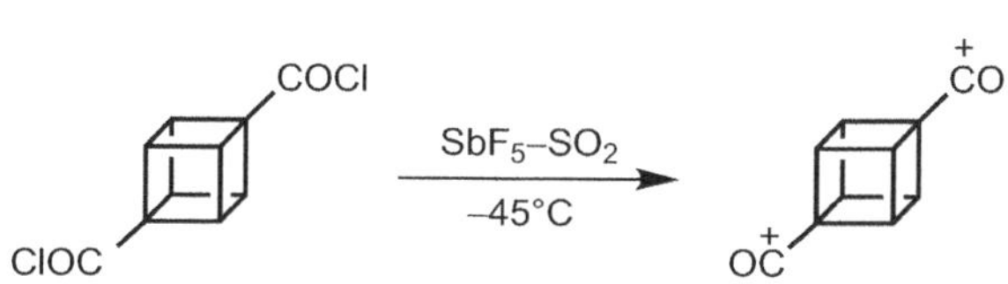

345

(3.96)

Prakash, Olah, and co-workers have shown that ketenes can also be transformed to acylium ions. Both the protonation and silylation of ketenes take place exclusively at the β-carbon to yield stable acylium ions **346** [Eq. (3.97)][660] and **347** [Eq. (3.98)].[661] A comparison of ^{13}C NMR chemical shift data show that both C_α and C_β in cation **347** are more deshielded than in cation **346** (R = Ph) ($\Delta\delta^{13}C_\alpha = 32.4$ ppm, $\Delta\delta^{13}C_\beta = 10.4$ ppm). These data indicate that cation **346** is primarily stabilized by delocalization involving oxygen (oxonium ion character), whereas β-silyl hyperconjugation results in enhanced charge delocalization in cation **347**. Interestingly, DFT calculations (B3LYP/6-311+G* level) predict that *O*-silylation of diphenylketene is preferred over *C*-silylation by 5.4 kcal mol^{-1}, whereas for the parent ketene *C*-silylation is preferred by 8.2 kcal mol^{-1}.

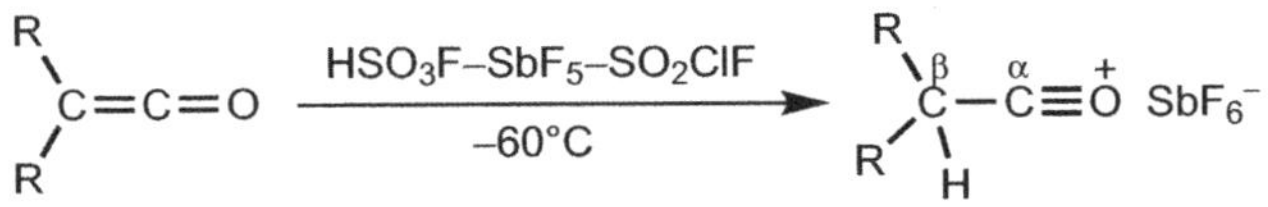

346 (3.97)

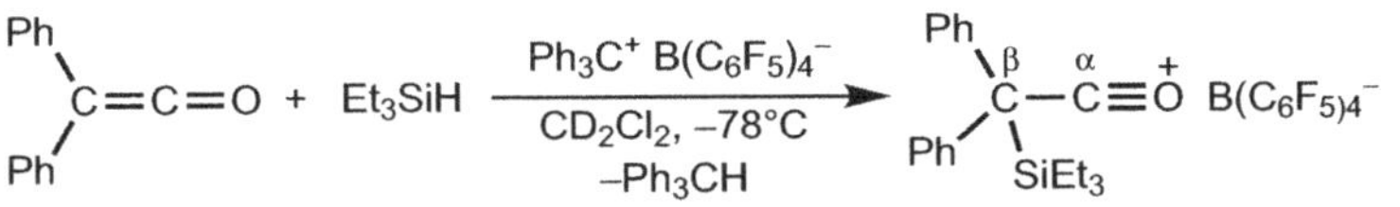

347 (3.98)

3.4.16.3. Sulfur as Heteroatom.

Thiols and sulfides are protonated on sulfur in superacid media and give mono- and dialkylsulfonium ions, respectively.[136] Thiocarboxylic acids, *S*-alkyl esters, thioesters, dithioesters, and thiocarbonates in similar media also form stable protonated ions[541,647] such as cations **348–353**.

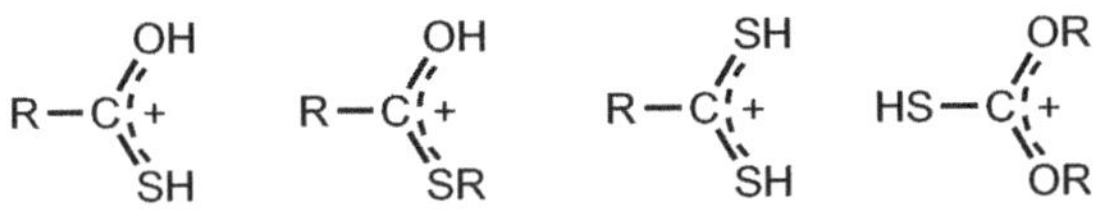

348 **349** **350** **351**

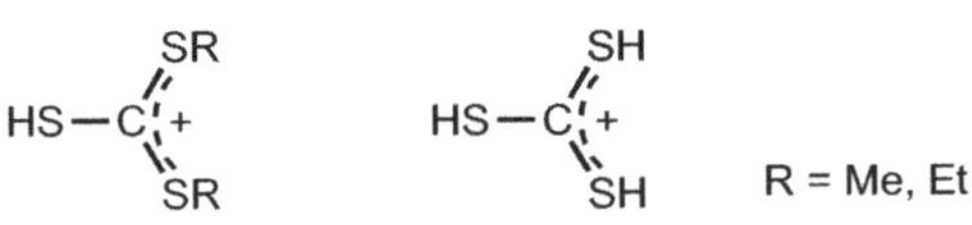

352 **353**

Minkwitz and co-workers have recently obtained the X-ray crystal structure of protonated thiocarbonic acid, protonated *O*,*S*-dimethyl dithiocarbonate, and protonated dimethyl trithiocarbonate. Protonated thiocarbonic acid **354** has C–S bond lengths of 1.683–1.708 Å.[662] The small differences imply a nonequal distribution of the positive charge over the S atoms. All hydrogen atoms have *trans* positions and they are not located in the CS_3 plane, resulting in a C_1 symmetry. The cations have contacts with the F atoms of the anions through H bonds with bond distances smaller than the sum of the van der Waals radii.

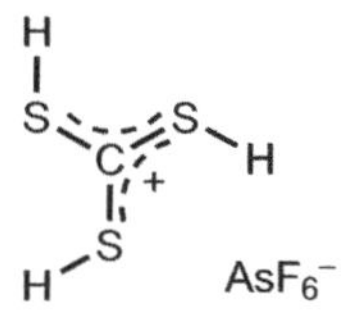

354

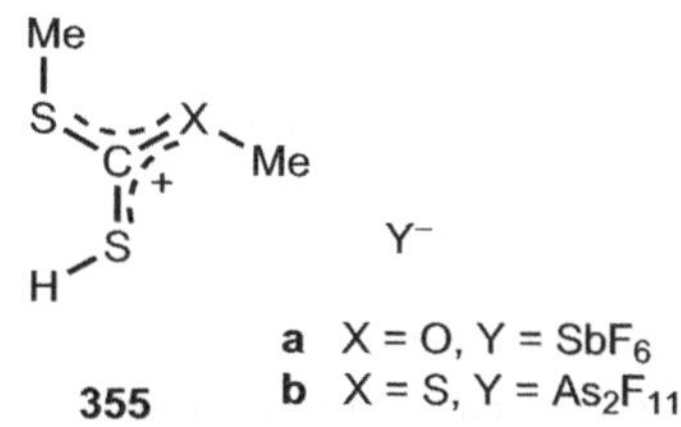

355

The C–S lengths (1.692 and 1.702 Å) and the C–O bond length (1.293 Å) of the protonated *O,S*-dimethyl dithiocarbonate cation **355a** are between those of typical single and double bonds.[663] A planar sp^2 hybridization is indicated by the sum of 360° of the corresponding angles around the central carbon atom with the methyl groups having out-of-plane arrangements. Two cations make contact via hydrogen bonds with two anions in the solid state. Strong anion–cation interactions were also detected in the protonated dimethyl trithiocarbonate salt **355b.**[664] The central carbon is planar and has one shorter C–S(H) bond (1.722 Å) and two longer C–S (Me) bonds (1.681 and 1.682 Å). The H and the methyl substituents are slightly out of the CS_3 plane.

Sulfur-stabilized heteroaromatic species such as **356** and **357** are also known.[566]

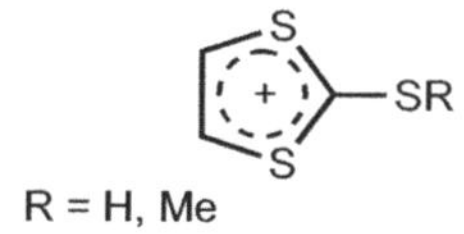

356

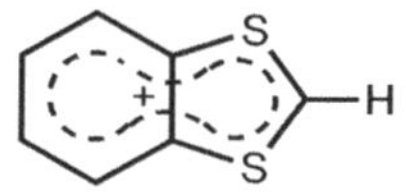

357

A series of thioketones have also been protonated to form carbosulfonium ions **358**[541,542] and studied by ^{13}C NMR spectroscopy.[665] It was found that deshielding of the thiocarbonyl carbon is much less pronounced than that of ketones. This shows that there is significant mercaptocarbenium **358b** contribution to the overall protonated thioketone structure. However, this is not the case in the case of protonated ketones.

358

Olah and coworkers have used *ab initio* methods (MP2/6-311+G** level) to calculate the structures of diprotonated superelectrophilic[3–5] formaldehyde and thioformaldehyde dications.[666] In both cases, the heteroatom-diprotonated structure

359 was found to be more stable (by 53.5 kcal mol^{-1} and 27.6 kcal mol^{-1}, respectively) than the 3*c*–2*e* bonded structure **360**.

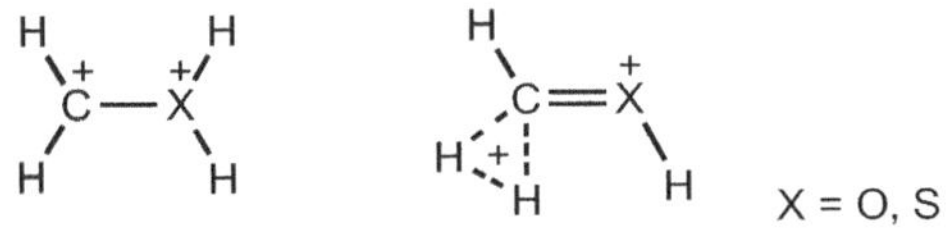

359 **360**

Tertiary carbosulfonium ions may be generated using the methods applied for the synthesis of tertiary carboxonium ions. The alkylation of thiocarbonyl compounds, preferably by trialkyloxonium ions,[667] is the most widely used procedure. Dehalogenation of α-halothioethers[668] [Eq. (3.99)] and hydride abstraction with trityl salts are also practiced[669] [Eq. (3.100)]. Minkwitz and Meckstroth[670] have obtained the $Cl_2C{=}SCl^+$ and $Cl_2C{=}SBr^+$ cations by the oxidative halogenation of thiophosgene in AsF_5 solution.

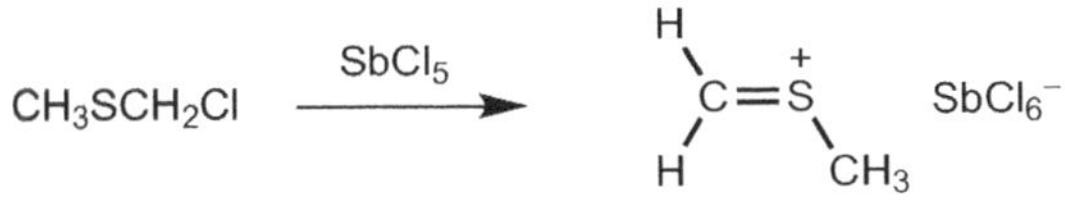

(3.99)

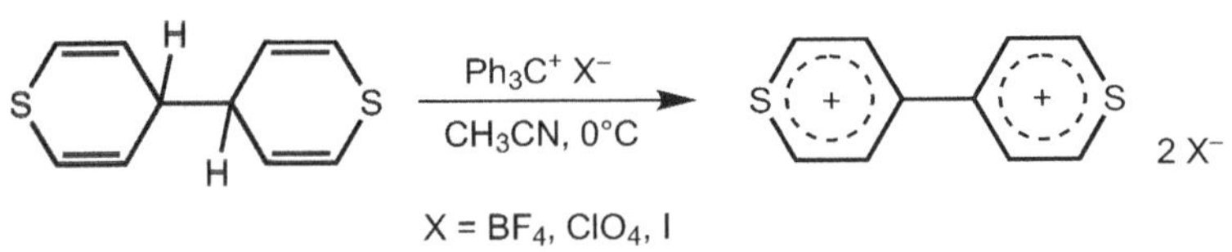

(3.100)

The cleavage of the sulfur–carbon bond of thioacetals and thioketals,[671] thioorthoformates[672] [Eq. (3.101)], and tetraalkyl thioorthocarbonates[673] [Eq. (3.102)] leads to the corresponding mono-, di-, and trithiocarbenium ions, respectively. Trimethylthiocarbenium ion **361** was also prepared by methylating dimethyl trithiocarbonate [Eq. (3.102)].

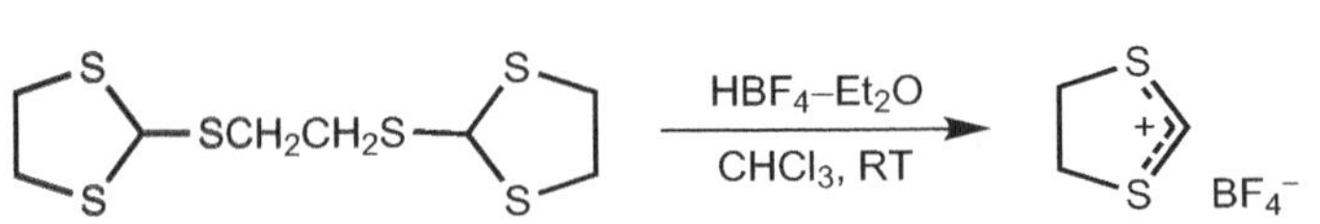

(3.101)

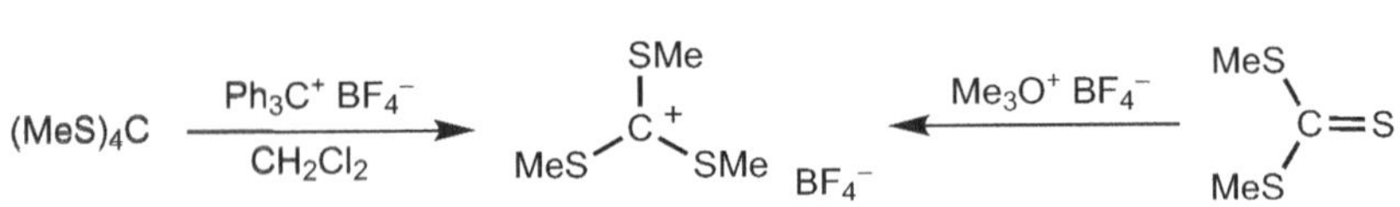

361

(3.102)

Miller et al.[674] generated the interesting long-lived dication **362** treating hexathia-1,3,5,7-tetramethyladamantane with acids (triflic acid, fuming H_2SO_4, or 25% Magic

Acid) at room temperature. The ^{13}C NMR chemical shift of dication **362** (δ^{13}C 230) closely matches that of the related cation **363** (δ^{13}C 229).[675]

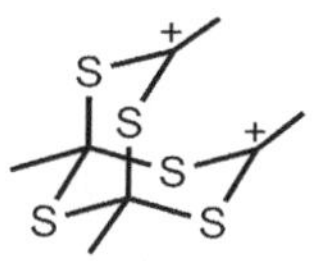

362

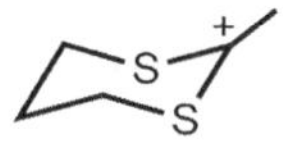

363

A series of thiobenzoyl cations **364** have been prepared and studied[676] using a metathetic silver salt reaction [Eq. (3.103)].

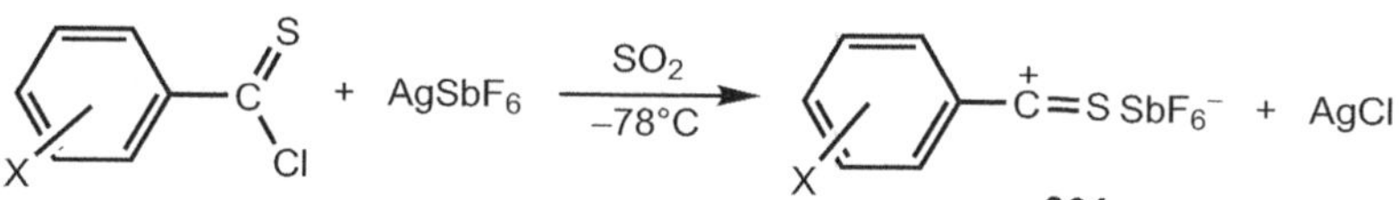

(3.103)

X = H, 2-Me, 4-Me, 4-MeO, 4-Cl, 4-F, 4-CF_3,

A controversial issue of heteroatom-stabilized cations is the relative stabilization of carbocationic centers adjacent to oxygen and sulfur.[541] In solution studies, α-*O*-substituted carbocations were found to be stabilized more than α-*S*-substituted carbocations.[677] Gas-phase studies reached an opposite conclusion,[678,679] whereas subsequent theoretical studies (high-level *ab initio* methods) supported the findings of solution chemistry. Recent results, namely, basicities of various vinylic compounds (**365**–**370**) measured in the gas phase also support this conclusion.[680] Although mono-heteroatom-substituted compounds **365** and **366** were found to have similar proton affinities, an additional α-methyl group increased the stability of the carbenium ion derived from **367** more than that of the sulfur counterpart **368**. Even larger differences were found between proton affinities of the bis-heteroatom-substituted compounds **369** and **370**.

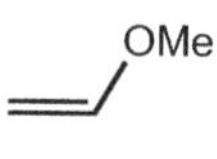

365

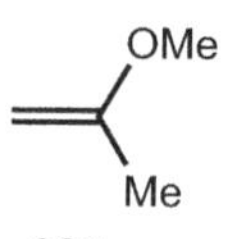

367

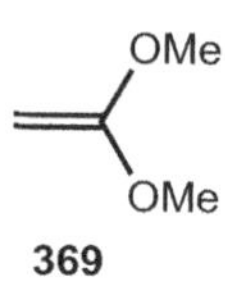

369

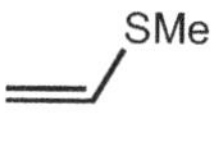

366

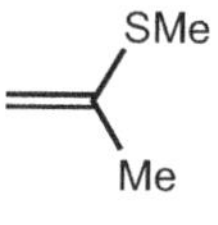

368

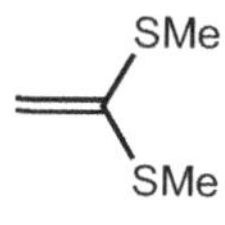

370

3.4.16.4. Nitrogen as Heteroatom. Amides are protonated on the carbonyl oxygen atom (cation **371**) in superacid media at low temperatures, as shown first by Gillespie and Birchall[681] [Eq. (3.104)]. The same conclusion has been arrived at in a recent multinuclear NMR characterization of the cation $CF_3C(OH)NH_2^+$.[682]

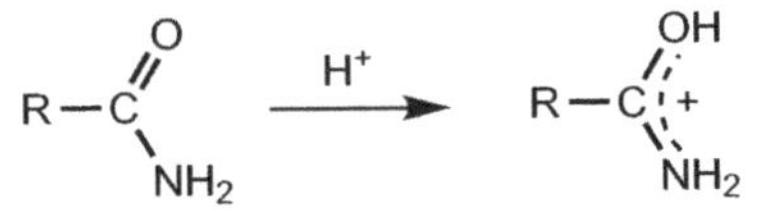

(3.104)

371

It was claimed that protonation of ethyl *N,N*-diisopropyl carbamate, a hindered amide, takes place on nitrogen and not on oxygen.[683] A reinvestigation, however, established that at low temperature initial *O*-protonation takes place (kinetic control) with the *O*-protonated amide subsequently rearranging to the more stable *N*-protonated form (thermodynamic control).[684] The possibility of observing the protonated amide linkage in strong acid media has particular relevance in the study of peptides and proteins.[685,686] There is, however, a notable exception. Since the bridgehead nitrogen of 2,2,-dimethyl-6-oxoquinuclidine **372** is more basic than nitrogen of usual amides, and resonance stabilization of the *O*-protonated form is not possible, exclusive *N*-protonation takes place[687] (**373**) [Eq. (3.105)].

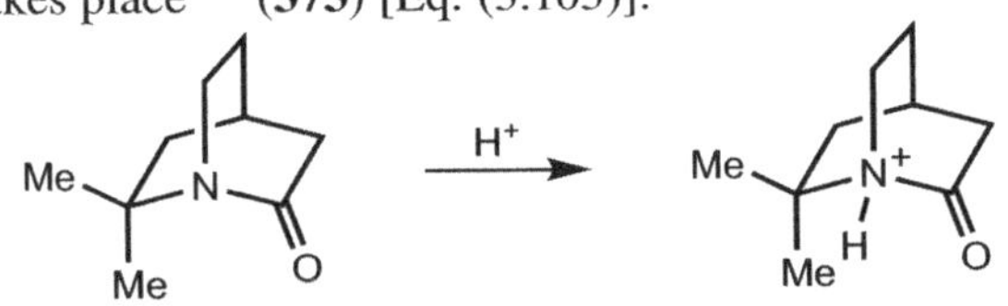

(3.105)

372 **373**

Protonation of aroyl azides and α-ketonitriles in superacid media ($HSO_3F–SbF_5$, −78°C) occurs exclusively on the oxygen to yield α-azidocarboxonium ions **374**[688] and α-cyanocarboxonium ions **375**,[689] respectively.

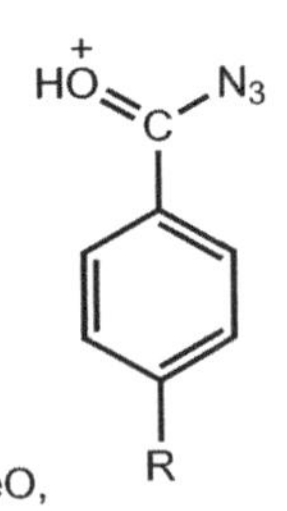

R = H, Me, MeO, F, Cl, Br

374

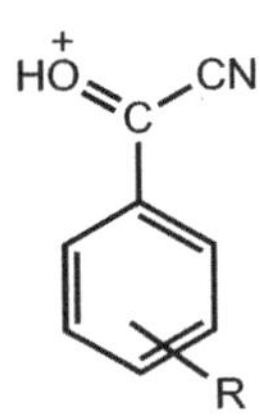

R = H, 2-Me, 3-Me, 4-Me, 4-MeO, 4-F, 4-Cl, 4-Br

375

The structures and ^{13}C and ^{15}N NMR chemical shifts of urea and its mono-, di-, and triprotonated forms have been calculated using high-level *ab initio* (MP2/6-311+G*) and IGLO methods.[690] Protonated urea (uronium ion) is a resonance-stabilized cation. The *O*-protonated structure uronium ion **376a** of C_1 symmetry was found to be more stable by 5.8 kcal mol^{-1} than the *N*-protonated isomer **376b**. The NMR spectra of urea solutions in H_2SO_4 and HNO_3 (0°C), as well as NMR spectra of $HSO_3F–SO_2ClF$

(−80°C), consist of a single peak (δ^{13}C ~160, δ^{15}N ~79). The equivalence of the ^{15}N NMR peak indicates that *O*- and *N*-protonated uronium ion forms undergo rapid exchange. Compared to chemical shifts of urea, the carbon in uronium ion is 1.5 ppm shielded and nitrogen is 3.2 ppm deshielded. These changes compare well with the IGLO calculated chemical shifts. Of the diprotonated ureas, the *O,N*-diprotonated form **377** of C_s symmetry is more stable by 22.3 kcal mol^{-1} than the *O,O*-diprotonated form. The calculated bond distances—that is, a longer C–O, a longer C–NH_3 and a shorter C–NH_2 bond compared to urea—clearly indicate that one of the positive charges is localized on N, and the other charge is delocalized among O–C–N. This was already described by Olah and White[691] when obtaining the ^{1}H and ^{15}N NMR spectra of ion **377**. In the case of triprotonated urea, which was not observed experimentally, the preferred structure is the *O,N,N*-triprotonated form **378** (X = O). This, however, is only 1.8 kcal mol^{-1} more stable than the *O,O,N*-triprotonated form.

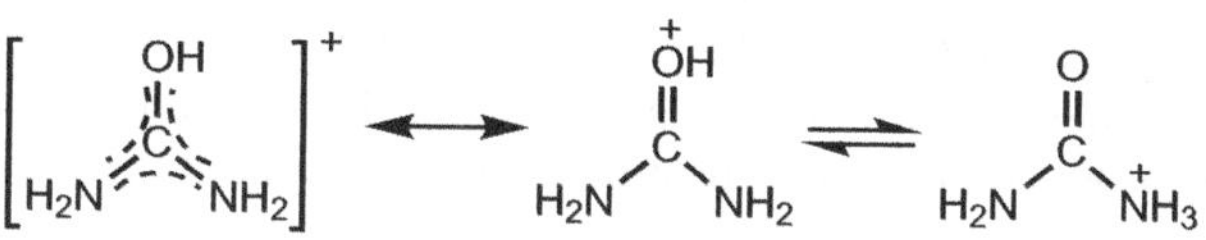

a b

376

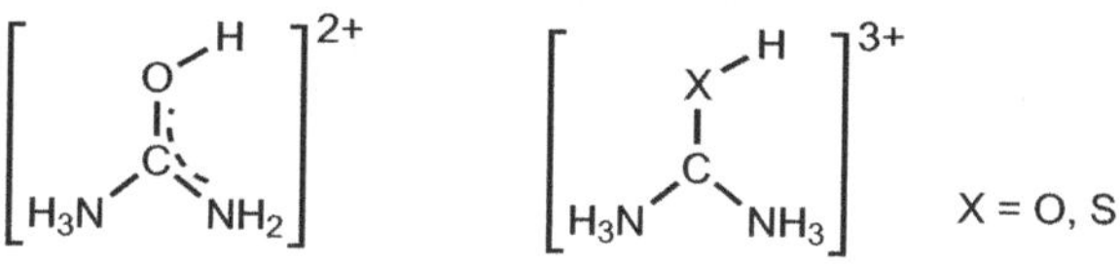

377 **378**

Similar studies have also been performed with thiourea.[692] Thiourea is monoprotonated in HSO_3F–SO_2ClF solution, and diprotonation occurs upon addition of SbF_5 [Eq. (3.106)]. Monoprotonation was calculated to be exothermic by 214.2 kcal mol^{-1}. Further protonation of **379** to form **380**, again, is exothermic by 60.1 kcal mol^{-1}. The changes in bond distances upon diprotonation of thiourea show similar tendencies found for diprotonation of urea, indicating a similar charge localization/delocalized pattern. Both cations were isolated as the AsF_6^- salts and characterized by low-temperature Raman spectroscopy (−110°C). With respect to the calculated structures (B3LYP/6-31G* level), the conclusions are very similar to those found for urea. Structure **379** of C_s symmetry is 20.0 kcal mol^{-1} more stable than the *N*-protonated thiouronium ion. This observation is in agreement with earlier X-ray structure analysis of thiouronium nitrate.[693] The *S,N*-diprotonated form **380** (C_s symmetry), the only structure observed by NMR spectroscopy, is the lowest in energy and 8.3 kcal mol^{-1} more stable than the *S,S*-diprotonated form. The preferred structure of triprotonated thiourea, not observed experimentally, is the *S,N,N*-triprotonated form **378** (X = S) being only 0.9 kcal mol^{-1} more preferred than the *S,S,N*-triprotonated form. However, including zero-point vibrational energy makes the latter to be more stable by 1.1 kcal mol^{-1}.

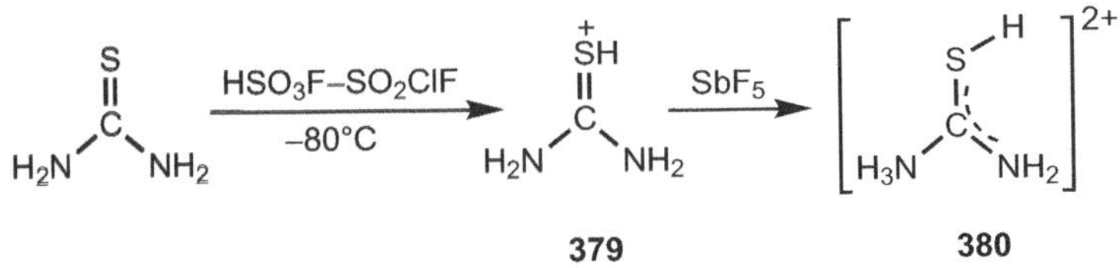

(3.106)

Mono-*O*-protonated carbamic acid **381** has been obtained by Olah, Prakash, and co-workers[694] by protolytic ionization of *tert*-butyl carbamate [Eq. (3.107)]. The NMR chemical shifts observed in $HSO_3F–SO_2ClF$ solution are at δ^1H 6.41 (2H,$J_{N–H}$ = 96.5 Hz) and $\delta^{15}N$ 69.5 (triplet, $J_{C–N}$ = 97.1 Hz), representing the NH_2 group, and at $\delta^{13}C$ 162.4 (doublet, $J_{C–N}$ = 28.4 Hz), representing the cationic carbon. This latter peak is 5.3 ppm deshielded relative to the carbonyl carbon of *tert*-butyl carbamate. The GIAO-MP2-calculated NMR shift values deviate significantly from the experimentally observed values. *O*-Protonated carbamic acid was calculated to be more stable than *N*-protonated carbamic acid (MP2/6-31G* level), by only 3.3 kcal mol^{-1}.

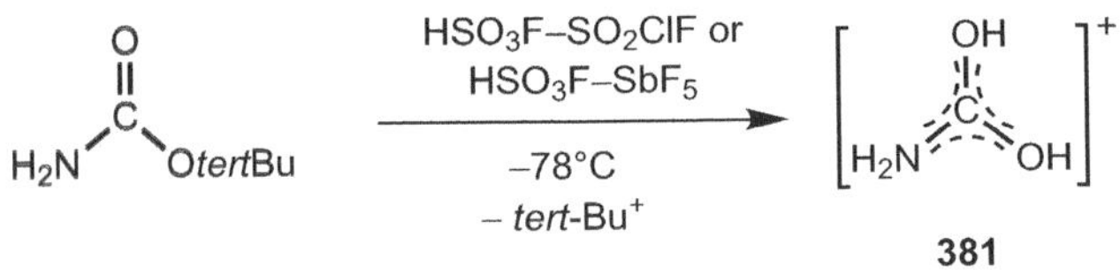

(3.107)

Further protonation of protonated carbamic acid to form isomeric diprotonated carbamic acids does not take place. Dication **382a** was calculated[694] to have a longer C–N bond compared to protonated carbamic acid (1.478 versus 1.312 Å). It is indicative of one positive charge localized on the NH_3 group, and the second charge is delocalized on the O–C–O unit. In contrast, dication **382b** is characterized by a shorter C–N and a longer C–O bond relative to protonated carbamic acid (1.288 versus 1.312 Å and 1.460 versus an average 1.303 Å, respectively) revealing a localized positive charge on oxygen and delocalization of the other positive charge on the O–C–N unit. Of the two isomeric dications, the *N*-protonated structure **382a** was calculated by 17.5 kcal mol^{-1} to be more stable than dication **382b**.

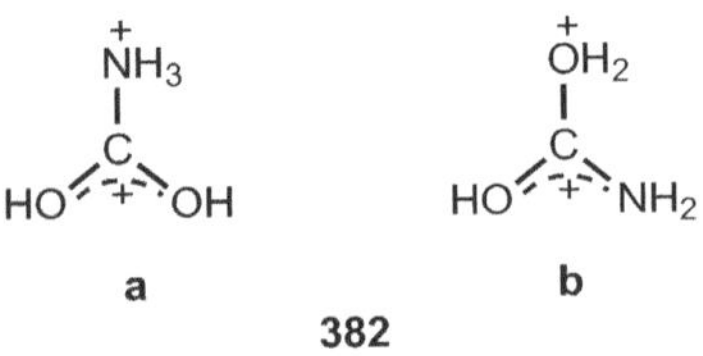

Diprotonated *N,N*-bis(carboxyl)-1,2-diaminoethane **383** was also generated by protolytic ionization[694] [Eq. (3.108)]. The carbocationic carbon resonance found at

$\delta^{13}C$ 161.5 in $HSO_3F–SO_2ClF$ solution is only 0.9 ppm shielded from that of protonated carbamic acid.

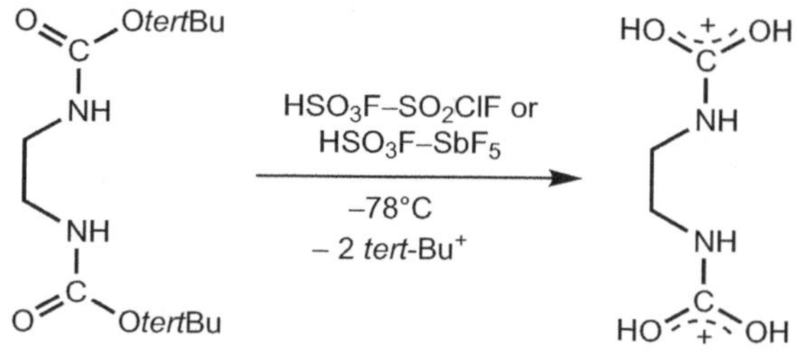

383

(3.108)

Protonation of ethyl *N*-methyl carbamate was also performed in $HSO_3F–SO_2ClF$ and $HSO_3F–SbF_5$ solutions. Due to the hindered rotation about the C–N bond, two rotamers are formed[694] [Eq. (3.109)]. The ^{13}C NMR spectra of the $HSO_3F–SO_2ClF$ solution showed resonances at $\delta^{13}C$ 160.4 assigned to the major *trans* isomer (**384a**) and at $\delta^{13}C$ 160.7 assigned to the minor *cis* isomer (**384b**). These resonances are shielded by 2.0 and 1.7 ppm, respectively, from that of protonated carbamic acid. Isomer **384a** was found to be more stable by 5.4 kcal mol^{-1} (MP2/6-31G* level). Again, the persistent diprotonated ethyl *N*-methyl carbamate dication was not detected. Likewise, only the *O*-monoprotonated cation was identified in the protonation of methyl carbamate.

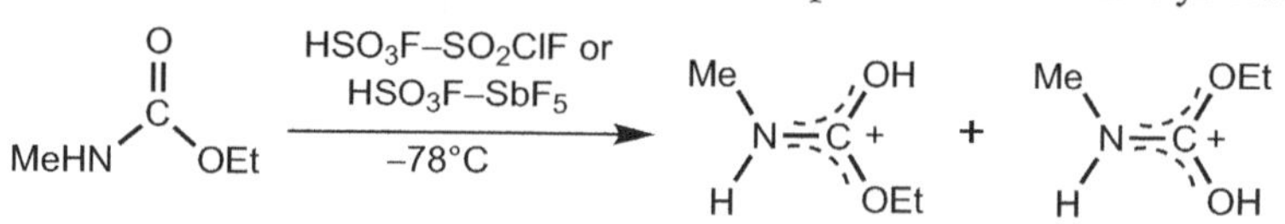

a b

384

(3.109)

Since nitrogen is a better electron donor than oxygen, in cationic derivatives of imines the contribution of the aminomethyl cation structure **385b**, compared with that of the iminium ion resonance form **385a**, is small.[691,695] The same also applies to amidines (**386**) and guanidines (**387**).

a b

385

R = H, alkyl, aryl, acyl

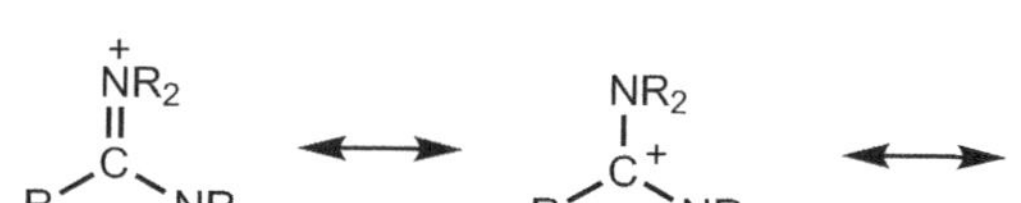

386

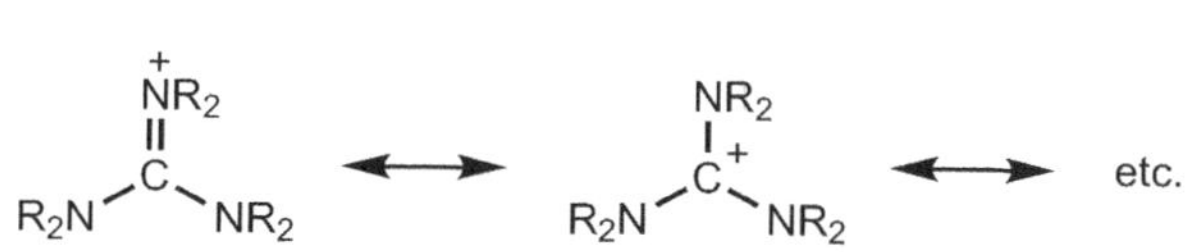

387

Acidic iminium ions (**385**, R = H) are usually prepared by the protonation of compounds with C–N double bond such as imines[542,696] [Eq. (3.110)] and ketoximes.[542] The nonequivalence of the two *C*-methyl groups below –20°C indicated that in cation **388** the rotation about the C=N bond is slow on the NMR time scale.[696]

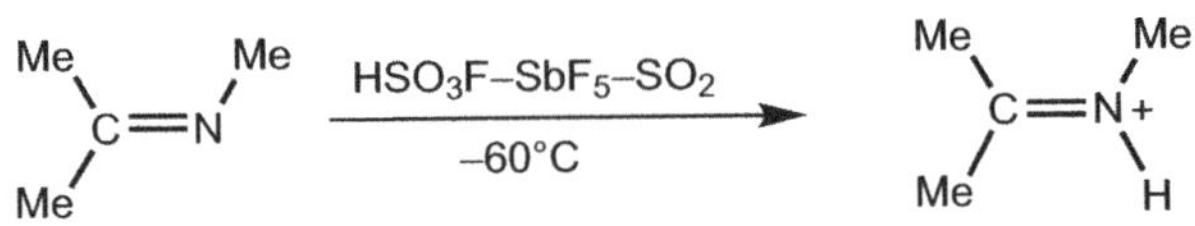

(3.110)

388

Olah and co-workers[666] also reported the calculated structures of protonated methyleniminium cations using *ab initio* methods (MP2/6-311+G** level). Both the *N*-protonated (**389**) and *C*-protonated (**390**) dications are minima on the potential energy surface, but the NH_3-substituted methyl cation **389** is more stable than the 3*c*–2*e* bonded structure **390**, by 27.6 kcal mol^{-1}. Structure **389** was detected by charge stripping mass spectrometry by Schwarz and co-workers.[697] Structures and energies of dications derived by protonation of methyl- and dimethyliminium ions have recently been calculated (MP2/6-311+G** level).[698] Both the *C*-protonated and *N*-protonated dications were identified as minimum structure. However, the *N*-protonated isopropyleniminium ion **391** (carbenium ammonium dication) was found to be more stable by 20.8 kcal mol^{-1}, even though the formal cationic centers are adjacent.

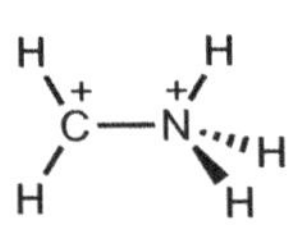

389

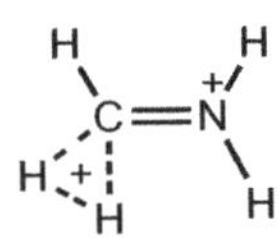

390

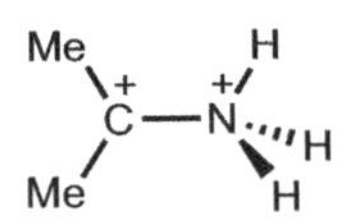

391

Minkwitz et al.[699] have prepared dichloromethyleneiminum salts **392** [Eq. (3.111)] and made vibrational and NMR spectroscopic characterization. They have also reported the X-ray crystal structure of the $Cl_2C{=}NH_2^+$ and $Cl_2C{=}NHCCl_3^+$ hexachloroantimonates. In the solid state the $Cl_2C{=}NHCCl_3^+$ cation and the $SbCl_6^-$ anion have one short H···Cl contact (2.313 Å).[700] The cation framework is planar with C–N bond lengths of 1.488 and 1.280 Å. The latter value compares well with the bond distance of 1.268 Å in cation $Cl_2C{=}NH_2^+$.[699]

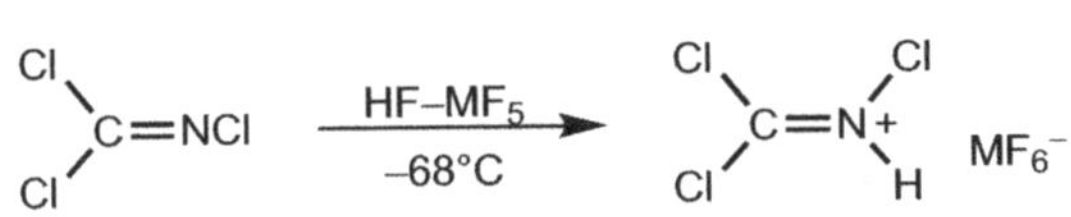

(3.111)

M = As, Sb

392

A number of methods, including alkylation,[701,702] acylation,[703] and halogenation,[704] are available for the synthesis of nonacidic iminium ions. *N*-Trifluoromethyl-substituted iminium ions have been obtained by Minkwitz and Lamek[704] using a variety of approaches (Scheme 3.12).

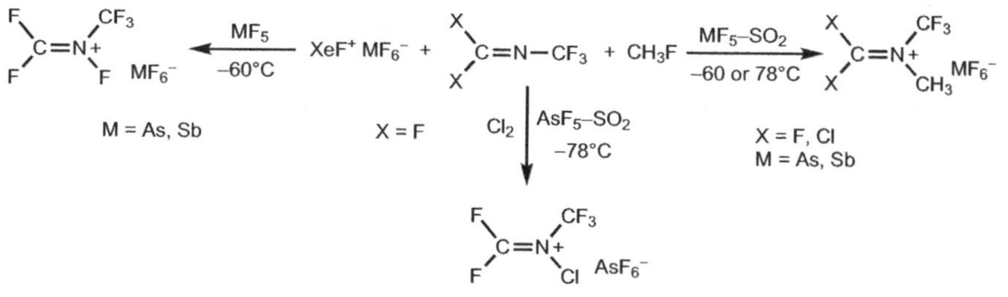

Scheme 3.12

^1H, ^{13}C, ^{15}N NMR and *ab initio*/IGLO/GIAO-MP2 study of mono-, di-, tri-, and tetraprotonated guanidine have been performed by Olah, Prakash, and co-workers.[705] Guanidinium ion **393** prepared according to a literature procedure[706] [Eq. (3.112)] and isolated as the sulfate salt shows a broad peak in its ^1H NMR spectrum (δ^1H 6.6), a peak in its ^{13}C NMR spectrum ($\delta^{13}C$ ~157), and a peak in its ^{15}N NMR spectrum ($\delta^{15}N$~74). In agreement with previous calculations,[707] guanidinium ion has a propeller-shaped structure of C_3 symmetry (MP2/6-31G* level). The equivalence of the ^{15}N NMR peaks indicates that no further protonation of ion **393** to form long-lived diprotonated guanidinium dication occurs at these acidities (H_2SO_4 and CF_3SO_3H at room temperature, and HSO_3F–SO_2ClF at −40°C). Cation **393** is, however, readily further protonated to diprotonated guanidinium dication **394** in substantially more acidic Magic Acid solution [Eq. (3.112)].

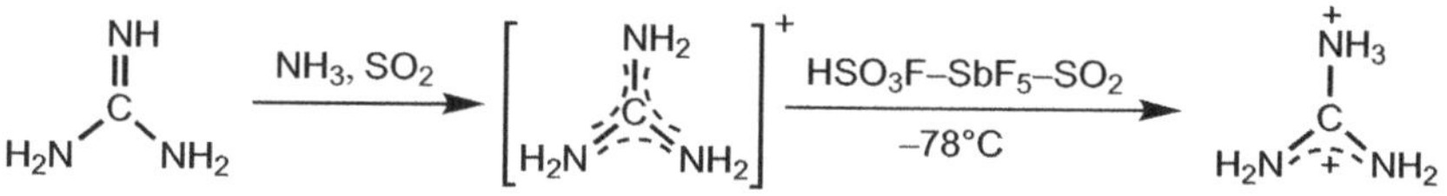

393 **394**

(3.112)

In its ^1H NMR spectrum, dication **394** exhibits an NH_3^+ peak (δ^1H 6.76) and two peaks at δ^1H 6.32 and 6.09 in agreement with data previously reported by Olah and White.[691] The ^{13}C NMR spectrum consists of a peak centered at $\delta^{13}C$ 148.9, and it is shielded by 8 ppm from that of the guanidinium cation. The calculated bond distances—that is, a longer C–NH_3 and a shorter C–NH_2 bond compared to guanidinium ion—indicate that one of the positive charges is localized on NH_3 group, and the second charge is delocalized among H_2N–C–NH_2. Previous calculations on diprotonated urea[690] and thiourea[692] showed a similar pattern of charge distribution. No further protonation of the dication could be achieved. The calculated structure of triprotonated guanidine indicates that two of the positive charges are localized on two NH_3 groups and the third charge is delocalized over the amiocarbenium moiety (**395**) resembling an ammonium–iminium structure. Formation of trication **395** is endothermic by 70.1 kcal mol^{-1}. GIAO-MP2 calculations indicate a deshielding effect of 6.3 in the ^{13}C NMR chemical shift of the trication compared to the dication

(δ^{13}C 155.3 versus 149). Deshielding was also found in the ^{15}N NMR chemical shifts. Structure **396** is the calculated structure of tetraprotonated guanidine. The four localized positive charges are, obviously, highly unfavorable.

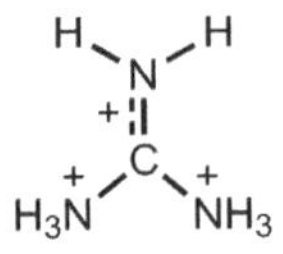

395

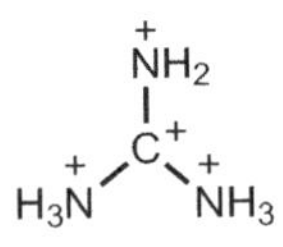

396

Even diazomethane has been protonated[708] in the superacid media. In Magic Acid media, both methyldiazonium ion **397** as well as *N*-protonated diazomethane **398** are formed [Eq. (3.113)].

$$CH_2N_2 + HSO_3F\text{–}SbF_5 \xrightarrow[-120°C]{} \underset{\mathbf{397}}{CH_3N_2^+} + \underset{\mathbf{398}}{CH_2{=}\overset{+}{N}{=}NH} \qquad (3.113)$$

A series of 2-azaallenium ions have been prepared and characterized including X-ray crystal structures.[709,710] Charge delocalization and, consequently, structural features depend significantly on substituents. In many cases, the contribution of the allenium ion form **399a** is more significant.

a **b**

399

Meerwein et al.[711] reported *N*-alkylnitrilium salts and Olah and Kiovsky[712] have studied protonated alkyl nitriles and hydrogen cyanide generated in $HSO_3F–SbF_5$ solution at low temperatures. ^{13}C and ^{15}N NMR characterization of the cations **400** showed that the nitrile carbon is *sp*-hybridized and, therefore, nitrilium ions are linear species. This conclusion is supported by theoretical studies[713] and X-ray crystal structure[714] of nitrilium ion **401**, indicating linear arrangement and triple-bond character of the C–N bond.

R = H, Me, Et, $MeOCH_2CH_2$, $NCCH_2$, MeCO
CH_2=CH, PhCH=CH, NCCH=CH

400

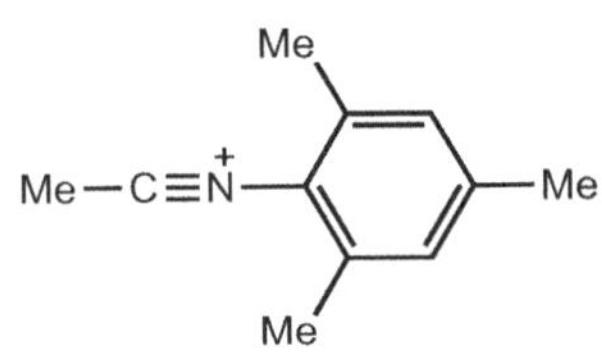

401

A variety of dications have recently been prepared and characterized by X-ray diffraction. The tetrakis(dimethylamino)ethylene dication **402** has been generated by two-electron oxidation of tetrakis(dimethylamino)ethylene[715] [Eq. (3.114)]. Theoretical calculations for the ethylene dication predict D_{2d} geometry.[716,717] In agreement with this prediction, X-ray characterization of dication **402** showed that the two halves of the ion rotate by up to 76°. Furthermore, the planar Me_2N groups are twisted out of the two CCN_2 planes. The C–C bond lengthens (1.51 Å), whereas the C–N bonds significantly shorten (1.30–1.33 Å).

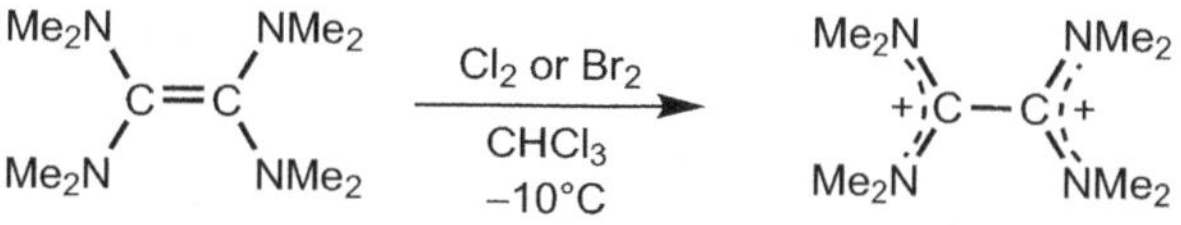

402

(3.114)

Cationic formamidinium derivatives have also been prepared and characterized.[718,719] The 1,1,3,3-tetrakis(alkylamino)allyl cation **403** formed according to Eq. (3.115) gives dication **404** after protonation. In cation **403**, both C–C and C–N bonds are significantly reduced compared with the corresponding single bonds. One of the C–N bonds is significantly longer than the others (1.41 Å versus 1.34 Å). This is presumably due to the unfavorable steric arrangement of one of the attached *tert*-butyl groups resulting in less favorable delocalization. The allyl bonds are, consequently, nonsymmetric (1.38 and 1.42 Å). The C–C bonds of dication **404** are considerably longer (1.50 and 1.52 Å), the C–C–C angle is close to tetrahedral (112.5°), and the CCN_2 planes are inclined at an angle of 82° compared with 28° in cation **403**. These changes indicate that delocalization is confined to the outer atoms. The barrier of rotation about the amidine C–N bond was estimated from variable temperature NMR data to be ~18 kcal mol^{-1}.

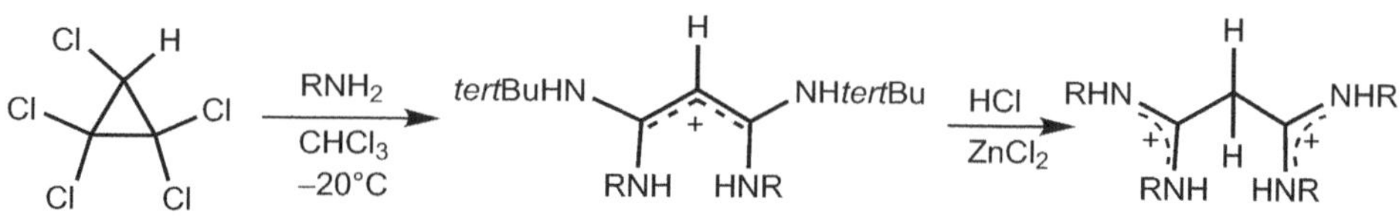

403 **404**

(3.115)

A series of diprotonated bisguanidinium dications **405** have been synthesized and characterized.[720,721] X-ray crystal structure analysis shows that the C=N and C–NMe_2 bonds have very similar bond lengths (1.326–1.341 Å and 1.331–1.343 Å, respectively). This indicates that each guanidinium moiety is highly symmetrical because of efficient charge delocalization. Crystal packing forces, in turn, result in significant differences in N–C–C–C torsion angles and angles between the planes of guanidinium units.

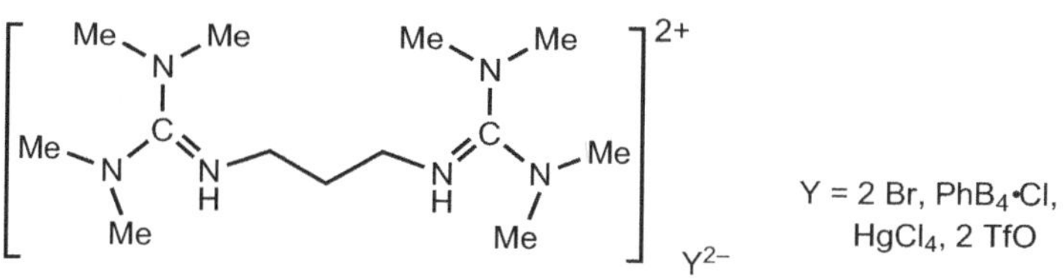

405

3.4.17. Carbocations Complexed to Metal Atoms

Organometallic cations, in which an organic ligand is coordinated to a metal atom bearing a unit positive charge, constitute a significant class of compounds.[722–726] Most common of these are the π-allylic **406** [Eq. (3.116)] and **407**, π-dienyl **408**, π-cycloheptatrienyl **409**, and π-cyclooctatetraenyl **410** systems.[722] Generally, metal atoms of low oxidation states are involved in complexing the electron-deficient ligands. The usual metals involved are transition metals such as Fe, Cr, Co, Rh, Ir, Pd, and Pt.

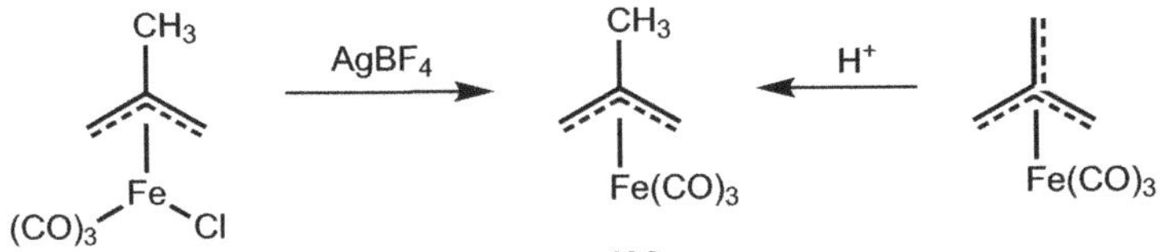

406

(3.116)

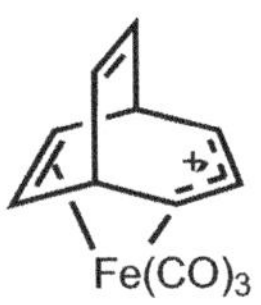

407

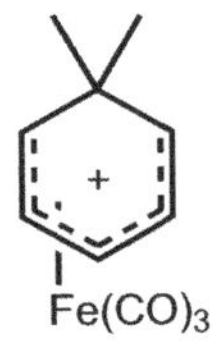

408

409

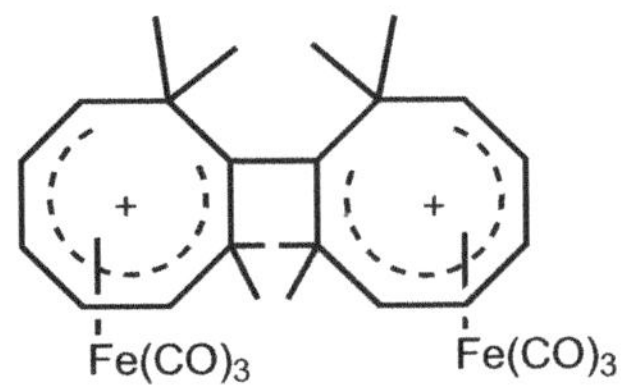

410

Since their discovery, cobalt and, in particular, cationic dicobalt propargyl complexes have played an important role in organic synthesis[725–727] The X-ray structure of such a complex has recently been reported.[728] Similar heterobimetallic complexes (**411**) have also been synthesized. IR and NMR spectroscopic data suggest that in complex **412** the stabilized $HC{\equiv}CCMe_2^+$ ions are π-coordinated to Mo or W atoms and not to nickel.[729] A recent example is the α-CF_3-substituted **413** cation stabilized by the bimetallic Co–Co cluster.[730] Unlike the uncomplexed 2-alkynylbornyl cations, bimetallic (Mo,Mo and Co,Mo) 2-propynylbornyl cation complexes related to cations **413** were found not to undergo Wagner–Meerwein rearrangement.[731] According to X-ray data, the 2-bornyl cation leans toward Mo in the mixed-metal Mo,Co complex. The synthesis and X-ray structure of a dicationic dimolybdenum complex was also reported.[732]

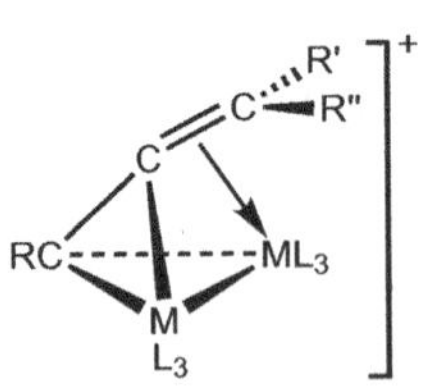

M = Co, Mo, W
L = CO, C_5H_5

411

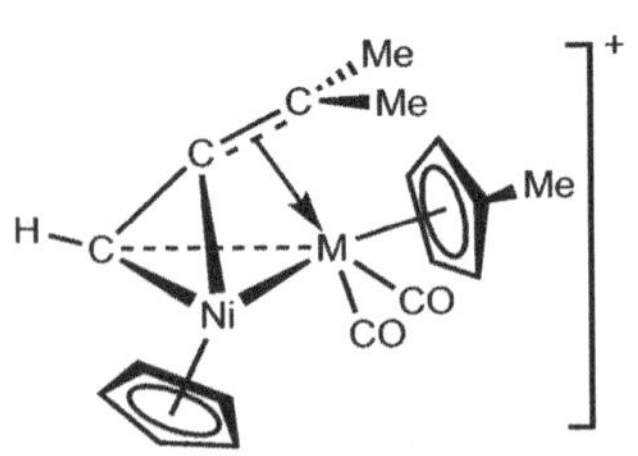

M = Mo, W

412

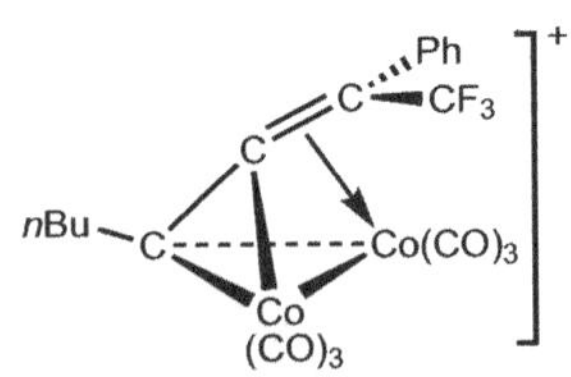

413

$Cr(CO)_3$ is also known to strongly stabilize carbocations. Theoretical studies for benzyl cations are abundant.[733] Hoffmann and co-workers[734] proposed that conformation **414a** would be more stable by approximately 6.8 kcal mol^{-1} than its rotamer. According to recent computations (B3LYP), however, the energy barrier is negligible (0.2 kcal mol^{-1}).[735] An interesting feature of complex **414** is that the benzylic carbon bends down to coordinate to Co; that is, the contribution of resonance form **414b** is significant.[735–737] Consequently, the complex may be better viewed as a pentadienyl cation with an exocyclic double bond. The bending angle, however, decreases with increasing substitution at the benzylic carbon.

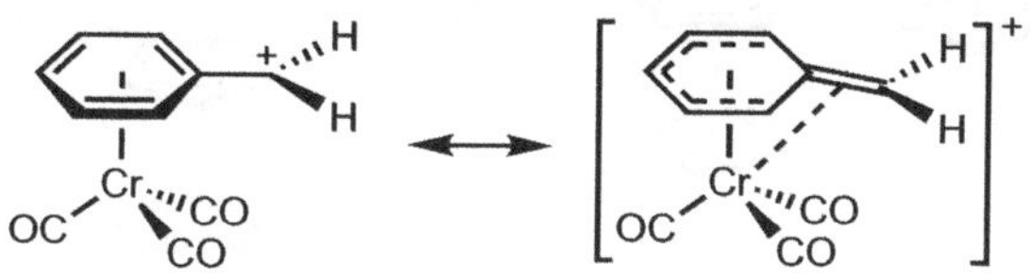

a **b**

414

Harman and co-workers[738] reported the synthesis of arenium cations stabilized by Os complexation (for example, **415**) via protonation with triflic acid of $[Os(NH_3)_5(Ar)](OTf)_2$ complexes (Ar = benzene, toluene, xylenes, naphthalene, anthracene).

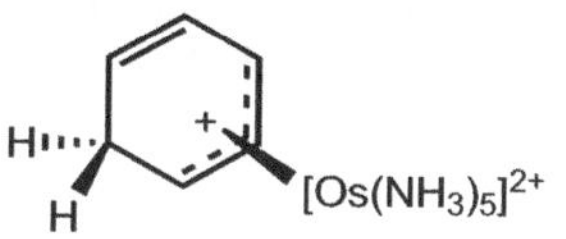

415

There are also several carbocations that have an α-π-complexed organometallic system. Cation **416** was easily prepared as the hexachloroantimonate salt[739] [Eq. (3.117)].

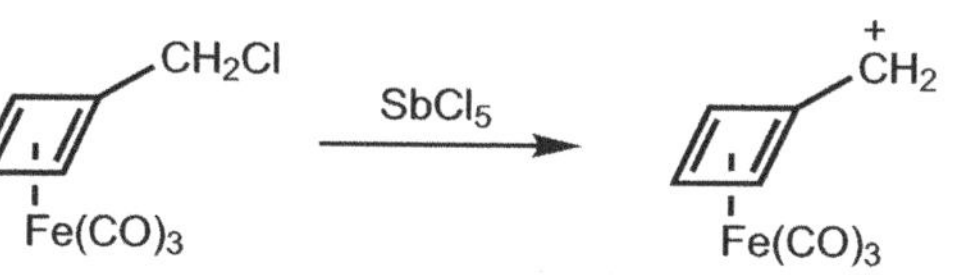

(3.117)

416

The most notable of these ions are α-ferrocenyl carbenium ions such as **417**, **418**, and **419**. As shown, cation **417** has a transoid conformation with a $C–C^+–C$ bond angle of 127°.[740] The ferrocenyl rings are bent at the carbocation center to reduce the distance between the Fe atoms and the carbocation carbon. ^{13}C NMR data for cation **418** indicate significant charge delocalization attributed to either *p*–π conjugation resulting in partial double bond character of the C^+–ferrocenyl bond or direct C^+–Fe interaction.[741] Allylic cation **419** was generated by protonation with HBF_4 of the corresponding alcohol and was fully characterized.[742] X-ray structure analysis of the tetrafluoroborate salt has found that the 1,3-ferrocenyl units are twisted (9.7–21.9°) and the cyclopentadienyl rings are planar but not parallel. Furthermore, similar to other α-ferrocenyl carbenium ions, the ferrocenyl units tilt toward the cationic allyl moiety (2.8–4.9 Å), indicating a weak interaction (Fe–α-C^+ bond distances = 2.83 and 3.04 Å, α-C^+–ferrocene bond angles = 4.8° and 15.7°).

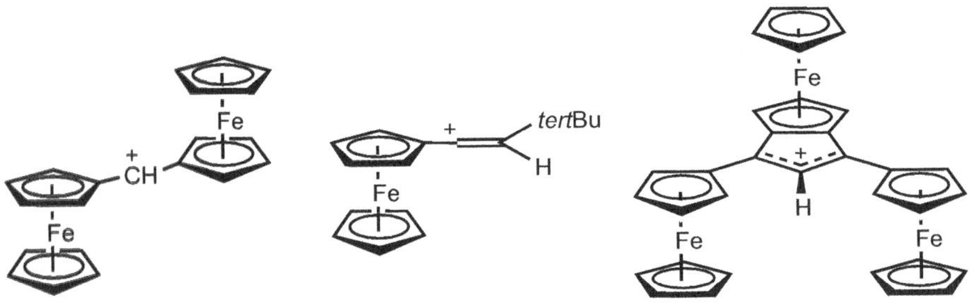

417 **418** **419**

As mentioned, the facile ring opening of the cyclopropyl cation **44** to the energetically more favorable allyl cation **45** prevented direct observation [see Eq. (3.29)]. Since the ferrocenyl group is a superstabilizing group for α-carbocationic centers, Olah, Prakash, and co-workers[743] ionized the **420** trimethylsilyl ether to find that upon heating, the intermediate protonated ether transformed to cation **421** [Eq. (3.118)]. The carbon signals of the methylene carbons of the cyclopropyl group (δ^{13}C 7.6) and the resonance of the carbocationic center at δ^{13}C 117.1 indicate extensive charge delocalization into the ferrocenyl moiety.

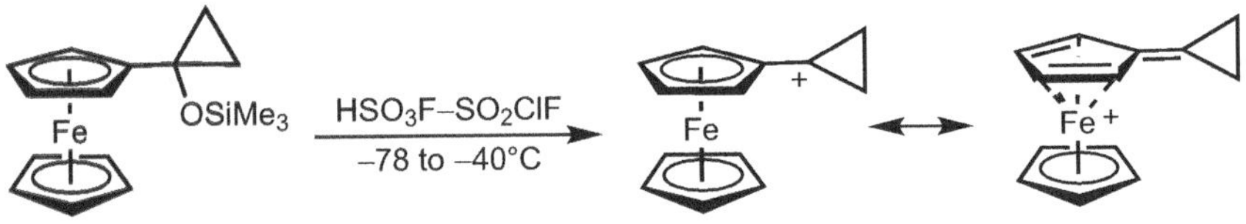

420 **421**

(3.118)

3.5. EQUILIBRATING (DEGENERATE) AND HIGHER (FIVE OR SIX) COORDINATE (NONCLASSICAL) CARBOCATIONS

3.5.1. Alkonium Ions (Protonated Alkanes $C_nH_{2n+3}^+$)

As recognized in the pioneering work of Meerwein, Ingold, and Whitmore,[10,12,18,19,744] trivalent alkyl cations ($C_nH_{2n+1}^+$) play important roles in the acid-catalyzed transformations of hydrocarbons as well as various electrophilic and Friedel–Crafts-type reactions. Trivalent alkyl cations can directly be formed by the ionization of lone-pair (nonbonded electron pair) containing precursors (*n*-bases) such as alkyl halides, alcohols, thiols, and so on, or by protonation of singlet carbenes or olefins.

Protonated alkanes ($C_nH_{2n+3}{}^+$) also play a significant role in alkane reactions. Saturated hydrocarbons can be protonated to alkonium ions, of which the methonium ion $CH_5{}^+$, **422** [Eq. (3.119)] is the parent, and formation of these pentacoordinate carbocations involves two-electron three-center (2*e*–3*c*) bonds. The dotted lines in the structure symbolize the bonding orbitals of the three-center bonds [Eq. (3.119)]; their point of junction does not represent an additional atom.

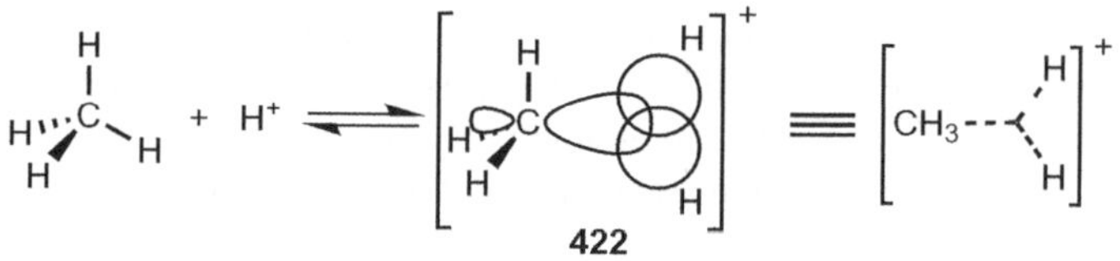

422 (3.119)

3.5.1.1. The Methonium Ion ($CH_5{}^+$). The existence of the methonium ion ($CH_5{}^+$, **422**)[745] was first indicated by mass spectrometric studies of methane[746] at relatively high source pressures—that is, molecular–ion reaction between a neutral CH_4 and a proton. Isotope exchange and collisional association in the reactions of $CH_3{}^+$ and its deuteriated analogs with H_2, HD, and D_2 have also been studied by mass spectrometry using a variable-temperature ion-flow method.[747] The chemistry of methane and homologous alkanes (e.g., hydrogen–deuterium exchange and varied electrophilic substitutions in superacidic media) pointed out the significance of alkonium ions in condensed state chemistry. The energy of dissociation of $CH_5{}^+$ to $CH_3{}^+$ and H_2 is known experimentally[748] to be $40\,\text{kcal mol}^{-1}$ and calculated as $34.7\,\text{kcal mol}^{-1}$ by *ab initio* MO theory [MP4(SDQ)/6-311G**+ZPE].[749]

Direct spectroscopic observation of $CH_5{}^+$ in the condensed state is difficult, because the concentration of the ion even in superacidic media at any time is extremely low. The matrices of superacids, such as $HSO_3F–SbF_5$ or $HF–SbF_5$ saturated with methane, were studied by ESCA[750] at $-180°C$, and the observed carbon 1*s* binding energy differing by less than 1 eV from that of methane is attributed to $CH_5{}^+$. Neutral methane has practically no solubility in the superacids at such low temperature of the experiment, and at the applied high vacuum (10^{-9} torr) it would be pumped out of the system. The relatively low 1*s* carbon binding energy in $CH_5{}^+$ is in good accord with theoretical calculations,[751] indicating that charge density is heavily on the hydrogen atoms and the five-coordinate carbon carries relatively little charge.

There are four possible structures of the methonium ion (C_s, C_{2v}, C_{4v}, or D_{3h} symmetry) (Figure 3.18). C_s has two structures: the eclipsed C_s(e) and the staggered

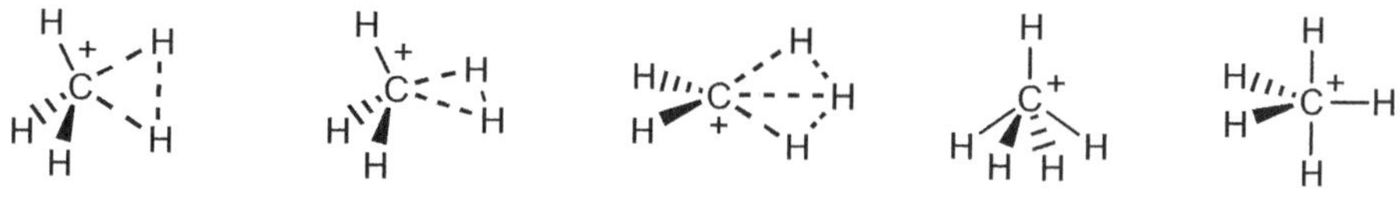

C_s(e) C_s(s) C_{2v} C_{4v} D_{3h}

Figure 3.18. Methonium ion structures of various symmetries.

C_s(s) conformation. The latter is the transition state for rotation of the H_2 moiety of $3c$–$2e$ bond. These structures resemble a complex between CH_3^+ and a hydrogen molecule, resulting in the formation of a $3c$–$2e$ bond. Of the possible structures, Olah, Klopman, and Schlosberg[751] suggested preference for the C_s front-side protonated form.

Preference for this form was based on consideration of the observed chemistry of methane in superacids (hydrogen–deuterium exchange and, more significantly, polycondensation indicating ease of cleavage to CH_3^+ and H_2) and also on the basis of self-consistent field (SCF) calculations.[751] More extensive calculations including *ab initio* methods[749,752–754] utilizing "all geometry" parameter search confirmed the favored structure of C_s symmetry. This structure is about 3.7 kcal mol^{-1} more stable than the structure of C_{4v} symmetry, which in turn is about 11.7 kcal mol^{-1} more stable than the trigonal bipyramidal D_{3h} symmetry structure.[749] Interconversion of stereoisomeric forms of CH_5^+ is obviously possible by a pseudorotation process. Muetterties suggested[755] that stereoisomerization processes of this type in pentacoordinated compounds could be termed "polytopal rearrangements." However, it is preferable to call intramolecular carbonium ion rearrangements as "bond-to-bond rearrangements" since these are not limited to equivalent bonds in the case of the higher homologs of CH_5^+ (see subsequent discussion).

More sophisticated quantum mechanical calculations have been performed on CH_5^+ during the last 20 years. The C_s symmetrical form is the energy minimum structure [MP2(FU)/6-31G** level of theory],[756] and frequency calculations at the same level showed that no other forms are minima on the potential energy surface. Structure C_{2v} was located as a transition structure for intramolecular hydrogen transfer. The C–H and H–H bond distances in the $3c$–$2e$ interactions [calculated at the MP2(FU)/6-311++G(2df,2pd) level] are 1.180 Å and 0.980 Å, respectively.[756] Single-point energies, however, showed that C_s is only 0.86 kcal mol^{-1} more stable than C_{2v} [QCISD(T)/6-311++G(3df,3pd)//MP2(FU)/6-311++G(2df,2pd) level]. After corrections for zero point vibrational energies, however, C_s and C_{2v} have essentially the same energies at absolute zero temperature. As a consequence, hydrogen scrambling in C_s, in accordance with experimental observations is a very facile process.

The C_s and C_{2v} structures have been reinvestigated by Schleyer and co-workers[757] at even higher-level *ab initio* theory. Changes in geometry were found to be very small, implying that the optimizations are essentially converged within theoretical limits. The differences in energies between the structures increasingly vanish at the most sophisticated levels and hydrogen scrambling becomes an essentially barrierless process. CH_5^+ was concluded to be a fluxional molecule and thus a rather unique species. Protonated methane is not the nonclassical carbocation prototype, but CH_5^+ is unique.[757–759] A more recent study has used quantum diffusion Monte Carlo techniques on an interpolated potential surface constructed from CCSD(T)/aug'-cc-pVTZ *ab initio* data.[760] The ground state of CH_5^+ was found to be significantly more symmetric than its global minimum energy structure. The zero-point motion of CH_5^+ renders all five protons equivalent in the ground state; that is, the ground state structure of CH_5^+ is more symmetric than the C_s global minimum energy configuration.

Marx and Parrinello performed an extensive *ab initio* electronic structure calculation (Car–Parrinello simulation[761]) also including the quantum effects and showed that a preference for the C_s quantum ground state does exist.[762] Furthermore, a small but definite barrier for hydrogen equilibration was found to exist. The preferred C_s symmetrical structure of CH_5^+ was also reconfirmed by Kutzelnigg and co-workers[763] [high-level CCSD(T)-R12 with large basis sets].

Attempts have been made to observe and experimentally determine the structure of CH_5^+ in the gas phase and study it in the condensed state using IR spectroscopy,[764,765] pulse electron-beam mass spectrometry,[766] and Fourier transform ion cyclotron resonance mass spectrometry (FT–ICR MS).[767] However, an unambiguous structure determination was unsuccessful. Retardation of the degenerate rearrangement was achieved by trapping the ion in clusters with H_2, CH_4, Ar, or N_2.

The IR spectrum of the CH_5^+ ion solvated by a hydrogen molecule [$CH_5^+(H_2)$, **423**] has been reported by Bo and Lee.[764] The frequency of the stretching band centered at 2966 cm^{-1} matches the calculated frequency of **423** containing the C_s(e) structure by the *ab initio* method.[768] IR spectra of solvated ions $CH_5^+(H_2)_n$ ($n = 1$–6) were also disclosed.[769] Data indicated the scrambling of CH_5^+ through large-amplitude motions such as the CH_3 internal rotation and in-plane wagging motion. Attachment of solvent H_2 molecules to the core ion slowed the scrambling, and binding three H_2 molecules to the CH_5^+ core ion ultimately resulted in the complete freezing of scrambling motions.

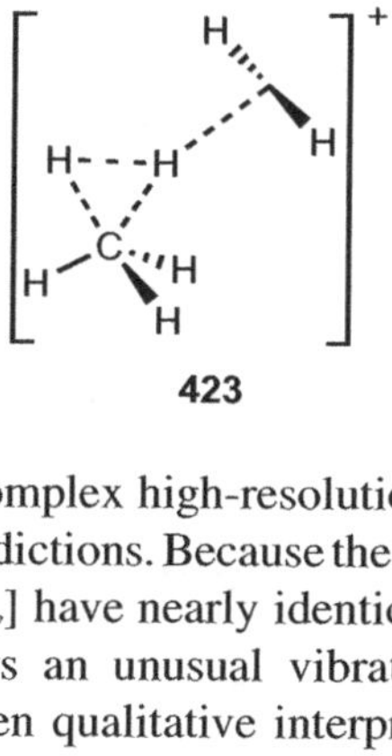

423

Protonated methane has a complex high-resolution IR spectrum,[765] which is not inconsistent with theoretical predictions. Because the three nonequivalent equilibrium structures [C_s(e), C_s(s), and C_{2v}] have nearly identical energies and its five protons scramble freely, the ion shows an unusual vibrational and rotational behavior. However, no assignment or even qualitative interpretation was offered. Marx and co-workers[770,771] have recently disclosed a comprehensive study of the properties of protonated methane using laser-induced reaction technique. A comparison of the experimental infrared spectrum of bare CH_5^+ at -110 K to finite-temperature spectra calculated by *ab initio* molecular dynamics supports the fluxionality of the protonated methane cation. Interestingly, hydrogen scrambling and internal rotation of the H_2 moiety could be computationally frozen out, which allowed the interpretation of the observed IR spectrum. Three distinct, well-separated peaks of C—H stretching modes involving the carbon nucleus and the protons that form the H_2 moiety, along with the hydrogens engaged in the CH_3 tripod, lend experimental support to three-center two-electron bonding. Hydrogen scrambling was suggested to result from softening of a mode that involves bending of the H_2 moiety relative to the CH_3 tripod.

Pulse electron-beam mass spectrometry was applied by Kebarle, Hiraoka, and co-workers[766,772] to study the existence and structure of $CH_5^+(CH_4)_n$ cluster ions in the gas phase. These $CH_5^+(CH_4)_n$ clusters were previously observed by mass spectrometry by Field and Beggs.[773] The enthalpy and free energy changes measured are compatible with the C_s symmetrical structure. Electron ionization mass spectrometry has been recently used by Jung and co-workers[774] to explore ion–molecule reactions within ionized methane clusters. The most abundant $CH_5^+(CH_4)_n$ cluster is supposed to be the product of the intracluster ion–molecule reaction depicted in Eq. (3.120) involving the methane dimer ion **424**.

$$\underset{\mathbf{424}}{(CH_4)_n\,(CH_4^{\bullet}\text{–}CH_4^+)} \longrightarrow (CH_4)_n\,(CH_3^{\bullet}\text{–}CH_5^+) \longrightarrow (CH_4)_n\,CH_5^+ + CH_3^{\bullet} \qquad (3.120)$$

Isotopic protonated (deuteriated) methanes were generated under various pressure conditions, and their reactivity toward ammonia was studied by FT–ICR mass spectrometry.[767] Competition between proton and deuterium transfer from protonated perdeuteriomethane and deuteriated methanes to ammonia shows chemically distinguishable hydrogens. The results allowed to conclude that the chemical behavior of protonated methane appears to be compatible with the theoretically predicted stable structure with C_s symmetry, involving a two-electron three-center bond. Interconversion of this structure due to exchange between one of the two associated hydrogens and one of the three remaining hydrogens appears to be fast, which is induced by interactions with the chemical ionization gas.

As early as 1972, Libby and co-workers[775] studied radiolysis by γ-rays of solid methane in liquid argon at 77 K and detected the formation of polymers. The CH_5^+ ion, most probably formed according to Eq. (3.121), may be involved in the process, whereas subsequent ion–molecule reactions followed by neutralization lead to heavier hydrocarbons [Eq. (3.122)].

$$CH_4^{\bullet+} + CH_4 \rightleftharpoons CH_5^+ + CH_3^{\bullet} \qquad (3.121)$$

$$CH_4 + C_n^+H_{2n+1} \rightleftharpoons H_2 + C_{n+1}^+H_{2n+3} \qquad (3.122)$$

Sommer and co-workers[776] have used experimental techniques (determining secondary kinetic deuterium isotope effects) and computational (DFT and high-level *ab initio*) methods to study methane activation in HF–SbF_5. The methonium ion solvated by HF (**425**) was found to be the potential energy minimum. The unsolvated superacid H_2F^+ is the only species strong enough to protonate methane in a barrierless process (B3LYP/BS2 and MP2/BS1 levels). The potential energy of the activated complex (**426**) for the exchange process was calculated to be only 1.9 kcal mol^{-1} higher than that of the strongly HF-bonded product CH_5^+ ion (**425**). The $(HF)_x$-solvated H_2F^+ superacid ($x = 1$–4) is not able to yield stable solvated methonium ions, but it participates in the transition states of the exchange in which CH_5^+ ions are solvated by $(HF)_x$.

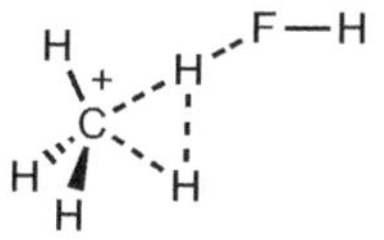

425

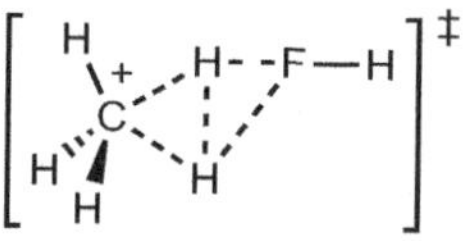

426

On the basis of the information discussed above, the CH_5^+ ion appears to be a fluxional species that undergoes rapid, very low-energy bond-to-bond rearrangements as postulated by Olah et al.[751] as early as 1969. The core protonated methane, however, is still best represented by the C_s symmetry ground-state structure and can be considered the parent carbonium ion.

Schleyer and co-workers[777,779] have calculated by *ab initio* methods the geometry of several lithiated analogs of CH_5^+, including structures **427** and **428**. Both $CH_3Li_2^+$ and CLi_5^+ have been observed by mass spectrometry.

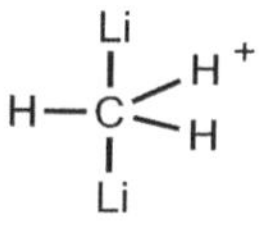

427

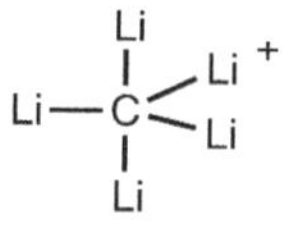

428

Schmidbaur and co-workers[780–783] have prepared the monopositively charged gold complexes $[(LAu)_5C]^+$**429**, which are the substituted analogs of CH_5^+. Considering the isolobal relationship (i.e., similarity in bonding) between LAu^+ and H^+, complexes **429** represent the isolobal analog of CH_5^+. The structure is of these compounds were determined by various experimental techniques (analytical and spectroscopic identifications, single-crystal X-ray crystallography). The X-ray data of the $[(Ph_3PAu)_5C]^+BF_4^-$ complex show a trigonal bipyramidal structure (C_3 symmetry). This was the first isolation and X-ray determination of an analog of CH_5^+, followed by the preparation and characterization of the pyramidal complex **430**.[784]

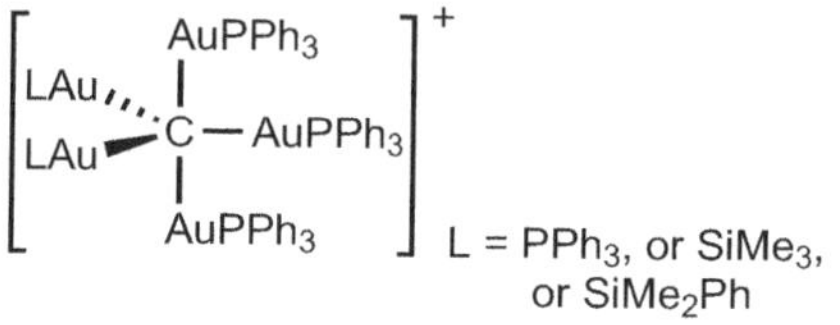

429

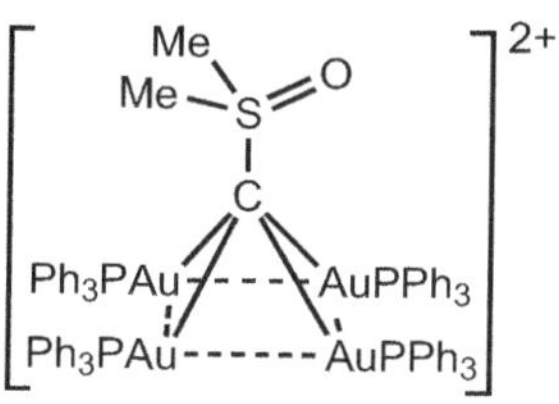

430

Boron and carbon are consecutive first-row elements. Consequently, pentacoordinate monopositively charged carbonium ions are isoelectronic with the corresponding neutral pentacoordinate boron compounds; that is, the BH_5 molecule is isoelectronic with CH_5^+. Within this context, it is interesting to refer to the five-coordinate square-pyramidal gold cation complex {$[Cy_3PB(AuPPh_3)_4]^+$, Cy = cyclohexyl} **431** reported and studied by Schmidbaur and co-workers,[781,785] which is the isolobal analog of BH_5. The symmetry of the $PB(AuP)_4$ skeleton of cation **431** is close to C_{4v}, and the four gold atoms are arranged in a slightly distorted and folded square capped by the boron atom with an apical Cy_3P ligand. The short Au–Au distances (2.82 Å average) indicate bonding interactions between neighboring Au atoms.

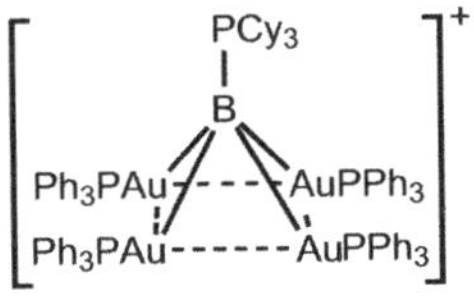

431

3.5.1.2. Multiply Protonated Methane Ions and Their Analogs. In addition to the involvement in a single 2e–3c bond formation, carbon is also capable of simultaneously participating in two 2e–3c bonds in some carbodications. Diprotonated methane (protiomethonium dication, CH_6^{2+}) is the parent of such carbodications.

The C_{2v} symmetrical structure (**432**) of diprotonated methane with two stabilizing orthogonal 2e–3c interactions was shown by quantum mechanical calculations (*ab initio* methods)[786,787] to be the only minimum on the potential energy surface. The other structures (**433**–**436**), in turn, were calculated to be saddle points on the potential energy surface. These were found to be 5.7, 12.0, 17.4, 41.5, and 35.2 kcal mol^{-1}, respectively, higher in energy than structure **432** (HF/6-31G**//HF/6-31G* level). At the HF/6-31G* level, the C–H bond length of each of the 2e–3c interactions is considerably longer than that of 2e–2c bonds (1.208 Å and 1.123 Å, respectively). The H–H bond length in 2e–3c interactions is 0.972 Å. On the basis of these data, structure **432** can be visualized as a complex of H_2 with the *p*-orbital of the carbon atom of the planar C_{2v} symmetrical CH_4^{2+} dication. Olah et al.[788] have recently recalculated CH_6^{2+} at even higher theoretical levels [MP4(SDTQ)/6-311G**//MP2/6-31G**]. C–H and H–H bond lengths in 2e–3c interactions were found to be 1.202 and 1.018 Å, respectively.

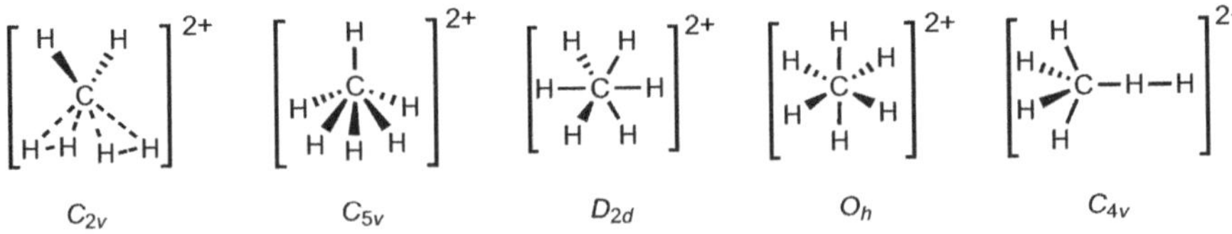

432 **433** **434** **435** **436**

The dication $(C_6Me_6)^{2+}$ (**437**) and the CLi_6^{2+} dication (**438**) are two additional six-coordinate carbon-containing species. Ion **437** was observed by Hogeveen and

Kant[41,789] and has a pyramidal structure (see Section 3.5.4.2). Calculations carried out for CLi_6^{2+} (**438**) show a preference for the octahedral structure.[777] Furthermore, similar to the fivecoordinate CH_5^+ analog gold complexes, the dipositively charged $[(LAu)_6C]^{2+}$ gold complexes **439**, containing a six-coordinate carbon atom have also been prepared and characterized by Schmidbaur and co-workers.[781,790–792] According to X-ray data, complex **439** (L = PPh_3), which is an isolobal mimic of CH_6^{2+}, has a distorted octahedral structure with Au–Au edges in the range 2.887–3.226 Å. Relativistic electronic structure calculations (LCGTO-LDF method) for the corresponding PH_3- and PMe_3-ligated models showed that the complexes are stable with respect to the loss of an $AuPR_3^+$ ligand.[793,794]

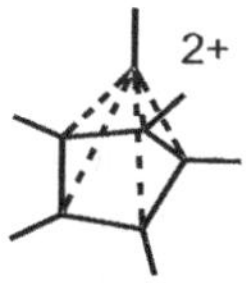

437

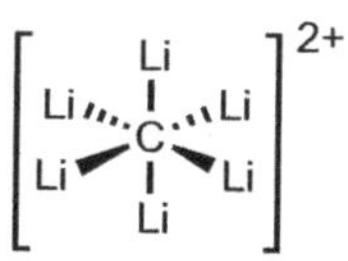

438

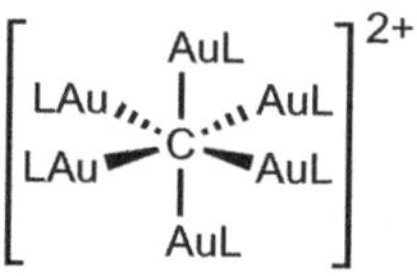

L = PPh_3,
$PPh_2(4\text{-}NH_2C_6H_4)$

439

Six-coordinate dipositively charged carbocations are isoelectronic with the corresponding monopositively charged hexavalent boron compounds. The parent six-coordinate boronium ion BH_6^+, isoelectronic with CH_6^{2+}, was generated in the gas phase by DePuy et al.,[795,796] and Olah and coworkers[797,798] studied simultaneously its structure and energetics. The C_{2v} symmetric form isostructural with CH_6^{2+}, containing two 2*e*–3*c* interactions and two 2*e*–2*c* bonds (see structure **432**), was found to be the only stable minimum for BH_6^+ [MP2/6-31G** level[797] and B3LYP/6-31++G (3df,2pd) level[798]].

Olah and Rasul[799] have reported *ab initio* calculations for triprotonated methane (diprotiomethonium trication, CH_7^{3+}). For the parent seven-coordinate CH_7^{3+} carbocation the C_{3v} symmetrical form (**440**) was found to be the only stable minimum (MP2/6-31G** level). It is a propeller-shaped molecule and can be visualized as a complex between CH^{3+} and three hydrogen molecules, resulting in the formation of three 2*e*–3*c* interactions with the hydrogen atoms bearing the positive charges (NBO charge calculation). It appears that strong bonding interactions in the cation can counter the substantial charge–charge repulsion. The C—H bonds in the 2*e*–3*c* interactions are slightly longer (1.282 and 1.277 Å) than bond distances in the CH_5^+ C_s structure (1.181 Å) at the same level of theory. Similarly, the H—H bond distance in the 2*e*–3*c* interactions (1.085 Å) is longer than that in the free hydrogen molecule (0.734 Å) at the same theoretical level and is also slightly longer than that found in H_2^+ (1.031 Å). The transition structure for intramolecular hydrogen transfer (**441**) lies only 0.7 kcal mol^{-1} higher in energy than structure **440**, indicating very facile intramolecular proton transfer. In accordance with expectations, various dissociation reactions of CH_7^{3+} are highly exothermic processes (data calculated at G2 theory) (Scheme 3.13).

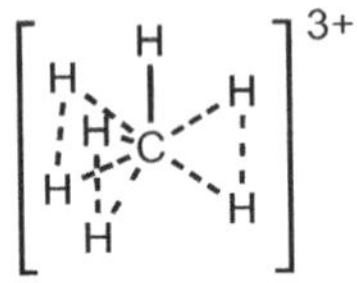

440

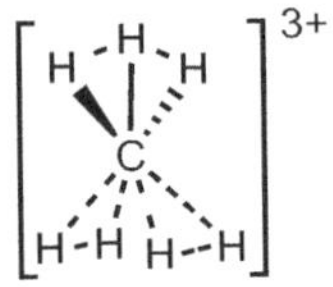

441

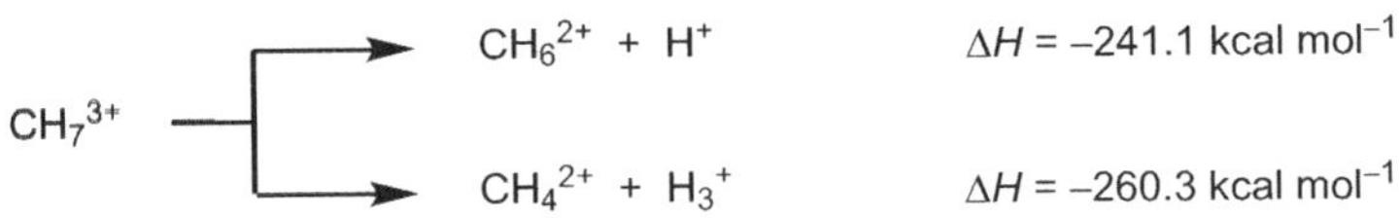

Scheme 3.13

A search was also made for the minimum-energy structure for the octacoordinate carbocation complex tetraprotonated methane (CH_8^{4+}).[799] This can be viewed as four dihydrogen units complexed with the tetrapositively charged C^{4+} cation (**442**) and can be compared with the $CH^{3+} + 3H_2$, $CH_2^{2+} + 2H_2$, $CH_3^+ + H_2$ complexes in CH_7^{3+}, CH_6^{2+}, CH_5^+ carbocations, respectively. It appears, however, that charge–charge repulsions have reached a prohibitive limit, since efforts at MP2/6-31G** level proved to be futile; that is, CH_8^{4+} remains even computationally elusive.

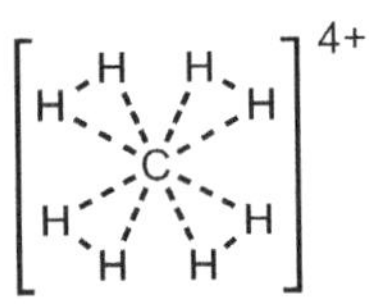

442

3.5.1.3. Varied Methane Cations. The methane molecular ion (methane radical cation, $CH_4^{+\bullet}$), the parent ion in mass spectrometry, and the methane dication (CH_4^{2+}) are of great significance and have been studied both experimentally and theoretically.[800–802] Recent advanced studies have shown that the methane radical cation, $CH_4^{+\bullet}$ has a fivecoordinate planar structure as suggested in early calculations by Olah and Klopman.[800]

The methane dication (CH_4^{2+}) with a lifetime of only ~3 μs[803] has been detected experimentally in the gas phase.[804–807] A square-planar D_{4h} symmetrical structure was predicted by early calculations.[808,809] However, the planar but not square C_{2v} symmetrical form (**443**) was shown to be the preferred structure by Wong and Radom.[810] The cation contains an sp^2-hybridized carbon, an empty *p*-orbital perpendicular to the plane of the molecule, and a 2*e*–3*c* interaction. It can be visualized as a complex of a hydrogen molecule weakly bounded by a 2*e*–3*c* bond to the CH_2^{2+} core as shown by *ab initio* calculations [MP2 and B3LYP levels with 6-311++G(2df,2pd) basis set].[811] The pentacoordinate $CH_5^{2+\bullet}$ radical dication was studied by Stahl et al.[807] by charge stripping mass spectrometry and the C_s symmetric geometry (**444**) was rationalized by theoretical calculations.

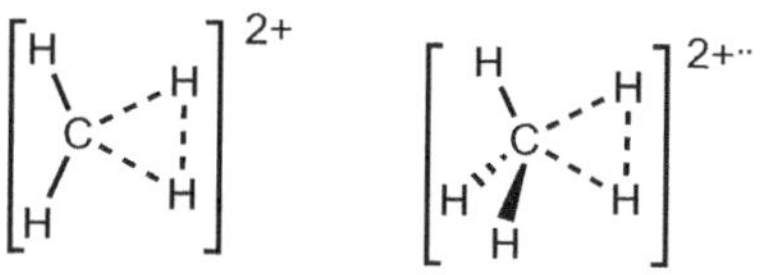

443 **444**

The CH_5^{3+} trication, which is formed by the removal of one electron from the $CH_5^{2+\bullet}$ radical dication, has been studied theoretically by Olah and Rasul.[812] Only a single minimum was found on the potential energy surface (MP2/6-31G** level). This is the planar D_{5h} symmetrical structure (**445**) with the five hydrogens bonded to the carbon atom sharing only six valence electrons and a vacant *p*-orbital perpendicular to the plane of the molecule. Another possible structure (**446**) with two 2*e*–3*c* bonds is not a minimum and converges to structure **445** upon optimization. The C–H bonds in trication **446** are electron-deficient and, consequently, weaker. Therefore, the bond lengths in structure **446** are longer (1.317 Å) than those in the 2*e*–3*c* C–H bonds of the C_s structure of CH_5^+ (1.181 Å). Bonding interactions between adjacent hydrogens are negligible, since hydrogens are separated by 1.549 Å. The dissociation of CH_5^{3+} to CH_4^{2+} and H^+ is highly exothermic (274.6 kcal mol^{-1}), indicating that the CH_5^{3+} trication is highly unstable. The search for the minimum energy structure of the CH_6^{4+} tetracation was also unsuccessful.

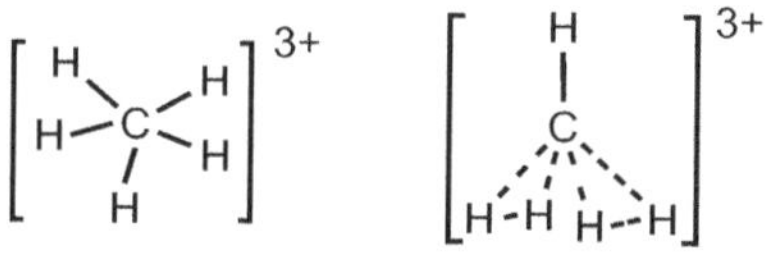

445 **446**

Calculations on the structure and stabilities of helium-containing polyatomic ions[813] were reported in the 1980s by Wilson and co-workers,[814,815] Schleyer,[816] and Radom and co-workers[817] and more recently by Koch, Frenking, and co-workers.[818,819] The quadruply charged tetraheliomethane tetracation CHe_4^{4+} shown by Radom and co-workers[817] to be a remarkably stable species was first calculated by Schleyer.[816] More recently, Olah et al.[820] studied the CH_4He^{2+} heliomethonium dication, which is isoelectronic with CH_5^+. The C_s symmetric form **447** is the only stable minimum, whereas the C_{4v} symmetrical structure (**448**) is the transition state for hydrogen scrambling [MP2/6-31G* and MP2/6-311+G(2d,p) levels]. Structure **447**, which may be considered as a complex between a neutral helium atom and methane dication CH_4^{2+}, contains one 2*e*–3*c* bond. Structure **447** is only 1.7 kcal mol^{-1} more stable than structure **448** (MP2/6-31G**//MP2/6-31G** level). The energy difference between the two structures, however, disappears at the highest level of theory applied, allowing very facile hydrogen scrambling.

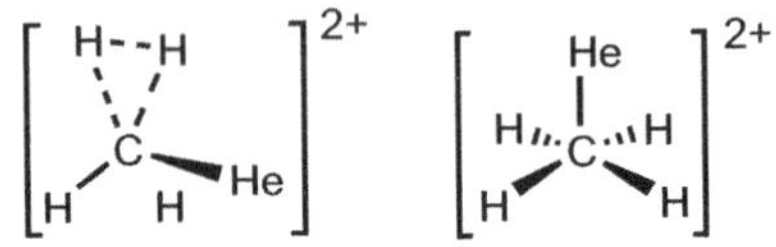

447 **448**

The energetics of several dissociation paths of the CH_4He^{2+} dication was also

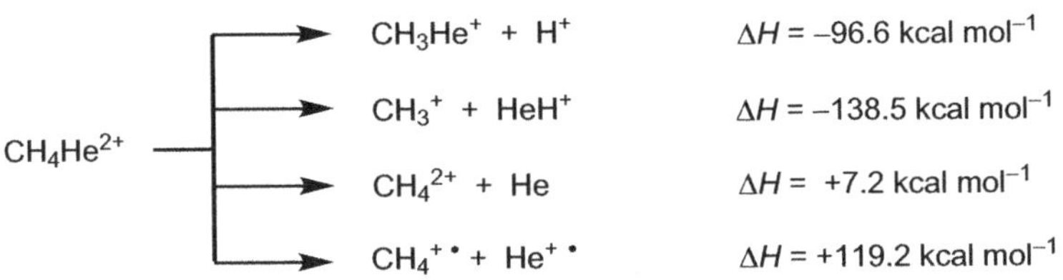

Scheme 3.14

calculated by the G2 method (Scheme 3.14). The first two processes are highly exothermic, but deprotonation has a significant kinetic barrier. The dissociation of CH_4He^{2+} into CH_4^{2+} and a neutral helium atom is endothermic. This provides a possibility to generate CH_4He^{2+} by reacting CH_4^{2+} with He in the gas phase.

3.5.1.4. Ethonium Ion ($C_2H_7^+$) and Analogs. The next higher alkonium ion, the ethonium ion (protonated ethane, $C_2H_7^+$) **449a** and **449b** (Scheme 3.15), is

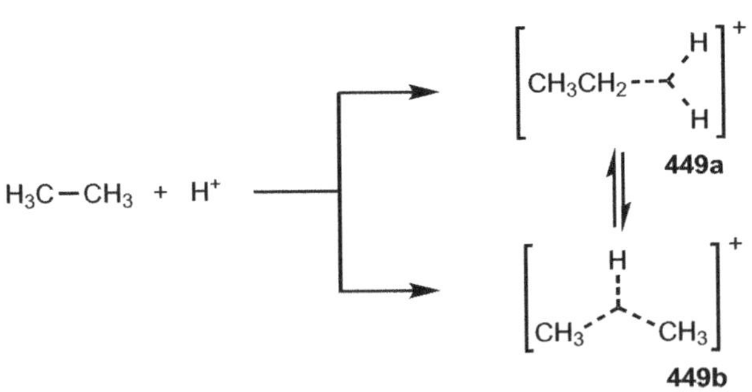

Scheme 3.15

analogous to its parent CH_5^+ ion. Protonation of ethane can take place either at a C–H bond or at the C–C bond, but the interconversion of the resulting ions is a facile low-energy process. Experimental studies by Hiraoka and Kebarle (pulsed electron beam MS)[748] and later by Lee et al. (IR spectroscopy)[821] have shown the existence of isomeric $C_2H_7^+$ cations in the gas phase with an energy difference of 7–8 kcal mol^{-1}. The **449b** C–C protonated form is preferred over the **449a** C–H protonated form by about 4.4 kcal mol^{-1} (*ab initio* calculations).[822] Such a protonation process would be consistent with the observed H–D exchange in labeled systems as well as with the formation of methane as a byproduct in the protolytic cleavage of ethane in superacids.[751] FT–ICR MS experiments showed that intramolecular hydrogen randomization is a very fast process in the gas-phase preceding the decomposition or any other transformations of $C_2H_7^+$.[823]

Subsequent theoretical investigations, however, indicated that a third isomer also exists. This may be considered to be an ethyl cation solvated by H_2[824] or a complex between bridged protonated ethylene and a hydrogen molecule (**450**).[822] Recent FT–ICR experiments[825] for the reaction of CH_3^+ and CH_4 leading to the intermediacy of the $C_2H_7^+$ ion lent support to the existence of structure **450**, which is lower in energy than structure **449**. It was also suggested that the isomer of higher energy identified in previous experiments[748,821] as structure **449** may, in fact, correspond to structure **450**. Interestingly, Hiraoka and Kebarle also considered but discounted the possibility of complexes between $C_2H_5^+$ and H_2.

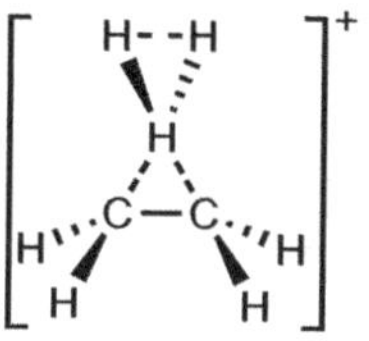

450

Various transformations of ethane on zeolites were also interpreted by invoking the involvement of protonated ethane. The hydrogen exchange has been studied using local density calculations (DFT with double-ξ plus polarization basis set),[826] applying the B3LYP and MP2 methods (6-31G** basis set)[827] or using DFT and an analysis with the atoms-in-molecules method.[828] The T3 zeolite cluster model (a cluster formed by three tetrahedral units) is generally used in most theoretical works. Structure **451** is suggested to be the transition state of hydrogen exchange characterized by a slightly ionic interaction between a distorted $C_2H_7^+$ structure and a negatively charged zeolite cluster.

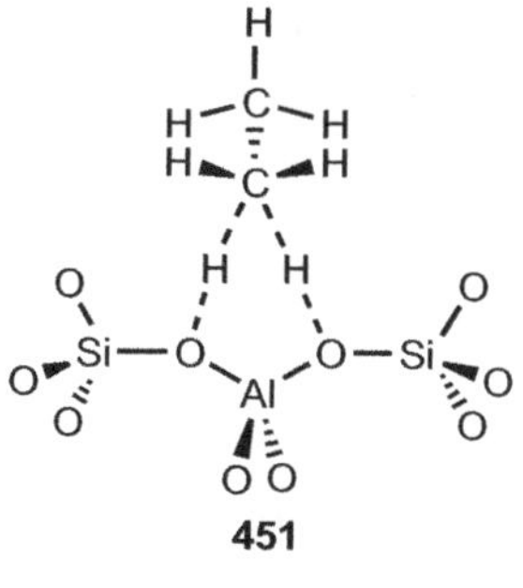

451

The preferred structure of the diprotonated $C_2H_8^{2+}$ ethane dication (**452**) considered theoretically[786] incorporates two pentacoordinate carbons and an unprotonated C–C bond (HF/6-31G* level). A new calculation by Olah et al.[788] (MP2/6-31G**) has shown that both the C_2 symmetry structure **452** and the C_{2h} symmetry structure **453** are the minima on the potential energy surface, with **453** being more stable by 0.4 kcal mol^{-1} than **452**.

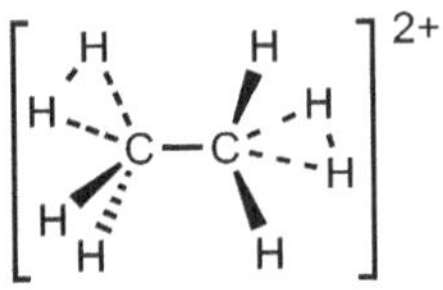

452

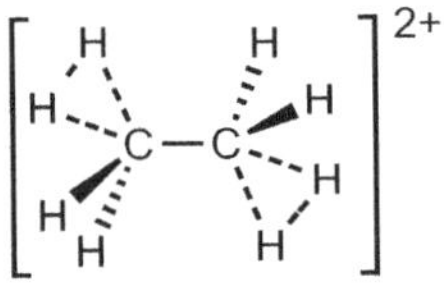

453

Schmidbaur and co-workers[829] have prepared and characterized the tetraaurated ethonium cations **454**. The $MeC(AuP)_4$ unit has a square-pyramidal geometry with the methyl group in apical position. The bonding interactions between neighboring Au atoms indicated by the short Au–Au distances (2.8 Å average) along the edges strongly contribute to stability.

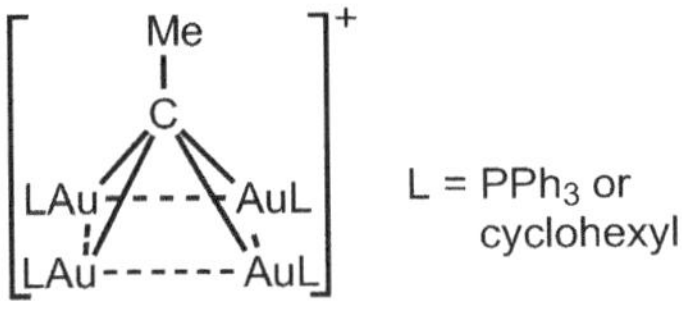

454

3.5.1.5. Proponium Ions and Analogs. The higher alkonium ions, including protonated propane ($C_3H_9^+$), have been observed in the gas phase by high-pressure mass spectrometry.[746,748,749] In solution, the higher hydrocarbons show an increasing tendency to form C–H–C three-center, two-electron bonds on protonation as evidenced by the increasing tendency to form C–C bond cleavage products.

Hiraoka and Kebarle[748,830] studied the $C_3H_9^+$ protonated propane cation in the gas phase and found two isomers (cations **455** and **456**). The *C*-protonated cation **455** was shown to be of lower energy and also much more stable toward dissociation to $C_2H_5^+ + CH_4$ or *sec*-$C_3H_7^+ + H_2$ than the 2-*H*-protonated (**456**) cation. The energy barrier of the interconversion of the ions was calculated to be 9.1 kcal mol^{-1}.

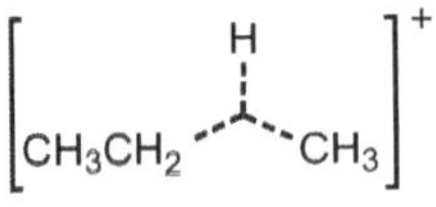

455

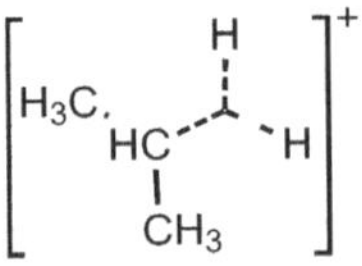

456

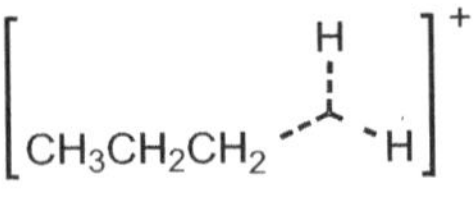

457

Two isomeric proponium ions, cation **455** and **457** were computed by Collins and O'Malley[831] (DFT and MP2 calculations). The 2-*H*-protonated cation **456**, in turn, was suggested to be rather the transition state for the interconversion of the ions. Structure **456** is better described as the van der Waals complex of *sec*-$C_3H_7^+$ and H_2, which was already suggested by Hiraoka and Kebarle.[748] Mota and co-workers[832] have recently calculated the potential energy surface of the $C_3H_9^+$ cation [MP4(SDTQ)/6-311++G**/MP2(full)/6-31G** level]. In accordance with previous results, the *C*-proponium cation (**455**) was found to be of lowest energy, but the complex *sec*-$C_3H_7^+ + H_2$ lies only 0.3 kcal mol^{-1} above structure **455**. The C–H bond lengths in the 3*c*–2*e* interactions are 1.272 and 1.188 Å, whereas the C–C bond distance is 2.099 Å. The interconversion between the 1-*H*-protonated cation (**457**) and the *C*-proponium cation (**455**) was shown to have no energy barrier. This result has relevance to zeolite chemistry. On zeolites where steric effects are important, the primary—that is, more accessible—hydrogens are protonated initially to yield the 1-*H*-proponium cation (**457**) (kinetic control), which then easily rearranges to give the thermodynamically more stable *C*-proponium ion.

On the potential energy surface of diprotonated propane ($C_3H_{10}^{2+}$), three structures were located as minima (MP2/6-31G** level).[788] The 1-*H*,2-*H*-diprotonated and 1-*H*,3-*H*-diprotonated forms (**458** and **459**) are of C_1 and C_{2v} symmetry, respectively, whereas the 1-*H*,*C*-diprotonated structure (**460**) has C_1 symmetry. Structure **459**, a *distonic* dication with the two charge-bearing centers separated by one carbon, is significantly more stable (by 22.6 kcal mol^{-1}) than structure **458** (a *gitonic* dication with adjacent charges) and also more stable than structure **460** (by 6.9 kcal mol^{-1}).

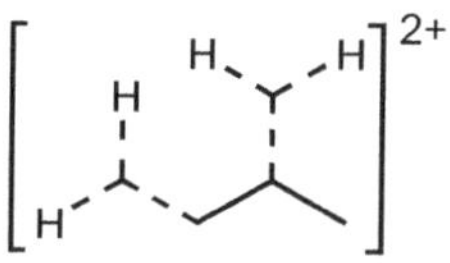

458

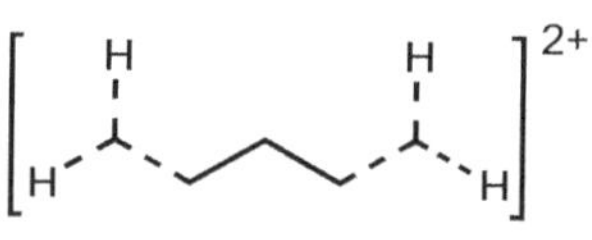

459

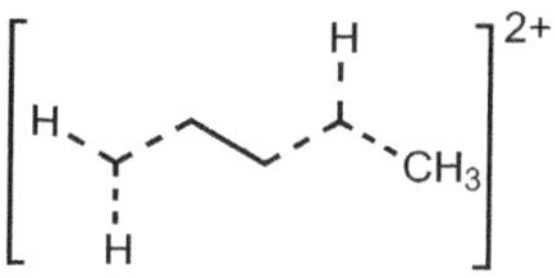

460

3.5.1.6. Higher Alkonium Ions. The acid-induced H–D exchange of isobutane (**461**) at conventional acidities (e.g., with deuterosulfuric acid) was studied by Otvos and co-workers.[833] All nine methyl hydrogens are readily exchanged, but not the methine hydrogen. The mechanistic explanation for this observation must involve formation of trivalent *tert*-butyl cation (**1**), probably in an oxidative ionization step (Scheme 3.16). The *tert*-butyl cation then undergoes reversible deuteriation

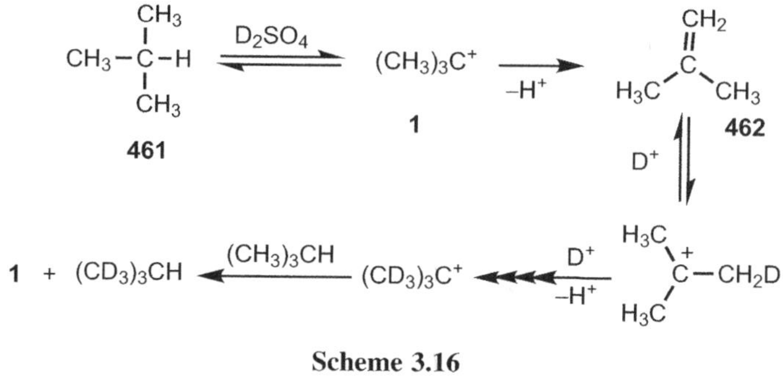

Scheme 3.16

involving isobutylene, with the process repeating itself and thus accounting for exchange of the methyl hydrogens with deuteriated sulfuric acid.

The *tert*-butyl cation (**1**) reforms isobutane via hydride abstraction from isobutane according to Bartlett et al.,[834] Nenitzescu et al.,[835] and Schmerling[836] involving the tertiary C–H bond only through intermediate structure **463**, and thus not exchanging the methine hydrogen.

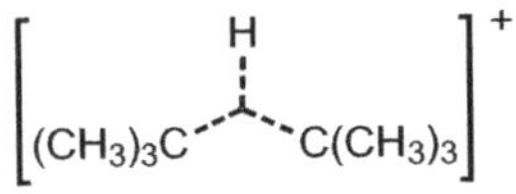

463

In contrast, Olah et al.[837] showed that in deuteriated superacidic media (e.g., DSO_3F–SbF_5 or DF–SbF_5 at low temperatures), initially only the methine hydrogen is exchanged (with the acid), indicating no deprotonation–protonation equilibria involving species such as **1** and **462**. The latter reaction proceeds through the intermediary of the **464** arising by protonation (deuteriation) of the tertiary C–H bond of isobutane [Eq. (3.123)]. Subsequent reactions exchange the methyl hydrogens through protonation (deuteriation) of the methyl groups.

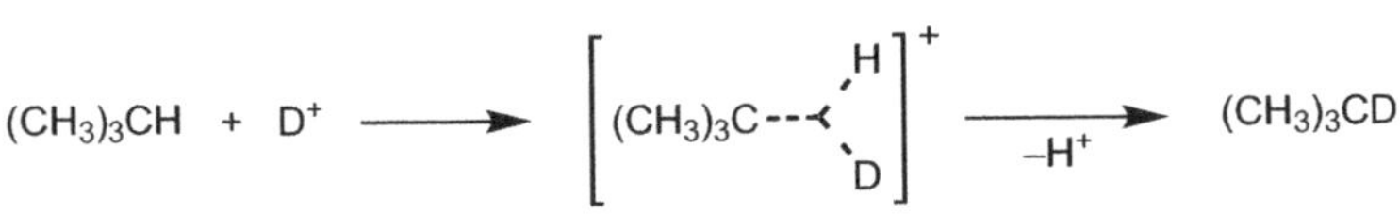

464 (3.123)

Similar regioselectivity of exchange was found in triflic acid.[838] Moreover, a fast deuterium exchange at all C–H bonds was observed in isobutane recovered after short contact times with DF–SbF_5 at 0°C (in contrast to −78°C).[839] Isomeric carbonium ions **465a**, **465b**, and **465c** are formed in equilibrium. The relative concentration of these pentacoordinate reaction intermediates, in accord with the Olah σ-basicity concept, depends only on the relative basicity of the proton-accepting bonds. An extended study with various C_3–C_5 alkanes resulted in similar observations.[840] The amount of hydrogen formed up to about 20 mol% of SbF_5 roughly parallels conversion, in agreement with the protolytic ionization.[839]

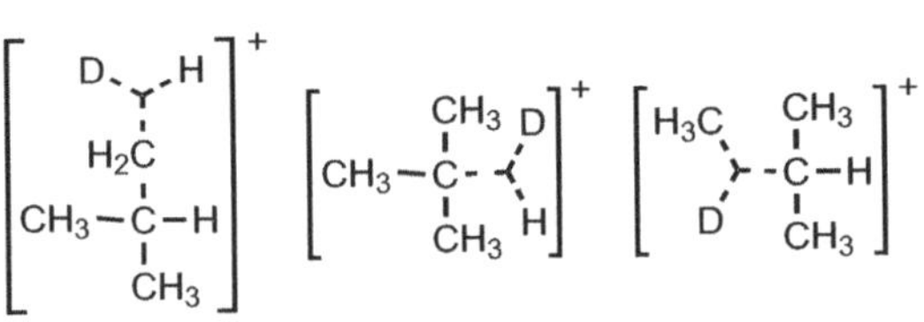

a b c
465

Evidence for a C–C protonated $C_4H_{11}^+$ ion (**466**) resembling cation **463** was obtained by Siskin.[841] When studying the HF–TaF_5-catalyzed ethylation of a large excess of ethane with ethylene in a flow system, *n*-butane (**467**) was obtained as the only four-carbon product free from isobutane [Eq. (3.124)]. This remarkable result can only be explained by C–H bond ethylation of ethane, through the five-coordinate carbocation intermediate **466** which subsequently yields *n*-butane (**467**) by proton elimination. Use of a flow system that limits the contact of the product *n*-butane with the acid catalyst is essential, because on more prolonged contact, isomerization of *n*-butane to isobutane occurs.

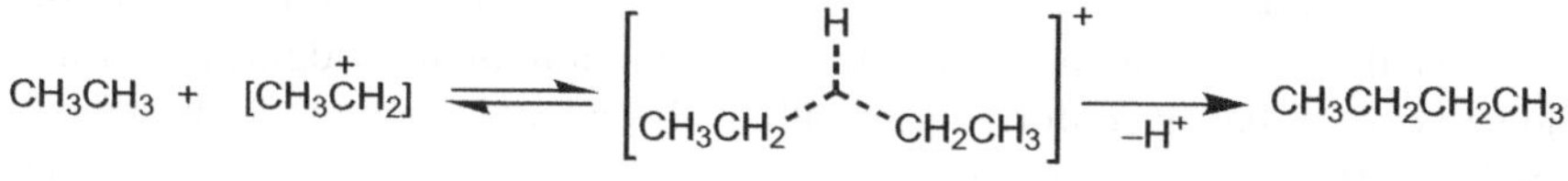

466 **467**

(3.124)

Alternatively, if the reaction involved trivalent *n*-butyl cation **468** (from ethylation of ethylene) the ion would inevitably rearrange via 1,2-hydrogen shift to *sec*-butyl cation **19**, which in turn would isomerize into the *tert*-butyl cation (**1**) and thus give isobutane (**461**) [Eq. (3.125)].

$$H_2C{=}CH_2 + [CH_3CH_2]^+ \longrightarrow \underset{\textbf{468}}{[CH_3CH_2CH_2CH_2]^+} \xrightarrow{1,2\text{ ~H}} \underset{\textbf{19}}{CH_3CH_2\overset{+}{C}HCH_3} \longrightarrow \underset{\textbf{1}}{(CH_3)_3C^+} \longrightarrow \underset{\textbf{461}}{(CH_3)_3CH} \qquad (3.125)$$

The pulsed electron beam MS technique was also used by Hiraoka and Kebarle[842] to study the $C_4H_{11}{}^+$ cations. In the ion–molecule reaction of ethane and the ethyl cation, two species were observed and identified as the 2-*H*-*n*-butonium cation **469** and the 2-*C*-*n*-butonium cation **470**. C–C protonated ion **470** formed first rearranges to C–H protonated ion **469** (energy barrier = 9.6 kcal mol^{-1}) and then dissociation to *sec*-$C_4H_9{}^+ + H_2$ takes place.

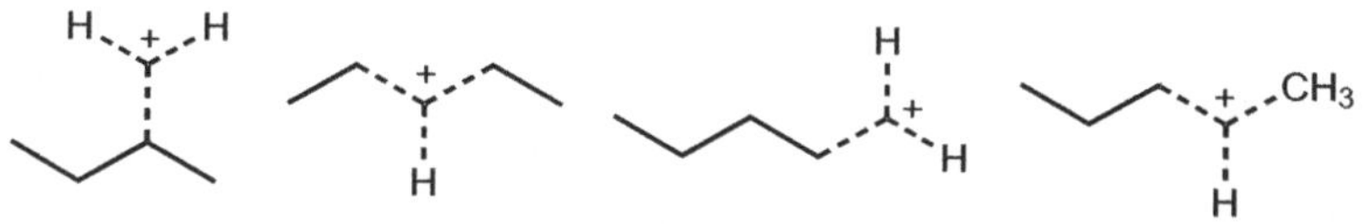

469 **470** **471** **472**

Three cations, an *H*-protonated (**471**) and two *C*-protonated (**470** and **472**) isomers were found by Collins and O'Malley[831] by DFT and MP2 calculations. Ion **472** is more stable than ion **470** by about 5 kcal mol^{-1}. Eleven stable isomeric protonated butane cations including rotamers were found by extensive studies by Mota and co-workers[843] [MP4SDTQ(fc)/6-311++G**//MP2(full)6-31G** level]. The stability order 2-*C*-*n*-butonium (**470**) > 1-*C*-*n*-butonium (**472**) > 2-*H*-*n*-butonium (**469**) > 1-*H*-*n*-butonium (**471**) was interpreted in terms of charge delocalization in the involved 3*c*–2*e* bonds. Cation **471** prefers to rearrange to cation **472**, whereas cation

469 decomposes to *sec*-$C_4H_9^+ + H_2$. Both cations **471** and **472** are higher in energy than the corresponding van der Waals complexes ($C_2H_5^+ + C_2H_6$, *sec*-$C_4H_9^+ + H_2$, and *n*-$C_3H_7^+ + CH_4$). Good agreement was found between experimental and computed proton affinity values (153.7 kcal mol^{-1} versus 156.7 kcal mol^{-1}). An exploratory topological study of the Laplacian of the electronic charge density by Jubert and co-workers[844] was performed on the 11 butonium carbocations.

Geometries and energies of the conformers of the symmetrically protonated $C_4H_{11}^+$ cation **470** have recently been studied by *ab initio* methods.[845] The *trans*-$C_4H_{11}^+$ and the *gauche* rotamer with a staggered dihedral for the bridging proton are of the lowest energy, but all conformers lie within a 1-kcal mol^{-1} range. The best method [CCS(T)/cc-pVTZ] used for optimization of the geometry of the C_2-symmetry minimum for *trans*-$C_4H_{11}^+$ gave the following bond geometry values: C—H bond length = 1.2424 Å, C—C bond distance = 2.177 Å, C—H—C bond angle = 122.4°.

The reaction of the isopropyl cation with methane to give the isobutonium ion was also studied by Hiraoka and Kebarle.[748] The *C*-protonated cation **473** has a heat of formation of 170.7 kcal mol^{-1}. Mota et al.[846] have characterized by *ab initio* studies [MP2(full)/6-31G** level] three structures (**473**–**475**) and two van der Waals complexes with increasing energies in the order **473** < **474** < **475**. The C—H bond lengths in the 3*c*–2*e* interactions are 1.470 and 1.137 Å, whereas the C—C bond distance is 2.470 Å. Decomposition of the cations into the corresponding van der Waals complexes (**474** to *sec*-$C_3H_7^+ + CH_4$ and **475** to *tert*-$C_4H_9^+ + H_2$) was found to have low or no activation energy. This is due to the high stability of the carbenium ions they collapse to. Using DFT calculations, Collins and O'Malley have arrived at the same conclusion for both structure **473** and C—C protonated neopentane.[831]

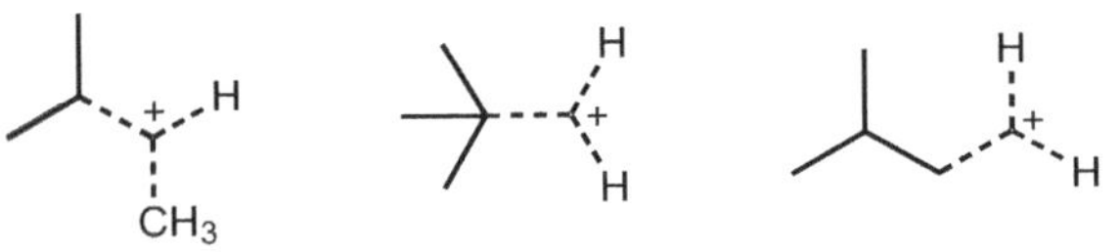

473 **474** **475**

Seitz and East selected five isomeric protonated octane isomers ($C_8H_{19}^+$), all featuring C—H—C or C—H—H 3*c*–2*e* bonds for their theoretical studies (*ab initio* calculations at the MP2/6-31G(d) level of theory).[847] In most cases, dissociation into ion–molecule complexes was found to be again barrierless. Proton affinities of C—C and C—H bonds are in the range 154–187 kcal mol^{-1} and 139–150 kcal mol^{-1}, respectively.

There is a general pattern having emerged from the theoretical works performed for higher protonated alkanes.[847] (i) Protonation of alkanes always produces a C—H—C or C—H—H 3*c*–2*e* bond through the attack of C—C or C—H bonds, respectively. (ii) The *C*-protonated structures are always lower in energy than the *H*-protonated structures with an equivalent carbon skeleton. (iii) Of the *C*-protonated structures, cations with more substituted carbon atoms participating in the 3*c*–2*e* bond have the higher stability.

Diprotonated *n*-butane and isobutane cations ($C_4H_{12}^{2+}$) have been computed by Olah et al.[788] Two *distonic* dications [the 1-*H*,4-*H*-diprotonated (**476**) and the terminal C–H and C–C diprotonated (**477**) forms] were found to have energy minima (MP2/6-31G** level), with structure **476** being more stable by only 1.3 kcal mol^{-1}.

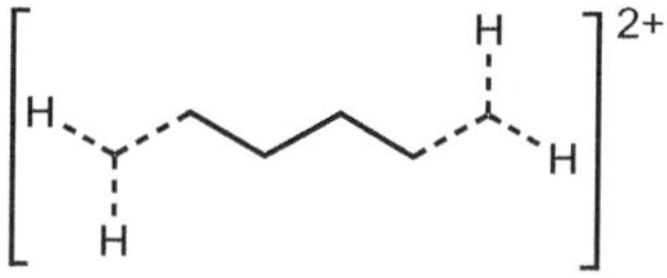

476

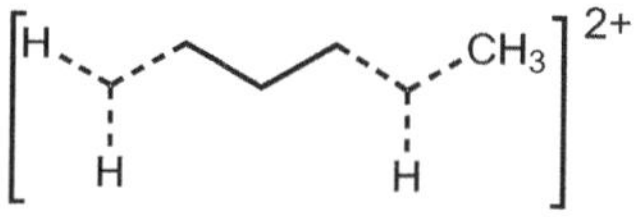

477

For diprotonated isobutane ($C_4H_{12}^{2+}$) the structures found as stable minima are analogous to those computed for diprotonated propane ($C_3H_{10}^{2+}$). Again, structure **478** (a *distonic* dication) is 17.7 kcal mol^{-1} more stable than structure **479** (a *gitonic* dication), whereas structure **480** is only slightly less stable than **478** (2.6 kcal mol^{-1}).[788]

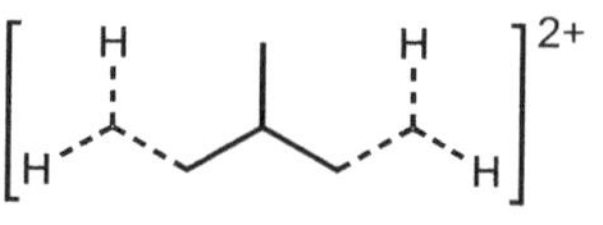

478

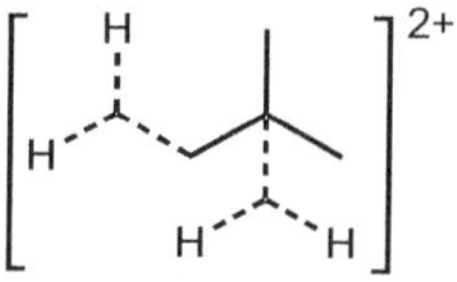

479

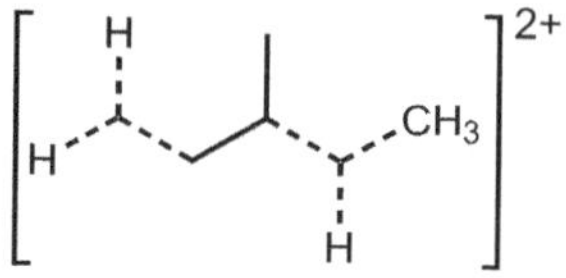

480

Olah, Prakash, and Rasul[125] have recently reported the structure of the protonated *tert*-pentyl cation ($C_5H_{12}^{2+}$, diprotonated isopentane dication). Of the five minima located on the potential energy surface, structure **481** was found to be the global minimum, being even more stable than the *tert*-pentyl cation (**23** and **24**). Structure **481** has a trivalent carbocationic center and a pentacoordinate carbonium ion center (involving a 3*c*–2*e* bond) separated by a carbon atom. The ion can be considered a carbenium–carbonium dication. One of the hydrogens of each methyl group attached to the carbocationic center is aligned in plane with the empty *p* orbital. These bonds are elongated (1.11 Å and 1.12 Å) and the corresponding H–C–C angles are significantly smaller (101° and 102°) than the other bond angles (113°–116°), which indicates significant hyperconjugation from two C–H σ bonds. In other structures with higher energy, the positive charges are closer to each other, generating larger intramolecular coulombic repulsion.

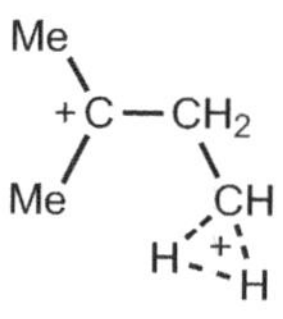

481

Protonation of spiro[2.2]pentane **482** yields the $C_5H_9^+$ ions. Cation **483** was shown to be the initial protonated spiropentane by early experimental (MS) and theoretical

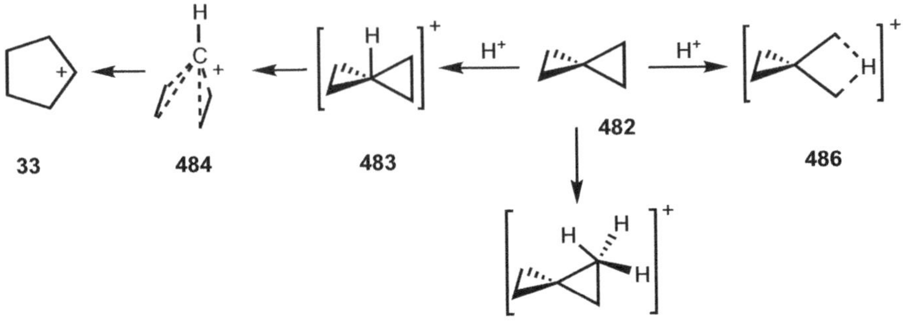

33 484 483 482 486 485

Scheme 3.17

(semiempirical) studies (Scheme 3.17). Once formed, cation **483** rearranges to pyramidal intermediate **484**, which yields cyclopentyl cation **33** upon ring opening. A combination of experimental methods (nuclear decay, radiolysis, FT–ICR mass spectrometry) and *ab initio* calculations (HF/6-31G* and MP2/6-31G* levels) performed by Schleyer and co-workers[848] showed that the corner-protonated cation **485** is a relatively long-lived intermediate in the gas-phase. Edge-protonated isomeric cation **486**, in turn, is less stable than ion **485** by 35.5 kcal mol^{-1}.

In a subsequent calculational study [HF and MP2 levels of theory with 6-31G(d) basis set and MP4(SDQ)], Szabó and Cremer[849] explored the he $C_7H_{11}^+$ potential energy surface. Cation **487** (tricyclo[4.1.0.$0^{1,3}$]heptyl cation), the protonated ethano-bridged derivative of spirocyclopentyl cation, was considered to be the missing link between the bicyclo[3.2.0]hept-3-yl cation **488** and the 7-norbornyl cation **489**. It is a kinetically stabilized species separated from cations **488** and **489** by 18.9 and 15.9 kcal mol^{-1}, respectively.

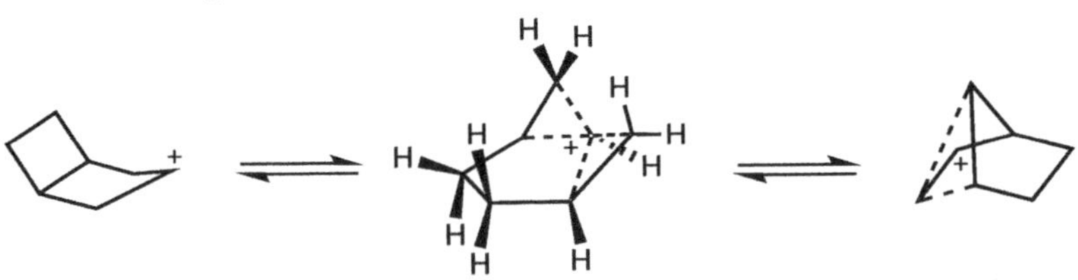

488 487 489

3.5.1.7. Adamantonium Ions. There exists a single theoretical study for adamantonium ions.[850] Mota and co-workers have found three isomeric structures (**490**–**492**) and two van der Waals complexes (1-adamantyl cation + H_2 and 2-adamantyl cation + H_2) [MP2(full)/6-31G** level]. The C–H bond lengths in the 3*c*–2*e* interactions in ion **490** are 1.276 and 1.266 Å and in ion **491** are 1.266 and 1.280 Å. The *C*-adamantonium ion (**492**) has nonequivalent C–H bond distances (1.191 and 1.294 Å) and the C–C bond distance is 2.348 Å. Interestingly, the 2-adamantyl cation + H_2 complex shows a nonclassical bonding nature similar to the

2-norbornyl cation. Cation **492** is the most stable, but it is 9.30 kcal mol^{-1} less stable than the 1-adamantyl cation + H_2 complex.

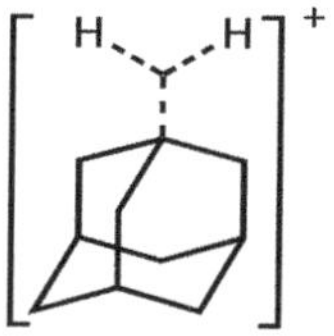

490

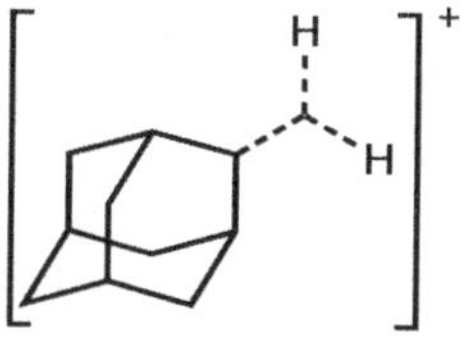

491

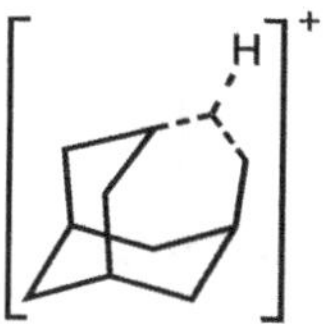

492

3.5.2. Equilibrating and Bridged Carbocations

Some carbocations, because of their flat potential energy surfaces, show great tendency to undergo fast degenerate rearrangements, through intermolecular hydrogen or alkyl shifts leading to the corresponding identical structures.[17–19,851] The question arises whether these processes are true equilibria between the limiting trivalent carbocations ("classical ion intermediates") separated by low-energy level transition states or whether they are hydrogen- or alkyl-bridged higher-coordinate (nonclassical) carbocations. Extensive discussion of the kinetic and stereochemical results in these systems has been made, and it is not considered to be within the scope of this chapter to recapitulate the arguments. The reader is referred to reviews[17–19,851] and the original literature.

3.5.2.1. Degenerate 1,2-Shifts in Carbocations. Many acyclic and monocyclic tertiary and secondary cations undergoing degenerate 1,2-hydrogen (and alkyl) shift give average proton and carbon absorptions in their NMR spectra, even at low temperatures. If the exchange rate process is rapid on the NMR time scale, single sharp resonances will appear for the exchanging nuclei at frequencies that are weighted averages of the frequencies being exchanged. It has been possible to freeze exchange processes in some equilibrating ions by observing ^{13}C NMR spectra at low temperature (ca. −160°C) at high magnetic field strength (to enhance signal separation). Typically, the barriers for such migrations range from 2.4 to 10 kcal mol^{-1}. Another important technique that has been immensely useful is the low-temperature solid-state ^{13}C NMR spectroscopy.

The *sec*-butyl cation (**19**) has been prepared from 2-chlorobutane in SbF_5–SO_2ClF at −100°C in a vacuum line by Saunders and Hagen[852,853] with very little contamination from the *tert*-butyl cation (**1**) (Scheme 3.18). Even at −110°C, only two peaks from 2,3 and 1,4 protons are observed in the ^{1}H NMR spectrum of **19** (at δ^1H 6.7 and 3.2). This is consistent with a *sec*-butyl cation averaged by very rapid 1,2-hydride shifts ($\Delta G^{\ddagger} \approx 6$ kcal mol^{-1}). Warming the sample from −110 to −40°C first causes line broadening and then coalescence of the two peaks, revealing a rearrangement process making all protons equal on the ^{1}H NMR time scale (indicating the formation of **1**). Line-shape analysis gave an activation barrier of 7.5 ± 0.1 kcal mol^{-1} for the process. This low barrier is not compatible with a mechanism involving primary cations as suggested for the corresponding rearrangement of the isopropyl cation. It appears

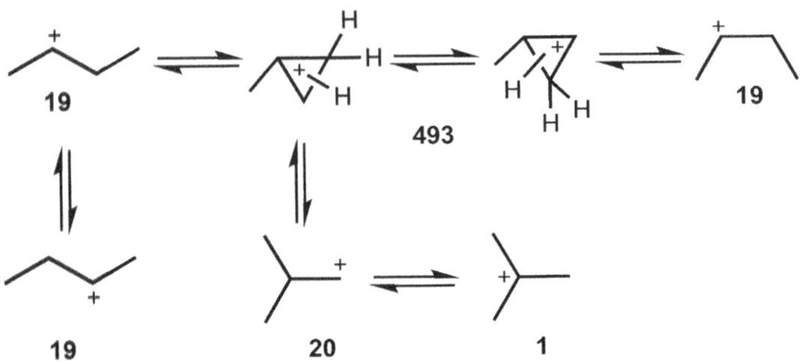

Scheme 3.18

necessary to invoke protonated methylcyclopropanes **493** as intermediates. The barrier for the irreversible rearrangement to **1** was measured to be about 18 kcal mol^{-1}, indicating that this rearrangement probably involves primary cationic structures as intermediates.

Olah and White[94] obtained an early ^{13}C NMR INDOR spectrum of **19** that showed a single peak from the two central carbon atoms in reasonable agreement with values calculated from model equilibrating ions. Therefore, it was concluded that **19** is a classical equilibrating ion rather than being bridged as in **494**.

494

In a comprehensive ^{13}C NMR spectroscopic study of alkyl cations, Olah and Donovan[95] applied the constancy of ^{13}C methyl substituent effects to the study of equilibrating cations and their rearrangements. They calculated the chemical shifts of the 2-butyl cation (**19**) from both the isopropyl cation and *tert*-pentyl cation using methyl group substituent effects and reached practically the same result in both cases. The observed chemical shifts deviate from the calculated ones by 9.2 and 19.8 ppm for the equilibrating methyl and carbocation carbons, respectively. Therefore, a hydrogen-bridged intermediate (**494**) was suggested to be involved. A static hydrogen-bridged 2-butyl cation was excluded by the observation of two quartets in the fully ^{1}H-coupled ^{13}C NMR spectrum. Comparison with bridged halonium ions indicates that equilibrating hydrogen-bridged ions have more shielded carbons [C(2), C(3)] than are observed experimentally for the 2-butyl cation. Therefore, it was suggested that the open-chain 2-butyl cation is of similar thermodynamic stability as the hydrogen-bridged **494** and that these intermediates in equilibrium may contribute to the observed average ^{13}C NMR shifts. However, the percentage of different structures could not be calculated, owing to lack of accurate models to estimate ^{13}C chemical shifts of hydrogen-bridged structures.

In a study of rates of degenerate 1,2-shifts in tertiary carbocations, Saunders and Kates[854] used higher-field (67.9 MHz) ^{13}C NMR line broadening in the fast-exchange limit. The 2-butyl cation showed no broadening at -140°C. Assuming the hypothetical "frozen out" chemical shift difference between C(2) and C(3) to be 227 ppm, an upper limit for $\Delta G^{\ddagger}$ was calculated to be 2.4 kcal mol^{-1}.

Application of the isotopic perturbation technique by Saunders et al.[56] to the 2-butyl cation [Eq. (3.126)] showed it to be a mixture of equilibrating open-chain ions since a large splitting of the ^{13}C resonance [C(2), C(3)] was obtained upon deuterium substitution.[855]

(3.126)

The cross-polarization, magic-angle spinning method (CP MAS) has been applied by Myhre and Yannoni[50] to cation **19** in the solid state at very low temperatures using ^{13}C NMR spectroscopy. In the initial study, no convincing evidence for a frozen 2-butyl cation was obtained even at −190°C. However, subsequently they managed to freeze out the equilibration of the 2-butyl cation (**19**) at −223°C.[62] It behaves like a normal secondary trivalent carbocation.

As mentioned earlier (Section 3.4.2), the cyclopentyl cation **33** shows a single peak in the ^{1}H NMR spectrum of δ^1H 4.75 even at −150°C.[143] In the ^{13}C NMR spectrum,[856] a 10-line multiplet centered around 95.4 ppm with $J_{C\text{–}H} = 28.5$ Hz was observed. This is in excellent agreement with values calculated for simple alkyl cations and cyclopentane and supports the complete hydrogen equilibration by rapid 1,2-shifts [Eq. (3.127)].

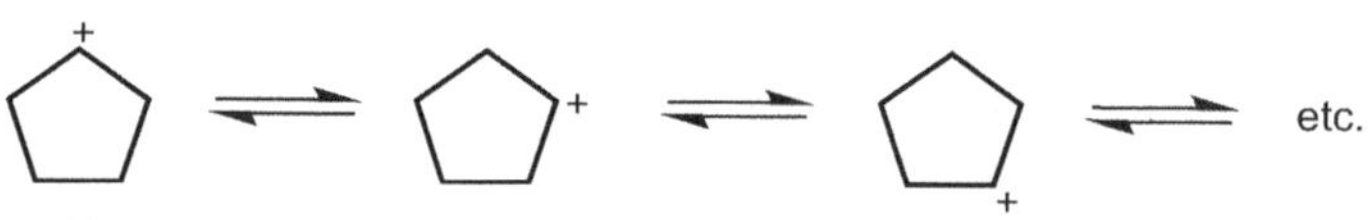

33

(3.127)

Subsequently, Yannoni and co-workers[857] succeeded in freezing out the degenerate hydride shift in **33** in the solid state at −203°C. The observed ^{13}C chemical shifts at δ^{13}C 320.0, 71.0, and 28.0 indicate the regular trivalent nature of the ion and are in good agreement with the estimated shifts in solution based on the average shift data.

The NMR spectrum of 2,2,3-trimethyl-2-butyl cation (triptyl cation, **495**) [Eq. (3.128)] consists of a single proton signal at δ^1H 2.90 for all the methyl groups.[134] This indicates that all five methyl groups undergo rapid interchange through 1,2-methyl shifts. The chemical shift of the singlet methyl is similar to that of 2,3-dimethyl-2-butyl cation **496** [Eq. (3.129)], another equilibrating ion that undergoes rapid 1,2-hydride shifts.[134]

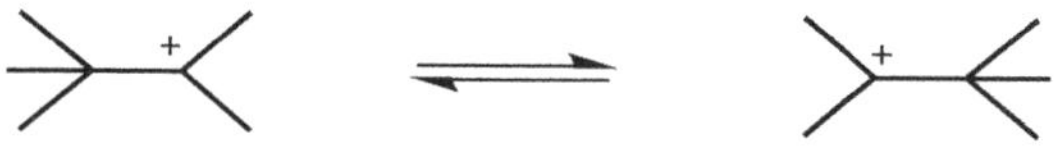

495

(3.128)

496

(3.129)

The ^{13}C NMR spectroscopic data of the average cationic center in **495** and **496** were found to be at δ^{13}C 205 and 197 ($J_{C\text{–}H} \approx 65$ Hz), respectively, indicating their regular trivalent carbenium nature. From studies of methyl substituents effects, Olah and Donovan[95] reached the same conclusions and these are supported by laser Raman and

ESCA studies.[72,858] Saunders and Vogel[58] have introduced deuteriums into a methyl group of **495** (cation **497**) and thereby perturbed the statistical distribution of the otherwise degenerate methyl groups and split the singlet into a doublet. The CD_3 group prefers to be attached to the tertiary carbon (**498**) [Eq. (3.130)].

497 **498** (3.130)

Saunders and Kates[854] have been successful in measuring the rates of degenerate 1,2-hydride and 1,2-methide shifts of simple tertiary alkyl cations employing higher-field (67.9 MHz) ^{13}C NMR spectroscopy. From line broadening in the fast-exchange limit, the free energies of activation ($\Delta G^{\ddagger}$) were determined to be 3.5 ± 0.1 kcal mol^{-1} at $-136°C$ for **495** and 3.1 ± 0.1 kcal mol^{-1} at $-138°C$ for **496**. The rapid equilibrium in cations **495** and **496** has been frozen out in the solid state at $-165°C$ and $-160°C$, respectively, by Yannoni and co-workers.[857]

Many more cyclic and polycyclic equilibrating carbocations have been reported. Some representative examples, namely, the bisadamantyl (**499**),[859] 2-norbornyl (**500**),[40] 7-perhydropentalenyl (**501**),[188] 9-decalyl (**502**),[188] and pentacyclopropylethyl (**503**)[860] cations, are given in Scheme 3.19. All these systems again involve hypercoordinate high-lying intermediates or transition states.

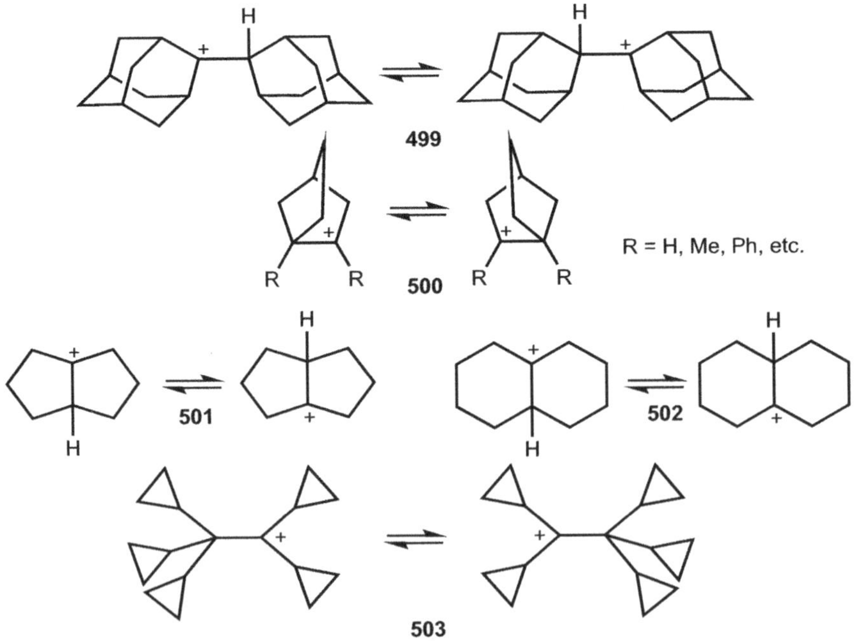

Scheme 3.19

3.5.2.2. The 2-Norbornyl Cation. The 2-norbornyl cation ($C_7H_{11}^+$) holds a unique position in the history of organic chemistry because of the important role it has played in the bonding theory of carbon compounds. Since Winstein's early solvolytic work in 1949[861] the 2-norbornyl cation was at the heart of the so-called nonclassical ion problem, and no other system has been studied so much by various physical and chemical methods and by so many investigators. The controversy[27–37,40,862–866] about it is well known, and the question has been whether the ion has a symmetrically bridged nonclassical structure **504** with a pentacoordinate carbon atom or is a rapidly equilibrating pair of classical trivalent ions **505a** and **505b**.

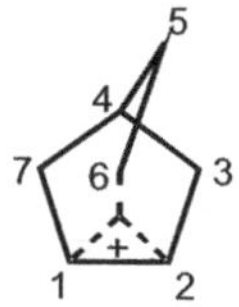

504

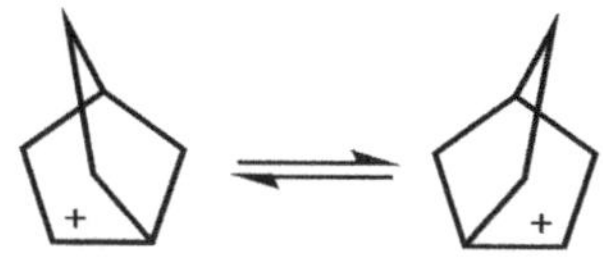

a **505** **b**

This controversy has been instrumental in the development of important structural methods in physical organic chemistry with respect to critical evaluation of results as well as to concepts behind the methods.

The methods that were developed in the early 1960s to generate and observe stable carbocations in low-nucleophilicity solutions[18] were successfully applied to direct observation of the norbornyl cation ($C_7H_{11}^+$). Preparation of the ion by the "σ route" from 2-norbornyl halides, by the "π route" from 4-(2-haloethyl)-cyclopentenes, and by the protonation of nortricyclene ("bent σ route") all led to the same 2-norbornyl cation.

The method of choice for the preparation of the norbornyl cation (giving the best-resolved NMR spectra, free of dinorbornylhalonium ion equilibration) is from *exo*-2-fluoronorbornane in SbF_5–SO_2 (or SO_2ClF) solution (Scheme 3.20).

Scheme 3.20

In a joint effort, Saunders, Schleyer, and Olah[867] first investigated the 60-MHz ^1H NMR spectrum of the 2-norbornyl cation in the early 1960s. Subsequently, Olah and co-workers[38,39] carried out detailed 100-MHz ^1H and 25-MHz ^{13}C NMR spectroscopic studies in the early 1970s at successively lower temperatures. From the detailed ^1H NMR investigations at various temperatures (RT to −154°C), the barrier for the

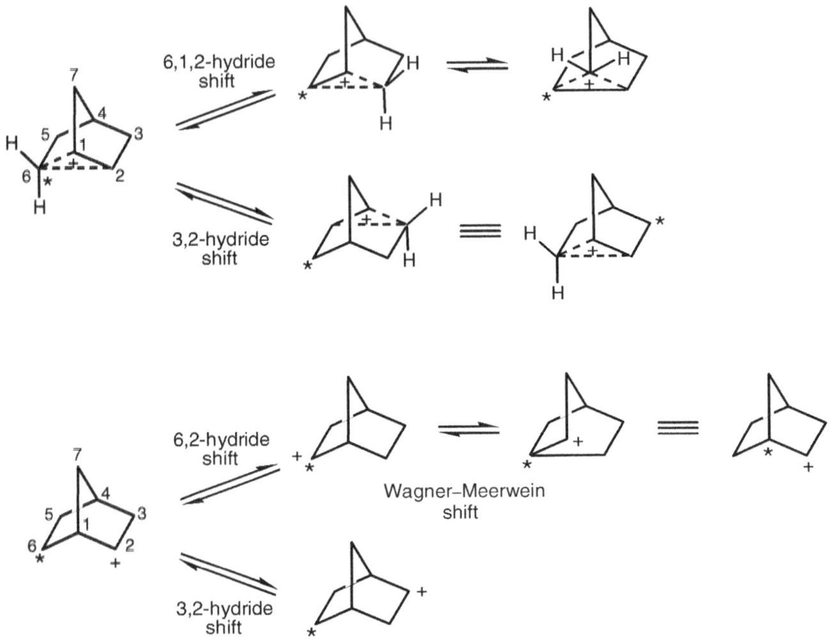

Figure 3.19. Degenerate shifts in the 2-norbornyl cation (one of the carbons is labeled for clarity).

2,3-hydrogen shift, as well as the 6,1,2-hydrogen shift, was determined by line-shape analysis and found to be 10.8 ± 0.6 kcal mol^{-1} and 5.9 kcal ± 0.2 mol^{-1}, respectively (Figure 3.19).[38]

The 60-MHz ^{1}H NMR spectrum of the 2-norbornyl cation at room temperature shows a single peak at δ^1H 3.10 for all protons, indicating fast 2,3-hydrogen, 6,1,2-hydrogen, and Wagner–Meerwein shifts.[867] Cooling the solution of the 2-norbornyl cation in the SbF_5–SO_2ClF–SO_2F_2 solvent system down to −100°C at 395 MHz (Figure 3.20) results in three peaks at δ^1H 4.92 (4 protons), 2.82 (1 proton), and 1.93 (6 protons), indicating that the 2,3-hydrogen shift is fully frozen, whereas the 6,1,2-hydrogen and Wagner–Meerwein shifts are still fast on the NMR time scale.[868]

Cooling the solution down further to −158°C results in significant changes in the spectrum. The peak at δ^1H 4.92 splits into two peaks at δ^1H 6.75 and 3.17 with a ratio of 2:2. The high-field peak broadens and splits into two peaks at δ^1H 2.13 and 1.37 and in the ratio 4:2. The peak at δ^1H 2.82 remains unchanged. The line width (~60 Hz) observed at 395 MHz was found to be rather small as compared to the one obtained[868] at 100 MHz (~30 Hz). This has some implications. If the line width were due to any slow exchange process occurring at this temperature, the line width should have broadened 15.6 times at 395 MHz over the one observed at 100 MHz. The observation of comparably narrow line widths at 395 MHz indicates that either the 6,1,2-hydrogen shift and the Wagner–Meerwein shift (σ-bond shift) are completely frozen and the 2-norbornyl cation has the symmetrically bridged structure **504** or the 6,1,2-hydrogen

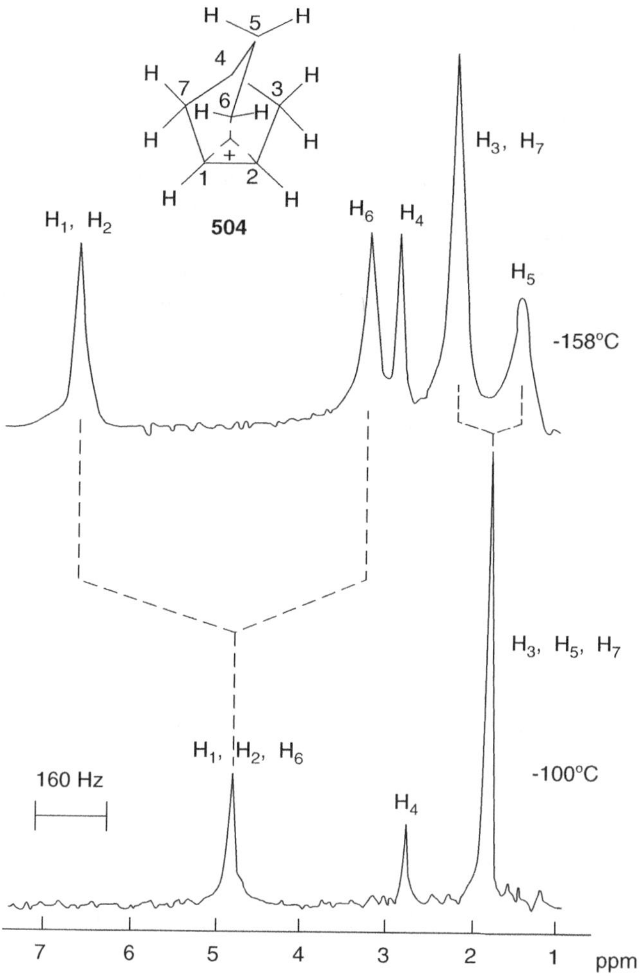

Figure 3.20. 395-MHz ^{1}H NMR spectra of 2-norbornyl cation in SbF_5–SO_2ClF–SO_2 solution.

shift is frozen and the so-called Wagner–Meerwein shift (if any) is still fast on the NMR time scale through a very shallow activation energy barrier (less than 3 kcal mol^{-1}). The second possibility raises the question as to the nature of the ion still undergoing equilibration through an extremely low-activation energy barrier. It has been pointed out[33] that if such a process occurs, it must be exclusively between unsymmetrically bridged ions **506** equilibrating through the intermediacy of the symmetrically bridged species **504**.

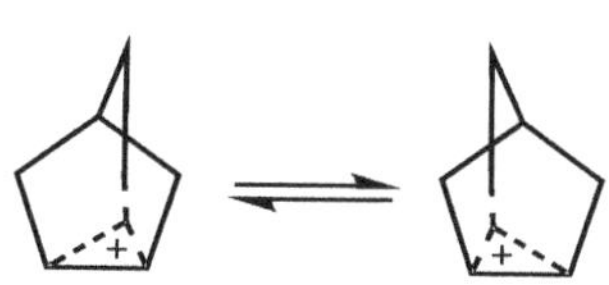

506

The unsymmetrically bridged ions **506** would be undistinguishable from the symmetrically bridged system **504** in solution NMR experiments. (However, see subsequent discussion of solid-state low-temperature ^{13}C NMR as well as ESCA studies.) It is important to recognize that equilibrating open classical cations **505** cannot explain the NMR data and thus cannot be involved as populated species.

The 50-MHz ^{13}C NMR spectrum of the 2-norbornyl cation ($C_7H_{11}^+$) has also been obtained in the mixed SbF_5–SO_2ClF–SO_2F_2 solvent system at $-159°C$.[868] To obtain a well-resolved ^{13}C NMR spectrum, the cation was generated from 15% ^{13}C-enriched *exo*-2-chloronorbornane [the label corresponds to one carbon per molecule randomly distributed over the C(1), C(2), and C(6) centers]. The ionization of the ^{13}C-enriched *exo*-2-chloronorbornane in SbF_5–SO_2ClF–SO_2F_2 solution at $-78°C$ results in the 2-norbornyl cation wherein the ^{13}C label is distributed evenly over all the seven carbons as a result of slow 2,3-hydrogen and fast 6,1,2-hydrogen and Wagner–Meerwein shifts.[868]

At $-80°C$, the 50-MHz ^{13}C NMR spectrum of the cation (Figure 3.21) showed three absorptions at δ^1H 91.7 (quintet, $J_{C-H} = 55.1$ Hz), 37.7 (doublet, $J_{C-H} = 153.1$ Hz),

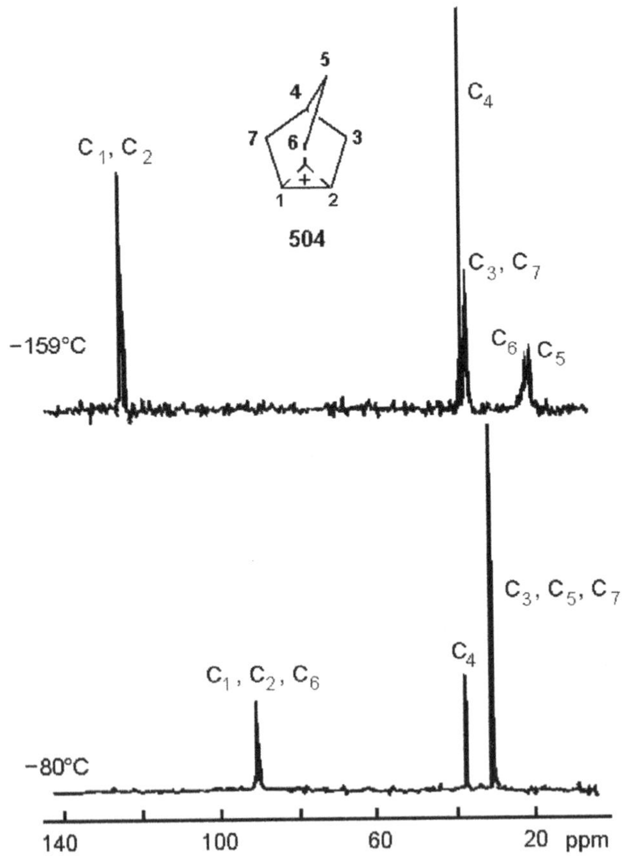

Figure 3.21. 50-MHz ^{13}C NMR spectra of the 2-norbornyl cation in SbF_5–SO_2ClF–SO_2F_2 solution.

and 30.8 (triplet, $J_{C\text{–}H} = 139.1$ Hz), indicating that the 2,3-hydrogen shift is frozen, but the 6,1,2-hydrogen and the Wagner–Meerwein shift is still fast on the NMR time scale. Cooling the solution down results in broadening and slow merger into the baseline of the peaks at δ^{13}C 91.7 and 30.8, but the peak at δ^{13}C 37.7 remains relatively sharp. At −159°C, the peaks at δ^{13}C 91.7 and 30.8 separate into two sets of two peaks at δ^{13}C 124.5 (doublet, $J_{C\text{–}H} = 187.7$ Hz), 21.2 (triplet, $J_{C\text{–}H} = 147.1$ Hz), and 36.3 (triplet, $J_{C\text{–}H} = 131.2$ Hz), 20.4 (triplet, $J_{C\text{–}H} = 153.2$ Hz), respectively. The observed ^{13}C NMR spectral data at −159°C complement well the 395 MHz ^{1}H NMR data at −158°C. The observation of the C(1) and C(2) carbons at δ^{13}C 124.5 and the C(6) carbon at δ^{13}C 21.2 clearly supports the bridged structure for the ion. Five (or higher) coordinate carbons generally show shielded (upfield) ^{13}C NMR shifts.[55]

Applying the additivity of chemical shift analysis[55] to the 2-norbornyl cation also supports the bridged nature of the ion. A chemical shift difference of 168 ppm is observed between the ion ($C_7H_{11}^+$) and its parent hydrocarbon [i.e., norbornane (**507**)], whereas ordinary trivalent carbocations such as the cyclopentyl cation (**33**) reveal a chemical shift difference of ~360 ppm.[55]

507 $\Delta\Sigma\delta = 168$ $\Delta\Sigma\delta = 360$ **33**

Yannoni, Macho, and Myhre[869] obtained magic-angle spinning cross-polarization ^{13}C NMR spectra of the ^{13}C-enriched 2-norbornyl cation in SbF_5 solid matrix down to −196°C. The solid-state chemical shifts and measured barriers for the 6,1,2-hydrogen shift of 6.1 kcal mol^{-1} correlate well with the discussed solution data. Subsequently, they even obtained ^{13}C NMR spectra in the solid state at −268°C (5 K),[870] a remarkable achievement indeed.

A fortuitous combination of large isotropic chemical shifts and small chemical shift anisotropies permitted them to obtain reasonable resolution of the positively charged carbon resonance without the need for magic-angle spinning. Comparison with their previous MAS spectra[869] down to −196°C shows that the nonspinning spectra reflect slowing of 6,2,1-hydride shift. Since no changes were observed in the positively charged carbon resonance (at δ^{13}C ~125) between −173°C and −268°C (Figure 3.22), the authors concluded[870] that if the hypothetical 1,2-Wagner–Meerwein shift is still occurring, then it should be rapid, and an upper limit for the barrier for such a process (involving structures **505**) can be estimated to be no greater than 0.2 kcal mol^{-1}. This can be taken as the most definitive evidence besides ESCA studies for the symmetrical σ-bridged structure of the 2-norbornyl cation. Subsequently, Dewar and Merz[871] raised the possibility of low-energy carbon tunneling between unsymmetrically delocalized nonclassical ions such as **506** based on MINDO/3 calculations. Such carbon tunneling, however, is unlikely.

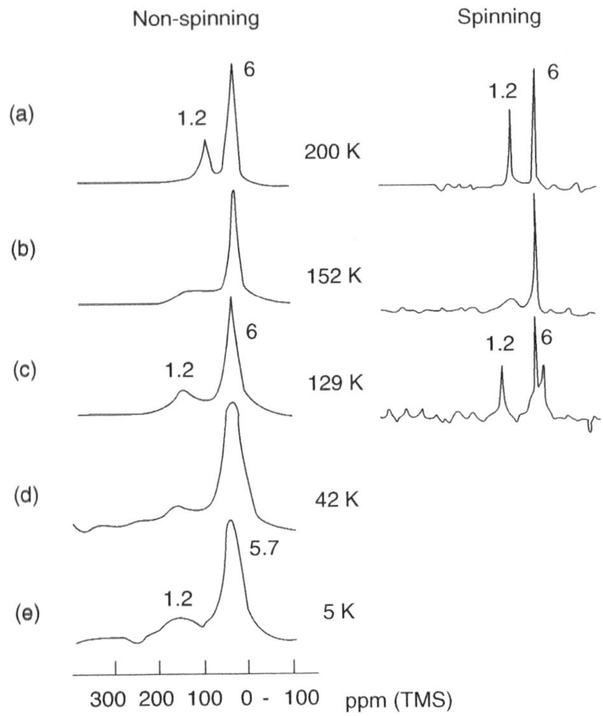

Figure 3.22. Solid-state ^{13}C NMR spectra of the 2-norbornyl cation according to Yannoni and Myhre.[870]

As discussed earlier, the method of observing changes in NMR spectra produced by asymmetric introduction of isotopes (isotopic perturbation) as a means of distinguishing systems involving equilibrating species passing rapidly over a low barrier from molecules with single-energy minima, intermediate between the presumed equilibrating structures, has been developed by Saunders et al.[56] Applying this method to the 2-norbornyl cation further supports its bridged nature.[872] In the ^{13}C NMR spectrum of the 2-norbornyl cation, even at low temperatures, besides Wagner–Meerwein rearrangement, the 6,1,2-hydrogen shift has a barrier of only 5.9 kcal mol^{-1} and results in a certain amount of line broadening of the lowest field signal observed. Even in the ion with no deuterium, the downfield signal at $\delta^{13}C$ ~124.5 [C(2) and C(6) cyclopropane-like carbons] is found to be 2 ppm wide. Nevertheless, no additional isotopic splitting or broadening was observed with either 2-monodeutero or 3,3-dideutero cations, and therefore the isotopic splitting can be no more than 2 ppm. This is true even if a slow 6,2-hydride shift converts part of the latter ions to a symmetrical 5,5-dideutero system that lacks an equilibrium isotope effect. This result, when compared with the significantly larger splitting observed for deuteriated dimethylcyclopentyl (**508**) and dimethylnorbornyl (**509**) cations[61,872] (known to be equilibrating ions) is in accordance with the nonclassical nature of the 2-norbornyl cation. A similar conclusion was reached[873,874] based on high-temperature deuterium isotopic perturbation effect in 2-norbornyl cation.

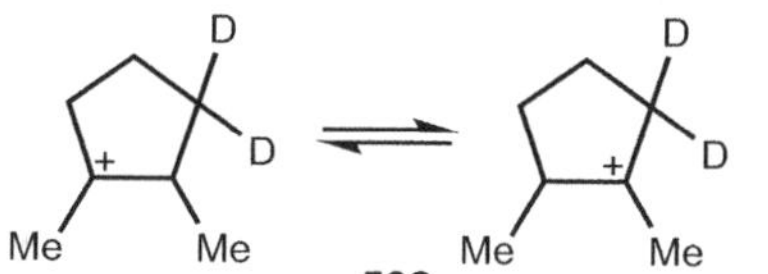

508

$\Delta\delta C$ = 104 ppm

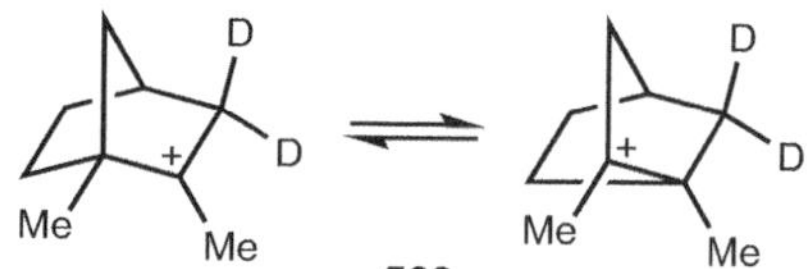

509

$\Delta\delta C$ = 24 ppm

$\Delta\delta C$ < 2 ppm

504

Farnum and Olah's groups, respectively, have extended the so-called Gassman–Fentiman tool of increasing electron demand coupled with ^{1}H and ^{13}C NMR spectroscopy as the structural probe under stable ion conditions to show the onset of π, πσ, and σ delocalization in a variety of systems.[153,154,873–883] The ^{13}C NMR chemical shifts of the cationic carbon of a series of regular trivalent 1-aryl-1-cyclopentyl, 1-aryl-1-cyclohexyl, 2-aryl-2-adamantyl, 6-aryl-6-bicyclo[3.2.1]octyl, and 7-aryl-7-norbornyl cations (so-called classical cations) correlate linearly with the observed cationic chemical shifts of substituted cumyl cations over a range of substituents[881–883] [generally from the most electron-releasing 4-MeO to the most electron-withdrawing 3,5-$(CF_3)_2$ groups].

However, 2-aryl-2-norbornyl cations **510** show deviations from linearity in such chemical shift plots with electron-withdrawing substituents indicative of the onset of nonclassical σ-delocalization fully supporting the nonclassical nature of the parent secondary cation (Figure 3.23). These conclusions were criticized by Brown et al.[883] In a subsequent paper, Olah et al.[67] have shown major flaws in such criticisms.

As mentioned earlier, since in electron spectroscopy the time scale of the ionization processes is on the order of 10^{-16} s, definite ionic species are characterized, regardless of their possible intra- and intermolecular rearrangements (e.g., Wagner–Meerwein rearrangements, hydride shifts, etc.) even at rates equaling or exceeding those of vibrational transitions. Thus, electron spectroscopy can give an unequivocal answer to the long-debated question of the "classical" or "nonclassical" nature of the norbornyl cation, regardless of the rate of any possible equilibration processes.

Olah et al.[39] succeeded in observing the ESCA spectrum of the norbornyl cation (**504**) and compared it with those of the 2-methyl-2-norbornyl cation (**511**) and other trivalent carbenium ions such as the *tert*-butyl (**1**), cyclopentyl (**33**), and 1-methylcyclopent-1-yl (**28**) cations. The 1*s* electron spectrum of the norbornyl cation shows no high-binding energy carbenium center, and a maximum separation of less than 1.5 eV is observed between the two "cyclopropyl"-type carbons, to which bridging takes place from the other carbon atoms (including the pentacoordinate bridging carbon). In contrast, the 2-methyl-2-norbornyl cation (**511**) shows a high-binding energy carbenium center, deshielded with the ΔE_b of 3.7 eV from the other carbon atoms. Typical ESCA shift differences are summarized in Table 3.6.

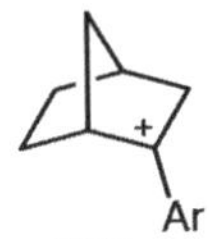

510

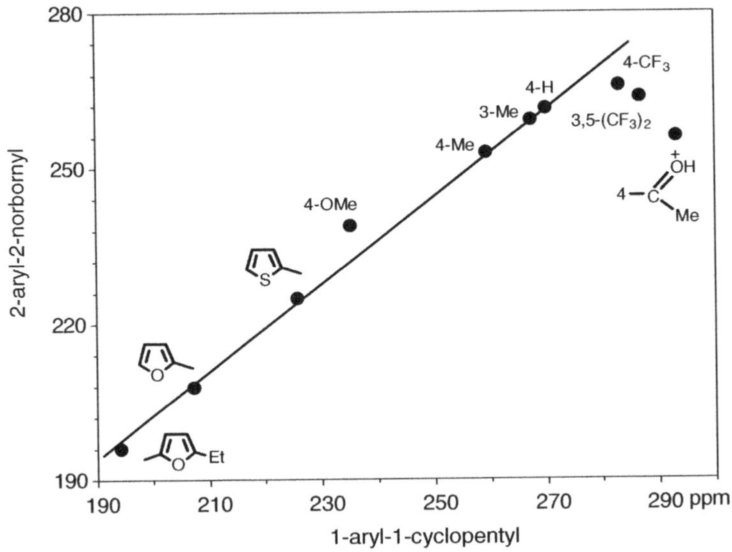

Figure 3.23. Plot of the ^{13}C NMR chemical shifts of the cationic center of 2-aryl-2-norbornyl cations versus those of model 1-aryl-1-cyclopentyl cations.[880]

Table 3.6. Binding Energy Differences of Carbocation Centers from Neighboring Carbon Atoms ΔE_{b+C-C}

Ion		ΔE_{b+C-C}	Approximate Relative C_+: C Intensity
Me_3C^+	**1**	3.9 ± 0.2	1:3
	**33**	4.3 ± 0.5	1:4
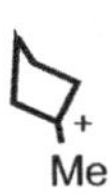	**28**	4.2 ± 0.2	1:5
	**511**	3.7 ± 0.2	1:7
	**504**	1.5 ± 0.2	2:5

Subsequently, Grunthaner reexamined the ESCA spectrum of the 2-norbornyl cation on a higher-resolution X-ray photoelectron spectrometer using highly efficient vacuum techniques.[884] The spectrum closely matches the previously published spectra. Furthermore, the reported ESCA spectral results are consistent with the theoretical studies of Allen and co-workers[885] on the classical and nonclassical norbornyl cation at the STO-3G and STO-4.31G levels. Using the parameters obtained by Allen and co-workers, Clark and co-workers[886,887] were able to carry out a detailed interpretation of the experimental ESCA data for the core-hole state spectra at SCF STO-4.31G level and calculated equivalent cores at STO-3G level. Agreement between experimentally obtained spectra and those calculated for the nonclassical cation are good and dramatically different from those for the classical cation.

If the classical structure were correct, the 2-norbornyl cation would be a usual secondary carbocation with no additional stabilization provided by σ-delocalization (such as the cyclopentyl cation). The facts, however, seem to be to the contrary. Direct experimental evidence for the unusual stability of the secondary 2-norbornyl cation comes from the low-temperature solution calorimetric studies of Arnett and Petro.[75] In a series of investigations, Arnett and Hofelich[76] determined the heats of ionization (ΔH_i) of secondary and tertiary chlorides in SbF_5–SO_2ClF [Eq. (3.131)] and subsequently alcohols in HSO_3F–SbF_5–SO_2ClF solutions [Eq. (3.132)].

$$\mathrm{R{-}Cl} \quad + \quad \mathrm{SbF_5} \xrightarrow[\mathrm{SO_2ClF}]{\Delta H_i} \mathrm{R^+\,SbF_5Cl^-} \qquad (3.131)$$

$$\mathrm{R{-}OH} \quad + \mathrm{HSO_3F{-}SbF_5} \xrightarrow[\mathrm{SO_2ClF}]{\Delta H_i} \mathrm{R^+\,SbF_5(FSO_3)^-} \quad + \quad \mathrm{H_3O^+} \qquad (3.132)$$

However, it was found that whereas the difference observed in the heats of ionization of 2-methyl-2-*exo*-norbornyl chloride and 2-*exo*-norbornyl chloride in SbF_5–SO_2ClF solution is 7.4 kcal mol^{-1}, the same difference between the corresponding alcohols in HSO_3F–SbF_5–SO_2ClF solution is only 2.5 kcal mol^{-1}. This indicates that the heats of ionization values (ΔH_i) seem to largely depend on the nature of the starting precursors (initial state effects). However, the observed differences are remarkably small for the corresponding secondary and tertiary cations, which generally is 10–15 kcal mol^{-1}. In the case of norbornyl, there seems to be at least 7.5 kcal mol^{-1} extra stabilization. A further compelling evidence for the nonclassical stabilization of the 2-norbornyl cation also comes from Arnett's measured heats of isomerization of secondary cations to tertiary cations.[888] The measured heat of isomerization of 4-methyl-2-norbornyl cation **512** (secondary system) to 2-methyl-2-norbornyl cation **511** is -6.6 kcal mol^{-1} [Eq. (3.133)]. In contrast, the related isomerization of *sec*-butyl cation **19** and *tert*-butyl cation **1** involved a difference in ΔH_i of -14.2 kcal mol^{-1} [Eq. (3.134)].

512 **511**

$\Delta H_i = -6.6$ kcal mol^{-1} (3.133)

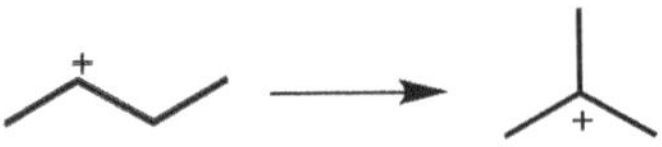

$\Delta H_i = -14.2$ kcal mol^{-1} (3.134)

19 1

Taking this latter value as characteristic of isomerization of secondary to tertiary ions, one must conclude that the secondary norbornyl ion **512** has an extra stabilization of at least 7.6 k cal mol^{-1}. Fărcaşiu[889] questioned these conclusions, arguing that they neglected to account for the extra stabilization by bridgehead methyl substitution as indicated by his molecular force field calculations. Schleyer and Chandrashekar[890] have subsequently pointed out that Fărcaşiu failed to include corrections of β-alkyl branching. Correcting for this effect, there is still 6 ± 1 kcal mol^{-1} extra stabilization in the 2-norbornyl cation for which no other reasonable explanation other than bridging was offered.

Gas-phase mass spectrometric studies[891–894] also indicate exceptional stability of the 2-norbornyl cation relative to other potentially related secondary cations. A study by Kebarle and co-workers[895] also suggests that the 2-norbornyl cation is more stable than the *tert*-butyl cation in the gas phase (based on hydride transfer equilibria from their respective hydrocarbons).

Subsequent experimental observations lent further strong support to the nonclassical structure of high stability. In 1987, Laube determined the crystal structure of the 1,2,4,7-*anti*-tetramethyl-2-norbornyl cation as the $Sb_2F_{11}^-$ salt[896] and redetermined the structure in 1994.[122,897] The geometry of the skeleton of the cation is significantly different than that of the norbornane skeleton and more similar to the geometry calculated for the symmetrical 2-norbornyl cation **504**. Forsyth and Panyachotipun[898] observed large isotope shifts for the ^{13}C resonance at the cationic center of the deuteriated 2-methyl-2-norbornyl cation, which was attributed to the bridging structure. The IR spectra of the matrix-generated 2-norbornyl cation were recorded by Koch, Sunko and co-workers[899] at 150–200 K. The experimental and calculated spectra (MP2/6-31G* level)[899,900] computed on the 2-norbornyl cation of C_s symmetry agree well and again support the bridged nonclassical structure. Finally, the stability of a range of bridgehead cations including the 2-norbornyl cation has been determined by the FT–ICR method on dissociative proton attachment of bromides and alcohols.[901] Good correlations were found for the stability of the ions with the solvolytic reactivity of bridgehead derivatives and theoretical calculations (MP2/6-311G** level) for hydride transfer of bridgehead hydrocarbons. The 2-norbornyl cation, a secondary carbocation, lies nicely on the correlation line of $\Delta G^0_{(exp)}$ versus $\Delta G^0_{(theor)}$, although this is defined by tertiary ions. The stability of the 2-norbornyl cation (**504**) is close to that of the 1-adamantyl cation, that is, **504** is much more stable than simple classical secondary or strained tertiary cations. The calculated C–C bond lengths agree within 0.04 Å with that reported for **504** in the gas phase and in solution by a high-level theoretical study.[902]

Theoretical quantum mechanical calculations[903–908] have also been performed on the 2-norbornyl cation at various levels. These calculations reveal a significant preference for the σ-delocalized nonclassical structure. An extensive calculation by Schaefer and co-workers[906] using full geometry optimization for symmetrically and unsymmetrically bridged systems showed a difference of only 1.0 kcal mol^{-1} between these structures. (Some confusion was introduced by Schaefer, who called the unsymmetrically bridged ion "classical.") Similar high-level calculations, including electron correlations (with a double zeta plus polarization basis set), by Schleyer, Schaefer, and co-workers[907] and Liu and co-workers,[908] indicate that the only minimum on the 2-norbornyl cation potential energy surface is the symmetrically bridged structure. The nonclassical symmetrically bridged 2-norbornyl cation (**504**) was calculated to be 24.8 kcal mol^{-1} more stable than the isopropyl cation (based on the hydride transfer reaction). The structure with "classical" 2-norbornyl-like geometry (**505**), however, did not correspond to a fixed point on the potential energy surface. The extra stabilization of the bridged structure (**504**) was roughly estimated to be 12–15 kcal mol^{-1} at this high level of *ab initio* theory.[907,908]

In further theoretical studies, Schleyer and co-workers[902,909] found that the nonclassical 2-norbornyl cation (**504**) is more stable than the classical structure **505**, by 13.6 kcal mol^{-1}. Cation **504** was shown to be the only stable form in the gas phase and in solution, and the classical form (**505**) is unlikely to be involved in solvolysis reactions [MP4(fc)SDQ/6-31G**/MP2(full)/6-31G*+ZPVE level]. A rigorous *ab initio* calculation led to the suggestion that the classical **505** structure is an artifact and a transition state in the rearrangement of **504** into the bridged 2-norpinyl cation. The computed MP2-GIAO ^{13}C chemical shifts for **504** are close to the experimental values, whereas those calculated by Schindler[910] (IGLO) for **505** deviate considerably. The comparison of the chemical shifts for the two structures with the experimental data proves once again the symmetrical nonclassical structure.

Werstiuk and coworkers published a series of articles[911–914] and showed that neither the 2-norbornyl cation nor the 1,2,4,7-*anti*-tetramethyl-2-norbornyl cation is an energy minimum on the potential energy surface and they are not nonclassical, σ-bridged species [AIM (atoms in molecules) method]. In a recent extensive interpretation, Mamantov[915] suggested a structure with H–C(1)–H σ-bond delocalization, that is, C(1) rather than C(6) being the hypercoordinated carbon atom. However, such a structure is highly unlikely based on the experimental results.

Nine nonclassical carbocations derived from cyclic hydrocarbons including the 2-norbornyl cation have recently been studied by means of *ab initio* calculations (MP2/6-311G**).[916] Electron density, chemical shifts, geometries, energetics and orbital interactions within the natural bond orbital (NBO) formalism were computed and analyzed. In sharp contrast to the findings by Werstiuk's group, two-electron, three-center bonding structures were found with the charge largely and similarly distributed over the three carbon centers. Bond lengths for two C–C bonds are 1.84 Å and the third C–C bond is 1.40 Å. The computed chemical shift values agree with the experimental values within a difference of 3.2 ppm.

3.5.2.3. The 7-Norbornyl Cation. 7-Norbornyl derivatives were found to be extremely unreactive in solvolysis studies and product formation was shown to occur with predominant retention of configuration.[917–920] These observations led to the suggestion by Winstein et al.[917] that the cationic intermediate is a nonclassical ion. Attempts to isolate the 7-norbornyl cation under stable ion conditions in superacid

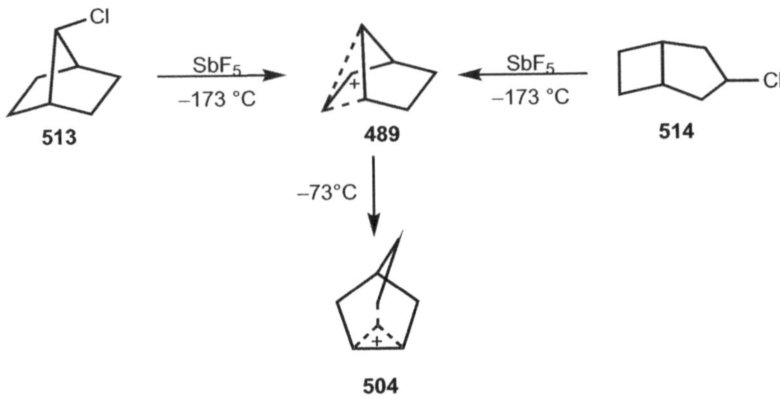

Scheme 3.21

media, however, failed: both 7-chloronorbornane (**513**) and the isomeric 3-chlorobicyclo[3.2.0]heptane (**514**) afforded the 2-norbornyl cation.

Subsequently, Sunko, Schleyer, and co-workers[921] succeded in observing the 7-norbornyl cation using the cryogenic matrix isolation technique. When 7-chloronorbornane **513** and a large excess of SbF_5 were co-deposited at −263°C and slowly warmed to −173°C, the IR signals indicated the formation of the 7-norbornyl cation **489** (Scheme 3.21). Further warming gave the 2-norbornyl cation **504** at −73°C. The IR spectrum was in very good agreement with the calculated frequencies for the 7-norbornyl cation [MP4(sdq,fc)/6-31G*//MP2(full)6-31G*+ZPE level]. The possibility of the formation of a complex of **513** with SbF_5 was ruled out when similar experiments with both **513** and **514** resulted in identical spectra. Recent *ab initio* calculations[916] support the nonclassical bent structure of the cation.

3.5.2.4. The 2-Bicyclo[2.1.1]hexyl Cation. The bicyclo[2.1.1]hexyl cation **515** was first observed by Wiberg and co-workers[922] in superacidic media by ^{1}H NMR spectroscopy. From the observed chemical shift data, they suggested a symmetrically bridged structure (**516**) for the ion, although they could not freeze out the degenerate equilibria. Similar conclusions were drawn from solvolytic studies.[923,924]

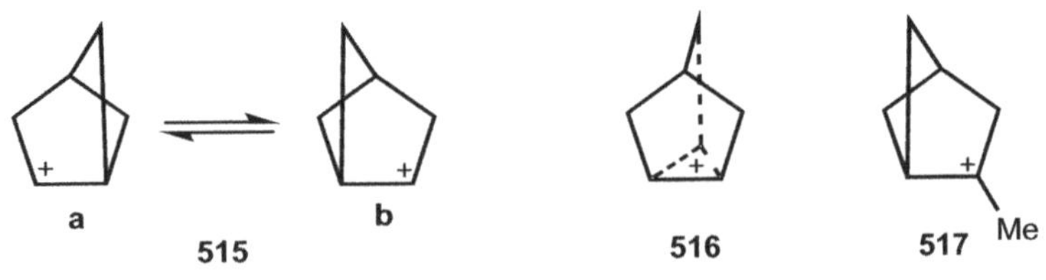

In a subsequent ^{13}C NMR study, Olah, Liang, and Jindal[925] concluded that there is very little σ-bridging in the rapidly equilibrating ion. The ^{1}H NMR spectrum of the ion **515** in SbF_5–SO_2ClF showed three resonances at δ^1H 8.32 (two protons), 3.70 (six protons), and 2.95 (one proton) with no significant line broadening down to −140°C.

The ^{13}C NMR spectrum of the ion **515** also shows three resonances at $\delta^{13}C$ 157.8 [doublet, $J_{C-H}=184.5$ Hz; C(1) and C(2)], 49.1 [triplet, $J_{C-H}=156.9$ Hz; C(3), C(5), and C(6)], and 43.4 [doublet, $J_{C-H}=164.6$ Hz; C(4)]. Above −90°C the ion irreversibly rearranges to the cyclohexenyl cation.

A study by Saunders, Wiberg, and co-workers[926] involving deuterium labeling at the exchanging sites indicates that there is significant σ-bridging in the ion **515**. Schmitz[927] and Sorensen have shown that the free-energy difference between cations **515** and **517** is 7–9.8 kcal mol^{-1} compared with 5.5 and 11.4 kcal mol^{-1} for the analogous 2-norbornyl and cyclopentyl cations substantiating the intermittent (partially bridged) nature of the ion **515**.

An unequivocal support for the existence of interconverting bridged ions has come from labeling experiments by Kirmse et al.[928] including the use of the double-labeled [2-^{2}H,3-^{13}C] derivative. Further support was provided by *ab initio* calculations[929] [MP(full)/6-31G* level] indicating that both **515** and **516** are energy-minimum structures but **516** is favored over **515** by 4 kcal mol^{-1}. Structure **516** is also supported by a comparison of experimental and calculated (IGLO) ^{13}C NMR chemical shifts values.[929,930]

Attempts to prepare[931] the analogous bicyclopentyl cation **518**, however, were unsuccessful and instead gave the rearranged cyclopentenyl cation **73**.[221]

518

73

3.5.2.5. Degenerate Cyclopropylmethyl and Cyclobutyl Cations. In contrast to the "classical" tertiary and secondary cyclopropylmethyl cations (showing substantial charge delocalization into cyclopropane ring but maintaining their identity), primary cyclopropylmethyl cations rearrange to cyclobutyl and homoallylic cations under both solvolytic and stable ion conditions.[18,195,196,932–934] The nonclassical nature of cyclopropylmethyl and 1-methylcyclopropylmethyl cations **519** and **520** is now firmly established.[171–176,935–940] Wide-ranging studies showed[201,212–214] that the cyclopropyl group is equal to or better than a phenyl group in stabilizing an adjacent carbocationic center. The direct observation of cyclopropylmethyl cations provides a clear example of positive charge delocalization into a saturated πσ-hydrocarbon system. The majority of the secondary cyclopropylmethyl cations, however, undergo degenerate equilibria.[941–943]

The cyclopropylmethyl cation **519** can be generated from allylic, cyclobutyl, and cyclopropylmethyl precursors (Scheme 3.22). The ^{1}H NMR spectrum is shown in Figure 3.24.

At the lowest temperatures studied (∼−140°C), ^{13}C NMR spectroscopy indicates that **519** is still an equilibrating mixture of bisected σ-delocalized cyclopropylcarbinyl cations **521** and bicyclobutonium ion **522**.[171,172] From the comparison of calculated NMR shifts, the low-lying species is considered to be the bicyclobutonium ion.[171,172]

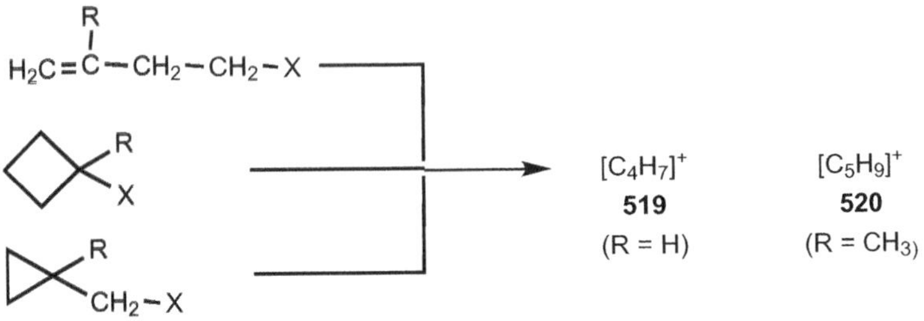

Scheme 3.22

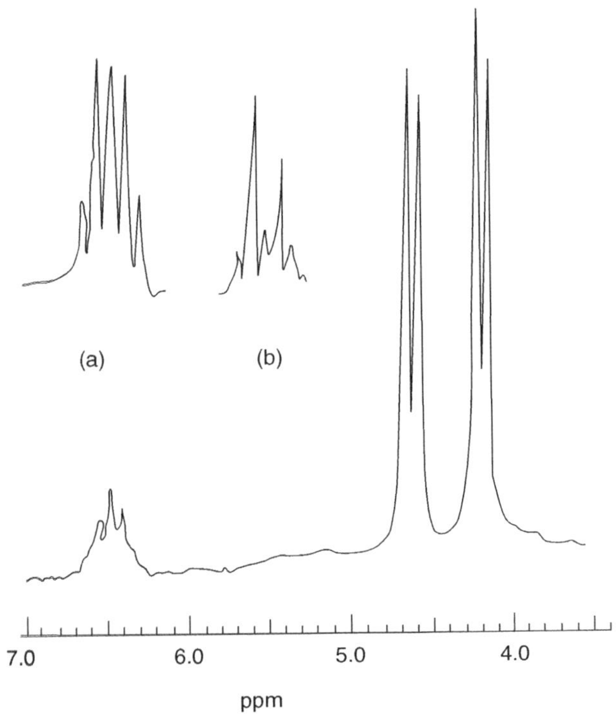

Figure 3.24. 100-MHz 1H NMR spectrum of the cyclopropylcarbinyl cation in SbF_5–SO_2ClF solution at −80°C. (*a*); 60-MHz spectrum of H_2 region; (*b*); 60-MHz spectrum of H_2 region from the α,α-dideuteriocyclopropylcarbinyl precursor.[171]

A similar conclusion has been reached by Saunders, Roberts, and co-workers[937,938] based on isotopic perturbation studies.

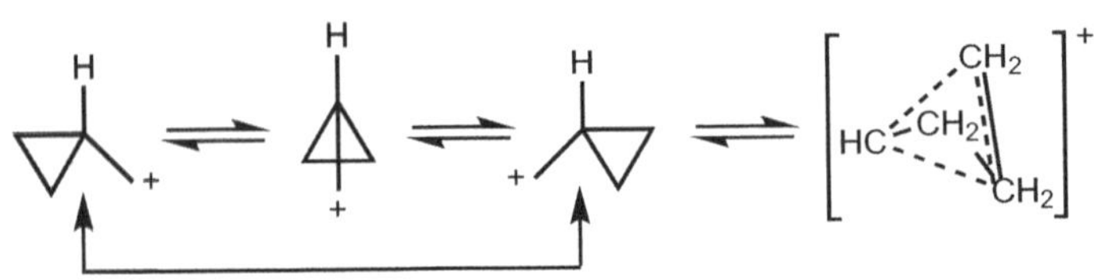

521 **522**

The same conclusion was reached by calculations using the IGLO method on the MP2/6-31G* optimized geometries.[944] Cations **521** and **522** were found to equally stable isomers lying 9.0 kcal mol^{-1} higher than the global minimum (1-methylallyl cation) reported in computational studies [MP4/6-311G**//MP2/6-31G*+ZPVE level].[945] Additional studies using ultralow-temperature CP-MAS NMR[946] and results of the IR spectra of $C_4H_7^+$ at 180 K in SbF_5 matrices in combination with MP2/6-31G* calculations also support these conclusions.[234] Cacace et al.[947] have recently found that in the gas phase and in a gaseous microsolvated environment the equilibrium ratio of **521/522a+522b** is very close to unity and equilibration occurs within a time interval of $\leq 10^{-10}$ s (FT–ICR mass spectrometry and high-pressure radiolytic techniques).

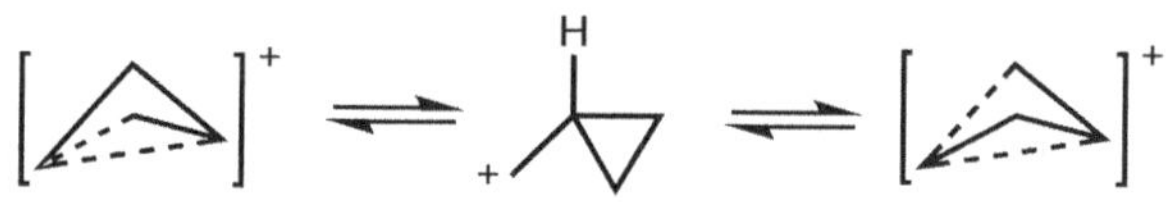

522a **521** **522b**

$C_4H_7^+$ ions were generated by collisionally activated dissociation (CAD) in the gas phase from various precursors.[216] Mass spectrometric analysis showed that homoallyl chloride and cyclopropylmethyl chloride generated primarily cation **521**, whereas cyclobutyl chloride gave a substantial amount of bicyclobutonium ion **522**.

However, in the case of $C_5H_9^+$ **520**, the low-lying species are the nonclassical methylbicyclobutonium ions **523** with no contribution from either the bisected 1-methyl-1-cyclopropylmethyl cation **524** or the 1-methylcyclobutonium cation **48a**. The highly shielded β-methylene resonance at $\delta^{13}C$ −2.81 in the ^{13}C NMR spectrum is particularly convincing evidence for the nonclassical bicyclobutonium structures.[176] Support for the bridged structure **525** comes from deuterium isotope perturbation studies.[939,940] The 1-ethyl and 1-propyl analogs of **520** are similarly nonclassical but rearrange irreversibly upon warming to cycloalkyl cations.[176]

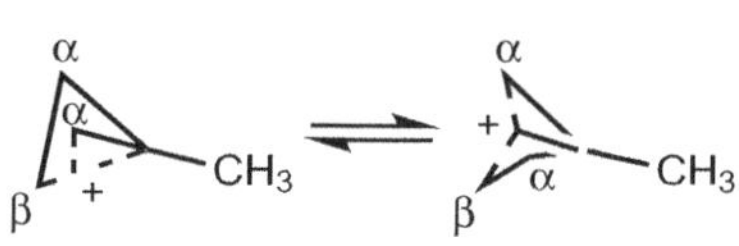

523

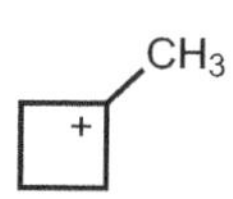

48a

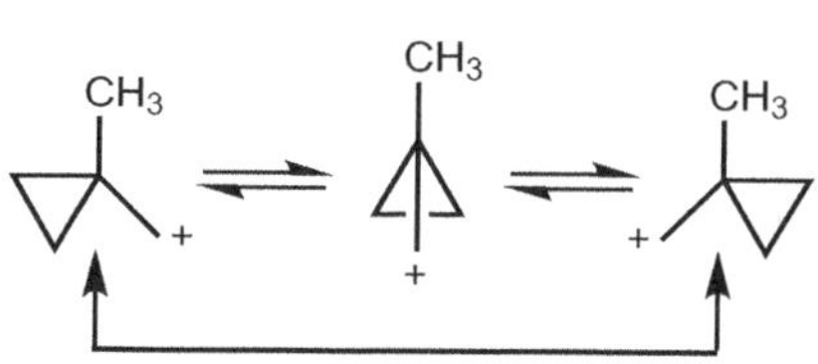

524

525

The potential energy surface of the analogous $C_7H_{11}^+$ cation with the built-in $C_4H_7^+$ bicyclobutonium subunit has recently been investigated[948] by *ab initio* (MP2/6-31G*) and DFT (B3LYP/6-31G*) calculations. The pentacoordinated unsymmetrical bicyclobutonium ion **526** was found to be the global minimum, but the boat conformer of cyclopropylcarbinyl cation **527** is only less stable by 0.8 kcal mol^{-1}. With the larger basis set or higher level of the MP method, the gap between the two cations decreases further.

526 **527**

Siehl and co-workers[949,950] have used the matrix co-condensation technique to generate the 1-(trimethylsilyl)bicyclobutonium ion **528** (Scheme 3.23). The ^1H and ^{13}C NMR spectra of ion **528** show averaged methylene signals, which is in accord with a fast threefold degenerate rearrangement and a puckered hypercoordinate structure.

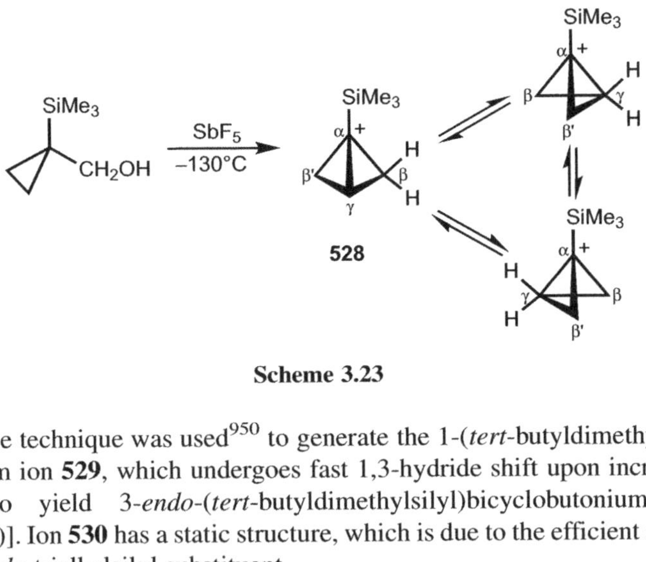

Scheme 3.23

The same technique was used[950] to generate the 1-(*tert*-butyldimethylsilyl)bicyclobutonium ion **529**, which undergoes fast 1,3-hydride shift upon increasing temperature to yield 3-*endo*-(*tert*-butyldimethylsilyl)bicyclobutonium ion **530** [Eq. (3.135)]. Ion **530** has a static structure, which is due to the efficient stabilization by the γ-*endo*-trialkylsilyl substituent.

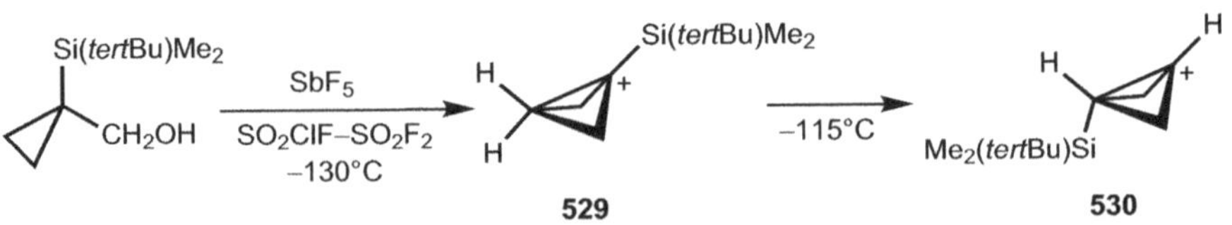

(3.135)

Schmitz and Sorensen[951] have prepared the primary cyclopropylmethyl cation **531** which shows static behavior. The nortricyclylmethyl cation **531** is regarded as a vinyl bridged 2-norbornyl cation **531b**. The support for the structure comes not only from ^{1}H and ^{13}C NMR studies but also from molecular orbital calculations.[952]

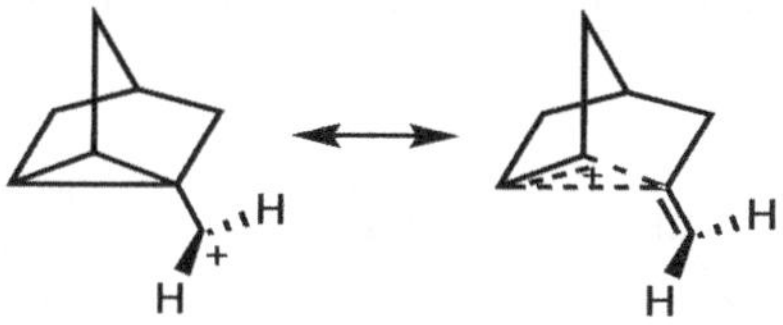

a b

531

Ionization of alcohol **532** (Scheme 3.24) and diol **536** [Eq. (3.136)] precursors to generate the corresponding triaxane-2-methyl cation and ditriaxane-2,2-dimethyl dication, respectively, have been explored by Olah, Prakash, and coworkers.[953,954] ^{13}C NMR characteristics of the ion generated from alcohol **532** (SbF_5–SO_2ClF, −78°C) are in accordance with a static classical, bisected cyclopropylmethyl cation (**533**), a fast equilibrium between nonclassical, unsymmetrically bridged bicyclobutonium ions (**534a** and **534b**), or a fast equilibrium of classical protoadamantyl cations (**535a** and **535b**) (Scheme 3.24). A comparison of calculated and experimental ^{13}C NMR shift values, however, excludes ion **533**. This and the application of the ^{13}C chemical shift additivity concept giving a $\Delta\delta^{13}$C value of 226 indicate that the most likely structure for the cation is a set of rapidly equilibrating bridged bicyclobutonium ions (**534a** and **534b**).

Because the geometry of **536** is similar to that of **532**, similar spectral features can be expected. The only exception is that the two distonic positive charges result in a higher atom-to-charge ratio and, consequently, a lower extent of σ delocalization into the strained cyclopropyl moieties. Ionization of **536** was performed under similar conditions[953,954] [Eq. (3.136)]. In accordance with expectations, the ^{13}C NMR spectroscopic features (deshielding of the methylenes by 71 ppm compared to that of **534a** and **534b**) indicate higher positive charge density and lower degree of charge delocalization. However, this value is still relatively shielded by 72.5 ppm when compared to the nortricyclylmethyl cation (**531**); that is, charge delocalization is still

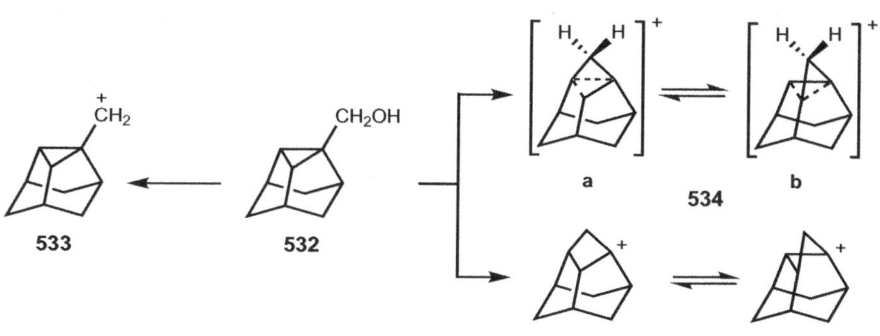

533 532 a b 534 a b 535

Scheme 3.24

substantial ruling out the static bisected structure **537**. The lack of temperature dependence of chemical shifts and the calculated ^{13}C chemical shift additivity value ($\Delta\delta^{13}C = 250$ per positive charge) support the bicyclobutonium-type nonclassical structures **538a–538c**.

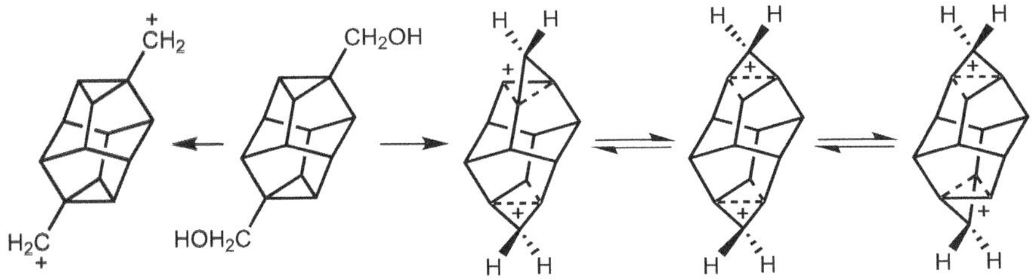

537 536 a b c
538

(3.136)

Prakash, Olah, and co-workers[209] succeeded in preparing the stable persistent tertiary cyclobutyldicyclopropylmethyl cation **69** by ionizing the corresponding alcohol (see Section 3.4.4). Primary and secondary cyclobutylmethyl cations, in turn, are nonclassical in nature and rearrange to thermodynamically more stable cyclopentyl cations. In a recent *ab initio* study (MP2/6-31G* and MP2/cc-pVTZ levels)[955] the authors have found that the primary cyclobutylmethyl cation with a σ-bridged nonclassical structure is an energy minimum on the potential energy surface. Charge delocalization into the cyclobutyl ring is evidenced by the significant elongation of the C(1)–C(2) bond (1.903 Å) which is comparable to that of the C(2)–C(Me) bond (1.738 Å). Two structures of energy minima were also identified for the secondary 1-cyclobutylethyl cation. Conformer **539a** with *exo* methyl group is more stable than conformer **539b** (*endo* methyl) (MP2/cc-pVTZ//MP2/cc-pVTZ+ZPE level), by 1.2 kcal mol^{-1}. The elongated, almost equal, bond distances of conformer **539a** clearly show its true nonclassical nature [C(1)–C(2) = 1.837 Å, C(2)–C(5) = 1.822 Å]. The shorter C(1)–C(2) bond distance of conformer **539b** (1.819 Å) indicates its relatively lower stability [C(2)–C(5) = 1.839 Å]. Calculated ^{13}C NMR shift data ($\delta^{13}C$ 154.1 and 142.2 for **539a** and **539b**, respectively)—that is, the larger deshielding of conformer **539a**—again reflect its relatively higher nonclassical stabilization.

539a 539b

3.5.2.6. *Shifts to Distant Carbons* Although there are many examples of 1,2-hydrogen and alkyl shifts, the occurrence of 1,3, 1,4, and 1,5 shifts must also be considered. A sequence of 1,2 shifts, however, can often yield the same results as a 1,3 or 1,4 shift, and the unambiguous demonstration of such can be difficult.

Hydrogen-Bridged Acyclic Ions. The 2,4-dimethylpent-2-yl cation **8** is able to undergo a degenerate 1,3-hydrogen shift[956,957] ($E_a = 8.5\,kcal\,mol^{-1}$). The alternative mechanism of the successive 1,2-hydrogen shift can be eliminated in this case, since line broadening of methyl peak but not methylene peak occurs (in the NMR spectrum) in the temperature range of −70°C to −100°C. A third possible mechanism involving a corner-protonated cyclopropane **540** is highly unlikely based on energy estimates.[97]

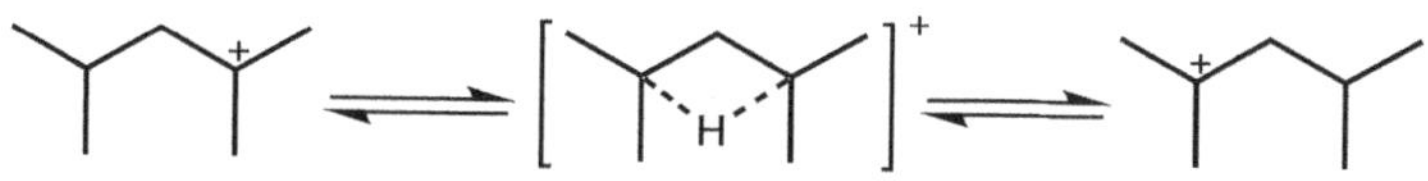

8

540

A similar activation energy barrier of 10.5 kcal mol^{-1} is found for 1,3-hydrogen shift in 1,3-dimethylcyclohexyl cation **541**[958]; incorporation of the six-membered ring constrains the transition state and raises the activation energy barrier.

541

An activation energy barrier of 12–13 kcal mol^{-1} was estimated for the 1,4-hydrogen shift in 2,5-dimethyl-2-hexyl cation **542** using magnetization transfer techniques.[957] The possibility of protonated cyclobutane intermediate similar to the previously considered protonated cyclopropane intermediate is highly unlikely. A similar degenerate 1,4-hydrogen shift is found to occur in the 1,4-dimethyl-1-cyclohexyl cation **543**. The occurrence of successive 1,2- or 1,3-hydrogen shifts was clearly ruled out from a variable temperature NMR study. The activation energy barrier for such a process was estimated at 13 kcal mol^{-1}.

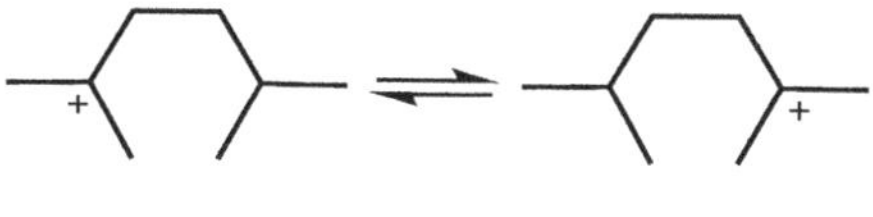

542

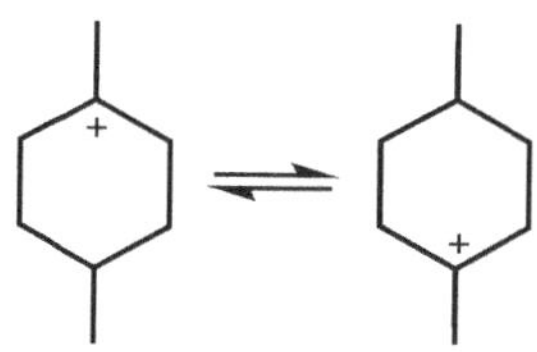

543

Intrigued by facile transannular hydride transfers in medium-sized rings,[959–963] Saunders et al.[56] examined the 2,6-dimethylheptyl cation **544**. Even at the lowest temperature studied (about $-100°C$), the ion exhibits a single averaged peak for the four methyl groups, implying that the 1,5-hydrogen shift occurs with an activation energy barrier of 5 kcal mol^{-1} or less or the ion could have a symmetrically structure such as **545**.

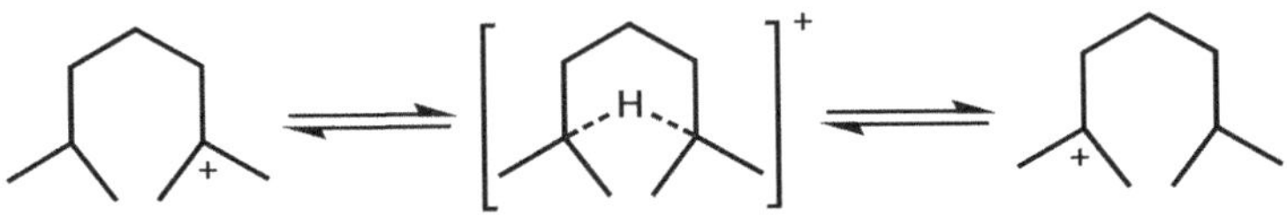

544 **545**

Sun and Sorensen[964] have successfully prepared mono- and di-μ-hydrido-bridged carbodications **546a–546c** by reacting the corresponding diols with $FSO_3H–SbF_5$ in SO_2ClF. NMR characterization data are shown in Table 3.7.

a R = H, R' = isoBu
b R = Me, R' = isoBu
c R, R' = isoBu

546

Dication **546a** showed an NMR spectrum with a single shielded ^1H signal (Table 3.7), although the signal is not as shielded as reported for the cyclic μ-hydrido structures. Furthermore, the signal is highly temperature-dependent, which is characteristic of rapidly equilibrating systems. It appears that the bulky isobutyl group at C(4) in the chain affects significantly the chain conformation allowing for an equilibrium between the μ-hydrogen bridged **546a** and the unbridged structure **546a′**. Disubstitution at C(4) (structure **546b**) results in a similar equilibrating dicationic mixture. The NMR characteristics of 4,4-diisobutyl-substituted dication **546c** (highest upfield shift, triplet for the bridging centers, $J_{^1H-^{13}C} = \sim 20$ Hz) indicate that **546c** exists as a di-μ-hydrido-bridged system, the first of its kind. The two hydride

Table 3.7. **^{1}H NMR Shifts of Carbodications 546a–546c**[964]

Dication Structure	δ^1H (ppm)	Temperature (K)
546a	−0.78	203
	−1.34	159
546b	−3.36	200
	−4.73	153
546c	−4.53[a]	200
	−5.10	160

[a]Two protons.

hydrogens undergo very rapid mutual exchange via unbridged species present in low concentration.

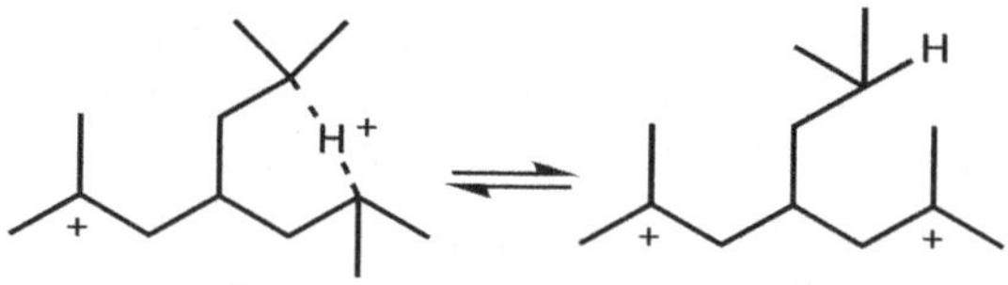

a a'

546

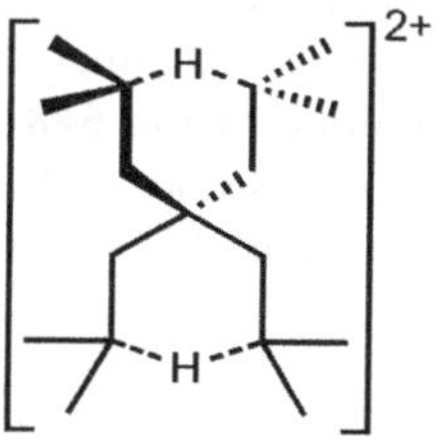

546c

Hydrogen-Bridged Cycloalkonium Ions. The studies of Prelog and Traynham[959] and Cope et al.[960] established that medium-sized cycloalkyl rings (C_8 to C_{11}) undergo direct transannular hydride shifts in reactions involving an electrophilic (i.e., carbocationic) intermediate.[965] Sorensen and co-workers[966] have shown that at very low temperature (−130°C) the cyclodecyl cation exists as a static 1,6- or 1,5-hydrido structure **547c** or **547e**, respectively. Similar behavior was also observed for the 1,6-dimethyl analog **548**.[967] The bridging hydrogen in ion **547c** is observed at an unusually high field of δ^1H = −6.85.

Stable hydrogen-bridged cycloalkyl cations **547a**, **547b**, and **547d** (8-, 9-, and 11-membered rings) have subsequently been observed.[968] The bridging hydrogen was found to be increasingly more shielded in the ^{1}H NMR spectra as the ring size was increased. This indicates increased negative charge on the bridged hydrogen (conversely increased positive charge on the terminal hydrogens) as the distance of separation between the bridged carbons is increased. The ^{1}H NMR shifts of the terminal and bridging hydrogens of various hydrogen-bridged carbocations are shown in Table 3.8.

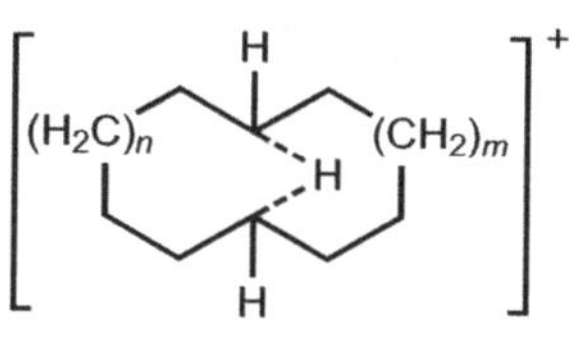

a $m = 0, n = 0$
b $m = 1, n = 0$
c $m = 1, n = 1$
d $m = 2, n = 1$ or $m = 3, n = 0$

547

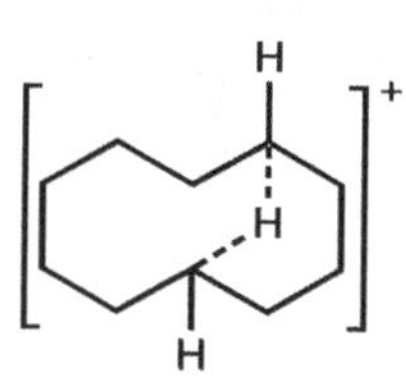

547e

Table 3.8. δ^1H in Hydrogen-Bridged Carbocations[968]

Cation Structure	H–C···H···C–H		
547a	+7.9	−7.7	+7.9
547b	+6.8	−6.6	+6.8
547c	+6.8	−6.85	+6.8
547d	+6.3	−6.0	+6.3
548	—	−3.9	—

Sorensen and co-workers[969] also obtained evidence for 1,5-μ-hydrido bridging between secondary and tertiary carbon sites in several substituted cyclooctyl cations. The μ-1,5-bridged 1,5-dimethylcyclodecyl cation **549** has also been obtained[970] and studied as a distinct stable species. Application of Saunder's isotopic perturbation technique to ion **548** confirmed the bridged structure. With one trideuteromethyl group, an isotopic splitting of only 0.6 ppm is observed in the ^{13}C resonance of bridged carbon, and this clearly supports the assigned hydrido-bridged structure.

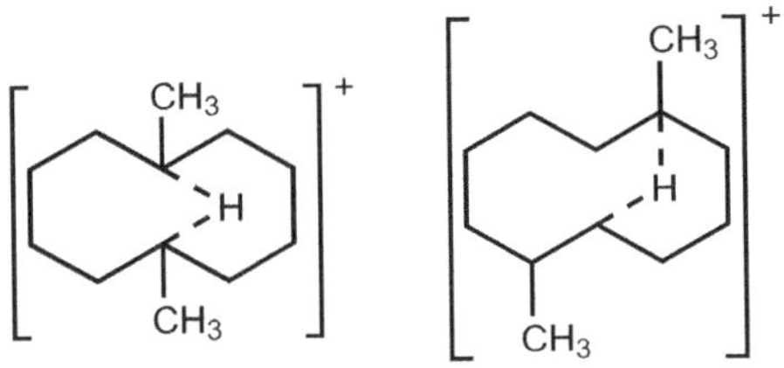

548 **549**

McMurry and co-workers have successfully prepared the unique μ-hydrido bridged cation **550** in *in,out*-bicyclo[4.4.4]tetradecane by protonating the bridgehead alkene in relatively weak trifluoroacetic acid[971] or reacting the corresponding saturated precursor with triflic acid.[972] [Eq. (3.137)]. This hydrogen-bridged propellane cation (**550**) is remarkably stable in acid solution even at room temperature. The bridgehead carbon appearing at δ^{13}C 139.3 in the ^{13}C NMR spectrum is far shielded for a classical cation.[973] The ^{13}C NMR of the 2-deuteriated cation showed a splitting of 0.8 ppm with a very small temperature dependence, whereas the ^{2}H NMR of the cation with bridgehead deuterium gave a chemical shift of δ^{2}H -3.36 (0.1 ppm deshielded from the unlabeled ion). These data are indicative of a nonclassical structure with a 3*c*–2*e* bond.

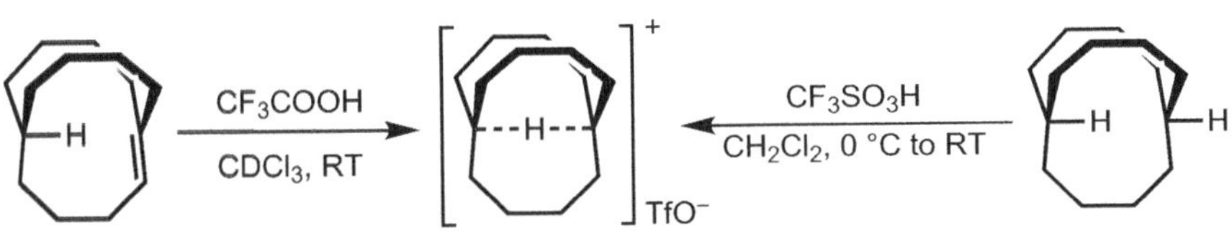

550

(3.137)

An extensive computational study [BLYP/6-31G(d,p) level, AIM and NBO theories] by DuPré has shown[974] that charge distribution in cation **550** prevents the development of unstable bridgehead carbocation. They also found that atoms across the C···H···C bonding are nearly neutral because of electron delocalization and the bridging hydrogen has essentially 1*s* electron configuration and thus highly shielded in the ^{1}H NMR spectrum.

Subsequently, cations with a range of different ring sizes (**551–553**) were prepared and studied both experimentally and computationally.[975,976] The expectation that smaller ring size would result in more bent three-center bond was fulfilled when calculated data (AM1 level) showed significant bending and changes in C–H bond

Table 3.9. NMR Shifts and Calculated Data for Hydrogen-Bridged Carbocations[976]

Cation	δ^1H (Inner H)	δ^{13}C (Bridgehead)	C–H–C Angle (Degrees)	C–H Bond Length (Å)
550	−3.46	139.3	180	1.31
551	−4.20	152.5	166	1.33
552	−6.42	178.8	135	1.34
553	−6.5	178.0	113	1.37

lengths (Table 3.9). Furthermore, the chemical shift of the bridging hydrogen becomes more shielded with increasing C–H–C bond angle indicative of an overall polarization of the bond. The deshielding in the ^{13}C NMR shifts indicates decreasing bridging with decreasing ring size. Ions **551** and **552** have stabilities similar to that of **550**. In contrast, ion **553** is stable only below 10°C and its NMR spectrum is temperature-dependent with two high-shielded absorptions at −50°C (δ^1H −6.4 and −6.5).

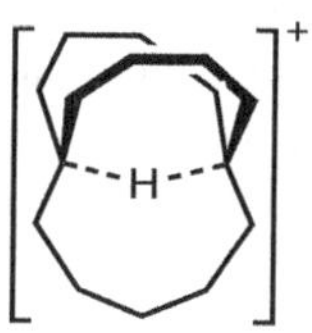

551

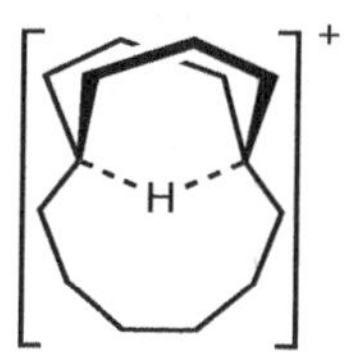

552

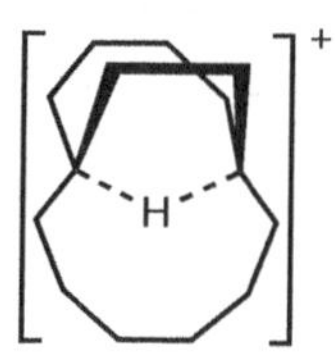

553

The effect of distance between the carbon centers on the nature of the 3*c*–2*e* bonds has been studied by Sorensen and Whitworth[977] in the tricyclic systems **554** prepared from the corresponding bridgehead alkenes. The NMR spectra of the unconstrained cation **555** studied for comparison showed the presence of two classical structures in equilibrium. The chemical shift of the interchanging hydrogen (δ^1H 0.30) indicates a shielded hydrogen but not sufficient enough for a μ-bridged hydrogen.

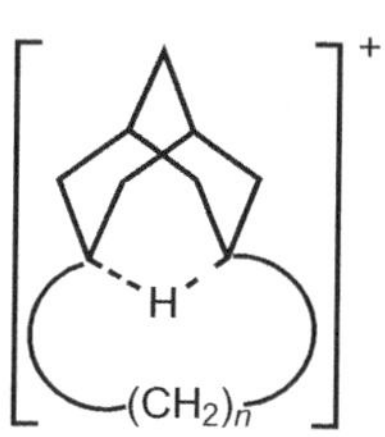

a *n* = 8
b *n* = 7
c *n* = 6
d *n* = 5

554

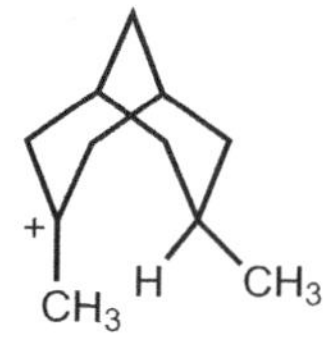

555

Ions **554a** and **554b** were generated by protonation in HSO_3F–SbF_5, and they were stable below −20°C. The NMR spectral data of cation **554a** (Table 3.10) are characteristic of a classical tertiary carbocation. Although the changes in the ^{1}H and

Table 3.10. NMR Shifts for Hydrogen-Bridged Carbocations[977]

Cation	δ^1H (Inner H)	δ^{13}C (Bridgehead)
554a	0.07	324.8
554b	−1.46	296.5
554c	−4.28^a (−4.56)^b	170.1
554d	−5.64	166.8

^{a}At −10°C.
^{b}At −105°C.

^{13}C NMR shifts of cation **554b** indicate the onset of bridging, MINDO/3 calculations for the lowest energy conformers gave classical structures for both cations. Cations **554c** and **554d** were generated in HSO_3F–TFA, and they were stable at room temperature. The NMR spectra of both cations show the features of nonclassical bridged structures. The temperature dependence of cation **554c** indicates the presence of two isomers: one symmetrical (minor) and the other (major) with less symmetrical bridging. Experimental data and calculations for these two ions do not agree well. The structure of lowest energy for **554c** was found to be a classical cation and the bridging structure of lowest energy is 5.1 kcal mol^{-1} less stable. The lowest energy structure for **554d**, in turn, is a μ-hydride bridged structure. However, an unsymmetrical μ-bridged ion lies higher in energy only by 0.45 kcal mol^{-1}.

The 5-pentacyclo[6.2.1.1^{3,6}.0^{2,7}.0^{4,10}]dodecyl cation has been suggested to be the intermediate in the solvolysis and rearrangement of pentacyclododecane derivatives.[978] Recent computational studies have shown that the μ-hydride bridged cation **556** is the structure of the lowest energy [MP2/6-31g(d,p)].[979] The C−H bond length is 1.261 Å, the C−C bond distance is 2.109 Å, and the C−H−C bond angle is 113.5°. A second minimum is the nonclassical cation **557** with the 3*c*–2*e* bond lying 3.3 kcal mol^{-1} higher than **556** [B3LYP/6-31g(d,p)].

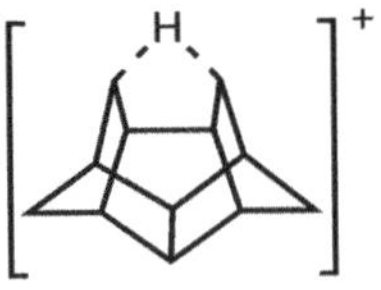

556

557

Five-Center Four-Electron Bonding Structures. The potential for the existence of 5-center 4-electron (5*c*–4*e*) bonding structures **558** have recently been surveyed by Tantillo and Hoffmann[980] [calculations at the B3LYP/6-31G(d) level]. A cation with three anthracenyl units joined around the C---H---C---H---C core with two approximately trigonal pyramidal carbon atoms and one five-coordinate trigonal bipyramidal carbon was found to have 5*c*–4*e* bonding. The anticipated existence of such cations has been strongly supported by an independent generalized population analysis.[981] The same method has been applied[982,983] to explore the nature of bonding in cation **559** with tetracoordinated proton sandwiched between two C—C double bonds with the 5*c*–4*e* bonding residing in the C---C---H---C---C fragment. The results further corroborate the possible occurrence of delocalized 5*c*–4*e* bonding.

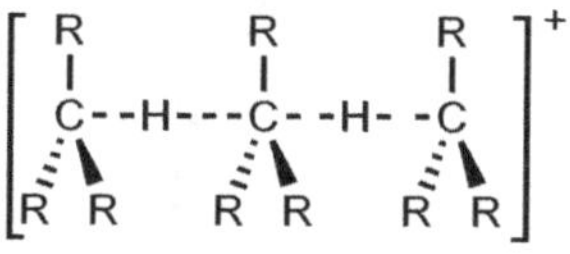

558

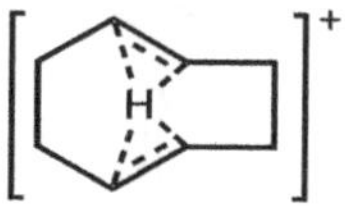

559

3.5.2.7. 9-Barbaralyl (Tricyclo[3.3.1.0^{2,8}]nona-3,6-dien-9-yl) Cations and Bicyclo[3.2.2]nona-3,6,8-trien-2-yl Cations.

The 9-barbaralyl cation **560** is the cationic counterpart of bullvalene **561**. The unique stereoelectronic composition of the structural elements of **560** suggests that it is very reactive in both degenerate (partial and total) and nondegenerate rearrangements. Bullvalene shows total degeneracy through a series of Cope rearrangements.[984] There are several intriguing structural and mechanistic questions connected with the barbaralyl cations; for example, what are their structures and how does the positive charge influence the degenerate Cope rearrangement in ion **560**? Ion **560** is closely related to the bicyclo [3.2.2.]nona-3,6,8-trien-2-yl cation **562**, which has been of interest in connection with the development of the concept of bicycloaromaticity.[985]

560

561

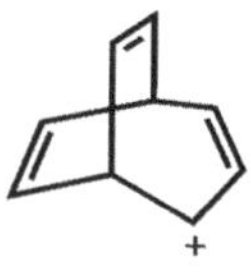

562

When bicyclo[3.2.2]nonatrien-2-ol **563** was treated with superacid at −135°C and observed at the same temperature by ^{1}H NMR spectroscopy, a sharp singlet at δ^1H 6.59 was obtained. A rapidly rearranging carbocation was inferred to be responsible for the observed singlet since there is no regular polyhedron with nine corners, that is, nine equivalent positions [Eq. (3.138)]. The chemical shift of the singlet compared with that estimated from appropriate reference compounds indicated that the ion that was rapidly exchanging all its nine CH groups was the 9-barbaralyl cation **560** rather than the bicyclo[3.2.2]nona-3,6,8-trien-2-yl cation **562**. A combination of mechanism (**IV**) (Scheme 3.25) and mechanism (**V**) (Scheme 3.26) or structure **565** itself was proposed for the total degeneracy, and the rearrangement barrier was estimated to be ~6 kcal mol^{-1}. Even at −125°C, the singlet disappeared rapidly, and a novel type of ion, a

(IV)

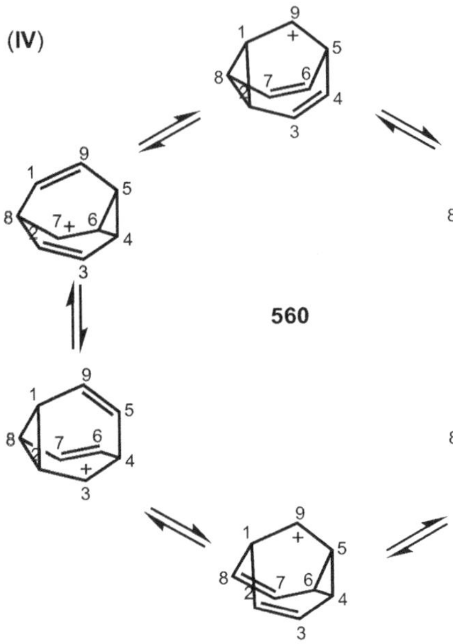

560

Scheme 3.25

(V)

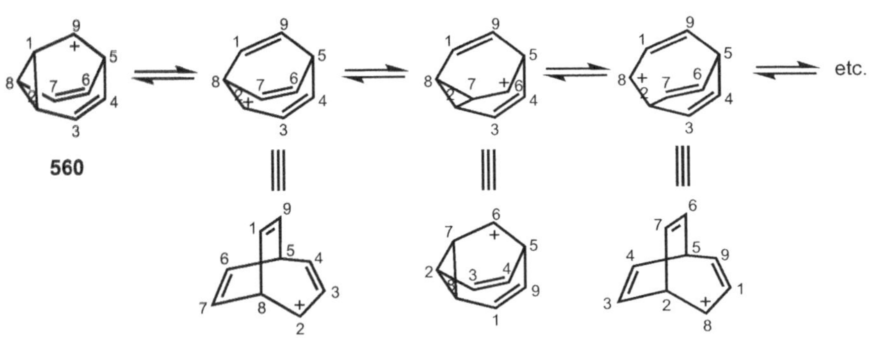

560

Scheme 3.26

1,4-bishomotropylium ion, bicyclo[4.3.0]nonatrienyl cation **564** was quantitatively formed.[45,986–988]

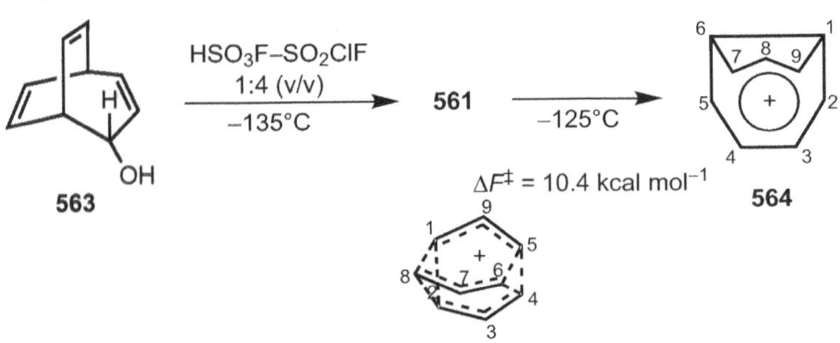

563 564 565

(3.138)

The total degeneracy was found to be slow on the ^{13}C NMR time scale. No signal was detected above noise level in the spectrum. Thus, the signals must be very broad as a consequence of slow rearrangements.[54] In the hope of being able to solve controversies concerning the structure and mechanisms of rearrangement of the barbaralyl cation, the ^{13}C-labeled precursor **566** was synthesized and the ^{13}C-labeled barbaralyl cation **567** was prepared[989] [Eq. (3.139)].

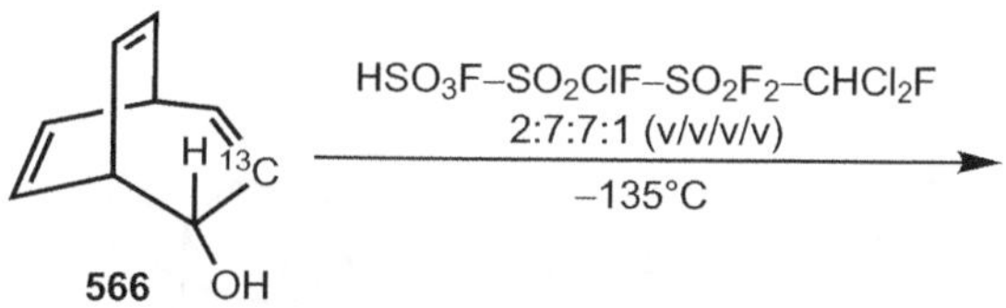

566 $(CH)_8{}^{13}CH^+$ **567** (3.139)

The ^{13}C NMR spectrum at $-135°C$ showed a broad band at $\delta^{13}C$ 118.5 conforming the fast scrambling of all nine carbon atoms. The barrier was found to be 5.5 kcal mol^{-1}. Upon lowering the temperature to $-150°C$, the signal broadened and split into two new signals at 101 ppm and 152 ppm with the area ratio 6:3. Further lowering of the temperature to $-152°C$ sharpened the signals. These results exclude ion **562** as the observed ion. However, the data do not allow discrimination between the two proposed ions **560** and **565**. If rearrangement (**IV**) gives the area ratio 6:3, the barrier for such a rearrangement is estimated to be $\leq$4 kcal mol^{-1}. The static ion **565** should also show the area ratio observed.

Subsequently, the controversy was solved by Ahlberg and co-workers[990,991] using a combination of ^{13}C labeling and isotopic perturbation. The specifically octadeuteriated and ^{13}C-labeled precursor **568** was synthesized, and reaction with superacid gave the $(CD)_8{}^{13}CH^+$ cation **569** [Eq. (3.140)]. The ^{13}C NMR spectrum of **569** is similar to that of **567** but has some important differences. The broad singlet observed from **569** at $-135°C$ appeared 4.5 ppm downfield of that of **567** and the two signals at $-151°C$ had also been shifted; the low-field one had been shifted to upfield by ~1 ppm and the high-field one downfield by 6 ppm. Furthermore, the area ratio had changed from 6 : 3 to 5 : 3. These changes caused by the isotopic perturbation are only consistent with the labeled perturbed ion being a 9-barbaralyl cation **560** and not the D_{3h} structure **565**.

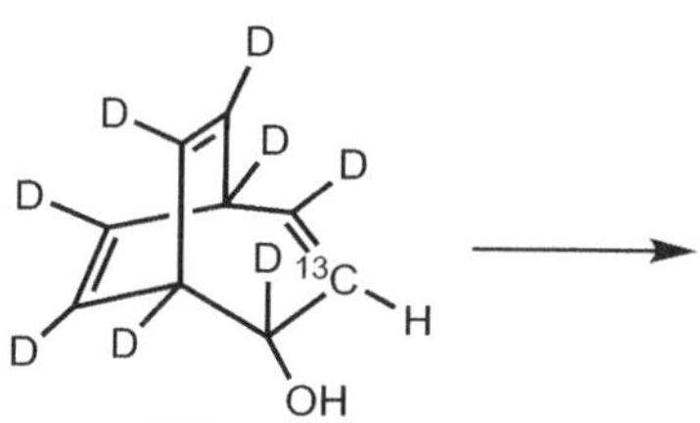

$(CD)_8{}^{13}CH^+$ **569** (3.140)

568

If the ion **569** has structure **565**, the following spectral changes would have been expected. Due to equilibrium (**VI**) (Scheme 3.27), which is likely to have an equilibrium constant smaller than 1 because of the difference in zero-point energy between a "cyclopropane" C–H and "olefinic" C–H, a downfield shift of the singlet might be observed. The two signals at $-151°C$, on the other hand, are not expected to

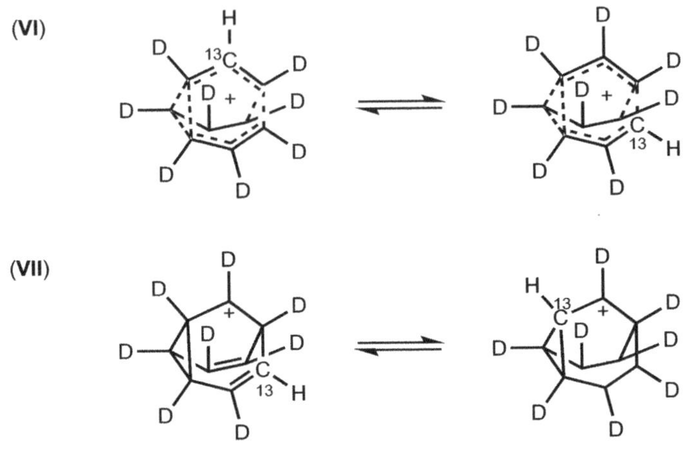

Scheme 3.27

shift relative to those of **567**, since the six "cyclopropane" carbon–hydrogen bonds are equivalent and so also are the three "olefinic" carbon–hydrogen bonds. However, an area ratio different from 6:3 is expected.

If, on the other hand, **569** has the 9-barbaralyl cation structure **560**, the observed shifts of the two signals at −151°C are as expected: The carbon–hydrogen bonds involving C(1), C(2), C(4), C(5), C(6), and C(8) are not equivalent and therefore ^{13}C–H will preferentially be found in olefinic position since olefinic C–H bonds have lower zero point energy than saturated C–H bonds. Therefore, the ^{13}C average chemical shift will be shifted downfield [by process (**VII**), Scheme 3.52]. By the same reasoning, only a minor shift is expected for the C(3), C(7), and C(9) group involving olefinic carbon–hydrogen bonds. The observed shift at −135°C of the broad singlet is also as expected. Thus, the barbaralyl cation has 9-barbaralyl cationic structure **560** and undergoes the sixfold degenerate rearrangement (**IV**) (Scheme 3.25), which has a barrier of only 4 kcal mol^{-1}. Structure **565** has been excluded as either transition state or intermediate in this rearrangement.[990]

Complete degeneracy of ion **560** is probably achieved through mechanism (**V**) (Scheme 3.26), where ion **562** is either an intermediate or transition state.[992]

Similar rearrangements are also observed in 9-substituted and 4,9-dimethyl-substituted 9-barbaralyl cations. The mechanisms of such rearrangements have been thoroughly investigated.[993–995]

Ab initio studies have been performed to explore the potential energy surface of the $C_9H_9^+$ cations.[996–998] Calculations by Ahlberg, Cremer, and co-workers[998] have shown [MP2, MP3, and MP4(SDQ) with 6-31G(d) basis set] that 9-barbaralyl cation **560** is more stable than nonclassical barbaralyl cation **565** and bicyclo[3.2.2]nona-3,6,8-trien-2-yl cation **562** by 6.9 and 4.6 kcal mol^{-1}, respectively. Both sit at a transition state or in a very shallow minimum surrounded by transition states of similar energies. The energetically most favorable transformations of **560** are sixfold degenerate rearrangements via transition state **570** of C_2 symmetry representing an activation energy barrier of 3.6 kcal mol^{-1}. A new reaction mechanism has been suggested to interpret the complete degenerate rearrangement of **560** leading to equilibration of all carbon atoms.

It is characterized by double-bifurcation reactions with three directly connected first-order transition states. These are two transition states **570** and transition state **562**. The chemical behavior of cation **560** is determined by the stereocomposition of a cyclopropylcarbynyl cation moiety in conjugation with two vinyl group. The 9-barbaralyl cation **560** is characterized by four weak C–C single bonds (bond orders = 0.54 and 0.84) and one strong nonbonded interaction that can be easily broken and closed.

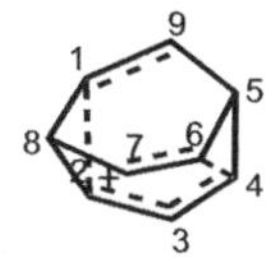

570

3.5.2.8. The 1,3,5,7-Tetramethyl- and 1,2,3,5,7-Pentamethyl-2-adamantyl Cations.

The nature of the 2-adamantyl cation **571** has been difficult to study under stable ion conditions since it undergoes facile rearrangements to the more stable 1-adamantyl cation **49**.[182] This difficulty was circumvented by Lenoir, Schleyer, Saunders, and co-workers[999] by blocking all four bridgehead positions by methyl groups in a study of 1,3,5,7-tetramethyl- and 1,2,3,5,7-pentamethyl-2-adamantyl cations **572** and **573**.

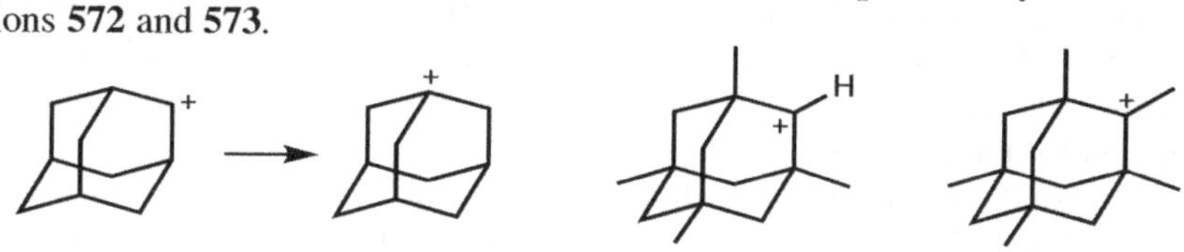

571 **49** **572** **573**

The ^{1}H NMR spectrum of **572** in superacid had the correct number of peaks to fit the symmetry of a static 2-adamantyl cation, but the chemical shift of the CH proton at the presumed carbocation center C(2) was δ^1H 5.1. This is 8 ppm to higher field than expected for a typical static secondary carbenium ion such as the isopropyl cation **2**.[17] Since the symmetry of the spectrum was incompatible with either a static bridged 2-adamantyl cation **574** or a static protoadamantyl cation **575**, two mechanisms were postulated involving sets of **574** or **575** undergoing rapid degenerate rearrangements at −47°C [Eqs. (3.141) and (3.142)].

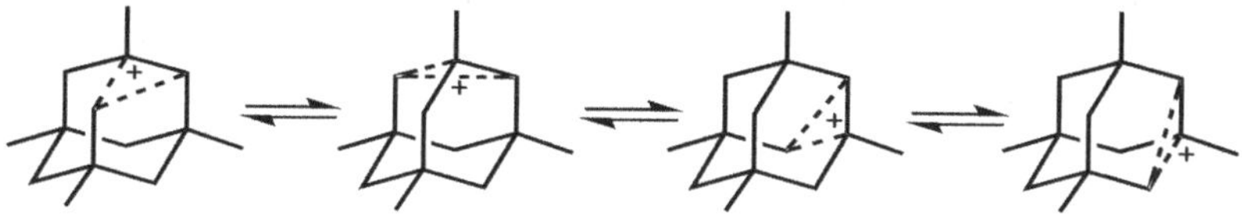

(3.141)

574

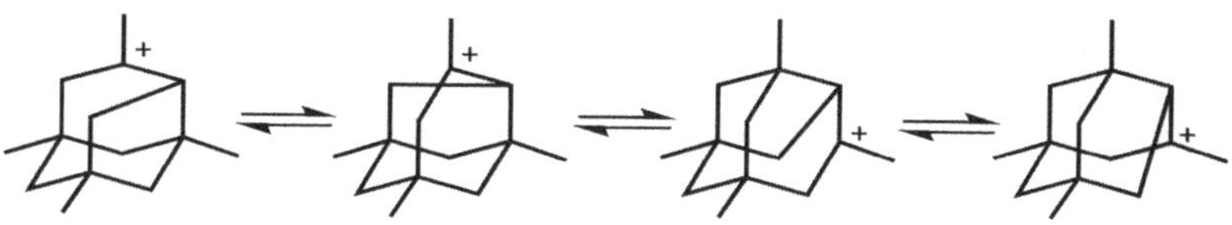

(3.142)

575

Apart from one of these degenerate rearrangements, **572** also underwent a non-degenerate rearrangement to the more stable tertiary 2-adamantyl cation **577** with a half-life about 1 h at −47°C [Eq. (3.143)]. The kinetics of this rearrangement, which involves protoadamantyl cations **576** as intermediates, was advantageously studied in the tertiary 2-adamantyl system **573**, where it is degenerate. Line-shape analysis for the degenerate rearrangement of **573** gave $E_a = 12.1 \pm 0.4$ kcal mol^{-1} in accord with molecular mechanics calculations.

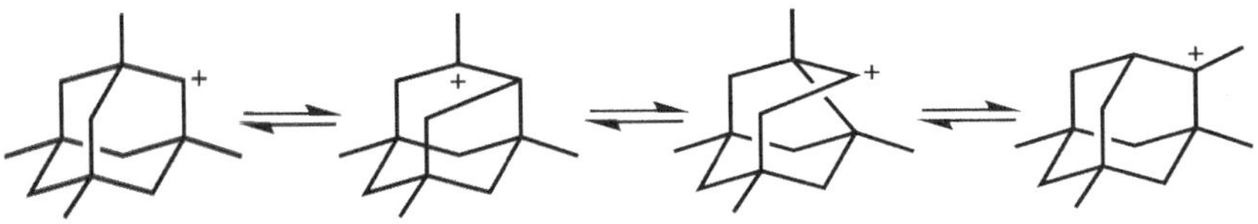

572 **576** **577**

(3.143)

Since it was difficult to make an exclusive choice between mechanisms for the degenerate rearrangement shown in Eqs. (3.141) and (3.142) and the average structure of **572** on the basis of ^{1}H NMR data only, further arguments were taken from a solvolytic study, and the mechanism involving **574** [Eq. (3.141)] was preferred as an explanation for the behavior of **572** in superacid.

Criticism of these conclusions by Fărcaşiu[1000] led Schleyer, Olah, and co-workers to study[55] **572** and **573** further by ^{13}C NMR spectroscopy. The spectra of **573** confirmed its classical static carbenium ion structure at low temperature. At 30°C an average of the C(1), C(2), and C(3) signals and those of the CH_3 groups attached to these positions, respectively, were observed due to the degenerate rearrangement via the mechanism in Eq. (3.142).

An entirely different spectrum was obtained for **572** with the C(2) ^{13}C resonance at δ^{13}C 92.3, more than 200 ppm removed from the position expected for a static classical cation. Since a static structure like **572** was incompatible with the observed spectrum, a chemical shift estimate was made for the protoadamantyl cation **575**. However, the discrepancy between these estimated chemical shifts and those observed was too large to explain the behavior of the 1,3,5,7-tetramethyl-2-adamantyl cation within the properties of an equilibrating set of ions **575** even with the partial contribution of **572**. This left the set of σ-bridged ions **574** equilibrating according to the mechanism in Eq. (3.141) as the only possible structure for this ion.

3.5.3. Homoaromatic Cations

The concept of homoaromaticity[29,1001–1006] was advanced by Winstein in 1960. It represented a challenge to experimental and theoretical chemists alike.[1007–1009] The question of homoaromatic overlap has been mainly studied in six-π-electron Hückeloid systems,[29,1001–1006] although several two-π-electron homoaromatic systems have been discovered subsequently.[230,1006,1010–1013]

Theoretical studies have recently been performed to find appropriate criteria for homoaromaticity of carbocations.[1014,1015] Using four criteria [aromatic ring-current shielding (ARCS), nucleus-independent chemical shifts (NICS), bond-length alterations, and ^{1}H NMR shielding], cations $C_8H_9^+$, $C_9H_{10}^{2+}$, and $C_{12}H_{13}^{3+}$ were found to be the most homoaromatic. No other molecules are ranked as homoaromatic by all four criteria. For homoaromatic molecules of small ring size, the ARCS method is the method of choice to assess homoaromaticity. Interestingly, a C(1)–C($n-1$) distance of about 2 Å seems to be a good indication of homoaromaticity.

3.5.3.1. Monohomoaromatic Cations. The simplest 2π monohomoaromatic cation, the homocyclopropenyl cation ($C_4H_5^+$, cyclobutenyl cation, **578**), along with its analogs, has been prepared and studied by Olah et al.[230,1010] The experimental evidence for the existence of 1,3-overlap has been derived from ^{1}H and ^{13}C NMR data.[230,1010] At low temperature, the methylene protons of the ion **578** exhibit nonequivalence, indicating ring puckering. The experimentally determined barrier for ring flipping of 8.4 kcal mol^{-1} compares closely with the theoretical estimates.[1016] Subsequent computational studies by Schleyer and co-workers[1017] have confirmed the bent structure of **578**. Ring inversion energies calculated at various levels of theory were found to be 9.0–9.3 kcal mol^{-1}, which agree well with experiment. The homoaromatic character of ion **578** is shown by the nearly equal charges on C(1), C(2), and C(3), the considerable bond order, the short C(1)–C(3) bond distance (1.74 Å), and the large stabilization energy relative to the methylallyl cation. Long-lived polysubstituted cations **579** have recently been generated from the corresponding methylenecyclobutenes in HSO_3F–SbF_5–SO_2ClF solution at −130°C and characterized by ^{1}H and ^{13}C NMR spectroscopy.[1018,1019]

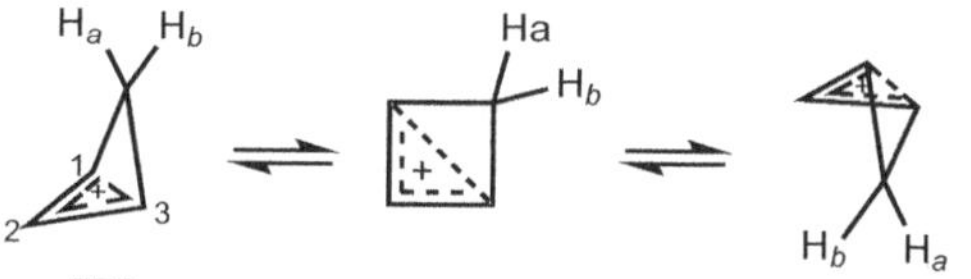

578

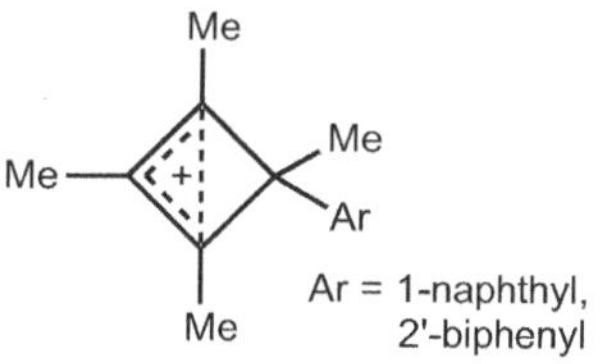

579

Other examples of monohomoaromatic cations are the 6π homotropylium ion **179** of Pettit[1020–1022] and Winstein[1023–1025] and the previously discussed 2π homoaromatic 11-methyltricyclo[4.4.1.0^{1,6}]undec-11-yl cation **46**.[163]

The structure and properties of the homotropylium ion **179** of C_s symmetry have been addressed theoretically.[346,1026] Both MP2(full)/6-31G* and MMP2 geometries give almost perfect C–C bond equalization in the seven-membered ring moiety. This observation, the strong stabilization (−22.8 kcal mol^{-1}) by the π overlap, and the short C(1)–C(7) bond distance (1.957–2.149 Å) indicate efficient electron delocalization involving through-space 1,7 interaction. Therefore, the homotropylium ion **179** is a truly homoaromatic six-π-electron system. The planar structure **580** (C_{2v} symmetry) participating in ring inversion of cation **179**, in turn, has a stabilization energy of only −2.1 kcal mol^{-1} and, consequently, cannot be regarded as an aromatic system. Alkorta et al.[1027] have found in a recent study that fluorination has not much effect on the homoaromaticity, that is, homoaromaticity is a persistent property of the parent ion **179**.

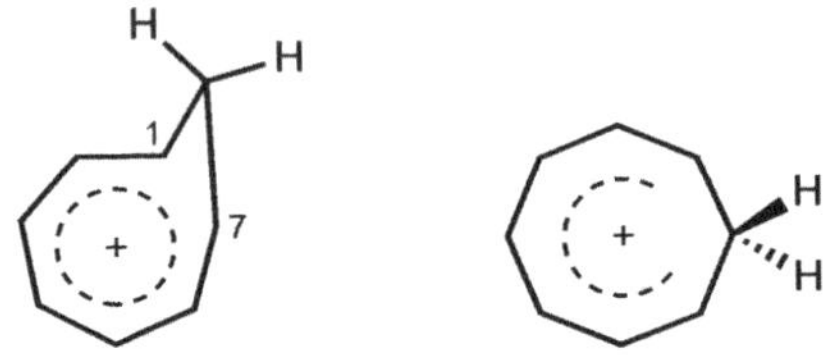

179 **580**

3.5.3.2. Bishomoaromatic Cations. The magnitude of homoaromatic stabilization is expected to decrease with increasing interruption by methylene groups of the otherwise conjugated π-framework in neutral molecules. However, in an ionic species, additional driving force is present for charge delocalization. Two of the most widely studied bishomoaromatic cations are the 7-norbornenyl and 7-norbornadienyl cations **581** and **582**.[1011,1028–1034]

581 **582** **583**

The ^{13}C NMR spectrum of the cation **581** shows substantial shielding of both the C(7) cationic and vinylic carbon chemical shifts (δ^{13}C 34.0 and 125.9, respectively).[1011] A similar shielding phenomenon is observed for ion **582**. Interestingly, ion **582** undergoes a bridge-flipping rearrangement[1029,1032,1033] as well as ring contraction–expansion through the intermediacy of bicyclo[3.2.0]heptadienyl cation **583**. These two processes can result in scrambling of all seven carbon atoms.

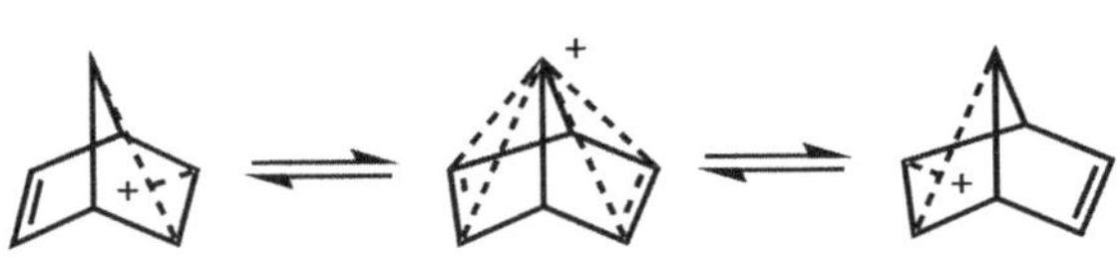

582 bridge flipping

Several studies, including the application of the tool of increasing electron demand,[691,1030–1032] best describe ion **581** as a symmetrical bridged π-bishomocyclopropenium cation and not as a rapidly equilibrating pair of cyclopropylmethyl cations, such as **584**. The observed unusually large ^{13}C–^{1}H coupling constants at the C(7) position of **581** and **582** (218.9 and 126.4 Hz, respectively) demonstrate the higher coordination of the carbocationic carbon. X-ray crystal structure analysis of the 7-phenyl-2,3-dimethyl derivative of the 7-norbornenyl ion showed short C–C distances of the carbons in the 3*c*–2*e* bonding (both are 1.86 Å).[78] The X-ray crystal structure, the optimized geometry[1035] (MP2FC/6-31G*), and calculated ^{13}C NMR shift data[78] agree well.

584

Olah, Prakash, and co-workers[1036] have generated and characterized the sandwiched bishomoaromatic dication **585** [Eq. (3.144)]. The ^{1}H NMR spectrum of the solution at $-80°C$ indicated that the species shows the same symmetry as the starting diol with chemical shifts of some of protons being more shielded. The ^{13}C NMR spectra with significantly shielded absorptions [δ^{13}C 131.7 (doublet, $J_{C\text{–}H} = 199.3$ Hz), 52.9 (doublet, $J_{C-H} = 219.2$ Hz), and 38.1 (doublet, $J_{C-H} = 169.6$ Hz)] indicate the highly symmetric dicationic structure **585**. Dication **585** can be considered as a four-π-electron bicyclo(polycyclo)aromatic system. The magnitude of the observed chemical shifts and the lack of temperature-dependent behavior rule out, respectively, rearranged structures and equilibrium processes.

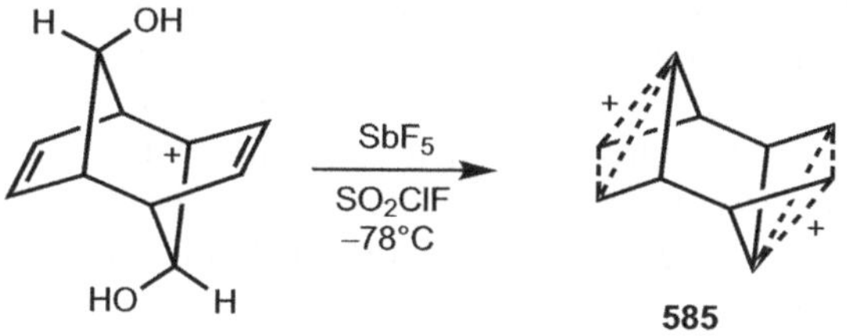

585 (3.144)

Several studies[1037–1041] on the hexamethylbicyclo[2.1.1]hexenyl cation have shown that the ion is best represented as a bishomoaromatic species **586** analogous to **581** and **582**. Laube and Lohse[1042] succeeded in acquiring the X-ray crystal structure of the *endo*-**586** ion and its 6-chloro derivative. An unusually strong $3c$–$2e$ bond [C(2)–C(3) = 1.406 Å, C(2)–C(5) and C(3)–C(5) = 1.741 Å] was found in the latter. The optimized geometry[78] of **586** (B3LYP/6-31G*) agrees well with the experimental values. These observations unequivocally indicate that the hexamethylbicyclo[2.1.1]hexenyl cation is a nonclassical ion and the equilibrating ions can be excluded.

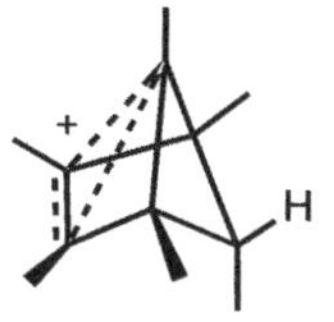

586

The extent of bishomoaromatic delocalization, as expected, is critically dependent upon structural geometry. Attempts to prepare the parent bishomoaromatic 4-cyclopentenyl cation **588** from 4-halocyclopentene **587** were unsuccessful

[Eq. (3.145)].[222,1043] They gave instead the cyclopentenyl cation. The lack of formation of bishomoaromatic ions from cyclopentenyl derivatives is mainly due to steric reasons. The planar cyclopentene skeleton has to bend into the "chair" conformation to achieve any significant overlap between the empty *p* orbital and the π–*p* lobe of the olefinic bond, which is sterically unfavorable. However, such conformation already exists in ions **581** and **582**.

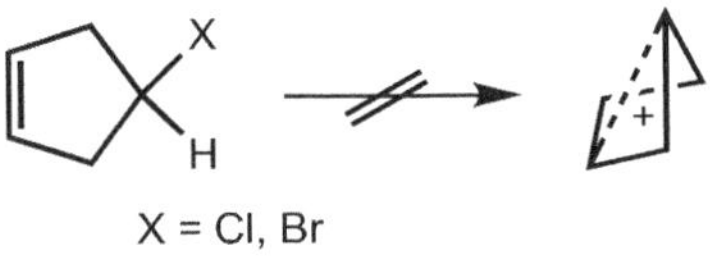

(3.145)

587 **588**

Bishomotropylium cations are 6π homoaromatic cations. The first bishomotropylium cations, the 1-methyl-1,4-bishomotropylium ion and the parent ion **589**, were generated and observed by Winstein and co-workers under superacid conditions.[986,1044,1045] The intermediate identified[991] as the barbaralyl cation **560** rearranged exclusively to cation **589** at −116°C. Strong support for the homoaromatic character of **589** comes from the observed ^{13}C NMR shifts and $^{13}C-^{13}C$ coupling constants indicating that all nine carbons are sp^2-hybridized. Solvolytic experiments, in turn, favored the *cis*-8,9-dihydro-1-indenyl structure **590**.[1046] Subsequent calculations[1047,1048] [HF/6-31G(d), MP2/6-31G//HF/6-31G, and MP2, MP3, MP4(SQD)/6-31G(d) levels] showed, however, that **589** has a folded structure. The folding angle and C(2)–C(9) and C(5)–C(7) interaction distances (93° and 2.1 Å, respectively) differ significantly from those in the classical structure **590** (114° and 2.34 Å). Further observations,—namely, large degrees of bond equalization (1.40 ± 0.01 Å) and charge delocalization, a similar strong equalization of ^{13}C chemical shifts in the seven-membered ring [IGLO/6-31G(d)], and a bond order characteristic of an aromatic system—lend additional strong support to the nonclassical nature of ion **589**.

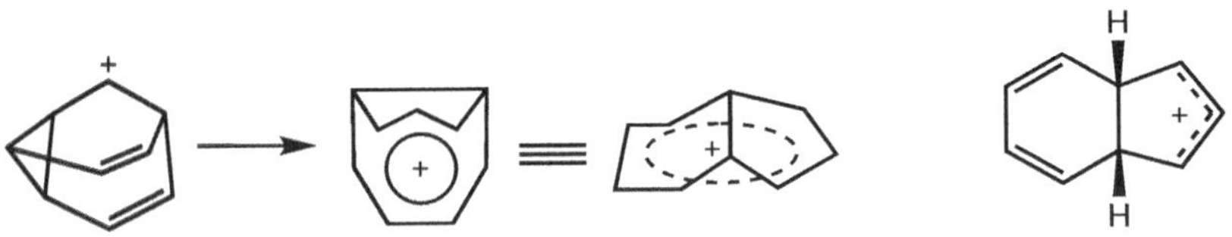

560 **589** **590**

Olah, Prakash, and co-workers in collaboration with Prinzbach have studied bishomoaromatic cage dications derived from pagodane (an undecacyclic hydrocarbon) and dodecahedrane cage hydrocarbons.[385,1049] Oxidation of **591** and its valence isomer **592** by two-electron removal from the central cyclobutane ring resulted in the formation of paramagnetic solutions which, upon standing at room temperature, gave diamagnetic solutions of a unique, surprisingly stable species[1050] [Eq. (3.146)]. The symmetry of the ^1H and ^{13}C NMR spectra (four peaks in each spectrum) and the extent of deshielding indicated that the species is ionic and has the same D_{2h} symmetry of the parent pagodane. A crystalline material was isolated and fully characterized by X-ray, which showed that the product retained the integrity of the hydrocarbon skeleton. The NMR spectra, which were found to be temperature-independent down to −130°C, along with additional evidence as well as computational studies (SCF–MO calculations), firmly established the nature of dication **593** excluding the possibility of rapidly equilibrating processes. The bishomologous dication **594**, in contrast, rearranges to the bisallylic dication **595** even at −130°C. Dication **593** is a novel four-center two-electron delocalized σ-bishomoaromatic species, which resembles two-electron Woodward–Hoffmann transition states and owes its unprecedented character to its specific rigid framework.

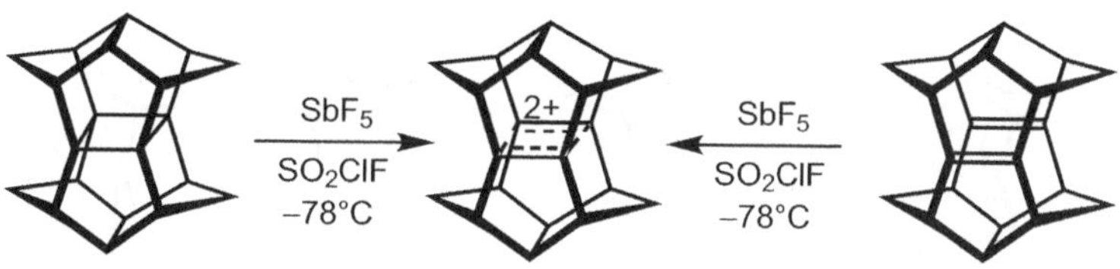

591 **593** **592**

(3.146)

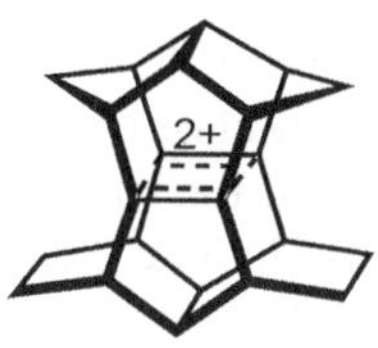

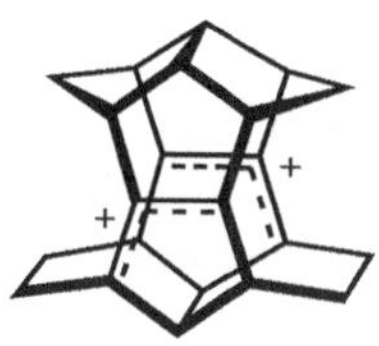

594 **595**

To further explore the limiting structural and energetic prerequisites for the unique σ-bishomoconjugation the structurally related seco-1,16-dodecahedradiene **596** and 1,16-dodecahedradiene **597** were also studied. For **596**, only the corresponding bisallylic dication could be observed,[1051,1052] whereas for **597** the dication could not be observed at all.[1053] The σ-bishomoaromatic species **599**, however, could be generated via electrochemical oxidation of alkene **598**.[1054]

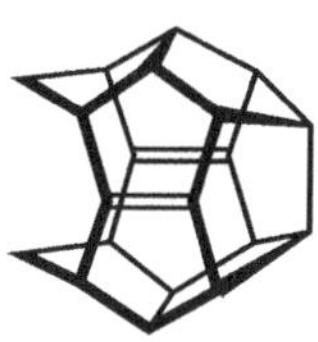

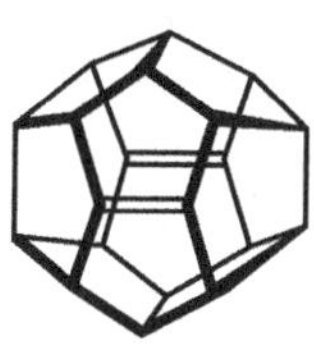

596 **597** **598** **599**

Isopagodanes **600** and **602** were also studied under the usual oxidative conditions (SbF_5–SO_2ClF, −130°C, prolonged stirring) and both gave diamagnetic solutions as a result of two-electron oxidation.[1055] The number of absorptions of the ^1H and ^{13}C

NMR spectra (seven peaks in each spectrum) and the observed deshielding of protons and carbons relative to the neutral **600** established the C_{2v} symmetrical σ-bishomoaromatic structure **601a** and excluded a rapid equilibration with the degenerate **601b**. The bishomoaromatic structure of **603** was established in an analogous way. *Ab initio* calculations (B3LYP/6-31G* level of theory) gave only one minimum on the corresponding hypersurfaces and also indicated the C_{2v} symmetrical structure with the rectangular geometries of the central C_4 units.[1055] Computations by Schleyer and co-workers[1056] [MINDO/3, MNDO, AM1, and MC-SCF] gave further strong evidence for the nonclassical nature of pagodane dications excluding two rapidly equilibrating classical C_{2v} structures and a diradical of D_{2h} symmetry.

600 a b 602 603
601

The novel four-center two-electron delocalized σ-bishomoaromatic species **593**, **594**, **599**, **601a**, and **603** are representatives of a new class of 2π-aromatic pericyclic systems. These may be considered as the transition state of the Woodward–Hoffmann allowed cycloaddition of ethylene to ethylene dication or dimerization of two ethylene radical cations[985] (Figure 3.25, **604**). Delocalization takes place among the orbitals in the plane of the conjugated system, which is in sharp contrast to cyclobutadiene dication **605** having a conventional *p*-type delocalized electron structure (Figure 3.25).

Precedents of this type of bonding pattern are the assumed bicyclo[2.2.2]octane-1,4-diyl dication **606**,[393] which was found to be the monocation monodonor–acceptor

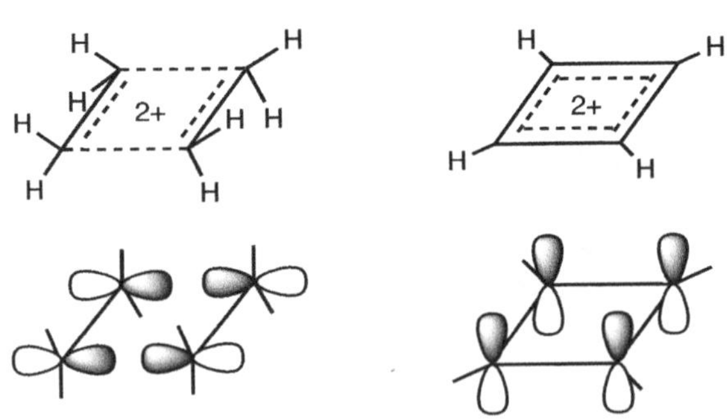

604 605

Figure 3.25. Molecular orbital representations of the bonding structure of the parent bishomoaromatic dication $C_4H_8^{2+}$ (**604**) and cyclobutadiene dication $C_4H_4^{2+}$ (**605**).

complex in a subsequent reinvestigation,[394] and the octamethylnorbornadienediyl dication **607** prepared and characterized by Hogeveen and co-workers.[1057]

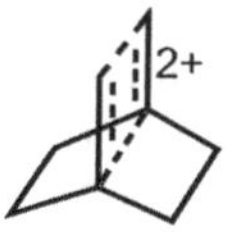

606

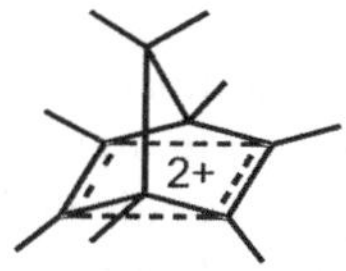

607

The nature of bonding in the parent $C_4H_8^{2+}$ dication [tetramethylene dication $(CH_2)_4^{2+}$] was analyzed by various levels of theory.[1050,1056] Of the various species selected, the open form **608** with C_i symmetry was found to be the global minimum at all levels of theory.[1056] However, optimization with MINDO/3, AM1, and 3-21G for the classical pagodane dications **609a** and **609b** led to ring closure to the parent bishomoaromatic pagodane dication **593** of D_{2h} symmetry. IGLO ^{13}C NMR chemical shift calculations lend additional support for the bishomoaromatic stabilization.[1056] Apparently, despite the strain imposed by the planar cyclobutane moiety in the polycyclic cage structure, the strong Coulombic repulsion that could lead to geometric distortion and, consequently, cancellation of bishomoaromaticity is successfully counterbalanced by the rigid molecular framework thus contributing to stabilization.

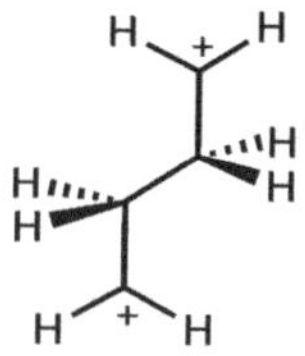

608

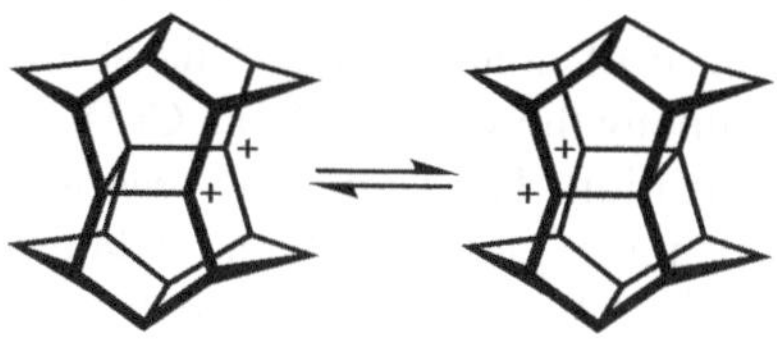

a b

609

3.5.3.3. Trishomoaromatic Cations. Following Winstein's proposal[1058–1061] of the formation of a trishomoaromatic cation in the solvolysis of *cis*-bicyclo[3.1.0] hexyl tosylate, extensive effort was directed toward its generation under stable ion conditions.[222] Masamune et al.[1012] were first able to prepare the ion **610** by the ionization of *cis*-3-chlorobicyclo[3.1.0]hexane in the superacidic media. Subsequently, it has also been generated from the corresponding *cis*-bicyclo[3.1.0]hexan-3-ol in protic acid-free SbF_5–SO_2ClF solution. The observed NMR spectral data agree very well with the C_{3v} symmetry of the ion **610**. The ^{13}C NMR shifts are highly shielded for the three equivalent five-coordinate carbons (δ^{13}C 4.9 with a ^{13}C–H coupling constant of 195.4 Hz), which is indicative of the nonclassical nature of the ion. Attempts to prepare methyl- and aryl-substituted trishomocyclopropenyl cations were, however, unsuccessful,[1043] which is consistent with Jorgensen's calculations.[1062]

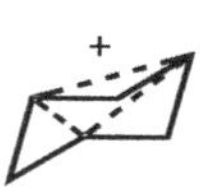

610

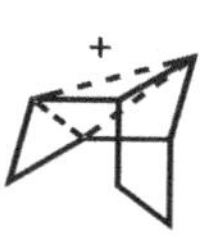

611

612

The ethano-bridged analog **611** was also prepared from the 8-chlorotricyclo [3.2.1.0^{2,4}]octane precursor.[1063] Ion **611** closely resembles the trishomoaromatic ion **610** in its spectral properties.

Another interesting trishomoaromatic system which has C_{3v} symmetry is the 9-pentacyclo[4.3.0.0^{2,4}0^{2,8}0^{5,7}]nonyl cation **612** prepared by Coates and Fretz.[1013] Ion **612** has been thoroughly investigated in solvolysis.[1064,1065] A rate enhancement of 10^{10} to 10^{12} compared with ordinary systems uncovered its highly delocalized nature (strain relief is not the reason for the degenerate rearrangement). Also, analysis of remote and proximate substituent effects upon ionization[1066] and application of the tool of increasing electron demand[879,1067] fully reinforced its nonclassical nature. A deuterium labeling study also confirms this conclusion.[872,1068]

Olah, Rasul, and Prakash[1069] have recalculated the trishomocyclopropenium ion **610** and calculated the structure of the hitherto unknown 1,3-dehydro-5-adamantyl cation **613** [DFT/IGLO method]. The B3LYP/6-31G* optimized geometry of cation **610** agrees well with that of the *ab initio* structure (MP2/6-31G*) reported previously.[1070] Both ions have C_{3v} symmetry, and the relevant C(1)–C(3) bond distance of ion **613** is only slightly longer than that of ion **610** (1.911 Å versus 1.859 Å), which indicates similar trishomoaromatic stabilization. The *p* orbitals of the three bridgehead sp^2 carbon atoms [C(1), C(3), C(5)] of cation **613** overlap, involving two electrons. The C(1)–C(8) and C(8)–C(7) bond lengths differ only slightly from those of adamantane. This shows extensive delocalization of the positive charge among the sp^2-type C(1), C(3), and C(5) carbons without additional C–C hyperconjugation. The isodesmic reaction of the adamantyl cation and 1,3-dehydro-5-adamantane yields **613** and adamantane in an exothermic reaction ($-20.0\,\text{kcal mol}^{-1}$). This is again indicative of significant stabilization due to trishomoaromatic interaction.

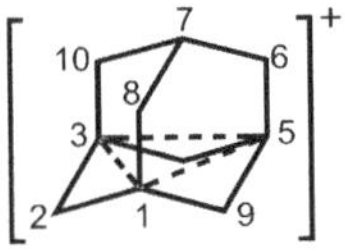

613

Recent computational studies gave conflicting results with respect to the nonclassical nature of the trishomocyclopropenium ion. According to the study of Werstiuk and Wang[1071] [B3PW91/6-311G(d,p) and MP4(SDQ)/6-311G(d,p) levels of calculations], the ion does not have pentacoordinate carbons, although it is stabilized by σ, σ-bond homoconjugation. Molecular properties calculated by Cremer and coworkers[1072] [HF and MP2/6-31G(d) level], in contrast, indicate a nonclassical structure with two-electron delocalization in the cycle as first predicted by Winstein.[1060]

3.5.3.4. Three-Dimensional Homoaromaticity. Schleyer and co-workers[1073] generated and observed the intriguing 1,3-dehydro-5,7-adamantanediyl dication **614** of T_d symmetry with four-center two-electron bonding corresponding to a three-dimensional homoaromaticity [Eq. (3.147)]. In this system the *p*-orbitals of the four bridgehead carbon atoms overlap inward in the cage in a tetrahedral fashion. The chemical shifts of the bridgehead and the CH_2 carbons (δ^{13}C 6.6 and 35.6, respectively), the unusual single sharp proton chemical shift (δ^1H 3.8), and the strong deformation (C–CH$_2$–C bond angle = 87.3°) predicted by calculations (3-21G *ab initio* geometry) are indicative of the nonclassical structure of ion **614**. Population analysis at various levels (MNDO, MINDO/3, STO-3G, 3-21G) showed that the positive charge resides not only on the bridgehead carbons but also on all 12 hydrogens and the methylene carbons are neutral. The three-dimensional homoaromatic character of ion **614** has been further confirmed by more recent high-level theoretical studies by Chan and Arnold[1074] (*ab initio* HF/6-31G* level), Prakash et al.[1075] (B3LYP/6-31G*), and Schleyer and co-workers[1076] (B3LYP/6-31G*).

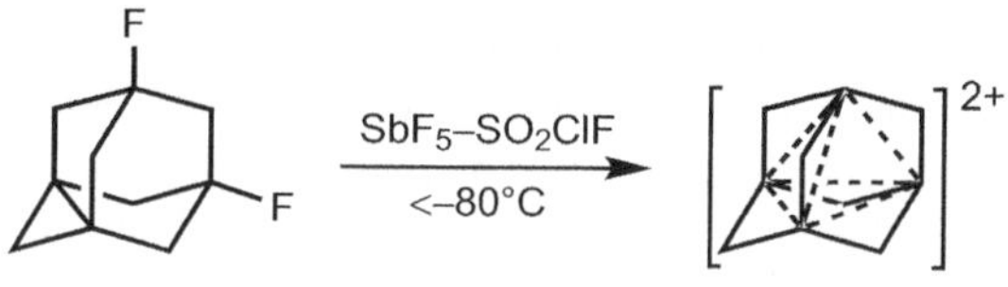

(3.147)

614

3.5.4. Pyramidal Cations

3.5.4.1. (CH)$_5^+$-Type Cations. The close relationship between carbocations and boranes led Williams[1077] to suggest the square-pyramidal structure **615** for the (CH)$_5^+$ cation based on the square pyramidal structure of pentaborane. Stohrer and Hoffmann[1078] subsequently came to the same conclusion concerning the preferred square-pyramidal structure for the (CH)$_5^+$ cation using extended Hückel MO calculations.

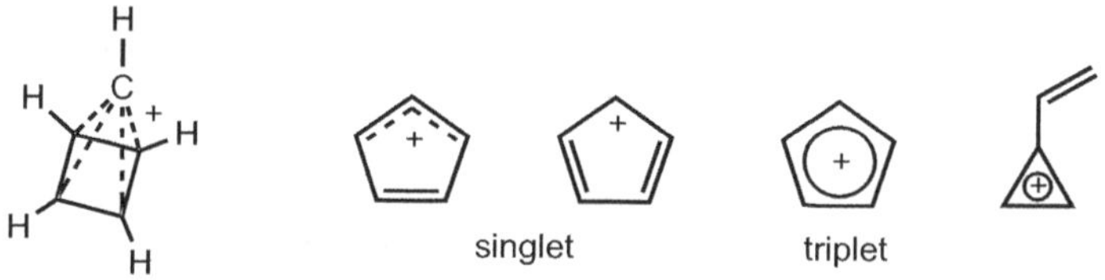

615 **616** **617** **618** **619**

Cation **615** with C_{4v} symmetry can be viewed[1078] as square cyclobutadiene capped by CH$^+$. Since then, several calculations of (CH)$_5^+$ at more sophisticated levels have appeared.[905,1079–1083] The MINDO/3 method[1084] indicated that the pyramidal cation **615** is less stable by 14.4 kcal mol^{-1} compared with the isomeric singlet cyclopentadienyl cations **616** and **617**. The triplet ion **618** of D_{5h} symmetry was found to be only 1.6 kcal mol^{-1} more stable than the singlet ions. The triplet ion **618** has been prepared[1085] by molecular beam codeposition of 1-bromocyclopentadiene with SbF$_5$ at 78 K and was characterized by ESR spectroscopy.

The fifth isomer, the vinylcyclopropenyl cation **619**, has been suggested by Zerner and co-workers[1086] to be the most stable of the isomeric (CH)$_5^+$ cations (MP2/6-31G**//6-31G* level). Subsequent high-level *ab initio* calculations (G2 level of theory),[1087] however, found almost identical energies for **618** and **619**. Indeed, recent *ab initio* [MP2(full)/6-31G*] and force-field calculations have shown[1088] that **618** is lower in energy than **619** by 1.2 kcal mol^{-1}. This study also showed[1088] that of the parent singlet cyclopentadienyl cations the allylic structure **616** is slightly favored over the localized cyclopentadienyl cation structure **617**. The stability of the alkyl-substituted analogs, however, depends on the substitution patterns.

Lambert et al.[1089] have claimed in a recent report the synthesis of the stable pentamethylcyclopentadienyl cation **621** by hydride transfer from pentamethylcyclopentadiene (**620**) with the trityl cation [Eq. (3.148)]. The NMR data, however, are characteristic of allyl cations, and it became obvious that the product is not cation **621** but the known *trans*-1,2,3,4,5-pentamethyl-1-cyclopentenyl allylic cation **622**.[1090–1094] Particularly revealing is the C(4)–C(5) bond length (1.51 Å) of the reported X-ray structure, which is too long for a carbon–carbon double bond. The calculated bond distance for **621** is 1.36 Å and 1.351 Å (reported by Lambert et al.[1089] and by Müller,[1091] respectively). The carbon–carbon single bond lengths for cation **622** observed experimentally (1.517 Å[1093]) are practically identical with computed bond lengths (1.545 Å[1091] and 1.535 Å[1093]). In the synthesis performed by Lambert, cation **622** was most probably formed by protonation of some acid impurity. In fact, ion **622** could be synthesized by reacting **620** with triflic acid.[1090] Consequently, cation **621** remains elusive.

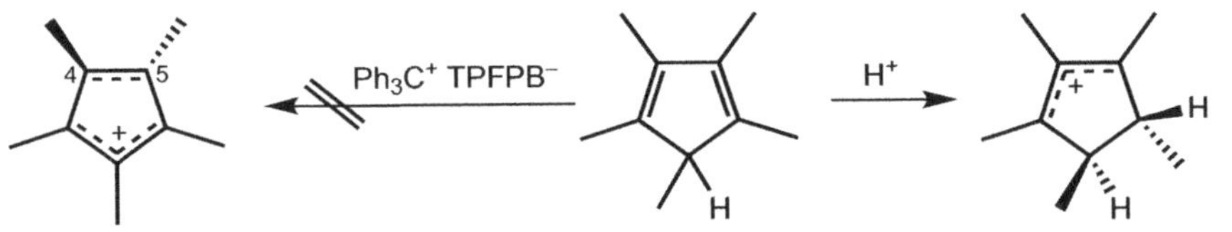

621 TPFPB = tetrakis(pentafluorophenyl)borate **620** **622**

(3.148)

Although experimental work on the parent square pyramidal $(CH)_5^+$ ion has not been reported, a dimethyl-substituted derivative **623** of the pyramidal ion **615**, however, has been prepared by Masamune et al.[1095,1096] and studied by ^{1}H and ^{13}C NMR spectroscopy. Ionization of the dimethylhomotetrahedranol under superacidic conditions gave the pyramidal ion **623** [Eq. (3.149)]. The alternative singlet cyclopentadienyl structures (**616** and **617**) for the species was eliminated based on ^{1}H and ^{13}C NMR data. The highly shielded C(5) carbon chemical shift (δ^{13}C −23.0) supports the structure **623** over a set of rapidly equilibrating structures such as **624a** and **624b**. The quenching of ion **623** at low temperature affords cyclopentenes.

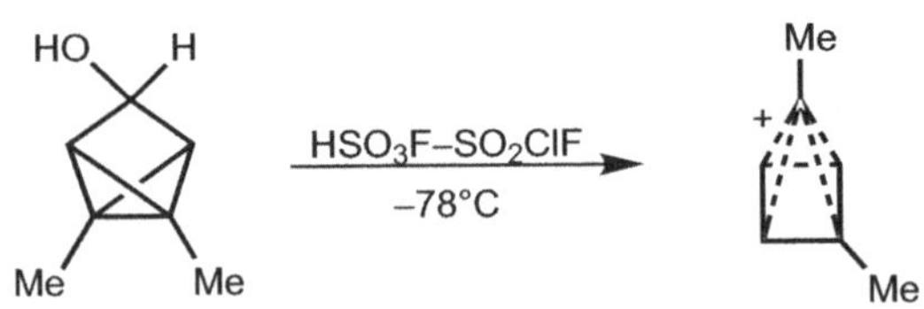

623

(3.149)

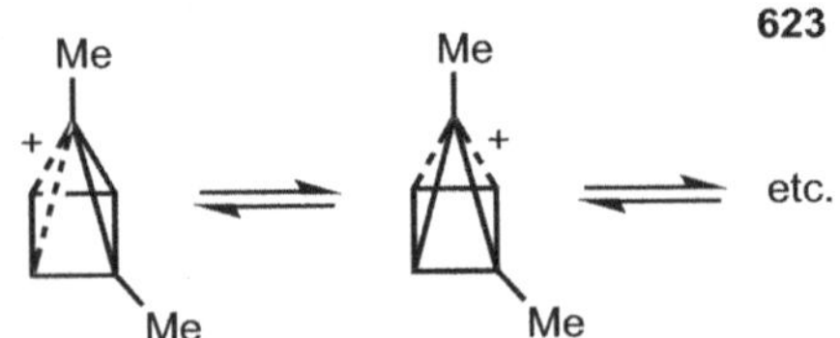

a **b**

624

Attempts have been made to observe[1084] the assumed interconversion of **625** to **626** in fluorenyl cations. Such intramolecular interconversion through the capped pyramidal ion **627** was not observed [Eq. (3.150)]. MINDO/3 calculations[1084] on the isomeric structures of cyclopentadienyl, indenyl, and fluorenyl cations indicated strongly decreasing relative stabilities of pyramidal forms due to bezannulation.

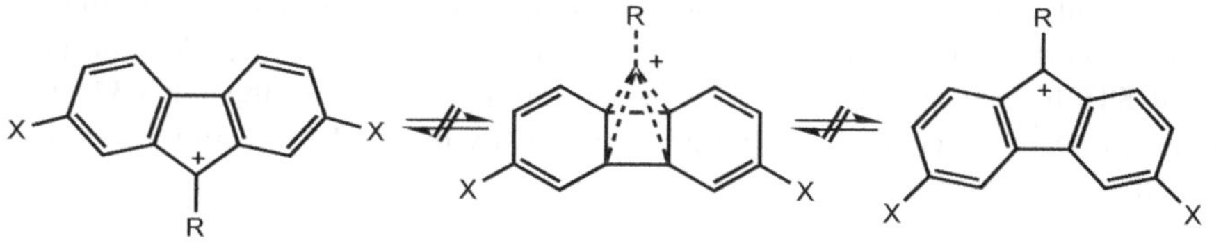

625 **627** **626**

(3.150)

Insertion of a methylene group into the four-membered ring of the pyramidal $(CH)_5^+$ cation **615** would give rise to the homo derivative **628**. As discussed earlier,[41,1037–1041] hexamethylbicyclo[2.1.1]hexenyl cation is best represented as a bishomocyclopropenyl cation **586** and not pyramidal-type ion **629**.

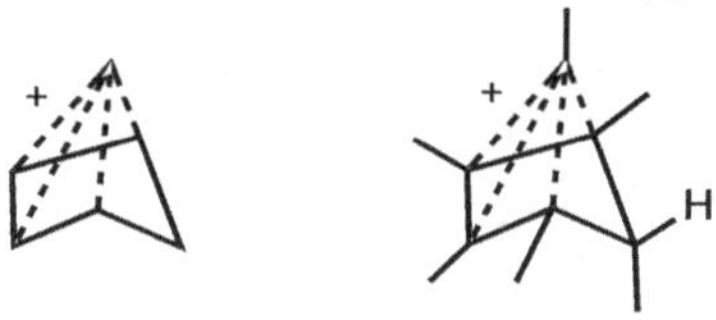

628 **629**

The trishomocyclopropenyl cation **630** has been investigated by both solvolytic[1097–1100] and stable ion studies.[1096] The obtained ^{1}H and ^{13}C NMR data could be explained with the intermediacy of the bishomo square pyramidal ion **631**, although no conclusive distinction could be made between rapidly equilibrating systems of **630** and **631** structures. However, the ion **631** of C_{2v} symmetry was preferred based on related MINDO/2 calculations.[1101] Subsequent high-level computational studies have confirmed,[1102] that cation **630** is not an energy minimum on the potential energy surface and converges upon optimization into structure **631** (MP2/cc-pVTZ level). Recent *ab initio* calculations [GIAO-CCSD(T) method][1103] have arrived at the same conclusion. The classical structure **632** (C_s symmetry), a dicyclopropylcarbinyl cation, was also shown to be energy minimum on the potential energy surface being 5.1 kcal mol^{-1} less stable than cation **631**. The calculated ^{13}C NMR chemical shifts agree very well with the experimental values.

630 **631** **632**

The other two substituted bishomo-$(CH)_5^+$ cations that have been investigated in superacid media are the octamethylated ion **633** and its methano-bridged analog **634**.[41] The observed ^{13}C NMR data of both ions[1097,1104–1106] are consistent with their highly symmetrical structures **633** and **634**, but the data could also be explained with degenerate rapidly equilibrating systems of lesser symmetry. Hart and Willer[1106] determined the apical $^{13}C-^{1}H$ coupling constant of ion **633** ($J_{C-H} = 220$ Hz), which is consistent with the single pyramidal structure **633** with nearly *sp* hybridization of the apical carbon atom. Surprisingly, the bishomoaromatic 7-norbornenyl cation **581** has nearly identical $^{13}C-^{1}H$ coupling constant as the ion **633** at the C(7) carbon.[1011] Recently, rearrangements involving $C_8H_8F^+$ cation have been studied.[1107]

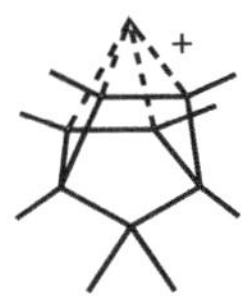

633

Both the bishomo square pyramidal structure **634** of C_{2v} symmetry and the trishomocyclopropenium-type structure **635a/635b** (C_s symmetry) were found to be minima on the potential energy surface of $C_8H_9^+$, but they are almost isoenergetic: Cation **635** is higher in energy than ion **634** only by 0.2 kcal mol^{-1} [MP4(SDQ)/6-31G*//MP2/6-31G*+ZPE level[1102]] or 0.7 kcal mol^{-1} (MP2/cc-pVTZ//MP2/cc-pVTZ+ZPE level[1103]). The observed spectrum of $C_8H_9^+$ could be reproduced quite well with a weighted average of the experimental and calculated NMR spectra. These observations indicate that an equilibrium mixture undergoing rapid exchange on the NMR time scale involving ion **634** and **635** (in a 1:2 ratio) can best represent the structure of $C_8H_9^+$.

635a **634** **635b**

3.5.4.2. $(CH)_6^{2+}$-Type Dications.

The first known representative of the $(CH)_6^{2+}$-type hexacoordinate pyramidal dications is Hogeveen's hexamethyl cation **437**.[41,789,1108] This dication **437** can be prepared from a variety of precursors in superacidic media (HSO_3F, HSO_3F–SbF_5) at low temperature (Scheme 3.28).

The observed fivefold symmetry in the ^{1}H and ^{13}C NMR spectra even at very low temperature (−150°C) with no line broadening leaves only two alternatives for the structure of the dication: the nonclassical fivefold symmetrical, static structure **437** or

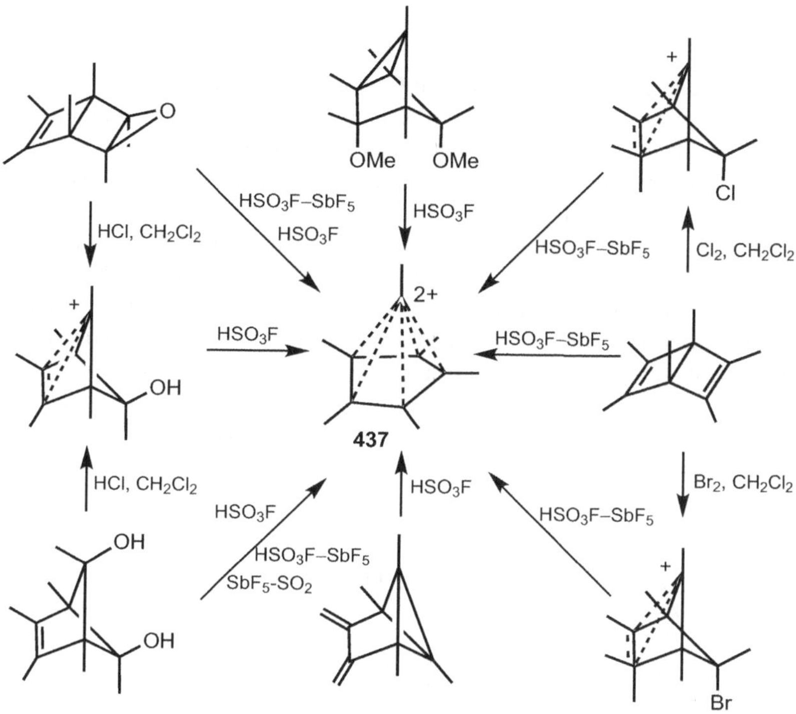

Scheme 3.28

a set of rapidly equilibrating degenerate dications **636** (Scheme 3.29) with an activation energy barrier of less than 5 kcal mol^{-1}.

The ^1H and ^{13}C NMR data, the rates of deuterium exchange, the rate of carbonylation, and the thermal stability of **437** provide strong evidence for the nonclassical

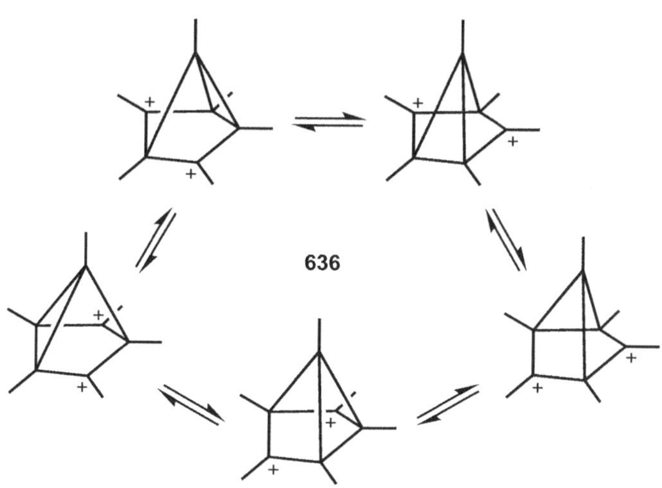

Scheme 3.29

nature of this carbodication. However, the structure can be also be considered in terms of the rapidly equilibrating set of structures **636** that have only one plane of symmetry during the degenerate process; the apical C-atom describes a circle.

Although the observed ^1H and ^{13}C NMR data strongly support the bridged nature of the ion **437**, further definitive evidence comes form isotope perturbation studies.[1109] Very little isotopic perturbation (the degeneracy is not lifted due to CD_3 substitution) on the basal carbon signal is observed in the ^{13}C NMR spectrum of the dication $[C_6(CH_3)_4(CD_3)_2{}^{2+}]$, indicating that the ion is symmetrically bridged. Other supporting evidence comes from a comparison of the ^{13}C and ^{11}B NMR shifts of the isoelectronic borane B_6H_{10} **637**. Williams and Field have shown[936] that the ^{13}C NMR chemical shifts of a series of nonclassical cations may be compared with the ^{11}B NMR chemical shifts of the isoelectronic polyhedral polyboranes.

Hogeveen, Heldeweg, and co-workers[1110,1111] also prepared the ethyl- and isopropyl-substituted analogs **638a–638d** of the dipositive ion $(CH)_6{}^{2+}$ and their bridged structures have been confirmed by ^1H and ^{13}C NMR spectroscopic and quenching studies.

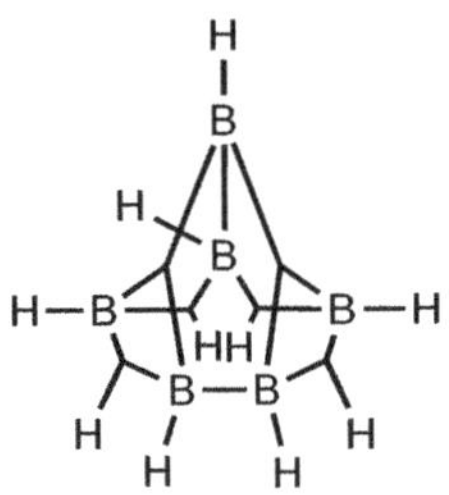

637

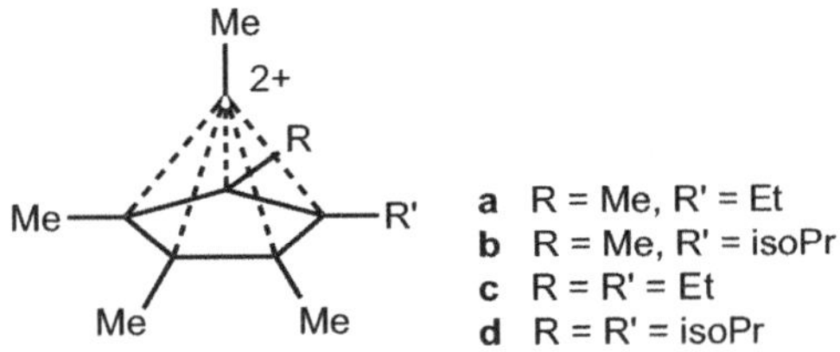

638

Lambert et al.[1112] have performed a study to learn if the $C_5SiMe_7{}^+$ cation **639**, the monosila analog of the $C_6Me_7{}^+$ cation, can be accessed [Eq. (3.151)]. Cation **639** is related to Hogeveen's hexamethyl cation **437**. Either a static structure with fivefold symmetry or a dynamic structure with the dimethylsilyl group equilibrating rapidly among positions is in agreement with the ^1H and ^{13}C NMR spectra (a single resonance for the ring methyl groups and a single resonance for the methyl groups on silicon). DFT calculations [B3LYP/6-31G(d,p) level] indicated the unsymmetrical silabicyclo [3.1.0]hexenyl structure **640** to be the global minimum. GIAO calculated and experimentally observed ^1H, ^{13}C, and ^{29}Si chemical shift values are in good agreement.

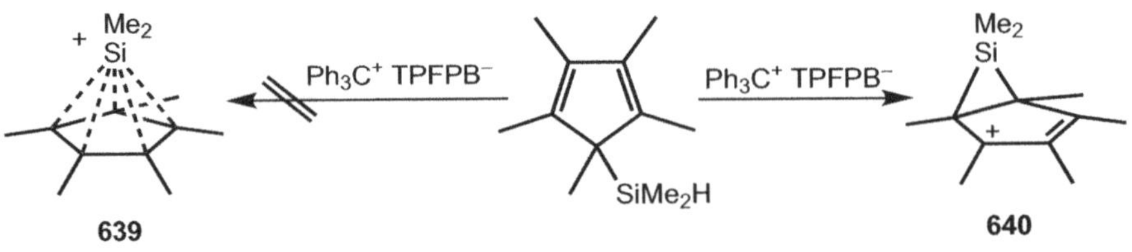

639 **640**

(3.151)

REFERENCES

1. G. A. Olah, A. Germain, H. C. Lin, and D. A. Forsyth, *J. Am. Chem. Soc.* **97**, 2928 (1975).
2. G. A. Olah, G. K. S. Prakash, and K. Lammertsma, *Res. Chem. Intermed.* **12**, 141 (1989).
3. G. A. Olah, *Angew. Chem. Int. Ed. Engl.* **32**, 767 (1993).
4. G. A. Olah, and D. A. Klumpp, *Acc. Chem. Res.* **37**, 211 (2004).
5. G. A. Olah and D. A. Klumpp, *Superelectrophiles and Their Chemistry*, Wiley-Interscience, Hoboken, NJ, 2008.
6. J. F. Norris, and W. W. Saunders, *J. Am. Chem. Soc.* **23**, R85 (1901).
7. K. Kehrmann and F. Wentzel, *Ber. Dtsch. Chem. Ges.* **34**, 3815 (1901).
8. M. Gomberg, *Ber. Dtsch. Chem. Ges.* **35**, 1822 (1902).
9. A. Baeyer and V. Villiger, *Ber. Dtsch. Chem. Ges.* **35**, 1189, 3013 (1902).
10. H. Meerwein and K. van Emster *Ber. Dtsch. Chem. Ges.* **55**, 2500 (1922).
11. For a summary, see C. K. Ingold, *Structure and Mechanism in Organic Chemistry*, Cornell University Press, Ithaca, NY, 1953.
12. F. C. Whitmore, *J. Am. Chem. Soc.* **54**, 3274, 3276 (1932); *Ann. Rep. Prog. Chem. (Chem. Soc. London)*, 177 (1943).
13. H. Meerwein, in *Methoden der Organischen Chemie (Houben-Weyl)*, Vol. VI/3, 4th ed. E. Müller,ed., Thieme, Stuttgart, 1965, p. 329 and references cited therein.
14. F. Seel, *Z. Anorg. Allg. Chem.* **250**, 331 (1943); **252**, 24 (1943).
15. For earlier reviews, see G. A. Olah, *Chem. Eng. News* **45**, 76 (1967); *Science* **168**, 1298 (1970).
16. Magic Acid is a registered trade name.
17. G. A. Olah, *Angew. Chem. Int. Ed. Engl.* **12**, 173 (1973).
18. For comprehensive reviews, see *Carbonium Ions*, Vols. 1–5, G. A. Olah and P. v. R. Schleyer, eds., Wiley-Interscience, New York, London, Vol. 1, (1968); Vol. 2, (1970); Vol. 3, (1972); Vol. 4, (1973); Vol. 5, 1976.
19. For recent reviews, see G. A. Olah, *Topics Curr. Chem.* **80**, 19 (1979); *Chem. Scr.* **18**, 97 (1981); *Angew. Chem. Int. Ed. Engl.* **34**, 1393 (1995); *J. Org. Chem.* **66**, 5943 (2001); *J. Org. Chem.* **70**, 2413 (2005).
20. G. A. Olah, *J. Am. Chem. Soc.* **94**, 808 (1972).
21. "Existence" is defined by Webster as: "The state of fact of being especially as considered independently of human consciousness."
22. R. J. Gillespie, *J. Chem. Soc.* 1002 (1952); and for rebuttal M. J. S. Dewar, *J. Chem. Soc.* 2885 (1953).
23. R. J. Gillespie, *J. Chem. Phys.* **21**, 1893 (1953).
24. E. L. Muetterties and R. A. Schunn, *Quart. Rev.* **20**, 245 (1966).
25. For a review critical discussion see F. R. Jensen and B. Rickborn, *Electrophilic Substitution of Organomercurials*, McGraw-Hill, New York, 1968.
26. J. D. Roberts,seems to have first used the term "nonclassical ion" when he proposed the tricyclobutonium ion structure for the cyclopropylcarbinyl cation. See J. D. Roberts and R. H. Mazur, *J. Am. Chem. Soc.* **73**, 3542 (1951). Winstein referred to the nonclassical structure of norbornyl, cholesteryl, and 3-phenyl-2-butyl cations. See S. Winstein and D.

Trifan, *J. Am. Chem. Soc.* **74**, 1154 (1952). Bartlett's definition is widely used for nonclassical ions; it states "an ion is nonclassical if its ground state has delocalized bonding σ electrons " Also see ref. 18.

27. For leading references, see P. D. Barlett, *Nonclassical Ions*, reprints, and commentary, W. A. Benjamin, New York, 1965.
28. G. D. Sargent, *Quart. Rev.* **20**, 301 (1966).
29. S. Winstein, *Quart. Rev.* **23**, 141 (1969).
30. For an opposing view, see H. C. Brown, *Chem. Eng. News* **45** (Feb. 13), 86–97 (1967) and references given therein.
31. H. C. Brown, *Acc. Chem. Res.* **6**, 377 (1973).
32. H. C. Brown, *Tetrahedron* **32**, 179 (1976).
33. G. A. Olah, *Acc. Chem. Res.* **9**, 41 (1976).
34. H. C. Brown, *The Nonclassical Ion Problem*, Commentary by P. v. R. Schleyer, Plenum Press, New York, 1977.
35. H. C. Brown, *Acc. Chem. Res.* **16**, 432 (1983).
36. C. A. Grob, *Acc. Chem. Res.* **16**, 426 (1983).
37. C. Walling, *Acc. Chem. Res.* **16**, 448 (1983).
38. G. A. Olah, A. M. White, J. R. DeMember, A. Commeyras, and C. Y. Lui, *J. Am. Chem. Soc.* **92**, 4627 (1970).
39. G. A. Olah, G. Liang, G. D. Mateescu, and J. G. Riemenschneider, *J. Am. Chem. Soc.* **95**, 8698 (1973).
40. G. A. Olah, G. K. S. Prakash, and M. Saunders, *Acc. Chem. Res.* **16**, 440 (1983).
41. H. Hogeveen and P. W. Kant, *Acc. Chem. Res.* **8**, 413 (1975).
42. For a review discussion, see J. J. Jensen, *Chimia* **20**, 309 (1966); however, for a contrasting view on general use of "carbonium ions," see H. Volz, *Chem. Unserer Zeit* **4**, 101 (1970).
43. V. Gold, *Pure Appl. Chem.* **51**, 1725 (1979); **55**, 1281 (1983).
44. This method has been employed in most of his work. For the procedure, see ref. 95.
45. P. Ahlberg, *Chem. Scr.* **2**, 81, 231 (1972).
46. M. Saunders, D. Cox, and W. Ohlmstead, *J. Am. Chem. Soc.* **95**, 3018 (1973).
47. P. Ahlberg and C. Engdahl, *Chem. Scr.* **11**, 95 (1977).
48. P. Ahlberg and M. Ek, *J. Chem. Soc., Chem. Commun.* 624 (1979).
49. M. Saunders, D. Cox, and J. R. Lloyd, *J. Am. Chem. Soc.* **101**, 6656 (1979).
50. P. C. Myhre and C. S. Yannoni, *J. Am. Chem. Soc.* **103**, 230 (1981).
51. D. P. Kelly and H. C. Brown, *Aust. J. Chem.* **29**, 957 (1976).
52. S. Forsén and R. A. Hoffman, *Acta Chem. Scand.* **17**, 1787 (1963).
53. S. Forsén and R. A. Hoffman, *J. Chem. Phys.* **39**, 2892 (1963); **40**, 1189 (1964).
54. C. Engdahl and P. Ahlberg, *J. Am. Chem. Soc.* **101**, 3940 (1979).
55. P. v. R. Schleyer, D. Lenoir, P. Mison, G. Liang, G. K. S. Prakash, and G. A. Olah, *J. Am. Chem. Soc.* **102**, 683 (1980).
56. M. Saunders, J. Chandrasekhar, and P. v. R. Schleyer, *Rearrangements in Ground and Excited States*, Vol. 1, P. de Mayo, ed., Academic Press, New York, 1980, p. 1.
57. M. Saunders, M. H. Jaffe, and P. Vogel, *J. Am. Chem. Soc.* **93**, 2558 (1971).

58. M. Saunders and P. Vogel, *J. Am. Chem. Soc.* **93**, 2559, 2561 (1971).
59. M. Saunders, H. A. Jiméney-Vázquez, and O. Kronja, in *Stable Carbocation Chemistry*, G. K. S. Prakash and P. v. R. Schleyer, eds., Wiley, New York, 1997, Chapter 9, p. 297.
60. M. Saunders and O. Kronja,in *Carbocation Chemistry*, G. A. Olah and G. K. S. Prakash, eds., Wiley, Hoboken, NJ, 2004, Chapter 8, p. 213
61. M. Saunders, L. A. Telkowski, and M. R. Kates, *J. Am. Chem. Soc.* **99**, 8070 (1977).
62. M. Saunders and M. R. Kates, *J. Am. Chem. Soc.* **99**, 8071 (1977).
63. C. S. Yannoni, *Acc. Chem. Res.* **15**, 201 (1982).
64. J. R. Lyerla, C. S. Yannoni, and C. A. Fyfe, *Acc. Chem. Res.* **15**, 208 (1982).
65. M. Sundaralingam, and A. K. Chwang in *Carbonium Ions*, Vol. 3, G. A. Olah and P. v. R. Schleyer, eds., Wiley-Interscience, New York, 1976, Chapter 39, p. 2427.
66. T. Laube, in *Stable Carbocation Chemistry*, G. K. S. Prakash and P. v. R. Schleyer, eds., Wiley, New York, 1997, Chapter 14, p. 453.
67. For a detailed account of the method, see G. A. Olah, G. K. S. Prakash, D. G. Farnum, and T. P. Clausen, *J. Org. Chem.* **48**, 2146 (1983).
68. J. M. Hollander and W. L. Jolly, *Acc. Chem. Res.* **3**, 193 (1970).
69. D. Betteridge and A. D. Baker, *Anal. Chem.* **42**,, 43A. (1970).
70. G. A. Olah, G. D. Mateescu, L. A. Wilson, and M. H. Goss, *J. Am. Chem. Soc.* **92**, 7231 (1970).
71. G. A. Olah, E. B. Baker, J. C. Evans, W. S. Tolgyesi, J. S. McIntyre, and I. J. Bastien, *J. Am. Chem. Soc.* **86**, 1360 (1964).
72. G. A. Olah, J. R. DeMember, A. Commeyras, and J. L. Bribes, *J. Am. Chem. Soc.* **93**, 459 (1971).
73. G. A. Olah, C. U. Pittman, Jr. R. Waack, and M. Doran, *J. Am. Chem. Soc.* **88**, 1488 (1966).
74. G. A. Olah, C. U. Pittman, Jr., and M. C. R. Symons,in *Carbonium Ions*, Vol. 1, G. A. Olah and P. v. R. Schleyer,eds., Wiley-Interscience, New York, London, 1968, Chapter 5, p. 153.
75. E. M. Arnett and C. Petro, *J. Am. Chem. Soc.* **100**, 2563 (1978).
76. E. M. Arnett and T. C. Hofelich, *J. Am. Chem. Soc.* **105**, 2889 (1983).
77. W. J. Hehre, L. Radom, and P. v. R. Schleyer, *Ab Initio Molecular Orbital Theory*, Wiley, New York, 1986.
78. P. v. R. Schleyer and C. Maerker, *Pure Appl. Chem.* **67**, 755 (1995).
79. P. v. R. Schleyer, C. Maerker, S. Buzek, S. Sieber, in *Stable Carbocation Chemistry*, G. K. S. Prakash and P. v. R. Schleyer, eds., Wiley, New York, 1997, Chapter 2, p. 19.
80. T. Helgaker, M. Jaszuński, and K. Ruud, *Chem. Rev.* **99**, 293 (1999).
81. For an extensive review, see *Mass Spectrometry of Organic Ions*, F. W. McLafferty, ed., Academic Press, New York, 1963.
82. H. C. Brown, H. W. Pearsall, and L. P. Eddy, *J. Am. Chem. Soc.* **72**, 5347 (1950).
83. E. Wertyporoch and T. Firla, *Ann. Chim.* **500**, 287 (1933).
84. G. Olah, S. J. Kuhn, and J. Olah, *J. Chem. Soc.* 2174 (1957).
85. F. Fairbrother, *J. Chem. Soc.* 503 (1945).

86. J. Rosenbaum and M. C. R. Symons, *Proc. Chem. Soc. (London)* 92 (1959); *Mol. Phys.* **3**, 205 (1960); *J. Chem. Soc.* 1 (1961).
87. A. C. M. Finch and M. C. R. Symons, *J. Chem. Soc.* 378 (1965).
88. N. C. Deno, *Prog. Phys. Org. Chem.* **2**, 129 (1964).
89. For preliminary communications and lecture, see G. A. Olah, Conference Lecture at the 9th Reaction Mechanism Conference, Brookhaven, New York, August 1962; Abstracts, 142nd National Meeting of the American Chemical Society (Atlantic City, NJ, 1962), p. 45; W. S. Tolgyesi, J. S. McIntyre, J. J. Bastien, and M. W. Meyer, p. 121;G. A. Olah, *Angew. Chem.* **75**, 800 (1963); *Rev. Chim. Rep. Populaire Roumaine* **7**, 1139 (1962); Preprints, *Div. Petr. Chem., ACS* **9**(7), 21 (1064); Organic Reaction Mechanism Conference, Cork, Ireland, June 1964, Special Publications No. 19, Chemical Society, London, 1965.
90. G. A. Olah and C. U. Pittman, Jr., *Adv. Phys. Org. Chem.* **4**, 305 (1966).
91. G. A. Olah, W. S. Tolgyesi, S. J. Kuhn, M. E. Moffatt, I. J. Bastien, and E. B. Baker, *J. Am. Chem. Soc.* **85**, 1328 (1963).
92. J. Bacon, P. A. W. Dean, and R. J. Gillespie, *Can. J. Chem.* **47**, 1655 (1969).
93. A. Commeyras and G. A. Olah, *J. Am. Chem. Soc.* **91**, 2929 (1969).
94. For a comprehensive report, see G. A. Olah and A. M. White, *J. Am. Chem. Soc.* **91**, 5801 (1969).
95. G. A. Olah and D. J. Donovan, *J. Am. Chem. Soc.* **99**, 5026 (1977).
96. M. Saunders and E. L. Hagen, *J. Am. Chem. Soc.* **90**, 6881 (1968).
97. M. Saunders, P. Vogel, E. L. Hagen, and J. Rosenfeld, *Acc. Chem. Res.* **6**, 53 (1973).
98. M. Saunders, A. P. Hewett, and O. Kronja, *Croat. Chem. Acta* **65**, 673 (1992).
99. C. C. Lee and D. J. Woodcock, *J. Am. Chem. Soc.* **92**, 5992 (1970).
100. G. J. Karabatsos, C. Zioudrou, and S. Meyerson, *J. Am. Chem. Soc.* **92**, 5996 (1970).
101. L. Radom, J. A. Pople, V. Buss, and P. v. R. Schleyer, *J. Am. Chem. Soc.* **93**, 1813 (1971); **94**, 311 (1972).
102. N. Bodor and M. J. S. Dewar, *J. Am. Chem. Soc.* **93**, 6685 (1971).
103. N. Bodor, M. J. S. Dewar, and D. H. Lo, *J. Am. Chem. Soc.* **94**, 5303 (1972).
104. H. Lischka and H.-J. Köhler, *J. Am. Chem. Soc.* **100**, 5297 (1978).
105. M. S. Dewar, E. F. Healy, and J. M. Ruiz, *J. Chem. Soc., Chem. Commun.* 943 1987
106. W. Koch, P. v. R. Schleyer, P. Buzek, and B. W. Liu, *Croat. Chem. Acta* **65**, 655 (1992).
107. W. Koch and B. Liu, *J. Am. Chem. Soc.* **111**, 3479 (1989).
108. P. v. R. Schleyer, W. Koch, B. Liu, and U. Fleischer, *J. Chem. Soc., Chem. Commun.* 1098 (1098).
109. B. Chiavarino, M. E. Crestoni, A. A. Fokin, and S. Fornarini, *Chem. Eur. J.* **7**, 2916 (2001).
110. B. Chiavarino, M. E. Crestoni, S. Fornarini, J. Lemaire, L. Mac Aleese, and P. Maître, *Chem Phys Chem* **5**, 1679 (2004).
111. H. Hogeveen and C. J. Gaasbeek, *Recl. Trav. Chim. Pays-Bas* **89**, 857 (1970).
112. M. Saunders and S. P. Budiansky, *Tetrahedron* **35**, 929 (1979).
113. G. K. S. Prakash, A. Husain, and G. A. Olah, *Angew. Chem. Int. Ed.* **22**, 50 (1983).
114. G. E. Walker, O. Kronja, and M. Saunders, *J. Org. Chem.* **69**, 3598 (2004).

115. D. Frcaşiu, P. Lukinskas, and S. V. Pamidighantam, *J. Phys. Chem. A* **106**, 11672 (2002).
116. G. E. Douberly, A. M. Ricks, B. W. Ticknor, P. v. R. Schleyer, and M. A. Duncan, *J. Am. Chem. Soc.* **129**, 13782 (2007).
117. Gh. D. Mateescu and J. L. Riemenschneider, in *Electron Spectroscopy*, D. A. Shirley, ed., North Holland, Amsterdam, London, 1972, p. 661.
118. G. A. Olah, G. K. S. Prakash, and R. Krishnamurti, *J. Am. Chem. Soc.* **112**, 6422 (1990).
119. C. S. Yannoni, R. D. Kendrick, P. C. Myhre, D. C. Bebout, and B. L. Petersen, *J. Am. Chem. Soc.* **111**, 6440 (1989).
120. S. Sieber, P. Buzek, P. v. R. Schleyer, W. Koch, J. W. de, and M. Carneiro, *J. Am. Chem. Soc.* **115**, 259 (1993).
121. S. Hollenstein and T. Laube, *J. Am. Chem. Soc.* **115**, 7240 (1993).
122. T. Laube, *Acc. Chem. Res.* **28**, 399 (1995).
123. T. Kato and C. A. Reed, *Angew. Chem. Int. Ed.* **43**, 2908 (2004).
124. P. v. R. Schleyer, J. W. de, M. Carneiro, W. Koch, and D. A. Forsyth, *J. Am. Chem. Soc.* **113**, 3990 (1991).
125. G. A. Olah, G. K. S. Prakash, and G. Rasul, *Dalton Trans.* 521 (2008).
126. G. A. Olah, M. B. Comisarow, C. A. Cupas, and C. U. Pittman, Jr., *J. Am. Chem. Soc.* **87**, 2997 (1965).
127. G. A. Olah and E. Namanworth, *J. Am. Chem. Soc.* **88**, 5327 (1966).
128. G. A. Olah, J. Sommer, and E. Namanworth, *J. Am. Chem. Soc.* **89**, 3576 (1967).
129. G. A. Olah, P. Shilling, J. M. Bollinger, and J. Nishimura, *J. Am. Chem. Soc.* **96**, 2221 (1974).
130. P. D. Bartlett and M. Stiles, *J. Am. Chem. Soc.* **77**, 2806 (1955).
131. G. A. Olah, A. Wu, O. Farooq, and G. K. S. Prakash, *J. Org. Chem.* **55**, 1792 (1990).
132. G. A. Olah, G. K. S. Prakash, G. Liang, P.v. R. Schleyer, and W. D. Graham, *J. Org. Chem.* **47**, 1040 (1982).
133. G. A. Olah and O'Brien, D. H. *J. Am. Chem. Soc.* **89**, 1725 (1967).
134. G. A. Olah, and L. Lukas, *J. Am. Chem. Soc.* **89**, 2227, 4739 (1967).
135. H. Hogeveen and C. J. Gaasbeek, *Recl. Trav. Chim. Pays-Bas* **89**, 395 (1970) and references cited therein.
136. G. A. Olah, D. H. O'Brien, and C. U. Pittman, Jr. *J. Am. Chem. Soc.* **89**, 2996 (1967).
137. G. A. Olah, and P. Schilling, *Justus Liebigs Ann. Chem.* **761**, 77 (1972).
138. G. A. Olah, N. Friedman, J. M. Bollinger, and J. Lukas, *J. Am. Chem. Soc.* **88**, 5328 (1966).
139. J. B. Lambert, Y. Zhao, and H. Wu, *J. Org. Chem.* **64**, 2729 (1999).
140. J. B. Lambert, *Tetrahedron* **46**, 2677 (1990).
141. H.-U. Siehl and T. Müller, in *The Chemistry of Organosilicon Compounds*, Vol. 2, Part 1, Y. Apeliog and Z. Rappoport, eds., Wiley, Chichester, 1998, p. 595.
142. J. B. Lambert, Y. Zhao, R. W. Emblidge, L. A. Salvador, X. Liu, J.-H. So, and E. C. Chelius, *Acc. Chem. Res.* **32**, 183 (1999).
143. G. A. Olah, J. M. Bollinger, C. A. Cupas, and J. Lukas, *J. Am. Chem. Soc.* **89**, 2692 (1967).
144. G. A. Olah and J. Lukas, *J. Am. Chem. Soc.* **90**, 933 (1968).

145. M. Saunders and J. Rosenfeld, *J. Am. Chem. Soc.* **91**, 7756 (1969).
146. D. M. Brouwer and E. L. Mackor, *Proc. Chem. Soc. (London)* 147 (1964).
147. D. M. Brouwer, *Recl. Trav. Chim. Pays-Bas* **87**, 210 (1968).
148. G. A. Olah, A. P. Fung, T. N. Rawdah, and G. K. S. Prakash, *J. Am. Chem. Soc.* **103**, 4646 (1981).
149. G. K. S. Prakash, A. P. Fung, G. A. Olah, and T. N. Rawdah, *Proc. Natl. Acad. Sci. USA* **84**, 5092 (1987).
150. R. P. Kirchen, T. S. Sorensen, and K. E. Wagstaff, *J. Am. Chem. Soc.* **100**, 5134 (1978).
151. H.-U. Siehl, V. Vrček, and O. Kronja, *J. Chem. Soc., Perkin Trans.* **2**, 106 (2002).
152. A. Rauk, T. S. Sorensen, and P. v. R. Schleyer, *J. Chem. Soc., Perkin Trans.* **2**, 869 (2001).
153. G. A. Olah, A. L. Berrier, and G. K. S. Prakash, *Proc. Natl. Acad. Sci. USA* **78**, 1998 (1981).
154. G. A. Olah, A. L. Berrier, M. Arvanaghi, and G. K. S. Prakash, *J. Am. Chem. Soc.* **103**, 1122 (1981).
155. G. K. S. Prakash and P. Iyer, *Rev. Chem. Intermed.* **9**, 65 (1988).
156. D. P. Kelly, M. Periasamy, and P. T. Perumal, *J. Org. Chem.* **50**, 4566 (1985).
157. T. Laube and S. Hollenstein, *Helv. Chim. Acta* **77**, 1773 (1994).
158. J. D. Roberts and V. C. Chambers, *J. Am. Chem. Soc.* **73**, 3176, 5034 (1951).
159. G. A. Olah and J. M. Bollinger, *J. Am. Chem. Soc.* **90**, 6082 (1968).
160. G. Biale, A. J. Parker, S. G. Smith, I. D. R. Stevens, and S. Winstein, *J. Am. Chem. Soc.* **92**, 115 (1970).
161. P. v. R. Schleyer, T. M. Su, M. Saunders, and J. C. Rosenfeld, *J. Am. Chem. Soc.* **91**, 5174 (1969).
162. J. M. Bollinger, J. M. Brinich, and G. A. Olah, *J. Am. Chem. Soc.* **92**, 4025 (1970).
163. G. A. Olah, G. Liang, D. B. Ledlie, and M. G. Costopoulos, *J. Am. Chem. Soc.* **99**, 4196 (1977).
164. L. Radom, P. C. Hariharan, J. A. Pople, and P. v. R. Schleyer, *J. Am. Chem. Soc.* **95**, 6531 (1973) and references cited therein.
165. V. Buss, P. v. R. Schleyer, and L. C. Allen, *Topics Stereochem.* **7**, 253 (1973).
166. L. Radom, J. A. Pople, and P. v. R. Schleyer, *J. Am. Chem. Soc.* **95**, 8193 (1973).
167. D. H. Aue, W. R. Davidson, and M. T. Bowers, *J. Am. Chem. Soc.* **98**, 6700 (1976). This reference gives the heats of formation of $C_3H_5^+$ cations in the gas phase.
168. U. Schöllkopf, K. Fellenberger, M. Patsch, P. v. R. Schleyer, T. Su, and G. W. van Dine *Tetrahedron Lett.* 3639 (1967).
169. W. Kutzelnigg, *Tetrahedron Lett.* 4965 (1967).
170. U. Schöllkopf, *Angew. Chem. Int. Ed. Engl.* **7**, 588 (1968).
171. G. A. Olah, C. L. Jeuell, D. P. Kelly, and R. D. Porter, *J. Am. Chem. Soc.* **94**, 146 (1972).
172. J. S. Staral, I. Yavari, J. D. Roberts, G. K. S. Prakash, D. J. Donovan, and G. A. Olah, *J. Am. Chem. Soc.* **100**, 8016 (1978).
173. M. Saunders and J. C. Rosenfeld, *J. Am. Chem. Soc.* **92**, 2548 (1970).
174. G. A. Olah, R. J. Spear, P. C. Hiberty, and W. J. Hehre, *J. Am. Chem. Soc.* **98**, 7470 (1976).
175. R. P. Kirchen and T. S. Sorensen, *J. Am. Chem. Soc.* **99**, 6687 (1977).

176. G. A. Olah, G. K. S. Prakash, D. J. Donovan, and I. Yavari, *J. Am. Chem. Soc.* **100**, 7085 (1978).

177. J. Bredt, J. Houben, and P. Levy, *Ber. Dtsch. Chem. Ges.* **35**, 1286 (1902).

178. J. Bredt, H. Thouet, and J. Schmitz, *Justus Liebig's Ann. Chem.* **437**, 1 (1924).

179. P. D. Bartlett and L. H. Knox, *J. Am. Chem. Soc.* **61**, 3184 (1939).

180. H. Koch and W. Haaf, *Angew. Chem.* **72**, 628 (1960).

181. P. v. R. Schleyer, R. C. Fort, Jr. W. E. Watts, M. B. Comisarow, and G. A. Olah, *J. Am. Chem. Soc.* **86**, 4195 (1964).

182. G. A. Olah, G. K. S. Prakash, J. G. Shih, V. V. Krishnamurthy, G. D. Mateescu, G. Liang, G. Sipos, V. Buss, T. M. Gund, and P. v. R. Schleyer, *J. Am. Chem. Soc.* **107**, 2764 (1985).

183. G. Yan, N. R. Brinkmann, and H. F. Schaefer, III, *J. Phys. Chem. A* **107**, 9479 (2003).

184. N. Polfer, B. G. Sartakov, and J. Oomens, *Chem. Phys. Lett.* **400**, 201 (2004).

185. T. Laube, *Angew. Chem. Int. Ed. Engl.* **25**, 349 (1986).

186. T. Laube and E. Schaller, *Acta Crystallogr. B* **51**, 177 (1995).

187. G. A. Olah and G. Liang, *J. Am. Chem. Soc.* **95**, 194 (1973).

188. G. A. Olah, G. Liang, and P. W. Westerman, *J. Org. Chem.* **39**, 367 (1974).

189. G. A. Olah, G. Liang, P. v. R. Schleyer, W. Parker, and C. I. F. Watt, *J. Am. Chem. Soc.* **99**, 966 (1977).

190. G. A. Olah, G. K. S. Prakash, and M. A. S. Stephenson, *Abstr. Papers ACS National Meeting* **201**, Part 2, 351-Orgn. (1991).

191. A. A. Fokin, B. A. Tkachenko, P. A. Gunchenko, D. V. Gusev, and P. R. Schreiner, *Chem. Eur. J.* **11**, 7091 (2005).

192. P. v. R. Schleyer, W. E. Watts, R. C. Fort, Jr. M. B. Comisarow, and G. A. Olah, *J. Am. Chem. Soc.* **86**, 5679 (1964).

193. G. A. Olah, G. K. S. Prakash, T. Kobayashi, and L. A. Paquette, *J. Am. Chem. Soc.* **110**, 1304 (1988).

194. G. A. Olah, G. K. S. Prakash, W.-D. Fessner, T. Kobayashi, and L. A. Paquette, *J. Am. Chem. Soc.* **110**, 8599 (1988).

195. J. D. Roberts and R. H. Mazur, *J. Am. Chem. Soc.* **73**, 2509 (1951).

196. H. Hart and J. M. Sandri, *J. Am. Chem. Soc.* **81**, 320 (1959).

197. N. C. Deno, H. G. Richey, Jr. J. S. Liu, J. D. Hodge, J. J. Houser, and M. J. Wisotsky, *J. Am. Chem. Soc.* **84**, 2016 (1962).

198. H. G. Richey, Jr. in *Carbonium Ions*, Vol. 3, G. A. Olah and P. v. R. Schleyer, eds., Wiley-Interscience, New York, London, 1972, Chapter 25, p. 1201.

199. G. A. Olah, G. K. S. Prakash, and V. P. Reddy, *Chem. Rev.* **92**, 69 (1992).

200. N. C. Deno, H. G. Richey, Jr. J. S. Liu, D. N. Lincoln, and J. O. Turner, *J. Am. Chem. Soc.* **87**, 4533 (1965).

201. G. A. Olah, P. W. Westerman, and J. Nishimura, *J. Am. Chem. Soc.* **96**, 3548 (1974).

202. C. U. Pittman, Jr. and G. A. Olah, *J. Am. Chem. Soc.* **87**, 2998 (1965).

203. G. A. Olah, G. Liang, K. A. Babiak, T. M. Ford, D. L. Goff, T. K. Morgan, Jr., and R. K. Murray, Jr. *J. Am. Chem. Soc.* **100**, 1494 (1978) and references cited therein.

204. D. S. Kabakoff and E. Namanworth, *J. Am. Chem. Soc.* **92**, 3234 (1970).

205. G. A. Olah, G. K. S. Prakash, and T. N. Rawdah, *J. Org. Chem.* **45**, 965 (1980).
206. G. A. Olah and G. Liang, *J. Am. Chem. Soc.* **97**, 1920 (1975).
207. G. A. Olah and G. Liang, *J. Am. Chem. Soc.* **98**, 7026 (1976).
208. G. A. Olah, G. Liang, K. A. Babiak, T. K. Morgan, Jr., and R. K. Murray, Jr., *J. Am. Chem. Soc.* **98**, 576 (1976).
209. G. K. S. Prakash, V. P. Reddy, G. Rasul, J. Casanova, and G. A. Olah, *J. Am. Chem. Soc.* **120**, 13362 (1998).
210. V. P. Reddy, G. K. S. Prakash, and G. Rasul, in *ACS Symposium Series*, Vol. 965, K. K. Laali, ed., American Chemical Society, Washington, DC, 2007, Chapter 6, p. 106.
211. G. A. Olah, V. P. Reddy, G. Rasul, and G. K. S. Prakash, *J. Org. Chem.* **57**, 1114 (1992).
212. G. A. Olah and G. Liang, *J. Org. Chem.* **40**, 2108 (1975).
213. G. A. Olah, G. K. S. Prakash, and G. Liang, *J. Org. Chem.* **42**, 2666 (1977).
214. G. K. S. Prakash, Ph. D. Dissertation, University of Southern California, 1978.
215. Y. Mo, P. v. R. Schleyer, H. Jiao, and Z. Lin, *Chem. Phys. Lett.* **280**, 439 (1997).
216. R. W. Holman, J. Plocica, L. Blair, D. Giblin, and M. L. Gross, *J. Phys. Org. Chem.* **14**, 17 (2001).
217. N. C. Deno, H. G. Richey, Jr., N. Friedman, J. D. Hodge, J. J. Houser, and C. U. Pittman, Jr., *J. Am. Chem. Soc.* **85**, 2991 (1963).
218. N. C. Deno, J. Bollinger, N. Friedman, K. Hafer, J. D. Hodge, and J. J. Houser, *J. Am. Chem. Soc.* **85**, 2998 (1963).
219. T. S. Sorensen, *Can. J. Chem.* **42**, 2768 (1964).
220. G. A. Olah and M. B. Comisarow, *J. Am. Chem. Soc.* **86**, 5682 (1964).
221. G. A. Olah and G. Liang, *J. Am. Chem. Soc.* **94**, 6434 (1972) and references cited therein.
222. G. A. Olah, G. Liang, and Y. K. Mo, *J. Am. Chem. Soc.* **94**, 3544 (1972).
223. G. A. Olah, G. Liang, and S. P. Jindal, *J. Org. Chem.* **40**, 3259 (1975).
224. G. A. Olah and R. J. Spear, *J. Am. Chem. Soc.* **97**, 1539 (1975).
225. G. A. Olah, G. Asensio, and H. Mayr, *J. Org. Chem.* **43**, 1518 (1978).
226. G. A. Olah, G. K. S. Prakash, and G. Liang, *J. Org. Chem.* **41**, 2820 (1976).
227. G. A. Olah, V. P. Reddy, G. Rasul, and G. K. S. Prakash, *J. Am. Chem. Soc.* **121**, 9994 (1999).
228. B. K. Carpenter, *J. Chem. Soc., Perkin Trans.* **2**, 1 (1974).
229. N. C. Deno, in *Carbonium Ions*, Vol. 2, G. A. Olah and P. v. R. Schleyer, eds., Wiley-Interscience, New York, London, 1970, Chapter 18, p. 783.
230. G. A. Olah, J. S. Stratal, R. J. Spear, and G. Liang, *J. Am. Chem. Soc.* **97**, 5489 (1975).
231. P. Buzek, P. v. R. Schleyer, H. Vančik, Z. Mihalic, and J. Gauss, *Angew. Chem. Int. Ed. Engl.* **33**, 448 (1994).
232. A. I. Biaglow, R. J. Gorte, and D. White, *J. Chem. Soc., Chem. Commun.* 1164 (1993).
233. G. E. Douberly, A. M. Ricks, P. v. R. Schleyer, and M. A. Duncan, *J. Chem. Phys.* **128**, 021102 (2008).
234. H. Vančik, V. Gabelica, D. E. Sunko, P. Buzek, and P. v. R. Schleyer, *J. Phys. Org. Chem.* **6**, 427 (1993).
235. D. Kidemet, Z. Mihalić, I. Novak, and H. Vančik, *J. Org. Chem.* **64**, 4931 (1999).

236. H. Vančik, *Croat. Chem. Acta* **73**, 595 (2000).

237. G. A. Olah, V. P. Reddy, J. Casanova, and G. K. S. Prakash, *J. Org. Chem.* **57**, 6431 (1992).

238. C. U. Pittman, Jr. *J. Chem. Soc. D: Chem. Commun.* 122 (1969).

239. T. S. Sorensen, *J. Am. Chem. Soc.* **87**, 5075 (1965); *Can. J. Chem.* **43**, 2744 (1965).

240. G. A. Olah, C. U. Pittman, Jr., and T. S. Sorensen, *J. Am. Chem. Soc.* **88**, 2331 (1966).

241. G. A. Olah, G. K. S. Prakash, and G. Liang, *J. Org. Chem.* **42**, 661 (1977).

242. D. M. Dytnerski, K. Ranganayakulu, B. P. Singh, and T. S. Sorensen, *Can. J. Chem.* **60**, 2993 (1982); *J. Org. Chem.* **48**, 309 (1983).

243. W. Adam, I. Casades, V. Fornés, H. García, and O. Weichold, *J. Org. Chem.* **65**, 3947 (2000).

244. A. A. V. Stuart and E. L. Mackor, *J. Chem. Phys.* **27**, 826 (1957).

245. G. Dallinga, E. L. Mackor, and A. A. V. Stuart, *Mol. Phys.* **1**, 123 (1958).

246. G. MacLean and E. L. Mackor, *Discuss. Faraday Soc.* **34**, 165 (1962).

247. T. Birchall and R. J. Gillespie, *Can. J. Chem.* **42**, 502 (1964).

248. V. A. Koptyug, *Arenium Ion—Structure and Reactivity*, Ch. Rees, ed., Springer, Berlin, 1984.

249. G. A. Olah and S. J. Kuhn, *Nature* **178**, 693 (1956); *Naturwissenschaften* **43**, 59 (1956); *J. Am. Chem. Soc.* **80**, 6535 (1958).

250. G. A. Olah, *Abstracts, 138th National Meeting of the ACS*, New York, 1960, p. 3P; *J. Am. Chem. Soc.* **87**, 1103 (1965).

251. G. A. Olah, A. Germain, and A. M. White, in *Carbonium ions*, Vol. 5, G. A. Olah and P. v. R. Schleyer, eds., Wiley-Interscience, New York, 1976, Chapter 35, p. 209.

252. G. A. Olah, R. H. Schlosberg, R. D. Porter, Y. K. Mo, D. P. Kelly, and G. D. Mateescu, *J. Am. Chem. Soc.* **94**, 2034 (1972).

253. G. A. Olah, J. S. Stratal, G. Asencio, G. Liang, D. A. Forsyth, and G. D. Mateescu, *J. Am. Chem. Soc.* **100**, 6299 (1978).

254. G. A. Olah, G. D. Mateescu, and Y. K. Mo, *J. Am. Chem. Soc.* **95**, 1865 (1973).

255. G. D. Mateescu, Ph.D. dissertation, Case Western Reserve University, 1971.

256. W. v. E. Doering, M. Saunders, H. G. Boyton, H. W. Earhart, E. F. Wadley, W. R. Edwards, and G. Laber, *Tetrahedron* **4**, 178 (1958).

257. J. R. Lyerla, C. S. Yannoni, D. Bruck, and C. A. Fyfe, *J. Am. Chem. Soc.* **101**, 4770 (1979).

258. R. Rathore, J. Hecht, and J. K. Kochi, *J. Am. Chem. Soc.* **120**, 13278 (1998).

259. C. A. Reed, N. L. P. Fackler, K.-C. Kim, D. Stasko, D. R. Evans, P. D. W. Boyd, and C. E. F. Rickard, *J. Am. Chem. Soc.* **121**, 6314 (1999).

260. C. A. Reed, K.-C. Kim, E. S. Stoyanov, D. Stasko, F. S. Tham, L. J. Mueller, and P. D. W. Boyd, *J. Am. Chem. Soc.* **125**, 1796 (2003).

261. J. B. Lambert, S. Zhang, C. L. Stern, and J. C. Huffman, *Science* **260**, 1917 (1993).

262. G. A. Olah, G. Rasul, X.-Y. Li, H. A. Buchholz, G. Sandford, and G. K. S. Prakash, *Science* **263**, 983 (1994).

263. G. A. Olah, G. Rasul, and G. K. S. Prakash, *J. Organomet. Chem.* **521**, 271 (1996).

264. R. Meyer, K. Werner, and T. Müller, *Chem. Eur. J.* **8**, 1163 (2002).
265. N. Choi, P. D. Lickiss, M. McPartlin, P. C. Masangane, and G. L. Veneziani, *Chem. Commun.* 6023 (2005).
266. T. Nishinaga, R. Inoue, A. Matsuura, and K. Komatsu, *Org. Lett.* **4**, 1435 (2002).
267. K. Komatsu, in *ACS Symposium Series*, Vol. 965, K. K. Laali,ed., American Chemical Society, Washington, DC, 2007, Chapter 2, p. 32.
268. R. Rathore, S. H. Loyd, and J. K. Kochi, *J. Am. Chem. Soc.* **116**, 8414 (1994).
269. K. Lammertsma and H. Cerfontain, *J. Am. Chem. Soc.* **101**, 3618 (1979).
270. N. V. Bodoev, V. I. Mamatyuk, A. P. Krysin, and V. A. Koptyug, *Russ. J. Org. Chem. (Engl. Transl.)* **14**, 1789 (1978).
271. K. K. Laali, *Chem. Rev.* **96**, 1873 (1996).
272. K. K. Laali, in *Carbocation Chemistry*, G. A. Olah and G. K. S. Prakash, eds., Wiley, Hoboken, NJ, 2004, Chapter 9, p. 237 and references cited therein.
273. K. Laali and H. Cerfontain, *J. Org. Chem.* **48**, 1092 (1983).
274. G. I. Borodkin, M. M. Shakirov, and V. G. Shubin, *Russ. J. Org. Chem. (Engl. Transl.)* **26**, 1944 (1990).
275. F. van de Griendt and H. Cerfontain, *Tetrahedron* **35**, 2563 (1979).
276. K. K. Laali and P. E. Hansen, *J. Org. Chem.* **56**, 6795 (1991).
277. K. K. Laali and M. Tanaka, *J. Org. Chem.* **63**, 7280 (1998).
278. K. K. Laali, S. Hollenstein, S. E. Galembeck, and M. M. Coombs, *J. Chem. Soc., Perkin Trans.* **2**, 211 (2000).
279. T. Okazaki and K. K. Laali, *J. Org. Chem.* **69**, 510 (2004).
280. K. K. Laali, M. A. Arrica, T. Okazaki, and R. G. Harvey, *J. Org. Chem.* **72**, 6768 (2007).
281. K. K. Laali, T. Okazaki, R. H. Mitchell, and T. R. Ward, *J. Org. Chem.* **66**, 5329 (2001).
282. K. K. Laali, T. Okazaki, R. H. Mitchell, K. Ayub, R. Zhang, and S. G. Robinson, *J. Org. Chem.* **73**, 457 (2008).
283. K. K. Laali and J. J. Houser, *J. Chem. Soc., Perkin Trans.* **2**, 1303 (1994).
284. P. Warner and S. Winstein, *J. Am. Chem. Soc.* **91**, 7785 (1969).
285. D. J. Cram and J. M. Cram, *Acc. Chem. Res.* **4**, 204 (1971).
286. G. M. Wallraff, E. Vogel, and J. Michl, *J. Org. Chem.* **53**, 5807 (1988).
287. D. J. Cram, *J. Am. Chem. Soc.* **71**, 3863, 3871, 3875 (1949); for a critical review, see D. J. Cram, *J. Am. Chem. Soc.* **86**, 3767 (1964).
288. G. A. Olah, M. B. Comisarow, E. Namanworth, and B. Ramsey, *J. Am. Chem. Soc.* **89**, 5259 (1967).
289. G. A. Olah, M. B. Comisarow, and C. J. Kim, *J. Am. Chem. Soc.* **91**, 1458 (1969).
290. G. A. Olah and R. D. Porter, *J. Am. Chem. Soc.* **92**, 7627 (1970); **93**, 6877 (1971).
291. G. A. Olah, R. J. Spear, and D. A. Forsyth, *J. Am. Chem. Soc.* **99**, 2615 (1977).
292. S. Sieber, P. v. R. Schleyer, and J. Gauss, *J. Am. Chem. Soc.* **115**, 6987 (1993).
293. E. del Río, M. I. Menéndez, R. López, and T. L. Sordo, *J. Phys. Chem. A* **104**, 5568 (2000).
294. G. A. Olah, N. J. Head, G. Rasul, and G. K. S. Prakash, *J. Am. Chem. Soc.* **117**, 875 (1995).

295. G. A. Olah, R. J. Spear, and D. A. Forsyth, *J. Am. Chem. Soc.* **98**, 6284 (1976).
296. E. del Río, M. I. Menéndez, R. López, and T. L. Sordo, *J. Am. Chem. Soc.* **123**, 5064 (2001).
297. G. A. Olah and B. P. Singh, *J. Am. Chem. Soc.* **104**, 5168 (1982).
298. G. A. Olah, B. P. Singh, and G. Liang, *J. Org. Chem.* **49**, 2922 (1984).
299. L. Eberson and S. Winstein, *J. Am. Chem. Soc.* **87**, 3506 (1965).
300. G. A. Olah and B. P. Singh, *J. Am. Chem. Soc.* **106**, 3265 (1984).
301. G. A. Olah and G. Liang, *J. Am. Chem. Soc.* **98**, 6304 (1976).
302. H. Volz, J.-H. Shin, and R. Miess, *J. Chem. Soc., Chem. Commun.* 543 (1993).
303. T. Laube, *J. Am. Chem. Soc.* **126**, 10904 (2004).
304. H. G. Richey, Jr. and J. M. Richey, in *Carbonium Ions*, Vol. 2, G. A. Olah and P. v. R. Schleyer, eds., Wiley-Interscience, New York, London, 1970, Chapter 21, p. 899.
305. M. Hanack, *Acc. Chem. Res.* **3**, 209 (1970).
306. G. Modena and U. Tonellato, *Adv. Phys. Org. Chem.* **9**, 185 (1971).
307. P. J. Stang, *Prog. Phys. Org. Chem.* **10**, 205 (1973).
308. P. J. Stang, Z. Rappoport, M. Hanack, and L. R. Subramanian, *Vinyl Cations*, Academic Press, New York, 1979.
309. *Dicoordinated Carbocations*, Z. Rappoport and P. J. Stang, eds., Wiley, Chichester, 1997.
310. M. W. Crofton, M.-F. Jagod, B. D. Rehfuss, and T. Oda, *J. Chem. Phys.* **91**, 5139 (1989). and references cited therein.
311. J. A. Pople, *Chem. Phys. Lett.* **137**, 10 (1987).
312. R. Lindh, J. E. Rice, and T. J. Lee, *J. Chem. Phys.* **94**, 8008 (1991).
313. Z. Vager, D. Zajfman, T. Graber, and E. P. Kanter, *Phys. Rev. Lett.* **71**, 4319 (1993).
314. L. Knoll, Z. Vager, and D. Marx, *Phys. Rev.* **67**, 022506 (2003) and references cited therein.
315. C. U. Pittman, Jr. and G. A. Olah, *J. Am. Chem. Soc.* **87**, 5632 (1965).
316. G. A. Olah, R. J. Spear, P. W. Westerman, and J.-M. Denis, *J. Am. Chem. Soc.* **96**, 5855 (1974).
317. G. A. Olah, A. L. Berrier, L. D. Field, and G. K. S. Prakash, *J. Am. Chem. Soc.* **104**, 1349 (1982).
318. K. Komatsu, T. Takai, S. Aonuma, and K. Takeuchi, *Tetrahedron Lett.* **29**, 5157 (1988).
319. H.-U. Siehl and H. Mayr, *J. Am. Chem. Soc.* **104**, 909 (1982).
320. H.-U. Siehl and E.-W. Koch, *J. Org. Chem.* **49**, 575 (1984).
321. H.-U. Siehl, in *Stable Carbocation Chemistry*, G. K. S. Prakash and P. v. R. Schleyer, eds., Wiley, New York, 1997, Chapter 5, p. 165.
322. H.-U. Siehl and F.-P. Kaufmann, *J. Am. Chem. Soc.* **114**, 4937 (1992).
323. H.-U. Siehl, F.-P. Kaufmann, and K. Hori, *J. Am. Chem. Soc.* **114**, 9343 (1992).
324. H.-U. Siehl, T. Müller, J. Gauss, P. Buzek, and P. v. R. Schleyer, *J. Am. Chem. Soc.* **116**, 6384 (1994).
325. H.-U. Siehl, T. Müller, and J. Gauss, *J. Phys. Org. Chem.* **16**, 577 (2003).
326. H.-U. Siehl and S. Brixner, *J. Phys. Org. Chem.* **17**, 1039 (2004).

327. H.-U. Siehl, in *ACS Symposium Series*, Vol. 965, K. K. Laali, ed., American Chemical Society, Washington, DC, 2007, Chapter 1, p. 1.
328. T. Müller, D. Margraf, Y. Syha, H. R. Nasiri, C. Kaiser, R. Maier, B. Boltres, M. Juhasz, and C. A. Reed, in *ACS Symposium Series*, Vol. 965, K. K. Laali, American Chemical Society, Washington, DC, 2007, Chapter 3, p. 51.
329. H.-U. Siehl, F.-P. Kaufmann, Y. Apeloig, V. Braude, D. Danovich, A. Berndt, and N. Stamatis, *Angew. Chem. Int. Ed. Engl.* **30**, 1479 (1991).
330. T. Müller, R. Meyer, D. Lennartz, and H.-U. Siehl, *Angew. Chem. Int. Ed.* **39**, 3074 (2000).
331. T. Müller, M. Juhasz, and C. A. Reed, *Angew. Chem. Int. Ed.* **43**, 1543 (2004).
332. T. Müller, D. Margraf, and Y. Syha, *J. Am. Chem. Soc.* **127**, 10852 (2005).
333. H.-U. Siehl, *Pure Appl. Chem.* **67**, 769 (1995).
334. G. K. S. Prakash, V. P. Reddy, G. Rasul, J. Casanova, and G. A. Olah, *J. Am. Chem. Soc.* **114**, 3076 (1992).
335. W. A. Waters, *J. Chem. Soc.* 266 (1942).
336. P. J. Stang, in *Dicoordinated Carbocations*, Z. Rappoport and P. J. Stang, eds., Wiley, Chichester, 1997, Chapter 10, p. 453.
337. K. Laali, I. Szele, and K. Yoshida, *Helv. Chim. Acta* **66**, 1710 (1983).
338. K. Laali, I. Szele, and H. Zollinger, *Helv. Chim. Acta* **66**, 1737 (1983).
339. Y. Himeshima, H. Kobayashi, and T. Sonoda, *J. Am. Chem. Soc.* **107**, 5286 (1985).
340. K. K. Laali, G. Rasul, G. K. S. Prakash, and G. A. Olah, *J. Org. Chem.* **67**, 2913 (2002) and references cited therein.
341. V. Dichiarante, A. Salvaneschi, S. Protti, D. Dondi, M. Fagnoni, and A. Albini, *J. Am. Chem. Soc.* **129**, 15919 (2007).
342. M. Slegt, H. S. Overkleeft, and G. Lodder, *Eur. J. Org. Chem.* 5364 (2007).
343. M. Winkler and W. Sander, *Angew. Chem. Int. Ed.* **39**, 2014 (2000); *J. Org. Chem.* **71**, 6357 (2006).
344. I. Manet, S. Monti, G. Grabner, S. Protti, D. Dondi, V. Dichiarante, M. Fagnoni, and A. Albini, *Chem. Eur. J.* **14**, 1029 (2008).
345. A. H. Gomes de Mesquita, C. H. MacGillavry, and K. Eriks, *Acta Crystallogr.* **18**, 437 (1965).
346. B. Reindl, T. Clark, and P. v. R. Schleyer, *J. Phys. Chem. A* **102**, 8953 (1998).
347. V. Gold and F. L. Tye, *J. Chem. Soc.* 2172 (1952).
348. D. G. Farnum, *J. Am. Chem. Soc.* **86**, 934 (1964).
349. G. A. Olah, *J. Am. Chem. Soc.* **86**, 932 (1964).
350. G. A. Olah and C. U. Pittman, Jr., *J. Am. Chem. Soc.* **87**, 3507 (1965).
351. M. Okajima, K. Soga, T. Nokami, S. Suga, and J. Yoshida, *Org. Lett.* **8**, 5005 (2006).
352. D. P. Kelly and M. J. Jenkins, *J. Org. Chem.* **49**, 409 (1984).
353. G. A. Olah, C. U. Pittman, Jr. E. Namanworth, and M. B. Comisarow, *J. Am. Chem. Soc.* **88**, 5571 (1966).
354. G. A. Olah, R. D. Porter, and D. P. Kelly, *J. Am. Chem. Soc.* **93**, 464 (1971).
355. C. A. Cupas, M. B. Comisarow, and G. A. Olah, *J. Am. Chem. Soc.* **88**, 361 (1966).

356. J. M. Bollinger, M. B. Comisarow, C. A. Cupas, and G. A. Olah, *J. Am. Chem. Soc.* **89**, 5687 (1967).

357. G. A. Olah, T. Shamma, A. Burrichter, G. Rasul, and G. K. S. Prakash, *J. Am. Chem. Soc.* **119**, 3407, 12923 (1997).

358. A. S. Siegel, *J. Am. Chem. Soc.* **92**, 5277 (1970).

359. C. Lifshitz, *Acc. Chem. Res.* **27**, 138 (1994).

360. T. Laube, G. A. Olah, and R. Bau, *J. Am. Chem. Soc.* **119**, 3087 (1997).

361. G. A. Olah, M. D. Heagy, and G. K. S. Prakash, *J. Org. Chem.* **58**, 4851 (1993).

362. M. D. Heagy, G. A. Olah, G. K. S. Prakash, and J. S. Lomas, *J. Org. Chem.* **60**, 7355 (1995).

363. G. A. Olah, Q. Liao, J. Casanova, R. Bau, G. Rasul, and G. K. S. Prakash, *J. Chem. Soc., Perkin Trans.* **2**, 2239 (1998).

364. H. Kawai, T. Nagasu, T. Takeda, K. Fujiwara, T. Tsuji, M. Ohkita, J. Nishida, and T. Suzuki, *Tetrahedron Lett.* **45**, 4553 (2004).

365. H. Kawai, T. Takeda, K. Fujiwara, and T. Suzuki, *J. Am. Chem. Soc.* **127**, 12172 (2005).

366. F. C. Krebs, B. W. Laursen, I. Johannsen, A. Faldt, K. Bechgaard, C. S. Jacobsen, N. Thorup, and K. Boubekeur, *Acta Crystallogr. B* **55**, 410 (1999).

367. B. W. Laursen, F. C. Krebs, M. F. Nielsen, K. Bechgaard, J. B. Christensen, and N. Harrit, *J. Am. Chem. Soc.* **120**, 12255 (1998).

368. B. W. Laursen and F. C. Krebs, *Angew. Chem. Int. Ed.* **39**, 3432 (2000); *Chem. Eur. J.***7**, 1773 (2001).

369. C. Herse, D. Bas, F. C. Krebs, T. Bürgi, J. Weber, T. Wesolowski, B. W. Laursen, and J. Lacour, *Angew. Chem. Int. Ed.* **42**, 3162 (2003).

370. G. I. Borodkin, S. M. Nagy, Yu.V. Gatilov, V. G. Shubin, *Dokl. Chem. (Engl. Transl.)* **280**, 33 (1985).

371. E. Solari, C. Floriani, A. Chiesi-Villa, and C. Rizzoli, *J. Chem. Soc., Chem. Commun.* 841 (1991).

372. K. K. Laali, T. Okazaki, and M. M. Coombs, *J. Org. Chem.* **65**, 7399 (2000).

373. K. K. Laali, M. Tanaka, S. Hollenstein, and M. Cheng, *J. Org. Chem.* **62**, 7752 (1997).

374. S. Takekuma, M. Tanizawa, M. Sasaki, T. Matsumoto, and H. Takekuma, *Tetrahedron Lett.* **43**, 2073 (2002).

375. M. Sasaki, M. Nakamura, T. Uriu, H. Takekuma, T. Minematsu, M. Yoshihara, and S. Takekuma, *Tetrahedron* **59**, 505 (2003).

376. S. Takekuma, Y. Hata, T. Nishimoto, E. Nomura, M. Sasaki, T. Minematsu, and H. Takekuma, *Tetrahedron* **61**, 6892 (2005).

377. S. Takekuma, K. Takahashi, A. Sakaguchi, Y. Shibata, M. Sasaki, T. Minematsu, and H. Takekuma, *Tetrahedron* **61**, 10349 (2005).

378. F. Mühlthau, D. Stadler, A. Goeppert, G. A. Olah, G. K. S. Prakash, and T. Bach, *J. Am. Chem. Soc.* **128**, 9668 (2006).

379. G. K. S. Prakash, T. N. Rawdah, and G. A. Olah, *Angew. Chem. Int. Ed. Engl.* **22**, 390 (1983).

380. V. G. Nenajdenko, N. E. Sevchenko, E. S. Balenkova, and I. V. Alabugin, *Chem. Rev.* **103**, 229 (2003).

381. H. Hart and R. W. Fish, *J. Am. Chem. Soc.* **80**, 5894 (1958); **82**, 5419 (1960); **83**, 4460 (1961).
382. R. J. Gillespie and E. A. Robinson, *J. Am. Chem. Soc.* **86**, 5676 (1964).
383. N. C. Deno, N. Friedman, and J. Mockus, *J. Am. Chem. Soc.* **86**, 5676 (1964).
384. E. A. Robinson and J. A. Ciruna, *J. Am. Chem. Soc.* **86**, 5677 (1964).
385. G. K. S. Prakash, *Pure Appl. Chem.* **70**, 2001 (1998).
386. M. J. S. Dewar, and M. K. Holloway, *J. Am. Chem. Soc.* **106**, 6619 (1984).
387. K. Lammertsma, P. v. R. Schleyer, and H. Schwarz, *Angew. Chem. Int. Ed. Engl.* **28**, 1321 (1989).
388. K. Krogh-Jespersen, *J. Am. Chem. Soc.* **113**, 417 (1991).
389. H. Hart, T. Sulzberg, and R. R. Rafos, *J. Am. Chem. Soc.* **85**, 1800 (1963).
390. H. Volz and M. J. Volz de Lecea, *Tetrahedron Lett.* 1871 (1964); 4683 (1966).
391. G. A. Olah, J. L. Grant, R. J. Spear, J. M. Bollinger, A. Serianz, and G. Sipos, *J. Am. Chem. Soc.* **98**, 2501 (1976).
392. J. M. Bollinger, C. A. Cupas, K. J. Friday, M. L. Woolfe, and G. A. Olah, *J. Am. Chem. Soc.* **89**, 156 (1967).
393. G. A. Olah, G. Liang, P. v. R. Schleyer, E. M. Engler, M. J. S. Dewar, and R. C. Bingham, *J. Am. Chem. Soc.* **95**, 6829 (1973).
394. A. de Meijere, O. Schallner, P. Gölitz, W. Weber, P. v. R. Schleyer, G. K. S. Prakash, and G. A. Olah, *J. Org. Chem.* **50**, 5255 (1985).
395. C. Taeschler and T. S. Sorensen, *Tetrahedron Lett.* **42**, 5339 (2001).
396. C. Taeschler, M. Parvez, and T. S. Sorensen, *J. Phys. Org. Chem.* **15**, 36 (2002).
397. G. K. S. Prakash, A. P. Fung, T. N. Rawdah, and G. A. Olah, *J. Am. Chem. Soc.* **107**, 2920 (1985).
398. G. K. S. Prakash, V. V. Krishnamurthy, M. Arvanaghi, and G. A. Olah, *J. Org. Chem.* **50**, 3985 (1985).
399. G. Rasul, G. A. Olah, and G. K. S. Prakash, *Proc. Natl. Acad. Sci. USA* **101**, 10868 (2004).
400. M. D. Heagy, Q. Wang, G. A. Olah, and G. K. S. Prakash, *J. Org. Chem.* **60**, 7351 (1995).
401. N. J. Head, G. K. S. Prakash, A. Bashir-Hashemi, and G. A. Olah, *J. Am. Chem. Soc.* **117**, 12005 (1995).
402. T. Suzuki, J. Nishida, and T. Tsuji, *Angew. Chem. Int. Ed. Engl.* **36**, 1329 (1997); *Chem. Commun.* 2193 (2193).
403. N. J. Head, G. A. Olah, and G. K. S. Prakash, *J. Am. Chem. Soc.* **117**, 11205 (1995).
404. R. Rathore, C. L. Burns, and I. A. Guzei, *J. Org. Chem.* **69**, 1524 (2004).
405. G. A. Olah, V. P. Reddy, G. Lee, J. Casanova, and G. K. S. Prakash, *J. Org. Chem.* **58**, 1639 (1993).
406. G. A. Olah, G. K. S. Prakash, and T. N. Rawdah, *J. Am. Chem. Soc.* **102**, 6127 (1980).
407. G. A. Olah, J. S. Staral, G. Liang, L. A. Paquette, W. P. Melega, and M. J. Carmody, *J. Am. Chem. Soc.* **99**, 3349 (1977).
408. G. A. Olah, G. Liang, L. A. Paquette, and W. P. Melega, *J. Am. Chem. Soc.* **98**, 4327 (1976).

409. G. A. Olah, J. S. Staral, and L. A. Paquette, *J. Am. Chem. Soc.* **98**, 1267 (1976).

410. G. A. Olah, M. Arvanaghi, and G. K. S. Prakash, *Angew. Chem. Int. Ed. Engl.* **22**, 712 (1983).

411. K. Lammertsma and H. Cerfontain, *J. Am. Chem. Soc.* **102**, 3257 (1980).

412. K. Lammertsma, *J. Am. Chem. Soc.* **103**, 2062 (1981).

413. V. I. Mamatyuk, A. P. Krysin, N. V. Bodoev, and V. A. Koptyug, *Russ. Chem. Bull. (Engl. Transl.)* **23**, 2312 (1974).

414. T. Suzuki, H. Higuchi, M. Ohkita, and T. Tsuji, *Chem. Commun.* 1574 (2001).

415. H. Higuchi, E. Ohta, H. Kawai, K. Fujiwara, T. Tsuji, and T. Suzuki, *J. Org. Chem.* **68**, 6605 (2003).

416. H. Wang and F. P. Gabbaï, *Angew. Chem. Int. Ed.* **43**, 184 (2004).

417. H. Wang, C. E. Webster, L. M. Pérez, M. B. Hall, and F. P. Gabbaï, *J. Am. Chem. Soc.* **126**, 8189 (2004).

418. N. C. Baenziger, R. E. Buckles, and T. D. Simpson, *J. Am. Chem. Soc.* **89**, 3405 (1967).

419. R. Rathore, S. V. Lindeman, A. S. Kumar, and J. K. Kochi, *J. Am. Chem. Soc.* **120**, 6931 (1998).

420. T. Mori and Y. Inoue, *J. Phys. Chem. A* **109**, 2728 (2005).

421. S. Takekuma, S. Takata, M. Sasaki, and H. Takekuma, *Tetrahedron Lett.* **42**, 5921 (2001).

422. M. Sasaki, M. Nakamura, G. Hannita, H. Takekuma, T. Minematsu, M. Yoshihara, and S. Takekuma, *Tetrahedron Lett.* **44**, 275 (2003).

423. S. Takekuma, K. Tone, M. Sasaki, T. Minematsu, and H. Takekuma, *Tetrahedron* **63**, 2490 (2007).

424. A. D. Allen and T. T. Tidwell, *Chem. Rev.* **101**, 1333 (2001) and references cited therein.

425. T. L. Amyes, J. P. Richard, and M. Novak, *J. Am. Chem. Soc.* **114**, 8032 (1992).

426. C. F. Rodriquez, D.Lj. Vucković, and A. C. Hopkinson, *J. Mol. Struct. (Theochem)* **363**, 131 (1996).

427. N. S. Mills, K. B. Llagostera, C. Tirla, S. M. Gordon, and D. Carpenetti, *J. Org. Chem.* **71**, 7940 (2006) and references cited therein.

428. J. L. Malandra, N. S. Mills, D. E. Kadlecek, and J. A. Lowery, *J. Am. Chem. Soc.* **116**, 11622 (1994).

429. N. S. Mills, M. A. Benish, and C. Ybarra, *J. Org. Chem.* **67**, 2003 (2002).

430. N. S. Mills, *J. Am. Chem. Soc.* **121**, 11690 (1999).

431. W. C. Hendron and N. S. Mills, *J. Org. Chem.* **70**, 8492 (2005).

432. N. S. Mills, in *ACS Symposium Series*, Vol. 965, K. K. Laali, ed., American Chemical Society, Washington, DC, 2007, Chapter 11, p. 210.

433. S. Pogodin and I. Agranat, *J. Org. Chem.* **72**, 10096 (2007).

434. R. Breslow and C. Yuan, *J. Am. Chem. Soc.* **80**, 5991 (1958).

435. R. Breslow, J. T. Groves, and G. Ryan, *J. Am. Chem. Soc.* **89**, 5048 (1967).

436. J. Ciabattoni and E. C. Nathan, III, *J. Am. Chem. Soc.* **91**, 4766 (1969).

437. J. Ciabattoni, E. C. Nathan, A. E. Feiring, and P. J. Kocienski, *Org. Synth.* **54**, 97 (1974).

438. A. W. Krebs, *Angew. Chem. Int. Ed. Engl.* **4**, 10 (1965).
439. W. Billups and A. W. Moorehead, in *The Chemistry of the Cyclopropyl Group*. Z. Rappoport, ed., Wiley Chichester 1987, Chapter 24, p. 1533.
440. M. Wyss, E. Riaplov, and J. P. Maier, *J. Chem. Phys.* **114**, 10355 (2001).
441. A. de Meijere, D. Faber, M. Noltemeyer, R. Boese, T. Haumann, T. Müller, M. Bendikov, E. Matzner, and Y. Apeloig, *J. Org. Chem.* **61**, 8564 (1996).
442. R. D. Gilbertson, T. J. R. Weakley, and M. M. Haley, *J. Org. Chem.* **65**, 1422 (2000).
443. U. Burger, P. Müller, and L. Zuidema, *Helv. Chim. Acta* **57**, 1881 (1974).
444. B. Halton, H. M. Hügel, D. P. Kelly, P. Müller, and U. Burger, *J. Chem. Soc., Perkin Trans.* **2**, 258 (1976).
445. P. Müller, J. Pfyffer, E. Wentrup-Byrne, and U. Burger, *Helv. Chim. Acta* **61**, 2081 (1978).
446. P. Müller, R. Etienne, J. Pfyffer, N. Pineda, and M. Schipoff, *Helv. Chim. Acta* **61**, 2482 (1978).
447. G. A. Olah and J. S. Stratal, *J. Am. Chem. Soc.* **98**, 6290 (1976).
448. G. A. Olah, J. M. Bollinger, and A. M. White, *J. Am. Chem. Soc.* **91**, 3667 (1969).
449. G. A. Olah and G. D. Mateescu, *J. Am. Chem. Soc.* **92**, 1430 (1970).
450. K. Krogh-Jespersen, D. Cremer, J. D. Dill, J. A. Pople, and P. v. R. Schleyer, *J. Am. Chem. Soc.* **103**, 2589 (1981).
451. B. A. Hess, Jr. C. S. Ewig, and L. J. Schaad, *J. Org. Chem.* **50**, 5869 (1985).
452. M. Bremer, P. v. R. Schleyer, and U. Fleischer, *J. Am. Chem. Soc.* **111**, 1147 (1989).
453. K. Komatsu, M. Arai, Y. Hattori, K. Fukuyama, Y. Katsube, and K. Okamoto, *J. Org. Chem.* **52**, 2183 (1987).
454. T. Eicher and H. Berneth, *Tetrahedron Lett.* 2039 (2039).
455. T. Nozoe, in *Non-Benzenoid Aromatic Compounds*, D. Ginsburg, ed., Interscience, New York, Chapter 7, 1959.
456. K. M. Harmon, in *Carbonium Ions*, Vol. 4, G. A. Olah and P. v. R. Schleyer, eds., Wiley-Interscience, New York, London, 1973, Chapter 29, p. 1579
457. I. S. Akhrem, E. I. Fedin, B. A. Kvasov, and M. E. Vol'pin *Tetrahedron Lett.* 5265 (1967).
458. M. A. Battiste, *J. Am. Chem. Soc.* **83**, 4101 (1961).
459. R. W. Murray and M. L. Kaplan, *Tetrahedron Lett.* 2903 (1965); 1307 (1307).
460. S. Brydges, J. F. Britten, L. C. F. Chao, H. K. Gupta, M. J. McGlinchey, and D. L. Pole, *Chem. Eur. J.* **4**, 1201 (1998).
461. K. Komatsu, H. Akamatsu, S. Aonuma, Y. Jinbu, N. Maekawa, and K. Takeuchi, *Tetrahedron* **47**, 6951 (1991).
462. M. Oda, H. Kainuma, T. Uchiyama, R. Miyatake, and S. Kuroda, *Tetrahedron* **59**, 2831 (2003).
463. A. Kagayama, K. Komatsu, T. Nishinaga, K. Takeuchi, and C. Kabuto, *J. Org. Chem.* **59**, 4999 (1994).
464. de Wit P. and H. Cerfontain, *Recl. Trav. Chim. Pays-Bas* **104**, 25 (1985).
465. L. T. Scott, C. A. Sumpter, M. Oda, and I. Erden, *Tetrahedron Lett.* **30**, 305 (1989).

466. M. Oda, N. Nakajima, N. C. Thanh, T. Kajioka, and S. Kuroda, *Tetrahedron* **62**, 8177 (2006) and references cited therein.

467. M. Oda, T. Kajioka, T. Uchiyama, K. Nagara, T. Okujima, S. Ito, N. Morita, T. Sato, R. Miyatake, and S. Kuroda, *Tetrahedron* **55**, 6081 (1999).

468. S. Ito, N. Morita, and T. Asao, *Tetrahedron Lett.* **32**, 773 (1991); **35**, 751 (1994).

469. S. Ito, S. Kikuchi, N. Morita, and T. Asao, *J. Org. Chem.* **64**, 5815 (1999).

470. K. S. Hassall and J. M. White, *Org. Lett.* **6**, 1737 (2004).

471. G. A. Olah and G. Liang, *J. Am. Chem. Soc.* **98**, 3033 (1976).

472. G. A. Olah and G. Liang, *J. Am. Chem. Soc.* **99**, 6045 (1977).

473. J. W. Bausch, P. S. Gregory, G. A. Olah, G. K. S. Prakash, P. v. R. Schleyer, and G. A. Segal, *J. Am. Chem. Soc.* **111**, 3633 (1989).

474. J. F. M. Oth, D. M. Smith, U. Prange, and G. Schröder, *Angew. Chem. Int. Ed. Engl.* **12**, 327 (1973).

475. D. H. O'Brien, A. J. Hart, and C. R. Russel, *J. Am. Chem. Soc.* **97**, 4410 (1975).

476. J. B. Stothers, *Carbon-13 NMR Spectroscopy*, Academic Press, New York, 1972.

477. G. C. Levy and G. L. Nelson, *Carbon-13 Nuclear Magnetic Resonance for Organic Chemists*, Wiley-Interscience, New York, 1972.

478. H. Spiesecke and W. G. Schneider, *J. Chem. Phys.* **35**, 731 (1961); *Tetrahedron Lett.* 468 (1961).

479. T. Schaefer and W. G. Schneider, *Can. J. Chem.* **41**, 966 (1963).

480. P. v. R. Schleyer, C. Maerker, A. Dransfeld, H. Jiao, and N. J. R. v. Eikema Hommes, *J. Am. Chem. Soc.* **118**, 6317 (1996).

481. F. Borchers and K. Levsen, *Org. Mass Spectr.* **10**, 584 (1975) and references cited therein.

482. O. Hammerich and V. D. Parker, *Electrochim. Acta* **18**, 537 (1973).

483. H. J. Shine and Y. Murata, *J. Am. Chem. Soc.* **91**, 1872 (1969).

484. Y. Murata and H. J. Shine, *J. Org. Chem.* **34**, 3368 (1969).

485. L. S. Marcoux, *J. Am. Chem. Soc.* **93**, 537 (1971).

486. I. C. Lewis and L. S. Singer, *J. Chem. Phys.* **43**, 2712 (1965).

487. W. I. Aalbersberg, G. J. Hoijtink, E. L. Mackor, and W. P. Weijland, *J. Chem. Soc.* 3055 (1959).

488. W. Th. A. M. van der Lugt, H. M. Buck, and L. J. Oosterhoff, *Tetrahedron* **24**, 4941 (1968).

489. D. M. Brouwer and van Doorn J. A. *Recl. Trav. Chim. Pays-Bas* **91**, 1110 (1972).

490. S. I. Weissman, E. de Boer, and J. J. Conradi, *J. Chem. Phys.* **26**, 963 (1957).

491. G. J. Hoijtink, and W. P. Weijland, *Recl. Trav. Chim. Pays-Bas* **76**, 836 (1957).

492. D. M. Brouwer, *Chem. Ind. (London)*, 177 (1961); *J. Catal.* **1**, 372 (1962).

493. J. J. Rooney and R. C. Pink, *Proc. Chem. Soc.* 70 (1961).

494. D. A. Forsyth and G. A. Olah, *J. Am. Chem. Soc.* **98**, 4086 (1976).

495. K. Lammertsma, G. A. Olah, C. M. Berke, and A. Streitwieser, Jr., *J. Am. Chem. Soc.* **101**, 6658 (1979).

496. G. A. Olah and B. P. Singh, *J. Org. Chem.* **48**, 4830 (1983).

497. I. Willner and M. Rabinovitz, *J. Am. Chem. Soc.* **100**, 337 (1978).

498. I. Willner, J. Y. Becker, and M. Rabinovitz, *J. Am. Chem. Soc.* **101**, 395 (1979).
499. K. K. Laali and M. Tanaka, *J. Chem. Soc., Perkin Trans.* **2**, 2509 (1998).
500. R. D. Bolskar, R. S. Mathur, and C. A. Reed, *J. Am. Chem. Soc.* **118**, 13093 (1996).
501. C. A. Reed, K.-C. Kim, R. D. Bolskar, and L. J. Mueller, *Science* **289**, 101 (2000).
502. L. J. Mueller, D. W. Elliott, K.-C. Kim, C. A. Reed, and P. D. W. Boyd, *J. Am. Chem. Soc.* **124**, 9360 (2002).
503. T. Kitagawa and K. Takeuchi, *Bull. Chem. Soc. Jpn.* **74**, 785 (2001).
504. T. Kitagawa, in *ACS Symposium Series*, Vol. 965, K. K. Laali, ed., American Chemical Society, Washington, DC, 2007, Chapter 12, p. 234.
505. T. Kitagawa, H. Sakamoto, and K. Takeuchi, *J. Am. Chem. Soc.* **121**, 4298 (1999).
506. A. G. Avent, P. R. Birkett, H. W. Kroto, R. Taylor, and D. R. M. Walton, *Chem. Commun.* 2153 (1998).
507. Y. Murata, F. Cheng, T. Kitagawa, and K. Komatsu, *J. Am. Chem. Soc.* **126**, 8874 (2004).
508. T. Kitagawa, Y. Lee, N. Masaoka, and K. Komatsu, *Angew. Chem. Int. Ed.* **44**, 1398 (2005).
509. H. Grützmacher and C. M. Marchand, *Coord. Chem. Rev.* **163**, 287 (1997).
510. G. A. Olah, G. K. S. Prakash, M. Arvanaghi, V. V. Krishnamurthy, and S. C. Narang, *J. Am. Chem. Soc.* **106**, 2378 (1984).
511. G. A. Olah, C. A. Cupas, and M. B. Comisarow, *J. Am. Chem. Soc.* **88**, 362 (1966).
512. For a summary, see G. A. Olah and Y. K. Mo, *Adv. Fluorine Chem.* **7**, 69 (1972).
513. G. A. Olah and Y. K. Mo, in *Carbonium Ions*, Vol. 5, G. A. Olah and P. v. R. Schleyer, ed., Wiley-Interscience, New York, London, 1976, Chapter 36, p. 2135.
514. G. A. Olah, R. D. Chambers, and M. B. Comisarow, *J. Am. Chem. Soc.* **89**, 1268 (1967).
515. K. O. Christe, X. Zhang, R. Bau, J. Hegge, G. A. Olah, G. K. S. Prakash, and J. A. Sheehy, *J. Am. Chem. Soc.* **122**, 481 (2000).
516. T. Laube, E. Bannwart, and S. Hollenstein, *J. Am. Chem. Soc.* **115**, 1731 (1993).
517. G. A. Olah and C. U. Pittman, Jr. *J. Am. Chem. Soc.* **88**, 3310 (1966).
518. G. A. Olah and M. B. Comisarow, *J. Am. Chem. Soc.* **89**, 1027 (1967).
519. R. Filler, C.-S Wang, M. A. McKinney, and F. N. Miller, *J. Am. Chem. Soc.* **89**, 1026 (1967).
520. P. B. Sargeant and C. D. Krespan, *J. Am. Chem. Soc.* **91**, 415 (1969).
521. G. G. Belen'kii *J. Fluorine Chem.* **77**, 107 (1996).
522. V. D. Shteingarts, in *Carbocation Chemistry*, G. A. Olah and G. K. S. Prakash, eds., Wiley, Hoboken, NJ, 2004, Chapter 7, p. 159.
523. G. K. S. Prakash, G. Rasul, A. Burrichter, K. K. Laali, and G. A. Olah, *J. Org. Chem.* **61**, 9253 (1996).
524. G. A. Olah and Y. K. Mo, *J. Org. Chem.* **37**, 1028 (1972).
525. H. Volz and M. J. Volz de Lecca, *Tetrahedron Lett.* 3413 (1965).
526. H. Volz and W. D. Mayer, *Tetrahedron Lett.* 5249 (1966).
527. M. Ballester, J. Riera-Figueras, and A. Rodríguez-Siurana, *Tetrahedron Lett.* 3615 (1970).

528. R. West and P. T. Kwitowski, *J. Am. Chem. Soc.* **88**, 5280 (1966).

529. G. A. Olah and M. B. Comisarow, *J. Am. Chem. Soc.* **91**, 2955 (1969).

530. G. A. Olah, Y. K. Mo, and Y. Halpern, *J. Am. Chem. Soc.* **94**, 3551 (1972).

531. G. A. Olah, G. Liang, and Y. K. Mo, *J. Org. Chem.* **39**, 2394 (1974).

532. G. A. Olah, L. Heiliger, and G. K. S. Prakash, *J. Am. Chem. Soc.* **111**, 8020 (1989).

533. G. A. Olah, G. Rasul, L. Heiliger, and G. K. S. Prakash, *J. Am. Chem. Soc.* **118**, 3580 (1996).

534. K. O. Christe, B. Hoge, J. A. Boatz, G. K. S. Prakash, G. A. Olah, and J. A. Sheehy, *Inorg. Chem.* **38**, 3132 (1999).

535. G. A. Olah, G. Rasul, A. K. Yudin, A. Burrichter, G. K. S. Prakash, A. L. Chistyakov, I. V. Stankevich, I. S. Akhrem, N. P. Gambaryan, and M. E. Vol'pin, *J. Am. Chem. Soc.* **118**, 1446 (1996).

536. G. A. Olah and Y. K. Mo, *J. Org. Chem.* **38**, 2682 (1973).

537. G. Frenking, S. Fau, C. M. Marchand, and H. Grützmacher, *J. Am. Chem. Soc.* **119**, 6648 (1997).

538. H. P. A. Mercier, M. D. Moran, G. J. Schrobilgen, C. Steinberg, and R. J. Suontamo, *J. Am. Chem. Soc.* **126**, 5533 (2004).

539. H. P. A. Mercier, M. D. Moran, and G. J. Schrobilgen, in *ACS Symposium Series*, Vol. 965, K. K. Laali, ed., American Chemical Society, Washington, DC, 2007, Chapter 19, p. 394.

540. I. Krossing, A. Bihlmeier, I. Raabe, and N. Trapp, *Angew. Chem. Int. Ed.* **42**, 1531 (2003).

541. G. A. Olah, K. K. Laali, Q. Wang, and G. K. S. Prakash, *Onium Ions*, Wiley, New York, 1998, Chapter 7, p. 269.

542. G. A. Olah, A. M. White, and D. H. O'Brien, *Chem. Rev.* **70**, 561 (1970) and references cited therein.

543. C. MacLean and E. L. Mackor, *J. Chem. Phys.* **34**, 2207 (1961).

544. T. Birchall and R. J. Gillespie, *Can. J. Chem.* **43**, 1045 (1965).

545. H. Hogeveen, *Recl. Trav. Chim. Pays-Bas* **86**, 696 (1967).

546. D. M. Brouwer, *Recl. Trav. Chim. Pays-Bas* **86**, 879 (1967).

547. M. Brookhart, G. C. Levy, and S. Winstein, *J. Am. Chem. Soc.* **89**, 1735 (1967).

548. G. A. Olah, D. H. O'Brien and M. Calin, *J. Am. Chem. Soc.* **89**, 3582 (1967).

549. G. A. Olah, M. Calin, and D. H. O'Brien, *J. Am. Chem. Soc.* **89**, 3586 (1967).

550. G. A. Olah and M. Calin, *J. Am. Chem. Soc.* **89**, 4736 (1967).

551. G. Liang, Ph.D. dissertation, Case Western Reserve University, 1973.

552. G. A. Olah, A. L. Berrier, and G. K. S. Prakash, *J. Am. Chem. Soc.* **104**, 2373 (1982).

553. G. A. Olah, A. Burrichter, G. Rasul, R. Gnann, K. O. Christe, and G. K. S. Prakash, *J. Am. Chem. Soc.* **119**, 8035 (1997).

554. P. V. Huong and G. Noel, *Spectrochim. Acta* **32A**, 831 (1976).

555. O. Dopfer, R. V. Olkhov, D. Roth, and J. P. Maier, *Chem. Phys. Lett.* **296**, 585 (1998).

556. G. E. Douberly, A. M. Ricks, B. W. Ticknor, and M. A. Duncan, *Phys. Chem. Chem. Phys.* **10**, 77 (2008).

557. V. V. Krishnamurthy, G. K. S. Prakash, P. S. Iyer, and G. A. Olah, *J. Am. Chem. Soc.* **106**, 7068 (1984).

558. N. Hartz, G. Rasul, and G. A. Olah, *J. Am. Chem. Soc.* **115**, 1277 (1993).

559. L. B. Krivdin, S. V. Zinchenko, G. A. Kalabin, J. C. Facelli, M. F. Tufró, R. H. Contreras, A.Yu. Denisov, O. A. Gavrilyuk, and V. I. Mamatyuk, *J. Chem. Soc., Faraday Trans. 1*, **88**, 2459 (1992).

560. R. F. Childs, R. Faggiani, C. J. L. Lock, M. Mahendran, and S. D. Zweep, *J. Am. Chem. Soc.* **108**, 1692 (1986).

561. S. K. Chadda, R. F. Childs, R. Faggiani, and C. J. L. Lock, *J. Am. Chem. Soc.* **108**, 1694 (1986).

562. G. Maier, K. Euler, and R. Emrich, *Tetrahedron Lett.* **27**, 3607 (1986).

563. R. West and D. L. J. Powell, *J. Am. Chem. Soc.* **85**, 2577 (1963).

564. G. A. Olah, J. Bausch, G. Rad, H. George, and G. K. S. Prakash, *J. Am. Chem. Soc.* **115**, 8060 (1993).

565. G. K. S. Prakash, G. Rasul, G. A. Olah, R. Liu, and T. T. Tidwell, *Can. J. Chem.* **77**, 525 (1999).

566. G. A. Olah and J. L. Grant, *J. Org. Chem.* **42**, 2237 (1977).

567. H. Hogeveen, A. F. Bickel, C. W. Hilbers, E. L. Mackor, and C. MacLean, *Chem. Commun.* 898 (1966); *Recl. Trav. Chim. Pays-Bas* **86**, 687 (1967).

568. G. A. Olah and A. M. White, *J. Am. Chem. Soc.* **89**, 3591 (1967).

569. G. A. Olah and A. M. White, *J. Am. Chem. Soc.* **89**, 4752 (1967).

570. H. Hogeveen, *Recl. Trav. Chim. Pays-Bas* **86**, 289 (1967); *Adv. Phys. Org. Chem.* **10**, 29 (1973).

571. G. A. Olah, D. H. O'Brien, and A. M. White, *J. Am. Chem. Soc.* **89**, 5694 (1967).

572. H. Hogeveen, *Recl. Trav. Chim. Pays-Bas* **86**, 816 (1967).

573. R. Minkwitz, S. Schneider, M. Seifert, and H. Hartl, *Z. Anorg. Allg. Chem.* **622**, 1404 (1996).

574. R. Minkwitz and T. Hertel, *Z. Naturforsch.* **52b**, 1283 (1997).

575. R. Minkwitz, F. Neikes, and D. Rüttershoff, *Z. Anorg. Allg. Chem.* **626**, 2203 (2000).

576. R. Minkwitz, N. Hartfeld, and C. Hirsch, *Z. Anorg. Allg. Chem.* **625**, 1479 (1999).

577. G. K. S. Prakash, G. Rasul, G. Liang, and G. A. Olah, *J. Phys. Chem.* **100**, 15805 (1996).

578. G. A. Olah, N. Hartz, G. Rasul, A. Burrichter, and G. K. S. Prakash, *J. Am. Chem. Soc.* **117**, 6421 (1995).

579. N. J. Head, G. Rasul, A. Mitra, A. Bashir-Heshemi, G. K. S. Prakash, and G. A. Olah, *J. Am. Chem. Soc.* **117**, 12107 (1995).

580. G. K. S. Prakash, G. Rasul, N. J. Head, A. Mitra, and G. A. Olah, in *Advances in Strained and Interesting Organic Molecules*, K. K. Laali, ed., JAI Press, Stamford, 1999, Suppl. 1, p. 109.

581. G. A. Olah, A. Burrichter, G. Rasul, A. K. Yudin, and G. K. S. Prakash, *J. Org. Chem.* **61**, 1934 (1996).

582. G. A. Olah and A. T. Ku, *J. Org. Chem.* **35**, 3916 (1970).

583. G. Carr and D. Whittaker, *J. Chem. Soc., Perkin Trans.* **2**, 1877 (1987).

584. G. Asensio, M. A. Miranda, M. J. Sabater, J. Pérez-Prieto, A. Simón-Fuentes, and R. Mello, *J. Org. Chem.* **65**, 964 (2000).

585. H. Hogeveen, *Recl. Trav. Chim. Pays-Bas* **89**, 1303 (1968).

586. G. A. Olah and A. M. White, *J. Am. Chem. Soc.* **90**, 1884 (1968).

587. G. A. Olah and J. Shen, *J. Am. Chem. Soc.* **95**, 3582 (1973).

588. G. Rasul, V. P. Reddy, L. Z. Zdunek, G. K. S. Prakash, and G. A. Olah, *J. Am. Chem. Soc.* **115**, 2236 (1993).

589. R. Minkwitz and S. Schneider, *Angew. Chem. Int. Ed.* **38**, 714 (1999).

590. R. Minkwitz and F. Neikes, *Eur. J. Inorg. Chem.* 1261 (2000).

591. H.-S. Andrei, S. A. Nizkorodov, and O. Dopfer, *Angew. Chem. Int. Ed.* **46**, 4754 (2007).

592. A. Hantzsch, *Ber. Dtsch. Chem. Ges.* **55**, 953 (1922); **58**, 612, 941 (1925).

593. J. Nishida, T. Suzuki, M. Ohkita, and T. Tsuji, *Angew. Chem. Int. Ed.* **40**, 3251 (2001).

594. H. Meerwein, K. Bodenbenner, P. Borner, F. Kunert, and K. Wunderlich, *Justus Liebigs Ann. Chem.* **632**, 38 (1960).

595. H. Meerwein, V. Hederich, H. Morschel, and K. Wunderlich, *Justus Liebigs Ann. Chem.* **635**, 1 (1960).

596. H. Perst, in *Carbonium Ions*, Vol. 5, G. A. Olah and P. v. R. Schleyer, eds., Wiley-Interscience, New York, 1976, Chapter 34, p. 1961.

597. B. G. Ramsey and R. W. Taft, *J. Am. Chem. Soc.* **88**, 3058 (1966).

598. R. H. Martin, F. W. Lampe, and R. W. Taft, *J. Am. Chem. Soc.* **88**, 1353 (1966).

599. R. A. Burt, C. A. Chambers, Y. Chiang, C. S. Hillock, A. J. Kresge, and J. W. Larsen, *J. Org. Chem.* **49**, 2622 (1984).

600. P. Kandanarachchi and M. L. Sinnott, *J. Chem. Soc., Chem. Commun.* 777 (1992).

601. S. P. Miller, S. Krasutsky, and P. Kiprof, *J. Mol. Struct. (Theochem)* **674**, 43 (2004).

602. G. A. Olah and J. M. Bollinger, *J. Am. Chem. Soc.* **89**, 2993 (1967).

603. G. A. Olah, S. Yu, G. Liang, G. D. Mateescu, M. R. Bruce, D. J. Donovan, and M. Arvanaghi, *J. Org. Chem.* **46**, 571 (1981).

604. R. Minkwitz, S. Reinemann, O. Blecher, H. Hartl, and I. Brüdgam, *Inorg. Chem.* **38**, 844 (1999).

605. R. Minkwitz, S. Schneider, and H. Preut, *Angew. Chem. Int. Ed.* **37**, 494 (1998).

606. G. A. Olah, D. G. Parker, and N. Yoneda, *Angew. Chem. Int. Ed. Engl.* **17**, 909 (1978).

607. O. V. Lubinskaya, A. S. Shashkov, V. A. Chertkov, and W. A. Smit, *Synthesis*, 742 (1976).

608. W. A. Smit, A. V. Semenovskii, O. V. Lubinskaya, and V. F. Kucherov, *Dokl. Akad. Nauk SSSR* **203**, 604 (1972); *Chem. Abstr.* **77**, 87955 (1972).

609. D. L. Rakhmankulov, R. T. Akhmatdinov, and E. A. Kantor, *Russ. Chem. Rev.* **53**, 888 (1984).

610. J.-P. Bégué and M. Charpentier-Morize, *Acc. Chem. Res.* **13**, 207 (1980).

611. J.-P. Bégué, D. Bonnet-Delpon, and M. Charpentier-Morize, *New J. Chem.* **16**, 243 (1992).

612. G. A. Olah, Y. Halpern, Y. K. Mo, and G. Liang, *J. Am. Chem. Soc.* **94**, 3554 (1972).

613. J. W. Larsen and S. Ewing, *J. Am. Chem. Soc.* **93**, 5107 (1971).

614. C. U. Pittman, Jr., and S. P. McManus, *J. Am. Chem. Soc.* **91**, 5915 (1969).

615. T. Nakai, E. Wada, and M. Okawara, *Tetrahedron Lett.* 1531 (1975).
616. H. Brinkmann and C. Rüchardt, *Tetrahedron Lett.* 5221 (1972).
617. C. Rüchardt, G. Hamprecht, and H. Brinkmann, *Chem. Ber.* **108**, 3210 (1975).
618. H. Meerwein, H. Allendörfer, P. Beekmann, Fr. Kunert, H. Morschel, F. Pawellek, and K. Wunderlich, *Angew. Chem.* **70**, 211 (1958).
619. H. Hart and D. A. Tomalia, *Tetrahedron Lett.* 3383 (1966).
620. D. A. Tomalia and H. Hart, *Tetrahedron Lett.* 3389 (1966).
621. H. Hart and D. A. Tomalia, *Tetrahedron Lett.* 1347 (1967).
622. U. Pindur, J. Müller, C. Flo, and H. Witzel, *Chem. Soc. Rev.* **16**, 75 (1987).
623. V. P. Reddy, G. Rasul, G. K. S. Prakash, and G. A. Olah, *J. Org. Chem.* **68**, 3507 (2003).
624. G. A. Olah and S. H. Yu, *J. Am. Chem. Soc.* **97**, 2293 (1975).
625. G. A. Olah and J. J. Svoboda, *Synthesis*, 52 (1973).
626. H. A. Bates, *J. Am. Chem. Soc.* **104**, 2490 (1982).
627. H. Meerwein, P. Borner, O. Fuchs, H. J. Sasse, H. Schrodt, and J. Spille, *Chem. Ber.* **89**, 2060 (1956).
628. V. Petrov, A. A. Marchione, and W. Marshall, *J. Fluorine Chem.* **127**, 68 (2006).
629. Y. Yang and Y. Okamoto, *J. Fluorine Chem.* **125**, 1825 (2004).
630. L. K. Montgomery, M. P. Grendze, and J. C. Huffman, *J. Am. Chem. Soc.* **109**, 4749 (1987).
631. M. Kira, T. Hino, and H. Sakurai, *Chem. Lett.* 555 (1992).
632. G. K. S. Prakash, Q. Wang, G. Rasul, and G. A. Olah, *J. Organomet. Chem.* **550**, 119 (1998).
633. G. K. S. Prakash, C. Bae, G. Rasul, and G. A. Olah, *J. Org. Chem.* **67**, 1297 (2002).
634. P. J. F. de Rege, J. A. Gladysz, and I. T. Horváth, *Science* **276**, 776 (1997).
635. P. Cremaschi and M. Simonetta, *Theor. Chim. Acta* **43**, 351 (1977).
636. N. L. Summers and J. Tyrrell, *J. Am. Chem. Soc.* **99**, 3960 (1977).
637. W. J. Bouma and L. Radom, *J. Am. Chem. Soc.* **105**, 5484 (1983).
638. M. W. Wong, B. F. Yates, R. H. Nobes, and L. Radom, *J. Am. Chem. Soc.* **109**, 3181 (1987).
639. S. Raugei and M. L. Klein, *J. Phys. Chem. AB* **105**, 8212 (2001).
640. G. K. S. Prakash, J. W. Bausch, and G. A. Olah, *J. Am. Chem. Soc.* **113**, 3203 (1991).
641. G. A. Olah, A. Burrichter, T. Mathew, Y. D. Vankar, G. Rasul, and G. K. S. Prakash, *Angew. Chem. Int. Ed. Engl.* **36**, 1875 (1997).
642. E. Bernhardt, H. Willner, and F. Aubke, *Angew. Chem. Int. Ed.* **38**, 823 (1999).
643. D. Cook, *Can. J. Chem.* **37**, 48 (1959). For a review, see also D. Cook, in *Friedel–Crafts and Related Reactions*, Vol. 1, G. A. Olah, ed; Interscience, New York, 1963, Chapter 9, pp. 790–812.
644. B. P. Susz and J.-J. Wuhrmann, *Helv. Chim. Acta* **40**, 971 (1957) and subsequent publications.
645. G. Olah, S. Kuhn, and S. Beke, *Chem. Ber.* **89**, 862 (1956).
646. G. A. Olah, S. J. Kuhn, W. S. Tolgyesi, and E. B. Baker, *J. Am. Chem. Soc.* **84**, 2733 (1962).

647. G. A. Olah, A. Germain, and A. M. White, in *Carbonium Ions*, Vol. 5, G. A. Olah and P. v. R. Schleyer, eds., Wiley-Interscience, New York, London, 1976, Chapter 35, p. 2049.

648. N. C. Deno, C. U. Pittman, Jr., and M. J. Wisotsky, *J. Am. Chem. Soc.* **86**, 4370 (1964).

649. G. A. Olah and A. M. White, *J. Am. Chem. Soc.* **89**, 7072 (1967).

650. G. A. Olah, P. S. Iyer, G. K. S. Prakash, and V. V. Krishnamurthy, *J. Org. Chem.* **49**, 4317 (1984).

651. F. P. Boer, *J. Am. Chem. Soc.* **88**, 1572 (1966).

652. G. A. Olah and M. B. Comisarow, *J. Am. Chem. Soc.* **88**, 4442 (1966).

653. G. A. Olah and M. B. Comisarow, *J. Am. Chem. Soc.* **88**, 3313 (1966).

654. G. A. Olah and M. B. Comisarow, *J. Am. Chem. Soc.* **89**, 2694 (1967).

655. I. V. Stankevich, A. L. Chistyakov, I. S. Akhrem, A. V. Orlinkov, and M. E. Vol'pin, *Russ. Chem. Bull.* **42**, 80 (1993).

656. T. Xu, D. H. Barich, P. D. Torres, J. B. Nicholas, and J. F. Haw, *J. Am. Chem. Soc.* **119**, 396 (1997).

657. T. M. Krygowski, M. K. Cyrański, D. D. Sung, and B. T. Stepień, *J. Phys. Org. Chem.* **17**, 699 (2004).

658. G. A. Olah, J.-M. Denis, and P. W. Westerman, *J. Org. Chem.* **39**, 1206 (1974).

659. G. Rasul, G. K. S. Prakash, and G. A. Olah, *J. Phys. Chem. A* **110**, 1041 (2006).

660. G. A. Olah, M. Alemayehu, A.-H. Wu, O. Farooq, and G. K. S. Prakash, *J. Am. Chem. Soc.* **114**, 8042 (1992).

661. G. K. S. Prakash, C. Bae, G. Rasul, and G. A. Olah, *Proc. Natl. Acad. Sci. USA* **102**, 6251 (2005).

662. R. Minkwitz, F. Neikes, and U. Lohmann, *Eur. J. Inorg. Chem.* 27 (2002).

663. R. Minkwitz and F. Neikes, *Eur. J. Inorg. Chem.* 2283 (2000).

664. R. Minkwitz and F. Neikes, *Inorg. Chem.* **38**, 5960 (1999).

665. G. A. Olah, T. Nakajima, and G. K. S. Prakash, *Angew. Chem. Int. Ed. Engl.* **19**, 811 (1980).

666. G. Rasul, G. K. S. Prakash, and G. A. Olah, *J. Mol. Struct. (Theochem)* **466**, 245 (1999).

667. G. A. Olah, K. K. Laali, Q. Wang, and G. K. S. Prakash, *Onium Ions*, Wiley, New York, 1998, Chapter 3, p. 140 and references cited therein.

668. H. Meerwein, K.-F. Zenner, and R. Gipp, *Justus Liebigs Ann Chem.* **688**, 67 (1965).

669. Z. Yoshida, S. Yoneda, T. Sugimoto, and O. Kikukawa, *Tetrahedron Lett.* 3999 (1971).

670. R. Minkwitz and W. Meckstroth, *Z. Anorg. Allg. Chem.* **619**, 583 (1993).

671. B. Pruski, R. Antkowiak, and W. Z. Antkowiak, *Tetrahedron* **46**, 6133 (1990).

672. G. A. Olah, V. P. Reddy, G. Rasul, and G. K. S. Prakash, *Chem. Croat Acta* **65**, 721 (1992).

673. W. P. Tucker and G. L. Roof, *Tetrahedron Lett.* 2747 (1967).

674. G. P. Miller, I. Jeon, G. Wilson, and A. J. Athans, *Org. Lett.* **1**, 1771 (1999).

675. J. Klaveness and K. Undheim, *Chem. Acta Scand B* **37**, 687 (1983).

676. G. A. Olah, G. K. S. Prakash, and T. Nakajima, *Angew. Chem. Int. Ed. Engl.* **19**, 812 (1980).

677. L. Hevesi, *Bull. Soc. Chim. Fr.* **127**, 697 (1990).

678. R. W. Taft, *Prog. Phys. Org. Chem.* **14**, 247 (1983).
679. J. K. Pau, M. B. Ruggera, J. K. Kim, and M. C. Caserio, *J. Am. Chem. Soc.* **100**, 4242 (1978).
680. K. Ösapay, J. Delhalle, K. M. Nsunda, E. Rolli, R. Houriet, and L. Hevesi, *J. Am. Chem. Soc.* **111**, 5028 (1989).
681. R. J. Gillespie and T. Birchall, *Can. J. Chem.* **41**, 148 (1963).
682. G. J. Schrobilgen and J. M. Whalen, *Inorg. Chem.* **33**, 5207 (1994).
683. V. C. Armstrong, D. W. Farlow, and R. B. Moodie, *Chem. Commun.* 1362 (1968).
684. G. A. Olah, A. M. White, and A. T. Ku, *J. Org. Chem.* **36**, 3585 (1971).
685. G. A. Olah, D. L. Brydon, and R. D. Porter, *J. Org. Chem.* **35**, 317 (1970).
686. J. L. Sudmeier and K. E. Schwartz, *Chem. Commun.* 1646 (1968).
687. H. Pracejus, *Chem. Ber.* **92**, 988 (1959).
688. A. Mertens, M. Avarnaghi, and G. A. Olah, *Chem. Ber.* **116**, 3926 (1983).
689. A. Mertens and G. A. Olah, *Chem. Ber.* **116**, 103 (1983).
690. G. Rasul, G. K. S. Prakash, and G. A. Olah, *J. Org. Chem.* **59**, 2552 (1994).
691. G. A. Olah and A. M. White, *J. Am. Chem. Soc.* **90**, 6087 (1968).
692. G. A. Olah, A. Burrichter, G. Rasul, K. O. Christe, and G. K. S. Prakash, *J. Am. Chem. Soc.* **119**, 4345 (1997).
693. D. Feil and W. S. Loong, *Acta Crystallogr. B* **24**, 1334 (1968).
694. G. A. Olah, T. Heiner, G. Rasul, and G. K. S. Prakash, *J. Org. Chem.* **63**, 7993 (1998).
695. G. A. Olah and D. J. Donovan, *J. Org. Chem.* **43**, 860 (1978).
696. G. A. Olah and P. Kreienbühl, *J. Am. Chem. Soc.* **89**, 4756 (1967).
697. F. Maquin, D. Stahl, A. Sawaryn, P. v. R. Schleyer, W. Koch, G. Frenking, and H. Schwarz, *J. Chem. Soc., Chem. Commun.* 504 (1984).
698. G. A. Olah, G. K. S. Prakash, and G. Rasul, *J. Org. Chem.* **67**, 8547 (2002).
699. R. Minkwitz, W. Meckstroth, and H. Preut, *Z. Anorg. Allg. Chem.* **617**, 136 (1992).
700. R. Minkwitz, D. Lamek, and H. Preut, *Z. Naturforsch.* **48b**, 1075 (1993).
701. J. V. Paukstelis and A. G. Cook, in *Enamines—Synthesis, Structure and Reactions*, 2nd ed., A. G. Cook, ed., Marcel Dekker, New York, 1988, Chapter 6, p. 275
702. G. A. Olah, K. K. Laali, Q. Wang, and G. K. S. Prakash, *Onium Ions*, Wiley, New York, 1998, Chapter 3, p. 110 and references cited therein.
703. I. Malassa and D. Matthies, *Chem.-Zeit.* **111**(181), 253 (1987).
704. R. Minkwitz and D. Lamek, *Z. Anorg. Allg. Chem.* **620**, 514 (1994).
705. G. A. Olah, A. Burrichter, G. Rasul, M. Hachoumy, and G. K. S. Prakash, *J. Am. Chem. Soc.* **119**, 12929 (1997).
706. J. L. Boivin, *Can. J. Chem.* **34**, 827 (1956).
707. A. Gobbi and G. Frenking, *J. Am. Chem. Soc.* **115**, 2362 (1993).
708. J. F. McGarrity and D. P. Cox, *J. Am. Chem. Soc.* **105**, 3961 (1983).
709. M. Al-Tabib, I. Jibril, E.-U. Würthwein, J. C. Jochims, and G. Huttner, *Chem. Ber.* **117**, 3365 (1984).
710. E.-U. Würthwein, R. Kupfer, P. H. M. Budzelaar, C. Strobel, and H. P. Beck, *Angew. Chem. Int. Ed. Engl.* **24**, 340 (1985).

711. H. Meerwein, P. Laasch, R. Mersch, and J. Spille, *Chem. Ber.* **89**, 209 (1956).

712. G. A. Olah and T. E. Kiovsky, *J. Am. Chem. Soc.* **90**, 4666 (1968).

713. G. Leroy, M. T. Nguyen, M. Sana, K. J. Dignam, and A. F. Hegarty, *J. Am. Chem. Soc.* **101**, 1988 (1979).

714. A. K. Gjoystdal and C. Romming, *Acta Chem. Scand.* **31B**, 56 (1977).

715. H. Bock, K. Ruppert, K. Merzweiler, D. Fenske, and H. Goesmann, *Angew. Chem. Int. Ed.* **28**, 1684 (1989).

716. K. Lammertsma, M. Barzaghi, G. A. Olah, J. A. Pople, A. J. Kos, and P. v. R. Schleyer, *J. Am. Chem. Soc.* **105**, 5252 (1983).

717. K. Lammertsma and T. Ohwada, *J. Am. Chem. Soc.* **118**, 7247 (1996).

718. M. J. Taylor, P. W. J. Surman, and G. R. Clark, *J. Chem. Soc., Chem. Commun.* 2517 (1994).

719. G. R. Clark, C. E. F. Rickard, P. W. J. Surman, and M. J. Taylor, *J. Chem. Soc., Faraday Trans.* **93**, 2503 (1997).

720. S. Pohl, M. Harmjanz, J. Schneider, W. Saak, and G. Henkel, *J. Chem. Soc. Dalton Trans* 3473 (2000).

721. U. Flörke, S. Herres-Pawlis, A. Heuwing, A. Neuba, O. Seewald, and G. Henkel, *Acta Crystallogr. C* **62**, m234 (2006).

722. R. Pettit and L. W. Haynes, in *Carbonium Ions*, Vol. 5, G. A. Olah and P. v. R. Schleyer, eds., Wiley-Interscience, New York, 1976, Chapter 37, p. 2263

723. M. J. McGlinchey, L. Girard, and R. Ruffolo, *Coord. Chem. Rev.* **143**, 331 (1995).

724. H. El Amouri and M. Gruselle, *Chem. Rev.* **96**, 1077 (1996).

725. A. J. M. Caffyn and K. M. Nicholas, in *Comprehensive Organometallic Chemistry II*, E. W. Abel, F. G. A. Stone, and G. Wilkinson, eds., Vol. 12: *Transition Metal Organometallics in Organic Synthesis*, L. S. Hegedus, ed., Pergamon Press, Oxford, 1995, Chapter 7.1, p. 685.

726. K. M. Nicholas, *Acc. Chem. Res.* **20**, 207 (1987).

727. K. M. Nicholas and R. Pettit, *Tetrahedron Lett.* 3475 (1971).

728. G. G. Melikyan, S. Bright, T. Monroe, K. I. Hardcastle, and J. Ciurash, *Angew. Chem. Int. Ed.* **37**, 161 (1998).

729. M. J. Chetcuti and S. R. McDonald, *Organometallics* **21**, 3162 (2002).

730. H. Amouri, J.-P. Bégué, A. Chennoufi, D. Bonnet-Delpon, M. Gruselle, and B. Malézieux, *Org. Lett.* **6**, 807 (2000).

731. M. Gruselle, H. El Hafa, M. Nikolski, G. Jaouen, J. Vaissermann, L. Li, and M. J. McGlinchey, *Organometallics* **12**, 4917 (1993).

732. M. D. McClain, M. S. Hay, M. D. Curtis, and J. W. Kampf, *Organometallics* **13**, 4377 (1994).

733. B. N. Hietbrink, D. J. Tantillo, K. N. Houk, and C. A. Merlic,in *Carbocation Chemistry*, G. A. Olah and G. K. S. Prakash, eds., Wiley, Hoboken, NJ, 2004, Chapter 10, p. 279.

734. T. A. Albright, R. Hoffmann, and P. Hofmann, *Chem. Ber.* **111**, 1591 (1978).

735. A. Pfletschinger, T. K. Dargel, J. W. Bats, H.-G. Schmalz, and W. Koch, *Chem. Eur. J.* **5**, 537 (1999).

736. C. A. Merlic, J. C. Walsh, D. J. Tantillo, and K. N. Houk, *J. Am. Chem. Soc.* **121**, 3596 (1999).

737. C. A. Merlic, B. N. Hietbrink, and K. N. Houk, *J. Org. Chem.* **66**, 6738 (2001).
738. M. D. Winemiller, M. E. Kopach, and W. D. Harman, *J. Am. Chem. Soc.* **119**, 2096 (1997).
739. J. D. Fitzpatrick, L. Watts, and R. Pettit, *Tetrahedron Lett.* 1299 (1966).
740. M. Cais, S. Dani, F. H. Herbstein, and M. Kapon, *J. Am. Chem. Soc.* **100**, 5554 (1978).
741. E.-W. Koch, H.-U. Siehl, and M. Hanack, *Tetrahedron Lett.* **26**, 1493 (1985).
742. J. Lukasser, H. Angleitner, H. Schottenberger, H. Kopacka, M. Schweiger, B. Bildstein, K.-H. Ongania, and K. Wurst, *Organometallics* **14**, 5566 (1995).
743. G. K. S. Prakash, H. Buchholz, V. P. Reddy, A. de Meijere, and G. A. Olah, *J. Am. Chem. Soc.* **114**, 1097 (1992).
744. D. Bethell and V. Gold, *Carbonium Ions, An Introduction*, Academic Press, London, 1967.
745. G. A. Olah and G. Rasul, *Acc. Chem. Res.* **30**, 245 (1997).
746. F. H. Field and M. S. B. Munson, *J. Am. Chem. Soc.* **87**, 3289 (1965) and references cited therein.
747. D. Smith, N. G. Adams, and E. Alge, *J. Chem. Phys.* **77**, 1261 (1982).
748. K. Hiraoka and P. Kebarle, *J. Am. Chem. Soc.* **98**, 6119 (1976).
749. K. Raghavachari, R. A. Whiteside, J. A. Pople, and P. v. R. Schleyr, *J. Am. Chem. Soc.* **103**, 5649 (1981).
750. G. A. Olah, unpublished results.
751. G. A. Olah, G. Klopman, and R. H. Schlosberg, *J. Am. Chem. Soc.* **91**, 3261 (1969).
752. V. Dyczmons, V. Staemmler, and W. Kutzelnigg, *Chem. Phys. Lett.* **5**, 361 (1970).
753. A. Dedieu and A. Veillard,Lecture at 21st Annual Meeting at the Société Chimique Physique, Paris, September 1970; A. Veillard, private communication.
754. W. A. Lathan, W. J. Hehre, and J. A. Pople, *Tetrahedron Lett.* 2699 (1970); *J. Am. Chem. Soc.* **93**, 808 (1971).
755. E. L. Muetterties, *J. Am. Chem. Soc.* **91**, 1636 (1969).
756. P. v. R. Schleyer, J. W. de, and M. Carneiro, *J. Comput. Chem.* **13**, 997 (1992).
757. P. R. Schreiner, S.-J. Kim, H. F. Schaefer III, and P. v. R. Schleyer, *J. Chem. Phys.* **99**, 3716 (1993).
758. G. E. Scuseria, *Nature* **366**, 512 (1993).
759. P. R. Schreiner, *Angew. Chem. Int. Ed.* **39**, 3239 (2000).
760. K. C. Thompson, D. L. Crittenden, and M. J. T. Jordan, *J. Am. Chem. Soc.* **127**, 4954 (2005).
761. R. Car and M. Parrinello, *Phys. Rev. Lett.* **55**, 2471 (1985).
762. D. Marx and M. Parrinello, *Nature* **375**, 216 (1995).
763. H. Müller, W. Kutzelnigg, J. Noga, and W. Klopper, *J. Chem. Phys.* **106**, 1863 (1997).
764. D. W. Boo and Y. T. Lee, *Chem. Phys. Lett.* **211**, 358 (1993); *J. Chem. Phys.* **103**, 520 (1995).
765. E. T. White, J. Tang, and T. Oka, *Science* **284**, 135 (1999).
766. K. Hiraoka, I. Kudaka, and S. Yamabe, *Chem. Phys. Lett.* **184**, 271 (1991).
767. A. J. R. Heck, L. J. de Koning, and N. M. M. Nibbering, *J. Am. Soc. Mass Spectrom.* **2**, 453 (1991).

768. S.-J. Kim, P. R. Schreiner, P. v. R. Schleyer, and H. F. Schaefer III *J. Phys. Chem.* **97**, 12232 (1993).

769. D. W. Boo, Z. F. Liu, A. G. Suits, and Y. T. Lee, *Science* **269**, 57 (1995).

770. O. Asvany, P. Kumar P., B. Redlich, I. Hegemann, S. Schlemmer, and D. Marx, *Science* **309**, 1219 (2005).

771. P. Kumar and D. Marx, *Phys. Chem. Chem. Phys.* **8**, 573 (2006).

772. K. Hiraoka and P. Kebarle, *J. Am. Chem. Soc.* **97**, 4179 (1975).

773. F. H. Field and D. P. Beggs, *J. Am. Chem. Soc.* **93**, 1585 (1971).

774. H. J. Yi, Y. S. Kim, C. J. Choi, and K.-H. Jung, *J. Mass. Spectrom.* **33**, 599 (1998).

775. M. E. Sheridan, E. Greer, and W. F. Libby, *J. Am. Chem. Soc.* **94**, 2614 (1972).

776. P. Ahlberg, A. Karlsson, A. Goeppert, S. O. Nilsson Lill, P. Dinér, and J. Sommer, *Chem. Eur. J.* **7**, 1936 (2001).

777. E. D. Jemmis, J. Chandrasekhar, E.-U. Würthwein, P. v. R. Schleyer, J. W. Chinn, Jr., F. J. Landro, R. J. Lagow, B. Luke, and J. A. Pople, *J. Am. Chem. Soc.* **104**, 4275 (1982).

778. P. v. R. Schleyer, B. Tidor, E. D. Jemmis, J. Chandrasekhar, E.-U. Würthwein, A. J. Kos, B. T. Luke, and J. A. Pople, *J. Am. Chem. Soc.* **105**, 484 (1983).

779. W. E. Rhine, G. Stucky, and S. W. Peterson, *J. Am. Chem. Soc.* **97**, 6401 (1975).

780. F. Scherbaum, A. Grohmann, G. Müller, and H. Schmidbaur, *Angew. Chem. Int. Ed. Engl.* **28**, 463 (1989).

781. H. Schmidbaur, *Chem. Soc. Rev.* **24**, 391 (1995).

782. N. Dufour, A. Schier, and H. Schmidbaur, *Organometallics* **12**, 2408 (1993).

783. S. Bommers, H. Beruda, N. Dufour, M. Paul, A. Schier, and H. Schmidbaur, *Chem. Ber.* **128**, 137 (1995).

784. J. Vicente, M. T. Chicote, R. Guerrero, and P. G. Jones, *J. Am. Chem. Soc.* **118**, 699 (1996).

785. A. Blumenthal, H. Beruda, and H. Schmidbaur, *J. Chem. Soc., Chem. Commun.* 1005 (1993).

786. K. Lammertsma, G. A. Olah, M. Barzaghi, and M. Simonetta, *J. Am. Chem. Soc.* **104**, 6851 (1982).

787. K. Lammertsma, M. Barzaghi, G. A. Olah, J. A. Pople, P. v. R. Schleyer, and M. Simonetta, *J. Am. Chem. Soc.* **105**, 5258 (1983).

788. G. A. Olah, G. K. S. Prakash, and G. Rasul, *J. Org. Chem.* **66**, 2907 (2001).

789. H. Hogeveen and P. W. Kant, *Tetrahedron Lett.* 1665 (1973).

790. F. Scherbaum, A. Grohmann, B. Huber, C. Krüger, and H. Schmidbaur, *Angew. Chem. Int. Ed. Engl.* **27**, 1544 (1988).

791. H. Schmidbaur, B. Brachthäuser, and O. Steigelmann, *Angew. Chem. Int. Ed. Engl.* **30**, 1488 (1991).

792. F. P. Gabbaï, A. Schier, J. Riede, and H. Schmidbaur, *Chem. Ber./Receuil* **130**, 111 (1997).

793. A. Görling, N. Rösch, D. E. Ellis, and H. Schmidbaur, *Inorg. Chem.* **30**, 3986 (1991).

794. O. D. Häberlen, H. Schmidbaur, and N. Rösch, *J. Am. Chem. Soc.* **116**, 8241 (1994).

795. C. H. DePuy, R. Gareyev, J. Hankin, and G. E. Davico, *J. Am. Chem. Soc.* **119**, 427 (1997).

796. C. H. DePuy, R. Gareyev, J. Hankin, G. E. Davico, M. Krempp, and R. Damrauer, *J. Am. Chem. Soc.* **120**, 5086 (1998).

797. G. Rasul and G. A. Olah, *Inorg. Chem.* **36**, 1278 (1997).

798. S. Salzbrunn, G. Rasul, G. K. S. Prakash, and G. A. Olah, *J. Mol. Model.* **6**, 213 (2000).

799. G. A. Olah and G. Rasul, *J. Am. Chem. Soc.* **118**, 8503 (1996).

800. G. A. Olah and G. Klopman, *Chem. Phys. Lett.* **11**, 604 (1971).

801. W. Koch and H. Schwarz, in *Structure/Reactivity and Thermochemistry of Ions*, P. Ausloosand S. G. Lias, eds., Reidel, Dordrecht, 1987, p. 413.

802. K. Lammertsma, *Rev. Chem. Intermed.* **9**, 141 (1988).

803. K. Negesha, V. R. Marathe, and D. Mathur, *Rapid Commun. Mass Spectrosc.* **5**, 15 (1991).

804. C. J. Proctor, C. J. Porter, A. T. Ast, P. D. Bolton, and J. H. Beynon, *Org. Mass Spectrom.* **16**, 454 (1981).

805. M. Rabrenović, C. J. Proctor, C. G. Herbert, A. G. Brenton, and J. H. Benyon, *J. Phys. Chem.* **87**, 3305 (1983).

806. S. Singh, R. K. Boyd, F. M. Harris, and J. H. Beynon, *Proc. R. Soc. London, Ser. A* **402**, 373 (1985).

807. D. Stahl, F. Maquin, T. Gäumann, H. Schwarz, P.-A. Carrupt, and P. Vogel, *J. Am. Chem. Soc.* **107**, 5049 (1985).

808. J. A. Pople, B. Tidor, and P. v. R. Schleyer, *Chem. Phys. Lett.* **88**, 533 (1982).

809. P. E. M. Siegbahn, *Chem. Phys.* **66**, 443 (1982).

810. M. W. Wong and L. Radom, *J. Am. Chem. Soc.* **111**, 1155 (1989).

811. P. Babinec and J. Leszczynski, *Int. J. Quantum Chem.* **72**, 319 (1999).

812. G. A. Olah and G. Rasul, *J. Am. Chem. Soc.* **118**, 12922 (1996).

813. G. Frenking and D. Cremer, in *Structure and Bonding*, Vol.73, Springer, Heidelberg, 1990, Chapter 2, p. 17.

814. S. Wilson and S. Green, *J. Chem. Phys.* **73**, 419 (1980).

815. D. L. Cooper and S. Wilson, *Mol. Phys.* **44**, 161 (1981).

816. P. v. R. Schleyer, in *Advances in Mass Spectrometry 1985*, J. F. J. Todd, ed., Wiley, Chichester, 1986, p. 287

817. M. W. Wong, R. H. Nobes, and L. Radom, *J. Chem. Soc., Chem. Commun.* 233 (1987).

818. W. Koch and G. Frenking, *J. Chem. Soc., Chem. Commun.*, 1095 (1986).

819. G. Frenking, W. Koch, F. Reichel, and D. Cremer, *J. Am. Chem. Soc.* **112**, 4240 (1990).

820. G. A. Olah, G. K. S. Prakash, and G. Rasul, *J. Mol. Struct. (Theochem)* **489**, 209 (1999).

821. L. I. Yeh, J. M. Price, and Y. T. Lee, *J. Am. Chem. Soc.* **111**, 5597 (1989).

822. J. W. de, M. Carneiro, P. v. R. Schleyer, M. Saunders, R. Remington, H. F. Schaefer III A. Rauk, and T. S. Sorensen, *J. Am. Chem. Soc.* **116**, 3483 (1994).

823. A. J. R. Heck, L. J. de Koning, and N. M. M. Nibbering, *Int. J. Mass Spectrom. Ion Process.* **117**, 145 (1992).

824. A. L. L. East, Z. F. Liu, C. McCague, K. Cheng, and J. S. Tse, *J. Phys. Chem. A* **102**, 10903 (1998).

825. J. J. Fischer, G. K. Koyagani, and T. B. McMahon, *Int. J. Mass Spetrom.* **195/196**, 491 (2000).

826. S. R. Blaszkowski, M. A. C. Nascimento, and R. A. van Santen, *J. Phys. Chem.* **100**, 3463 (1996).

827. P. M. Esteves, M. A. C. Nascimento, and C. J. A. Mota, *J. Phys. Chem. B* **103**, 10417 (1999).

828. N. B. Okulik, R. P. Diez, and A. H. Jubert, *J. Phys. Chem. A* **108**, 2469 (2004).

829. O. Steigelmann, P. Bissinger, and H. Schmidbaur, *Z. Naturforsch.* **48B**, 72 (1993).

830. K. Hiraoka and P. Kebarle, *J. Chem. Phys.* **63**, 394 (1975).

831. S. J. Collins and P. J. O'Malley, *J. Chem. Soc., Faraday Trans.* **92**, 4347 (1996).

832. P. M. Esteves, C. J. A. Mota, A. Ramírez-Solís, and R. Hernández-Lamoneda, *J. Am. Chem. Soc.* **120**, 3213 (1998).

833. D. P. Stevenson, C. D. Wagner, O. Beeck, and J. W. Otvos, *J. Am. Chem. Soc.* **74**, 3269 (1952).

834. P. D. Bartlett, F. E. Condon, and A. Schneider, *J. Am. Chem. Soc.* **66**, 1531 (1944).

835. C. D. Nenitzescu, M. Avram, and E. Sliam, *Bull. Soc., Chim. Fr.* 1266 (1955) and references therein.

836. L. Schmerling, *J. Am. Chem. Soc.* **66**, 1422 (1944); *J. Am. Chem. Soc.* **67**, 1778 (1945).

837. G. A. Olah, Y. Halpern, J. Shen, and Y. K. Mo, *J. Am. Chem. Soc.* **93**, 1251 (1971).

838. A. Goeppert, B. Louis, and J. Sommer, *Catal. Lett.* **56**, 43 (1998).

839. J. Sommer, J. Bukala, M. Hachoumy, and R. Jost, *J. Am. Chem. Soc.* **119**, 3274 (1997).

840. A. Goeppert and J. Sommer, *New J. Chem.* **26**, 1335 (2002).

841. M. Siskin, *J. Am. Chem. Soc.* **98**, 5413 (1976).

842. K. Hiraoka and P. Kebarle, *Can. J. Chem.* **58**, 2262 (1980).

843. P. M. Esteves, G. G. P. Alberto, A. Ramírez-Solís, and C. J. A. Mota, *J. Phys. Chem. A* **104**, 6233 (2000).

844. R. M. Lobayan, G. L. Sosa, A. H. Jubert, and N. M. Peruchena, *J. Phys. Chem. A* **108**, 4347 (2004).

845. Q. Li, K. C. Hunter, C. Steitz, and A. L. L. East, *J. Chem. Phys.* **119**, 7148 (2003).

846. C. J. A. Mota, P. M. Esteves, A. Ramírez-Solís, and R. Hernández-Lamoneda, *J. Am. Chem. Soc.* **119**, 5193 (1997).

847. C. Seitz and A. L. L. East, *J. Phys. Chem. A* **106**, 11653 (2002).

848. P. Cecchi, A. Pizzabiocca, G. Renzi, F. Grandinetti, C. Sparapani, P. Buzek, P. v. R. Schleyer, and M. Speranza, *J. Am. Chem. Soc.* **115**, 10338 (1993).

849. K. J. Szabó and D. Cremer, *J. Org. Chem.* **60**, 2257 (1995).

850. P. M. Esteves, G. G. P. Alberto, A. Ramírez-Solís, and C. J. A. Mota, *J. Phys. Chem. A* **105**, 4308 (2001).

851. P. Ahlberg, G. Jonsäll, and C. Engdahl, *Adv. Phys. Org. Chem.* **19**, 223 (1983).

852. M. Saunders, E. L. Hagen, and J. Rosenfeld, *J. Am. Chem. Soc.* **90**, 6882 (1968).

853. M. Saunders and E. L. Hagen, *J. Am. Chem. Soc.* **90**, 2436 (1968).

854. M. Saunders and M. R. Kates, *J. Am. Chem. Soc.* **100**, 7082 (1978).

855. M. R. Kates, Ph.D. thesis Yale University (1978).

856. G. A. Olah and A. M. White, *J. Am. Chem. Soc.* **91**, 3954 (1969).

857. P. C. Myhre, J. D. Kruger, B. L. Hammond, S. M. Lok, C. S. Yannoni, V. Macho, H. H. Limbach, and H. M. Vieth, *J. Am. Chem. Soc.* **106**, 6079 (1984).

858. G. D. Mateescu and J. L. Riemenschneider, *Application of Electron Spectroscopy in Organic Chemistry*, D. A. Shirley, ed., North-Holland, Amsterdam, 1972.
859. G. A. Olah, P. Schilling, P. W. Westerman, and H. C. Lin, *J. Am. Chem. Soc.* **96**, 3581 (1974).
860. Y. M. Jun and J. W. Timberlake, *Tetrahedron Lett.* **23**, 1761 (1982).
861. S. Winstein and D. S. Trifan, *J. Am. Chem. Soc.* **71**, 2953 (1949). 74, 1147 1952
862. H. C. Brown, Lectures presented at The International Symposium on Carbocation Chemistry, Bangor, England, 1981; at The Sixth IUPAC Conference on Physical Organic Chemistry, Louvain-la-Neuve, Belgium, 1982; and at The Symposium on Carbocation Chemistry, ACS National Meeting, Seattle Washington, 1983.
863. G. M. Kramer, *Adv. Phys. Org. Chem.* **11**, 177 (1975).
864. H. C. Brown, *Topics Curr. Chem.* **80**, 1 (1979).
865. C. A. Grob, *Angew. Chem. Int. Ed. Engl.* **21**, 87 (1982).
866. G. A. Olah and G. K. S. Prakash, *Preprints Div. Petr. Chem. ACS* **28**(2), 366 (1983).
867. M. Saunders, P. v. R. Schleyer, and G. A. Olah, *J. Am. Chem. Soc.* **86**, 5680 (1964).
868. G. A. Olah, G. K. S. Prakash, M. Arvanaghi, and F. A. L. Anet, *J. Am. Chem. Soc.* **104**, 7105 (1982).
869. C. S. Yannoni, V. Macho, and P. C. Myhre, *J. Am. Chem. Soc.* **104**, 907 (1982).
870. C. S. Yannoni, V. Macho, and P. C. Myhre, *J. Am. Chem. Soc.* **104**, 7380 (1982).
871. M. J. S. Dewar and K. M. Merz, Jr., *J. Am. Chem. Soc.* **108**, 5634 (1986).
872. M. Saunders and M. R. Kates, *J. Am. Chem. Soc.* **102**, 6867 (1980).
873. M. Saunders and M. R. Kates, *J. Am. Chem. Soc.* **105**, 3571 (1983).
874. K. Ranganayakalu and T. S. Sorensen, *J. Org. Chem.* **49**, 4310 (1984).
875. G. A. Olah, G. K. S. Prakash, and G. Liang, *J. Am. Chem. Soc.* **99**, 5683 (1977).
876. D. G. Farnum and R. E. Botto, *Tetrahedron Lett.* 4013 (1975).
877. D. G. Farnum, R. E. Botto, W. T. Chambers, and B. Lam, *J. Am. Chem. Soc.* **100**, 3847 (1978).
878. D. G. Farnum and A. D. Wolf, *J. Am. Chem. Soc.* **96**, 5166 (1974).
879. D. G. Farnum and T. P. Clausen, *Tetrahedron Lett.* **22**, 549 (1981).
880. G. A. Olah, A. L. Berrier, and G. K. S. Prakash, *J. Org. Chem.* **47**, 3903 (1982).
881. G. A. Olah, R. D. Porter, C. L. Jeuell, and A. M. White, *J. Am. Chem. Soc.* **94**, 2044 (1972).
882. H. C. Brown, D. P. Kelly, and M. Periasamy, *Proc. Natl. Acad. Sci. USA* **77**, 6956 (1980).
883. H. C. Brown, M. Periasamy, D. P. Kelly, and J. J. Giansiracusa, *J. Org. Chem.* **47**, 2089 (1982) and references cited therein.
884. The ESCA spectrum was obtained by Dr. Grunthaner on HP 7950A ESCA spectrometer at Jet Propulsion Labs in Pasadena, CA.
885. D. W. Goetz, H. B. Schlegel, and L. C. Allen, *J. Am. Chem. Soc.* **99**, 8118 (1977).
886. D. T. Clark, B. J. Cromarty, and L. Colling, *J. Am. Chem. Soc.* **99**, 8120 (1977).
887. S. A. Johnson and D. T. Clark, *J. Am. Chem. Soc.* **110**, 4112 (1988).
888. E. M. Arnett, N. Pienta, and C. Petro, *J. Am. Chem. Soc.* **102**, 398 (1980).

889. D. Fărcaşiu, *J. Org. Chem.* **46**, 223 (1981).
890. P. v. R. Schleyer and J. Chandrashekar, *J. Org. Chem.* **46**, 225 (1981).
891. F. Kaplan, P. Cross, and R. Prinstein, *J. Am. Chem. Soc.* **92**, 1445 (1970).
892. J. J. Solomon and F. H. Field, *J. Am. Chem. Soc.* **98**, 1567 (1976).
893. R. H. Staley, R. D. Wieting, and L. J. Beuchamp, *J. Am. Chem. Soc.* **99**, 5964 (1977).
894. P. P. S. Saluja and P. Kebarle, *J. Am. Chem. Soc.* **101**, 1084 (1979).
895. R. B. Sharma, D. K. Sen Sharma, K. Hiraoka, and P. Kebarle, *J. Am. Chem. Soc.* **107**, 3747 (1985).
896. T. Laube, *Angew. Chem. Int. Ed. Engl.* **26**, 560 (1987).
897. T. Laube, *Helv. Chim. Acta* **77**, 943 (1994).
898. D. A. Forsyth and C. Panyachotipun, *J. Chem. Soc., Chem. Commun.* 1564 (1988).
899. W. Koch, B. Liu, D. J. DeFrees, D. E. Sunko, and H. Vanćik, *Angew. Chem. Int. Ed. Engl.* **29**, 183 (1990).
900. W. Koch, B. Liu, and D. J. DeFrees, *J. Am. Chem. Soc.* **111**, 1527 (1989).
901. J.-L. M. Abboud, O. Castaño, E. W. Della, M. Herreros, P. Müller, R. Notario, and J.-C. Rossier, *J. Am. Chem. Soc.* **119**, 2262 (1997).
902. P. R. Schreiner, D. L. Severance, W. L. Jorgensen, P. v. R. Schleyer, and H. F. Schaefer III, *J. Am. Chem. Soc.* **117**, 2663 (1995).
903. M. J. S. Dewar, R. C. Haddon, A. Komornicki, and H. Rzepa, *J. Am. Chem. Soc.* **99**, 377 (1977).
904. G. Wenke and D. Lenoir, *Tetrahedron* **35**, 489 (1979).
905. H. J. Köhler and H. Lischka, *J. Am. Chem. Soc.* **101**, 3479 (1979).
906. J. D. Goddard, Y. Osamura, and H. Schaefer III *J. Am. Chem. Soc.* **104**, 3258 (1982).
907. K. Raghavachari, R. C. Haddon, P. v. R. Schleyer, and H. Schaefer III *J. Am. Chem. Soc.* **105**, 5915 (1983).
908. M. Yoshimine, A. D. McLean, B. Liu, D. J. DeFrees, and J. Binkley, *J. Am. Chem. Soc.* **105**, 6185 (1983).
909. P. v. R. Schleyer and S. Sieber, *Angew. Chem. Int. Ed. Engl.* **32**, 1606 (1993).
910. M. Schindler, *J. Am. Chem. Soc.* **109**, 1020 (1987).
911. N. H. Werstiuk and H. M. Muchall, *J. Mol. Struct. (Theochem)* **463**, 225 (1999).
912. H. M. Muchall and N. H. Werstiuk, *J. Phys. Chem. A* **103**, 6599 (1999).
913. N. H. Werstiuk and H. M. Muchall, *J. Phys. Chem. A* **104**, 2054 (2000).
914. N. H. Werstiuk, H. M. Muchall, and S. Noury, *J. Phys. Chem. A* **104**, 11601 (2000).
915. A. Mamantov, *Prog. React. Kinet. Mech.* **29**, 243 (2004).
916. I. Alkorta, J. L. M. Abboud, E. Quintanilla, and J. Z. Dávalos, *J. Phys. Org. Chem.* **16**, 546 (2003).
917. S. Winstein, F. Gadient, E. T. Stafford, and P. E. Klinedinst, Jr., *J. Am. Chem. Soc.* **80**, 5895 (1958).
918. P. G. Gassman and J. M. Hornback, *J. Am. Chem. Soc.* **89**, 2487 (1967).
919. W. Krimse and J. Streu, *J. Org. Chem.* **50**, 4187 (1985).
920. D. E. Sunko, H. Vancik, V. Deljac, and M. Milun, *J. Am. Chem. Soc.* **105**, 5364 (1983).

921. S. Sieber, P. v. R. Schleyer, H. Vančik, M. Mesić, and D. E. Sunko, *Angew. Chem. Int. Ed.* **32**, 1604 (1993).
922. G. Seybold, P. Vogel, M. Saunders, and K. B. Wiberg, *J. Am. Chem. Soc.* **95**, 2045 (1973).
923. J. Meinwald and P. G. Gassman, *J. Am. Chem. Soc.* **85**, 57 (1963).
924. J. Meinwald and J. K. Crandall, *J. Am. Chem. Soc.* **88**, 1292 (1966).
925. G. A. Olah, G. Liang, and S. P. Jindal, *J. Am. Chem. Soc.* **98**, 2508 (1976).
926. M. Saunders, M. R. Kates, K. B. Wiberg, and W. Pratt, *J. Am. Chem. Soc.* **99**, 8072 (1977).
927. L. R. Schmitz and T. S. Sorensen, *J. Am. Chem. Soc.* **102**, 1645 (1980).
928. W. Kirmse, V. Zellmer, and B. Goer, *J. Am. Chem. Soc.* **108**, 4912 (1986).
929. P. v. R. Schleyer, K. Laidig, K. B. Wiberg, M. Saunders, and M. Schindler, *J. Am. Chem. Soc.* **110**, 300 (1988).
930. M. Saunders, K. E. Laidig, and M. Wolfsberg, *J. Am. Chem. Soc.* **111**, 8989 (1989).
931. G. A. Olah, G. K. S. Prakash, and G. Liang, *J. Am. Chem. Soc.* **101**, 3932 (1979).
932. R. H. Mazur, W. N. White, D. A. Semenov, C. C. Lee, M. S. Silver, and J. D. Roberts, *J. Am. Chem. Soc.* **81**, 4390 (1959).
933. K. B. Wiberg, *Tetrahedron*, **24**, 1083 (1968).
934. K. B. Wiberg and G. Szeimies, *J. Am. Chem. Soc.* **90**, 4195 (1968).
935. R. E. Williams, IMEBORON-III, Munchen-Ettal, (1976).
936. R. E. Williams and L. D. Field,in *Boron Chemistry*, R. H. Parry and G. Kodoma, eds., Pergamon Press, New York, 1980.
937. M. Saunders and H.-U. Siehl, *J. Am. Chem. Soc.* **102**, 6868 (1980).
938. W. J. Brittain, M. E. Squillacote, and J. D. Roberts, *J. Am. Chem. Soc.* **106**, 7280 (1984).
939. G. K. S. Prakash, M. Arvanaghi, and G. A. Olah, *J. Am. Chem. Soc.* **107**, 6017 (1985).
940. H.-U. Siehl, *J. Am. Chem. Soc.* **107**, 3390 (1985).
941. G. A. Olah, G. K. S. Prakash, and T. Nakajima, *J. Am. Chem. Soc.* **104**, 1031 (1982).
942. R. K. Murray, Jr., T. M. Ford, G. K. S. Prakash, and G. A. Olah, *J. Am. Chem. Soc.* **102**, 1865 (1980).
943. G. A. Olah, D. J. Donovan, and G. K. S. Prakash, *Tetrahedron Lett.* 4779 (1978).
944. M. Saunders, K. E. Laidig, K. B. Wiberg, and P. v. R. Schleyer, *J. Am. Chem. Soc.* **110**, 7652 (1988).
945. W. Koch, B. Liu, and D. J. DeFrees, *J. Am. Chem. Soc.* **110**, 7325 (1988); for an ab initio/GIAO-CCSD(T) study see, G.A. Olah, G.K.S. Prakash and G. Rasul, *J. Am. Chem. Soc.* **130,** 9168 (2008).
946. P. C. Myhre, G. G. Webb, and C. S. Yannoni, *J. Am. Chem. Soc.* **112**, 8992 (1990).
947. F. Cacace, B. Chiavarino, and M. E. Crestoni, *Chem. Eur. J.* **6**, 2024 (2000).
948. J.-F. Fuchs and J. Mareda, *J. Mol. Struct. (Theochem)* **718**, 93 (2005).
949. H.-U. Siehl, M. Fuss, and J. Gauss, *J. Am. Chem. Soc.* **117**, 5983 (1995).
950. H.-U. Siehl and M. Fuss, *Pure Appl. Chem.* **70**, 2015 (1998).
951. L. R. Schmitz and T. S. Sorensen, *J. Am. Chem. Soc.* **104**, 2600 (1982).
952. L. R. Schmitz and T. S. Sorensen, *J. Am. Chem. Soc.* **104**, 2605 (1982).
953. G. A. Olah, H. A. Buchholz, G. K. S. Prakash, G. Rasul, J. J. Sosnowski, R. K. Murray, Jr., M. A. Kusnetsov, S. Liang, and A. de Meiejre, *Angew. Chem. Int. Ed. Engl.* **35**, 1499 (1996).

954. G. K. S. Prakash and V. Reddy, in *Carbocation Chemistry*, G. A. Olah and G. K. S. Prakash,eds., Wiley, Hoboken, NJ, 2004, Chapter 4, p. 73.
955. V. P. Reddy, G. Rasul, G. K. S. Prakash, and G. A. Olah, *J. Org. Chem.* **72**, 3076 (2007).
956. D. M. Brouwer and J. A. Van Doorn, *Recl. Trav. Chim. Pays-Bas* **88**, 573 (1969).
957. M. Saunders and J. J. Stofko, Jr., *J. Am. Chem. Soc.* **95**, 252 (1973).
958. A. P. Hewett, Ph.D. dissertation, Yale University, (1975).
959. V. Prelog and T. J. Traynham,in *Molecular Rearrangements*, Part 1, P. de Mayo, Interscience, New York, 1963, Chapter 9, p. 593
960. A. C. Cope, M. M. Martin, and M. A. McKervey, *Quart. Rev.* **20**, 119 (1966).
961. L. Stehglin, J. Lhomme, and G. Ourisson, *J. Am. Chem. Soc.* **93**, 1650 (1971).
962. L. Stehglin, K. Kannellias, and G. Ourisson, *J. Org. Chem.* **38**,847 851 (1973).
963. W. Parker and C. I. F. Watt, *J. Chem. Soc. Perkin* **2**, 1642 (1975).
964. F. Sun and T. S. Sorensen, *J. Am. Chem. Soc.* **115**, 77 (1993).
965. T. S. Sorensen, in *Stable Carbocation Chemistry*, G. K. S. Prakash and P. v. R. Schleyer, eds., Wiley, New York, 1997, Chapter 3, p. 75
966. R. P. Kirchen, T. S. Sorensen, and K. Wagstaff, *J. Am. Chem. Soc.* **100**, 6761 (1978).
967. R. P. Kirchen and T. S. Sorensen, *J. Chem. Soc., Chem. Commun.* 769 (1978).
968. R. P. Kirchen and T. S. Sorensen, *J. Am. Chem. Soc.* **101**, 3240 (1979).
969. R. P. Kirchen, N. Okazawa, K. Ranganayakulu, A. Rauk, and T. S. Sorensen, *J. Am. Chem. Soc.* **103**, 597 (1981).
970. R. P. Kirchen, K. Ranganayakulu, A. Rauk, B. P. Singh, and T. S. Sorensen, *J. Am. Chem. Soc.* **103**, 588 (1981).
971. J. E. McMurry and C. N. Hodge, *J. Am. Chem. Soc.* **106**, 6450 (1984).
972. J. E. McMurry and T. Lectka, *J. Am. Chem. Soc.* **112**, 869 (1990).
973. J. E. McMurry, T. Lectka, and C. N. Hodge, *J. Am. Chem. Soc.* **111**, 8867 (1989).
974. D. B. DuPré, *J. Phys. Chem. A* **109**, 622 (2005).
975. J. E. McMurry and T. Lectka, *Acc. Chem. Res.* **25**, 47 (1992).
976. J. E. McMurry and T. Lectka, *J. Am. Chem. Soc.* **115**, 10167 (1993).
977. T. S. Sorensen and S. M. Whitworth, *J. Am. Chem. Soc.* **112**, 8135 (1990).
978. L. de Vries and S. Winstein, *J. Am. Chem. Soc.* **82**, 5363 (1960).
979. J. W. de M. Carneiro, C. A. Taft, C. H. T. de Paula e Silva, J. G. R. Tostes, P. R. Seidl, P. S. da S. Pinto, V. E. U. Costa, and J. Alifantes, *Chem. Phys. Lett.* **345**, 189 (2001).
980. D. J. Tantillo and R. Hoffmann, *J. Am. Chem. Soc.* **125**, 4042 (2003).
981. R. Ponec and G. Yuzhakov, *J. Org. Chem.* **68**, 8284 (2003).
982. P. Gutta and D. J. Tantillo, *Angew. Chem. Int. Ed. Engl.* **44**, 2719 (2005).
983. R. Ponec, P. Bultinck, P. Gutta, and D. J. Tantillo, *J. Phys. Chem. A* **110**, 3785 (2006).
984. G. Schröder, *Angew. Chem. Int. Ed. Engl.* **2**, 481 (1963).
985. M. J. Goldstein and R. Hoffmann, *J. Am. Chem. Soc.* **93**, 6193 (1971) and references cited therein.
986. P. Ahlberg, D. L. Harris, and S. Winstein, *J. Am. Chem. Soc.* **92**, 4454 (1970).

987. P. Ahlberg, D. L. Harris, M. Roberts, P. Warner, P. Seidl, M. Sakai, D. Cook, A. Diaz, J. P. Dirlam, H. Hamberger, and S. Winstein, *J. Am. Chem. Soc.* **94**, 7063 (1972).
988. C. Engdahl and P. Ahlberg, *J. Chem. Res. (S)* 342 (1977).
989. C. Engdahl, G. Jonsäll, and P. Ahlberg, *J. Chem. Soc. D: Chem. Commun.* 626 (1979).
990. P. Ahlberg, C. Engdahl, and G. Jonsäll, *J. Am. Chem. Soc.* **103**, 1583 (1981).
991. C. Engdahl, G. Jonsäll, and P. Ahlberg, *J. Am. Chem. Soc.* **105**, 891 (1983).
992. T. D. Bouman and C. Trindle, *Theor. Chim. Acta* **37**, 217 (1975).
993. J. W. McIver, Jr., *Acc. Chem. Res.* **7**, 72 (1974).
994. P. Ahlberg, J. B. Grutzner, D. L. Harris, and S. Winstein, *J. Am. Chem. Soc.* **92**, 3478 (1970).
995. C. Engdahl and P. Ahlberg, *J. Phys. Org. Chem.* **3**, 349 (1990).
996. M. B. Huang and G. Jonsäll, *Tetrahedron* **41**, 6055 (1985).
997. J. Bella, J. M. Poblet, A. Demoulliens, and F. Volatron, *J. Chem. Soc., Perkin Trans.* **2**, 37 (1989).
998. D. Cremer, P. Svensson, E. Kraka, and P. Ahlberg, *J. Am. Chem. Soc.* **115**, 7445 (1993).
999. D. Lenoir, P. Mison, E. Hyson, P. v. R. Schleyer, M. Saunders, P. Vogel, and L. A. Telkowski, *J. Am. Chem. Soc.* **96**, 2157 (1974).
1000. D. Fărcaşiu, *J. Am. Chem. Soc.* **98**, 5301 (1976); *J. Org. Chem.* **43**, 3878 (1978).
1001. S. Winstein, *Chem. Soc. Spec. Publ. No.21* 5 (1967).
1002. P. R. Story and B. C. Clark, Jr., in *Carbonium Ions*, Vol. 3, G. A. Olah and P. v. R. Schleyer,ed., Wiley-Interscience, New York, 1972, Chapter 23, p. 1007.
1003. P. J. Garratt and M. V. Sargent, in *Nonbenzoid Aromatics*, Vol. 2, J. P. Snyder, ed., Academic Press, New York, 1971, Chapter 4, p. 207.
1004. L. A. Paquette, *Angew. Chem. Int. Ed. Engl.* **17**, 106 (1978).
1005. R. V. Williams and H. A. Kurtz, *Adv. Phys. Org. Chem.* **29**, 273 (1994).
1006. R. V. Williams, *Chem. Rev.* **101**, 1185 (2001).
1007. W. J. Hehre, *J. Am. Chem. Soc.* **95**, 5807 (1973).
1008. R. C. Haddon, *Tetrahedron Lett.* **2797**, 4303 (1974).
1009. R. S. Brown and T. G. Taylor, *J. Am. Chem. Soc.* **95**, 8025 (1973).
1010. G. A. Olah, J. S. Staral, and G. Liang, *J. Am. Chem. Soc.* **96**, 6233 (1974).
1011. G. A. Olah and G. Liang, *J. Am. Chem. Soc.* **97**, 6803 (1975) and references cited therein.
1012. S. Masamune, M. Sakai, and A. V. Kemp-Jones, *Can. J. Chem.* **52**, 855 (1974).
1013. R. M. Coates and E. R. Fretz, *J. Am. Chem. Soc.* **97**, 2538 (1975).
1014. P. K. Freeman and J. E. Dacres, *J. Org. Chem.* **68**, 1386 (2003).
1015. J. Jusélius, M. Patzschke, and D. Sundholm, *J. Mol. Struct. (Theochem)* **633**, 123 (2003).
1016. W. L. Jorgensen, *J. Am. Chem. Soc.* **97**, 3082 (1975): **98**, 6784 1976.
1017. S. Sieber, P. v. R. Schleyer, A. H. Otto, J. Gauss, F. Reichel, and D. Cremer, *J. Phys. Org. Chem.* **6**, 445 (1993).
1018. S. A. Osadchii, N. V. Mikushova, and V. G. Shubin, *Russ. J. Org. Chem. (Engl. Transl.)* **35**, 1777 (1999).
1019. V. A. Bushmelev, A. M. Ganaev, S. A. Osadchii, M. M. Shakirov, and V. G. Shubin, *Russ. J. Org. Chem. (Engl. Transl.)* **39**, 1301 (2003).

1020. J. L. Rosenber, J. E. Mahler, and R. Pettit, *J. Am. Chem. Soc.* **84**, 2842 (1962).

1021. C. E. Keller and R. Pettit, *J. Am. Chem. Soc.* **88**, 604 (1966).

1022. J. D. Holmes and R. Pettit, *J. Am. Chem. Soc.* **85**, 2531 (1963).

1023. S. Winstein, H. D. Kaesz, C. G. Kreiter, and E. C. Friearich, *J. Am. Chem. Soc.* **87**, 3267 (1965).

1024. P. Warner, D. L. Harris, C. H. Bradley, and S. Winstein, *Tetrahedron Lett.* 4013 (1970).

1025. S. Winstein, C. G. Kreiter, and J. A. Bauman, *J. Am. Chem. Soc.* **88**, 2047 (1966).

1026. D. Cremer, F. Reichel, and E. Kraka, *J. Am. Chem. Soc.* **113**, 9459 (1991).

1027. I. Alkorta, J. Elguero, M. Eckert-Maksić, and Z. B. Maksić, *Tetrahedron*, **60**, 2259 (2004).

1028. R. K. Lustgarten, M. Brookhart, S. Winstein, P. G. Gassman, D. S. Patton, H. G. Richey, Jr., and J. D. Nichols, *Tetrahedron Lett.* 1699 (1970).

1029. R. K. Lustgarten, M. Brookhart, and S. Winstein, *J. Am. Chem. Soc.* **89**, 6350 (1967).

1030. P. G. Gassman, J. Zeller, and J. T. Lumb, *J. Chem. Soc., Chem. Commun.* 69 (1968).

1031. P. G. Gassman and A. F. Fentiman, Jr., *J. Am. Chem. Soc.* **91**, 1545 (1969); 92, 2549 (1970).

1032. H. G. Richey, Jr., J. D. Nichols, P. G. Gassman, A. F. Fentiman, Jr., S. Winstein, M. Brookhart, and R. K. Lustgarten, *J. Am. Chem. Soc.* **92**, 3783 (1970).

1033. R. K. Lustgarten, M. Brookhart, and S. Winstein, *J. Am. Chem. Soc.* **94**, 2347 (1972).

1034. M. Brookhart, R. K. Lustgarten, and S. Winstein, *J. Am. Chem. Soc.* **89**, 6352 (1967).

1035. P. v. R. Schleyer and C. Maerker, *Pure Appl. Chem.* **67**, 755 (1995).

1036. G. K. S. Prakash, M. Farnia, S. Keyanian, G. A. Olah, H. K. Kuhn, and K. Schaffner, *J. Am. Chem. Soc.* **109**, 911 (1987).

1037. H. Hogeveen and H. C. Volger, *Recl. Trav. Chim. Pays-Bas* **87**, 385, 1042 (1968); 88, 353 (1969).

1038. L. A. Paquette, G. R. Krow, J. M. Bollinger, and G. A. Olah, *J. Am. Chem. Soc.* **90**, 7147 (1968).

1039. H. Hogeveen and P. W. Kwant, *J. Am. Chem. Soc.* **95**, 7315 (1973).

1040. P. W. Kwant, Ph.D. dissertation, Universitat Groningen, (1974).

1041. H. Hogeveen, P. W. Kwant, E. P. Schudde, and P. A. Wade, *J. Am. Chem. Soc.* **96**, 7518 (1974).

1042. T. Laube and C. Lohse, *J. Am. Chem. Soc.* **116**, 9001 (1994).

1043. G. A. Olah, G. K. S. Prakash, T. N. Rawdah, D. Whittaker, and J. C. Rees, *J. Am. Chem. Soc.* **101**, 3935 (1979).

1044. P. Ahlberg, D. L. Harris, and S. Winstein, *J. Am. Chem. Soc.* **92**, 2146 (1970).

1045. P. Ahlberg, in *Stable Carbocation Chemistry*, G. K. S. Prakash, and P. v. R. Schleyer, ed., Wiley, New York, 1997, Chapter 6, p. 197

1046. D. Cook, A. Diaz, J. P. Dirlam, D. L. Harris, M. Sakai, S. Winstein, J. C. Barborak, P. v. R. Schleyer, *Tetrahedron Lett.* 1405 (1971).

1047. P. Svensson, F. Reichel, P. Ahlberg, and D. Cremer, *J. Chem. Soc., Perkin Trans. 2* 1463 (1991).

1048. D. Cremer, P. Svensson, E. Kraka, Z. Konkoli, and P. Ahlberg, *J. Am. Chem. Soc.* **115**, 7457 (1993).

1049. G. K. S. Prakash, in *Stable Carbocation Chemistry*, G. K. S. Prakash and P. v. R. Schleyer, eds., Wiley, New York, 1997, Chapter 4, p. 137

1050. G. K. S. Prakash, V. V. Kishnamurthy, R. Herges, R. Bau, H. Yuan, G. A. Olah, W.-D. Fessner, and H. Prinzbach, *J. Am. Chem. Soc.* **108**, 836(1986); 7764 (1988).

1051. H. Prinzbach, J. Reinbold, M. Bertau, T. Voss, H.-D. Martin, B. Mayer, J. Heinze, D. Neschchadin, G. Gescheidt, G. K. S. Prakash, and G. A. Olah, *Angew. Chem. Int. Ed.* **40**, 911 (2001).

1052. J. Reinbold, M. Bertau, T. Voss, D. Hunkler, L. Knothe, H. Prinzbach, D. Neschchadin, G. Gescheidt, B. Mayer, H.-D. Martin, J. Heinze, G. K. S. Prakash, and G. A. Olah, *Helv. Chim. Acta* **84**, 1518 (2001).

1053. K. Weber, H. Prinzbach, R. Schmidlin, F. Gerson, and G. Gescheidt, *Angew. Chem. Int. Ed. Engl.* **32**, 875 (1993).

1054. A. Weiler, E. Quennet, M. Keller, K. Exner, and H. Prinzbach, *Tetrahedron Lett.* **41**, 4763 (2000).

1055. G. K. S. Prakash, K. Weber, G. A. Olah, H. Prinzbach, M. Wollenweber, M. Etzkorn, T. Voss, and R. Herges, *Chem. Commun.* 1029 (1999).

1056. R. Herges, P. v. R. Schleyer, M. Schindler, and W.-D. Fessner, *J. Am. Chem. Soc.* **113**, 3649 (1991).

1057. H. Carnadi, C. Giordano, R. F. Heldeweg, and H. Hogeveen, *Isr. J. Chem.* **21**, 229 (1981).

1058. S. Winstein, *J. Am. Chem. Soc.* **81**, 6524 (1959).

1059. S. Winstein, J. Sonnenberg, and L. deVries, *J. Am. Chem. Soc.* **81**, 6523 (1959).

1060. S. Winstein and J. Sonnenberg, *J. Am. Chem. Soc.* **83** 3235, 3244 (1961).

1061. S. Winstein, E. C. Friedrich, R. Baker, and Y. I. Liu, *Tetrahedron*, **22**, suppl. 8, 621 (1966).

1062. W. L. Jorgensen, *Tetrahedron Lett.* 3029, 3033 (1976).

1063. S. Masamune, M. Sakai, A. V. Kemp-Jones, and T. Nakashima, *Can. J. Chem.* **52**, 858 (1974).

1064. R. M. Coates and J. L. Kirkpatrick, *J. Am. Chem. Soc.* **90**, 4162 (1968).

1065. R. M. Coates and J. L. Kirkpatrick, *J. Am. Chem. Soc.* **92**, 4883 (1970).

1066. R. M. Coates and E. R. Fretz, *J. Am. Chem. Soc.* **99**, 297 (1977).

1067. H. C. Brown and M. Ravindranathan, *J. Am. Chem. Soc.* **99**, 299 (1977).

1068. D. P. Kelly, J. J. Giansiracusa, D. R. Leslie, I. D. McKern, and G. C. Sinclair, *J. Org. Chem.* **53**, 2497 (1988).

1069. G. A. Olah, G. Rasul, and G. K. S. Prakash, *J. Org. Chem.* **65**, 5956 (2000).

1070. G. K. S. Prakash, G. Rasul, A. K. Yudin, and G. A. Olah, *Gazz. Chim. Ital.* **126**, 1 (1996).

1071. N. H. Werstiuk and Y.-G. Wang, *J. Phys. Chem. A* **107**, 9434 (2003).

1072. K. J. Szabo, E. Kraka, and D. Cremer, *J. Org. Chem.* **61**, 2783 (1996).

1073. M. Bremer, P. v. R. Schleyer, K. Schötz, M. Kausch, and M. Schindler, *Angew. Chem. Int. Ed. Engl.* **26**, 761 (1987).

1074. M. S. W. Chan and D. R. Arnold, *Can. J. Chem.* **75**, 192 (1997).

1075. G. K. S. Prakash, G. Rasul, A. K. Yudin, and R. E. Williams,in *The Borane, Carborane, Carbocation Continuum*, J. Casanova, ed. Wiley, 1998, Chapter 6, in 152–153.
1076. A. A. Fokin, B. Kiran, M. Bremer, X. Yang, H. Jiao, P. v. R. Schleyer and P. R. Schreiner, *Chem. Eur. J.* **6**, 1615 (2000).
1077. R. E. Williams, *Inorg. Chem.* **10**, 210 (1971).
1078. W. D. Stohrer and R. Hoffmann, *J. Am. Chem. Soc.* **94**, 1661 (1972).
1079. S. Yoneda and Z. Yoshida, *Chem. Lett.* 607 (1972).
1080. H. Kollmar, H. O. Smith, and P. v. R. Schleyer, *J. Am. Chem. Soc.* **95**, 5834 (1973).
1081. M. J. S. Dewar and R. C. Haddon, *J. Am. Chem. Soc.* **95**, 5836 (1972).
1082. W. J. Hehre and P. v R. Schleyer, *J. Am. Chem. Soc.* **95**, 5837 (1973).
1083. M. J. S. Dewar and R. C. Haddon, *J. Am. Chem. Soc.* **96**, 255 (1974).
1084. G. A. Olah, G. K. S. Paraksh, G. Liang, P. W. Westerman, K. Kunde, J. Chandrasekhar, and P. v. R. Schleyer, *J. Am. Chem. Soc.* **102**, 4485 (1980).
1085. M. Saunders, R. Berger, A. Jaffe, J. M. McBride, J. O'Neill R. Breslow, J. M. Hoffman, Jr., C. Perchonock, E. Wassermann, R. S. Hutton, and V. J. Kuck, *J. Am. Chem. Soc.* **95**, 3017 (1973).
1086. J. Feng, J. Leszczynski, B. Weiner, and M. C. Zerner, *J. Am. Chem. Soc.* **111**, 4648 (1989).
1087. M. N. Glukhovtsev, R. D. Bach, and S. Laiter, *J. Phys. Chem.* **100**, 10952 (1996).
1088. B. Reindl and P. v. R. Schleyer, *J. Comput. Chem.* **19**, 1402 (1998).
1089. J. B. Lambert, J. Lin, and V. Rassolov, *Angew. Chem. Int. Ed.* **41**, 1429 (2002).
1090. M. Otto, D. Scheschkewitz, T. Kato, M. M. Midland, J. B. Lambert, and G. Bertrand, *Angew. Chem. Int. Ed.* **41**, 2275 (2002).
1091. T. Müller, *Angew. Chem. Int. Ed.* **41**, 2276 (2002).
1092. J. B. Lambert, *Angew. Chem. Int. Ed.* **41**, 2278 (2002).
1093. J. N. Jones, A. H. Cowley, and C. L. B. Macdonald, *Chem. Commun.* 1520 (2002).
1094. G. K. S. Prakash and G. A. Olah, *Chem. Eng. News* **80**(21), 8 (2002).
1095. S. Masamune, M. Sakai, H. Ona, and A. L. Jones, *J. Am. Chem. Soc.* **94**, 8956 (1972).
1096. S. Masamune, M. Sakai, A. V. Kemp-Jones, H. Ona, A. Venot, and T. Nakashima, *Angew. Chem. Int. Ed. Engl.* **12**, 769 (1973).
1097. A. V. Kemp-Jones, N. Nakamura, and S. Masamune, *J. Chem. Soc., Chem. Commun.* 109 (1974).
1098. R. K. Lustgarten, *J. Am. Chem. Soc.* **94**, 7602 (1972).
1099. S. Masamune, R. Vukov, M. J. Bennett, and J. T. Purdham, *J. Am. Chem. Soc.* **94**, 8239 (1972).
1100. P. G. Gassman and X. Creary, *J. Am. Chem. Soc.* **95**, 2729 (1973).
1101. K. Morio and S. Masamune, *Chem. Lett.* 1107 (1974).
1102. G. K. S. Prakash, G. Rasul, and G. A. Olah, *J. Phys. Chem. A* **102**, 2579 (1998).
1103. G. Rasul, G. K. S. Prakash, and G. A. Olah, *J. Phys. Chem. A* **110**, 11320 (2006).
1104. H. Hart and M. Kuzuya, *J. Am. Chem. Soc.* **94**, 8958 (1972); 95, 4096 (1973); 96, 6436 (1974)
1105. H. Hart and M. Kuzuya, *Tetrahedron Lett.* 4123 (1973).
1106. H. Hart and R. Willer, *Tetrahedron Lett.* 4189 (1978).

1107. C. W. Jefford, J. Mareda, J. P. Blaudzun, and U. Burger, *Helv. Chim. Acta* **65**, 2476 (1982).

1108. H. Hogeveen and P. W. Kant, *J. Am. Chem. Soc.* **96**, 2208 (1974) and references cited therein.

1109. H. Hogeveen and E. M. G. A. van Kruchten, *J. Org. Chem.* **46**, 1350 (1981).

1110. C. Giordano, R. F. Heldeweg, and H. Hogeveen, *J. Am. Chem. Soc.* **99**, 5181 (1977).

1111. R. F. Heldeweg,Ph.D. dissertation, Universitat Groningen, (1977).

1112. J. B. Lambert, L. Lin, and S. Keinan, *Org. Biomol. Chem.* **1**, 2559 (2003).

CHAPTER 4

Heterocations in Superacid Systems

4.1. INTRODUCTION

This chapter deals with cations generated in superacid media wherein the positive charge is located on an atom other than carbon. The heteroatoms include oxygen, sulfur, selenium, tellurium, nitrogen, chlorine, bromine, iodine, hydrogen, xenon, and krypton. This discussion is divided for convenience into the following sections: (i) higher valency onium ions such as oxonium, sulfonium, selenonium, telluronium, halonium, diazonium, and other species; (ii) enium ions of boron (borenium) and aluminum (alumenium), silicon (silicenium), germanium (germenium), tin (stannylium), and lead (plumbylium); nitrogen (nitronium), phosphorus (phosphenium), arsenic (arsenium), and antimony (stibenium); (iii) homo- and heteropolyatomic cations of halogens (interhalogen cations), polyhomoatomic cations of oxygen, sulfur, selenium, and tellurium, and other polyatomic and polyheteroatomic cations; (iv) cations of group 6–12 elements; and (v) miscellaneous cations of hydrogen, xenon, and krypton.

4.2. ONIUM IONS

4.2.1. Oxonium Ions

The synthesis, structural characteristics, and chemistry of oxonium ions have been extensively investigated and reviewed.[1]

4.2.1.1. Hydronium Ion (H_3O^+). The hydronium ion (H_3O^+, **1**) is the parent of saturated oxonium ions. It was first postulated in 1907[2] and gained wider acceptance with the acid–base theory of Brønsted[3] and Lowry.[4] The lifetime of **1** in aqueous solution has been estimated to be about 10 times longer than the time scale of molecular vibrations.[5,6] The hydronium ion **1** has been studied in the gas phase as well as in solution (IR,[7–10] Raman,[11,12] NMR,[13–16] MS,[17,18] and neutron diffraction[19,20]).

Superacid Chemistry, Second Edition, George A. Olah, G. K. Surya Prakash, Árpád Molnár, and Jean Sommer

In fact, Christe et al.[21,22] have isolated hydronium salts with a variety of counterions such as SbF_6^-, AsF_6^-, and BF_4^-. Complete vibrational[7–10] and neutron diffraction studies[19,20] support the pyramidal nature of the hydronium ion **1**. Extensive evidence for the long-lived hydronium ion has been obtained in superacid solution.[13,14,16] In the mass spectrum, evidence has also been obtained for water-solvated hydronium ions such as $H_5O_2^+$ and $H_7O_3^+$. Isolation and X-ray crystal structure characterization of $H_3O^+SbF_6^-$,[22,23] $H_3O^+Sb_2F_{11}^-$,[24] $H_3O^+SbCl_6^-$ stabilized by acetonitrile,[25] and $H_5O_2^+SbX_6^-$ salts (X = F, Cl)[26,27] have been reported.

The ^{1}H NMR spectrum of H_3O^+ (**1**) in superacids such as $HSO_3F–SbF_5$ (Magic Acid) and HF–SbF_5 is observed at δ ^{1}H 11.0 (from trimethylsilane).[13,14,16] ^{17}O NMR spectroscopic studies[28,29] in HF–SbF_5 with ^{17}O-enriched hydronium ion have indicated strong ^{17}O–^{1}H coupling (Figure 4.1). In the proton-coupled spectrum, a quartet is observed at δ^{17}O 9±0.2 (with reference to SO_2 at 505 ppm) with $J_{^{17}O-^1H} = 106 \pm 1.5$ Hz.

Mateescu and Benedikt[28] suggested that **1** is planar and not pyramidal, on the basis of a 33% increase in ^{17}O–^{1}H coupling constant upon going from H_2O to H_3O^+. They assumed a linear relationship between $J_{^{17}O-^1H}$ and the hybridization state of oxygen. Symons[30] subsequently showed that the assumed sp^3 hybridization for H_2O by Mateescu and Benedikt was not true (H_2O has a bond angle of 104.5°[31]). Redoing the calculation, Symons showed that the hydronium ion **1** is pyramidal with an H–O–H bond angle of 111.3°. Similar conclusions have been reached on the basis of quantum mechanical calculations.[32]

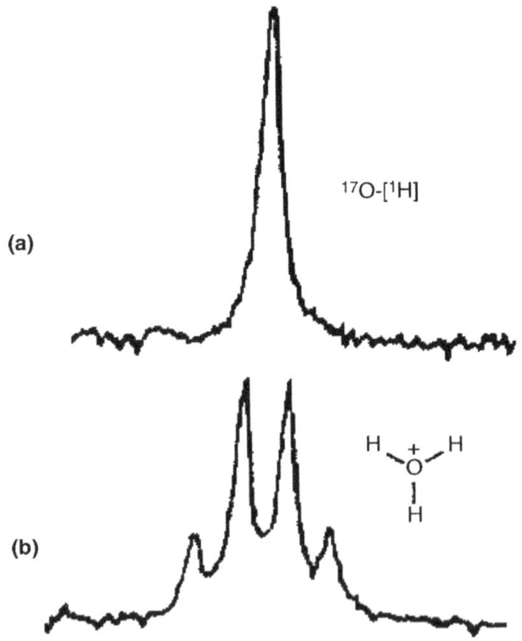

Figure 4.1. ^{17}O NMR spectrum of H_3O^+ (1) in $HSO_3F–SbF_5–SO_2$ at −20°C. (***a***) Proton decoupled, (***b***) proton coupled.

The hydronium ion forms clusters with electron pair donor molecules such as water[33] and crown ethers.[34–36] Since the pioneering work of Kebarle[37] on $H_3O^+ \cdot (H_2O)_n$ clusters by mass spectrometry, their structures and energetics have been thoroughly studied.[33,38,39] Two main structural models have emerged. Eigen proposed[40] the formation of a $H_9O_4^+$ complex in which the H_3O^+ core is strongly hydrogen-bonded to three H_2O molecules. Zundel,[41] in turn, suggested the existence of a $H_5O_2^+$ complex with the proton being shared between two H_2O molecules. $H_3O^+ \cdot (H_2O)_{20}$ is considered a magic number cluster due to its enhanced stability originating from the closed cage structure. This cluster has been shown to form a pentagonal dodecahedral cage with 11 non-hydrogen-bonded hydrogens extending out from the ion.[42,43] Recent theoretical studies such as Car–Parrinello molecular dynamics indicate[44] that Eigen and Zundel structures may be considered as limiting, equilibrating cases. By the use of a novel approach (multiconfigurational time-dependent Hartree), Meyer and co-workers[45] were able to simulate the experimental gas-phase spectrum of $H_5O_2^+$ with surprising accuracy.

The oxonium dication H_4O^{2+}, which may play a role in superacid-catalyzed H–D exchange reactions via an associative mechanism, has been studied theoretically. The tetrahedral symmetry structure was found to be a minimum (HF/6-31G* level).[46] Treatment with GAUSSIAN-2 method[47] gives a dissociation energy of -61.9 kcal mol^{-1} and a kinetic barrier of 38.2 kcal mol^{-1} for deprotonation. Practically the same values (-61.3 and 38.1 kcal mol^{-1}) have been computed at the QCISD(T)/6-311G (2df,2p)/MP2(full)/TZP + ZPE/MP2(full)/6-31G** level.[48]

4.2.1.2. Primary Oxonium Ions [ROH$_2^+$]. It has been shown[49] that methyl and ethyl alcohol in sulfuric acid give stable solutions of the corresponding alkyl hydrogen sulfates **2** [Eq. (4.1)]. Many other alcohols show similar initial behavior, but the solutions are not stable at room temperature.

$$\mathrm{R{-}OH} + 2\mathrm{H_2SO_4} \longrightarrow \underset{\mathbf{2}}{\mathrm{R{-}OSO_3H}} + \underset{\mathbf{1}}{\mathrm{H_3O^+}} + \mathrm{HSO_4^-} \qquad (4.1)$$

The first direct NMR spectroscopic evidence for the existence of primary alkyloxonium ions (protonated alcohols) in superacid solutions was found in 1961 by MacLean and Mackor.[50] The NMR spectrum of ethanol in HF–BF_3 solution at -70°C gave a well-resolved triplet at about δ^1H 9.90 for the protons on oxygen coupled to the methylene protons. In HSO_3F this fine structure is not observed, even at -95°C, due to the fast proton exchange.[51]

The NMR spectra of a series of aliphatic alcohols have been investigated in the stronger acid system HSO_3F–SbF_5 using sulfur dioxide as diluent[52,53] [Eq. (4.2)]. Methyl, ethyl, *n*-propyl, isopropyl, *n*-butyl, isobutyl, *sec*-butyl, *n*-amyl, neopentyl, *n*-hexyl, and neohexyl alcohol all give well-resolved NMR spectra at -60°C, under these conditions (Figure 4.2).

$$\mathrm{R{-}OH} \xrightarrow[-60^\circ\mathrm{C}]{\mathrm{HSO_3F{-}SbF_5{-}SO_2}} \mathrm{R{-}\overset{+}{O}H_2} \qquad (4.2)$$

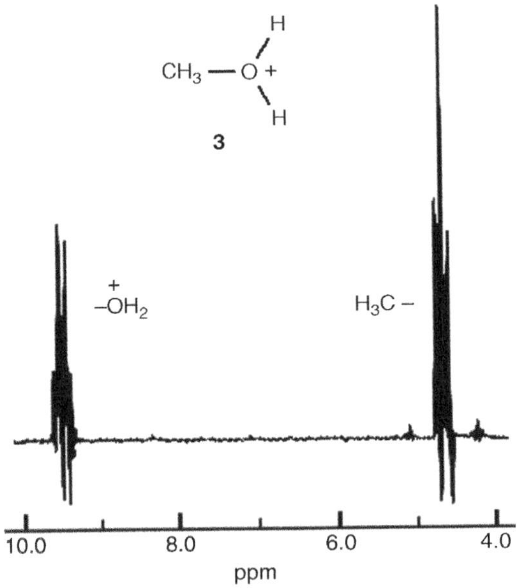

Figure 4.2. ^1H NMR spectrum of protonated methyl alcohol in $HSO_3F–SbF_5–SO_2$ at $-60°C$.

The strength of the $HSO_3F–SbF_5$ acid system is reflected by the fact that even at 25°C, solutions of primary alcohols show fine structure for the proton on oxygen (Figure 4.2). This indicates that at this relatively high temperature the exchange rate is still slow on the NMR time scale. In ^1H NMR, the OH_2^+ protons of the primary alcohols appear at lower field (δ^1H 9.3–9.5) than the isomeric secondary alcohols (δ^1H 9.1). This is due to a different charge distribution, as confirmed by the C(1) protons of the primary alcohols appearing at higher field (δ^1H 5.0–4.7) compared with the C(1) methine proton of the secondary alcohols (δ^1H 5.4 and 5.5). The ^{17}O chemical shift of protonated methyl alcohol **3** is deshielded by approximately 25 ppm from the neutral methyl alcohol.[29] The $^{17}O–^1H$ coupling constant of 107.6 Hz is similar in magnitude to that of the parent oxonium ion H_3O^+ **1**.

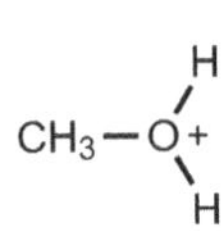

3

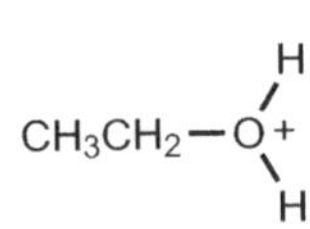

4

Minkwitz and Schneider[54] have been able to obtain protonated methanol in the form of hexafluorometalate salts ($MeOH_2^+MF_6^-$, M = As, Sb), and these were characterized by X-ray diffraction. Protonated propanol and proton-bound dimers of methanol, ethanol, and propanol with the proton shared between two alcohol molecules (**5**) have also been characterized by infrared multiphoton dissociation spectroscopy.[55] The synthesis and X-ray crystal structure of the oxonium salt $[H(OEt_2)_2]^+(C_6F_5)_4B^-$ have recently been reported.[56] Unfortunately, the proton could not be located. Thin molecular films of ethanol and *tert*-butyl alcohol on Ru(001) surface have been prepared and reacted with HBr.[57] Only protonated ethanol $EtOH_2^+$ was detected on the ethanol surface by Cs^+ reactive ion scattering and low-energy sputtering. In contrast, on the *tert*-butyl alcohol surface, both protonated alcohol $Me_3COH_2^+$ and *tert*-butyl cation, Me_3C^+ were formed.

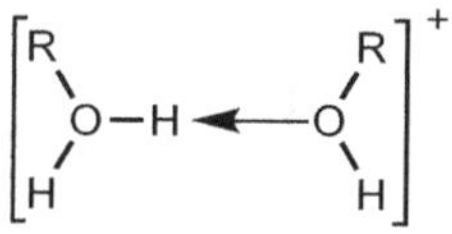

R = Me, Et, Pr

5

Aliphatic glycols in $HSO_3F–SbF_5–SO_2$ solution give diprotonated species at low temperatures.[58] In diprotonated diols, the protons on oxygen are found at lower fields than in protonated alcohols reflecting the presence of two positive charges. This is especially true for ethylene glycol (δ^1H 11.2), where the positive charges are adjacent. As the separation of the positive charges becomes greater with increasing chain length, the chemical shift of the protons on oxygen of protonated diols approaches that of protonated alcohols.

The reactivity of protonated alcohols[53] and protonated diols[58] in strong acids has been studied by NMR spectroscopy. Protonated methyl alcohol shows surprising stability in $HSO_3F–SbF_5$ and can be heated to 50°C without significant decomposition, although surprising condensation to *tert*-butyl cation was occasionally observed. Protonated ethyl alcohol **4** is somewhat less stable and begins to decompose at 30°C. The cleavage of protonated *n*-propyl alcohol **6** has been followed in the temperature range 5–25°C, giving a mixture of *tert*-butyl and *tert*-hexyl cations [Eq. (4.3)].

$$\underset{\mathbf{6}}{CH_3CH_3CH_2\overset{+}{O}H_2} \xrightarrow[5\text{–}25°C]{HSO_3F\text{–}SbF_5} (CH_3)_3C^+ + (CH_3)_2CH\overset{+}{C}(CH_3)_2 + H_3O^+ \qquad (4.3)$$

Higher protonated alcohols cleave to stable tertiary alkyl cations. For protonated primary and secondary alcohols, the initially formed primary and secondary carbocations rapidly rearrange to the more stable tertiary carbenium ions under the conditions of the reaction. For example, protonated *n*-butyl alcohol **7** cleaves to *n*-butyl cation which rapidly rearranges to *tert*-butyl cation ($k_2 >> k_1$) [Eq. (4.4)].

$$\underset{\mathbf{7}}{CH_3CH_2CH_2CH_2\overset{+}{O}H_2} \xrightarrow[-H_2O]{k_1} [CH_3CH_2CH_2CH_2]^+ \xrightarrow{k_2} (CH_3)_3C^+ \qquad (4.4)$$

The cleavage of carbocations, shown to be first order, is enhanced by branching of the chain: Protonated 1-pentanol is stable up to 0°C, isopentyl alcohol is stable up to

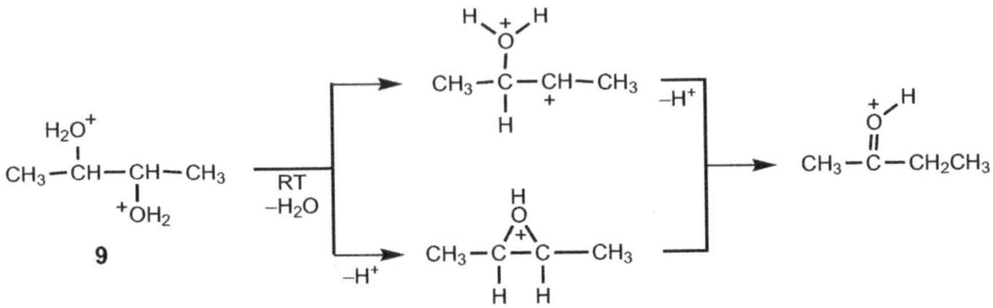

Scheme 4.1

$-30°C$, and neopentyl alcohol cleaves at $-50°C$. The stability of the protonated primary alcohols also decreases as the chain length is increased.

When the $HSO_3F–SbF_5–SO_2$ solutions of diprotonated glycols are allowed to warm up,[58] pinacolone rearrangements, formation of allylic carbenium ions, and cyclization reactions of diprotonated glycols can be directly observed by NMR spectroscopy. Diprotonated ethylene glycol **8** rearranges to protonated acetaldehyde in about 24 h at room temperature [Eq. (4.5)]. Protonated 1,2-propanediol undergoes a pinacolone rearrangement to protonated propionaldehyde, probably through the initial cleavage of water from the secondary position. Diprotonated 2,3-butanediol **9** rearranges to protonated ethyl methyl ketone either through direct hydride shift or through a bridged intermediate (Scheme 4.1).

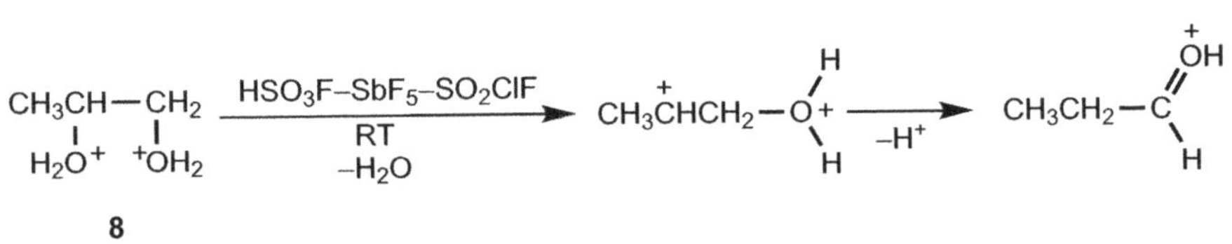

(4.5)

Diprotonated 2,4-pentanediol **10** loses water and rearranges to form 1,3-dimethyl allyl cation [Eq. (4.6)]. Diprotonated 2,5-hexanediol **11**, above $-30°C$, rearranges to a mixture of protonated *cis*- and *trans*-2,5-dimethyltetrahydrofurans [Eq. (4.7)]. This would seem to indicate that there is a significant amount of the monoprotonated form present or that the carbocation formed can easily lose a proton before ring formation occurs.

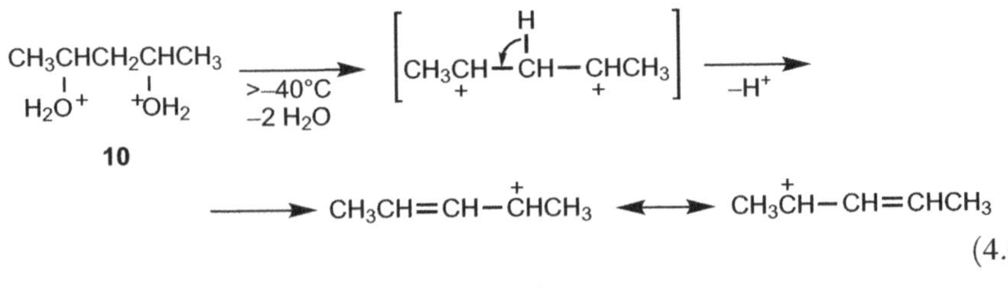

(4.6)

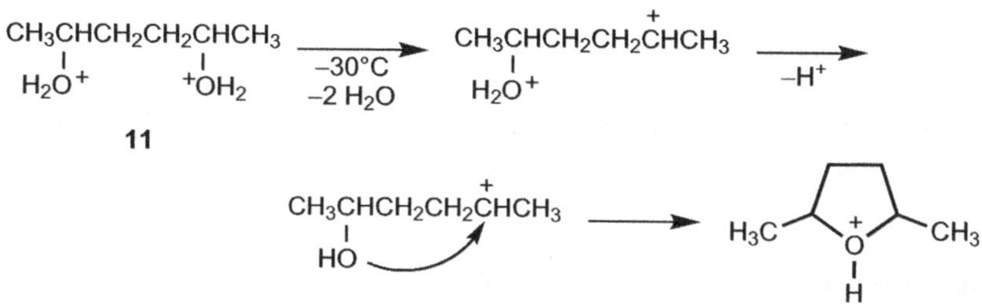

(4.7)

According to protonation thermochemistry of simple α,ω-diols (1,2-ethanediol, 1,3-propanediol, 1,4-butanediol), the diols show enhanced proton affinities in the gas phase compared with primary alcohols attributed to the formation of a strong intramolecular hydrogen bond.[59,60]

In the case of phenols, either *C*- or *O*-protonated ion **12** or **13** are formed, depending upon the acidity of the medium. In relatively weak acid medium such as H_2SO_4[61] only *O*-protonation is observed. However, in HSO_3F only the *C*-protonated species is formed.[62] In a much stronger acid system such as $HSO_3F–SbF_5$ (Magic Acid), both *C*- or *O*-protonated dication **14** can be observed. Both the *C*- and *O*-protonated ions were shown to exist in the gas phase by IR photodissociation spectroscopy.[63]

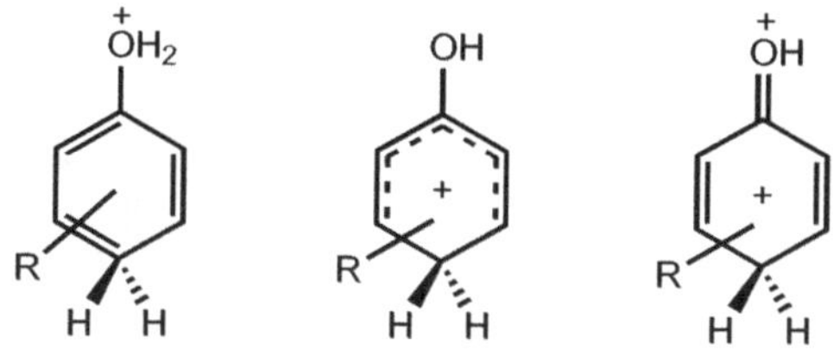

In addition to the direct protonation of alcohols or phenols, haloalkyloxonium ions can be generated from carbonyl compounds. Olah et al.[64] prepared halomethyloxonium ions $XCH_2OH_2^+$ ($X = F, Cl$) by the protonation of formaldehyde with $HF–SbF_5$ or using a mixture of HCl and $HSO_3F–SbF_5–SO_2$. Minkwitz and Reinemann[65] reacted hexafluoroacetone with $HF–SbF_5$ to prepare an analogous primary oxonium ion [Eq. (4.8)].

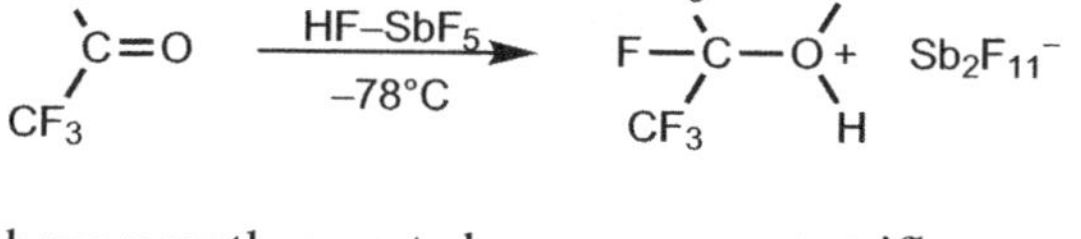

(4.8)

Christe et al.[66] have recently reported an easy access to trifluoromethanol (CF_3OH) by reacting carbonyl dilfuoride with anhydrous HF. In fact, trifluoromethanol exists in equilibrium with CF_2O and HF. It was found during their study that CF_3OH is not protonated by HF; however, it could be converted to trifluoromethyloxonium MF_6^- salts ($M = Sb$, As).

The synthesis under stable ion conditions and characterization by ^{1}H, ^{13}C, and ^{29}Si NMR spectroscopy of the [(trimethylsilyl)methyl]oxonium ion **15** have been reported by Olah et al.[67] Protonation of the corresponding alcohol with Magic Acid gives ion **15**, which is stable and can be isolated at room temperature [Eq. (4.9)].

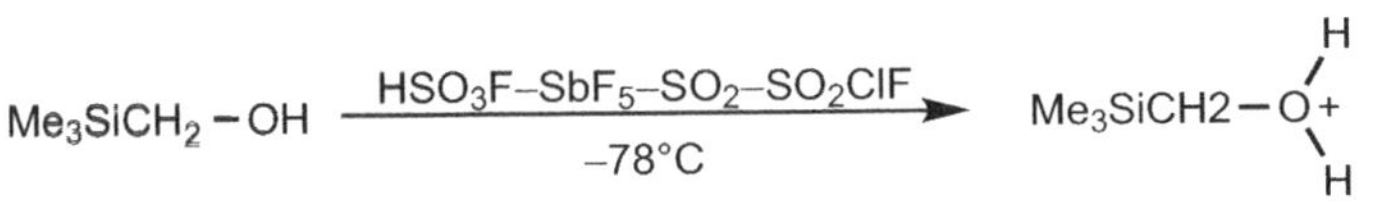

15

(4.9)

Silyloxonium ions (protonated silanols) **16** have also been prepared,[67] and the crystal structure of the *tert*-butyl derivative **16c** as the $Br_6CB_{11}H_6^-$ salt has been reported.[68] The most notable features of ion **16c** are the flattening of the silicon center (average C–Si–C angle = 116.0°) and the long Si–O bond (1.779 Å). The Si–O bond distance of the corresponding methyl derivative **16a** (1.929 Å) calculated by Olah et al.[69] (B3LYP/6-31G* level) is only 0.1 Å longer than that found in **16c**. The IGLO II′ calculated ^{29}Si NMR chemical shift of **16a** is considerably deshielded from that of the neutral precursor Me_3SiOH ($\delta^{29}Si$ 101.9 versus 15.1). In contrast, the ^{29}Si NMR chemical resonance of *tert*-butyl derivative **16c** is even shielded compared with that of **16a** ($\delta^{29}Si$ 46.7).[68]

a R = Me
b R = Et
c R = *tert*-Bu

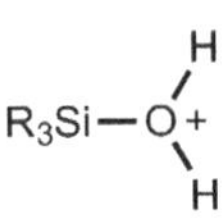

16

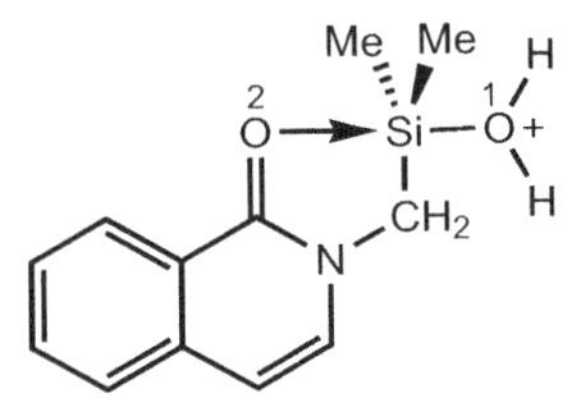

17

In protonated (amidomethyl)dimethylsilanols, such as **17**, the Si–O(1) distance is even longer (1.9114 Å) due to the additional Si–O(2) interaction.[70] According to AIM calculations[71] the O–Si–O moiety is a 3*c*–4*e* bond and the positive charge is largely localized on silicon. The Si–O(1) bond length (1.9604 Å) is similar to that of Si–O(2) and the sum of the three C–Si–C bond angles is 359.9°. These data indicate that the structure may be described as an intermediate of aqueous hydrolysis of silanes.

Monofluorooxonium hexafluorometalates have been synthesized by Minkwitz and co-workers[72,73] by fluorinating H_2O with $XeF^+MF_6^-$ [Eq. (4.10)]. The salts can be stored at room temperature for 2 h without decomposition.

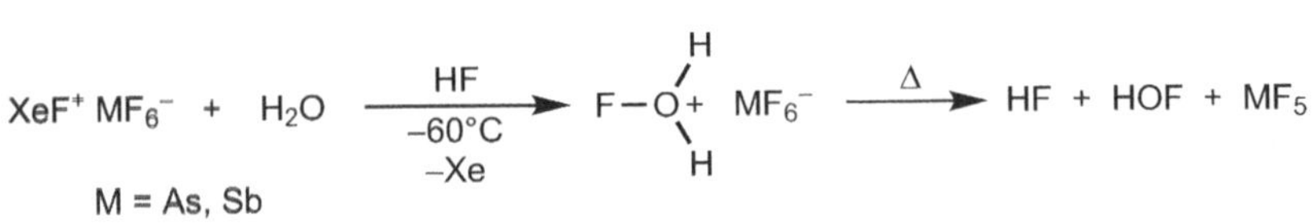

(4.10)

Superelectrophilic fluorooxonium dications FOH_3^{2+} and $F_2OH_2^{2+}$ derived from HOF and F_2O, respectively, have been studied theoretically [QCISD(T)/6-311G**].[74] Both the *O,O*-diprotonated structure **18** and the *O,F*-diprotonated structure **19** were found to be energy minima with the oxonium dication **18** being 14.9 kcal mol^{-1} less stable than the oxonium–fluoronium dication **19**. The *O,O*-diprotonated dication **20**, in turn, is the only minimum on the potential energy surface for the $F_2OH_2^{2+}$ system. All of these species have substantial kinetic barriers for deprotonation to the corresponding monocations.

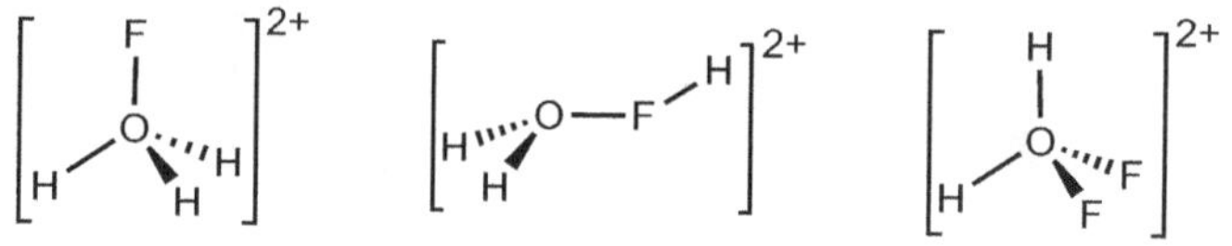

C_{3v} **18** C_s **19** C_{2v} **20**

4.2.1.3. Secondary Oxonium Ions [RR′OH$^+$].

^{1}H NMR and IR investigation[75] of protonated ether salts (hexachloroantimonates) in dichloromethane solutions showed the formation of both (a) dialkyloxonium ions in which the proton is bound to only one oxygen (**21**) and (b) bidentate complexes in which the proton is shared between two ether molecules (**22**). Structural analysis of such a bidentate complex of diethyl ether with a complex anion shows a broad H^+ resonance at δ^1H 16.3 and unequal O–H bond distances (1.39 and 1.11 Å).[76]

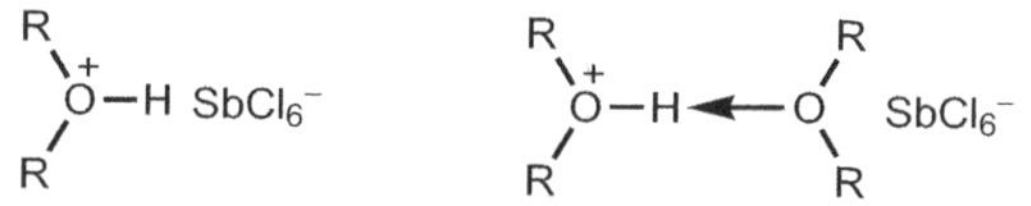

21 **22**

In the ^{1}H NMR, the proton on oxygen showed up as a singlet around δ^1H 7–9 with no coupling with adjacent alkyl protons, indicating rapid proton exchange under the reaction conditions. Investigations using the superacid system HSO_3F–SbF_5–SO_2 at low temperatures lead to comparable values for the chemical shift to the proton on oxygen for a variety of aliphatic ethers at δ^1H 7.88–9.03. Because of the stronger acid system and the low temperature, however, the exchange rate is slowed down sufficiently so that the expected splitting of the proton resonance on oxygen by the adjacent hydrogens is observed[77] (Figure 4.3). Diprotonated dimethyl ether $Me_2OH_2^{2+}$ has been computed (MP2/6-31G*/MP2/6-31G* level) and found to be a stable *O,O*-diprotonated isomer.[47]

The cleavage of protonated ethers in strong acid systems has not been studied extensively. Kinetic investigation of the cleavage of ethers in 99.6% sulfuric acid using cryoscopic methods showed that cleavage takes place by unimolecular fission of the conjugated acid of the ether to form the most stable carbocation and an alcohol. The carbocation and alcohol formed rapidly unite with the hydrogen sulfate anion.

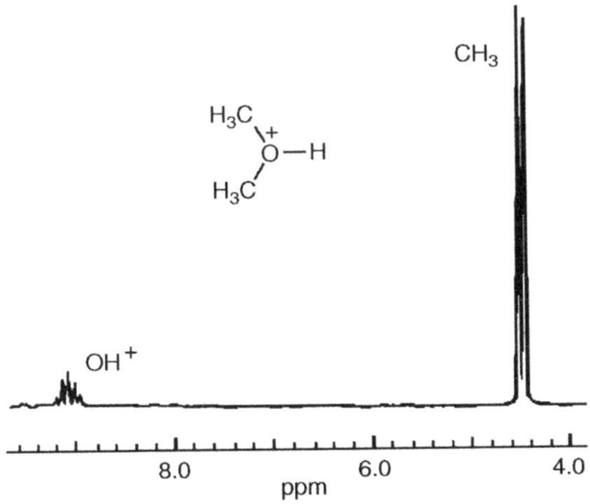

Figure 4.3. ¹H NMR spectrum of protonated dimethyl ether in $HSO_3F–SbF_5–SO_2$ solution.

The overall rate in sulfuric acid, however, appears to be dependent upon the concentration of sulfur trioxide. To rationalize these observations, the mechanism outlined in Scheme 4.2 has been proposed.[78]

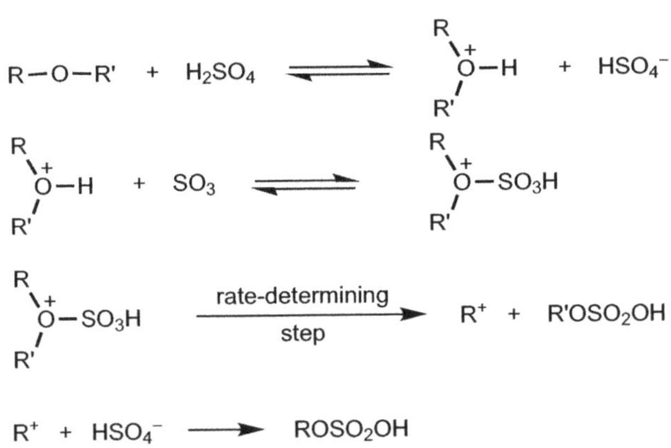

Scheme 4.2

In a solution of $HSO_3F–SbF_5$, protonated *n*-butyl methyl ether **23** does not show any significant change, neither cleavage nor rapid exchange, as indicated by the NMR spectrum up to +40°C. Above +40°C, however, it cleaves and a sharp singlet appears at δ¹H 4.0. This can be attributed to the rearrangement of the *n*-butyl cation, formed in the cleavage, to *tert*-butyl cation [Eq. (4.11)].

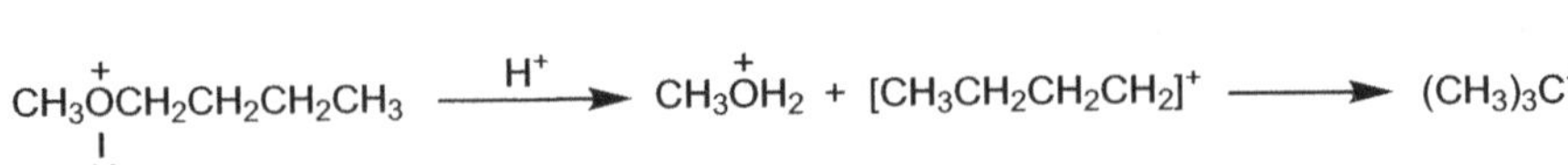

23

(4.11)

Ethers in which one of the groups is secondary begin to show appreciable cleavage at −30°C. Protonated *sec*-butyl methyl ether **24** cleaves cleanly at −30°C to protonated methanol and *tert*-butyl cation [Eq. (4.12)]. Ethers in which one of the alkyl groups is tertiary cleave rapidly even at −70°C.

$$\underset{\mathbf{24}}{CH_3CH_2CH(CH_3)\overset{+}{O}(H)CH_3} \xrightarrow[k_1]{H^+} CH_3\overset{+}{O}H_2 + [CH_3CH_2\overset{+}{C}HCH_3] \xrightarrow{k_2} (CH_3)_3C^+ \qquad (4.12)$$

It was found possible to measure the kinetics of cleavage of protonated *sec*-butyl methyl ether **24** by following the disappearance of the methoxy doublet in the NMR spectrum with simultaneous formation of protonated methanol and *tert*-butyl cation. The cleavage shows pseudo-first-order kinetics. Presumably, the rate-determining step is the formation of *sec*-butyl cation followed by rapid rearrangement to the more stable *tert*-butyl cation ($k_2 \gg k_1$).

Olah and Szilagyi[79] have studied the protonation of three-, four-, five-, and six-membered cyclic ethers in HSO_3F–SbF_5–SO_2 solution at −60 and −78°C. The ^{1}H NMR spectra of ethylene oxide and propylene oxide showed a broad signal, which could be due to protonated oxirane moiety. No evidence of any long-lived intermediate was found for more highly substituted oxiranes at the temperature range studied. This, however, is not unexpected, considering the strained structures and that ring opening leads to more stable tertiary carbonium ions. In contrast, the homologous cyclic ethers of larger rings gave well-resolved spectra with negligible exchange rates.

Protonation and cleavage of cyclic ethers, particularly that of oxiranes, have been extensively studied by calculations.[80–84] Protonated ethylene oxide and propylene oxide were found to be energy minima on the potential energy surface, being more stable than the corresponding carbenium ions. In contrast, benzene oxide and styrene oxide give carbenium ions as energy minima upon protonation. Protonation of propylene oxide can yield both *syn* and *anti* oxonium ions with the *syn* stereoisomer having marginally higher stability at all levels of theory (0.2 kcal mol^{-1} at the MP2/6-31G* level). High-level *ab initio* calculations show that a concerted, asynchronous pathway exists between protonated propylene oxide and ring-opened product protonated propanal.[81] The reaction shows a 20:1 preference for migration of the proton *trans* to the methyl group. AIM theory employed for a series of protonated oxacycloalkanes $(CH_2)_nOH^+$ ($n = 2$–6) indicates that the positive charge is partially concentrated on hydrogen. This led the authors to suggest that protonated ethers are better described by an O–H$^+$ structure.[83]

Carlier et al.[85] have carried out a comparative study of protonated 2,2-dimethyloxirane using a range of density functional methods. It was found that relative to CCSD, all methods except MP2 overestimated the C(2)–O bond length of protonated 2,2-dimethyloxirane by about 0.2 Å. The difficulty lies in the extremely weak C(2)–O bond, which is shown by the highly asymmetric charge distribution between the two ring carbons [Mulliken charges of C(2) and C(3) are +0.419 and −0.460, respectively]. Data for protonated oxiranes with more symmetrical charge distributions and cyclic homologues with less ring strain (protonated 2,2-dimethyloxetane and 2,2-dimethyloxolane) can be calculated by B3LYP with greater accuracy (deviations in bond lengths are +0.073 Å and +0.061 Å, respectively).

Olah et al.[67] also prepared secondary [(trimethylsilyl)methyl]oxonium and trimethylsilyloxonium ions (**25** and **26**). ^{13}C NMR characterization of the ions showed that the α- or β-trialkylsilyl substituents induce little charge delocalization. The ^{29}Si NMR chemical shift of the bistrimethylsilyl derivative **26** (R = Me_3Si) calculated by Olah, Rasul, and Prakash[69] (IGLO II′) is considerably deshielded from that of the neutral precursor Me_3SiOH (δ^{29}Si 67.6 vs. 15.1). Secondary oxonium ions with halogen heteroatom ligands (**27**) were synthesized by Minkwitz and Konikowski.[86] Proton resonances were found at δ^1H 5.3 ppm (X = F) and 4.1 ppm (X = Cl), respectively.

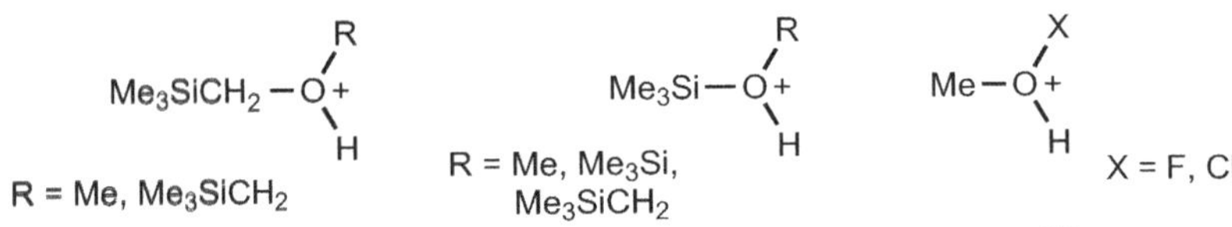

25 **26** **27**

Alkylaryl ethers and diaryl ethers undergo protonation on either oxygen or carbon, depending upon the acidity of the medium[87] (Scheme 4.3). Both the *C*-protonated species **28** and the *O*-protonated species **29** have been observed. The evidence mainly comes from NMR and UV data. Sommer et al.[88] have used *para*-anisaldehyde as an indicator in acidity measurements in the superacidity range. The barrier of rotation around the C_{ipso}–O bond upon *O*-protonation has been used as a criterion in such studies. The torsional barrier around the phenyl–alkoxy bond in the *C*-protonated forms of aromatic ethers has also been measured by spin-saturation transfer measurements.[89]

4.2.1.4. Tertiary Oxonium Ions. The synthesis, structural characterization, and chemistry of compounds containing trivalent, positively charged oxygen (R_3O^+) have

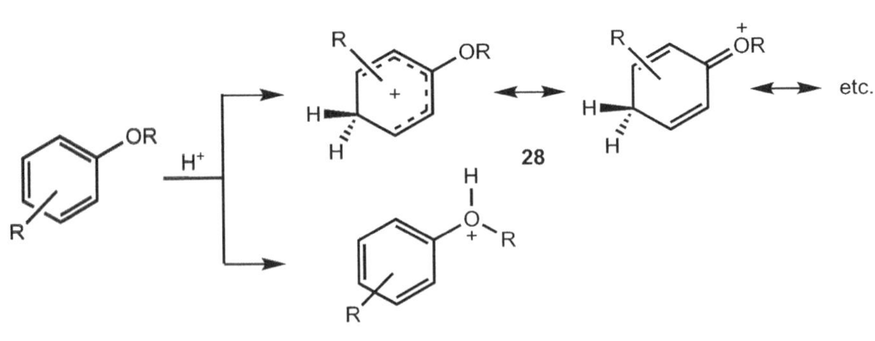

Scheme 4.3

been extensively investigated and reviewed[1,90] since Meerwein pioneering work on tertiary oxonium ions in 1937.[91]

Trialkyloxonium salts are generally prepared by the alkylation of dialkyl ethers.[1,90] Alkyl halides can be used as alkylating agents in the presence of strong halide acceptors such as BF_3, PF_5, $SbCl_3$, and SbF_5 [Eq. (4.13)].[91]

$$R_2O + SbCl_5 + R{-}Cl \longrightarrow R_3O^+ SbCl_6^- \qquad (4.13)$$

A metathetic silver salt reaction[92] has been employed to prepare bicyclo[2.2.2]octyl-1-oxonium hexafluoroantimonate **30** [Eq. (4.14)].

(4.14)

30

The analogous reaction, the alkylation of methyl (trimethylsilyl)methyl ether with alkyl iodide/$AgBF_4$, and the reaction in Eq. (4.15) have been employed to prepare tertiary [(trimethylsilyl)methyl]oxonium salts.

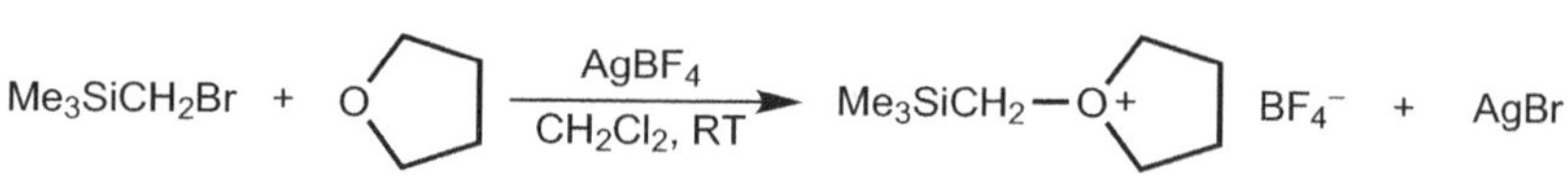

(4.15)

Trialkyloxonium salts can also be prepared by the reaction of secondary oxonium ion salts with diazoalkanes[93] [Eq. (4.16)].

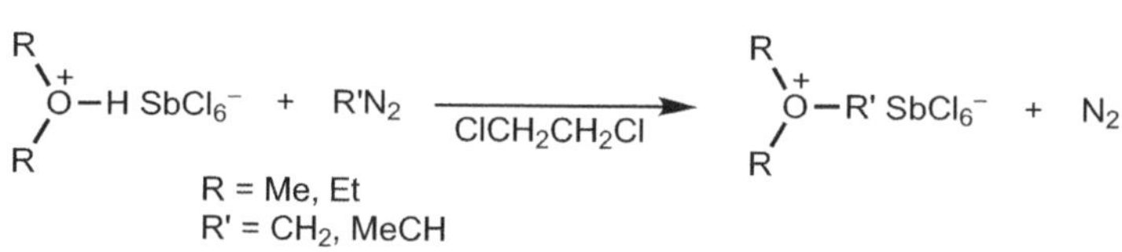

(4.16)

Disproportionation of dialkyl ether–PF_5 adducts also give trialkyloxonium ions[94] [Eq. (4.17)].

$$3R_2O \rightarrow PF_5 \xrightarrow{\text{RT}} 2R_3O^+ PF_6^- + F_3P{=}O \qquad (4.17)$$

Another important synthesis method is transalkylation reactions with other oxonium ions.[95] However, these reactions are reversible to a certain extent. The equilibrium between two oxonium salts and the corresponding ethers in solution is established by both differences in solubility and stability of the oxonium salts (Scheme 4.4). By an exchange reaction of this type, the trimethyloxonium salt **31** can be

$$R_3O^+ BF_4^- + R'_2O \rightleftharpoons R_2O + R'_2RO^+ BF_4^-$$

$$R_2O + R'_2RO^+ BF_4^- \rightleftharpoons R_2R'O^+ BF_4^- + ROR'$$

$$R_2R'O^+ BF_4^- + R'_2O \rightleftharpoons R'_3O^+ BF_4^- + ROR'$$

Scheme 4.4

prepared in excellent yield (86–94%) from the readily available triethyloxonium salt[95,96] [Eq. (4.18)].

$$2(CH_3CH_2)_3O^+ BF_4^- + 3(CH_3)_2O \longrightarrow \underset{\mathbf{31}}{2(CH_3)_3O^+ BF_4^-} + 3(CH_3CH_2)_2O \quad (4.18)$$

In many cases, the main step in the syntheses of trialkyloxonium salts is the alkylation of a dialkyl ether with a reactive intermediate oxonium ion formed in situ. Thus, the most widely used method for preparing trialkyloxonium tetrafluoroborates by the reaction of epichlorohydrin and BF_3 is based on the intermediacy of the inner oxonium salt **32**[91,95,97] [Eq. (4.19)].

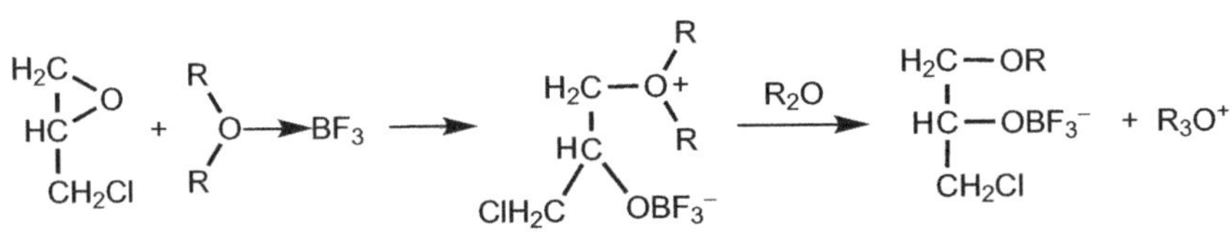

32

(4.19)

This method has been used by Minkwitz, Christe, and co-workers[98] to synthesize perfluoroalkyl-substituted trialkyloxonium salts **33**. X-ray crystal structure analysis of **33** ($R_F = CF_3$) shows that replacement of a methyl group in trimethyloxonium cation by the bulkier and more electronegative CF_3 group results in increases in both the Me–O bond lengths and the C–O–C bond angles (by 0.037 Å and 4.5°, respectively). Whereas pronounced deshielding of both the α-methyl protons and α-methyl carbons are observed in trialkyloxonium ions compared with the parent ethers, the shielding effect of the perfluoroalkyl substituent is small.

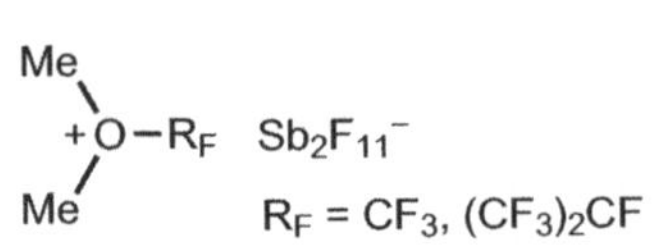

33

The method has been extended by the use of reactive intermediate oxonium ions such as sulfonyl cation–ether adducts, dialkyloxycarbenium salts with ether, and so on. The more reactive dialkyl halonium ions also readily alkylate dialkyl ethers to the corresponding oxonium ions.[99]

Olah and co-workers[47,100] have recently reported experimental and theoretical studies of the trimethyloxonium ion Me_3O^+. An early study[29] resulting in inconclusive results has been repeated and the ^{17}O NMR spectra of $Me_3O^+PF_6^-$ salt was observed.[100] A sharp peak at $\delta^{17}O - 32.1$ was recorded, which is deshielded by 20.4 ppm compared to the experimental ^{17}O NMR shift of the parent Me_2O ($\delta^{17}O$ -52.5), which agrees well with the deshielding effect of 21 ppm calculated by the GIAO-MP2 method. Although not yet observed experimentally, the protonated trimethyloxonium dication Me_3OH^{2+} is a stable species both thermodynamically and kinetically (MP2/6-31G*//HF/6-31G* level).[47,100] The tetramethyloxonium dication Me_4O^{2+} was found to be a stable minimum (HF/6-31G*).[47]

In contrast to trialkyloxonium salts, the triphenyloxonium salt **34** can be obtained by the reaction of benzenediazonium tetrafluoroborate with diphenyl ether in poor (2%) yield [Eq. (4.20)] and is extremely inert toward nucleophiles. This fact is clearly demonstrated by the remarkable stability of triphenyloxonium halides (Cl^-, Br^-, I^-).[101–103]

$$Ph_2O + PhN_2^+ BF_4^- \xrightarrow[-N_2]{} \underset{\mathbf{34}}{Ph_3O^+ BF_4^-} \tag{4.20}$$

The surprising stability of triaryloxonium ions has been demonstrated by the fact that reduction of the tris(4-nitrophenyl)oxonium tetrafluoroborate **35** can be performed to give the tris(4-aminophenyl)oxonium salt **36** without cleavage of aryl oxygen bonds [Eq. (4.21)]. Diazotation of this amino derivative and reaction with iodide leads to the tris(*para*-iodophenyl)oxonium salt **37** in excellent yields.[104]

$$\underset{\mathbf{35}}{(4\text{-}NO_2C_6H_4)_3O^+} \xrightarrow[HCl]{Sn} \underset{\mathbf{36}}{\underset{78\%}{(4\text{-}NH_2C_6H_4)_3O^+}} \xrightarrow[2.\ NaI]{1.\ NaNO_2,\ H_2SO_4} \underset{\mathbf{37}}{\underset{87\%}{(4\text{-}IC_6H_4)_3O^+}} \tag{4.21}$$

Dialkylaryl- or alkyldiaryl-oxonium ions are less stable and are prepared by alkylation of alkylaryl ethers or diaryl ethers with either methyl or ethyl hexafluoroantimonate in SO_2ClF at low temperatures.[105,106] Dialkylaryloxonium ions are stable below $-70°C$; above this temperature they are transformed into ring-alkylated alkoxybenzenes. These onium ions are stronger alkylating agents than trialkyloxonium ions, as shown by the immediate formation of trimethyloxonium ion from dimethylphenyloxonium ion and dimethyl ether (the reverse alkylation of anisole fails with trimethyloxonium salt).

The protons of methyl and methylene groups in the α-position of the oxonium center of trialkyloxonium ions exhibit a deshielding effect of 1 ppm in the ^1H NMR spectra compared with protons of the corresponding dialkyl ethers.[107,108]

A deshielding effect of 15–20 ppm is also observed in the ^{13}C NMR spectra.[109] The spectra generally reflect the expected result of replacing a neutral center by a positively charged one.

The geometry of trialkyloxonium salts has been presumed to be pyramidal rather than planar. This assumption has been proved by an NMR experiment in which pyramidal oxygen inversion in the case of the *O*-alkylethyleneoxonium ion **38** was unambiguously demonstrated.[110]

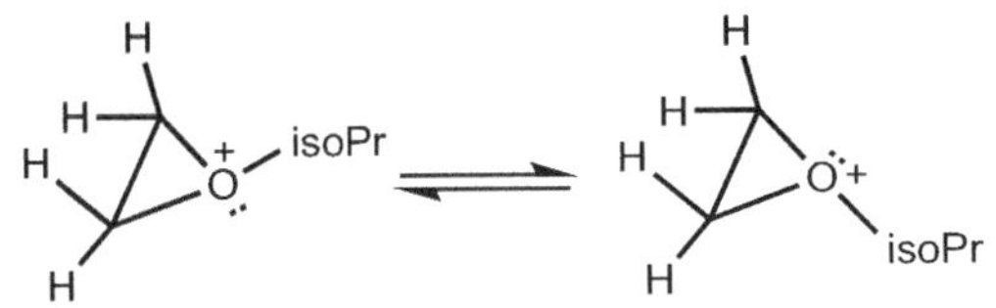

38

At +40°C, the ^{1}H NMR spectrum of **38** contains a very sharp singlet due to the ring protons. As the temperature is lowered, the ring proton resonance broadens to coalescence at −50°C. At lower temperatures (−70°C) the signal sharpens to a closely coupled ($\nu_{AB} \sim 3$ Hz) AA′BB′ spectrum. Thus, at the lowest temperature, the slowing of the rate of oxygen inversion results in the nonequivalence of the ring protons. The activation energy of the pyramidal inversion has been determined[110] to be 10 ± 2 kcal mol^{-1}. The NMR spectra of six-membered cyclic oxonium ions are temperature-independent; therefore, even at −70°C, rapid oxygen inversion must be assumed.[110] Further evidence for the pyramidal nature of trialkyloxonium salts comes from X-ray crystallographic studies on triethyloxonium ion.[111] On the contrary, the triphenyloxonium ion has a planar structure.[111]

In a recent theoretical study [MP2, B3LYP, CCSD(T)][112] the inversion at the oxygen of the *O*-methylated oxirane cation **39** was calculated to be 15.7 kcal mol^{-1}, which agrees well with the experimentally determined value for cation **38**. Cation **39** is kinetically stable because a high-energy barrier (35.3 kcal mol^{-1}) separates it from the ring-opened, significantly more stable isomeric oxonium ions (Scheme 4.5). In

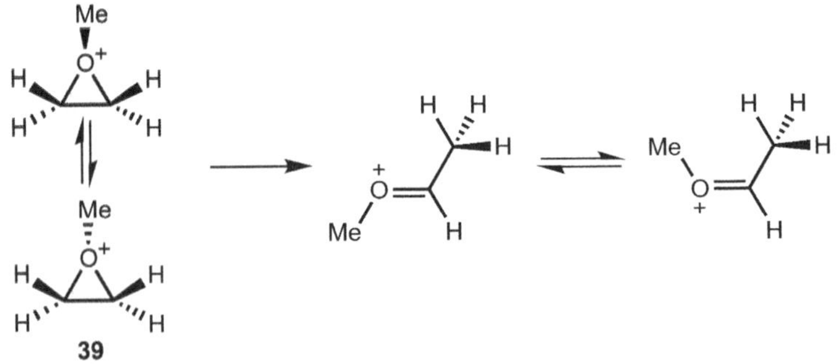

Scheme 4.5

comparison, there are two major differences in the behavior of the *O*-methylated tetramethyloxirane cation. (i) The potential energy surface is much shallower and, in particular, the barrier to ring opening is much smaller. In fact, it is essentially identical to that of the oxygen inversion [$\sim$15 kcal mol^{-1} (MP2) or $\sim$11 kcal mol^{-1} (B3LYP)]. (ii) The hydride migration associated with ring opening of cation **39** is a concerted process, as opposed to the stepwise methyl shift for the tetramethyl analog.

The unusual tertiary cage cation **40** (1-oxoniaadamantane) has recently been synthesized and characterized by Olah, Prakash, and co-workers[113a] using various approaches [Eq. (4.22)]. The main characteristics of the ^{13}C NMR spectrum are triplets of the methylene groups (OCH_2 at δ^{13}C 94.6 and CH_2 at δ^{13}C 31.7) and a doublet of the bridgehead carbon at δ^{13}C 28.2. X-ray analysis of single crystals of the unsubstituted cation with the low nucleophilicity carborane anion $CB_{11}H_6Cl_6^-$ reveals no unusual bond features. A highly stable bridgehead oxoniatriquinane cation has been prepared and characterized.[113b]

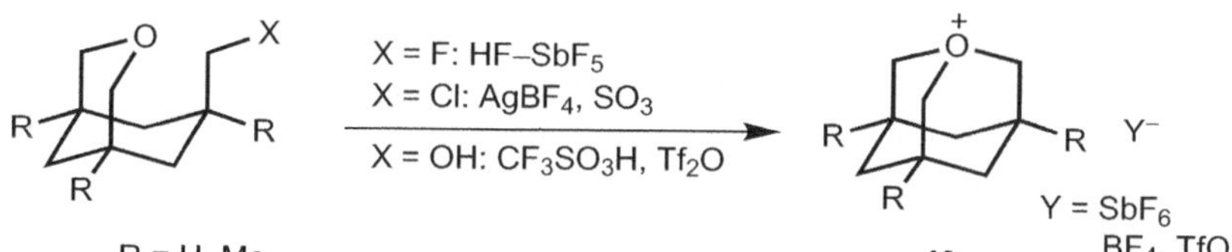

R = H, Me **40** (4.22)

Trialkyloxonium ions have been extensively used to carry out *O*-alkylation of lactams, amides, sulfoxides, oxo-sulfonium ylides, *N*-alkylation of *N*-heterocycles, *S*-alkylation of thioethers, thioacetals, thioamides, and thiolactams.[1,90] Although trialkyloxonium ions do not alkylate benzene or toluene directly, alkylations readily occur in combination with highly ionizing superacids such as HSO_3F–SbF_5. This may indicate protosolvation of a nonbonded electron pair on oxygen in the superacid medium.

Silyloxonium ions, which are key intermediates in cationic polymerization of cyclosiloxanes, have been prepared by treating diethyl ether according to Scheme 4.6.[114] A similar route has been applied to synthesize cyclic silyloxonium ions **41**. Trisilyloxonium ions **42** and **43** have been generated by reacting Me_3SiH with $Ph_3C^+(C_6F_5)_4B^-$ in the presence of siloxanes.[115,116]

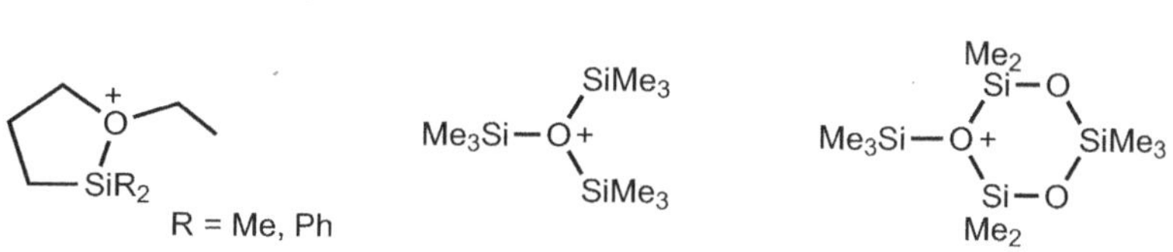

41 **42** **43**

The intermediacy of trimethyloxonium ion has been proposed in the first step of the acid–base-catalyzed conversion of methyl alcohol into gasoline range hydrocarbons over WO_3–Al_2O_3.[117] The crucial step is the base-catalyzed subsequent deprotonation of the trimethyloxonium ion to the very reactive surface-bound

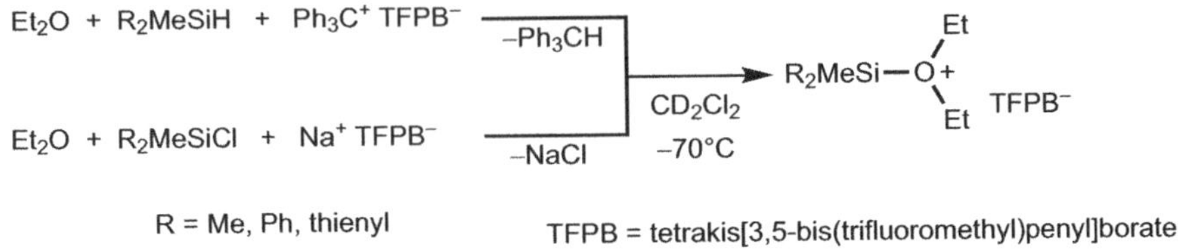

Scheme 4.6

methylene-dimethyloxonium ylide. The ylide subsequently undergoes alkylation at its negative pole, and the resulting ethyldimethyloxonium ion then cleaves to give ethylene and dimethyl ether. The key to the success of such reactions is the use of bifunctional catalysts that have both acidic and basic sites.

4.2.1.5. Aurated Oxonium Ions. A variety of oxonium gold complexes with triaryl-, trialkyl-, alkylaryl-, and trialkoxyphosphine ligands (**44**) have been synthesized[118–122] according to the general synthesis methods in Eq. (4.23). Since LAu^+ is isolobal with H^+—that is, they have similar bondings—complexes **44** represent the isolobal analog of H_3O^+. Some of the complexes were characterized by X-ray crystallography and shown to have trigonal pyramidal structure in the solid state.[118,121] The cations are associated with edge-to-edge contact (Au···Au distances about 3.10 Å) to form a square, but this may be disrupted with a sufficiently large phosphine ligand such as (*ortho*-tolyl)$_3$P.[121] The smallest phosphine ligand (Me_3P), however, allows the formation of a new structural unit $[(Me_3PAu)_6O_2]^{2+}(BF_4^-)_2$, with aggregation through crossed edges to give a tetrahedron.[122] Factors influencing aggregation behavior were addressed in subsequent theoretical calculations (LCGTO-DF approach).[123]

$$3LAuCl + Ag_2O \xrightarrow{NaBF_4} \underset{\mathbf{44}}{(LAu)_3O^+\,BF_4^-} \xleftarrow{KOH} LAu^+\,BF_4^- \qquad (4.23)$$

L = Me_3P, Et_3P, *tert*-Bu_3P, cyclohexyl$_3$P,
Ph_3P, $(2\text{-}MeC_6H_4)_3P$, Mes_3P, $(4\text{-}ClC_6H_4)_3P$,
Ph_2MeP, Ph_2EtP, Ph_2isoPrP, $PhMe_2P$,
$Ph_2(OEt)P$, $(OMe)_3P$

Schmidbaur et al.[124] have succeded in synthesizing cations **45**, which are gold analogs of the elusive double protonated water molecule H_4O^{2+} [Eq. (4.24)]. The central oxygen atom in the solid state is tetrahedrally coordinated to four gold atoms (characteristic data for the salt **45b**: O–Au bond lengths = 2.0571 Å; Au–O–Au bond angles = 109.5°; Au···Au contacts = 3.3593 Å). The three *ortho*-tolyl groups at phosphorus form propellers of C_3 symmetry and are directionally disordered (right- or left-handed). The products are thermally very stable and show only one resonance in the ^{31}P NMR spectrum at ambient temperature. At very low temperature (–90°C) the ^{31}P NMR singlet of cation **45b** but not **45a** is split into a closely separated set of signals. This change arises from the relative orientation of the four (*ortho*-tolyl)$_3$P propellers, which are not able to change the sign of rotation by concerted *ortho*-tolyl rotation on the NMR time scale. Since the hindrance of rotation of the phenyl groups in cation **45a** is much less, no splitting is observed.

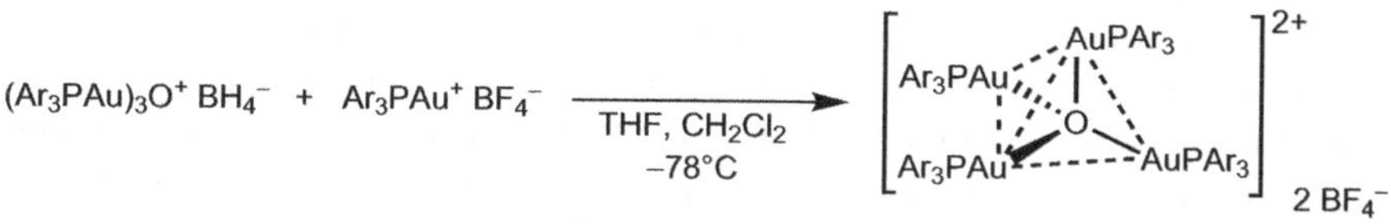

45 **a** Ar = Ph
b Ar = *o*-tolyl

(4.24)

4.2.1.6. Hydrogen Peroxonium Ion ($H_3O_2^+$) and Derivatives.

The use of hydrogen peroxide as a source of electrophilic oxygen under acidic conditions is gaining increasing importance.[125] The hydrogen peroxonium ion $H_3O_2^+$ **46** has been invoked as an intermediate in the electrophilic hydroxylation of aromatics, oxygenation of alkanes, and so on.[126] The hydrogen peroxonium ion $H_3O_2^+$ **46** has been isolated under superacid conditions.[127] H_2O_2 is also protonated in HF–MF_5 (M = Sb, As) solutions [Eq. (4.25)].

$$H_2O_2 \; + \; HF \; + \; MF_5 \longrightarrow H_3O_2^+\,MF_6^- \qquad (4.25)$$

M = Sb, As **46**

The $H_3O_2^+MF_6^-$ **46** salts are white crystalline solids with marginal stability. They exothermically decompose at room temperature to hydronium ion salt and molecular oxygen[127] [Eq. (4.26)].

$$H_3O_2^+\,MF_6^- \longrightarrow H_3O^+\,MF_6^- \; + \; 0.5\,O_2 \qquad (4.26)$$

46 M = Sb, As **1**

A detailed infrared analysis[127] of $H_3O_2^+SbF_6^-$ **46** shows that the ion possesses C_s symmetry similar to its isoelectronic analog NH_2OH. The ^1H NMR spectrum of **46** in HF–AsF_5 solution at −80°C shows only one broad signal at δ^1H 11.06 indicating rapid proton exchange between **46** and the HF solution. This was subsequently confirmed by ^{17}O NMR spectroscopy. The $H_3O_2^+$ ion **46** has a ^{17}O NMR chemical shift of $\delta^{17}O$ 151 (with respect to SO_2 at 505 ppm) and shielded by 36 ppm with respect to hydrogen peroxide.[29]

The crystal structure of $H_3O_2^+SbF_6^-$ has been determined by Minkwitz et al.[128] An interesting feature of the cation **46** is that the O–O bond distance (1.443 Å) is slightly shorter than the bond length in free hydrogen peroxide (1.461 Å by X-ray diffraction and 1.458 Å by neutron diffraction[129]). This indicates that the conversion of the lone pair of electrons into a bonding electron pair strengthens the weak O–O bond in hydrogen peroxide.

The synthetic utility of hydrogen peroxonium ion reagent is discussed in Chapter 5. Similarly, protonation and cleavage reactions of hydroperoxides and peroxides have been investigated extensively[125] and are also discussed in Chapter 5.

Mitchell and co-workers[130–132] have studied substituted derivatives of hydrogen peroxide intermediates in organic transformations. Ions **47** and **48** were prepared from the corresponding bromoperoxides with $AgBF_4$ and trapped in appropriate reactions. When ionization was performed in $SbCl_5$, the NMR spectra of the reaction mixture (downfield singlets at δ^1H 6.29 and 6.14 assigned to the methylene protons) gave compelling evidence of the existence of cation **48**.[131] Further support comes from a similar study of the tetradeutero derivative.

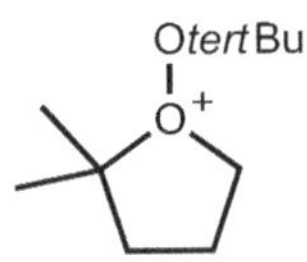

47

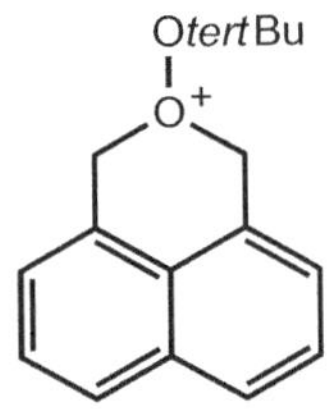

48

Minkwitz and Gerhard[133] have isolated protonated dimethyl peroxide $MeOO(H)Me^+$ prepared by protonating dimethyl peroxide in HF–AsF_5 and reported characterization by ^{1}H NMR and vibrational spectroscopy. Their efforts to obtain the trimethylperoxonium ion, however, were unsuccessful. Subsequently, however, Olah, Prakash, Christe, and co-workers[134] reported the preparation and NMR spectroscopic characterization of the $MeOOMe_2^+$ ion **49** as a long-lived species [Eq. (4.27)]. The ^{1}H NMR spectrum showed two singlets [δ^1H 5.26 (6H) and 4.95 (3H)]. The ^{13}C NMR chemical shift at δ^{13}C 85.6 (s) assigned to the methyl groups attached to the oxonium center is deshielded by 24.2 ppm compared to that of the methyl group in MeOOMe. The carbon chemical shift at δ^{13}C 70.8 (s) assigned to the methyl group on the uncharged oxygen, in turn, is only 9.4 ppm deshielded. This indicates that the positive charge is mostly localized on the tricoordinate oxygen atom.

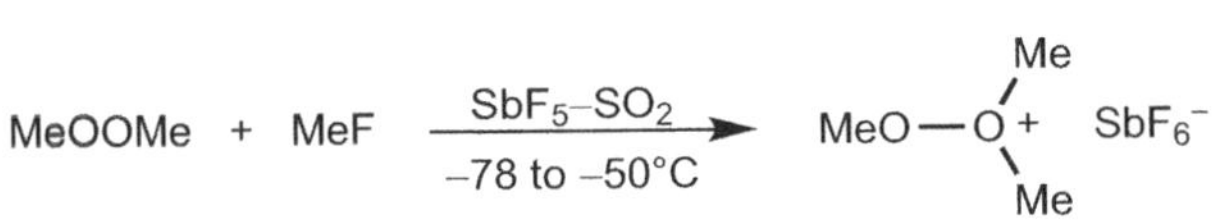

(4.27)

49

4.2.1.7. Ozonium Ion (HO_3^+).

Ozone is a resonance hybrid of canonical structures **50a–50d**.[135] Ozone does in fact act as a 1,3-dipole—that is, either as an electrophile or a nucleophile. The electrophilic nature of ozone has been recognized for a long time in its reactions with alkenes, alkynes, arenes, amines, phosphines, sulfides, and so on.[136–142] However, its reactivity as a nucleophile has not been well-recognized.[143]

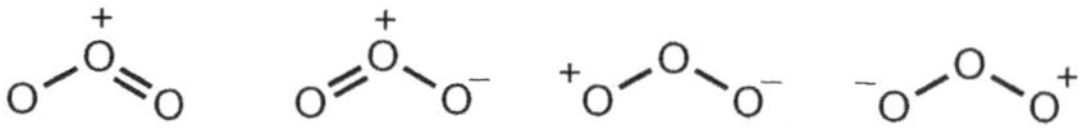

a b c d

50

Ozone has been shown to be protonated in the superacid media to ozonium ion HO_3^+ **51** [Eq. (4.28)], which reacts with alkanes as a powerful electrophilic oxygenating agent.[125] Similarly, ozone reacts with carbocations, giving alkylated ozonium ion that undergoes further cleavage reactions. These reactions are well-covered in Chapter 5.

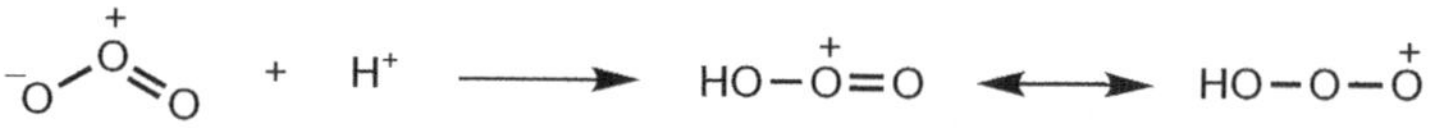

51

(4.28)

The long elusive ozonium ion HO_3^+ **51** has been experimentally detected in the gas phase with the use of Fourier transform ion cyclotron resonance (FT–ICR) generated by reacting ozone with strong Brønsted acids (H_3^+, KrH^+, XeH^+, CH_5^+).[144,145] The proton affinity of ion **51** determined with the bracketing technique is 148 ± 3 kcal mol^{-1} at 298 K. This value is in excellent agreement with those calculated at various levels of theory (CCSDT-1 level[146] = ~149.5 kcal mol^{-1}; BP, BPRO, PP91, PW98 levels[147] = 146.8–149.5 kcal mol^{-1}). Varied values have recently been computed at more advanced levels including temperature and entropy effects.[148] The absolute minimum on the potential energy surface is a structure with terminally attached proton in *trans* arrangement to the O(2)–O(3) bond.[146,147] The *cis* isomer is 3.6 kcal mol^{-1} higher in energy.

4.2.2. Sulfonium Ions

4.2.2.1. Hydrosulfonium Ion (H_3S^+). The parent of sulfonium ions is the protonated H_2S, the hydrosulfonium ion H_3S^+ **52**. Olah et al.[149] observed ion **52** for the first time in $HSO_3F–SbF_5–SO_2$ media at low temperature by ^1H NMR spectroscopy. H_3S^+ appeared as a singlet at δ^1H 6.6 from tetramethylsilane. Christe[150] has isolated $H_3S^+SbF_6^-$ salt by treating H_2S with HF–SbF_5 [Eq. (4.29)].

$$H_2S + HF + SbF_5 \longrightarrow H_3S^+ SbF_6^- \quad (4.29)$$

52

The hexafluoroantimonate salt of **52** is a stable white solid that reacts with water to produce H_2S. This reaction can be conveniently used to generate H_2S. The ion **52** has been thoroughly characterized by vibrational spectroscopy and normal coordinate analysis.[150]

Cation **52** has also been studied in the gas phase by mass spectrometry.[151,152] In addition, SH_3^+ and its isotopomers (H_3S^+, HD_2S^+, D_3S^+) were also observed by various spectroscopic methods.[153–155] The structure and energy of ion **52** have been calculated by *ab initio* method [MP2/6-311(2df,2pd) level].[156] Ion **52** is of C_{3v} symmetry and has a bond angle of 94.2°, which is slightly larger than that of H_2S (92.1°). This is attributed to the positive charge resulting in the contraction of the lone electron pair of sulfur, thus leaving more room for the bonding electrons of the S–H bonds.

Olah et al.[157] have observed that the sulfonium ion H_3S^+ undergoes hydrogen–deuterium exchange in superacid media, indicating the involvement of diprotonated hydrogen sulfide (tetrahydridosulfonium ion) H_4S^{2+}. The structure of T_d symmetry was found to be the minimum energy structure at the HF/6-31G* level. Dication H_4S^{2+} is thermodynamically unstable (dissociation energy $= -25.2\,\text{kcal mol}^{-1}$), but has a significant kinetic barrier ($59.2\,\text{kcal mol}^{-1}$) for deprotonation. The values calculated by Boldyrev and Simons[48] are $-91.4\,\text{kcal mol}^{-1}$ and $19.7\,\text{kcal mol}^{-1}$, respectively (QCISD (T)/6-311G(2df,2p)/MP2(full)/TZP + ZPE/MP2(full)/6-31G** level). Olah, Rasul, and Prakash[158] have calculated the structure of the trication H_5S^{3+} (MP2/6-31G** and QCISD(T)/6-311G** levels). The minimum-energy structure of C_s symmetry resembles a complex between H_3S^{3+} and a hydrogen molecule forming a two-electron three-center bond. The S–H bond lengths in the 2*e*–3*c* system (1.622 and 1.624 Å) are about 0.2 Å longer than normal S–H bonds. Similarly, the H–H bond distance in the 2*e*–3*c* bond (1.028 Å) is longer by about 0.29 Å than that in the H_2 molecule.

4.2.2.2. Primary Sulfonium Ions. Aliphatic thiols are completely protonated in $HSO_3F–SbF_5$ diluted with SO_2 at −60°C [Eq. (4.30)].[149]

$$\text{RSH} \xrightarrow[-60^\circ\text{C}]{\text{HSO}_3\text{F–SbF}_5\text{–SO}_2} \text{RSH}_2^+\ \text{SbF}_5\text{FSO}_3^- \tag{4.30}$$

The proton at sulfur is at considerably higher field in the ^1H NMR (δ^1H 5.93–6.45) than the corresponding proton on oxygen in protonated alcohols (δ^1H 9.1–9.5), reflecting the larger size of the sulfur atom compared with oxygen. The protonated thiols are considerably more stable than protonated alcohols. Protonated *tert*-butyl thiol **53** shows no appreciable decomposition at −60°C in $HSO_3F–SbF_5–SO_2$, whereas protonated tertiary alcohols could not be observed under similar conditions and even protonated secondary alcohols cleave at a significant rate. Protonated thiols cleave at higher temperature to give protonated hydrogen sulfide (singlet, δ^1H 6.60) and stable alkyl cations. For example, protonated 2-methylpropane-2-thiol **53** slowly cleaves to *tert*-butyl cation and protonated hydrogen sulfide when the temperature is increased to −30°C ($t_{1/2} \sim 15$ min) [Eq. (4.31)]. Protonated 2-methylbutane-2-thiol **54** also cleaves at this temperature to the *tert*-amyl cation [Eq. (4.32)].

$$\underset{\mathbf{53}}{(\text{CH}_3)_3\text{CSH}_2^+} \xrightarrow[-30^\circ\text{C}]{\text{H}^+} (\text{CH}_3)_3\text{C}^+ + \text{H}_3\text{S}^+ \tag{4.31}$$

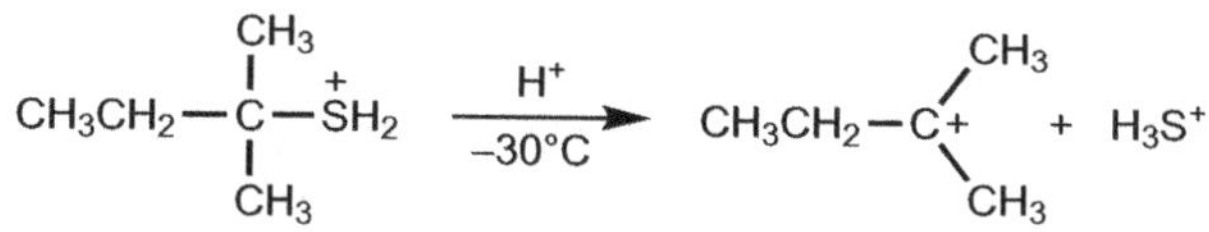

54

(4.32)

Protonated secondary thiols are stable even at higher temperatures. Protonated isopropyl thiol cleaves slowly at 0°C in HSO_3F–SbF_5 (1:1 *M*) solution. No well-identified carbocations were found in the NMR spectra due to the instability of the isopropyl cation under these conditions. Protonated *sec*-butyl thiol **55** cleaves to *tert*-butyl cation at this temperature [Eq. (4.33)].

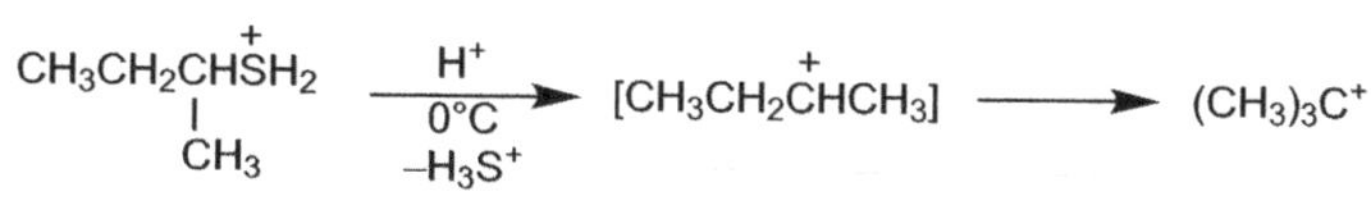

55

(4.33)

Protonated primary thiols are stable at much higher temperatures. Protonated *n*-butyl thiol **56** cleaves to *tert*-butyl cation only at +25°C [Eq. (4.34)].

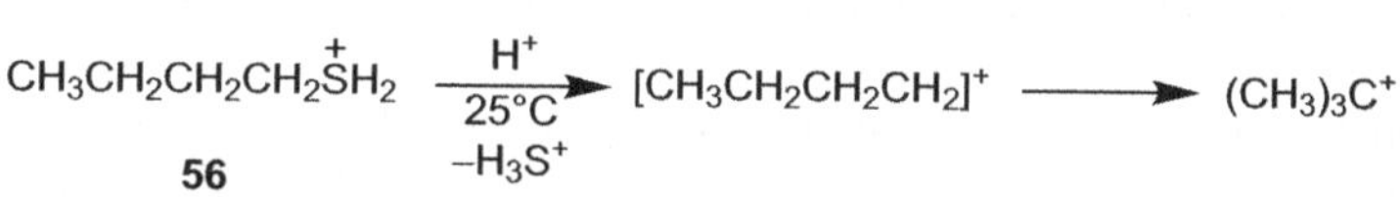

56

(4.34)

Minkwitz et al.[159] have prepared monoalkylsulfonium salts **57** by protonation of the corresponding thiols in superacids [Eq. (4.35)]. Monofluorosulfonium hexafluoroantimonate **58**, in turn, was generated by oxidative fluorination[73,160] of H_2S with equimolar amount of $XeF^+SbF_6^-$ [Eq. (4.36)]. Excess of either reagent results in decomposition into $S_8^{2+}(SbF_6^-)_2$. Crystal structure analysis of the isoPr$SH_2^+SbF_6^-$ salt gave the first experimentally determined S–H bond distance (1.12 Å).

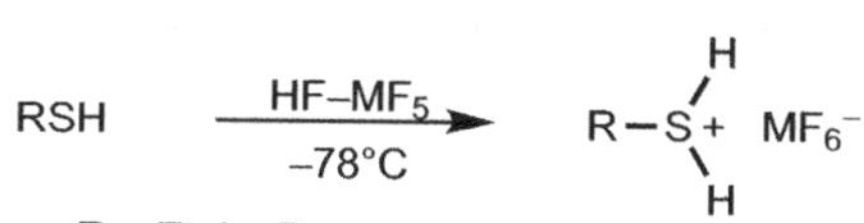

R = Et, isoPr
M = As, Sb

57

(4.35)

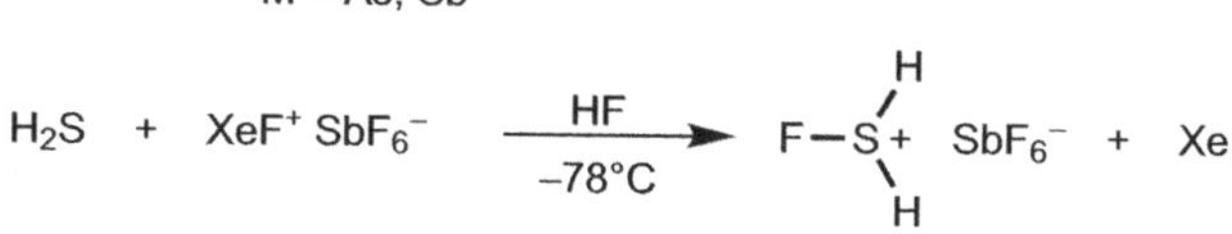

58

(4.36)

Mercaptosulfonium salts $H_3S_2^+MF_6^-$ (M = As, Sb), analogs of protonated hydrogen peroxide, have been prepared [Eq. (4.37)] and characterized by vibrational spectroscopy by Minkwitz et al.[161] According to calculations (*ab initio*, general force field), the cation **59** has a conformation with the lone pair of H_2S^+ antiperiplanar to the S–H bond.

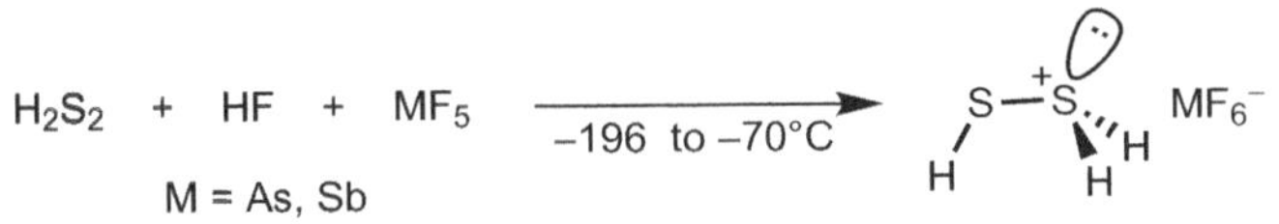

59

(4.37)

4.2.2.3. Secondary Sulfonium Ions.

Protonated aliphatic sulfides have been studied at low temperatures by NMR spectroscopy in strong acid systems[149] [Eq. (4.38)]. They show well-resolved NMR spectra, with the proton on sulfur being observed at about δ^1H 6.0.

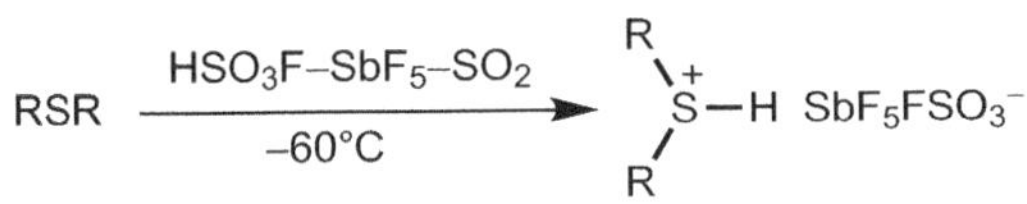

(4.38)

The NMR spectrum of protonated thiane-3,3,5,5-d_4 has also been studied in HSO_3F–SbF_5 to determine the conformational position of the proton on sulfur in the six-membered ring and to study the ring inversion process.[162] The proton on sulfur resides exclusively in the axial position.

The protonated sulfides are less susceptible to cleavage than the corresponding protonated ethers and also more stable than the protonated thiols.[149] Protonated *tert*-butyl methyl ether is completely cleaved to *tert*-butyl cation and protonated methanol even at −70°C. On the other hand, protonated *tert*-butyl methyl sulfide **60** is stable even at −60°C. When the temperature is increased to −15°C, protonated *tert*-butyl methyl sulfide very slowly cleaves to *tert*-butyl cation and protonated methyl thiol[149] [Eq. (4.39)].

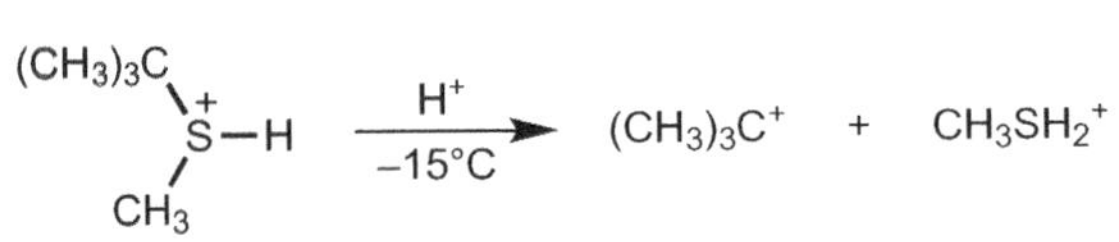

60

(4.39)

Protonated di-*tert*-butyl sulfide **61** shows very little cleavage at −60°C. At −35°C it cleaves slowly ($t_{1/2} \sim 1$ h) to *tert*-butyl cation and protonated hydrogen sulfide [Eq. (4.40)], with the latter showing the ^{1}H NMR peak at δ^1H 6.60.

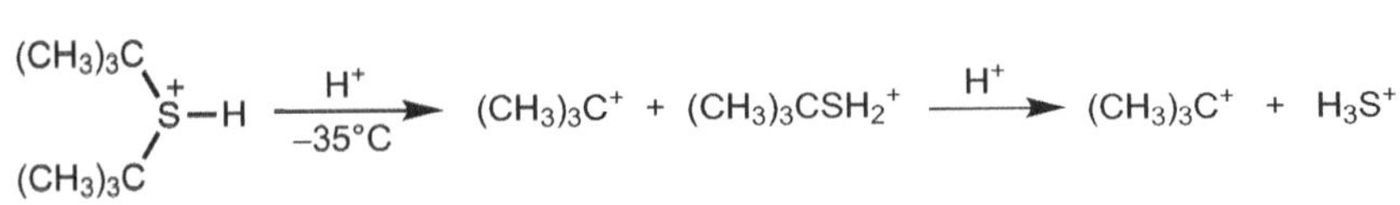

61

(4.40)

Protonated secondary sulfides show extraordinary stability toward the strongly acidic medium. Protonated isopropyl sulfide shows no appreciable cleavage up to +70°C in a solution of $HSO_3F–SbF_5$ (1:1).

Protonation of dimethyl sulfide and di-*tert*-butyl sulfide with $HF–MF_5$ (M = As, Sb) has been used to prepare the corresponding secondary dialkylsulfonium salts.[163] Oxidative fluorination with $XeF^+SbF_6^-$ has also been used to generate alkyl fluorosulfonium hexafluorometalates[164] [Eq. (4.41)]. The salts are thermally labile (decomposition is observed above −40°C) and exhibit ^{19}F NMR chemical shifts at $\delta^{19}F$ −177.0 (R = CH_3) and −133.2 (R = CF_3), which are downfield from those of the corresponding dialkyl-substituted derivatives [$(CH_3)_2SF^+$ $\delta^{19}F$ −190.7[165] and $(CF_3)_2SF^+$ $\delta^{19}F$ −159.4[166]].

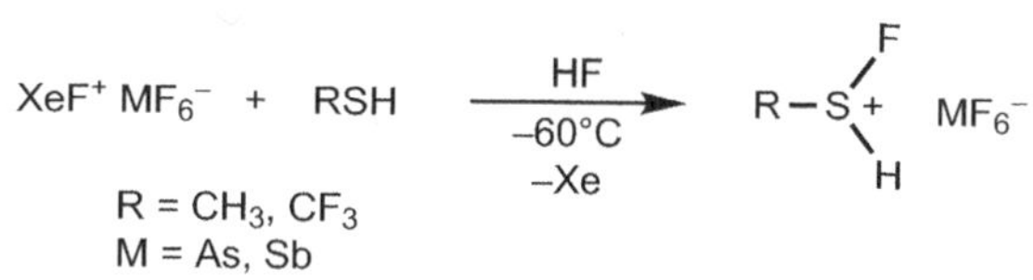

(4.41)

The protonation of cyclic sulfides have also been studied by Olah and Szilagyi.[79] Both protonated thiirane and methylthiirane could be observed by ^1H NMR spectroscopy ($HSO_3F–SbF_5–SO_2$ solution, −78°C). Protonated thiirane gives a complex symmetrical spectrum (A_2B_2X) centered at δ^1H 3.63 deshielded by 1.30 ppm from the precursor but the SH proton could not be observed. Protonation of thiirane-1-oxide studied for comparison, occurred at the sulfur atom and the ion exhibited ring protons at δ^1H 3.87 deshielded from δ^1H 2.47 of the precursor. The SH proton was observed at δ^1H 5.27. Protonation of the larger homologs were also successful. The SH proton resonances appear at consistently higher field (average 6.30 ppm) than those of the corresponding acyclic sulfides (average 7.61 ppm). Furthermore, ^1H NMR chemical shifts of thiiranes are more shielded than those of oxiranes, and both are more shielded when compared with those found for cyclic iodonium and bromonium ions. The same tendency was observed for the corresponding five-membered cyclic protonated derivatives.

4.2.2.4. Tertiary Alkyl(Aryl)Sulfonium Ions. In contrast to tertiary oxonium ions, the tertiary sulfonium ions are stable and are prepared rather readily.[167–169] They are even stable in aqueous solutions. Trialkylsulfonium ions are obtained by alkylation of dialkyl sulfides with alkyl halides. Trialkyloxonium and dialkylhalonium ions readily transalkylate basic dialkylsulfides.

In contrast, alkylation of diaryl sulfides and thiophene requires rather drastic conditions due to poor nucleophilicity of sulfur. The alkylations have been achieved using alkyl triflates[170] [Eq. (4.42)] or with alkyl halide and silver tetrafluoroborate[171] [Eq. (4.43)].

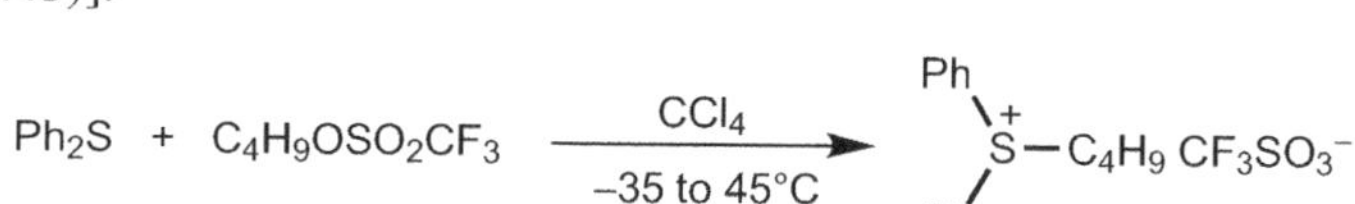

(4.42)

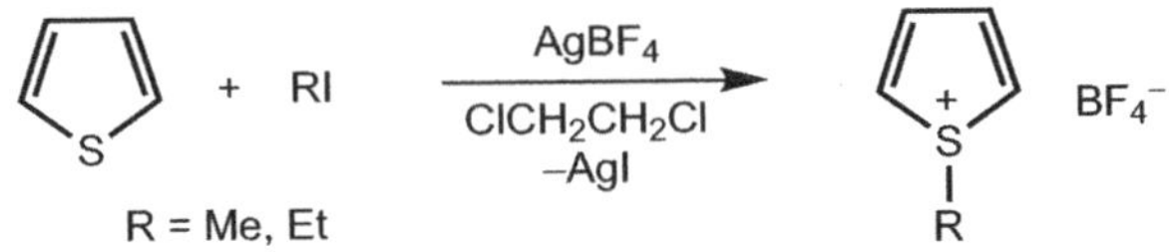

(4.43)

Another method is the use of esters as alkylating agents in triflic acid[172] [Eq. (4.44)]. This method has been used to synthesize *S*-methylated phenylene sulfide oligomers (methyl triflate, triflic acid, 25°C, 10 h) and to solubilize poly(phenylenesulfide) by transforming it to poly(arylenesulfonium) salts **62** (methyl triflate, methyl or ethyl formate, triflic acid, 110°C, 10 h). In a similar way, alkylation of thianthrene to yield the corresponding sulfonium salts (**63**) has been achieved using alkyl formates.[173]

(4.44)

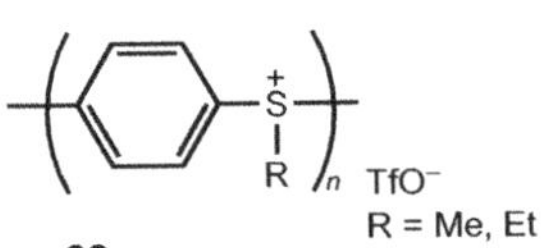

62

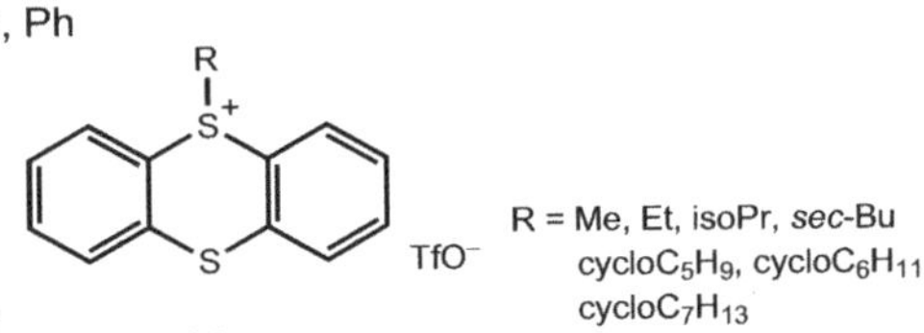

63

Cyclizations shown in Eq. (4.45) have been used to obtain *S*-perfluoroalkylbenzothiophenium ion **64** and ring-substituted derivatives.[174,175] Salts **64** and the *S*-(difluoromethyl)- and *S*-(monofluoromethyl)-diarylthiophenium salts **65** reported recently by Prakash, Olah, and co-workers[176,177] are used as fluoromethylating agents.

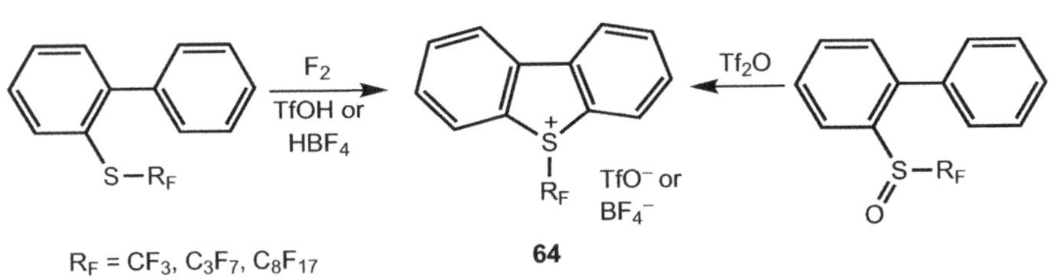

(4.45)

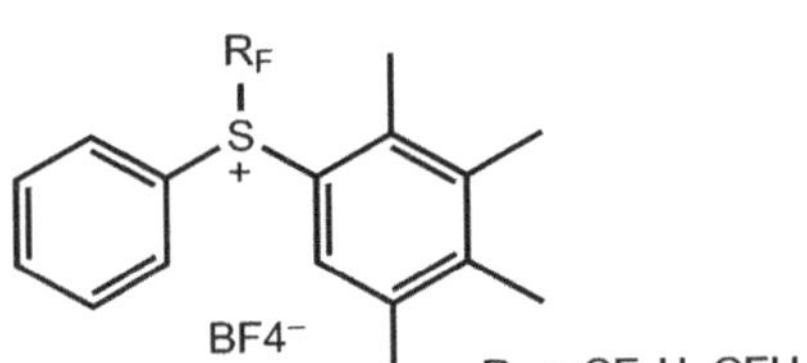

65

The absolute configuration of ethylmethylpropylsulfonium ion, the simplest chiral sulfonium ion has been determined.[178]

The most important application of trisubstituted sulfonium ions is in the generation of sulfur ylides.[179]

Thiiranium ions (episulfonium ions) were first synthesized by Lucchini and co-workers[180] and subsequently characterized by X-ray crystallography by Simonetta and co-workers.[181] A number of thiiranium ions with bulky substituents capable of hindering the attack of nucleophiles have been isolated and characterized. The molecular structure of ion **66** shows two longer S–C bonds in the thiiranium ring with some asymmetry[182] (1.970 and 1.987 Å). Ion **67** is dimeric in the solid state with shorter bonds and one ion exhibiting significant asymmetry[183] (1.914 and 1.937 Å, 1.909 and 1.913 Å). The S–C_{ipso} bonds are significantly shorter (**66** = 1.784 Å, **67** = 1.789 Å). One of the adamantyl rings is twisted away from the plane of the thiiranium ring.

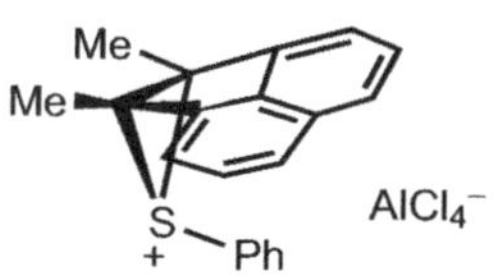

66

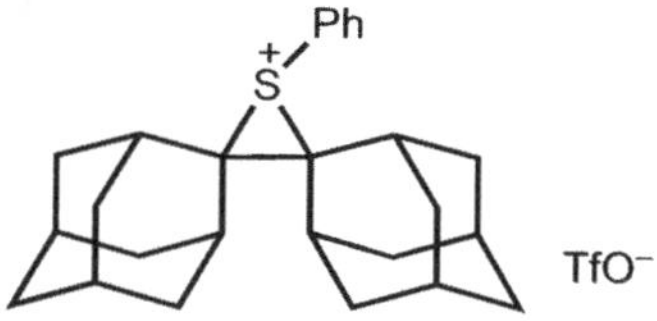

67

Lucchini, Pasquato, and co-workers have obtained stereoisomeric 2,3-disubstituted 1-methylthiiranium ions, reported characterization (^{1}H and ^{13}C NMR, X-ray crystallography), and studied their anionotropic rearrangements. The synthesis was performed by reacting methylbis(methylthio)sulfonium salts with alkenes[184,185] (Scheme 4.7).

Because of the steric interactions of the two *tert*-butyl groups,[186] ion ***trans*, *trans*-68** features a longer C–C bond in the ring (1.500 versus 1.452 Å) and increased bond angles [C(3)–C(2)–C(4) = 135.4 versus 127.0°, C(2)–C(3)–C(5) = 135.4 versus 125.5°] as compared to ion ***cis*,*trans*-68**. The interaction of S–Me with the nearest Me group of the *tert*-butyl group in ion ***cis*,*trans*-68** brings about an enlarged S–C(2)–C(4) angle with respect to S–C(3)–C(5) (125.1 versus 117.1°) and asymmetric S–C bonds (1.876 and 1.860 Å).

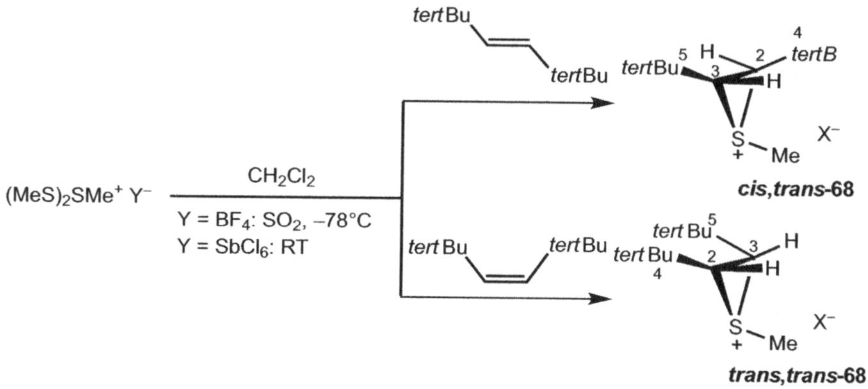

Scheme 4.7

Ion ***cis*, *trans*-68** undergoes rearrangement at room temperature in CD_2Cl_2 to yield quantitatively the ring-enlarged product thiethanium ion **69**. X-ray crystal structure characterization shows[187] that the exocyclic S–C bond (1.797 Å) is significantly shorter than the endocyclic S–C bonds (1.864 and 1.896 Å). The difference (about 0.03 Å) between the latter two bonds is related to the steric hindrance between the two *cis* Me groups. The lengthening of the ring S–C bonds and the torsion angle [C(1)–S–C(2)–C(3) = 12.2°] allows the separation of the two Me groups by 3.100 Å. This value, however, is still smaller than the sum of the van der Waals radii of the C atoms (3.4 Å). The four-membered ring has a puckering angle of 145.1°.

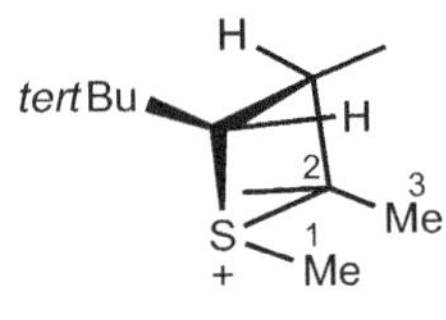

69

Thiirenium ion **70**, synthesized as described in Scheme 4.7 in a reaction with the corresponding acetylene, exhibits an analogous behavior to yield the thiethium ion **71**.[188] Phenyl-substituted thiiranium ion **72a** was shown to undergo rapid isomerization presumably through benzyl cations at −50°C to yield the *trans* compound.[189] In contrast, **72b** is configurationally stable at 0°C. Stereoisomeric ions **73a** and **73b** prepared by methylation of the corresponding thiiranes (Scheme 4.8) undergo isomerization at room temperature and above.[190]

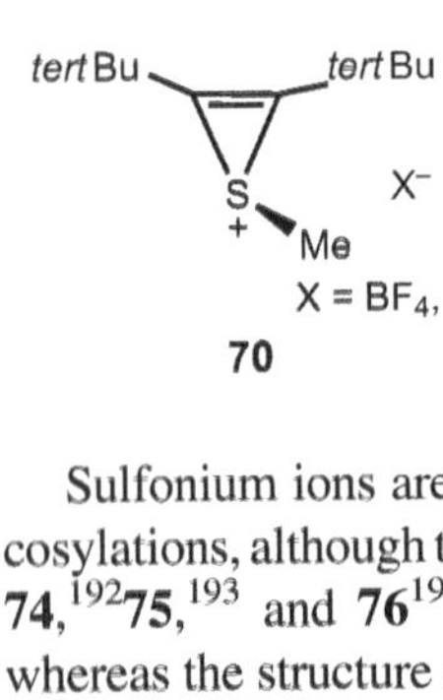

X = BF_4, $SbCl_6$, PF_6

70

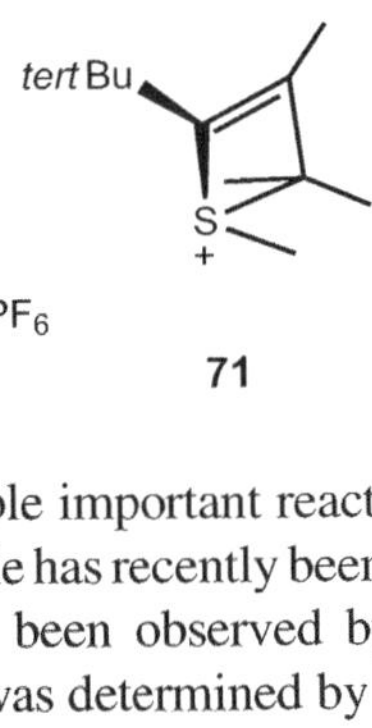

71

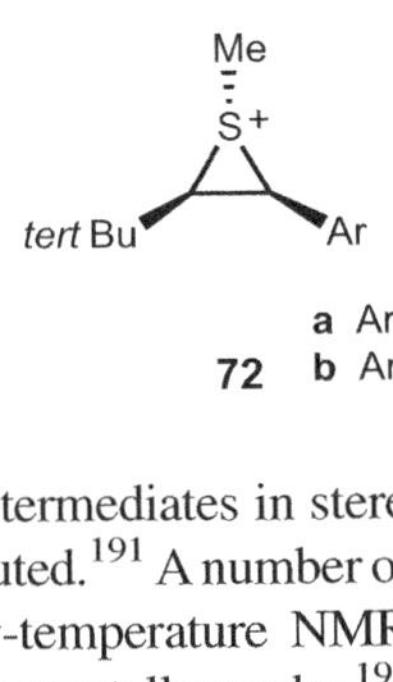

72 **a** Ar = Ph
b Ar = 3,5-diClC_6H_3

Sulfonium ions are possible important reactive intermediates in stereoselective glycosylations, although their role has recently been disputed.[191] A number of ions including **74**,[192] **75**,[193] and **76**[194] have been observed by low-temperature NMR spectroscopy, whereas the structure of **77** was determined by X-ray crystallography.[195]

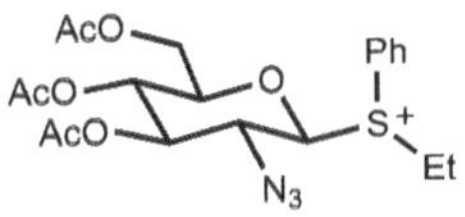

74

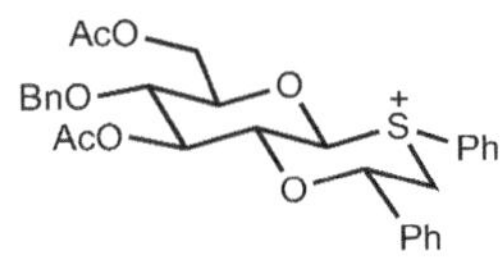

75

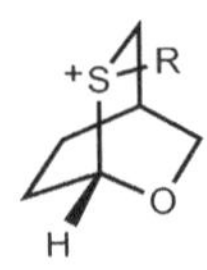

R = Et, Ph

76

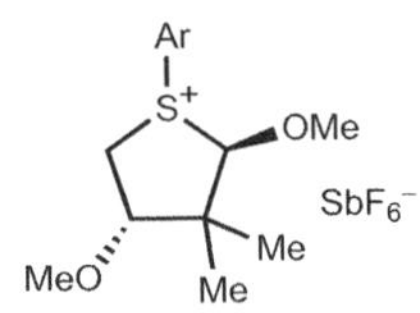

Ar = 4-MeC_6H_4

77

The tertiary cage sulfonium cation **78** was generated and characterized by NMR spectroscopy as early as 1963[196] [Eq. (4.46)].

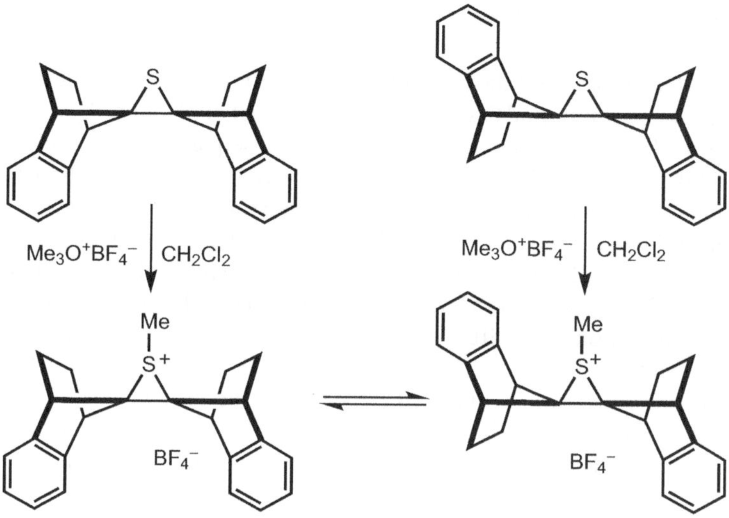

a **b**

73

Scheme 4.8

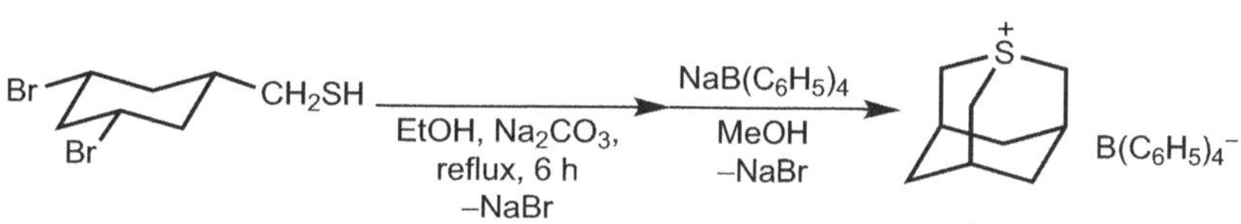

78

(4.46)

Sulfonium dication **79** with unusual bonds and charges has been prepared by Furukawa, Sato, and co-workers[197,198] [Eq. (4.47)]. Both salts were characterized by NMR spectroscopy and X-ray crystallography. The data indicate that the dications have distorted tetrahedral configuration (bond angles of the tetrafluororborate salt are 94.9°, 115.9°, 117.5°, and 118.3°).[197] Bond lengths are within those of normal tetrahedral sulfur compounds.

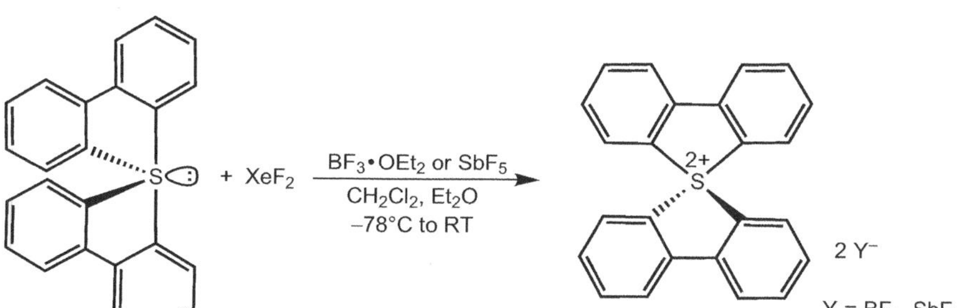

79

(4.47)

4.2.2.5. Halosulfonium Ions

Trihalosulfonium Ions. Of the parent trihalosulfonium ions, X_3S^+ (X = F, Cl, Br), F_3S^+ and Cl_3S^+ were reported first. The reaction product of SF_4 and BF_3 was described by Bartlett as the $SF_4{\cdot}BF_3$ adduct,[199] but it was subsequently proved by structural studies (NMR, Raman, IR)[200,201] to be the $F_3S^+BF_4^-$ salt. According to crystal structure analysis,[202] the ion F_3S^+ is of C_{3v} symmetry and has very short S—F bonds (1.495 and 1.499 Å) indicative of a substantial positive charge on sulfur. Other $F_3S^+MF_6^-$ (M = As, P, Sb) salts were studied by Gillespie and co-workers.[200]

Various $Cl_3S^+M_n^-$ salts were prepared by Kolditz and Schäfer[203] ($M_n = AsF_6$, oxidative chlorination of sulfur with $AsCl_3$–AsF_3), Minkwitz and Gerhard[204] [Eq. (4.48)], and Passmore et al.[205] [Eq. (4.49)]. The cation Cl_3S^+ in $Cl_3S^+AsF_6^-$ was found to be pyramidal (C_{3v} symmetry) by X-ray structure analysis.[206] The cations in $Br_3S^+SbF_6^-$ and $Cl_3S^+SbCl_6^-$, in turn, were shown to have significantly shortened sulfur–halogen bonds and distorted octahedral arrangement around sulfur, when cation–anion secondary bonding interactions are taken into account.[207–210]

$$H_3S^+\,SbF_6^- + 2Cl_2 \xrightarrow{SO_2} Cl_3S^+\,SbF_6^- + 3HCl \quad (4.48)$$

$$0.25S_8 + 3Br_2 + 3MF_5 \xrightarrow{SO_2} 2Br_3S^+\,MF_6^- + MF_3 \quad (4.49)$$

M = As, Sb

Olah, Rasul, and Prakash have calculated the structure of the trication SF_5^{3+} (MP2/6-31G** level).[158] In contrast to SH_5^{3+}, it has a halonium-type structure $(F_3SFF)^{3+}$; that is, the F_3S group acts as a pesudohalogen.

Dihalo-monoalkyl(aryl)sulfonium Ions. A general synthetic route to produce dihalo-monoalkyl(aryl)sulfonium salts **80** is oxidative halogenations of RSX[169,211–213] [Eq. (4.50)]. Other approaches may also be used for the synthesis of $CF_3SCl_2^+AsF_6^-$ salts [Eq. (4.51)]. Ionization of CF_3SF_3 with Lewis superacids yields the corresponding $CF_3SF_2^+$ salts.[214] The crystal structure of $CF_3SF_2^+AsF_6^-$[215,216] and $CF_3SCl_2^+SbF_6^-$[217] have been reported.

$$2RSX + X_2 + 3MF_5 \xrightarrow{SO_2} \underset{\mathbf{80}}{2RSX_2^+\,MF_6^-} + MF_3 \quad (4.50)$$

R = CH_3, CF_3
M = As, Sb
X = Cl, Br

$$CF_3SCl + AsF_5 + ClF \xrightarrow{SO_2} CF_3SCl_2^+\,AsF_6^- \longleftarrow CF_3SCl + AsF_5 + Cl_2 \quad (4.51)$$

Pentafluorophenyl derivatives $C_6F_5SX_2^+MF_6^-$ (X = F, Cl, Br; M = As, Sb) can also be prepared by oxidative halogenation[218] [Eq. (4.50)] or by the ionization of

$C_6F_5SF_3$ with Lewis superacids.[169] Significant deshielding of fluorines in *ortho* and *para* positions indicates charge dispersion into the aromatic ring. C_6F_5 derivatives were found to be more stable than their CF_3 counterparts.

Mixed chlorofluorosulfonium salts are prepared by oxidative halogenation of acidic secondary sulfonium salts[164,219] [Eq. (4.52)] or oxidative fluorination[73,217] with XeF^+ [Eq. (4.53)]. The products were characterized by spectroscopic methods (NMR, IR, Raman). $CF_3S(Cl)F^+SbF_6^-$ undergoes symmetrization in SO_2 solution to yield $CF_3SF_2^+SbF_6^-$ and $CF_3SCl_2^+SbF_6^-$.[217] The X-ray structure of the latter compound was also reported.

$$RS(H)F^+\ MF_6^- + X_2 \xrightarrow{HF} RS(X)F^+\ MF_6^- + HX \qquad (4.52)$$

$R = CH_3, CF_3$
$M = As, Sb$
$X = Cl, Br$

$$CF_3SX + XeF^+\ MF_6^- \xrightarrow[-Xe]{HF} CF_3S(X)F^+\ MF_6^- \qquad (4.53)$$

$M = As, Sb$
$X = Cl, Br$

Dialkyl(diaryl)monohalosulfonium Ions. These ions can readily be prepared by reacting dialkylsulfides with appropriate halogenating agents[169] [Eq. (4.54)]. Meerwein used this method to synthesize $Me_2SCl^+BF_4^-$ by anion exchange and also reacted Me_2S–$SbCl_5$ to obtain the corresponding hexachloroantimonate.[220] Chlorination of Et_2S and isoPr_2S was also accomplished with excess $SbCl_5$.[221]

$$R_2S + \begin{matrix}Cl_2,\text{ or NCIS,}\\ \text{or } SO_2Cl_2\end{matrix} \longrightarrow R_2ClS^+\ Cl^- \xrightarrow{HBF_4} R_2ClS^+\ BF_4^- \qquad (4.54)$$

$R = CH_3, CF_3$

Dialkylsulfoxides can also be transformed to R_2XS^+ ($X = Cl, Br$) ions [Eq. (4.55)].

$$R_2SO + SOCl_2 \xrightarrow{HBF_4 \text{ or } SbF_5} R_2ClS^+\ Y^- \qquad (4.55)$$

$R = CH_3, CF_3$ $\qquad$ $Y = BF_4, SbF_6$

In addition to the procedure developed by Meerwein [see Eq. (4.54)], Me_2XS^+ salts have also been obtained according to the method shown in Eq. (4.56)[222] and by using XeF_2 as a reagent [Eqs. (4.57) and (4.58)].[165] Minkwitz and co-workers reported the synthesis of $(CF_3)_2FS^+$ salts[166] [Eq. (4.59)] and the crystal structure of $Me_2ClS^+SbF_6^-$ and $Et_2ClS^+SbCl_6^-$.[221,222]

$$Me_2SH^+\ SbF_6^- + Cl_2 \xrightarrow[SO_2]{} Me_2ClS^+\ SbF_6^- + HCl \qquad (4.56)$$

$$Me_2S + XeF_2 \xrightarrow[HF]{} Me_2FS^+ [F(HF)_n]^- \qquad (4.57)$$

$$Me_2S{\cdot}MF_n + XeF_2 \xrightarrow[HF]{} Me_2FS^+ MF_{n+1}^- \qquad (4.58)$$

$$MF_n = BF_3, AsF_5$$

$$(CF_3)_2SF_2 + MF_5 \xrightarrow[-40°C]{HF\text{–}SO_2} (CF_3)_2FS^+ MF_6^- \xleftarrow[-40°C]{HF} (CF_3)_2S + XeF^+ MF_6^-$$

$$M = As, Sb \qquad (4.59)$$

A comparison of ^{19}F NMR chemical shifts of $(CF_3)_2FS^+$, $CF_3F_2S^+$, and F_3S^+ shows an increasing shielding with replacement of CF_3 by F ($\delta^{19}F$ −159.0, −54.7, and 30.5, respectively).[223] All three ions react with acetonitrile to form ψ-pentacoordinated ions [Eq. (4.60)]. NSF_3 forms similar but weaker donor–acceptor complexes.

$$[(CF_3)_nSF_{3-n}]^+ AsF_6^- + CH_3CN \rightleftharpoons [(CF_3)_nSF_{3-n}{\cdot}NCCH_3]^+ AsF_6^- \qquad (4.60)$$

$$n = 0\text{–}2$$

Fluorination of $(C_6F_5)_2S$ with XeF^+ has been applied to obtain $(C_6F_5)_2FS^+MF_6^-$ (M = As, Sb) salts.[224] An S–F bond distance of 1.584 Å was found by crystal structure analysis of $(C_6F_5)_2FS^+SbF_6^-$. Minkwitz et al.[225,226] have reported the synthesis and structural characterization of the cyclic monofluorinated cations **81** and **82**. The S–F bond lengths (1.51 and 1.522 Å) are significantly shorter than that in $(C_6F_5)_2FS^+$ (1.584 Å).

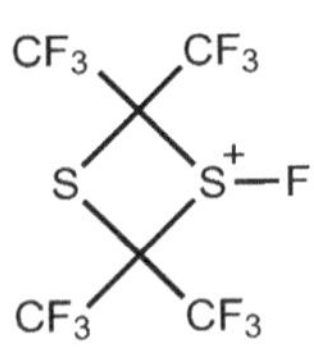

81

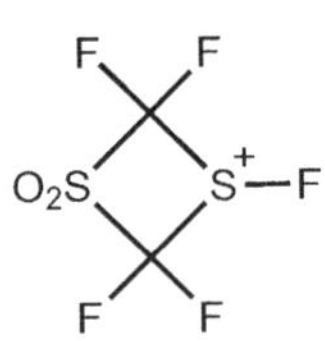

82

4.2.2.6. *Sulfonium Ions with Other Heteroligands*

Hydroxy- and Alkoxysulfonium Ions. In addition to the well-documented formation of the SO_2–SbF_5 complex in a mixture of the two components,[227] Minkwitz and co-workers[228] have recently detected the formation of the fluorodihydroxysulfonium cation **83** [Eq. (4.61)]. Crystal structure analysis of the isolated salt showed bond distances typical of S–O single bonds (1.537 and 1.522 Å) and S–F single bonds (1.547 Å). The steric demand of the lone electron pair of sulfur results in the compression of bond angles from the ideal tetrahedral arrangement

(F—S—O = 99.0 and 100.1°, O—S—O = 98.6°). A Raman line at 830 cm^{-1} may be assigned to the ion **83**.

$$2HF + SbF_5 \xrightarrow[-90^\circ C]{SO_2} H_2F^+ SbF_6^- \xrightarrow[-25^\circ C]{SO_2} SO_2SbF_5 + 2HF \rightleftharpoons \underset{\textbf{83}}{(HO)_2FS^+ SbF_6^-} \quad (4.61)$$

The existence of the trihydroxysulfonium ion $H_3SO_4^+$ formed by autoprotolysis of sulfuric acid was confirmed by various experimental techniques. Minkwitz and colleagues[229] have been able to prepare the ion using the technique, which was successfully applied in the generation of protonated hydrogen peroxide and carbonic acid [Eq. (4.62)]. The molecular structure of the colorless crystals of $D_3SO_4^+SbF_6$ shows three S—O bonds with approximately the same bond lengths (1.499–1.512 Å). This bond length is intermediate between the S—O single and S=O double bond of sulfuric acid indicative of dispersion of the positive charge. The S=O double bond length is close to that in sulfuric acid (1.413 Å versus 1.426 Å). The cations and anions are linked by three hydrogen bondings.

$$(Me_3SiO)_2SO_2 + XF + SbF_5 \xrightarrow[-196\text{ to }-78^\circ C]{} X_3SO_4^+ SbF_6^- + 2Me_3SiF \qquad X = H, D \quad (4.62)$$

Olah and co-workers studied protonated dimethyl sulfoxide in SbF_5–SO_2ClF (or SO_2) solution and identified the *O*-monoprotonated Me_2SOH^+ species by ^1H and ^{13}C NMR spectroscopy[230] also indicated by a Raman study.[231] The ^{13}C, ^{17}O, and ^{33}S NMR chemical shifts calculated in a recent theoretical study[232] (DFT/GIAO-MP2 method) match well with the experimental values. According to DFT calculations (B3LYP/6-311 + G** level), the *O*-protonated form is 37.0 kcal mol^{-1} more stable than the *S*-protonated form.[232] The *O,O*-diprotonated dication was calculated to be the global minimum being more stable than the *O,S*-diprotonated ion by 20.8 kcal mol^{-1}. Interestingly, the *O,O*- and *S,O*-dimethylated dications were found to be isoenergetic. Diaryl sulfoxides were also shown to be protonated at the oxygen in Magic Acid with the exception of the parent diphenyl sulfoxide and bis(2,4,6-trifluorophenyl) sulfoxide.[233,234] *O*-Protonation was calculated to be favored over *S*-protonation by about 17 kcal mol^{-1}.[235] Low-temperature NMR studies indicated significant π-electron delocalization and, consequently, increased double-bond character of the C—S bond and, in some cases, two conformations were observed.[234] The conformers of diphenyl sulfoxide with *syn* or *anti* orientation of the OH proton relative to the aromatic rings are separated by a barrier of 1.3 kcal mol^{-1}.

Dimethoxyfluorosulfonium ion **84** has been generated from methylfluorosulfite by methylation with CH_3F under superacidic conditions[236] [Eq. (4.63)]. A similar

O-methylation of thionyl fluoride yields ion $F_2(MeO)S^+$.[237] Both ions show a single deshielded methoxy resonance (δ^1H 4.89 and 5.27, respectively).

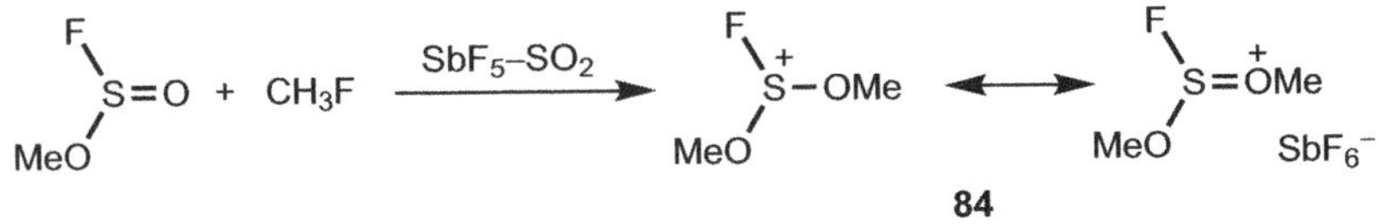

84

(4.63)

Minkwitz and Molsbeck[238] have performed *O*-methylation of various sulfoxides to obtain the corresponding methoxysulfonium ions [Eq. (4.64)] and also reported their spectroscopic characterization (Raman and ^1H, ^{13}C, and ^{19}F NMR).

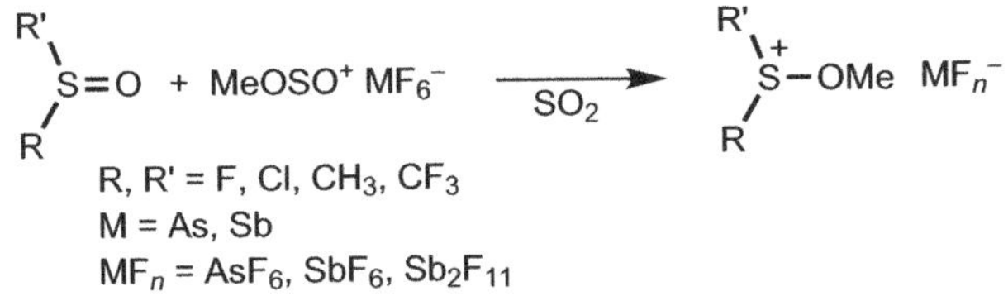

(4.64)

Shine and co-workers[239,240] have reported the synthesis and full NMR characterization of a series of alkoxysulfonium ion perchlorates including ions **85** derived from *cis*- and *trans*-substituted cyclohexanols. The X-ray structural study of four salts showed that the orientation of the S—O bond is always pseudoaxial. The optically active (—)-**86** ion and similar systems have been synthesized.[241] ^1H and ^{13}C NMR measurements indicate that sulfur has a trigonal bipyramidal geometry due to the S---O intramolecular interaction in the axial position.

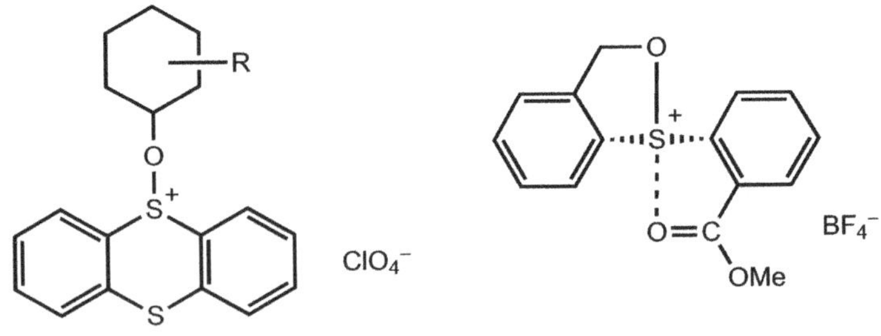

R = 2-, 3-, 4-Me, 3,5-diMe, 2-OH **85** (–)-**86**

Sulfonium Ions with Nitrogen Ligands. Long-lived aminofluorosulfonium ions have been prepared by ionization under superacidic conditions[242,243] [Eq. (4.65)]. The stereochemistry of ion $Me_2NSF_2^+$ was studied by dynamic ^1H NMR spectroscopy and structure **87** was established with a rotation barrier of 14.7 kcal mol^{-1} about the N–S bond.[243]

$$(R_2N)_mSF_{4-m} + MF_n \xrightarrow[SO_2]{} (R_2N)_xSF_{3-m}^{+}\, MF_{n+1}^{-} \tag{4.65}$$

R_2N = Me_2N, Et_2N, piperidono, morpholino
m = 1–3
MF_n = BF_3, PF_5, AsF_5, SbF_5

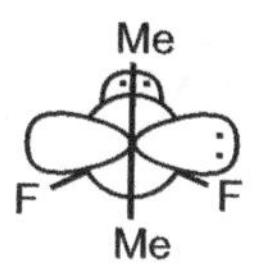

87

Nucleophilic substitution of fluorine in perfluorinated sulfonium ions yields the corresponding aminosulfonium ions when reacted with Me_3SiNMe_2 [Eq. (4.66)]. Crystal structure analysis of $CF_3S(NMe_2)_2^+$, $(CF_3)_2SNMe_2^+$, and $F_2SNMe_2^+$ hexafluoroarsenates gives similar S–N bond distances (1.605, 1.578, and 1.535 Å, respectively).[215,244]

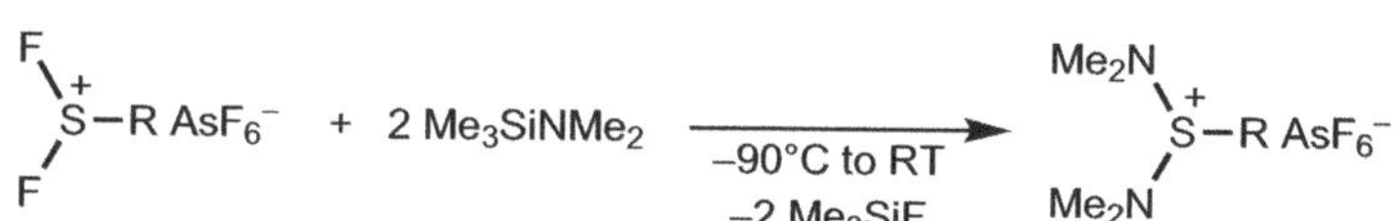

(4.66)

R = CF_3, $(CF_3)_2CF$

Studies by Olah et al.[245] have shown that due to its ambident character the NO_2^+ ion when reacting with sulfides gives both *S*-nitro- and *S*-nitritiosulfonium ions (**88** and **89**). Ions **88** are irreversibly transformed into **89** upon raising the temperature. The corresponding dimethyl species were detected by ^{13}C NMR spectroscopy (two methyl resonances at $\delta^{13}C$ 25.2 and 34.5). Dialkyl- and diarylsulfoxides also yield two ions reacting with NO_2^+ (**90** and **91**).[246] Again, two methyl resonances were observed in a controlled reaction of dimethylsulfoxide. The intensity of the resonance at $\delta^{13}C$ 35.4 slowly decreased upon raising the temperature from −60 to −20°C and disappeared after 4–5 h, whereas the resonance at $\delta^{13}C$ 42.4 remained unchanged.

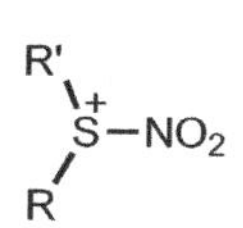

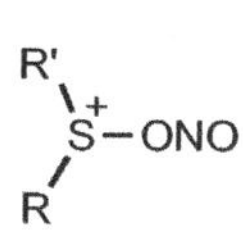

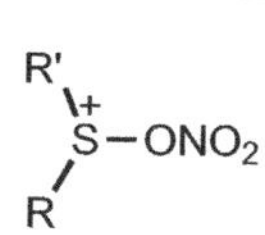

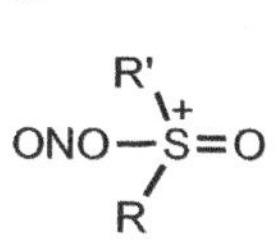

R, R' = Me, Et, Ph, 4-ClC_6H_4 — R, R' = Me, Et, *n*-Pr, Ph, 4-MeC_6H_4

88 **89** **90** **91**

Sulfonium Ions with Sulfur Ligands. Steudel and co-workers[247] prepared and characterized the $Me_2SSMe^+SbCl_6^-$ salt, the sulfur analog of ion **49**. The ion has C_s symmetry and the sole methyl group on S(2) bisects the angle formed by the other two methyl groups on S(1). They also determined the crystal structure of cation $(MeS)_3^+$ (**92**). The chain-end methyl groups were shown to occupy *trans* position with respect to the plane defined by the three sulfur atoms. The S–S–S–C torsion angle is 90.5° and the S–S bond distances (2.041 and 2.056 Å) are close to the single bond lengths.

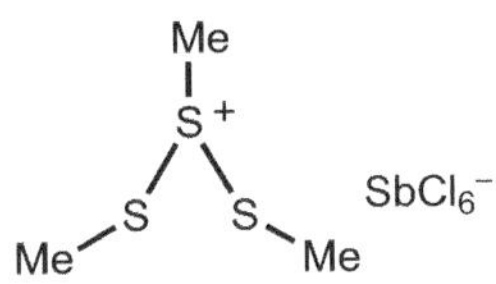

92

Trimercaptosulfonium salts were prepared by reacting Cl_3S^+ salts with excess H_2S[248] [Eq. (4.67)]. According to calculation, the pyramidal structure of C_{3v} symmetry is the energy minimum structure with the S–H bonds pointing toward the top of the pyramid. The corresponding dimercapto(methyl)sulfonium salts were also synthesized in an analogous way.[249] The products were transformed into chlorothio derivatives [Eq. (4.68)].

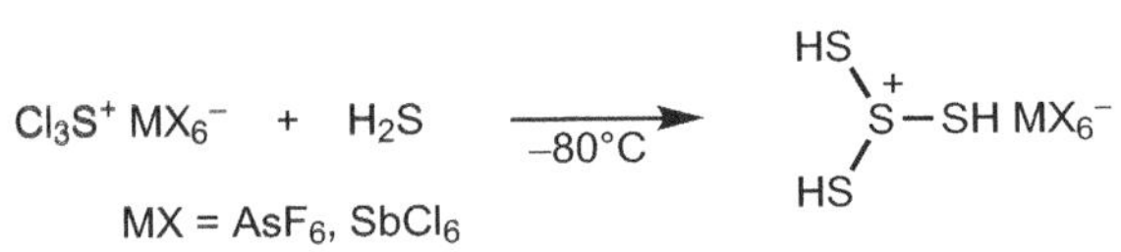

(4.67)

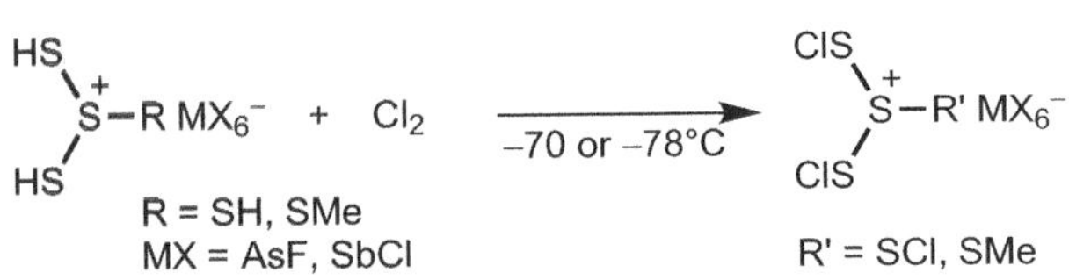

(4.68)

Minkwitz et al.[250] have obtained tris(methylthio)sulfonium hexafluoroantimonate in a solventless reaction at low temperature [Eq. (4.69)] and characterized the salt by Raman, IR, ^1H and ^{13}C NMR to spectroscopy. The salt is stable below 10°C but decomposes in SO_2 solution at −45°C with rapid elimination of sulfur. It possesses C_3 symmetry with a pyramidal trithiosulfonium unit and the methyl groups point to the central sulfur atom (*ab initio* calculations).

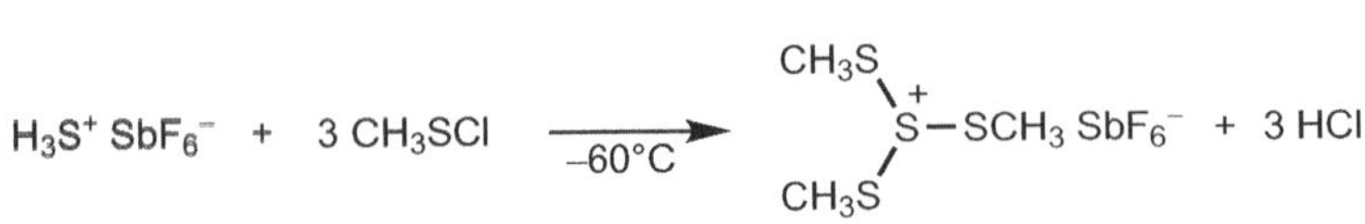

(4.69)

Substitution of chlorine in ion $MeSCl_2^+$ results in the formation of dithio-substituted methylsulfonium ions[251] [Eq. (4.70)]. X-ray structure determination of the $(MeS)_2SMe^+AsF_6^-$ salt gives S—S bond distances of 2.048 and 2.054 Å and a S—S—S bond angle of 107.6°.

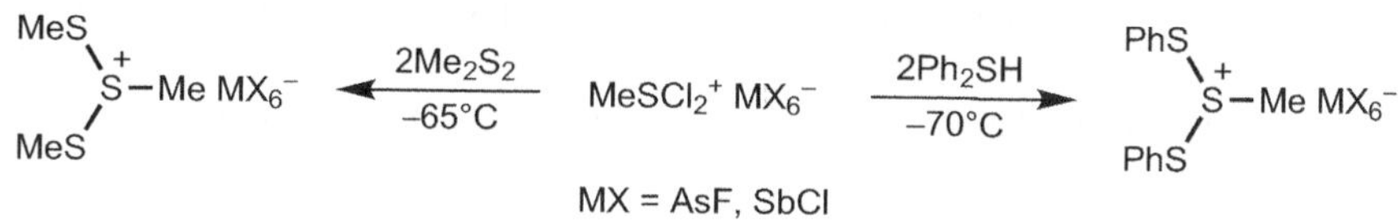

(4.70)

Sulfonium ion **93** has been prepared by treatment of the starting disulfur compound with $NO^+PF_6^-$ and the transformation was interpreted with the involvement of a dicationic species[252] [Eq. (4.71)]. Treatment of the corresponding sulfoxide with Tf_2O resulted in the formation of a cyclic product ion[253] [Eq. (4.72)].

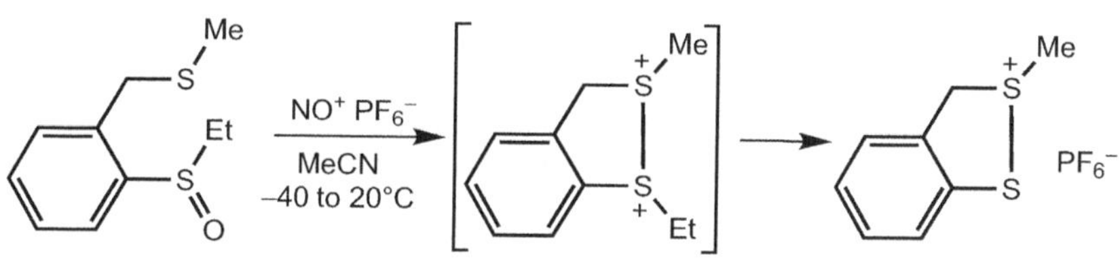

93

(4.71)

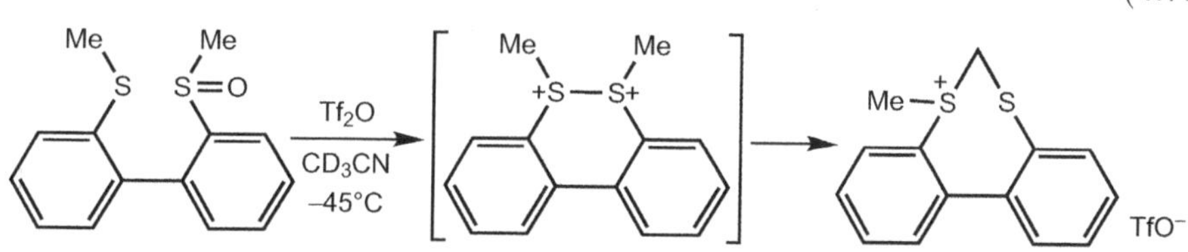

(4.72)

Thiosulfonium ions derived from methylation of substituted 1,2-dithianes (**94a**) were found to exist in undistorted chair conformation with the methyl group in axial position.[254] The phenomenon was interpreted to result from minimizing electron repulsion in orthogonal position (**94b**) (*ab initio* calculation at STO-3G*//STO-3G level). Unlike 1,2-dithianes, methylation of 1,2-dithiolanes and substituted thianes to form the corresponding *S*-methylsulfonium hexafluorophosphate and perchlorate salts[255] is non-stereoselective.

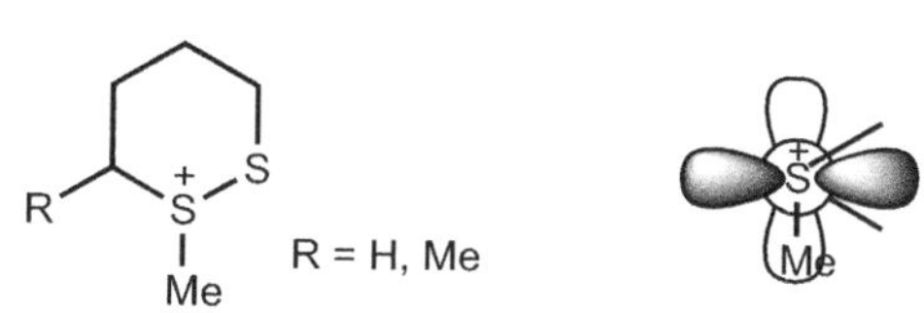

a **b**

94

Chemical two-electron oxidation of the corresponding 1,2-dithiin resulted in the formation of a stable solution of dication **95** (no decomposition was observed at −18°C for months)[256] [Eq. (4.73)]. ^{1}H and ^{13}C NMR spectra of the dication exhibited resonances at considerably lower fields compared with those of the neutral starting material suggesting aromatic stabilization. Calculated geometries (B3LYP/6-31G*) indicate a planar ring and calculated chemical shifts are in fair agreement with observed values. The related 1,4-dithiin dication was also characterized.[257]

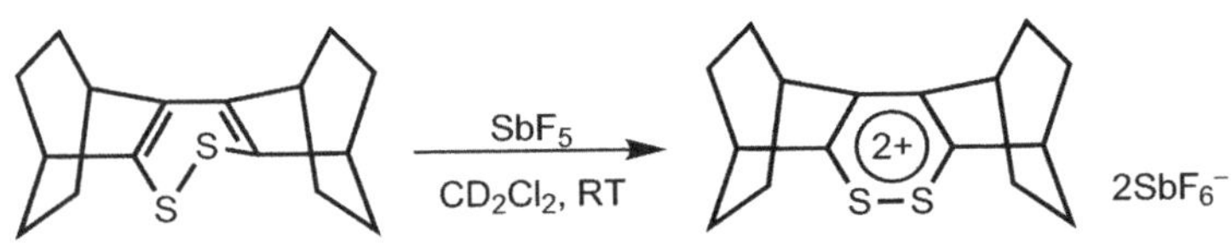

95

(4.73)

Silylated Sulfonium Ions. Olah, Prakash, and co-workers[258] have prepared silyl-substituted sulfonium ions [Eq. (4.74)] and characterized the products by multinuclear NMR spectroscopy. Calculated (DFT/IGLO) ^{13}C and ^{29}Si chemical shifts agree well with experimental data. The minimum energy structures of ions **96a** and **96b** are of C_1 and C_3 symmetry, respectively, with pyramidalization levels of 19° and 16° (pyramidalization level = out-of-plane bending angle of the central sulfur atom relative to the plane defined by its three bonding groups). A solution of ion **96a** is more stable than that of ion **96b**. The preparation of dimethyl(trimethylsilyl)sulfonium ion under similar conditions failed and the more stable $Me_3S^+OTf^-$ was formed instead.

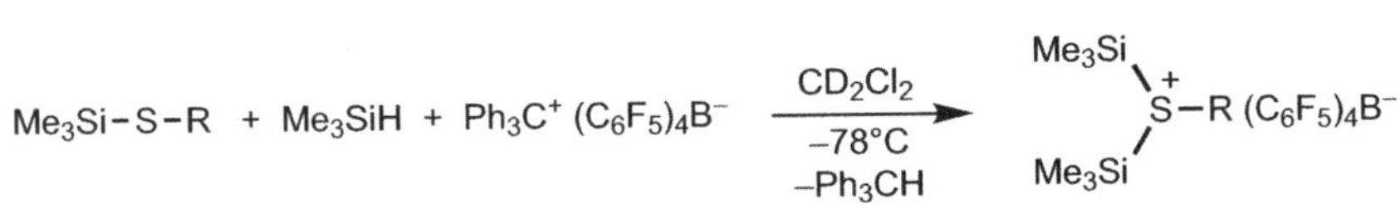

96 **a** R = Me
b R = Me3Si

(4.74)

Aurated Sulfonium Ions. Sheldrick and co-workers[259] were the first to report the synthesis and X-ray characterization of the aurated sulfonium salt $(Ph_3PAu)_3S^+PF_6^-$ and subsequently Schmidbaur et al.[119,260] synthesized similar tetrafluoroborates using two approaches [Eq. (4.75)]. The aurated sulfonium cations in the solid state are pyramidal and form dimeric units (**97a**) or strings of dimers (**97b**). In contrast, **97c** and **97d** do not build supracationic aggregates. In cation **97c** the Au−S−Au bond angle is 90.8°, the Au−S bond distance is 3.253 Å, and the intramolecular Au···Au contact is 2.285 Å. The supramolecular structure of ions **97a** and **97b** results in significant distortions of the cationic structures: much smaller (as low as 80.8°) bond angles are detected and the Au–S bonds are generally longer. The inter- and intramolecular Au···Au contacts are similar but shorter than in complex **97c**.

$$(R_3PAu)_3O^+ BF_4^- + (Me_3Si)_2S \longrightarrow (R_3PAu)_3S^+ BF_4^- \longleftarrow R_3PAu^+ BF_4^- + (Me_3Si)_2S$$

97 **a** R = Me
b R_3 = $MePh_2$
c R = isoPr
d R = Ph

(4.75)

The partially aurated ion **98** has been obtained by Sladek and Schmidbaur,[261] who reacted $(Ph_3PAu)_3O^+BF_4^-$ with 2-methylpropane-2-thiol. Cation **98** is dimeric in the crystal with very similar Au–Au distances within and between the monomeric units indicating significant aurophilic interaction between cations.

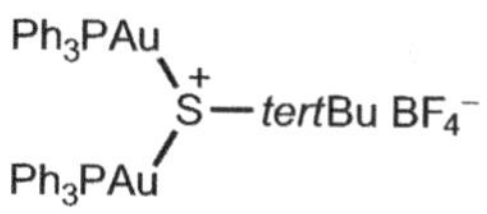

98

A unique mixed-valence aurated sulfonium cation **99** has been synthesized by Laguna and co-workers[262] by reacting the Au(III) precursor $[(C_6F_5)_2Au(OEt_2)_2]^+TfO^-$ with $[S(Au_2dppf)]$ [dppf = 1,1′-bis(diphenylphosphino)ferrocene]. The P–Au–S arrangements are close to linear (170.70–175.62°) and the S(1)–Au(III)–S(2) angle is 95.19°. The most significant feature of the cation **99**, in addition to three Au(I)···Au(I) contacts, is the Au(I)···Au(III) interactions of 3.2195 and 3.3661 Å. Results of DFT calculations agree only qualitatively with experimental data.

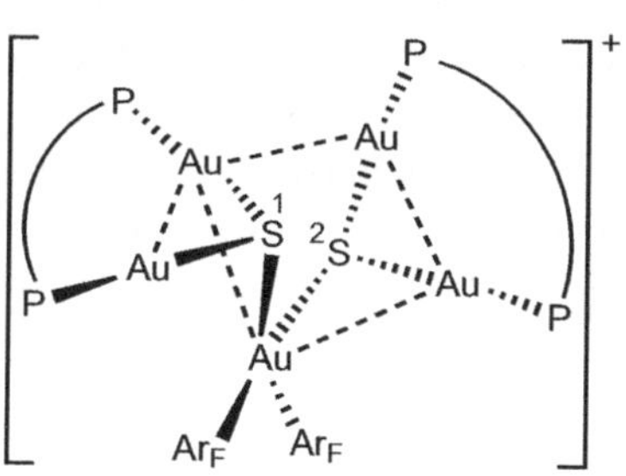

(P P = 1,1′-bis(diphenylphosphino)-ferrocene

$Ar_F = C_6F_5$

99

Laguna and co-workers[263] have reported the synthesis and X-ray characterization of the aurated sulfonium dication $[(Ph_3PAu)_4S]^{2+}(TfO^-)_2$, which also forms dimers in the crystal. It possesses a tetragonal pyramidal framework, with the sulfur atom occupying the apical position. The Au–S–Au bond angles are significantly smaller than those in the corresponding monocation $[(Ph_3PAu)_3S]^+$ (73.5–75.6° versus 80.82–87.8°[119,259]), resulting in short Au–Au distances between adjacent and longer ones between opposite Au atoms (2.883–2.938 and 3.4 Å, respectively). There are two distinctive Au–Au–Au angles: about 75° and about 100°. Subsequently, they synthesized a variety of highly aurated sulfonium complexes $[(Ar_3PAu)_nS]^{(n-2)+}$ and reported the X-ray structure of dication $[(Ph_3PAu)_4S]^{2+}(ClO_4^-)_2$ with structural features similar to that of the triflate salt.[264]

Aurated dications of a different type, bissulfonium salts **100** formed from dithiols, have been reported[265,266] including the systematic study by Sladek and Schmidbaur.[267] Single-crystal X-ray diffraction study indicated that they form complicated aggregated structures.

$$(Ph_3PAu)_2\overset{+}{S}-(CH_2)_n-\overset{+}{S}(AuPPh_3)_2\ (BF_4^-)_2$$

100 $n = 4, 5$

4.2.3. Selenonium and Telluronium Ions

A great number of trialkyl(aryl) selenonium and telluronium ions are known, and their synthesis does not require the use of strong electrophilic alkylating or arylating agents.[268,269] The synthesis and transformations of triorganotellurium ions have been treated in recent reviews.[270,271] However, acidic selenonium and telluronium ions can be obtained only under superacidic conditions.

4.2.3.1. Hydridoselenonium and Hydridotelluronium Ions.

Hydrogen selenide is considerably less basic than hydrogen sulfide and, consequently, more difficult to protonate and the resulting hydridoselenonium salts are expected to be more thermolabile. Nevertheless, hydrogen selenide can be protonated by $HF–BF_3$ in excess HF solution[272] [Eq. (4.76)]. The hydridoselenonium ion **101** formed in this way at -70°C shows a singlet ^{1}H NMR resonance at δ^1H 5.8, deshielded by 6.1 ppm from the ^{1}H NMR resonance of parent H_2Se. In the ^{77}Se NMR spectrum, **101** is observed at δ^{77}Se -142 (from dimethyl selenide) with no resolvable ^{77}Se–^{1}H coupling even at -90°C.[273] Methyl selenide also undergoes protonation in $HF–BF_3$ media to methylselenonium ion $MeSeH_2^+$. The methylselenonium ion, however, shows ^{77}Se–^{1}H coupling in the ^{77}Se NMR spectrum ($J_{^{77}Se-^{1}H} = 129.5$ Hz). A long-range ^{77}Se–methyl proton coupling of 8.9 Hz is also observed.[273]

$$H_2Se + HF + BF_3 \longrightarrow H_3Se^+\ BF_4^- \qquad (4.76)$$

101

The $HF–SbF_5$ superacid has been used to prepare the $H_3Se^+SbF_6^-$ salt [Eq. (4.77)] and the perdeuteriated derivative.[274] The salt is stable at 195 K but decomposes within minutes at 213 K to reform the starting materials.

$$H_2Se + HF + SbF_6 \xrightarrow[195\ K]{} H_3Se^+\ SbF_6^- \qquad (4.77)$$

The parent hydrotelluronium ion H_3Te^+ could not be observed in superacid solution of hydrogen telluride, under conditions where the selenonium ion is observed.

4.2.3.2. Acidic Selenonium and Telluronium Ions. Alkyl selenides are much more stable to oxidation than hydrogen selenide, and they can be protonated in $HSO_3F–SbF_5–SO_2$ solution[272] [Eq. (4.78)]. The dimethylselenonium ion Me_2SeH^+ (protonated dimethyl selenide) shows in its ^{1}H NMR spectrum the methyl double at δ^1H 2.96 ($J = 7.0$ Hz) and the SeH septet at δ^1H 4.50 ($J = 7.0$ Hz). A double irradiation experiment showed that the doublet and septet are coupled. The ^{1}H NMR spectrum also shows an unidentified small doublet at δ^1H 3.50 and a singlet at δ^1H 3.80 for the $Me_2Se–SbF_5$ complex. The diethylselenonium ion shows the methyl triplet at δ^1H 2.00, the methylene quintet at δ^1H 3.77, and the SeH quintet at δ^1H 4.40. The ^{13}C and ^{77}Se NMR data of these two ions agree well with ^{1}H NMR results. The acidic, secondary alkylselenonium ions are remarkably stable. The ^{1}H NMR spectra show no significant change from -60°C to $+65$°C.

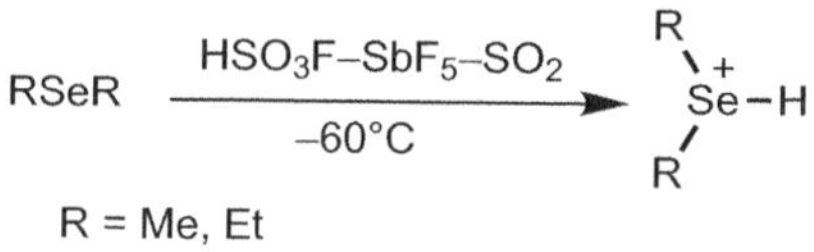

(4.78)

Alkyl tellurides in $HSO_3F–SbF_5–SO_2$ solution at -60°C show deshielded alkyl proton chemical shifts (with no long-range proton coupling) as compared with the corresponding dialkyl tellurides themselves in SO_2. This indicates that in this medium the tellurides are probably oxidized. However, using $HF–BF_3$ in excess HF solution [Eq. (4.79)], both the TeH^+ proton and its coupling to secondary alkyl groups can be observed.[272]

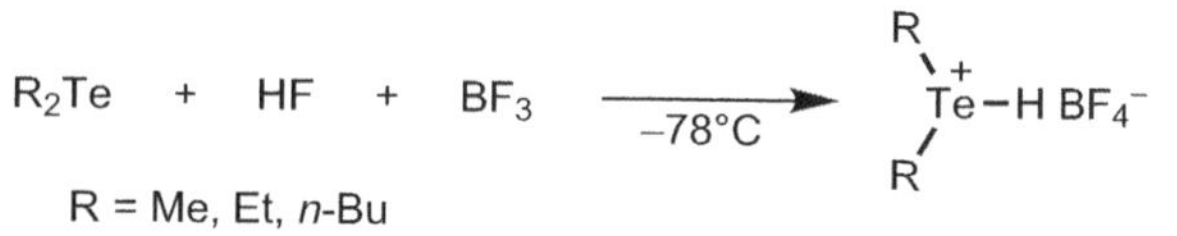

(4.79)

The dimethyltelluronium ion Me_2TeH^+ (protonated dimethyl telluride) shows the methyl doublet at δ^1H 2.7 ($J = 7$ Hz) and the septet at δ^1H 1.6. Similarly, the diethyltelluronium ion Et_2TeH^+ (protonated diethyl telluride) shows the methyl triplet at δ^1H 1.9, the methylene quintet at δ^1H 3.4, and the multiplet, partially overlapping the methyl triplet at δ^1H 1.6.

The proton on selenium in selenonium ions and on tellurium in telluronium ions is considerably more shielded than the proton on oxygen in the related oxonium ions (δ 7.88–9.21) and the proton on sulfur in the corresponding sulfonium ions (δ 5.80–6.52). There is a consistent trend of increasing shielding going from related oxonium to sulfonium to selenonium to telluronium ions (which is particularly significant when considering the directly observed protons on heteroatoms). Charge delocalization and shielding by increasingly heavier atoms is thus indicated.

Laali et al.[275] also observed that the ^{77}Se chemical shifts in the long-lived acidic selenonium ions R_2SeH^+ (Me = δ^{77}Se 343, Et = δ^{77}Se 399) are more deshielded than that in the corresponding $R_2Se \rightarrow SbF_5$ donor–acceptor complexes (Me = δ^{77}Se 258, Et = δ^{77}Se 317). The ^{13}C chemical shifts, in turn, show an opposite trend; that is, the α-methyl carbon is more deshielded in the donor–acceptor complexes than in the ions.

4.2.3.3. Tertiary Selenonium and Telluronium Ions. Trialkylselenonium fluorosulfates are conveniently prepared by the reaction of dialkyl selenide and alkyl fluorosulfates, using 1,1,2-trichlorotrifluoroethane as the reaction medium[272] [Eq. (4.80)]. Trimethylselenonium fluorosulfate **102a** thus prepared is a stable, white solid (mp = 83–85°C), which, when dissolved in liquid sulfur dioxide, exhibits a singlet proton NMR absorption at δ^1H 2.7. Triethylselenonium fluorosulfate **102b** has also been prepared in the same way. It is also a stable, white solid (mp = 25–28°C). When dissolved in liquid sulfur dioxide, **102b** shows the methylene protons at δ^1H 3.2 (quartet) and the methyl protons at δ^1H 1.4 (triplet).

$$RSeR + ROSO_2F \xrightarrow[\text{RT}]{Cl_2FCCF_2Cl} R_3Se^+ FSO_3^- \qquad (4.80)$$

102 **a** R = Me
b R = Et

Further examples of alkylation of selenides are shown in Eqs. (4.81)[276] and (4.82).[277]

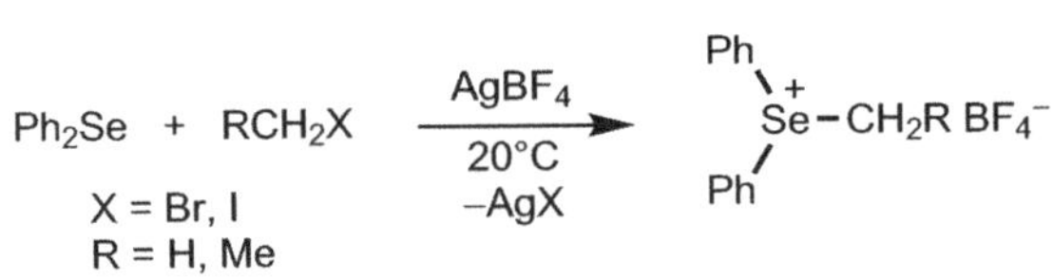

(4.81)

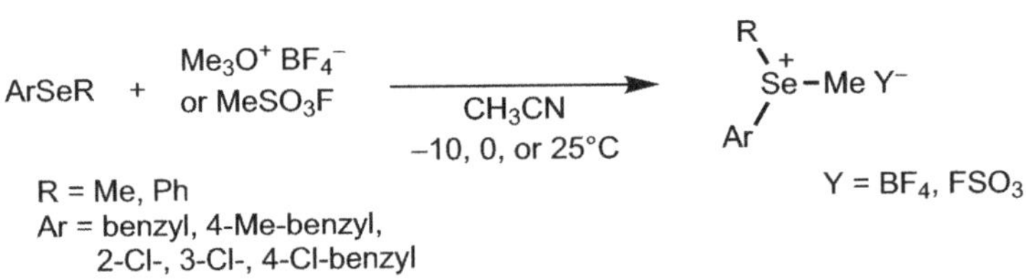

(4.82)

Methylation of trifluoromethylphenylselenide under conditions shown in Eq. (4.81) was successfully employed in the synthesis of $Me(CF_3)PhSe^+BF_4^-$.[278] The cyclization methods described in Eq. 4.45 have also been used to prepare analogous Se and Te derivatives[175] [Eq. (4.83)].

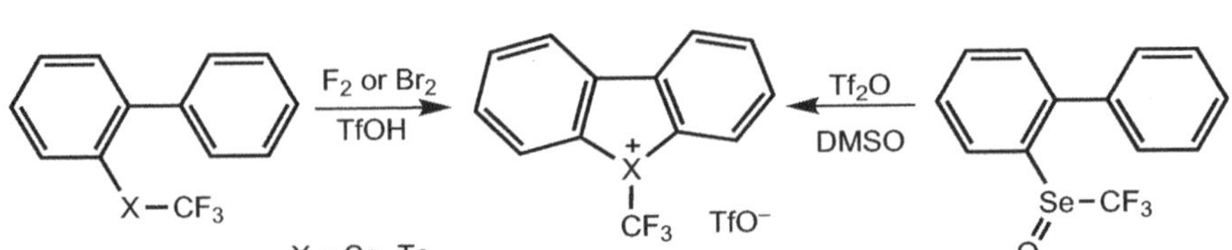

(4.83)

Kataoka and co-workers have developed methods to obtain alkynyl- and alkenylselenonium ions[279,280] [Eq. (4.84)] and allenylselenonium triflate derivatives.[281]

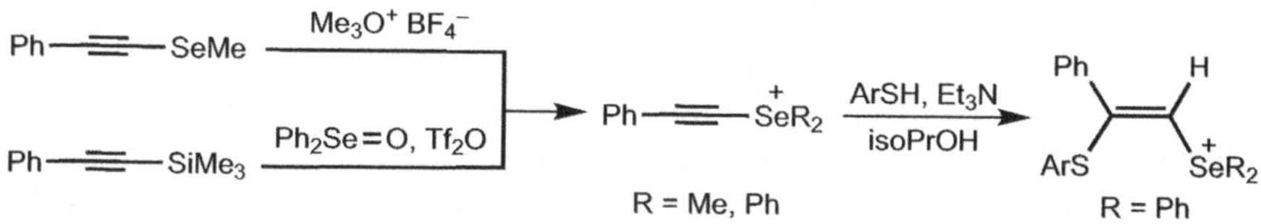

(4.84)

Various organoselenium compounds are able to add in stereospecific *anti* fashion to alkenes[282] [Eq. (4.85)][283] or alkynes[282] [Eq. (4.86)] to form stable cyclic selenonium ions.

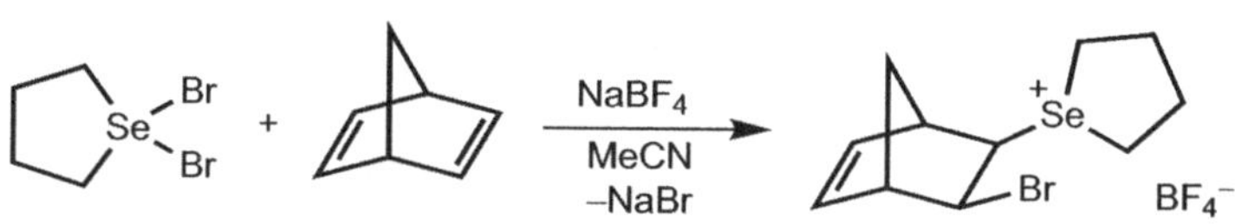

(4.85)

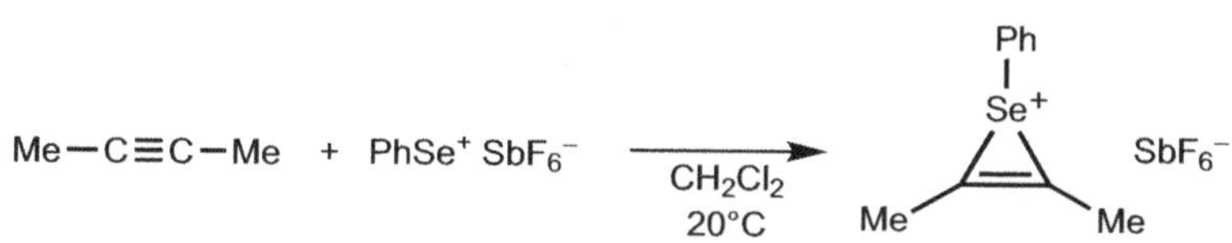

(4.86)

In the molecular structure of the selenium analog of cation **66**, the phenyl and acenaphthyl rings have a similar *cis* arrangement.[284] The Se–C bond lengths (2.16 and 2.17 Å) are significantly longer than those in Me_3Se^+ (1.946 and 1.962 Å). The X-ray crystal structure of the cation $(MeSe)_3{}^+$ the selenium analog of **92** has also been determined.[247] The two cations have very similar structures with a notable difference: The Se–Se–Se–C torsion angle of the ion $(MeSe)_3{}^+$ is significantly smaller (only 47.1°), that is, ion $(MeS)_3{}^+$ has a more open structure. The Se–Se bond distances (2.354 and 2.372 Å) are close to the single bond lengths.

Trialkyltelluronium fluorosulfates were prepared similarly to the trialkylselenonium salts from dialkyl tellurides and alkyl fluorosulfates.[272] Trimethyltelluronium fluorosulfate **103** prepared in this way [Eq. (4.87)] is a stable, light-yellow salt (mp 130°C), which, dissolved in liquid sulfur dioxide, exhibits a singlet ^1H NMR signal at δ^1H 2.3. The corresponding triethyltelluronium salt Et_3Te^+ could not be isolated, although prepared in solution it is also quite stable. No cleavage of the ions in solution is observed up to 65°C.

$$(CH_3)_2Te \ + \ CH_3OSO_2F \longrightarrow \underset{\mathbf{103}}{(CH_3)_3Te^+ \ FSO_3^-} \qquad (4.87)$$

The synthesis and detailed ^1H, ^{13}C, ^{77}Se, and ^{125}Te NMR spectroscopic characterization of the corresponding triflates have been reported by Laali et al.[275] The ^{77}Se

and ^{125}Te chemical shifts were shown to increase with the increasing size of the alkyl groups. The ^{13}C NMR chemical shifts of carbons directly attached to the heavy atom are significantly shielded, showing the tendency of increasing shielding Me_3S^+ (δ^{13}C 27.20) < Me_3Se^+ (δ^{13}C 21.78) < Me_3Te^+ (δ^{13}C 3.76).

A wide variety of synthetic procedures have been developed to synthesize telluronium ions from disubstituted tellurides.[271] Naumann and Tyrra[285,286] have prepared ion **104** using various approaches [Eq. (4.88)] whereas du Mont and coworkers[287] obtained the trimesityltelluronium cation **105** [Eq. (4.89)]. Cation **105** has a propeller-like conformation with C–Te–C angles close to tetrahedral (104.4–109.3°). The triflate salt of ion **104** has a distorted ψ-trigonal bipyramidal geometry and the C–Te–C angles (91.1–106.1°) are somewhat different from those in ion **105**.

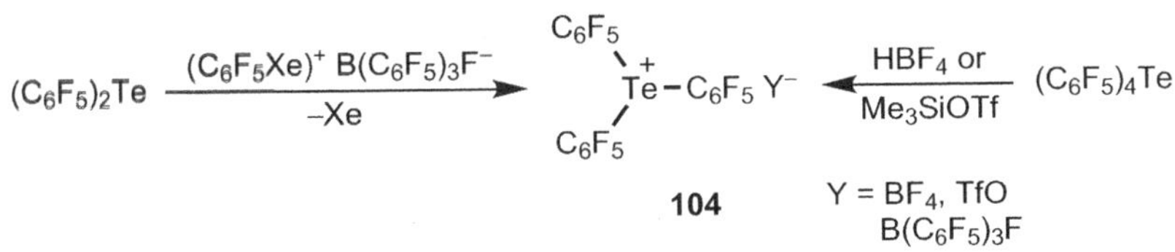

(4.88)

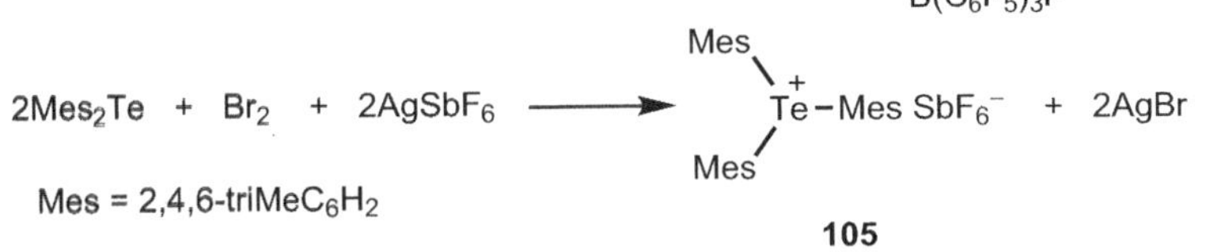

(4.89)

Diaryltellurium difluorides also serve as convenient and versatile starting materials to obtain various telluronium tetrafluoroborates[288] (Scheme 4.9). The tellurium atom in cation **106** (Ar = Ph, R = *tert*-Bu) has a distorted pyramidal geometry. The observed Te–C bond distances (2.108–2.129 Å) are close to those reported for other telluronium ions. There is a weak nonbonding interaction between Te and O manifested in the lower IR absorption of C=O (1678 cm^{-1}). The ^{125}Te NMR chemical shifts in the range δ^{125}Te 645–755 indicate the onium ion character of the central tellurium atom. Ph_2TeF_2 also reacts with aryl- and alkenylboronic acids in the presence of $BF_3 \cdot OEt_2$ to yield the corresponding aryl- or vinyl-substituted telluronium tetrafluoroborates.[289]

1-Alkynylphenyliodonium tetrafluoroborates selectively transfer the unsaturated group to diphenyl chalcogens to form 1-alkynyldiphenylonium ions[290]

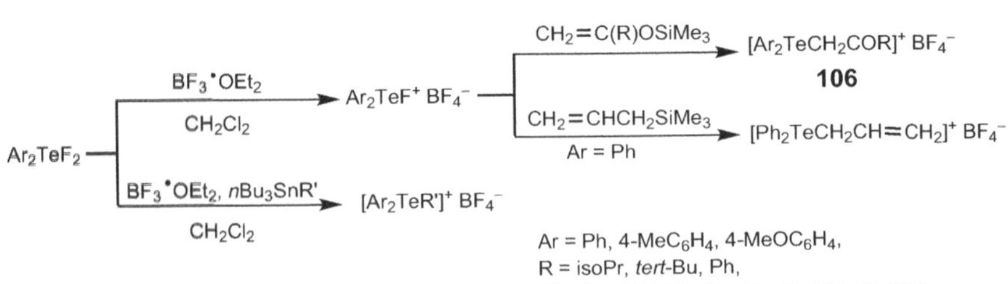

Scheme 4.9

[Eq. (4.90)]. Analogous reactions can be performed with vinyl derivatives to obtain 1-vinyldiphenylonium ions.[291]

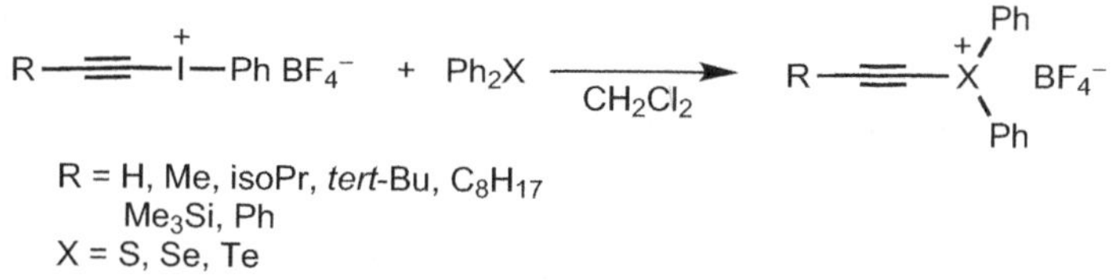

(4.90)

Hypervalent pentaphenyltelluronium salts obtained by Akiba and co-workers[292] [Eq. (4.91)] were shown by X-ray crystal structure analysis to have unusual square pyramidal geometry. The Te$-$C$_{apical}$ bond lengths (2.101–2.125 Å) are significantly shorter than the four basal Te$-$C bond distances (2.174–2.214 Å). The ^{125}Te NMR spectra of the three salts show almost identical chemical shifts (δ^{125}Te 659.0–659.9) which are upfield from that of Ph_3Te^+ (δ^{125}Te 753.0).

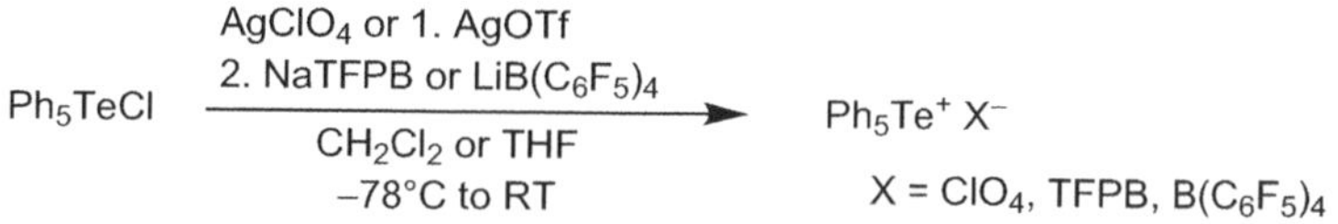

(4.91)

The unusual telluronium dication analogous to **79** [see Eq. (4.47)] has also been prepared by Furukawa and Sato.[197,198] The ^{1}H and ^{13}C NMR spectra are consistent with those of dication **79**. The X-ray crystallographic analysis of the triflate salt indicates, however, a nearly hexagonal structure for the telluronium dication. This results from the strong interaction of the anions with the Te center forming a hexavalent structure.

Gillespie and coworkers obtained the compound $Ph_2Se_6{}^+AsF_6{}^-\cdot SO_2$.[293] The cation **107** consists of a six-membered selenium ring in a boat conformation with the phenyl rings at opposite corners. The average Se$-$Se bond length of 2.416 Å is typical for selenium dications. The distance across the top of the boat is 3.550 Å, which is smaller than the sum of the van der Waals radii (4.00 Å) indicating a weak intracationic interaction.

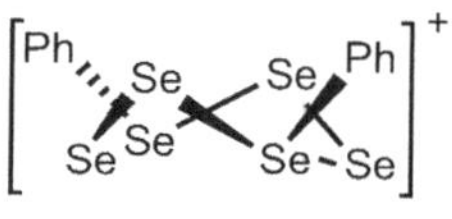

107

du Mont and co-workers[287] have prepared cations **108** containing Te$-$Se (**108a**) or Te$-$Te (**108b**) moiety [Eq. (4.92)]. The X-ray crystal structure of the ions shows that the coordination geometry of all chalcogen atoms is pseudo-trigonal-bipyramidal, when secondary M---F contacts are taken into consideration. The ions are dimeric in the solid state with a bridging SbF_6^- unit. Te---F and Se---F interactions weaken the Te–Se and Te–Te bonds (Te–Se bond length = 2.5755 Å, Te–Te bond length = 2.7645 Å). The ions, therefore, can be regarded as $MesTe^+$ and $C_6F_5Se^+$ cations stabilized by the coordinating Lewis base Mes_2Te.

$$(C_6F_5)_2Se_2 + Mes_2Te + Br_2 + 2AgSbF_6 \xrightarrow{CH_2Cl_2} [Mes_2TeXAr]^+ \, SbF_6^- \quad (4.92)$$

108 **a** X = Se, Ar = C_6F_5
b X = Te, Ar = Mes

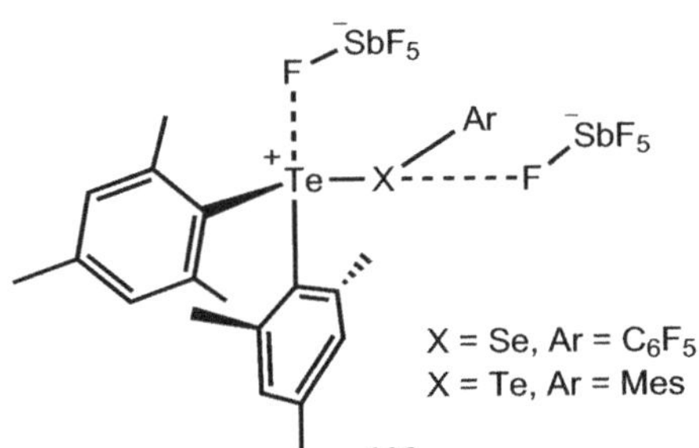

108

4.2.3.4. Haloselenonium and Halotelluronium Ions.

Oxidative chlorination of Se and Te in the presence of $AsF_3/AlCl_3$ was used by Kolditz and Schäfer to synthesize $SeCl_3^+AsF_6^-$ and $TeCl_3^+AsF_6^-$.[203] F_3Se^+ salts with a variety of anions were prepared by Peacock[294] and Bartlett and Robinson.[199] The ionic nature of these salts was first demonstrated by Edwards and Jones,[295] who performed X-ray crystal structure analysis of $TeF_3^+Nb_2F_{11}^-$ to show substantial interaction between the ions through fluorine bridging. Later, Gillespie and Whitla[296] arrived at a similar conclusion on the basis of conductometric, cryoscopic, vibrational, and NMR spectroscopic measurements with $SeF_3^+MF_6^-$ (M = Sb, As, Nb, Ta) and $SeF_3^+BF_4^-$ adducts.

Gillespie and Whitla[297] also studied the $SF_4{\cdot}SO_3$ system with a variety of techniques (IR, Raman, ^{19}F NMR, conductivity and cryoscopic measurements). In the solid and molten states the polymeric fluorosulfate bridged structure **109** exists, whereas in dilute solution the cyclic dimer **110** is the predominant species. A comparison of the ^{19}F NMR chemical shifts of F_3S^+ and F_3Se^+ shows that fluorines bonded to Se^+ ($\delta^{19}F$ +9.6 from $CFCl_3$) are about 21 ppm shielded compared to fluorines on S^+ ($\delta^{19}F$ +30.5) indicating charge delocalization and shielding by the heavier atom.[169] This observation is similar to that found for trialkylonium salts.

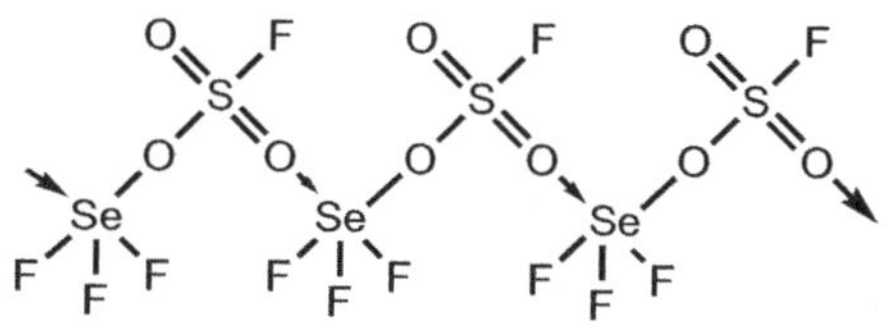

109

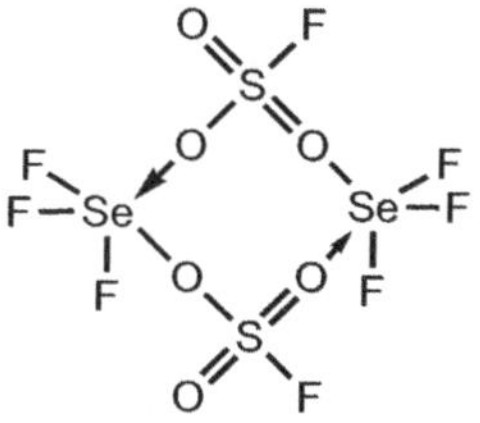

110

Subsequently, all Se and Te trihalide cations were prepared. X-ray single-crystal characterization of hexafluoroarsenate and hexafluoroantimonate salts shows that these cations, similar to the corresponding SX_3^+ cations, have trigonal pyramidal geometry and the MX_3^+ units are linked to the anions via M···F interactions.[207,210,298–300] An interesting feature of the crystal structure of the hemisolvate $TeI_3^+AsF_6^- \cdot SO_2$ is the pairwise association of the TeI_3^+ ions facing each other with the I atoms and forming large voids.[301] This results in significantly lower density in comparison to the unsolvated salt. A computational study of SeX_3^+ ions (X = Cl, Br, I) and ^{77}Se chemical shifts, along with the first solid-state FT–Raman spectrum of $SeI_3^+AsF_6^-$, has been reported.[302] The interesting salt $[Cl_3Te{-}F{-}TeCl_3]^+[Sb(OTeF_5)_6]^-$ contains isolated cations and anions.[303] The cation has two pseudo-bipyramidal Cl_3TeF(lone pair) units bridged centrosymmetrically by fluorine. Each tellurium has two shorter and one longer additional fluorine contacts.

4.2.3.5. Aurated Selenonium and Telluronium Ions. The crystal structures of $[(Ph_3PAu)_3Se]^+PF_6^-$ prepared by Sheldrick and co-workers[304] and $[(Ph_3PAu)_3Te]^+BF_4^-$ obtained by Angermaier and Schmidbaur[305] according to Eq. (4.93) have been determined. The Au–Te–Au angles in the Au_3Te pyramids are all much smaller than 90° (72.6–84.7°), resulting in Au–Au distances of 3.074–3.515 Å, which are considered to be bonding contacts. The cations are associated to dimers through short intermolecular Au···Au contacts in the same range. When the tellurium reagent was reacted with four equivalents of $Ph_3PAu^+BF_4^-$, the tetranuclear dication $[(Ph_3PAu)_4Te]^{2+}$ was obtained and assumed to have a pyramidal geometry.

$$(tert\text{-}BuMe_2Si)_2Te + (Ph_3PAu)_3O^+ BF_4^- \xrightarrow[\substack{-78°C \\ -(tert\text{-}BuMe_2Si)_2O}]{CH_2Cl_2,\ THF} [(Ph_3PAu)_3Te]^+ BF_4^- \qquad (4.93)$$

Laguna and co-workers[306] have succeeded in preparing the $[(Ph_3PAu)_4Se]^{2+}$ dication **111**, which crystallizes with two moles of dichloromethane and isostructural with the corresponding sulfur salts $[(Ph_3PAu)_4S]^{2+}(Y^-)_2$ (Y = ClO_4, TfO) discussed above. In the tetragonal pyramidal framework, Se occupies the apical position. The base of the pyramid is not exactly coplanar but has a flattened butterfly form with a hinge angle of 27° about diagonal Au atoms. Au–Au distances between adjacent Au atoms (2.8959–2.9605 Å) and opposite Au atoms (3.6 and 4.5 Å) are longer than those in the sulfur analogs.

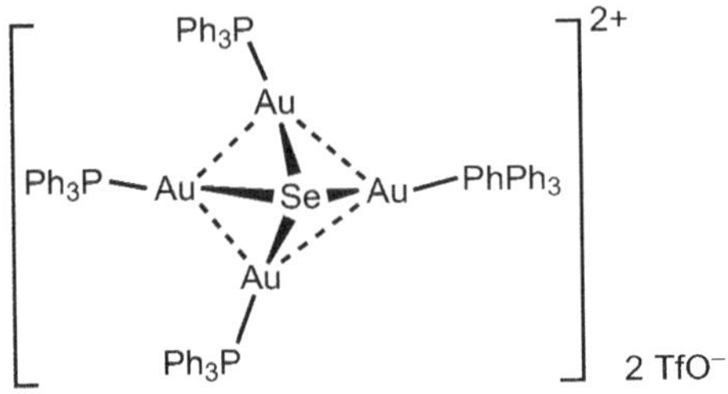

111

4.2.3.6. Polychalcogen Dications. The first disulfonium dication derived from 1,5-dithiacyclooctane was prepared by Musker et al.[307,308] [Eq. (4.94)] and subsequently Furukawa and co-workers[309,310] reported the synthesis and crystal structure of dication **112**. Dication **112** was prepared by reacting the corresponding monosulfoxide with $(TfO)_2O$, which is a general method for the synthesis of analogous dichalcogen dications.[311] In the crystals, there are two independent dications with slightly different characteristics. The S–C bonds (1.828–1.842 Å) are slightly longer, whereas the C–C bonds are shorter (1.509–1.527 Å) than the corresponding normal single bonds. The C–S–C bond angles are 104.1° and 104.6°. The eight-membered ring has a distorted C_2 chair–chair conformation. Four energy-minimum structures were located by *ab initio* studies (RFH/3-21G*).[312] In three structures, the lone electron pairs of the sulfur atoms have *cis* configuration and all three are much more stable (by about 27–28 kcal mol^{-1}) than the structure with the *trans* arrangement. The related dications **113** could be characterized by NMR spectroscopy.[313,314] An *ab initio* molecular orbital study was performed on dimeric dication **114**.[315] Disulfonium dications are presumed intermediates in organic transformations.[316]

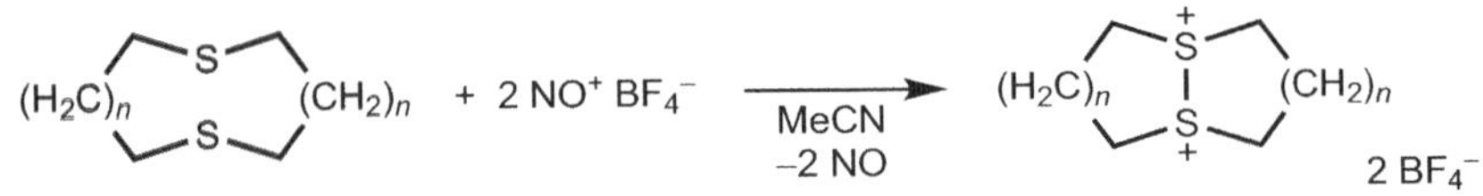

n = 1, 2

(4.94)

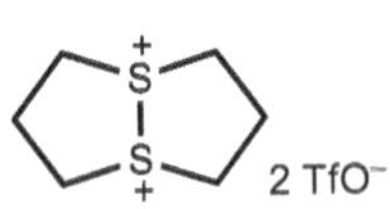

112

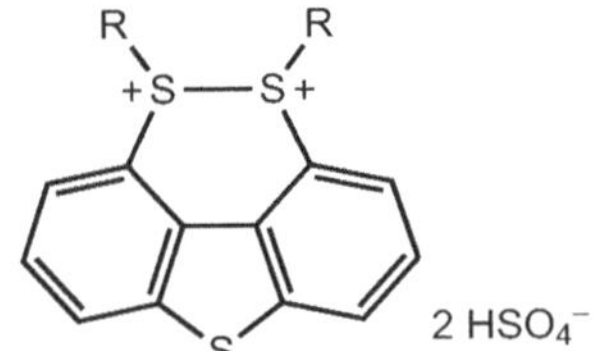

R = Me, Ph

113

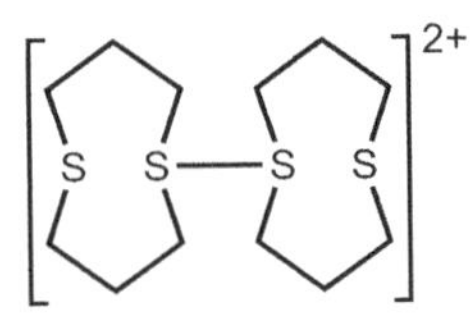

114

Furukawa and coworkers have also studied dications with a Se–Se unit. Dication **115** was generated by two-electron oxidation with $NO^+PF_6^-$ [Eq. (4.95)].[317] Computational studies of the isomeric structures have led to the same conclusions as already discussed above for disulfonium dication **112**. The analogous ditellurium dication (BF_4^- and PF_6^- salts) has also been prepared.[318]

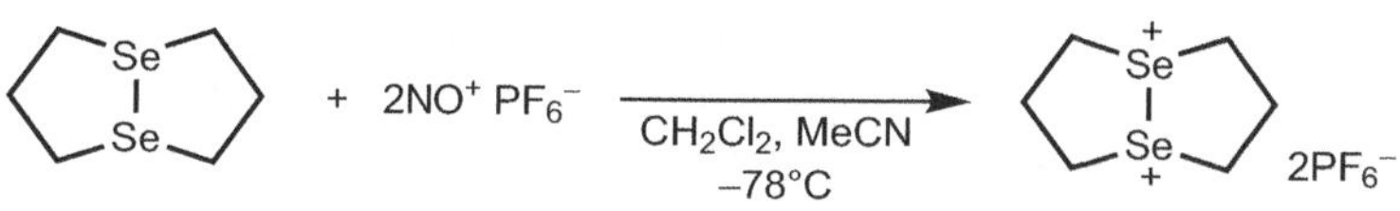

(4.95)

115

Minkwitz et al.[319] have obtained dications **116** in the reactions of Me_2SH^+ and Me_2SSH^+ with SCl_2 and S_2Cl_2, respectively, and reported spectroscopic characterization (IR, X-ray, and ^1H and ^{13}C NMR).

$$Me_2\overset{+}{S}-S_n-\overset{+}{S}Me_2$$

n = 1–3

116

Disulfonium dicationic species with the participation of hypervalent Se (**117**) or Te (**118**) atom have been described.[316,320] The ^{77}Se NMR spectrum of the dihexafluorophosphate salt of dication **117** shows a single peak at $\delta^{77}Se$ 946.7 characteristic of selenuranes.[321] The ^{125}Te NMR chemical shifts for ions **118** are in the range $\delta^{125}Te$ 1327.3–1343.[322,323] X-ray structure determination of the ditriflate salt of **118** (R = Ph) showed that the Te atom has a distorted trigonal bipyramidal geometry with two apical Te–S bonds and two equatorial Te–C bonds, with the lone electron pair occupying the third equatorial position. NMR spectral characteristics indicate that the ions have *cis–trans* or *trans–cis* configuration [the phenyl group attached to S(1) or S(3) is *trans* to the other two phenyl groups]; that is, a racemic mixture exists. According to *ab initio* calculations (RHF/3-21G* level), the total positive charge, located exclusively on the chalcogen atoms, is +2.578 [S(1) = +0.483, S(2) = +0.495, Te = +1.600], that is, larger than 2, which is attributed to the polarization of the Te–C and S–C bonds.

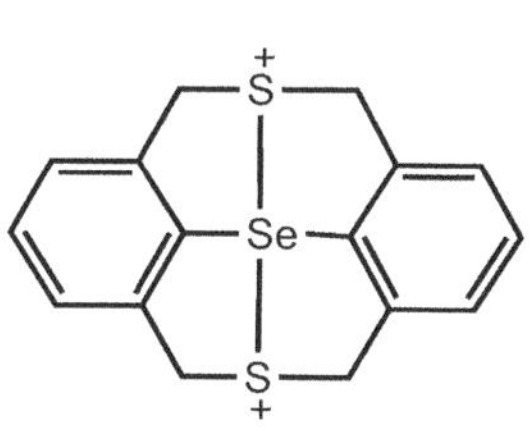

117

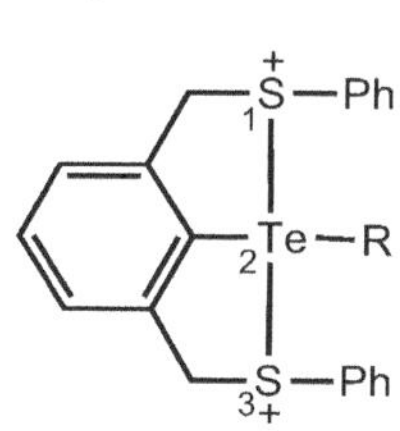

118

2 Y^-

R = Ph, $(C_6H_4S)_nC_6H_4SMe$

n = 0, 1

Y = BF_4, TfO

Other dications with three chalcogen atoms have also been studied.[316] When the tris-selenide precursor was oxidized with $NO^+PF_6^-$, the twin-chair conformer changed to the more rigid twin boat conformer **119** [Eq. (4.96)].[321] This was evidenced by a change in the ^{1}H NMR resonance of the benzylic protons (δ^1H 3.88 and 5.33 versus δ^1H 4.07 and 4.59). The crystal structure analysis also supports this conclusion.[324] Dication **119** has an almost linear Se–Se–Se bond (Se–Se–Se bond angle = 170.21°) with a C–Se(2)–C bond angle of 95.6°. The Se(2) atom has a distorted trigonal bipyramidal configuration with two apical Se–Se bonds and two equatorial Se–C bonds, with the lone pair occupying the third equatorial position. Se(2) carries a charge of +1.020, whereas the charges on the other two Se atoms are +0.701. The analogous O–Se–Se dication, the only example with participation of oxygen, has very similar structural characteristics.[325]

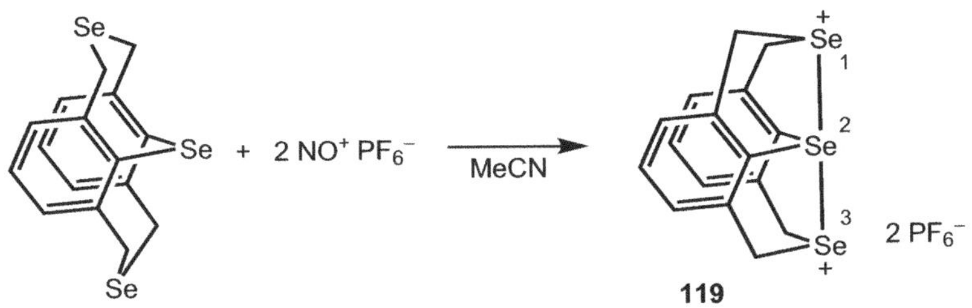

119

(4.96)

The X-ray structure of dication **120** exhibits features very similar to those of the analogous dication **118** (R = Ph).[322] Treatment of bis(4-methylphenyl)telluride with $NO^+BF_4^-$ or TfO_2 has afforded dication **121** with NO or O_2 as the oxygen source.[326] The ions are dimeric in the solid state and the anions interact with the Te atoms through their O atoms. The Te---O contacts are considerably shorter than the sum of the van der Waals radii of the two atoms (2.64–2.98 Å versus 3.60 Å). The Te atoms have pseudo-octahedral geometry.

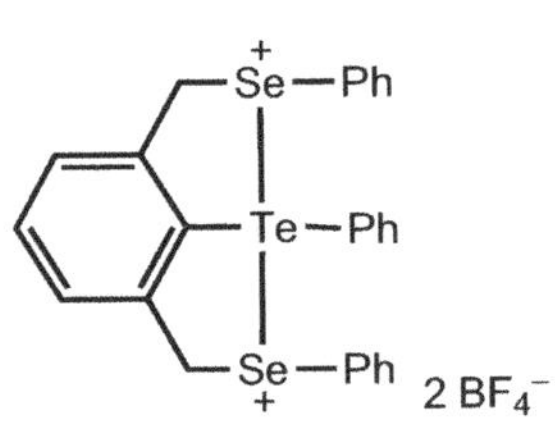

120

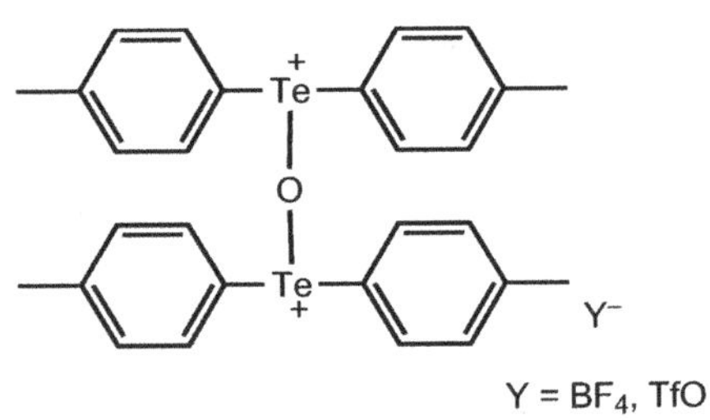

121

4.2.4. Halonium Ions

Organic halogen cations—that is, halonium ions of acyclic (open-chain) (**122**) or cyclic (**123**) nature—have gained increasing significance, both as reaction intermediates and as preparative reagents. They are related to oxonium ions in reactivity but offer much more selectivity. Halogen atoms that form organic halonium ions are chlorine, bromine, and iodine. To date, no stable organic fluoronium ion has been prepared or characterized. A comprehensive monograph[99] pertaining to the chemistry of halonium ions that covers more than 200 research publications and a review[327] have been published.

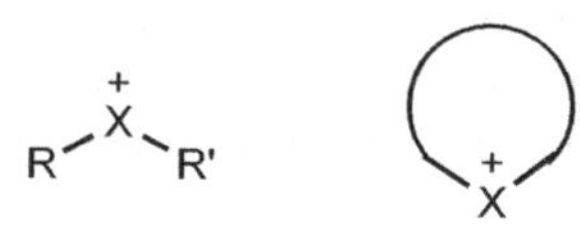

122 123

In 1894, Hartmann and Meyer[328a] were the first to prepare a diphenyliodonium ion salt when they reacted iodosobenzene in concentrated H_2SO_4 and obtained *para*-iodophenylphenyliodonium bisulfate. Diphenyliodonium ions have since been studied by various research groups, most notably by Nesmeyanov[103] and Beringer[328b] since 1950. Diarylbromonium and -chloronium ions, although considerably less stable and much less investigated, were also prepared in the 1950s by Nesmeyanov and co-workers.[103] Iodonium ions, which are useful reagents and synthons in organic synthesis and frequently characterized as organic polyvalent iodine(III) compounds, have been extensively studied.[329–332]

Open-chain (acyclic) dialkylhalonium ions of the type R_2X^+ ($X = Cl, Br, I$) were unknown until the 1960s, as were alkylarylhalonium ions ($ArRX^+$). Realization of their possible role as intermediates in alkylation reactions of haloalkanes and -arenes has followed their preparation and study.[333]

One of the most daring proposals of an organic reaction mechanism of its time was made in 1937 by Roberts and Kimball,[334] who suggested that the observed *trans* stereospecificity of bromine addition to alkenes is a consequence of intermediate bridged bromonium ion formation. The brief original publication suggested the actual structure of the ion is undoubtedly intermediate between **124** and **125**. Structure **124** was not intended to represent a conventional free carbocation, however. Since the two carbons in either structure are joined by a single bond and by a halogen bridge, free rotation is not to be expected. A clear description of the difference in bonding between carbon and bromine in **124** and **125** was not given.

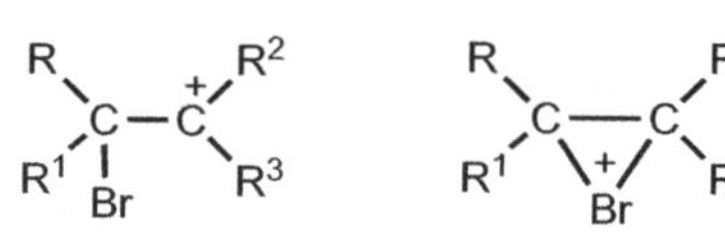

124 125

The bromonium ion concept was quickly used by other investigators to account for stereospecific transformation of alkenes,[335,336] notably by Winstein and Lucas,[337] but was not unanimously accepted.[338] For example, in discussing the mechanistic concepts of bromonium ion formation, Gould, in his still popular text,[339] wrote in 1959, "Although a number of additions are discussed in terms of the halogenonium-ion mechanism, the reader should bear in mind that few organic mechanisms have been accepted so widely while supported with such limited data."

In 1967, Olah and Bollinger[340] reported the first preparation and spectroscopic characterization of stable, long-lived bridged alkylenehalonium ions. This was followed by Olah and DeMember's[333] preparation in 1969 of the first dialkylhalonium ions. Since then the field of organic halonium ions has undergone rapid development through substantial contributions from an increasing number of investigators, notably by Peterson.[341]

The halogen atom in organic halonium ions is generally bound to two carbon atoms, although in the case of acidic halonium ions—that is, protonated alkyl halides—one ligand is hydrogen.

Of the dihydrohalonium ions—that is, acidic halonium ions—only chloronium ion is characterized. HCl has been protonated to dihydrochloronium ion in HSO_3F–SbF_5 media and studied by ^1H NMR spectroscopy.[342] Christe[150] managed to isolate the $H_2Cl^+SbF_6^-$ salt at low temperature, and the equilibrium molecular structure of H_2Cl^+ has been determined in the gas phase.[343] Under these strongly acidic conditions, HBr and HI are readily oxidized. H_2F^+ in the condensed state cannot be considered as dihydrofluoronium ion,[344] because the very electronegative fluorine atom resists acquiring positive charge. Instead, the HF solvated proton can have linear or more probably 2*e*–3*c* bonding. Mootz and Bartmann,[345] however, succeeded in determining the crystal structure of the $H_2F^+SbF_6^-$ salt without being able to locate the hydrogens. Subsequently, they isolated the $H_7F_6^+SbF_6^-$ salt and found that the cation forms an unbranched chain held together by very strong hydrogen bonds, which become weaker from center to the ends.[346] Both this study and an *ab initio* molecular dynamic calculation[347] on the formation of H_2F^+ in the HF–SbF_5 superacid system have concluded that the cationic species is a protonated HF chain mediating a fast proton jump process.

Both *ab initio* [MP2(FU)/6-31G** level][48] and DFT (B3LYP/6-31-G** level)[348] calculations have been reported for the H_3X^{2+} dications. The H_3Cl^{2+}, H_3Br^{2+}, and H_3I^{2+} dications have C_{3v} symmetry and considerable kinetic barriers for deprotonation (27.4–37.9 kcal mol^{-1}), although deprotonation is exothermic (66.5–35.0 kcal mol^{-1}). In contrast, the H_3F^{2+} dication has a planar D_{3h} structure and the dissociation energy is only 8.6 kcal mol^{-1}.

4.2.4.1. Acyclic (Open-Chain) Halonium Ions

Alkyl- and Arylhydridohalonium Ions. The self-condensation of alkyl halides in strongly acidic media represents a convenient preparative route to symmetric dialkylhalonium ions **127** [Eq. (4.97)]. This reaction involves hydridohalonium ions **126** as intermediates that subsequently undergo nucleophilic attack by excess alkyl halide.

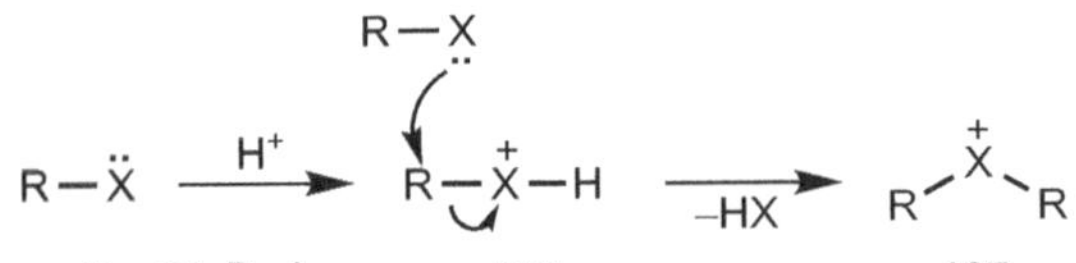

126 127

(4.97)

Methyl bromide and methyl iodide in $HSO_3F–SbF_5–SO_2ClF$ at −78°C give both dimethylhalonium ions (**127**, R = CH_3, X = Br, I) as well methylhydridohalonium ions (**126**, R = CH_3, X = Br, I).[349] These hydridohalonium ions have been characterized by ^{13}C NMR spectroscopy.[349] Methyl chloride does not give methylhydridochloronium ion under similar conditions.

Protonation of methylhalonium ions to form diprotonated dications $CH_3XH_2^{2+}$ have been calculated (B3LYP/6-31-G** level)[348] to occur primarily on the C−H bond to form a pentacoordinated carbon with 2*e*–3*c* bond. The halogen atoms in dications, as expected, carry more charge than in the corresponding monocations and the dications are less stable than the monocations (49.5–13.0 kcal mol^{-1}).

Halobenzenes do not form diarylhalonium ions under superacidic conditions. The protonation occurs either on the halogen or on the aromatic ring.[350] Indeed, chloro- and bromobenzenes quantitatively yield the corresponding 4-halobenzenium ions[349] (Wheland intermediates) on protonation with $HSO_3F–SbF_5–SO_2ClF$ at −78°C. Iodobenzene under similar conditions gives the hydridoiodonium ion **128** [Eq. (4.98)].[349] The ion **128** is rather stable and does not rearrange to ring-protonated $C_6H_6I^+$ even when the temperature is raised to −20°C.

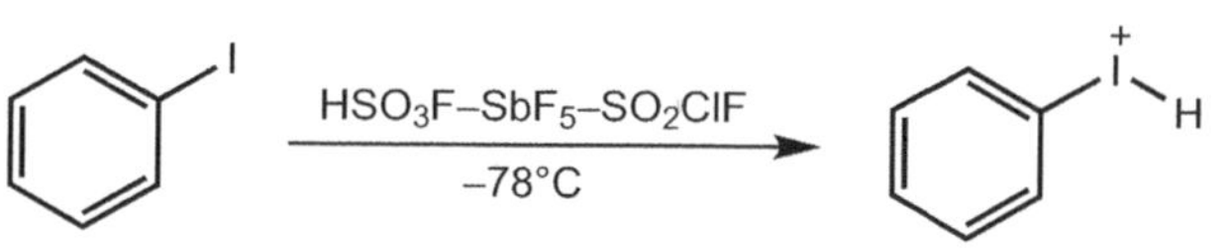

128

(4.98)

Protonation of fluorobenzene in the gas phase has been studied by infrared photodissociation (IRPD) spectroscopy by Solcà and Dopfer.[351] F-protonated fluorobenzene was formed in significant amount when protonation was carried out with CH_5^+. It was found to be the most stable isomer in the gas phase by quantum mechanical calculations [B3LYP/6-311G(2df,2pd) level] separated by a large energy barrier from the four Wheland intermediates. F-protonated fluorobenzene is best described as a weakly bound ion–dipole complex between the phenyl cation and HF.

Dialkylhalonium Ions. Dialkylhalonium ions were first observed as stable fluoroantimonates and characterized by Olah and DeMember[333] in 1969. Since then, a large number of unsymmetrical and symmetrical halonium ions have been prepared. The alkylating ability as well as intermolecular exchange reactions of dialkylhalonium ions were also studied.[107,352–355]

There are two general methods for the preparation of dialkylhalonium ions. (i) The reaction of excess primary and secondary alkyl halides with $SbF_5–SO_2$ [Eq. (4.99)], anhydrous fluoroantimonic acid ($HF–SbF_5$) [Eq. (4.100)], or anhydrous silver hexafluoroantimonate (or related complex fluoro silver salts) in SO_2 solution [Eq. (4.101)]. (ii) The alkylation of alkyl halides with methyl or ethyl fluoroantimonate (or alkylcarbenium fluoroantimonates) in SO_2 solution[107,333,352,353] [Eq. (4.102)]. The first method is only suitable for the preparation of symmetrical dialkylhalonium ions, whereas the second method can be used for both symmetrical and unsymmetrical dialkylhalonium ions. Additional methods for the preparation of dialkylhalonium ions are also available, but these methods generally have less practical value.

$$2RX \xrightarrow[-60°C]{SbF_5\text{–}SO_2} R\overset{+}{X}R\ SbF_5X^- \tag{4.99}$$

$$2RX + H^+SbF_6^- \longrightarrow R\overset{+}{X}R\ SbF_6^- + HX \tag{4.100}$$

$$2RX + Ag^+SbF_6^- \xrightarrow[SO_2]{} R\overset{+}{X}R\ SbF_6^- + AgX \tag{4.101}$$

$$R'X + R^+SbF_6^- \xrightarrow[-60°C]{SO_2} R'\overset{+}{X}R\ SbF_6^- \tag{4.102}$$

R = Me, Et
R′ = Me, Et, Pr
X = Cl, Br, I

Ions **129–133** are some of the representative dialkylhalonium ions that have been prepared and characterized.[99] (Trimethylsilyl)methylhalonium ions **134** have also been obtained under stable ion conditions and characterized by ^{1}H, ^{13}C, and ^{29}Si NMR spectroscopy.[67] The Raman and IR spectroscopic studies of dimethylhalonium ions **129** seem to indicate that these species exist in a bent conformation.

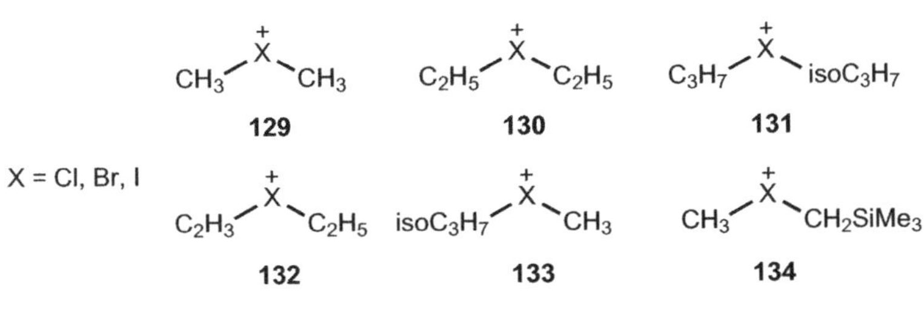

Minkwitz and Gerhard[355] have reported the synthesis of the AsF_6^- and $Sb_2F_{11}^-$ salts of cation **129** (X = Br) by reacting bromine with $CH_3OSO^+MF_6^-$ (M = As, Sb). In the analogous reaction with iodine, the salts $(CH_3I_2)_n^{n+}(MF_6^-)_n$ (M = As, Sb) of unknown structure were isolated. They also succeeded in methylating CF_3I to obtain cation **135** [Eq. (4.103)] but failed to prepare the analogous bromonium compound.[356]

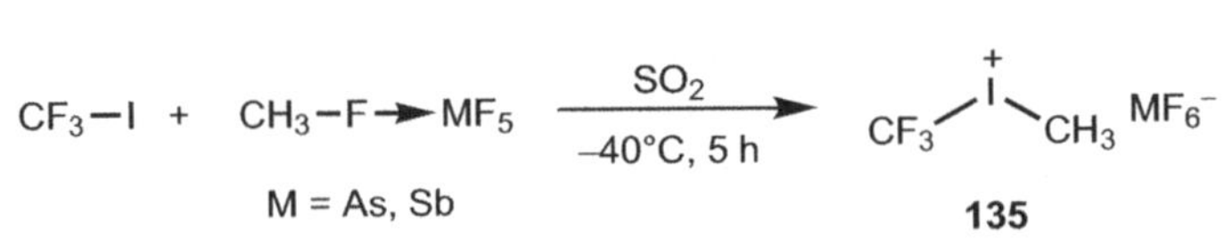

(4.103)

Even dicyclopropylbromonium ion **136** has been prepared[357] by the ionization of cyclopropyl bromide in SbF_5–SO_2ClF solution at low temperature. Also a series of alkylcyclopropylhalonium ions **137** and **138** has been studied.[357]

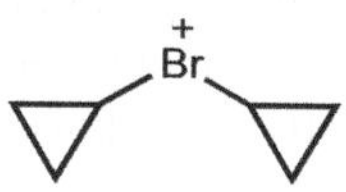

136

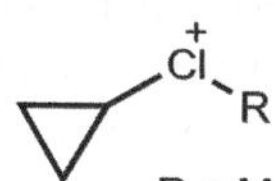

R = Me, isoPr

137

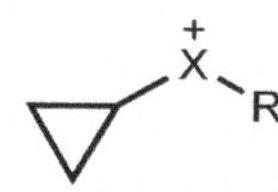

X = Br, I
R = Me, Et, isoPr

138

Alkylvinylhalonium ions **139**, which are stable only below −90°C, have also been investigated.[358]

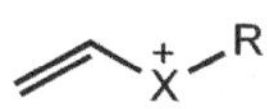

X = Cl, Br
R = Me, Et

139

Cubylhalonium ions, the first stable acyclic tertiary halonium ions, have been prepared by Olah, Prakash, and co-workers.[359,360] Methylation of 1,4-dihalocubanes with CH_3F–SbF_5 at −60°C gave the ions **140a** and **140b**, which could be identified unequivocally on the basis of their ^{13}C NMR spectra. Assignments for **140c**, however, were inconclusive due to the formation of unidentified products. Attempts to prepare the corresponding monosubstituted analogs (**141**) in a similar way were unsuccessful. This, however, is not surprising considering the fact that halonium ions are known to localize most of their charge on the halogen atoms.

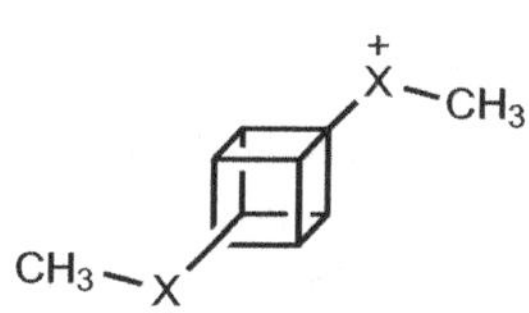

140 **a** X = I
b X = Br
c X = Cl

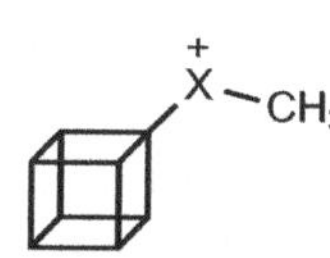

141

Stang and co-workers[361–363] have reported the synthesis of the triflate salts of dialkynyliodonium ions (**142**), the phenyl(cyano)iodonium ion **143** and the dicyanoiodonium ion **144**.

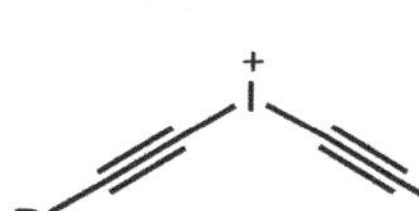

R = *tert*-Bu, Me_3Si, $isoPr_3Si$

142

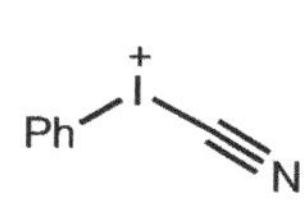

143

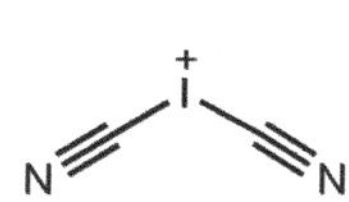

144

In search for nonvolatile and, therefore, safer chloromethylating agents, even chloromethylhalonium ions have been synthesized.[364] Bis(chloromethyl) chloronium ion is formed by the ionization of dichloromethane in SbF_5–SO_2ClF at −130°C.

Generally, the preparation of symmetrical dialkylhalonium ions is simpler and the reactions are clean. Unsymmetrical dialkylhalonium ions undergo disproportionation and alkylation (self-condensation) reactions, even at low temperatures (about −30°C).

A qualitative measure of the stability and the sequence of reactivity of dialkylhalonium ions and their deshielding characteristics are shown in Scheme 4.10.

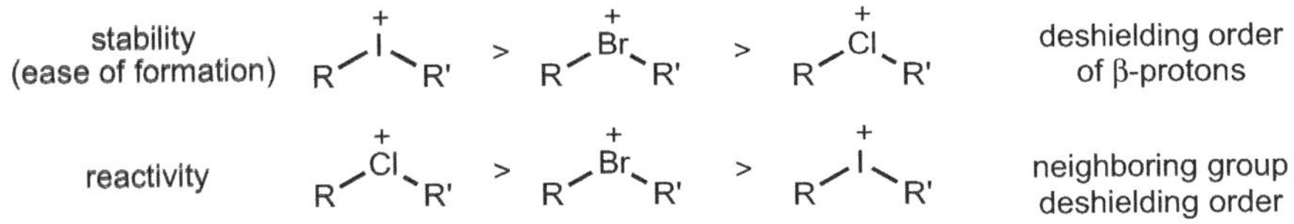

Scheme 4.10

Dimethylhalonium fluoroantimonates such as dimethylbromonium and -iodonium fluoroantimonates can be isolated as crystalline salts. They are stable in a dry atmosphere at room temperature, and some are now commercially available. Dimethylhalonium fluoroantimonate salts are very hygroscopic, and exposure to atmospheric moisture leads to immediate hydrolysis.

The ^1H and ^{13}C NMR data[107,352,353] on dialkylhalonium ions seem to indicate the neighboring group deshielding order shown in Scheme 4.10, indicating inductive effect of the halogen atoms. Chlorine, being the smallest halogen atom in halonium ions (fluoronium ions are not known in solution), can accommodate the least amount of positive charge, whereas iodine, the largest of the halogen atoms, can accept essentially most of the charge. However, β-protons in the ^1H NMR spectra of related homologous dialkylhalonium ions show an opposite deshielding order (Scheme 4.10) indicating that the inductive effect of the positively charged halogen atoms diminishes and the anisotropy effect of halogen atoms causes an opposite trend.

Dialkylhalonium ions are reactive alkylating agents. The alkylation of π-donor (aromatic and olefinic) and *n*-donor bases with dialkylhalonium ions has been studied.[353] Alkylation of aromatics with dialkylhalonium ions was found to be not significantly different from conventional Friedel–Crafts alkylations, showing particular similarities in the case of alkylation with alkyl iodides. Alkylation of *n*-donor bases with dialkylhalonium salts provides a simple synthetic route to a wide variety of onium ions.

Alkylation of alkylene dihalides with methyl and ethyl fluoroantimonate (CH_3F–SbF_5–SO_2 and C_2H_5F–SbF_5–SO_2) gives monoalkylated halonium ions and/or dialkylated dihalonium ions, depending on the reaction conditions.[365] Iodine shows an unusual ability to stabilize positive charge, as demonstrated by the formation of dialkyl alkylenediiodonium ions [$RI^+(CH_2)_nI^+R$, $n = 1$ to 6, R = Me, Et]. In the extreme case ($n = 1$), the two positive iodonium cation sites are separated only by a single methylene group. However, dialkyl alkylenedibromonium ions were formed only when the two positive bromines were separated by three methylene groups. In the case with four methylene groups, rearrangement to the more stable five-membered ring tetramethylenebromonium ion takes place. Dialkyl alkylenedichloronium ions have not yet been directly observed. Consequently, the ease of formation of dialkyl alkylenedihalonium ions is similar to that of dialkylhalonium ions (Scheme 4.10).

Olah, Rasul, Prakash, and co-workers[348] have calculated the structure and stability of protonated dimethylhalonium dications $(CH_3)_2XH^{2+}$ (B3LYP/6-31-G** level). Both the halogen-protonated and C–H bond protonated Br and I dications, all of C_2 symmetry, were found to be energy minima on the potential energy surface. Both C–H bond protonated dications possess a pentacoordinated carbon with $2e$–$3c$ bond. The Br- and I-protonated forms are more stable than the C–H bond protonated species by 21.7 kcal mol^{-1} and 19.0 kcal mol^{-1}, respectively. Of the corresponding fluorine and chlorine analogs, only the C–H bond protonated chloronium dication is an energy minimum structure.

Similar calculations[348] of the methylated dimethylhalonium dications $(CH_3)_3X^{2+}$ have found the trimethylchloronium, trimethylbromonium, and trimethyliodonium dications of C_{3v} symmetry to be the stable minima. The fluoro analog is also an energy minimum structure but has a planar D_{3h} symmetry, a relatively long C–F bond, and a bond order of 0.53, indicating a very weak C–F bond. The order of increasing atomic charges of the halogen atoms (F = –0.27, Cl = 0.54, Br = 0.87, and I = 1.16) corresponds to the order of electronegativities (F > Cl > Br > I) and the order of sizes (I > Br > Cl > F) of the halogen atoms.

Alkyl(aryl)halonium Ions. Dence and Roberts[366] attempted to prepare the cyclopropylphenyliodonium ion from phenyliodoso chloride and cyclopropyllithium [Eq. (4.104)]. However, they were unable to obtain the corresponding iodonium ion or any cyclopropylbenzene from the reaction mixture. Thus, the iodonium ion was not formed, even as an unstable reaction intermediate.

$$C_6H_5ICl_2 + RLi \xrightarrow[-LiCl]{} [C_6H_5\overset{+}{I}R]\,Cl^- \longrightarrow C_6H_5I + RI + C_6H_5Cl + RCl \qquad (4.104)$$

$R = \text{cyclo}C_3H_5$

Perfluoroalkyliodoso trifluoroacetates react with aromatic compounds to give perfluoroalkylaryliodonium ions **145** [Eq. (4.105)].[367]

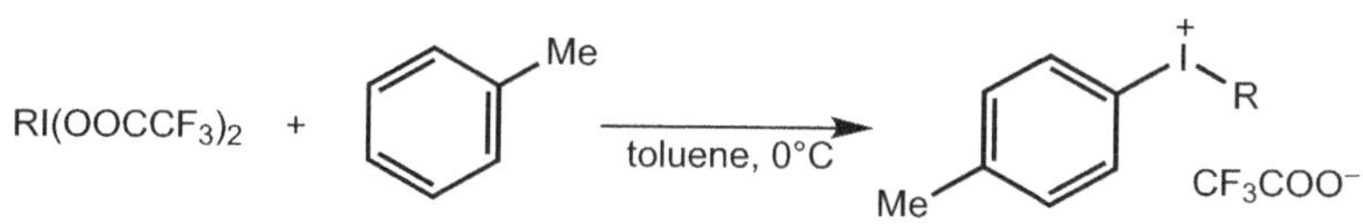

(4.105)

$R = n\text{-}C_3F_7,\ n\text{-}C_6F_{13},\ C_6F_5$

145

Alkyl(aryl)halonium ions (other than perfluorinated derivatives) were first prepared by Olah and Melby.[368] When a SO_2 solution of iodobenzene was added to a SO_2 solution of the CH_3F–SbF_5 complex (methyl fluoroantimonate) at –78°C, a clear, slightly colored solution resulted. The ^{1}H NMR spectrum of this solution at $-80°$C showed in addition to the excess methyl fluoroantimonate a methyl singlet at δ^1H 3.80 and a multiplet aromatic region (7.7–8.3) with a peak ratio of 3:5. The aromatic signals showed the same coupling pattern as that of iodobenzene in SO_2 but were deshielded by approximately 0.5 ppm. The species that accounts for the NMR data is the methylphenyliodonium ion **146-I**. When bromobenzene and other aryl bromides or iodides were added in the same manner to methyl fluoroantimonate in SO_2, analogous spectra were obtained, indicating the formation of the corresponding methylarylbromonium ions [Eq. (4.106)].

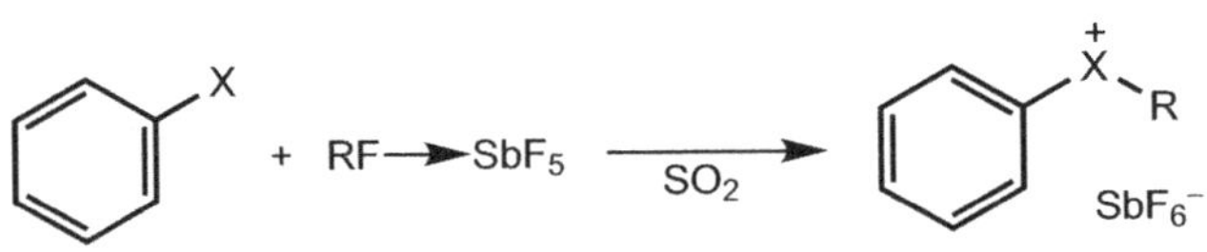

(4.106)

X = Br, I
R = Me, Et

146
R = Me, X = I **146-I**

Likewise, the reaction of bromo- and iodoarenes with ethyl fluoroantimonate in SO_2 gave the corresponding ethylarylhalonium ions. The structures of all aklylarylhalonium ions prepared have been characterized by ^{1}H and ^{13}C NMR spectroscopy.

Similarly, a series of dialkylphenylenedihalonium ions such as **147**–**150** have been prepared and characterized.[365] Even trihalonium ions such as **151** have been prepared. Many of the mentioned halonium ions are stable only below $-20°$C and above which they undergo ring alkylation.

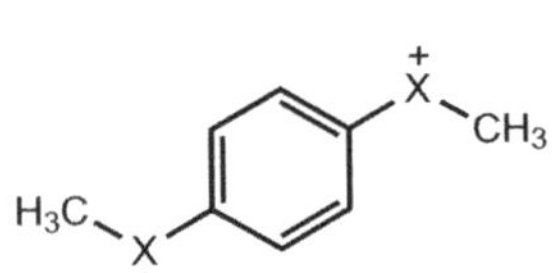

X = Br, I
147

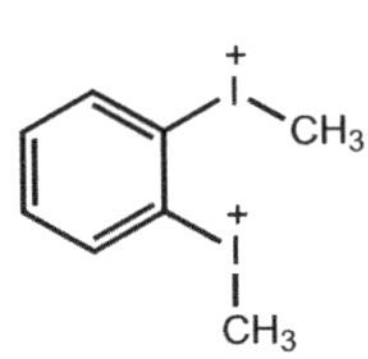

148

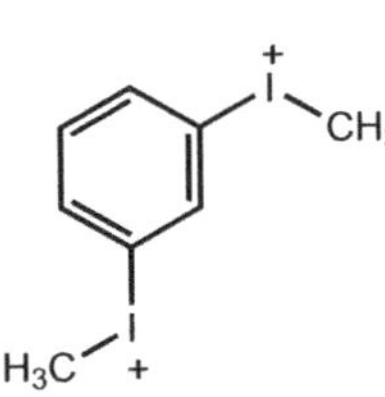

149

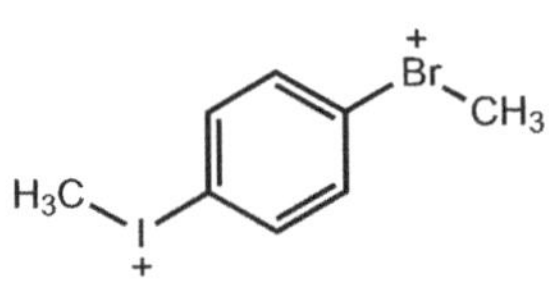

150

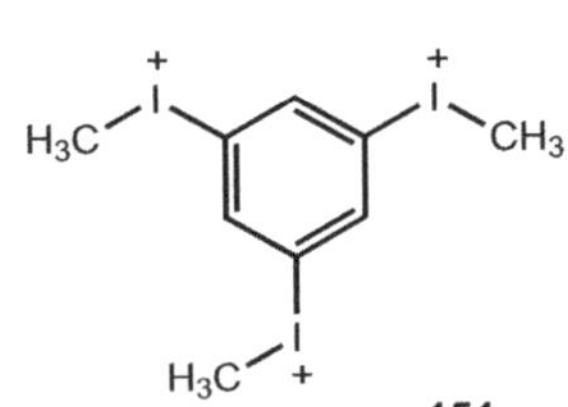

151

Ochiai et al.[369] have developed a method for the synthesis of alkenyl(aryl)iodonium ions with the use of trimethylvinylsilanes and idosobenzene [Eq. (4.107)]. Product formation is stereospecific with retention of configuration.

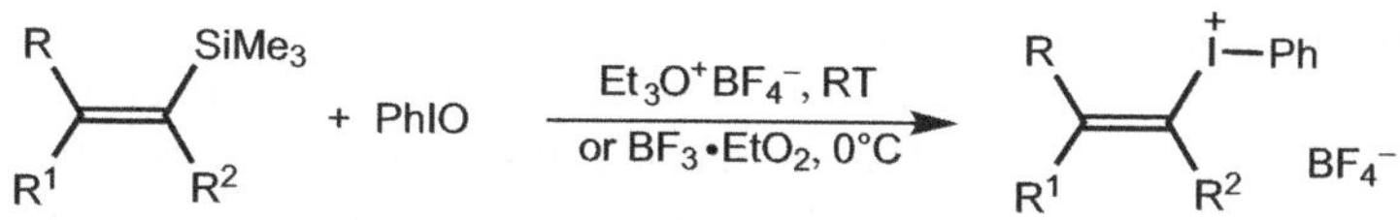

(4.107)

R, R^1, R^2 = H, alkyl, Ph,
R = H, R^1–R^2 = C_4H_8

Tributylstannanes have proved to be versatile reagents in the preparation of various aryliodonium ions with unsaturated organic substituents. Stang and coworkers[370,371] have shown that iodonium ion transfer from phenyl(cyano)iodonium triflate requiring Pd(II) and Cu(I) catalysis is a stereospecific process [Eq. (4.108)]. Stang and Zhdankin[372] also prepared the bisalkenyliodonium salt **153** using the bisphenyl(alkynyl)iodonium triflate **152** [Eq. (4.109)] and have provided full spectroscopic characterization (^{1}H, ^{13}C, ^{19}F NMR, and X-ray).

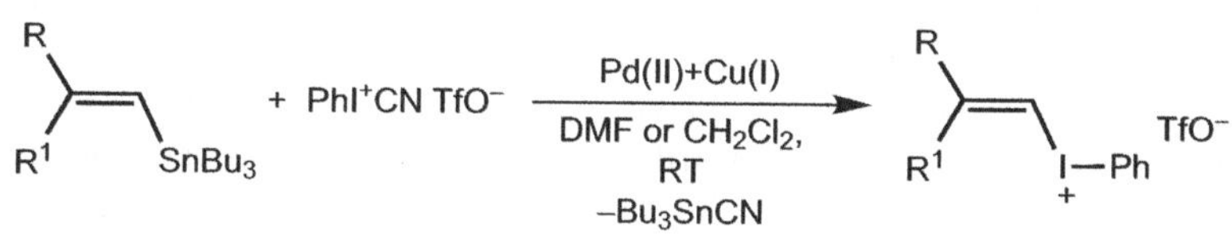

(4.108)

R, R^1 = Me, Et, *n*-Bu

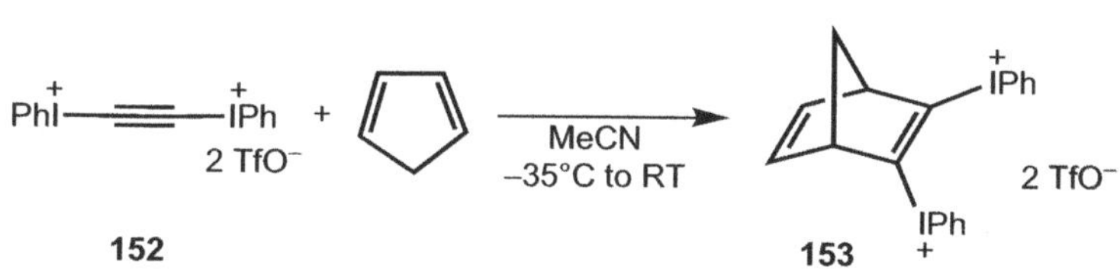

(4.109)

Tributylstannanes have also been used in the synthesis of alkynyl(aryl)iodonium salts, including 1,3-diynyl derivatives[373] and the parent member of the family, $HC{\equiv}CIPh^{+}TfO^{-}$, which was characterized by X-ray structure analysis.[374] The bisphenyliodonium triflate reagent **152** [Eq. (4.110)][372] and analogs[375] were synthesized in a similar way. Alkynyltrimethylsilanes may also serve as similar useful starting materials.[376,377] X-ray characterization of a variety of alkynyl(aryl)iodonium ion has been reported.[332]

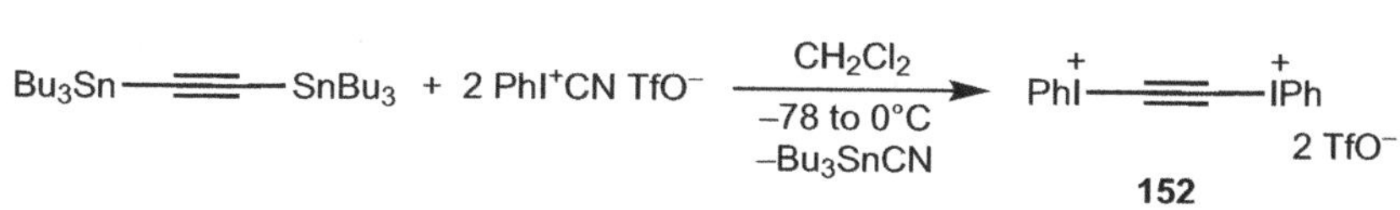

(4.110)

Diarylhalonium Ions. In contrast to dialkylhalonium ions and alkylarylhalonium ions, diarylhalonium ions are considerably more stable. This is particularly the case for diaryliodonium ions, which have been known for 110 years.[328a] It is interesting, however, to compare the discovery and assumed significance of these ions with that of triarylmethyl cations. The latter were discovered early in the twentieth century, but were considered only as a specific class of organic cations limited exclusively to the highly stabilized triarylmethyl systems. The general significance of carbocations as intermediates in electrophilic reactions was not recognized until many years later, when it became evident that they are intermediates in all electrophilic organic reactions. Subsequently, methods were developed to prepare practically any conceivable type of carbocations under stable ion conditions. Halonium ions represent a somewhat similar class. Diaryliodonium (or, to a lesser extent, diarylbromonium and -chloronium) ions were also for long considered a specific class of highly stabilized halonium ions. No relationship or significance was attached to these ions until many years later, when it was realized that many other types of halonium ions (such as dialkyl-, alkylaryl,- and alkylenehalonium ions) can exist, and the significance of dialkylhalonium ions in electrophilic alkylation with alkyl halides was pointed out.

Diaryliodonium salts (diaryl-λ^3-iodanes) are widely used as arylating agents. There are a number of methods available for their synthesis typically involving two or three steps.[378,379] A recent one-pot approach, however, offers a simple and high-yielding access to unsymmetrical diaryliodonium triflates using *meta*-chloroperbenzoic acid (*m*CPBA) as the oxidant[380] [Eq. (4.111)]. Moreover, symmetrical diaryliodonium salts can directly be prepared from iodine and arenes without the use of expensive aryl iodides [Eq. (4.112)].

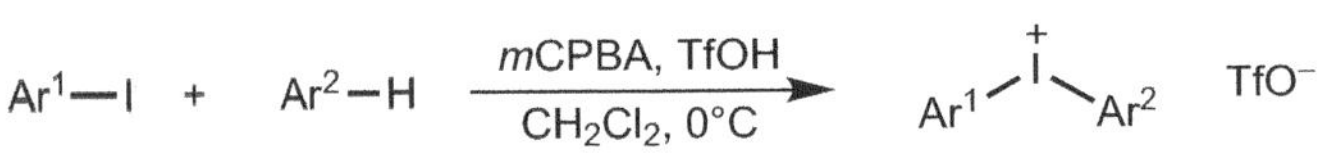

(4.111)

51–94% yield

Ar^1 = Ph, 4-ClC_6H_4, 4-BrC_6H_4, 2-MeC_6H_4, 4-MeC_6H_4, 4-*tert*-BuC_6H_4, 4-$NO_2C_6H_4$, 3-$CF_3C_6H_4$, 4-$CF_3C_6H_4$, 4-$CO_2HC_6H_4$

Ar^2 = Ph, ClC_6H_4, BrC_6H_4, IC_6H_4, MeC_6H_4, $MeOC_6H_4$, *tert*-BuC_6H_4, $AcNHC_6H_4$, 1,3-diMeC_6H_3, 1,4-diMeC_6H_3, 1,3,5-triMeC_6H_2

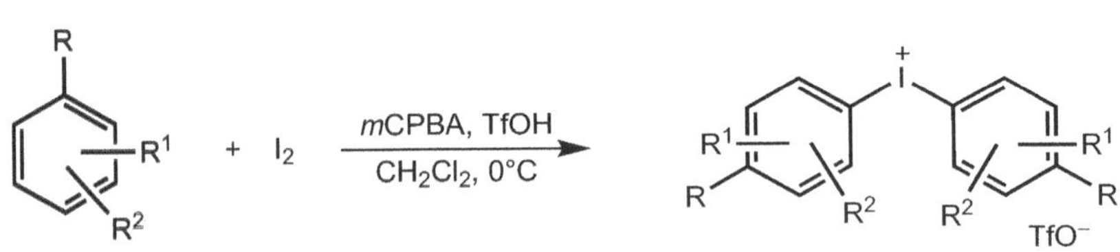

24–93% yield

R, R^1, R^2 = H
R = F, Cl, Br, Me, *tert*-Bu, R^1, R^2 = H
R = Me, R^1 = H, R^2 = 4-Me
R = Me, R^1 = 3-Me, R^2 = 5-Me
R = NO_2, R^1 = 3-Me, R^2 = 5-Me

(4.112)

Olah et al.[381,382] have studied the ^{13}C NMR spectra of a series of diarylhalonium ions. Some of the representative diarylhalonium ions are ions **154–157**.[99] The synthesis of these diarylhalonium ions does not require strongly acidic conditions and thus will not be discussed here. The use of arylstannane and arylsilane precursors in the synthesis of diarylhalonium ions has been reviewed[332] and the X-ray structure analysis of a number of these ions has also been reported.[332] Frohn and co-workers[383] have recently prepared and characterized aryl(pentafluorophenyl)iodonium tetrafluoroborates. Significant cation–anion interactions were shown to result in the formation of two arrangements in the solid state: dimers with an eight-membered ring or polymers with a zigzag chain.

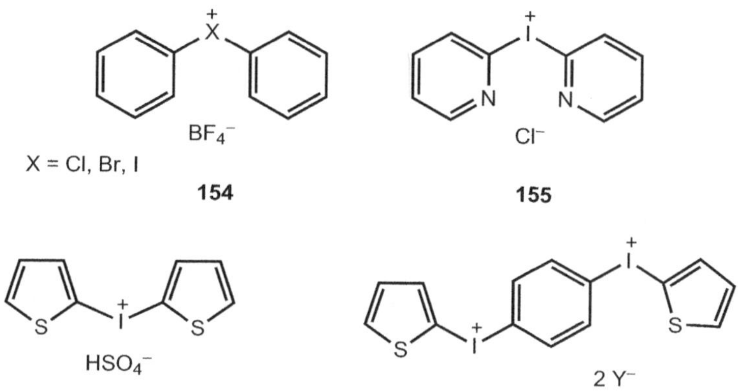

154 **155** **156** **157**

High-yield preparation and characterization (multinuclear NMR, MS) of iodonium-containing macrocycles such as rhomboid [Eq. (4.113)], square, and pentagon have been reported by Stang and co-workers.[384,385]

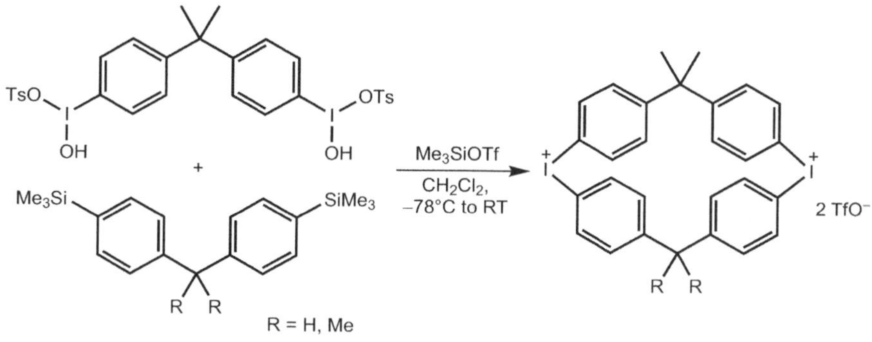

(4.113)

4.2.4.2. Cyclic Halonium Ions

Ethylenehalonium Ions. The parent ethylenehalonium ions **158** were obtained when 1-halo-2-fluoroethanes were ionized in SbF_5–SO_2 solution at −60°C[386] [Eq. (4.114)]. The ^1H NMR spectra of the ethylenebromonium and -iodonium ions show a singlet at δ^1H 5.53 and 5.77, respectively. Under similar experimental conditions, when 1,2-dichloroethane and 1-chloro-2-fluoroethane were treated with SbF_5–SO_2 solution, only donor–acceptor complexes **159** and **160** were formed instead of the ethylenechloronium ion.

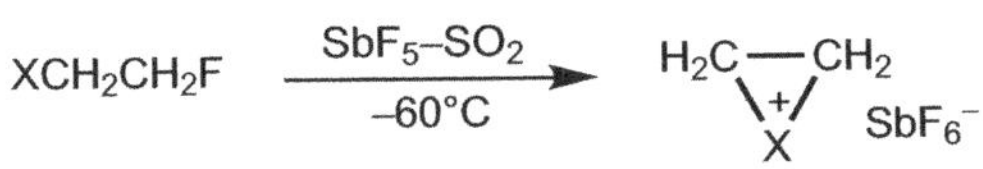

(4.114)

X = Br **158-Br**
X = I **158-I**

$ClCH_2CH_2Cl \rightarrow SbF_5$ **159** $\quad$ $ClCH_2CH_2F \rightarrow SbF_5$ **160**

Subsequently, the ethylenechloronium ion **158-Cl** was obtained.[387] When, instead of SO_2, SO_2ClF was used as the solvent in the reaction of antimony pentafluoride with 1-chloro-2-fluoroethane at −80°C, a solution was obtained whose ^1H NMR spectrum consisted of these absorptions: a doublet (δ^1H 4.6, 3H, $J=6$ Hz), a quartet (δ^1H 13.3, ^1H, $J=6$ Hz), and a singlet (δ^1H 5.9), consistent with the formation of ethylenechloronium ion **158-Cl** and methylchlorocarbenium ion **161**, respectively (Scheme 4.11).

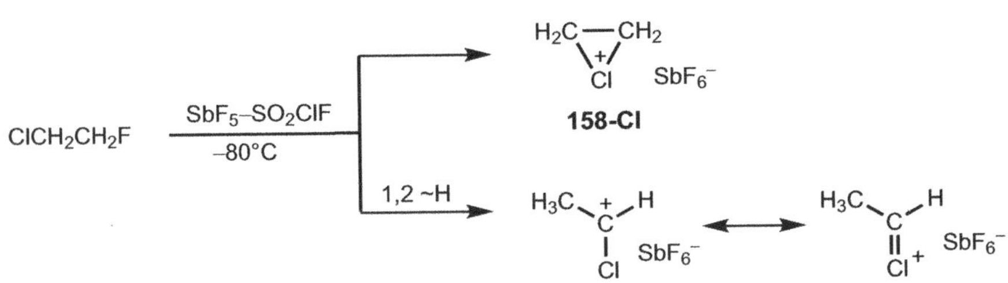

158-Cl

161

Scheme 4.11

The ethyleneiodonium ion **158-I** has also been prepared by the direct iodination of ethylene using ICN–SbF_5–SO_2ClF solution[388] [Eq. (4.115)]. However, similar reaction with either BrCN or ClCN did not give the corresponding halonium ions **158-Br** and **158-Cl**.

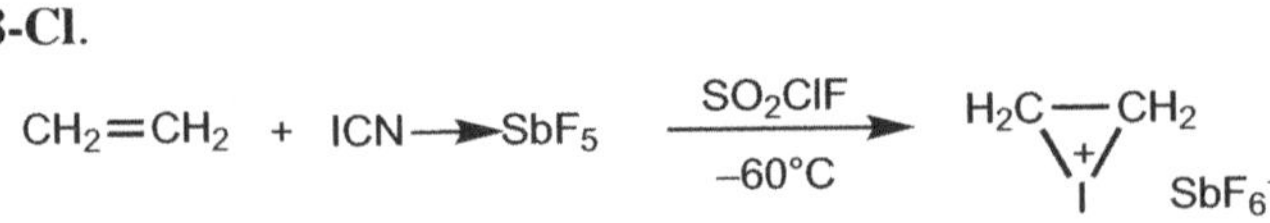

(4.115)

158-I

Propyleneiodonium ion **162-I** and propylenebromonium ion **162-Br** were also obtained by the ionization of 2-fluoro-1-iodopropane and 2-fluoro-1-bromopropane in SbF_5–SO_2ClF solution at −78°C[386] [Eq. (4.116)]. The propylenechloronium ion is also known.

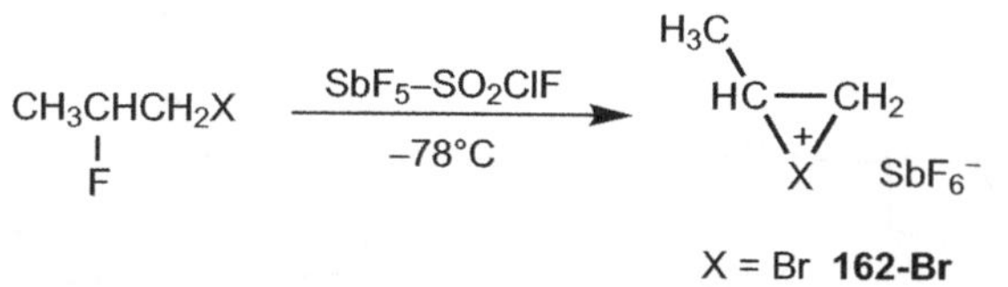

(4.116)

X = Br **162-Br**
X = I **162-I**

Similarly, a variety of dimethylethylene, trimethylethylene, and tetramethylethylene halonium ions have been prepared and studied by ^1H and ^{13}C NMR spectroscopy.[99,386,389] Some of the representative examples are ions **163–166**.

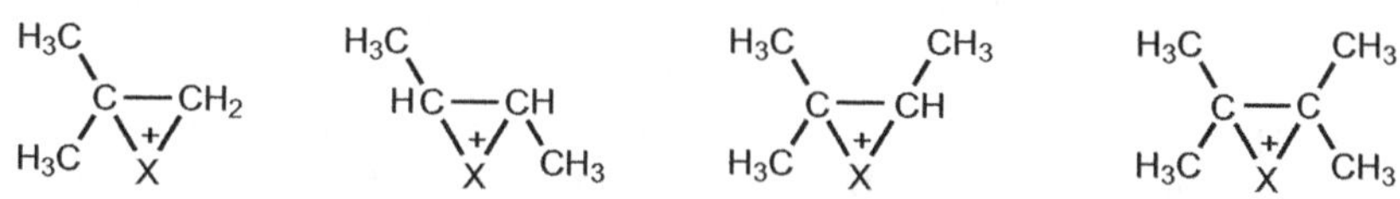

X = Cl, Br — **163**; X = Br, I — **164**; X = Cl, Br, I — **165**; X = Cl, Br, I — **166**

The unsymmetrical halonium ions such as **165-Br** are found in equilibrium with open-chain β-bromocarbenium ion. Many of the above-mentioned halonium ions have been prepared by the protonation of the appropriate cyclopropyl halides in superacids.[390]

Attempts to obtain the tetramethylfluoronium ion **166-F** has, however, been unsuccessful.[340] Ionization of 2,3-difluoro-2,3-dimethylbutane in SbF_5–SO_2 at −60°C gave an equilibrating β-fluorocarbenium ion **167**. The equilibration was shown to occur through the intermediacy of 1-*tert*-butyl-1-fluoroethyl cation **168** by a recent ^{13}C NMR spectroscopic study.[391]

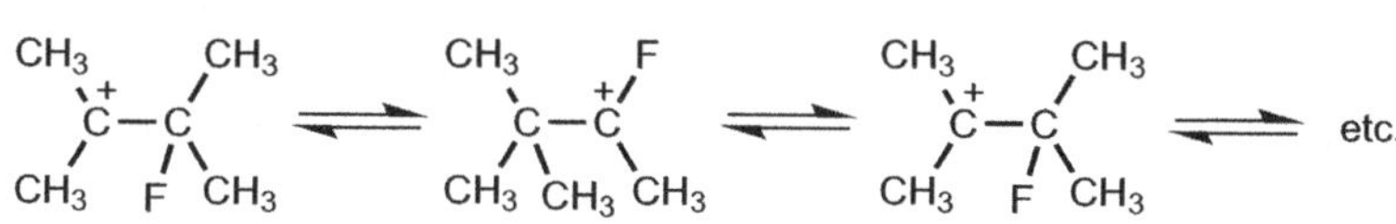

167 **168**

Strating, Wieringa, and Wynberg[392,393] reported that adamantylideneadmantane, a highly sterically hindered olefin, reacted with chlorine in hexane and bromine in CCl_4 solution to give the corresponding chloronium and bromonium salts **169**. They proposed the structure of these rather insoluble salts (on which no spectroscopic study was conducted) to be that of three-membered-ring ethylenehalonium ions (σ-complexes). Olah et al.[388] subsequently investigated these stable salts by ^1H and

^{13}C NMR spectroscopy, and concluded that they are not three-membered-ring halonium ions but molecularly bound π-complexes **170**.

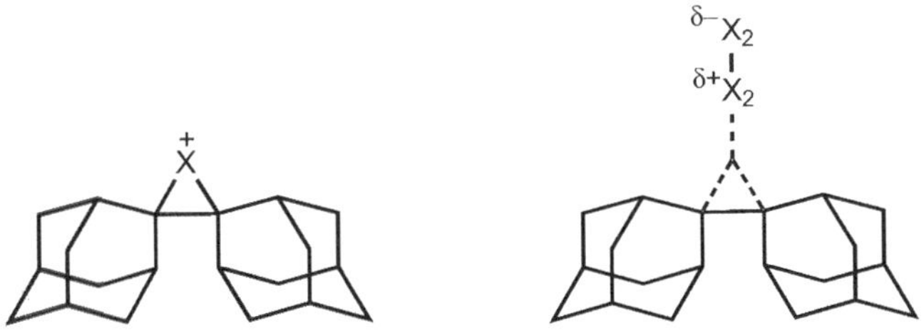

X = Cl, Br, I

169 **170**

Nugent[394] has isolated the hexafluoroantimonate salt of **169-Cl** and subsequently Brown and co-workers[395,396] established the molecular structure of the Br_3^- and triflate salts of **169-Br** and the triflate of **169-I** by X-ray crystal structure analysis. The halonium portion of the triflate salts were found to be essentially symmetric (averaged structural parameters for **169-Br**: Br–C bond length = 2.11 Å, C–C bond length 1.49 Å, Br–C–C angle = 69.4°, C–Br–C angle = 41.3°). It was also observed that addition of the parent olefin results in translocation of Br^+ from the top side of the halonium ion to its bottom side in an equilibrium process. Kochi and co-workers[397] obtained the hexachloroantimonate salt of **169-Cl** and showed to have an unsymmetrical cyclopropane ring system. Chlorine is ascribed to be σ-bonded to a single carbon center, and the cationic charge on the adjacent carbon is stabilized by the chlorine lone pair acting as *n*-donor. The 1-methyl- and 4-chloro-susbtituted derivatives of **169-Br** have also been characterized by ^1H and ^{13}C NMR spectroscopy.[398,399]

Trimethylenehalonium Ions. Attempts to prepare trimethylenehalonium ions by ionizing the appropriate 1,3-dihaloalkanes with SbF_5–SO_2 or with 1:1 HSO_3F–SbF_5–SO_2 solution at low temperature have been unsuccessful.[400] The ions obtained were either three- or five-membered-ring halonium ions, formed through ring contraction or ring expansion, respectively. For example, when 1-halo-3-iodopropanes were reacted with SbF_5–SO_2ClF solution at −78°C, the propyleneiodonium ion **162-I** is formed (Scheme 4.12). This ion could be formed through the trimethyleneiodonoium ion **171**,

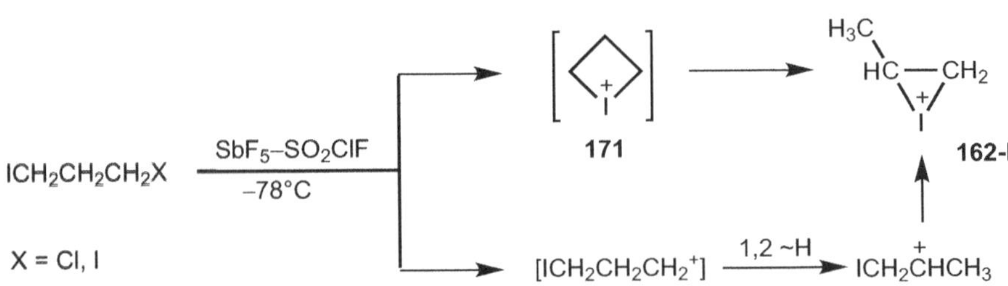

Scheme 4.12

which was, however, not observed as an intermediate. Alternatively, the nonassisted primary ion could undergo rapid 1,2-hydrogen shift to the secondary ion, which then would form the propyleneiodonium ion via iodine participation. 1-Halo-3-bromopropanes behave similarly yielding **162-Br**.

The ionization of 1,3-dihalo-2-methylpropanes with SbF_5–SO_2ClF gave both three-membered and five-membered ring halonium ions and open-chain halocarbenium ions.

The only reported preparation of long-lived four-membered-ring halonium ions is that reported by Exner et al.[401] 3,3-Bis(halomethyl)trimethylenebromonium ions **172** were prepared by treating tetrahaloneopentanes with SbF_5–SO_2ClF solution at low temperature [Eq. (4.117)]. Seemingly, halogen substitution stabilizes the four-membered ring. The ^1H NMR spectrum of fluorinated ion **172-F** shows a broad singlet at δ^1H 5.28 and a doublet at δ^1H 4.68 ($J_{H\text{–}F} = 47$ Hz). In contrast, the brominated ion **172-Br** displays a temperature-dependent proton absorption at δ^1H 5.17 (from internal $Me_4N^+BF_4^-$), indicating the occurrence of the exchange process shown.

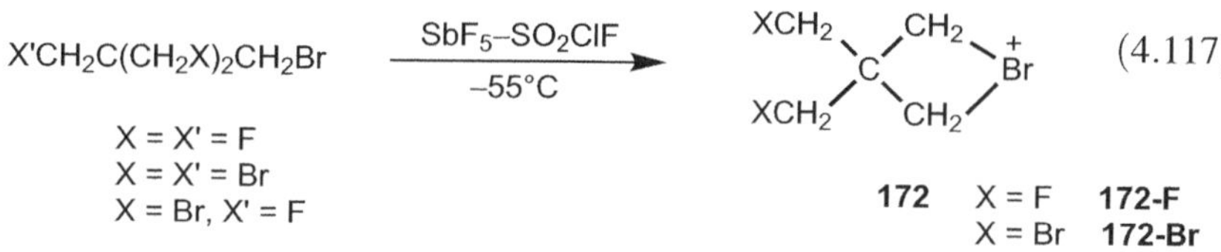

(4.117)

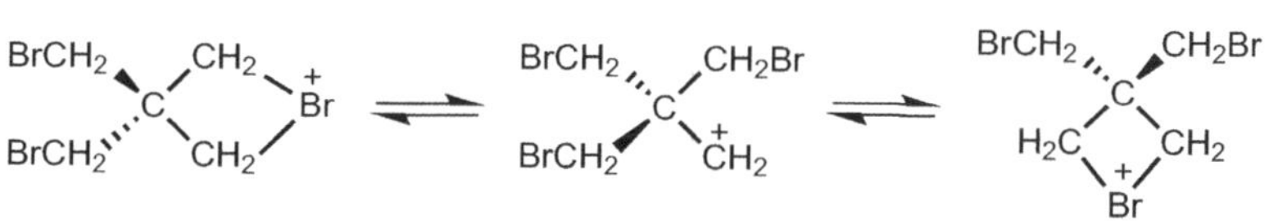

172-Br

Tetramethylenehalonium Ions. 1,4-Halogen participation was first postulated to occur in the acetolysis of 4-iodo- and 4-bromo-1-butyl tosylates.[402] In subsequent studies, Peterson and coworkers found anomalous rates in the addition of trifluoroacetic acid to 5-halo-1-hexenes[403,404] and 5-halo-1-pentynes.[405,406] Such observations were recognized as due to 1,4-halogen participation.[341] Also, a study of the solvolysis of δ-chloroalkyl tosylates indicated rate-accelerating 1,4-chlorine participation effects up to 99-fold.[407]

Direct experimental evidence for 1,4-halogen participation comes from the direct observation (by NMR spectroscopy) of five-membered-ring tetramethylenehalonium ions by Olah and Peterson[408] and by Olah et al.[400]

The ionization of 1,4-dihalobutanes in SbF_5–SO_2 solution gave the parent tetramethyenehalonium ions **173** [Eq. (4.118)].[408,409] The ^1H NMR spectra of these ions are similar to each other. Subsequently it was found that even 1,2- and

1,3-dihalobutanes when reacted with SbF_5–SO_2 solution[400] give the same tetramethylenehalonium ions[365] [Eq. (4.119)].

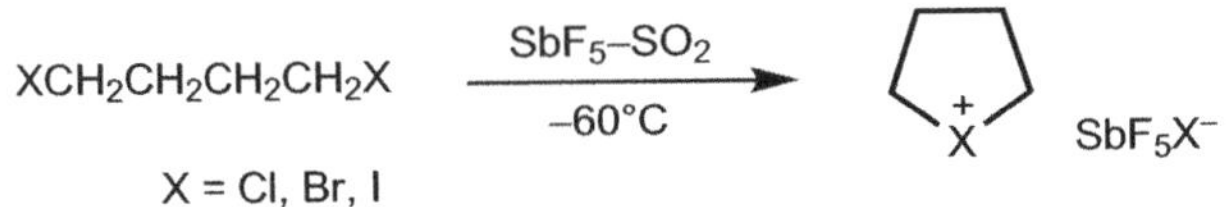

173

(4.118)

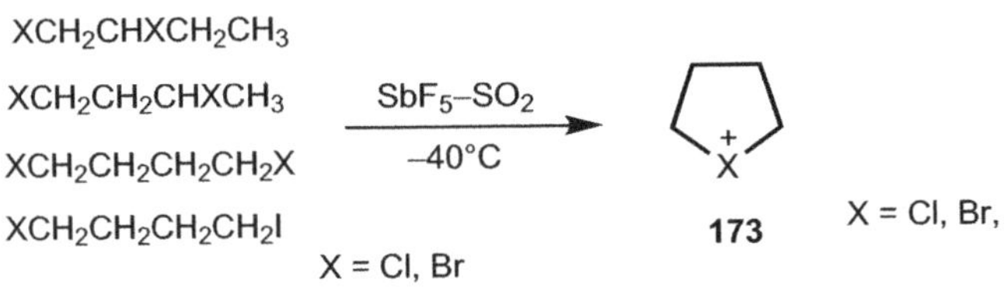

173

(4.119)

Similarly, several 2- and 2,5-substituted tetramethylenehalonium ions (**174–178**) have been prepared and studied by both ^{1}H and ^{13}C NMR spectroscopy.[99,410,411]

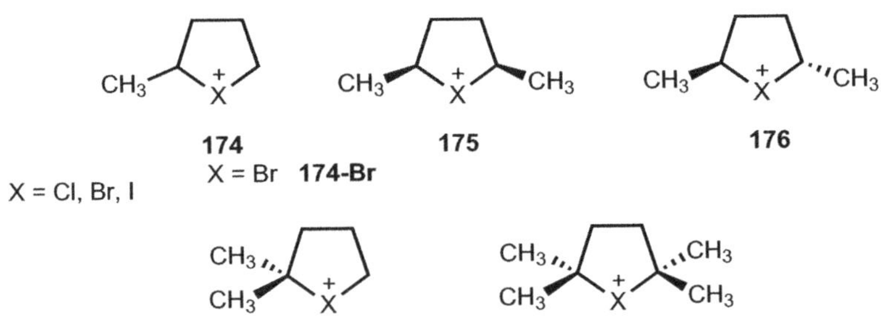

174 X = Br **174-Br**

175

176

X = Cl, Br, I

177

178 X = Cl **178-Cl**

The 2,2,5,5-tetramethyltetramethylenechloronium ion **178-Cl** shows only one ^{1}H and ^{13}C NMR signal for the nonequivalent methyl groups indicating that some kind of equilibrium exists with the open-chain carbenium ion.[411]

Even 2-methylenetetramethyleneiodonoium ion **179** has been prepared by the protonation of 5-iodopentyne[405,406] [Eq. (4.120)].

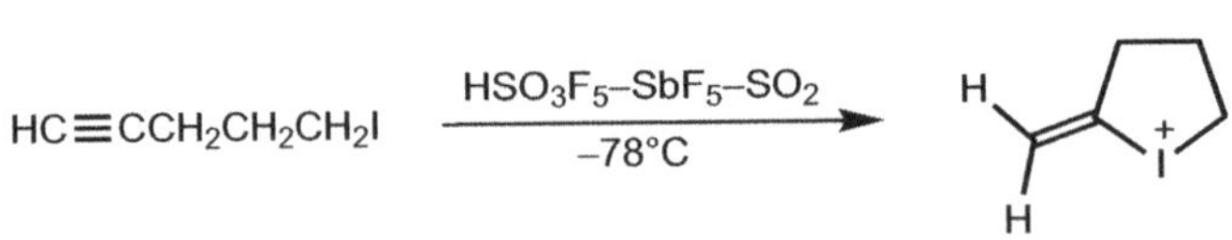

179

(4.120)

Pentamethylenehalonium Ions Attempts to prepare six-membered-ring halonium ions by treating 1,5-dihalopentanes with SbF_5–SO_2 gave exclusive rearrangement to five-membered-ring halonium ions.[99,408] Peterson et al.[412], however, were able to prepare six-membered-ring halonium ions **180** by the methylation of 1,5-dihalopentanes with methyl fluoroantimonate (CH_3F–SbF_5) in SO_2 solution[412]

[Eq. (4.121)]. However, some rearrangement to five-membered-ring halonium ions **174** was also observed.

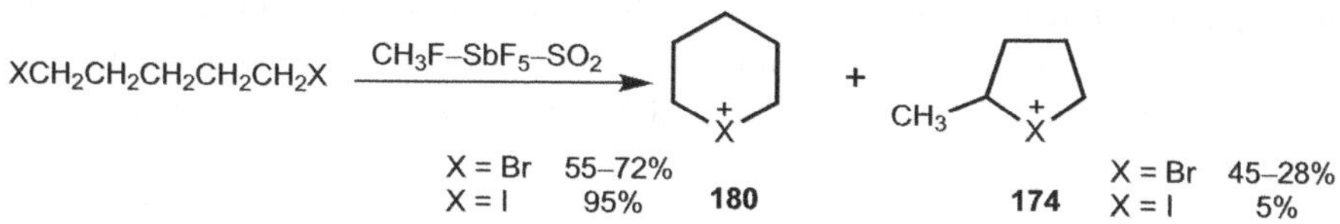

(4.121)

Alternatively, six-membered-ring halonium ions were also formed when equimolar amount of 1,5-dihalopentane was added to dihalonium ions [Eq. (4.122)]. The dihalonium ions were prepared from 1,5-dihalopentane and 2 mol of methyl fluoroantimonate. Furthermore, the dimethylbromonium ion **129-Br** is also a sufficiently active methylating agent to form cyclic pentamethylenebromonium ion **180-Br** from 1,5-dibromopentane [Eq. (4.123)].

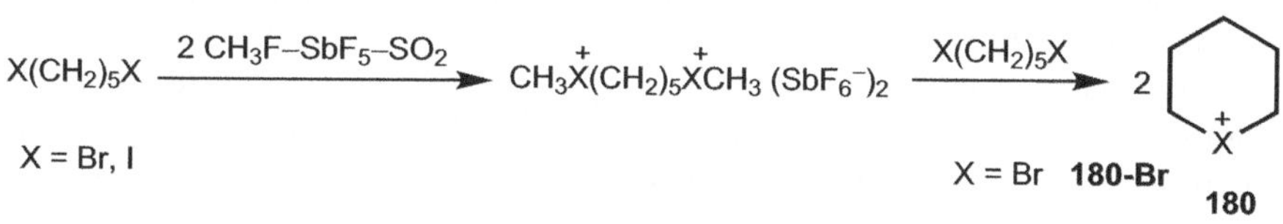

(4.122)

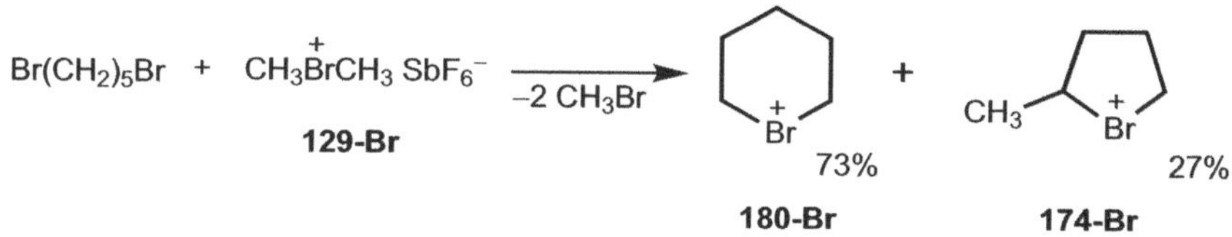

(4.123)

Bicyclic Halonium Ions. Although halogen addition to cycloalkenes are assumed to proceed through the corresponding bicyclic halonium ions, these ions are quite elusive. Olah, Liang, and Staral[413] were able to prepare the cyclopentenebromonium ion **181** by the ionization of *trans*-1,2-dibromocyclopentane in SbF_5–SO_2ClF solution at $-120°C$ [Eq. (4.124)]. The ^1H NMR (60 MHz) spectrum of the ion solution showed a broadened peak at δ^1H 7.32 (two protons) and two broad peaks centered at δ^1H 3.14 (four protons), and δ^1H 2.50 (two protons). The ^1H NMR spectrum of the solution also showed the presence of the related cyclopentenyl cation. When the solution was slowly warmed up to $-80°C$, the cyclopentenebromonium ion gradually and cleanly transformed into the allylic ion. The cyclopentenyl ion present initially in solution might be formed as a result of local overheating during preparation.

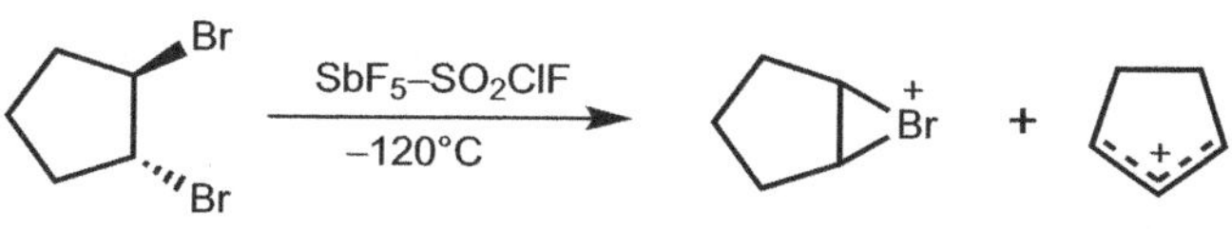

(4.124)

181

The cyclopentenebromonium ion **181** was also obtained via protonation of 4-bromocyclopentene in $HSO_3F–SbF_5–SO_2ClF$ solution at $-120°C$, through the reaction sequence shown in Eq. (4.125).[413] The proton noise-decoupled ^{13}C NMR spectrum of the cyclopentenebromonium ion **181** shows three carbon resonances at $\delta^{13}C$ 114.6 (doublet, $J_{C–H} = 190.6$ Hz), 31.8 (triplet, $J_{C–H} = 137.6$ Hz), and 18.7 (triplet, $J_{C–H} = 134.0$ Hz).

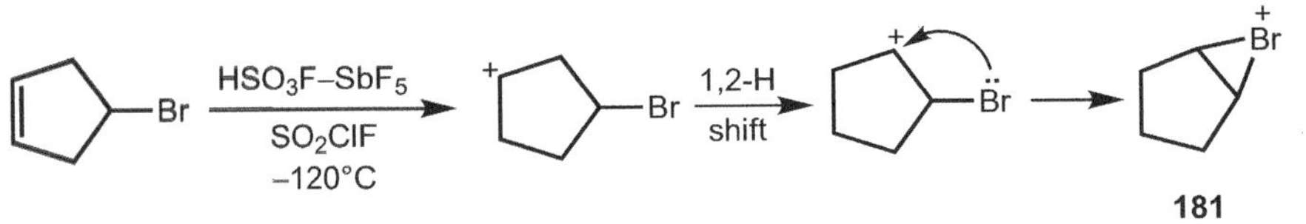

181

(4.125)

Attempts were also made to prepare the cyclopentenechloronium ion **182** via ionization of *trans*-1,2-dichlorocyclopentane in $SbF_5–SO_2ClF$ solution at $-120°C$. However, instead of the cyclopentenechloronium ion **182** only the 1-chloro-1-cyclopentyl cation **183** was obtained[413] [Eq. (4.126)]. Apparently, the participation of the smaller chlorine atom could not effectively compete with the fast 1,2-hydride shift forming the tertiary ion. The larger bromine atom, however, preferentially participates with the neighboring electron-deficient center, forming the bicyclic bridged ion.

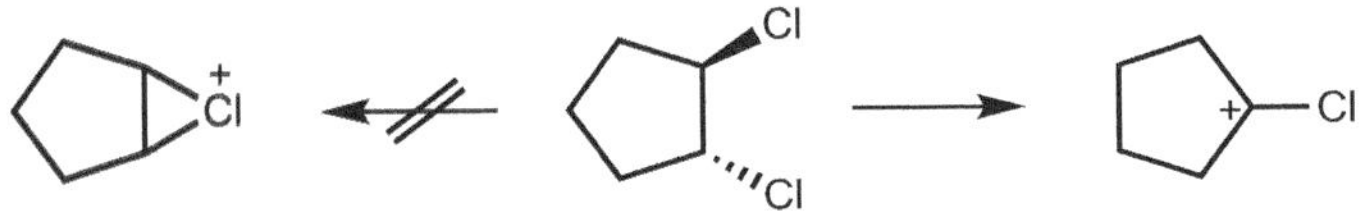

182 183

(4.126)

Olah, Prakash, and co-workers[414] have isolated an unusual fluorinated product in the transformation of bromocyclohexane and suggested the involvement of 7-bromoniabicyclo[4.1.0]heptane **184** [Eq. (4.127)].

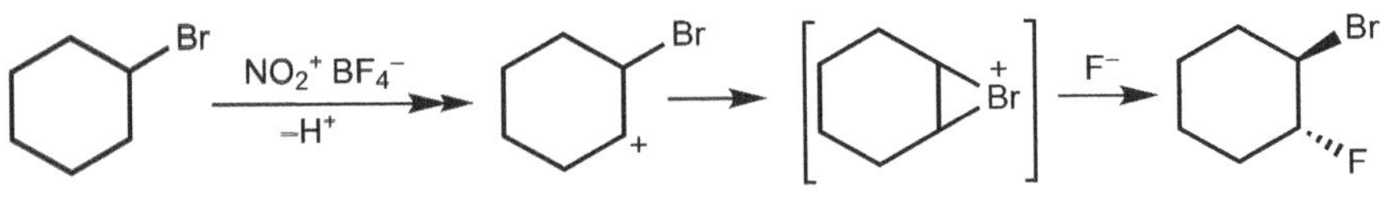

184

(4.127)

The attempted generation of ion **184** in a subsequent study[415] resulted in the exclusive formation of 7-bromoniabicyclo[2.2.0]heptane **185** (Scheme 4.13) identified unequivocally on the basis of its NMR characteristics [$\delta^{13}C$ 118.7 ($J_{C–H} = 172.1$ Hz) and 37.4 ($J_{C–H} = 135.5$ Hz)]. Calculations have shown (MP2/BB2//MP2/SB level)[416]

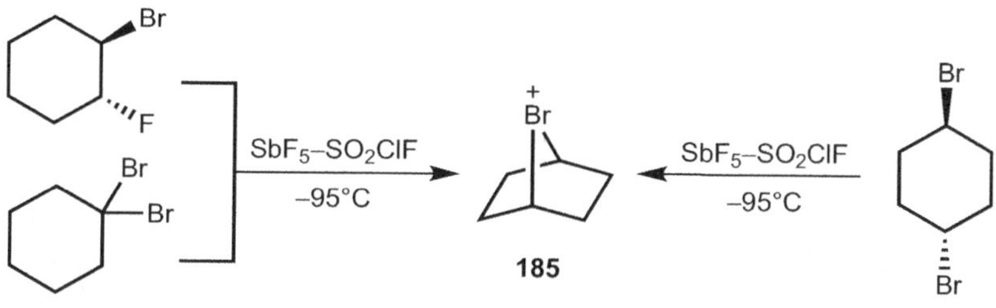

Scheme 4.13

that on the potential energy surface of the $C_6H_{10}X^+$ cations (X = F, Cl, Br), the 1,4-bridged bromonium ion **185** is the most stable species being more stable than the 1-bromocyclohexylium cation and the 1,2-bridged ion by about 5 kcal mol^{-1} and 10 kcal mol^{-1}, respectively. The corresponding chloro derivatives are of equal energy, whereas the 1-fluorocyclohexylium cation is considerably more stable than the 1,4-fluoro-bridged ion. On the basis of these results, the unsuccessful attempt to prepare the 1,2-bridged ion **184** is not surprising. Similar studies of the bromocyclopentyl cations found the 1,2-bridged ion more stable than its isomers. Electrostatic and size effects seem to dominate the stability of these cations.

Peterson and Bonazza[417] have reported that ionization of *cis*-1,2-bis(chloromethyl) cyclohexane in SbF_5–SO_2 solution at −78°C gives the bicyclic five-member-rig chloronium ion **186** along with smaller amounts of other species [Eq. (4.128)]. Warming the solution containing the ion to −10°C leads to the formation of the open-chain tertiary carbenium ion **187**.

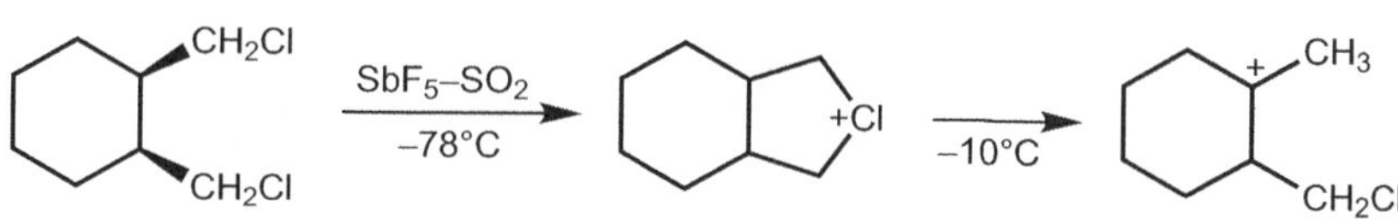

(4.128)

Hall, Gabbai, and co-workers[418] have obtained the tetrafluoroborate salt of cation **188** from the 1,8-bis(diphenylmethylium)naphthalenediyl dication [Eq. (4.129)]. The molecular structure shows that the C(1) atom is tetrahedral, C(2) has a trigonal planar arrangement, and the C−F bond is a regular bond (1.424 Å). The fluorine forms a long interaction with the methylium center (2.444 Å) and the C(1)–F–C(2) angle is 111.11° characteristic of a formally sp^3-hybridized F atom. DFT calculations, AIM analysis, and Boys localized orbitals indicate that the long C(2)–F interaction is a dative bond and the unsymmetrical **188b** structure must also contribute. Indeed, variable-temperature ^1H NMR measurements show that **188** is a fluxional ion and the fluorine atom oscillates between the two carbon centers involving a symmetrical fluoronium ion as low-energy transition state.

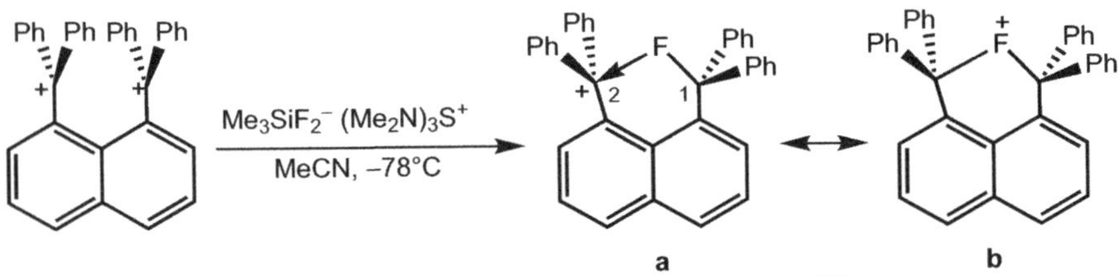

188

(4.129)

Heteroaromatic Halophenium Ions. Halophenium ions are a class of halonium ions analogous to thiophene, furan, and pyrrole. To date, no parent halophenium ions **189** are known, but many stabilized tetraphenyl iodophenium, benzoiodophenium, and dibenzoiodophenium ions have been prepared by Beringer and co-workers.[419,420] Some of them have been analyzed by X-ray crystal structure investigations. Representative examples are ions **190–192**.

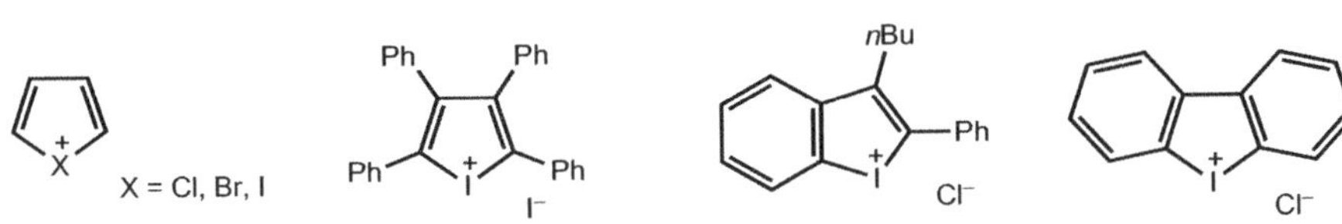

189 **190** **191** **192**

Olah and Yamada[421] have shown that thermal decomposition of *ortho*-(β,β-dichloroethenyl)phenyl diazoniumfluorophosphate yielding *ortho*-(β-chloroethynyl) chlorobenzene involves benzochloronium ion **193** as the intermediate [Eq. (4.130)].

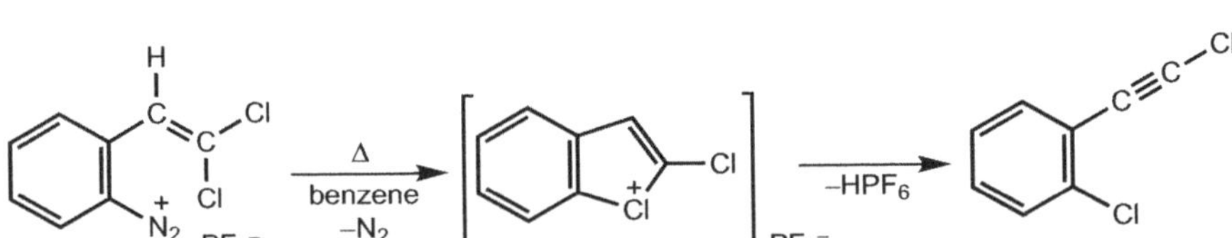

193

(4.130)

Miscellaneous Halonium Ions. The Si-containing ring system **194** analogous to **188** has been reported by Müller and co-workers.[422] The silyl cation is stabilized by intramolecular interactions with the F atoms increasing the coordination number of Si and, consequently, Si becomes considerably shielded. In the ^{29}Si NMR spectrum, the resonance at $\delta^{29}Si$ 77 indicate that cation **194** has only a small silicenium ion character and better be described as a fluoronium ion. Characteristic data for cation **194** are as follows: Si–F bond lengths = 1.755 and 1.763 Å, Si–F–Si angle = 129.9°. This is in sharp contrast to the structure of ion **188**. Cations **195** have similar characteristics (δ^{29}Si 90.2–90.8).[423] X-ray crystal analysis shows that the Si–X–Si bridge in these cations is symmetric.

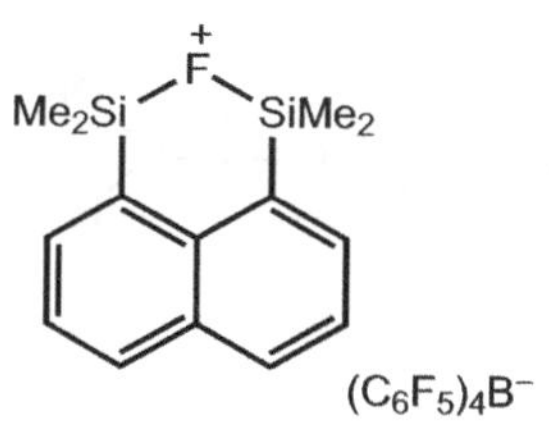

194

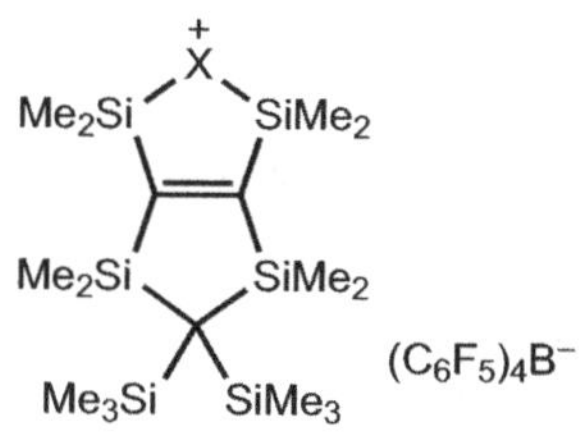

Y = F, Cl, Br

195

4.2.5. Onium Ions of Group 15 Elements

4.2.5.1. 2-Azoniaallene and Derived Cations. Various monocations can be derived from allenes by replacing the carbons by nitrogen atoms. Whereas only a few examples of ketiminium salts **196** are known,[424] 2-azoniaallene salts (**197**) can be prepared by a number of methods.[425–431] They are well-characterized and used widely in dipolar cycloadditions. Spectroscopic studies of hexachloroantimonate salts of phenyl-substituted cations **197** first synthesized by Würthwein[432] and X-ray crystallographic analysis of various salts[426,431,433] including **198** indicate a C=N=C unit, that is, the allene geometry (**197a**, D_{2d} symmetry). Experimental and theoretical studies reveal, however, that depending on the substitution pattern and particularly with amino and alkoxy substituents, the ions may exist as bent 2-azaallyl cations **197b** (C_{2v} symmetry) or adopt structures in between.[428,433,434]

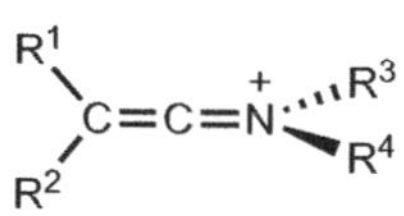

196

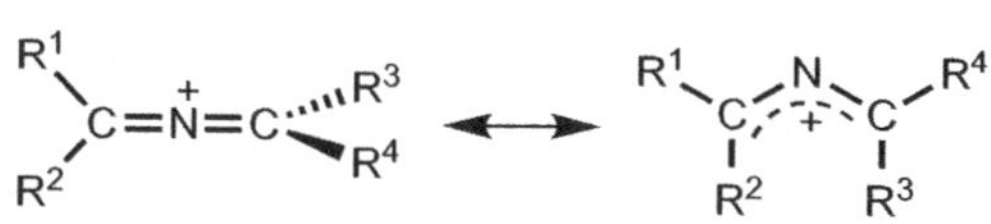

a **b**

197

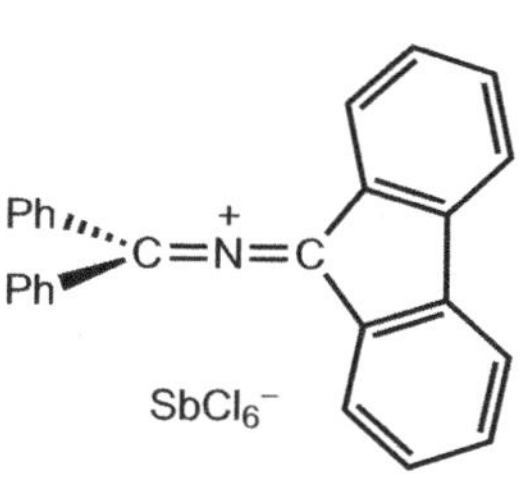

198

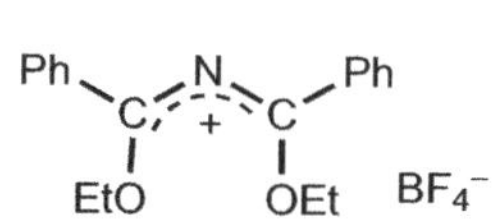

199

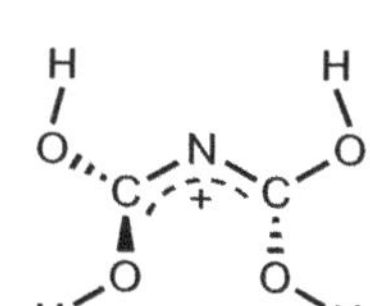

200

Cation **199** synthesized according to Eq. (4.131)[428] has, for example, a chiral, bent geometry with the ethoxy substituents in *exo* position in *s-cis* conformation (C–N–C angle = 133.0°, torsional angle = 70.4°).[427] The 2-azaallyl structure **200** of the parent tetrahydroxy model compound (C_2 symmetry) was shown by quantum mechanical calculations to be 20 kcal mol^{-1} lower in energy than the corresponding 2-azoniaallene.[435]

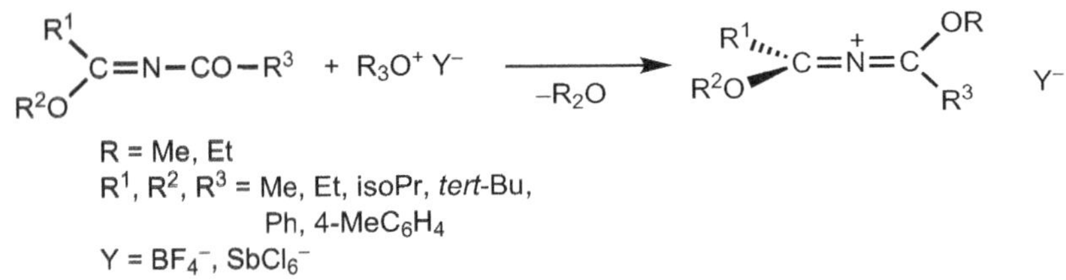

(4.131)

A variety of 1-aza-2-azoniaallene salts **201** have been developed and are in use as reactive intermediates in cycloadditions.[436] According to X-ray analysis, methylation of di-*tert*-butylcarbodiimide yielded cyanamidium salt **202a** (N–C bond lengths 1.15 and 1.25 Å, C–N–C bond angle = 177°) instead of the expected carbodiimidium salt **202b**.[437]

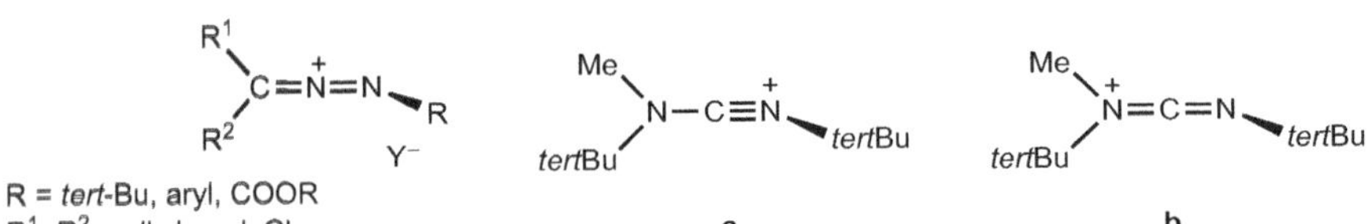

201 **a** **202** **b**

Jochims and co-workers[438] synthesized 1-oxa-3-azabutatrienium and 1-thia-3-azabutatrienium chloroantimonates [Eq. (4.132)]. X-ray structure analysis of the phenyl-4-bromophenyl-substituted derivatives shows bent structures (C=N=C angles for X = O and S, respectively, are 129° and 139°). Bond distances indicate that all possible mesomeric structures contribute to the structure of the O analog. In contrast, the mesomeric form with a C≡S$^+$ moiety appears to be less important. ^{13}C NMR data suggest a certain delocalization of the positive charge into the aromatic rings.

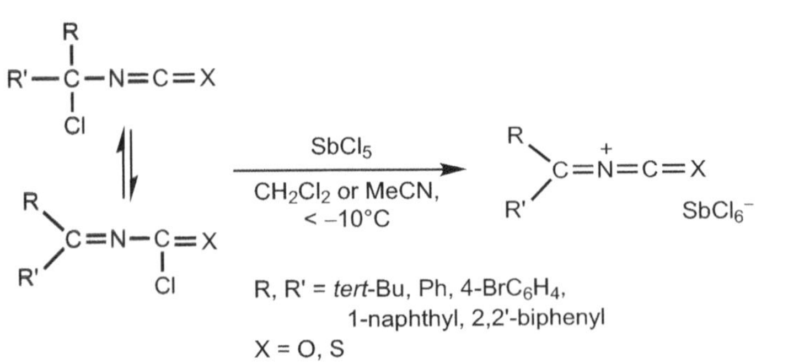

(4.132)

Seppelt and co-workers[439] isolated various salts of the cation $OCNCO^+$ [Eq. 4.133)]. The molecular structure of $OCNCO^+Sb_3F_{16}^-$ comprises a strongly bent cation (N–C–O angles = 173.1 and 173.6°, C–N–C angle = 130.7°). This finding is in good agreement with the structure of the ion calculated by Pyykkö and Runeberg[440] using *ab initio* methods (C–N–C angle = 138.6 and 139.4°). It is to be noted that the N_5^+ cation (**217**, vide supra) is also strongly bent at the central nitrogen atom.

$$\underset{\text{M = Sb, As}}{\text{F–CO–NCO} + MF_5} \xrightarrow[\text{−196°C to RT}]{CF_3CH_2CF_3 \text{ or } CF_2Cl_2,} \underset{X = Sb_3F_{16},\ AsF_6,\ As_2F_{11}}{OCNCO^+\ X^-} \quad (4.133)$$

4.2.5.2. Diazonium Ions

Parent Diazonium Ion. The simplest diazonium ion, protonated dinitrogen **203**, is still elusive. In fact, molecular nitrogen is a very weak base. *Ab initio* calculations of protonated dinitrogen have shown[441] that the end-protonated linear structure **203** of $C_{\infty v}$ symmetry is energetically preferred over the edge-protonated cyclic ion **204** of C_{2v} symmetry. Subsequent high-level calculations [HF/6-31G** and MP2(FU)/6-31G** levels] by Rasul, Prakash, and Olah[442] have found that the cyclic structure is a transition state for N_α–N_β rearrangement with an energy barrier of 47.1 kcal mol^{-1}. The N–N bond distance in **203** is slightly shorter than that in dinitrogen (1.124 Å versus 1.130 Å). Further protonation leads to linear diprotonated dinitrogen (protodiazonium dication), which is less stable by 60.5 kcal mol^{-1} than ion **203** but dissociation to **203** and H^+ has a kinetic barrier of 28.7 kcal mol^{-1}.

203 **204**

Using an indirect approach, Olah et al.[443] were able to diazotize ammonia [Eq. (4.134)], bis(trimethylsilyl)amine, and isocyanic acid with $^{15}NO^+BF_4^-$, resulting in the formation of $^{14}N^{15}N$, which requires the intermediacy of the parent diazonium ion HN_2^+.

$$^{14}NH_3 + {}^{15}NO^+\ BF_4^- \xrightarrow[-H_2O]{CH_2Cl_2,\ -80°C} [{}^{15}N\equiv{}^{14}NH]^+\ BF_4^- \longrightarrow {}^{15}N\equiv{}^{14}N + HBF_4 \quad (4.134)$$

Protonated (deuterated) nitrogen has been studied in the gas phase,[444,445] but attempts to observe it in the condensed phase were unsuccessful.[443] An analog of **203**, fluorodiazonium ion **205**, is, however, known. Ionization of *cis*-difluorodiazine with arsenic pentafluoride below ambient temperature gives fluorodiazonium hexafluoroarsenate **205** as a white solid.[446] The same reaction does not occur with *trans*-difluorodiazine. The salts of ion **205** have been studied by X-ray diffraction, vibrational spectroscopy, and NMR spectroscopy.[447–449] Both nitrogens are more shielded compared with benzenediazonium ion, and N_β is deshielded relative to N_α (δ^{14}N −166.1 versus −191.2 relative to CD_3NO_2).[448] The J_{14N-F} value measured for **205** at 339.0 Hz is the largest known $^{19}F-^{14}N$ coupling constant, indicating the high *s*-character of the nitrogen–nitrogen bond.[448] In a subsequent study of the hexafluoroarsenate salt of ion **205** the F–N and N≡N bond lengths were determined (1.217 and 1.099 Å, respectively).[449] The F–N bond distance is the shortest F–N bond length known, whereas the N≡N bond length is close to the value measured for N_2 (1.0976 Å). The shortness of both bonds is interpreted as resulting from the high σ-character of the N molecular orbital and the formal positive charge on the cation.

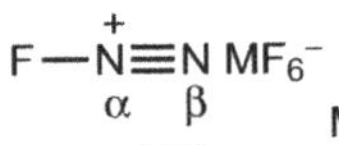

M = As, Sb

205

Alkyldiazonium Ions. The intermediacy of alkyldiazonium ions in a variety of organic reactions is well established.[450–452] They are common intermediates in the acid-catalyzed decomposition of diazo compounds and the nitrous acid-induced deamination of aliphatic primary amines. The evidence for RN_2^+ (R = alkyl) intermediates come from both rate data and product analysis studies. However, direct investigation of alkyldiazonium ions has been difficult due to their instability.

The first direct observation[453] of an aliphatic diazonium ion was achieved by protonation of trifluoromethyldiazomethane in HSO_3F at −60°C. The 2,2,2-trifluoroethyldiazonium ion **206** is stable for 1 h at −60°C. Similarly bis(trifluoromethyl)methane diazonium ion **207** has been prepared and characterized.[454] These ions were studied by ^1H NMR spectroscopy. Similar diazonium structures have been assigned to protonated 2-diazo-5α-cholestan-3-one.[455] None of these studies, however, showed nitrogen protonation.

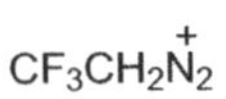

206

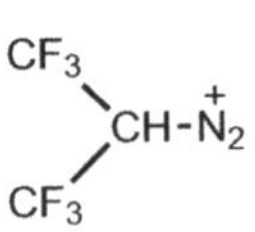

207

Subsequently, McGarrity and Cox[456] have succeeded in protonating diazomethane in $HSO_3F–SbF_5–SO_2ClF$ at −120°C. In this acid media, exclusive formation of methyldiazonium ion $CH_3N_2^+$ **208**, the thermodynamically more stable product, is observed. With a more acidic $HSO_3F–SbF_5$ system, both methyldiazonium ion **208** and methylenediazenium ion **209** are observed in a ratio corresponding to their gas-phase stabilities. The two ions **208** and **209** have been characterized by ^1H, ^{13}C, and ^{15}N NMR spectroscopy.

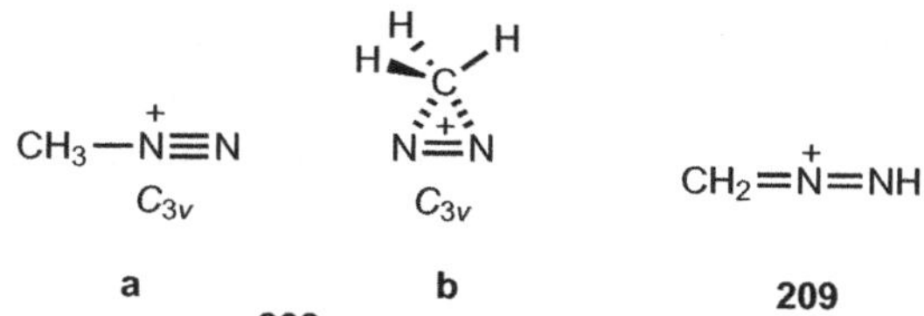

208 **209**

Early *ab initio* calculations predicted[457,458] that the open structure **208a** is significantly more stable than the bridged cation **208b**. High-level studies[442] [MP2 (FU)/6-31G**] have recently found that *C*-protonated diazomethane (**208a**) is more stable than the *N*-protonated $CH_2N_2H^+$ isomer (**209**) by 39.3 kcal mol^{-1}. Glaser and co-workers[459,460] studied the methyldiazonium cation and ethyldiazonium cation and found a remarkable difference of 30.7 kcal mol^{-1} for the dediazoniation enthalpy for the ions (42.2 versus 11.5 kcal mol^{-1}). A small part (5.6 kcal mol^{-1}) of the difference may be accounted for by the formation of the nonclassical ethyl cation. It was also concluded that alkyldiazonium ions are best described as carbocations and Lewis structures do not adequately describe the structure of these ions. Of the diprotonated diazomethanes,[442] the *N,N*-diprotonated structure was found to be the global minimum, which is more stable than the *C,N*-diprotonated and *C,C*-diprotonated forms by 11.3 kcal mol^{-1} and 34.3 kcal mol^{-1}, respectively.

The stability of alkyldiazonium ions may be greatly increased by an appropriate molecular structure. Bott has reported the generation of a variety of stable vinyldiazonium ions[461] [Eqs. (4.135) and (4.136)]. The results are summarized in reviews.[462,463]

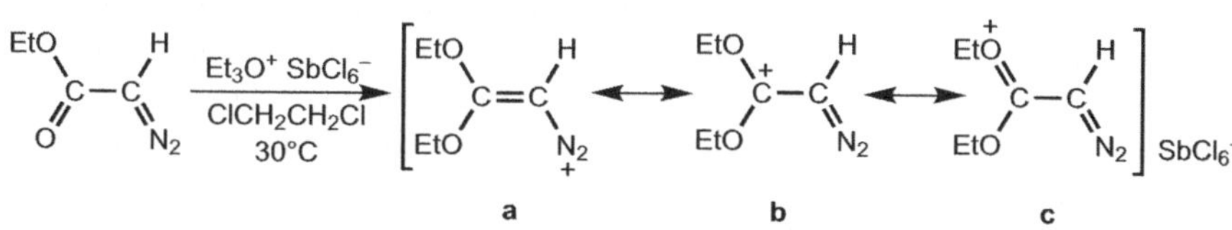

210

(4.135)

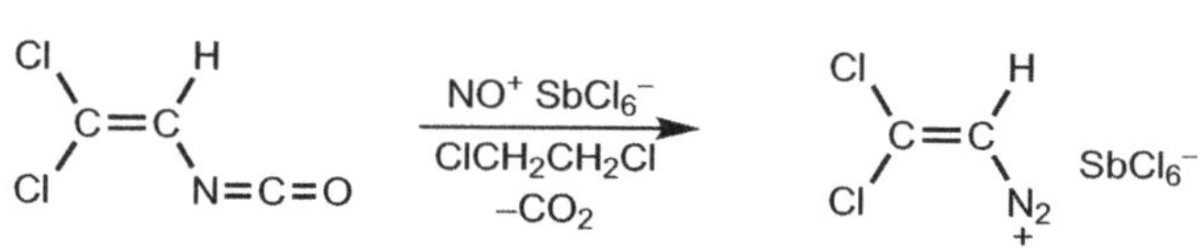

211

(4.136)

Glaser and co-workers[464,465] determined the X-ray crystal structure of β,β-disubstituted ions **210** and **211**. Structural parameters indicate that carbenium ion resonance forms stabilized by heteroatoms (O and Cl), particularly the type **210b**, contribute most to the overall stability of the ions.

Levisalles and co-workers[466] and Wentrup and Dahn[467] studied enoldiazonium ions generated under superacidic conditions. More recently Laali and coworkers[468] have shown that the stability of enoldiazonium ions can be greatly enhanced by an α-silyl functionality. Equation (4.137) shows the characteristic diazonium ion products.

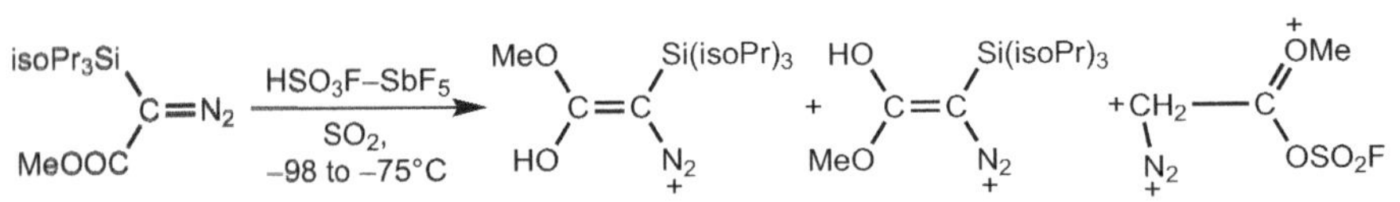

(4.137)

Aromatic Diazonium Ions. In contrast to alkyldiazonium ions, aryldiazonium ions are well-studied.[469–477] They were known as early as 1894. They are isolable as ionic salts with a variety of counterions such as BF_4^-, PF_6^-, $SbCl_6^-$, SbF_6^-, AsF_6^-, and ClO_4^-. They undergo a variety of nucleophilic reactions and an excellent review is available on the subject.[478]

The ambident reactivity of aryldiazonium ions has also been established.[479] The diazonium group is an interesting substituent on the aryl ring and by far the most strongly electron-withdrawing substituent known ($\sigma_p = 1.8$).[480] A ^{13}C NMR spectroscopic investigation on a series of aryldiazonium ions[481] seems to support the above fact and also indicates their ambident electrophilic character.[478] Aryldiazonium ions undergo interesting reaction of N_α–N_β inversion catalyzed by metals that complex molecular nitrogen (Scheme 4.14), probably through the intermediacy of a phenyl cation.[482–488] Such inversions have been observed by Zollinger and co-workers[489,490] in dediazoniation reaction of β-^{15}N-labeled diazonium ions with nitrogen gas under pressure (300 atmosphere) (Scheme 4.14). Such exchange reactions have been further studied with sterically hindered 2,6-disubstituted diazonium ions.[491,492]

The study of 2,6-disubstituted benzenediazonium ions[492] did not show the presence of any persistent *C*-protonated benzenediazonium dication even in Magic Acid [Eq. (4.138)], reinforcing the notion that charge delocalization into the aromatic ring plays a significant role.

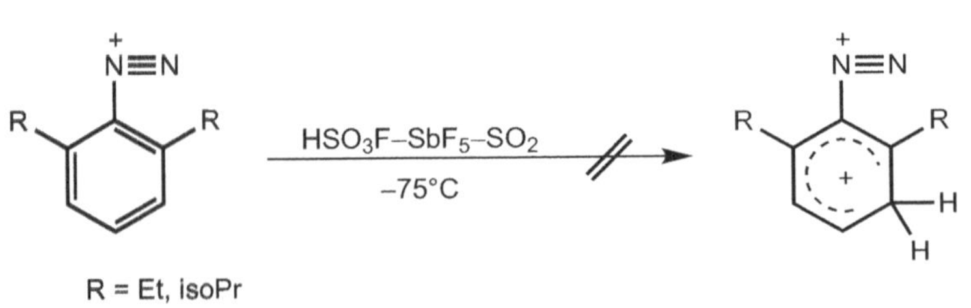

(4.138)

R = Et, isoPr

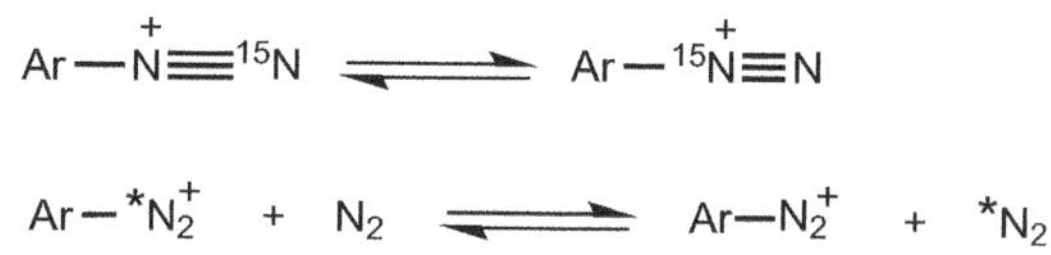

Scheme 4.14

In contrast, Laali and Olah[493] were able to *O*-protonate *para*-methoxybenzenediazonium ion in HSO_3F–SbF_5–SO_2ClF solution [Eq. (4.139)]. The C(4) carbon in the dication is slightly shielded relative to the precursor whereas all other carbons are deshielded. Unlike the *para*-methoxybenzenediazonium ion, the *ortho*-methoxybenzenediazonium ion is not *O*-protonated under similar conditions due to the proximity of the developing positive charge of the oxonium ion with the N_2^+. Attempted *O*-methylation with MeF–SbF_5–SO_2 was unsuccessful.

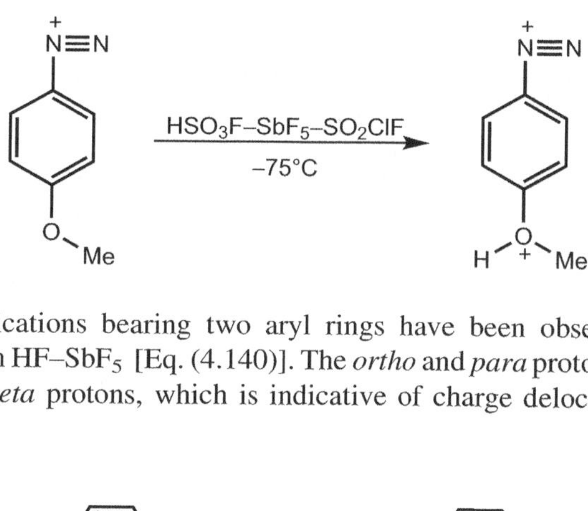

(4.139)

Diazonium dications bearing two aryl rings have been observed by treating azoxybenzenes in HF–SbF_5 [Eq. (4.140)]. The *ortho* and *para* protons are deshielded relative to the *meta* protons, which is indicative of charge delocalization into the aromatic rings.

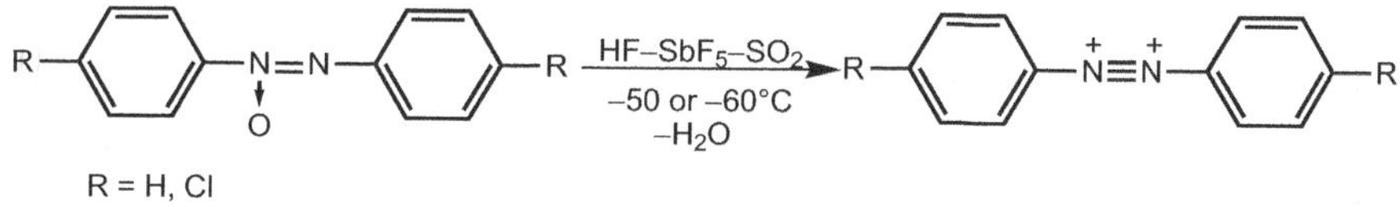

(4.140)

Other Substituted Diazonium Ions. A series of aminodiazonium ions have been prepared under superacidic conditions [Eq. (4.141)]. Schmidt[495] described the preparation and IR spectra of protonated hydrazoic acid **212** and methylazide as their hexachloroantimonate salts. Olah and co-workers[496] have carried out a comprehensive study on aminodiazonium ions (protonated azides) by ^1H, ^{13}C, and ^{15}N NMR spectroscopy. Even the electrophilic aminating ability of aromatics of **212** has been explored.[496] The tetrachloroaluminate salt of **212** has also been prepared[496]

[Eq. (4.142)] and the tetrafluroborate was also generated by reacting Me_3SiN_3 with $HF–BF_3–SO_2ClF$.[496]

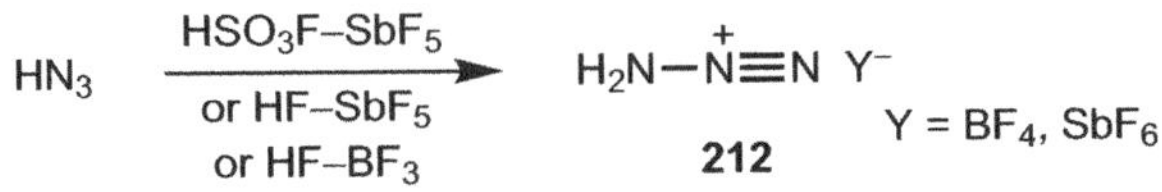

(4.141)

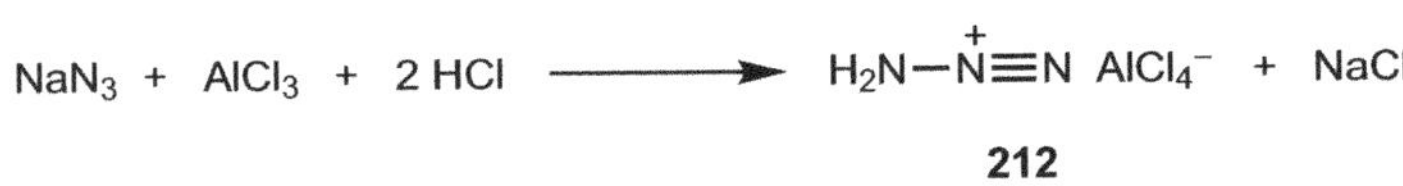

(4.142)

The evidence for the aminodiazonium structure **212** for the protonated hydrazoic acid comes from ^{15}N NMR spectroscopy and molecular structure of the hexafluoroantimonate salt.[497] These studies showed that both hydrogens are bonded to the same nitrogen, which has a pyramidal structure (**212**). The N–N single and triple bond lengths are quite different (1.295 versus 1.101 Å) and the N–N–N bond is slightly distorted (bond angle = 175.3°). Gas-phase protonation of hydrazoic acid has been studied by Cacace et al.[498,499] Gas-phase proton affinity was found experimentally to be $180 \pm 2\,kcal\,mol^{-1}$, whereas the calculated value is $179.4 \pm 1\,kcal\,mol^{-1}$. The iminodiazenium (1,3-diaza-2-azoniaallene) structure **213** is not observed in the case of either hydrazoic acid or the alkyl azides. Various salts ($AlCl_4^-$, $SbCl_6^-$, PF_6^-) of substituted 1,3-diaza-2-azoniaallene cations have recently been synthesized.[500,501] The strong absorption at $2012\,cm^{-1}$ was assigned to the symmetric stretching vibration of the $N{=}N^+{=}N$ unit, which was shown by AM1 calculations to have a bent structure (155°).

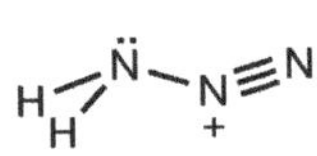

212

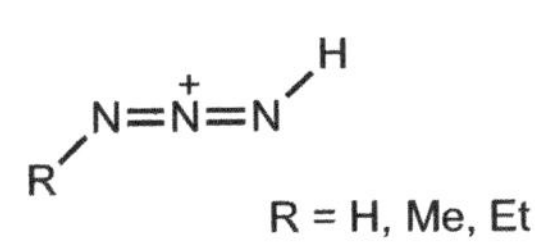

213

Alkyl azides are also protonated to alkylaminodiazonium ions **214** (Scheme 4.15). Multinuclear NMR studies and calculations for the parent aminodiazonium ion

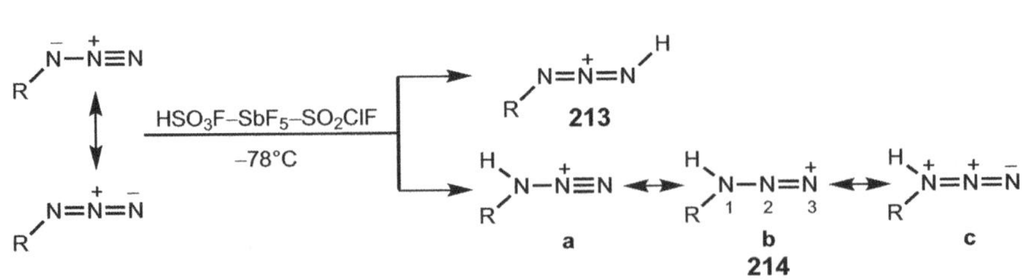

Scheme 4.15

indicate[496] that N(3) has significant positive charge and, therefore, the mesomeric structure **214b** is an important contributor to the overall structure.

Rasul, Prakash, and Olah[442] have attempted to generate bisdiazonium dication **215** by the diazotization of aminodiazonium ion **212** with $NO^+BF_4^-$ (HSO_3F–SbF_5–SO_2ClF, -78°C), but no further diazotization was observed. The linear $D_{\infty h}$ structure was found to be a minimum on the potential energy surface at various levels of theory.[442,502] Bond distances represent bisdiazonium character.

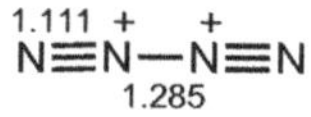

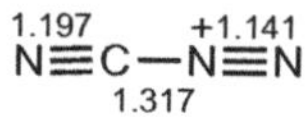

215 **216**

The cyanodiazonium ion NCN_2^+ proposed as a transient intermediate in solution,[503] has been observed in the gas phase by Cacace et al.[504] It was generated by ionization of NF_3 and cyanamide and characterized by collisionally activated dissociation. The linear structure of $C_{\infty v}$ symmetry (**216**) has been found[442] to be the global minimum by high-level *ab initio* calculations [MP2(FU)/6-31G**]. Of the protonated structures the dication protonated on the cyano nitrogen is 55.0 kcal mol^{-1} more stable.

Christe and co-workers[505] have reported the synthesis [Eq. (4.143)] and full characterization of the homoleptic N_5^+ polynitrogen cation. Both salts are surprisingly stable and decompose at −70°C. Crystal structure of the fluoroantimonate salt **217** is in excellent agreement with calculated values (MP2/6-31G*[440] and B3LYP[506]). The N_5^+ cation has a V-shaped structure of C_{2v} symmetry and its exceptional stability is largely due to resonance stabilization.

$$N_2F^+\,SbF_6^- + HN_3 \xrightarrow[\substack{-78\text{ to }25^\circ C\\ -HF}]{HF} N_5^+\,SbF_6^- \xrightarrow[25^\circ C]{SbF_5} \underset{\mathbf{217}}{N_5^+\,Sb_2F_{11}^-} \quad (4.143)$$

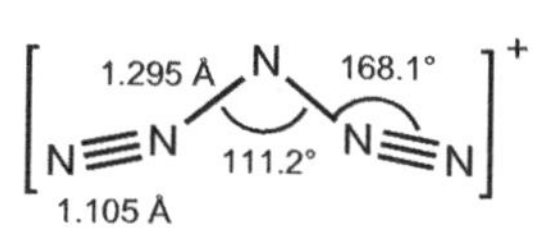

217

Olah et al.[507] have generated long-lived methoxydiazonium ion **218** by *O*-methylation or transmethylation of N_2O [Eq. (4.144)]. Multinuclear NMR data indicate that the nature of the N–N bond is quite different from that in benzenediazonium ion. Particularly revealing are the upfield shift of N_β and the N–N coupling ($\delta^{15}N$ 86.8 versus 316.8, $J_{N-N} = 12$ Hz versus <0.7). *Ab initio* calculations (4-31G and 4-31G* levels) show that ion **218** is planar and slightly distorted (N–N–O bond angle 173.3°) and has a considerably elongated N–O bond (1.425 Å) compared with that of N_2O (1.176 Å) and a weak C–O bond (1.489 Å). According to spectroscopic and computational results, both structures **218b** and **218c** are significant contributors to the overall structure of the ion.

$$N_2O + CH_3F\rightarrow SbF_5 \xrightarrow[-78^\circ C]{SO_2F_2} \underset{\mathbf{218}}{CH_3ON_2^+} \xleftarrow[-78^\circ C]{SO_2ClF} N_2O + CH_3O^+\ SO_2ClF\ SbF_6^- \qquad (4.144)$$

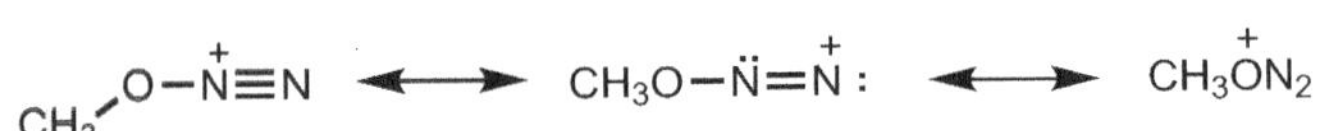

a b c

218

Generation of the hydroxydiazonium ion HON_2^+ as long-lived species under superacidic conditions was not successful.[507] According to an early theoretical study, the ion has a structure similar to that of methoxydiazonium ion **218**. Recent high-level calculations[442] [MP2(FU)/6-31G** level] have found the *O*-protonated structure of C_s symmetry to be the global minimum but it is only 2.5 kcal mol^{-1} more stable than *N*-protonated nitrous oxide. Of the dications formed by a second protonation, the *O,N*-diprotonated nitrous oxide is more stable than the *O*, *O*-diprotonated nitrous oxide by 7.5 kcal mol^{-1}.

Eberlin, Laali, and co-workers[508] have studied the reaction of N_2O with a variety of cations in the gas phase using collision-induced dissociation (CID) for detection and identification of products. Methoxydiazonium ion, $MeON_2^+$, is formed by the reaction of Me^+ with N_2O. This observation contradicts earlier findings[509] reporting the preferential formation of the isomeric nitroso-onium ion $Me–N_2O^+$. The reaction of $PhCH_2^+$, in turn, does produce the corresponding *N*-nitroso-onium ion $PhCH_2–N_2O^+$.

4.2.5.3. *Nitronium Ion (NO_2^+).* Nitration is one of the most studied and best understood organic reactions.[510–512] The species responsible for electrophilic aromatic nitration was shown to be the nitronium ion (NO_2^+) **219**. Since the early 1900s, extensive efforts have been directed toward the identification of this ion, whose existence was first shown by Hantzsch and later firmly established by Ingold and Hughes.[510]

Raman spectroscopic studies in the 1930s on $HNO_3–H_2SO_4$ mixtures by Médard[513] showed the presence of nitronium ion **219** in the media. A sharp band at 1400 cm^{-1} was assigned to the symmetric stretching mode of the species. Since then the nitronium ion has been characterized in a variety of Brønsted and Lewis acid mixtures of nitric acid.[514]

There are more than 15 crystalline nitronium salts that have been isolated and characterized with a variety of counterions. The most important salts are of BF_4^-,

PF_6^-, SbF_6^-, ClO_4^-, FSO_3^- counterions. The X-ray crystal structure of nitronium ion is known with hydrosulfate anion.[515] The most widely used tetrafluoroborate nitronium salt ($NO_2^+BF_4^-$) is prepared by treating a mixture of nitric acid and anhydrous hydrogen fluoride wit boron trifluoride[516] [Eq. (4.145)].

$$HNO_3 + HF + 2BF_3 \longrightarrow \underset{\mathbf{219}}{NO_2^+\,BF_4^-} + BF_3\text{–}H_2O \qquad (4.145)$$

Nitronium salts are also available from the reaction of NO_2F with $SbCl_5$ followed by anion exchange[517] [Eq. (4.146)], and methods have also been developed for the synthesis of the triflate salt.[517]

$$NO_2F + SbCl_5 \longrightarrow NO_2^+\,SbCl_5F^- \xrightarrow{Me_4N^+\,X^-} NO_2^+\,X^- \qquad (4.146)$$

$$X = ClO_4,\ BF_4$$

Ingold[518] was the first to propose that nitric acid anhydride N_2O_5 has an ionic structure $NO_2^+NO_3^-$. This was later confirmed by Wilson and Christe,[519] who showed a linear symmetric NO_2^+ with an N–O bond length of about 1.154 Å. The structure of reaction products formed between nitrogen oxides and BF_3, including $NO_2^+BF_4^-$ was established by Olah and co-workers[520,521] on the basis of Raman, IR, and X-ray diffraction data.

In the ^{15}N NMR spectrum,[245] the nitronium ion is about 130 ppm deshielded from NO_3^- of aqueous sodium nitrate solution.[245,448] It has also been shown by ^{14}N NMR spectroscopy that a mixture of nitric acid in 88% sulfuric acid contains both nitronium ion and free nitric acid.[522] At 95% H_2SO_4 only the NO_2^+ ion was detected (a single peak at $\delta^{14}N$ 251 relative to external liquid NH_3). Prakash, Heiliger, and Olah[523] have recently arrived at similar conclusions through ^{15}N NMR studies of the nitric acid–nitronium ion system. The ^{14}N NMR spectrum of $NO_2^+BF_4^-$ in SO_2 solution shows a single peak at $\delta^{14}N$ 248.6,[448] whereas in 92% H_2SO_4 it is detected at $\delta^{15}N$ 251.[523] The IGLO calculated value is $\delta^{15}N$ 268.3.[524]

Nitronium ion, which is linear with *sp*-hybridized nitrogen, is not very reactive in aprotic media and not capable of nitrating deactivated aromatics. The reactivity, however, can be increased in superacid solution (HF, HSO_3F). The enhanced reactivity is attributed to protosolvation,[525,526] that is, to the formation of protonitronium dication NO_2H^{2+} (**220**) suggested as early as 1975 by Olah et al.[527] The interaction of the superacid with the nitronium ion weakens the N–O π-bond character resulting in the bending of the linear ion and rehybridization of the N from *sp* to sp^2 (**220b**). Early *ab initio* calculations did not find a minimum for NO_2H^{2+}, but subsequent high-level studies by Olah et al.[524] (HF/6-31G* and MP2/6-31G**) showed it to be a minimum. It is characterized by a shorter N–O(1) (1.055 Å) and a longer N–O(2) (1.165 Å) bond, and an O–N–O bond angle of 172.7°. Studies by Schwarz and co-workers[528]

[RHF/6-311G** and MP2(fc)/6-311G**] resulted in similar values (1.125 Å, 1.175 Å, and 169.1°, respectively).

220

Attempts to observe ion **220** in the condensed phase under superacidic conditions were unsuccessful, but it was generated in the gas phase.[528] However, Prakash, Olah, and colleagues[529] made an attempt to identify ion **220** by means of ^{17}O NMR spectroscopy. ^{17}O enriched nitronium tetrafluoroborate, when dissolved in HSO_3F, exhibited a sharp peak at $\delta^{17}O$ 196.6 assigned for nitronium ion NO_2^+ shielded by 217.4 ppm with respect to nitric acid. Inceasing the acidity by adding SbF_5 to the solution (HSO_3F–SbF_5 = 1: 3), the NO_2^+ signal moved upfield by 5 ppm and the line broadened significantly (line width = 930 Hz). This line broadening could be brought about by possible proton exchange between the acid and NO_2^+ via the protonitronium dication **220**, which is present in small equilibrium concentration.

The applications of nitronium salts as a synthetic reagent[530] are discussed in Chapter 5. Until recently, the nitronium ion was recognized only as a nitrating agent. However, it has been found that it possesses significant ambident reactivity. This has been recently shown in the oxidation of sulfides, selenides, and phosphines. In fact, the sulfide reaction has been monitored by ^{15}N NMR spectroscopy wherein both nitrosulfonium and nitritosulfonium ions **88** and **89** were detected as distinct intermediates [Eq. (4.147)].

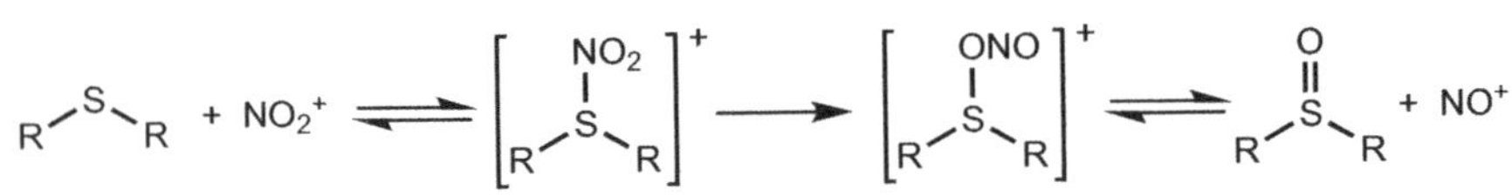

88 **89**

(4.147)

4.2.5.4. Nitrosonium Ion (NO^+).

Nitrosonium ion (NO^+) **221** is an important species that is generally present in nitrous acid media. It acts as a powerful nitrosating agent of amines (both aromatic and aliphatic) resulting in the diazotization reaction [Eq. (4.148)].

$$R{-}NH_2 \quad + \quad \underset{\mathbf{221}}{NO^+} \longrightarrow R{-}N_2^+ \quad + \quad H_2O \qquad (4.148)$$

The first isolation of nitrosonium ion **221** as a distinct species was in the reaction of dinitrogen trioxide and dinitrogen tetroxide with boron trifluoride[531] [Eqs. (4.149) and (4.150)].

$$3N_2O_3 + 8BF_3 \longrightarrow \underset{\textbf{221}}{6NO^+ BF_4^-} + B_2O_3 \quad (4.149)$$

$$3N_2O_4 + 8BF_3 \longrightarrow \underset{\textbf{221}}{3NO^+ BF_4^-} + 3NO_2^+ BF_4^- + B_2O_3 \quad (4.150)$$

Since then, a variety of nitrosonium salts have been isolated. The important ones are with the following counterions: BF_4^-, PF_6^-, FSO_3^-, HSO_4^-, BCl_4^-, and $SbCl_6^-$. The ion has been characterized by ^{15}N NMR, IR, and X-ray analysis.[245,448,514,531,532] A detailed NMR study by Mason and Christe have showed[448] that conditions (solvent, counterion, temperature) have minor effects on observed ^{14}N NMR shifts ($\delta^{14}N$ 372.5–376.8).

Similar to nitronium ion, attempt was also made to identify protonated nitrosonium ion (protonitrosonium dication, HNO^{2+}) by means of ^{17}O NMR spectroscopy.[529] The sharp peak at $\delta^{17}O$ 461.5 observed in HSO_3F and identified for NO^+ shifted upfield by 5 ppm upon addition of SbF_5. This, again, can be attributed to the presence of protonitrosonium dication in small equilibrium concentration.

The nitrosonium ion does not react toward aromatics except in activated systems. It forms a π-complex with aromatics with deep color.[533,534] However, it is a powerful hydride-abstracting agent in the case of activated benzylic or allylic positions. Olah and Friedman[535] have demonstrated that isopropylbenzenes undergo hydride abstraction to cumyl cations **222** [Eq. (4.151)] which further reacts to give various condensation products. The reaction has been employed to prepare a variety of stable carbocations.[536]

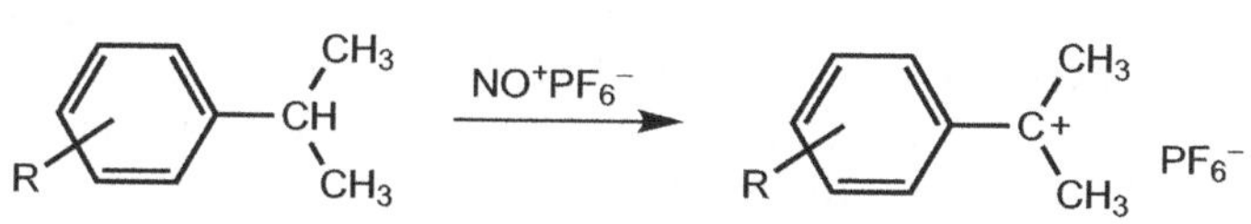

(4.151)

222

The unique hydride abstraction property has been gainfully employed in developing novel synthetic reactions.[530] Reactive hydrocarbons such as triphenylmethane, adamantane, and diamantane are readily fluorinated in the presence of nitrosonium ion in HF–pyridine media.[537] In the presence of a suitable oxygen donor such as dimethyl sulfoxide, the nitrosonium ion can act as a nitrating agent[538] [Eq. (4.152)]. The initially formed nitrito onium ion **223** transfer nitrates aromatics rather readily.[245] The NO^+-induced reactions are further reviewed in Chapter 5.

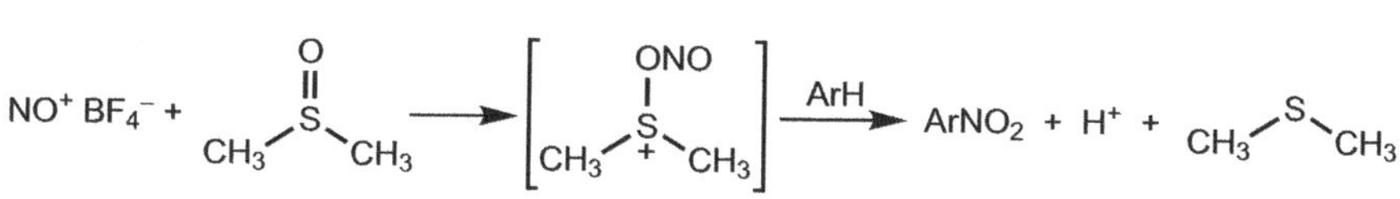

223

(4.152)

4.2.5.5. Ammonium, Phosphonium, Arsonium, and Stibonium Ions

Tetrahaloonium Ions. Whereas only tetrafluoro and tetrachloro cations of N and tetrachloro and tetrabromo cations of Sb are known, all four tetrahalophosphonium and arsonium cations, along with mixed tetrahaloonium cations, have been prepared and characterized mostly by vibrational and NMR spectroscopy.[539,540] In addition, X-ray crystal structure analyses have been reported for $NF_4^+BF_4^-$,[541] $PCl_4^+SbF_6^-$,[542] $AsCl_4^+AsF_6^-$,[543] and $SbCl_4^+Sb_2F_{11}^-$.[544,545] Schrobilgen and co-workers have recently reported spectroscopic characterization and X-ray crystal structures of $AsX_4^+[As(OTeF_5)_6]^-$ (X = Cl, Br)[539] and $SbX_4^+[Sb(OTeF_5)_6]^-$ (X = Cl, Br).[540] The method of preparation is the oxidation of AsX_3 with $XOTeF_5$ in the presence of $As(OTeF_5)_n$ ($n = 4, 5$) at room temperature or oxidation of $Sb(OTeF_5)_3$ with X_2. In agreement with Raman spectroscopy data, all four cations have undistorted tetrahedral geometry in the crystal state and the anion–cation interactions are weaker than in previously known tetrahaloarsonium and tetrahalostibonium salts. By reexamining the XRD structure of $NF_4^+BF_4^-$, Christe *et al.* have recently shown that, in contrast to earlier incorrect structural analysis, the NF_4^+ cation is really tetrahedral.[546] The hexafluoroarsenate of the mixed cation $AsFCl_3^+$ was also prepared and characterized.[547]

Acidic Onium Ions. Relatively little is known about halogenated acidic onium ions. A few fluoroammonium salts including $FNH_3^+HF_2^- \cdot nHF$[548] and $F_2NH_2^+MF_6^-$ (M = As, Sb)[549] are known. In addition, Christe and coworkers have been able to obtain the salts $ClNH_3^+M^-$ (M = BF_4, AsF_6, SbF_6) by protonating NH_2Cl formed in situ from $(Me_3Si)_2NCl$[550] [Eq. (4.153)]. All salts are stable at room temperature and contain small amounts of NH_4^+ impurities, which prevented crystallization. IR, Raman, and NMR spectroscopy and theoretical calculations provide evidence for the existence of ion $ClNH_3^+$. For example, the ^{14}N NMR shift at $\delta^{14}N$ –364 is only slightly deshielded in comparison with that in NH_4^+ ($\delta^{14}N$ –367) but considerably more shielded than that in FNH_3^+ ($\delta^{14}N$ –252.1).

$$(Me_3Si)_2NCl + HF + M \xrightarrow[-196 \text{ to } 25^\circ C]{} ClNH_3^+ \, MF^- \qquad (4.153)$$

$$M = BF_3, AsF_5, SbF_5$$

The fluorinated salts $F_{4-n}PH_n^+Y^-$ ($n = 1$–4, Y = AsF_6, SbF_6, Sb_2F_{11}, Sb_3F_{16}) of phosphorus have been characterized by vibrational and NMR spectroscopy.[551–554] The crystal structure of the salt $PF_3H^+SbF_6^- \cdot HF$ and trihalogenphosphonium salts $X_3PH^+As_2F_{11}^-$ (X = Cl, Br) prepared by protonation with the corresponding conjugated superacids HF–MF_6 (M = As, Sb) have recently been reported by Seppelt and co-workers[554] and Minkwitz and Dzyk,[555] respectively. The structure of the PF_3H^+ cation is pseudo-tetrahedral and, surprisingly, the cation has no contact with either the anion or HF. The cations of the $X_3PH^+As_2F_{11}^-$ salts have trigonal pyramidal structure and interionic F–Cl, P–Br, and P–F contacts but F–H interactions were not detected.

Superacidic protonation of $MePF_2$ was also used to prepare the salts $MePF_2H^+MF_6^-$ (M = As, Sb).[556] The cations show distorted tetrahedral structure and weak P---F interionic contacts.

Protonation of PH_3 and AsH_3 with HF was studied by conductometric measurements with the conclusion that the less basic AsH_3 is not fully protonated.[557] After treatment in HF–TaF_5, crystalline $PH_4^+TaF_6^-$ and a mixture of $AsH_4^+TaF_6^-$ and $AsH_4^+Ta_2F_{11}^-$ could be isolated. $AsH_4^+MF_6^-$ (M = As, Sb) and $SbH_4^+SbF_6^-$ were obtained by protonation of the corresponding hydrides with HF–MF_5.[558] Me_3AsH^+ ions were generated in a similar manner[559] [Eq. (4.154)]. The cation in the $Me_3AsH^+As_2F_{11}^-$ salt has a trigonal pyramidal shape with weak hydrogen bonding between the hydrogen atoms of the methyl groups and fluorine atoms.

$$\underset{M\,=\,As,\ Sb}{Me_3As + HF + MF_5} \xrightarrow[-78°C]{} \underset{M'\,=\,As_2F_{10},\ SbF_5}{Me_3As^+\ M'F^-} \tag{4.154}$$

Hydroxy(alkoxy)phosphonium Ions. Olah and McFarland[560] studied the protonation in HSO_3F or HSO_3F–SbF_5 solution of varied phosphorus oxyacids and derivatives. Treatment of tetravalent phosphorus compounds (phosphorus, phosphonic, and phosphinic acid and their trialkyl and triaryl derivatives) results in *O*-protonation and the formation of hydroxyphosphonium ions. Trivalent phosphites, in turn, are protonated at the phosphorus atom. The ^{31}P shifts observed for the latter ions are significantly deshielded, which was attributed to significant oxonium ion character.

Minkwitz and Schneider[561] obtained tetrahydroxyphosphonium salts by protonation of phosphorus acid with HF–AsF_5 and HF–SbF_5 [Eq. (4.155)]. XRD structure study shows that the cation of the hexafluoroantimonate salt has almost S_4 symmetry and equal P–O bonds (1.529–1.536 Å). Cations and anions in the crystals are linked three-dimensionally by hydrogen bonds (each cation is bonded to six anions).

$$\underset{M\,=\,As,\ Sb}{H_3PO_4 + (Me_3SiO)_3PO + HF + MF_5} \xrightarrow[-Me_3SiF]{-196\ \text{to}\ -60\ °C} P(OH)_4^+\ MF_6^- \tag{4.155}$$

Aurated Onium Ions. A variety of gold ammonium cations $[(AuPPh_3)_3NR]^+$ (R = Me, Et, *n*-Pr, isoPr, *tert*-Bu, cyclohexyl, Ph, Bn, 4-FC_6H_4, 4-BrC_6H_4, 4-$NO_2C_6H_4$) have been obtained as tetrafluoroborate salts by treating $[(AuPPh_3)_3O]^+BF_4^-$ with excess RNH_2 or RNCO.[562,563] X-ray crystal structure analysis show a distorted tetrahedral geometry of the cations with the Au–PPh_3 ligands linearly coordinated to nitrogen.[562,563] The cations of $[(AuPPh_3)_3NR]^+BF_4^-$ (R = *tert*-Bu, cyclohexyl) exhibit Au–N–Au angles smaller than the tetrahedral value (average 102°) and short Au---Au distances (~3.1 Å), indicating attractive forces between gold atoms. The cation in $[(AuPPh_3)_4N]^+BF_4^-$ shows similar features.[564]

The powerful aurating agent $[(AuPPh_3)_3O]^+BF_4^-$ has been used by Schmidbaur et al.[565] to generate triaurated phosphonium ions [Eq. (4.156)]. In contrast to aurated ammonium ions and as a result of the larger size of phosphorus, the Au–P–Au angle of cation **224** deviates only slightly (average 106°) from the ideal tetrahedral angle and the Au···Au distances (~3.7 Å) are significantly longer. This latter feature indicates the lack of Au···Au interactions. Cation **225** with the five-coordinate, electron-deficient central phosphorus atom was characterized by ^1H and ^{31}P NMR spectroscopy. Characteristic features are a doublet ($\delta^{31}P$ 39.6) and a sextet ($\delta^{31}P$ 122) in the area ratio 5:1. Furthermore, the coupling constant is significantly reduced compared to that in **224** (J_{P-P} = 186 Hz versus 249 Hz), indicative of diminished *s* character of sp^3d-hybridized P. The trigonal bipyramidal structure is suggested on the basis of C and N analogs.

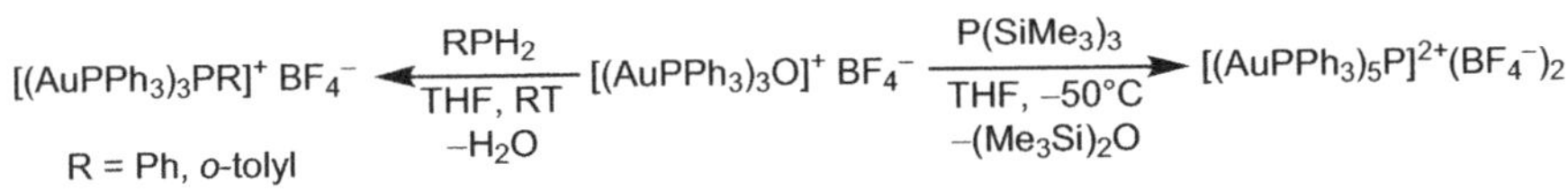

224 **225**

(4.156)

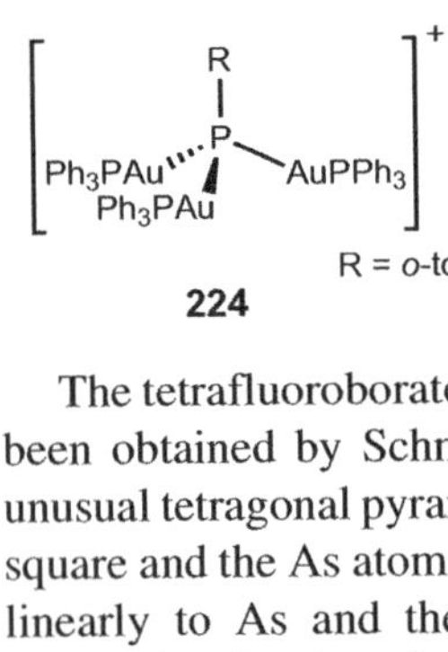

R = *o*-tolyl

224

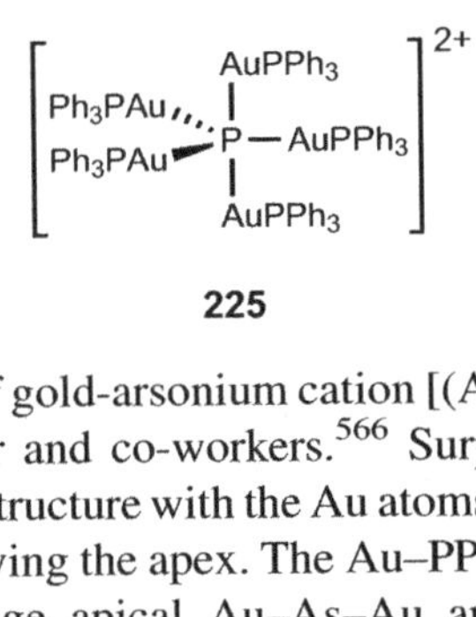

225

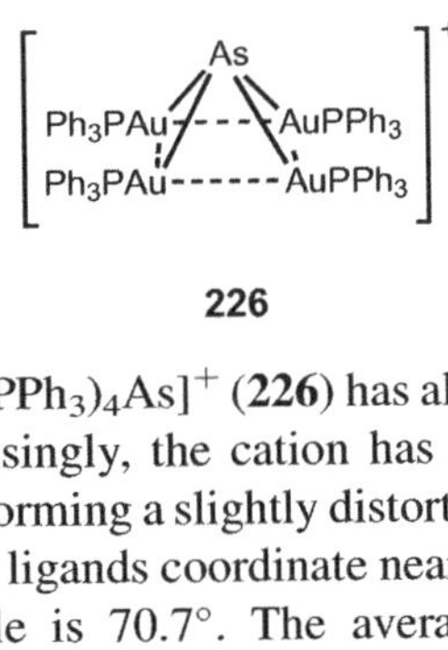

226

The tetrafluoroborate salt of gold-arsonium cation $[(AuPPh_3)_4As]^+$ (**226**) has also been obtained by Schmidbaur and co-workers.[566] Surprisingly, the cation has an unusual tetragonal pyramidal structure with the Au atoms forming a slightly distorted square and the As atom occupying the apex. The Au–PPh_3 ligands coordinate nearly linearly to As and the average apical Au–As–Au angle is 70.7°. The average intramolecular Au···Au contact is 2.90 Å and intermolecular Au···Au and As···Au contacts also exist.

Other Onium Ions. Minkwitz and co-workers have obtained a variety of halophosphonium ions with a sulfur ligand including $Cl_nBr_{3-n}PSMe^+MF_6^-$ (n = 0–3, M = As, Sb),[567] $X_2FPSMe^+MF_6^-$ (X = Cl, Br, M = As, Sb), and $XF_2PSMe^+SbF_6^-$ (X = F, Cl, Br).[568] These were prepared by methylation of the corresponding thiophosphorylhalides. Oxidative bromination, in turn, furnished $Cl_{n-3}Br_nPSBr^+AsF_6^-$ (n = 0–3) and $Cl_3PSBr^+SbF_6^-$.[569] Characterization by spectroscopic methods (vibrational, NMR) have also been reported. In addition, they have isolated the methylfluoroarsonium salts $MeAsF_3^+AsF_6^-$[570] and $MeAsF_3^+SbF_6^-$.[571]

Laali et al. have studied the protonation of tetra-*tert*-butyltetraphosphacubane (**227a**) and tetra-*tert*-butyltetraarsacubane (**227b**). In the case of tetra-*tert*-butyltetraphosphacubane, monoprotonation was observed in FSO_3H or CF_3SO_3H to furnish cation **228**,[572,573] whereas diprotonation occurred in the stronger superacid $HSO_3F–SbF_5$ (1:1) to give dication **229**. The formation of a mixture of the corresponding mono- and diprotonated cations was detected when tetra-*tert*-butyltetraarsacubane (**227b**) was treated in $HSO_3F–SO_2$.[574] The unique tetraphosphatricyclodiene **230** also underwent similar mono- and diprotonation with FSO_3H and $HSO_3F–SbF_5$ (1 : 1), respectively.[575]

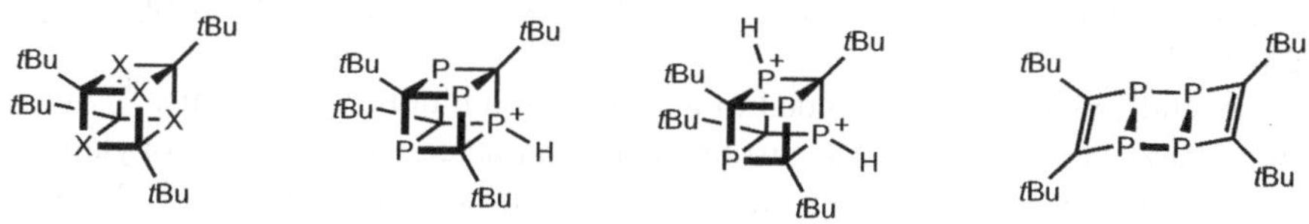

227 a X = P
227 b X = As
228
229
230

4.3. ENIUM IONS

4.3.1. Enium Ions of Group 13 Elements

4.3.1.1. Borenium Ions. Boron cations with coordination number four such as $[BH_2(NH_3)_2]^+$ and many others of the type $[H_2BL_2]^+$, $[HXBL_2]$, and $[X_2BL_2]$ are well known[576–581] (X = halogen, L = electron-donating ligand). Even doubly and triply charged tetracoordinate boron cations are known. However, dicoordinate borenium ions, wich are elusive and highly electrophilic species, are much less known. To date, no dicoordinate borenium ion with either only alkyl and/or aryl substituent is known. Developments in the area have been reported in reviews.[582,583]

It is well recognized that electron deficiency of boron compounds can be considerably compensated by π-back bonding. Exploiting this principle, Nöth and Staudigl[584] have succeeded in obtaining borenium ions.

Reaction of anhydrous aluminum bromide with a series of 2,2,6,6-tetramethylpiperidinoaminoboron bromides in dichloromethane leads to specific displacement of bromide, which is trapped as tetrabromoaluminate. By formation of this less nucleophilic anion, and owing to the steric and electronic shielding of the β atom by the bulky 2,2,6,6-tetramethylpiperidino moiety, dicoordinate borenium ions **231** are generated [Eq. (4.157)].

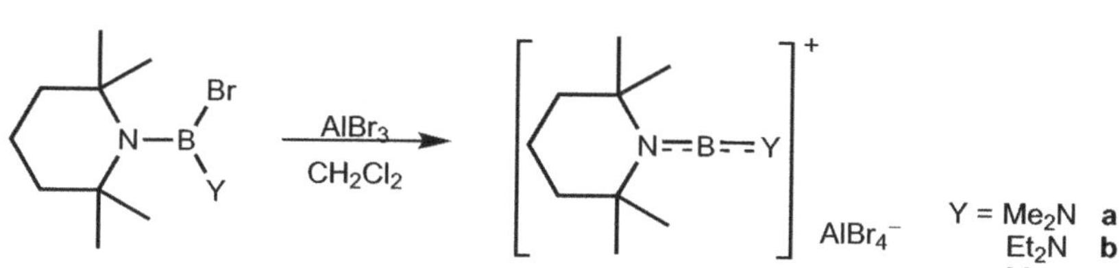

231

(4.157)

^{27}Al NMR spectra confirm exclusive formation of the $AlBr_4^-$ anion, which, compared to Al_2Br_6, is characterized by its substantially sharper signal; the linewidth of $\nu_{1/2} \sim 20$ Hz in **231a** and **231b** corresponds to an undistorted tetrahedral $AlBr_4^-$. The ^{11}B NMR signals of **231** are shifted 6–18 ppm downfield relative to those of the starting compounds; their linewidths, which are greater by a factor of about 5, are consistent with a linear heteroallene structure. Also consistent with such a structure are the isotopically split IR bands at 1850–1900 cm^{-1}, which are assigned to an antisymmetric BN_2 vibration. The heteroallene structure has been confirmed by X-ray analysis of **231a**.[585]

Such studies have been extended to a variety of amidoborenium ions.[585] Subsequently, Parry and coworkers have isolated and characterized bis(diisopropylamino) borenium ion **232** at low temperature as tetrachloroaluminate salt.[586] Attempts to prepare the analogous bis(dimethylamino)borenium ion **233** was, however, unsuccessful.

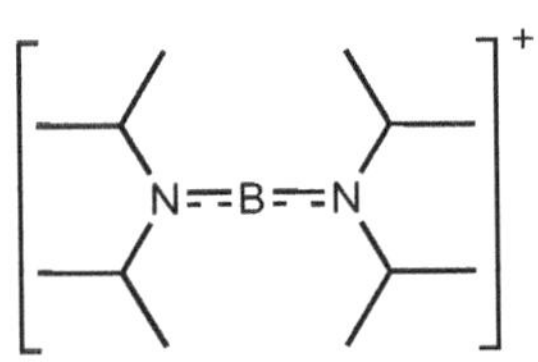

232

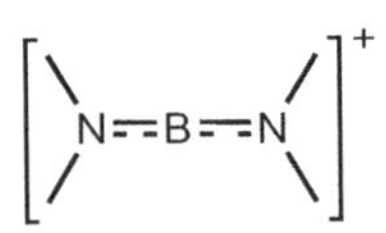

233

Cations with sterically demanding *tert*-Bu and Me_3Si groups on the nitrogen atoms have been obtained by Kölle and Nöth.[587] Because of the protective bulky amino groups, the cations are stable in solution. ^{11}B NMR spectroscopy gives resonances in the region expected for boron cations ($\delta^{11}B$ 32–36), whereas IR bands around 1800 cm^{-1} were assigned to B–N stretching and N–B–N bending vibrations.

Stephan and co-workers[588] have synthesized cation **234** with an extended structure containing a P=N–B–N=B moiety [Eq. (4.158)]. The crystal structure reveals a linear arrangement of the five atoms (N–B–N bond angle = 180.0°) with a B–N bond length of 1.236 Å. This bond distance is considerably shorter than those in cation **231a** (1.30 and 1.42 Å) indicative of strong electron donation from the phosphinimide ligand.

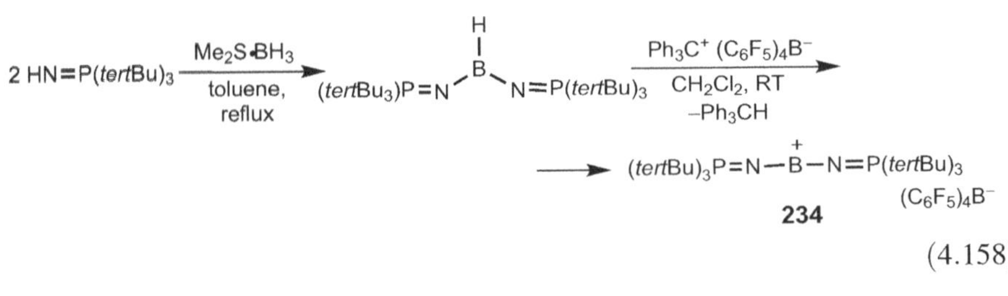

(4.158)

The chemistry of dicoordinate boron cations with additional coordinating donor ligands has also been explored.[583] Three-coordinate cations **235** have been reported by Jutzi et al.[589] and cations **236** have been synthesized by Kuhn et al.[590] The X-ray crystal structure of cation **237a** reported by Cowley et al.[591] indicates that the 1,3,2-diazaborenium ring is planar with trigonal boron geometry (sum of angles = 359.9°) and longer B–N bond lengths (average 1.449 Å). ^{11}B NMR spectroscopy gives a resonance at δ^{11}B 72, whereas calculations (B3LYP/3-21G* level) show π-interactions in the ring. Cation **237b** prepared by methyl abstraction shows a resonance at δ^{11}B 37.1 shifted downfield substantially and a line width typical for three-coordinate boron.[592] The downfield shift of the methine proton at δ^{1}H 6.73 coupled with the fact that cation **237b** coordinates only with the strong Lewis base pyridine is indicative of a significant aromatic stabilization.

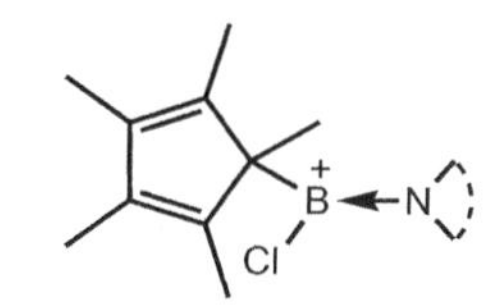

N = phenanthridine, acridine

235

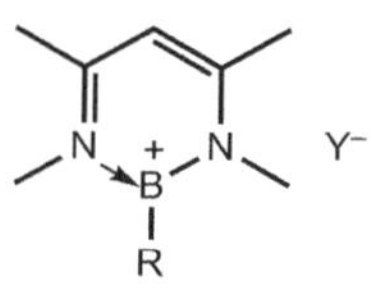

R = F, Cl, Et, Ph
Y = BF_4, $AlCl_4$

236

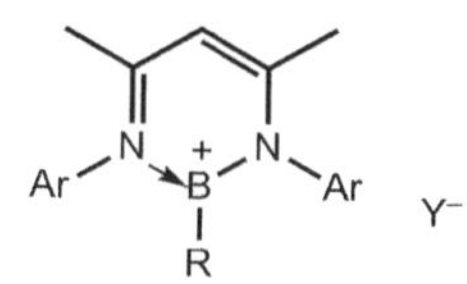

a R = Ph, Ar = 2,6-diisoPrC_6H_3, Y = Al_2Cl_7
b R = Me, Ar = 4-MeC_6H_4, Y = $MeB(C_6F_5)_3$

237

Further examples are cations **238**[593] and **239**.[594] Cation **239** is particularly interesting, since the two boron substituents (phenyl and H) are barely able to contribute to π-stabilization. A broad resonance at δ^{11}B 38.7 in the low-temperature ^{11}B NMR spectrum, which disappears at warming, was assigned to the trivalent B atom.

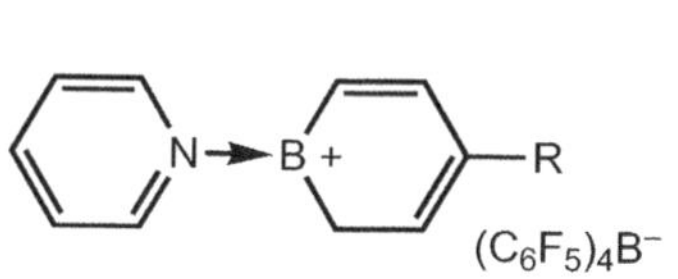

R = H, Me, isoPr

238

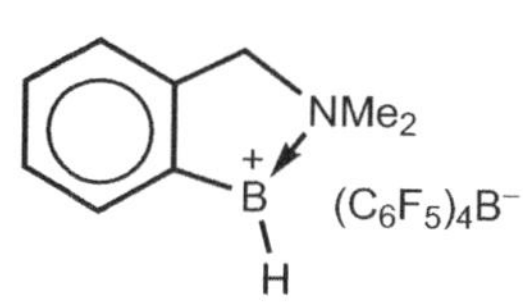

239

An additional *n*-donor ligand results in the most stable boron cations because of the filled octet of boron and the complete coordination sphere.[583] Boron in cation **240** has a distorted tetrahedral geometry and a N(1)–B–C angle of 99.5°.[583] In cation **241** the boron atom is complexed to [Cp*Fe(CO)$_2$].[595] The Cp* group is bonded to boron in an η^5 fashion and the Cp–B–Fe vector is essentially linear (bond angle = 177.86°). Both spectroscopic data and DFT calculations (B3LYP with LANL2DZ and 6-31 + G* basis sets) show that the Fe–B bond order is 1.

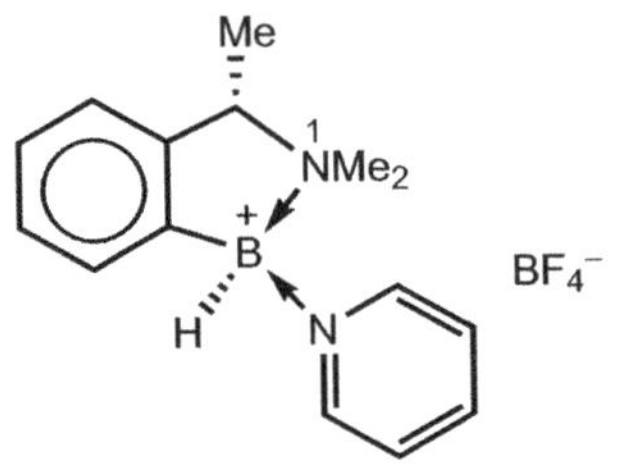

240

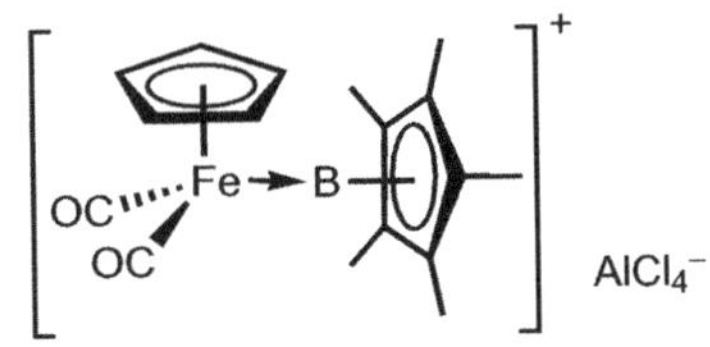

241

Examples of boron dications are also known. The tris(amine) dications $[X_3BH]^{2+}(PF_6^-)_2$ (X = pyridine, 2-Me- and 4-Me-pyridine, 3,5-diMe-pyridine) were synthesized by Mathur and Ryschkewitsch.[596] Cowley and co-workers have recently succeeded in generating boron dication **242** with the coordinated bidentate base 2,2′-bipyridine[597] [Eq. (4.159)]. The ^{11}B NMR chemical shift (δ^{11}B 6.44) is in the range of boronium cations.[583] The BN_2C_3 ring is planar and orthogonal to the plane of the bipyridine ligand. The average bond distances (N–C = 1.358 Å, C–C = 1.381 Å) are identical to those of cation **237a**. Since boron in **242** has a tetragonal geometry, the average B–N bond length (1.514 Å) is longer than that in the trigonal planar cation **237a** (1.450 Å).

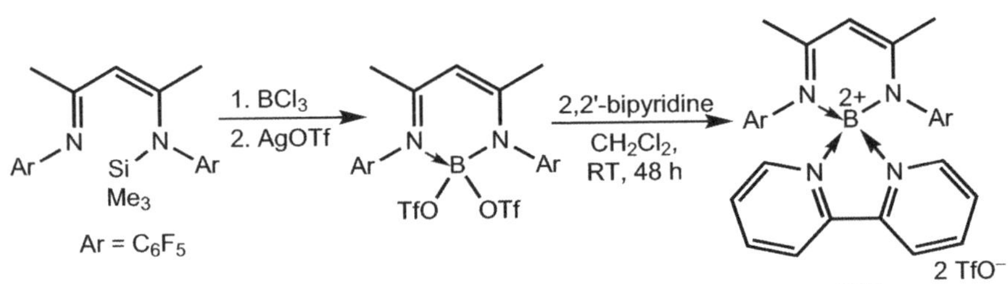

(4.159)

4.3.1.2. Alumenium Ions. A few examples of alumenium cations are known. Cation **243**[598] and the pentamethylcyclopentadienyl analog[599] most likely possess η^5-bound metallocene-type structure. Cation **244** has been prepared from the reaction of Et_3Al with carborane trityl salts.[600] The crystal structure of the salts **244** reveal weak bidentate interaction between the Et_2Al^+ fragment and two halogen substituents of the anions. The Al···Br interactions (2.54 and 2.58 Å) are longer than typical Al–Br bonds (2.25–2.30 Å). The C–Al–C bond angles (130.0° and 136.6° for the bromo and chloro derivatives, respectively), which are larger than the ideal tetrahedral and trigonal angles suggest considerable alumenium ion character. Salts of the cation $isoBu_2Al^+$ with the weakly coordinating anions $(C_6F_5)_4M^-$ (M = B, Al) and $[Al\{OC(CF_3)_3\}_4]^-$ have recently been generated.[601] The cation in $[isoBu_2Al]^+(C_6F_5)_4B^-$ is stabilized by two tetrahydrofuran donor ligands.

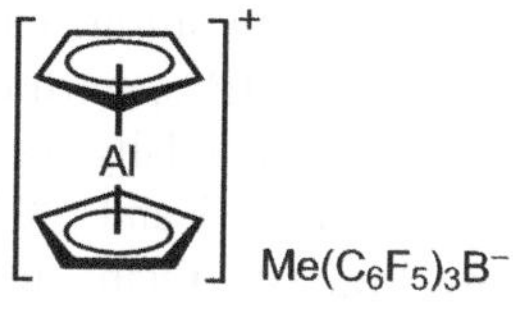

243

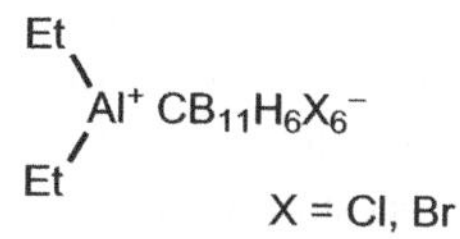

244

4.3.2. Enium Ions of Group 14 Elements

4.3.2.1. Silicenium Ions. One of the key intermediates that was not observed in solution for a long time is trivalent positively charged silicon, the silicenium ion R_3Si^+, the analog of a carbocation.[602–604] On the contrary, silicenium ions are well known in the gas phase as high-abundance fragments in the mass spectra of organosilicon compounds.[605,606] The failure to observe them in solution is due to the poor ability of silicon to undergo $p\pi$–$p\pi$ bonding. Whereas carbocations are readily stabilized by $2p$–$2p$ resonance, the silicenium ion is more weakly stabilized through $2p$–$3p$ overlap over longer bonds with lone pairs or π-electrons on carbon, nitrogen, or oxygen.[607] Furthermore, silicon, unlike carbon, has the ability to increase its coordination number and valency, and, consequently, is prone to complexation with various nucleophiles including counterions and solvents. Moreover, the very large bond strength of silicon with oxygen, nitrogen, and most halogens make common leaving groups unavailable. This is the main reason, why attempts to prepare silicenium ions under superacidic conditions have failed (due to nucleophilic fluorosulfate or fluoride ions which strongly bond electrophilic silicon).[608–611]

Olah and Field[612] were able to obtain only a polarized complex **245** from methylsilyl bromide and aluminum bromide in methylene bromide solution. However, they were able to correlate ^{29}Si NMR chemical shifts with ^{13}C NMR chemical shifts of analogous compounds. Based on such an empirical relationship, they have been able to predict ^{29}Si chemical shift of trivalent silicenium ion.

$$\overset{\delta+}{Me_3Si}-Br \rightarrow \overset{\delta-}{AlBr_3}$$

245

Later, Lambert and Schulz[613] have prepared triisopropylthiosilicenium ion **246** by hydride abstraction from triisopropylthiosilane using trityl perchlorate in dichloromethane solution [Eq. (4.160)].

$$(isoPr_2S)_3Si-H \;+\; Ph_3C^+\,ClO_4^- \longrightarrow \underset{\mathbf{246}}{(isoPrS)_3Si^+\,ClO_4^-} \;+\; Ph_3CH \qquad (4.160)$$

The evidence for **246** comes from both electrical conductivity measurements and ^1H and ^{13}C NMR spectra. The IR spectrum of **246** clearly showed the presence of perchlorate anion. The ^1H and ^{13}C NMR data were interpreted to indicate that there is not much positive charge delocalization from silicon to sulfur. Unfortunately, the authors were not able to obtain a satisfactory ^{29}Si NMR spectrum. Subsequently, Lambert and coworkers reported the synthesis of triphenylsilyl perchlorate[614] and trimethylsilyl perchlorate[615] using the same method. ^{29}Si and ^{35}Cl NMR spectroscopic and X-ray crystallographic studies by Prakash, Olah, and co-workers[616] showed, however, that triphenylsilyl perchlorate exists as a covalent perchloryl ester in both the solid state and solution. They also pointed out that existence of a silicenium ion can be proved by observing a significantly downfield shifted ^{29}Si NMR signal. Such proof, however, did not exist.

Since the late 1980s a variety of approaches has been employed to generate silicenium ions.[606,617–621] Most of the results, however, have been challenged.[606,620,622–628]

Lambert and co-workers[629–631] have made continued efforts to generate silicenium ions in high-polarity, low-nucleophilicity solvents (sulfolane, dichloromethane) using the hydride abstraction method with trityl perchlorate. In a subsequent study, Olah, Prakash, and co-workers[622] compared experimental and computed (*ab initio*/IGLO) ^1H, ^{13}C, and ^{29}Si NMR chemical shifts of trimethylsilyl perchlorate and concluded that no long-lived persistent trimethylsilicenium ion was observed in solution. The ^{13}C chemical shifts at various concentrations in sulfolane–CD_2Cl_2 reported by Lambert[631] correspond closely to the calculated values of the covalent trimethylsilyl perchlorate. Furthermore, the experimentally observed ^{29}Si chemical shift ($\delta^{29}Si$ 47.0) is fundamentally different from the IGLO-calculated value of $\delta^{29}Si$ 355.7.

In 1993 Lambert and co-workers[632,633] reported the synthesis of the $Et_3Si^+(C_6F_5)_4B^-$ salt [Eq. (4.161)]. It was identified as the cation **247a** on the basis of a long $Si-C_{tolyl}$ bond distance of 2.18 Å. However, the Et_3Si moiety is not planar (average C–Si–C bond angle = 114°), and calculated (carbocation chemistry) bond distances and bond angles (HP/6-31G* level) are in close agreement with those found experimentally.[625] Similarly, there is a particularly good agreement between the observed ($\delta^{29}Si$ 81.8 in toluene[632] and $\delta^{29}Si$ 92.3 in benzene[633]) and calculated ($\delta^{29}Si$ 82.1; IGLO" at HP/6-31G* level) ^{29}Si NMR chemical shifts, in contrast with the calculated highly deshielded shift of 355.7 ppm of the still elusive Me_3Si^+ ion. These observations and additional calculations[626,634,635] showed that there is a substantial covalent bonding and substantial transfer of positive charge to toluene. Consequently, the actual species is the *para*-triethylsilyltoluenium ion **247b** with contribution from the resonance form **247a**. According to the interpretation of Reed,[636] the actual structure is not a π-arene complex nor a σ-bonded arenium ion (Wheland intermediate), but a structure in between. Additional reports about the generation of silicenium ions with a variety of weakly coordinating anions such as $(C_6F_5)_4B^-$,[637] $CB_9H_5Br_5^-$,[638] $CB_{11}H_6Br_6^-$,[639,640] and 1-Me-$CB_{11}F_{11}^-$[641] similarly resulted in the formation of species with limited silicenium ion-like character as indicated by ^{29}Si NMR chemical shift values in the range $\delta^{29}Si$ 100–120.

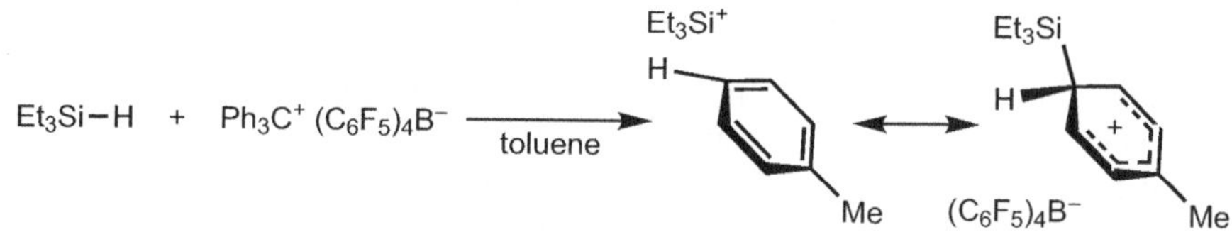

247

(4.161)

A report by Jørgensen and co-workers[642] about the synthesis of the first chiral tertiary alkylsilicenium ion **248** was also questioned. DFT/IGLO NMR studies showed[627] that it is a silylated acetonitrilium ion. Calculations (B3LYP/6-31G* level) to find a minimum energy structure for the free silicenium ion failed. Instead, the intramolecularly silylated spirocyclopropylarenium ion was found.

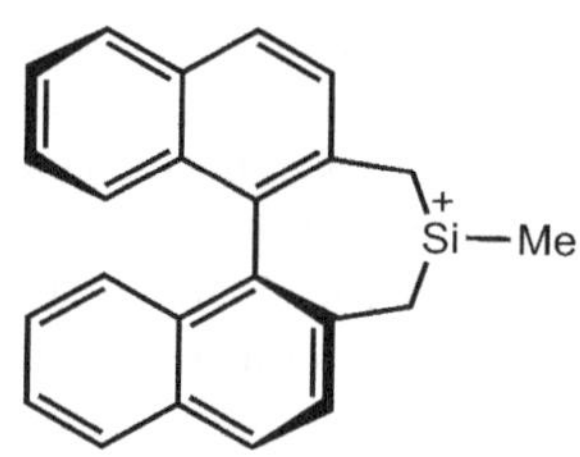

248

It became clear after these efforts that the successful synthesis of free silicenium ion requires the use of a counterion as inert as possible. Furthermore, it is necessary to hinder the Si^+ environment with bulky, sterically demanding groups and electronically stabilize the cation by appropriate substituents to suppress any interaction with the anion or the solvent.[643,644]

Recently, Lambert and co-workers[645,646] have reported the preparation of the trimesityl-substituted Mes_3Si^+ cation **249** using the so-called allyl leaving group approach [Eq. (4.162)] instead of the hydride transfer reaction applied in earlier studies. The electrophile attacks the double bond and the intermediate carbenium decomposes, yielding the cation.

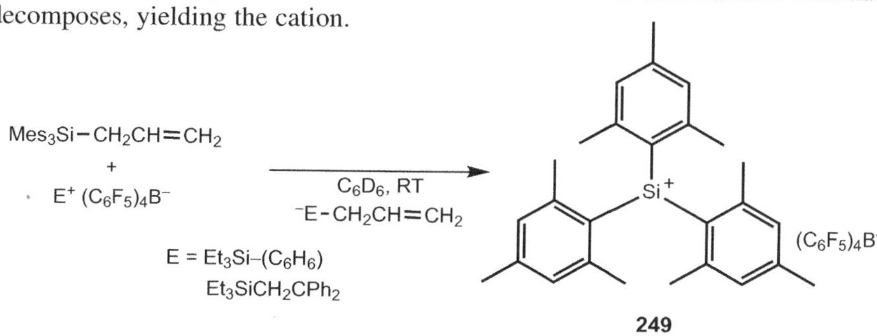

249

(4.162)

Due to the steric crowding of the mesityl groups, ion **249** is not able to react with nucleophiles or interact with the counterion or solvent. The ^{29}Si NMR chemical shift of the ion is practically independent from the solvent (δ^{29}Si 225.5–225.7). Calculations for the chemical shifts of arylsilicenium ions give values in the range δ^{29}Si 226–243.9 depending on the basis sets used.[647,648] X-ray crystal structure analysis of the related $Mes_3Si^+(1\text{-}H\text{-}CB_{11}Me_5Br_5)^-$ species[649] reveals well-separated anions, cations, and solvent molecules. The silicon center is three-coordinate and planar as indicated by the sum of the three C–Si–C bond angles (120.5°+121.7°+117.7° = 359.9°). As expected, the Si–C bond lengths (average 1.817 Å) are shorter than those in the neutral precursor (average 1.91 Å). The mesityl groups adopt a propeller-like arrangement with twist angles 51.3, 54.5, and 41.9°, which are close to the calculated values of 47.3–49°. This allows still considerable conjugation between the aryl rings and the empty orbital of silicon. The ^{29}Si NMR chemical shift measured in the solid sate is δ^{29}Si 226.7. A new silicenium ion, tris(2,3,5,6-tetramethylphenyl)Si$^+$ exhibits a similar ^{29}Si NMR chemical shift (δ^{29}Si 226.8).[650]

Olah et al.[622] made an unsuccessful attempt to prepare the dibenzosilatropylium ion **250** and only the corresponding covalent perchlorate ester was isolated. However, Komatsu and coworkers[651] have recently synthesized the stable silatropylium ion **251** by hydride transfer reaction under conditions suitable for the formation of stable silicenium ions. The observed ^{29}Si NMR chemical shift of δ^{29}Si 142.9 deshielded from that of the precursor by 192.2 ppm, compares favorably with δ^{29}Si 159.9, which is the calculated value [GIAO/HF/6-311 + G(2df,p)(Si), 6-31 + G*//B3LYP/6-31G* level] for the idealized gas-phase structure. This indicates only small interaction between the cation and CD_2Cl_2 the solvent for NMR measurements.

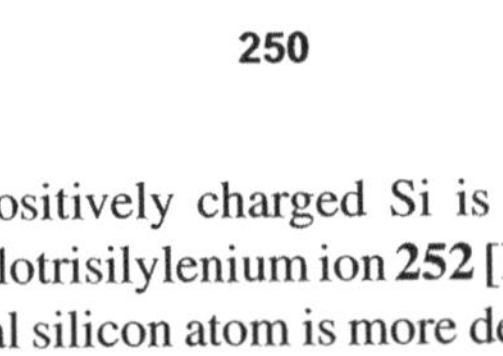

250

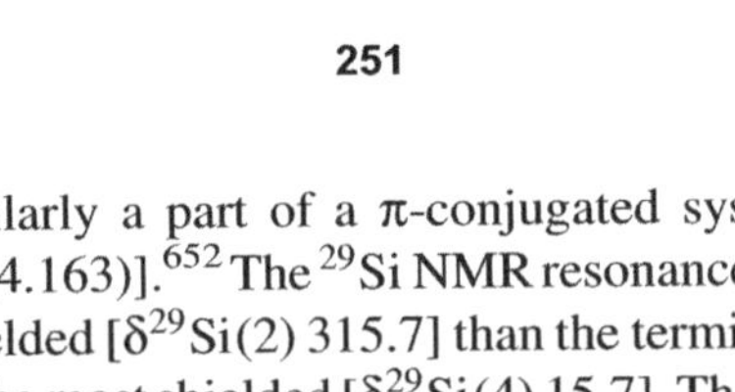

251

The positively charged Si is similarly a part of a π-conjugated system in the homocyclotrisilylenium ion **252** [Eq. (4.163)].[652] The ^{29}Si NMR resonances show that the central silicon atom is more deshielded [δ^{29}Si(2) 315.7] than the terminal silicons [δ^{29}Si(1) and Si(3) 77.3] and Si(4) is the most shielded [δ^{29}Si(4) 15.7]. The molecular structure in the solid state shows no interaction between the cation and the benzene molecule present as solvent of crystallization. The four-membered ring is folded [dihedral angle between the planes Si(1)–Si(2)–Si(3) and Si(1)–Si(4)–Si(3) is 46.6°]. The cationic part is completely planar and Si(4) has a distorted sp^3 environment and the Si–Si bond lengths are between the Si–Si single and double bonds of the precursor. These features and the observation that the most deshielded Si is the central tricoordinated Si(2) indicate homoaromatic character with charge delocalization at Si(2).

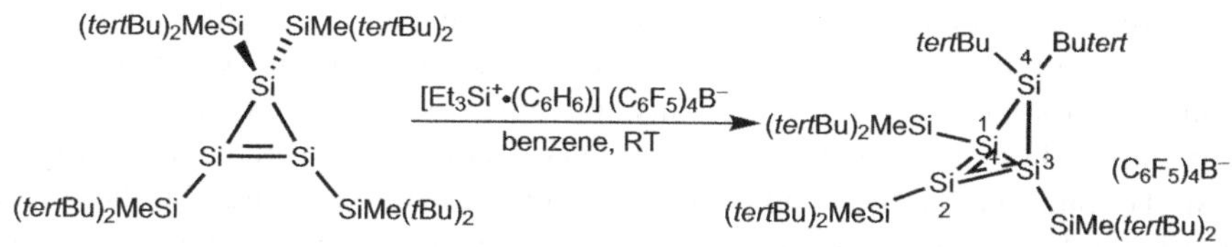

252

(4.163)

Lickiss and co-workers[653] have performed the hydride abstraction reaction shown in Eq. (4.164). On the basis of multinuclear NMR characterization, including the single sharp resonance at δ^{29}Si 17.8 observed at 60°C, the species is fluxional with rapidly equilibrating ions but with little benzenium character **253a**. $(Me_3Si)_4C$ as a byproduct (about 20%) was also detected indicative of intermolecular Me and Ph exchange processes and intramolecular migrations. X-ray crystal studies allowed to identify the salt as the ion **253b**. The phenyl group forms an almost symmetrical bridge [C_{ipso}–Si bond lengths = 2.104 and 2.021 Å]. The four-membered ring is folded about the Si–Si vector (dihedral angle = 13.5°). The bond distances in the aromatic ring (C_{ipso}–C_{ortho} = 1.408 and 1.400 Å, C_{ortho}–C_{meta} = 1.372 and 1.363 Å, C_{meta}–C_{para} = 1.358 and 1.369 Å) show small differences, are closely similar to those of the neutral species $(Me_3Si)_3CSiMe_2Ph$, and do not show the typical long, short, long characteristic of cyclohexadienyl resonance structures. Consequently, the ion has little **253a** character and better represented as structure **253b**.

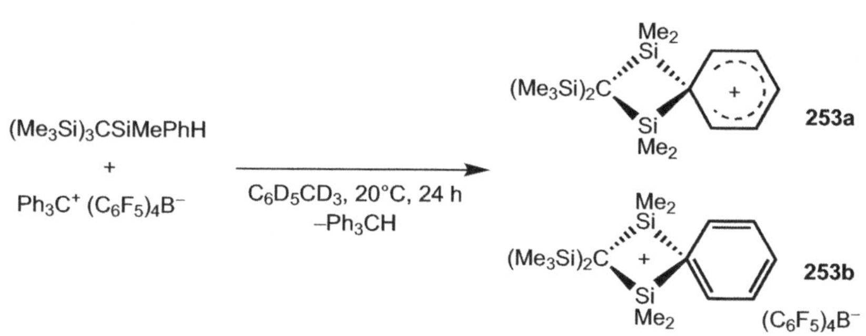

(4.164)

Ions with bridging hydrogen have also been prepared and studied.[422,423,654] Common characteristics of ions **254**–**256** are the strong dependence of the ^{29}Si NMR chemical shift on the system (δ^{29}Si 99.1, 76.7, 54.4, respectively) and the markedly reduced J_{Si-H} coupling constants (two doublets of heptets with J_{Si-H} = 26.0 and 6.6 Hz for ion **254**, J_{Si-H} = 39 Hz for ion **255**, and J_{Si-H} = 46 Hz for ion **256**) compared to neutral silanes ($J_{Si\text{-}H} = \sim$180–200 Hz). Furthermore, the hydrogen in the Si–H–Si two-electron three-center bond is unusually shielded in comparison to the precursor silanes ($\Delta\delta^1$H −2.43 and −1.85 for **255** and **256**, respectively). Calculations at the MP2/6-31G(d,p) level[422,654] predict a geometry that is very similar to the X-ray structure found experimentally, and computed δ^{29}Si chemical shifts and $J_{Si\text{-}H}$ coupling constants agree well with experimental data. Furthermore, theoretical analysis suggests very small 3*s*(Si)-orbital contribution to the Si–H bond, that is, the bond is almost completely of 3*p* nature. All these data suggest that these cations have larger silicenium ion character than the corresponding halonium ions (see Section 4.2.4).

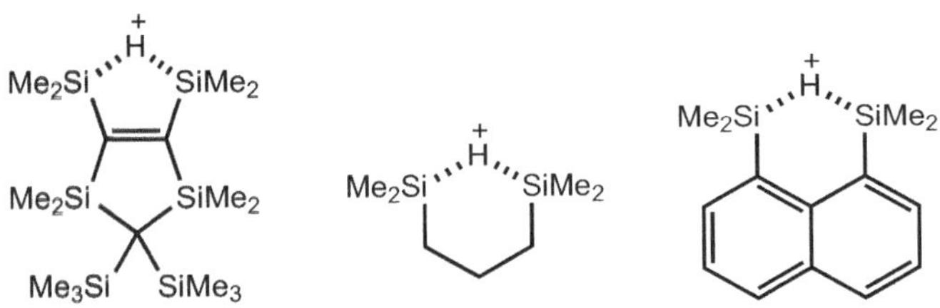

254 **255** **256**

Reed and co-workers[655] have obtained carborane salts of simple hydrogen-bridged disilyl cations **257**. The X-ray structure of **257** (R = Me) confirms an essentially symmetrical cation well-separated from the anion with the following parameters: Si–H bond lengths = 1.60 and 1.62 Å, Si···Si distance = 3.1732 Å, Si–H–Si bond angle = 160°, average C–Si–C bond angle = 116.7°. The corresponding calculated data [B3LYP/6-311 + G(d,p)] are Si···Si = 3.82 Å, Si–H–Si = 179.7°, C–Si–C 116.0°. The ^{29}Si NMR chemical shifts (δ^{29}Si 85.4 and 82.2) are reproduced well by DFT/IGLO calculations (δ^{29}Si 107).

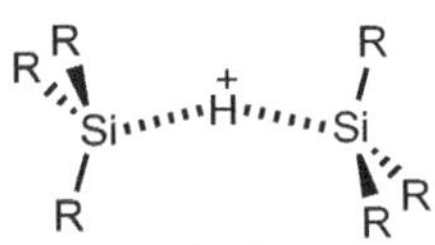

257 R = Me, Et

Silylium ion **258** generated from perdimethylsilyl-substituted benzene by hydride abstraction is stabilized by two agostic Si–H···Si interactions.[656] The highly decreased $J_{Si\text{-}H}$ coupling constant of the α-Si (46.3 Hz), again, is indicative of the hydrogen-bridged silylium ion, whereas the upfield shifts of the α-SiH and β-SiH resonances (δ^1H 4.26 and 4.41, respectively) are characteristic of agostic bondings. Ion **259** made from perdimethylsilyl-substituted toluene, in contrast, has no bridging hydrogen and exhibits resonances at δ^{29}Si 34.3 (α-Si) and 33.5 (β-Si). The agostic interactions are also supported by DFT calculations.

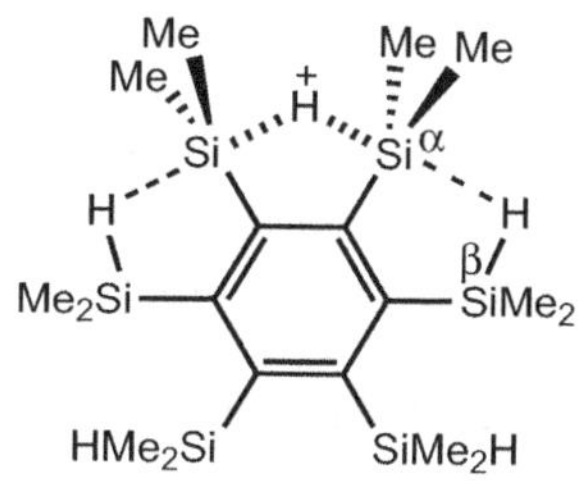

258

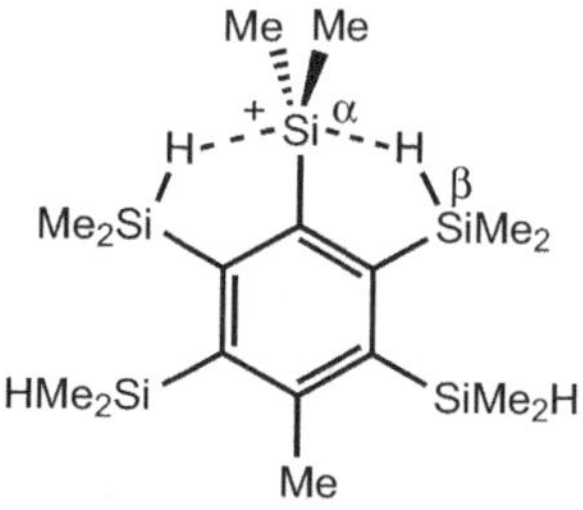

259

Schleyer and co-workers[657] and Müller et al.[658] have prepared silanorbornyl cations **260** by the π-route via the intramolecular addition of the transient silicenium ion [Eq. (4.165)]. All three cations **260** exhibit strong downfield shift upon ionization ($\Delta\delta^{29}$Si = 86.4–102.1). The ^{29}Si NMR chemical shifts of the cations are δ^{29}Si 80.2–87.2 and agree well with calculated values (δ^{29}Si 77–93). These values are similar to those found for silabenzenium ions (δ^{29}Si 70–100), but markedly smaller than the chemical shifts for trivalent cations (δ^{29}Si 225 for Mes_3Si^+). The ^{13}C NMR chemical shift pattern of the C(3)–C(4)–C(5) saturated backbone is characteristic of the norbornyl cage. The chemical shifts for the C(6) and C(7) vinylic carbons (δ^{13}C 149.6–150.6) show a marked downfield shift relative to the precursors ($\Delta\delta^{13}$C = 19.3–20.3) indicative of the intramolecular coordination and charge transfer from Si to the vinylic carbons. The calculated structures [GIAO/B3LYP/6-311G(3d,p)//MP2/6-311G(p,d) level] support the bridged norbornyl structure of the ions.

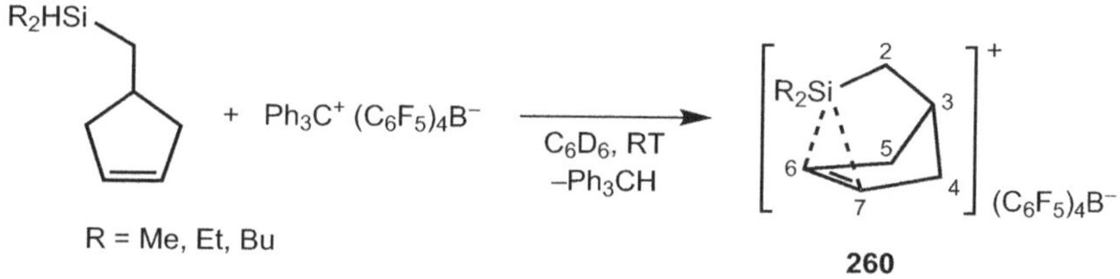

260

(4.165)

Müller and co-workers[659] have also attempted to generate the 7-silabenzonorbornadien-7-ylium cation **261**. However, only the corresponding nitrilum and oxonium ions, prepared in acetonitrile and diethyl ether, respectively, could be isolated, and the solvent-free ion **261** could not be detected.

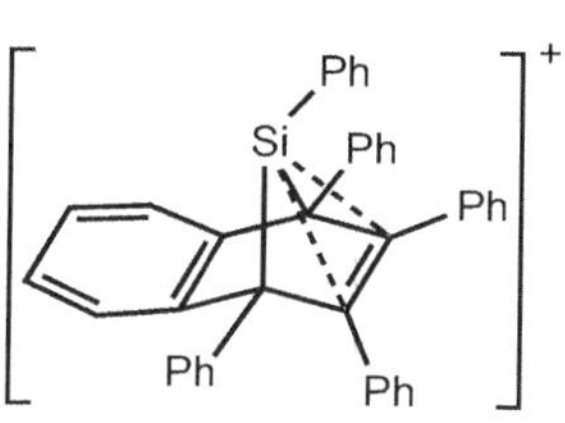

261

The first synthesis and characterization of a persilaaromatic compound, the cyclotrisilenylium ion **262** have been accomplished by Sekiguchi and co-workers[660] [Eq. (4.166)]. In the ^{29}Si NMR spectrum the signals $\delta^{29}Si$ 284.6 and 288.1 (relative intensity = 2:1) were assigned to the cationic ring Si atoms bearing the (*tert*-Bu)$_3$Si and (*tert*-Bu)$_2$MeSi substituents, respectively. The significant downfield shifts were well reproduced by calculations [GIAO/B3LYP/6-311 + G(2df,p)//B3LYP/6-31G(d) level]. In the solid state, the crystals contain two molecules of toluene and the cations are well-separated from both the solvent and the anion clearly showing a free silicenium ion. The ring atoms form an almost regular triangle (internal bond angles = 59.76–60.20°) with the substituent Si atoms being in the same plane within 0.39 Å. The ring Si–Si bond distances are in the range 2.211–2.221 Å, which are intermediate between the single and double bond lengths of the precursor and agree well with the calculated value of the $(HSi)_3^+$ analog of D_{3h} symmetry (2.203 Å).[661]

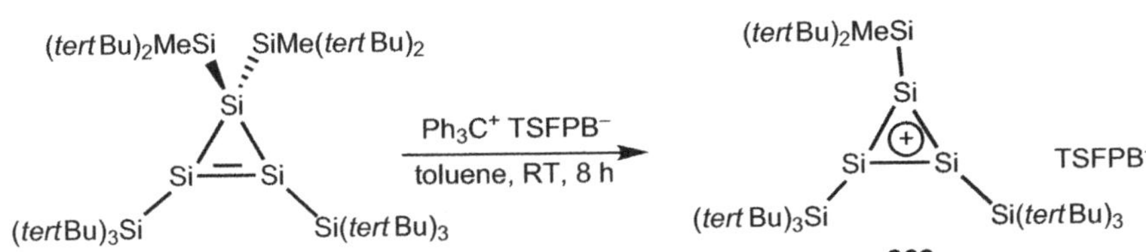

262

TSFPB = [4-(*tert*-BuMe$_2$Si)C$_6$F$_4$]$_4$B

(4.166)

Sekiguchi and co-workers[662] have performed one-electron oxidation of the precursor radical **263** to generate silicenium ion **264** [Eq. (4.167)]. In the presence of acetonitrile, the corresponding nitrile adduct was isolated, whereas in CD_2Cl_2, the isomeric silicenium ion **265** was formed as a results of rapid 1,2-methyl migration from the peripheral Si to the central cationic Si atom. The observed ^{29}Si NMR resonances are at $\delta^{29}Si$ 29.1 and −69.7, and for the cationic Si atom at $\delta^{29}Si$ 303. This value indicates that ion **265** may exist as a free silicenium ion. A temperature-dependent NMR study shows that there is a rapid exchange of Me groups through 1,3-methyl migration between the peripheral Si atoms with an activation energy of $13.1 \pm 0.4\ kcal\ mol^{-1}$.

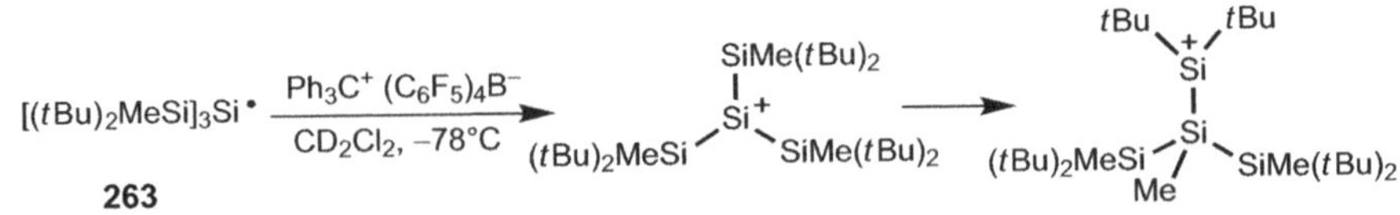

263 **264** **265**

(4.167)

Jutzi and Bunte[663] have reported the synthesis and characterization of the unusual cation **266** with divalent Si. In the ^{13}C NMR spectra, one signal for the Me and one signal for the ring carbon atoms are observed. A doublet at δ^{29}Si -12.1 in the coupled ^{29}Si NMR spectrum corresponds to a downfield shift of $\Delta\delta$ 386 compared to the precursor. Note that this signal is still more than 230 ppm less deshielded that the one for the Mes_3Si^+ ion (δ^{29}Si 225). The large difference of the coupling constant of ion **266** ($J_{Si-H} = 302$ Hz) when compared to that of $Cp^*{}_2SiH_2$ ($J_{Si-H} = 194$ Hz) indicates an increase in *s*-character in the Si–H bond as expected for the sp^2-hybridized silicon. Theoretical analysis of the parent Cp_2SiH^+ cation [B3LYP/6-311G(2d,p)//B3LYP/6-31G(d) + ΔZPV level] reveals[664] that ion **266a** of C_2 symmetry ($\eta^{1.5}:\eta^{1.5}$ coordination) is the most stable but ion **266b** of C_s symmetry ($\eta^2:\eta^3$ bonding) is very close in energy ($\Delta E < 1$ kcal mol^{-1}). Because the positive charge is effectively transferred to the cyclopentadienyl rings cation **266** has high thermodynamic stability. The ion is 18.7 kcal mol^{-1} more stable than Me_3Si^+.

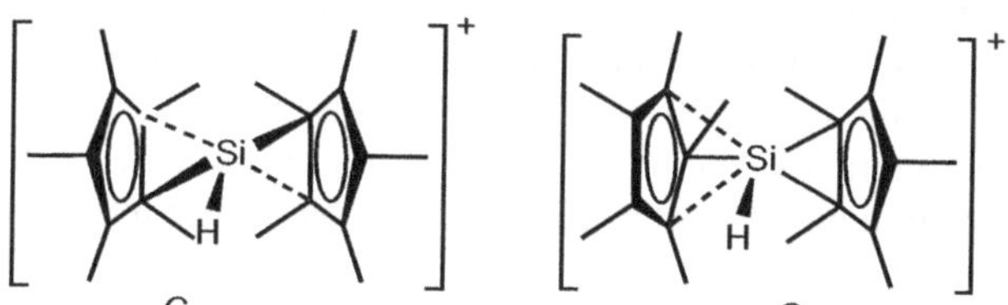

a b

266

Another unusual cation reported by Jutzi et al.[665,666] is the half-sandwich cation **267** [Eq. (4.168)]. Both salts show equivalency of all ring substituents and ring carbon atoms, respectively, in the ^{1}H and ^{13}C NMR spectra. Resonances in the ^{29}Si NMR spectra for salts **267a** and **267b** appear at δ^{29}Si -400.2 and -397.4, respectively. Such high-field shifts are characteristic of π-complexes of divalent Si. In the solid-state ion, **267a** has an almost ideal pentagonal–pyramidal structure with weak interactions with the anion but no contact between Si and incorporated solvent molecules (CH_2Cl_2) is observed. The distance between Si and the center of the Cp* ring is 1.76 Å and the Si–C(Cp*) distances are in the range 2.14–2.16 Å. These distances are significantly shorter than those in the starting $Cp^*{}_2Si$ molecule.

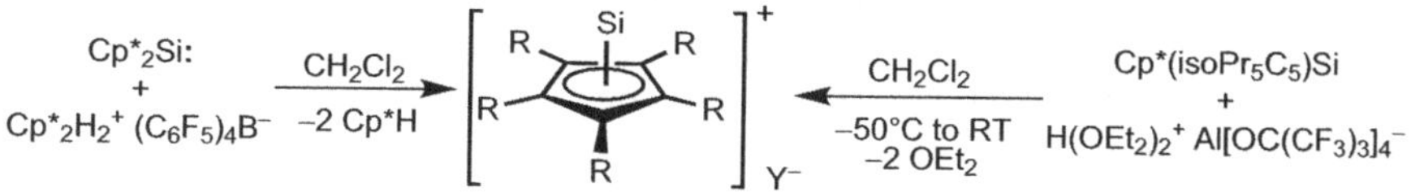

267 **a** R = Me, Y = $(C_6F_5)_4B$
b R = isoPr, Y = $Al[OC(CF_3)_3]_4$

(4.168)

A related species with divalent Si is cation **268**.[667] The Si resonance in the ^{29}Si NMR spectrum (δ^{29}Si 69.3) is indicative of strong donor–acceptor interaction of Si with the N ligand. The ^{1}H NMR chemical shift of the γ-CH proton (δ^1H 6.92) and the molecular structure of the salt suggest aromatic 6π-electron delocalization (**268b**).

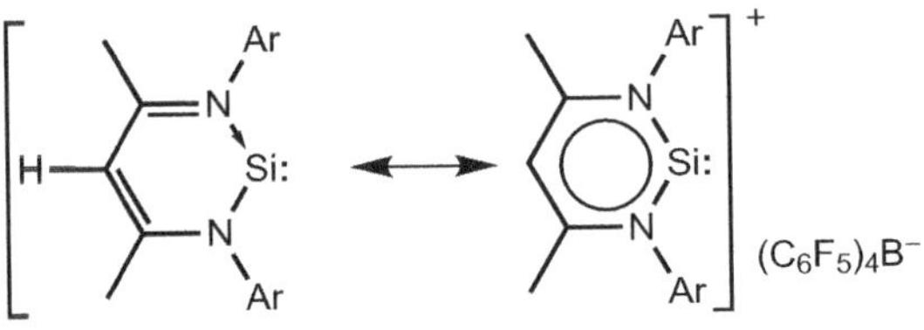

a **268** b Ar = 2,6-diisoPrC$_6$H$_3$

The high electrophilicity of the silicenium ion can be modified by intramolecular electron donation from remote substituents. Corriu and co-workers[668] were the first to synthesize species **269** with the bidentate nitrogen ligand. Cations with other structural units and donor atoms (O, S, P) were subsequently generated.[669–674] Due to the interaction called internal solvation,[675] the silicon atom becomes pentacoordinated and largely loses its ionic character, that is, such ions differ in principle from the trivalent silicenium ions.

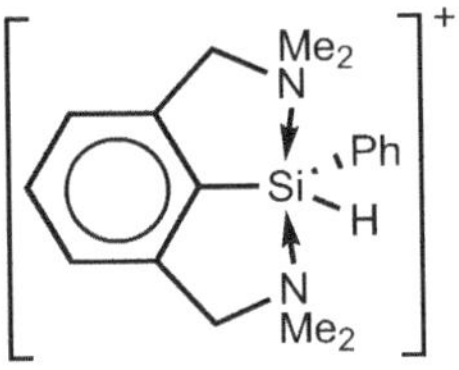

269

Finally, it is appropriate to make a short note about five-coordinate siliconium ions. The parent ion SiH_5^+, a penta-coordinate, tetravalent species, has been studied theoretically[676–679] and by cyclotron resonance[680] and IR spectroscopy.[681] According to the latest computational study[682] the global minimum structure is of C_s symmetry. Ion SiH_5^+ can be described as a complex between SiH_3^+ and H_2 with the hydrogen molecule bound sideways to the SiH_3^+ fragment. This conclusion is in agreement with spectroscopic observations.

Penta-coordinate siliconium species have been referred to as possible reactive intermediates in organic transformations. Such claims, however, were unsubstantiated.[683] Olah, Prakash, and colleagues,[684] in turn, reported that trialkylsilanes undergo H/D exchange of the tertiary Si–H(D) bond when treated with HI–AlI_3 at 0°C. This is in contrast to other HX Brønstead acids (X = F, Cl, Br, OSO_3H, OSO_3F) and HX–MX_n conjugate acids (X = $AlCl_3$, $AlBr_3$, BF_3, SbF_5), which give halogen-exchanged R_3SiX products. The suggested mechanism of the exchange process, which includes the 2*e*–3*c* intermediate **270**, is depicted in Eq. (4.169).

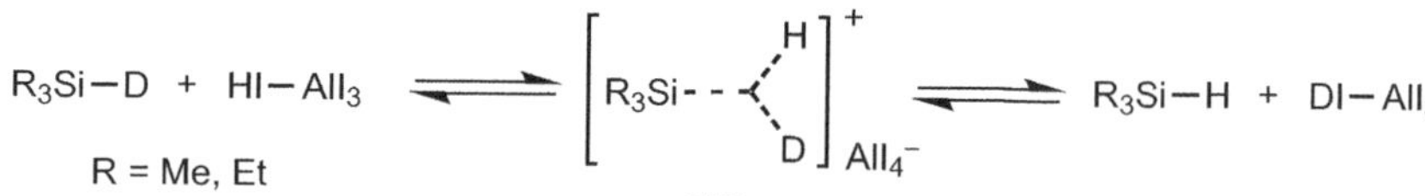

270

(4.169)

4.3.2.2. Germenium Ions. The allyl leaving group approach [see Eq. (4.162)] has been successfully applied to the synthesis of the ion Mes_3Ge^+, the germanium analog of ion **249**.[646] Since Ge lacks a sensitive and convenient nuclide, there is no easy and direct NMR method to assess the germenium ion character of the species isolated. Nevertheless, a comparison of the ^{13}C NMR chemical shifts with those of the analogous Si and Sn ions indicates comparable cationic character.

Sekiguchi and co-workers have recently generated numerous germenium ions of varied structures.[621] Cation **271** is formed as a result of an unexpected rearrangement of the initially formed intermediate germenium ion[685] [Eq. (4.170)]. According to the ^{1}H NMR spectrum, the three *tert*-butyl groups are nonequivalent and one shows hindered rotation. The ^{13}C NMR chemical shift of C(2) of one of the phenyl groups (δ^{13}C 102.8) is upfield by about 25 ppm relative to the C(2) carbons of the other two aromatic rings (δ^{13}C 130.3 and 126.3). In contrast, the other carbons are shifted downfield. These changes are close to those calculated for the Me_3Si^+–benzene complex. Consequently, in cation **271** the Ge^+ ion is intramolecularly coordinated to the C(2) atom of the aromatic ring which has a weak arenium ion character.

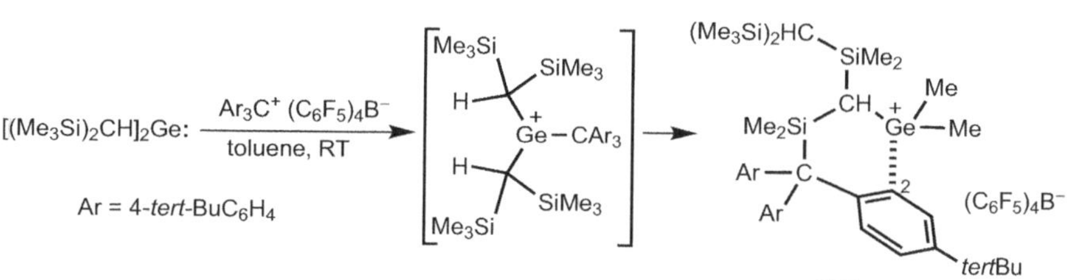

271

(4.170)

One-electron oxidation of the isolable Ge-centered radical results in the formation of the corresponding germenium cation **272** [Eq. (4.171)].[686] Because of its highly symmetric nature, cation **272** exhibits simple NMR spectra. The resonances of Me groups attached to Si (δ^1H 1.17) and the Si signals (δ^{29}Si 49.9) are highly deshielded, which are indicative of the positive charge transferred from Ge to the electropositive Si atom. Indeed, DFT calculations [B3LYP/6-31G(d) level] on the compound $(H_3Si)_3Ge^+$ show the following charge distribution: Ge = –0.12, Si = +0.37. In the solid state, the Ge center has a completely planar geometry with very long Ge–Si bonds (average bond distance = 2.5195 Å).

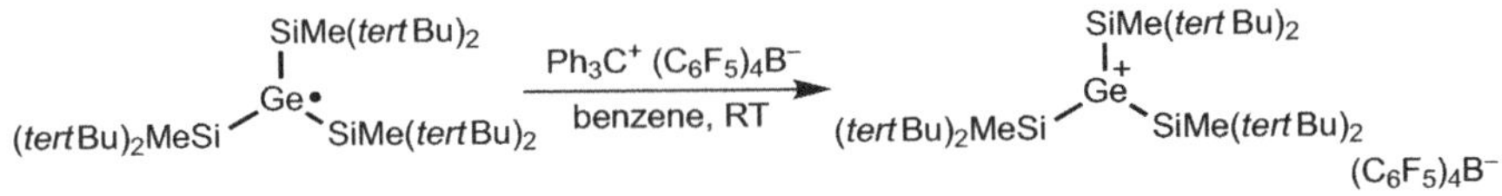

272

(4.171)

Using the method applied for the preparation of the persilaaromatic ion **262** shown in Equation (4.166), Sekiguchi and co-workers[687,688] have succeded in obtaining the corresponding Ge analogs **273**. According to X-ray crystal structure analysis of **273a** and **273b** (Y = TFPB or TSFPB), the Ge atoms form an equilateral triangle (Ge–Ge bond lengths = 2.321–2.335 Å, Ge–Ge–Ge bond angles = 59.8–60.3°) with the Si atoms lying approximately in the same plane. The Ge–Ge bond distances are intermediate between Ge–Ge single (2.522 Å) and Ge=Ge double bonds (2.239 Å). The Ge–Ge bond lengths agree well with that calculated for the $(HGe)_3{}^+$ analog of D_{3h} symmetry (2.361 Å).[661] There is no significant interaction between the cation and the counterion. The characteristic NMR features are practically the same for cation **273b** with all three anions and independent from the solvents used. The ^{29}Si NMR resonances are shifted downfield (δ^{29}Si 37.2 for the *tert*-Bu_3Si attached to the saturated Ge atom and 50.1 for the *tert*-Bu_3Si attached to the Ge=Ge double bond) relative to the neutral precursor, which indicates significant charge transfer from Ge to Si. The Mulliken charges (Ge = –0.07, Si = +0.64) are in harmony with this observation.

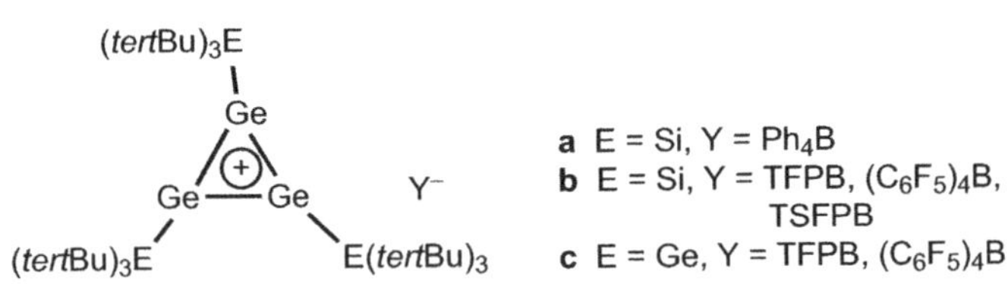

273

Two other germenium ions obtained by Sekiguchi and co-workers are the bishomocyclopropenylium ion **274**[689] and the cation cluster **275** with trishomoaromaticity.[690] Cation **274** is free in the solid state and forms a Ge_3 equilateral triangle and a Ge_2C_2 four-membered ring. The Ge(3)–C(1) and Ge(3)–C(2) distances are short (2.415 and 2.254 Å), whereas the C = C double bond is long (1.411 Å) because of through-space interaction and effective homoconjugation. This results in a significant bending of Ge(3) toward the C = C double bond [dihedral angle between the Ge(1)–Ge(2)–Ge(3) and Ge(1)–Ge(2)–C(1)–C(2) planes = 67°] and a slight pyramidalization of the Ge(3), C(1), and C(2) atoms. Cation **274** shows the characteristic deshielded ^{29}Si NMR resonances (δ^{29}Si 56.1 and 67.2) and considerably shielded ^{1}H chemical shift of the proton attached to C(2) and ^{13}C NMR shifts of the C(1) and C(2) atoms relative to the precursor.

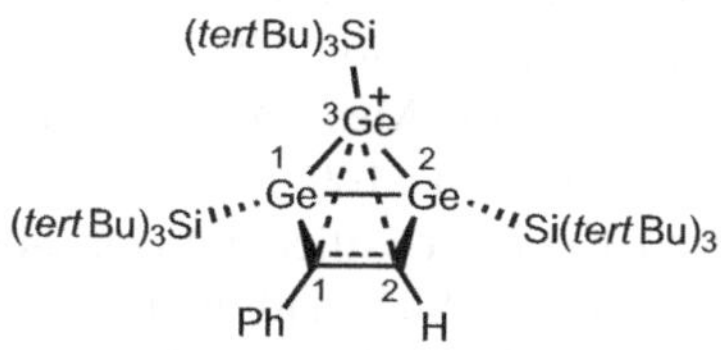

274

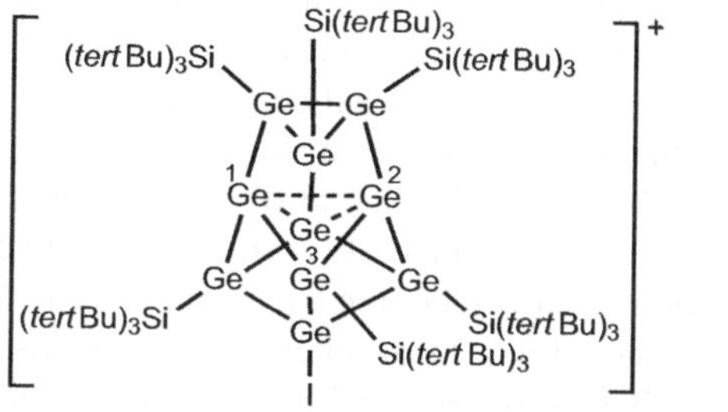

275

The structure of cation **275** was characterized by X-ray crystallography and multinuclear NMR spectroscopy.[690] The cluster of 10 Ge atoms has an approximate C_{3v} symmetry. The Ge–Ge bond distances of unsubstituted Ge atoms [G(1), G(2), and G(3)] in the central Ge_3 core (3.2542–3.2642 Å) are longer than the other Ge–Ge bonds in the cluster (2.4711–2.5449 Å), but they are in the range of metallic Ge–Ge bond lengths. A DFT calculation [B3LYP/6-31G(d) level] on the compound $Ge_{10}H_7^+$ gives the cluster skeleton of cation **275** as the global minimum structure. Molecular orbital calculations indicate bonding interactions between the three Ge atoms in the core and show that the positive charge is evenly distributed over these atoms, indicating the formation of a $2e$–$3c$ bond. The calculated aromatic stabilization energy of $-19.2\,kcal\,mol^{-1}$ indicates a large homoaromatic stabilization.

The characteristic structural features of two-coordinate low-valent germanium cation **276** are similar to those of the analogous Si cation **268** discussed above. The average C–N and C–C bond distances (1.348 Å and 1.392 Å, respectively) and the essentially planar ring indicate aromatic delocalization.[691] In contrast, however, the γ-CH proton (δ^1H 4.23) is more shielded. The aromatic substituents are perpendicular to the GeN_2C_3 plane.

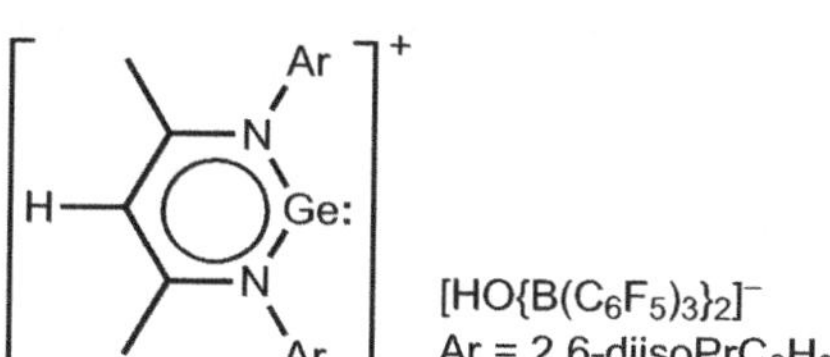

276

4.3.2.3. Enium Ions of Other Group 14 Elements. Stannylium ions obtained in early studies by Birchall and Manivannan ($Me_3Sn^+FSO_3^-$),[692] Lambert and Kuhlmann[693] [$Bu_3Sn^+H(C_6F_5)_3B^-$], and Kira et al.[694] (Bu_3Sn^+TFPB$^-$) exhibit ^{119}Sn NMR chemical shifts in the region δ^{119}Sn 322–360. Using the empirical correlation between ^{29}Si and ^{119}Sn chemical shifts, Arshadi et al.[695] concluded that these claimed stannylium ions have, in fact, little ionic character (assuming linear relationship between ^{29}Si and ^{119}Sn chemical shifts, the chemcial shift of δ^{29}Si 225 corresponds to δ^{119}Sn 1100). However, using the approach, which finally led to the preparation of the first long-lived bona fide silicenium cations, Lambert et al.[646] succeeded in synthesizing the stannylium cation (Mes_3Sn^+) analogous to silicenium ion **249**. The ^{119}Sn NMR chemical shift value at δ^{119}Sn 806 observed for the ion $Mes_3Sn^+(C_6F_5)_4B^-$, although compares favorably with the calculated value (δ^{119}Sn 1100), indicates that the cationic character of the ion is smaller (about 75%) than that of silicenium ion **249**. Other stannylium ions, phenylbis(2,4,6-triisopropylphenyl)Sn^+ and tris(2,3,5,6-tetramethylphenyl)Sn^+, exhibit even lower shift values (δ^{119}Sn 697 and 720).[646,650]

Subsequently, Michl prepared the cations Me_3Sn^+ and Bu_3Sn^+ and obtained the crystal structure of $Bu_3Sn^+CB_{11}Me_{12}^-$.[696,697] The geometry around the tin, however, was found to be nonplanar (sum of C–Sn–C bond angles = 353.1°) and the cation weakly coordinated to the anion. Furthermore, the ^{119}Sn NMR chemical shifts of these cations are much smaller (δ^{119}Sn 335.9 for Me_3Sn^+ and 454.3 for Bu_3Sn^+) than those mentioned above. Lambert et al.[698] succeeded in obtaining the crystal structure of tris(2,4,6-triisopropylphenyl)$Sn^+(C_6F_5)_4B^-$. Although the chemical shift value is also low (δ^{119}Sn 714; the value calculated by the GIAO method is δ^{119}Sn 763), the cation has a practically planar geometry (sum of C–Sn–C bond angles = 359.9°).

Sekiguchi et al.[621,699] have prepared the cation $[(tert\text{-}Bu)_2MeSi]_3Sn^+$ the tin analog of cation **272** using the one-electron oxidation method [see Eq. (4.171)]. X-ray analysis shows that the Sn ion has a perfectly planar geometry (sum of the bond angles around Sn is 360.0°) and the Me substituents of Si atoms lie in the same plane. An unexpectedly high ^{119}Sn NMR chemical shift at δ^{119}Sn 2653 was observed. This value greatly exceeds those observed earlier for any stannylium ion and the estimate given by Arshadi using the empirical correlation with ^{29}Si. However, it agrees well with the value of δ^{119}Sn 2841 calulated for the model parent ion $(H_3Si)_3Sn^+$.

In addition to the silanorbornyl cations **260** [see Eq. (4.165)], Müller et al.[658] obtained the corresponding Ge and Sn (R = *n*-Bu), and Pb (R = Et) analogs. The downfield shift of the C(6) and C(7) vinylic carbons is also observed for the Ge, Sn, and Pb cations, but the values are smaller and they decrease from Si to Sn ($\Delta\delta^{13}$C 19.3–20.5 for Si cations, 14.6 for Ge, 11.1 for Sn, and 7.8 for Pb). This shows that electron transfer from the C = C double bond to the metal decreases from Si to Sn. The Pb center is less deshielded (δ^{207}Pb 1049) than those in other triorganolead ions (for example, δ^{207}Pb for $[Et_3Pb{\cdot}C_6D_6]^+(C_6F_5)_4B^-$ is 1432[620,658]] and the same is observed for the Sn ion (δ^{119}Sn 334 versus 806 for Mes_3Sn^+). Additionally, the coupling constants between vinylic carbons and Sn or Pb are reduced by about an order of magnitude ($J_{C\text{–}Sn}$ = 26 Hz, $J_{C\text{–}Pb}$ = 16 Hz), suggesting a direct bonding.

The plumbylium cation **277** is an intramolecular bisalkene complex and can also be considered as a spironorbornyl cation[700] [Eq. (4.172)]. The ^{13}C NMR spectrum (only six resonances) suggests that the cyclopentenemethyl substituents are equivalent and symmetrically situated. Other spectral characteristics are similar to those of related norbornyl cations of Group 14 elements already discussed. The vinylic carbons are downfield-shifted ($\Delta\delta^{13}$C 5.9), the C–Pb coupling constant is significantly reduced

(J_{C-Pb} = 14.4 Hz), and the ^{207}Pb NMR resonance is strongly downfield-shifted relative to the precursor (δ^{207}Pb 807, $\Delta\delta^{207}$Pb 777) but less deshielded than in other triorganolead ions. Anions and cations are well-separated in the solid state and the trigonal base is planar (sum of C–Pb–C bond angles = 360.0°). The Pb has a distorted trigonal–bipyramidal geometry with the double bonds in apical position but tilted by 14° toward the C_3Pb plane. According to calculations [MP2/6-311G(d,p)(C,H),SDD (Pb)//MP2/6-31G(d)SDD(Pb)] the intramolecular interactions stabilizes the ion by about 28.3 kcal mol^{-1}.

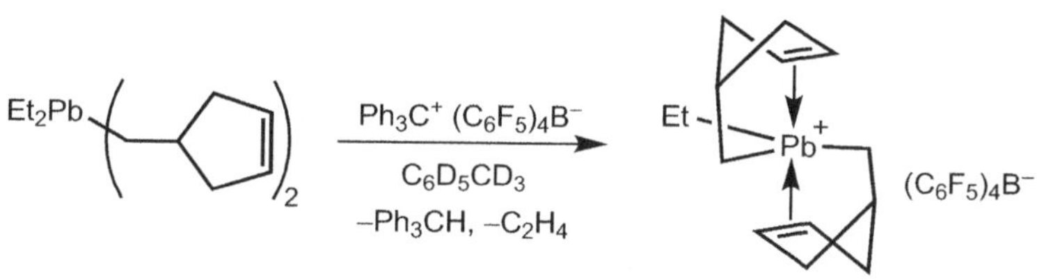

(4.172)

277

In the synthesis of cation **277** the cation **278** was also isolated as a byproduct. ^{13}C NMR data show that all three cyclopentenemethyl substituents are equivalent at the NMR time scale, suggesting a dynamic equilibrium between bisalkene complexes.[620]

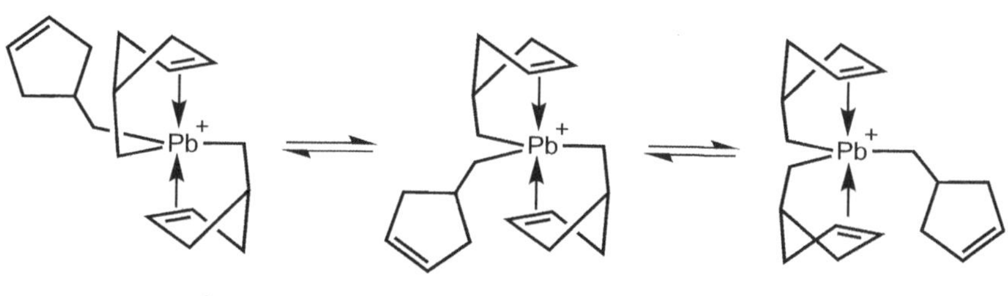

278

The tin analog of ion **267a** [see Eq. (4.168)] has been prepared and a ^{119}Sn NMR chemical shift at δ^{119}Sn −2219 was reported.[701]

4.3.3. Enium Ions of Group 15 Elements

4.3.3.1. Nitrenium Ions. Nitrenium ions containing positive nitrogen have been postulated as intermediates in rearrangement, synthesis, and cleavage of nitrogen-containing organic compounds.[702–704] Arylnitrenium ions have recently attracted considerable interest because of their suspected role in chemical carcinogenesis. In contrast to a trivalent carbocation, the nitrenium ion **279** is unusual in that it has both a positive charge and a nonbonding pair of electrons [Eq. (4.173)]. Hence, the nitrenium ion could exist both as a singlet (**279a**) and a triplet (**279b**).[704] The singlet would resemble a carbocation and the triplet a radical cation with great tendency for hydride abstraction.

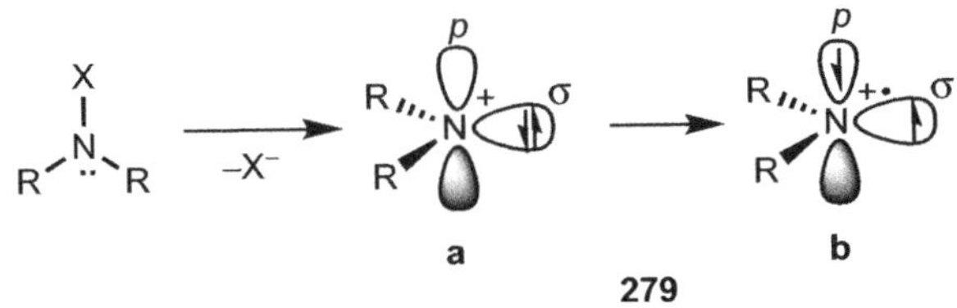

(4.173)

Quantum-chemical calculations for a wide variety of nitrenium ions are abundant.[704] For the parent H_2N^+ ion the triplet state was found experimentally to be more stable by 30 kcal mol^{-1}.[704] The H–N–H angle in the triplet is calculated to be considerably greater (149.4 versus 107.1°).[705] The infrared spectrum of the complex He–H_2N^+ has also been observed.[706]

Attempted generation of nitrenium ions as distinct species under long-lived stable ion conditions has thus far been unsuccessful.[707–710] Protonation of nitrosobenzenes in superacid media, for example, has led only to benzenium-iminium dications **280**.[707–710] [Eq. (4.174)].

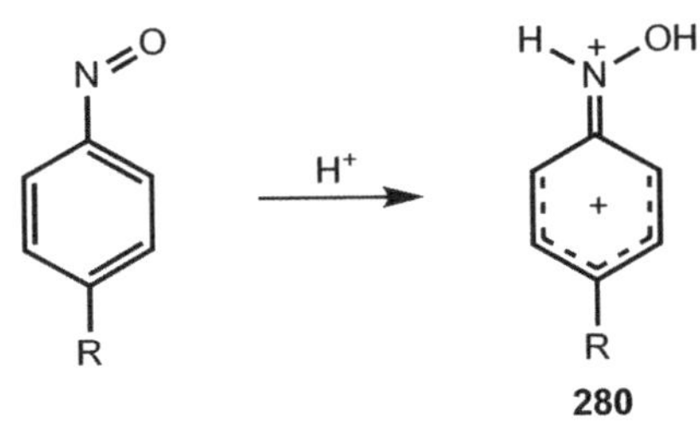

(4.174)

One exception is the trifluorodiazenium ion $N_2F_3^+$ **281**, which could be considered as a potential nitrenium ion. Christe and Schack[711] have obtained $N_2F_3^+$ **281** by the ionization of tetrafluorohidrazine in HF–SbF_5 solutions [Eq. (4.175)]. Similarly, they have been successful in preparing pentafluorostannate and hexafluorostannate salts. The vibrational and ^{19}F and ^{15}N NMR spectroscopic data[448,711] are consistent with planar structure **281a** with C_s symmetry with very little nonplanar nitrenium ion (**281b**) character of C_1 symmetry.

(4.175)

Recently, persistent nitrenium ions—including the triazolinium ion **282**, which is considered a nitrenium ion on the basis of the resonance form **282b**—have been isolated and structurally characterized by X-ray crystallography.[712] The N–N bond distances were found to be equal (1.286 Å) and distinctly shorter than those in an 1,2,3-triazole studied for comparison (1.375 and 1.302 Å).

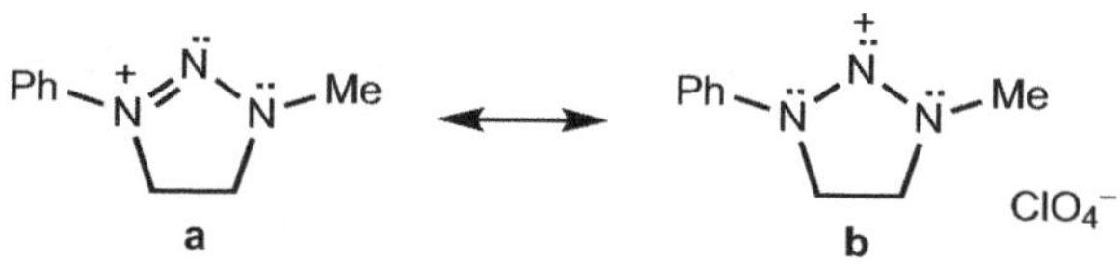

a b

282

The ambivalent nature of an α-cyano group on a carbocationic center has been demonstrated by the solvolytic work of Gassman and co-workers.[713,714] Inductively, the cyano group strongly destabilizing. However, the major portion of this effect is offset by the mesomeric nitrenium ion structure **283b**.

a b

283

Olah et al.[715,716] have prepared a series of α-cyanodiarylcarbenium ions **283** (R = Ar) under superacidic conditions and have evaluated their mesomeric nitrenium ion character by ^{1}H, ^{13}C, and ^{15}N NMR spectroscopy. A subsequent one-bond $^{13}C-^{13}C$ coupling constant measurement[717] also indicates significant mesomeric nitrenium ion character of **283** (R = Ar). Protonated aroyl cyanides **284**, however, exist predominantly in the carboxonium ion form of **284a** over the nitrenium ion form **284b**.[718]

a b

284

4.3.3.2. Phosphenium Ions. In contrast to widely studied phosphonium ions (R_4P^+), the chemistry of the dicoordinate phosphenium ions was little known,[719] but there have recently been significant developments.[720–722] It has been recognized that phosphenium ions can only be generated if one of the substituents is a dialkylamino group[723] [Eq. (4.176)]. Obviously, these ions are stabilized by the strong electron-donating amino function directly attached to phosphorus. On the other hand, the iminium ion character (**285b**) compromises their phosphenium ion character (**285a**).

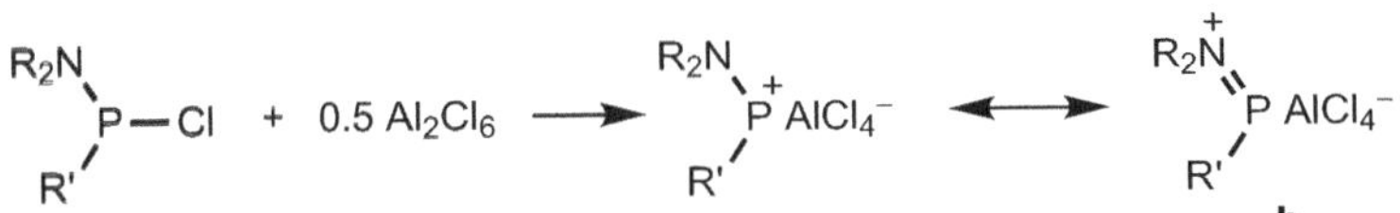

a b

285

R = alkyl, Me_3Si
R' = Me_2N, $(Me_3Si)_2N$, Me_3C, Cl

(4.176)

Parry and co-workers[724,725] have carried out detailed ^{31}P NMR spectroscopic study of a series of phosphenium ions. Their study indicates that the chemical shift of the P^+ center depends upon the steric crowding around phosphorus and the extent of back-donation from the nitrogen lone pair. In the case of *tert*-butyldimethylaminophosphenium ion **286a** the ^{31}P chemical shift of the P^+ center is at δ^{31}P 510 (from 85% H_3PO_4), and it appears to be the largest-ever ^{31}P downfield shift to be measured. On the other hand, bis(dimethylamino)phosphenium ion **286b** shows a chemical shift of δ^{31}P 264. Two-coordinate geometry of phosphenium ions has been confirmed by X-ray diffraction studies on the tetrachloroaluminate salt of cation **286b** with a N–P–N angle of 114.8°.[726]

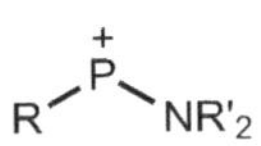

286

a R = *tert*-Bu, R' = isoPr
b R = Me_2N, R' = Me
c R = $(isoPr)_2N$, R' = isoPr

Structurally related cations are cyclic phosphenium ions **287**,[727,728] **288**,[728–731] **289**,[732–734] and **290**[735] with an NPN moiety. No interionic contacts were observed in the crystals of hexafluorophosphate, tetrafluoroborate, and triflate salts of cations **288**. The planar ring structure and the ring bond distances being intermediate between single and double bonds lent evidence for π electron delocalization. This, however, is only a weak delocalization and the cations cannot be described as genuine aromatic systems.[730]

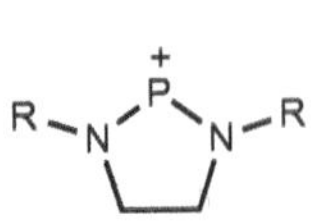

R = *tert*-Bu
4-$MeOC_6H_4$
2,4,6-triMeC_6H_2
2,6-$(isoPr)_2C_6H_3$

287

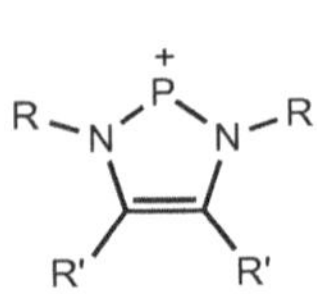

R = *tert*-Bu
2,6-diMeC_6H_3
2,4,6-triMeC_6H_2
2,6-$(isoPr)_2C_6H_3$
R' = H, Me, Cl

288

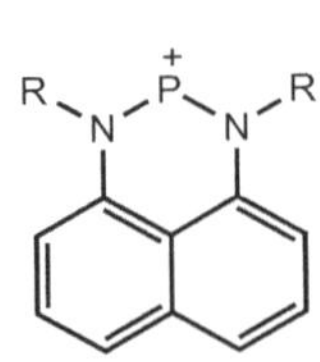

R = isoPr, Ph

289

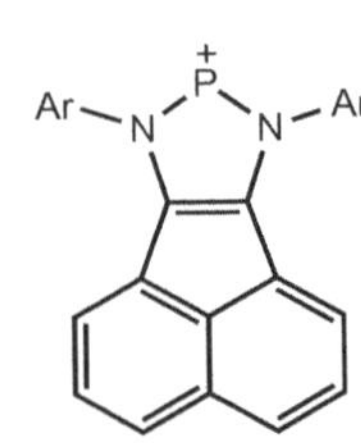

Ar = 2,6-$(isoPr)_2C_6H_3$

290

Veith et al.[736] have prepared the unique four-membered cyclic phosphenium ion **291**. X-ray crystal structure analysis of the tetrachloroaluminate salt clearly indicates the intramolecular backbonding from ring nitrogen atoms (average N–P bond distance = 1.633 Å).

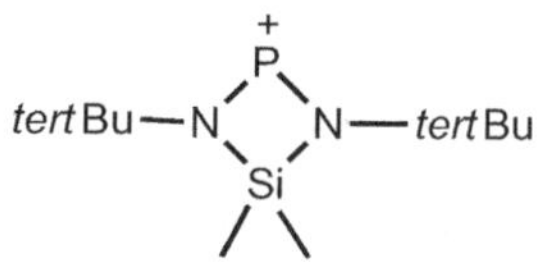

291

The phosphenium ion **292** with a P–C σ-bond has been isolated and structurally characterized by X-ray crystallography.[737] The loss of electronic stabilization upon the replacement of a dialkylamino group is countered by kinetic stabilization by the bulky mesityl substituent in ion **292**. The ^{31}P NMR resonance of the ion is at δ^{31}P 500 (368 ppm downfield from that of the parent chlorophosphine). As a rare example, the ^{31}P signal is a triplet due to coupling to the quadrupolar ^{14}N ($J_{P\text{–}N}$ = 65 Hz). A notable feature of the X-ray structure is the N–P bond distance (1.617 Å), which is, surprisingly, not different from those found in ion **286c** (1.611 and 1.615 Å), where the double-bond character is distributed over two dialkylamido substituents. The P–C bond length (1.787 Å) is of single-bond character, although it is relatively short for P(III) (the P–C bond length in the neutral parent chlorophosphine is 1.848 Å, the P–C double bond in a phosphaalkene is 1.684 Å). The mesityl group is rotated out of the NPC(1)C(2) plane by 69°, which prevents any conjugation with the aromatic π system. Interestingly, the N–P–C(3) angle is only 107.0°, that is, significantly smaller than the N–P–N angle in ion **286b** (114.8°). The unique ferrocenyl-stabilized two-coordinate phosphenium ion **293** was also prepared and analyzed.[738]

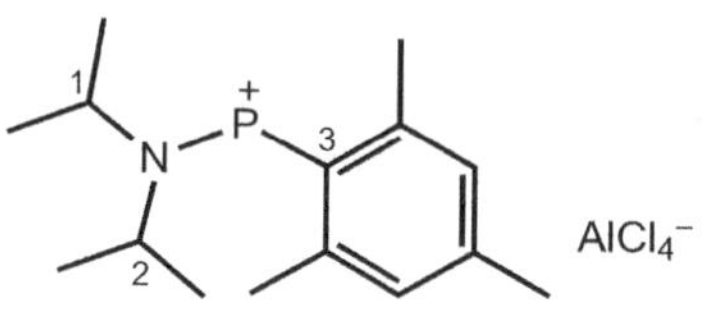

292

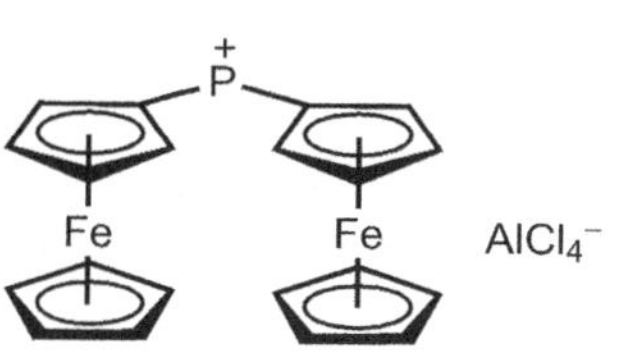

293

Phosphenium ions can also be stabilized by intramolecular coordination to a Lewis base. In ion **294**, which was prepared by hydride abstraction[739] [Eq. (4.177)], intramolecular bis-coordination exists. The ^{31}P NMR spectrum of the corresponding P–H derivative[740] exhibits one doublet at δ^{31}P 37.6 ($J_{P\text{–}N}$ = 243 Hz). The X-ray structure, which shows no interaction between the cation and the hexafluorophosphate anion, clearly demonstrates an ionic structure. Both arms are coordinated to the phosphorus center. The N–P distances are 2.082 and 2.068 Å, which are longer than the N–P σ-bond (1.769 Å) but significantly shorter than the sum of the two van der

Waals radii (3.4 Å). The P–H bond is almost orthogonal to the plane of the benzene ring (C_{ipso}–P–H bond angle = 99.4°). The phosphorus atom has a distorted trigonal bipyramid with the amino groups occupying the axial sites and C_{ipso}, H, and the phosphorus lone pair occupying the equatorial sites.

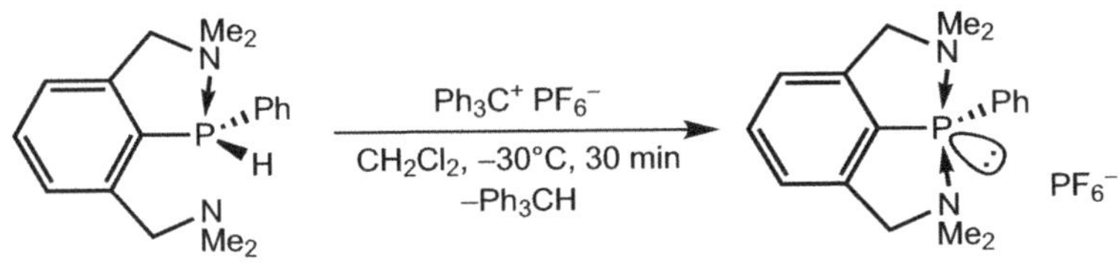

294

(4.177)

In cation **295** analogous to borenium ion **236**, both an N directly attached to P and a donor nitrogen ligand contribute to stabilization.[741] Ion **296**, in contrast, is N, C-bonded.[742] A single-crystal X-ray diffraction study of **296** (R = Cl, Y = TfO) reveals a hydrogen bond between the triflate oxygen and the nonligated nitrogen. The six-membered ring adopts an envelope conformation with a nearly planar P–N (1)–C(2)–C(3)–C(4) moiety and a flip of C(5) by 45.99°. Bond distances are indicative of considerable electron delocalization.

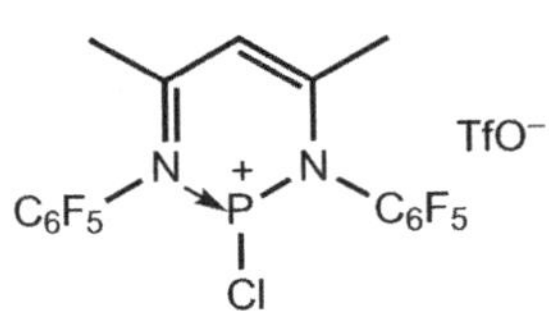

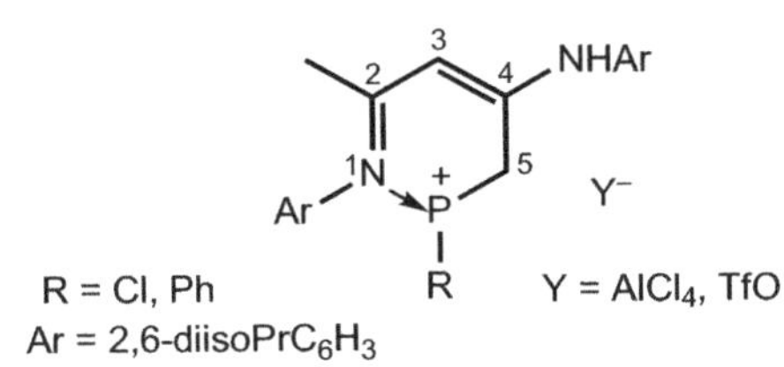

295 **296**

Considering the observation[743] that the stabilizing effect of donor substituents decreases in the order NH_2 > SH, OH > Cl, it is not surprising that examples of cations with oxygen as directly attached donor atoms, such as cation **297**, are scarce.[744] Cation **298** with sulfur as an intramolecular *n*-donor is the only example of its kind.[745]

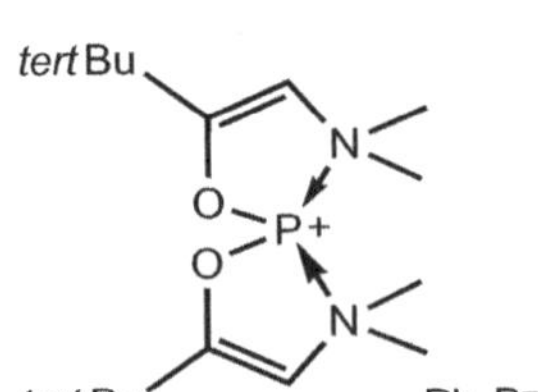

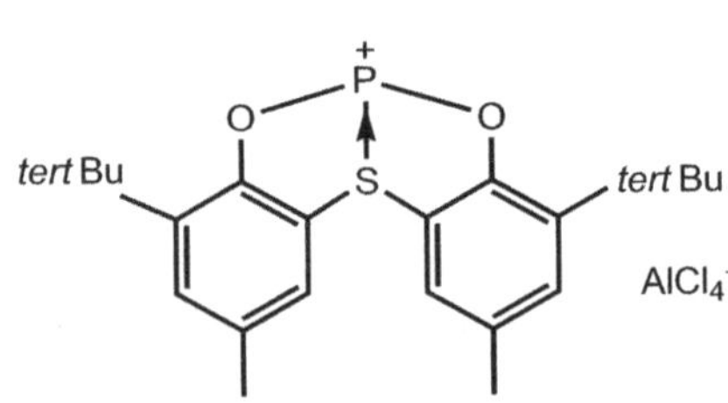

297 **298**

Di- and triphosphenium ions have also been synthesized and characterized. The **299** phosphanyl phosphonium ion was generated by a large excess of methyl triflate[746]

[Eq. (4.178)]. Creation of the positive charge brings about a considerable low-frequency shift of both ^{31}P NMR resonances (from $\delta^{31}P$ 495 of the precursor to two doublets at $\delta^{31}P$ 237 and 332.2). The phosphorus center P(1) is trigonal planar, and there is no twist around the P=P double bond. The only notable change in the X-ray characteristics upon methylation is the enlargement of the P(1)–P(2)–C_{ipso} bond angle (102.8 versus 123.1°). These features indicate that the ion has a low phosphenium ion (**299a**) character.

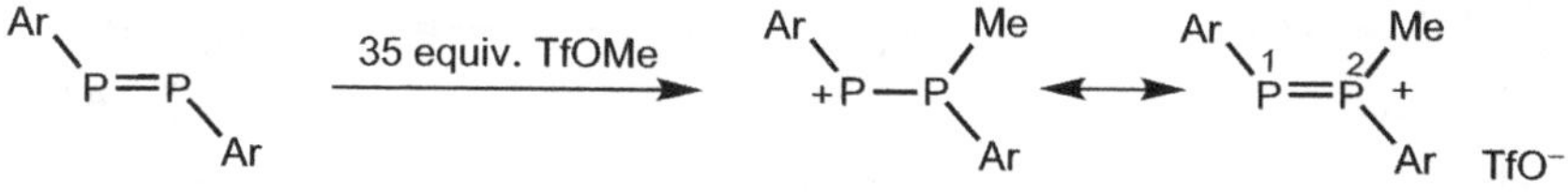

Ar = 2,4,6-tri-*tert*-BuC_6H_2 **a** **b** **299**

(4.178)

The mixture of Ph_3P and $AlCl_3$ reacts with PCl_3 to yield the **300** triphosphenium ion[747] [Eq. (4.179)]. There are two independent, yet not significantly different cations in the crystal. The P–P distances (2.128 and 2.137 Å, and 2.124 and 2.141 Å) lie between that of a P–P single bond (2.20–2.25 Å) and a P=P double bond (2.00–2.05 Å).

$$PCl_3 + 3Ph_3P + 2AlCl_3 \xrightarrow[CHCl_3,\ 3°C]{} [Ph_3P{-}P{-}PPh_3]^+ AlCl_4^- + Ph_3PCl^+ AlCl_4^-$$

300

(4.179)

The cation **301** with homoatomic P→P coordination has been shown to be better represented by resonance structure **301b**.[727] In contrast, cations **302** and **303**, as well as dications **304** obtained by Burford et al.,[748,749] have been characterized as diphosphines linked to one or two R_2P^+ Lewis acid units. The P of phosphine ligands has distorted tetragonal geometry and the geometry of P^+ is distinctly pyramidal with standard phosphine-like bond angles (90–105°). Furthermore, all cations exhibit strikingly uniform P–P bond distances (2.187–2.230 Å for monocations and 2.163–2.233 Å for dications). The salts are R_2P^+ cation donors in ligand exchange reactions.

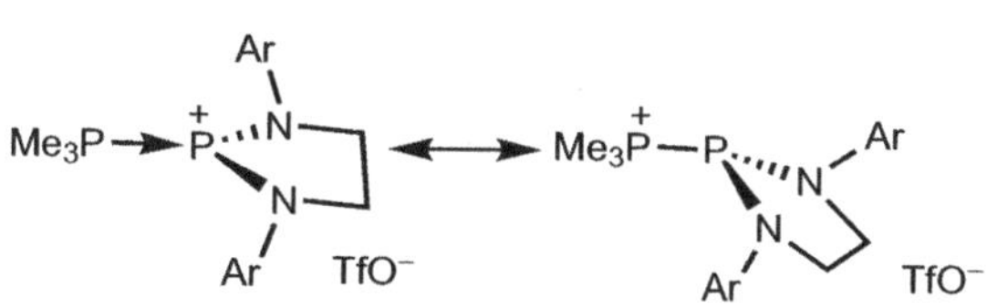

301

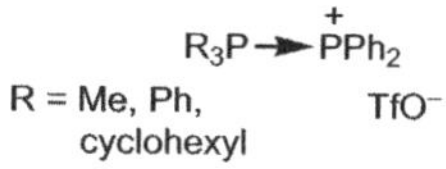

302

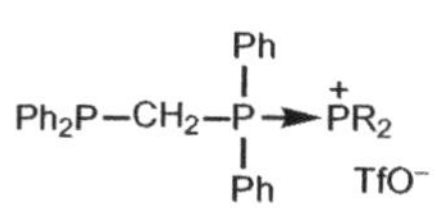

303

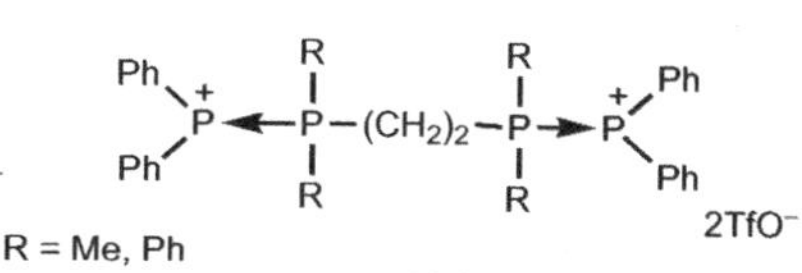

304

Attempts have been made to observe bisphosphenium dications [Eq. (4.180)] and trisphosphenium trications[723] [Eq. (4.181)]. In all cases, however, only monoionization takes place.

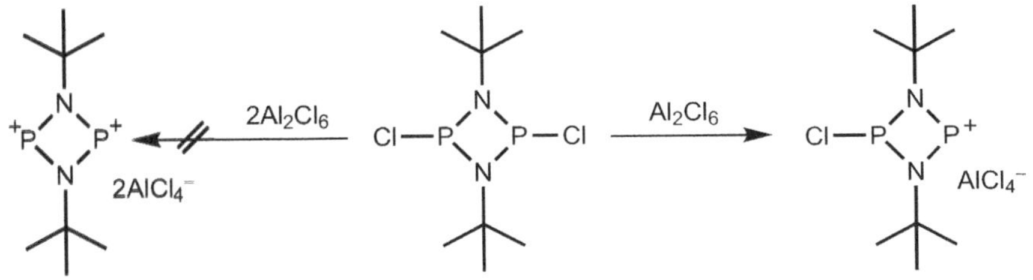

(4.180)

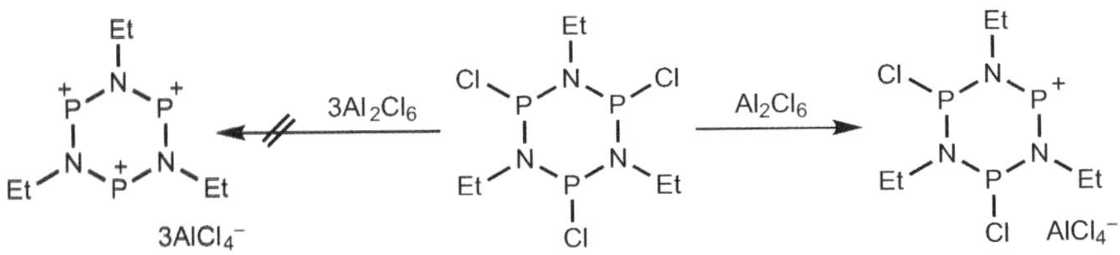

(4.181)

Laali, Regitz, and co-workers[750] have reported the first synthesis and characterization of a phosphirenylium ion. Cation **305** was obtained by exploiting the super Lewis acidity of $B(OTf)_3$ [Eq. (4.182)]. The single phosphorus resonance in the ^{31}P NMR spectrum at $\delta^{31}P$ 309.7 is deshielded by 313 ppm from the precursor, indicating the cationic character of the P atom. Both ^{1}H and ^{13}C NMR spectral data of all ring atoms show downfield shifts indicative of charge delocalization over the ring. The ^{13}C NMR phenyl resonances are also deshielded, suggesting that charge delocalization to the aromatic ring also takes place. *Ab initio* calculations (HF/6-31 + G* level) show that bonds are delocalized (P–C and C–C bond orders are 0.980 and 1.776, respectively).

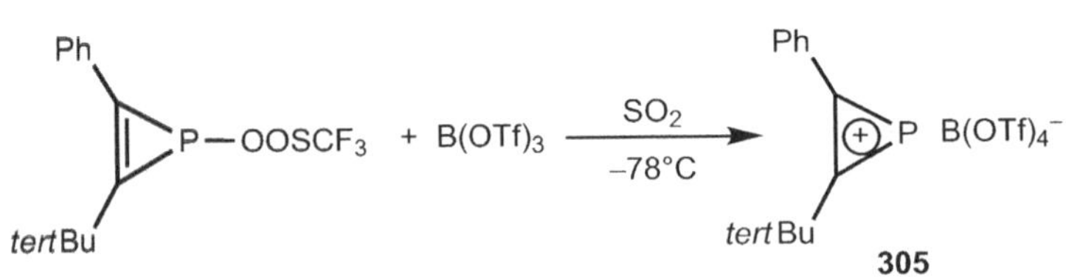

305

(4.182)

Regitz and co-workers[751] have recently reported the phosphirenylium ion **306** complexed to $W(CO)_5$. Similar to ion **305**, the formation of cation **306** is characterized by a significant downfield shift of the ^{31}P NMR resonance (about 300 ppm). NMR spectroscopic data, in general, exhibit close resemblance to the values of the non-complexed ion **305**. The prototype aromatic phosphirenylium ion cyclo-$C_2H_2P^+$ **307** has been generated in the gas phase and characterized using collision-induced dissociation (CID) spectra.[752] DFT calculations (B3LYP/6-31G* and G3//B3LYP/6-31G* levels) have found that the ion **307** is the lowest energy structure of 10 possible isomers.

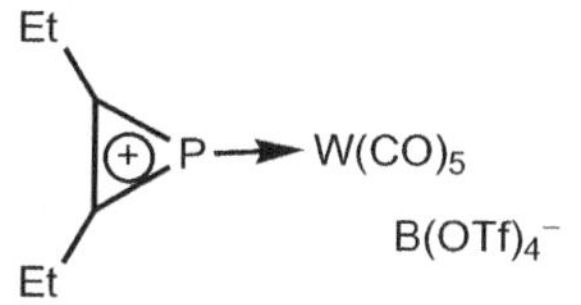

306

307

4.3.3.3. Enium Ions of Other Group 15 Elements. Similar to phosphenium ions, amino group(s) directly attached to arsenic or antimony provide stabilization through strong electron donation (**308**–**311**).[729,753–757] Interestingly, however, in the $AlCl_4^-$ salts of the stibium and bismuth analogs of ion **291**, the origin of stabilization is a strong interionic interaction between Sb^+ and Bi^+, respectively, and the chlorine atoms of Al (Sb---Cl = 3.05 Å, Bi---Cl = 3.09 Å).[736] The As and Sb analogs of **289** and **290** have also been prepared and characterized.[734,735] Diphosphane donors may also stabilize arsenium cations as shown by the synthesis of $[(Me_2N)_3PAsP(NMe_2)_3]^+BPh_4^-$.[758]

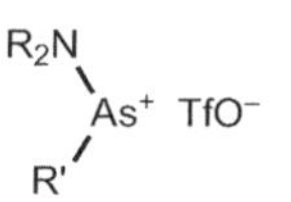

R = Me, Et
R' = Cl, NR_2

308

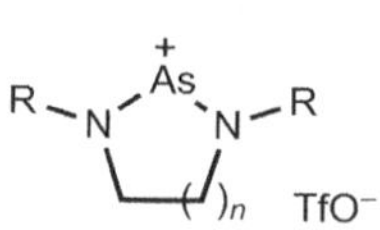

R = Me, Et
n = 1, 2

309

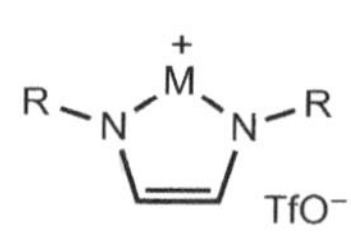

M = As, Sb
R = *tert*-Bu, Mes

310

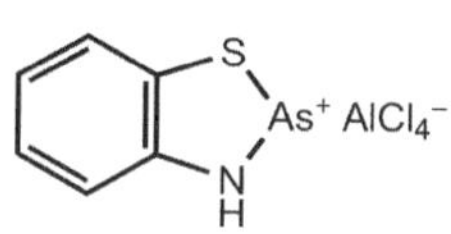

311

Chloride abstraction from the corresponding precursors of identical structure has resulted in the formation of the amidocyclopentadienyl As and Sb cations **312** and **313** with different structures.[759] In both cases ^{31}P NMR spectra show a downfield shift. Single crystal structure of the arsenium cation **312** reveals a modest increase in the N–As–C angle and decrease in the As–N and As–C bond lengths. In contrast to the η^1 (σ) attachment found in cation **312**, the stibenium cation **313** has a η^3 bonding to the cyclopentadienyl ring. An additional noteworthy feature is the shortening of the N–Sb bond by about 0.2 Å indicating a more significant electron donation.

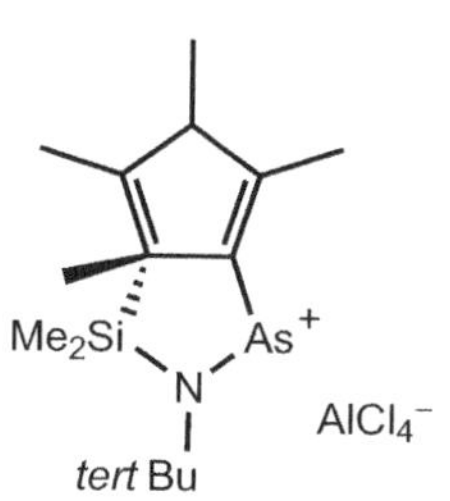

312

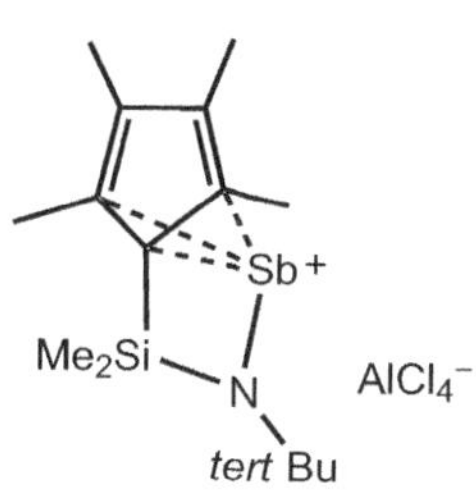

313

Intramolecular coordination to a Lewis base such as in cation **314** likewise contributes to the stabilization of arsenium ion.[760] Examples for intermolecular complexation are more numerous. X-ray crystal structure analysis of cations **315** and **316** reveals similar features to those of the corresponding phosphenium ions discussed above.[761,762] The dimeric cation **317** is unique since it is stabilized by both the amino group attached directly and through intermolecular coordination.[763] In the stibenium cation **318**,[764] the stereochemistry around antimony is trigonal pyramidal with the antimony at the center of the trigonal plane containing the two phenyl groups and the lone pair of electrons of the six-electron Ph_2Sb^+ group. The Me_3P moiety lies directly above the antimony atom and is orthogonal to the trigonal plane. In the solid state, four cationic units surround a central chloride or bromide ion in a centrosymmetrical, square-planar arrangement of C_{4h} symmetry. The Sb---Cl and Sb---Br contacts are 3.1362 and 3.2236 Å (sums of the van der Waals radii for the two elements are 3.87 and 4.24 Å, respectively).

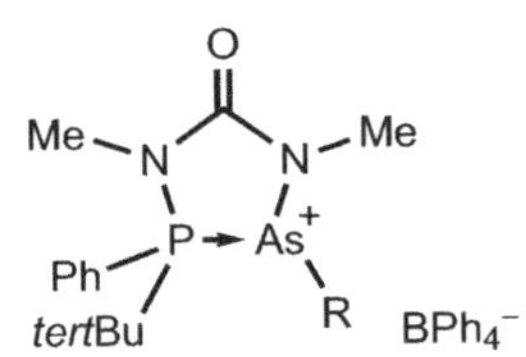

R = Me, Ph

314

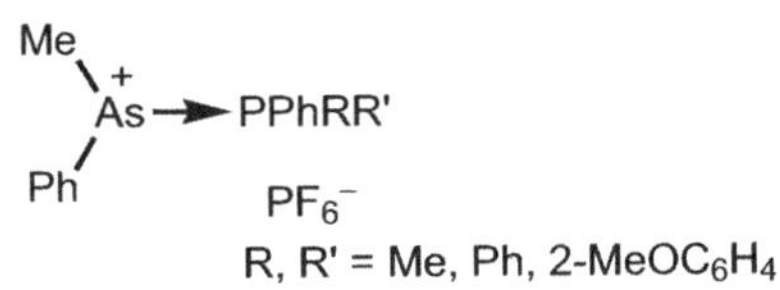

315

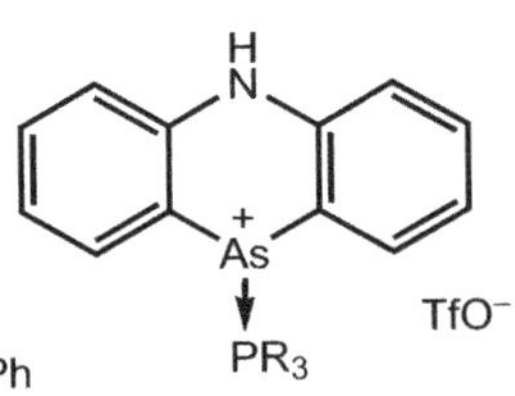

R = Me, Ph

316

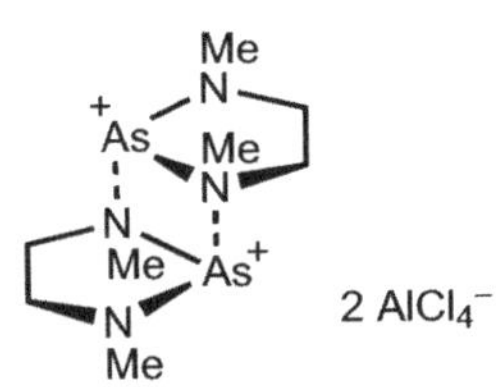

317

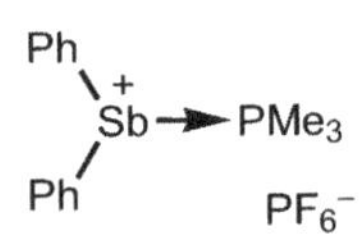

318

4.3.4. Enium Ions of Group 16 Elements

4.3.4.1. Oxenium Ions. Oxenium ions similar to nitronium ions are, in general, too reactive to be observed. The parent ion—that is, the hydroxyl cation HO^+—is elusive, and it is improbable that it can be observed in its "free" form in the condensed state. However, the incipient hydroxyl cation is involved in acid-catalyzed electrophilic hydroxylation with protonated (or Lewis acid complexed) hydrogen peroxide ($HO{-}OH_2^+$) or ozone ($HO{-}O{-}O^+$).[125] Nitrous oxide is also a potential precursor for the hydroxyl cation (in its protonated form). The hydroxy diazonium ion HON_2^+ has not yet been observed.

Alkyl- and aryloxenium ions (RO^+) are similarly too reactive to be observed; however, they may be involved in the oxidation of alkanes under superacidic conditions with ozone (see Section 5.12.2). Of all the possible alkyloxenium ions, the *tert*-butyloxenium ion is the most significant. Superacid cleavage of *tert*-butyl hydroperoxide in a Hock-type reaction gives acetone and methyl alcohol, indicative of the intermediacy of the *tert*-butyloxenium ion **319** [Eq. (4.183)]. Under stable ion conditions even at low temperatures, only the rearranged carboxonium ion could, however, be observed.[125]

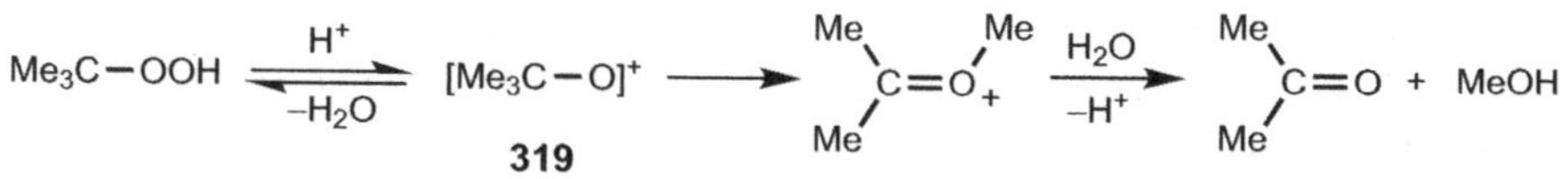

319

(4.183)

Similarly, the cumyloxenium ion **320** is involved in the acid-catalyzed cleavage rearrangement reaction of cumene hydroperoxide to phenol and acetone [Eq. (4.184)].

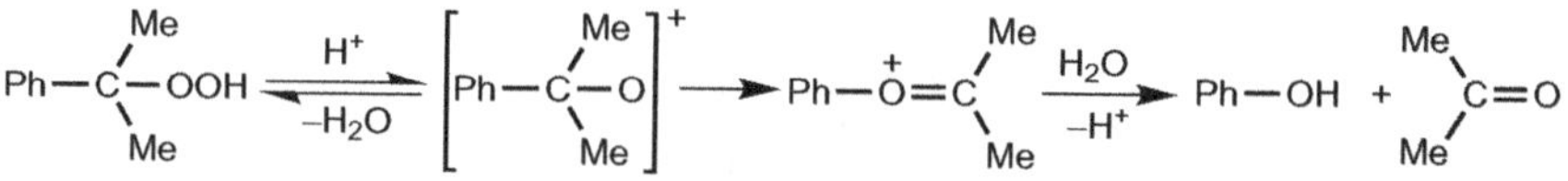

320

(4.184)

Aryloxenium cations, however, can be best regarded as cyclohexadienyl ions because of strong delocalization (**321b**). The lifetime of cation **321** in water was estimated to be 0.55 μs.[765] Properties of phenyloxenium and phenylnitrenium ions have been calculated (HF/6-31G* and pBP/DN*//HF/6-31G* levels).[766] Both ions are ground-state singlets and stabilized by 4-methyl and 4-phenyl substituents. The phenyloxenium ion has much greater charge localization on the ring primarily at the *para* position.

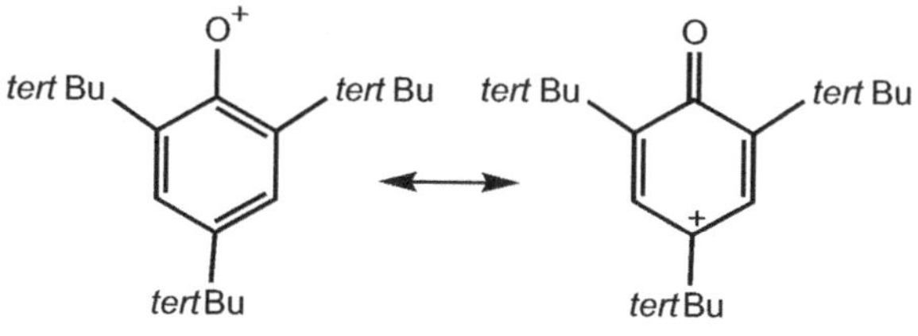

a **b**

321

4.3.4.2. Enium Ions of Other Group 16 Elements. Furukawa et al.[316,320] have obtained enium ions of the heavier chalcogen elements stabilized by intramolecular complexation with dimethylamino groups (**322**). Resonances of the benzylic and methyl protons in the ^1H NMR spectrum of cation **322a** are shifted downfield indicative of the coordination of nitrogen to sulfur.[767] There is no cation–anion contact in the solid state. The S–N distances are longer than the covalent single bond (2.063 versus 1.74 Å) and the molecular structure reveals a significant distortion of the N–S–N angle from linearity (168.54°). The charge on the sulfur atom was calculated to be +0.550 with the two nitrogen atoms negatively charged (–0.535). Bond orders were found to be 0.379 (Mulliken) and 0.539 (Löwdin). The singlet peak in the ^{77}Se NMR spectrum of cation **322b** is significantly deshielded relative to the neutral precursor (δ^{77}Se 1208 and 90, respectively)[768] and a similar phenomenon is observed in the ^{125}Te NMR spectrum of cation **322c** (δ^{125}Te 1950 versus 287).[768] These changes are characteristic of cationic species.

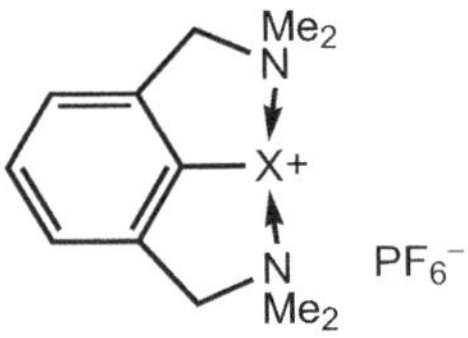

322 **a** X = S
b X = Se
c X = Te

The **323** pentamesityltritellurium cation has been prepared and characterized.[287,769] The two Mes_2Te groups in the cation are NMR equivalent. Their signal (δ^{125}Te +854) and that of the central MesTe unit (δ^{125}Te +388) exhibit different broadening indicative of dynamic exchange equilibria. The Te_3 unit is not linear (Te–Te–Te bond angle = 159.51°), and the Te–Te bonds are unsymmetric (2.979 and 3.049 Å) and significantly longer than the single bond in **108b**. The valence shell of the central Te atom contains 10 electrons and, consequently, the Te_3 unit is a nonclassical hypervalent moiety.

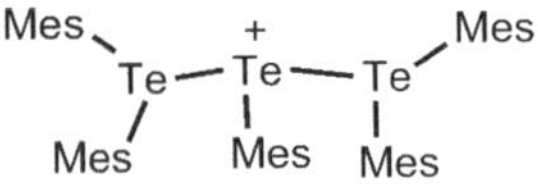

323

The energies and geometries of the phenylsulfenium cation have been optimized by *ab initio* calculations (MP2/6-31G*//MP2/6-31G* level)[770] and the singlet state has been found to be 63.0 kcal mol^{-1} more stable than the triplet state.

4.4. HOMO- AND HETEROPOLYATOMIC CATIONS

In this section, the polyatomic cations of group 17 elements (halogen and interhalogen cations and polycations), cations and polycations of group 16 elements (O, S, Se, and

Te), and mixed polyheteroatom cations of group 15, 16, and 17 elements obtained in superacid media will be discussed.

4.4.1. Halogen Cations

The existence of many well-known compounds in which chlorine, bromine, and iodine are found in the +1 oxidation state led to the assumption that the cations Cl^+, Br^+, and I^+ are important and stable entities or at best as reaction intermediates. However, no evidence exists for monoatomic Cl^+, Br^+, and I^+ as stable species.[771,772] In contrast, a whole series of polyatomic halogen cations are known.[773–776]

4.4.1.1. Iodine Cations. The existence of I_3^+ and I_5^+ ions, deduced by Masson[777] 70 years ago in aromatic iodination reactions, has been confirmed now by physical measurements. The controversy over the nature of blue solutions of iodine in strong acid media has now been settled. It has been shown conclusively that these solutions contain I_2^+ ions[778–780] and not I^+ as suggested earlier.[781]

I_3^+ and I_5^+ Ions. The first evidence for stable iodine cations was obtained by Masson.[777] He postulated the presence of I_3^+ (**324**) and I_5^+ (**325**) ions in solutions of iodine and iodic acid in sulfuric acid to explain the stoichiometry of the reaction of such solutions with chlorobenzene to form both iodo and iodoso derivatives. Later, Symons and co-workers[782] gave conductometric evidence for I_3^+ formed from iodic acid and iodine in 100% sulfuric acid and suggested that I_5^+ may be formed on the basis of changes in the UV and visible spectra when iodine is added to I_3^+ solutions. Gillespie and co-workers,[783] on the basis of detailed conductometric and cryoscopic measurements, confirmed that I_3^+ is formed from HIO_3 and I_2 in 100% sulfuric acid [Eq. (4.185)]. The I_3^+ cation may also be prepared in fluorosulfuric acid.[778] Solutions of red brown I_3^+ **324** in H_2SO_4 or HSO_3F have characteristic absorption maxima at 303 and 470 nm, with a molar extinction coefficient of 5200 at 305 nm.

$$HIO_3 + 7I_2 + 8H_2SO_4 \longrightarrow \underset{\mathbf{324}}{5I_3^+} + 3H_3O^+ + 8HSO_4^- \qquad (4.185)$$

Solutions of I_3^+ in 100% sulfuric acid[783] or fluorosulfuric acid[778] dissolve at least 1 mol of iodine per mol of I_3^+, and a new absorption spectrum is obtained which has bands at 240, 270, 345, and 450 nm. At the same time, there is no change in either the conductivity or the freezing point of the solutions. This leads to the conclusion that ion I_5^+ **325** is formed [Eq. (4.186)]. Some further iodine also dissolves in solutions of I_3^+, indicating possible formation of I_7^+.

$$\underset{\mathbf{324}}{I_3^+} + I_2 \longrightarrow \underset{\mathbf{325}}{I_5^+} \qquad (4.186)$$

Solutions of I_3^+ **324** in H_2SO_4 give Raman spectra[784] that have three bands, in addition to the solvent peaks at 114, 207, and 233 cm^{-1} which may be assigned as the ν_2, ν_1, and ν_3 vibrations of an angular molecule. The average stretching frequency of 220 cm^{-1} in the I_3^+ molecule is appreciably lower than the stretching frequency of 238 cm^{-1} for the I_2^+ molecule and, in fact, closer to the frequency of 213 cm^{-1} for the neutral molecule I_2. This is consistent with I_3^+ having a formal I–I bond order of 1.0 for both bonds as in the simple valence bond formation **324**, whereas that in I_2^+ is 1.5.

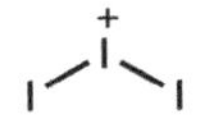

324

On the basis of ^{127}I nuclear quadrupole resonance (NQR) studies of $I_3^+AlCl_4^-$, Corbett and co-workers[785] have predicted a bond angle of 97° between the two bonding orbitals on the central atom. X-ray crystal structures of $I_3^+AlCl_4^-$ and $I_5^+SbF_6^-$ have been obtained.[786,787] In the latter salt, two cations are weakly bound to a central I_5^+ ion and the moiety can be regarded as $I_{15}^{3+}(SbF_6^-)_3$. In contrast, in the salt $I_5^+AsF_6^-$ prepared subsequently by Passmore, White, and co-workers[788] by oxidizing iodine with AsF_5 in liquid SO_2, the cation I_5^+ (**325**) exists. It is planar, has C_{2h} symmetry and a Z-shaped (*trans*) structure (**325**), and features the following characteristics: I(1)–I(2) and I(2)–I(3) bond lengths = 2.645 and 2.895 Å, respectively, I–I–I bond angles = 97.0° and 180°.

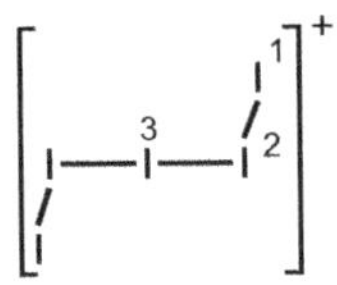

325

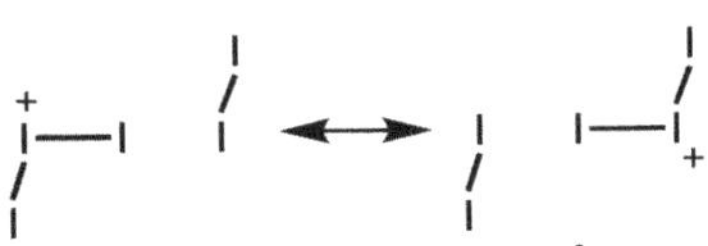

a **b**

325

The bonding in I_5^+ (and in Br_5^+) can be described in terms of valence bond structures **325a** and **325b** showing a formal bond order of 1 for the terminal I–I bonds.[788] The bond order of the central bonds is 0.5, and these bonds may be considered as three-center four-electron bonds.

I_2^+ Ion. Gillespie and Milne have shown,[778] by conductometric, spectrophotometric, and magnetic susceptibility measurements, that the blue iodine species observed in strong acids is I_2^+ **326**. When iodine was oxidized by peroxodisulfuryl difluoride in fluorosulfuric acid [Eqs. (4.187) and (4.188)], the concentration of the blue iodine species reached a maximum at 2:1 I_2:$S_2O_6F_2$ mole ratio and not at the 1:1 mole ratio as would be anticipated for the formation of I^+. The conductivities of 2:1 solutions of iodine:$S_2O_6F_2$ at low concentrations were found to be very similar to solutions of KSO_3F at the same concentration, showing that 1 mol of FSO_3^- had been formed per mole of iodine. The magnetic moment of the blue species in fluorosulfuric acid was found to be 2.0 ± 0.1 D, which agreed with the value expected for the $^3\pi_{3/2}$ ground state of the I_2^+ cation **326**. The I_2^+ has characteristic peaks in its UV absorption spectrum at

640, 490, and 410 nm and has a molar extinction coefficient of 2560 at 640 nm.

$$2I_2 + S_2O_6F_2 \longrightarrow \underset{\mathbf{326}}{2I_2^+} + 2FSO_3^- \quad (4.187)$$

$$I_2 + S_2O_6F_2 \longrightarrow 2I^+ + 2FSO_3^- \quad (4.188)$$

The I_2^+ cation is not completely stable in fluorosulfuric acid and undergoes some disproportionation to the more stable I_3^+ ion **324** and $I(SO_3F)_3$ [Eq. (4.189)]. This disproportionation is largely prevented in a 1:1 iodine:$S_2O_6F_2$ solution in which $I(SO_3F)_3$ is also formed [Eq. (4.190)].

$$\underset{\mathbf{326}}{8I_2^+} + 3FSO_3^- \rightleftharpoons \underset{\mathbf{324}}{5I_3^+} + I(SO_3F)_3 \quad (4.189)$$

$$5I_2 + 5S_2O_6F_2 \longrightarrow \underset{\mathbf{326}}{4I_2^+} + 4FSO_3^- + 2I(SO_3F)_3 \quad (4.190)$$

The disproportionation can also be prevented if the fluorosulfate ion concentration in fluorosulfuric acid is lowered by addition of antimony pentafluoride [Eq. (4.191)] or by using the less basic solvent, 65% oleum.

$$SbF_5 + FSO_3^- \longrightarrow (SbF_5SO_3F)^- \quad (4.191)$$

In 100% H_2SO_4, the disproportionation of I_2^+ to I_3^+ and an iodine(III) species, probably $I(SO_4H)_3$, is essentially complete, and only traces of I_2^+ can be detected by means of its resonance Raman spectrum.

Solution of the blue iodine cation in oleum has been reinvestigated[779] by conductometric, spectrophotometric, and cryoscopic methods confirming the formation of I_2^+. In 65% oleum, iodine is oxidized to I_2^+ [Eq. (4.192)].

$$2I_2 + 5SO_3 + H_2S_4O_{13} \longrightarrow \underset{\mathbf{326}}{2I_2^+} + 2HS_4O_{13}^- + SO_2 \quad (4.192)$$

Adhami and Herlem[789] have carried out a coulometric titration at controlled potential of iodine in fluorosulfuric acid and have shown that iodine is quantitatively oxidized to I_2^+ by removal of one electron per mole of iodine.

Pure crystalline $I_2^+Sb_2F_{11}^-$ has been prepared by the reaction of iodine with antimony pentafluoride in liquid sulfur dioxide as solvent.[789] After removal of insoluble SbF_3, deep blue crystals of $I_2^+Sb_2F_{11}^-$ were obtained from the solution. An X-ray crystallographic structure determination showed the presence of the discrete

I_2^+ and $Sb_2F_{11}^-$ ions.[790] Crystalline solids that can be formulated as $I_2^+Sb_2F_{11}^-$ have also been prepared by Kemmitt et al.[780] by the reaction of iodine with antimony or tantalum pentafluoride in iodine pentafluoride solutions.

Besida and O'Donnell[791] have studied the existence and stabilization of iodine cations in hydrogen fluoride using Lewis acids (NbF_5, TaF_5, SbF_5) and NaF to control acidity and basicity. Excess F^- causes disproportionation of the cations I_5^+, I_3^+, and I_2^+ to I_2 and IF_5. When these products are dissolved in HF and the acidity level is adjusted appropriately, the individual cations can be generated.

I_4^{2+} Dication. In an early study of the I_2^+ cation, Gillespie et al.[792] observed that on cooling the HSO_3F solution the color of the solution changed from an intense blue to a deep red-brown at about −80°C with the concomitant considerable decrease of the paramagnetism. This was suggested to result from the dimerization of the paramagnetic I_2^+ ion to the diamagnetic I_4^{2+} cation. Subsequently, they were able to prepare the I_4^{2+} dication **327** by the reaction of iodine with either AsF_5 or SbF_5 solution[793] [Eq. (4.193)].

$$2\,I_2 + 4\,MF_5 \xrightarrow[RT]{SO_2} \underset{\mathbf{327}}{I_4^{2+}\,Y^-} \qquad M = As, Sb \qquad Y = AsF_6,\ (Sb_3F_{14}^-)SbF_6^- \tag{4.193}$$

The X-ray crystal structure of salts $I_4^{2+}(AsF_6)_2^{2-}$ and $I_4^{2+}(Sb_3F_{14})^-SbF_6^-$ have been determined.[793] The I_4^{2+} dication is described as two I_2^+ cations bonded together by two relatively weak bonds. The interaction between the two I_2^+ ions may be described as a four-center two-electron π^*–π^* bond formed between the singly occupied antibonding π^* molecular orbitals of the two I_2^+ monomers. Each of the long iodine–iodine bonds has a bond order of 0.5. This model is consistent with the long I–I distances and the diamagnetism of the dication.

4.4.1.2. Bromine Cations

Br_3^+ Ion. In 1906, Ruff et al.[794] prepared a compound by the reaction of Br_2 and SbF_5, which was formulated as SbF_5Br. Later McRae[795] showed that Br_3^+ **328** is formed in the system. Subsequently, Gillespie and Morton[796,797] found that Br_3^+ **328** is formed quantitatively in the superacid medium HSO_3F–SbF_5–SO_2 (mainly be the reaction of $S_2O_6F_2$) [Eq. (4.194)].

$$3\,Br_2 + S_2O_6F_2 \longrightarrow \underset{\mathbf{328}}{2\,Br_3^+\,FSO_3^-} \tag{4.194}$$

These solutions are brown in color and have a strong absorption at 300 nm with a shoulder at 375 nm. Solutions of Br_3^+ can also be obtained in a similar way in fluorosulfuric acid; however, they are not completely stable in this solvent and undergo some disproportionation [Eq. (4.195)].

$$\underset{\mathbf{328}}{Br_3^+} + FSO_3^- \longrightarrow Br_2 + BrOSO_2F \quad (4.195)$$

Glemser and Šmalc[798] have prepared the compound $Br_3^+AsF_6^-$ by the displacement of oxygen in dioxygenyl hexafluoroarsenate by bromine [Eq. (4.196)] and by the reaction of bromine pentafluoride, bromine, and arsenic pentafluoride [Eq. (4.197)]. The compound is chocolate-brown and in solution has absorption bands at 310 nm and 375 nm; it has fair thermal stability and can be sublimed at 30–50°C under nitrogen atmosphere.

$$2O_2^+AsF_6^- + 3Br_2 \longrightarrow \underset{\mathbf{328}}{2Br_3^+AsF_6^-} + 2O_2 \quad (4.196)$$

$$7Br_2 + BrF_5 + 5AsF_5 \longrightarrow \underset{\mathbf{328}}{5Br_3^+AsF_6^-} \quad (4.197)$$

Christe et al.[799] have obtained the crystal structure of the $Br_3^+AsF_6^-$ salt and showed that it contains discrete ions with some cation–anion interactions. The Br_3^+ cation is symmetric with a bond distance of 2.270 Å and a bond angle of 102.5°.

Br_2^+ Ion. The Br_2^+ cation **329** can be prepared[797] by oxidation of bromine by $S_2O_6F_2$ in the superacid $HSO_3F–SbF_5–3SO_3$; however, even in this very weakly basic medium, the Br_2^+ ion is not completely stable because it undergoes appreciable disproportionation [Eq. (4.198)].

$$\underset{\mathbf{329}}{2Br_2^+} + 2HSO_3F \rightleftharpoons \underset{\mathbf{328}}{Br_3^+} + BrOSO_2F + H_2SO_3F^+ \quad (4.198)$$

Moreover, the $BrOSO_2F$ that is formed also undergoes some disproportionation by itself to Br_2^+, Br_3^+, and $Br(OSO_2F_3)_3$, [Eqs. (4.199) and (4.200)], and the equilibria in these solutions are quite complex.

$$5BrOSO_2F + 2H_2SO_3F^+ \rightleftharpoons \underset{\mathbf{329}}{2Br_2^+} + Br(OSO_2F)_3 + 4HSO_3F \quad (4.199)$$

$$4BrOSO_2F + H_2SO_3F^+ \rightleftharpoons \underset{\mathbf{328}}{Br_3^+} + Br(OSO_2F)_3 + 2HSO_3F \quad (4.200)$$

A solution of Br_2^+ **329** in superacid has a characteristic cherry red color with maximum absorption at 510 nm and a single band in the Raman spectrum at 360 cm^{-1}. The paramagnetic scarlet crystalline compound $Br_2^+Sb_3F_{16}^-$ has been prepared[800,801] according to Eq. (4.201).

$$9Br_2 + 2BrF_5 + 30SbF_5 \longrightarrow \underset{\mathbf{329}}{10Br_2^+ Sb_3F_{16}^-} \quad (4.201)$$

It is a stable salt and can be sublimed at 200°C. The X-ray crystal structure of $Br_2^+Sb_3F_{16}^-$ **329** shows a bromine–bromine bond distance of 2.13 Å. The shorter bond distance of **329** compared to neutral bromine is in accord with increase in bond order resulting from the loss of an antibonding electron from the neutral molecule.[800,801]

Br_5^+ Ion. The Br_5^+ cation was first isolated[799] as a mixture of $Br_3^+AsF_6^-$ and $Br_5^+AsF_6^-$. On the basis of the Raman spectrum, it was shown to have three colinear central bromine atoms with a semi-ionic, three-center four-electron bond and two terminal, perpendicular bromine atoms with covalent bonds. Minkwitz and co-workers[802] have obtained the **330** salts by oxidation with XeF^+ [Eq. (4.202)]. The planar *trans* geometry of the cation is similar to that of I_5^+ (**325**). It is centrosymmetric in the solid state with two short bonds between the external bromine atoms (2.275 Å in the AsF_6^- and 2.268 Å in the SbF_6^- salt) and two longer bonds to the central atom (2.512 and 2.514 Å in the AsF_6^- and SbF_6^- salts, respectively). The Br(1)–Br(2)–Br(3) bond angles are 96.91 and 97.69°, respectively. According to local density functional calculations,[803] in the optimized structure the three central bromine atoms deviate from linearity by 11.4° and the terminal bromine atoms form a dihedral angle of 82° about the Br(2)–Br(4) axis.

$$\underset{M = As, Sb}{3XeF^+ MF_6^- + 8Br_2} \xrightarrow{RT} \underset{\mathbf{330}}{3Br_5^+ MF_6^-} + 3Xe + BrF_3 \quad (4.202)$$

4.4.1.3. Chlorine Cations

Cl_3^+ Ion. There is no evidence for either Cl^+ or Cl_2^+ in superacid media.[804] However, Cl_2, ClF, and AsF_5 react at −70°C to form $Cl_3^+AsF_6^-$ **331** [Eq. (4.203)].[772]

$$Cl_2 + ClF + AsF_5 \longrightarrow \underset{\mathbf{331}}{Cl_3^+ AsF_6^-} \quad (4.203)$$

The Cl_3^+ cation **331** has been identified by its Raman spectrum in the yellow solid which precipitates from solution of Cl_2 and ClF in HF–SbF_5 at −76°C or formed by treating $ClF_2^+AsF_6^-$ with chlorine.[772] At room temperature the Cl_3^+ cation completely disproportionates in the HF–SbF_5 media to chlorine and ClF_2^+ salts. There is no evidence for the formation of $Cl_3^+BF_4^-$ salt from mixtures of Cl_2, ClF, and BF_3 at temperatures ranging from ambient to −120°C.

The Raman spectrum of $Cl_3^+AsF_6^-$ shows bands due to the AsF_6^- ion, together with three relatively intense bands at 490 (split to 485 and 493), 225, and 508 cm^{-1}.[805] These frequencies are very close to the vibrational frequencies of neutral SCl_2 molecule.[806] Subsequently, Clegg and Downs[807] isolated the bright-orange $Cl_3^+Sb_2F_{11}^-$ salt at room temperature and Minkwitz et al.[808] performed Raman characterization of $Cl_3^+SbF_6^-$.

Seppelt and co-workers[809] have synthesized Cl_3^+ salts by warming $Cl_2O_2^+$ salts to room temperature [Eq. (4.204)] and isolated yellow crystals. The Cl_3^+ cation in all salts exists as a symmetric, bent entity with Cl−Cl bond distances of 1.972–1.994 Å and bond angles of 104.51–105.62°.

$$Cl_2O_2^+\ M^- \xrightarrow[\text{RT, 1 day}]{\text{HF–SbF}_5} Cl_3^+\ M'^- \qquad (4.204)$$

$$M = SbF_6,\ Sb_2F_{11} \qquad M' = SbF_6,\ Sb_2F_{11},\ Sb_3F_{16}$$

A DZP calculation by Burdett and Marsden[810] predicted a bent structure for Cl_3^+ (Cl−Cl bond length = 2.010, bond angle = 105.6°). Calculations by Olah and co-workers[811] found the Cl_3^+ ion a bent ground-state singlet (1.998 Å and 107°) with a singlet–triplet gap of only 2.5 kcal mol^{-1}. Further *ab initio* calculations for all four X_3^+ ions were performed by Schwarz and co-workers[812] (RHF and MP2 levels with different basis sets). Cacace et al.[813] have prepared the Cl_3^+ ion in the gas phase and investigated with FT–ICR mass spectrometry and DFT method. The theoretical study [B3LYP/6-311++G(3df)] shows that the global minimum on the Cl_3^+ energy surface is a singlet of C_{2v} symmetry with a Cl−Cl bond distance of 1.996 Å and a bond angle of 110.0°.

Cl_4^+ Ion. In an attempt to generate the elusive Cl_2^+ cation, Seidel and Seppelt[814] have oxidized chlorine with IrF_6. Instead, however, they isolated the blue salt $Cl_4^+IrF_6^-$. The compound contains a rectangular Cl_4^+ ion with one short (1.941 Å) and one long Cl−Cl bond (2.936 Å) and without any significant contact with fluorine atoms.

4.4.2. Interhalogen Cations

Interhalogen cations form a class of polycations containing at least two different halogen atoms.[771] Cations containing one or more halogens and another element such as oxygen, nitrogen, or xenon will not be considered here. The class of interhalogen cations includes triatomic, pentaatomic, and heptaatomic systems. Many of these interhalogen cation salts, which are strong oxidants, have been found useful for collecting radioactive noble gases such as ^{222}Rn and ^{133}Xe.[815,816]

4.4.2.1. Triatomic Interhalogen Cations. Of all the possible 16 triatomic interhalogen cations ClF_2^+, BrF_2^+, IF_2^+, Cl_2F^+, Br_2F^+, I_2F^+, $ClBrF^+$, $ClIF^+$, $BrIF^+$,

$BrCl_2^+$, ICl_2^+, Br_2Cl^+, I_2Cl^+, $BrICl^+$, IBr_2^+, and I_2Br^+, only five are known. They are ClF_2^+, BrF_2^+, IF_2^+, Cl_2F^+, and ICl_2^+. It seems reasonable to predict that the least electronegative halogen occupies the central position, where it carries a formal positive charge.

ClF_2^+ Ion. Adducts of ClF_3 with Lewis acids such as AsF_5, SbF_5, and BF_3 have been known for some time,[817] and it has been established by IR, Raman, and ^{19}F NMR spectroscopic studies that these compounds are best formulated as salts of ClF_2^+ cation **332-Cl** (e.g., $ClF_2^+AsF_6^-$).[818,819] The spectroscopic data indicate a bent structure. Additional support for the bent structure of **332-Cl** comes from X-ray crystallographic studies[820] on $ClF_2^+SbF_6^-$ salt. The ClF_2^+ ion **332-Cl** has a bond angle of 95.9° and a bond length of 1.58 Å. There is a good evidence for the fluorine bridging between the anion and the cation and the two fluorine bridges formed by each ClF_2^+ give rise to a very approximately square coordination of fluorine around chlorine, which is the geometry predicted by the valence shell electron pair-repulsion theory for AX_4E_2 coordination (where X is a ligand and E a lone pair). It is interesting to note that the SbF_6^- ion in this structure forms *trans* bridges rather than the *cis* bridges that have been observed in other related systems.

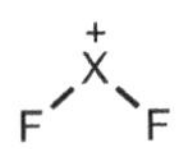

332 X = Cl **332-Cl**
X = Br **332-Br**

BrF_2^+ Ion. The 1:1 adduct of BrF_3 with SbF_5 has been shown by X-ray crystallography[821] to contain BrF_2^+ **332-Br** and SbF_6^- ions held together by fluorine bridging in such a way that bromine acquires a very approximately square-planar configuration. Each bromine atom has two fluorine atoms at 1.69 Å, making an angle of 93.5° at bromine, and two other neighboring fluorine atoms at 2.29 Å which form part of the distorted octahedral coordination of the antimony atoms. The two fluorine bridges formed by SbF_6^- are *cis* rather than *trans*, as in the unusual structure $ClF_2^+SbF_6^-$.

The IR and Raman spectra of BrF_3–SbF_5, BrF_3–AsF_3, and $(BrF_3)_2GeF_4$ have been reported.[822,823] The electrical conductivity of liquid bromine trifluoride[824] (specific conductance = 8 10^{-3} ohm^{-1} cm^{-1}) may be attributed to self-ionization [Eq. (4.205)].

$$2\,BrF_3 \rightleftharpoons \underset{\textbf{332-Br}}{BrF_2^+}\,BrF_4^- \qquad (4.205)$$

IF_2^+ Ion. The salts of IF_2^+ with AsF_6^- and SbF_6^- anions have been prepared from IF_3 and AsF_5, and from IF_3 and SbF_5 in AsF_5 as solvent at −78°C.[825] The compound $IF_2^+SbF_6^-$ is stable up to 45°C and the solid gives two broad overlapping ^{19}F NMR

signals whose relative intensities were estimated to be 1:2.6 and which were assumed to arise, therefore, from fluorine on iodine and fluorine on antimony, respectively. $IF_2^+AsF_6^-$ was found to be stable only up to –20°C.

Cl_2F^+ Ion. Raman spectra of the adducts AsF_5–2ClF and BF_3–2ClF have established that these compounds contain the unsymmetrical $ClClF^+$ cation[805] and not the symmetrical $ClFCl^+$ cation previously reported on the basis of the IR spectrum alone.[826] The observed vibrational frequencies indicate that there is a strong fluorine bridging between the cation and the anion in $Cl_2F^+AsF_6^-$ **333** salt. Theoretical calculations confirm that the $ClClF^+$ cation is more stable by 43.0 kcal mol^{-1} [MP4(SDTQ)/6-311G(2df)//M3/6-311G(2df) + ZPE level][827] and 44.3 kcal mol^{-1} [CCSD(T)/cc-pVQZ level].[813] Cl—F bond strength in the $ClClF^+$ cation is twice as strong as in cation $ClFCl^+$ (41.6 kcal mol^{-1} versus 21.2 kcal mol^{-1}).[827]

The Cl_2F^+ ion **333** appears to be unstable in solution and was found to be completely disproportionated in HF—SbF_5 even at –78°C[805] [Eq. (4.206)]. As mentioned, Cl_3^+ in the media disproportionates further at room temperature to give chlorine and **332-Cl**.

$$2\,Cl_2F^+ \rightleftharpoons ClF_2^+ + Cl_3^+ \qquad (4.206)$$

333 **332-Cl**

ICl_2^+ Ion. X-ray crystallographic investigations[828] of the adducts of ICl_3 with SbF_5 and $AlCl_3$ have shown that these may be regarded as ionic compounds, that is, $ICl_2^+SbCl_6^-$ and $ICl_2^+AlCl_4^-$. However, there is considerable interaction between the two ions via two bridging chlorines, which give an approximately square-planar arrangement of four chlorines around the iodine atom, similar to the arrangement of fluorines around bromine and chlorine in $BrF_2^+SbF_6^-$ and $ClF_2^+SbF_6^-$, respectively. The bond angle and bond length for ICl_2^+ **334** were found to be 92.5° and 2.31 Å in $ICl_2^+SbCl_6^-$ and 96.7° and 2.28 Å in $ICl_2^+AlCl_4^-$.

The electrical conductivity of ICl_3 (specific conductance = 9.85 10^{-2} ohm^{-1} cm^{-1}) can be attributed to the self-ionization.[828]

I_2Cl^+ Ion. There is no certain evidence for the I_2Cl^+ cation **335**, but presumably the electrical conductivity of liquid ICl (specific conductance = 4.60 10^{-3} ohm^{-1} cm^{-1} at 35°C) which has previously been ascribed to the self-ionization[829] according to Eq. (4.207) is in fact due to a self-ionization that produces I_2Cl^+ ion **335** [Eq. (4.208)].

$$2ICl \rightleftharpoons I^+ + ICl_2^- \qquad (4.207)$$

$$3ICl \rightleftharpoons I_2Cl^+ + ICl_2^- \qquad (4.208)$$

335

The I_2Cl^+ cation **335**, however, is possibly extensively disproportionated to give the known I_3^+ cation **324** and ICl_2^+ cation **334** [Eq. (4.209)].

$$\underset{\mathbf{335}}{2I_2Cl^+} \rightleftharpoons \underset{\mathbf{324}}{I_3^+} + \underset{\mathbf{334}}{ICl_2^+} \quad (4.209)$$

4.4.2.2. Pentaatomic Interhalogen Cations. Chlorine pentafluoride forms 1:1 adducts with AsF_5 and SbF_5. The interpretation of Raman spectra of these adducts indicates the formation of ClF_4^+ cation.[830] The ^{19}F NMR spectrum of ClF_4^+ (a doublet with equal intensity) suggests two pairs of nonequivalent fluorine ligands and a structure of C_{2v} symmetry.[818] Bromine pentafluoride forms the adducts BrF_5–$2SbF_5$ and BrF_5–SO_3.[831] These may, presumably be formulated as BrF_4^+ ion salts, although the latter compound might be the covalent BrF_4SO_3F. Iodine pentafluoride also forms adducts with SbF_5[832] and PtF_5.[833] The electrical conductivity of liquid IF_5 (specific conductance = 2.30 10^{-5} ohm^{-1} cm^{-1}) has been attributed to the self-ionization [Eq. (4.210)]. In the Raman spectrum of IF_4^+, the observed nine lines have been assigned to cation IF_4^+ **336**,[834] which is consistent with its C_{2v} structure found by X-ray crystallography.

$$2IF_5 \rightleftharpoons \underset{\mathbf{336}}{IF_4^+\ IF_6^-} \quad (4.210)$$

The X-ray crystal structure of the $ClF_4^+SbF_6^-$, $BrF_4^+Sb_2F_{11}^-$, and $IF_4^+Sb_2F_{11}^-$ salts have been obtained.[835–837] All three salts contain discrete ions with the cation having pseudotrigonal bipyramidal structure like SF_4 with two fluorines occupying the axial positions and two fluorines and a lone pair occupying the equatorial positions (**337**). The axial X–F bonds are longer and of more ionic character (mean bond distances for ClF_4^+, BrF_4^+, and IF_4^+ are 1.618, 1.81, and 1.84 Å, respectively), whereas the equatorial bonds are shorter and more covalent (1.53, 1.77, and 1.77 Å). Each X atom forms two fluorine bridges with two different anions, resulting in pseudooctahedral coordination of the central X atom and the formation of infinite zigzag chains. The minimum-energy structure of the cation ClF_4^+ was calculated to be of C_{2v} symmetry at all levels of theory.[837–840] Because of discrepancies between the observed and calculated values, Christe and co-workers[841] have redetermined the structure of $BrF_4^+Sb_2F_{11}^-$ and $IF_4^+Y^-$ ($Y = SbF_6$, Sb_2F_{11}).

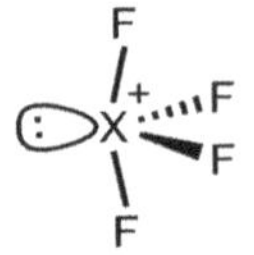

X = Cl, Br, I

337

4.4.2.3. Heptaatomic Interhalogen Cations

ClF_6^+ Ion. The PtF_6^- salt of cation ClF_6^+ (**338**) has been prepared by the reaction of PtF_6 with chlorine fluorides[842] [Eq. (4.211)] or oxyfluorides [Eq. (4.212)].[843]

$$2ClF_5 + 2PtF_6 \xrightarrow[\text{RT, 8 days}]{\text{sapphire reactor}} \underset{\mathbf{338}}{ClF_6^+ PtF_6^-} + ClF_4^+ PtF_6^- \quad (4.211)$$

$$6FClO_2 + 6PtF_6 \xrightarrow[-78°C]{} \underset{\mathbf{338}}{ClF_6^+ PtF_6^-} + 5ClO_2^+ PtF_6^- + CO_2 \quad (4.212)$$

Christe and co-workers[844] have recently developed a new synthesis applying the highly reactive and thermally unstable NiF_3^+ cation [Eq. (4.213)].

$$\underset{X = Cl, Br}{Cs_2NiF_6 + 5AsF_5 + XF_5} \xrightarrow[-60 \text{ to } 10°C]{HF} XF_6^+ AsF_6^- + Ni(AsF_6)_2 + 2CsAsF_6 \quad (4.213)$$

The structure of ClF_6^+ cation **338** has been established beyond any reasonable doubt by ^{19}F NMR spectroscopy.[818,843] The ClF_6^+ cation **338**, except for the $ClO_2F_2^+$ cation,[845] is the only known heptacoordinate chlorine cation. Besides the well-known NF_4^+ cation,[846,847] it is the only known example of a fluorocation derived from the hitherto unknown compounds (i.e., NF_5 and ClF_7). Complete vibrational analysis of ClF_6^+ cation **338** has been reported,[848] and it indicates the octahedral symmetry of the ion.

The $ClF_6^+PtF_6^-$ salt is canary yellow in color and is quite stable at 25°C when stored in Teflon-FEP containers. The ClF_6^+ salts are very powerful oxidizers and react explosively with organic compounds and water.

BrF_6^+ Ion. In 1974 Gillespie and Schrobilgen reported the direct oxidation of bromine pentafluoride to BrF_6^+ cation **339** by $Kr_2F_3^+$ cation[849,850] [Eq. (4.214)].

$$BrF_5 + \underset{M = As, Sb}{Kr_2F_3^+ MF_6^-} \xrightarrow[0°C]{} \underset{\mathbf{339}}{BrF_6^+ MF_6^-} + KrF_2 + Kr \quad (4.214)$$

The ^{19}F NMR and Raman spectroscopic studies[850,851] on BrF_6^+ **339** indicate the octahedral symmetry of the species. The ion **339** is a powerful oxidizing agent and rapidly oxidizes oxygen and xenon to O_2^+ and XeF^+ cations, respectively, under ambient conditions.

IF_6^+ Ion. Iodine heptafluoride has been shown to form the adduct IF_7AsF_5 and IF_7–3 SbF_5.[852] The latter complex was postulated to have the ionic structure $IF_4^{3+}(SbF_6^-)_3$. These findings were questioned by Christe and Sawodny,[853] who indicated that the adduct may contain IF_6^+ cation. Indeed they showed that the IF_7AsF_5 adduct is

actually the $IF_6^+AsF_6^-$ salt. Hohorst, Stein, and Gebert have been successful in preparing the $IF_6^+SbF_6^-$ salt.[854]

The Raman and IR spectral analyses of the salts indicate the octahedral nature of the IF_6^+ cation.[854,855] The $IF_6^+SbF_6^-$ salt rapidly reacts with radon gas at room temperature forming a nonvolatile radon compound. The salt is claimed to have potential application in purifying radon-contaminated air and in the analysis of radon in air.[854]

Schrobilgen, Christe, and co-workers have recently carried out the first detailed structural characterization of the XF_6^+ cations (X = Cl, Br, I) as their $Sb_2F_{11}^-$ salts.[856] The XF_6^+ cations have octahedral geometries with average bond lengths of 1.550 Å (Cl–F), 1.666 Å (Br–F), and 1.779 Å (I–F) measured at −130°C and −173°C. The cations have 13–16 interionic F---F contacts to the neighboring anions and six F---X contacts between the fluorine atoms of the anions and the central halogen atoms. Additional studies (NMR characterization of the central X atoms, Raman spectroscopy, and calculations) were also performed.

4.4.3. Polyatomic Cations of Group 16 Elements

4.4.3.1. The O_2^+ Cation. The existence of O_2^+ cation **340** in the gas phase at low pressures has been well established.[772,857] However, it was not until 1962 that a compound containing O_2^+ was identified.[858] It was discovered as a reaction product during the fluorination of platinum in a silica apparatus. The product was first thought to be $PtOF_4$,[859] but later it was shown to be $O_2^+PtF_6^-$.[858] It was subsequently prepared by direct oxidation of molecular oxygen using platinum hexafluoride at room temperature. It now appears that the dioxygenyl salt $O_2^+BF_4^-$ may have been prepared prior to 1962.[860] There are at least nine O_2^+ salts known with a variety of anions. The anions are PtF_6^-, AsF_6^-, SbF_6^-, $Sb_2F_{11}^-$, PF_6^-, BF_4^-, VF_6^-, BiF_6^-, and SnF_6^-. The most convenient route to O_2^+ salts appears to be the photochemical synthesis of $O_2^+Sb_2F_{11}^-$ from oxygen, fluorine, and antimony pentafluoride.[861] Most O_2^+ preparations involve the reaction of fluoride ion acceptors with O_2F_2 or O_4F_2 at low temperatures or with O_2 and F_2 mixtures under conditions favoring synthesis of the long-lived radical O_2F [Eq. (4.215)].

$$O_2 + F_2 \xrightarrow[-F]{h\nu} O_2F \xrightarrow{SbF_5} \underset{\mathbf{340}}{O_2^+\,Sb_2F_{11}^-} \qquad (4.215)$$

Compounds containing O_2^+ cation **340** are colorless with the exception of $O_2^+PtF_6^-$, which is red due to the PtF_6^- ion. The compound $O_2^+PF_6^-$ decomposes slowly at −80°C[862] and rapidly at room temperature, giving oxygen, fluorine, and phosphorous pentafluoride. ^{18}F tracer studies on $O_2^+BF_4^-$ have led to the conclusion that the mechanism of the decomposition involves the equilibrium [Eq. (4.216)] followed by a bimolecular decomposition of O_2F.[863]

$$O_2^+\,BF_4^- \rightleftharpoons O_2F(g) + BF_3(g) \qquad (4.216)$$

Dioxygenyl hexafluoroantimonate has been studied by differential thermal analysis.[864] Decomposition of $O_2^+SbF_6^-$ proceeds in two stages, according to the mechanism given in Eqs. (4.217) and (4.218).

$$2O_2^+\,SbF_6^- \xrightarrow{\sim 245^\circ C} O_2 + 0.5F_2 + O_2^+\,Sb_2F_{11}^- \quad (4.217)$$

$$O_2^+\,Sb_2F_{11}^- \xrightarrow{\sim 280^\circ C} O_2 + 0.5F_2 + 2\,SbF_5 \quad (4.218)$$

The $O_2^+Sb_2F_{11}^-$ was converted to $O_2^+SbF_6^-$ by heating at 130°C in vacuo; conversely, $O_2^+Sb_2F_{11}^-$ was prepared by reaction of $O_2^+SbF_6^-$ and SbF_5 at 180–200°C. Dioxygenyl hexafluoroarsenate is markedly less stable than the hexafluoroantimonate salts; it decomposes rapidly at 130–180°C. $O_2^+PtF_6^-$ can be sublimed above 90°C in vacuo and melts with some decomposition at 219°C in a sealed tube.[863]

X-ray powder data obtained from the cubic form of O_2PtF_6 were consistent with the presence of O_2^+ and PtF_6^- ions.[863] The structure was refined using neutron diffraction powder data. The O_2^+ cation **340** resembles nitrosonium ion.[864] The IR, Raman, and ESR spectra of O_2^+ salts have been studied in detail.[865–868]

Recently, the X-ray crystal structures of $O_2^+MF_6^-$ salts (M = Sb, Ru, Pt, Au) have been determined.[869] For the O_2^+ cation in the $O_2^+RuF_6^-$ salt, an interatomic O—O bond distance of 1.125 Å was found at −127°C[870] in agreement with the gas-phase value of 1.1227 Å.[857] Values obtained for the O—O bond length in the $O_2^+AuF_6^-$ salt are 1.068 Å (–122°C) and 1.079 Å.[871,872]

The $O_2^+AsF_6^-$ salt readily oxidizes molecular bromine to Br_3^+ ion **328**. The $O_2^+BF_4^-$ salt also reacts with xenon.[873] The major application of O_2^+ salts is as an oxidant for collecting ^{222}Rn in uranium mines, since it has negligible dissociation pressure at ambient temperature and releases oxygen as a gaseous product. Reactions of the salt with radon and components of diesel exhausts (CO, CO_2, CH_4, SO_2, NO, and NO_2) have therefore been studied in some detail.[815]

4.4.3.2. Polysulfur Cations. The nature of colored solutions obtained by dissolving elemental sulfur in oleum[874] remained a mystery for a long time, since their discovery by Bucholz[875] in 1804.[776,876] Red, yellow, and blue solutions have been prepared; however, it appears that particular attention has been given to the blue solutions. The species responsible for blue color has been identified by various workers as S_2O_3, S_2, radical ion $(X_2S–SX_2)^+$, and the species designated as S_n.[877] The various colors have now been shown to be due to cations S_{16}^{2+}, S_8^{2+}, and S_4^{2+}.[878–880] Subsequently, it was shown by X-ray crystallography[881] that the originally assigned S_{16}^{2+} cation is indeed S_{19}^{2+}.

Sulfur (and selenium) can be oxidized by a variety of oxidizing agents including AsF_5, SbF_5, and $S_2O_6F_2$ and by protic acids such as $H_2S_2O_7$ and HSO_3F. It has been shown[882] that a trace amount of halogen facilitates the formation of S_4^{2+} and Se_4^{2+} dications. Along with previously mentioned doubly charged species, several singly charged radical cations have been claimed, of which only $S_5^{+\cdot}$ has been positively identified.[883] So far these singly charged species have been observed only in solution. They do not appear to be as stable as the doubly charged species and hence are not well-characterized. In contrast, all the dipositive ions have been obtained in the form of relatively stable salts such as $S_4^{2+}(FSO_3^-)_2$ and $S_8^{2+}(AsF_6^-)_2$.[878] The vibrational spectra of the $S_4^{2+}(MF_6^-)_2$ (M = As, Sb) salts solvated by SO_2 have been reported.[884]

In the case of $S_8^{2+}(AsF_6^-)_2$ a single-crystal X-ray crystallographic analysis showed unequivocally that this cation has an *exo–endo* cyclic structure with a long transannular bond (**341**).[880] Recently, the crystal structure of $S_8^{2+}(AsF_6^-)_2$ have been redetermined.[885] The S_8^{2+} dication consists of a folded eight-membered ring of approximate C_s symmetry. These findings are similar to those reported in the original study[880] and reported for the $S_8^{2+}(SbF_6^-)(Sb_3F_{14}^-)$ salt.[886] The average S–S single bond length in the cation is 2.051 Å and all sulfur atoms exhibit at least one S–F contact. The unusually long transannular bonds of 2.8–2.9 Å [S(2)–S(8), S(3)–S(7), S(4)–S(6)] were reproduced computationally [B3PW91 level of theory with 6-311G* and 6-311G(2df) basis sets].

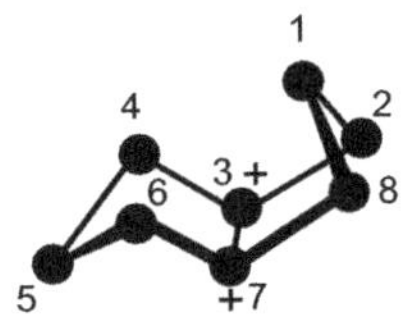

341

The question of the blue color has recently been raised again by Passmore and Krossing.[887] In fact, the blue color was assigned by Gillespie and Passmore to S_8^{2+},[888] but $S_8^{2+}(AsF_6^-)_2$ has been shown to be red[885]; that is, the blue color was not due to S_8^{2+}. However, the $S_8^{2+}(AsF_6^-)_2$ salt, when dissolved in HSO_3F or SO_2, gives a blue solution, which contains the S_5^+ cation and an unknown component. A comparison of calculated and experimental UV–vis spectra indicates that the unknown species is the S_6^{2+} dication, which is a 10π-electron Hückel system, and the blue color is likely due to the HOMO–LUMO π*–π* transition.[887] The transition into the first excited state was calculated to be at 564 nm (CIS) and 555 nm (TD-DFT). The experimentally found maximum in the UV–vis spectra is at 585 nm.

By comparison with analogous selenium and tellurium ions,[889,890] the S_4^{2+} ion is proposed to have a square-planar structure. Additional support comes from infrared spectra,[891] Raman and electronic spectra,[891,892] magnetic circular dichroism,[893] and molecular orbital calculations.[894] Indeed, the X-ray analysis of $S_4^{2+}(AsF_6^-)_2$, which crystallizes with at least 0.62 SO_2, shows that the cation is square-planar (S–S bond distance = 2.014 Å, bond angle = 90°).[895] The corresponding values for the $S_4^{2+}(Sb_2F_4^{2+})(Sb_2F_5^+)(SbF_6^-)_5$ salt are 1.988 Å and 90°.[886] According to recent observations,[896] the crystal structure of the salt $S_4^{2+}(AsF_6^-)_2 \cdot AsF_3$ consists of discrete square-planar S_4^{2+} cations, octahedral AsF_6^-, and AsF_3 molecules, weakly linked by cation–fluorine interactions. The average S–S bond distance (1.964 Å) and angle (90°) are not significantly different from those observed for other salts.

Calculations showed that the S_4^{2+} cation is lattice-stabilized in the solid state, as compared to the S_2^+ cation, and presently available anions are unlikely to stabilize S_2^+ (and Se_2^+ and Te_2^+) in the solid phase.[896,897]

Early evidence for the dipositive ion S_{16}^{2+} was not so conclusive. The existence of cation S_{16}^{2+} was first reported as a result of isolation of a red solid from the reaction of sulfur and AsF_5 in anhydrous HF solution.[888] From the analytical data, the solid appeared to have a composition corresponding to $S_{16}(AsF_6)_2$. In a study of the progressive oxidation of sulfur by $S_2O_6F_2$ in HSO_3F, it was observed that some unreacted sulfur always remained until sufficient oxidant had been added to give a ratio of sulfur to $S_2O_6F_2$ of 16:1. The cryoscopic measurements were also consistent with the formulation of this ion as S_{16}^{2+}.[878] However, attempts to obtain a single crystal of the ion from various acid solutions were unsuccessful. Subsequently, Gillespie and co-workers[881] obtained both needle- and plate-like crystals originally believed to be $S_{16}^{2+}(AsF_6^-)_2$ in 2:1 mixture of SO_2 and SO_2ClF at −25°C. An X-ray crystallographic study of these crystals led to the surprising and the unexpected result that the compound had the composition $S_{19}^{2+}(AsF_6^-)_2$,—that is, that it is a salt of S_{19}^+ rather than S_{16}^+.

The structure of $S_{19}^{2+}(AsF_6^-)_2$ contains discrete AsF_6^- anions and S_{19}^+ cations, the latter consisting of two seven-membered rings joined by a five-atom chain (**342**). In contrast to $S_{19}^{2+}(AsF_6^-)_2$, the two rings in $S_{19}^{2+}(SbF_6^-)_2$ have similar chair conformations.[886] In both systems, there is a very noticeable short–long bond alternation. Furthermore, two of the S−S bonds are significantly elongated [S(1)–S(7) = 2.203 Å, S(13)–S(14) = 2.202 Å] since bonding and lone pairs of electrons on these atoms are virtually eclipsed.

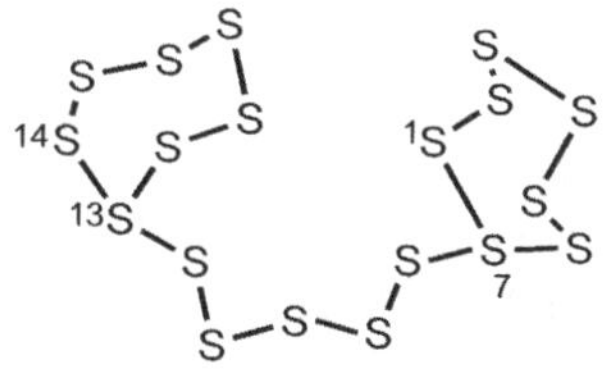

342

Further information of the chemistry of polysulfur, polyselenium, and polytellurium cations can be found in reviews.[773,776,876,898,899]

4.4.3.3. *Polyselenium Cations.* The colored solutions produced by dissolving elemental selenium in sulfuric acid were first observed by Magnus in 1827.[900] Since then a number of workers have investigated the nature of selenium solutions in sulfuric acid, oleum, and sulfur trioxide, providing a substantial amount of data[901] but with little understanding of the system. Subsequently, it was shown that these solutions contain the yellow Se_4^{2+} and green Se_8^{2+} polyatomic cations.[902,903]

Polyselenium cations are less electrophilic than their sulfur analogs and give stable solutions in various strong acids.[902] In fluorosulfuric acid, selenium can be oxidized quantitatively by $S_2O_6F_2$ to give yellow Se_4^{2+} **343** [Eq. (4.219)].

$$4Se + S_2O_6F_2 \xrightarrow{HSO_3F} \underset{\mathbf{343}}{Se_4^{2+}} + 2FSO_3^- \tag{4.219}$$

The addition of selenium to the yellow solution up to a 8:1 ratio of Se:$S_2O_6F_2$ did not appreciably affect the conductivity. This indicated that the FSO_3^- ion concentration remained unchanged and that the Se_4^{2+} ion is reduced by selenium to Se_8^{2+} **344** [Eq. (4.220)].

$$Se_4^{2+} + 4SeS \longrightarrow \underset{\mathbf{344}}{e_8^{2+}} \tag{4.220}$$

Conductivity measurements of selenium in pure fluorosulfuric acid were also consistent with the formation of Se_8^{2+} **344**.

Solutions of Se_8^+ in 100% H_2SO_4 may be prepared by heating selenium in the acid at 50–60°C; the element is oxidized by sulfuric acid [Eq. (4.221)]. The Se_4^{2+} **343** can also be oxidized to Se_8^{2+} **344** by selenium dioxide.

$$8Se + 5H_2SO_4 \longrightarrow \underset{\mathbf{344}}{Se_8^{2+}} + 2H_3O^+ + 4HSO_4^- + SO_2 \tag{4.221}$$

Various compounds containing cations Se_4^{2+} and Se_8^{2+} have been prepared by oxidizing selenium with $SeCl_4$ in the presence of acids ($AlCl_3$, SO_3, oleum, SbF_5, or AsF_5),[772] or reacting Se with AsF_5 or SbF_5.[904,905] All polyatomic selenium cations are diamagnetic, and so far no evidence has been reported for radicals analogous to $S_5^{+\cdot}$. The crystal structure of $Se_4(HS_2O_7)_2$ has shown[889,906] Se_4^{2+} **343** to be square planar with an Se–Se bond distance of 2.283 Å. The average bond length in $Se_4^{2+}(Sb_2F_4^{2+})(Sb_2F_5^+)(SbF_6^-)_5$ is slightly smaller (2.260 Å),[907] whereas in $Se_4^{2+}(AlCl_4^-)_2$ it is practically identical (2.285 Å). These values are significantly less than that of 2.34 Å found in the Se_8 molecule,[908] indicating some degree of multiple bonding. Such a result is consistent with a valence bond description of the molecule involving four-membered ring structures of the type **343a**. Alternatively, the structure can be understood in terms of molecular orbital theory. The circle in structure **343b** denotes a closed-shell (aromatic) 6π-electron system. Of the four π molecular orbitals, the two almost nonbonding (e_g) orbitals and the lower-energy (b_{2u}) bonding orbital are occupied by the six π electrons, leaving the upper antibonding (a_{1g}) orbitals empty. The intense yellow-orange color of Se_4^{2+} has been attributed to the dipole-allowed excitation of an electron from an e_g orbital to the lowest empty π orbital (b_{2u}). Stephens[893] has shown that the magnetic circular dichroism results are consistent with such a model. The structure is also consistent with a vibrational spectroscopic study and molecular orbital calculations.[891,909]

343

The structure of Se_8^{2+} **344** in $Se_8(AlCl_4)_2$[910] is similar to that of S_8^{2+} (see also structure **341**) except that the cross-ring distance Se(3)–Se(7) is relatively shorter (2.84 Å) than that found in the sulfur cation (2.859 Å). However, the distance in cation S_8^{2+} in the mixed $(Te_6^{4+})(Se_8^{2+})(AsF_6^-)_6(SO_2)$ system is 2.905 Å.[911] The cross-ring distances Se(4)–Se(6) and Se(2)–Se(8), are relatively long (3.29 and 3.35 Å) compared to those in S_8^{2+}. All bond angles in Se_8^{2+} are more acute than those in S_8^{2+}. The cation Se_8^{2+} is, therefore, reasonably well-described by valence bond structure **344′**. In contrast to the S_8^{2+} cation, calculations could not reproduce the transannular bonds of the Se_8^{2+} cation.[885]

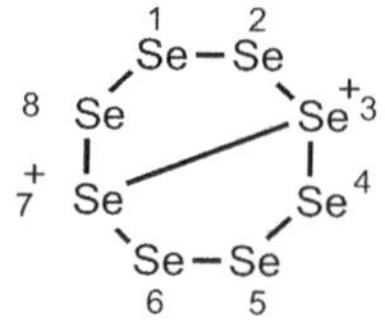

344'

The Se_{10}^{2+} cation **345** has also been prepared and characterized.[912,913] The salts include $Se_{10}^{2+}(AsF_6^-)_2$, $Se_{10}^{2+}(AlCl_4^-)_2$, $Se_{10}^{2+}(SbF_6^-)_2$, and $Se_{10}^{2+}(FSO_3^-)_2$. The X-ray crystal structure of $Se_{10}^{2+}(SbF_6^-)_2$ indicates that the Se_{10}^{2+} cation can be described as a six-membered boat-shaped ring linked across the middle by a chain of four selenium atoms; that is, it has a bicyclo[4.2.2]decane-type structure (**345**). The selenium–selenium bonds in the cation vary greatly in length from 2.24 to 2.44 Å. According to a ^{77}Se NMR study, the Se_{10}^{2+} cation undergoes structural isomerization in SO_2 at ambient temperature.[914] One of the isomers disproportionates below 273 K to give Se_8^{2+} and a higher nuclearity species, which was identified as Se_{17}^{2+} or, less likely, Se_{18}^{2+}. The $Se_{17}^{2+}(NbCl_6^-)_2$ and $Se_{17}^{2+}(TaBr_6^-)_2$ salts have recently been isolated.[915] The cation Se_{17}^{2+} consists of two seven-membered rings connected by a Se_3 chain; that is, it contains two three-coordinate Se atoms, which formally carry the positive charge.

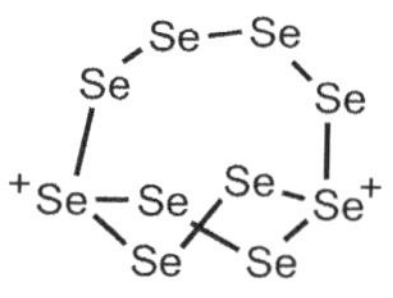

345

4.4.3.4. Polytellurium Cations. The red color produced when tellurium dissolves in concentrated sulfuric acid was first observed as long ago as 1798,[916] but the origin of this color for long remained a mystery. Subsequently, Bjerrum and Smith[917] and Bjerrum[918] have studied the reaction of tellurium chloride with tellurium in molten $AlCl_3$–NaCl. They obtained a purple melt that they concluded contained the species $Te_{2n}{}^{n+}$, probably $Te_4{}^{2+}$ **346**, formed by the reaction shown in Eq. (4.222).

$$7Te \quad + \quad Te^{4+} \longrightarrow \underset{\mathbf{346}}{2Te_4{}^{2+}} \tag{4.222}$$

About the same time, solutions of tellurium in various acids were investigated in detail.[919,920] It was found that the red solutions are produced when tellurium is dissolved in sulfuric acid, fluorosulfuric acid, or oleum with the simultaneous production of SO_2, indicating that tellurium is oxidized. The electronic spectra of the solutions were found to be identical with those obtained by Bjerrum and Smith from their melts. Conductometric and cryoscopic measurements of the acid solutions led to the conclusion that they contain a species $Te_{2n}{}^{n+}$, which was certainly not Te^{2+} but probably $Te_4{}^{2+}$ **346**.

Reaction of tellurium with disulfuric acid, $S_2O_6F_2$, SbF_5, and AsF_5 in SO_2 gave[920–922] the compounds $Te_4{}^{2+}S_3O_{10}{}^{2-}$, $Te_4{}^{2+}(FSO_3{}^-)_2$, $Te_4{}^{2+}(Sb_2F_{11}{}^-)_2$, and $Te_4{}^{2+}(AsF_6{}^-)_2$; and, from Te–($TeCl_4$–$AlCl_3$) melts,[923] compounds $Te_4{}^{2+}(AlCl_4{}^-)_2$, and $Te_4{}^{2+}(Al_2Cl_7{}^-)_2$ were obtained. Gillespie and co-workers[907] also prepared $Te_4{}^{2+}(SbF_6{}^-)_2$. The formulation of the red species as $Te_4{}^{2+}$ **346** was confirmed by the determination of the crystal structures of the chloroaluminate salts.[890] In both cases, the $Te_4{}^{2+}$ **346** ion lies on a center of symmetry and is almost exactly square planar. The average tellurium–tellurium distances of 2.661–2.673 Å are significantly shorter than the tellurium–tellurium distance of 2.864 Å within the spiral chain in elemental tellurium.[924] This is consistent with a structure exactly analogous to that for $Se_4{}^{2+}$ in which each bond has 25% double-bond character. The Raman spectra of $Te_4{}^{2+}$ in solution and the solid state are analogous to those of $Se_4{}^{2+}$ and $S_4{}^{2+}$ but shifted to lower frequency.[891] The magnetic circular dichroism[893] and visible and UV spectrum of solutions of $Te_4{}^{2+}$ **346** are also similar to those of $Se_4{}^{2+}$ **343**, as expected on the basis of their structural similarity and a molecular orbital study.[909] Thermochemical measurements have shown that $Te_4{}^{2+}$ in $Te_4{}^{2+}(AsF_6{}^-)_2$, but not $Se_4{}^{2+}$ in $Se_4{}^{2+}(AsF_6{}^-)_2$, is lattice-stabilized in the solid state.[925] The $Te_4{}^{2+}$ cation in $Te_4{}^{2+}(SbF_6{}^-)_2$ has bond lengths slightly longer (2.669 and 2.676 Å) than those in the chloroaluminate salts. The cation is significantly distorted and has D_{2h} symmetry compared with the expected D_{4h} symmetry.

If the acid solutions of tellurium described earlier are warmed or if the oleum is sufficiently strong, then the color of the solutions changes from red to orange-yellow.[919] The same change may also be produced by addition of an oxidizing agent such as peroxydisulfate to the sulfuric acid solutions or $S_2O_6F_2$ to the HSO_3F solutions. Absorption spectra and conductometric, cryoscopic, and magnetic measurements on the solutions in HSO_3F suggested that the yellow species was tellurium in +1 oxidation state and it was formulated as "Te_n^{n+}" where n is even as the cation was found to be diamagnetic. Furthermore, these studies also established that "Te_n^{n+}" could not be Te_2^{2+} and is probably Te_4^{4+}, although higher-molecular-weight species such as Te_6^{6+} and Te_8^{8+} could not be ruled out.[920,921]

On the contrary, Bjerrum[926] concluded, from spectrophotometric measurements on solutions formed by the reduction of $TeCl_4$ with tellurium metal in $KAlCl_4$ melts buffered with KCl–$ZnCl_2$, that the species was Te_2^{2+}. Subsequently, Gillespie and co-workers[927,928] were able to show that the Te_n^{n+} species is indeed a cluster cation Te_6^{4+}. The single-crystal X-ray crystallographic study on $Te_6(AsF_6)_4$–$2AsF_3$ and $Te_6(AsF_6)_4$–2 SO_2 indicate that the Te_6^{4+} cation has a trigonal prismatic arrangement **347**. In the $(Te_6^{4+})(Se_8^{2+})(AsF_6^-)_6(SO_2)$ system containing two different homopolyatomic cations the Te_6^{4+} cation has the same prismatic structure.[911]

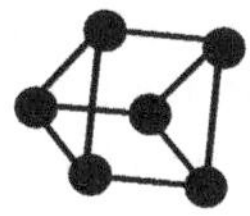

347

Beck[929] prepared a polymeric Te_7WOBr_5 material, and subsequently the $Te_7^{2+}(AsF_6^-)_2$ polymeric product was isolated by Kolis and coworkers.[930] The infinite chains of the Te_7^{2+} polycation in $Te_7^{2+}(AsF_6^-)_2$ consist of six-membered rings of chair conformation connected through Te atoms bridging between the Te(1) and Te(4) atoms (**348**). The shortest bond distance (2.688 Å) is between Te(2) and Te (3), whereas the longest one is the Te(4′)–Te(1) distance (2.859 Å). There is also a long contact between Te(4′) and Te(3) (3.221 Å), which is much less than the sum of the van der Waals radii (4.40 Å).

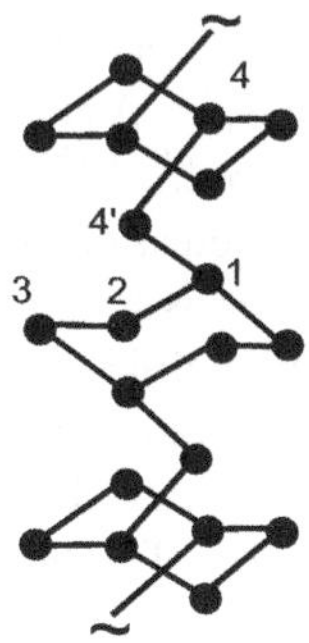

348

4.4.3.5. Polyheteroatom Cations. So far, only studies on homopolyatomic cations of oxygen, sulfur, selenium, and tellurium have been discussed. Most of them were of the type X_n^{2+}. There is a new class of polyatomic cations that comprise two or more heteroatoms.

When powdered tellurium metal (0.002 mol) was allowed to react with $Se_8^{2+}(AsF_6^-)_2$ (0.001 mol) in SO_2 at $-78°C$ and the mixture was allowed to warm up to room temperature under stirring,[904,905] the dark green color of Se_8^{2+} **344** slowly diminished (over a 12-h period) resulting in the formation of $Te_2Se_8(AsF_6)_2SO_2$.[931] The X-ray crystal structure of this product indicates that the adduct has $Te_2Se_8^{2+}$ cation with a bicyclo[4.2.2]decane-type structure similar to that of the Se_{10}^{2+} cation (**345**). By changing the ratio of tellurium metal to Se_8^{2+} **344**, Gillespie and co-workers[931] have managed to prepare $Te_{3.7}Se_{6.3}^{2+}(AsF_6^-)_2$. The structure of $Te_{3.7}Se_{6.3}^{2+}$ is similar to that of $Te_2Se_8^{2+}$.

Preparation of adducts $Te_2Se_6^{2+}(MF_6^-)_2$ (M = As, Sb), $Te_{4.5}Se_{5.5}^{2+}(AsF_6^-)_2$, and $(Te_2Se_6)^{2+}(Te_2Se_8)^{2+}(AsF_6^-)_4(SO_2)_2$ was accomplished by reacting powdered Te and Se with AsF_5 or SbF_5 in SO_2.[932] The novel $Te_2Se_6^{2+}$ cation in $(Te_2Se_6)^+(SbF_6^-)_2$ is not isostructural with previously known cations S_8^{2+} and Se_8^{2+}. Instead, it has a bicyclo[2.2.2]octane structure with the two Te atoms occupying bridgehead positions (**349**) resembling a distorted cube with three long (nonbonding) edges. The average Te–Se bond is 2.603 Å, significantly longer than the sum of the covalent radii (2.54 Å) and the average Te–Se distance of 2.57 Å in $Te_2Se_6^{2+}(AsF_6^-)_2$. Furthermore, the Se(1)–Se(2) bond is longer than the other Se–Se bonds. Interestingly, the three transannular Se–Se contacts at the cube edges (3.34–3.42 Å) are shorter than the sum of the van der Waals radii, indicating weak cross-ring bonding.

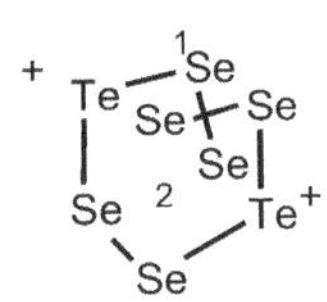

349

Based on the same principle, adducts $Te_3S_3^+(AsF_6^-)_2$, $Te_2Se_4^{2+}(SbF_6^-)_2$, $Te_2Se_4^{2+}(Sb_3F_{14}^-)(SbF_6^-)$, $Te_2Se_4^{2+}(AsF_6^-)_2$, $Te_{2.7}Se_{3.3}^{2+}(SbF_6^-)_2$, and $Te_{3.4}Se_{2.6}^{2+}(SbF_6^-)_2$ have been prepared.[933,934] The $Te_3S_3^{2+}$ and $Te_2Se_4^{2+}$ cations have novel structures that can be described as consisting of a three-membered ring and a five-membered ring fused together or as a boat-shaped six-membered ring with a cross-ring bond corresponding to a bicyclo[3.1.0]hexane-type arrangement.

The *trans*-$Te_2Se_2^{2+}$ dication[891,935] and the $Te_{3.0}Se_{1.0}^{2+}$ dication[935] have been also prepared and studied by Raman and IR spectroscopy. The *trans*-$Te_2Se_2^{2+}$ cation in $Te_2Se_2^{2+}(Sb_3F_{14}^-)(SbF_6^-)$ has Te–Se bond lengths of 2.446 and 2.481 Å and has Se–Te–Se′ and Te–Se–Te′ bond angles of 89.1 and 90.9°, respectively. Consequently, it is significantly distorted because of the different sizes of Te and Se. Furthermore, the Te–Se bond lengths are considerably shorter than bond distances in cations $Te_2Se_4^{2+}$ (2.525–2.539 Å) and $Te_2Se_8^{2+}$ (2.579–2.644 Å). This is consistent with the bonds in *trans*-$Te_2Se_2^{2+}$ having some double bond character. The bonding in the *trans*-$Te_2Se_2^{2+}$ cation is similar to the resonance structures **343** found for the Se_4^{2+} cation. The $Te_{3.0}Se_{1.0}^{2+}$ dication was found to be a disordered mixture of cations Te_3Se^{2+}, Te_4^{2+}, and *trans*-$Te_2Se_2^{2+}$. A disordered mixture of $S_nSe_{4-n}^{2+}$

dications was also shown to give a cation with an average composition of $S_{3.0}Se_{1.0}{}^{2+}$.[936]

Selenium and tellurium multinuclear NMR studies provided useful structural information for Te–Se polyheteroatom cations.[937–939]

4.4.4. Mixed Polyheteroatom Cations of Group 15, 16, and 17 Elements

4.4.4.1. Polyheteroatom Cations of Nitrogen and Sulfur. The simplest stable S–N species is the NS^+ thiazyl cation.[940] A convenient synthesis is the reaction of excess $AgAsF_6$ with $S_3N_3Cl_3$ in SO_2.[941] The bond length of the NS^+ cation shows a pronounced temperature dependence (corrected value = 1.42 Å), and of the two resonance structures $N{\equiv}S^+$ is stabilized by ionic interactions.[942]

Thionitronium ion $NS_2{}^+$,[943–945] the sulfur analog of nitronium ion $NO_2{}^+$ was first obtained by Gillespie and co-workers[946] [Eq. (4.223)] and an improved synthesis was subsequently reported by Passmore and co-workers[947] [Eq. (4.224)]. The X-ray crystal structure analysis of the salt $NS_2{}^+SbCl_6{}^-$ shows a linear centrosymmetric cation with short N–S π-bondings, nevertheless, the positive charge is delocalized to the S atoms.[946,948] The experimental bond length of 1.49 Å is well reproduced by calculations.

$$S_7NH + BCl_3 \xrightarrow[\text{-HCl}]{SO_2,\ -196\text{ to }-48^\circ C} S_7NBCl_2 \xrightarrow[-S_8]{SbCl_5 \atop SO_2,\ -48^\circ C} S_2N^+\,SbCl_6{}^- \qquad (4.223)$$

$$1/8\,S_8 + S_4N_4 + 6\,AsF_5 \xrightarrow[SO_2,\ RT]{Br_2} 4S_2N^+\,AsF_6{}^- + 2AsF_3 \qquad (4.224)$$

Several other cations containing both sulfur and nitrogen have cyclic structure. These include the $S_3N_2{}^+$ cation,[949–951] $S_3N_2{}^{2+}$ dication,[951,952] $S_6N_4{}^{2+}$ dication,[944,950] $S_4N_3{}^+$ cation,[953] $S_4N_4{}^{2+}$ dication,[954] and $S_5N_5{}^+$ cation.[955,956]

The $S_3N_2{}^+$ radical cation has a planar ring and exists only in the $S_3N_2{}^+AsF_6{}^-$ salt,[950] whereas in all other salts it is dimeric. The $S_6N_4{}^{2+}$ dication (**350a**) consists of two centrosymmetric units with delocalized 6π electrons held together through strong S···S interactions (2.971 Å).[944,950] The mixed $Se_4S_2N_4{}^{2+}$ dication (**350b**) analogous to $S_6N_4{}^{2+}$ has also been obtained and characterized.[957] The length of the diselenide bonds in the $AsF_6{}^-$ and $SbF_6{}^-$ salts (2.345–2.369 Å) are close to the single-bond length (2.335 Å), whereas the Se···Se contacts between the two rings are much longer (3.111–3.171 Å).

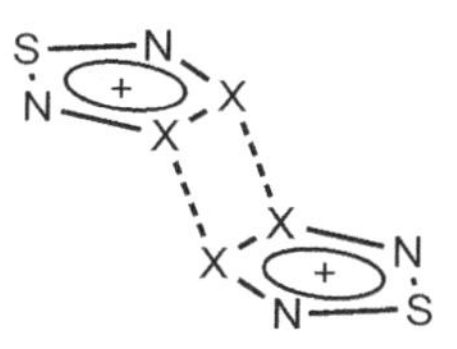

350 **a** X = S **b** X = Se

Passmore and co-workers[952] have prepared the $S_3N_2^{2+}$ dication by the cycloaddition of SN^+ to SNS^+. The ion, predicted to be a stable 6π-electron-rich aromatic, dissociates reversibly in SO_2 solution, but it is stable in the solid state. The S–N and S–S bond distances (1.525–1.584 Å and 2.093 Å, respectively) are close to those found in other S_3N_2 ring systems. The analogous $SeS_2N_2^{2+}$, $Se_2SN_2^{2+}$, and $Se_3N_2^{2+}$ dications have also been prepared and characterized.[958–960]

In the planar ring of the $S_4N_3^+$ cation, the S–S bond distance is 2.066 Å and the S–N bond lengths show small variations (1.542–1.569 Å).[953] The eight-membered ring of the $S_4N_4^{2+}$ dication in the $S_4N_4^{2+}(AlCl_4^-)_2$ and $S_4N_4^{2+}(FSO_3^-)_2$ salts has been shown to be planar with D_{4h} symmetry and equal bond lengths, whereas a nonplanar boat-shaped structure has been found for the dication in the $S_4N_4^{2+}(SbCl_6^-)_2$ salt.[954] For the $S_5N_5^+$ cation, which is a 14π-electron system, heart-shaped[955] and azulene-like[956] structures were found and the latter was predicted by calculations.[961] According to Gillespie et al.[962] the cation in the $S_5N_5^+(SbCl_6^-)$ salt is planar and its geometry is intermediate between the structures suggested earlier (**351**).

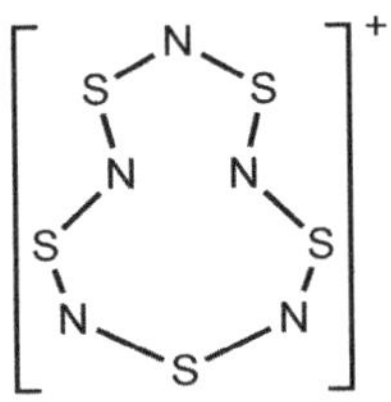

351

The $(S_3N_2)_2N^+AsF_6^-$ salt with two identical $(S_3N_2)_2N^+$ cations and AsF_6^- units in the crystal has been prepared by Passmore and co-workers.[963] In the unique cation **352** the S_3N_2 rings are connected by a bridging nitrogen attached to a sulfur atom of each ring (S–N–S bond angle = 111.6°). The average bridging S–N bond length is 1.624 Å corresponding to a bond order of 1.47. The two rings are eclipsed with an average S–N bond length of 1.586 Å.

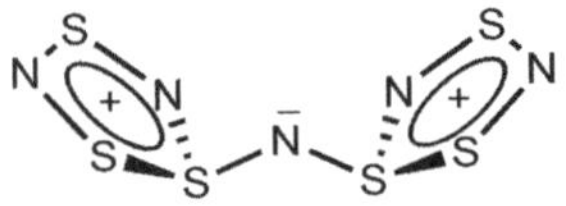

352

4.4.4.2. Polyheteroatom Cations of Halogens with Oxygen or Nitrogen.

Lewis acid adducts of chlorine trifluoride oxide (ClF_3O),[964,965] chlorine trifluoride dioxide (ClF_3O_2),[845,966] and nitrogen trifluoride oxide (NF_3O)[967] are all known. They are best described as ClF_2O^+, $ClF_2O_2^+$, and NF_2O^+, respectively. The spectroscopic properties of these cations have been investigated extensively.[448,818,845,964–966,968,969]

A new stable nitrogen fluoride oxide, N_3NFO^+ has recently been reported by Christe and co-workers[970] [Eq. (4.225)]. Spectroscopic investigation of the fluoroantimonate salt ($^{14/15}N$ and ^{19}F NMR, Raman) showed the presence of two stereoisomers in agreement with theoretical studies [B3LYP/G(2df), MP2, CCSD(T) levels] giving two structures of minimum energy ($\Delta E \leq 0.6\,kcal\,mol^{-1}$). Both structures are planar (C_s symmetry) and differ in the relative position of the azido moiety and fluorine. The nitrogen chemical shifts indicate a covalent azido group attached to a highly electronegative ligand and a nitrogen bonded to a F and an O. The calculated bond distances and bond angles are given for the *E*-[N_3NFO^+] isomer **353**. The ^{19}F NMR chemical shifts and the N–F bond lengths of the isomers (1.377 Å for the *Z*-isomer) differ significantly.

$$NF_2O^+\,SbF_6^- + HN_3 \xrightarrow[-45\text{ to }25^\circ C,\ -HF]{HF} N_3NFO^+\,SbF_6^- \qquad (4.225)$$

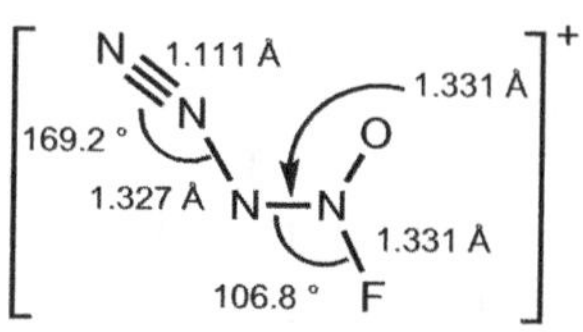

353

Strähle and co-workers[971] were the first to report the synthesis of the dichloronitronium ion [Eq. (4.226)]. According to X-ray crystal structure analysis of the hexachloroantimonate salt, which shows a remarkable stability up to 145°C, the cation is almost planar (maximum deviation from the plane is 0.038 Å) and has C_s symmetry. The N–O bond distance is 1.31 Å, that is, lies between the length of a NO single bond (1.151 Å) and a NO double bond (1.47 Å). This is indicative of a decreased π-bond contribution, which is also reflected in the IR spectrum (stretching vibration at $1650\,cm^{-1}$ as compared to $1827\,cm^{-1}$ of phosgene).

$$NCl_3 + SbCl_5 + SOCl_2 \xrightarrow[CCl_4]{} Cl_2NO^+\,SbCl_6^- + SCl_2 \qquad (4.226)$$

An improved method developed by Minkwitz et al.[972] allowed to obtain other salts [Eq. (4.227)]. The structure of the cation was characterized by vibrational spectroscopy, ^{14}N NMR (a single peak at $\delta^{14}N$ -282 from CH_3NO_2), and by *ab initio* theory.

$$2ONCl + Cl_2 + 3MF_5 \xrightarrow[-70^\circ C]{SO_2} 2Cl_2NO^+\,MF_6^- + MF_3 \qquad (4.227)$$

M = As, Sb

An ionic structure for the complex between NF_3O and AsF_5 was suggested by Bartlett et al.,[973] which was subsequently proved by Christe and Maya[967] by preparing and characterizing (IR, X-ray) stable salts [Eq. (4.228)]. The ^{19}F NMR spectrum of ion F_2NO^+ shows a triplet with equal intensity and line width ($\delta^{19}F$ -331, $J_{N-F} = 250$ Hz), and the ^{14}N NMR shows a singlet ($\delta^{14}N$ 281).[445,974] Ion F_2NO^+ has also been prepared[975] by fluorination with XeF^+ [Eq. (4.229)].

$$NF_3O + HF + SbF_5 \xrightarrow{-78°C} F_2NO^+ M^- \longleftarrow NF_3O + MF_n \qquad M = BF_4, B_2F_7, AsF_6, SbF_6 \qquad MF_n = BF_3, AsF_5 \tag{4.228}$$

$$ONF + XeF^+ MF_6^- \xrightarrow[-50°C]{HF} F_2NO^+ MF_6^- + Xe \qquad M = As, Sb \tag{4.229}$$

The calculated geometric parameters [B3LYP/6-311 + G(2d,p) level of theory] of the ion F_2NO^+ (N–F = 131.2 Å, N–O = 112.9 Å, F–N–F angle = 108.4°, F–N–O angle = 125.8°)[976] deviate considerably from those determined by Christe and coworkers[977] for the $F_2NO^+AsF_6^-$ salt (N–F = 124.5 Å, N–O = 119.0 Å, F–N–F angle = 115.9°, F–N–O angle = 122.1°). This suggests O/F disorder and refinement of the structure with variable occupancy factors resulted in values with good agreement with those predicted by theory (N–F = 128.4 Å, N–O = 111.4 Å, F–N–F angle = 107.9°, F–N–O angle = 126.0°).

The mixed halonitronium ion $ClFNO^+$ has been obtained by Minkwitz et al.[975] by oxidative fluorination of ONCl [Eq. (4.230)]. The ion has much lower stability than the difluoro analog. Subsequently, however, the validity of the structure determination of both Cl_2NO^+ and $ClFNO^+$ was questioned on the basis of large discrepancies between experimental and theoretical vibrational spectra.[978,979]

$$ONCl + N_2F^+ AsF_6^- \xrightarrow[-78°C]{CCl_3F} ClFNO^+ AsF_6^- + N_2 \tag{4.230}$$

Fluorination and methylation have been used to synthesize[975] related fluorinated nitronium ions [Eq. (4.231)]. The trifluoromethyl(methyl)nitronium ion ON(Me)CF_3^+ exits in the keto form in solution, but X-ray crystal structure data indicate that the enol form HON(CH_2)CF_3^+ exists in the solid state stabilized by a hydrogen bond between the enolic OH group and one of the fluorines of the counterion.

$$CF_3NO \begin{cases} \xrightarrow[CCl_3F,\ -35°C]{N_2F^+ AsF_6^-} ON(CF_3)F^+ AsF_6^- + N_2 \\ \xrightarrow[SO_2,\ -40°C]{MeF,\ MF_5} ON(Me)CF_3^+ MF_6^- \qquad M = As, Sb \end{cases} \tag{4.231}$$

4.4.4.3. *Polyheteroatom Cations of Chalcogens with Halogens.* A number of binary polyheteroatom cations of Group 6 and 7 elements have been prepared and characterized.

The $S_2I_4^{2+}$ and $Se_2I_4^{2+}$ dications have been studied extensively.[980] The $S_2I_4^{2+}$ dication has been prepared by Passmore and co-workers[981] [Eq. (4.232)]. The dication in $S_2I_4^{2+}(SbF_6^-)_2$ consists of two planar quadrilateral S_2I_2 units joined at the common S–S bond (I–S–I bond angle = 90.38°) (**354**).[981] It has C_{2v} symmetry and a distorted trigonal-prism-like shape. The dication in $S_2I_4^{2+}(AsF_6^-)_2$ shows a significant difference: The S–S bond is not parallel with the I–I bonds and, consequently, has C_2 symmetry. The S–S, I–I, and S–I bond distances in $S_2I_4^{2+}(SbF_6^-)_2$ (1.818, 2.571, and 2.993 Å, respectively) correspond to bond orders of 2.7, 1.4, and 0.1, respectively. The $S_2I_4^{2+}$ cation, therefore, can be considered to contain an S_2 unit of bond order 2.33 and two I_2^+ units of bond order 1.33 with a total π-bond order of 2, held together by a weak interaction.[982,983] The positive charge is delocalized evenly over all atoms.

$$0.25S_8 + 2I_2 + 3MF_5 \xrightarrow{SO_2} S_2I_4^{2+}(MF_6^-)_2 + MF_3 \qquad (4.232)$$

M = As, Sb

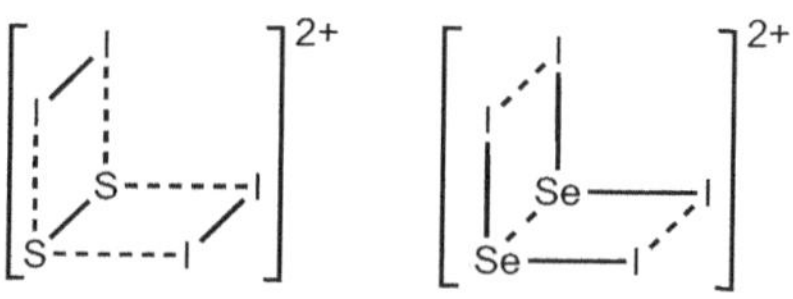

354 **355**

The $Se_2I_4^{2+}$ dication, synthesized in a similar way,[984] has a shape similar to that of the $S_2I_4^{2+}$ dication and a long Se–Se bond (2.841 Å). In sharp contrast, however, the bonding of $Se_2I_4^{2+}$ can be described as a weak dimer of two SeI_2^+ units (**355**) with evenly delocalized positive charge.[983] According to recent quantum chemical calculations (MPW1PW91 method),[985] the $Se_2I_4^{2+}$ dication has aromaticity resulting from through-space conjugation, analogous to transition states of some pericyclic reactions.

$Y_3X_3^+MF_6^-$ salts (Y = S, Se, X = Cl, Br, M = As, Sb) have been synthesized and characterized by Raman and ^{77}Se NMR spectroscopy, and X-ray crystallography[986,987] [Eq. (4.233)]. The $Se_3Cl_3^+$ cation adopts a structure with an intracationic Se···Cl contact (3.289 Å) and substantial Se–Se bond alternation (2.191 and 2.551 Å) resulting from the delocalization of the positive charge with resonance structure **356b**. The crystal structure of $S_3Cl_3^+AsF_6^-$ contains two crystallographically different disordered cations. The disorder arises from the superimposition of the two ordered cations **357a** and **357b**.

$$SeCl_3AsF_6 + 2Se \xrightarrow{SO_2,\ RT} Se_3Cl_3^+\ AsF_6^- \qquad (4.233)$$

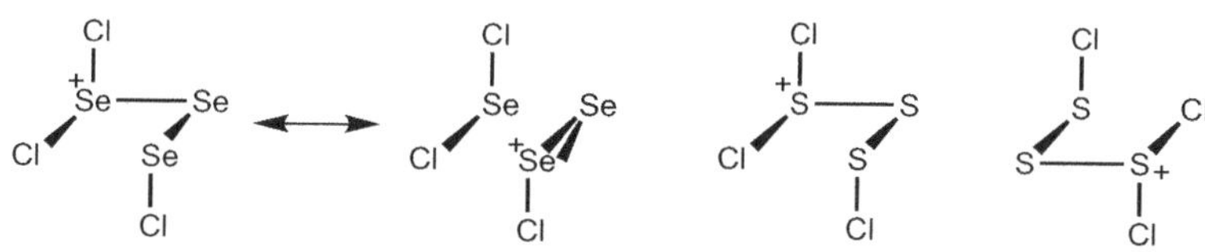

a **b** **a** **b**

356 **357**

The $Se_3Br_3^+$ cation gives three resonances of equal intensity in the ^{77}Se NMR spectra at $-70°C$ ($\delta^{77}Se$ 1065, 1263, 1735).[987] At $-40°C$ the two peaks at $\delta^{77}Se$ 1065 and 1263 coalesce into a broad peak ($\delta^{77}Se$ 1184) consistent with an intercationic exchange process rendering Se(1) and Se(3) equivalent (**358a** and **358b**). The short S–S and Se–Se bond distances of these cations are indicative of $3p_\pi$–$3p_\pi$ and $4p_\pi$–$4p_\pi$ bonds, respectively.

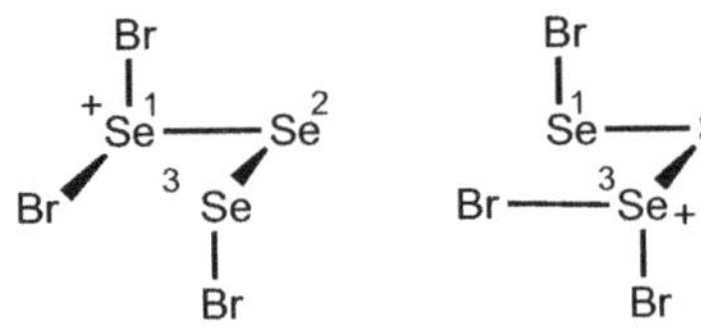

a **b**

358

Passmore and co-workers[988] have prepared the $Se_6I_2^{2+}$ dication **359** [Eq. (4.234)]. X-ray crystal structure analysis showed that the dication has C_i symmetry and contains a hexaselenium ring in the chair conformation with iodine atoms in *endo* position (**359**). The Se–Se bonds of di- and tricoordinate selenium atoms alternate significantly [average Se(1)–Se(2) and Se(2)–Se(3) bond lengths are 2.475 and 2.227 Å, respectively]. The reaction of Se, I_2, and AsF_5 in a molar ratio of 12:1:3 results in the formation of the polymeric $(Se_6I^+)_n \cdot n(AsF_6^-)$ salt, and the reaction of Se with $I_2Sb_2F_{11}$ yields $(Se_6I^+)_n \cdot n(SbF_6^-)$.[988] Both salts contain polymeric strands of $(Se_6I^+)_n$ species. The geometry of the cationic unit is similar to that of the $Se_6I_2^{2+}$ dication with iodine atoms connecting the Se_6 rings (**360**). The Se–I–Se bond is almost linear (bond angles are 173.7° and 174.2° for the AsF_6^- and SbF_6^- salts, respectively) and can be considered a three-center four-electron bond. The $(Se_6I^+)_n$ strands are joined together by weak intercationic interactions.

$$6Se + I_2 + 3AsF_5 \xrightarrow[SO_2,\ -AsF_3]{} \underset{\mathbf{359}}{Se_6I_2^{2+}(AsF_6^-)_2} \xleftarrow[SO_2]{} Se_8(AsF_6)_2 + SeI_3AsF_6 \quad (4.234)$$

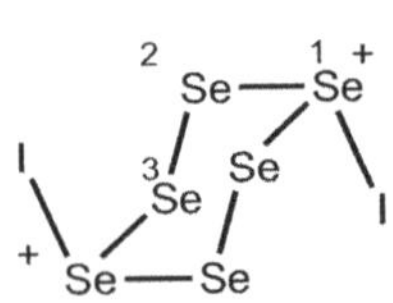

359

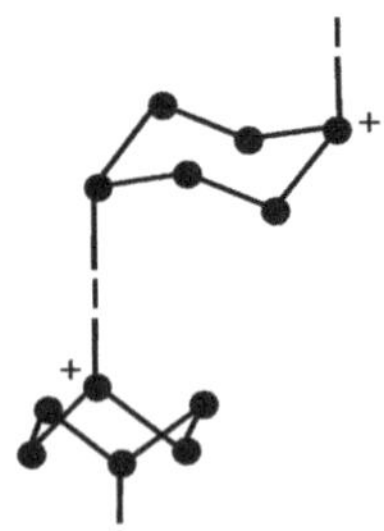

360

The general synthetic route shown in Eq. (4.235) has been used to prepare $S_7X^+MF_6^-$ (M = As, Sb) salts.[989–991] The stability of the cations decreases in the order I > Br > Cl > F. The X-ray structure of the S_7I^+ and S_7Br^+ cations was obtained.[980,989,990] Both cations were shown to have a seven-membered sulfur ring in a slightly distorted chair conformation with markedly varying S−S bond distances (1.900–2.389 Å for $S_7I^+SbF_6^-$).[980] The S−I (2.30–2.37 Å) and S−Br (2.11 Å) bond lengths are slightly shorter then the sum of the van der Waals radii (2.37 and 2.18 Å, respectively). All S_7I^+ cations have I---F contacts with the SbF_6^- anions in the solid state.

$$S_8^{2+}(MF_6^-)_2 + EX \xrightarrow[SO_2]{} S_7X^+MF_6^- + EMF_6 + 1/8S_8 \qquad (4.235)$$

M = As, Sb
E = Na, K
X = F, Cl, Br, I

4.5. CATIONS OF GROUP 6–12 ELEMENTS

4.5.1. Homoleptic Metal Carbonyl Cations

The first homoleptic carbonyl cations of Group 7 metals with the general formula $[M(CO)_6]^+$ (M = Mn, Tc, Re) were generated by Fischer and co-workers[992,993] and Hieber and co-workers[994,995] in the 1960s[996,997] applying halide abstraction by Lewis acids [Eq. (4.236)].

$$M(CO)_5X + AlX_3 + CO \xrightarrow[85–100°C,\ 300\ atm\ CO]{} [M(CO)_6]^+AlX_4^- \qquad (4.236)$$

M = Mn, Tc, Re
X = Cl, Br

The method, however, could not be extended to the synthesis of other metal carbonyl cations. Thus, Group 7 metal carbonyl cations remained isolated examples until 1990, when Aubke and co-workers generated $[Au(CO)_2]^+$ in HSO_3F.[998] This was followed by the isolation and characterization of the salts $[Au(CO)_2]^+Sb_2F_{11}^-$[999] and $[Hg(CO)_2]^{2+}(Sb_2F_{11}^-)_2$.[1000] Since then, stable salts of Group 6–12 metal carbonyl cations with the SbF_6^- or $Sb_2F_{11}^-$ anion have been isolated and characterized mainly by Aubke's group (Figure 4.4).[1001,1002] These cations with the general formula $[M(CO)_n]^{m+}$ ($n = 2, 4, 6$; $m = 1–3$) have mainly σ-bonded CO ligands. In addition, thermally unstable polycarbonyl complexes $[M(CO)_n]$Y of Cu and Ag are also known ($n = 1–4$); however, the nature of the M−Y bond, in most cases, is covalent.

Versatile synthetic routes are available to prepare metal carbonyl cations in superacidic media. The most successful methods include reductive carbonylation of metal fluorosulfates or fluorides [Eqs. (4.237)[1003] and (4.238)[1004]].[1002,1005] This

6	7	8	9	10	11	12
24 Cr	25 Mn	26 Fe	27 Co		29 Cu	
42 Mo	43 Tc	44 Ru	45 Rh	46 Pd	47 Ag	
74 W	75 Re	76 Os	77 1s	78 Pt	79 Au	80 Hg

Figure 4.4. Metals with known homoleptic carbonyl cations (shading indicates structural characterization).

method was also applied in the synthesis of $[Au(CO)_2]^+Sb_2F_{11}^-$ from $Au(SO_3F)_3$ and AuF_3,[999] and the corresponding metal chlorides could also be used as starting materials to prepare Au, Pd, and Pt fluoroantimonates.[1006] In contrast to reductive carbonylation, the oxidation state of the metal remains unchanged in solvolytic carbonylation [Eqs. (4.239)[1000] and (4.240)[1003]].

$$Pt(SO_3F)_4 + 5CO + 8SbF_5 \xrightarrow[\substack{25^\circ C,\ 1\ day \\ -CO_2}]{HF\text{–}SbF_5} [Pt(CO)_4]^{2+}(Sb_2F_{11}^-)_2 + 2Sb_2F_9(SO_3F) + S_2O_5F_2 \quad (4.237)$$

$$2IrF_6 + 15CO + 6SbF_5 + 8HF \xrightarrow[25^\circ C,\ 28\ h]{HF\text{–}SbF_5} [Ir(CO)_6]^{3+}(SbF_6^-)_3 \cdot 4\,HF + 3COF_2 \quad (4.238)$$

$$Hg(SO_3F)_2 + 2CO + 8SbF_5 \xrightarrow[100^\circ C,\ 1\ h]{SbF_5} [Hg(CO)_2]^{2+}(Sb_2F_{11}^-)_2 + 2Sb_2F_9(SO_3F) \quad (4.239)$$

$$cis\text{-}M(CO)_2(SO_3F)_2 + 2CO + 8SbF_5 \xrightarrow[25^\circ C,\ 1\ day]{HF\text{–}SbF_5} [M(CO)_4]^{2+}(Sb_2F_{11}^-)_2 + 2Sb_2F_9(SO_3F) \quad (4.240)$$

M = Pt, Pd

Oxidative reactions are the method of choice in the generation of cations of Group 6, 8, and 9 metals [Eq. (4.241)].[1007] The *fac*-$Rh(CO)_3(FSO_3)_3$ complex used in the synthesis of the Rh(I) salt **361** [Eq. (4.242)] was prepared from the solvated $[Rh(CO)_4]^+$ cation generated from $[Rh(CO)_2Cl]_2$.[1008] All procedures are performed under a CO atmosphere (1–2 atm).[1002]

$$2Fe(CO)_5 + 2CO + XeF_2 + 4SbF_5 \xrightarrow[50^\circ C,\ 2\ days]{SbF_5} 2[Fe(CO)_6]^{2+}(Sb_2F_{11}^-)_2 + Xe \quad (4.241)$$

$$fac\text{-}Rh(CO)_4(FSO_3)_3 \xrightarrow[\substack{2.\ CO,\ HF\text{–}HSO_3F, \\ -196^\circ C\ to\ -40^\circ C}]{1.\ CO,\ HSO_3F,\ 25^\circ C} [Rh(CO)_4]^+ Sb_2F_{11}^- \quad (4.242)$$

361

The stable, isolable salts have been thoroughly characterized by vibrational and ^{13}C NMR spectroscopy, and molecular structures have also been elucidated.[1001,1002] The geometry of the cationic complexes depends on the coordination number and electron configuration of the metal ion: Metal ions with coordination number 2 (d^{10} electron configuration) and 4 (d^8 electron configuration), respectively, form linear ($D_{\infty h}$ symmetry) and square-planar (D_{4h} symmetry) cationic complexes, whereas complexes of metals with d^6 electron configuration (coordination number 6) are octahedral (O_h symmetry). Illustrative of each class is the XRD structure of linear $[Au(CO)_2]^+Sb_2F_{11}^-$[999] and $[Hg(CO)_2]^{2+}(Sb_2F_{11}^-)_2$,[1000] square-planar $[M(CO)_4]^{2+}(Sb_2F_{11}^-)_2$ (M = Pd, Pt),[1003] and octahedral $[M(CO)_6]^{2+}(Sb_2F_{11}^-)_2$ (M Fe, Ru, Os)[1007,1009] and $[Ir(CO)_6]^{3+}(Sb_2F_{11}^-)_3$[1004] metal carbonyl cations. The molecular geometries are very regular and only slight angular distortions are detected. In the $[Fe(CO)_6]^+$ cation, for example, the C–Fe–C angles deviate from 90° by 0.2–1.1° and the Fe–C–O angles are between177.5° and 179.4°.

The C–O bond lengths (about 1.1 Å) are unusually short, which is a common feature of all σ-bonded metal carbonyl cations. This is related to the fact that M→CO π backbonding is almost absent, resulting also in increased M–C bond lengths. For so-called superelectrophilic cations (metals with oxidation states +2 and +3) the M–C bond distances are about 2.0 Å. As a result, very high $\nu(CO)_{av}$ values are detected (up to about 2280 cm^{-1}), indicating high polarization of the C–O bond and electrophilic nature of carbon. The strength of the C–O bond in $[M(CO)_n]^{m+}$ homoleptic carbonyl cations is found to be proportional to the charge m and inversely proportional to the coordination number n.[1002] In many salts, significant intercationic contacts are observed between the electrophilic carbon and fluorine, whereas M···F interactions are negligible. These observations are supported by DFT calculations.

In addition to the varied homoleptic carbonyl cations isolated as undecafluorodiantimonates (V), salts with other counteranions are also known. Two examples are $[Co(CO)_5]^+[(CF_3)_3BF]^-$ and $[Rh(CO)_4]^+$1-Et-$CB_{11}F_{11}^-$. The Co salt contains the univalent metal carbonyl cation, which is the first example of a trigonal-bipyramidal structure (D_{3h} symmetry).[1010] It is synthesized by oxidation of the $Co_2(CO)_8$ cluster with H_2F^+ [Eq. (4.243)]. There are no significant interactions between cations and anions, and the Co–C_{ax} bond is somewhat shorter than the Co–C_{eq} bond (1.826 Å versus 1.853 Å). The formation of the Rh salt was first observed by differential total reflectance FT–IR in a rather unexpected reaction between the solid complex **362** and CO [Eq. (4.244)], and then it was independently prepared according to Eq. (4.245).[1011] The cation has D_{4h} symmetry and, interestingly, there is a contact between Rh and one of the hydrogens of the methyl group of the carborane anion.

$$[Co_2(CO)_8] + 2\ (CF_3)_3BCO + 2\ HF \xrightarrow[\text{−196°C to RT, overnight}]{\text{HF, CO (2 bar)}} 2\ [Co(CO)_5]^+[(CF_3)_3BF]^- + H_2 \qquad (4.243)$$

$$\underset{\mathbf{362}}{[(\eta^6\text{-}C_6H_6)Rh(CO)_2]^+ \ 1\text{-Et-}CB_{11}F_{11}^-} + 2\,CO \xrightarrow[-C_6H_6]{25^\circ C} [Rh(CO)_4]^+ \ 1\text{-Et-}CB_{11}F_{11}^- \quad (4.244)$$

$$[Rh(CO)_2Cl]_2 + Ag(1\text{-Et-}CB_{11}F_{11}) \xrightarrow[\text{16 h, }-AgCl]{CH_2Cl_2,\text{ RT,}} \xrightarrow[\text{50°C to RT, overnight}]{CO} [Rh(CO)_4]^+ \ 1\text{-Et-}CB_{11}F_{11}^- \quad (4.245)$$

Despite the fact that CO is reversibly bound to Ag(I), Strauss and co-workers[1012] succeeded in isolating crystalline silver carbonyl complexes at low temperature under a CO atmosphere and studying the crystal structure of $[Ag(CO)]^+[B(OTeF_5)_4]^-$ and $[Ag(CO)_2]^+[B(OTeF_5)_4]^-$.

The synthesis and characterization of heptachloroaluminate and -gallate complexes of Rh(I) [Eq. (4.246)][1008] and the first tetrafluoroborate salts have recently been recently reported. Oxidative decarbonylation with XeF_2 was used for the synthesis of the Fe salt as shown in Equation (4.241), whereas the $[M(CO)_6]^+BF_4^-$ (M = Ru, Os) derivatives were obtained by oxidative decarbonylation with F_2 followed by solvolytic carbonylation [Eq. (4.247)].[1013] Experimental studies (XRD, FT–IR) show that structural and spectroscopic properties of the cations $[M(CO)_6]^+$ (M = Fe, Ru, Os) are independent of the anion.

$$[Rh(CO)_2Cl]_2 + MCl_3 \xrightarrow[\text{25°C, several weeks}]{CO} [Rh(CO)_4]^+ \ M_2Cl_7^- \qquad \text{M = Al, Ga} \quad (4.246)$$

$$M_3(CO)_{12} + F_2 \xrightarrow[\text{2. HF–}BF_3\text{, CO, RT}]{\text{1. HF, –78°C to RT}} [M(CO)_6]^+ \ BF_4^- \qquad \text{M = Ru, Os} \quad (4.247)$$

4.5.2. Other Cations of Group 6–12 Elements

A group of metal carbonyl cations with $Sb_2F_{11}^-$ as counteranion generated in superacids include $[\eta^6\text{-}(C_6H_6)Rh(CO)_2]^+$, $[M(CO)_5Cl]^+$ (M = Rh, Ir), $[W(CO)_6(FSbF_5)]^+$, and the polymeric $[\{Mo(CO)_4\}_2(cis\text{-}\mu\text{-}F_2SbF_4)_3]^+$.[1014] The PF_6^- salt of the hexamethylbenzene analog of cation **362** was first reported in 1982 without structural characterization.[1015] Strauss and co-workers[1011] have isolated the carborane anion salt [Eq. (4.248)] and found that it has two nearly identical cations of C_{2v} symmetry in the crystal state. The C–Rh–C angle in the $Rh(CO)_2$ moiety is ~90° deduced from the equal intensities of the two ν(CO) bands. The molecular structure of the isostructural Rh and Ir cations show C_{4v} symmetry and nearly identical bond angles and bond lengths.[1008,1016] The unusually long M–C and short C–O bonds and high CO stretching frequencies are characteristic features observed for the corresponding homoleptic metal carbonyl cations as well.

$$[Rh(CO)_2Cl]_2 + [Ag(C_6H_6)]^+ \text{ 1-Et-CB}_{11}F_{11}^- \xrightarrow[\text{-AgCl}]{CH_2Cl_2} \underset{\mathbf{362}}{[(\eta^6\text{-}C_6H_6)Rh(CO)_2]^+ \text{ 1-Et-CB}_{11}F_{11}^-} \quad (4.248)$$

Two-electron oxidation of $W(CO)_6$ with SbF_5 in HF–SbF_5 to give $[W(CO)_6(FSbF_5)]^+Sb_2F_{11}^-$ in quantitative yield without loss of CO is unprecedented.[1017] The cation is seven-coordinated with a distorted C_{2v} capped trigonal prismatic structure. $FSbF_5$ is tightly coordinated to W with nearly equal W---F and Sb---F bond lengths. $Mo(CO)_6$ reacts with SbF_5 in a very similar manner to form the fluoro-bridged product $[Mo(CO)_6(FSbF_5)]^+Sb_2F_{11}^-$.[1014] This, however, undergoes condensation with partial elimination of CO and SbF_5, and *cis*-μ-F_2SbF_4 bridges are formed to yield polymeric $[\{Mo(CO)_4\}_2(cis\text{-}\mu\text{-}F_2SbF_4)_3{}^+]_x(Sb_2F_{11}{}^-)_x$. The polymerization of the tungsten analog, in turn, requires elevated temperature (>100°C). A comparison of XRD data shows a rather wide range of bond parameters for the cations, but the differences are less pronounced for the $Mo(CO)_4$ moiety.

During the synthesis of the $[Os(CO)_6]^{2+}(Sb_2F_{11}{}^-)_2$ salt by the reductive carbonylation of $Os(SO_3F)$, OsF_6, or OsO_4, Aubke and co-workers[1018] observed a band at 2253 cm^{-1} in the IR spectra of the products. They used a modified synthesis [Eq. (4.249)] to isolate the extremely moisture-sensitive compound **363**. The cation identified by vibrational spectroscopy has octahedral geometry with *trans* oxygen atoms. Experimental frequencies agree well with calculated data (gradient-corrected DFT calculations at the BP86/ECP2 level).

$$OsO_4 + 5\,CO \xrightarrow[25°C]{SbF_5,\ 1.5\text{ atm CO}} \underset{\mathbf{363}}{[OsO_2(CO)_4]^+ Sb_2F_{11}^-} + SbOF_3 + CO_2 \quad (4.249)$$

Schrobilgen and co-workers[1019] have prepared oxafluoro cations of Tc, Re, and Os as $Sb_2F_{11}^-$ and AsF_6^- salts and characterized them by X-ray diffraction and Raman and NMR spectroscopy. The $Tc_2O_2F_9^+$ cation contains two fluorine-bridged square pyramidal $TeOF_4$ groups with the F bridge *trans* to the oxygen atoms (Tc---F---Tc angle = 158.2°, bridging Tc---F bond distances = 2.061 Å and 2.075 Å).[1019] Geometric and spectral characteristics (vibrational frequencies, ^{19}F and ^{99}Tc NMR data) calculated by local DFT are in good agreement with experimental values. According to the ^{19}F NMR spectrum, the $Re_2O_4F_5^+$ cation consists of two ReO_2F_2 moiety with a bridging fluorine.[1020] In the crystal structure of $[\mu\text{-}F(OsO_2F_3)_2]^+Sb_2F_{11}^-$ the dinuclear fluorine-bridged (F_3O_2Os---F---OsO_2F_3) cation has a distorted octahedral *cis*-dioxo arrangement and the fluorine bridge is *trans* to one of the oxygen atoms of each OsO_2F_3 unit (Os---F---Os angle = 155.2°, Os---F bond distance = 2.086 Å).[1021]

Varied salts of the OsO_3F^+ cation were generated by reacting osmium trioxide difluoride with AsF_5 and SbF_5 in HF solvent.[1022] The OsO_3F^+ cations in $OsO_3F^+Y^-$ (Y = AsF_6, SbF_6) have distorted C_{3v} symmetry with one Os–O bond significantly longer (1.711 Å and 1.708 Å for the AsF_6^- and SbF_6^- salts, respectively). The cations are bridged through a fluorine of the anion and two such cation–anion pairs form a cyclic dimer by additional fluorine bridges. The cation–anion pairs in the cyclic dimer of the solvated salt $OsO_3F^+(HF)_2AsF_6^-$ are linked together by two hydrogen-bonded $(HF)_2$ moieties. One Os–O bond of the cation is much shorter (1.666 Å). The X-ray structure of the $OsO_3F^+(HF)SbF_6^-$ salt contains helical $(FO_3Os\cdots FH\cdots FSbF_5)_\infty$ chains. In sharp contrast, the cations are well-separated from the *cis*-fluorine-bridged $Sb_3F_{16}^-$ anion in the crystals of the $OsO_3F^+Sb_3F_{16}^-$ salt.

Acetonitrile complexes of Pd, Pt, and Au were prepared by complete ligand exchange of the corresponding homoleptic metal carbonyl salts under solvolysis conditions with the concomitant conversion of the dioctahedral anion $Sb_2F_{11}^-$ into SbF_6^-. The $[Au(MeCN)_2]^+SbF_6^-$ salt formed is not solvated,[999] in contrast to the resulting Pd and Pt salts [Eq. (4.250)].[1023] The Pd and Pt cations are square planar with a slight deviation of the N–C–C group form linearity (177°), whereas the molecular structure of the $[Au(MeCN)_2]^+$ cation is highly regular.

$$[M(CO)_4]^{2+}(Sb_2F_{11}^-)_2 + 7MeCN \xrightarrow[-CO]{50°C,} [M(MeCN)_4]^{2+}(SbF_6^-)_2 \cdot MeCN + 2MeCN \cdot SbF_5 \quad (4.250)$$

M = Pd, Pt

The Re oxafluoro acetonitrile complex prepared by Schrobilgen and co-workers[1020] [Eq. (4.251)] and characterized by spectroscopic methods (^1H, ^{13}C, ^{19}F NMR and Raman) has a pseudooctahedral *cis*-dioxo arrangement. The acetonitrile ligands are *trans* to the oxygens and the fluorines are *trans* to each other.

$$ReO_2F_3 \cdot SbF_5 + 2MeCN \xrightarrow{-196°C \text{ to RT}} [ReO_2F_2(MeCN)_2]^+ SbF_6^- \quad (4.251)$$

The homoleptic carbonyl salt $[Hg(CO)_2]^+(Sb_2F_{11}^-)_2$ reacts with traces of water in the HF–SbF_5 superacid to yield the bis-aqua solvate $[Hg_2(OH_2)_2]^{2+}(SbF_6^-)_2 \cdot 4\,HF$.[1024] The cation is almost linear with nearly C_{2h} symmetry. The Hg–Hg and Hg–O bond distances are 2.4917 Å and 2.148 Å, respectively, and the O–Hg–Hg bond angle is 177.6°.

The $[Au(PF_3)_2]^+Sb_2F_{11}^-$ salt was synthesized from the corresponding homoleptic dicarbonyl complex by ligand exchange with PF_3 and reductive phosphorylation of $Au(SO_3F)_3$ with excess PF_3 in HSO_3F.[1025] Spectroscopic data (^{19}F and ^{31}P NMR, FT-IR, Raman) suggest a linear structure, predominantly covalent Au···P bonding, and highly reduced π backdonation. The crystal structure of $[(PF_3)_2Au]^+SbF_6^-$ supports the above conclusions.[1026]

The first metal–xenon compound with direct Au–Xe bonds has been reported by Seidel and Seppelt[1027–1029] [Eq. (4.252)]. The salt crystallizes in two crystallographic modifications differing in cation–anion interactions. In each form the cation is a regular square. In the triclinic modification the Au–Xe bond lengths are between 2.7330 Å and 2.7779 Å and there are three weak interionic contacts (Au···F distances = 2.671 Å, 2.950 Å, 3.153 Å). The interionic contacts in the tetragonal modification are weaker (Au---F distance = 2.928 Å).

$$AuF_3 + 6Xe \xrightarrow[-196°C \text{ to RT}]{HF–SbF_5 (2:1)} [AuXe_4]^+ (Sb_2F_{11}^-)_2 + Xe_2^+ \quad (4.252)$$

When excess Xe was pumped off from the reaction mixture at −78°C, the [*cis*-$AuXe_2]^{2+}(Sb_2F_{11}^-)_2$ was isolated in the form of violet-black crystals.[1028] The cation has two Xe and two F contacts with slightly shorter Au–Xe bond lengths (2.658 Å and 2.671 Å) and rather short Au---F contacts (2.181 and 2.238 Å). Seppelt and coworkers were also successful in isolating [*trans*-$AuXe_2]^{2+}(SbF_6^-)_2$ and $[Au_2Xe_2F]^{3+}(SbF_6^-)_3$ with the Au^{2+} center residing in a square-planar environment. In the latter case, the cation has a Z-shaped $[Xe–Au–F–Au–Xe]^{3+}$ ion. In addition to these gold salts, Hwang and Seppelt[1030] also isolated the unique $[Au(HF)_2]^{2+}(SbF_6^-)_2 \cdot 2$ HF by reducing AuF_3 in HF–SbF_5 upon UV irradiation. It is a rare example of compounds with HF as ligand with square-planar AuF_4 units in the crystal structure.

In all salts discussed above the cation has a gold(II) center. The only example of a cation with a gold(III) center is [*trans*-$AuXe_2F]^{2+}(SbF_6^-)(Sb_2F_{11}^-)$.[1028] The Au–Xe distances (2.593 Å and 2.619 Å) are markedly shorter than those in the Au^{2+} complexes. In further studies Seppelt and co-workers prepared two gold(I) complexes. The very strong cation–anion contacts in $[(F_3As)Au]^+SbF_6^-$ through one fluorine atom allow the possible description as $F_3As–Au–F$---SbF_5.[1026] This was then transformed to $[(F_3As)AuXe]^+Sb_2F_{11}^-$ [Eq. (4.253)], which is stable at room temperature.[1031] The As–Au–Xe moiety is almost linear (173.26°), the Au^+–Xe bond is similar to the Au^{3+}–Xe distance but much shorter than the Au^{2+}–Xe bonds, and there are only weak interionic interactions (the shortest Au---F separation is 2.848 Å).

$$[(F_3As)Au]^+ SbF_6^- + Xe \xrightarrow[-196°C \text{ to RT then to } -50°C]{HF–SbF_5} [(F_3As)AuXe]^+ Sb_2F_{11}^- \quad (4.253)$$

Additional Au(I) cation complexes with phosphane ligands are the binuclear halogen-bridged complexes $\{[(R_3P)Au]_2X\}^+BF_4^-$ (R = Et, Ph, *ortho*-tolyl, Bn, mesityl, X = Cl, Br, I).[1032] The molecular structure of $\{[(Ph_3P)Au]_2Br\}^+BF_4^-$ shows a V-shape structure (Au–Br–Au bond angle = 96.83°) and a quasi-linear coordination of the gold atoms (P–Au–Br = 177.87°). The Au–Au distance (3.6477 Å) indicates negligible interactions between the gold atoms. These data differ considerably from those observed earlier for $\{[(Ph_3P)Au]_2Cl\}^+ClO_4^-$[1033]: The Au–Br–Au bond angle is much smaller (81.7°) and the Au–Au distance is significantly shorter (3.06 Å), indicating intramolecular bonding interactions.

Hg^{2+} ions, which are isoelectronic with Au^+, reacting with SbF_5 in the presence of xenon above room temperature yield the stable $[HgXe]^{2+}(SbF_6^-)(Sb_2F_{11}^-)$ salt.[1031] The salt has a highly distorted capped structure with Xe in the capping position. The Hg–Xe bond length (2.7693 Å) is similar to that of the Au–Xe bond distance in $AuXe_4^{2+}$ and there are six Hg---F contacts with distances between 2.279 Å and 2.594 Å.

4.6. MISCELLANEOUS CATIONS

4.6.1. Hydrogen Cations

4.6.1.1. H^+ Ion. The naked proton "H^+" exists only in the gas phase. In the condensed state, the proton is always solvated, thus no free proton is capable of existence. It is customary, however, as short hand notation to depict "H^+" as the solvated proton.

4.6.1.2. H_3^+ Ion. The H_3^+ ion **364** was discovered by Thompson[1034] in 1912 in hydrogen discharge studies. Actually, it was the first observed gaseous ion–molecule reaction product [Eq. (4.254)] and the reaction sequence was established in 1925 by Hogness and Lunn.[1035,1036] Since then, extensive mass spectrometric studies of H_2, D_2, and their mixtures have been carried out in an effort to study thermodynamic and kinetic aspects of ion–molecule reactions of $(H,D)_3^+$ cations.[1037]

$$H_2^+ + H_2 \longrightarrow \underset{\mathbf{364}}{H_3^+} + H \qquad (4.254)$$

Despite numerous studies in the gas phase on H_3^+**364**, not much solution chemistry has been reported until the 1960s. Gillespie and Pez[1038] reported that according to their solubility, cryoscopic, and ^{1}H NMR spectroscopic measurements, $HSO_3F–SbF_5–$(Magic Acid)–SO_2 is not sufficiently strong to protonate a series of weak bases, including molecular hydrogen. Their investigation pertained, however, to observe H_3^+**364** as a stable, detectable intermediate with a long life.

Olah, Shen, and Schlosberg[1039] were subsequently able to observe the hydrogen–deuterium exchange of molecular H_2 and D_2, respectively, with 1:1 HF–SbF_5 and $HSO_3F–SbF_5$ at room temperature. The facile formation of HD does indicate that protonation or deuteriation occurs, involving H_3^+ **364** at least as transition states in the kinetic exchange process. The H_3^+ ion **364** is the simplest two-electron three-center-bonded entity. The H_3^+ ion **364** has also been observed by IR spectroscopy.[1040]

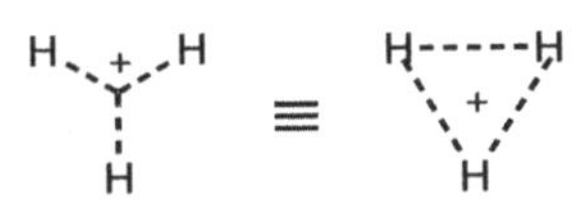

364

4.6.2. Cations of Noble Gases

Reviews about the chemistry of noble gases including the cations of xenon and krypton are available.[1041–1047]

A large number of Xe_n^+ cations ($n \leq 30$) have been detected by mass spectrometry,[1048] and the Xe_2^+ ion was first generated in the solution phase in 1978.[1049] Drews and Seppelt[1050] were the first, however, to report the isolation and molecular structure of $Xe_2^+Sb_4F_{21}^-$ (**365**) [Eq. (4.255)]. In the crystals of the dark green compound, weak multiple contacts exist between xenon and fluorine atoms. The Xe–Xe bond is surprisingly long (3.087 Å) but much shorter than the theoretically predicted values (3.17–3.27 Å).

$$XeF^+ Sb_2F_{11}^- + Xe \xrightarrow[-196 \text{ to } 0°C]{HF-SbF_5} \underset{\mathbf{365}}{Xe_2^+ Sb_4F_{21}^-} \qquad (4.255)$$

The solution of Xe_2^+ has recently been shown to undergo a color change from green to dark blue upon increasing the xenon pressure.[1051] Of the possible cationic species (Xe_n^{m+}, $n = 2$–4; $m = 1, 2$), calculated vibrational spectra and energies agree well with those observed for Xe_4^+, but the presence of higher xenon aggregates ($Xe_4^+ \cdot Xe_n$) cannot be ruled out. According to calculations, the ion has a symmetric linear structure ($D_{\infty h}$, Xe–Xe bond lengths = 3.529 Å and 3.190 Å).

Both xenon and krypton are known to undergo ion–molecule reaction with H^+ to give the corresponding onium ions. The XeH^+ and KrH^+ ions are well-recognized in mass spectroscopic studies.[1052–1060] and they have been also characterized by IR[1061] and microwave spectroscopy.[1062] The Xe_2H^+ cation has been observed in low-temperature matrices,[1063] whereas $Xe_2H_3^+$ has been studied by *ab initio* methods.[1064] Homogeneous and mixed noble gas species Ar_2H^+, KrH^+, Xe_2H^+, $ArKrH^+$, and $ArXeH^+$ have also been observed.[1063,1065] Even methylated xenon and krypton, CH_3Xe^+ and CH_3Kr^+, have been observed by Holtz and Beauchamp [1066] in the gas phase. The carbon–inert gas atom bond strengths in these cations are estimated to be 43 ± 8 and 21 ± 15 kcal mol^{-1}, respectively. The C–Xe bond strength in CH_3Xe^+ was determined by ion cyclotron resonance to be 55.2 ± 2.5 kcal mol^{-1}.[1067]

In solution chemistry too attempts have been made to protonate xenon to XeH^+ in the superacid media.[150,1038] Evidence for the protonation comes from suppression of proton–deuterium exchange rates of deuterium gas in the presence of xenon in strong acid medium.[342]

In 1989 the synthesis and isolation of the first compounds with stable Xe–C bond were reported by Naumann and Tyrra[285] and Frohn and Jakobs[1068] by the introduction of the organic group through nucleophilic substitution called xenodeborylation [Eq. (4.256)]. The boron as Lewis acid center polarizes the Xe–F bond and thereby allows it to overcome the low electrophilicity of the Xe center to form the fluoroarylxenonium fluoroborates **366**. When the reaction is carried out in anhydrous HF, all C_6F_5 groups of the borane are transferred to xenon [Eq. (4.256)].[1069]

$$XeF_2 + (C_6F_5)_3B \longrightarrow \underset{\mathbf{366}}{C_6F_5Xe^+ Y^-} \quad \begin{array}{l} \text{MeCN, 0°C, } Y = (C_6F_5)_3BF \\ CH_2Cl_2\text{, −30°C, } Y = C_6F_5BF_3 \\ \text{HF, −30°C, } Y = BF_4 \text{ and } F(HF)_n \end{array} \qquad (4.256)$$

Frohn et al.[1070] have obtained the X-ray structure of the acetonitrile complex of the salt $C_6F_5Xe^+(C_6F_5)_2BF_2^-$. The Xe–C bond distance is 2.092 Å and the Xe–N contact is 2.681 Å. The ^{129}Xe NMR spectrum contains a triplet at δ^{129}Xe −1956, and the coupling with the *ortho* F atoms ($J_{Xe-F} = 68.8$ Hz) indicates the presence of the Xe–C bond. The ^{129}Xe NMR resonance (−1784.5) is strikingly high-field compared with that of XeF_2, which is attributed to the solvation of the cation by acetonitrile. At the same time, the signals in the ^{19}F NMR spectrum are shifted significantly downfield relative to the isoelectronic C_6F_5I molecule, indicating charge delocalization to the aromatic ring. Subsequently, the X-ray structure of **366** ($Y = AsF_6$) prepared by metathesis from **366** [$Y = (C_6F_5)_2BF_2$] was also reported.[1071] In the crystal cell, there are two independent molecules. In each, one fluorine atom of the anion forms an almost linear but asymmetric C–Xe---F contact (C–Xe bond lengths = 2.079 and 2.082 Å, Xe---F distances = 2.714 and 2.672 Å, C–Xe---F angles = 170.5 and 174.2°). There are also additional Xe–Fe contacts. Schrobilgen, Frohn, and co-workers[1072] have recently synthesized new examples of $C_6F_5Xe^+$ salts with a range of weakly coordinating anions BY_4^- [$Y = CF_3$, C_6F_5, CN, $OTeF_5$].

The xenodeborylation method has been extended to hydrogen-containing aryl compounds, in which the aryl group bears at least one electron-withdrawing substituent (F or CF_3).[1073,1074] The resulting salts, however, are unstable in CH_2Cl_2. When the synthesis is carried out in the presence of $BF_3{\cdot}OMe_2$, in turn, stable tetrafluoroborates are isolated.[1075–1077] Recently, Frohn et al.[1078] have developed a more convenient route to xenonium salts by the use of aryldifluoroboranes generated in situ [Eq. (4.257)].

$$\mathrm{KArBF_3 + BF_3 \xrightarrow[\substack{CH_2Cl_2, -40\ to\ -50^\circ C \\ -KBF_4}]{} ArBF_2 \xrightarrow[CH_2Cl_2]{XeF_2} ArXe^+\,BF_4^-} \tag{4.257}$$

$$\mathrm{Ar = C_6F_5,\ 2,3,4,5\text{-}F_4C_6H,\ 3,4,5\text{-}F_3C_6H_2,\ 3,5\text{-}F_2C_6H_3}$$

Naumann et al.[1079,1080] developed the xenodeprotonation method using the unique electrophilic Xe(II) reagent to generate arylxenononium triflates from highly deactivated benzene derivatives [Eq. (4.258)]. Stable derivatives were isolated in low yields (10–15%), whereas other products were observed by NMR spectroscopy. The cations and anions in the crystals of $[Xe(2,6\text{-}F_2C_6H_3)]^+TfO^-$ are weakly coordinated through short Xe–O contacts.

$$\mathrm{XeF_2 \xrightarrow[CCl_3F, -40^\circ C]{\substack{1.\ CF_3COOH \\ 2.\ CF_3SO_3H}} CF_3CO_2XeO_3SCF_3 \xrightarrow[CCl_3F, -40^\circ C]{Ar-H} ArXe^+\,TfO^-} \tag{4.258}$$

$$\mathrm{Ar = C_6F_5,\ ClC_6H_4,\ 1,3\text{-}F_2C_6H_3,\ 1,3,5\text{-}F_3C_6H_2,\ 1,3,5\text{-}Cl_3C_6H_2,\ NO_2C_6H_4,\ 1,3\text{-}(NO_2)_2C_6H_4,\ 1\text{-}F,4\text{-}NO_2C_6H_3,\ 1\text{-}F,4\text{-}CF_3C_6H_3,\ 1,3\text{-}(CF_3)_2C_6H_3}$$

Fluorinated vinyl derivatives **367a** and **368a** have also been prepared by Frohn and Bardin,[1081] and the synthesis of the partially fluorinated **367b** and **368b** and oxygenated compounds **369** and **370** was also reported.[1082] They utilized the method described in Equation (4.257) for the synthesis of the first acyclic vinylxenon compound **371**.[1083] Alkynylxenonium salts have been obtained by Stang and co-workers[1084] (**372**) and Frohn and Bardin[1085,1086] (**373**).

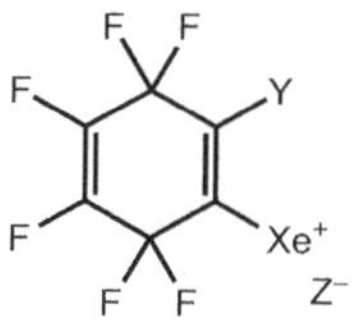

367

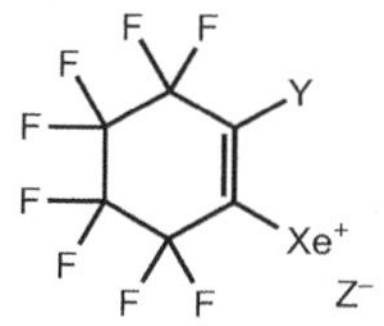

368

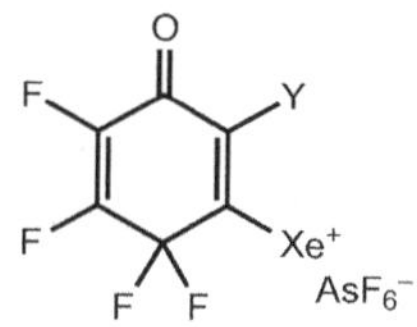

369

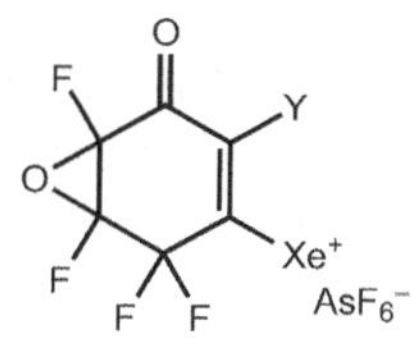

370

a Y = F, Z = AsF_6
b Y = H, Z = BF_4

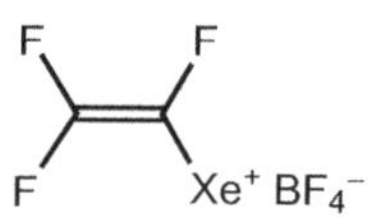

371

R—C≡CXe$^+$ BF_4^-

R = Et, Pr, *tert*-Bu, $SiMe_3$

372

F_3C—C≡CXe$^+$ BF_4^-

373

Frohn et al.[1087] have recently reported the first organoxenon(IV) compound [Eq. (4.259)]. The ^{129}Xe resonance is significantly shielded compared to that of XeF_4 ($\delta^{129}Xe$ −1706.5 versus 316.9). The three substituents and the two nonbonding electron pairs of Xe predict a pseudo-trigonal–bipyramidal arrangement and a T-shaped molecular geometry.

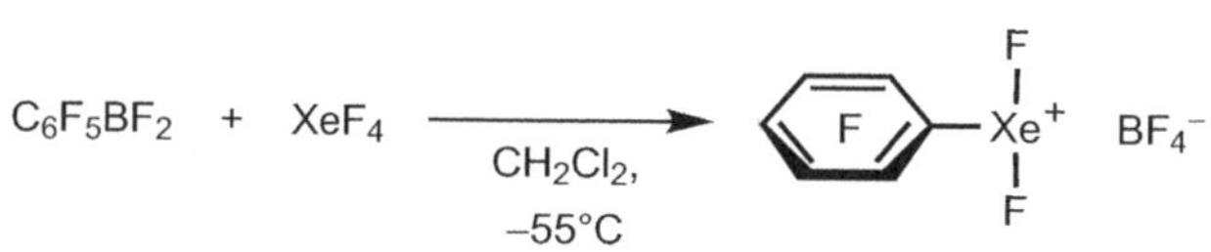

(4.259)

The fluorides and oxyfluorides of xenon are well recognized.[1088–1090] Most of the xenon fluorides and oxyfluorides react with Lewis acids to give the corresponding cations. The following cations have been prepared and characterized spectroscopically: XeF^+,[1091] XeF_3^+,[1092] $Xe_2F_3^+$,[1041] XeF_5^+,[1041] $Xe_2F_{11}^+$,[1093] $XeOF_3^+$,[1092] and XeO_2F^+,[1092,1094] and $[FO_2XeFXeO_2F]^+$.[1094] XeF^+ salts have widely been used for oxyfluorination.[73] The X-ray structure of XeF^+,[1050,1095] $Xe_2F_3^+$,[1096–1098] XeF_5^+,[1097] $XeOF_3^+$,[1099] XeO_2F^+,[1100] and $[FO_2XeFXeO_2F]^+$[1100] has been determined. Cations containing the highly electronegative fluoro analog $OTeF_5$ ligand, such as $XeOTeF_5^+$,[1101–1106] $Xe_2(OTeF_5)_3^+$,[1104] $O_2XeOTeF_5^+$,[1107] $[F_nXe(OTeF_5)_{3-n}]^+$ and $[OXeF_n(OTeF_5)_{3-n}]^+$ ($n = 0$–2)[1107] have also been prepared and studied experimentally and theoretically. Additional examples are $[F_5TeN(H)Xe]^+AsF_6^-$,[1108] $[CF_3C(OXeF)NH_2]^+AsF_6^-$,[1109] and various adducts such as $[RC{\equiv}NXeF]^+AsF_6^-$.[1110] Recently, the XeO^+ and $XeOH^+$ ions have been observed in the gas phase,[1111] and the $XeOO^+$ ion has been prepared and characterized in solid argon matrix.[1112]

Cations of Xe bonded to nitrogen are $[XeN(SO_2F)_2]^+MF^-$ ($MF = AsF_6$, Sb_3F_{16}),[1113] $F[XeN(SO_2F)_2]_2^+AsF_6^-$,[1113] $[F_5TeN(H)Xe]^+AsF_6^-$,[1114] $[CF_3C(OXeF)NH_2]^+AsF_6^-$,[1115] and various adducts such as $[RC{\equiv}NXeF]^+AsF_6^-$ (R = H, Me, Et, CH_2F, CF_3, C_2F_5, C_3F_7, C_6F_5),[1116,1117] and $[F_3S{\equiv}NXeF]^+AsF_6^-$.[1118] The anion of $[F_5TeN(H)Xe]^+AsF_6^-$ in the solid state forms a fluorine bridge with the Xe of the cation $[F_5TeN(H)Xe]^+$ (F–Xe–N = 171.6°).[1114] The arangement of the Xe, Te, and H atoms around the nitrogen is nearly tetrahedral. The salt $[F_3S{\equiv}NXeF]^+AsF_6^-$ synthesized by reacting $XeF + AsF_6^-$ with $N{\equiv}SF_3$ was extensively characterized.[1118] Calculations predict a linear geometry of the cation; however, the actual geometry in the crystal state is nonlinear (Xe–N–S bond angle = 142.6°), which is attributed to close N···F contacts in the unit cell and crystal packing. Calculations also reveal a very weak Xe–N donor–acceptor interaction.

In contrast to the numerous examples discussed, only two chlorine-containing xenon cations are known. Seidel and Seppelt[1119] observed the formation of a blue solution in the reaction shown in Eq. (4.260) containing Cl_4^+ formed by oxidation. Slow cooling of this solution resulted in the deposition of $XeCl^+Sb_2F_{11}^-$ as orange crystals formed in the Cl/F exchange reaction. As predicted by *ab initio* calculations, the Xe–Cl bond in the $XeCl^+$ cation corresponds to a Xe–Cl single bond (2.307 Å) and is much shorter than in any XeCl compounds. There is a cation–anion contact forming an almost linear Cl–Xe···F array with a contact distance of 2.628 Å.

$$XeF^+\,SbF_6^- + SbCl_5 \xrightarrow[\text{-196°C to RT then slow cooling to -30°C}]{HF\text{-}SbF_5} XeCl^+\,Sb_2F_{11}^- \qquad (4.260)$$

Frohn and coworkers have obtained the unique chloronium cation **374**,[1120] which is the first unambiguously characterized xenon(II) chlorine compound [Eq. (4.261)]. There is no strong contact between cation **374** and the anion in the solid state. The two C–Xe–Cl contacts are linear (bond angles = 176.0 and 178.8°) with shorter C–Xe (2.111 and 2.116 Å) and longer Xe–Cl (2.784 and 2.847 Å) bond lengths. The Xe–Cl–Xe bond angle is 116.96°.

$$2C_6F_5Xe^+\,AsF_6^- \xrightarrow[CH_2Cl_2,\ -78°C]{} (C_6F_5Xe)_2Cl^+\,AsF_6^- + 6Me_3SiF + AsCl_3 + Cl_2 \qquad (4.261)$$

374

Similarly to xenon fluorides, krypton fluorides give cations KrF^+ and $Kr_2F_3^+$ with Lewis acids.[1047,1121,1122] As mentioned earlier, cation salts $Kr_2F_3^+SbF_6^-$ and $Kr_2F_3^+AsF_6^-$ are capable of oxidizing bromine pentafluoride to BrF_6^+ cation **339**. The crystal structure of both cations has been obtained.[872,1123] In the solid state, the KrF^+ ion has an interesting feature that is not observed for XeF^+ salts: there is a small, but significant deviation from linearity of the F–Kr---F moiety (175.4–177.9°).[1047] The KrO_n^+ ($n = 1, 2$) ion has been shown to be a stable species in the gas phase.[1124] $[RC{\equiv}NKrF]^+AsF_6^-$ adducts[1117] as well as $[HC{\equiv}NKrF]^+AsF_6^-$[1125] with Kr–N bond in the cations were also prepared and characterized.

Cations of the lighter noble gases (He, Ne, Ar) have been observed in the gas phase and studied theoretically.[1126–1128] Koch and Frenking[1129] have calculated the structures and stabilities of He_2O^{2+}, He_2N^{2+}, and He_2C^{2+} dications, and Radom and coworkers[1130] found (*ab initio* studies at the MP4/6-311G** level) that the triheliomethyl trication (He_3C^{3+}) and tetraheliomethane tetracation (He_4C^{4+}) should be experimentally observable. Olah, Prakash, and Rasul[1131] have found that both the C_s symmetry structure of the helionitronium trication ($HeNO_2^{3+}$) and the $C_{\infty v}$ structure of helionitrosonium trication ($HeNO^{3+}$) are minima on the potential energy surface (*ab initio* MP2/6-31G** level). In the trication $HeNO_2^{3+}$, oxygen is strongly bonded to helium. Dissociation of the trication to NO^+ and OHe^{2+} is thermodynamically preferred by 183.1 kcal mol^{-1}, but dissociation has a kinetic barrier of 12.4 kcal mol^{-1}. Recent multireference configuration interaction (MRCI) studies[1132] have shown that the triplet state of C_{2v} symmetry with He binding to the N atom is the ground state for the helionitronium trication ($HeNO_2^{3+}$).

REFERENCES

1. G. A. Olah, K. K. Laali, Q. Wang, and G. K. S. Prakash, *Onium Ions*, Wiley, New York, 1998, Chapter 3, p. 96.
2. A. Hantzsch and K. S. Caldwell, *Z. Phys. Chem.* **58**, 575 (1907).
3. J. N. Brönsted, *Recl. Trav. Chim. Pays-Bas* **42**, 718 (1923); *J. Phys. Chem.* **30**, 777 (1926).
4. T. M. Lowry, *Trans. Faraday Soc.* **20**, 13 (1924); *Chem. Ind.* (London), 1048 (1923).
5. B. E. Conway, J. O'M. Bockris, and H. J. Linton, *J. Chem. Phys.* **24**, 834 (1956).
6. M. Eigen and L. DeMaeyer, *Proc. Roy. Soc. London, Ser. A* **247**, 505 (1958).
7. D. E. Bethell and N. Sheppard, *J. Chem. Phys.* **21**, 1421 (1953).
8. C. C. Ferrisso and D. F. Hornig, *Chem. Phys.* **23**, 1464 (1955); *J. Am. Chem. Soc.* **75**, 4113 (1953).
9. P. A. Giguère, *Rev. Chim. Miner.* **3**, 627 (1966).
10. M. Fournier and J. Roziere, *C.R. Acad. Sci. Ser. C* **270**, 729 (1970).
11. R. C. Taylor and G. L. Vidale, *J. Am. Chem. Soc.* **78**, 5999 (1956).
12. P. A. Giguère and C. Madec, *Chem. Phys. Lett.* **37**, 569 (1976) and references cited therein.
13. R. E. Richards and J. A. S. Smith, *Trans. Faraday Soc.* **47**, 1261 (1951).
14. Y. Kakiuchi, H. Shono, H. Komatsu, and K. Kigoshi, *J. Chem. Phys.* **19**, 1069 (1951).

15. A. Commeyras and G. A. Olah, *J. Am. Chem. Soc.* **91**, 2929 (1969).
16. V. Gold, L. J. Grant, and K. P. Morris, *J. Chem. Soc., Chem. Commun.* 397 (1976) and references cited therein.
17. P. Kebarle, R. N. Haynes, and J. G. Collins, *J. Am. Chem. Soc.* **89**, 5753 (1967).
18. A. W. Castelman Jr., I. N. Tang, and H. R. Munkelwitz, *Science* **173**, 1025 (1971).
19. J. O. Lundgren and J. M. Williams, *J. Chem. Phys.* **58**, 788 (1973).
20. J.-O. Lundgren, R. Tellgren, and I. Olovsson, *Acta Crystallogr. Sect. B* **B34** 2945 (1978).
21. K. O. Christe, C. J. Schack, and R. D. Wilson, *Inorg. Chem.* **14**, 2224 (1975).
22. K. O. Christe, P. Charpin, E. Soulie, R. Bougon, J. Fawcett, and D. R. Russel, *Inorg. Chem.* **23**, 3756 (1984).
23. E. M. Larson, K. D. Abney, A. C. Larson, and P. G. Eller, *Acta Crystallogr. Sect. B* **B47** 206 (1991).
24. D. Zhang, S. J. Rettig, J. Trotter, and F. Aubke, *Inorg. Chem.* **35**, 6113 (1996).
25. R. Minkwitz, C. Hirsch, and *Z. Naturforsch.* **54b**, 377 (1999).
26. R. Minkwitz, S. Schneider, and A. Kornath, *Inorg. Chem.* **37**, 4662 (1998).
27. R. Minkwitz and C. Hirsch, *Acta Crystallogr.* **C55**, 703 (1999).
28. G. D. Mateescu and G. M. Benedikt, *J. Am. Chem. Soc.* **101**, 3959 (1979).
29. G. A. Olah, A. L. Berrier, and G. K. S. Prakash, *J. Am. Chem. Soc.* **104**, 2373 (1982).
30. M. C. R. Symons, *J. Am. Chem. Soc.* **102**, 3982 (1980).
31. C. A. Coulson, Volume Commemoratif Victore Henri, Contributions a l' Étude de la Structure Moleculare, Liége, 1948, p. 15.
32. W. R. Rodwell and L. Radom, *J. Am. Chem. Soc.* **103**, 2865 (1981).
33. X. Huang, B. J. Braams, and J. M. Bowman, *J. Chem. Phys.* **122**, 044308 (2005) and references cited therein.
34. R. M. Izatt, K. Pawlak, and J. S. Bradshaw, *Chem. Rev.* **91**, 1721 (1991).
35. J. S. Bradshaw and R. M. Izatt, *Acc. Chem. Res.* **30**, 338 (1997).
36. J. W. Steed, *Coord. Chem. Rev.* **215**, 171 (2001).
37. P. Kebarle, *Ann. Rev. Phys. Chem.* **28**, 45 (1977).
38. A. Khan, *Chem. Phys. Lett.* **319**, 440 (2000).
39. J.-C. Jiang, Y.-S. Wang, H.-C. Chang, S. H. Lin, Y. T. Lee, G. Niedner-Schatteburg, and H.-C. Chang, *J. Am. Chem. Soc.* **122**, 1398 (2000).
40. M. Eigen, *Angew. Chem. Int. Ed. Engl.* **3**, 1 (1964).
41. G. Zundel, in *The Hydrogen Bond—Recent Developments in Theory and Experiments II. Structure and Spectroscopy*, P. Schuster, G. Zundel,and C. Sandorfy, eds., North-Holland, Amsterdam, 1976, p. 683.
42. U. Nagashima, H. Shinohara, N. Nishi, and H. Tanaka, *J. Chem. Phys.* **84**, 209 (1986).
43. S. Wei, Z. Shi, and A. W. Castleman Jr., *J. Chem. Phys.* **94**, 3268 (1991).
44. G. Niedner-Schatteburg, *Angew. Chem. Int. Ed.* **47**, 1008 (2008).
45. O. Vendrell, F. Gatti, and H.-D. Meyer, *Angew. Chem. Int. Ed.* **46**, 6918 (2007); *J. Chem. Phys.* **127**, 184303 (2007).
46. G. A. Olah, G. K. S. Prakash, M. Barzaghi, K. Lammertsma, P. v. R. Schleyer, and J. A. Pople, *J. Am. Chem. Soc.* **108**, 1032 (1986).
47. N. Hartz, G. Rasul, and G. A. Olah, *J. Am. Chem. Soc.* **115**, 1277 (1993).

48. A. I. Boldyrev and J. Simons, *J. Chem. Phys.* **97**, 4272 (1992); **99**, 769 (1993).
49. R. J. Gillespie and J. A. Leisten, *Quart. Rev. Chem. Soc.* **8**, 40 (1954).
50. C. MacLean and E. L. Mackor, *J. Chem. Phys.* **34**, 2207 (1961).
51. T. Birchall and R. J. Gillespie, *Can. J. Chem.* **43**, 1045 (1965).
52. G. A. Olah and E. Namanworth, *J. Am. Chem. Soc.* **88**, 5327 (1966).
53. G. A. Olah, J. Sommer, and E. Namanworth, *J. Am. Chem. Soc.* **89**, 3576 (1967).
54. R. Minkwitz and S. Schneider, *Z. Anorg. Allg. Chem.* **624**, 1989 (1998).
55. T. D. Fridgen, L. MacAleese, T. B. McMahon, J. Lemaire, and P. Maitre, *Phys. Chem. Chem. Phys.* **8**, 955 (2006).
56. P. Jutzi, C. Müller, A. Stammler, and H.-G. Stammler, *Organometallics* **19**, 1442 (2000).
57. S.-C. Park, K.-W. Maeng, and H. Kang, *Chem. Eur. J.* **9**, 1706 (2003).
58. G. A. Olah and J. Sommer, *J. Am. Chem. Soc.* **90**, 927 (1968).
59. Q.-F. Cheng and J. A. Stone, *J. Phys. Chem.* **99**, 1442 (1995).
60. G. Bouchoux and F. Berruyer-Penaud, *J. Phys. Chem. A* **107**, 7931 (2003).
61. E. M. Arnett and C. Y. Wu, *J. Am. Chem. Soc.* **82**, 5660 (1960).
62. T. Birchall, A. N. Bourns, R. J. Gillespie, and P. J. Smith, *Can. J. Chem.* **42**, 1433 (1964).
63. N. Solcà and O. Dopfer, *Chem. Phys. Lett.* **342**, 191 (2001); *J. Am. Chem. Soc.* **126**, 1716 (2004).
64. G. A. Olah, S. Yu, G. Liang, G. D. Mateescu, M. R. Bruce, D. J. Donovan, and M. Arvanaghi, *J. Org. Chem.* **46**, 571 (1981).
65. R. Minkwitz and S. Reinemann, *Z. Anorg. Allg. Chem.* **625**, 121 (1999).
66. K. O. Christe, J. Hegge, B. Hoge, and R. Haiges, *Angew. Chem. Int. Ed.* **46**, 6155 (2007).
67. G. A. Olah, H. Doggweiler, J. D. Felberg, and S. Frohlich, *J. Org. Chem.* **50**, 4847 (1985).
68. Z. Xie, R. Bau, and C. A. Reed, *J. Chem. Soc., Chem. Commun.* 2519 (1994).
69. G. A. Olah, G. Rasul, and G. K. S. Prakash, *J. Organomet. Chem.* **521**, 271 (1996).
70. A. R. Bassindale, D. J. Parker, P. G. Taylor, N. Auner, and B. Herrschaft, *Chem. Commun.* 565 (2000).
71. A. A. Korlyukov, S. A. Pogozhikh, Y. E. Ovchinnikov, K. A. Lyssenko, M. Yu. Antipin, A. G. Shipov, O. A. Zamyshlyaeva, E. P. Kramarova, V. V. Negrebetsky, I. P. Yakovlev, and Y. I. Baukov, *J. Organomet. Chem.* **691**, 3962 (2006).
72. R. Minkwitz and G. Nowicki, *Angew. Chem. Int. Ed. Engl.* **29**, 688 (1990).
73. R. Minkwitz and B. Bäck, *in ACS Symposium Series,* Vol. 555, J. S. Thrasher and S. H. Strauss, eds., American Chemical Society, Washington, DC, 1994, Chapter 6, p. 90
74. V. P. Reddy, E. Sinn, G. A. Olah, G. K. S. Prakash, and G. Rasul, *J. Phys. Chem. A* **108**, 4036 (2004).
75. F. Klages, J. E. Gordon, and H. A. Jung, *Chem. Ber.* **98**, 3748 (1965).
76. D. Vagedes, G. Erker, and R. Föhlich, *J. Organomet. Chem.* **641**, 148 (2002).
77. G. A. Olah and D. H. O'Brien, *J. Am. Chem. Soc.* **89**, 1725 (1967).
78. D. Jaques and J. A. Leisten, *J. Chem. Soc.* 4963 (1961).
79. G. A. Olah and P. J. Szilagyi, *J. Org. Chem.* **36**, 1121 (1971).
80. P. George, C. W. Bock, and J. P. Glusker, *J. Phys. Chem.* **96**, 3702 (1992).
81. J. M. Coxon, R. G. A. R. Maclagan, A. Rauk, A. J. Thorpe, and D. Whalen, *J. Am. Chem. Soc.* **119**, 4712 (1997).

82. B. Ganguly and B. Fuchs, *J. Org. Chem.* **65**, 558 (2000).
83. A. Vila and R. A. Mosquera, *Tetrahedron* **57**, 9415 (2001).
84. A. Vila and R. A. Mosquera, *Chem. Phys. Lett.* **371**, 540 (2003).
85. P. R. Carlier, N. Deora, and T. D. Crawford, *J. Org. Chem.* **71**, 1592 (2006).
86. R. Minkwitz and D. Konikowski, *Z. Anorg. Allg. Chem.* **621**, 2055 (1995).
87. G. A. Olah, A. M. White, and D. H. O'Brien, *Chem. Rev.* **70**, 561 (1970).
88. J. Sommer, P. Canivet, S. Schwartz, and P. Rimmelin, *Nouv. J. Chim.* **5**, 45 (1981).
89. R. Jost, J. Sommer, and C. Engdahl, *Acta Chem. Scand.* **43**, 888 (1989).
90. H. Perst,in *Carbonium Ions*, G. A. Olahand P. v. R. Schleyer, eds., Wiley-Interscience, New York, London, 1976, Vol. 5, Chapter 34, p. 1961.
91. H. Meerwein, G. Hinz, P. Hoffmann, E. Kroning, and E. Pfeil, *J. Prakt. Chem.,* [2] **147**, 257 (1937).
92. F. Klages and H. A. Jung, *Chem. Ber.* **98**, 3757 (1965).
93. F. Klages and H. Meuresch, *Chem. Ber.* **85**, 863 (1952).
94. R. A. Goodrich and P. M. Treichel, *J. Am. Chem. Soc.* **88**, 3509 (1966).
95. H. Meerwein, E. Battenberg, H. Gold, E. Pfeil, and G. Willfang, *J. Prakt. Chem.,* [2] **154**, 83 (1939).
96. H. Meerwein, *Org. Synth.* **46**, 120 (1966).
97. H. Meerwein, *Org. Synth.* **46**, 113 (1966).
98. R. Minkwitz, S. Reinemann, R. Seelbinder, D. Konikowski, H. Hartl, I. Brüdgam, J. Hegge, B. Hoge, J. A. Sheehy, and K. O. Christe, *Inorg. Chem.* **40**, 4404 (2001).
99. G. A. Olah, *Halonium Ions*, Wiley-Interscience, New York, 1975.
100. G. A. Olah, A. Burrichter, G. Rasul, R. Gnann, K. O. Christe, and G. K. S. Prakash, *J. Am. Chem. Soc.* **119**, 8035 (1997).
101. A. N. Nesmeyanov, L. G. Makarova, and T. P. Tolstaya, *Tetrahedron* **1**, 145 (1957).
102. A. N. Nesmeyanov and T. P. Tolstaya, *Russ. Chem. Bull. (Engl. Transl.)* **8**, 620 (1959).
103. A. N. Nesmeyanov, *Selected Works in Organic Chemistry*, Pergamon Press, Oxford, 1963. p. 772.
104. A. N. Nesmeyanov, T. P. Tolstaya, and A. V. Grib, *Dokl. Akad. Nauk SSSR* **139**, 114 (1961).
105. G. A. Olah and E. G. Melby, *J. Am. Chem. Soc.* **95**, 4971 (1973).
106. G. A. Olah, J. A. Olah, and T. Ohyama, *J. Am. Chem. Soc.* **106**, 5284 (1984).
107. G. A. Olah and J. R. DeMember, *J. Am. Chem. Soc.* **92**, 2562 (1970).
108. B. G. Ramsey and R. W. Taft, *J. Am. Chem. Soc.* **88**, 3058 (1966).
109. G. A. Olah, unpublished results.
110. J. B. Lambert and D. H. Johnson, *J. Am. Chem. Soc.* **90**, 1349 (1968).
111. M. I. Watkins, W. M. Ip, G. A. Olah and R. Bau, *J. Am. Chem. Soc.* **104**, 2365 (1982).
112. J. W. Schubert, T. J. Doodly and B. K. Ohta, *J. Org. Chem.* **72**, 2452 (2007).
113a. M. Etzkorn, R. Aniszfeld, T. Li, H. Buchholz, G. Rasul, G. K. S. Prakash, and G. A. Olah, *Eur. J. Org. Chem.*, 4555 (2008).
113b. M. Maskal, N. Hafezi, N. K. Maher and J. C. Ferringer, *J. Am. Chem. Soc.*, **130** 13532 (2008).
114. M. Kira, T. Hino, and H. Sakurai, *J. Am. Chem. Soc.* **114**, 6697 (1992).

115. G. A. Olah, X.-Y. Li, Q. Wang, G. Rasul, and G. K. S. Prakash, *J. Am. Chem. Soc.* **117**, 8962 (1995).

116. G. A. Olah, Q. Wang, X.-Y. Li, G. Rasul, and G. K. S. Prakash, *Macromolecules* **29**, 1857 (1996).

117. G. A. Olah, H. Doggweiler, J. D. Felberg, S. Frohlich, M. J. Grdina, R. Karpeles, T. Keumi, S. Inaba, W. M. Ip, K. Lammertsma, G. Salem, and D. C. Tabor, *J. Am. Chem. Soc.* **106**, 2143 (1984).

118. A. N. Nesmeyanov, E. G. Perevalova, Yu. T. Struchkov, M. Yu. Antipin, K. I. Grandberg, and V. P. Dyadchenko, *J. Organomet. Chem.* **210**, 343 (1980).

119. H. Schmidbaur, A. Kolb, E. Zeller, A. Schier, and H. Beruda, *Z. Anorg. Allg. Chem.* **619**, 1575 (1993).

120. A. Kolb, P. Bissinger, and H. Schmidbaur, *Z. Anorg. Allg. Chem.* **619**, 1580 (1993).

121. Y. Yang, V. Ramamoorthy, and P. R. Sharp, *Inorg. Chem.* **32**, 1946 (1993).

122. K. Angermaier and H. Schmidbaur, *Inorg. Chem.* **33**, 2069 (1994).

123. S.-C. Chung, S. Krüger, H. Schmidbaur, and N. Rösch, *Inorg. Chem.* **35**, 5387 (1996).

124. H. Schmidbaur, S. Hofreiter, and M. Paul, *Nature* **377**, 503 (1995).

125. G. A. Olah, D. G. Parker, and N. Yoneda, *Angew. Chem. Int. Ed. Engl.* **17**, 909 (1978).

126. R. W. Alder and M. C. Whiting, *J. Chem. Soc.* 4707 (1964) and references cited therein.

127. K. O. Christe, W. W. Wilson, and E. C. Curtis, *Inorg. Chem.* **18**, 2578 (1979).

128. R. Minkwitz, C. Hirsch, and H. Hartl, *Angew. Chem. Int. Ed.* **37**, 1681 (1998).

129. J.-M. Savariault and M. S. Lehmann, *J. Am. Chem. Soc.* **102**, 1298 (1980).

130. J. C. Mitchell, *Chem. Soc. Rev.* **14**, 399 (1985).

131. J. C. Mitchell, S. Heaton, and N. A. Porter, *Tetrahedron Lett.* **25**, 3769 (1984).

132. A. J. Bloodworth, T. Melvin, and J. C. Mitchell, *J. Org. Chem.* **51**, 2612 (1986). **53**, 1078 (1988).

133. R. Minkwitz and V. Gerhard, *Z. Anorg. Allg. Chem.* **603**, 95 (1991).

134. G. A. Olah, G. Rasul, A. Burrichter, M. Hachoumy, G. K. S. Prakash, R. I. Wagner, and K. O. Christe, *J. Am. Chem. Soc.* **119**, 9572 (1997).

135. R. Trambarulo, S. N. Ghosh, C. A. Barrus, Jr., and W. Gordy, *J. Chem. Phys.* **21**, 851 (1953).

136. P. D. Bartlett and M. Stiles, *J. Am. Chem. Soc.* **77**, 2806 (1955).

137. J. P. Wibaut, F. L. J. Sixma, L. W. F. Kampschmidt, and H. Boer, *Recl. Trav. Chim. Pays-Bas* **69**, 1355 (1950).

138. J. P. Wibaut and F. L. J. Sixma, *Recl. Trav. Chim. Pays-Bas* **71**, 761 (1951).

139. P. S. Bailey, *Chem. Rev.* **58**, 925 (1958).

140. L. Horner, H. Schaefer, and W. Ludwig, *Chem. Ber.* **91**, 75 (1958).

141. *Ozone Reactions with Organic Compounds, Adv. Chem. Ser.*, Vol. 112, P. S. Bailey, ed., American Chemical Society, Washington, DC, 1972.

142. S. D. Razumovskii and G. E. Zaikov, *Ozone and Its Reactions with Organic Compounds*, Elsevier, Amsterdam, 1984.

143. P. S. Bailey, J. W. Ward, F. E. Potts, and R. E. Hornish, *Adv. Chem. Ser.* **112**, 1 (1972).

144. F. Cacace and M. Speranza, *Science* **265**, 208 (1994).

145. M. Speranza, *Inorg. Chem.* **35**, 6140 (1996).

146. C. Meredith, G. E. Quelch, and H. F. Schaefer III, *J. Am. Chem. Soc.* **113**, 1186 (1991).

147. J. Rubio, N. Russo, and E. Sicilia, *Int. J. Quantum Chem.* **61**, 415 (1997).
148. M. Ceotto, F. A. Gianturco, and D. M. Hirst, *J. Phys. Chem. A* **103**, 9984 (1999).
149. G. A. Olah, D. H. O'Brien, and C. J. Pittman, Jr., *J. Am. Chem. Soc.* **89**, 2996 (1967).
150. K. O. Christe, *Inorg. Chem.* **14**, 2230 (1975).
151. A. G. Harrison and J. C. Thynne, *Trans. Faraday Soc.* **62**, 3345 (1966).
152. J. L. Beauchamp and S. E. Butrill, Jr., *J. Chem. Phys.* **48**, 1783 (1968).
153. M. Araki, H. Ozeki, and S. Saito, *J. Mol. Spectr.* **192**, 228 (1998).
154. L. Dore, C. D. Esposti, A. Mazzavillani, and G. Cazzoli, *Chem. Phys. Lett.* **300**, 489 (1999).
155. F. Tinti, L. Bizzocchi, C. D. Esposti, and L. Dore, *J. Mol. Spectr.* **240**, 202 (2006).
156. A. H. Otto and R. Steudel, *Z. Anorg. Allg. Chem.* **626**, 2063 (2000).
157. G. A. Olah, G. K. S. Prakash, M. Marcelli, and K. Lammertsma, *J. Phys. Chem.* **92**, 878 (1988).
158. G. A. Olah, G. Rasul, and G. K. S. Prakash, *Chem. Eur. J.* **3**, 1039 (1997).
159. R. Minkwitz, U. Lohmann, and H. Preut, *Z. Naturforsch.* **51b**, 277 (1996).
160. R. Minkwitz, G. Nowicki, B. Bäck, and W. Sawodny, *Inorg. Chem.* **32**, 787 (1993).
161. R. Minkwitz, A. Kornath, W. Sawodny, and J. Hahn, *Inorg. Chem.* **35**, 3622 (1996).
162. J. B. Lambert, R. G. Keske, and D. K. Weary, *J. Am. Chem. Soc.* **89**, 5921 (1967).
163. R. Minkwitz, V. Gerhard, and T. Norkat, *Z. Naturforsch.* **44b**, 1337 (1989).
164. R. Minkwitz and G. Nowicki, *Inorg. Chem.* **31**, 225 (1992).
165. A. M. Forster and A. J. Downs, *J. Chem. Soc., Dalton Trans.* 2827 (1984).
166. R. Minkwitz and A. Werner, *J. Fluorine Chem.* **39**, 141 (1988).
167. J. Goerdeler, in *Methoden der Organischen Chemie (Houben-Weyl)*, Vol. 9, E. Müller, ed., Thieme, Stuttgart, 1955, pp. 175–194 and references cited therein.
168. E. Vilsmaier, in *Methoden der Organischen Chemie (Houben-Weyl)*, Vol. E11, D. Klamann, ed., Thieme, Stuttgart, 1985, pp. 341–502 and references therein.
169. G. A. Olah, K. K. Laali, Q. Wang, and G. K. S. Prakash, *Onium Ions*, Wiley, New York, 1998, Chapter 4, p. 167.
170. C. S. F. Tang and H. Rapoport, *J. Org. Chem.* **38**, 2806 (1973).
171. R. M. Acheson and D. R. Harrison, *J. Chem. Soc. D: Chem. Commun.* 724 (1969).
172. K. Miyatake, K. Yamamoto, K. Endo, and E. Tsuchida, *J. Org. Chem.* **63**, 7522 (1998).
173. B. Liu and H. J. Shine, *J. Phys. Org. Chem.* **14**, 81 (2001).
174. T. Umemoto and S. Ishihara, *Tetrahedron Lett.* **31**, 3579 (1990).
175. T. Umemoto and S. Ishihara, *J. Am. Chem. Soc.* **115**, 2156 (1993).
176. G. K. S. Prakash, Cs. Weber, S. Chacko, and G. A. Olah, *Org. Lett.* **9**, 1863 (2007).
177. G. K. S. Prakash, I. Ledneczki, S. Chacko, and G. A. Olah, *Org. Lett.* **10**, 557 (2008).
178. E. Kelstrup, A. Kjìr, S. Abrahamsson, and B. Dahlén, *J. Chem. Soc., Chem. Commun.* 629 (1975).
179. B. M. Trost and L. S. Melvin, Jr., in *Organic Chemistry, A Series of Monographs*, Vol. 31, A. T. Blomquistand H. H. Wasserman, eds., Academic Press, New York, 1975.
180. G. Capozzi, V. Lucchini, G. Modena, and P. Scrimin, *Tetrahedron Lett.* 911 (1977).
181. R. Destro, T. Pilati, and M. Simonetta, *J. Chem. Soc., Chem. Commun.* 576 (1977).
182. G. I. Borodkin, Yu. V. Gatilov, E. I. Chernyak, and V. G. Shubin, *Russ. Chem. Bull. (Engl. Transl.)* **34**, 2622 (1985); **36**, 2065 (1987).

183. X. Huang, R. J. Batchelor, F. W. B. Einstein, and A. W. Bennet, *J. Org. Chem.* **59**, 7108 (1994).
184. V. Lucchini, G. Modena, and L. Pasquato, *J. Am. Chem. Soc.* **113**, 6600 (1991).
185. V. Lucchini, G. Modena, M. Pasi, and L. Pasquato, *J. Org. Chem.* **62**, 7018 (1997).
186. R. Destro, V. Lucchini, G. Modena, and L. Pasquato, *J. Org. Chem.* **65**, 3367 (2000).
187. R. Destro, E. Ortoleva, G. Modena, L. Pasquato, and V. Lucchini, *Helv. Chim. Acta* **84**, 860 (2001).
188. V. Lucchini, G. Modena, and L. Pasquato, *J. Am. Chem. Soc.* **115**, 4527 (1993).
189. L. Pasquato and G. Modena, *Chem. Commun.* 1469 (1999).
190. Y. Sugihara, Y. Aoyama, and J. Nakayama, *Chem. Lett.* 980 (2001).
191. M. G. Beaver, S. B. Billings, and K. A. Woerpel, *Eur. J. Org. Chem.* 771 (2008).
192. J. Park, S. Kawatkar, J.-H. Kim, and G.-J. Boons, *Org. Lett.* **9**, 1959 (2007).
193. J.-H. Kim, H. Yang, J. Park, and G.-J. Boons, *J. Am. Chem. Soc.* **127**, 12090 (2005).
194. M. G. Beaver, S. B. Billings, and K. A. Woerpel, *J. Am. Chem. Soc.* **130**, 2082 (2008).
195. M. I. Lazareva, Y. K. Kryschenko, R. Caple, D. Wakefield, A. Hayford, W. A. Smit, and A. S. Shashkov, *Tetrahedron Lett.* **39**, 8787 (1998).
196. F. Klages and H.-U. Schmidt, *Chem. Ber.* **96**, 2063 (1963).
197. S. Sato, H. Ameta, E. Horn, O. Takahashi, and N. Furukawa, *J. Am. Chem. Soc.* **119**, 12374 (1997).
198. N. Furukawa and S. Sato, *Heteroatom Chem.* **13**, 406 (2002).
199. N. Bartlett and P. L. Robinson, *J. Chem. Soc.* 3417 (1961).
200. M. Azeem, M. Brownstein, and R. J. Gillespie, *Can. J. Chem.* **47**, 4159 (1969).
201. M. R. Barr and B. A. Dunnel, *Can. J. Chem.* **48**, 895 (1970).
202. D. D. Gibler, C. J. Adams, M. Fischer, A. Zalkin, and N. Bartlett, *Inorg. Chem.* **11**, 2325 (1972).
203. L. Kolditz and W. Schäfer, *Z. Anorg. Allg. Chem.* **315**, 35 (1962).
204. R. Minkwitz and V. Gerhard, *Z. Naturforsch.* **44b**, 364 (1989).
205. J. Passmore, E. K. Richardson, and P. Taylor, *Inorg. Chem.* **17**, 1681 (1971).
206. R. Minkwitz, K. Jänichen, H. Prenzel, and V. Wölfel, *Z. Naturforsch.* **40b**, 53 (1985).
207. B. H. Christian, M. J. Collins, R. J. Gillespie, and J. F. Sawyer, *Inorg. Chem.* **25**, 777 (1986).
208. R. Minkwitz, R. Lekies, and H. Preut, *Z. Naturforsch.* **42b**, 1227 (1987).
209. R. Minkwitz, A. Kornath, and H. Preut, *Z. Naturforsch.* **47b**, 594 (1992).
210. J. P. Johnson, M. Murchie, J. Passmore, M. Tajik, P. S. White, and C.-M. Wong, *Can. J. Chem.* **65**, 2744 (1987).
211. R. Minkwitz, U. Nass, A. Radünz, and H. Preut, *Z. Naturforsch.* **40b**, 1123 (1985).
212. R. Minkwitz, R. Lekies, and A. Werner, *Z. Anorg. Allg. Chem.* **559**, 163 (1988).
213. R. Minkwitz, H. Prenzel, A. Werner, and H. Preut, *Z. Anorg. Allg. Chem.* **562**, 42 (1988).
214. M. Kramar and L. C. Duncan, *Inorg. Chem.* **10**, 647 (1971).
215. F. Pauer, M. Erhart, R. Mews, and D. Stalke, *Z. Naturforsch.* **45b**, 271 (1990).
216. R. Minkwitz, W. Molsbeck, B. Bäck, and H. Preut, *Z. Anorg. Allg. Chem.* **619**, 174 (1993).

217. R. Minkwitz, G. Nowicki, and H. Preut, *Z. Anorg. Allg. Chem.* **573**, 185 (1989).
218. R. Minkwitz, H. Preut, and J. Sawatzki, *Z. Anorg. Allg. Chem.* **569**, 158 (1989).
219. R. Minkwitz and B. Bäck, *Z. Naturforsch.* **48b**, 694 (1993).
220. H. Meerwein, K.-F. Zenner, and R. Gipp, Justus Liebigs Ann. Chem. **688**, 67 (1965).
221. R. Minkwitz, U. Lohmann, and H. Preut, *Z. Anorg. Allg. Chem.* **621**, 1547 (1995).
222. R. Minkwitz, V. Gerhard, and H. Preut, *Z. Anorg. Allg. Chem.* **596**, 99 (1991).
223. M. Erhart and R. Mews, *Z. Anorg. Allg. Chem.* **615**, 117 (1992).
224. R. Minkwitz, G. Nowicki, and H. Preut, *Z. Anorg. Allg. Chem.* **611**, 23 (1992).
225. R. Minkwitz, B. Bäck, and H. Preut, *Z. Naturforsch.* **49b**, 881 (1994).
226. R. Minkwitz, B. Bäck, and H. Preut, *Z. Anorg. Allg. Chem.* **620**, 1125 (1994).
227. A. Commeyras and G. A. Olah, *J. Am. Chem. Soc.* **91**, 2929 (1969).
228. A. Kornath, R. Seelbinder, and R. Minkwitz, *Angew. Chem. Int. Ed.* **44**, 973 (2005).
229. R. Minkwitz, R. Seelbinder, and R. Schöbel, *Angew. Chem. Int. Ed.* **41**, 111 (2002).
230. G. A. Olah, D. J. Donovan, H. C. Lin, H. Mayr, P. Andreozzi, and G. Klopman, *J. Org. Chem.* **43**, 2268 (1978).
231. M. Spiekermann and B. Schrader, *Angew. Chem. Int. Ed. Engl.* **16**, 197 (1977).
232. G. Rasul, G. K. S. Prakash, and G. A. Olah, *J. Org. Chem.* **65**, 8786 (2000).
233. G. A. Olah, A. T. Ku, and J. A. Olah, *J. Org. Chem.* **35**, 3904 (1970).
234. K. K. Laali and D. S. Nagvekar, *J. Org. Chem.* **56**, 1867 (1991).
235. K. K. Laali and J. J. Houser, *J. Phys. Org. Chem.* **5**, 244 (1992).
236. G. A. Olah, P. Schilling, J. M. Bollinger and J. Nishimura, *J. Am. Chem. Soc.* **96**, 2221 (1974).
237. J. Y. Calves and R. J. Gillespie, *J. Am. Chem. Soc.* **99**, 1788 (1977).
238. R. Minkwitz and W. Molsbeck, *Z. Naturforsch.* **47b**, 22 (1992).
239. W. Zhao, H. J. Shine, and B. R. Whittlesey, *J. Org. Chem.* **62**, 8693 (1997).
240. W. Zhao and H. J. Shine, *Can. J. Chem.* **76**, 695 (1998).
241. D. Szabó, J. Varga, A. Csámpai, and I. Kapovits, *Tetrahedron: Asymmetry* **11**, 1303 (2000).
242. R. Mews and H. Henle, *J. Fluorine Chem.* **14**, 495 (1979).
243. A. H. Cowley, D. J. Pagel, and M. L. Walker, *J. Am. Chem. Soc.* **100**, 7065 (1978).
244. M. Erhart, R. Mews, F. Pauer, D. Stalke, and G. M. Sheldrick, *Chem. Ber.* **124**, 31 (1991).
245. G. A. Olah, B. G. B. Gupta, and S. C. Narang, *J. Am. Chem. Soc.* **101**, 5317 (1979).
246. G. A. Olah and B. G. B. Gupta, *J. Org. Chem.* **48**, 3585 (1983).
247. R. Laitinen, R. Steudel, and R. Weiss, *J. Chem. Soc., Dalton Trans.* 1095 (1986).
248. R. Minkwitz, R. Krause, H. Härtner, and W. Sawodny, *Z. Anorg. Allg. Chem.* **593**, 137 (1991).
249. R. Minkwitz and A. Kornath, *Z. Anorg. Allg. Chem.* **605**, 101 (1991).
250. R. Minkwitz, A. Kornath, W. Sawodny, V. Typke, and J. A. Boatz, *J. Am. Chem. Soc.* **122**, 1073 (2000).
251. R. Minkwitz, V. Gerhard, R. Krause, H. Prenzel, and H. Preut, *Z. Anorg. Allg. Chem.* **559**, 154 (1988).
252. H. Naka, T. Maruyama, S. Sato, and N. Furukawa, *Tetrahedron Lett.* **40**, 345 (1999).

253. H. Shima, R. Kobayashi, T. Nabeshima, and N. Furukawa, *Tetrahedron Lett.* **37**, 667 (1996).

254. M. J. Caserio, J. K. Kim, S. D. Kahn, and W. J. Hehre, *J. Am. Chem. Soc.* **107**, 2807 (1985).

255. E. L. Eliel and R. L. Willer, *J. Am. Chem. Soc.* **99**, 1936 (1977).

256. A. Wakamiya, T. Nishinaga, and K. Komatsu, *J. Am. Chem. Soc.* **124**, 15038 (2002).

257. T. Nishinaga, A. Wakamiya, and K. Komatsu, *Chem. Commun.*, 777 (1999).

258. G. K. S. Prakash, C. Bae, Q. Wang, G. Rasul, and G. A. Olah, *J. Org. Chem.* **65**, 7646 (2000).

259. P. G. Jones, G. M. Sheldrick, and E. Hädicke, *Acta Crystallogr. B* **36**, 2777 (1980).

260. K. Angermaier and H. Schmidbaur, *Chem. Ber.* **127**, 2387 (1994).

261. A. Sladek and H. Schmidbaur, *Chem. Ber.* **128**, 907 (1995).

262. M. J. Calhorda, F. Canales, M. C. Gimeno, J. Jiménez, P. G. Jones, A. Laguna, and L. F. Veiros, *Organometallics* **16**, 3837 (1997).

263. F. Canales, M. C. Gimeno, P. G. Jones, and A. Laguna, *Angew. Chem. Int. Ed. Engl.* **33**, 769 (1994).

264. F. Canales, C. Gimeno, A. Laguna, and M. D. Villacampa, *Inorg. Chim. Acta* **244**, 95 (1996).

265. R. M. Dávila, A. Elduque, T. Grant, R. J. Staples, and J. P. Fackler, Jr., *Inorg. Chem.* **32**, 1749 (1993).

266. M. C. Gimeno, P. G. Jones, A. Laguna, M. Laguna, and R. Terroba, *Inorg. Chem.* **33**, 3932 (1994).

267. A. Sladek and H. Schmidbaur, *Inorg. Chem.* **35**, 3268 (1996).

268. H. Reinboldt,in *Methoden der Organischen Chemie (Houben-Weyl)*, Vol. 9, E. Müller, ed., Thieme, Stuttgart, 1955, pp. 1034–1043, 1075–1082 and references cited therein.

269. R. J. Shine,in *Organic Selenium Compounds: Their Chemistry and Biology*, D. L. Klaymanand W. H. H. Günther, eds., Wiley-Interscience, New York, 1973, Chapter 6, p. 223.

270. K. J. Irgolic, in *Methoden der Organischen Chemie (Houben-Weyl)*, Vol. E12b, D. Klamann,ed., Thieme, Stuttgart, 1990, pp. 676–706 and references therein.

271. A. V. Zakharov, I. D. Sadekov, and V. I. Minkin, *Russ. Chem. Rev.* **75**, 207 (2006).

272. G. A. Olah, J. J. Svoboda, and A. T. Ku, *J. Org. Chem.* **38**, 4447 (1973).

273. G. A. Olah, A. Howard, and G. K. S. Prakash, unpublished results.

274. R. Minkwitz, A. Kornath, and W. Sawodny, *Angew. Chem. Int. Ed. Engl.* **31**, 643 (1992).

275. K. Laali, H. Y. Chen, and R. J. Gerzina, *J. Org. Chem.* **52**, 4126 (1987).

276. W. Dumont, P. Bayet, and A. Krief, *Angew. Chem. Int. Ed. Engl.* **13**, 274 (1974).

277. P. G. Gassman, T. Miura, and A. Mossman, *J. Org. Chem.* **47**, 954 (1982).

278. N. V. Kondratenko, G. N. Timofeeva, V. I. Popov, and L. M. Yagupol'skii, *Zhur. Org. Khim.* **15**, 2432 (1979).

279. T. Kataoka, Y. Banno, S. Watanabe, T. Iwamura, and H. Shimizu, *Tetrahedron Lett.* **38**, 1809 (1997).

280. S. Watanabe, E. Mori, H. Nagai, T. Iwamura, T. Iwama, and T. Kataoka, *J. Org. Chem.* **65**, 8893 (2000).

281. S. Watanabe, Y. Miura, T. Iwamura, H. Nagasawa, and T. Kataoka, *Tetrahedron Lett.* **48**, 813 (2007).

282. G. H. Schmid and D. G. Garratt, *Tetrahedron Lett.* 3991 (1975).
283. N. N. Magdesieva and M. F. Gordeev, *Zhur. Org. Khim.* **23**, 949 (1987).
284. G. I. Borodkin, Yu. V. Gatilov, T. V. Rybalova, E. I. Chernyak, and V. G. Shubin, *Russ. Chem. Bull. (Engl. Transl.)* **35**, 2602 (1986).
285. D. Naumann and W. Tyrra, *J. Chem. Soc., Chem. Commun.* 47 (1989).
286. D. Naumann, W. Tyrra, R. Herrmann, I. Pantenburg, and M. S. Wickleder, *Z. Anorg. Allg. Chem.* **628**, 833 (2002).
287. J. Jeske W.-W. du Mont F. Ruthe P.G. Jones L.M. Mercuri and P. Deplano *Eur. J. Inorg. Chem.* 1591, (2000).
288. Y. Matano, H. Suzuki, and N. Azuma, *Organometallics* **15**, 3760 (1996).
289. Y. Matano, S. A. Begum, T. Miyamatsu, and H. Suzuki, *Organometallics* **17**, 4332 (1998).
290. M. Ochiai, T. Nagaoka, T. Sueda, J. Yan, D.-W. Chen, and K. Miyamoto, *Org. Biomol. Chem.* **1**, 1517 (2003).
291. M. Ochiai, T. Sueda, R. Noda, and M. Shiro, *J. Org. Chem.* **64**, 8563 (1999).
292. M. Minoura, T. Mukuda, T. Sagami, and K. Akiba, *J. Am. Chem. Soc.* **121**, 10852 (1999); *Heteroatom Chem.* **12**, 380 (2001).
293. R. Faggiani, R. J. Gillespie, and J. W. Kolis, *J. Chem. Soc., Chem. Commun.* 592 (1987).
294. R. D. Peacock, *Prog. Inorg. Chem.* **2**, 193 (1960).
295. A. J. Edwards and G. R. Jones, *Chem. Commun.* 346 (1968).
296. R. J. Gillespie and W. A. Whitla, *Can. J. Chem.* **48**, 657 (1970).
297. R. J. Gillespie and W. A. Whitla, *Can. J. Chem.* **47**, 4153 (1969).
298. J. Passmore, E. K. Richardson, T. K. Whidden, and P. S. White, *Can. J. Chem.* **58**, 851 (1980).
299. W. V. F. Brooks, J. Passmore, and K. Richardson, *Can. J. Chem.* **57**, 3230 (1979).
300. J. Passmore, T. S. Cameron, P. D. Boyle, G. Schatte, and T. Way, *Can. J. Chem.* **74**, 1671 (1996).
301. J. Beck and F. Steden, *Z. Naturforsch.* **58b**, 711 (2003).
302. J. M. Rautiainen, T. Way, G. Schatte, J. Passmore, R. S. Laitinen, R. S. Suontamo, and J. Valkonen, *Inorg. Chem.* **44**, 1904 (2005).
303. T. S. Cameron, I. Dionne, I. Krossing, and J. Passmore, *Solid State Sci.* **4**, 1435 (2002).
304. C. Lensch, P. G. Jones, and Sheldrick, *Z. Naturforsch.* **7b**, 944 (1982).
305. K. Angermaier and H. Schmidbaur, *Z. Naturforsch.* **51b**, 879 (1996).
306. S. Canales, O. Crespo, M. C. Gimeno, P. G. Jones, and A. Laguna, *Chem. Commun.* 679 (1999).
307. W. K. Musker, T. L. Wolford, and P. B. Roush, *J. Am. Chem. Soc.* **100**, 6416 (1978).
308. W. K. Musker, *Acc. Chem. Res.* **13**, 200 (1980).
309. H. Fujihara, R. Akaishi, and N. Furukawa, *J. Chem. Soc., Chem. Commun.* 930 (1987).
310. H. Fujihara and N. Furukawa, *J. Mol. Struct. (Theochem)* **186**, 261 (1989).
311. V. G. Nenajdenko, N. E. Sevchenko, E. S. Balenkova, and I. V. Alabugin, *Chem. Rev.* **103**, 229 (2003).
312. N. Nakayama, O. Takahashi, O. Kikuchi, and N. Furukawa, *Heteroatom Chem.* **11**, 31 (2000).

313. T. Kimura, Y. Horie, S. Ogawa, N. Furukawa, and F. Iwasaki, *Heteroatom Chem.* **4**, 243 (1993).
314. T. Kimura, Y. Izumi, E. Horn, and N. Furukawa, *Heteroatom Chem.* **7**, 313 (1996).
315. N. Nakayama, O. Takahashi, O. Kikuchi, and N. Furukawa, *J. Mol. Struct. (Theochem)* **542**, 215 (2001).
316. N. Furukawa, K. Kobayashi, and S. Sato, *J. Organomet. Chem.* **611**, 116 (2000).
317. H. Fujihara, R. Akaishi, T. Erata, and N. Furukawa, *J. Chem. Soc., Chem. Commun.* 1789 (1989).
318. H. Fujihara, T. Ninoi, R. Akaishi, T. Erata, and N. Furukawa, *Tetrahedron Lett.* **32**, 4537 (1991).
319. R. Minkwitz, A. Kornath, R. Krause, and H. Preut, *Z. Anorg. Allg. Chem.* **620**, 632 (1994).
320. N. Furukawa, *Bull. Chem. Soc. Jpn.* **70**, 2571 (1997).
321. H. Fujihara, H. Mima, T. Erata, and N. Furukawa, *J. Am. Chem. Soc.* **114**, 3117 (1992).
322. A. B. Bergholdt, K. Kobayashi, E. Horn, O. Takahashi, S. Sato, N. Furukawa, M. Yokoyama, and K. Yamaguchi, *J. Am. Chem. Soc.* **120**, 1230 (1998).
323. K. Kobayashi, S. Sato, E. Horn, and N. Furukawa, *Angew. Chem. Int. Ed.* **39**, 1318 (2000).
324. T. Nakahodo, O. Takahashi, E. Horn, and N. Furukawa, *Chem. Commun.* 1767 (1997).
325. H. Fujihara, T. Nakahodo, and N. Furukawa, *Chem. Commun.* 311 (1996).
326. K. Kobayashi, N. Deguchi, E. Horn, and N. Furukawa, *Angew. Chem. Int. Ed.* **37**, 984 (1998).
327. G. A. Olah, K. K. Laali, Q. Wang, and G. K. S. Prakash, *Onium Ions*, Wiley, New York, 1998, Chapter 6, p. 246.
328a. C. Hartmann and V. Meyer, *Ber. Dtsch. Chem. Ges.* **27**, 426 (1894).
328b. F. M. Beringer, P. Ganis, G. Avitabile, H. Jaffe, *J. Org. Chem.* **37**, 879 (1972) and references cited therein.
329. A. Varvoglis, *Chem. Soc. Rev.* **10**, 377 (1981).
330. A. Varvoglis, *The Organic Chemistry of Polycoordinated Iodine*, VCH, New York, 1992.
331. P. J. Stang, *Angew. Chem. Int. Ed. Engl.* **31**, 274 (1992).
332. P. J. Stang and V. V. Zhdankin, *Chem. Rev.* **96**, 1123 (1996).
333. G. A. Olah and J. R. DeMember, *J. Am. Chem. Soc.* **91**, 2113 (1969).
334. I. Roberts and G. E. Kimball, *J. Am. Chem. Soc.* **59**, 947 (1937).
335. B. Capon, *Quart. Rev.* **18**, 45 (1964).
336. R. C. Fay, *Topics Stereochem.* **3**, 237 (1968).
337. S. Winstein and H. J. Lucas, *J. Am. Chem. Soc.* **61**, 1576, 2845 (1939).
338. J. G. Traynham, *J. Chem. Ed.* **40**, 392 (1963).
339. E. S. Gould, *Mechanism and Structure in Organic Chemistry*, Holt, Reinhart & Winston, New York, 1959, p. 523.
340. G. A. Olah and J. M. Bollinger, *J. Am. Chem. Soc.* **89**, 4744 (1967).
341. P. E. Peterson, *Acc. Chem. Res.* **4**, 407 (1971) and references cited therein.
342. G. A. Olah and J. Shen, *J. Am. Chem. Soc.* **95**, 3582 (1973).
343. S. K. Lee, T. Amano, K. Kawaguchi, and M. Oldani, *J. Mol. Spectrosc.* **130**, 1(1988).

344. K. O. Christe, *J. Fluorine Chem.* **25**, 269 (1984).
345. D. Mootz and K. Bartmann, *Angew. Chem. Int. Ed. Engl.* **27**, 391 (1988).
346. D. Mootz and K. Bartmann, *Z. Naturforsch.* **46b**, 1659 (1991).
347. D. Kim and M. L. Klein, *J. Phys. Chem. B* **104**, 10074 (2000).
348. G. A. Olah, G. Rasul, M. Hachoumy, A. Burrichter, and G. K. S. Prakash, *J. Am. Chem. Soc.* **122**, 2737 (2000).
349. G. A. Olah, Y. Yamada, and R. J. Spear, *J. Am. Chem. Soc.* **97**, 680 (1975).
350. G. A. Olah and T. E. Kiovsky, *J. Am. Chem. Soc.* **89**, 5692 (1967).
351. N. Solcà and O. Dopfer, *J. Am. Chem. Soc.* **125**, 1421 (2003); *Angew. Chem. Int. Ed.* **42**, 1537 (2003).
352. G. A. Olah and J. R. DeMember, *J. Am. Chem. Soc.* **92**, 718 (1970).
353. G. A. Olah, J. R. DeMember, Y. K. Mo, J. J. Svoboda, P. Schilling, and J. A. Olah, *J. Am. Chem. Soc.* **96**, 884 (1974).
354. G. A. Olah and Y. K. Mo, *J. Am. Chem. Soc.* **96**, 3560 (1974).
355. R. Minkwitz and V. Gerhard, *Z. Naturforsch.* **46b**, 1470 (1991).
356. R. Minkwitz and V. Gerhard, *Z. Naturforsch.* **46b**, 884 (1991).
357. G. A. Olah, G. K. S. Prakash, and M. R. Bruce, *J. Am. Chem. Soc.* **101**, 6463 (1979).
358. G. K. S. Prakash, M. R. Bruce, and G. A. Olah, *J. Org. Chem.* **50**, 2405 (1985).
359. N. J. Head, G. Rasul, A. Mitra, A. Bashir-Heshemi, G. K. S. Prakash, and G. A. Olah, *J. Am. Chem. Soc.* **117**, 12107 (1995).
360. G. K. S. Prakash, G. Rasul, N. J. Head, A. Mitra, and G. A. Olah, in *Advances in Strained and Interesting Organic Molecules*, K. K. Laali, ed., JAI Press, Stamford, 1999, Suppl. 1, p. 109.
361. P. J. Stang, V. V. Zhdankin, and A. M. Arif, *J. Am. Chem. Soc.* **113**, 8997 (1991).
362. V. V. Zhdankin, C. M. Crittell, P. J. Stang, and N. S. Zefirov, *Tetrahedron Lett.* **31**, 4821 (1990).
363. P. J. Stang, V. V. Zhdankin, R. Tykwinski, and N. S. Zefirov, *Tetrahedron Lett.* **33**, 1419 (1992).
364. G. A. Olah and M. R. Bruce, *J. Am. Chem. Soc.* **101**, 4765 (1979).
365. G. A. Olah, Y. K. Mo, E. G. Melby, and H. C. Lin, *J. Org. Chem.* **38**, 367 (1973).
366. J. B. Dence and J. D. Roberts, *J. Org. Chem.* **33**, 1251 (1968).
367. V. V. Lyalin, V. V. Orda, L. A. Alekseeva, and L. M. Yagupol'skii, *Z. Org. Khim.* **7**, 1473 (1971).
368. G. A. Olah and E. G. Melby, *J. Am. Chem. Soc.* **94**, 6220 (1972).
369. M. Ochiai, K. Sumi, Y. Takaoka, M. Kunishima, Y. Nagao, M. Shiro, and E. Fujita, *Tetrahedron* **44**, 4095 (1988).
370. R. J. Hinkle, G. T. Poulter, and P. J. Stang, *J. Am. Chem. Soc.* **115**, 11626 (1993).
371. R. J. Hinkle and P. J. Stang, *Synthesis* 313 (1994).
372. P. J. Stang and V. V. Zhdankin, *J. Am. Chem. Soc.* **113**, 4571 (1991).
373. P. J. Stang and J. Ullmann, *Synthesis* 1073 (1991).
374. P. J. Stang, A. M. Arif, and C. M. Crittell, *Angew. Chem. Int. Ed. Engl.* **29**, 287 (1990).
375. P. J. Stang, R. Tykwinski, and V. V. Zhdankin, *J. Org. Chem.* **57**, 1861 (1992).
376. M. Ochiai, T. Ito, Y. Takaoka, Y. Masaki, M. Kunishima, S. Tani, and Y. Nagao, *J. Chem. Soc., Chem. Commun.* 118 (1990).

377. M. D. Bachi, N. Bar-Ner, C. M. Crittell, P. J. Stang, and B. L. Williamson, *J. Org. Chem.* **56**, 3912 (1991).

378. A. Varvoglis, *Hypervalent Iodine in Organic Synthesis*, Academic Press, San Diego, 1997, Chapter 8, pp. 133–137.

379. V. V. Zhdankin and P. J. Stang, in *Chemistry of Hypervalent Compounds*, K. Akiba, ed., Wiley-VCH, New York, 1999, Chapter 11, pp. 341–346.

380. M. Bielawski, M. Zhu, and B. Olofsson, *Adv. Synth. Catal.* **349**, 2610 (2007).

381. G. A. Olah, P. W. Westerman, E. G. Melby, and Y. K. Mo, *J. Am. Chem. Soc.* **96**, 3565 (1974).

382. G. A. Olah, T. Sakakibara, and G. Asensio, *J. Org. Chem.* **43**, 463 (1978).

383. F. Bailly, P. Barthen, H.-J. Frohn, and M. Köckerling, *Z. Anorg. Allg. Chem.* **626**, 2419 (2000).

384. P. J. Stang and V. V. Zhdankin, *J. Am. Chem. Soc.* **115**, 9808 (1993).

385. U. Radhakrishnan and P. J. Stang, *J. Org. Chem.* **68**, 9209 (2003).

386. G. A. Olah, J. M. Bollinger, and J. Brinich, *J. Am. Chem. Soc.* **90**, 2587 (1968).

387. G. A. Olah, D. A. Beal, and P. W. Westerman, *J. Am. Chem. Soc.* **95**, 3387 (1973).

388. G. A. Olah, P. Schilling, P. W. Westerman, and H. C. Lin, *J. Am. Chem. Soc.* **96**, 3581 (1974).

389. G. A. Olah and J. M. Bollinger, *J. Am. Chem. Soc.* **90**, 947 (1968).

390. G. A. Olah and J. M. Bollinger, *J. Am. Chem. Soc.* **90**, 6082 (1968).

391. G. A. Olah, G. K. S. Prakash, and V. V. Krishnamuthy, *J. Org. Chem.* **48**, 5116 (1983).

392. J. Strating, J. H. Wieringa, and H. Wynberg, *J. Chem. Soc. D: Chem. Commun.* 907 (1969).

393. J. H. Wieringa, J. Strating, and H. Wynberg, *Tetrahedron Lett.* 4579 (1970).

394. W. A. Nugent, *J. Org. Chem.* **45**, 4533 (1980).

395. H. Slebocka-Tilk, R. G. Ball, and R. S. Brown, *J. Am. Chem. Soc.* **107**, 4504 (1985).

396. R. S. Brown, R. W. Nagorski, A. J. Bennet, R. E. D. McClung, G. H. M. Aarts, M. Klobukowski, R. McDonald, and B. D. Santarsiero, *J. Am. Chem. Soc.* **116**, 2448 (1994).

397. T. Mori, R. Rathore, S. V. Lindeman, and J. K. Kochi, *Chem. Commun.* 927 (1998).

398. C. Chiappe, A. De Rubertis, P. Lemmen, and D. Lenoir, *J. Org. Chem.* **65**, 1273 (2000).

399. D. Lenoir, N. Hertkorn, and C. Chiappe, *Tetrahedron Lett.* **45**, 3003 (2004).

400. G. A. Olah, J. M. Bollinger, Y. K. Mo, and J. M. Brinich, *J. Am. Chem. Soc.* **94**, 1164 (1972).

401. J. H. Exner, L. D. Kershner, and T. E. Evans, *J. Chem. Soc., Chem. Commun.* 361 (1973).

402. R. E. Glick,Ph.D. thesis, University of California at Los Angeles, Los Angeles, CA, 1954.

403. P. E. Peterson and G. Allen, *J. Am. Chem. Soc.* **85**, 3608 (1963).

404. P. E. Peterson and E. V. P. Tao, *J. Am. Chem. Soc.* **86**, 4503 (1964).

405. P. E. Peterson and J. E. Duddy, *J. Am. Chem. Soc.* **88**, 4990 (1966).

406. P. E. Peterson and R. J. Bopp,Abstracts, 152nd National Meeting of the American Chemical Society, New York, 1966, Paper S3.

407. P. E. Peterson, R. J. Bopp, D. M. Chevli, E. Curran, D. E. Dillard, and R. J. Kamat, *J. Am. Chem. Soc.* **89**, 5902 (1967).

408. G. A. Olah and P. E. Peterson, *J. Am. Chem. Soc.* **90**, 4675 (1968).
409. G. A. Olah and J. J. Svoboda, *Synthesis* 203 (1973).
410. P. E. Peterson, P. R. Clifford, and F. J. Slama, *J. Am. Chem. Soc.* **92**, 2840 (1970).
411. P. M. Henrichs and P. E. Peterson, *J. Am. Chem. Soc.* **95**, 7449 (1973).
412. P. E. Peterson, B. R. Bonazza, and P. M. Henrichs, *J. Am. Chem. Soc.* **95**, 2222 (1973).
413. G. A. Olah, G. Liang, and L. Staral, *J. Am. Chem. Soc.* **96**, 8112 (1974).
414. T. Hashimoto, G. K. S. Prakash, J. G. Shih, and G. A. Olah, *J. Org. Chem.* **52**, 931 (1987).
415. G. K. S. Prakash, R. Aniszfeld, T. Hashimoto, J. W. Bausch, and G. A. Olah, *J. Am. Chem. Soc.* **111**, 8726 (1989).
416. R. Damrauer, M. D. Leavell, and C. M. Hadad, *J. Org. Chem.* **63**, 9476 (1998).
417. P. E. Peterson and B. R. Bonazza, *J. Am. Chem. Soc.* **94**, 5017 (1972).
418. H. Wang, C. E. Webster, L. M. Pérez, M. B. Hall, and F. P. Gabbai, *J. Am. Chem. Soc.* **126**, 8189 (2004).
419. F. M. Beringer, P. Ganis, G. Avitabile, and H. Jaffe, *J. Org. Chem.* **37**, 879 (1972).
420. F. M. Beringer and R. A. Nathane, *J. Org. Chem.* **34**, 685 (1969).
421. G. A. Olah and Y. Yamada, *J. Org. Chem.* **40**, 1107 (1975).
422. R. Panisch, M. Bolte, and T. Müller, *J. Am. Chem. Soc.* **128**, 9676 (2006).
423. A. Sekiguchi, Y. Murakami, N. Fukaya, and Y. Kabe, *Chem. Lett.* **33**, 530 (2004).
424. S. Minakata, A. Takemiya, K. Nakamura, I. Ryu, and M. Komatsu, *Synlett* 1810 (2000) and references cited therein.
425. M. Al-Talib and J. C. Jochims, *Chem. Ber.* **117**, 3222 (1984).
426. E.-U. Würthwein, R. Kupfer, P. H. M. Budzelaar, C. Strobel, and H. P. Beck, *Angew. Chem. Int. Ed. Engl.* **24**, 340 (1985).
427. E.-U. Würthwein, R. Kupfer, R. Allmann, and M. Nagel, *Chem. Ber.* **118**, 3632 (1985).
428. E.-U. Würthwein and R. Kupfer, *Chem. Ber.* **119**, 1557 (1986).
429. J. C. Jochims, A. Hamed, T. Huu-Phuoc, J. Hofmann, and H. Fischer, *Synthesis* 918 (1989).
430. A. El-Hamid Ismail, A. Hamed, I. Zeid, and J. C. Jochims, *Tetrahedron* **38**, 8271 (1992).
431. G. Böttger, A. Geisler, R. Fröhlich, and E.-U. Würthwein, *J. Org. Chem.* **62**, 6407 (1997).
432. E.-U. Würthwein, *Angew. Chem. Int. Ed. Engl.* **20**, 99 (1981).
433. M. Al-Talib, I. Jibril, E.-U. Würthwein, J. C. Jochims, and G. Huttner, *Chem. Ber.* **117**, 3365 (1984).
434. E.-U. Würthwein, *J. Org. Chem.* **49**, 2971 (1984).
435. R. Kupfer, S. Meier, and E.-U. Würthwein, *Chem. Ber.* **125**, 2487 (1992).
436. W. G. Wirschun, Y. A. Al-Soud, K. A. Nusser, O. Orama, G.-M. Maier, and J. C. Jochims, *J. Chem. Soc., Perkin Trans.* **1**, 4356 (2000).
437. J. Lambrecht, L. Zsolnai, G. Hutter, and J. C. Jochims, *Chem. Ber.* **114**, 3655 (1981).
438. M. Al-Talib, I. Jibril, J. C. Jochims, and G. Huttner, *Chem. Ber.* **117**, 3211 (1984).
439. I. Bernhardi, T. Drews, and K. Seppelt, *Angew. Chem. Int. Ed.* **38**, 2232 (1999).
440. P. Pyykkö and N. Runeberg, *J. Mol. Struct. (Theochem)* **234**, 279 (1991).
441. N. L. Summers and J. Tyrrell, *Theor. Chim. Acta* **47**, 223 (1978).
442. G. Rasul, G. K. S. Prakash, and G. A. Olah, *J. Am. Chem. Soc.* **116**, 8985 (1994).

443. G. A. Olah, R. Herges, J. D. Feldberg, and G. K. S. Prakash, *J. Am. Chem. Soc.* **107**, 5282 (1985).

444. Y.-H. Li and A. G. Harrison, *Int. Mass Spectrom. Ion. Phys.* **28**, 289 (1978).

445. K. Hiraoka, *Bull. Chem. Soc. Jpn.* **52**, 1578 (1979).

446. D. Moy and A. R. Young, II, *J. Am. Chem. Soc.* **87**, 1889 (1965).

447. K. O. Christe, R. D. Wilson, and W. Sawodny, *J. Mol. Struct.* **8**, 245 (1971).

448. J. Mason and K. O. Christe, *Inorg. Chem.* **22**, 1849 (1983).

449. K. O. Christe, R. D. Wilson, W. W. Wilson, R. Bau, S. Sukumar, and D. A. Dixon, *J. Am. Chem. Soc.* **113**, 3795 (1991).

450. W. Kirmse, *Angew. Chem. Int. Ed. Engl.* **15**, 251 (1976).

451. R. Huisgen, *Angew. Chem.* **67**, 439 (1955).

452. O. P. Studzinskii and K. Korobitsyna, *Russ. Chem. Rev. (Engl. Transl.)* **39**, 834 (1970).

453. J. R. Mohrig and K. Keegstra, *J. Am. Chem. Soc.* **89**, 5492 (1967).

454. J. R. Mohrig, K. Keegstra, A. Maverick, R. Roberts, and S. Wells, *J. Chem. Soc., Chem. Commun.* 780 (1974).

455. M. Avaro, J. Levisalles, and J. M. Sommer, *Chem. Commun.* 410 (1968).

456. J. F. McGarrity and D. P. Cox, *J. Am. Chem. Soc.* **105**, 3961 (1983).

457. M. A. Vincent and L. Radom, *J. Am. Chem. Soc.* **100**, 3306 (1978).

458. P. Demontis, R. Ercoli, A. Gamba, G. B. Suffritti, and M. Simonetta, *J. Chem. Soc., Perkin Trans.* **2**, 488 (1981).

459. R. Glaser, G. S.-C. Choy, and M. K. Hall, *J. Am. Chem. Soc.* **113**, 1109 (1991).

460. R. Glaser and G. S.-C. Choy, *J. Am. Chem. Soc.* **115**, 2340 (1993).

461. K. Bott, *Tetrahedron* **22**, 1251 (1966); *Tetrahedron Lett.* 2227 (1971).

462. K. Bott, *Angew. Chem. Int. Ed. Engl.* **18**, 259 (1979).

463. K. Bott, in *The Chemistry of Triple-Bonded Functional Groups*, S. Pataiand Z. Rappoport,eds., Wiley, New York, 1983, Supplement C, Chapter 16, p. 671.

464. R. Glaser, G. S. Chen, and C. L. Barnes, *Angew. Chem. Int. Ed. Engl.* **31**, 740 (1992).

465. G. S. Chen, R. Glaser, and C. L. Barnes, *J. Chem. Soc., Chem. Commun.* 1530 (1993).

466. M. Allard, J. Levisalles, and J. Sommer, *J. Chem. Soc. D: Chem. Commun.* 1515 (1969).

467. C. Wentrup and H. Dahn, *Helv. Chim. Acta* **53**, 1637 (1970).

468. K. K. Laali, G. Maas, and M. Gimmy, *J. Chem. Soc., Perkin Trans. 2*, 1387 (1993).

469. H. Zollinger, *Azo and Diazo Chemistry, Aliphatic and Aromatic Compounds*, Interscience, New York, 1961.

470. C. G. Overberger, J.-P. Anselme, and J. G. Lombardino, *Organic Compounds with Nitrogen–Nitrogen Bonds*, Ronald Press, New York, 1966.

471. P. A. S. Smith, *The Chemistry of Open-Chain Nitrogen Compounds*, W. A. Benjamin, New York, Vol. 1, 1965, Vol. 2, 1966.

472. B. A. Porai-Koshits, *Russ. Chem. Rev. (Engl. Transl.)* **39**, 283 (1970).

473. J. H. Ridd, *Quart. Rev.* **15**, 418 (1961).

474. W. E. Bachmann and R. A. Hoffman, *Org. React.* **2**, 224 (1944).

475. C. S. Rondestvedt, Jr., *Org. React.* **11**, 189 (1960).

476. D. F. DeTar, *Org. React.* **9**, 409 (1957).

477. A. Roe, *Org. React.* **5**, 193 (1949).
478. H. Zollinger, *Acc. Chem. Res.* **6**, 335 (1973).
479. G. A. Olah and J. Welch, *J. Am. Chem. Soc.* **97**, 208 (1975).
480. E. S. Lewis and M. D. Johnson, *J. Am. Chem. Soc.* **81**, 2070 (1959).
481. G. A. Olah and J. L. Grant, *J. Am. Chem. Soc.* **97**, 1546 (1975).
482. J. M. Insole and E. S. Lewis, *J. Am. Chem. Soc.* **85**, 122 (1963).
483. E. S. Lewis and J. M. Insole, *J. Am. Chem. Soc.* **86**, 32, 34 (1964).
484. E. S. Lewis and R. E. Holliday, *J. Am. Chem. Soc.* **88**, 5043 (1966); **91**, 426 (1969).
485. E. S. Lewis and P. G. Kotcher, *Tetrahedron* **25**, 4873 (1969).
486. G. Henrici-Olivé and S. Olivé, *Angew. Chem. Int. Ed. Engl.* **8**, 650 (1969).
487. M. E. Volpin, *Pure Appl. Chem.* **30**, 607 (1972).
488. W. Preetz, *Angew. Chem. Int. Ed. Engl.* **11**, 243 (1972).
489. R. G. Bergstrom, R. G. M. Landells, G. H. Wahl, Jr., and H. Zollinger, *J. Am. Chem. Soc.* **98**, 3301 (1976).
490. I. Szele and H. Zollinger, *J. Am. Chem. Soc.* **100**, 2811 (1978).
491. W. Maurer, I. Szele, and H. Zollinger, *Helv. Chim. Acta* **62**, 1079 (1979).
492. K. K. Laali, I. Szele, and H. Zollinger, *Helv. Chim. Acta* **66**, 1737 (1983).
493. K. Laali and G. A. Olah, *J. Org. Chem.* **50**, 3006 (1985).
494. G. A. Olah, K. Dunne, D. P. Kelly, and Y. K. Mo, *J. Am. Chem. Soc.* **94**, 7438 (1972).
495. A. Schmidt, *Chem. Ber.* **99**, 2976 (1966).
496. A. Mertens, K. Lammertsma, M. Arvanaghi, and G. A. Olah, *J. Am. Chem. Soc.* **105**, 5657 (1983).
497. K. O. Christe, W. W. Wilson, D. A. Dixon, S. I. Khan, R. Bau, T. Metzenthin, and R. Lu, *J. Am. Chem. Soc.* **115**, 1836 (1993).
498. F. Cacace, M. Attinà, G. de Petris, F. Grandinetti, and M. Speranza, *Gazz. Chim. Ital.* **120**, 691 (1990).
499. F. Cacace, M. Attinà, M. Speranza, G. de Petris, and F. Grandinetti, *J. Org. Chem.* **58**, 3639 (1993).
500. W. Wirschun and J. C. Jochims, *Synthesis* 233 (1997).
501. W. Wirschun, *J. Prakt. Chem.* **340**, 300 (1998).
502. G. A. Olah, G. K. S. Prakash, and G. Rasul, *J. Am. Chem. Soc.* **123**, 3308 (2001).
503. G. A. Olah, K. K. Laali, Q. Wang, and G. K. S. Prakash, *Onium Ions*, Wiley, New York, 1998, Chapter 2, p. 6.
504. F. Cacace, F. Grandinetti, and F. Pepi, *J. Chem. Soc., Chem. Commun.* 2173 (1994).
505. A. Vij, W. W. Wilson, V. Vij, F. S. Tham, J. A. Sheehy, and K. O. Christe, *J. Am. Chem. Soc.* **123**, 6308 (2001).
506. K. O. Christe, W. W. Wilson, J. A. Sheehy, and J. A. Boatz, *Angew. Chem. Int. Ed.* **38**, 2004 (1999).
507. G. A. Olah, R. Herges, K. Laali, and G. A. Segal, *J. Am. Chem. Soc.* **108**, 2054 (1986).
508. L. G. Cabrini, M. Benassi, M. N. Eberlin, T. Okazaki, and K. K. Laali, *Eur. J. Org. Chem.* 70 (2007).
509. T. B. McMahon, T. Heinis, G. Nicol, J. K. Hovey, and P. Kebarle, *J. Am. Chem. Soc.* **110**, 7591 (1988).

510. K. Schofield, *Aromatic Nitration*, Cambridge University Press, New York, 1980.
511. G. A. Olah, S. C. Narang, and J. A. Olah, *Proc. Natl. Acad. Sci. USA* **79**, 4487 (1982).
512. G. A. Olah, R. Malhotra, and S. C. Narang, *Nitration: Methods and Mechanisms,* VCH, New York, 1989.
513. L. Médard, *Compt. Rend.* **199**, 1615 (1934).
514. D. Cook, in *Friedel–Crafts and Related Reactions*, Vol. 1, G. A. Olah, ed., Interscience, New York, 1963, Chapter 9, pp. 769, 770.
515. J. W. M. Steeman and C. H. MacGillavry, *Acta Crystallogr.* **7**, 402 (1954).
516. G. A. Olah and S. J. Kuhn, *Org. Synth.* **47**, 56 (1967).
517. G. A. Olah, R. Malhotra, and S. C. Narang, *Nitration: Methods and Mechanisms*. VCH, New York, 1989, Chapter 2, p. 9.
518. C. K. Ingold, D. J. Millen, and H. G. Poole, *Nature* **158**, 480 (1946).
519. W. W. Wilson and K. O. Christe, *Inorg. Chem.* **26**, 1631 (1987).
520. D. Cook, S. J. Kuhn, and G. A. Olah, *J. Chem. Phys.* **33**, 1669 (1960).
521. J. C. Evans, H. W. Rinn, S. J. Kuhn, and G. A. Olah, *Inorg. Chem.* **3**, 857 (1964).
522. D. S. Ross, R. F. Kuhlmann, and R. Malhotra, *J. Am. Chem. Soc.* **105**, 4299 (1983).
523. G. K. S. Prakash, L. Heiliger, and G. A. Olah, *Inorg. Chem.* **29**, 4965 (1990).
524. G. A. Olah, G. Rasul, R. Aniszfeld, and G. K. S. Prakash, *J. Am. Chem. Soc.* **114**, 5608 (1992).
525. G. A. Olah, *Angew. Chem. Int. Ed. Engl.* **32**, 767 (1993).
526. G. A. Olah and D. A. Klumpp, *Acc. Chem. Res.* **37**, 211 (2004).
527. G. A. Olah, A. Germain, H. C. Lin, and D. A. Forsyth, *J. Am. Chem. Soc.* **97**, 2928 (1975).
528. T. Weiske, W. Koch, and H. Schwarz, *J. Am. Chem. Soc.* **115**, 6312 (1993).
529. G. K. S. Prakash, G. Rasul, A. Burrichter, and G. A. Olah, in *ACS Symposium Series*, Vol. 623, L. F. Albright, R. V. C. Carr, and R. J. Scmitt, eds., American Chemical Society, Washington, DC, 1996, Chapter 2, p. 10.
530. G. A. Olah, *Acc. Chem. Res.* **13**, 330 (1980) and references cited therein.
531. G. A. Olah and M. W. Meyer, in *Friedel–Crafts and Related Reactions*, Vol. 1, G. A. Olah, ed., Interscience, New York, 1963, Chapter 8, p. 684.
532. G. E. Toogood and C. Chieh, *Can. J. Chem.* **53**, 831 (1975).
533. C. K. Ingold, *Structure and Mechanism in Organic Chemistry*, Cornell University Press, Ithaca, NY, 1953.
534. Z. J. Allan, J. Podstata, D. Šnobl, and J. Jarkovský, *Tetrahedron Lett.* 3565 (1965).
535. G. A. Olah and N. Friedman, *J. Am. Chem. Soc.* **88**, 5330 (1966).
536. G. A. Olah, G. Salem, J. S. Staral, and T. L. Ho, *J. Org. Chem.* **43**, 173 (1978).
537. G.A. Olah, J.G. Shih, B.P. Singh, and B.G.B. Gupta, *J. Org. Chem.* **48**, 3356 (1983).
538. G. A. Olah, H. C. Lin, J. Olah, and S. C. Narang, *Proc. Natl. Acad. Sci. USA* **75**, 1045 (1978).
539. M. Gerken, P. Kolb, A. Wegner, H. P. A. Mercier, H. Borrmann, D. A. Dixon, and G. J. Schrobilgen, *Inorg. Chem.* **39**, 2813 (2000) and references cited therein.
540. W. J. Casteel, Jr., P. Kolb, N. LeBlond, H. P. A. Mercier, and G. J. Schrobilgen, *Inorg. Chem.* **35**, 929 (1996) and references cited therein.

541. K. O. Christe, M. D. Lind, N. Thorup, D. R. Russell, J. Fawcett, and R. Bau, *Inorg. Chem.* **27**, 2450 (1988).
542. H. Preut, D. Lennhoff, and R. Minkwitz, *Acta Crystallogr. C* **48**, 1648 (1992).
543. R. Minkwitz, J. Nowicki, and H. Bormann, *Z. Anorg. Allg. Chem.* **596**, 93 (1991).
544. H. B. Miller, H. W. Baird, C. L. Bramlett, and W. K. Templeton, *J. Chem. Soc., Chem. Commun.* 262 (1972).
545. U. Müller, *Z. Naturforsch.* **34b**, 681 (1979).
546. R. Haiges, M. Gerken, A. Iuga, R. Bau, and K. O. Christe, *Inorg. Chem.* **45**, 7981 (2006).
547. R. Minkwitz and W. Molsbeck, *Z. Anorg. Allg. Chem.* **607**, 175 (1992).
548. R. Minkwitz, A. Liedtke, and R. Nass, *J. Fluorine Chem.* **35**, 307 (1987).
549. K. O. Christe, *Inorg. Chem.* **14**, 2821 (1975).
550. S. Schneider, R. Haiges, T. Schroer, J. Boatz, and K. O. Christe, *Angew. Chem. Int. Ed.* **43**, 5213 (2004).
551. G. S. H. Chen, J. Passmore, *J. Chem. Soc., Chem. Commun.* 559 (1973).
552. H. W. Roesky, K.-L. Weber, and J. Schimkowiak, *Angew. Chem. Int. Ed. Engl.* **20**, 973 (1981).
553. R. Minkwitz and A. Liedke, *Z. Naturforsch.* **44b**, 679 (1989).
554. R. Küster, T. Drews, and K. Seppelt, *Inorg. Chem.* **39**, 2784 (2000) and references cited therein.
555. R. Minkwitz and M. Dzyk, *Eur. J. Inorg. Chem.* 569 (2002).
556. R. Minkwitz, A. Kornath, D. Lennhoff, and H. Preut, *Z. Anorg. Allg. Chem.* **620**, 509 (1994).
557. R. Gut, *Inorg. Nucl. Chem. Lett.* **12**, 149 (1976).
558. R. Minkwitz, A. Kornath, W. Sawodny, and H. Härtner, *Z. Anorg. Allg. Chem.* **620**, 753 (1994).
559. R. Minkwitz, C. Hirsch, and T. Berends, *Eur. J. Inorg. Chem.* 2249 (1999).
560. G. A. Olah and C. W. McFarland, *J. Org. Chem.* **36**, 1374 (1971).
561. R. Minkwitz and S. Schneider, *Angew. Chem. Int. Ed.* **38**, 210 (1999).
562. V. Ramamoorthy and P. R. Sharp, *Inorg. Chem.* **29**, 3336 (1990).
563. A. Grohmann, J. Riede, and H. Schmidbaur, *J. Chem. Soc., Dalton Trans.* 783 (1991).
564. Yu. L. Slovokhotov and Yu. T. Struchkov, *J. Organomet. Chem.* **277**, 143 (1984).
565. H. Schmidbaur, G. Weidenhiller, and O. Steigelmann, *Angew. Chem. Int. Ed. Engl.* **30**, 433 (1991).
566. E. Zeller, H. Beruda, A. Kolb, P. Bissinger, J. Riede, and H. Schmidbaur, *Nature* **352**, 141 (1991).
567. R. Minkwitz and G. Medger, *Z. Anorg. Allg. Chem.* **604**, 105 (1991).
568. R. Minkwitz, P. Garzarek, and G. Medger, *Z. Anorg. Allg. Chem.* **612**, 40 (1992).
569. R. Minkwitz, T. Hertel, and G. Medger, *Z. Anorg. Allg. Chem.* **615**, 114 (1992).
570. R. Minkwitz and C. Hirsch, *Z. Anorg. Allg. Chem.* **625**, 1362 (1999).
571. R. Minkwitz, T. Hertel, and U. Zachwieja, *Z. Naturforsch.* **52b**, 1185 (1997).
572. K. K. Laali, M. Regitz, M. Birkel, P. J. Stang, and C. M. Crittell, *J. Org. Chem.* **58**, 4105 (1993).

573. K. K. Laali, B. Geissler, M. Regitz, and J. Houser, *J. Org. Chem.* **60**, 47 (1995).
574. K. K. Laali, J. F. Nixon, and J. Johnson, *Heteroatom Chem.* **5**, 503 (1994).
575. K. K. Laali, B. Geissler, and M. Regitz, *J. Org. Chem.* **60**, 3149 (1995).
576. O. P. Shitov, S. L. Ioffe, V. A. Tarkovskii, and S. S. Novikov, *Russ. Chem. Rev. (Engl. Transl.)* **39**, 905 (1970).
577. S. G. Shore and R. W. Parry, *J. Am. Chem. Soc.* **77**, 6084 (1955).
578. R. D. Schultz and R. W. Parry, *J. Am. Chem. Soc.* **80**, 4 (1958).
579. H. Nöeth, H. Beyer, and H.-J. Vetter, *Chem. Ber.* **97**, 110 (1964).
580. N. E. Miller and E. L. Muetterties, *J. Am. Chem. Soc.* **86**, 1033 (1964).
581. H. Nöth and S. Lukas, *Chem. Ber.* **95**, 1505 (1962).
582. P. Kölle and H. Nöth, *Chem. Rev.* **85**, 399 (1985).
583. W. E. Piers, S. C. Bourke, and K. D. Conroy, *Angew. Chem. Int. Ed.* **44**, 5016 (2005).
584. H. Nöth and R. Staudigl, *Angew. Chem. Int. Ed. Engl.* **20**, 794 (1981).
585. H. Nöth, R. Staudigl, and H.-U. Wagner, *Inorg. Chem.* **21**, 706 (1982).
586. J. Higashi, A. D. Eastman, and R. W. Parry, *Inorg. Chem.* **21**, 716 (1982).
587. P. Kölle and H. Nöth, *Chem. Ber.* **119**, 3849 (1986).
588. S. Courtenay, J. Y. Mutus, R. W. Schurko, and D. W. Stephan, *Angew. Chem. Int. Ed.* **41**, 498 (2002).
589. P. Jutzi, B. Krato, M. Hursthouse, and A. J. Howes, *Chem. Ber.* **120**, 1091 (1987).
590. N. Kuhn, A. Kuhn, J. Lewandowski, and M. Speis, *Chem. Ber.* **124**, 2197 (1991).
591. A. H. Cowley, Z. Lu, J. N. Jones, and J. A. Moore, *J. Organomet. Chem.* **689**, 2562 (2004).
592. B. Qian, S. W. Baek, and M. R. Smith III, *Polyhedron* **18**, 2405 (1999).
593. I. Ghesner, W. E. Piers, M. Parvez, and R. McDonald, *Chem. Commun.* 2480 (2005).
594. E. Vedejs, T. Nguyen, D. R. Powell, and M. R. Schrimpf, *Chem. Commun.* 2721 (1996).
595. D. Vidovic, M. Findlater, G. Reeske, and A. H. Cowley, *Chem. Commun.* 3786 (2006).
596. M. A. Mathur and G. E. Ryschkewitsch, *Inorg. Chem.* **19**, 887 (1980).
597. D. Vidovic, M. Findlater, and A. H. Cowley, *J. Am. Chem. Soc.* **129**, 8436 (2007).
598. M. Bochmann and D. M. Dawson, *Angew. Chem. Int. Ed. Engl.* **35**, 2226 (1996).
599. C. Dohmeier, H. Schnöckel, C. Robl, U. Schneider, and R. Ahlrichs, *Angew. Chem. Int. Ed. Engl.* **32**, 1655 (1993).
600. K.-C. Kim, C. A. Reed, G. S. Long, and A. Sen, *J. Am. Chem. Soc.* **124**, 7662 (2002).
601. M. Klahn, C. Fischer, A. Spannenberg, U. Rosenthal, and I. Krossing, *Tetrahedron Lett.* **48**, 8900 (2007).
602. R. J. P. Corriu and M. Henner, *J. Organomet. Chem.* **74**, 1 (1974).
603. B. Bøe, *J. Organomet. Chem.* **107**, 139 (1976).
604. R. West and T. J. Barton, *J. Chem. Ed.* **57**, 334 (1980).
605. W. P. Weber, R. A. Felix, and A. K. Willard, *Tetrahedron Lett.* 907 (1970).
606. T. A. Kochina, D. V. Vrazhnov, E. N. Sinotova, and M. G. Voronkov, *Russ. Chem. Rev. (Engl. Transl.)* **75**, 95 (2006) and references cited therein.
607. W. J. Pietro and W. J. Hehre, *J. Am. Chem. Soc.* **104**, 4329 (1982).
608. G. A. Olah, D. H. O'Brien, and C. Y. Lui, *J. Am. Chem. Soc.* **91**, 701 (1969).

609. A. G. Brook and A. K. Pannell, *Can. J. Chem.* **48**, 3679 (1970).
610. G. A. Olah and Y. K. Mo, *J. Am. Chem. Soc.* **93**, 4942 (1971).
611. A. H. Cowley, M. C. Cushner, and P. E. Riley, *J. Am. Chem. Soc.* **102**, 624 (1980).
612. G. A. Olah and L. D. Field, *Organometallics* **1**, 1485 (1982).
613. J. B. Lambert and W. J. Schulz, Jr., *J. Am. Chem. Soc.* **105**, 1671 (1983).
614. J. B. Lambert, J. A. McConnell, and W. J. Schulz, Jr., *J. Am. Chem. Soc.* **108**, 2482 (1986).
615. J. B. Lambert, J. A. McConnell, W. Schilf, and W. J. Schulz, Jr., *J. Chem. Soc., Chem. Commun.* 455 (1988).
616. G. K. S. Prakash, S. Keyaniyan, R. Aniszfeld, L. Heiliger, G. A. Olah, R. C. Stevens, H.-K. Choi, and R. Bau, *J. Am. Chem. Soc.* **109**, 5123 (1987).
617. J. B. Lambert, L. Kania, and S. Zhang, *Chem. Rev.* **95**, 1191 (1995).
618. C. A. Reed, *Acc. Chem. Res.* **31**, 325 (1998) and references cited therein.
619. J. B. Lambert, Y. Zhao, and S. M. Zhang, *J. Phys. Org. Chem.* **14**, 370 (2001) and references cited therein.
620. T. Müller, *Adv. Organomet. Chem.* **53**, 155 (2005) and references cited therein.
621. W. Ya. Lee and A. Sekiguchi, *Acc. Chem. Res.* **40**, 410 (2007).
622. G. A. Olah, G. Rasul, L. Heiliger, J. Bausch, and G. K. S. Prakash, *J. Am. Chem. Soc.* **114**, 7737 (1992).
623. C. Eaborn, *J. Organomet. Chem.* **405**, 173 (1991).
624. P. D. Lickiss, *J. Chem. Soc., Dalton Trans.* 1333 (1992).
625. G. A. Olah, G. Rasul, X.-Y. Li, H. A. Buchholz, G. Sandford, and G. K. S. Prakash, *Science* **263**, 983 (1994).
626. P. v. R. Schleyer, P. Buzek, T. Müller, Y. Apeloig, and H.-U. Siehl, *Angew. Chem. Int. Ed.* **32**, 1471 (1993).
627. G. A. Olah, G. Rasul, H. A. Buchholz, X.-Y. Li, and G. K. S. Prakash, *Bull. Soc. Chim. Fr.* **132**, 569 (1995).
628. G. A. Olah, G. Rasul, and G. K. S. Prakash, *J. Am. Chem. Soc.* **121**, 9615 (1999).
629. J. B. Lambert, W. J. Schulz, Jr., J. A. McConnell, and W. Schilf, *J. Am. Chem. Soc.* **110**, 2201 (1988).
630. J. B. Lambert and W. Schilf, *J. Am. Chem. Soc.* **110**, 6364 (1988).
631. J. B. Lambert, L. Kania, W. Schilf, and J. A. McConnell, *Organometallics* **10**, 2578 (1991).
632. J. B. Lambert and S. Zhang, *J. Chem. Soc., Chem. Commun.* 383 (1993).
633. J. B. Lambert, S. Zhang, C. L. Stern, and J. C. Huffman, *Science* **260**, 1917 (1993).
634. L. Pauling, *Science* **260**, 983 (1993).
635. L. Olsson and D. Cremer, *Chem. Phys. Lett.* **215**, 433 (1993).
636. C. A. Reed and Z. Xie, *Science* **260**, 985 (1993).
637. J. B. Lambert, S. Zhang, and S. M. Ciro, *Organometallics* **13**, 2430 (1994).
638. Z. Xie, D. J. Liston, T. Jelínek, V. Mitro, R. Bau, and C. A. Reed, *J. Chem. Soc., Chem. Commun.* 384 (1993).
639. C. A. Reed, Z. Xie, R. Bau, and A. Benesi, *Science* **262**, 402 (1993).
640. Z. Xie, R. Bau, A. Benesi, and C. A. Reed, *Organometallics* **14**, 3933 (1995).

641. S. V. Ivanov, J. J. Rockwell, O. G. Polyakov, C. M. Gaudinski, O. P. Anderson, K. A. Solntsev, and S. H. Strauss, *J. Am. Chem. Soc.* **120**, 4224 (1998).

642. M. Johannsen, K. A. Jørgensen, and G. Helmchen, *J. Am. Chem. Soc.* **120**, 7637 (1998).

643. J. Belzner, *Angew. Chem. Int. Ed. Engl.* **36**, 1277 (1997).

644. P. v. R. Schleyer, *Science* **275**, 39 (1997).

645. J. B. Lambert and Y. Zhao, *Angew. Chem. Int. Ed. Engl.* **36**, 400 (1997).

646. J. B. Lambert, Y. Zhao, H. Wu, W. C. Tse, and B. Kuhlmann, *J. Am. Chem. Soc.* **121**, 5001 (1999).

647. E. Kraka, C. P. Sosa, J. Gräfenstein, and D. Cremer, *Chem. Phys. Lett.* **279**, 9 (1997).

648. T. Müller, Y. Zhao, and J. B. Lambert, *Organometallics* **17**, 278 (1998).

649. K.-C. Kim, C. A. Reed, D. W. Elliott, L. J. Mueller, F. Tham, L. Lin, and J. B. Lambert, *Science* **297**, 825 (2002).

650. J. B. Lambert and L. Lin, *J. Org. Chem.* **66**, 8537 (2001).

651. T. Nishinaga, Y. Izukawa, and K. Komatsu, *J. Am. Chem. Soc.* **122**, 9312 (2000); *Tetrahedron* **57**, 3645 (2001).

652. A. Sekiguchi, T. Matsuno, and M. Ichinohe, *J. Am. Chem. Soc.* **122**, 11250 (2000).

653. N. Choi, P. D. Lickiss, M. McPartlin, P. C. Masangane, and G. L. Veneziani, *Chem. Commun.* 6023 (2005).

654. T. Müller, *Angew. Chem. Int. Ed.* **40**, 3033 (2001).

655. S. P. Hoffmann, T. Kato, F. S. Tham, and C. A. Reed, *Chem. Commun.* 767 (2006).

656. A. Y. Khalimon, Z. H. Lin, R. Simionescu, S. F. Vyboishchikov, and G. I. Nikonov, *Angew. Chem. Int. Ed. Engl.* **46**, 4530 (2007).

657. H.-U. Steinberger, T. Müller, N. Auner, C. Maerker, and P. v. R. Schleyer, *Angew. Chem. Int. Ed. Engl.* **36**, 626 (1997).

658. T. Müller, C. Bauch, M. Ostermeier, M. Bolte, and N. Auner, *J. Am. Chem. Soc.* **125**, 2158 (2003).

659. J. Schuppan, B. Herrschaft, and T. Müller, *Organometallics* **20**, 4584 (2001).

660. M. Ichinohe, M. Igarashi, K. Sanuki, and A. Sekiguchi, *J. Am. Chem. Soc.* **127**, 9978 (2005).

661. E. D. Jemmis, G. N. Srinivas, J. Leszczynski, J. Kapp, A. A. Korkin, and P. v. R. Schleyer, *J. Am. Chem. Soc.* **117**, 11361 (1995).

662. M. Nakamoto, T. Fukawa, and A. Sekiguchi, *Chem. Lett.* **33**, 38 (2004).

663. P. Jutzi and E.-A. Bunte, *Angew. Chem. Int. Ed. Engl.* **31**, 1605 (1992).

664. T. Müller, P. Jutzi, and T. Kühler, *Organometallics* **20**, 5619 (2001).

665. P. Jutzi, A. Mix, B. Rummel, W. W. Schoeller, B. Neumann, and H.-G. Stammler, *Science* **305**, 849 (2004).

666. P. Jutzi, A. Mix, B. Neumann, B. Rummel, and H.-G. Stammler, *Chem. Commun.* 3519 (2006).

667. M. Driess, S. Yao, M. Brym, and C. van Wüllen, *Angew. Chem. Int. Ed. Engl.* **45**, 6730 (2006).

668. C. Chuit, R. J. P. Corriu, A. Mehdi, and C. Reyé, *Angew. Chem. Int. Ed. Engl.* **32**, 1311 (1993).

669. J. Belzner, D. Schär, B. O. Kneisel, and R. Herbst-Irmer, *Organometallics* **14**, 1840 (1995).

670. M. Chauhan, C. Chuit, R. J. P. Corriu, A. Mehdi, and C. Reyé, *Organometallics* **15**, 4326 (1996).
671. U.-W. Berlekamp, P. Jutzi, A. Mix, B. Neumann, H.-G. Stammler, and W. W. Schoeller, *Angew. Chem. Int. Ed. Engl.* **38**, 2048 (1999).
672. I. Kalikhman, S. Krivonos, L. Lameyer, D. Stalke, and D. Kost, *Organometallics* **20**, 1053 (2001).
673. D. Kost, V. Kingston, B. Gostevskii, A. Ellern, D. Stalke, B. Walfort, and I. Kalikhman, *Organometallics* **21**, 2293 (2002).
674. I. Kalikhman, B. Gostevskii, O. Girshberg, S. Krivonos, and D. Kost, *Organometallics* **21**, 2551 (2002).
675. C.-H. Ottosson and D. Cremer, *Organometallics* **15**, 5309 (1996).
676. P. v. R. Schleyer, Y. Apeloig, D. Arad, B. T. Luke, and J. A. Pople, *Chem. Phys. Lett.* **95**, 477 (1983).
677. A. E. Reed and P. v. R. Schleyer, *Chem. Phys. Lett.* **133**, 553 (1987).
678. C.-H. Hu, M. Shen, and H. F. Schaefer III, *Chem. Phys. Lett.* **190**, 543 (1992).
679. X. G. Gong, D. Guenzenburger, and E. B. Saitovich, *Chem. Phys. Lett.* **275**, 392 (1997).
680. M. D. Sefcik, J. M. S. Henis, and P. P. Gaspar, *J. Chem. Phys.* **61**, 4329 (1974).
681. B. W. Boo and Y. T. Lee, *Chem. Phys. Lett.* **103**, 514 (1995).
682. E. del Río, M. I. Menéndez, R. López, and T. L. Sordo, *Chem. Commun.* 1779 (1997).
683. R. J. P. Corriu and M. Henner, *J. Organomet. Chem.* **74**, 1 (1974).
684. G. A. Olah, L. Heiliger, R. Aniszfeld, and G. K. S. Prakash, *New J. Chem.* **14**, 877 (1990).
685. M. Ichinohe, Y. Hayata, and A. Sekiguchi, *Chem. Lett.* 1054 (2002).
686. A. Sekiguchi, T. Fukawa, V. Ya. Lee, M. Nakamoto, and M. Ichinohe, *Angew. Chem. Int. Ed.* **42**, 1143 (2003).
687. A. Sekiguchi, M. Tsukamoto, and M. Ichinohe, *Science* **275**, 60 (1997).
688. A. Sekiguchi, N. Fukaya, M. Ichinohe, and Y. Ishida, *Eur. J. Inorg. Chem.* 1155 (2000).
689. Y. Ishida, A. Sekiguchi, and Y. Kabe, *J. Am. Chem. Soc.* **125**, 11468 (2003).
690. A. Sekiguchi, Y. Ishida, Y. Kabe, and M. Ichinohe, *J. Am. Chem. Soc.* **124**, 8776 (2002).
691. M. Stender, A. D. Phillips, and P. P. Power, *Inorg. Chem.* **40**, 5314 (2001).
692. T. Birchall and V. Manivannan, *J. Chem. Soc., Dalton Trans.* 2671 (1985).
693. J. B. Lambert and B. Kuhlmann, *J. Chem. Soc., Chem. Commun.* 931 (1992).
694. M. Kira, T. Oyamada, and H. Sakurai, *J. Organomet. Chem.* **471**, C4 (1994).
695. M. Arshadi, D. Johnels, and U. Edlund, *Chem. Commun.* 1279 (1996).
696. I. Zharov, B. T. King, Z. Havlas, A. Pardi, and J. Michl, *J. Am. Chem. Soc.* **122**, 10253 (2000).
697. I. Zharov, T.-C. Weng, A. M. Orendt, D. H. Barich, J. Penner-Hahn, D. M. Grant, Z. Havlas, and J. Michl, *J. Am. Chem. Soc.* **126**, 12033 (2004).
698. J. B. Lambert, L. Lin, S. Keinan, and T. Müller, *J. Am. Chem. Soc.* **125**, 6022 (2003).
699. A. Sekiguchi, T. Fukawa, V. Ya. Lee, and M. Nakamoto, *J. Am. Chem. Soc.* **125**, 9250 (2003).
700. T. Müller, C. Bauch, M. Bolte, and N. Auner, *Chem. Eur. J.* **9**, 1746 (2003).
701. B. Rhodes, J. C. W. Chien, and M. D. Rausch, *Organometallics* **17**, 1931 (1998).
702. P. G. Gassman, *Acc. Chem. Res.* **3**, 26 (1970).

703. P. G. Gassman and G. D. Hartman, *J. Am. Chem. Soc.* **95**, 449 (1973).
704. G. I. Borodkin and V. G. Shubin, *Russ. J. Org. Chem. (Engl. Transl.)* **41**, 473 (2005) and references cited therein.
705. G. P. Ford and P. S. Herman, *J. Am. Chem. Soc.* **111**, 3987 (1989).
706. O. Dopfer, D. Roth, and J. P. Maier, *Chem. Phys. Lett.* **310**, 201 (1999).
707. G. A. Olah and D. J. Donovan, *J. Org. Chem.* **43**, 1743 (1978).
708. T. Okamoto, K. Shudo, and T. Ohta, *J. Am. Chem. Soc.* **97**, 7184 (1975).
709. T. Ohta, K. Shudo, and T. Okamoto, *Tetrahedron Lett.* 101 (1977).
710. K. Shudo, T. Ohta, Y. Endo, and T. Okamoto, *Tetrahedron Lett.* 105 (1977).
711. K. O. Christe and C. J. Schack, *Inorg. Chem.* **17**, 2749 (1978).
712. G. Boche, P. Andrews, K. Harms, M. Marsch, K. S. Rangappa, M. Schimeczek, and C. Willeke, *J. Am. Chem. Soc.* **118**, 4925 (1996).
713. P. G. Gassman and J. J. Talley, *J. Am. Chem. Soc.* **102**, 1214, 4138 (1980).
714. P. G. Gassman, K. Saito, and J. J. Talley, *J. Am. Chem. Soc.* **102**, 7613 (1980).
715. G. A. Olah, G. K. S. Prakash and M. Arvanaghi, *J. Am. Chem. Soc.* **102**, 6640 (1980).
716. G. A. Olah, M. Arvanaghi, and G. K. S. Prakash, *J. Am. Chem. Soc.* **104**, 1628 (1982).
717. V. V. Krishnamurthy, G. K. S. Prakash, P. S. Iyer, and G. A. Olah, *J. Am. Chem. Soc.* **108**, 1575 (1986).
718. A. Mertens and G. A. Olah, *Chem. Ber.* **116**, 103 (1983).
719. P. Beck,in *Organic Phosphorus Compounds*, Vol. 2, G. M. Kosolapoffand L. Meier, eds., Interscience, New York, 1972, Chapter 4, p. 189.
720. A. H. Cowley and R. A. Kemp, *Chem. Rev.* **85**, 367 (1985).
721. D. Gudat, *Coord. Chem. Rev.* **163**, 71 (1997).
722. C. Chuit and C. Reyé, *Eur. J. Inorg. Chem.* 1847 (1998).
723. A. H. Cowley, M. C. Cushner, M. Lattman, M. L. McKee, J. S. Szobota, and J. C. Wilburn, *Pure Appl. Chem.* **52**, 789 (1980).
724. C. W. Schultz and R. W. Parry, *Inorg. Chem.* **15**, 3046 (1976).
725. M. G. Thomas, C. W. Schultz, and R. W. Parry, *Inorg. Chem.* **16**, 994 (1977).
726. A. H. Cowley, M. C. Cushner, and J. S. Szobota, *J. Am. Chem. Soc.* **100**, 7784 (1978).
727. M. B. Abrams, B. L. Scott, and R. T. Baker, *Organometallics* **19**, 4944 (2000).
728. M. K. Denk, S. Gupta, and A. J. Lough, *Eur. J. Inorg. Chem.* 41 (1999).
729. C. J. Carmalt, V. Lomeli, B. G. McBurnett, and A. H. Cowley, *Chem. Commun.* 2095 (1997).
730. D. Gudat, A. Haghverdi, H. Hupfer, and M. Nieger, *Chem. Eur. J.* **6**, 3414 (2000).
731. S. Burck, D. Gudat, K. Nättinen, M. Nieger, M. Niemeyer, and D. Schmid, *Eur. J. Inorg. Chem.* 5112 (2007).
732. H. A. Spinney, G. P. A. Yap, I. Korobkov, G. DiLabio, and D. S. Richeson, *Organometallics* **25**, 3541 (2006).
733. H. A. Spinney, I. Korobkov, and D. S. Richeson, *Chem. Commun.* 1647 (2007).
734. H. A. Spinney, I. Korobkov, G. A. DiLabio, G. P. A. Yap, and D. S. Richeson, *Organometallics* **26**, 4972 (2007).
735. G. Reeske, C. R. Hoberg, N. J. Hill, and A. H. Cowley, *J. Am. Chem. Soc.* **128**, 2800 (2006).

736. M. Veith, B. Bertsch, and V. Huch, *Z. Anorg. Allg. Chem.* **559**, 73 (1988).
737. R. W. Reed, Z. Xie, and C. A. Reed, *Organometallics* **14**, 5002 (1995).
738. S. G. Baxter, R. L. Collins, A. H. Cowley, and S. F. Sena, *J. Am. Chem. Soc.* **103**, 714 (1981).
739. F. Carré, C. Chuit, R. J. P. Corriu, A. Mehdi, and C. Reyé, *J. Organomet. Chem.* **529**, 59 (1997).
740. J.-P. Bezombes, F. Carré, C. Chuit, R. J. P. Corriu, A. Mehdi, and C. Reyé, *J. Organomet. Chem.* **535**, 81 (1997).
741. D. Vidovic, Z. Lu, G. Reeske, J. A. Moore, and A. H. Cowley, *Chem. Commun.* 3501 (2006).
742. Z. Lu, G. Reeske, J. A. Moore, and A. H. Cowley, *Chem. Commun.* 5060 (2006).
743. D. Gudat, *Eur. J. Inorg. Chem.* 1087 (1998).
744. Y. V. Balitzky, S. E. Pipko, A. D. Sinitsa, A. N. Chernega, Y. G. Gololobov, and A. N. Nesmeyanov, Phosphorous, Sulfur, Silicon **75**, 167 (1993).
745. S. E. Solovieva, M. Gruner, I. S. Antipin, W. D. Habicher, and A. I. Konovalov, *Org. Lett.* **3**, 1299 (2001).
746. S. Loss, C. Widauer, and H. Grützmacher, *Angew. Chem. Int. Ed. Engl.* **38**, 3329 (1999).
747. A. Schmidpeter, S. Lochschmidt, and W. S. Sheldrick, *Angew. Chem. Int. Ed. Engl.* **24**, 226 (1985).
748. N. Burford, P. J. Ragogna, R. McDonald, and M. J. Ferguson, *Chem. Commun.* 2066 (2003); *J. Am. Chem. Soc.* **125**, 14404 (2003).
749. N. Burford, D. E. Herbert, P. J. Ragogna, R. McDonald, and M. J. Ferguson, *J. Am. Chem. Soc.* **126**, 17067 (2004).
750. K. K. Laali, B. Geissler, O. Wagner, J. Hoffmann, R. Armbrust, W. Eisfeld, and M. Regitz, *J. Am. Chem. Soc.* **116**, 9407 (1994).
751. J. Simon, U. Bergsträsser, M. Regitz, and K. K. Laali, *Organometallics* **18**, 817 (1999).
752. X. Liu, D. M. Ivanova, D. Giblin, M. L. Gross, and P. P. Gaspar, *Organometallics* **24**, 3125 (2005).
753. C. Payrastre, Y. Madaule, and J.-G. Wolf, *Tetrahedron Lett.* **33**, 1273 (1992).
754. C. Payrastre, Y. Madaule, J. G. Wolf, T. C. Kim, M.-R. Mazières, R. Wolf, and M. Sanchez, *Heteroatom Chem.* **3**, 157 (1992) and references cited therein.
755. N. Burford, T. M. Parks, B. W. Royan, J. F. Richardson, and P. S. White, *Can. J. Chem.* **70**, 703 (1992).
756. D. Gudat, T. Gans-Eichler, and M. Nieger, *Chem. Commun.* 2434 (2004).
757. S. Burck, J. Daniels, T. Gans-Eichler, D. Gudat, K. Nättinen, and M. Nieger, *Z. Anorg. Allg. Chem.* **631**, 1403 (2005).
758. M. Driess, H. Ackermann, J. Aust, K. Merz, and C. v. Wüllen, *Angew. Chem. Int. Ed.* **41**, 450 (2002).
759. R. J. Wiacek, C. L. B. Macdonald, J. N. Jones, J. M. Pietryga, and A. H. Cowley, *Chem. Commun.* 430 (2003).
760. R. Vogt, P. G. Jones, and R. Schmutzler, *Chem. Ber.* **126**, 1271 (1993).
761. K. A. Porter, A. C. Willis, J. Zank, and S. B. Wild, *Inorg. Chem.* **41**, 6380 (2002).
762. N. Burford, P. J. Ragogna, K. Sharp, R. McDonald, and M. J. Ferguson, *Inorg. Chem.* **44**, 9453 (2005).

763. N. Burford, T. M. Parks, B. W. Royan, B. Borecka, T. S. Cameron, J. F. Richardson, E. J. Gabe, and R. Hynes, *J. Am. Chem. Soc.* **114**, 8147 (1992).

764. J. W. Wieland, N. L. Kilah, A. C. Willis, and S. B. Wild, *Chem. Commun.* 3679 (2006).

765. A. F. Hegarty and J. P. Keogh, *J. Chem. Soc., Perkin Trans. 2*, 758 (2001).

766. S. A. Glover and M. Novak, *Can. J. Chem.* **83**, 1372 (2005).

767. K. Kobayashi, S. Sato, E. Horn, and N. Furukawa, *Tetrahedron Lett.* **39**, 2593 (1998).

768. H. Fujihara, H. Mima, and N. Furukawa, *J. Am. Chem. Soc.* **117**, 10153 (1995).

769. J. Jeske, W.-W. du Mont, and P. G. Jones, *Angew. Chem. Int. Ed. Engl.* **36**, 2219 (1997).

770. O. Bortolini, A. Guerrini, V. Lucchini, G. Modena, and L. Pasquato, *Tetrahedron Lett.* **40**, 6073 (1999).

771. R. J. Gillespie and M. J. Morton, *Quart. Rev.* **25**, 553 (1971).

772. R. J. Gillespie and J. Passmore, *Adv. Inorg. Chem. Radiochem.* **17**, 49 (1975).

773. T. A. O'Donnell, *Chem. Soc. Rev.* **16**, 1 (1987).

774. I. C. Tornieporth-Oetting and T. Klapötke, *Heteroatom Chem.* **4**, 543 (1993).

775. T. A. O'Donnell, *Superacids and Acidic Melts as Inorganic Chemical Reaction Media*, VCH, New York, 1993, Chapter 4, p. 82.

776. S. Brownridge, I. Krossing, J. Passmore, H. D. B. Jenkins, and H. K. Roobottom, *Coord. Chem. Rev.* **197**, 397 (2000).

777. I. Masson, *J. Chem. Soc.* 1708 (1938).

778. R. J. Gillespie and J. B. Milne, *Inorg. Chem.* **5**, 1577 (1966).

779. R. J. Gillespie and K. C. Malhotra, *Inorg. Chem.* **8**, 1751 (1969).

780. R. D. W. Kemmitt, M. Murray, V. M. McRae, R. D. Peacock, M. C. R. Symons, and T. A. O'Donnell, *J. Chem. Soc. A* 862 (1968).

781. J. Arotsky, H. C. Mishra, and M. C. R. Symons, *J. Chem. Soc.* 12 (1961).

782. J. Arotsky, H. C. Mishra, and M. C. R. Symons, *J. Chem. Soc.* 2582 (1962).

783. R. A. Garett, R. J. Gillespie, and J. B. Senior, *Inorg. Chem.* **4**, 563 (1965).

784. R. J. Gillespie, M. J. Morton, and J. M. Sowa,in *Advances in Raman Spectroscopy*, Vol. 1, J. P. Mathieu, ed., Heyden, London, 1973, Chapter 68, p. 539.

785. D. J. Merryman, P. A. Edwards, J. D. Corbett, and R. E. McCarley, *J. Chem. Soc., Chem. Commun.* 779 (1972).

786. J. Passmore, G. Sutherland, and P. S. White, *Inorg. Chem.* **20**, 2169 (1981).

787. J. Passmore, P. Taylor, T. Whidden, and P. S. White, *Can. J. Chem.* **57**, 968 (1979).

788. A. Apblett, F. Grein, J. P. Johnson, J. Passmore, and P. S. White, *Inorg. Chem.* **25**, 422 (1986).

789. G. Adhami and M. Herlem, *J. Electroanal. Chem.* **26**, 363 (1970).

790. C. G. Davies, R. J. Gillespie, P. R. Ireland, and J. M. Sowa, *Can. J. Chem.* **52**, 2048 (1974).

791. J. Besida and T. A. O'Donnell, *Inorg. Chem.* **28**, 1669 (1989).

792. R. J. Gillespie, J. B. Milne, and M. J. Morton, *Inorg. Chem.* **7**, 2221 (1968).

793. R. J. Gillespie, R. Kapoor, R. Faggiani, C. J. L. Lock, M. Murchie, and J. Passmore, *J. Chem. Soc., Chem. Commun.* 8 (1983).

794. O. Ruff, H. Gray, W. Heller, and Knock, *Ber. Dtsch. Chem. Ges.* **9**, 4310 (1906).

795. V. M. McRae,Ph.D. thesis, University of Melbourne (1966).
796. R. J. Gillespie and M. J. Morton, *Chem. Commun.* 1565 (1968).
797. R. J. Gillespie and M. J. Morton, *Inorg. Chem.* **11**, 586 (1972).
798. O. Glemser and A. Šmalc, *Angew. Chem. Int. Ed.* **8**, 517 (1969).
799. K. O. Christe, R. Bau, and D. Zhao, *Z. Anorg. Allg. Chem.* **593**, 46 (1991).
800. A. J. Edwards, G. R. Jones, and R. J. C. Sills, *Chem. Commun.* 1527 (1968).
801. A. J. Edwards and G. R. Jones, *J. Chem. Soc.* 2318 (1971).
802. H. Hartl, J. Nowicki, and R. Minkwitz, *Angew. Chem. Int. Ed. Engl.* **30**, 328 (1991).
803. K. O. Christe, D. A. Dixon, and R. Minkwitz, *Z. Anorg. Allg. Chem.* **612**, 51 (1992).
804. R. J. Gillespie and M. J. Morton, *Inorg. Chem.* **11**, 591 (1972).
805. R. J. Gillespie and M. J. Morton, *Inorg. Chem.* **9**, 811 (1970).
806. H. Siebert, *Anwendungen der Schwingungs Spektroscopic in der Anorganischen Chemie*, Springer, Berlin, 1966.
807. M. J. Clegg and A. J. Downs, *J. Fluorine Chem.* **45**, 13 (1989).
808. R. Minkwitz, J. Nowicki, H. Härtner, and W. Sawody, *Spectrochim. Acta* **47A**, 1673 (1991).
809. T. Drews, W. Koch, and K. Seppelt, *J. Am. Chem. Soc.* **121**, 4379 (1999).
810. J. K. Burdett and C. J. Marsden, *New J. Chem.* **12**, 797 (1988).
811. Y. Li, X. Wang, F. Jensen, K. N. Houk, and G. A. Olah, *J. Am. Chem. Soc.* **112**, 3922 (1990).
812. J. Li and S. Irle, W. H. E. Schwarz, *Inorg. Chem.* **35**, 100 (1996).
813. F. Cacace, G. de Petris, F. Pepi, M. Rosi, and A. Troiani, *J. Phys. Chem. A* **103**, 2128 (1999).
814. S. Seidel and K. Seppelt, *Angew. Chem. Int. Ed.* **39**, 3923 (2000).
815. L. Stein, *Chemistry* **47**, 15 (1974); *Nature* **30**, 243 (1973).
816. L. Stein, *Science* **175**, 1463 (1972).
817. *Halogen Chemistry,* V. Gutman, ed., Academic Press, New York, 1967.
818. K. O. Christe, J. F. Hon, and D. Pilipovich, *Inorg. Chem.* **12**, 84 (1973).
819. K. O. Christe and W. Sawodny, *Inorg. Chem.* **6**, 313 (1967).
820. A. J. Edwards and R. J. C. Sills, *J. Chem. Soc. A* 2697 (1970).
821. A. J. Edwards and G. R. Jones, *J. Chem. Soc. A* 1467 (1969).
822. K. O. Christe and C. J. Schack, *Inorg. Chem.* **9**, 2296 (1970).
823. T. Surles, H. H. Hyman, L. A. Quarterman, and A. I. Popov, *Inorg. Chem.* **9**, 2726 (1970).
824. A. A. Banks, H. J. Emeléus, and A. A. Wolf, *J. Chem. Soc.* 2861 (1949).
825. M. Schmeisser, W. Ludovici, D. Naumann, P. Sartori, and E. Scharf, *Chem. Ber.* **101**, 4214 (1968).
826. K. O. Christe and W. Sawodny, *Inorg. Chem.* **8**, 212 (1969).
827. G. Frenking and W. Koch, *Inorg. Chem.* **29**, 4513 (1990).
828. C. G. Vonk and E. H. Wiebenga, *Acta Crystallogr.* **12**, 859 (1959).
829. N. N. Greenwood and H. J. Emeléus, *J. Chem. Soc.* 987 (1950).
830. K. O. Christe and D. Pilipovich, *Inorg. Chem.* **8**, 391 (1969).
831. M. Schmeisser and E. Pammer, *Angew. Chem.* **69**, 781 (1957).

832. A. A. Woolf, *J. Chem. Soc.* 3678 (1950).
833. N. Bartlett and D. H. Lohmann, *J. Chem. Soc.* 619 (1964).
834. J. Shamir and I. Yaroslavsky, *Israel J. Chem.* **7**, 495 (1969).
835. M. D. Lind and K. O. Christe, *Inorg. Chem.* **11**, 608 (1972).
836. A. J. Edwards and P. Taylor, *J. Chem. Soc., Dalton Trans.* 2174 (1975).
837. K. O. Christe, X. Zhang, J. A. Sheehy, and R. Bau, *J. Am. Chem. Soc.* **123**, 6338 (2001).
838. S. R. Ungemach and H. F. Schaefer III, *Chem. Phys. Lett.* **38**, 407 (1976).
839. S.-P. So, *J. Chem. Soc., Faraday Trans. 2* **77**, 213 (1981).
840. R. M. Minyaev and D. J. Wales, *J. Chem. Soc., Faraday Trans.* **90**, 1831 (1994).
841. A. Vij, F. S. Tham, V. Vij, W. W. Wilson, and K. O. Christe, *Inorg. Chem.* **41**, 6397 (2002).
842. F. Q. Roberto, *Inorg. Nucl. Chem. Lett.* **8**, 737 (1972).
843. K. O. Christe, *Inorg. Nucl. Chem. Lett.* **8**, 741 (1972).
844. T. Schroer and K. O. Christe, *Inorg. Chem.* **40**, 2415 (2001).
845. K. O. Christe, *Inorg. Nucl. Chem. Lett.* **8**, 453 (1972).
846. K. O. Christe, W. W. Wilson, and R. D. Wilson, *Inorg. Chem.* **19**, 1494 (1980).
847. K. O. Christe, R. D. Wilson, and C. J. Schack, *Inorg. Chem.* **19**, 3046 (1980) and references cited therein.
848. K. O. Christe, *Inorg. Chem.* **12**, 1580 (1973).
849. R. J. Gillespie and G. J. Schrobilgen, *J. Chem. Soc., Chem. Commun.* 90 (1974).
850. R. J. Gillespie and G. J. Schrobilgen, *Inorg. Chem.* **13**, 1230 (1974).
851. K. O. Christe and R. D. Wilson, *Inorg. Chem.* **14**, 694 (1975).
852. F. Seel and O. Detmer, *Angew. Chem.* **70**, 163 (1958); *Z. Anorg. Chem.* **301**, 113 (1959).
853. K. O. Christe and W. Sawodny, *Inorg. Chem.* **6**, 1783 (1967).
854. F. A. Hohorst, L. Stein, and E. Gebert, *Inorg. Chem.* **14**, 2233 (1975).
855. K. O. Christe, *Inorg. Chem.* **9**, 2801 (1970).
856. J. F. Lehmann, G. J. Schrobilgen, K. O. Christe, A. Kornath, and R. J. Suontamo, *Inorg. Chem.* **43**, 6905 (2004).
857. G. Herzberg, *Spectra of Diatomic Molecules*, 2nd ed., Van Nostrand, Princeton, 1950.
858. N. Bartlett and D. H. Lohmann, *J. Chem. Soc.* 5253 (1962); *Proc. Chem. Soc. (London)* 115 (1962).
859. N. Bartlett and D. H. Lohmann, *Proc. Chem. Soc. (London)* 14 (1960).
860. E. W. Lawless and I. C. Smith, *Inorganic High-Energy Oxidizers*, Marcel Dekker, New York, 1968, and references cited therein.
861. D. E. McKee and N. Bartlett, *Inorg. Chem.* **12**, 2738 (1973).
862. A. R. Young II, T. Hirata and S. I. Morrow, *J. Am. Chem. Soc.* **86**, 20 (1964).
863. J. N. Keith, I. J. Solomon, I. Sheft, and H. H. Hyman, *Inorg. Chem.* **7**, 230 (1968).
864. Z. K. Nikitina and V. Ya. Rosolovskii, *Izv. Akad. Nauk SSSR, Ser. Khim.* 2173 (1970).
865. N. Bartlett, *Angew. Chem. Int. Ed. Engl.* **7**, 433 (1968).
866. J. Shamir, J. Binenboym, and H. H. Claassen, *J. Am. Chem. Soc.* **90**, 6223 (1968).
867. K. R. Loos, V. A. Campanile, and C. T. Goetschel, *Spectrochim. Acta A* **26**, 365 (1970).
868. I. B. Goldberg, K. O. Christe, and R. D. Wilson, *Inorg. Chem.* **14**, 152 (1975).

869. O. Graudejus and B. G. Müller, *Z. Anorg. Allg. Chem.* **622**, 1076 (1996).
870. P. Botkovitz, G. M. Lucier, R. P. Rao, and N. Bartlett, *Acta. Chim. Slov.* **46**, 141 (1999).
871. I.-C. Hwang and K. Seppelt, *Angew. Chem. Int. Ed.* **40**, 3690 (2001).
872. J. F. Lehmann and G. J. Schrobilgen, *J. Fluorine Chem.* **119**, 109 (2003).
873. C. T. Goetschel and K. R. Loos, *J. Am. Chem. Soc.* **94**, 3018 (1972).
874. J. W. Mellor, *Comprehensive Treatise on Inorganic and Theoretical Chemistry*, Vol. 10, Longmans, Green, London, 1930, pp. 184–186.
875. C. F. Bucholz, *Gehlen's Neues. J. Chem.* **3**, 7 (1804).
876. I. Krossing, *Top. Curr. Chem.* **230**, 135 (2003).
877. G. Nickless, ed., *Inorganic Sulfur Chemistry*, Elsevier, Amsterdam, 1968. p. 412.
878. R. J. Gillespie, J. Passmore, P. K. Ummat, and O. C. Vaidya, *Inorg. Chem.* **10**, 1327 (1971).
879. R. J. Gillespie and P. K. Ummat, *Inorg. Chem.* **11**, 1674 (1972).
880. C. G. Davies, R. J. Gillespie, J. J. Park, and J. Passmore, *Inorg. Chem.* **10**, 2781 (1971).
881. R. C. Burns, R. J. Gillespie, and J. F. Sawyer, *Inorg. Chem.* **19**, 1423 (1980).
882. M. P. Murchie, J. Passmore, G. W. Sutherland, and R. Kapoor, *J. Chem. Soc., Dalton Trans.* 503 (1992).
883. H. S. Low and R. A. Beudet, *J. Am. Chem. Soc.* **98**, 3849 (1976).
884. R. Minkwitz, J. Nowicki, W. Sadowny, and K. Härtner, *Spectrochim. Acta* **47A**, 151 (1991).
885. T. S. Cameron, R. J. Deeth, I. Dionne, H. Du, H. D. B. Jenkins, I. Krossing, J. Passmore, and H. K. Roobottom, *Inorg. Chem.* **39**, 5614 (2000).
886. R. Faggiani, R. J. Gillespie, J. F. Sawyer, and J. E. Vekris, *Acta Crystallogr.* **45C**, 1847 (1989).
887. I. Krossing and J. Passmore, *Inorg. Chem.* **43**, 1000 (2004).
888. R. J. Gillespie and J. Passmore, *J. Chem. Soc. D: Chem. Commun.* 1333 (1969).
889. I. D. Brown, D. B. Crump, and R. J. Gillespie, *Inorg. Chem.* **10**, 2319 (1971).
890. T. W. Couch, D. A. Lokken, and J. D. Corbett, *Inorg. Chem.* **11**, 357 (1972).
891. R. C. Burns and R. J. Gillespie, *Inorg. Chem.* **21**, 3877 (1982).
892. J. Barr, R. J. Gillespie, and P. K. Ummat, *J. Chem. Soc. D: Chem. Commun.* 264 (1970).
893. P. J. Stephens, *J. Chem. Soc. D: Chem. Commun.* 1496 (1969).
894. K. Tanaka, T. Yamabe, H. Terama-e, and K. Fukui, *Inorg. Chem.* **18**, 3591 (1979).
895. J. Passmore, G. Sutherland, and P. S. White, *J. Chem. Soc., Chem. Commun.* 330 (1980).
896. T. S. Cameron, I. Dionne, H. D. B. Jenkins, S. Parsons, J. Passmore, and H. K. Roobottom, *Inorg. Chem.* **39**, 2042 (2000).
897. I. Krossing and J. Passmore, *Inorg. Chem.* **38**, 5203 (1999).
898. T. A. O'Donnell, *Superacids and Acidic Melts as Inorganic Chemical Reaction Media*, VCH, New York, 1993, Chapter 4, p. 100.
899. R. J. Gillespie, *Chem. Soc. Rev.* **8**, 315 (1979).
900. G. Magnus, *Ann. Phys. (Leipzig)* [2] **10**, 491 (1827); **14**, 328 (1828).
901. J. W. Mellor, *Comprehensive Treatise on Inorganic and Theoretical Chemistry*, Vol. 10, Longmans, Green, London, 1930, pp. 922–923.
902. J. Barr, R. J. Gillespie, R. Kapoor, and K. C. Malhotra, *Can. J. Chem.* **46**, 149 (1968).

903. R. J. Gillespie and G. P. Prez, *Inorg. Chem.* **8**, 1229 (1969).
904. R. J. Gillespie and P. K. Ummat, *Can. J. Chem.* **48**, 1239 (1970).
905. P. A. W. Dean, R. J. Gillespie, and P. K. Ummat, *Inorg. Synth.* **15**, 213 (1974).
906. I. D. Brown, D. B. Crump, R. J. Gillespie, and D. P. Santry, *Chem. Commun.* 853 (1968).
907. G. Cardinal, R. J. Gillespie, J. F. Sawyer, and J. E. Vekris, *J. Chem. Soc., Dalton Trans.* 765 (1982).
908. R. E. March, L. Pauling, and J. D. McCullogh, *Acta Crystallogr. B* **6**, 71 (1953).
909. R. C. Burns, R. J. Gillespie, J. A. Barnes, and M. J. McGlinchey, *Inorg. Chem.* **21**, 799 (1982).
910. R. K. McMullan, D. J. Prince, and J. D. Corbett, *J. Chem. Soc. D: Chem. Commun.* 1438 (1969); *Inorg. Chem.* **10**, 1749 (1971).
911. M. J. Collins, R. J. Gillespie, and J. F. Sawyer, *Acta Crystallogr.* **44C**, 405 (1988).
912. R. C. Burns, W.-L. Chan, R. J. Gillespie, W.-C. Luk, J. F. Sawyer, and D. R. Slim, *Inorg. Chem.* **19**, 1432 (1980).
913. M. J. Collins, R. J. Gillespie, J. F. Sawyer, and G. J. Schrobilgen, *Acta Crystallogr.* **42C**, 13 (1986).
914. R. C. Burns, M. J. Collins, R. J. Gillespie, and G. J. Schrobilgen, *Inorg. Chem.* **25**, 4465 (1986).
915. J. Beck and A. Fischer, *Z. Anorg. Allg. Chem.* **623**, 780 (1997).
916. M. H. Klaproth, *Philos. Mag.* **1**, 78 (1798).
917. N. J. Bjerrum and G. P. Smith, *J. Am. Chem. Soc.* **90**, 4472 (1968).
918. N. J. Bjerrum, *Inorg. Chem.* **9**, 1965 (1970).
919. J. Barr, R. J. Gillespie, R. Kapoor, and G. P. Pez, *J. Am. Chem. Soc.* **90**, 6855 (1968).
920. J. Barr, R. J. Gillespie, G. P. Pez, P. K. Ummat, and O. C. Vaidya, *Inorg. Chem.* **10**, 362 (1971).
921. J. Barr, R. J. Gillespie, G. P. Pez, P. K. Ummat, and O. C. Vaidya, *J. Am. Chem. Soc.* **92**, 1081 (1970).
922. R. C. Paul, C. L. Arora, J. K. Puri, R. N. Virmani, and K. C. Malhotra, *J. Chem. Soc., Dalton Trans.* **781**, (1972).
923. D. J. Prince, J. D. Corbett, and B. Garbisch, *Inorg. Chem.* **9**, 2731 (1970).
924. M. Z. Straumanis, *Z. Kristallogr. Kristallogeometrie, Kristallophys. Kristallochem.* **102**, 432 (1946).
925. I. Tomaszkiewicz, J. Passmore, G. Schatte, and P. A. G. O'Hare, *J. Chem. Thermodyn.* **28**, 1019 (1996).
926. N. J. Bjerrum, *Inorg. Chem.* **11**, 2648 (1972).
927. R. J. Gillespie, W. Luk, and D. R. Slim, *J. Chem. Soc., Chem. Commun.* 791 (1976).
928. R. C. Burns, R. J. Gillespie, W.-C. Luk, and D. R. Slim, *Inorg. Chem.* **18**, 3086 (1979).
929. J. Beck, *Angew. Chem. Int. Ed. Engl.* **30**, 1128 (1991).
930. G. W. Drake, G. L. Schimek, and J. W. Kolis, *Inorg. Chem.* **35**, 1740 (1996).
931. P. Boldrini, I. D. Brown, R. J. Gillespie, P. R. Ireland, W. Luk, D. R. Slim, and J. E. Vekris, *Inorg. Chem.* **15**, 765 (1976).
932. M. J. Collins, R. J. Gillespie, and J. F. Sawyer, *Inorg. Chem.* **26**, 1476 (1987).
933. R. J. Gillespie, W. Luk, E. Maharajh, and D. R. Slim, *Inorg. Chem.* **16**, 892 (1977).

934. R. C. Burns, M. J. Collins, S. M. Eicher, R. J. Gillespie, and J. F. Sawyer, *Inorg. Chem.* **27**, 1807 (1988).
935. P. Boldrini, I. D. Brown, M. J. Collins, R. J. Gillespie, E. Maharajh, D. R. Slim, and J. F. Sawyer, *Inorg. Chem.* **24**, 4302 (1985).
936. M. J. Collins, R. J. Gillespie, J. F. Sawyer, and G. J. Schrobilgen, *Inorg. Chem.* **25**, 2053 (1986).
937. C. R. Lassigne and E. J. Wells, *J. Chem. Soc., Chem. Commun.* 956 (1978).
938. G. J. Schrobilgen, R. C. Burns, and P. Granger, *J. Chem. Soc., Chem. Commun.* 957 (1978).
939. M. J. Collins and R. J. Gillespie, *Inorg. Chem.* **23**, 1975 (1984).
940. O. Glemser and W. Koch, *Angew. Chem. Int. Ed. Engl.* **10**, 127 (1971).
941. A. Apblett, A. J. Banister, D. Biron, A. G. Kendrick, J. Passmore, M. Schriver, and M. Stojanac, *Inorg. Chem.* **25**, 4451 (1986).
942. W. Clegg, O. Glemser, K. Harms, G. Hartmann, R. Mews, M. Noltemeyer, and G. M. Sheldrick, *Acta Crystallogr.* **37B**, 548 (1981).
943. E. G. Awere and J. Passmore, *J. Chem. Soc., Dalton Trans.* 1343 (1992).
944. B. Ayres, A. J. Banister, P. D. Coates, M. I. Hansford, J. M. Rawson, C. E. F. Rickard, M. B. Hursthouse, K. M. A. Malik, and M. Motevalli, *J. Chem. Soc., Dalton Trans.* 3097 (1992).
945. S. Parsons and J. Passmore, *Acc. Chem. Res.* **27**, 101 (1994).
946. R. Faggiani, R. J. Gillespie, C. J. L. Lock, and J. D. Tyrer, *Inorg. Chem.* **17**, 2975 (1978).
947. A. J. Banister, R. G. Hey, G. K. MacLean, and J. Passmore, *Inorg. Chem.* **21**, 1679 (1982).
948. T. S. Cameron, A. Mailman, J. Passmore, and K. V. Shuvaev, *Inorg. Chem.* **44**, 6524 (2005).
949. R. J. Gillespie, P. R. Ireland, and J. E. Vekris, *Can. J. Chem.* **53**, 3147 (1975).
950. R. J. Gillespie, J. P. Kent, and J. F. Sawyer, *Inorg. Chem.* **20**, 3784 (1981).
951. B. M. Gimarc and D. S. Warren, *Inorg. Chem.* **30**, 3276 (1991).
952. W. V. F. Brooks, T. S. Cameron, S. Parsons, J. Passmore, and M. J. Schriver, *Inorg. Chem.* **33**, 6230 (1994).
953. B. Kruss and M. L. Ziegler, *Z. Anorg. Allg. Chem.* **388**, 158 (1972).
954. R. J. Gillespie, J. P. Kent, J. F. Sawyer, D. R. Slim, and J. D. Tyrer, *Inorg. Chem.* **20**, 3799 (1981).
955. A. J. Banister, P. J. Dainty, A. C. Hazell, R. G. Hazell, and J. G. Lomborg, *J. Chem. Soc. D: Chem. Commun.* 1187 (1969).
956. H. W. Roesky, W. G. Böwing, I. Rayment, and H. M. M. Schearer, *J. Chem. Soc., Chem. Commun.* 735 (1975).
957. R. J. Gillespie, J. P. Kent, and J. F. Sawyer, *Inorg. Chem.* **20**, 4053 (1981).
958. E. G. Awere, W. V. F. Brooks, J. Passmore, P. S. White, X. Sun, and T. S. Cameron, *J. Chem. Soc., Dalton Trans.* 2439 (1993).
959. E. G. Awere, J. Passmore, and P. S. White, *J. Chem. Soc., Dalton Trans.* 1267 (1992); 299 (1993).
960. A. Haas, J. Kasprowski, K. Angermund, P. Betz, C. Krüger, Y.-H. Tsay, and S. Werner, *Chem. Ber.* **124**, 1895 (1991).

961. R. Bartetzko and R. Gleiter, *Inorg. Chem.* **17**, 995 (1978).

962. R. J. Gillespie, J. F. Sawyer, D. R. Slim, and J. D. Tyrer, *Inorg. Chem.* **21**, 1296 (1982).

963. G. K. MacLean, J. Passmore, and P. S. White, *J. Chem. Soc., Dalton Trans.* 211 (1984).

964. K. O. Christe, C. J. Schack, and D. Pilipovich, *Inorg. Chem.* **11**, 2205 (1972).

965. C. J. Schack, C. B. Lindahl, D. Pilipovich, and K. O. Christe, *Inorg. Chem.* **11**, 2201 (1972).

966. K. O. Christe, R. D. Wilson, and E. C. Curtis, *Inorg. Chem.* **12**, 1358 (1973).

967. K. O. Christe and W. Maya, *Inorg. Chem.* **8**, 1253 (1968).

968. K. O. Christe and R. D. Wilson, *Inorg. Chem.* **12**, 1356 (1973).

969. C. A. Wamser, W. B. Fox, B. Sukornick, J. R. Holmes, B. B. Stewart, R. Juurik, N. Vanderkooi, and D. Gould, *Inorg. Chem.* **8**, 1249 (1969).

970. W. W. Wilson, R. Haiges, J. A. Boatz, and K. O. Christe, *Angew. Chem. Int. Ed. Engl.* **46**, 3023 (2007).

971. K. Dehnicke, H. Aeissen, M. Kölmel, and J. Strähle, *Angew. Chem. Int. Ed. Engl.* **16**, 545 (1977).

972. R. Minkwitz, D. Bernstein, W. Sawodny, and H. Härtner, *Z. Anorg. Allg. Chem.* **580**, 109 (1990).

973. N. Bartlett, J. Passmore, and E. J. Wells, *Chem. Commun.* 213 (1966).

974. K. O. Christe, J. F. Hon, and D. Pilipovich, *Inorg. Chem.* **12**, 84 (1973).

975. R. Minkwitz, D. Bernstein, H. Preut, and P. Sartori, *Inorg. Chem.* **30**, 2157 (1991).

976. R. J. Gillespie, E. A. Robinson, and G. L. Heard, *Inorg. Chem.* **37**, 6884 (1998).

977. A. Vij, X. Zhang, and K. O. Christe, *Inorg. Chem.* **40**, 416 (2001).

978. M. Brumm, G. Frenking, and W. Koch, *Chem. Phys. Lett.* **182**, 310 (1991).

979. M. Brumm, G. Frenking, J. Breidung, and W. Thiel, *Chem. Phys. Lett.* **197**, 330 (1992).

980. T. Klapötke and J. Passmore, *Acc. Chem. Res.* **22**, 234 (1989).

981. M. P. Murchie, J. P. Johnson, J. Passmore, G. W. Sutherland, M. Tajik, T. K. Whidden, P. S. White, and F. Grein, *Inorg. Chem.* **31**, 273 (1992).

982. S. Brownridge, T. S. Cameron, H. Du, C. Knapp, R. Köppe, J. Passmore, J. M. Rautiainen, and H. Schnöckel, *Inorg. Chem.* **44**, 1660 (2005).

983. S. Brownridge, M.-J. Crawford, H. Du, R. D. Harcourt, C. Knapp, R. S. Laitinen, J. Passmore, J. M. Rautiainen, R. S. Suontamo, and J. Valkonen, *Inorg. Chem.* **46**, 681 (2007).

984. W. A. S. Nandana, J. Passmore, P. S. White, and C.-M. Wong, *Inorg. Chem.* **29**, 3529 (1990).

985. Q. Zhang, X. Lu, R.-B. Huang, and L.-S. Zheng, *Inorg. Chem.* **45**, 2457 (2006).

986. P. Bakshi, P. D. Boyle, T. S. Cameron, J. Passmore, G. Schatte, and G. W. Sutherland, *Inorg. Chem.* **33**, 3849 (1994).

987. S. Brownridge, T. S. Cameron, J. Passmore, G. Schatte, and T. C. Way, *J. Chem. Soc., Dalton Trans.* 2553 (1996).

988. W. A. S. Nandana, J. Passmore, P. S. White, and C.-M. Wong, *Inorg. Chem.* **28**, 3320 (1989).

989. J. Passmore, G. W. Sutherland, P. Taylor, T. K. Whidden, and P. S. White, *Inorg. Chem.* **20**, 3839 (1981).

990. J. Passmore, G. W. Sutherland, T. K. Whidden, P. S. White, and C.-M. Wong, *Can. J. Chem.* **63**, 1209 (1985).
991. R. Minkwitz and J. Nowicki, *Inorg. Chem.* **29**, 2361 (1990).
992. E. O. Fischer, K. Fichtel, and K. Öfele, *Chem. Ber.* **95**, 249 (1962).
993. E. O. Fischer and K. Öfele, *Angew. Chem.* **73**, 581 (1961).
994. W. Hieber and Th. Kruck, *Z. Naturforsch.* **16b**, 709 (1961); *Angew. Chem.* **73**, 580 (1961).
995. W. Hieber, F. Lux, and C. Herget, *Z. Naturforsch.* **20b**, 1159 (1965).
996. E. W. Abel and S. P. Tyfield, *Adv. Organomet. Chem.* **8**, 117 (1970).
997. W. Beck and K. Sünkel, *Chem. Rev.* **88**, 1405 (1988).
998. H. Willner and F. Aubke, *Inorg. Chem.* **29**, 2195 (1990).
999. H. Willner, J. Schaebs, G. Hwang, F. Mistry, R. Jones, J. Trotter, and F. Aubke, *J. Am. Chem. Soc.* **114**, 8972 (1992).
1000. M. Bodenbinder, G. Balzer-Jöllenbeck, H. Willner, R. Batchelor, F. W. B. Einstein, C. Wang, and F. Aubke, *Inorg. Chem.* **35**, 82 (1996).
1001. H. Willner and F. Aubke, *Angew. Chem. Int. Ed. Engl.* **36**, 2403 (1997).
1002. H. Willner and F. Aubke, *Chem. Eur. J.* **9**, 1669 (2003); *Organometallics* **22**, 3612 (2003).
1003. H. Willner, M. Bodenbinder, R. Bröchler, G. Hwang, S. J. Rettig, J. Trotter, B. v. Ahsen, U. Westphal, V. Jonas, W. Thiel, and F. Aubke, *J. Am. Chem. Soc.* **123**, 588 (2001).
1004. B. v. Ahsen, M. Berkei, G. Henkel, H. Willner, and F. Aubke, *J. Am. Chem. Soc.* **124**, 8371 (2002).
1005. B. v. Ahsen, C. Bach, H. Pernice, H. Willner, and F. Aubke, *J. Fluorine Chem.* **102**, 243 (2000).
1006. C. Wang, S. C. Siu, G. Hwang, C. Bach, B. Bley, M. Bodenbinder, H. Willner, and F. Aubke, *Can. J. Chem.* **74**, 1952 (1996).
1007. E. Bernhardt, B. Bley, R. Wartchow, H. Willner, E. Bill, P. Kuhn, I. H. T. Sham, M. Bodenbinder, R. Bröchler, and F. Aubke, *J. Am. Chem. Soc.* **121**, 7188 (1999).
1008. B. v. Ahsen, C. Bach, M. Berkei, M. Köckerling, H. Willner, G. Hägele, and F. Aubke, *Inorg. Chem.* **42**, 3801 (2003).
1009. E. Bernhardt, C. Bach, B. Bley, R. Wartchow, U. Westphal, I. H. T. Sham, B. v. Ahsen, C. Wang, H. Willner, R. C. Thompson, and F. Aubke, *Inorg. Chem.* **44**, 4189 (2005).
1010. E. Bernhardt, M. Finze, H. Willner, C. W. Lehmann, and F. Aubke, *Angew. Chem. Int. Ed.* **42**, 2077 (2003).
1011. A. J. Lupinetti, M. D. Havighurst, S. M. Miller, O. P. Anderson, and S. H. Strauss, *J. Am. Chem. Soc.* **121**, 11920 (1999).
1012. P. K. Hurlburt, J. J. Rack, J. S. Luck, S. F. Dec, J. D. Webb, O. P. Anderson, and S. H. Strauss, *J. Am. Chem. Soc.* **116**, 10003 (1994).
1013. M. Finze, E. Bernhardt, H. Willner, C. W. Lehmann, and F. Aubke, *Inorg. Chem.* **44**, 4206 (2005).
1014. R. Bröchler, D. Freidank, M. Bodenbinder, I. H. T. Sham, H. Willner, S. J. Rettig, J. Trotter, and F. Aubke, *Inorg. Chem.* **38**, 3684 (1999).
1015. M. Valderrama and L. A. Oro, *Can. J. Chem.* **60**, 1044 (1982).

1016. H. Willner, C. Bach, R. Wartchow, C. Wang, S. J. Rettig, J. Trotter, V. Jonas, W. Thiel, and F. Aubke, *Inorg. Chem.* **39**, 1933 (2000).

1017. R. Bröchler, I. H. T. Sham, M. Bodenbinder, V. Schmitz, S. J. Rettig, J. Trotter, H. Willner, and F. Aubke, *Inorg. Chem.* **39**, 2172 (2000).

1018. E. Bernhardt, H. Willner, V. Jonas, W. Thiel, and F. Aubke, *Angew. Chem. Int. Ed.* **39**, 168 (2000).

1019. N. LeBlond, H. P. A. Mercier, D. A. Dixon, and G. J. Schrobilgen, *Inorg. Chem.* **39**, 4494 (2000).

1020. N. LeBlond, D. A. Dixon, G. J. Schrobilgen, *Inorg. Chem.* **39**, 2473 (2000).

1021. W. J. Casteel, Jr., D. A. Dixon, H. P. A. Mercier, and G. J. Schrobilgen, *Inorg. Chem.* **35**, 4310 (1996).

1022. M. Gerken, D. A. Dixon, and G. J. Schrobilgen, *Inorg. Chem.* **41**, 259 (2002).

1023. B. v. Ahsen, B. Bley, S. Proemmel, R. Wartchow, H. Willner, and F. Aubke, *Z. Anorg. Allg. Chem.* **624**, 1225 (1998).

1024. M. Berkei, M. Schürmann, E. Bernhardt, and H. Willner, *Z. Naturforsch.* **57b**, 615 (2002).

1025. B. Bley, M. Bodenbinder, G. Balzer, H. Willner, G. Hägele, F. Mistry, and F. Aubke, *Can. J. Chem.* **74**, 2392 (1996).

1026. R. Küster and K. Seppelt, *Z. Anorg. Allg. Chem.* **626**, 236 (2000).

1027. S. Seidel and K. Seppelt, *Science* **290**, 117 (2000).

1028. T. Drews, S. Seidel, and K. Seppelt, *Angew. Chem. Int. Ed.* **41**, 454 (2002).

1029. K. Seppelt, *Z. Anorg. Allg. Chem.* **629**, 2427 (2003).

1030. I.-C. Hwang and K. Seppelt, *Z. Anorg. Allg. Chem.* **628**, 765 (2002).

1031. I.-C. Hwang, S. Seidel, and K. Seppelt, *Angew. Chem. Int. Ed.* **42**, 4392 (2003).

1032. A. Bayler, A. Bauer, and H. Schmidbaur, *Chem. Ber./Receuil* **130**, 115 (1997).

1033. R. Uson, A. Laguna, and M. V. Castrillo, *Synth. React. Inorg. Metal-Org. Chem.* **9**, 317 (1979).

1034. J. J. Thompson, *Philos. Mag.* **24**, 209 (1912).

1035. T. R. Hogness and E. G. Lunn, *Phys. Rev.* **26**, 44 (1925).

1036. H. D. Smith, *Phys. Rev.* **25**, 45 (1925).

1037. L. Friedman and B. G. Reuben, *Adv. Chem. Phys.* **19**, 59 (1971).

1038. R. J. Gillespie and G. P. Pez, *Inorg. Chem.* **8**, 1233 (1969).

1039. G. A. Olah, J. Shen, and R. H. Schlosberg, *J. Am. Chem. Soc.* **95**, 4957 (1973).

1040. T. Oka, *Phys. Rev. Lett.* **45**, 531 (1980).

1041. N. Bartlettand F. O. Sladky, in *Comprehensive Inorganic Chemistry*, Vol. 1, J. C. Bailar, Jr., H. J. Emeléus, R. Nyholm,and A. F. Trotman-Dickenson, eds., Pergamon Press, Oxford, 1973, Chapter 6, p. 213.

1042. J. H. Holloway and E. G. Hope, *Adv. Inorg. Chem.* **46**, 51 (1998).

1043. M. Gerken and G. J. Schrobilgen, *Coord. Chem. Rev.* **197**, 335 (2000).

1044. H.-J. Frohn and V. V. Bardin, *Organometallics* **20**, 4750 (2001).

1045. K. O. Christe, *Angew. Chem. Int. Ed.* **40**, 1419 (2001).

1046. W. Tyrra,and D. Naumann, in *Inorganic Chemistry Highlights*, G. Meyer, D. Naumann, and L. Wesemann, eds., Wiley-VCH, Weinheim, 2002, Chapter 19, p. 297.

1047. J. F. Lehmann, H. P. A. Mercier, and G. J. Schrobilgen, *Coord. Chem. Rev.* **233–234**, 1 (2002).

1048. J. A. Gascón, R. W. Hall, C. Ludewigt, and H. Haberland, *J. Chem. Phys.* **117**, 8391 (2002) and references cited therein.

1049. L. Stein, J. R. Norris, A. J. Downs, and A. R. Minihan, *J. Chem. Soc., Chem. Commun.* 502 (1978).

1050. T. Drews and K. Seppelt, *Angew. Chem. Int. Ed. Engl.* **36**, 273 (1997).

1051. S. Seidel, K. Seppelt, C. v. Wüllen, and X. Y. Sun, *Angew. Chem. Int. Ed. Engl.* **46**, 6717 (2007).

1052. G. A. Sinnot, *University of Colorado Joint Institute Laboratory of Astrophysics Information Center* Report No. 9 (1969) and references cited therein.

1053. F. H. Field, H. H. Head, and J. L. Franklin, *J. Am. Chem. Soc.* **84**, 1118 (1962).

1054. T. O. Tiernan and P. S. Gill, *J. Chem. Phys.* **50**, 5042 (1969).

1055. G. R. Hertel and W. S. Koski, *J. Am. Chem. Soc.* **87**, 1686 (1965).

1056. S. Wexler, *J. Am. Chem. Soc.* **85**, 272 (1963).

1057. M. S. B. Munson and F. H. Field, *J. Am. Chem. Soc.* **87**, 4242 (1965).

1058. F. S. Klein and L. Friedman, *J. Chem. Phys.* **41**, 1789 (1964).

1059. V. Aquilanti, A. Galli, A. Giardini-Guidoni, and G. G. Volpi, *J. Chem. Phys.* **43**, 1969 (1965).

1060. W. A. Chupka and M. E. Russel, *J. Chem. Phys.* **49**, 5426 (1968).

1061. S. A. Rogers, C. R. Brazier, and P. F. Bernath, *J. Chem. Phys.* **87**, 159 (1987).

1062. K. A. Peterson, R. H. Petrmichl, R. L. McClain, and R. C. Woods, *J. Chem. Phys.* **95**, 2352 (1991).

1063. T. D. Fridgen and J. M. Parnis, *J. Chem. Phys.* **109**, 2155 (1998).

1064. J. Lundell and M. Petterson, *J. Mol. Struct.* **509**, 49 (1999).

1065. G. Hvistendahl and O. W. Saastad, E. Uggerud, *Int. J. Mass Spectrom. Ion Proces.* **98**, 167 (1990).

1066. D. Holtz and J. L. Beauchamp, *Science* **173**, 1237 (1971).

1067. J. K. Hovey and T. B. McMahon, *J. Am. Chem. Soc.* **108**, 528 (1986).

1068. H. J. Frohn and S. Jakobs, *J. Chem. Soc., Chem. Commun.* 625 (1989).

1069. H.-J. Frohn and T. Schroer, *J. Fluorine Chem.* **112**, 259 (2001).

1070. H. J. Frohn, S. Jakobs, and G. Henkel, *Angew. Chem. Int. Ed. Engl.* **28**, 1506 (1989).

1071. H.-J. Frohn, A. Klose, T. Schroer, G. Henkel, V. Buss, D. Opitz, and R. Vahrenhorst, *Inorg. Chem.* **37**, 4884 (1998).

1072. K. Koppe, V. Bilir, H.-J. Frohn, H. P. A. Mercier, and G. J. Schrobilgen, *Inorg. Chem.* **46**, 9425 (2007).

1073. H. J. Frohn, S. Jakobs, and C. Rossbach, *Eur. J. Solid State Inorg. Chem.* **29**, 729 (1992).

1074. H. J. Frohn and C. Rossbach, *Z. Anorg. Allg. Chem.* **619**, 1672 (1993).

1075. H. Butler, D. Naumann, and W. Tyrra, *Eur. J. Solid State Inorg. Chem.* **29**, 739 (1992).

1076. D. Naumann, H. Butler, R. Gnann, and W. Tyrra, *Inorg. Chem.* **32**, 861 (1993).

1077. D. Naumann, W. Tyrra, and D. Pfolk, *Z. Anorg. Allg. Chem.* **620**, 987 (1994).

1078. H.-J. Frohn, H. Franke, and V. V. Bardin, *Z. Naturforsch.* **54b**, 1495 (1999).

1079. D. Naumann, W. Tyrra, R. Gnann, and D. Pfolk, *J. Chem. Soc., Chem. Commun.* 2651 (1994).

1080. D. Naumann, W. Tyrra, R. Gnann, D. Pfolk, T. Gilles, and K.-F. Tebbe, *Z. Anorg. Allg. Chem.* **623**, 1821 (1997).

1081. H. J. Frohn and V. V. Bardin, *J. Chem. Soc., Chem. Commun.* 1072 (1993).

1082. H. J. Frohn and V. V. Bardin, *Z. Naturforsch.* **51b**, 1011 1996; *Z. Naturforsch.* **53b**, 562 (1998).

1083. H.-J. Frohn and V. V. Bardin, *Chem. Commun.* 919 (1999).

1084. V. V. Zhdankin, P. J. Stang, and N. S. Zefirov, *J. Chem. Soc., Chem. Commun.* 578 (1992).

1085. H.-J. Frohn and V. V. Bardin, *Chem. Commun.* 2352 (2003).

1086. H.-J. Frohn and V. V. Bardin, in *ACS Symposium Series* Vol. 965, K. K. Laali, ed., American Chemical Society, Washington, DC, 2007, Chapter 20, p. 428.

1087. H.-J. Frohn, N. LeBlond, K. Lutar, and B. Žemva, *Angew. Chem. Int. Ed.* **39**, 391 (2000).

1088. H. Selig, in *Halogen Chemistry*, Vol. 1, V. Gutman, ed., Academic Press, New York, 1967, p. 403.

1089. F. A. Cotton and G. Wilkinson, *Advanced Inorganic Chemistry*, Wiley, New York, 1980, p. 580.

1090. N. N. Greenwood, A. Earnshaw, *Chemistry of the Elements*, 2nd ed., Butterworth-Heinemann Oxford, Boston, 1997, p. 893.

1091. V. M. McRae, R. D. Peacock, and D. R. Russell, *J. Chem. Soc. D: Chem. Commun.* 62 (1969).

1092. R. J. Gillespie, B. Landa, and G. J. Schrobilgen, *Inorg. Chem.* **15**, 1256 (1976) and references cited therein.

1093. K. Leary, A. Zalkin, and N. Bartlett, *Inorg. Chem.* **13**, 775 (1974).

1094. K. O. Christe and W. W. Wilson, *Inorg. Chem.* **27**, 2714 (1988).

1095. A. Zalkin, D. L. Ward, R. A. Biagioni, D. H. Templeton, and N. Bartlett, *Inorg. Chem.* **17**, 1318 (1978).

1096. F. O. Sladky, P. A. Bulliner, and N. Bartlett, *Chem. Commun.* 1048 (1968).

1097. N. Bartlett, B. G. DeBoer, F. J. Hollander, F. O. Sladky, D. H. Templeton, and A. Zalkin, *Inorg. Chem.* **13**, 780 (1974).

1098. B. A. Fir, M. Gerken, B. E. Pointner, H. P. A. Mercier, D. A. Dixon, and G. J. Schrobilgen, *J. Fluorine Chem.* **105**, 159 (2000).

1099. H. P. A. Mercier, J. C. P. Sanders, G. J. Schrobilgen, and S. S. Tsai, *Inorg. Chem.* **32**, 386 (1993).

1100. B. E. Pointner, R. J. Suontamo, and G. J. Schrobilgen, *Inorg. Chem.* **45**, 1517 (2006).

1101. F. Sladky, *Angew. Chem. Int. Ed. Engl.* **9**, 375 (1970).

1102. N. Keller and G. J. Schrobilgen, *Inorg. Chem.* **20**, 2118 (1981).

1103. B. A. Fir, H. P. A. Mercier, J. C. P. Sanders, D. A. Dixon, and G. J. Schrobilgen, *J. Fluorine Chem.* **110**, 89 (2001).

1104. P. Ulferts and K. Seppelt, *Z. Anorg. Allg. Chem.* **630**, 1589 (2004).

1105. H. P. A. Mercier, M. D. Moran, J. C. P. Sanders, G. J. Schrobilgen, and R. J. Suontamo, *Inorg. Chem.* **44**, 49 (2005).

1106. H. P. A. Mercier, M. D. Moran, and G. J. Schrobilgen, in *ACS Symposium Series*, Vol. 965, K.K. Laali, ed., American Chemical Society, Washington, DC, 2007, Chapter 19, p. 394.

1107. R. G. Syvret, K. M. Mitchell, J. C. P. Sanders, and G. J. Schrobilgen, *Inorg. Chem.* **31**, 3381 (1992).
1108. B. Fir, J. M. Whalen, H. P. A. Mercier, D. A. Dixon, and G. J. Schrobilgen, *Inorg. Chem.* **45**, 1978 (2006).
1109. G. J. Schrobilgen and J. M. Whalen, *Inorg. Chem.* **33**, 5207 (1994).
1110. A. A. A. Emara and G. J. Schrobilgen, *J. Chem. Soc., Chem. Commun.* 1644 (1987).
1111. A. Filippi, A. Troiani, and M. Speranza, *J. Phys. Chem. A* **101**, 9344 (1997).
1112. M. Zhou, Y. Zhao, Y. Gong, and J. Li, *J. Am. Chem. Soc.* **128**, 2504 (2006).
1113. R. Faggiani, D. K. Kennepohl, C. J. L. Lock, and G. J. Schrobilgen, *Inorg. Chem.* **25**, 563 (1986).
1114. B. Fir, J. M. Whalen, H. P. A. Mercier, D. A. Dixon, and G. J. Schrobilgen, *Inorg. Chem.* **45**, 1978 (2006).
1115. G. J. Schrobilgen and J. M. Whalen, *Inorg. Chem.* **33**, 5207 (1994).
1116. A. A. A. Emara and G. J. Schrobilgen, *J. Chem. Soc., Chem. Commun.* 1644 (1987).
1117. G. J. Schrobilgen, *J. Chem. Soc., Chem. Commun.* 1506 (1988).
1118. G. L. Smith, H. P. A. Mercier, and G. J. Schrobilgen, *Inorg. Chem.* **46**, 1369 (2007).
1119. S. Seidel and K. Seppelt, *Angew. Chem. Int. Ed.* **40**, 4225 (2001).
1120. H.-J. Frohn, T. Schroer, and G. Henkel, *Angew. Chem. Int. Ed.* **38**, 2554 (1999).
1121. B. Frlec and J. H. Holloway, *J. Chem. Soc., Chem. Commun.*, 89 (1974).
1122. R. J. Gillespie, B. Landa, and G. J. Schrobilgen, *Inorg. Chem.* **15**, 22 (1976).
1123. J. F. Lehmann, D. A. Dixon, and G. J. Schrobilgen, *Inorg. Chem.* **40**, 3002 (2001).
1124. A. Filippi, A. Troiani, and M. Speranza, *Chem. Phys. Lett.* **278**, 202 (1997).
1125. G. J. Schrobilgen, *J. Chem. Soc., Chem. Commun.* 863 (1988).
1126. G. Frenking, W. Koch, F. Reichel, and D. Cremer, *J. Am. Chem. Soc.* **112**, 4240 (1990).
1127. G. Frenking and D. Cremer, in *Structure and Bonding*, Vol. 73, Springer, Heidelberg, 1990, p. 17.
1128. D. Schröder and H. Schwarz, *J. Phys. Chem. A* **103**, 7385 (1999).
1129. W. Koch and G. Frenking, *J. Chem. Soc., Chem. Commun.* 1095 (1986).
1130. M. W. Wong, R. H. Nobes, and L. Radom, *J. Chem. Soc., Chem. Commun.* 233 (1987).
1131. G. A. Olah, G. K. S. Prakash, and G. Rasul, *Proc. Natl. Acad. Sci. USA* **96**, 3494 (1999).
1132. W. Eisfeld and J. S. Francisco, *Proc. Natl. Acad. Sci. USA* **99**, 15303 (2002).

CHAPTER 5

Superacid-Catalyzed Reactions

In discussing superacids as catalysts for chemical reactions, we will review both liquid (Magic Acid, fluoroantimonic acid, etc.) and solid (Nafion-H, etc.) acid-catalyzed reactions, but not those of conventional Friedel–Crafts-type catalysts. The latter reactions have been extensively reviewed elsewhere (see G. A. Olah, *Friedel–Crafts Chemistry*, Wiley, New York, 1972; G. A. Olah, ed., *Friedel–Crafts and Related Reactions*, Vols. I–IV, Wiley-Interscience, New York, 1963–1965).

As already mentioned and shown, considerable experimental and theoretical evidence has been collected over the last decades, which supports the idea of superelectrophilic activation, that is, protosolvation[1,2] or *de facto* protonation of cationic intermediates.[3–5] Examples of superelectrophiles as highly reactive dicationic (doubly electron-deficient) and tricationic intermediates were discussed in Chapter 4.

The success of carbocation chemistry lies in the stabilization of carbocations in a medium of low nucleophilicity. Superelectrophiles, in turn, are reactive intermediates generated by further protonation (protosolvation). This second protonation increases electron deficiency, induces destabilization, and, consequently, results in a profound increase in reactivity. In particular, charge–charge repulsive interactions[6] play a crucial role in the enhanced reactivity of dicationic and tricationic superelectrophilic intermediates. As various examples of acidity dependence studies show, without an appropriate acidity level, transformations may occur at much lower rate or even do not take place at all. In addition to numerous examples of superacid catalyzed reactions, various organic transformations, in which the involvement of superelectrophilic intermediates is invoked or superelectrophiles are *de facto* observed in the condensed state, are also included in this chapter.

5.1. CONVERSION OF SATURATED HYDROCARBONS

Solid and liquid acid-catalyzed hydrocarbon conversions involve by far the largest amount of catalysts and largest economic efforts in oil refining and chemical

Superacid Chemistry, Second Edition, George A. Olah, G. K. Surya Prakash, Árpád Molnár, and Jean Sommer

industry.[7–9] Saturated hydrocarbons are the main components of natural gas and crude oil, which play presently an absolute key role in world economy as the number one source not only of energy but also of chemical feedstock covering our daily needs.

At the end of the 1800s, saturated hydrocarbons (paraffins) played only a minor role in industrial chemistry. They were mainly used as a source of paraffin wax as well as for heating and lighting oils. Aromatic compounds such as benzene, toluene, phenol, and naphthalene obtained from destructive distillation of coal were the main source of organic materials used in the preparation of dyestuffs, pharmaceutical products, and so on. Calcium carbide-based acetylene was the key starting material for the emerging synthetic organic industry. It was the ever-increasing demand for gasoline after the First World War that initiated study of isomerization and cracking reactions of petroleum fractions. After the Second World War, rapid economic expansion necessitated more and more abundant and cheap sources for chemicals and this resulted in the industry switching over to petroleum-based ethylene as the main source of chemical raw material. One of the major difficulties that had to be overcome is the low reactivity of some of the major components of the petroleum. The lower boiling components (up to 250°C) are mainly straight-chain saturated hydrocarbons or paraffins, which, as their name indicates (*parum affinis*: slight reactivity), have very little reactivity. Consequently, the lower paraffins were cracked to give olefins (mainly ethylene, propylene, and butenes). The straight-chain liquid hydrocarbons have also very low octane numbers, which make them less desirable as gasoline components. To transform these paraffins into useful components for gasoline and other chemical applications, they have to undergo diverse reactions such as isomerization, cracking, or alkylation. These reactions, which are used on a large scale in industrial processes, necessitate acidic catalysts (at temperatures around 100°C) or noble metal catalysts (at higher temperature, 200–500°C) capable of activating the strong covalent C–H or C–C bonds.[10]

Despite worldwide trends toward severe environmental legislation, the use of such acids as HF and H_2SO_4 is still in effect because of their high activity at low temperature. But simultaneously a huge effort has been and is still currently underway to develop solid or supported strong acids, which are easier to handle and to recycle. This trend is reflected by the impressive number of patents, special issues, and reviews devoted to this subject.[11–16]

The preparation of new solid acids, their characterization, mechanistic studies, and theoretical approaches to understand the fundamental aspects of acid-catalyzed hydrocarbon conversion constitute a very large fraction of the topics discussed in the last decade in all journals related to catalysis and physical chemistry. However, in contrast with liquid-acid-catalyzed activation processes, many fundamental questions concerning the initial step, the true nature of the reaction intermediates, and the number of active sites remain open for discussion. For this reason, the results obtained in liquid-superacid-catalyzed chemistry, which can be rationalized by classical reaction mechanisms, supported by the usual analytical tools of organic chemists, represent the fundamental basis to which scientist in the field refer.

Since the early 1960s, superacids are known to react with saturated hydrocarbons, even at temperatures much below 0°C. This discovery initiated extensive studies devoted to hydrocarbon conversions.

5.1.1. Sigma-Basicity: Reversible Protonation or Protolysis of C–H and C–C Bond

The fundamental step in acid-catalyzed hydrocarbon conversion processes is the formation of the intermediate carbocations. Whereas all studies involving isomerization, cracking, and alkylation reactions under acidic conditions (Scheme 5.1) agree that a trivalent carbocation (carbenium ion) is the key intermediate, the mode of their formation of this reactive species from the neutral hydrocarbon remained controversial for many years.

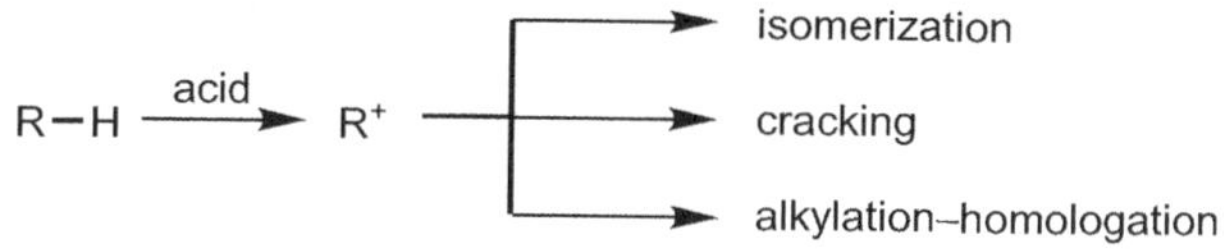

Scheme 5.1

Due to the unclear picture concerning the initial step in hydrocarbon conversion on solid acids, generally one of four pathways can be found in the literature: protolysis (1), hydride abstraction by an already existing carbenium ion (2), hydride abstraction by a Lewis acid (M) (3), and oxidation (4) (Scheme 5.2).

$$\mathrm{RH} + \text{"}\mathrm{H^+}\text{"} \longrightarrow \mathrm{R^+} + \mathrm{H_2} \qquad (1)$$

$$\mathrm{RH} + \mathrm{R'^+} \longrightarrow \mathrm{R^+} + \mathrm{R'H} \qquad (2)$$

$$\mathrm{M} + \mathrm{RH} \longrightarrow \mathrm{R^+} + \mathrm{MH} \qquad (3)$$

$$\mathrm{RH} \xrightarrow[-2\,e^-]{} \mathrm{R^+} + \text{"}\mathrm{H^+}\text{"} \qquad (4)$$

Scheme 5.2

In 1946 Bloch, Pines, and Schmerling[17] observed that *n*-butane (**1**) isomerizes to isobutane (**2**) under the influence of pure aluminum chloride only in the presence of HCl. They proposed that the ionization step takes place through initial protolysis of the alkane as evidenced by formation of minor amounts of hydrogen in the initial stage of the reaction [Eq. (5.1)].

$$\underset{\mathbf{1}}{\mathrm{CH_3CH_2CH_2CH_3}} + \mathrm{HCl} \xrightarrow{\mathrm{AlCl_3}} \underset{\mathbf{3}}{\mathrm{CH_3\overset{+}{C}HCH_2CH_3}}\ \mathrm{AlCl_4^-} + \mathrm{H_2} \qquad (5.1)$$

The first evidence of protonation of alkanes under superacid conditions has been reported by Olah and Lukas[18] as well as by Hogeveen and co-workers.[19,20]

When *n*-butane **1** or isobutane **2** was reacted with HSO_3F-SbF_5 (Magic Acid), *tert*-butyl cation **4** was formed exclusively [Eq. (5.2)] as evidenced by a sharp singlet at 4.5 ppm (from TMS) in the ^{1}H NMR spectrum. In excess Magic Acid, the stability of the ion is remarkable and the NMR spectrum of the solution remains unchanged even after having been heated to 110°C.

$$\underset{\mathbf{1}}{CH_3CH_2CH_2CH_3} \xrightarrow{HSO_3F-SbF_5} \underset{\mathbf{4}}{(CH_3)_3C^+\ SbF_5FSO_3^-} + H_2 \xleftarrow[RT]{HSO_3F-SbF_5} \underset{\mathbf{2}}{(CH_3)_3CH} \quad (5.2)$$

It was also shown[21] that the *tert*-butyl cation **4** undergoes degenerate carbon scrambling at higher temperatures (see Chapter 3). A lower limit of $E_a \sim 30$ kcal mol^{-1} was estimated for the scrambling process, which could correspond to the energy difference between *tert*-butyl cation **4** and primary isobutyl cation **5** (the latter being partially delocalized).

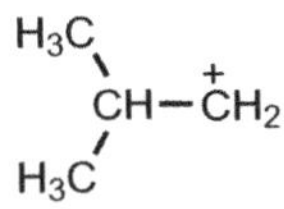

5

n-Pentane and isopentane are ionized under the same conditions to the *tert*-amyl cation. *n*-Hexane **6** and the branched C_6 isomers ionize in the same way to yield a mixture of the three tertiary hexyl ions [Eq. (5.3)] as shown by their ^{1}H NMR spectra.

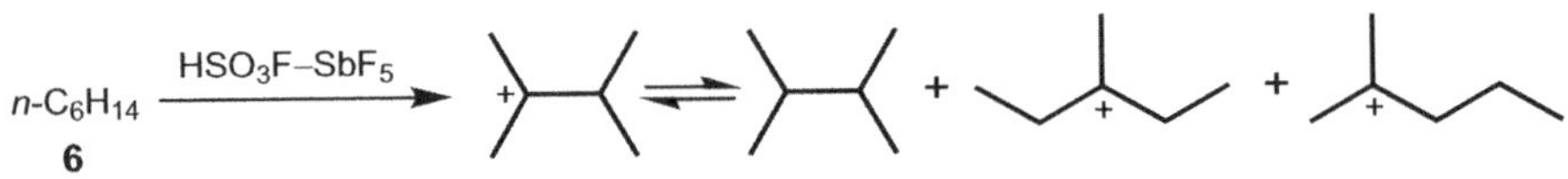

(5.3)

Both methylcyclopentane and cyclohexane were found to give the methylcyclopentyl ion, which is stable at low temperature, in excess superacid.[22] When alkanes with seven or more carbon atoms were used, cleavage was observed with formation of the stable *tert*-butyl cation **4**. Even paraffin wax (see Section 2.2.2.2 on Magic Acid) and polyethylene ultimately gave the *tert*-butyl cation **4** after complex fragmentation and ionization processes.

In compounds containing only primary hydrogen atoms such as neopentane **7** [Eq. (5.4)] and 2,2,3,3-tetramethylbutane, a carbon–carbon bond is broken rather than a carbon–hydrogen bond.[23]

$$\underset{\mathbf{7}}{C(CH_3)_4} \xrightarrow{H^+} CH_4 + \underset{\mathbf{4}}{(CH_3)_3C^+} \quad (5.4)$$

Results of protolytic reactions of hydrocarbons in superacid media were interpreted by Olah as indication for the general electrophilic reactivity of covalent C–H and C–C single bonds of alkanes and cycloalkanes. The reactivity is due to the σ-donor ability of a shared electron pair (of σ-bond) via two-electron, three-center bond formation. Consequently, the transition state of the reaction, is of three-center bound pentacoordinate carbonium ion nature [Eq. (5.5)].

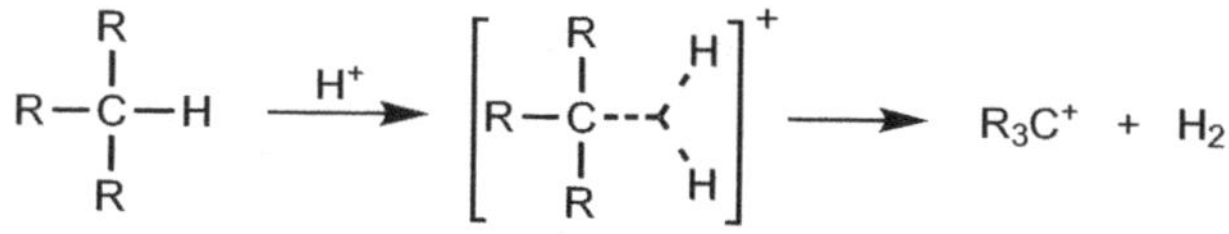

(5.5)

5.1.1.1. Deuterium–Hydrogen Exchange Studies. H/D exchange between alkanes and strong or superacids have proven very useful for mechanistic investigations. Already in the early 1970s, monodeuteromethane was reported to undergo H–D exchange without detectable side reactions in the HF–SbF_5 system[24] [Eq. (5.6)]. d_{12}-Neopentane, when treated with Magic Acid, was also reported to undergo H–D exchange before cleavage.[25]

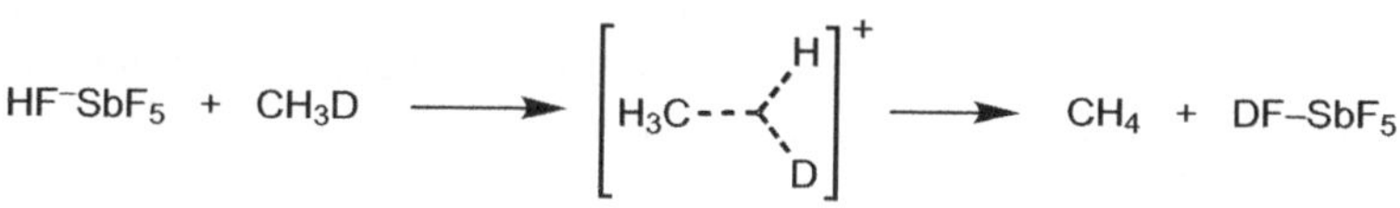

(5.6)

Based on the demonstration of H–D exchange of molecular hydrogen (and deuterium) in superacid solutions, Olah et al.[26] suggested that these reactions go through trigonal isotopic H_3^+ ions (**8**, **9**) in accordance with theoretical calculations and IR studies.[27]

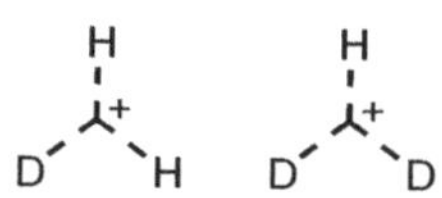

8 **9**

Consequently, the reverse reaction of protolytic ionization of hydrocarbons to carbenium ions—that is, the reduction of carbenium ion by molecular hydrogen[28,29]—can be considered as alkylation of H_2 by the electrophilic carbenium ion through a pentacoordinate carbonium ion. Indeed, Hogeveen and Bickel have experimentally proved this point by reacting stable alkyl cations in superacids with molecular hydrogen [Eq. (5.7)].

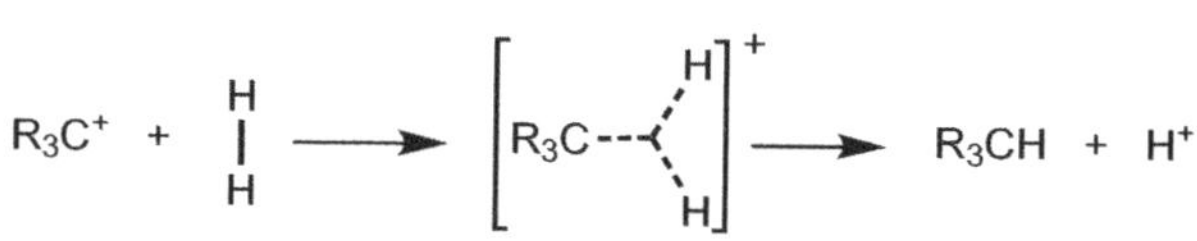

(5.7)

However, due to competitive reaction pathways the reaction of small alkanes yields a quite complex product distribution (Schemes 5.2 and 5.3). Especially hydride transfer to the initially formed carbenium ions [reactions (2) and (3) in Scheme 5.2] makes it difficult to study the initial steps.

$$\mathrm{RH} + \text{"}\mathrm{D}^{+}\text{"} \longrightarrow \mathrm{RD} + \mathrm{H}^{+} \quad (1)$$

$$\mathrm{R{-}R'} + \mathrm{H}^{+} \longrightarrow \mathrm{RH} + \mathrm{R'}^{+} \quad (2)$$

$$\mathrm{RH} + \mathrm{SbF_5} \longrightarrow \mathrm{R}^{+} + \text{"}\mathrm{H}^{+}\text{"} + 2\,\mathrm{F}^{-} + \mathrm{SbF_3} \quad (3)$$

Scheme 5.3

However, if the reactions are run in the presence of carbon monoxide the initially formed carbenium ions are trapped as oxocarbenium ions, which do not initiate hydride abstraction [Eq. (5.8)].

$$\mathrm{R}^{+} + \mathrm{R'}^{+} + 2\,\mathrm{CO} \longrightarrow \mathrm{RCO}^{+} + \mathrm{R'CO}^{+} \quad (5.8)$$

By using this technique, Sommer and his group reinvestigated the H–D exchange reaction occurring between small alkanes and deuteriated superacids and showed that the reaction depended not only on the structure of the alkane but also very much on the superacid system.[30,31]

Methane. Methane is abundant on earth. It is the major component of natural gas[8] and is produced on short time scale by biological conversion of biomass.[32] During the last decades much effort has been expended to the challenge of converting methane into useful products.[33–38] One of the approaches is the electrophilic activation, which relies on the σ-basicity of the C–H bond—that is, its ability to react with strong electrophiles.[39] In liquid superacids, Hogeveen and co-workers[24,40] and Olah and co-workers[23,25] have observed protium/deuterium exchange with methane and the methonium ion was suggested as an intermediate or transition state.

The methonium ion (CH_5^+) was first discovered in the mass spectrometer.[41,42] A number of theoretical studies have shown its shallow potential energy surface[43–47] (see also Section 3.5.1.1).

The eclipsed C_s conformer [C_s(e), Figure 5.1] is now accepted as being the lowest on the potential energy surface, although the barriers for rearrangements are extremely small. The barrier for methyl rotation to a staggered C_s conformation [C_s(s)] is about 0.1 kcal mol^{-1}, and the exchange between two eclipsed C_s structures via a flip-transition state with C_{2v} symmetry is about 0.8 kcal mol^{-1}. Addition of zero-point energy suggests that the energy difference is almost negligible.[43] A recent high-resolution infrared spectrum in the gas phase seems to confirm that the parent alkonium ion CH_5^+ is highly fluxional and that its structure is not described by a single nuclear configuration, although any assignment of peaks was not possible.[48]

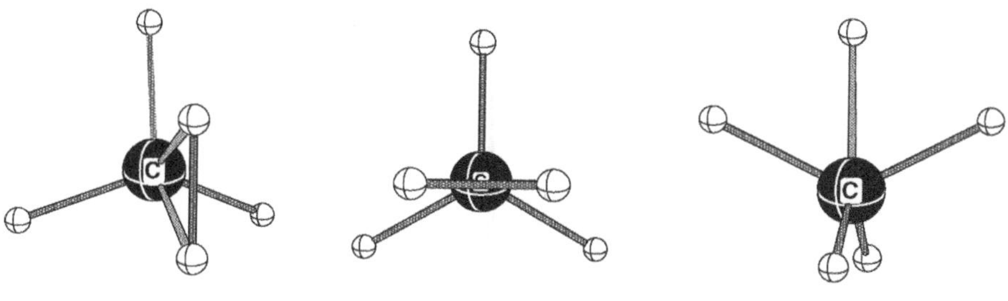

Figure 5.1. Calculated conformations of the methonium ion CH_5^+.

Methane and ethane, the weakest σ-bases, which are extremely difficult to ionize, will exchange hydrogens without side reaction in any acid having an H_0 value of −12 or below.

Methane is slightly soluble in HF–SbF_5 even at atmospheric pressure (0.005 *M*), which facilitates direct kinetic studies by NMR. Thus the transition states for methane activation in this medium have been studied experimentally by Ahlberg et al.[49] The first-order rate constants [Eqs. (5.9) and (5.10)], determined experimentally on the basis of ^2H-decoupled 600-MHz ^1H NMR time-dependent spectra (Figure 5.2), are on the order of $3.2 \times 10^{-4}\,s^{-1}$ at −20°C and show a secondary kinetic isotope effect (SKIE) of 1 ± 0 02.

$$^{13}C^1H_4 + {}^2HF{-}SbF_5 \xrightarrow{4\,k_a} {}^{13}C^1H_3{}^2H + {}^1HF{-}SbF_5 \tag{5.9}$$

$$^{12}C^1H_3{}^2H + {}^2HF{-}SbF_5 \xrightarrow{3\,k_b} {}^{12}C^1H_2{}^2H_2 + {}^1HF{-}SbF_5 \tag{5.10}$$

$$^{12}C^1H_2{}^2H_2 + {}^2HF{-}SbF_5 \xrightarrow{2\,k_c} {}^{12}C^1H^2H_3 + {}^1HF{-}SbF_5 \tag{5.11}$$

$$^{12}C^1H^2H_3 + {}^2HF{-}SbF_5 \xrightarrow{k_d} {}^{12}C^2H_4 + {}^1HF{-}SbF_5 \tag{5.12}$$

By using the reagent pairs $^{13}CH_4/^{12}CH_3D$ and $^{13}CH_4/^{12}CH_2D_2$ as starting materials, the first-order rate constants of all isotopologs were obtained and the secondary kinetic isotope effects were estimated to be around 1.00 ± 0.05. Because these results were surprising in view of the expected dramatic structural change in going from methane to the presumed methonium ion-like activated complex, DFT and *ab initio* methods were used to optimize the structures and energies of the intermediates and transition states at various levels of theory. Protonation of methane by H_2F^+ to yield the strongly HF-hydrogen-bonded CH_5^+ ion **10** was found to be barrierless and lowered the potential energy by 29.0 kcal mol^{-1}. The activated complex **11** for hydrogen exchange showed a potential energy of only 1.9 kcal mol^{-1}

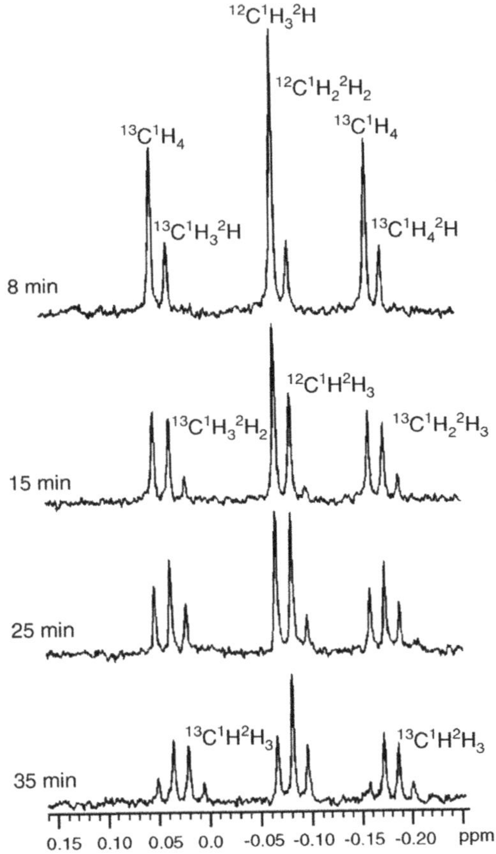

Figure 5.2. ^{1}H NMR spectra recorded at different reaction times for the $^{13}CH_4$–$^{12}CH_3D$ Deuterium-decoupled mixture (1:1 molar ratio) at −20°C.

higher than 1, the exchange taking place in the complex. The methonium ion structure in **10** and **11** is closely similar to the C_s structure in CH_5^+.

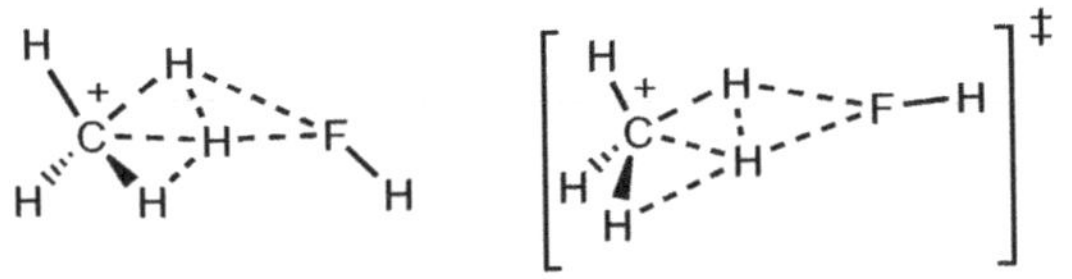

10 **11**

Only the unsolvated superacid H_2F^+ was found to be strong enough to protonate methane to yield the methonium ion solvated by HF as a potential energy minimum. Similar calculations at the B3LYP/6-31++G**+RECP (SB) level have been performed by Mota and colleagues[50] using the $H_2F^+Sb_2F_{11}^-$ cluster as an electrophile to mimic the liquid superacid HF–SbF$_5$ and C_1-C_4 alkanes.

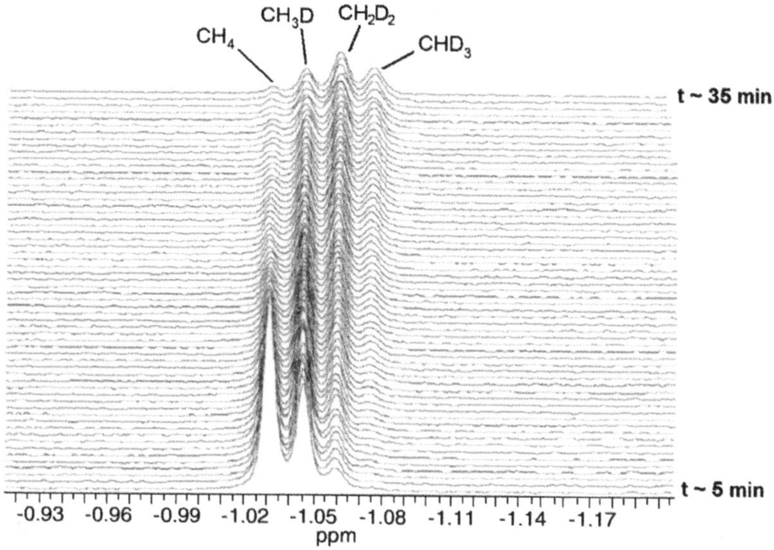

Figure 5.3. Deuterium-decoupled ^{1}H NMR spectra during the reaction between methane and a mixture of $DSO_3F–SbF_5$ (41 mol%) at 30°C.

As methane is also slightly soluble in Magic Acid, the hydrogen exchange rate between methane and a series of $DSO_3F–SbF_5$ superacids could be measured by in situ ^{2}H-decoupled ^{1}H NMR spectroscopy (Figure 5.3).[51] The rates of exchange, much lower than in $HF–SbF_5$, showed a strong dependence on antimony pentafluoride concentration, with the free energy of activation $\Delta G^{\ddagger}{}_{30°C}$ decreasing from 23 to 20 kcal mol^{-1} over the range of concentration 19 to 49 mol% SbF_5. The constant free enthalpy of activation $\Delta H^{\ddagger}$ (about 15.5 kcal mol^{-1}) and the decreasing entropy of activation $\Delta S^{\ddagger}$ seem to indicate that an increase in acidity of the superacid system does not substantially change the nature of the transition state but rather its solvation. If the rates of exchange were directly related to acidity, the activation energy measured in the temperature range of −25°C to +75°C appeared to be constant, which would indicate that not the nature but only the concentration of the protonating species was changing when the amount of SbF_5 was increasing in order to increase the acidity. No exchange was observed when pure HSO_3F was used in the same temperature range.

At temperatures above 270°C, H–D exchange occurs even in D_2SO_4.[52] Above 300°C, however, oxidation to CO takes place, which is then further oxidized to CO_2 with the formation of SO_2 and water [Eqs. (5.13) and (5.14)].

$$CH_4 + 3\,H_2SO_4 \longrightarrow CO + 3\,SO_2 + 5\,H_2O \qquad (5.13)$$

$$CO + H_2SO_4 \longrightarrow CO_2 + SO_2 + H_2O \qquad (5.14)$$

Small Alkanes with More than Two Carbon Atoms. Whereas methane and ethane show a very similar behavior towards superacidic media, alkanes with more than two carbon atoms undergo a more complex reaction scheme in which C–H and C–C bond cleavage compete with reversible protonation. In order to follow the initial steps, it is very useful to run the reaction in presence of carbon monoxide, which reacts rapidly with the initially formed carbenium ions yielding stable oxocarbenium ions unable to activate alkanes by hydride transfer.[53]

When isobutane (**2**) is contacted with HF–SbF_5 at -10°C in the presence of carbon monoxide, analysis of the reaction products both from the gas phase and from the liquid phase can be rationalized by the two pathways described in Scheme 5.4.[30,54,55]

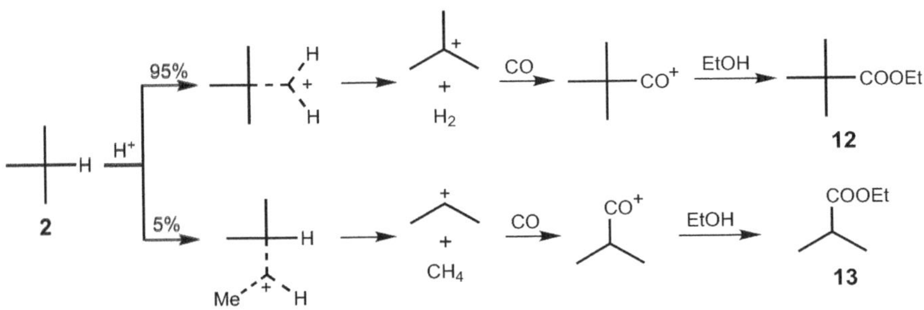

Scheme 5.4

The main pathway for ionization as expected is the protolytic cleavage of the tertiary C–H bond producing stoichiometric amounts of *tert*-butyl cation and hydrogen. The *tert*-butyl ion is converted into ethyl pivalate (**12**) after reaction with CO and neutralization of the superacid with ethanol–bicarbonate mixture. The minor pathway is C–C bond cleavage resulting in the formation of ethyl isobutyrate (**13**). The fact that in earlier work the purely protolytic pathway was questioned[23] has now been explained on the basis of the composition of the HF–SbF_5 system: When the concentration of SbF_5 exceeds 20% of HF, a small and increasing amount of uncomplexed SbF_5 is present, which participates in the activation by an oxidative process (*vide infra*).[56]

The protolytic activation of the alkane is, however, only the apparent part of the reaction as long as the alkane or the acid is not isotopically labeled. When HF is replaced by DF and the isobutane–CO mixture is bubbled through the DF–SbF_5 acid (6:1 molar ratio) at -10°C, the apparent conversion based on ester or H_2 formation is only 4% but the $^1H/^2H$ NMR analysis of the apparently unreacted isobutane (96%) shows extensive H–D exchange (18 atom% in the tertiary position and 9 atom% at each primary position).[30] The most plausible rationalization of hydrogen exchange is via the formation of carbonium ions (here pentacoordinate transition states or intermediates) as described in Eq. (5.15).

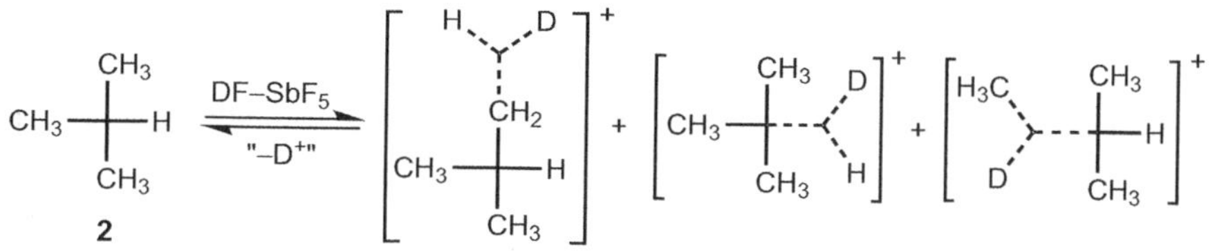

(5.15)

By selective labeling experiments, it has been shown that, in contrast with CH_5^+, the proton scrambling does not occur in the iso$C_4H_{11}^+$ and $C_3H_9^+$species. Rather, the exchange takes place between the alkane and the acid and not between the isotopolog structures of the iso$C_4H_{11\text{-}X}D_x^+$ and $C_3H_{9\text{-}X}D_x^+$species[57] [Eqs. (5.16) and (5.17)].

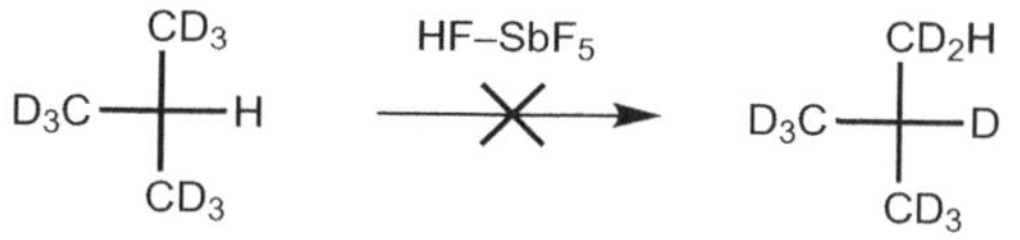

(5.16)

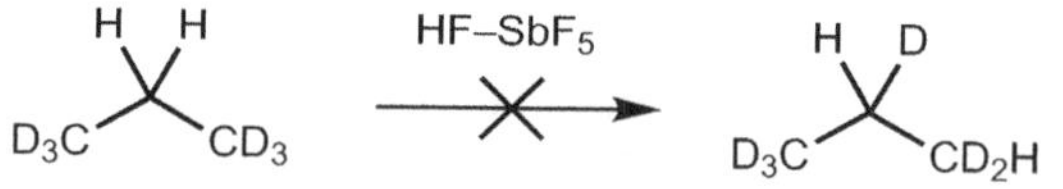

(5.17)

Under similar experimental conditions, propane is slightly ionized (2% conversion) but extensively deuteriated when bubbled through DF–SbF_5 in the presence of CO (12 atom% in the primary position and 17 atom% in the secondary position). Skeletal rearrangement in protonated propane via carbonium-ion-type intermediates has been suggested several times based on results obtained with zeolites.[58,59] This type of rearrangement could be excluded as ^{13}C-labeled propane, extensively deuteriated in DF–SbF_5, showed no ^{13}C label scrambling.[60]

Isopentane is the smallest alkane in the series, having primary, secondary, and tertiary hydrogens. Under similar conditions as propane and isobutane, isopentane was partially ionized in DF–SbF_5 at $-10°C$ (10% conversion); the recovered 90% of alkane showed extensive deuterium/proton exchange (12 atom% of the primary, 16 atom% of the secondary, 19 atom% of the tertiary hydrogens were exchanged as expected in accord with their relative σ-basicity).[61]

From these experiments, some general conclusions can be drawn concerning the behavior of small alkanes in the strongest HF–SbF_5 system. (1) The reversible protonation of the alkanes (i) is very fast in comparison with the ionization step; (ii) it takes place on all σ-bonds independently of the subsequent reactivity of the alkane; (iii) it involves carbonium ions (transition states), which do not undergo molecular rearrangements. (2) Protonation of an alkane is a typical acid–base reaction and carbon monoxide has no effect on this step.

The Oxidative Pathway. For a long time, one of the difficulties in understanding the mechanism of the superacid-catalyzed transformations of alkanes was that no stoichiometric amount of hydrogen gas evolution was observed from the reaction mixture. This led to a controversy, which was only solved in the early 1990s.

One mechanism that was proposed involves direct hydride abstraction by the Lewis acid[62] [Eq. (5.18)].

$$RH + 2\,SbF_5 \longrightarrow R^+\,SbF_6^- + SbF_3 + HF \quad (5.18)$$

Olah, however, pointed out that if SbF_5 would abstract H^-, it would need to form SbF_5H^- ion involving an extremely weak Sb–H bond compared to the strong C–H bond being broken.[63] Calculations based on thermodynamics[64] also seemed to indicate that the direct oxidation of alkanes by SbF_5 was not feasible. Hydrogen is generally assumed to be partially consumed in the reduction of one of the superacid components [Eqs. (5.19) and (5.20)].

$$2\,HSO_3F + H_2 \longrightarrow SO_2 + H_3O^+ + HF + SO_3F^- \qquad \Delta H = -49\ \text{kcal mol}^{-1} \quad (5.19)$$

$$SbF_5 + H_2 \longrightarrow SbF_3 + 2\,HF \qquad \Delta H = -33\ \text{kcal mol}^{-1} \quad (5.20)$$

It could, however, be shown[30] that the amounts of hydrogen and ester obtained in the presence of CO from isobutane in HF–SbF_5 are stoichiometric only as long as the SbF_5 concentration is lower then 20%. For higher concentrations the ester production increases steadily whereas hydrogen formation decreases. Competition between ionization and exchange is described in Scheme 5.5.

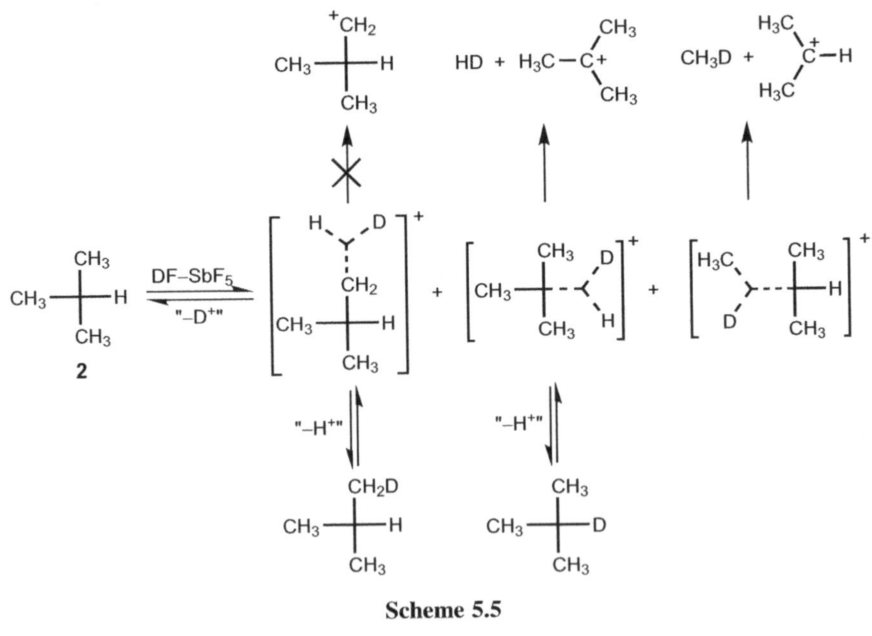

Scheme 5.5

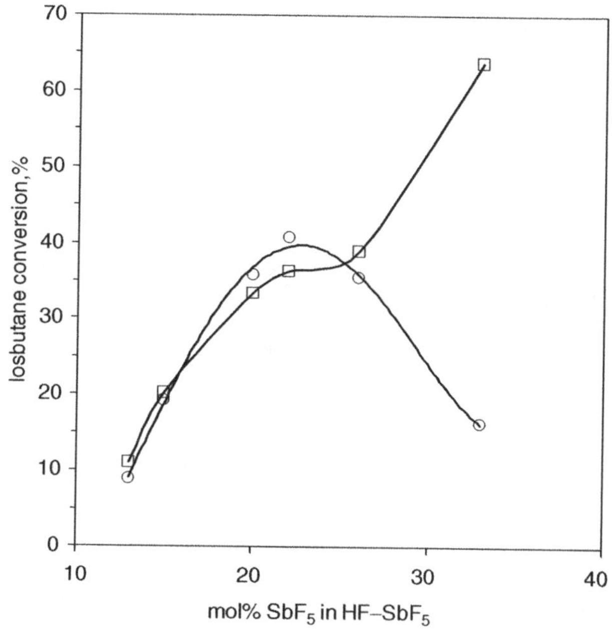

Figure 5.4. Isobutane conversion based on ethyl pivalate production (□) compared with hydrogen production (○) (60 min time-on-stream).[30]

An increasing amount of esters is accompanied by a decreasing amount of H_2 concomitant with reduction of SbF_5 to SbF_3 (Figure 5.4). The H–D exchange on tertiary and primary hydrogens as a function of the concentration of SbF_5 is shown in Figure 5.5.

The same phenomenon was observed[61] with propane (Figure 5.6) and isopentane (Figure 5.7): above 20 mol% of SbF_5, reversible protonation and protolytic ionization decrease rapidly whereas the conversion of the alkane with concomitant reduction of SbF_5 increases. H–D exchange data observed in small alkanes are collected in Table 5.1.

The direct reduction of SbF_5 in the absence of hydrocarbons by molecular hydrogen necessitates, however, more forcing conditions (50 atm, high temperature), which suggests that the protolytic ionization of alkanes proceeds probably via solvation of the protonated alkane by SbF_5 and concurrent ionization–reduction.[63] That SbF_5 could be reduced by alkanes even in the absence of protons was demonstrated by Culmann and Sommer[56] who showed that SbF_5 reacts with isobutane at 0° in the presence of a proton trap according to Eq. (5.21).

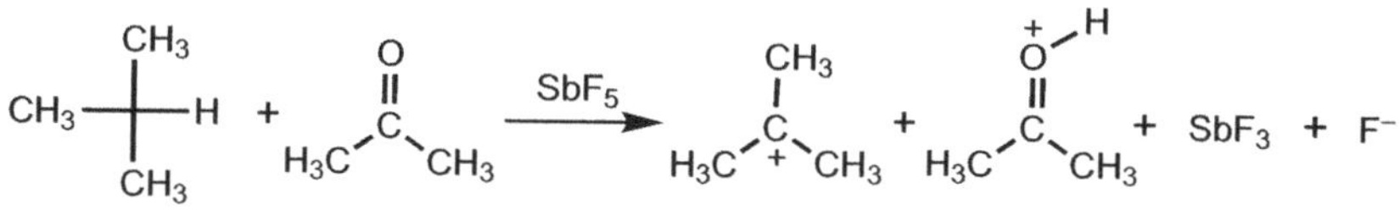

(5.21)

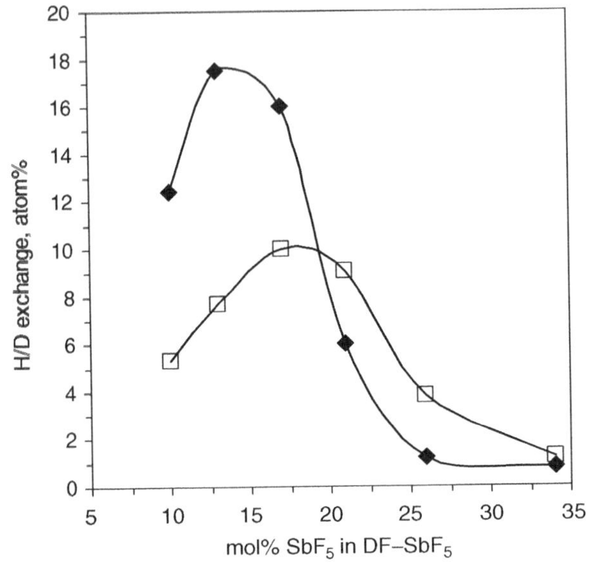

Figure 5.5. Importance of H–D exchange on tertiary (♦) and primary (□) hydrogens, depending on SbF_5 concentration.[30]

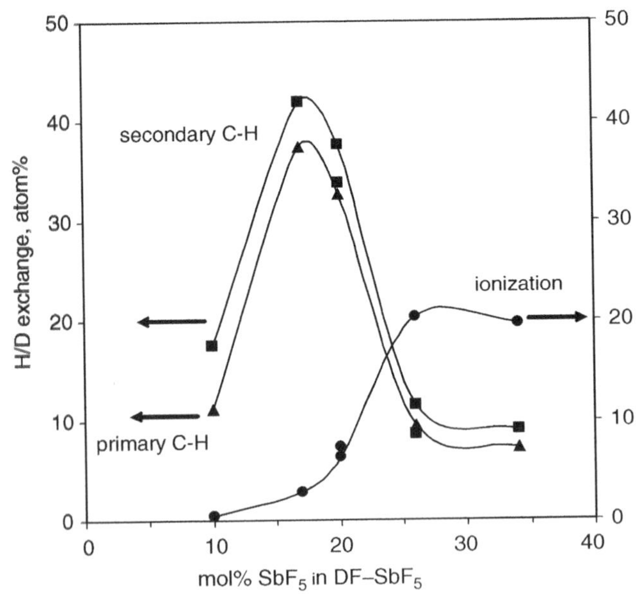

Figure 5.6. H–D exchange and conversion of propane reacted with $DF–SbF_5$ at various SbF_5 concentrations.[61]

Table 5.1. H–D Exchange Observed in Small Alkanes Upon Reaction with DF–SbF_5[61]

Alkane	SbF_5, mol% in DF[a]	Exchange (atom %)[b] on		
		Primary C–H	Secondary C–H	Tertiary C–H
Propane	12	12.0	16.5	
		1[d]	1.38	
n-Butane[c]	12.7	28.1	38.9	
		1	1.38	
Isobutane[e]	12.7	7.7		17.5
		1		*2.27*
Isopentane	12.4	12.1	16.2	19.3
		1	1.34	1.60

[a]Mol % determined by weight (± % of the indicated value).
[b]Determined by ^1H and ^2H NMR (± 3% of the indicated value).
[c]See ref. 65 for a complete study.
[d]All numbers in italic are normalized to the exchange in the primary position.
[e]See ref. 30 for a complete study.

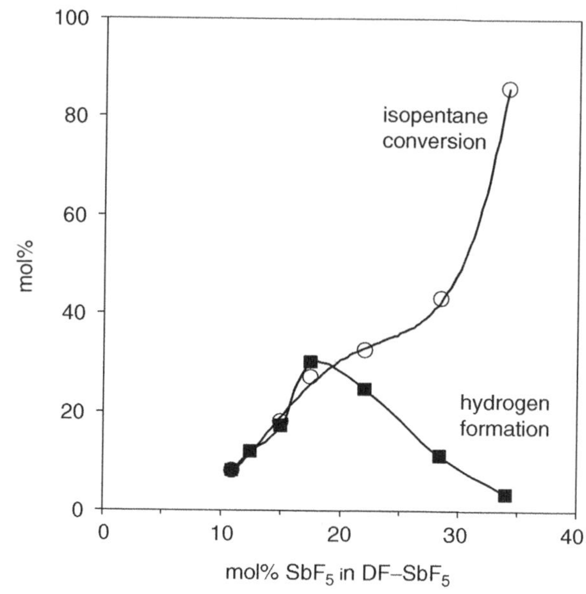

Figure 5.7. Isopentane conversion compared to hydrogen production.[61]

This direct oxidation generates itself a proton and is probably only a minor pathway under superacidic condition where, in the absence of the proton trap, protolysis of the C—H and C—C bond occurs very rapidly. The mechanism is most probably of electron transfer nature as suggested in Eq. (5.22) and (5.23)

$$SbF_5 + 2e^- \longrightarrow SbF_3^+ + 2F^- \tag{5.22}$$

$$RH \xrightarrow[-2e^-]{} R^+ + \text{"}H^+\text{"} \tag{5.23}$$

Small Alkanes with More than Two Carbon Atoms in Weaker Superacid Systems. As already observed for methane (*vide supra*), with decreasing acidity it becomes more and more difficult to protonate reversibly C—H bonds. Nevertheless, when alkanes with more than two carbon atoms are used as starting material, carbenium ions are generated by competitive protolytic and oxidative processes. Depending on the strength of the superacid system, proton exchange can take place by two competitive reactions: (i) directly via reversible protonation and (ii) via deprotonation of the carbenium ion and reprotonation of the alkene.

Whereas the importance of the exchange mechanism via route (i) is rapidly decreasing with decreasing acidity, route (ii) is facilitated by the increasing basicity of the superacid counterion.

For this reason when Magic Acid (HSO_3F–SbF_5, 1:1 molar ratio) is used under the same experimental conditions as above at room temperature, isobutane undergoes very slow ionization and the formation of the *tert*-butyl ion can be monitored. However, recovered isobutane shows no exchange because the reversible protonation via carbonium ion transition state does not take place and because the *tert*-butyl ion, stable in this solution at room temperature, does not deprotonate.

When still weaker superacids are used, such as HSO_3F ($H_0 = -15$) or CF_3SO_3H ($H_0 = -14$), linear alkanes do not react at room temperature, but branched alkanes are ionized via an oxidative process involving the reactive tertiary C—H bond. The tertiary cations that are generated undergo reversible deprotonation to alkenes that are reprotonated. This process ends when hydride transfer occurs from unreacted isoalkane; but, when deuteriated acids are used, it leads to isoalkanes extensively and regioselectively deuteriated on the carbons vicinal to the branching carbon. This exchange process is similar to the one observed by Otvos et al.[66] in the early 1950s while studying the deuteriation of isobutane in D_2SO_4 as described in Scheme 5.6.

Because the reaction is catalytic in *tert*-butyl cation and the deprotonation/reprotonation steps are very fast, extensive regioselective deuteriation of the isoalkane is observed at room temperature as shown by GC–MS analysis. The absence of mass 68 (d_{10}-isobutane) and the presence of mass 64 due to SO_2 formation in the oxidative process are typical features in accord with the oxidative activation of the alkane and the Markovnikov-type addition of deuterons on the intermediate isobutylene (**14**). However, the exchange process does not take place in the presence of carbon monoxide, which traps the *tert*-butyl cation and prevents deprotonation (Scheme 5.7).

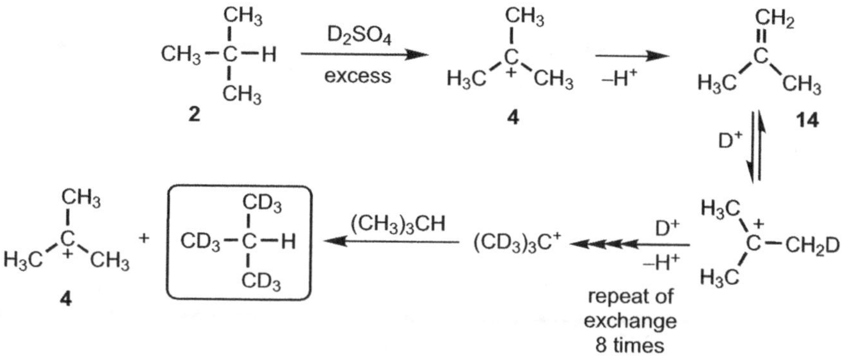

Scheme 5.6

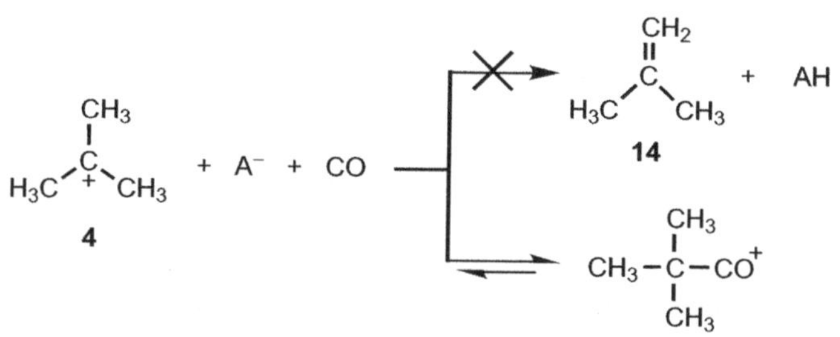

Scheme 5.7

Alkanes and Strong Solid Acids. Since the early reports by Nenizetscu and Dragăn[67] on alkane isomerization on wet aluminum chloride in 1933, all mechanistic studies have led to a general agreement on the carbenium-ion-type nature of the reaction intermediates involved in acid-catalyzed hydrocarbon conversions. In contrast with this statement, the nature of the initial step is still under discussion and a variety of suggestions can be found in the literature among which direct protolysis of C–H and C–C bonds, protonation of alkenes present as traces, and oxidative activation are the most often quoted.[54,55]

Solid acids are generally oxides containing various OH groups, which may function as proton donors. The deuteriation of these OH groups allows mechanistic isotope tracer studies. Monitoring the appearance and localization of the isotope in the alkane activated by the solid acids will provide interesting information on the reaction mechanism. H–D exchange of the OH groups occurs readily at moderate temperatures below 200°C when the solid is exposed to D_2O or at higher temperatures (500°C) in the presence of D_2.

When isoalkanes are contacted with D_2O-exchanged solid acids such as zeolites, sulfated zirconias, or heteropoly acids, proton–deuterium exchange takes place slowly even at room temperature; but above 100°C the catalysts are rapidly depleted of their deuterons, which are recovered in the alkane.[54,55] This exchange process is

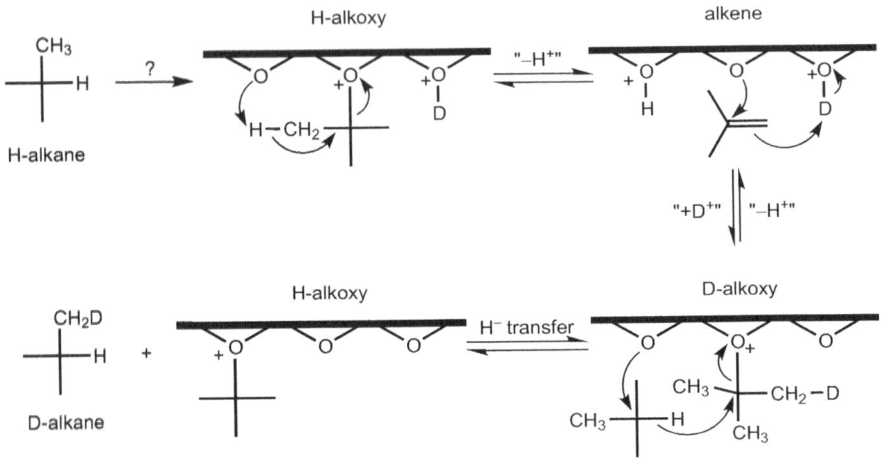

Scheme 5.8

regioselective because only hydrogens vicinal to the branched carbon are exchanged in accord with a mechanism very similar to the one observed in H_2SO_4 and in weak acids (Scheme 5.8).

Linear alkanes, which are known to be less reactive, also undergo H–D exchange by the same mechanistic scheme at slower rates at and above 150°C.[54,55] This exchange reaction occurs in a very clean way because no side products from cracking and isomerization are observed. The cations that are adsorbed on the surface are prone to deprotonation, but the alkenes that are formed are rapidly reprotonated before substantial oligomerization can take place.

The occurrence of carbenium ions as reaction intermediates is strongly supported by the observation that the isotopic exchange can be totally suppressed in the presence of carbon monoxide.[68] Furthermore, trapping of the intermediate carbenium ions by CO and water has been observed by insitu NMR spectroscopy when isobutane, water, and CO reacted on HZSM-5 zeolite to form pivalic acid.[69,70] Regarding the small conversion, only a limited number of acid sites are suggested to be strong enough for the initial protolytic activation to take place.

Methane does not react with D_2O-exchanged solid acids at temperatures below 400°C. The H–D exchange has, however, been observed on sulfated zirconia (SZ) at 400°C.[71] When the stronger acid SZA3 (sulfated zirconia doped with 3% Al_2O_3) is used,[72] the exchange rates are substantially higher.[71] In the presence of D_2O-exchanged zeolites, the H–D exchange process can only be measured at temperatures as high as 500°C or higher.[73]

Theoretical Approaches. Computer modeling is an increasingly fruitful tool in catalysis, and several research groups have attempted to rationalize high carbon conversion over zeolites from a theoretical point of view. The main problem to be

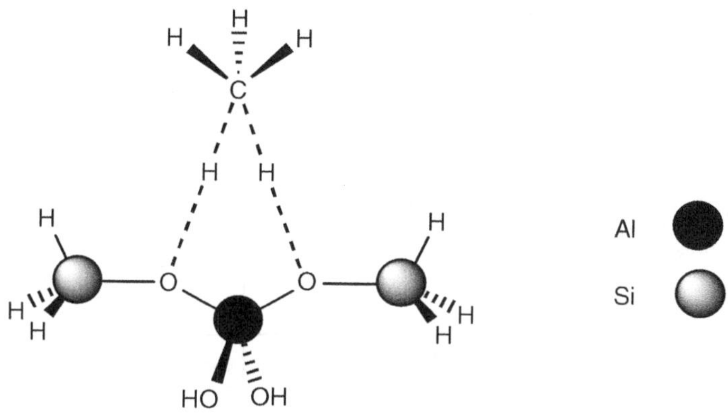

Figure 5.8. Calculated transition state geometry for H–D exchange between a zeolite cluster and methane.

solved is the choice of a model (generally a small cluster) representative of the zeolite framework.

Both the carbenium ion and carbonium ion transition states have been investigated. In the presence of the oxygen lone pairs, it is clear that the most stable reaction intermediates will not be free carbenium ion but surface alkoxy groups. The generation of these alkoxy species from alkenes has been demonstrated by Haw et al.[74] The structure of the transition state of the H–D exchange process in methane has been calculated by van Santen and Kramer[75] to be symmetrical with essentially covalent bond sharing character (Figure 5.8).

By modifying the proton affinity, the authors[76] could demonstrate the dependence of the exchange rate on the proton affinity of the zeolite cluster, relating the activity of the catalyst with its acidity. The activity of the catalyst was found to be determined by the acidity differences between the proton donating and proton accepting oxygen sites. The calculated energy of activation is close to the experimentally determined values of approximately 33.5 kcal mol^{-1} with an uncertainty of 5 kcal mol^{-1}.[76] Restudying the mechanism of hydrogen exchange between methane and H-USY and replacing the bare cluster by an accurate *ab initio* embedded cluster, including the Madelung potential, Vollmer and Truong[77] found that the transition state had carbonium-like characteristics from both geometric and electron density considerations. Using the CCSD(T)/6-31G(d,p) level of theory including zero-point energy and tunneling, the activation energy was predicted to be 30 ± 1 kcal mol^{-1} and 34 ± 1 kcal mol^{-1} for hydrogen exchange from two different binding sites.

It is interesting to compare this transition state in the solid with the one calculated from the HF–SbF_5 system. In the liquid superacid, the ionic character is very strong and it is easier to connect the reactivity with the unusual activity of the proton even when solvated by the HF solvent. In contrast, on the solid the theoretical calculated transition state is further away from the carbonium ion type and in line with the much higher temperatures needed to activate the alkane with weaker acids.

5.1.2. Electrochemical Oxidation in Strong Acids

The protolytic oxidation of alkanes is also strongly supported by electrochemical studies. In 1973, Fleischmann, Plechter, and co-workers[78] showed that the anodic oxidation potential of several alkanes in HSO_3F was dependent on the proton donor ability of the medium. This acidity dependence shows that there is a rapid protonation equilibrium before the electron transfer step and it is the protonated alkane that undergoes oxidation (Scheme 5.9).

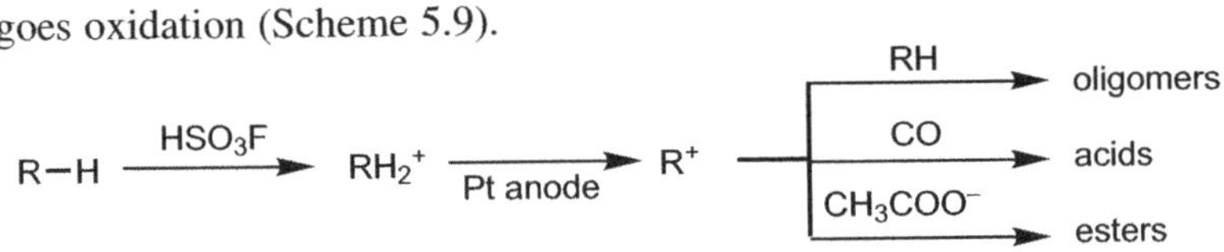

Scheme 5.9

The electrochemical oxidation of lower alkanes in the HF solvent system has been investigated by Devynck and co-workers[79] over the entire pH range. Classical and cyclic voltammetry show that the oxidation process depends largely on the acidity level. Isopentane (**15**, 2-methylbutane, 2MeBuH), for example, undergoes two-electron oxidation in HF–SbF_5 and HF–TaF_5 solutions[80] [Eq. (5.24)].

$$\underset{\mathbf{15}}{\text{2MeBuH}} \xrightarrow[-2\,e^-]{} \text{2MeBu}^+ + H^+ \tag{5.24}$$

In the higher-acidity region, the intensity–potential curve shows two peaks (at 0.9 V and 1.7 V, respectively, versus the Ag/Ag^+ system). The first peak corresponds to the oxidation of the protonated alkane and the second the oxidation of the alkane itself.

The chemical oxidation process in the acid solution [Eq. (5.25)] can be considered as a sum of two electrochemical reactions [Eqs. (5.26) and (5.27)].

$$RH + H^+ \rightleftharpoons R^+ + H_2 \tag{5.25}$$

$$2\,H^+ + 2\,e^- \rightleftharpoons H_2 \qquad (E^0_{H^+/H_2}) \tag{5.26}$$

$$RH \rightleftharpoons R^+ + H^+ + 2\,e^- \qquad (E^0_{R^+/RH}) \tag{5.27}$$

The chemical equilibrium [Eq. (5.25)] is characterized by the constant K_1 [Eq. (5.28)]. With the experimental conditions set to control p_{H_2}, the acidity level and R^+ concentration, [RH] can be evaluated from voltammetric results and K_1 can be determined.

$$K_1 = \frac{[R^+]p_{H_2}}{[RH][H^+]} \quad \text{and} \quad R_A = \frac{[R^+]}{[RH]} = \frac{K_1}{p_{H_2}} \cdot [H^+] \tag{5.28}$$

However, when log R_A is plotted as a function of pH, the authors noticed that in the case of 2MeBuH below pH = 1.25, the ratio becomes pH-independent (Figure 5.9).

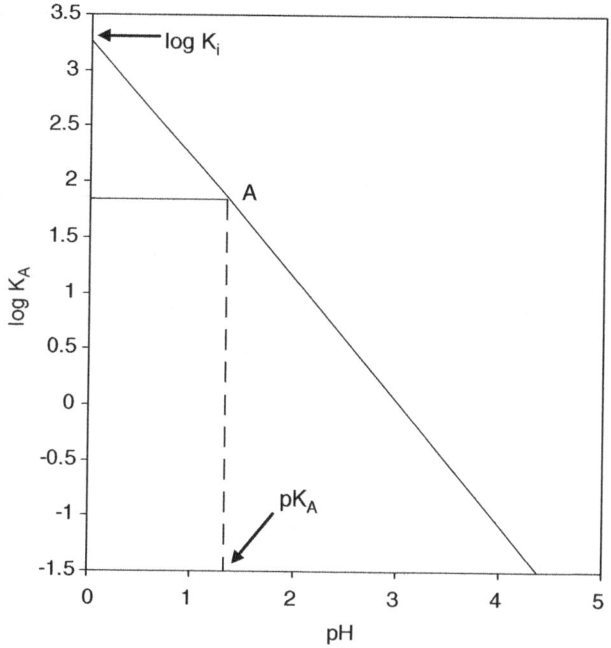

Figure 5.9. Variation of $2MeBu^+/2MeBuH$ as a function of pH in HF–SbF_5.[79]

Thus, at a higher acidity level, another equilibrium must be taken into account, which is the protonation of the alkane [Eq. (5.28)] with constant K_2 [Eq. (5.30)], and the oxidation reaction becomes pH-independent [Eq. (5.31)].

$$RH + H^+ \rightleftharpoons RH_2^+ \tag{5.29}$$

$$K_2 = \frac{[RH_2^+]}{[RH][H^+]} \tag{5.30}$$

$$RH_2^+ \longrightarrow R^+ + H_2 \tag{5.31}$$

The acidity level at point A is representative of the basicity of the hydrocarbon: If K_A is the acidity constant of the protonated alkane, then $pH_A = pK_A$. On the other hand, as the redox reactions [Eqs. 5.26 and 5.27] are combined in Eq. 5.25, the equilibrium constant K_1 is related to the standard potentials of redox couples R^+/RH and H^+/H_2 according to Eq. (5.32).

$$\frac{2.3RT}{2F} \log K_1 = E^0_{H^+/H_2} - E^0_{R^+/RH} \tag{5.32}$$

Because K_1 can be determined experimentally and the oxidation potential of H^+/H_2 is known in the acidity range, oxidation potential can be calculated. This allowed the

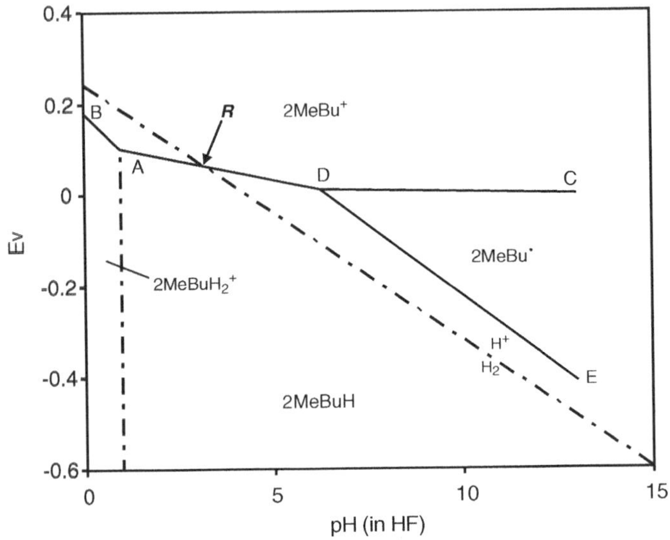

Figure 5.10. Potential–acidity diagram of isopentane **15** (2MeBuH) versus the hydrogen electrode system ($p_{H2} = 1$ atm).[79]

authors to plot the B–D part of the potential–acidity diagram versus the H^+/H_2 system as shown in Figure 5.10 for isopentane (**15**, 2MeBuH). The diagram shows that oxidation of the alkane (2MeBuH) by H^+ gives the carbocation only at pH values below 5.7. In the stronger acids ($pH < pH_A$), it is the protonated alkane which is oxidized. At pH values higher than 5.7, oxidation of isopentane gives the alkane radical that dimerizes [Eqs. (5.33) and 5.34] or is oxidized in a pH-independent process [Eq. 5.35].

$$\underset{\mathbf{15}}{\text{2MeBuH}} \xrightarrow[-e]{} \text{2MeBuH}^{+\bullet} \longrightarrow \text{2MeBu}^{\bullet} + H^+ \quad (5.33)$$

$$2\ \text{2MeBu}^{\bullet} \longrightarrow (\text{2MeBu})_2{}^{\bullet} \quad (5.34)$$

$$\text{2MeBu}^{\bullet} \xrightarrow[-e]{} \text{2MeBu}^+ \quad (5.35)$$

An interesting point is the intersection point ***R*** at which crossover of the H^+/H_2 and the R^+/RH systems occurs. From this acidity level onwards, RH is oxidized spontaneously into R^+.

This potential–acidity diagram (Pourbaix's type) has been determined for a large series of alkanes.[79] All of these results indicate two types of oxidation mechanism of the C−H bond: (i) oxidation of alkanes into carbenium ion at high acidity levels and (ii) oxidation of alkanes into radicals at low acidity levels.

Using isopentane **15** as a reference alkane, the authors calculated the Gibbs free energy between the redox couples of various alkanes and the $2MeBuH/2MeBu^+$ couple. This leads to the standard potential of the alkane redox couples in HF

Table 5.2. Standard Potential of Redox Couples of Alkanes in HF (at $H_0 \sim -22.1$) and Acidity Levels of Oxidation (pH_R in HF).[79]

Alkane	$E^0_{R^+/RH}$	pH_R
CH_4	1.15	−34
C_2H_6	0.70	−17
C_3H_8	0.34	−4.3
n-C_4H_{10}	0.38	−5.6
isoC_4H_{10}	0.17	2.0
n-C_5H_{12}	0.37	−5.2
isoC_5H_{12}	0.15	2.7
neoC_5H_{12}	0.78	−21
n-C_6H_{14}	0.36	−5.0
2-Methylpentane	0.14	3.0
3-Methylpentane	0.16	2.4
2,2-Dimethylbutane	0.26	−1.5
2,3-Dimethylbutane	0.13	3.5

(at $H_0 \sim -22.1$) (Table 5.2). The position of the redox couple R^+/RH versus H^+/H_2 system leads to the determination of the oxidation pH or acidity level at which reaction in Eq. (5.31) is quantitative.

These results are in agreement with the alkane behavior in superacid media and indicate the ease of oxidation of tertiary alkanes. However, high acidity levels are necessary for the oxidation of alkanes possessing only primary C−H bonds.

Once the alkane has been partly converted into the corresponding carbenium ion, the carbenium ion may undergo various reactions following (a) intramolecular routes such as skeletal rearrangement [Eq. (5.36)] and fragmentation [Eq. (5.37)] or (b) intermolecular routes such as hydride transfer [Eq. (5.38)] and alkylation of another alkane molecule [Eq. (5.39)]. The specificity of some of these reactions will be discussed in the following sections.

(5.36)

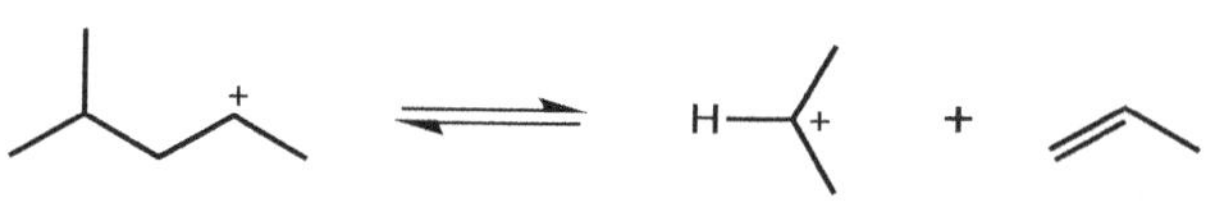

(5.37)

(5.38)

$$CH_3^+ + CH_3{-}CH_3 \longrightarrow CH_3{-}CH_2{-}CH_3 + H^+ \qquad (5.39)$$

In a subsequent study Devynck and co-workers[81,82] studied the electrochemical oxidation of alkanes and alkenes in triflic acid monohydrate. The acidity of $CF_3SO_3H \cdot H_2O$ was found to be intermediate between that of aqueous acid media and superacidity. Alkanes undergo two-electron oxidation, whereas alkenes are protonated to yield carbenium ions in this medium. In addition to various transformations characteristic of carbenium ions [Eqs. (5.36)–(5.38)], they undergo a reversible disproportionation to give an alkane and an aldehyde [Eqs. (5.40)].

$$2\,R^+ + H_2O \rightleftharpoons R'CHO + RH + 2\,H^+ \qquad (5.40)$$

5.1.3. Isomerization of Alkanes

Acid-catalyzed isomerization of saturated hydrocarbons was first reported in 1933 by Nenitzescu and Dragăn.[67] They found that when *n*-hexane was treted with aluminum chloride under reflux, it was converted into its branched isomers. This reaction is of major economic importance as the straight-chain C_5–C_8 alkanes are the main constituents of gasoline obtained by refining of the crude oil. Because the branched alkanes have a considerably higher octane number than their linear counterparts, the combustion properties of gasoline can be substantially improved by isomerization. Table 5.3 gives the octane number of a series of straight-chain and branched hydrocarbons.

The isomerization of *n*-butane **1** to isobutane **2** is of great importance because isobutane reacts under mild acidic conditions with olefins to give highly branched hydrocarbons in the gasoline range. A substantial number of investigations have been devoted to this isomerization reaction and a number of reviews are available.[83–89] The isomerization is an equilibrium reaction that can be catalyzed by various strong acids. In the industrial processes, aluminum chloride and chlorinated alumina are the most widely used catalysts. Whereas these catalysts become active only above 80–100°C, superacids are capable of isomerizing alkanes at room temperature and below. The major advantage (besides energy saving) is that lower temperatures thermodynamically favor the most branched isomers (Table 5.4).

During the isomerization process of pentanes, hexanes, and heptanes, cracking of the hydrocarbon is an undesirable side reaction. The discovery that cracking can be substantially suppressed by hydrogen gas under pressure was of significant importance. In our present-day understanding, the effect of hydrogen is to quench carbocationic sites through five-coordinate carbocations to the related hydrocarbons, thus decreasing the possibility of C–C bond cleavage reactions responsible for the acid-catalyzed cracking.

The nucleophilic nature of the alkanes is also shown by the influence of the acidity level on their solubility. Torck and co-workers[90] have investigated the composition of the catalytic phase obtained when *n*-pentane or *n*-hexane is thoroughly mixed with $HF–SbF_5$ in an autoclave under hydrogen pressure [Eq. (5.41)]. The total amount of hydrocarbon in the catalytic phase (dissolved ions and neutrals) was obtained by extraction with excess of methylcyclopentane. The amount of physically dissolved

Table 5.3. Comparison of the Octane Numbers of Straight-Chain and Branched Paraffinic Hydrocarbons[85]

Paraffinic Hydrocarbon	Octane Number
n-Butane	94
Isobutane	100
n-Pentane	63
2-Methylbutane	90
2,2-Dimethylpropane (neopentane)	116
n-Hexane	32
2-Methylpentane	66
3-Methylpentane	75
2,2-Dimethylbutane (neohexane)	94.6
2,3-Dimethylbutane (bis-isopropyl)	95
n-Heptane	0
2-Methylhexane	45
3-Methylhexane	65
2-Ethylpentane	68
2,2-Dimethylpentane	80
2,3-Dimethylpentane	82
2,4-Dimethylpentane	80
3,3-Dimethylpentane	98
2,2,3-Trimethylbutane (triptane)	116
n-Octane	−19
2-Methylheptane	23.8
3-Methylheptane	35
4-Methylheptane	39
2,2,3-Trimethylpentane	105
2,2,4-Trimethylpentane	100
2,3,3-Trimethylpentane	99
2,3,4-Trimethylpentane	97

hydrocarbons was obtained by extracting the catalytic phase with Freon 113. The amount of cations is calculated by difference. The results are shown in Figure 5.11.

$$RH + H^+ Sb_2F_{11}^- \longrightarrow R^+ Sb_2F_{11}^- + H_2 \qquad (5.41)$$

The total amount of hydrocarbons increases from 1.6% to 14.6% in weight when the SbF_5 concentration varies from 0 to 6.8 mol L^{-1}. The amount of carbenium ions increases linearly with the SbF_5 concentration, and the solubility of the hydrocarbon itself reaches a maximum for 5 mol L^{-1} of SbF_5. The apparent decrease in solubility of the hydrocarbon at higher SbF_5 concentration may be due to the rapid rate of hydrocarbon protolysis, along with the change in the composition of the acid, causes SbF_6^- anions to be transformed into $Sb_2F_{11}^-$ anions.

Isomerization of *n*-hexane **6** in superacid proceeds by three steps: formation of the carbenium ion (step 1, Scheme 5.10), isomerization of the carbenium ion via hydride shift, alkyde shift, and protonated cyclopropane (for the branching step) (step 2,

Table 5.4. Thermodynamic Isomerization Equilibria of Butanes, Pentanes, and Hexanes at Various Temperatures[85]

	Temperature (°C)			
	−6	38	66	93
Butanes				
Isobutane	85	75	65	57
n-Butane	15	25	35	43
Pentanes				
Isopentane	95	85	78	71
n-Pentane	5	15	22	29
2,2-Dimethylpropane (neopentane)	0	0	0	0
Hexanes				
2,2-Dimethylbutane (neohexane)	57	38	28	21
2,3-Dimethylbutane	11	10	9	9
2-Methylbpentane	20	28	34	36
3-Methylpentane	8	13	15	17
n-Hexane	4	11	14	17

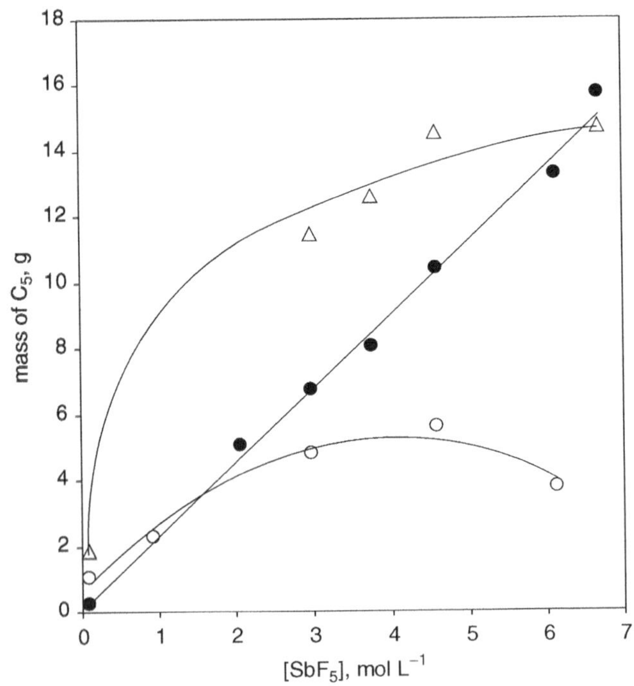

Figure 5.11. Variation of the composition of the catalytic phase as a function of the SbF_5 concentration in the *n*-pentane isomerization in HF–SbF_5.[90] $T = 15°C$, $p_{H2} = 5$ bars, volume of the catalytic phase = 57 ml. ●, Mass of C_5^+, ○, mass of C_5H (Freon-113 extract); △, % weight of $C_5^+ + C_5H$ (methylcyclopentane extract).

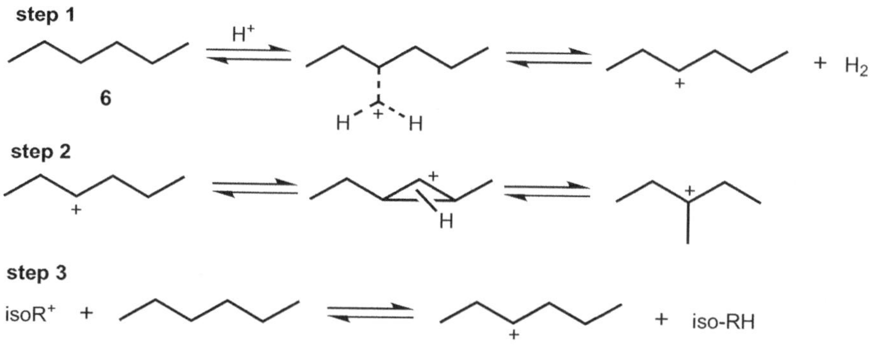

Scheme 5.10

Scheme 5.10), and hydride transfer from the alkane to the incipient carbenium ion (step 3, Scheme 5.10).

Whereas step 1 is stoichiometric, steps 2 and 3 form a catalytic cycle involving the continuous generation of carbenium ions via hydride transfer from a new hydrocarbon molecule (step 3) and isomerization of the corresponding carbenium ion (step 2). This catalytic cycle is controlled by two kinetic and two thermodynamic parameters that can help orient the isomer distribution, depending on the reaction conditions. Step 2 is kinetically controlled by the relative rates of hydrogen shifts, alkyl shifts, and protonated cyclopropane formation, and it is thermodynamically controlled by the relative stabilities of the secondary and tertiary ions. (This area is thoroughly studied; see Chapter 3.) Step 3, however, is kinetically controlled by the hydride transfer from excess of the starting hydrocarbon and by the relative thermodynamic stability of the various hydrocarbon isomers.

For these reasons, the outcome of reaction will be very different depending on which thermodynamic or kinetic factor will be favored. In the presence of excess hydrocarbon in equilibrium with the catalytic phase and long contact times, the thermodynamic hydrocarbon isomer distribution is attained. However, in the presence of a large excess of acid, the product will reflect the thermodynamic stability of the intermediate carbenium ions (which, of course, is different from that of hydrocarbons) if rapid hydride transfer or quenching can be achieved. Torck and co-workers[90–92] have shown that the limiting step in the isomerization of *n*-hexane **6** and *n*-pentane with the HF–SbF_5 acid catalyst is the hydride transfer with sufficient contact in a batch reactor, as indicated by the thermodynamic isomer distribution of C_6 isomers. Figure 5.12 shows the isomer distribution versus reaction time of *n*-hexane **6** (Hex) at 20°C.

It also shows that 2-methylpentane (**16**, 2MeP), 3-methylpentane (**17**, 3MeP), and 2,3-dimethylbutane (**18**, 2,3DiMeBu) appear simultaneously and their concentration reaches a maximum at the same time. Their relative ratio stays the same at all times and is identical to the thermodynamically calculated one. 2,2-Dimethylbutane (**19**, 2,2DiMeBu), however, appear much more slowly. The reaction network is depicted in Scheme 5.11.

The detailed kinetic study of this system shows that after the initial period during which the catalytic phase is formed, the experimental rate constant of isomerization K_1

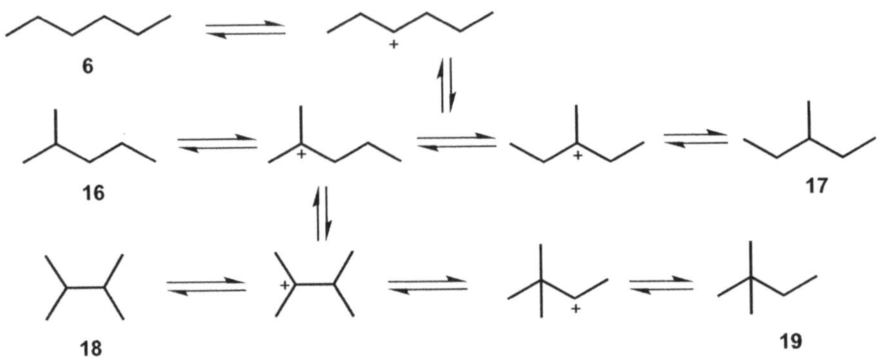

Scheme 5.11

is expressed as in Eq. 5.42, in which β is the partition coefficient of the paraffin between the hydrocarbon and the catalytic phase, N_c is the number of moles in the catalytic phase, N_0 is the total number of moles of paraffin at the start, $[R^+]$ is the concentration of carbenium ion, and K_{sp} is the temperature-dependent constant. Under external hydrogen pressure, the rate of isomerization slows down in agreement with the corresponding reduction of the carbenium ion into alkane (reverse step). This effect was found more pronounced for weaker superacids (SbF_5 content $< 2\,\mathrm{mol\,L^{-1}}$).

$$K_1 = K_{sp}[R^+]\beta\frac{N_c}{N_0} \qquad (5.42)$$

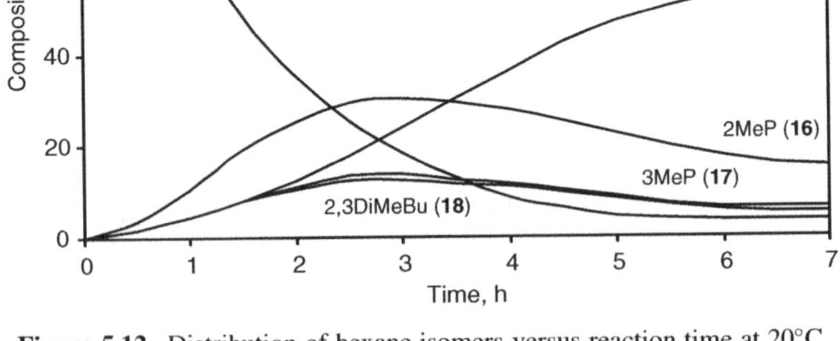

Figure 5.12. Distribution of hexane isomers versus reaction time at 20°C.

The isomerization of *n*-pentane in superacids of the type $R_FSO_3H{-}SbF_5$ ($R_F = C_nF_{2n+1}$) has been investigated by Commeyras and co-workers.[93] The influence of parameters such as acidity (A), hydrocarbon concentration (B), nature of the perfluoroalkyl group (C), total pressure (D), hydrogen pressure (E), temperature (F), and agitation has been studied. Only A, C, E, and F have been found to have an influence on the isomerization reaction in accordance with such reactions in the $HF{-}SbF_5$ system.

In weaker superacids such as neat CF_3SO_3H, alkanes that have no tertiary hydrogen are isomerized only very slowly, because the acid is not strong enough to hydride abstract to form the initial carbocation. This lack of reactivity can be overcome by introducing initiator carbenium ions in the medium to start the catalytic process. For this purpose, alkenes may be added, which are directly converted into their corresponding carbenium ions by protonation, or alternatively the alkane may be electrochemically oxidized (anodic oxidation) [Eq. (5.43)]. Both methods are useful to initiate isomerization and cracking reactions. The latter method has been studied by Commeyras and co-workers,[94,95] who came to the conclusion that it was much more favorable for side reactions such as condensation and cracking of alkanes than for simple isomerization reactions. Considering the low superacidity of the medium used, the result is in good agreement with the predictions of the acidity–potential diagrams described earlier.

$$\mathrm{RH} \xrightarrow[-e]{} \mathrm{RH}^{+\bullet} \xrightarrow[-\mathrm{H}^+]{} \mathrm{R}^{\bullet} \xrightarrow[-e]{} \mathrm{R}^+ \qquad (5.43)$$

Olah[96] has developed a method wherein natural gas liquids containing saturated straight-chain hydrocarbons can be conveniently upgraded to highly branched hydrocarbons (gasoline upgrading) using triflic acid, HF and a Lewis acid (BF_3, PF_5, AsF_5, SbF_5, TaF_5, NbF_5). The addition of small amount of olefins, preferably butenes, helped the reaction rate. This can be readily explained by the formation of alkyl fluorides (HF addition to olefins), whereby an equilibrium concentration of cations is maintained in the system during the upgrading reaction. The gasoline upgrading process is also improved in the presence of hydrogen gas, which helps to suppress side reactions such as cracking and disproportionation and minimize the amount of hydrocarbon products entering the catalyst phase of the reaction mixture. The advantage of the above method is that the catalyst $HF{-}BF_3$ (being gases at ambient temperatures) can easily be recovered and recycled.

Olah has also found HSO_3F and related superacids are efficient catalysts for the isomerization of *n*-butane to isobutane.[97] Isomerization carried out in a flow system with fluorosulfuric acid containing up to 5% HF as a co-acid at 21°C gave ~70% conversion to isobutane with generally less than 3% cracking.[98] The acidity of the $HF{-}HSO_3F$ system decreases significantly with higher amounts of HF; consequently, isomerization becomes less efficient. Under superacid conditions, butane isomerization is an intramolecular reaction proceeding through protonated methylcyclopropanes [Eq. (5.44)]. This process, however, differs from that of higher alkanes in that the secondary 2-butyl cation (**3**) cannot undergo facile β-scission.

This is why cracking is not significant in butane isomerization. In contrast, prolonged reaction of pentane and hexane in this acid system resulted in predominant protolytic cleavage.

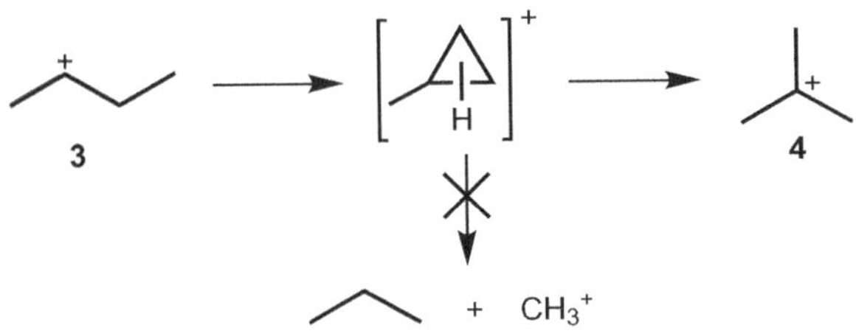

3 4 (5.44)

Fărcaşiu and Lukinskas have studied the isomerization of C_6 hydrocarbons (hexane, 3-methylpentane) in triflic acid (a two-phase system) under mild conditions (below 42°C).[99–101] It was observed that the immediate vicinity of the liquid–liquid interphase became yellow. If left undisturbed, dehydrogenation (formation of alkenyl cations) took place and then, after an induction period, rapid transformation was observed, resulting in the formation of mostly cracking and disproportionation products. The induction period could be reduced by the addition of one-electron oxidizers such as Fe^{3+} ions. When the reaction mixture was shaken periodically or a mild magnetic stirring was applied, isomerization became the main reaction with a much lower overall rate. Significant cracking was also observed in the homogenized reaction but only after a much longer period, when the concentration of the alkenyl cations reached a critical concentration throughout the acid phase. This was interpreted as the isomerization mode being an induction period to the cracking mode with features of a free-radical chain reaction.

Cracking and disproportionation in the reaction of hexane could be suppressed by the addition of cycloalkanes (cyclohexane, methylcyclopentane, cyclopentane).[101] Furthermore, 3-methylpentane and methylcyclopentane also reduced the induction period. These data indicate that reactions are initiated by an oxidative formation of alkene intermediates. These may be transformed into alkenyl cations, which undergo cracking and disproportionation. When there is intensive contact between the phases ensuring effective hydride transfer, protonated alkenes give isomerization products.

The relative reactivity of hexane and 3-methylpentane (about 1000) in the isomerization mode was shown to be the same as found for isomerization in HF–SbF_5.[102] In the cracking mode, however, the ratio is about 10, resulting from the dramatic acceleration of the reaction of hexane compared to that of 3-methylpentane. Further characteristics of the cracking mode are a large excess of branched isomers in the C_4–C_5 fractions, the absence of unsaturated cracking products, and formation of dibranched C_6 products, particularly 2,2-dimethylbutane. These features are very similar to those found in zeolite catalysis.[100,103]

The difficulties encountered in handling liquid superacids and the need for product separation from the catalyst in batch processes have stimulated research in the isomerization of alkanes over solid superacids. The isomerization of 2-methylpentane (**16**), 3-methylpentane (**17**), and 2,3-dimethylbutane (**18**), using SbF_5-intercalated graphite as a catalyst, has been studied in a continuous flow system.[104,105] The isomerization reaction carried out at −30°C, −17°C, and room temperature shows an unusually high activity of this catalyst. At −17°C, conversions over 50% with a selectivity over 99% in nonbranching isomerization could be achieved. For comparison, to obtain the same conversion with acidic zeolites, temperatures as high as 180°C are necessary. At room temperature, the activity of the catalyst is higher, but branching and cracking reactions compete. The study of the skeletal rearrangement of ^{13}C-labeled 2-methylpentane, 3-methylpentane, and 2,3-dimethylbutane has been carried out over the same catalyst under similar conditions.[106] Despite the high conversion, very little scrambling occurs. The isomerization process involves only intramolecular rearrangement of the hexyl ions and can be fully described in terms of 1,2-hydride, methide, or ethide shifts and protonated cyclopropane intermediates. The advantage of the ^{13}C tracer technique is to differentiate between mono- and multimolecular processes and to remove the degeneracy of otherwise indistinguishable pathways. Combined with a flow system, it allows the investigation of the initial steps under kinetic control. For example, in the isomerization process of 2-methylpentane to 3-methylpentane, it permits distinguishing between the methyl and ethyl shift and estimating the relative rates of these two alkyde shifts.

When 3-methyl[3-^{13}C]pentane (**17′**) is isomerized to 2-methylpentane, the label distribution shows that the isomerization cannot be explained by a simple 1,2-methide shift: 30% of the 2-methylpentane has the ^{13}C label in a position that can be best explained by the ethyl shift (Scheme 5.12). The recovered 3-methylpentane (96%) also shows a very large degree of internal shift [Eq. (5.45)].

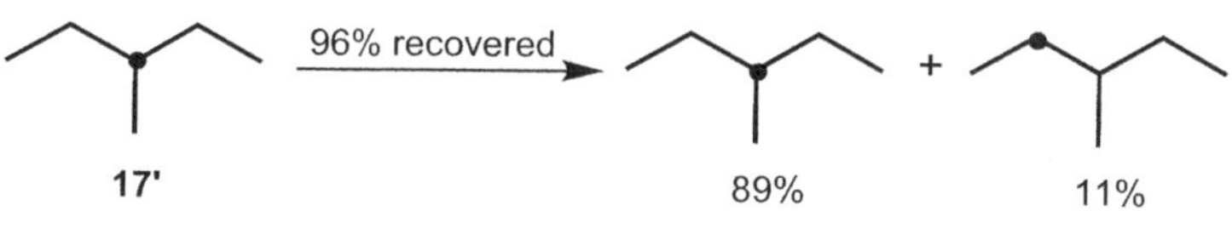

(5.45)

The rate of ethyl shift in this reaction is three times as high as the apparent rate of conversion of 3-methylpentane to 2-methylpentane. A simple calculation shows that

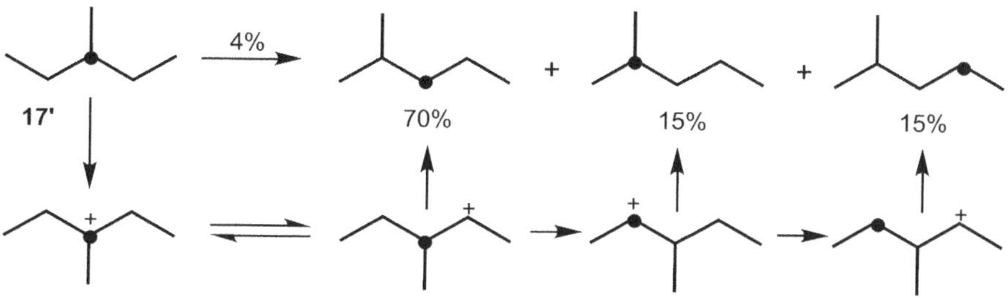

Scheme 5.12

the ethyl shift is about four times faster than the methyl shift in this isomerization process. The occurrence of a competitive ethyl shift in the 3-methyl-3-pentenyl ion has been studied by Olah and co-workers under stable ion condition. Subsequently, Sommer and co-workers[107] showed by using the deuterium-labeled 3-methyl-3-pentenyl ion that the ratio of ethyl:methyl shift was the same both in Magic Acid and in the presence of solid superacid. SbF_5-intercalated graphite has also been shown to exhibit a 99% selectivity in *n*-pentane skeletal isomerization at room temperature in a batch process[108] (48% conversion after 2 h). The selectivity in isomerization of *n*-hexane was much lower because of the competing cracking process described earlier. However, in the presence of HF, the selectivity of the catalyst substantially improves as shown by the skeletal isomerization of *n*-hexane (96% selectivity with 86% conversion after 2 h). The major drawback in the extended use of this catalyst system is its relatively rapid deactivation (studied by Heinerman and Gaaf[109]).

The isomerization of cycloalkanes over SbF_5-intercalated graphite can be achieved at room temperature without the usual ring opening and cracking reactions, which occur at higher temperatures and lower acidities.[110] In the presence of excess hydrocarbon after several hours, the thermodynamic equilibrium is reached for the isomers. Interconversion between cyclohexane (**20**) and methylcyclopentane (**21**) yields the thermodynamic equilibrium mixture [Eq. (5.46)].

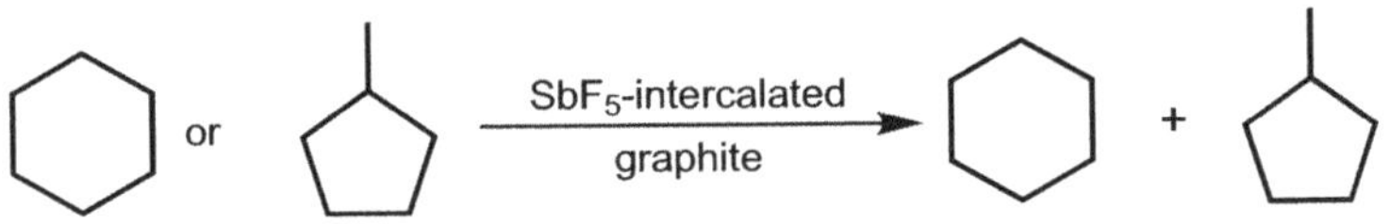

20 **21** **20** **21** (5.46)

The thermodynamic ratio for the neutral hydrocarbon isomerization is very different as compared with the isomerization of the corresponding ions. The large energy difference (> 10 kcal mol^{-1}) between secondary cyclohexyl cation **22** and the tertiary methylcyclopentyl ion **23** means that in the presence of excess superacid, only the latter can be observed [Eq. (5.47)].

22 **23** (5.47)

When a mixture of *cis*- and *trans*-decalins **24** (61.5:38.5) are treated with the same solid superacid at 0°C, the thermodynamic equilibrium is rapidly achieved[110] [Eq. (5.48)]. In the reaction, decalin **24** serves as the solvent, substrate, and the hydride donor. When the equilibrium is reached, the hydrocarbon can be separated from the catalyst by simple filtration. Perhydroindane **25** was also isomerized under mild condition at 0°C to the thermodynamic equilibrium mixture[110] [Eq. (5.48)].

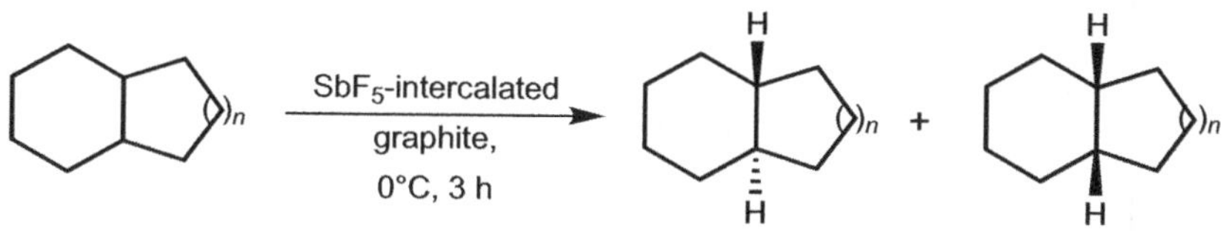

$n = 2$ **24** — **24**-*trans* 98% — **24**-*cis* 2%

$n = 1$ **25** — **25**-*trans* 62% — **25**-*cis* 38%

(5.48)

The potential of other solid superacid catalysts, such as Lewis-acid-treated metal oxides for skeletal isomerization of hydrocarbons, has been studied in a number of cases. The reaction of *n*-butane **1** with SbF_5–SiO_2–TiO_2 gave the highest conversion forming C_3, isoC_4, isoC_5, and traces of higher alkanes (Table 5.5). TiO_2–SbF_5, on the other hand, gave the highest selectivity for skeletal isomerization of *n*-butane. With SbF_5–Al_2O_3, however, the conversions were very low (Table 5.5).[111,112]

Similarly, isomerization of pentane and 2-methylbutane over a number of SbF_5-treated metal oxides has been investigated.[113] SbF_5–TiO_2–ZrO_2 system was the most reactive and at the maximum conversion the selectivity for skeletal isomerization was found to be about 100%. Table 5.6 summarizes the effect of temperature on desorption of SbF_5 out of the catalyst and shows that with increasing temperature, the weight % of the retained Lewis acid decreases.

A comparison of the reactivity of SbF_5-treated metal oxides with that of HSO_3F-treated catalysts showed that the former is by far the better catalyst for the reaction of alkanes at room temperature, although the HSO_3F-treated catalyst showed some potential for isomerization of 1-butene.[111]

SbF_5–SiO_2–Al_2O_3 has been used to isomerize a series of alkanes at or below room temperature. Methylcyclopentane, cyclohexane, propane, butane, 2-methylpropane, and pentane all reacted at room temperature, whereas methane, ethane, and 2,2-dimethylpropane could not be activated.[111]

An IR study of SbF_5–Al_2O_3 after addition of pyridine to the catalyst shows an absorption at 1460 cm^{-1}, which is assigned to pyridine coordinated with the Lewis acid site, and shows another absorption at 1540 cm^{-1}, which is attributed to the pyridinium ion resulting from the protonation of pyridine by the Brønsted acid sites. When the catalyst is heated up to 300°C, the IR band of the pyridinium ion disappears, whereas the absorption for the Lewis acid is still present. The fact that the catalyst is still active for *n*-butane isomerization suggests that the Lewis acid sites are the active sites for the catalysis.[111]

Methane produced in the reaction of butane with SbF_5–SiO_2–Al_2O_3 did not contain any deuterium when the surface OH groups of the catalyst were replaced by OD. Furthermore, no hydrogen evolution could be detected in these reactions.

Table 5.5. Reaction *n*-Butane and Isobutane over Solid Superacids at 20°C[111]

			Product Distribution (%)							
Reactant	Catalyst	Time (h)	Propane	Butane	Isobutane	Pentane	Isopentane	2,2-DiMeBu	Hexanes[b]	Heptanes
Butane (**1**)										
	$SbF_5-Al_2O_3$	720	0	100	0	0	0	0	0	0
	SbF_5-SiO_2	280	6.9	25.2	54.8	1.0	6.8	2.9	2.1	0
	SbF_5-TiO_2	280	6.4	17.1	59.1	1.1	13.4	2.0	1.0	0
	$SbF_5-Al_2O_3$	280	0.14	77.2	21.1	0.13	1.3	0.09	0	0
	$SbF_5-TiO_2-SiO_2$	280	21.9	13.1	58.1	0.6	4.8	1.0	0.5	0
	$SbF_5-SiO_2-Al_2O_3$	280	4.4	32.7	47.7	1.1	8.1	3.7	2.6	0
	$SbF_5-SiO_2-Al_2O_3$	20	1.8	49.7	41.0	0.6	4.8	1.4	0.7	0
Isobutane (**2**)										
	$SbF_5-Al_2O_3$	600	0	0	100	0	0	0	0	0
	$SbF_5-Al_2O_3$	4[a]	0	0	99.9	0	<0.1	0	0	0
	TiO_2-SiO_2	720	0	0	100	0	0	0	0	0
	SbF_5	280	2.9	0.16	96.4	0	0.49	0	0	0
	SbF_5-SiO_2	280	7.0	11.8	66.8	1.6	5.6	4.2	2.7	0.3
	$SbF_5-Al_2O_3$	280	0.17	0.93	97.9	0.09	0.86	0	0	0
	$SbF_5-SiO_2-Al_2O_3$	280	6.8	11.0	65.0	1.2	9.6	3.7	2.2	0.4
	$SbF_5-SiO_2-Al_2O_3$	2	0.07	0.4	94.7	3.5	0.4	0.4	0.5	0

[a]Reacted at 200°C.

[b]2,3-Dimethylbutane, 2-methylpentane, 3-methylpentane.

Table 5.6. Amount of SbF_5 Retained in Catalyst (wt%)[113]

Catalyst[a]	Adsorption Temperature of SbF_5 (°C)	Evacuation Temperature			
		0°C	30°C	50°C	100°C
$SbF_5-SiO_2-Al_2O_3$	0	49.7	37.6	27.7	
	10	53.2	40.9	34.5	30.7
	30		34.1	32.2	
	100				17.2
$SbF_5-TiO_2-Al_2O_3$	0	49.9	30.0	32.8	
	10		33.5	27.2	
	30			29.8	
	100				30.9
SbF_5-SiO_2	0	33.1	18.8	12.8	
	10			18.2	
	30			17.2	
$SbF_5-Al_2O_3$	0	37.1			
SbF_5-TiO_2	0	23.1			
SbF_5-ZrO_2	0	24.8			

[a]Adsorption–desorption cycle in SbF_5 was repeated four times.

Two alternative mechanisms have been suggested. (i) The reactions are initiated by hydride abstraction from the alkane by the Lewis acid to form a carbenium ion, and not by protonation of the C–C bond of *n*-butane. (ii) *n*-Butane is protonated by the Brønsted acid to form a carbonium ion intermediate; and either the hydrogen formed is used up to reduce SbF_5 or it loosely remains bound to the ion during the isomerization process.[111]

The isomerization of a large number of C_{10} hydrocarbons under strongly acidic conditions gives the unusually stable isomer adamantane **26**. The first such isomerization was reported by Schleyer in 1957.[114] During a study of the facile aluminum chloride-catalyzed *endo–exo* isomerization of tetrahydrodicyclopentadiene (trimethylenenorbornane) **27**, difficulty was often encountered with a highly crystalline material that often clogged distillation heads. This crystalline material was found to be adamantane **26**. Adamantane **26** can be prepared from a variety of C_{10} precursors and involves a series of hydride and alkyde shifts. Scheme 5.13 shows the rearrangement map of C_{10} hydrocarbons to adamantane. The mechanism of the reaction has been reviewed in detail.[115] Fluoroantimonic acid ($HF–SbF_5$) very effectively isomerizes tetrahydrodicyclopentadiene **27** into adamantane **26**.[116]

Boron tristriflate and the conjugate superacids triflic acid–SbF_5 and triflic acid–$B(OSO_2CF_3)_3$ were also found to be highly effective in subsequent studies.[117,118] They induce the isomerization of *endo*-trimethylenenorbornane to *exo*-trimethylenenorbornane in near-quantitative conversion followed by the isomerization of the latter to adamantane in high yields [Eq. (5.49)]. Sonication was shown to accelerate isomerization. Furthermore, the same catalysts could be

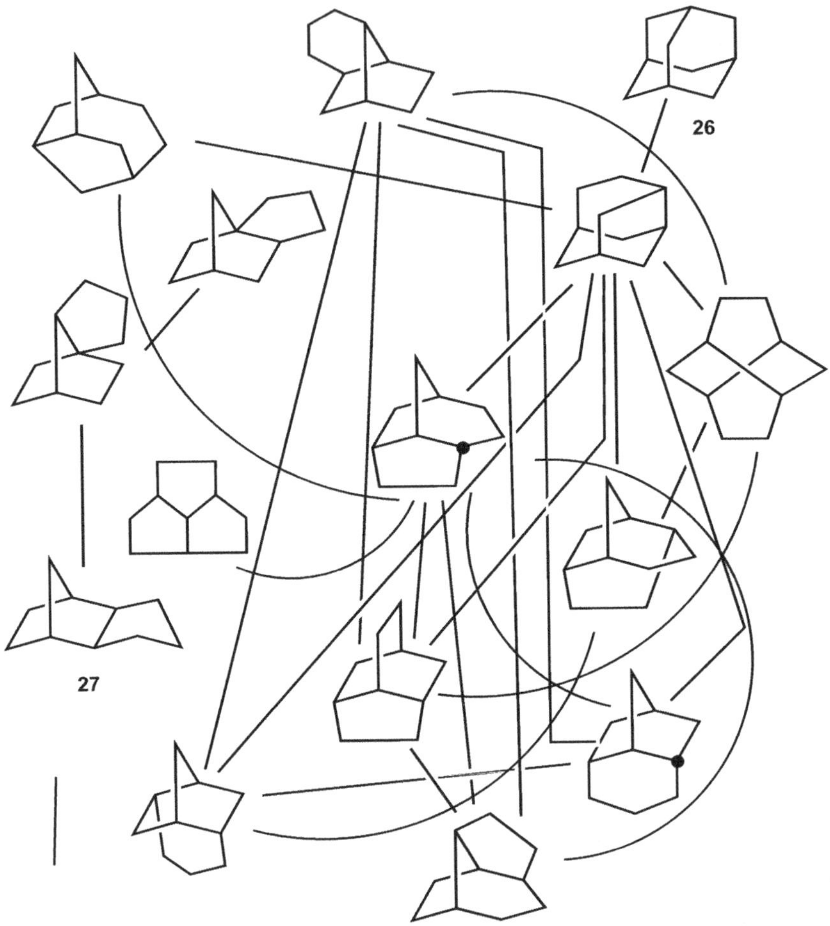

Scheme 5.13

successfully applied to transform $C_{14}H_{20}$ and $C_{18}H_{24}$ precursors to diamantane [Eq. (5.50)] and triamantane, respectively.[118]

(5.49)

$B(OSO_2CF_3)_3$, Freon-113, 0.25 h	98%
CF_3SO_3H–$B(OSO_2CF_3)_3$, 11 h	98%
CF_3SO_3H–SbF_5, 10 h	98%
sonication, 1.5 h	98%

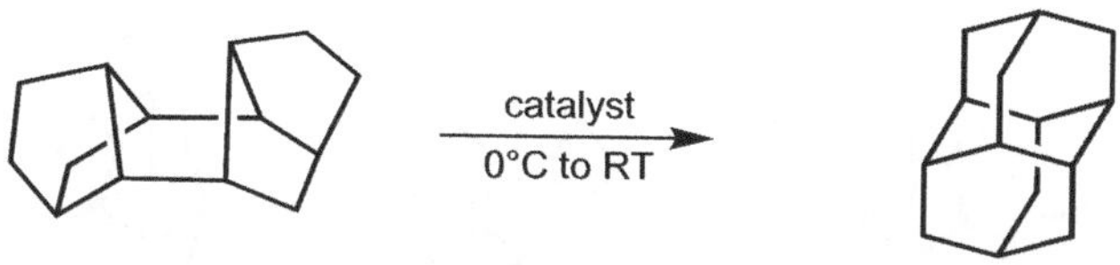

(5.50)

$B(OSO_2CF_3)_3$, Freon-113, 0.25 h	99%
CF_3SO_3H–$B(OSO_2CF_3)_3$, 11 h	98%
CF_3SO_3H–SbF_5, 10 h	98%
sonication, 1.75 h	98%

The superelectrophilic AcBr–nAlBr$_3$ complexes, called aprotic organic superacids developed by Vol'pin and co-workers,[119,120] were used in the isomerization of $C_{12}H_{20}$ tricyclanes with the main component being perhydroacenaphthene to dimethyl- and ethyladamantanes[121] [Eq. (5.51)].

n = 1.8–2

71–76% yield

(5.51)

Another example of superacid-catalyzed formation of an unusually stable highly symmetric hydrocarbon has been provided by Paquette and Balogh[122] in the synthesis of 1,16-dimethyldodecahedrane **28** [Eq. (5.52)].

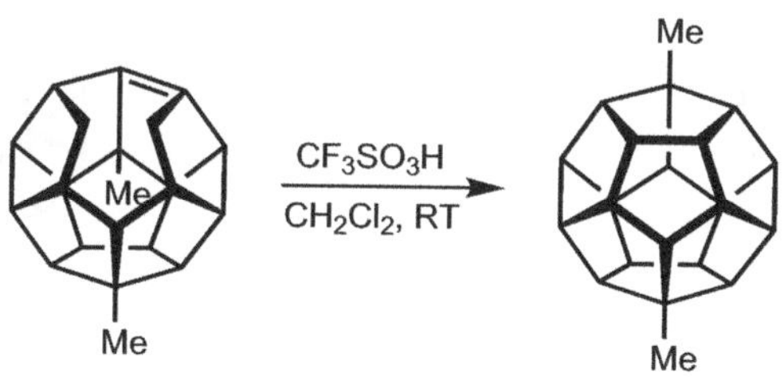

28

(5.52)

Two mechanistic schemes (Schemes 5.14 and 5.15) via carbocationic intermediates have been proposed by Paquette and Balogh.[122]

The reaction path depicted in Scheme 5.14 involves Wagner–Meerwein shifts of the methyl group prior to cyclization followed by hydride shift to a number of cationic intermediates. The second scheme (Scheme 5.15) depicts ring closure before methyl migration. The first step involves protolysis of the C–H bond next to the methyl-bearing carbon. The corresponding ion can then rearrange by a 1,2-methyl shift and yield 1,16-dimethyldodecahedrane **28** by hydride abstraction from a hydride donor.

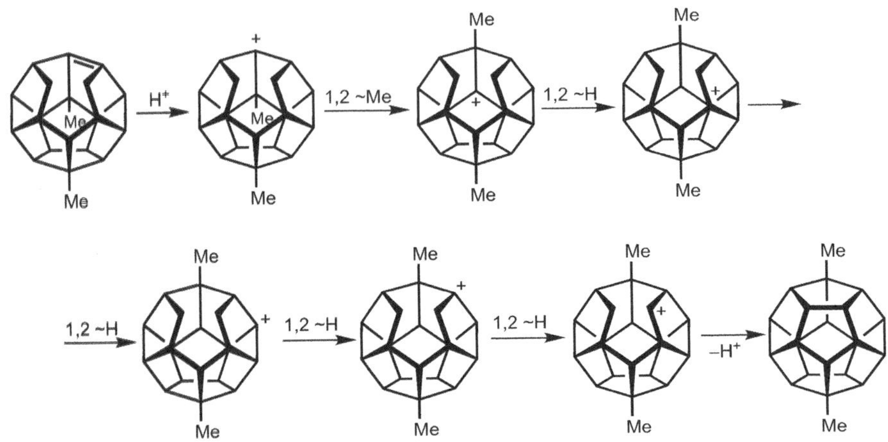

28

Scheme 5.14

An unusual isomerization of perhydro[2.2]paracyclophane **29**, an *out–in* hydrogen transfer process, was observed in triflic acid[123] [Eq. (5.53)]. Reaction in CF_3SO_3D resulted in the incorporation of up to five deuteriums without specific locations. The authors suggested the backside attack of the inside hydrogen (H_i) to the bridgehead

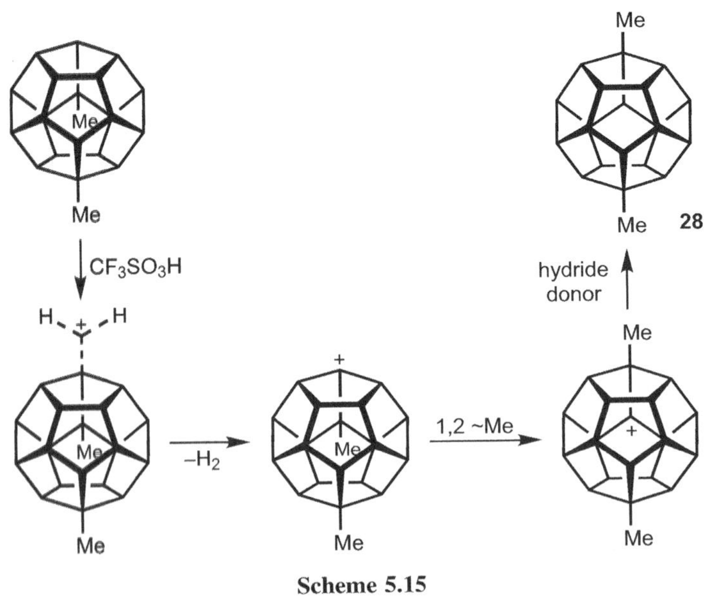

28

Scheme 5.15

carbon participating in 2*e*–3*c* bonding (**29**) as the key step in the isomerization.

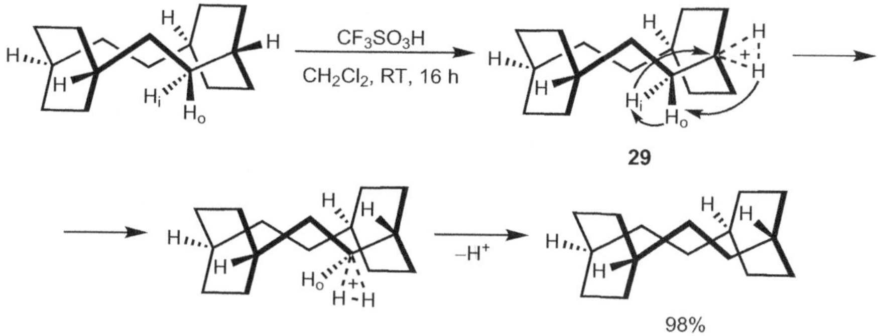

(5.53)

5.1.4. Cleavage Reactions (β-Cleavage versus C–C Bond Protolysis)

The reduction in molecular weight of various fractions of crude oil is an important operation in petroleum chemistry. The process is called cracking. Catalytic cracking is usually achieved by passing the hydrocarbons over a metallic or acidic catalyst, such as zeolites at about 400–600°C. The molecular weight reduction involves carbocationic intermediates and the mechanism is based on the β-scission of carbenium ions [Eq. (5.54)]. The main goal of catalytic cracking is to upgrade higher boiling oils, which, through this process, yield lower hydrocarbons in the gasoline range.[124–126]

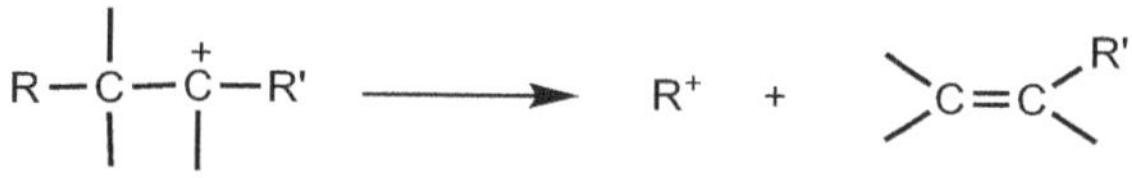

(5.54)

Historically, the first cracking catalyst used was aluminum trichloride. With the development of heterogeneous solids and supported catalysts, the use of $AlCl_3$ was soon superseded, since its activity was mainly due to the ability to bring about acid-catalyzed cleavage reactions.

The development of highly acidic superacid catalyst in the 1960s again focused attention on acid-catalyzed cracking reactions. $HSO_3F–SbF_5$, trade-named Magic Acid, derived its name (see Section 2.2.2.2) from its remarkable ability to cleave higher-molecular-weight hydrocarbons, such as paraffin wax, to lower-molecular-weight components, preferentially C_4 and other branched isomers.

As a model for cracking of alkanes, the reaction of 2-methylpentane (**16**, 2MeP) over SbF_5-intercalated graphite has been studied in a flow system, with the hydrocarbon being diluted in a hydrogen stream.[104,105] A careful study of the product

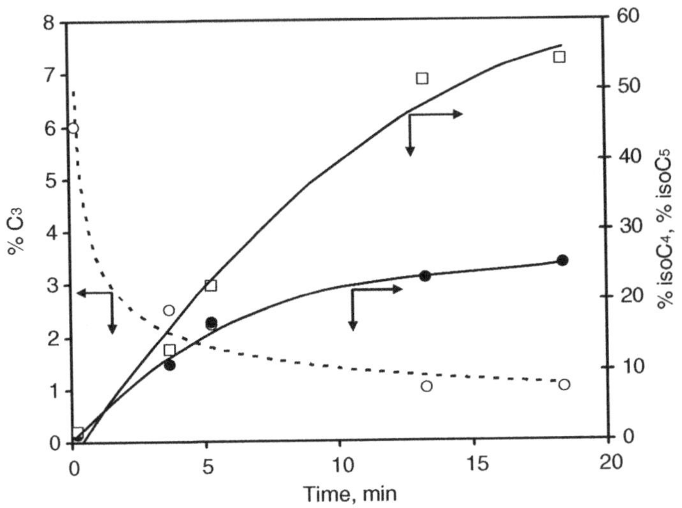

Figure 5.13. Distribution of cracking products (mass%) of 2-methylpentane (**16**) versus time on stream at 20°C.[104]

distribution versus time on stream showed that propane was the initial cracking product whereas isobutane and isopentane (as major cracking products) appeared only later (Figure 5.13).

This result can only be explained by the β-scission of the trivalent 4-methylpent-2-yl ion **30** as the initial step in the cracking process. Based on this and on the product distribution versus time profile, a general scheme for the isomerization and cracking process of the methylpentanes has been proposed[103,104] (Scheme 5.16).

Propene (**32**), which is formed in the β-scission step, never appears as a reaction product because it is alkylated immediately under the superacid condition by a C_6^+ carbenium ion (MeP^+), forming a C_9^+ carbocation, which is easily cracked to form an iso C_4^+ or iso C_5^+ ion and the corresponding C_4 or C_5 alkene ($C_4^=$, $C_5^=$). The alkenes are further alkylated by a C_6^+ carbenium ion in a cyclic process of alkylation and cracking reactions. The C_4^+ or C_5^+ ions also give the corresponding alkanes [isobutane, isoC_4H and isopentane, isoC_5H) by hydride transfer from the starting methylpentane. This scheme, which occurs under superacidic conditions, is at a variance with the scheme that was proposed for the cracking of C_6 alkanes under less acidic conditions (Scheme 5.17).[94,95]

Under superacidic conditions, it is known that the deprotonation equilibria (Scheme 5.17, first reaction) lie too far to the left ($K = 10^{-16}$ for isobutane[127]) to make this pathway plausible. On the other hand, among C_6 isomers, 2MeP is by far the easiest to cleave by β-scission. The 4-methylpent-2-yl ion **30** is the only species that does not give a primary cation by this process. For this reason, this ion is the key intermediate in the isomerization cracking reaction of C_6 alkanes.

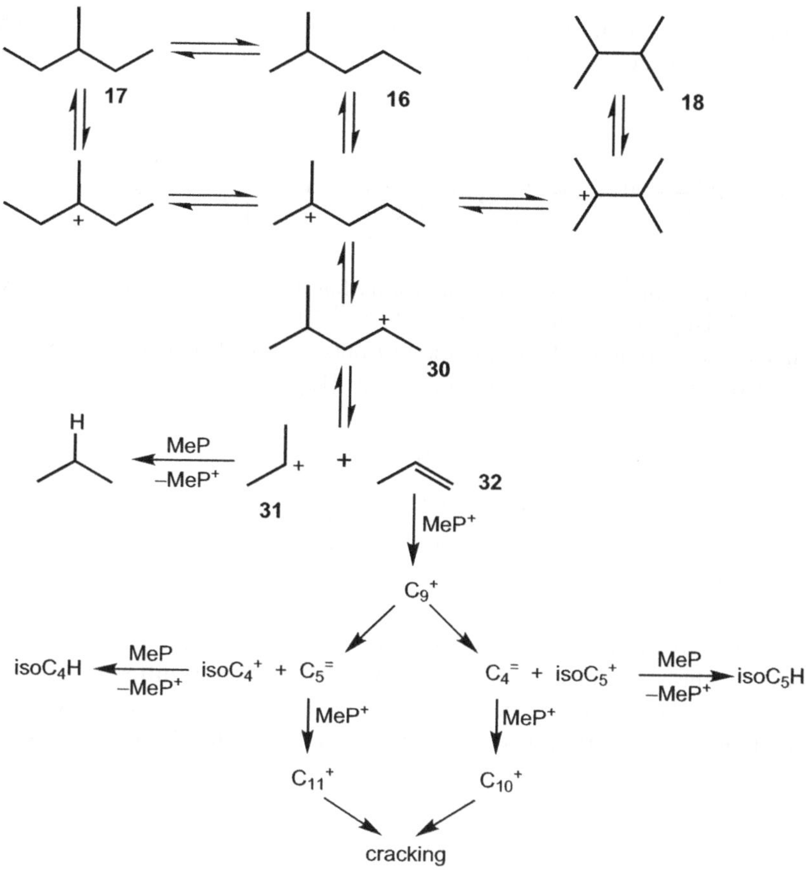

Scheme 5.16

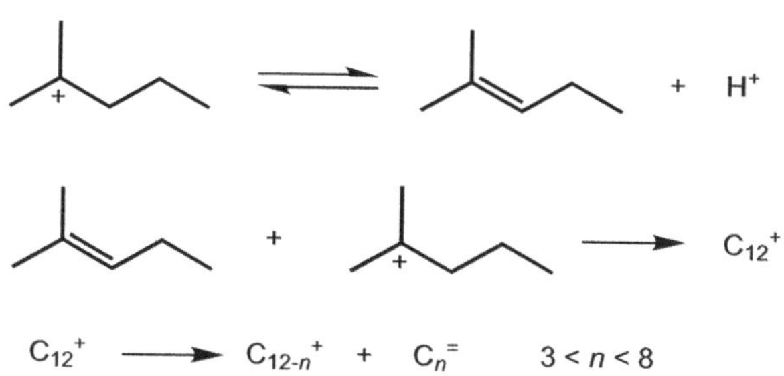

Scheme 5.17

Under superacid conditions, β-scission is not the only pathway by which hydrocarbons are cleaved. The C–C bond can also be cleaved by protolysis [Eq. (5.55)].

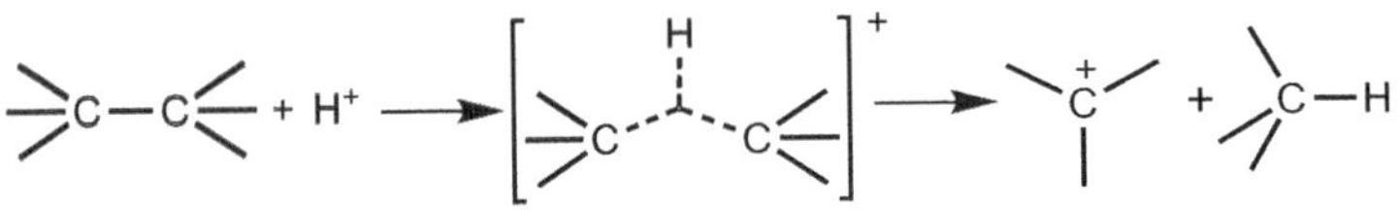

(5.55)

The protolysis under superacid conditions has been studied independently by Olah[128] and Hogeveen and co-workers[19,20,129] The carbon–carbon cleavage in neopentane (**7**) yielding methane and the *tert*-butyl cation **4** occurs by a mechanism different from the β-scission of carbenium ions [Eq. (5.56)].

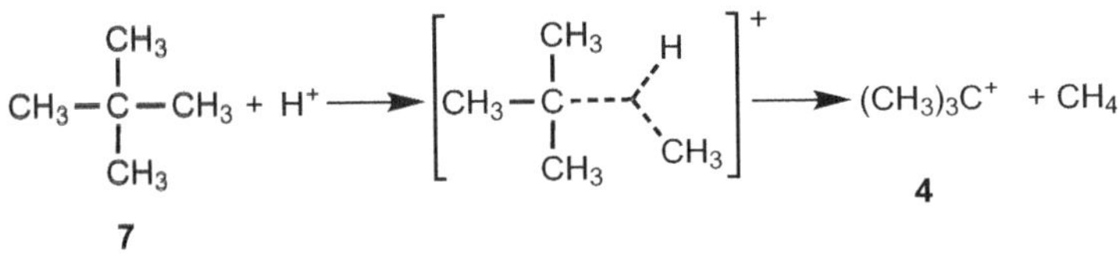

(5.56)

The protolysis occurs following the direct protonation of the σ-bond providing evidence for the σ-basicity of hydrocarbons. Under slightly different conditions, protolysis of a C–H bond occurs, yielding rearranged *tert*-amyl cation **33** (2-methylbut-2-yl cation) [Eq. (5.57)]. In cycloalkanes, the C–C bond cleavage leads to ring opening [Eq. (5.58)].

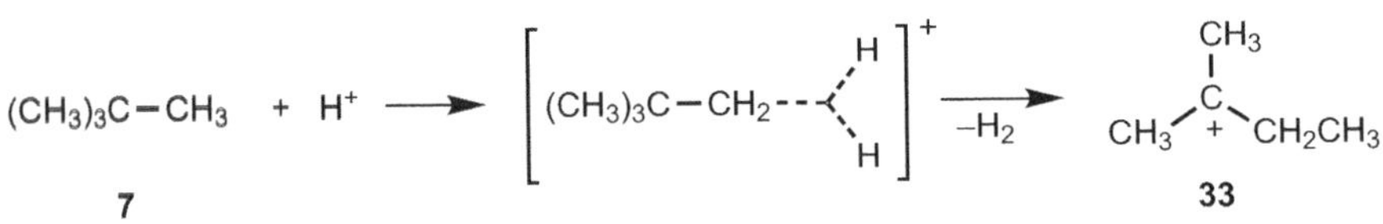

(5.57)

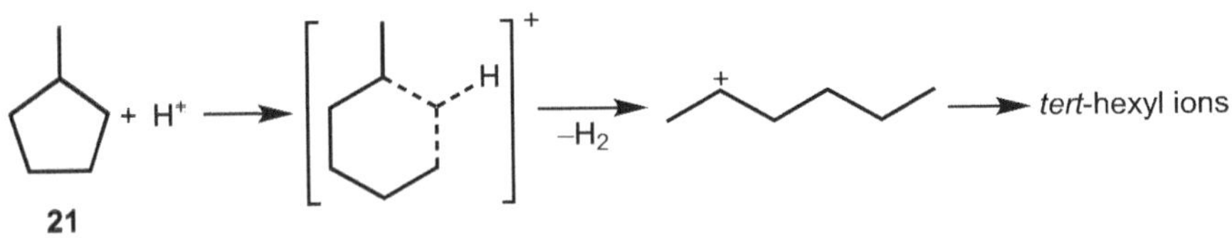

(5.58)

This reaction is much faster than the carbon–carbon cleavage in neopentane, despite the initial formation of secondary carbenium ions. Norbornane is also cleaved in a fast reaction, yielding substituted cyclopentyl ions. Thus, protonation of alkanes induces cleavage of the molecule by two competitive ways: (i) protolysis of a C–H bond followed by β-scission of the carbenium ions and (ii) direct protolysis of a C–C bond yielding a lower-molecular-weight alkane and a lower-molecular-weight carbenium ion.

This reaction, which is of economic importance in the upgrading of higher boiling petroleum fractions to gasoline, has also been shown applicable to coal depolymerization and hydroliquefaction processes.[130–132] The cleavage of selected model compounds representing coal structural units in the presence of HF and BF_3 and under hydrogen pressure has been studied by Olah et al.[130] Bituminous coal (Illinois No. 6) could be pyridine-solubilized to the extent of 90% by treating it with HF–BF_3 catalyst in the presence of hydrogen gas at 105°C for 4 h. Under somewhat more elevated temperatures (150–170°C), cyclohexane extractability of up to 22% and distillability of up to 28% was achieved. Addition of hydrogen donor solvents such as isopentane has been shown to improve the efficiency of coal conversion to cyclohexane-soluble products. The initial depolymerization of coal involves various protolytic cleavage reactions involving those of C—C bonds. HF–BF_3 was also shown to be a highly effective catalyst for the hydrocracking of oil sand bitumens evidenced by a high conversion of the bitumen to volatiles (56%).[133] When methylcyclohexane is applied as a hydrogen-donor solvent, new products are formed resulting from the oligomerization of methylcyclohexane. This points to possible complications arising in the application of hydrocarbons as stabilizers or hydrogen donors, since oligomerization consumes the hydrocarbons and the highly reactive polyenic oligomers formed may induce secondary reactions during product analyses.

Shimizu et al. have reported the use of triflic acid, HF, and BF_3 in the liquefaction of coals of various origins. Triflic acid was effective to convert lignite and Taiheiyo coal to pyridine-solubles almost completely in the presence of isopentane without hydrogen (150°C, 3 h).[134–136] Relatively low conversions with diphenyl ether and biphenyl as model compounds indicated that aromatic ether and aryl–aryl linkages are more difficult to cleave under these conditions. Solubilization was mainly attributed to cleavage of methylene bridges in the coal. Desulfurization of lignite (35.4–41.3%) was achieved with triflic acid in toluene originating mainly from sulfides.[137] Mixtures of HF and BF_3 were found to be highly effective in the depolymerization of Miike bituminous coal[138,139] and Taiheiyo subbituminous coal in toluene.[140] The high effectiveness of the HF–BF_3 system was ascribed to its ability to efficiently cleave both ether groups and methylene bridges. Since both deoxygenation and depolymerization are required for Yallourn lignite, it underwent solubilization under more forcing conditions (higher acidity, elevated temperature).

A combination of triflic acid with iodine was shown to be effective to liquefy three types of coal in toluene or tetralin under hydrogen pressure.[141] The major role of acid was found to enhance coal depolymerization to asphaltenes, whereas the main function of iodine was to hydrogenate and hydrocrack asphaltenes to oils. The combined catalytic system removed 50% of the nitrogen and 90% of the sulfur of the coals (Illinois No. 6 and Pittsburg seam samples).

5.1.5. Alkylation of Alkanes and Oligocondensation of Lower Alkanes

The alkylation of alkanes by olefins, from a mechanistic point of view, must be considered as the alkylation by the carbenium ion formed by the protonation of the olefin. The well-known acid-catalyzed isobutane–isobutylene reaction demonstrates the mechanism rather well (Scheme 5.18).

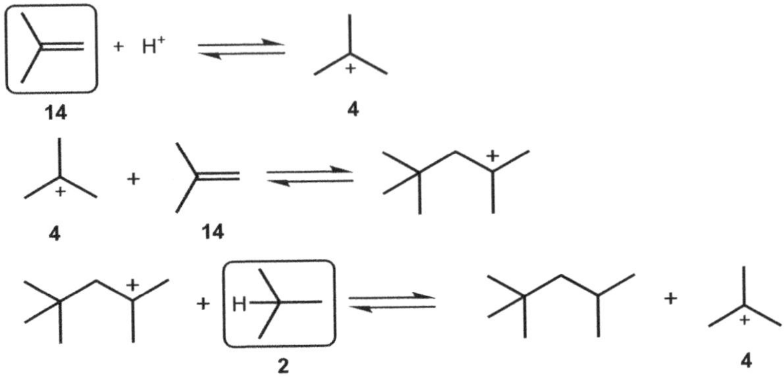

Scheme 5.18

As is apparent in the last step, isobutane is not alkylated but transfers a hydride to the C_8^+ carbocation before being used up in the middle step as the electrophilic reagent (*tert*-butyl cation **4**). The direct alkylation of isobutane by an incipient *tert*-butyl cation would yield 2,2,3,3-tetramethylbutane,[142] which indeed was observed in small amounts in the reaction of *tert*-butyl cation with isobutane under stable ion conditions at low temperatures (*vide infra*).

The alkylating ability of methyl and ethyl fluoride–antimony pentafluoride complexes has been investigated by Olah et al.,[143,144] who showed the extraordinary reactivity of these systems. Self-condensation was observed as well as alkane alkylation. When $CH_3F–SbF_5$ was reacted with excess of CH_3F at 0°C, at first only an exchanging complex was observed in the ^1H NMR spectrum. After 0.5 h, the starting material was converted into the *tert*-butyl cation **4** (Scheme 5.19).

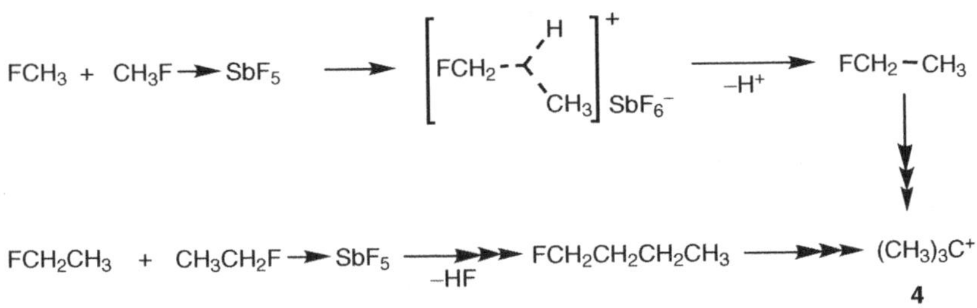

Scheme 5.19

Similar reactions were observed with the $CH_3CH_2F–SbF_5$ complex (Scheme 5.19). When the complex was treated with isobutane or isopentane, direct alkylation products were observed [Eq. (5.59)].

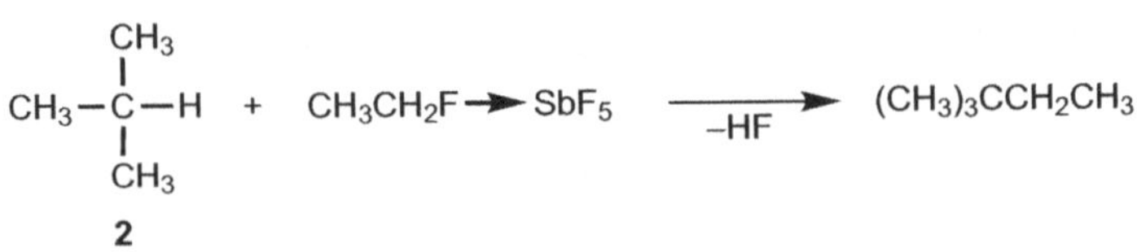

(5.59)

To improve the understanding of these alkane alkylation reactions, Olah and his group carried out experiments involving the alkylation of lower alkanes by stable carbenium ions under controlled superacidic stable ion conditions[128,145,146] [Eq. (5.60)].

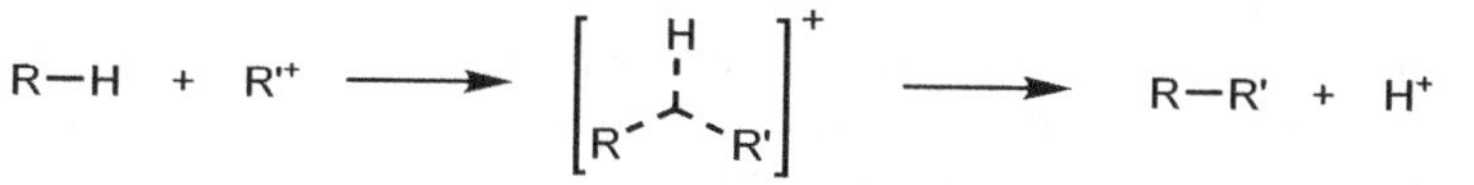

(5.60)

The σ-donor ability of the C—C and C—H bonds in alkanes was demonstrated from a variety of examples. The order of reactivity of single bonds was found to be tertiary C—H > C—C > secondary C—H >> primary C—H, although various specific factors such as steric hindrance can influence the relative reactivities.

Typical alkylation reactions are those of propane, isobutane, and *n*-butane by the *tert*-butyl or *sec*-butyl ion. These systems are somewhat interconvertible by competing hydride transfer and rearrangement of the carbenium ions. The reactions were carried out using alkyl carbenium ion hexafluoroantimonate salts prepared from the corresponding halides and antimony pentafluoride in sulfuryl chloride fluoride solution and treating them in the same solvent with alkanes. The reagents were mixed at −78°C warmed up to −20°C and quenched with ice water before analysis. The intermolecular hydride transfer between tertiary and secondary carbenium ions and alkanes is generally much faster than the alkylation reaction. Consequently, the alkylation products are also those derived from the new alkanes and carbenium ions formed in the hydride transfer reaction.

Propylation of propane by the isopropyl cation **31**, for example, gives a significant amount (26% of the C_6 fraction) of the primary alkylation product [Eq. (5.61)].

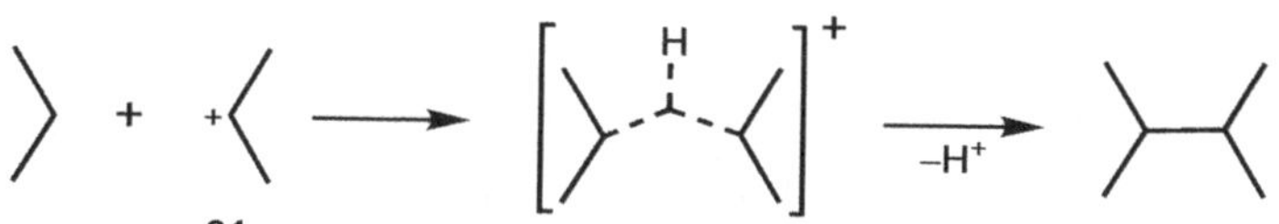

31

(5.61)

The C_6 isomer distribution—2-methylpentane (28%), 3-methylpentane (14%), and *n*-hexane (32%)—is very far from thermodynamic equilibrium, and the presence of these isomers indicates that both isopropyl cation **31** [Eq. (5.62)] and *n*-propyl cation **34** [Eq. (5.63)] are involved as intermediates [as shown by $^{13}C(2)-^{13}C(1)$ scrambling in the stable ion[147]].

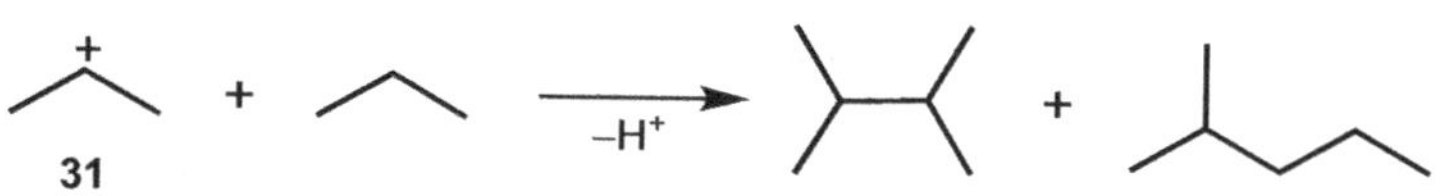

31

(5.62)

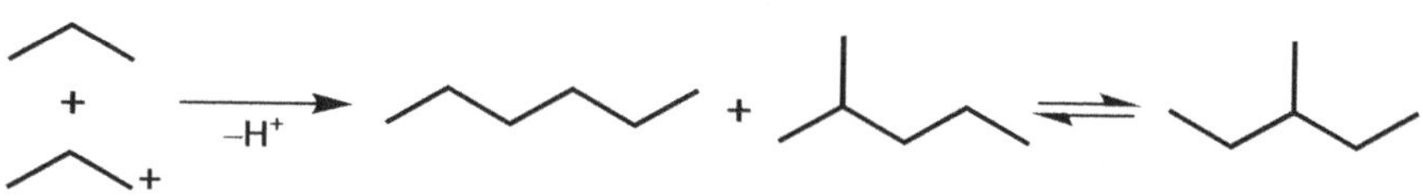

34

(5.63)

The strong competition between alkylation and hydride transfer appears in the alkylation reaction of propane by butyl cations, or butanes by the propyl cation. The amount of C_7 alkylation products is rather low. This point is particularly emphasized in the reaction of propane by the *tert*-butyl cation, which yields only 10% of heptanes. In the interaction of isopropyl cation **31** with isobutane **2** the main reaction is hydride transfer from the isobutane to the isopropyl ion followed by alkylation of propane by the isopropyl ions (Scheme 5.20).

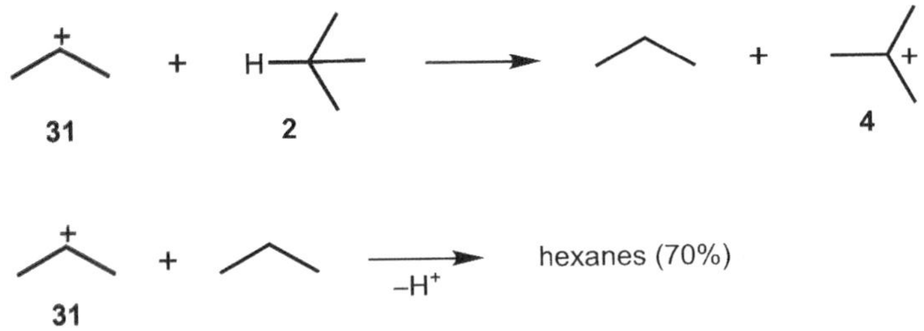

Scheme 5.20

Even the alkylation of isobutane by the *tert*-butyl cation **4** despite the highly unfavorable steric interaction has been demonstrated[142] by the formation of small amounts of 2,2,3,3-tetramethylbutane **36**. This result also indicates that the related five-coordinate carbocationic transition state (or high-lying intermediate) **35** of the degenerate isobutylene–*tert*-butyl cation hydride transfer reaction is not entirely linear, despite the highly crowded nature of the system (Scheme 5.21).

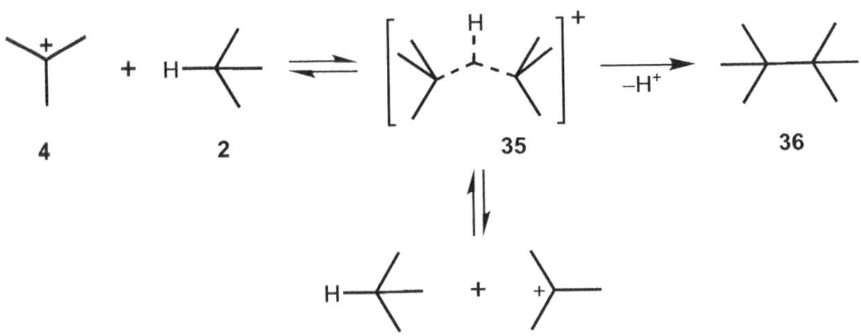

Scheme 5.21

2,2,3,3-Tetramethylbutane **36** was not formed when *n*-butane and *sec*-butyl cation were reacted. The isomer distribution of the octane isomers for typical butyl cation–butane alkylations is shown in Table 5.7.

Alkylation of methane, ethane, propane, and *n*-butane by the ethyl cation generated via protonation of ethylene in superacid media has been studied by Siskin,[148] Sommer et al.,[149] and Olah et al.[150] The difficulty lies in generating in a controlled way a very energetic primary carbenium ion in the presence of excess methane and at the same time avoiding oligocondensation of ethylene itself. Siskin carried out the reaction of

Table 5.7. Isomeric Octane Composition Obtained in Typical Alkylations of Butanes with Butyl Cations

	C_4–C_4			
Octane	$(CH_3)_3CH$ $(CH_3)_3C^+$	$CH_3CH_2CH_2CH_3$ $(CH_3)_3C^+$	$CH_3CH_2CH_2CH_3$ $MeEtCH^+$	$(CH_3)_3CH$ $MeEtCH^+$
2,2,4-Trimethylpentane	18.0	4.0	3.8	8.5
2,2-Dimethylhexane			0.4	
2,2,3,3-Tetramethylbutane	1–2		Trace	1–2
2,5-Dimethylhexane	43.0	0.6	1.6	29.0
2,4-Dimethylhexane	7.6	Trace		6.6
2,2,3-Trimethylpentane	3.0	73.6	40.6	3.2
3,3-Dimethylhexane			12.3	7.1
2,3,4-Trimethylpentane	1.5	7.2	15.5	6.2
2,3,3-Trimethylpentane	3.6		3.8	8.8
2,3-Dimethylhexane	4.2	6.9		12.8
2-Methylheptane		Trace	10.3	6.7
3-Methylheptane	19.3	7.6	6.8	9.5
n-Octane	0.2		4.8	Trace

methane–ethylene (86:14) gas mixture through a 10:1 HF–TaF_5 solution under pressure with strong mixing. Among the recovered reaction products 60% of C_3 was found (propane and propylene) [Eq. (5.64)]. Propylene is formed when propane, which is substantially a better hydride donor, reacts with ethyl cation [Eq. (5.65)].

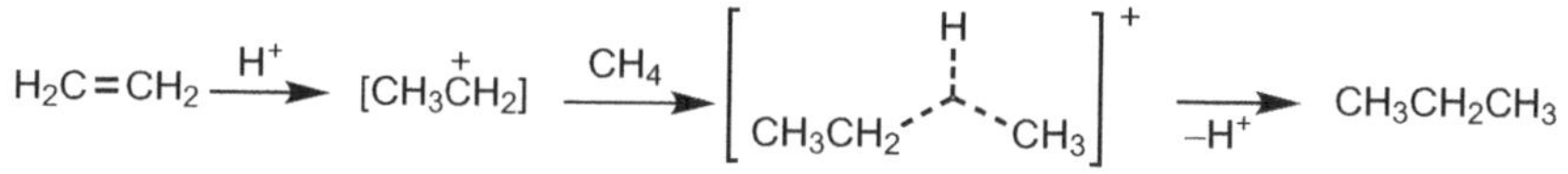

(5.64)

$$CH_3CH_2CH_3 + [CH_3\overset{+}{C}H_2] \longrightarrow CH_3CH_3 + CH_3-\overset{+}{C}H-CH_3 \xrightarrow[-H^+]{} CH_3-CH{=}CH_2 \tag{5.65}$$

Propane as a degradation product of polyethylene (a byproduct in the reaction) was ruled out because ethylene alone under the same conditions does not give any propane. Under similar conditions but under hydrogen pressure, polyethylene reacts quantitatively to form C_3 to C_6 alkanes, 85% of which are isobutane and isopentane. These results further substantiate the direct alkane alkylation reaction and the intermediacy of the pentacoordinate carbonium ion. Siskin also found that when ethylene was allowed to react with ethane in a flow system, *n*-butane was obtained as the sole product, indicating that the ethyl cation is alkylating the primary C—H bond through a five-coordinate carbonium ion [Eq. (5.66)].

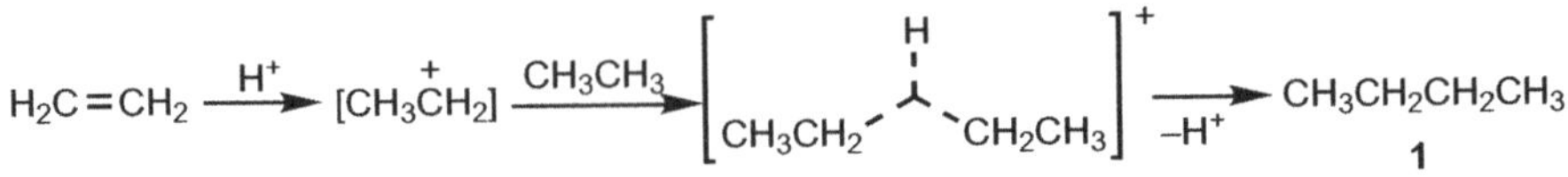

(5.66)

If the ethyl cation would have reacted with excess ethylene, primary 1-butyl cation would have been formed, which irreversibly would have rearranged to the more stable *sec*-butyl and subsequently *tert*-butyl cations giving isobutane as the end product.

Superacid-catalyzed alkylation of adamantane with lower alkenes (ethene, propene, isomeric butenes) has been investigated by Olah et al.[151] in triflic acid and triflic acid–$B(OSO_2CF_3)_3$. Only trace amounts of 1-*tert*-butyladamantane (**37**) were detected in alkylation with 1- and 2-butenes, whereas isobutylene gave consistently relatively good yield of **37**. Since isomerization of isomeric 1-butyladamantane under identical conditions did not give even traces of **37**, its formation can be accounted for by σ-alkylation, that is, through the insertion of the *tert*-butyl cation into the C–H bond (Scheme 5.22). This reaction is similar to that between *tert*-butyl cation and isobutane to form 2,2,3,3-tetramethylbutane discussed above (Scheme 5.21). In either case, the pentacoordinate carbocation intermediate, which may also lead to hydride transfer, does not attain a linear geometry, despite the unfavorable steric interactions.

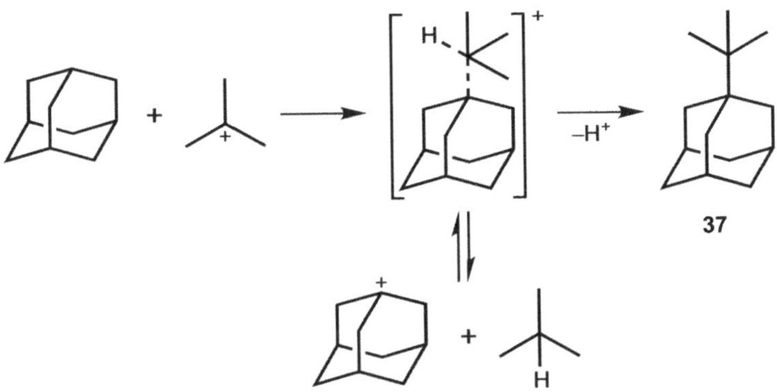

Scheme 5.22

The yield of the alkane–alkene alkylation in homogeneous HF–TaF_5 depending on the alkene–alkane ratio has been investigated by Sommer et al.[149] in a batch system with short reaction times. The results support direct alkylation of methane, ethane, and propane by the ethyl cation and the product distribution depends on the alkene–alkane ratio (Figure 5.14).

Despite the unfavorable experimental conditions in a batch system for kinetically controlled reactions, a selectivity of 80% in *n*-butane was achieved through ethylation of ethane. The results show, however, that to succeed in the direct alkylation the following conditions have to be met. (i) The olefin should be completely converted to

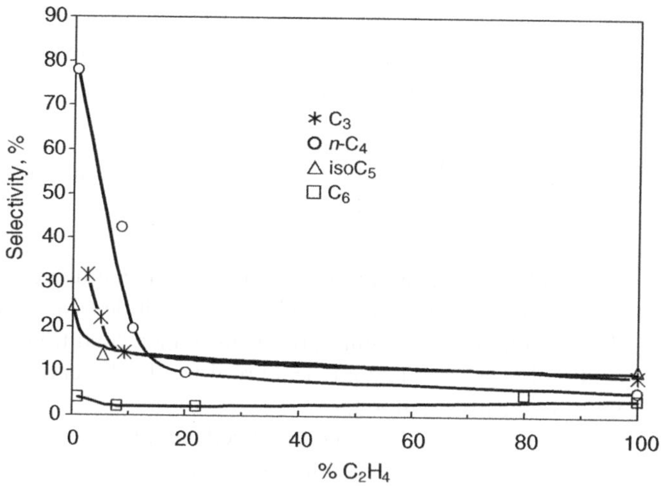

Figure 5.14. Selectivity dependence of alkylation on the ethylene–alkane ratio for the following reactions: ethylene + methane (*), ethylene + ethane (○), ethylene + propane (△), ethylene + *n*-butane (□).

the reactive cation (incomplete protonation favors the polymerization and cracking processes); this means the use of a large excess of acid and good mixing. (ii) The alkylation product must be removed from the reaction mixture before it transfers a hydride to the reactive cation, in which case the reduction of the alkene is achieved. (iii) The substrate to cation hydride transfer should not be easy; for this reason the reaction shows the best yield and selectivity when methane and ethane are used.

The direct ethylation of methane with ethylene was also investigated by Olah et al.[150] using ^{13}C-labeled methane (99.9 ^{13}C) over solid superacid catalysts such as TaF_5–AlF_3, TaF_5, and SbF_5–graphite. Product analyses by gas chromatography–mass spectrometry (GC–MS) are given in Table 5.8.

Table 5.8. Ethylation of $^{13}CH_4$ with C_2H_4[150]

			Products Normalized (%)[b,c]				Label Content of C_3 Fraction (%)[d]	
Run	$^{13}CH_4$:C_2H_4	Catalyst[a]	C_2H_6	C_3H_8	isoC_4H_{10}	C_2H_5F	$^{13}CC_2H_8$	C_3H_8
1	98.7:1.3	TaF_5–AlF_5	51.9	9.9	38.2		31	69
2	99.1:0.9	TaF_5		15.5	3.0	81.5	91	9
3	99.1:0.9	SbF_5–graphite	64.1	31.5		4.4	96	4

Values are in mol%.

[a]Catalysts pretreated with HF for 30 s.

[b]Excluding methane.

[c]Trace amounts of Me_2SiF_2 detected in all runs are probably due to SiF_4 impurity (from HF) reacting with methane.

[d]Isobutane contained no ^{13}C label; thus it is derived from ethylene.

These results show a high selectivity in monolabeled propane $^{13}CH_3CH_2CH_3$, which can only arise from direct electrophilic attack of the ethyl cation on methane via pentacoordinate carbonium ion [Eq. (5.67)].

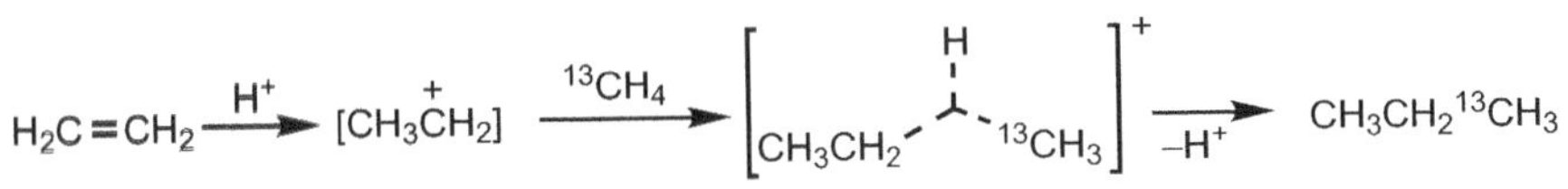

(5.67)

An increase in the alkene–alkane ratio results in a significant decrease in single-labeled propane; ethylene polymerization–cracking and hydride transfer become the main reaction. This labeling experiment carried out under conditions where side reactions were negligible is indeed unequivocal proof for the direct alkylation of an alkane by a very reactive carbenium ion.

Corma and co-workers[152] have performed a detailed theoretical study (B3PW91/6-31G* level) of the mechanism of the reactions between carbenium ions and alkanes (ethyl cation with ethane and propane and isopropyl cation with ethane, propane, and isopentane) including complete geometry optimization and characterization of the reactants, products, reaction intermediates, and transition states involved. Reaction enthalpies and activation energies for the various elemental steps and the equilibrium constants and reaction rate constants were also calculated. It was concluded that the interaction of a carbenium ion and an alkane always results in the formation of a carbonium cation, which is the intermediate not only in alkylation but also in other hydrocarbon transformations (hydride transfer, disproportionation, dehydrogenation).

Olah et al.[153] have made detailed studies of the use of triflic acid in the isobutane–isobutylene alkylation. A high-quality alkylate with about 70% branched C_8 isomers could be produced at −30°C with a catalyst/hydrocarbon ratio of 0.5. At higher temperatures and longer residence time, cracking becomes significant, resulting in the formation of increasing amounts of products of low molecular weight (light end) (Scheme 5.23). Triflic acid could be recycled several times without purification to produce alkylates of good quality and contained low amounts of acid-soluble oils (ASO) after four cycles. Acid-soluble oils are conjunct polymers and formed primarily from the decomposition of ester intermediates.[154,155] These are rich in heavy

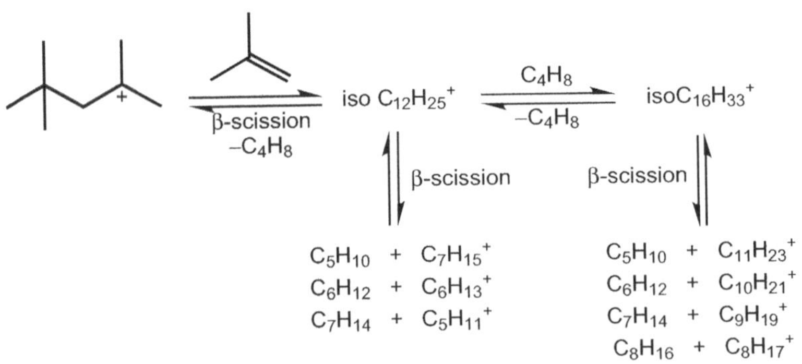

Scheme 5.23

isoalkanes and alkenes and dilute the acid, resulting in loss of effectiveness. It was shown that esters, including triflate esters, which are short-lived intermediates formed in the reaction of alkenes with triflic acid, play an important role in alkylate selectivity.[156] Intermediates such as surface alkoxides formed between an alkene and an acid site are believed to passivate solid acids.

Acidity-dependence studies were also made using triflic acid modified with trifluoroacetic acid (TFA) and water in the range of acidity between $H_0 = -10.1$ and -14.1.[157] The best alkylation conditions were found to be at an acid strength of about $H_0 = -10.7$, giving calculated research octane numbers of 89.1 (triflic acid–TFA) and 91.3 (triflic acid–water). Alkylation in liquid CO_2 with weak acids shows a decrease in alkylate quality because CO_2 acts as a competing weak base and decreases acidity.[158] The heavy ends—that is, C_9 and higher fractions—formed by oligomerization, in turn, increase (Scheme 5.23). With the stronger acids (triflic acid, HF), however, higher optimum research octane numbers were obtained than in the neat acids (95.6 for HF and 88.0 for triflic acid). Moreover, these reactions required less acid catalyst and afforded increased selectivities to trimethylpentanes.

Pyridinium poly(hydrogen fluoride) (PPHF), which serves as an HF equivalent catalyst with decreased volatility,[159] showed similar characteristics in liquid CO_2.[158] Other liquid amine poly(hydrogen fluoride) complexes with high (22:1) HF/amine ratios are also effective catalysts in the alkylation of isobutane with butenes and, at the same time, also act as ionic liquid solvents.[160] Likewise the solid poly(ethyleneimine)/HF and poly(4-vinylpyridinium)/HF (1:24) complexes have proved to be efficient catalysts affording excellent yields of high-octane alkylates with research octane numbers up to 94.

Silica-supported triflic acid catalysts were prepared by various methods (treatment of silica with triflic acid at 150°C or adsorption of the acid from solutions in trifluoroacetic acid or Freon-113) and tested in the isobutane–1-butene alkylation.[161] All catalysts showed high and stable activity (near-complete conversion at room temperature in a continuous flow reactor at 22 bar) and high selectivity to form saturated C_8 isomers (up to 99%) and isomeric trimethylpentanes (up to 86%). Selectivities to saturated C_8 isomers, however, decreased considerable with time-on-stream (79% and 80% after 24 h).

Various Nafion preparations have extensively been studied in alkane–alkene alkylations. The initial high activity of Nafion-H beads applied in isobutane–2-butene alkylation at 80°C decreased rapidly, which was attributed to coking.[162] Corma and coworkers[163] have studied Nafion–silica nanocomposites with 13%, 20%, and 53% loadings. They observed a lower initial activity and selectivity and faster deactivation for catalysts with the same Nafion content but with larger surface area. This was explained by the fact that in nanocomposites with higher surface area the interaction between silanol groups of silica and sulfonic acid groups of highly dispersed Nafion particles is more extensive. A charge transfer may occur, resulting in decreased acidity and, consequently, lower catalytic activity. This may be significant for a demanding reaction such as alkane–alkene alkylation. Because of this leveling effect,[164] the catalytic performance of a 16% Nafion–SiO_2 catalyst prepared by impregnation was similar to those of the nanocomposites.

The isobutane–1-butene alkylation was studied in dense CO_2 in both fixed-bed and slurry reactors.[165–167] Both Nafion SAC-13 and Nafion SAC-25 exhibited steady-state conversions and selectivities for 50 h. Enhanced C_8 alkylate selectivity could be achieved at near total butene conversion. The maximum value attained, however, was only about 40%. The higher effective alkylation rate constant for SAC-25 compared to SAC-13 indicates improved accessibility of the acid sites. Nafion SAC-13 and SAC-25 applied in a study to test the effect of supercritical fluids on alkylation exhibited only modest activities.[168]

A comparative study of nanocomposites (16% Nafion–silica and commercial SAC-13) has been performed by Hoelderich and co-workers[169] in the alkylation of isobutane and Raffinate II. Raffinate II, the remaining C_4 cut of a stream cracker effluent after removal of dienes, isobutane, propane, and propene, contains butane, isobutylene, and butenes as main components. High conversion with a selectivity of 62% to isooctane was found for Nafion SAC-13 (batch reactor, 80°C). Both the quality of the product and the activity of the catalysts, however, decrease rapidly due to isomerization and oligomerization. Treating under reflux, the deactivated catalysts in acetone followed by a further treatment with aqueous hydrogen peroxide (80°C, 2 h), however, restores the activity.

The protolytic condensation of methane in Magic Acid solution at 60°C is evidenced by the formation of higher alkyl cations such as *tert*-butyl and *tert*-hexyl cations[23,25,170] (Scheme 5.24).

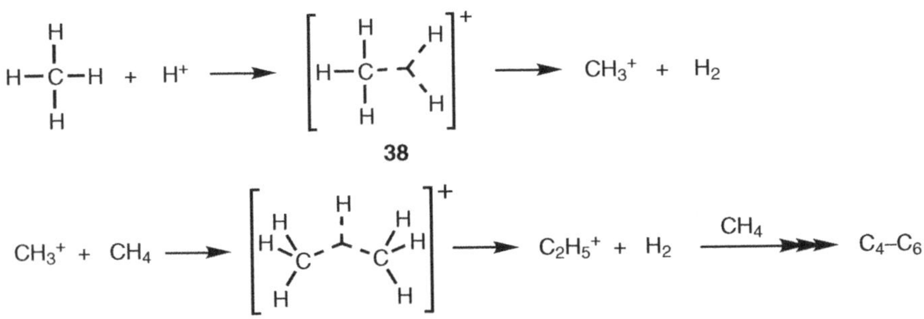

Scheme 5.24

It is not necessary to assume a complete cleavage of methonium ion $[CH_5]^+$ **38** to a free, energetically unfavorable methyl cation. The carbon–carbon bond formation can indeed be visualized as the C—H bond of methane reacting with the developing methyl cation [Eq. (5.68)].

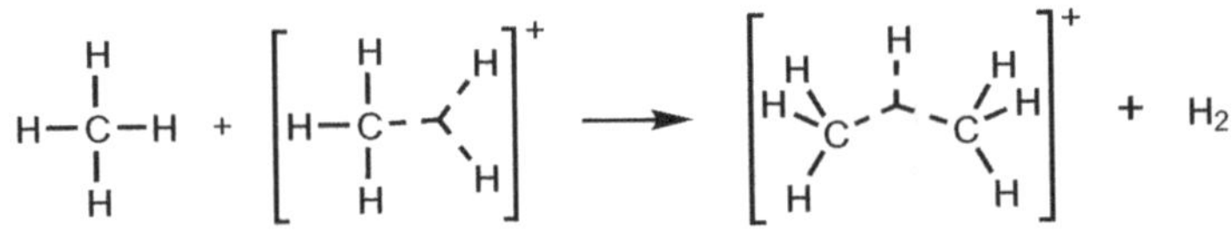

38

(5.68)

In order to overcome unfavorable thermodynamics, hydrogen must be oxidatively removed (either by superacid or added oxidant). Considering the abundance of methane in nature, the conversion of natural gas into branched liquid hydrocarbons in the gasoline range is of immense interest.

Polycondensation of alkanes over HSO_3F–SbF_5 has also been achieved by Roberts and Calihan.[171] Several low-molecular-weight alkanes such as methane, ethane, propane, *n*-butane, and isobutane were polymerized to highly branched oily oligomers with a molecular weight range from the molecular weight of monomers to around 700. These reactions again follow the same initial protolysis of the C–H or C–C bond, which results in a very reactive carbenium ion. Similarly, the same workers[172] were able to polycondense methane with a small amount of olefin such as ethylene, propylene, butadiene, and styrene to yield oily polymethylene oligomer with a molecular weight ranging form 100 to 700.

Herlem and Jobert-Perol[173] carried out the oligocondensation of methane, ethane, and propane in HF–HSO_3F (1:1) by generating carbenium ions by anodic oxidation using a Pt anode. Reactions were quenched by injecting hydrogen into the high-pressure electrolysis cell. Methane gave ethane with high selectivity (90%), which is attributed to the low electrooxidation rates to ethane. Propane was formed with similar selectivity from ethane the rest being isobutane, whereas propane was transformed to isobutane (74%) and isopentane (25%). The observed high electrical yields are likely due to high local acidity around the anode.

The superelectrophilic RCOX–$2AlBr_3$ (R = Me, Pr; X = Cl, Br, I) complexes induce rapid oxidative coupling of excess C_5–C_6 cycloalkanes without solvent to form a mixture of isomeric dimethyldecalins with high selectivities[174] [Eq. (5.69)]. Yields calculated on the superelectrophiles are near quantitative. Complexes of polyhalomethanes (CBr_4, CCl_4, $CHCl_3$) with AlX_3 (X = Cl, Br) have recently been shown to be equally effective.[175]

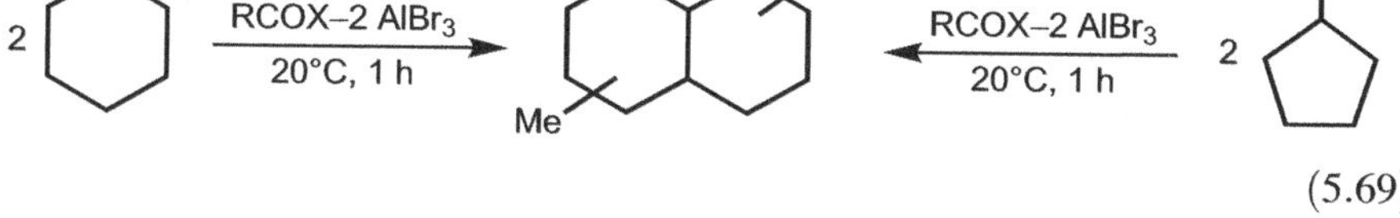

(5.69)

The AcBr–$2AlX_3$ (X = Cl, Br) complexes display high activity in the alkylation of adamantane with alkanes to form polyalkylated adamantanes ($C_{11} < C_n < C_{33}$) and bisadamantylalkanes ($C_{23} < C_n < C_{50}$)[119] [Eq. (5.70)]. The suggested pathway includes the 1-adamantyl cation and alkyl cations generated by hydride removal by the superacidic complexes. The 1-adamantyl cation then alkylates alkenes equilibrating with the alkyl cations. Various transformations may follow, resulting in the formation of additional products.

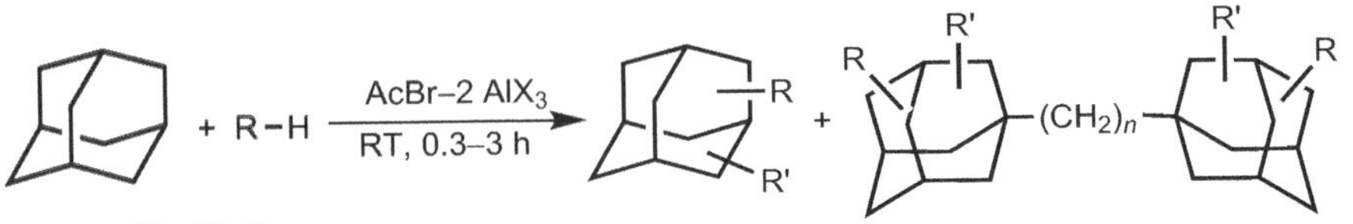

X = Cl, Br
R = C_6H_{13}, C_8H_{17}, $C_{12}H_{25}$

(5.70)

5.2. ALKYLATION OF AROMATIC HYDROCARBONS

5.2.1. Alkylation with Alkenes

Cumene is industrially produced by propylating benzene over supported acidic catalysts such as phosphoric acid. On the other hand, the largest-scale single industrial alkylation process—that is, ethylation of benzene with ethylene—is still carried out to a significant degree in the liquid phase using acid catalysts; since ethylene is less polar than propylene, it requires more forcing conditions in the protolytic initiation step [Eq. (5.71)].

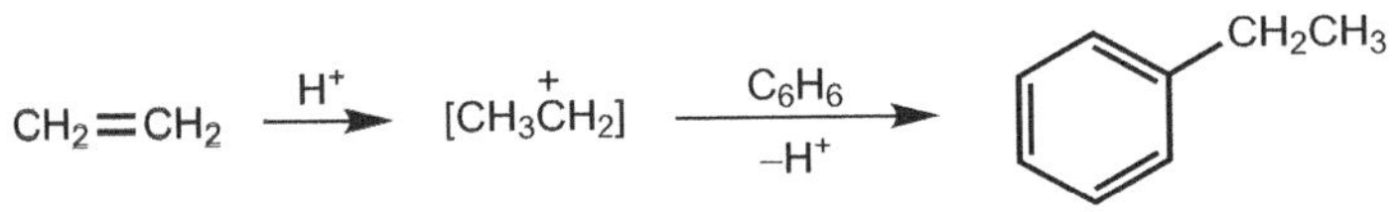

(5.71)

Consequently, stronger solid acids were needed to activate the ethylation reaction. The solid acids that were available earlier exhibited only limited acidity, which was sufficient to promote the propylation of benzene with propylene at reasonable temperatures and pressures, but it was not high enough to promote ethylation of benzene with ethylene under similar conditions.

The conventional resinsulfonic acids such as sulfonated polystyrenes (Dowex-50, Amberlite IR-112, and Permutit Q) are of moderate acidity with limited thermal stability. Therefore, they can be used only to catalyze alkylation of relatively reactive aromatic compounds (like phenol) with alkenes, alcohols, and alkyl halides. Nafion-H, however, has been found to be a suitable superacid catalyst in the 110–190°C temperature range to alkylate benzene with ethylene (*vide infra*).[176] Furthermore, various solid acid catalysts (ZSM-5, zeolite β, MCM-22) are applied in industrial ethylbenzene technologies in the vapor phase.[177]

In addition to alkenes, functionalized alkenes can also be used as alkylating agents. Koltunov, Walspurger, and Sommer,[178,179] have reported the alkylation of benzene with α,β-unsaturated carboxamides in the presence of excess aluminum chloride [Eq. (5.72)]. The reaction takes place under mild conditions and gives the products in near-quantitative yields. Results with *ortho*-dichlorobenzene and triflic acid are usually inferior. Triflic acid, however, can catalyze similar reactions of cyclic and open-chain unsaturated amines with benzene to give phenylalkylamines in excellent yields.[180] The transformations are interpreted by invoking the involvement of dicationic intermediates **39** and **40**.

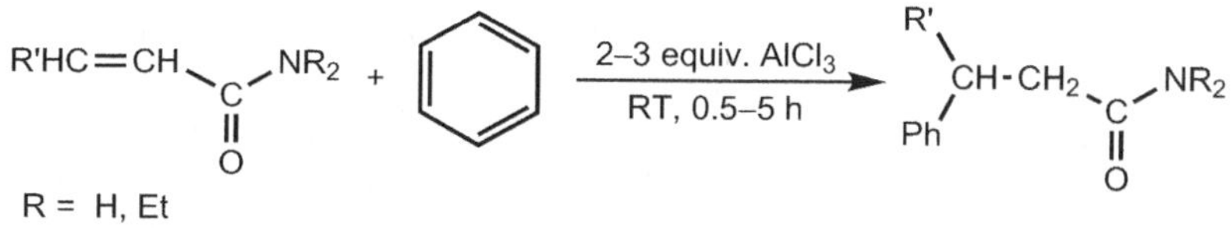

R = H, Et
R' = H, Me, Ph, 4-MeOC$_6$H$_4$

93–100% yield

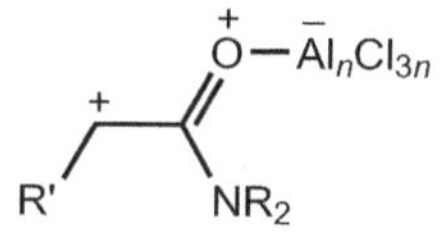

39

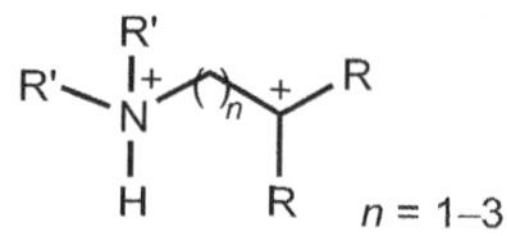

40

(5.72)

Shudo and co-workers[181] have studied the effect of various functional groups on the reactivity of the carbon–carbon double bond in alkylations. α,β-Unsaturated carbonyl compounds react with benzene under superacidic conditions to yield the corresponding ketones [Eq. (5.73)]. The primary alkylation products, however, may undergo cyclization (R = H, Me) to give phenylindene derivatives. The parent compound cinnamaldehyde is almost unreactive in triflic acid (6% of 3-phenylindene), but ketones react readily. Higher yields were found in the presence of triflic acid–SbF$_5$. The corresponding oximes are more reactive to give products in very high yields (65–99%).

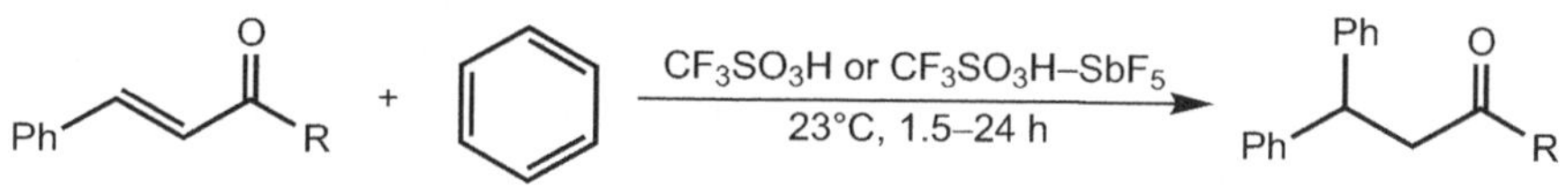

R = H, Me, Ph, PhCH=CH

6–98% yield

(5.73)

Superelectrophilic dicationic species **41**, **42**, and **43** were suggested to be involved in the alkylation steps of the above transformations.

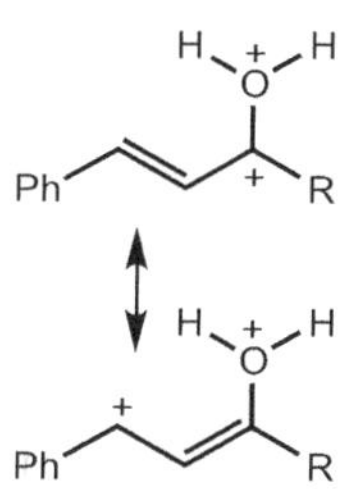

41

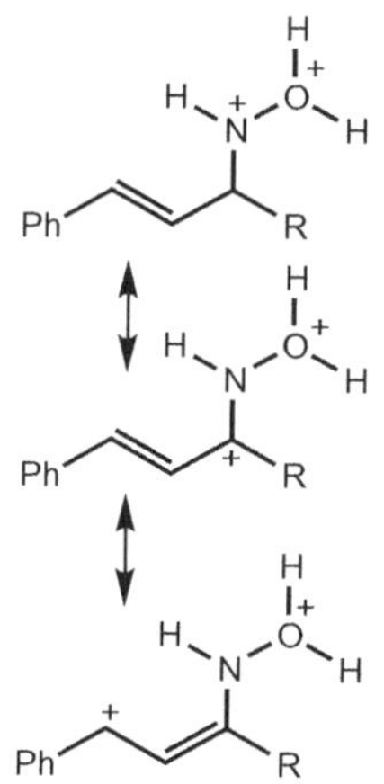

42

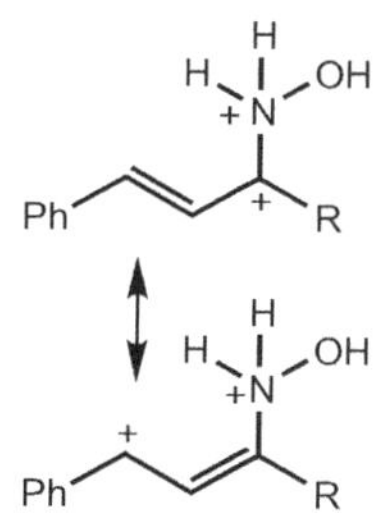

43

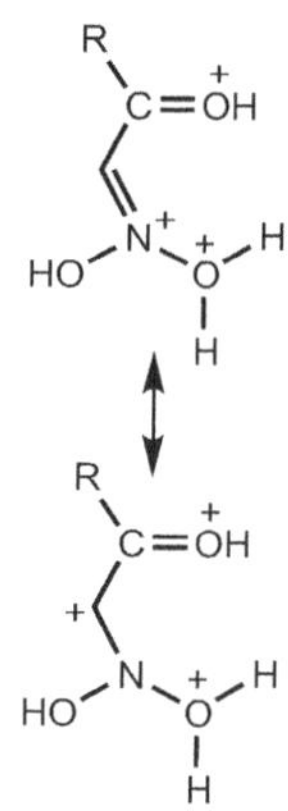

44

Nitroalkenes react with benzene derivatives at low temperature in triflic acid to afford α-aryl ketones after quenching with methanol[182,183] [Eq. (5.74)]. At higher temperature the *O*-protonated oxime intermediate may react further to yield 4*H*-1,2-benzoxazines (see Section 5.14.1.3). α-Nitrocarbonyl compounds show similar characteristics as alkylating agents to yield oximes with the involvement of the tricationic intermediate **44**.[181]

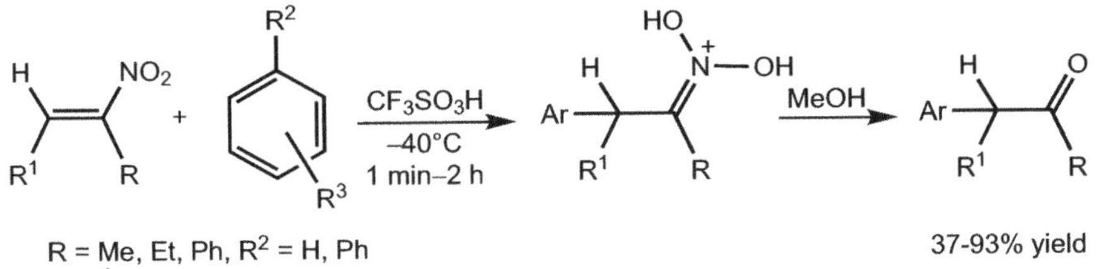

R = Me, Et, Ph, R^2 = H, Ph
R–R^1 = $(CH_2)_4$
R^2 = Me, MeO, Cl
R^3 = H, 4-Me, 4-MeO

(5.74)

Shudo and co-workers[181] have made the general observation that electron-withdrawing substituents on the cationic center increase reactivity toward benzene by increasing electrophilicity. Reactive carbenium ions may be described by the general formula **45**, where X represents genuine electron-withdrawing groups such as carbenium, oxonium, and ammonium.

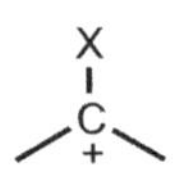

45

X =

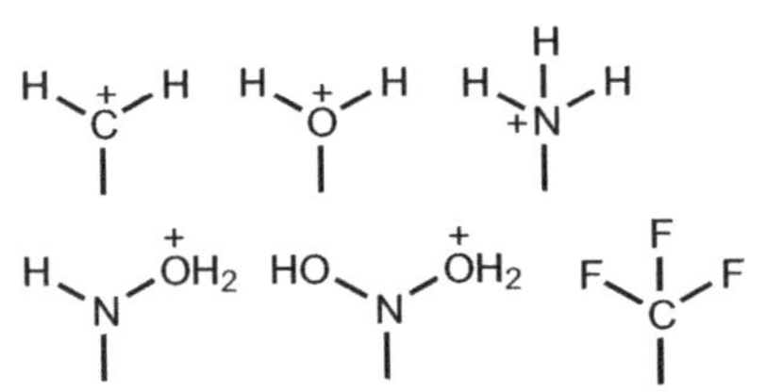

Klumpp and co-workers have shown that phosphonium[184] or ammonium[185] groups can also dramatically enhance the reactivities of adjacent electrophilic centers and, consequently, the alkylating ability of alkenes (and carbonyl compounds). For example, they have prepared aryl-substituted piperidines in high yields (60–99%) by reacting 1,2,3,6-tetrahydropyridines with benzene in triflic acid.[186] The alkylation is regioselective: the parent compound and *N*-methyl- and 2,2,6,6-tetramethyl-substituted derivatives give 4-phenyl-substituted products exclusively, whereas 3-substituted *N*-methyl compounds afford 3-phenyl-substituted derivatives. This observation and reactivities of substituted benzenes indicate the possible involvement of 1,4-dicationic and 1,3-dicationic species (**46**) and (**47**), respectively, as the most stable intermediates.

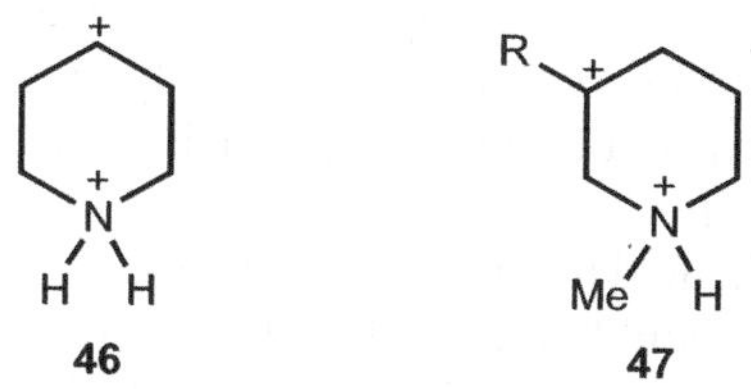

The $AlCl_3$–graphite and graphite intercalates of related Lewis acid halides have been tested as solid catalysts for the gas-phase alkylation of aromatic hydrocarbons.[112,187] The catalytic activity of intercalation compounds of $AlCl_3$, $AlBr_3$, and $FeCl_3$ in graphite was measured toward two model reactions, the alkylation of benzene with ethylene and propylene [Eq. (5.75)] and the transethylation of benzene with diethylbenzene. The ethylation of benzene with ethylene proceeded to give high initial yield at temperatures as low as 125°C (Table 5.9) when the intercalated Lewis acid halide was $AlCl_3$. The initial yield is lower when $AlBr_3$ is used and very low when SbF_5 or $FeCl_3$ are intercalated in the graphite layers. The same results were observed when these catalysts were used to induce transethylation of benzene with diethylbenzene (see Section 5.2.6).

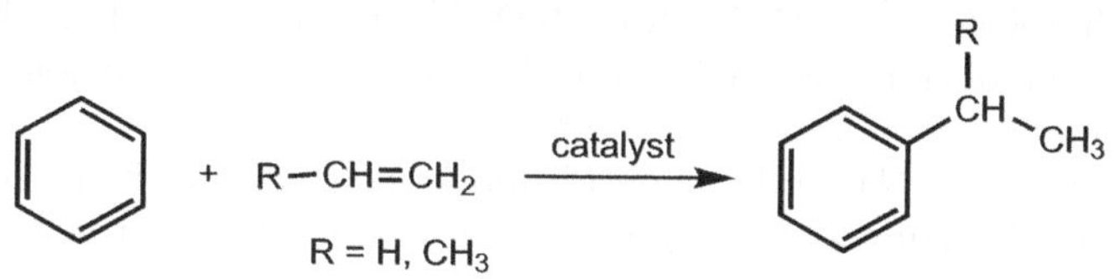

(5.75)

Although in the case of relatively active catalysts ($AlCl_3$–graphite and $AlBr_3$–graphite) the initial conversions were high, their activity declined rapidly with the

Table 5.9. Ethylation of Benzene with Ethylene over Graphite-Intercalated Metal Halides[187]

Metal Halide:	$AlCl_3$	$AlCl_3$	$AlCl_3$	$AlCl_3$	$AlBr_3$	SbF_5	$FeCl_3$
% Intercalated:	16.6	16.6	16.6	28.4	4.5	26.1	50.5
Temperature (°C):	125	160	200	160	160	160	185
$[C_6H_6]/[C_2H_4]$ Ratio:	3.3	3.4	3.2	3.5	3.3	3.6	3.2
Time-on-Stream (h)	% Conversion (Based on Ethylene)						
1	61.7	41.7	15.1	60.4	31.7	4.8	
2	10.3	12.0	5.7	62.9	15.1		
3	5.4	7.3	2.4	43.8			
4	4.3	5.3	1.7	33.7			
5	2.9	5.7	1.9	26.5			
6	1.9	1.4	0.7	6.8			
7				2.6			
8				2.2			
9				1.7			
10				1.5			

time-on-stream. The catalysts were completely deactivated after a period of 6–8 h. There are two possible reasons for the loss in catalytic activity of these systems: (i) possible hydrolysis by traces of moisture present in the feed and (ii) leaching of the metal halide from the graphite layers by the feed.

Lalancette et al.[188] indicated that intercalation process can be carried out in CCl_4 solution only if the Lewis acid halide is slightly soluble in the solvent. Therefore, $AlCl_3$ is probably eluted from the graphite by the feed. The rate of this process will increase with the increase in temperature, which increases the vapor pressure of Lewis acid halide. Hence, there are several drawbacks in using these graphite intercalates.

On the other hand, the Nafion resin in its acidic form (Nafion-H) shows high activity in a variety of electrophilic reactions. Gas-phase alkylation of benzene with ethylene and propylene in a flow system proceeds at temperatures as low as 110°C over Nafion-H (Table 5.10).

In the alkylation of ethylbenzene with ethylene, with conventional acid catalysts under usual conditions, *sec*-butylbenzene is a byproduct. *sec*-Butylbenzene was detected when the reaction was carried out over catalysts such as supported phosphoric acid,[189] ferric phosphate,[189] or $AlCl_3$–NiO–SiO_2.[190] When Nafion-H or $AlCl_3$ are used, no such byproduct is detected, probably due to fast dealkylation of *sec*-butylbenzene under the more acidic conditions.

The high acidity of the Nafion-H catalyst is further demonstrated by its ability to promote both polyalkylation and isomerization. In reaction between benzene and ethylene at 190°C, 20% of the alkylated products are diethylbenzenes.[187] The isomer distribution of the diethylbenzenes is 1% of the *ortho*, 75% of the *meta*, and 24% of the *para* isomers. This composition is very close to the equilibrium composition of diethylbenzenes determined in solution chemistry with $AlCl_3$ catalyst and indicates that the reaction is thermodynamically controlled.

Nafion–silica nanocomposite catalysts and silicas modified with superacidic surface functions (see Section 2.4.2.2) developed recently have been tested in

Table 5.10. Alkylation of Benzene with Alkenes over Nafion-H Catalyst[187]

	Temperature (°C)	$[C_6H_6]/[Alkene]$	Contact Time (s)	Alkene Conversion (%)
Ethylene	110	4	7	10
	150	4	6	24
	180	4	6	36
	190	3.4	3.5	44
Propylene	110	1.5	7	9
	150	1.5	6	16
	180	1.5	6	19
	180	3	4	21
	180	6	4	29

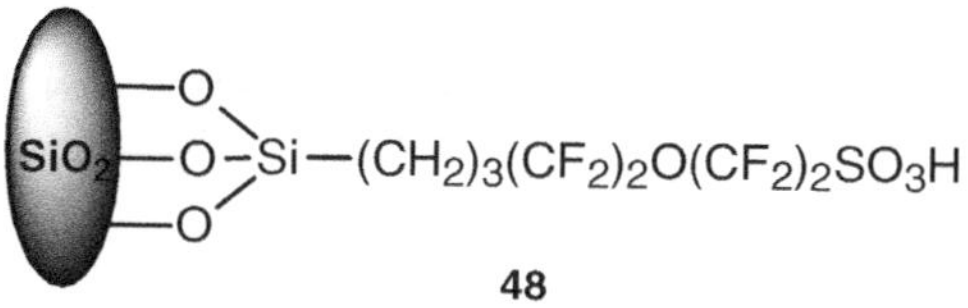

48

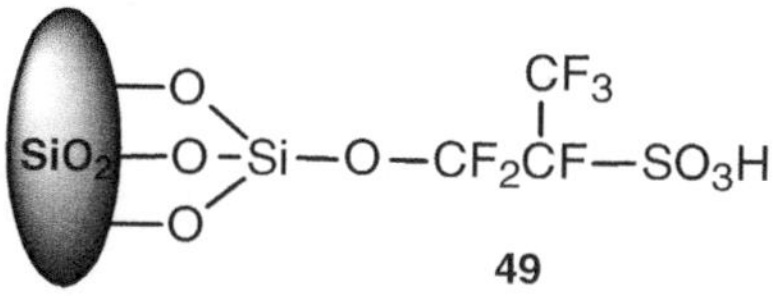

49

Figure 5.15. Superacidic perfluorinated sulfonic acid functions anchored to silica surfaces.

Friedel–Crafts alkylation of aromatics with various alkenes.[191,192] Alkylation of benzene with 1-dodecene to form linear alkylbenzenes (LABs) is of particular interest because LABs play a key role in the industrial production of surfactants. The preferred product is 2-phenyldodecane, and a high linearity of all alkylated products is also required. A nanocomposite sample made by the hydrolysis of sodium silicate gave better than 95% LAB selectivity at 99% conversion,[193,194] whereas lower conversion was measured over silica with anchored perfluorinated sulfonic acid sites[192] (structure **48**, Figure 5.15). At low conversion, selectivity of 2-phenyldodecane was 28%,[194] which is significantly higher than values for the industrial HF process (15–18%). Another significant advantage of nanocomposites is the very low selectivity for the undesirable branched alkylbenzenes, which is attributed to the low activity of Nafion–silica to catalyze skeletal isomerization. Superior catalyst performance was observed for the type 2 catalyst in transalkylation as well.[193,194]

An alkene mixture of industrial source (equal amounts of C_9–C_{13} alkenes and alkanes) was used in the alkylation of benzene on three Nafion–silica catalysts with 5%, 13%, and 20% loadings.[195] 20% Nafion–silica showed high and stable activity and its performance exceeded that of a Y-zeolite-based material. The selectivity to 2-phenylalkanes (25%) was higher than in the Detal process using fluorinated silica–alumina but decreased somewhat with increasing Nafion content.

Harmer et al.[196] used 1,1,2,2-tetrafluoroethanesulfonic acid in the alkylation of *para*-xylene with 1-dodecene. The silica-embedded catalyst prepared by the sol–gel method showed much higher activity than the neat acid (almost complete conversion in 15 min at 100°C over the sol–gel-derived material versus 10% conversion, using the same molar amounts of acid). Practically no leaching was detected and the catalyst could be recycled with a slight decrease in conversion. It is in sharp contrast with silica-supported triflic acid, which showed much lower activity due to the loss of volatile triflic acid.

Superelectrophilic dications **50** have been observed by Coustard,[197] who used ^1H and ^{13}C NMR spectroscopy in triflic acid at low temperature (Scheme 5.25). They

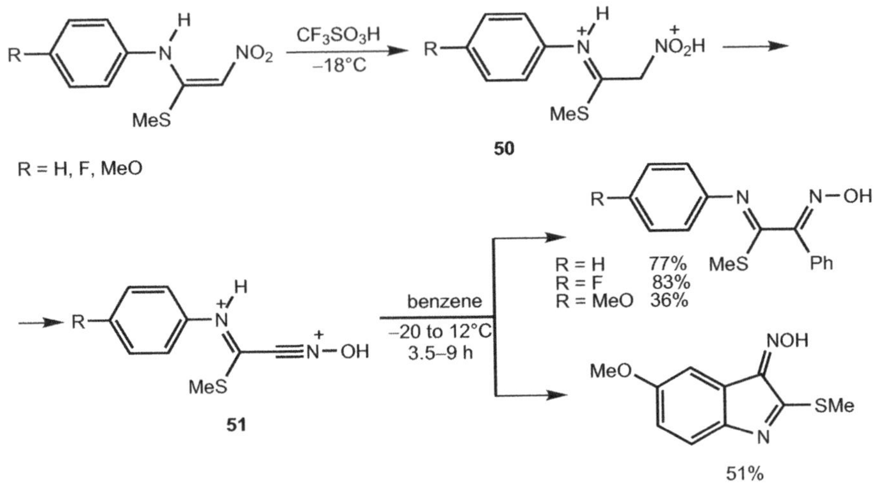

Scheme 5.25

were slowly transformed into ions **51**, which alkylated benzene to yield aryliminohydroxyimino derivatives. The 4-MeO-substituted compound gave the cyclialkylated product as well.

5.2.2. Alkylation with Alcohols and Cyclic Ethers

Data for the use of alcohols as alkylating agents in superacids are scarce. A study of the alkylation of phenol and naphthols with *tert*-butyl alcohol has shown[198] that triflic acid adsorbed on aminopropyl-modified silica is the most selective to yield monoalkylated products compared to solid acids (triflates immobilized in silica).

Sarca and Laali[199] have used triflic acid in butylmethylimidazolium hexafluorophosphate [BMIM][PF_6] ionic liquid for the benzylation of various arenes with benzyl alcohol [Eq. (5.76)]. When compared with $Yb(OTf)_3$, triflic acid proved to be a better catalyst showing higher selectivity (less dibenzyl ether byproduct) by exhibiting similar activity (typically complete conversion). Of the isomeric products, *para* isomers dominate. Experimental observations indicate that dibenzyl ether originates from less complete protonation of benzyl alcohol and, consequently, serves as a competing nucleophile. Both substrate selectivity (k_T/k_B) and positional selectivity (*ortho*/*para* ratio) found in competitive benzylation with a benzene–toluene mixture (1:1 molar ratio) are similar to those determined in earlier studies, indicating that the nature of the electrophile is not affected in the ionic liquid.

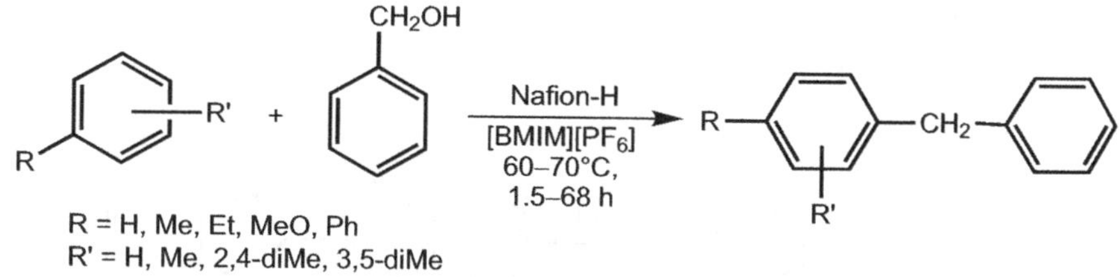

(5.76)

Prakash, Bach, and co-workers[200] have subjected a series of chiral benzylic alcohols to S_N1-type displacement reaction with aromatics catalyzed by triflic acid or $HBF_4{\cdot}OEt_2$. Under optimized conditions the alcohols give alkylated products in high yield with high facial diastereoselectivity [Eq. (5.77)]. The enantomerically pure alcohol (R = H) showed no significant racemization. NMR characterization of the long-lived benzylic cation **52** generated under superacidic conditions (SbF_5–SO_2ClF, −70°C) allowed the authors to demonstrate that such conformationally restricted carbocationic intermediates are responsible for the high stereoselectivity of alkylation observed.

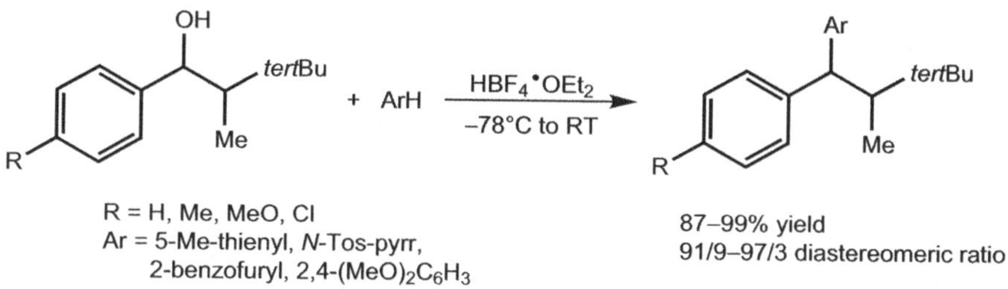

(5.77)

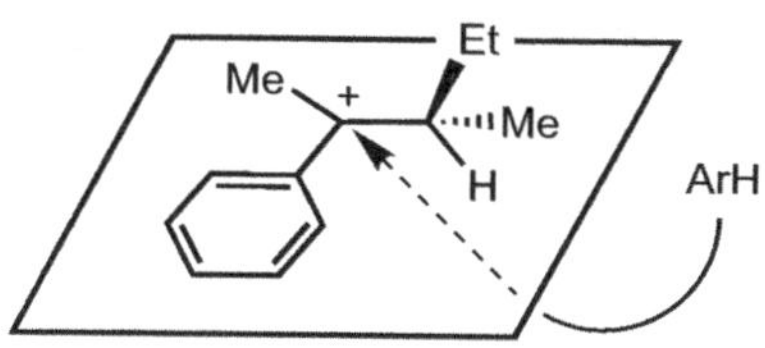

52

Triflic acid is also efficient in the alkylation of electron-rich aromatics (anisole, 1,3-dimethoxybenzene, 2-methylfurane, pyrrole, benzofurane, indole) with secondary benzylic alcohols and 3-phenylallyl alcohols (50°C, 1–9 h, 66–95% yield).[201] Benzene, toluene, and halobenzenes are also alkylated with hydroxybiindantetraone **53** in triflic acid within 1–2 h[202] [Eq. (5.78)]. Suprisingly, however, the primary products (with the exception of the 4-methylphenyl-substituted compound) undergo rearrangement upon prolonged treatment to yield alkenes selectively with the indicated stereochemistry [Eq. (5.78)]. The possible involvement of dicationic intermediates is suggested.

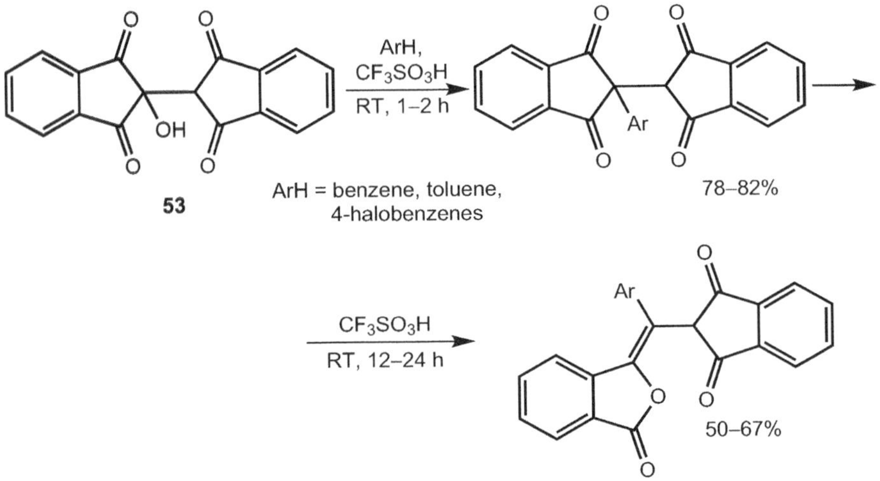

(5.78)

Kotsuki et al.[203] have reported high yields of alkylated products formed in the presence of triflic acid using cyclohexyl methanesulfonate [Eq. (5.79)]. Similar performance was found for mesitylene, durene, and naphthalene, as well as for other secondary alcohol methanesulfonates.

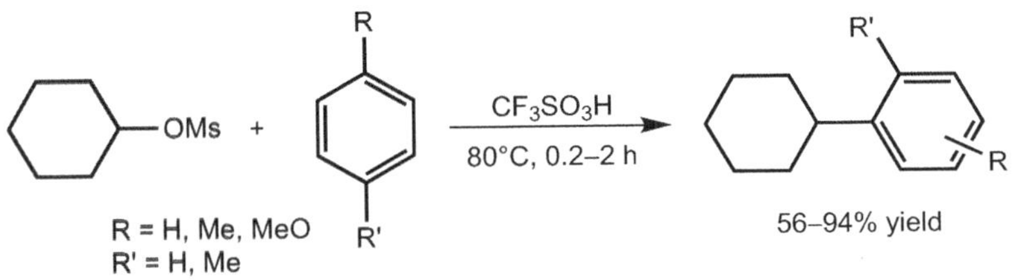

(5.79)

Epoxides are well-known carbon electrophiles and proved to be useful reagents in alkylations. A systematic study[204] with oxetanes, oxiranes, and oxolanes in the presence of triflic acid showed that various alkylated products are formed, but with low selectivities due to multiple alkylations and cyclialkylation. Methyl (*R*)-glycidate, in turn, has been shown to react with electron-rich arenes in triflic acid to give α-hydox-β-arylpropanoates with high stereospecificity[205] [Eq. (5.80)]. The high regioselectivity was attributed to the selective ring opening to yield the intermediate cation **54**, and dication **55** was also postulated. In a recent similar study,[206] the facile synthesis of 1-trifluoromethyl- and 1,1-bis(trifluoromethyl)-2-arylethanols has been reported [Eq. (5.81)]. Observations with respect to reactivities and selectivities were well demonstrated through DFT calculations (B3LYP/6-31G* level).

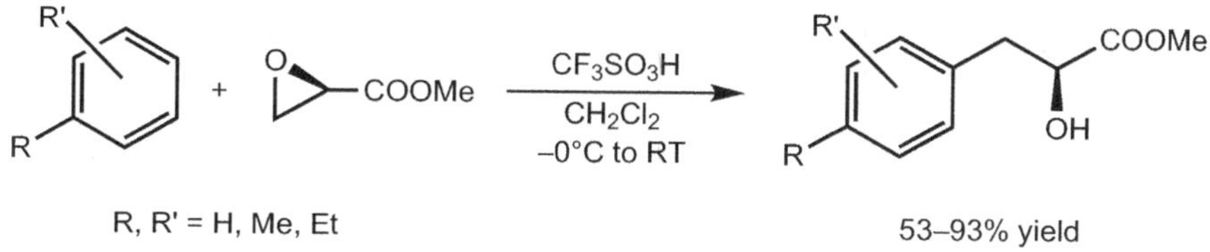

(5.80)

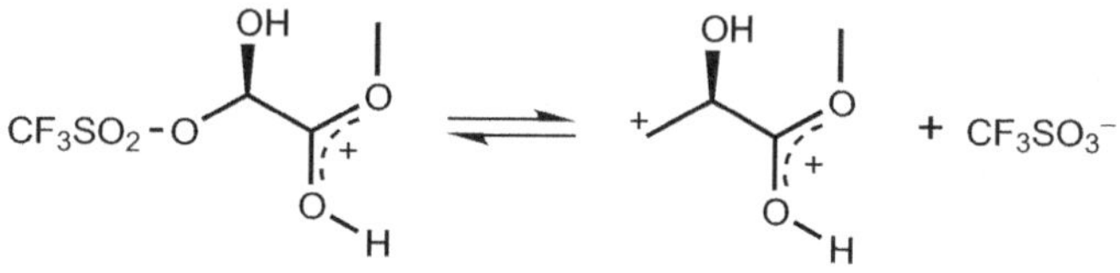

54 **55**

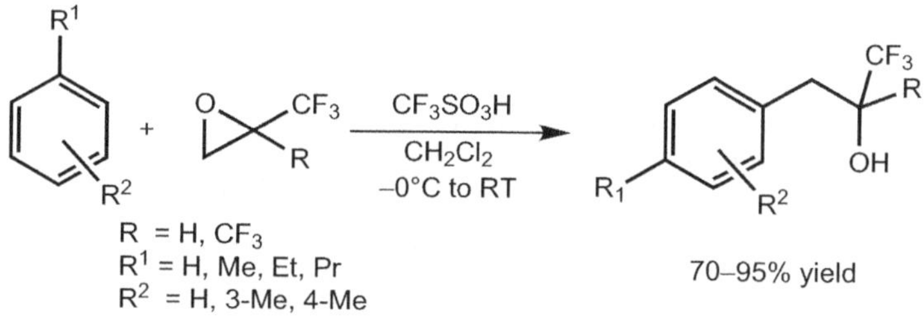

(5.81)

When olefins are used as alkylating agents, the catalytic activity of Nafion-H slowly decreases, most probably due to some polymerization on the surface, which deactivates the catalytic sites. The activity decreases faster when more reactive branched alkenes are used. The use of alcohols instead of olefins as the alkylating agents improves the lifetime of the catalyst. With alcohols, no ready polymerization takes place, since water formed as byproduct inhibits polymerization of any olefin formed (by dehydration) but does not affect the acidity of the catalyst at the reaction temperatures.

Reaction of alcohols with benzene over Nafion-H catalysts gave the corresponding alkylbenzenes (Table 5.11). When *n*-propyl alcohol was the alkylating agent, no *n*-propylbenzene was detected, and the only product obtained was cumene.[112,187] This indicates the intermediacy of the isopropyl cation in the alkylation process.

The alkylation of aromatic hydrocarbons with methyl alcohol over Nafion-H catalysts, including the mechanistic aspects, has been studied in detail. The degree of conversion of methyl alcohol was much dependent on the nucleophilic reactivity of the aromatic hydrocarbon. For example, the reactivity of isomeric xylenes was higher than that of toluene or benzene.

A study of the temperature dependence of the reaction between toluene and methyl alcohol showed a substantial increase in xylene formation as the temperature was increased.[207] However, the overall yield of xylene at 209°C was lower than that expected from extrapolation of data obtained at lower temperatures. This is probably

Table 5.11. Alkylation of Benzene with Alcohols over Nafion-H Catalyst

Alcohol	$[C_6H_6]/[ROH]$	Temperature (°C)	Contact time (s)	Alcohol Conversion (%)
EtOH	2.6	180	9	3.5
	2.6	210	8	6
n-PrOH	0.85	110	10	0
	0.85	175	9	5
	2	175	9	17
isoPrOH	2	170	9	11
	2	210	8	16

due to the increasing thermal instability of Nafion-H at temperatures around and above 200°C and hence its reduced activity.

Yields and conversion of methyl alcohol were much higher when the aromatic substrate was phenol or anisole and their derivatives[208] (Table 5.12). In recent studies, various degrees of selectivity were reported in the gas-phase methylation of phenol toward the *ortho* substitution, using catalysts such as Al_2O_3,[209–211] TiO_2,[212] ZnO, Fe_2O_3,[213,214] ZnO · MO (M = Cu, Ba, Ca, Co, Mn, Mg, Ni),[215] MgO alone,[216] or mixed with oxides of Mn, Cu, Sn, Bi, Pb or Cr[217] at temperatures ranging from 250°C to 400°C. Unlike these examples, the selectivity toward *ortho* methylation when using Nafion-H as catalyst is somewhat lower,[208] probably due to the absence of basic sites on this solid catalyst.

In gas-phase methylation reactions over Nafion-H using methyl alcohol as the alkylating agent, the consumption of methyl alcohol was higher than that calculated by product analysis.[207,208] This is due to the formation of dimethyl ether as the byproduct [Eq. (5.82)]. Indeed, when neat methyl alcohol is passed over Nafion-H catalyst at temperatures over 150°C, dimethyl ether is the only product formed quantitatively with water as the byproduct.[218]

Table 5.12. Methylation of Phenol, Cresols, and Anisole with Methyl Alcohol over Nafion-H Catalyst[208]

	Product Composition, (%)						
	Unreacted Starting Material	Anisole[a]	Me-Anisoles	DiMe-Anisoles	Phenol[a]	Cresols	Xylenols
Phenol	37.3	37.2	9.7	1.0		10.4	4.4
Anisole	58.6		13.9	3.0	18.1	4.7	1.7
ortho-Cresol	51.4	0.1	23.4	4.7	0.4	0.6	19.4
meta-Cresol	48.1	Trace	26.0	5.8	0.2	1.0	18.9
para-Cresol	39.2	3.3	23.4	6.4	8.4	4.6	14.8

[a]Excluding starting material.

$$2\ CH_3OH \xrightarrow[150°C]{\text{Nafion-H}} CH_3OCH_3 + H_2O \qquad (5.82)$$

Therefore, studies were also carried out using dimethyl ether as the methylating agent over Nafion-H catalyst. The initial conversions were similar to those obtained with methyl alcohol as the methylating agent. However, the reactivity of the catalyst was found to decrease sharply with time probably due to the increased esterification of the acidic sites of the solid catalyst.[207] The reactivity of the catalyst could be, however, readily regenerated by passing steam over it at 180°C for 1 h. In general, it appears that the presence of water vapor (i.e., steam) in the systems catalyzed by Nafion-H does not reduce the activity of the catalyst. In cases where polymerization and other possible side reactions would lead to the deactivation of the catalyst, the presence of water helps to maintain the catalytic activity of Nafion-H catalyst. For example, when benzene was alkylated by dimethyl ether over Nafion-H, the use of benzene saturated with water slowed down the deactivation of the catalyst considerably.

Olah and co-workers[219] have applied Nafion-H in the benzylation of benzene with benzyl alcohols [Eq. (5.83)] and also reported the reaction of benzyl alcohol with substituted aromatics (toluene, xylenes, mesitylene) to yield diphenylmethanes. The reaction is performed under mild conditions and produces the corresponding dibenzyl ethers as byproducts (2–22%). The substrate and positional selectivity in competitive benzylation of benzene and toluene (1:1 molar ratio) was found to be almost the same as observed in solution-phase Friedel–Crafts benzylation with benzyl chloride ($AlCl_3$–CH_3NO_2). Cyclic products **56** and **57** resulting from cyclialkylation were isolated when Nafion-H-catalyzed benzylation was applied to 2-(hydroxymethyl) diphenylmethane and 3,4-dimethoxybenzyl alcohol, respectively.

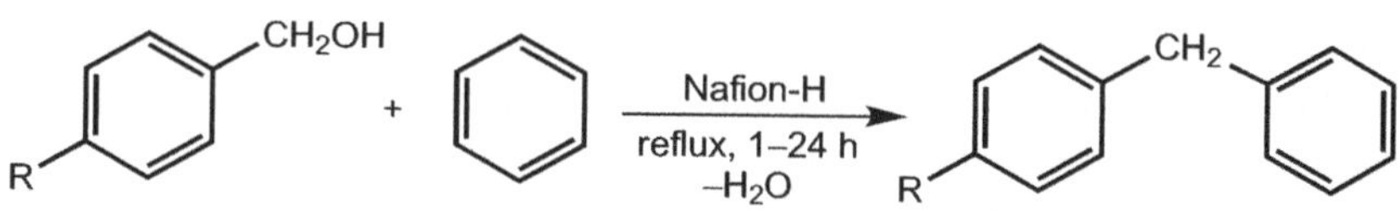

R = H, Me, *tert*-Bu, Br, NO_2 60–97% yield

(5.83)

56

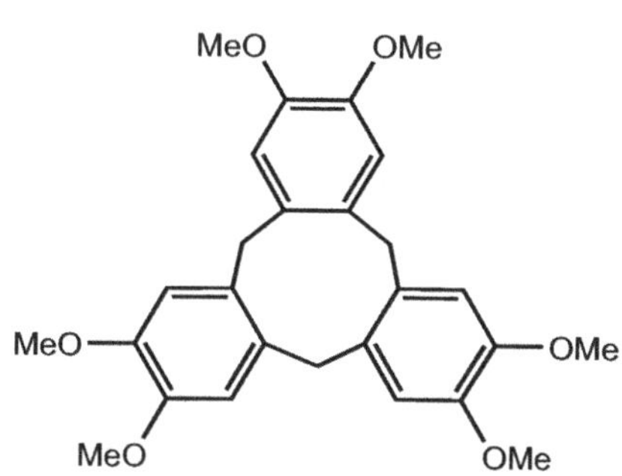

57

Friedel–Crafts alkylation of benzene,[220,221] toluene,[222] *para*-xylene,[220] and naphthalene[223] with benzyl alcohols have been studied over Nafion–silica nanocomposite catalysts, including the kinetics of alkylation.[221,223] In most cases, 13% Nafion–silica showed the highest activity, testifying again to the much higher accessibility of the active sites. Complete conversion of *para*-xylene was found in the presence of triflic acid, and it was the only reaction when ether formation as side reaction did not occur.

The isopropylation of *meta*-cresol with propylene and isopropyl alcohol[224,225] and anisole with propan-1-ol[226] [Eq. (5.84)] was studied in supercritical CO_2 in a continuous flow reactor over Nafion SAC-13 and, for comparison, other organic-based and inorganic solid acids. The optimal reaction temperature for Nafion SAC-13 was found to be 200–250°C. This temperature range is significantly higher than those for organic-based catalysts, which lost sulfonic acid groups at these high temperatures. Selectivity toward monoalkylated products, however, decreases with increasing temperature.

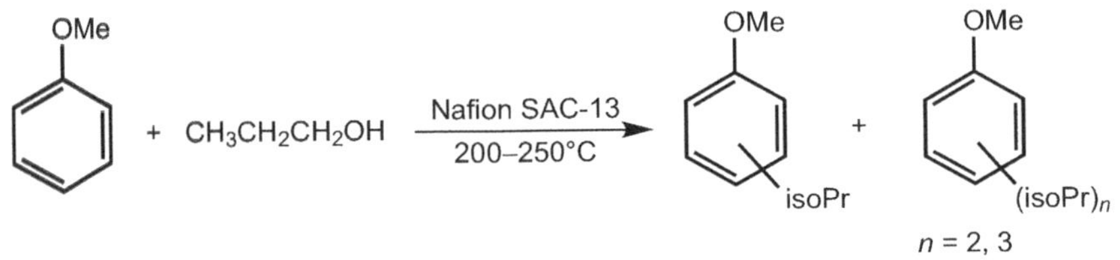

(5.84)

Fujiwara et al.[227] tested a nanocomposite material having Nafion immobilized in MCM-41 mesoporous silica in Friedel–Crafts alkylations with benzyl alcohol. Whereas Nafion–MCM-41 showed lower activity in the alkylation of toluene than 13% Nafion SAC-13 under identical conditions, it exhibited increased activity when used in the alkylation of *para*-xylene.

In 1976, Lalancette et al. studied[228] the catalytic activity of graphite intercalated $AlCl_3$ and compared it with neat $AlCl_3$ in solution-phase alkylations. Whereas the rate of alkylation slowed down using the intercalated catalyst, a higher selectivity toward monoalkylation was found (Table 5.13).

5.2.3. Alkylation with Alkyl Halides

Jouannetaud and co-workers[229] have explored electrophilic trifluoromethylation under superacidic conditions of aniline derivatives[229] and *N*-heterocycles. Methyl-substituted anilines and substituted acetanilides [Eq. (5.85)] react with the "CCl_3^+" cation generated from CCl_4 in HF–SbF_5 followed by fluorination to yield the corresponding trifluoromethyl derivatives. Under similar conditions, indolines are transformed to the 6-triluoromethyl derivatives, whereas substituted indoles yield 5-triluoromethyl derivatives.[230]

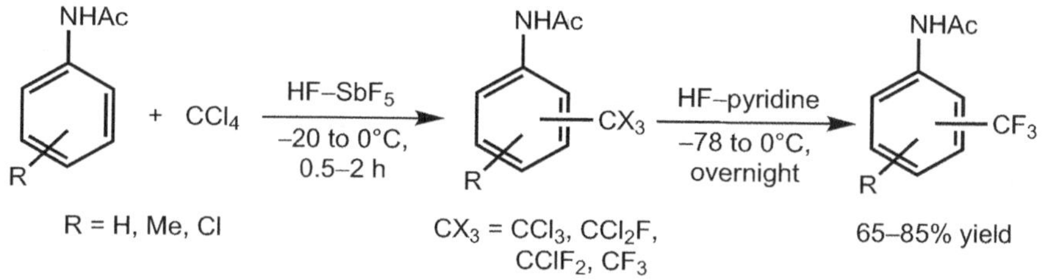

(5.85)

Olah et al.[231] have used boron tris(triflate) [$B(OSO_2CF_3)_3$ or $B(OTf)_3$] in Friedel–Crafts alkylation with alkyl halides to give alkylated arenes in low yields (Table 5.14). It is known that alkylation of alkylbenzenes promoted by Lewis acids, in general, leads to the formation of increased amounts of *meta* isomers. Indeed, high, sometimes exclusive formation of the *meta* isomer is observed with the use of $B(OTf)_3$. In the case of ethyl chloride, for example, high initial *meta* selectivity is found, which changes after prolonged reaction approaching the equilibrium distribution.

High *meta* selectivities can be explained by intramolecular isomerization within the arenium ion intermediates to the most stable 1,3-disubstituted isomeric ethyltoluenium ion [Eq. (5.86)]. Isomerization is induced by the very strong conjugate acid formed between $B(OTf)_3$ and HF or HCl generated during alkylation from the corresponding alkyl halides. Ethyltoluene isomers were found in independent experiments to readily undergo isomerization with $B(OTf)_3$ in dichloromethane to give equilibrium isomer composition. Alkylation with isopropyl chloride shows that isomerization is more pronounced in the less coordinating dichloromethane solvent than in nitromethane. In *tert*-butylation with *tert*-butyl fluoride the high initial *meta* isomer content rapidly decreases. This is accounted for by facile de-*tert*-butylation with excess toluene yielding predominantly para-*tert*-butyltoluene (kinetic product). After prolonged reaction the quantity of the *meta* product increases again due to the thermodynamically directed intramolecular isomerization.

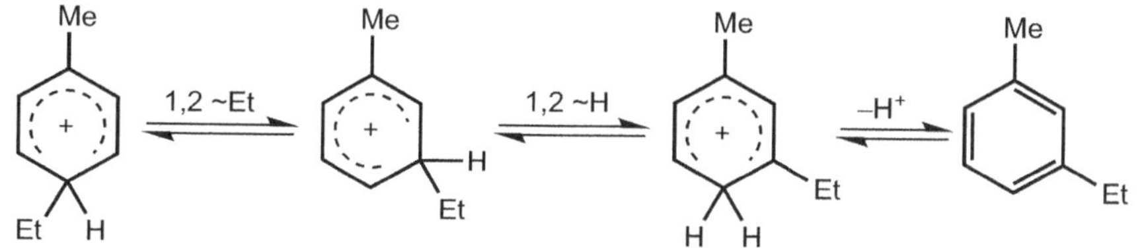

(5.86)

Boron tris(triflate) has also been tested in the adamantylation of benzene and toluene with 1-haloadamantanes [Eq. (5.87)] and 2-haloadamantanes.[232] $B(OTf)_3$ is a highly active catalyst to promote the transformation in very short time under mild conditions to yield isomeric aryladamantanes and adamantane byproduct (Table 5.15). Of the isomeric 1-tolyladamantanes, 1-*meta*-tolyladamantane predominates, whereas the *para* isomer is the main product of the 2-tolyladamantanes. The *ortho* isomers were

Table 5.13. Electrophilic Substitutions Catalyzed by $AlCl_3$–Graphite[228]

System	Substrate	Reagent	Temperature (°C)	Time (h)	Products	Composition	
						$AlCl_3$	$AlCl_3$:Graphite
Sealed tube	Benzene	Ethyl bromide[a]	25	48	Benzene	2	29
					Mono-Et	13	40
					Di-Et	27	23
					Tri-Et	54	8
					Tetra-Et	4	Traces
Sealed tube	Toluene	Ethyl bromide[a]	25	24	Toluene	0	17
					Mono-Et	Traces	52
					Di-Et	41	31
					Trie-Et	60	Traces
Atmospheric	Toluene	Ethyl Bromide[b]	−10	24	Toluene	5	5
					Mono-Et	Traces	33
					Di-Et	30	52
					Tri-Et	65	10
Sealed tube	Naphthalene	Ethyl bromide[a]	25	44	Naphthalene	26	77
					Mono-Et	39	21
					Di-Et	17	1
					Di-Et, tetrahydro	2	Traces
					Tri-Et	16	0.5
Sealed tube	Biphenyl	Ethyl bromide[a]	25	48	Biphenyl	26	81
					Di-Et	16	15
					Tetra–hexa-Et	33	4
					Octa-Et	25	0
Atmospheric	Benzene	Ethylene	75	24	Benzene	0	3
					Mono-Et	Traces	19
					Die-Et	1.1	47
					Tri-Et	25	22

					Tetra-Et	53	6
					Penta-Et	18	2
					Hexa-Et	4	1
Atmospheric	Benzene	Propylene	60	0.5	Benzene	23	44
					Mono-isoPr	56	54
					Di-isoPr	18	2
					Tri-isoPro	3	Traces
Atmospheric	Toluene	Ethylene	80	1.45	Toluene	8	17
					Mono-Et	49	51
					Di-Et	40	26
Atmospheric	Toluene	Propylene	80	2.50	Toluene	Traces	4
					Mono-isoPr	Traces	33
					Di-isoPr	42	59
					Tri-isoPr	46	Traces
Atmospheric	Toluene	Isobutylene	80	1.15	Toluene	20	35
					Mono-isoBu	66	63
					Di-isoBu	13	1

[a]Ethyl bromide: aromatics ratio = 3.
[b]Ethyl bromide: toluene ratio = 2.

Table 5.14. Alkylation of Toluene with Alkyl Halides in the Presence of $B(OTf)_3$[231]

Alkyl halide	Time (min)	Yield (%)	Isomer Distribution (%)		
			ortho	*meta*	*para*
Methyl fluoride[a]	1	18	41	24	35
	30	30	47	26	27
Methyl chloride[a]	1	2	46	22	32
	60	11	17	58	25
Ethyl fluoride[a]	1	23	17	70	13
	30	32	28	45	27
Ethyl chloride[a]	1	15	3	97	
	60	25	4	83	13
	150	20	7	64	29
Ethyl bromide[a]	10	15		100	
	60	18		100	
Isopropyl chloride[a]	5	25		83	17
	30	17		77	23
Isopropyl chloride[b]	5	10	47	24	29
	60	27	58	21	21
tert-Butyl fluoride[a]	5	42		60	40
	25	46		4	96
tert-Butyl chloride[a]	5	30		79	21
	30	36		75	25
tert-Butyl chloride[b]	1	30		45	55
	30	16		70	30

Reaction conditions: 10 mol% catalyst, 25°C.
[a]In dichloromethane as solvent.
[b]In nitromethane as solvent.

Table 5.15. Adamantylation of Toluene with 1-Haloadamantanes in the Presence of $B(OTf)_3$[232]

	Time (min)	Yield (%)		Tolyladamantanes, Isomer Distribution (%)			
		Adamantane	Tolyladamantanes	1-*meta*	1-*para*	2-*meta*	2-*para*
1-Ad-Cl	0.25	27	72	65	34		1
	1	33	66	72	25		3
	30	34	56	68	27	Trace	5
	120	35	54	49	29	6	16
1-Ad-Br	1	17	82	66	33		
	5	28	71	70	29		1
	30	32	66	67	28		5

Reaction conditions: toluene/1-haloadamantane/$B(OTf)_3$ molar ratio = 1:1:0.25, dichloromethane, room temperature.

not detected, which is attributed to the significant steric hindrance with the tertiary bridgehead system and possible fast isomerization.

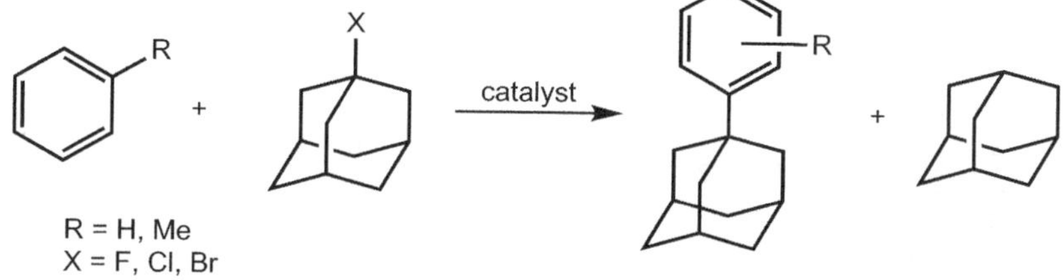

(5.87)

The relative reactivities of benzene and toluene were studied in competitive reactions with benzene–toluene mixture. Isomeric tolyladamantanes, in this case, was formed in much lower yield (Table 5.16); that is, low substrate selectivity ($k_T/k_B = 0.5$–4.8) prevails. These data, however, are obscured by significant disproportionation indicated by the high amounts of adamantane detected. Furthermore, various isomerizations resulting in rather varied isomer distributions were also observed. In the isomeric tertiary and secondary substituted phenyladamantanes, 1-phenyladamantane is highly favored. Interestingly, 1-adamantanoyl chloride was also found to give aryladamantane products. 1-Adamantanoyl chloride readily ionizes with the strong Lewis acid $B(OTf)_3$ and, then, the formed adamantanoyl cation loses CO to form the 1-adamantyl cation.

Formation of the 2-substituted aryldamantane products can be accounted for by an intermolecular isomerization process (Scheme 5.26). It requires the formation of

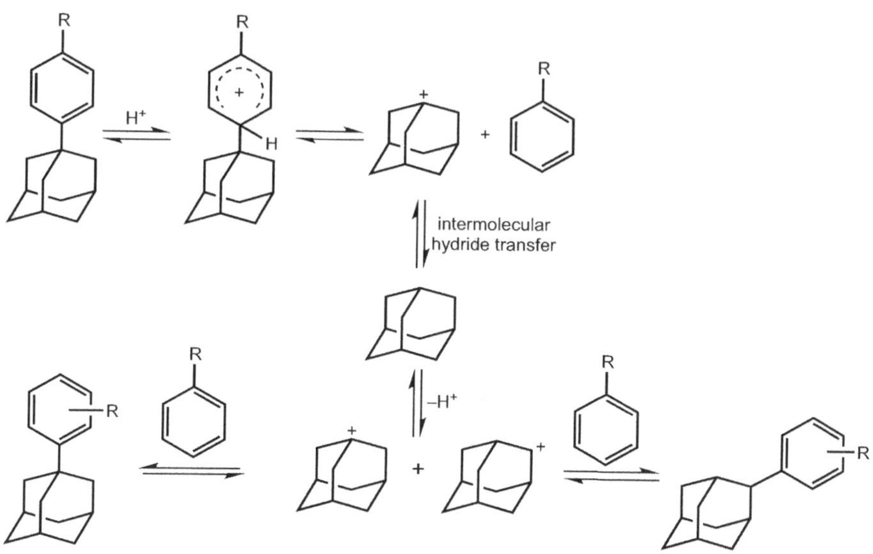

Scheme 5.26

Table 5.16. Adamantylation of Benzene–Toluene Mixture with 1-Haloadamantanes in the Presence of $B(OTf)_3$[232]

			Phenyladamantanes			Tolyladamantanes				
				Isomer Distribution[a] (%)			Isomer Distribution[a] (%)			
	Time (min)	Adamantane (%)	Yield (%)	1-Ph	2-Ph	Yield (%)	1-*meta*	1-*para*	2-*meta*	2-*para*
1-Ad-F[b]	1	20	44	99	1	5	76	22	Traces	Traces
	5	23	42	98	2	7	51	30	Traces	18
	15	35	40	98	2	6	56	30	1	13
	120	46	34	94	6	4	50	28	4	18
1-Ad-Cl[b]	1	37	39	96	4	20	59	33		8
	5	44	37	98	2	12	63	25		12
	15	50	36	98	2	9	61	27		12
	25	52	35	98	2	10	59	27		14
1-Ad-Cl[c]	1	Traces	42	100		2	84	16		
	5	1	51	100		7	74	26		
	15	3	52	100		8	70	30		
	30	2	51	100		8	68	32		

Reaction conditions: benzene/toluene/$B(OTf)_3$ molar ratio = 5:1:0.25, room temperature.

[a]Normalized data.

[b]In dichloromethane.

[c]In nitromethane.

adamantane through hydride transfer followed by the formation of isomeric 1- and 2-adamantyl cations, which alkylate the arene formed in the first step. Isomerization of *meta*- and *para*-tolyladamantes, in turn, is interpreted as intramolecular 1,2-migration processes (Scheme 5.27).

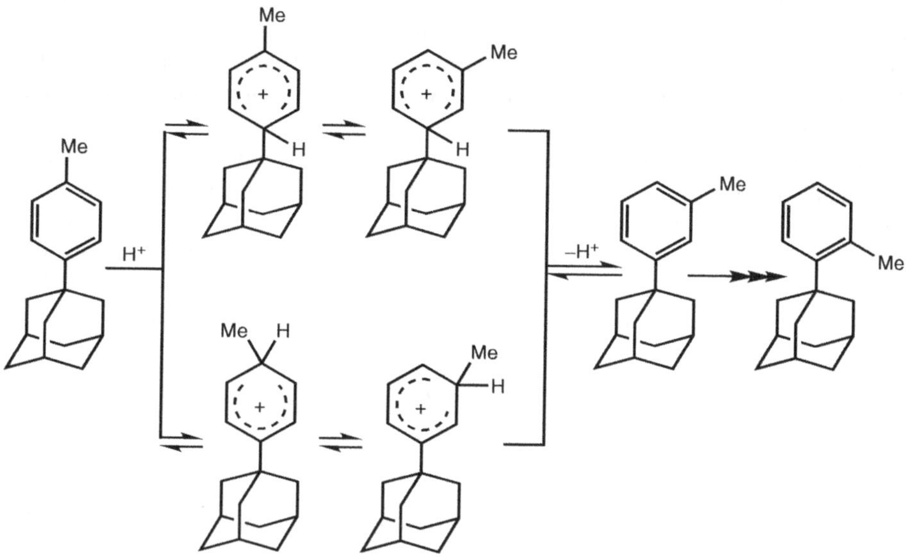

Scheme 5.27

Olah and et al.[233] have studied the alkylation of aromatics with 1-chloronorbornane, 3-halonoradamantane, and fluorocubane in the presence of $AlCl_3$ and BF_3. Backside S_N2-type displacement at bridgehead positions is not possible. Consequently, Friedel–Crafts alkylation with these halides must involve strongly polarized bridgehead halide–Lewis acid complexes in equilibrium with their energetic, reactive carbocations. Since the strained bridgehead centers cannot flatten out, the carbocationic center is sp^3-hybridized and the empty orbital of these carbocations is of sp^3 nature (**58**–**60**). The reactive 1-norbornyl, 3-noradamantyl, and cubyl cations could not be observed under stable ion conditions by NMR spectroscopy at low temperature (SbF_5-SO_2ClF solution, −78°C).

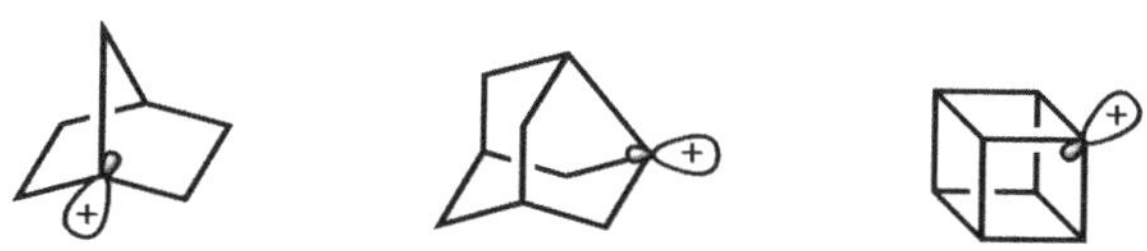

58 **59** **60**

Laali et al.[234] have developed a method to the highly selective *para*-adamantylation of arenes (toluene, ethylbenzene, anisole) with haloadamantanes (1-chloro- and 1-bromoadamantane, 1-bromo-3,5,7-trimethyladamantane) and 1-adamantanol promoted by triflic acid in butylmethylimidazolium triflate [BMIM][OTf] ionic liquid. In contrast to reactions run in 1,2-dichloroethane, little or no adamantane byproduct was detected in [BMIM][OTf]. Furthermore, no isomerization of *para*-tolyladamantane was observed supporting the intramolecular nature of the formation of *meta* isomers. In competitive experiments with benzene–toluene mixture (1:1 molar ratio), high substrate selectivities were found ($k_T/k_B = 16–17$) irrespective of the alkylating agent. This is in sharp contrast to values about unity measured in 1,2-dichloroethane.

Satisfactory results were obtained in the Nafion-H-catalyzed gas-phase alkylation of aromatic hydrocarbons with alkyl halides[235] [Eq. (5.88)]. Alkyl halides are reactive Friedel–Crafts alkylating agents and give high conversions when alkylating benzene in the gas phase over Nafion-H catalyst. For example, in the alkylation of benzene with isopropyl chloride, conversions as high as 87% were achieved (Table 5.17, run 11). Conversions, however, were temperature and contact time dependent (Table 5.17).

$$\underset{X = \text{halogen}}{\text{ArH} + \text{RX}} \xrightarrow[140\text{–}200^\circ\text{C}]{\text{Nafion-H}} \text{ArR} + \text{HX} \qquad (5.88)$$

The selectivity of the Nafion-H catalyst for monoalkylation has been found to be generally high. With a molar ratio of benzene:isopropyl chloride being 5:1, about 94% of the alkylate is monoalkylbenzene. This result is comparable to the highly selective monoalkylation reaction reported by Langlois.[236] They alkylated benzene with propylene (5.2:1 molar ratio) over H_3PO_4-quartz catalyst at ~200°C and obtained cumene in 95% yield.

The results obtained in the gas-phase isopropylation of various aromatic hydrocarbons with isopropyl chloride over Nafion-H catalyst showed only a relatively small variation of reactivity in going from fluorobenzene to xylenes.[235] Therefore, it has been assumed that the reaction rate is controlled by the formation of a reactive electrophilic intermediate (possibly, protonated alkyl halide **61**, or some form of incipient alkyl cation) rather than by σ-complex formation between the electrophile and the aromatic nucleus [Eq. (5.89)].

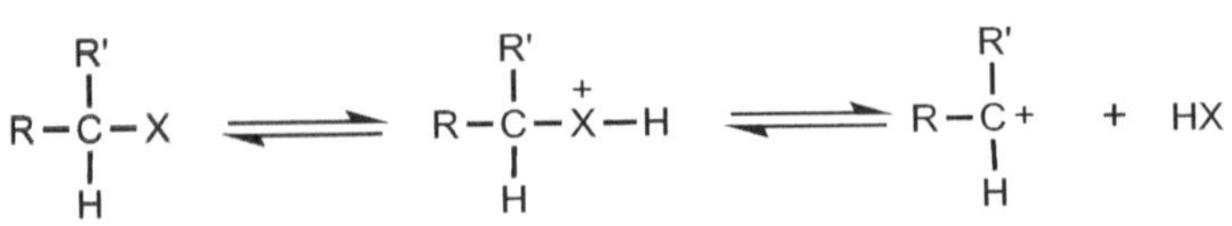

61

(5.89)

The relatively minor differences observed in the degree of conversion in the reaction of various aromatic hydrocarbons with isopropyl chloride over Nafion-H are

Table 5.17. Effect of Temperature and Contact Time on the Isopropylation of Benzene with Isopropyl Chloride over Nafion-H Catalyst[235]

Run	Temperature (°C)	Feed Rate ml/min	Yield of Isopropylbenzene (%)				Mono/Di Ratio	Conversion[a] (%)
			Mono	*meta*-Di	*para*-Di	Tri		
1	135	0.2	2.5	0.1	0.1	<0.1	12.5	3.0
2	150	0.2	6.4	0.4	0.3	<0.1	9.1	8.1
3	165	0.2	15.8	1.1	0.8	0.1	8.3	19.9
4	174	0.2	25.6	1.9	1.2	0.2	8.2	32.4
5	180	0.05	44.3	3.8	2.0	0.3	7.6	56.8
6	180	0.1	36.8	3.0	1.7	0.3	7.8	47.1
7	180	0.2	33.2	2.7	1.5	0.2	7.9	42.2
8	196	0.2	41.4	3.2	1.7	0.2	8.4	51.8
9	196	0.05	53.7	4.1	2.0	0.2	8.8	66.5
10	180	0.2	63.0	5.5	2.4	0.3	8.0	78.7
11	180	0.1[b]	69.5	5.8	2.4	0.3	8.5	86.8
12	180	0.2[c]	55.0	2.5	1.3	0.0	14.5	62.6
13	180	0.2[d]	19.7	3.3	1.6	0.5	4.0	31.0

Reaction conditions: 1 g Nafion-H, benzene:isoPrCl ratio = 5:2, unless otherwise indicated.
[a]Total conversion based on isoPrCl.
[b]2 g Nafion-H was used.
[c]Benzene:IsoPrCl = 5:1.
[d]Benzene:IsoPrCl = 5:4.

Table 5.18. Benzylation of Aromatics with Benzyl Chloride over 13% Nafion–Silica Catalyst[237]

Aromatic	Time (h)	Conversion (%)	Yield (%)	Isomer Ratio *ortho/meta/para*
Benzene	10	30	29	—
Toluene	1.5	98	92	45/8/47
para-Xylene	0.75	100	96	—
Anisole	8	96	94	47/<1/52
para-Methylanisole	8	98	98	82/18^a
Chlorobenzene	1	100	99	38/<1/61
Naphthalene	0.5	100	99	67/33^b

Reaction conditions: 1.2 mmol of benzyl chloride, 24 mmol of aromatic, 100 mg of catalyst, 120°C.
^{a}2-Benzyl-4-methylanisole/3-benzyl-4-methylanisole.
$^b\alpha/\beta$.

explained by the possibility of some dealkylation (i.e., reverse reaction). Dealkylation reactions occur as a competitive process to the alkylation process. The more nucleophilic is the alkylated aromatic product, the higher the rate of dealkylation reaction.

Besides the advantage of their high reactivity toward alkylation reactions, primary and secondary alkyl halides show little tendency for dehydrohalogenation in Nafion-H-catalyzed gas-phase reactions.[235] Although a minor amount of olefin is reported to be formed, no polymer formation was observed on the catalyst. As a result, the catalytic activity of Nafion-H stays constant over prolonged on-stream periods.

Benzylation of various aromatics with benzyl chloride also proceeds smoothly over 13% Nafion–silica to afford diphenylmethane derivatives in high yields[237] (Table 5.18). Although deactivated aromatics (nitrobenzene, methyl benzoate) gave low (<10%) yields, chlorobenzene reacted readily with complete conversion similar to naphthalene. Furthermore, the catalyst, after recovery, exhibited the same activity.

Olah et al.[238] have performed a comparative study of the adamantylation of substituted benzenes with 1-bromoadamantane [see Eq. (5.87)] using various Nafion preparations and higher perfluroalkanesulfonic acids supported on zeolite HY. Nafion-H and Nafion–silica nanocomposite exhibited high activity in the adamantylation of bromobenzene and toluene, resulting in the formation of the *para* isomer as the main alkylated product (Table 5.19). Exclusive formation of the *para* isomer was observed in the adamantylation of phenol, anisole, and fluorobenzene over Nafion-H at complete conversion. *tert*-Butylbenzene, in turn, gave only byproducts. Among these, 1-phenyladamantane was the only adamantylated product, which indicates the facile de-*tert*-butylation of the intermediate carbocation.

In a similar study, Nafion-H beads, 10% Nafion-H on silica, and 13% Nafion–silica were compared in the adamantylation of toluene with 1-bromoadamantane.[239] 13% Nafion–silica nanocomposite exhibited the highest activity and showed a change in the isomeric ratio of the two alkylated products: the *para/meta* ratio shifted to lower values in prolonged reaction (Table 5.20). Acidity is known to exert a significant influence on

Table 5.19. Adamantylation of Bromobenzene and Toluene over Various Solid Acids[238]

	Bromobenzene[a]			Toluene[b]		
Catalyst	Time (h)	Conversion (%)	Selectivity *para*/*meta*	Time (h)	Conversion (%)	Selectivity *para*/*meta*
Nafion-H	1	100	88/12(16)	1.5	100	70/30(9)[c]
Nafion-H–SiO_2	2	5	97/3	1	30	92/8
Nafion–silica nanocomposite	1.5	92	89/11(8)	1.5	100	71/29(7)
$C_8F_{17}SO_3H$–HY	1	76	84/16(14)	1	100	79/21
$C_{10}F_{21}SO_3H$–HY	2	56	89/11(21)	1.5	100	79/21

[a]Reaction temperature = 156°C.
[b]Reaction temperature = 111°C.
[c]In parentheses: % adamantane formed.

regioselectivity.[232,240] In this case, after complete conversion the kinetic distribution (high excess of the *para* isomer) changed to the thermodynamic distribution (increasing amount of the *meta* isomer) with increasing reaction time over the nanocomposite catalyst with highly accessible acid sites. As it was pointed out, such isomerization takes place within the *para*-adamantylated arenium ion intermediate (Scheme 5.27).[238]

5.2.4. Alkylation with Carbonyl Compounds and Derivatives

Carbonyl compounds, particularly aromatic aldehydes, when activated with electrophilic catalysts, can also react with aromatics.[241] The process is often called condensation or reductive alkylation, but it is actually a multistep Friedel–Crafts alkylation reaction.

Olah et al.[242] and Shudo and co-workers[243] have shown that benzaldehyde reacts with benzene in various superacidic systems [triflic acid, triflic acid–SbF_5,

Table 5.20. Adamantylation of Toluene over Various Nafion Catalysts[239]

Catalyst	Time (min)	Conversion (%)	Rate[a]	Selectivity *para*/*meta*
Nafion beads	5	5.4	5.02 10^{-2}	85/15
	35	37.5		82/18
10% Nafion–SiO_2	5	19	0.58	83/17
	35	34		86/14
13% Nafion–silica nanocomposite	5	47	3.38	80/20
	35	100		55/45

Reaction conditions: 5 ml of toluene, 1 mmol of 1-bromoadamantane, 0.2 g of catalyst, 111°C.
[a]Initial reaction rate, mmol $(g_{cat}\ min)^{-1}$.

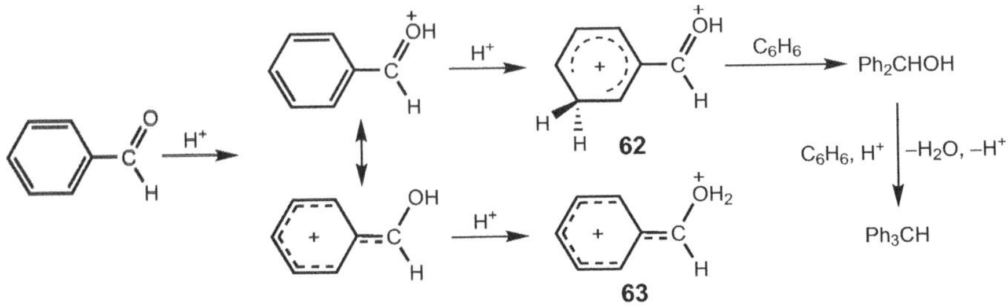

Scheme 5.28

$CF_3SO_3H_2^+B(OSO_2CF_3)_4^-$] to give triphenylmethane (Scheme 5.28). Experimental evidence supports the involvement diprotonated benzaldehyde in the reaction. Calculations showed (MP2/6-31G*//MP2/6-31G*+ZPE level) that the reactive intermediate is the *O*,*C*(aromatic)-diprotonated dication **62** and not the *O*,*O*-diprotonated **63** dication, which is less stable by 20.6 kcal mol^{-1}.

In a subsequent study, Shudo and co-workers[244] showed that benzaldehydes with electron-withdrawing groups (NO_2, CF_3) react with 2 equivalents of benzene in the presence of triflic acid to give substituted triphenylmethanes in good yields [Eq. (5.90)]. They also observed that *para*-fluorobenzaldehyde and biphenyl-4-carboxaldehyde yield diphenylmethane and triphenylmethanol under similar conditions, and the same products were also isolated in the reaction of triphenylmethane (Scheme 5.29).

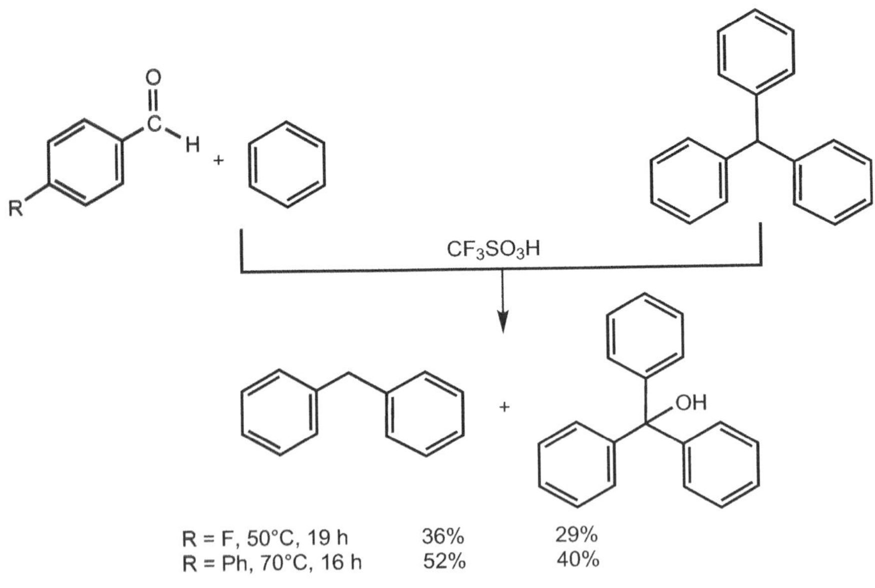

Scheme 5.29

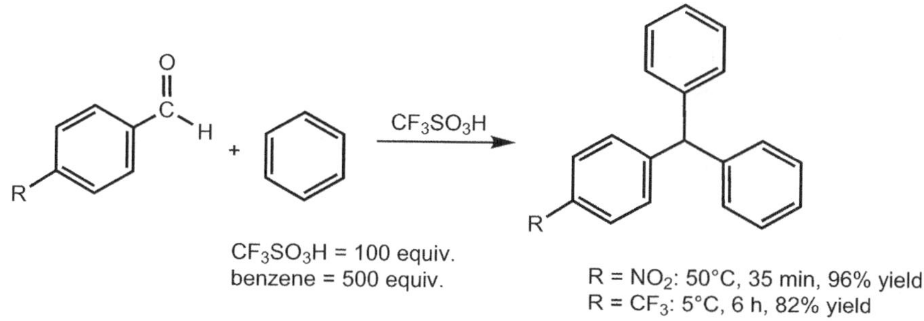

(5.90)

Product formation was interpreted in terms of transalkylation of substituted triphenylmethanes. Protonation at the *ipso* position of the substituted phenyl ring to form arenium ion **64** followed by the C–C bond breaking yields the diphenylmethyl cation, which alkylates benzene or is stabilized by hydride transfer (Scheme 5.30). The protonated intermediate **64** is highly unstable when the ring has an electron-withdrawing substituent. Consequently, its transformation is extremely slow and the primary product triphenylmethane can be isolated.

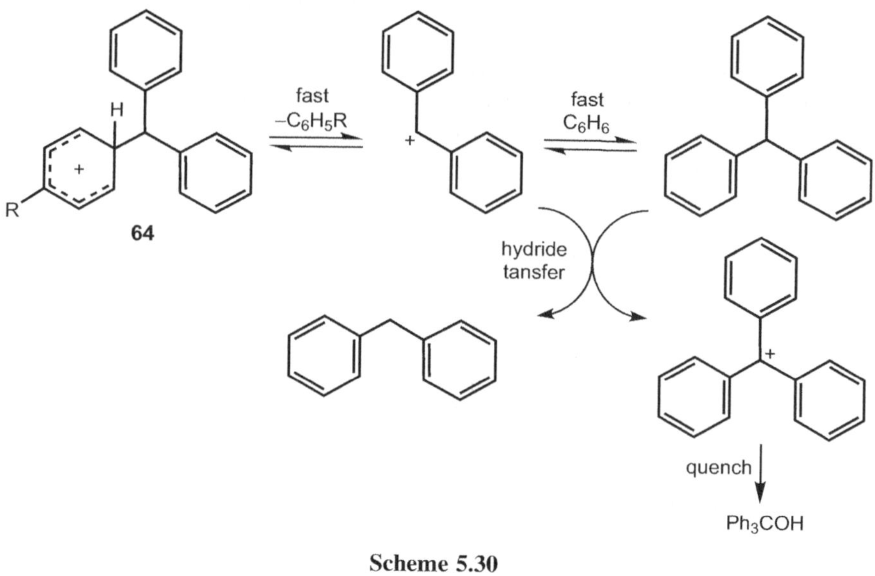

Scheme 5.30

Fukuzawa et al.[245] used 2-phenyl-1,3-dioxane to benzylate a variety of arenes [Eq. (5.91)]. Similar observations were made when substituted benzaldehydes were treated in the presence of 1,3-propanediol under identical conditions. Although 2-phenyl-1,3-dioxane gave similar results, benzaldehyde dialkylacetals in general were unreactive under similar conditions. Mechanistic studies including reaction of a labeled dioxane indicate the involvement of the alkylated intermediate **65** and product formation was interpreted via an 1,3-hydride shift.

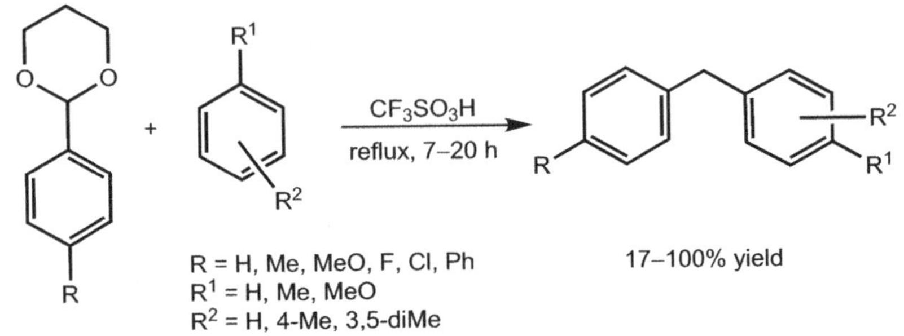

(5.91)

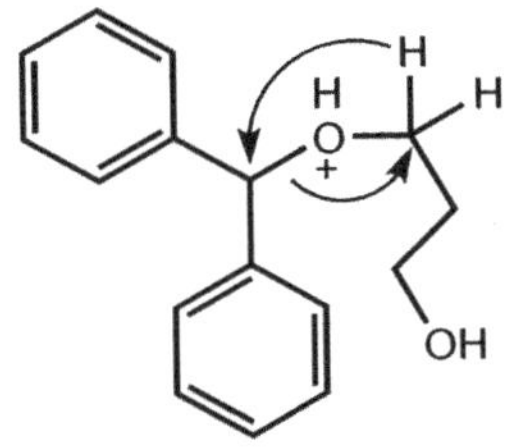

65

Aminoacetals proved to be more reactive in triflic acid reacting with benzene at room temperature to give diphenyl-substituted products of alkylation[246] [Eq. (5.92)]. Propanal acetals ($n = 1$), in general, give somewhat lower yields than acetaldehyde acetals ($n = 0$). In addition to benzene, halobenzenes also react, affording decreasing yields with decreasing nucleophilic character. Only monoprotonated species (ammonium cations) were detected in triflic acid by NMR spectroscopy. In turn, when the dioxolane derivative of 3-piperidin-1-yl-propionaldehyde was treated in triflic acid–SbF_5 (SO_2ClF solution, $-80°C$), the isomeric dicationic intermediates **66** were shown to exist. This allowed the suggestion of a mechanism involving diprotonated electrophiles (carboxonium–ammonium dications).

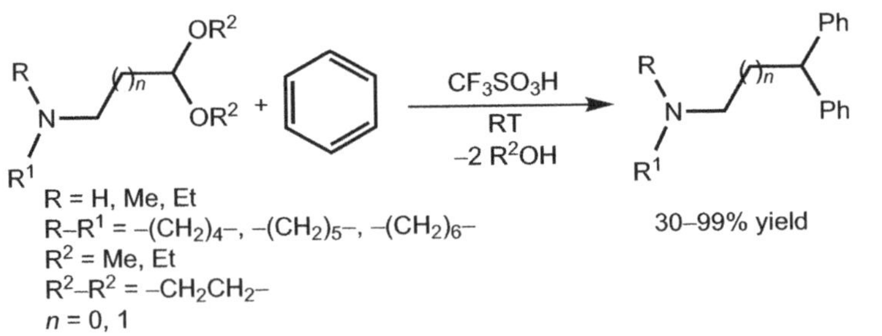

(5.92)

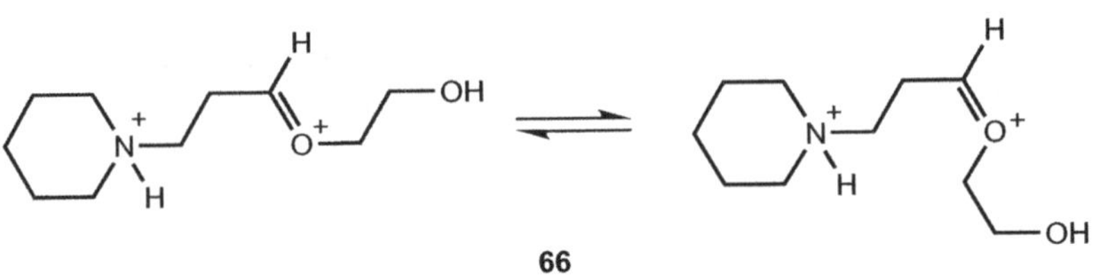

66

Klumpp et al. and Sommer and co-workers have studied the alkylating ability of a variety of heterocycles with carbonyl substituents in the presence of triflic acid. Pyrazolecarboxaldehydes [Eq. (5.93)],[247] pyridinecarboxaldehydes,[248–250] quinolinecarboxaldehydes,[251] piperidinones [Eq. (5.94)],[252,253] tropinone and quinuclidone, and acetyl-substituted heteroaromatic compounds (pyridines, thiazoles, pyrazine, quinoline, and isoquinoline)[254] proved to be highly reactive alkylating agents when reacted with benzene to give the corresponding diphenyl-substituted products in good to excellent yields.

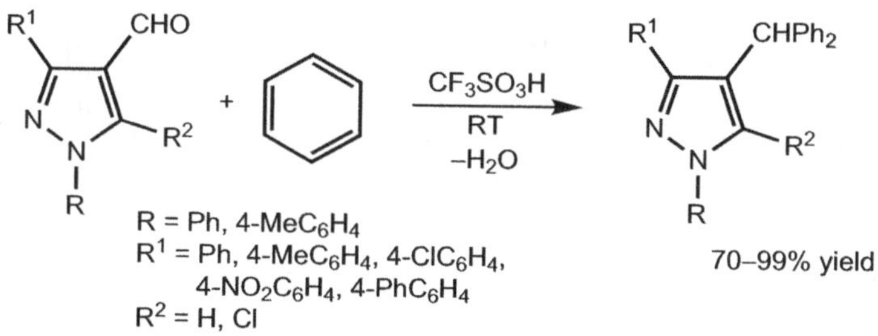

(5.93)

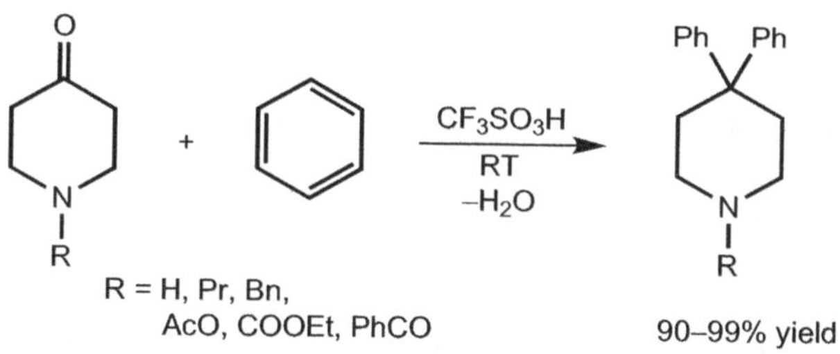

(5.94)

In all cases, superelectorophilic dicationic intermediates[3–5] were suggested to be involved in the activation of carbonyl compounds based on the observation that protonated *N*-heterocycles significantly enhance the reactivity of adjacent carbocationic centers. For example, cyclohexanone and acetophenone are unreactive toward benzene in triflic acid, whereas 4-piperidones[252] and acetylpyridines[254] react readily. Likewise, 3-pyridinecarboxaldehyde is able to alkylate deactivated aromatics such as chlorobenzene, *ortho*-dichlorobenzene, and nitrobenzene.[248] Furthermore, dicationic intermediates **67**, **68**, and **69** generated in FSO_3H–SbF_5–SO_2ClF solution have been directly observed by low-temperature NMR spectroscopy at −80°C.[247,248,253] Recent computational studies have shown[255] that the first protonation of 4-heterocyclohexanones at the carbonyl oxygen in triflic acid is exergonic (4-piperidone is an exception protonated at the N atom). The second protonation is slightly endergonic, but it is thermodynamically highly unfavorable for cyclohexanone.

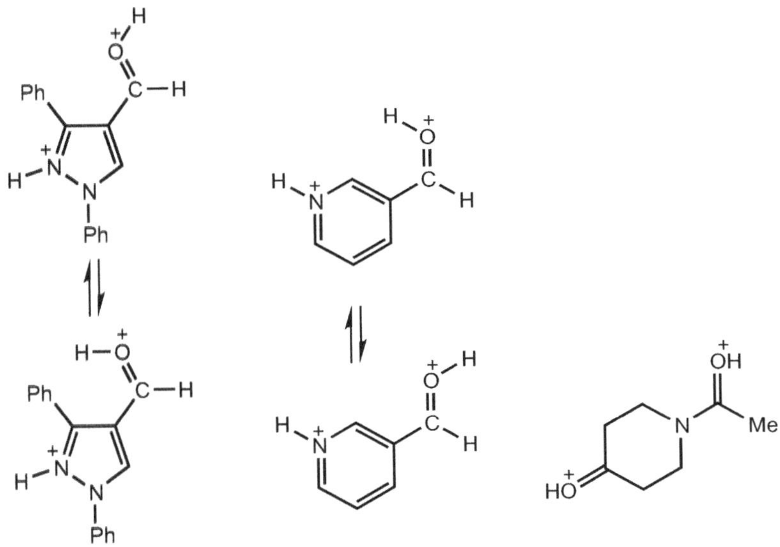

67 **68** **69**

Prakash, Olah, and co-workers[256] have prepared Mosher's acid analogs by the hydroxyalkylation of substituted benzenes with ethyl trifluoropyruvate [Eq. (5.95)]. Deactivated aromatics (fluorobenzene, chlorobenzene) required the use of excess triflic acid indicative of superelectrophilic activation.[3–5] In contrast to these observations, Shudo and co-workers[257] reported the formation *gem*-diphenyl-substituted ketones in the alkylation of benzene with 1,2-dicarbonyl compounds [Eq. (5.96)]. In weak acidic medium (6% trifluoroacetic acid–94% triflic acid), practically no reaction takes place. With increasing acidity the reaction accelerates and complete conversion is achieved in pure triflic acid, indicating the involvement of diprotonated intermediates.

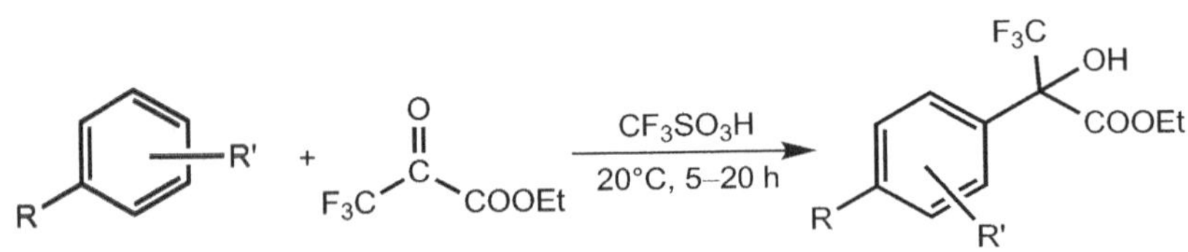

(5.95)

R = H, Me, *tert*-Bu, MeO, F, Cl
R' = H, Me, 3,5-diMe

46-86% yield

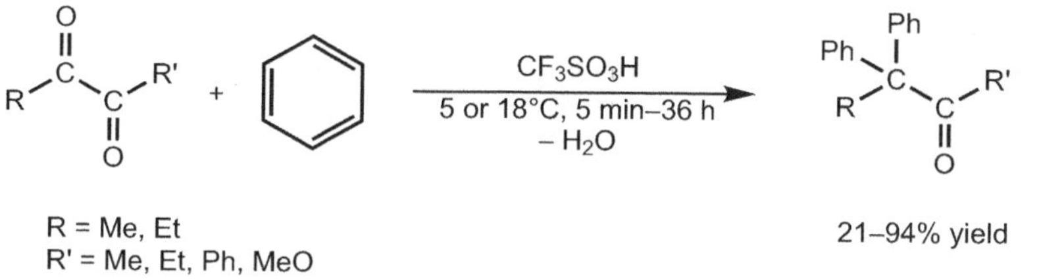

(5.96)

The use of triflic acid in the alkylation with other dicarbonyl compounds, such as isatins,[258] parabanic acid,[259] and ninhydrin,[260] has also been explored. A significant acidity dependence was found in the reaction of isatin with benzene.[258] Alkylation does not take place in the presence of trifluoroacetic acid ($H_0 = -2.7$) at 25°C in 12 h. Adding 22% of triflic acid ($H_0 = -10.6$) brings about a 90% yield of the 3,3-diphenyloxindole, whereas complete conversion is achieved in neat triflic acid ($H_0 = -14.1$) in 8 h. Ninhydrin undergoes facile reaction with various benzene derivatives in the presence of triflic acid[260] to yield 3-(diarylmethylene)isobenzofuranones [Eq. (5.97)], which is interpreted with the participation of the ring-opened intermediate **70**. In a similar manner, alkylation in the case of α-ketoacids[261] is usually followed by additional reactions to results in the formation various products. For example, α-ketosuccinic acid and α-ketoglutaric acid give 3,3-diphenylindanone and 4,4-diphenyltetralone, respectively, as a result of dehydrative decarbonylation. Two major products were isolated in the reaction of phenylpyruvic acid as a result of secondary transformations (Scheme 5.31).

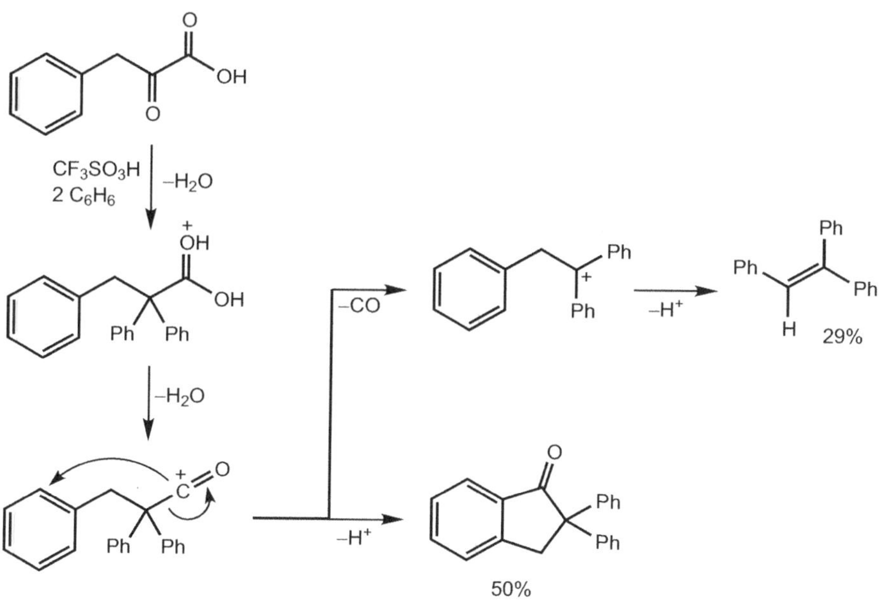

Scheme 5.31

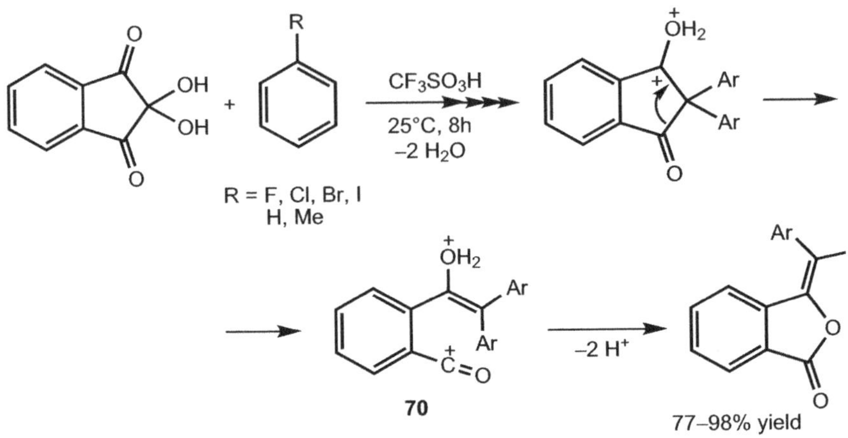

70

(5.97)

β-Ketoesters, β-ketophosphonates, and β-ketosulfones have been used to alkylate ferrocene to afford the corresponding β-ferrocenyl-α,β-unsaturated derivatives in excess triflic acid[262,263] [Eq. (5.98)]. The transformations are highly stereoselective, giving exclusively the (*E*)-isomers; this was explained by the *exo*-deprotonation of carbenium ion **71a** of more stable conformation. Acetals of formylphosphonates and formylsulfones react in a similar manner.

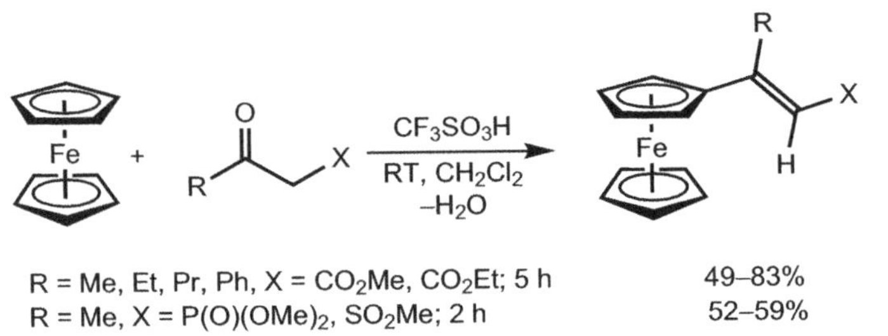

(5.98)

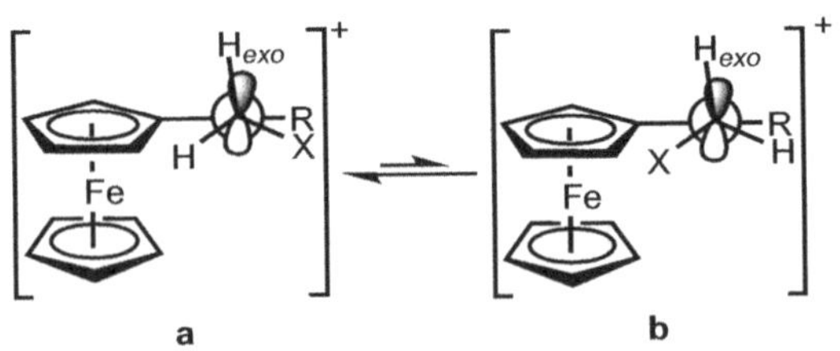

71

A variety of solid acids has been studied in the alkylation of benzene with formaldehyde to produce diphenylmethane[264] [Eq. (5.99)]. Aciplex–SiO_2 exhibited the highest activity and gave the highest yield. Nafion SAC-13, in turn, showed the highest specific activity and highest selectivity. The difference between these two catalysts is due to the amount of acid sites. In the reaction with formalin solution, all perfluorinated resinsulfonic acid catalysts including Nafion NR50 were superior to other solid acids studied for comparison.

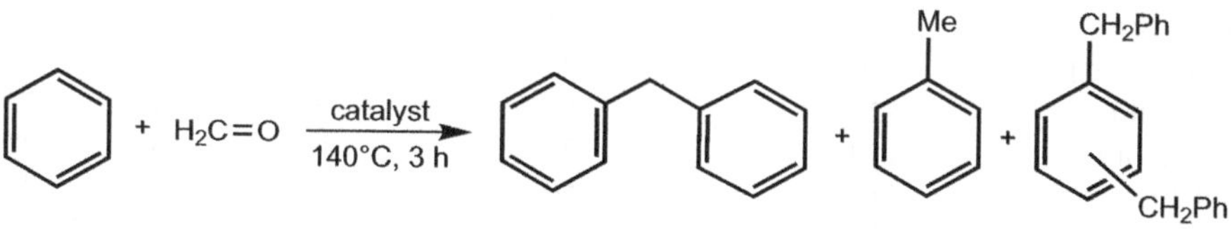

	yield %	selectivity %
Aciplex–SiO_2	66	81
Nafion–SiO_2	32	88
NafionNR50	16	81

(5.99)

5.2.5. Alkylation with Acid Derivatives

The versatility of the catalytic activity of Nafion-H is well-demonstrated in the alkylation of aromatic hydrocarbons by carboxylic acid alkyl esters both in the gas phase and in heterogeneous liquid-phase reactions.[265] Esters in the presence of conventional Lewis acid halide catalysts tend to give rise to acylation products along with alkylation products.[266,267] The heterogeneous liquid-phase alkylation reactions have been generally carried out under reflux conditions. Two types of alkylating agent have been studied: (i) alkyl esters of carboxylic acids [Eq. (5.100)], preferentially those of oxalic acid and (ii) alkyl chloroformates [Eq. (5.101)]. The advantage of alkyl chloroformates lies primarily in their volatile byproducts (HCl and CO_2). Diethyl oxalate shows particularly good alkylating ability even under mild conditions (Table 5.21).

$$\text{ArH} + \text{RCOOR'} \xrightarrow{\text{Nafion-H}} \text{ArR'} + \text{RCOOH} \qquad (5.100)$$

$$\text{ArH} + \text{ClCOOR} \xrightarrow{\text{Nafion-H}} \text{ArR} + \text{HCl} + \text{CO}_2 \qquad (5.101)$$

The gas-phase alkylation of toluene with alkyl chloroformates over Nafion-H was also reported to be an efficient transformation. It is interesting to compare the alkylating ability of methyl chloroformate with that of methyl alcohol on toluene under similar reaction conditions. For example, a 59% conversion with methyl chloroformate was observed,[265] compared with 15% conversion with methyl alcohol.

The gas-phase alkylation of toluene with dimethyl and diethyl oxalate over Nafion-H was also reported.[265] The alkylating ability of diethyl oxalate is

Table 5.21. Nafion-H Catalyzed Liquid-Phase Alkylation of Toluene with Esters and Haloesters[265]

			Isomer Distribution (%)		
Alkylating Agent	Temperature °C	Conversion (%)	*ortho*	*meta*	*para*
MeO(CO)Cl	70–72	2	48	26	26
EtO(CO)Cl	90	24	46	26	28
isoPrO(CO)Cl	110	80	42	21	37
CF_3COOEt	82	5.5	49	24	27
CCl_3COOEt	110	20	44	28	28
$(COOEt)_2$	110	50	48	24	28

Reaction time = 12 h.

comparable with that of ethyl chloroformate. However, the alkylating ability of dimethyl oxalate is lower than that of methyl chloroformate.

5.2.6. Isomerization and Transalkylation of Alkylbenzenes

When dialkylbenzenes are passed over Nafion-H at 160°C, both isomerization and disproportionation take place [Eqs. (5.102) and (5.103)]. Monoalkylbenzenes also disproportionate under these conditions[268–271] [Eq. (5.104)].

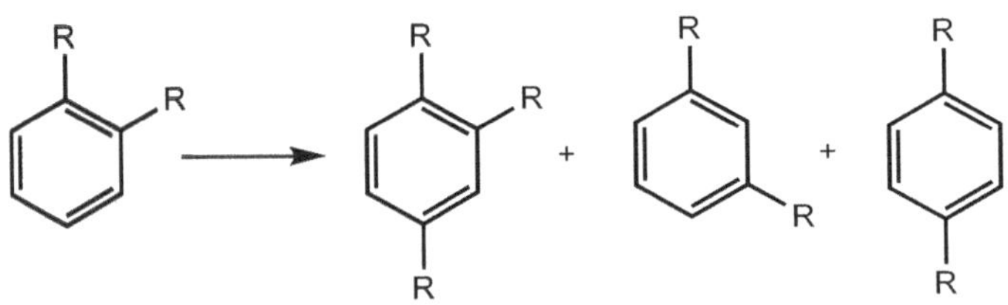

(5.102)

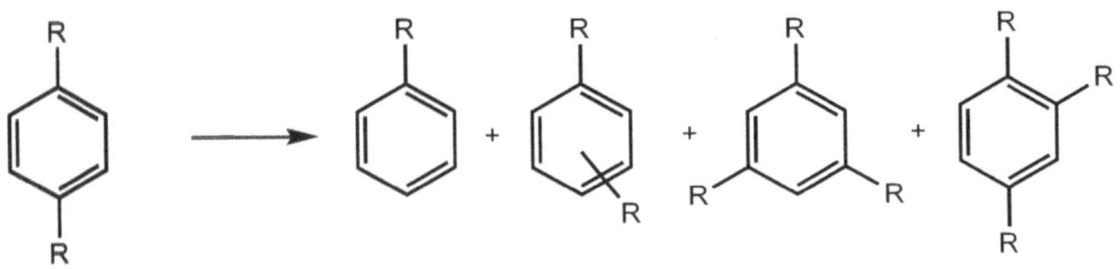

(5.103)

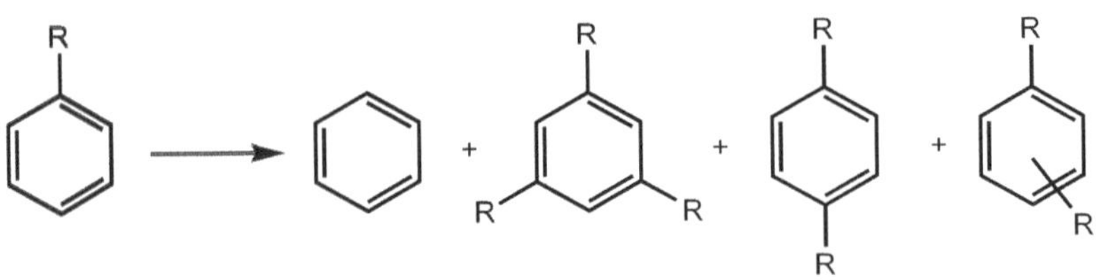

(5.104)

As expected, the aptitude for disproportionation of the aromatic compound depends upon the nature of the alkyl group, and the order of reactivity is isopropyl > ethyl > methyl. Due to their higher nucleophilicity, polyalkylbenzenes react faster than monoalkylbenzenes. This effect is pronounced in the case of methylbenzenes. Toluene itself shows little reactivity over Nafion-H at 193°C. Diethylbenzenes react much faster than dimethylbenzenes. The rate of conversion of diethylbenzenes over Nafion-H at 193°C is ~5 10^{-5} mol min^{-1} g^{-1} of catalyst.[269] This is a low rate when compared with that using $AlCl_3$−HCl in the liquid phase at room temperature (10^{-4} mol min^{-1} g^{-1} of catalyst).[272] However, one should bear in mind that Nafion-H is a truly insoluble heterogeneous catalyst, whereas in the case of $AlCl_3$−HCl a soluble complex is formed with the hydrocarbon and therefore the rates are not directly comparable. The equilibrium composition of the acid-catalyzed disproportionation of diethylbenzenes depends upon the nature of the catalyst.

The predominance of *meta*-diethylbenzene in the isomerization of diethylbenzenes is easily rationalized. Since isomerization reaction proceeds via arenium ion intermediates, the σ-complex derived from *meta*-diethylbenzene is the most stable one. Moreover, *meta*-diethylbenzene is also the most basic of the isomeric diethylbenzenes.[273,274] Therefore, more acidic catalysts increase the amount of the *meta* isomer at the expense of the *para* (and *ortho*) isomers, due to the increased stability of the substrate–catalyst complex.

The $AlCl_3$–graphite and graphite intercalates of related Lewis acid halides have also been tested as solid catalysts for the transethylation of benzene with diethylbenzene in the gas phase[112] [Eq. (5.105)]. The results were very similar to those observed in the ethylation of benzene (see Section 5.2.1); that is, high initial yields were found for intercalated $AlCl_3$, lower yields were found for $AlBr_3$, and very low yields were obtained when SbF_5 was intercalated (Table 5.22).

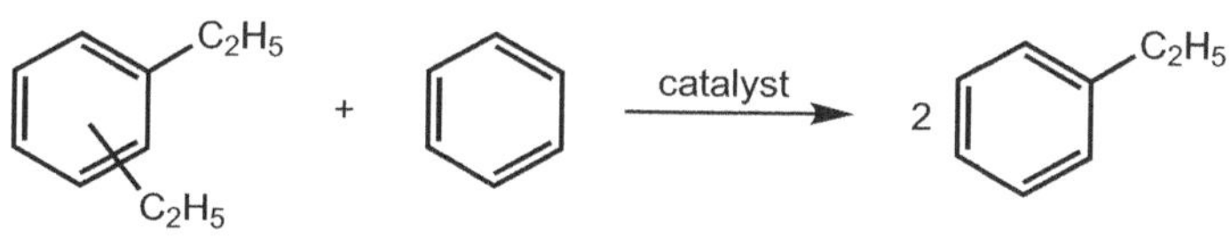

(5.105)

Nafion-H appears to be a very useful catalyst for transalkylation reaction as indicated in these studies. Transalkylation of benzene with diethylbenzenes, as well as with diisopropylbenzene, is efficiently catalyzed by Nafion-H in a flow system. The efficiency of the catalyst is, however, more limited when the transferring group is a methyl group.[268] Beltrame and co-workers have also carried out[269] detailed mechanistic studies on the isomerization of xylenes over Nafion-H.

The use of Nafion-H in de-*tert*-butylation—in fact, trans-*tert*-butylation—has been extensively studied by Olah, Yamato, and co-workers. An early study established[275] that *tert*-butyl-substituted aromatics, when treated in the presence of a suitable acceptor compound, are easily de-*tert*-butylated over Nafion-H used in catalytic amount [Eq. (5.106)]. Additional compounds including substituted biphenyls, bibenzyls, and cyclophanes, along with a range of substituted *tert*-butylphenols, all gave de-*tert*-butylated products in high (80–97%) yields.

Table 5.22. Transethylation of Benzene with Diethylbenzene over Graphite-Intercalated Metal Halides.[112]

Metal Halide:	$AlCl_3$	$AlBr_3$	SbF_5
% Intercalated:	28.4	4.5	26.1
Time-on-Stream (h)	% Conversion (Based on Diethylebenzene)		
1	45.0	66.2	2.4
2	70.4	51.8	
3	66.9	41.4	
4	70.3	3.1	
5	56.9		
6	16.2		
7	13.8		
8	15.2		
9	12.7		
10	10.4		

Reaction temperature = 180°C, $[C_6H_6]/[C_6H_4Et_2]$ ratio = 4.

2-Amino- and 2-acetamino-4-*tert*-butyltoluene, in turn, are unreactive. In the case of 2,2′,6,6′-tetramethyl-4,4′-di-*tert*-butyldiphenylmethane, trans-debenzylation competes with trans-*tert*-butylation. It was also observed in the ring closure of 2,2′-dihydroxybiphenyls[276] (see Section 5.14.1.1).

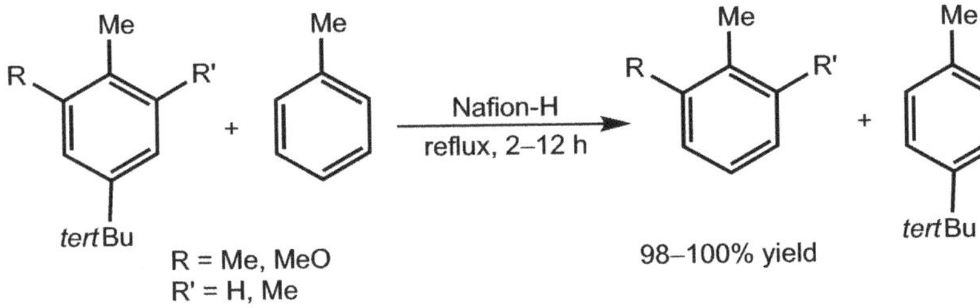

(5.106)

Subsequently, Yamato and co-workers[277–280] successfully applied trans-*tert*-butylation in the multi-step synthesis of various polycyclic aromatic compounds, where the *tert*-butyl group served as a positional protective group and could be easily removed in the final step. The method was also applied to introduce the *tert*-butyl group into a suitable position in the aromatic ring using 2,6-di-*tert*-butyl-*para*-cresol[281] [Eq. (5.107)]. The de-*tert*-butylation of *para*-*tert*-butylcalix[4]arene has also been reported[282] and the process (with toluene under reflux or *para*-xylene at 120°C) was found to be more convenient than the widely employed $AlCl_3$-catalyzed reaction. In addition, the use of Nafion-H allows the isolation of partially de-*tert*-butylated products.

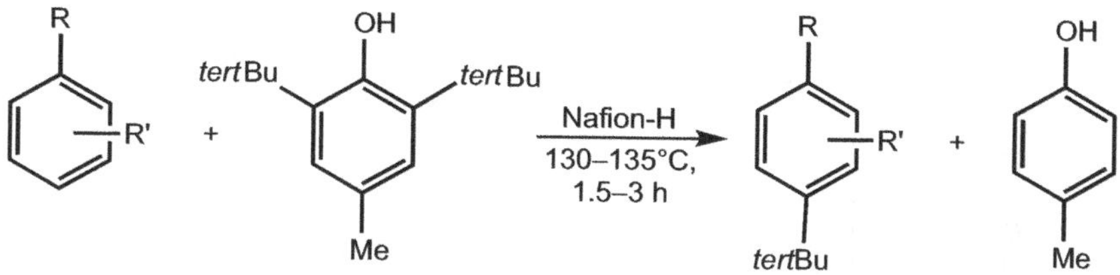

R = Me, Et, Pr, isoPr, *tert*-Bu, MeO
R' = H, 2-Me, 3-Me

67–93% yield
89–100% *para* selectivity

(5.107)

Nafion-H is also very efficiently catalyzes the rearrangement of anisole, methylanisoles, and phenetole to ring-alkylated phenols and products of transalkylation when vapors of the alkyl aryl ethers are passed over it at temperatures higher than 160°C.[208,268] At these reaction temperatures, some of the starting alkyl phenyl ethers undergo cleavage of the alkyl group to give phenol.

Recently, the results of the isomerization and transalkylation of isomeric diethylbenzenes with benzene in the presence of triflic acid have been reported. The aim is to find the best condition for the preparation of ethylbenzene.[283–285] *ortho*-Diethylbenzene and benzene reacting in 1:1 molar ratio at 35°C gave ethylbenzene in 49% yield in 6 h.[285] An even higher yield was obtained with *para*-diethylbenzene (51% at 22°C), whereas *meta*-diethylbenzene produced ethylbenzene only in 29% yield.[283] Both decreasing temperature and decreasing diethylbenzene/benzene ratio resulted in decreasing yields.

5.2.7. Alkylation with Miscellaneous Reagents

In liquid-phase alkylations besides conventional Friedel–Crafts systems, superacids that are capable of forming stable carbocations and onium ions have also found applications. Olah,[128] in his extensive studies, has shown that alkyl halides readily ionize in SbF_5 to the corresponding alkylcarbenium hexafluoroantimonates. Tertiary and some secondary carbocations are remarkably stable in solutions of SO_2–SbF_5 and SO_2ClF–SbF_5, respectively. These systems were also found to be highly efficient aromatic alkylating agents[286] (their ability to alkylate saturated aliphatic hydrocarbons was discussed in Section 5.1).

Methyl fluoride and ethyl fluoride form stable addition complexes with SbF_5, which are powerful methylating and ethylating agents, respectively.[143,144] In the study of alkyl halide–antimony pentafluoride systems, Olah and DeMember[287] found that dialkylhalonium ions R_2X^+ are formed when 2 mol (or excess) of an alkyl halide (except fluoride) are reacted with SbF_5 (see Section 4.2.4) in SO_2 or SO_2ClF solution. The alkylation of aromatic hydrocarbons, such as benzene, toluene, and ethylbenzene, has been studied with highly electrophilic dimethylhalonium and diethylhalonium hexafluoroantimonates (**72a** and **72b**) in SO_2ClF solution under superacidic conditions[288] [Eq. (5.108)]. Data for alkylations are summarized in Table 5.23.

Table 5.23. Alkylation of Benzene, Toluene, and Ethylbenzene with Dimethyl- and Diethylhalonium Fluoroantimonates in SO_2ClF[288]

Halonium Ion		Temperature (°C)	Time (min)	k_T/k_B	Isomer Distribution (%) *ortho*	*meta*	*para*	*ortho/para*
Me_2Cl^+	Toluene	25	10		46.6	27.2	26.7	1.75
		0	1		51.8	16.2	32.1	1.61
		−50	5		52.3	15.7	32.0	1.63
		−50	2		58.6	13.0	28.4	2.06
Me_2Br^+	Toluene	−50	5		57.8	9.5	32.7	1.76
		−50	2		59.0	8.6	32.4	1.82
Me_2I^+	Toluene	25	10		46.2	15.6	38.1	1.27
		0	10		53.9	11.8	34.3	1.57
		−20	60			No reaction		
Et_2Cl^+	Benzene–Toluene	−78	1	4.9	33.4	28.1	38.5	0.86
		−78	2.5	4.8	31.4	24.6	44.0	0.75
	Ethylbenzene–Toluene	−78	2.5	1.1	31.9	19.3	48.8	0.62
Et_2Br^+	Benzene–Toluene	−78	1	4.0	38.7	19.3	42.0	0.92
		−78	5	4.5	36.0	18.2	45.8	0.78
	Ethylbenzene–Toluene	−78	5	1.2	32.8	14.5	52.8	0.62
					(25.2)	(19.4)	(55.4)	(0.45)
Et_2I^+	Benzene–Toluene	−45	5	4.1	44.0	10.2	45.8	0.96

All data are the average of at least three parallel experiments.

$$\mathrm{ArH} + \overset{+}{\mathrm{R}}\mathrm{XR}\ \mathrm{SbF_5X^-} \longrightarrow \mathrm{ArR} + \mathrm{RX} + \mathrm{HSbF_5X} \qquad (5.108)$$

72a $R = CH_3$
72b $R = C_2H_5$ $X = Cl, Br, I$

Dimethyl-chloronium and -bromonium ions give similar methylation results and are quite reactive even at temperatures as low as −50°C. The dimethyliodonium ion is less reactive and alkylates benzene and toluene in SO_2ClF solution only when allowed to react (if necessary under pressure) at or above 0°C.

The ethylation of toluene by diethylhalonium ions gives ethyltoluenes with *ortho*: *para* isomer ratios between 0.60 and 0.96. The *ortho*: *para* isomer ratios obtained for the alkylation of toluene in conventional Friedel–Crafts ethylations range from 1.17 to 1.84 (average ~1.60). Such differences are considered to be due to the steric ortho effect caused by diethylhalonium ions, and are in accordance with the most probable displacement reaction on the bulky diethylhalonium ions by the aromatic substrate. This can be envisioned to proceed through an S_N2-type transition state involving no free alkyl cations [Eq. (5.109)].

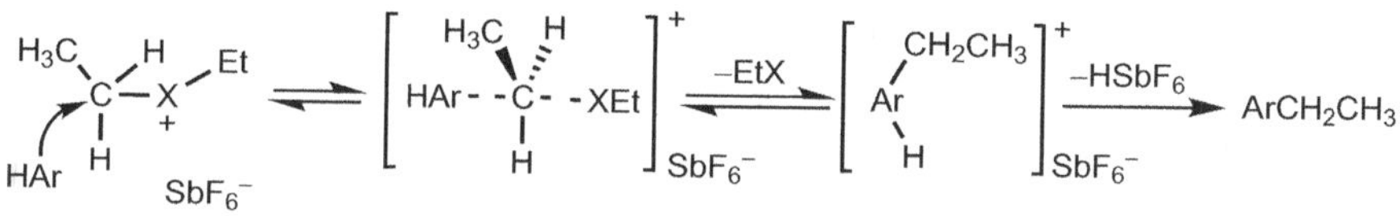

(5.109)

The alkylation data obtained from the reaction of dimethyl- and diethylhalonium ions provide evidence for direct alkylation of aromatics by dialkylhalonium ions. In addition, the data also indicate that dialkylhalonium ions are not necessarily involved as active alkylating agents in general Friedel–Crafts systems, although some of the reported anomalous alkylation results, particularly with alkyl iodides, could be attributed to reaction conditions favoring dialkylhalonium ion formation.

Vol'pin and co-workers[289] used the superelectrophilic reagent CBr_4–$n AlBr_3$ to alkylate pentafluorobenzene a deactivated arene with propane to form the isopropylated derivative in almost quantitative yield (Scheme 5.32). In the absence of propane, tribromomethylation takes place in a slow reaction. The suggested mechanism involves cation $C_6F_5^+$ formed by hydride abstraction induced by CBr_3^+. The formed intermediate cation then reacts with either propane or propene also formed by hydride abstraction to give the alkylated product. Pentafluorobenzene was also alkylated with perfluoroindane promoted by SbF_5 to yield perfluoro-1-phenylindane in 61% yield (C_6F_6 solvent, 22°C, 3.5 h).[290]

A variety of substituted aromatics have recently been found to be effective in alkylations. Phenols, naphthols, and their ethers,[291,292] 5-amino-1-naphthol,[293] hydroxyquinolines[294,295] and hydroxyisoquinolines[295,296] [Eq. (5.110)] are activated

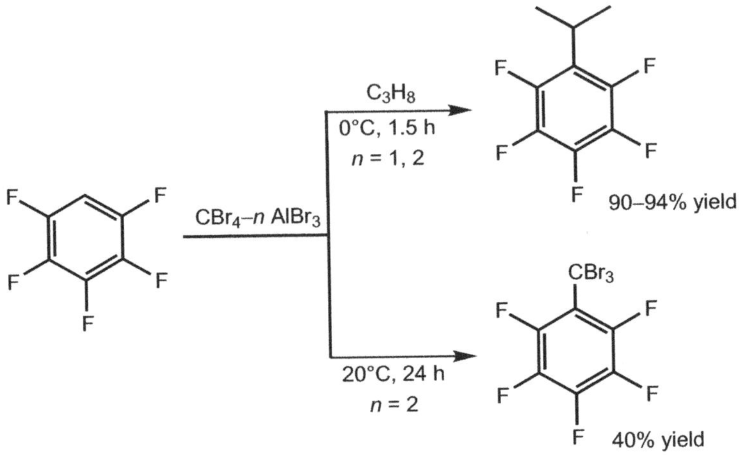

Scheme 5.32

under superacidic conditions (HF–SbF_5, $AlCl_3$ or $AlBr_3$) to form intermediate dications, which serve as the alkylating agents to produce partially saturated cyclic carbonyl compounds.

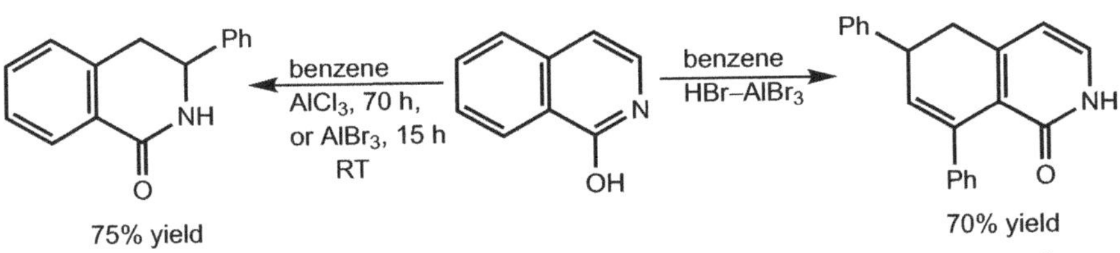

(5.110)

Quinoline and isoquinoline react in an analogous manner with benzene in conjugated superacids (HCl–$AlCl_3$, HBr–$AlBr_3$) to yield predominantly *cis*-5,7-diphenyl-5,6,7,8-tetrahydroquinoline and *cis*-6,8-diphenyl-5,6,7,8-tetrahydroisoquinoline, respectively (75–94% yield).[297] The regioselectivity observed corresponds to the most stable dicationic intermediates **73** and **74**. Unsaturated imides[298] exhibit similar behavior in triflic acid (Scheme 5.33).

Klumpp and co-workers have recently shown[299] that 2-oxazolines, which are prone to ring opening, when bearing a phenyl substituent at C(5) are capable of alkylating weak nucleophiles under superacidic conditions to yield phenyl-substituted amides [Eq. (5.111)]. Benzene and even *ortho*-dichlorobenzene could be alkylated in excellent yield. The diprotonated ring-opened intermediate **75** has been invoked to interpret the reactions.

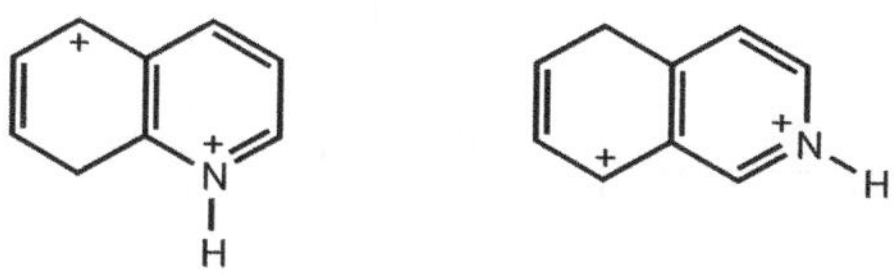

73 74

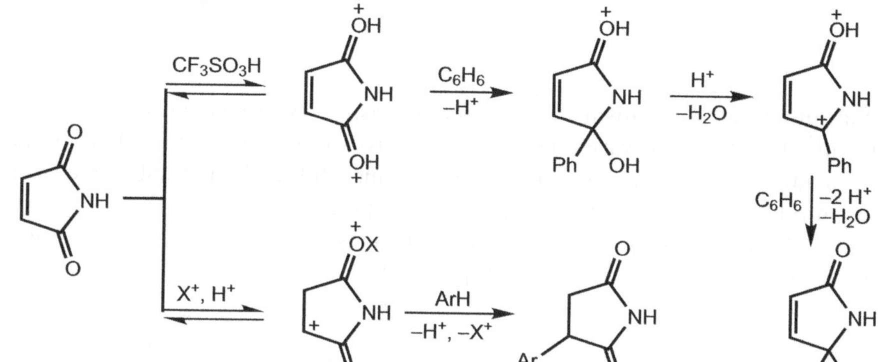

X = H, $Al_nCl_{3n}^-$, $Al_nBr_{3n}^-$
Ar = Ph, 4-MeC_6H_4, 3,4-diClC_6H_4

Scheme 5.33

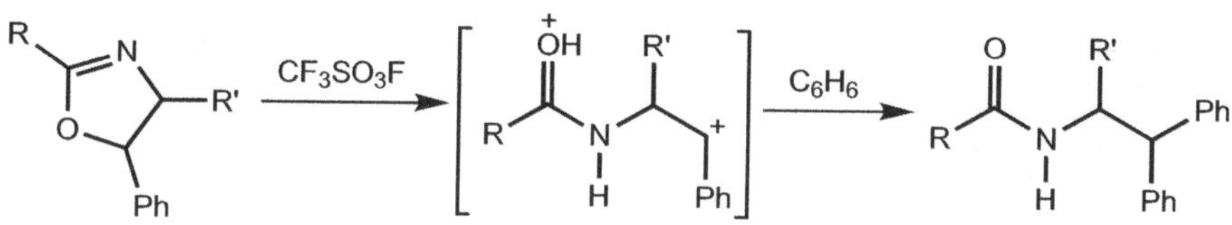

75

40–99% yield

R = Me, Et, Ph
R' = H, Me, CH_2OH, CH_2OMe, Ph

(5.111)

Protonated phenols and phenol ethers formed in superacids can be trapped by aromatics (benzene, naphthalene, tetrahydroquinoline).[292,300] The products are either cyclohexenone derivatives[301] [Eq. (5.112)] or aryl-substituted phenols. In the reaction of phloroglucinol with benzene, the diphenyl-substituted derivative is the main product [Eq. (5.113)], whereas 1,3,5-trimethoxybenzene gives selectively the monophenyl derivative (80% yield). Protonated dicationic species, such as **76**, detected by Olah and Mo[302] using NMR, were suggested to be intermediates in these processes.

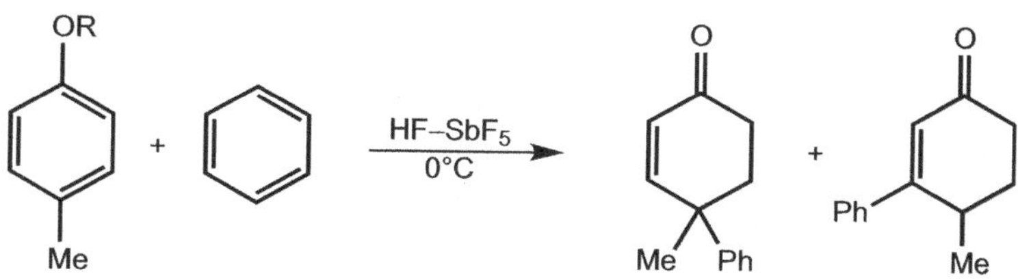

(5.112)

R = H, 1.5 min	29%	33%
R = Me, 15 min	2–3%	90%

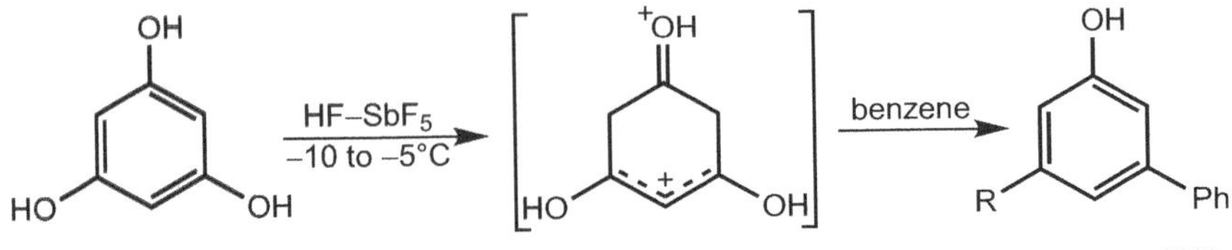

76

R = OH 10%
R = Ph 75%

(5.113)

Substituted alkynes (disubstituted propynones, 3-arylpropyonates) are used to perform alkenylation of varied benzene derivatives in the presence of HSO_3F and triflic acid.[303–305] It is known[306] that arylpropyonic acids and esters substituted with electron-releasing groups on the aromatic ring are protonated in HSO_3F at C(2) to give the unstable vinyl cation **77**, which, in turn, alkylates aromatic compounds to furnish isomeric alkenes in moderate yields [Eq. (5.114)].

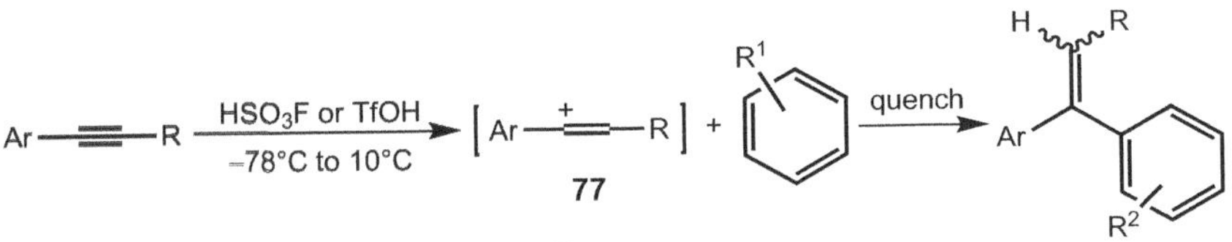

77

Ar = Ph, 4-HlgC_6H_4, 4-MeC_6H_4, 4-MeOC_6H_4, 2,4-diMeC_6H_3, 2,4,6-triMeC_6H_2, 2,3,5,6-tetraMeC_6H
R = Me, COOH, COOMe, COOEt, CN
R^1 = MeO, 3-MeO-1,2,4,5-tetraMeC_6H, 3-FSO_2-1,2,4,5-tetraMeC_6H, 3-CN-1,2,4,5-tetraMeC_6H

R^2 = 4-MeO, 4-MeO-2,3,5,6-tetraMeC_6H, 4-FSO_2-2,3,5,6-tetraMeC_6H, 4-CN-2,3,5,6-tetraMeC_6H

(5.114)

Depending on the substitution pattern and reaction conditions, however, dimerization may take place; that is, the intermediate vinyl cation alkenylates another acetylenic molecule (self-alkenylation)[305,307] [Eq. (5.115)].

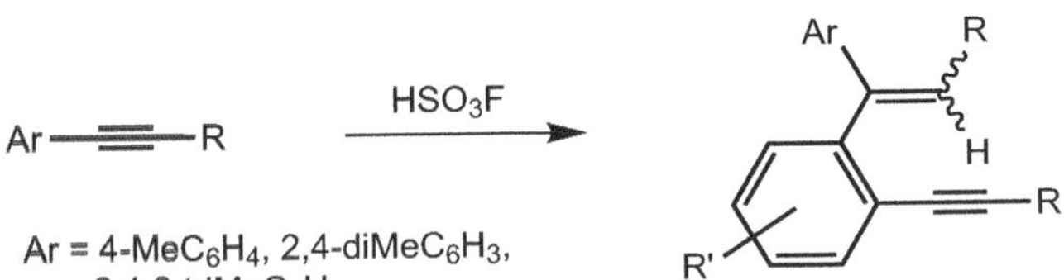

(5.115)

Ar = 4-MeC_6H_4, 2,4-diMeC_6H_3, 2,4,6-triMeC_6H_2,
R = COMe, COPh, COOMe, $PO(OEt)_2$
R' = 4-Me, 2,4-diMe, 2,4,6-triMe

Alkynes bearing an adjacent *N*-heterocycle, such as 5-ethynyl-1-methyl-imidazole and 1-propargylbenzotriazole, readily alkylate benzene in the presence of triflic acid to yield diphenyl-substituted products.[308] Isomeric ethynylpyridines exhibit distinct differences in reactivity: 3-ethynylpyridine exhibits the highest reactivity, whereas 2-ethynylpyridine is the least reactive [Eq. (5.116)]. This is consistent with the involvement of dicationic intermediates as the *de facto* alkylating agents. Indeed, calculations showed (B3LYP/6-311G** level), that of the three dications with vinyl cation moiety, **78** is the most stable formed most easily.

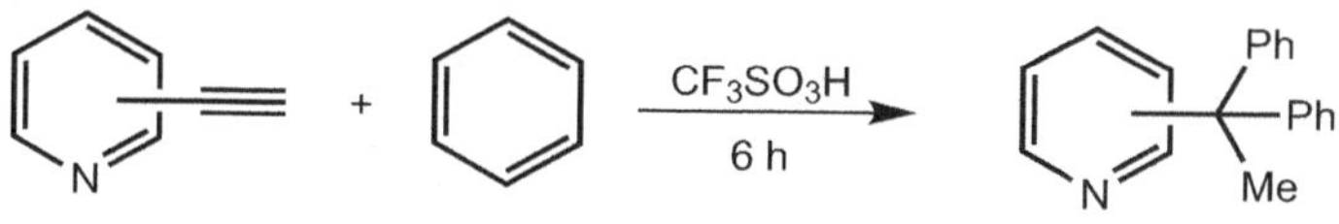

	25°C	80°C
2-ethynylpyridine	5%	5%
3-ethynylpyridine	84%	94%
4-ethynylpyridine	5%	75%

(5.116)

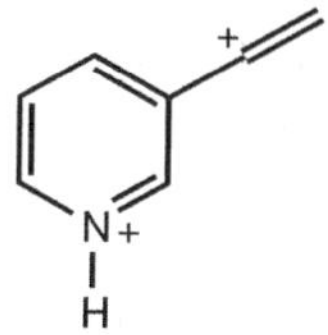

78

5.2.8. Cyclialkylation

Electrophilic ring closure of aryl-substituted compounds such as alkenes, halides, alcohols, and carbonyl compounds called cyclialkylation can be induced by conventional Friedel–Crafts catalysts[309] and by superacids. Examples are also known in which an intermolecular alkylation step is followed by intramolecular alkylation of the intermediate to furnish a cyclic product.

A range of diols and cyclic ethers were used to carry out alkylation of aromatics (benzene, toluene, xylenes, trimethylbenzenes, naphthalene) in the presence of triflic acid.[204,310] In a recent study,[311] various methyl-substituted benzene derivatives were alkylated with 1,4-diols [Eq. (5.117)] to form substituted tetralin derivatives in high yields. The transformations involve an intermolecular alkylation step followed by intramolecular alkylation (cyclialkylation). 2,2,5,5-Tetramethyltetrahydrofuran is similarly effective. For example, it alkylates benzene to give octamethyloctahydroanthracene (98% yield) and reacts with naphthalene to yield octamethyloctahydrotetracene [Eq. (5.118)].

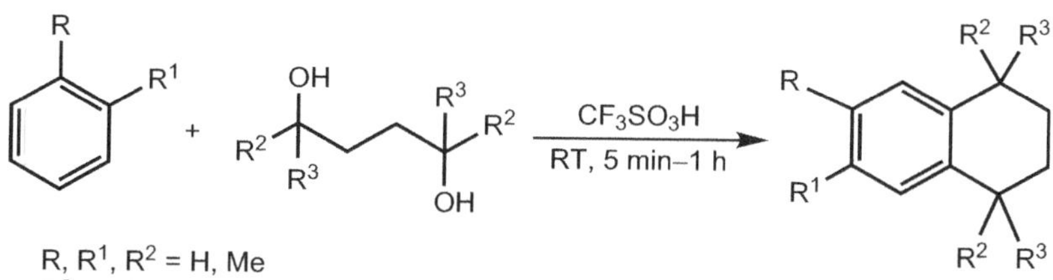

R, R^1, R^2 = H, Me
R^3 = Me

51–95% yield

(5.117)

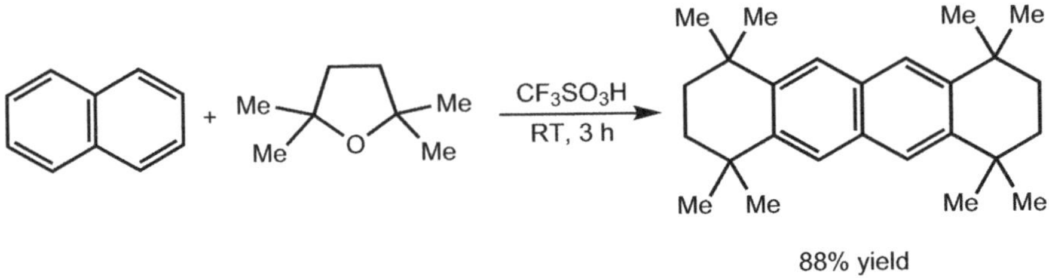

(5.118)

Török and co-workers[312] have reported the one-pot synthesis of *N*-arylsulfonyl heterocycles through the reaction of primary aromatic sulfonamides with 2,5-dimethoxytetrahydrofuran. When triflic acid is used in catalytic amount, *N*-arylsulfonylpyrroles are formed (Scheme 5.34). Equimolar amount of triflic acid results in the formation of *N*-arylsulfonylindoles, whereas *N*-arylsulfonylcarbazoles are isolated in excess acid (Scheme 5.34). In the reaction sequence 1,4-butanedial formed in situ from 2,5-dimethoxytetrahydrofurane reacts with the sulfonamide to give the pyrrole derivative (Paal–Knorr synthesis). Subsequently, one of the formyl groups of 1,4-butanal alkylates the pyrrole ring followed by a second, intramolecular alkylation (cyclialkylation) step.

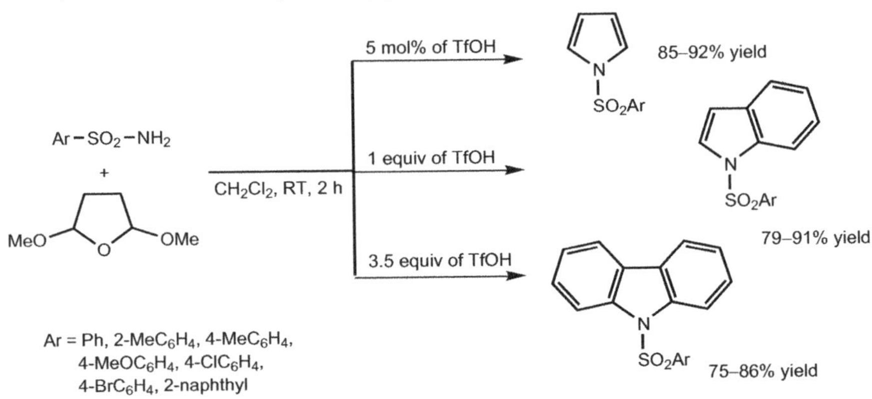

Scheme 5.34

Intramolecular alkylation of aryl ketones with the concomitant elimination of water, also called cyclodehydration,[313] has been studied by Shudo and co-workers.[314] Cyclodehydration of 1,3-diphenylpropanones to give 1-phenyl-1*H*-indenes [Eq. (5.119)] shows strong dependence on acidity.

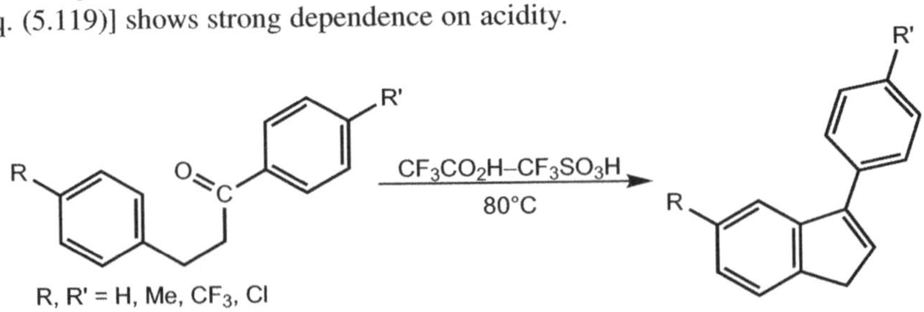

(5.119)

Table 5.24. Acidity Dependence of Cyclodehydration of 1,3-Diphenylpropanones[314]

Substrate (**79**)					
R	R′	Acids	$-H_0$	Time	Yield (%)
H	H	TFA	2.7	17 h	0
		95% TFA–5% TfOH	9	17 h	7
		TfOH	14	17 h	72
H	Me	95% TFA–5% TfOH	9	144 h	16
		TfOH	14	144 h	86
H	CF_3	95% TFA–5% TfOH	9	20 min	2
		TfOH	14	20 min	68
H	Cl	95% TFA–5% TfOH	9	18 h	22
		TfOH	14	18 h	84
Me	H	95% TFA–5% TfOH	−9	5.5 h	28
		TfOH	14	5.5 h	84

The reaction of the parent compound did not take place in trifluoroacetic acid (TFA), but did proceed on the addition of triflic acid to give the alkylated product in low yield (Table 5.24). In neat triflic acid the reaction is faster and the product is formed in much higher yield. Similar changes were observed with the other 1,3-diphenylpropanones. Substituents on the benzoyl group also induce significant differences in reactivities (compare the results for R′ = Me and CF_3 in Table 5.24). These observations suggest that the protosolvated dicationic species **79** is involved in the rate-determining step of cyclization (Scheme 5.35).

Cyclodehydration has been used to prepare the antihistamine precursor **80** under superacid conditions[315] [Eq. (5.120)].

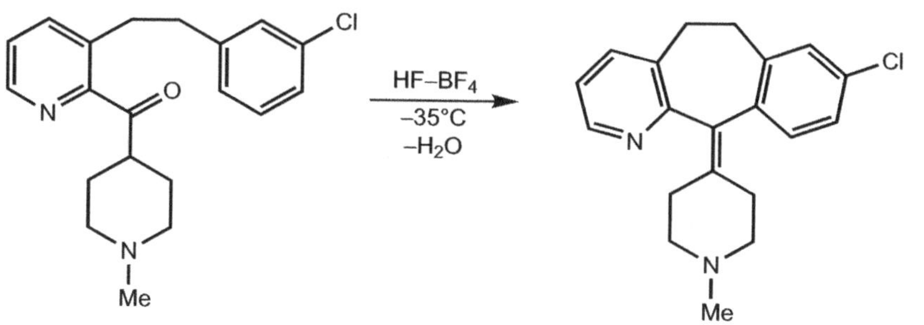

80 91% yield

(5.120)

The related transformation of 1-phenyl-2-propen-1-ones **81** (a Nazarov-type cyclization) [Eq. (5.121)] has also been studied by Shudo and co-workers.[316] The acidity dependence (Table 5.25) and, in particular, the linear relationship between rates and acidity values strongly suggest the involvement of *O*,*O*-diprotonated

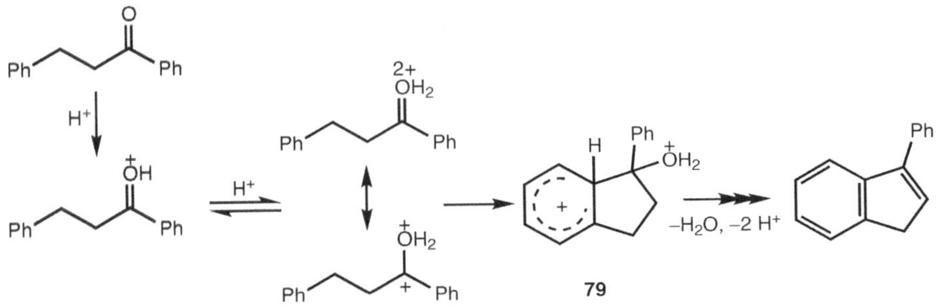

Scheme 5.35

intermediates. In agreement with earlier observations, an electrocyclization mechanism rather than Friedel–Crafts type reaction is suggested to be operative. According to calculations (B3LYP/6-31G*+ZPE), the energy barrier of the electrocyclization of the carbenium–oxonium dication of the parent compound through transition state **82** is much lower than that of the monocation through transition state **83** (12.4 kcal mol^{-1} versus 24.9 kcal mol^{-1}). Appropriate substitution ($R' = Me$) lowers the energy barrier even further (9.0 kcal mol^{-1}). A similar study[317] with respect to the cyclization of hydroxycarbonyl compounds **84** likewise indicated the involvement of diprotonated species **85** [Eq. (5.122)]. 3-Arylbenzofurans affording the same intermediate show the same transformation. Subsequently, the substituent effects of electrocyclization of similar systems were also studied to explore the synthetic utility of the reaction.[318]

Table 5.25. Acidity Dependence of Cyclization of 1-Phenyl-2-propen-1-ones[316]

Substrate (**81**) R	R′	Acids	$-H_0{}^a$	Time (h)	Temperature (°C)	Yield (%)
H	H	94% TFA–6% TfOH	8.7	120	25	51
		TfOH	12.7	120	25	63
H	Ph	TFA	2.7	5	0	0
		94% TFA–6% TfOH	8.7	5	0	59
		TfOH	12.7	5	0	98
CF_3	Ph	70% TFA–30% TfOH	10.6	5	0	61
		TfOH	12.7	5	0	93
Me	Ph	94% TFA–6% TfOH	8.7	5	0	48
		TfOH	12.7	5	0	91
H	4-$CF_3C_6H_4$	94% TFA–6% TfOH	8.7	5	0	43
		TfOH	12.7	5	0	99
H	4-MeC_6H_4	94% TFA–6% TfOH	8.7	5	0	78
		TfOH	12.7	5	0	100
H	Me	93% TFA–7% TfOH	8.9	2	25	49
		TfOH	12.7	2	25	97
H	Et	93% TFA–7% TfOH	8.9	3	25	54
		TfOH	12.7	3	25	97

[a] Acidity function values are corrected for the reaction media.

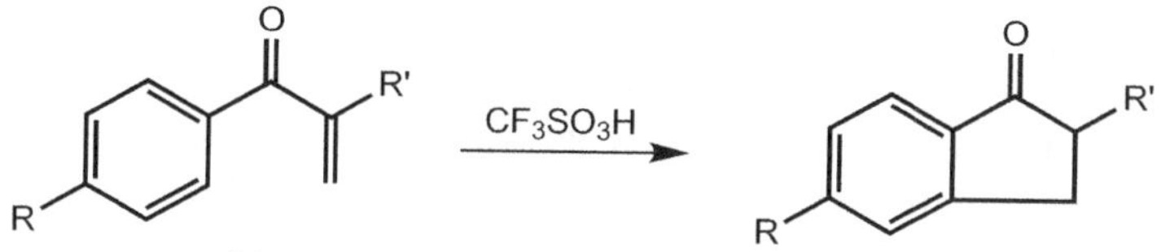

(5.121)

81

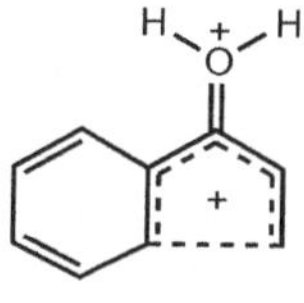

82

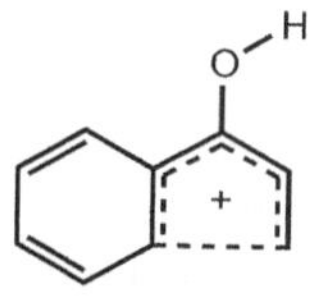

83

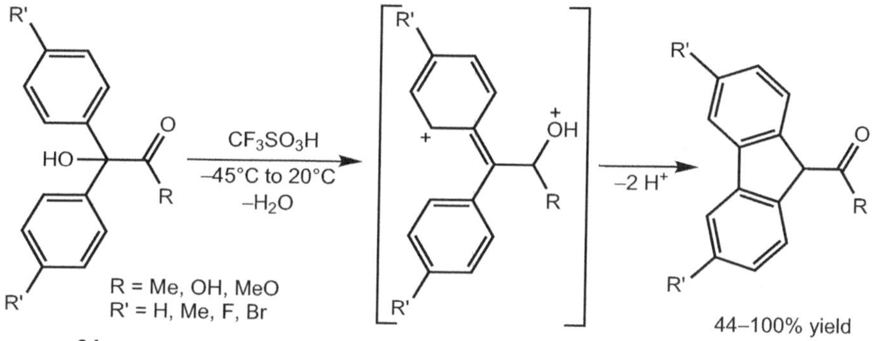

84

85

(5.122)

Coxon, Steel, and co-workers studied the transformations of a series of phenylalkanols in fluorosulfuric acid at low temperature to find a variety of reaction modes. Cyclization of 2-phenylethanols, in most of the cases, is accompanied by rearrangement to afford various polycyclic products.[319] The formation of propellane **86** was rationalized by the plausible mechanism shown in Eq. (5.123).

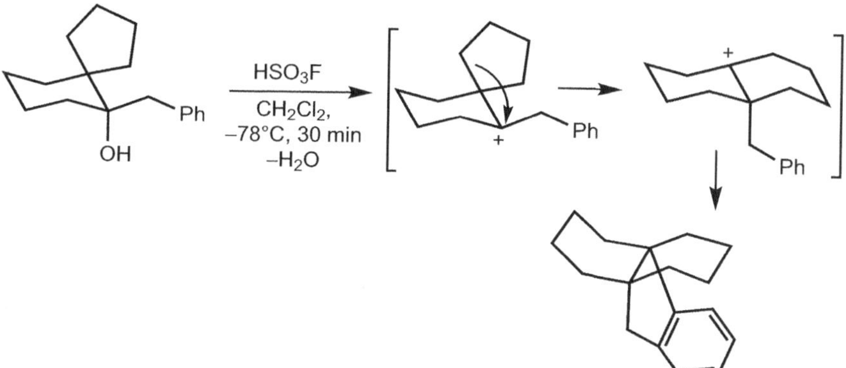

86 80% yield

(5.123)

3-Phenylpropanols generally undergo rearrangement prior to cyclization.[320] For example, the reaction of 1-(2-phenylethyl)cyclohexanol (**87**, R = H) in HSO_3F gives a mixture of two products (**88** and **89**) in a ratio of 1:3 in 40% yield (Scheme 5.36). The initially formed tertiary cation yields the spiro compound **88**. The 1,2-hydride shift to give the secondary cation, however, competes successfully with the first process and results in the preferential formation of the octahydrophenanthrene derivative **89**. This latter cyclization is stereoselective; this affords only the less stable *cis* isomer, indicating that it is a kinetically controlled product. In the transformation of the 2-methyl derivative (**87**, R = Me), only isomeric octahydrophenanthrenes were isolated (80% yield, **90a**/**90b** = 3:1). In this case, formation of the spiro compound is not competitive with hydride transfer to form the other tertiary carbocation. 4-Phenylbutanols undergo direct cyclization of the initially formed carbocation to give tetralin derivatives.

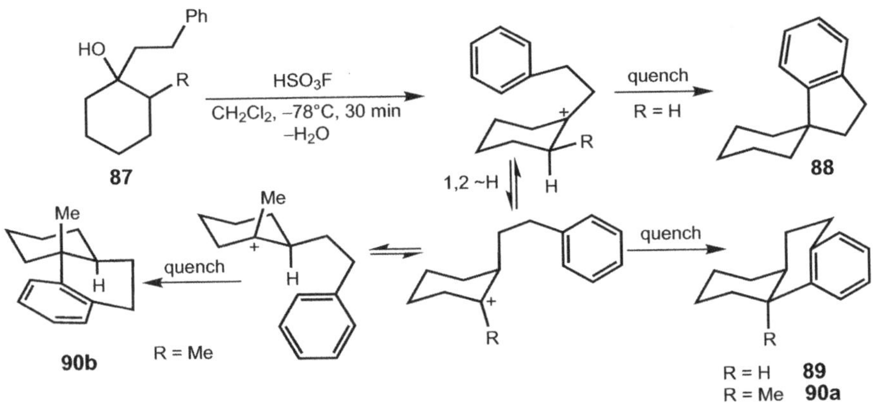

Scheme 5.36

Sommer and co-workers[321–323] have performed detailed studies of the cyclization of 1,3-diarylpropynones to form 3-arylindenones in various superacids (TfOH, TfOH–SbF_5, HF–SbF_5, HSO_3F; $-H_0 = 14$–20). In the superacid HSO_3F, the starting compounds with electron-withdrawing R substituents undergo protonation at the carbonyl oxygen to give stable carbocationic intermediates characterized by ^{1}H and ^{13}C NMR spectroscopy (−80°C and 0°C). Electron-releasing R groups, in turn, increase the electron-density of the ynone moiety and the intermediates undergo ring closure to 3-arylindenones [Eq. (5.124)]. *O*-Protonated and *C*-protonated intermediates **91a** and **91b** and *O*,*C*-diprotonated dication **92** have been proposed to participate in product formation.

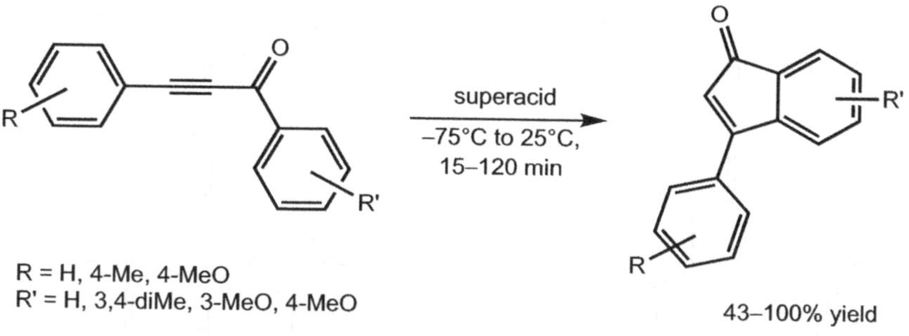

(5.124)

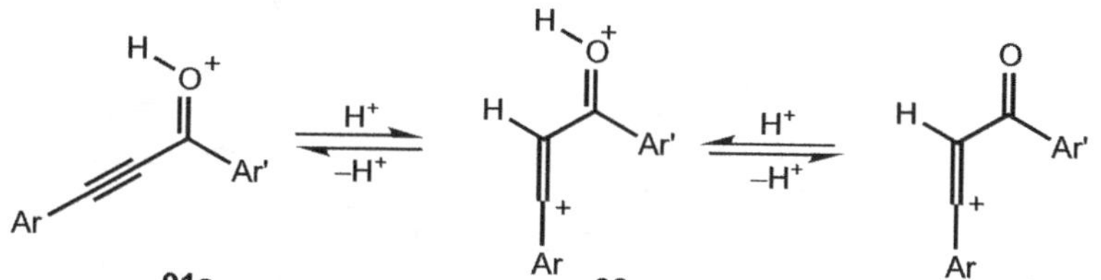

91a **92** **91b**

It has been reported in a follow-up study,[324] that the product 3-arylindenones underwent double protonation in strong superacids (HSO_3F, CF_3SO_3H) and the stability of formed species allowed their observation by NMR spectroscopy at room temperature. Although fast proton exchange prevented to observe *O*-protonation, ^{1}H and ^{13}C chemical shifts unequivocally showed the presence of dication **93**.

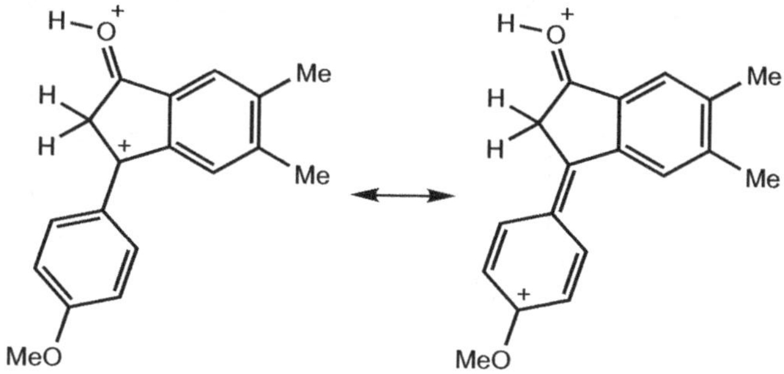

93

Shubin and co-workers[325] have generated long-lived cyclobutenyl cations and studied their varied rearrangements, which involve cyclialkylation steps. For example, cation **94** gives the isomeric pentacycle **95** under superacid conditions [Eq. (5.125)].

Further treatment of product **95** in triflic acids allows the isolation of the isomeric product via ring-opening–ring-closing reaction steps [Eq. (5.126)].

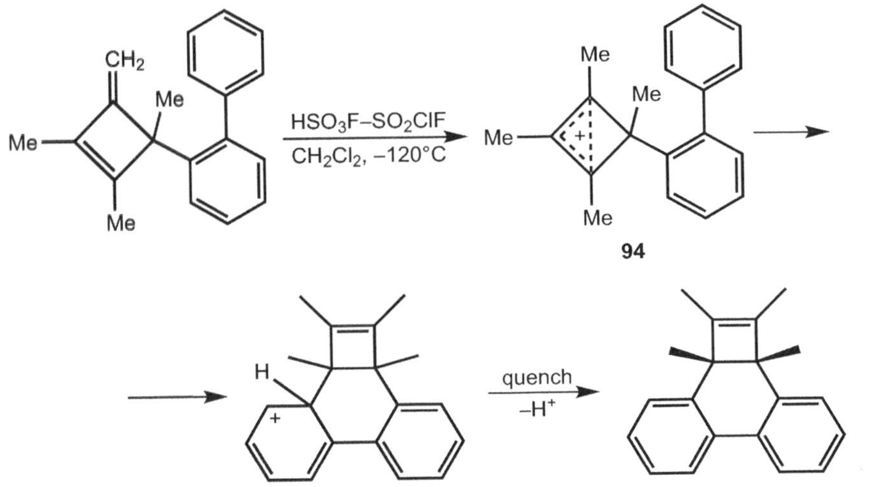

95 61% yield

(5.125)

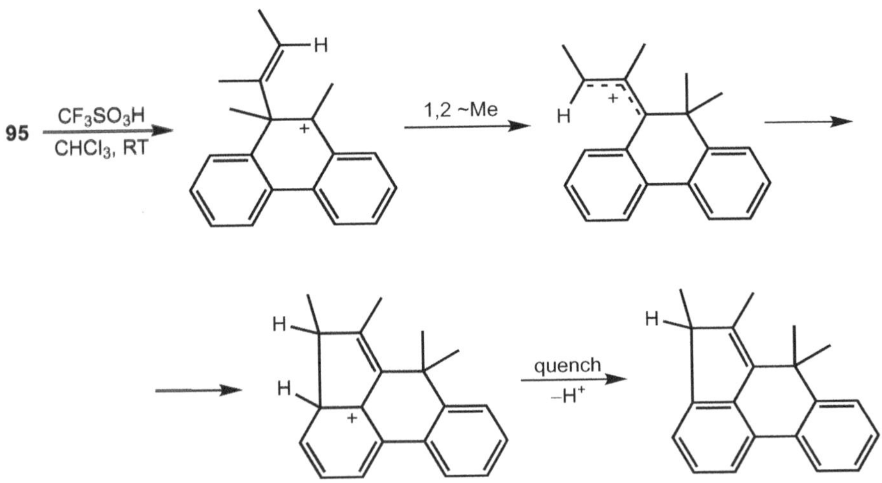

100% yield

(5.126)

Triflic acid has been used in the ring closure of allyl-substituted heterocycles to synthesize compounds **96** and **97**,[326] whereas isomeric compounds **98** was isolated in the reaction of propargyl-substituted benzylamines.[308]

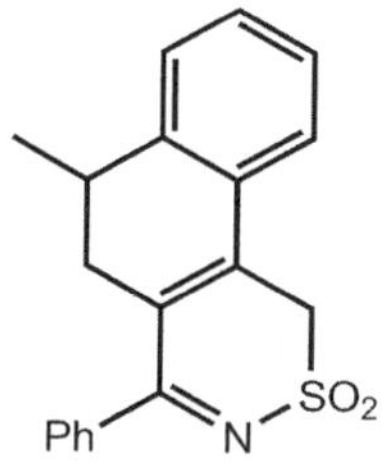

96

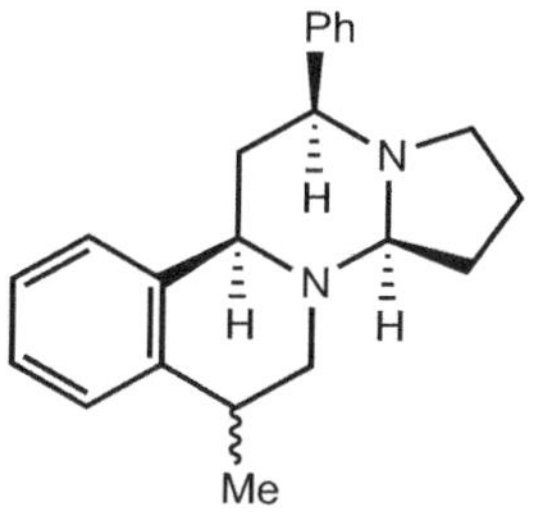

97

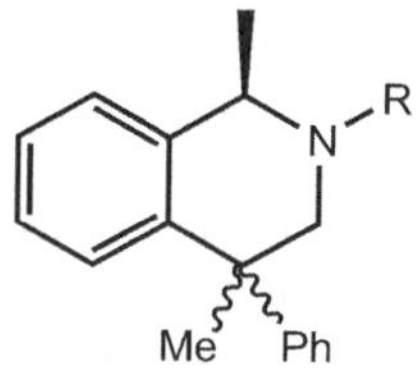

R = H, Me

98

Tricationic and dicationic intermediates were suggested to be involved in the intramolecular cyclization of 2-nitromethylene-1-phenylalkyl-substituted *N*-heterocycles with the participation of superelectorophilic hydroxynitrilium cation **99** to give tricyclic products in triflic acid[327] (Scheme 5.37). Likewise, the intramolecular trapping of the intermediate hydroxynitrilium cation affords the six- to nine-membered ring oximinoorthodithiolactones[328] [Eq. (5.127)].

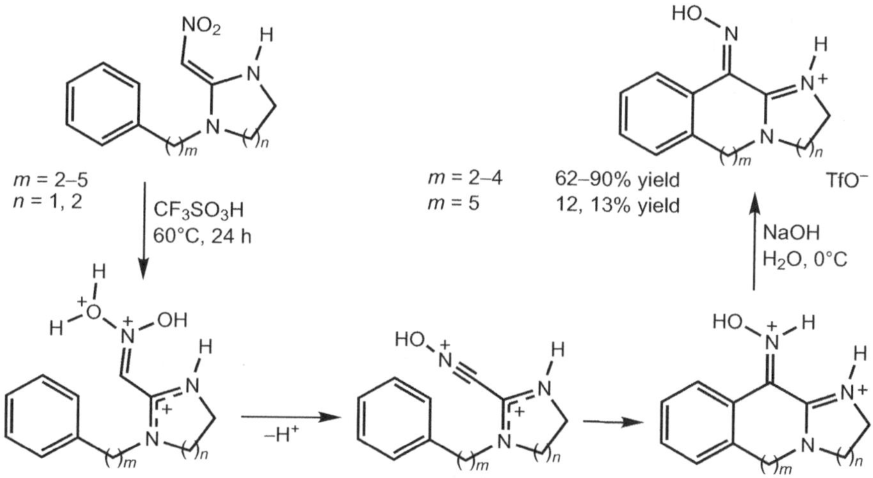

99

Scheme 5.37

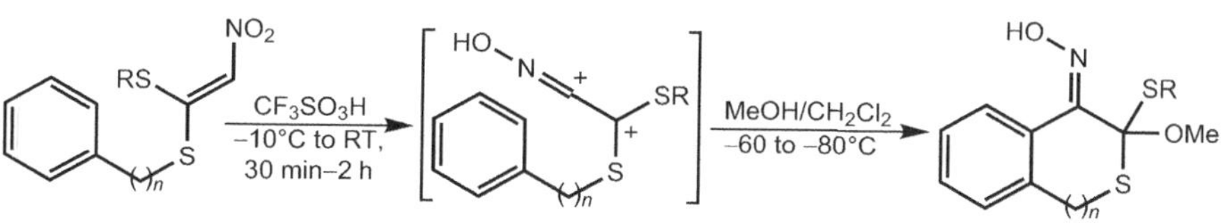

n =	1	2	3	4
R = Me	70	74	82	28% yield
R = $Ph(CH_2)_n$	83	73	71	12% yield

(5.127)

Construction of the 1,2,3,4-tetrahydroisoquinoline skeleton through cyclialkylation has also been achieved with triflic acid. Elevated temperature is required to transform aldimines of 2-arylethylamine (Pictet–Spengler cyclization) to substituted tetrahydroisoquinolines[329] [Eq. (5.128)]. Kinetic and acidity dependence studies and substituent effects led to postulation regarding the intervention of the dicationic species **100** in the rate-determining cyclization step. In a subsequent study,[330] high stereoselectivity was observed under superacidic conditions particularly for 2-alkyl-*N*-benzylidene derivatives as compared with the corresponding reactions in weak acid (TFA). *N*-Acyl enaminoketones were assumed to react with the involvement of an iminium ion intermediate to give the *N*-acyl tetrahydroisoquinoline derivatives in good yield[331] [Eq. (5.129)].

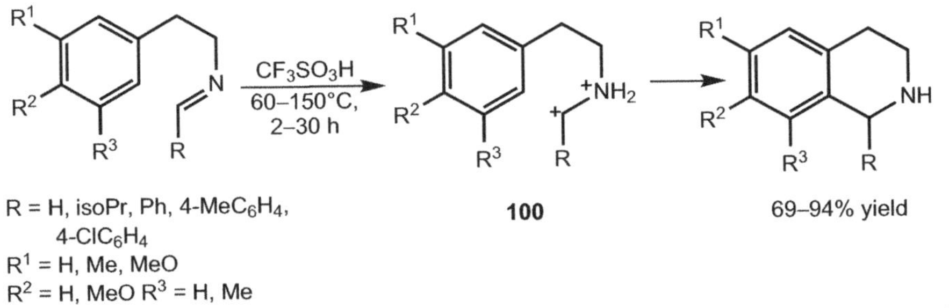

(5.128)

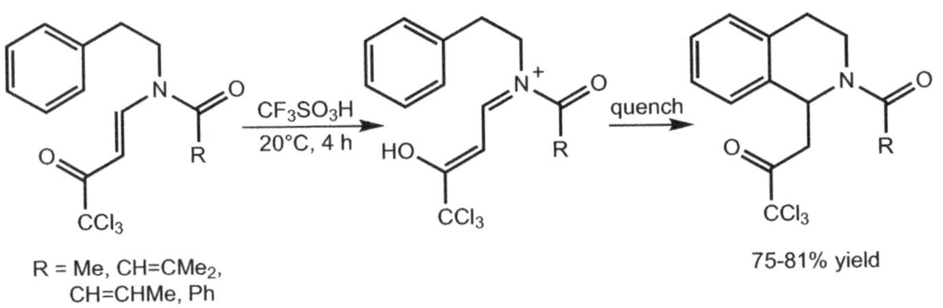

(5.129)

Cyclialkylation has been applied in the synthesis of *N*-heterotri- and tetracycles. Reimann et al.[332] have performed the stereoselective ring closing of the diastereomeric mixture of piperidinols **101** and isolated the product ergoline with a C/D *cis* configuration in low yield [Eq. (5.130)]. It was also observed, however, that one of the isomers afforded the product quantitatively under identical conditions, whereas the other slowly decomposed.

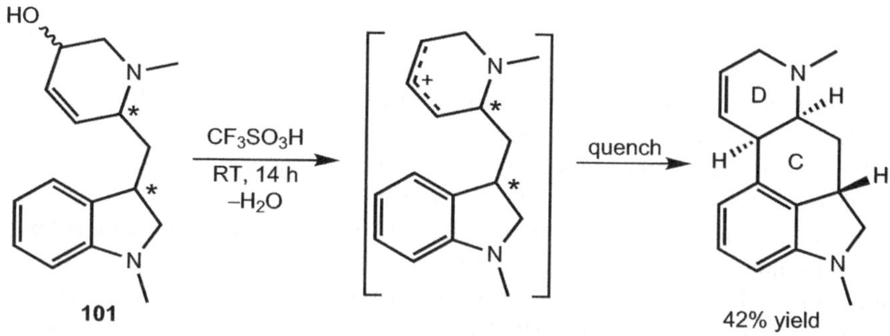

(5.130)

N-acyliminium ions, in general, are widely used in the formation of varied ring systems.[333] Triflic acid was applied in the synthesis of polycyclic compounds **102** and **103** starting from hydroxy lactam precursors,[334,335] whereas a lactone was transformed to compound **104**.[336]

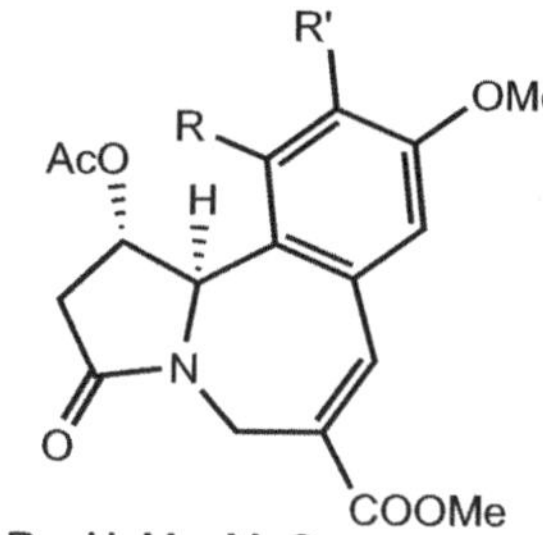

R = H, Me, MeO
R' = H, OAc

102

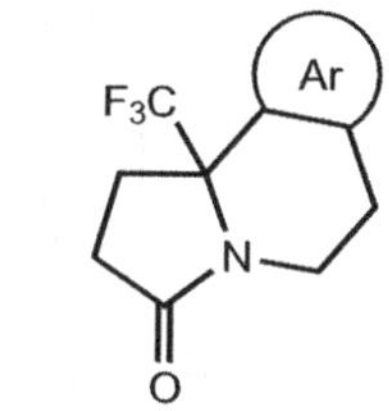

Ar = benzene, Cl-benzene, diMeO-benzene, naphthalene, indole, *N*-methylpyrrol, thiophene

103

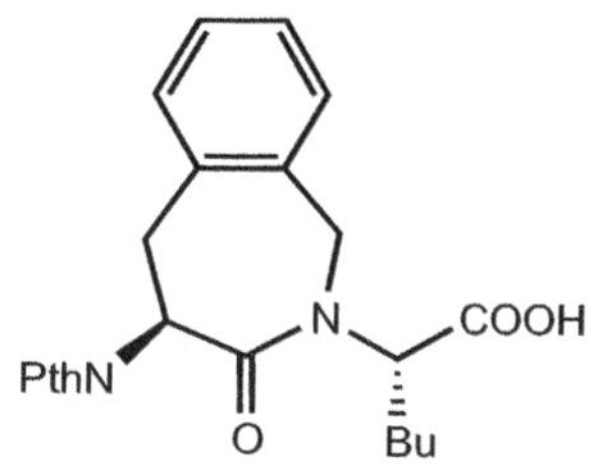

Pth = phthalyl

104

In a study by Klumpp and co-workers,[337] *N*-acyliminium ions were prepared in situ and ring closure was induced by the addition of triflic acid to the reaction mixture [Eq. (5.131)].

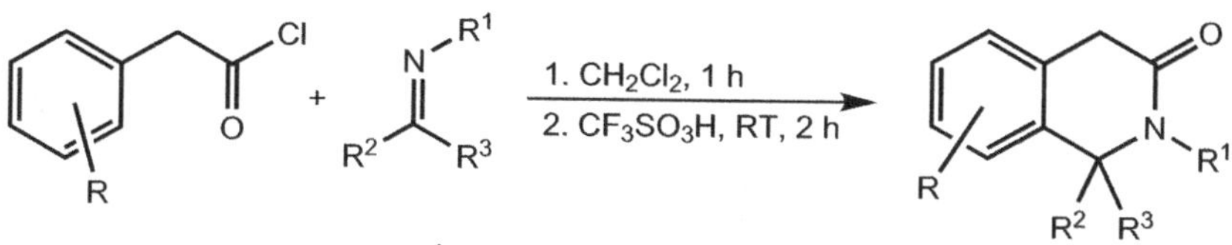

R = H, 4-F, 2,5-diMeO

R^1 = Me, Ph, Bn, 4-FC_6H_4, 1-naphthyl
R^2 = H, Ph
R^3 = Ph, 4-FC_6H_4, C_6F_5

37–97% yield

(5.131)

Detailed experimental and theoretical investigations were subsequently performed with 1-methyl-5-hydroxypyrrolidin-2-one.[338] It was shown that *para*-dimethoxybenzene, a stronger nucleophile, is able to react in CF_3COOH, whereas the deactivated aromatic *para*-dichlorobenzene is unreactive. However, it gives the alkylated product in 67% yield in the presence of excess triflic acid. These observations show that ion **105** formed in CF_3COOH can react with the activated aromatic, but further protolytic activation—that is, formation of protosolvated species **106a** or dication **106b**—is required to induce the transformation of the deactivated compound. Cation **105** was identified by ^1H and ^{13}C NMR spectroscopy under stable ion conditions (HSO_3F–SO_2ClF). In the even stronger acid HSO_3F–SbF_5 at −40°C, two sets of peaks appeared with all ^1H NMR resonances significantly deshielded. Notable is the chemical shift of the iminium proton at δ^1H 9.42. The carboxonium and iminium carbon signals of the fully protonated dication **106b** in the calculated gas-phase NMR spectrum are significantly deshielded from those in the experimental spectrum suggesting the formation of the protosolvated structure **106a**.

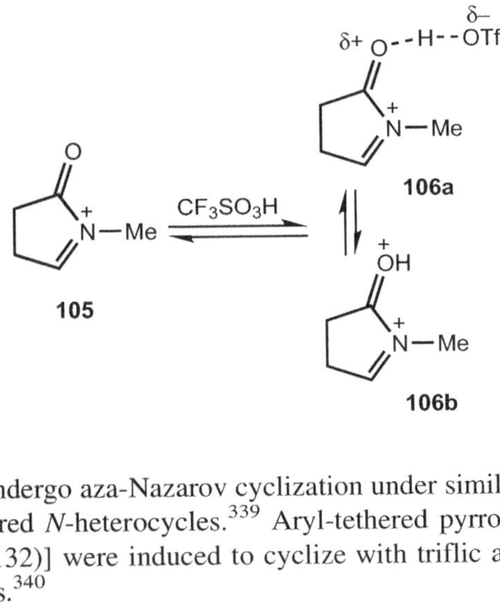

Acylium salts undergo aza-Nazarov cyclization under similar conditions to form varied five-membered *N*-heterocycles.[339] Aryl-tethered pyrrolinones and dihydropyridones [Eq. (5.132)] were induced to cyclize with triflic acid to afford tri- and tetracyclic products.[340]

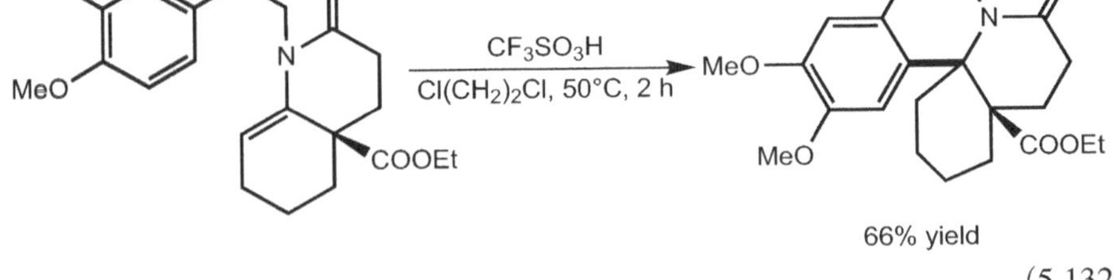

(5.132)

When 5- and 8-hydroxyquinolines and 5-hydroxyisoquinoline are used as alkylating agents (see Section 5.2.7) the primary alkylation products undergo a second, intramolecular alkylation to form methano-bridged compounds in low yields, which is due to the reversibility of the reactions[294,295] [Eq. (5.133)].

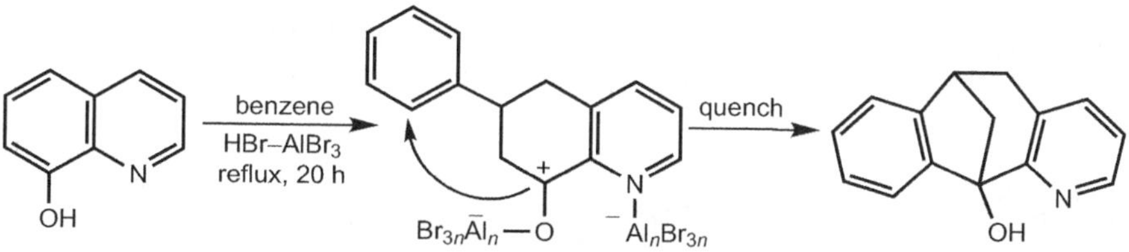

54% yield

(5.133)

The alkylation method developed by Fukuzawa et al.[245] (see Section 5.2.4) has been used in cyclialkylation to synthesize polycyclic aromatic hydrocarbons such as benzopentaphene **107** [Eq. (5.134)].[341]

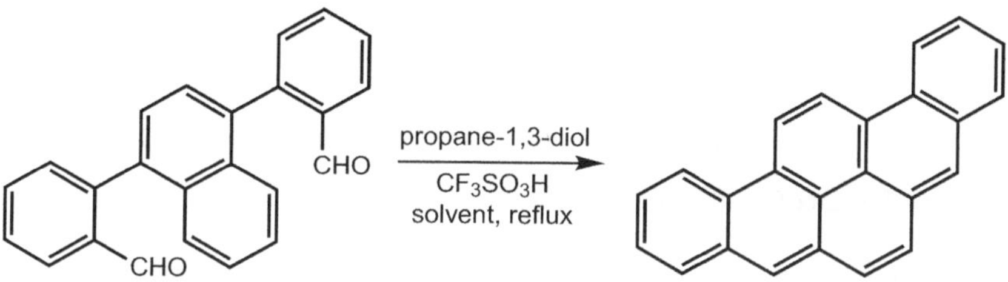

107

CH_2Cl_2	43%
1,4-dioxane	37%

(5.134)

Nafion-H has been shown to promote cyclialkylation of arylalkylepoxides to form tetralin derivatives[342] [Eq. (5.135)]. The reactions were performed by passing a solution of the epoxide in a solvent mixture (CH_2Cl_2, $CFCl_3$, 1,1,2,2-tetrafluoroethanol) through a 0.8-cm × 10-cm column packed with Nafion powder. Aryl-alkenes and -alkanols could also be used as starting materials.[343]

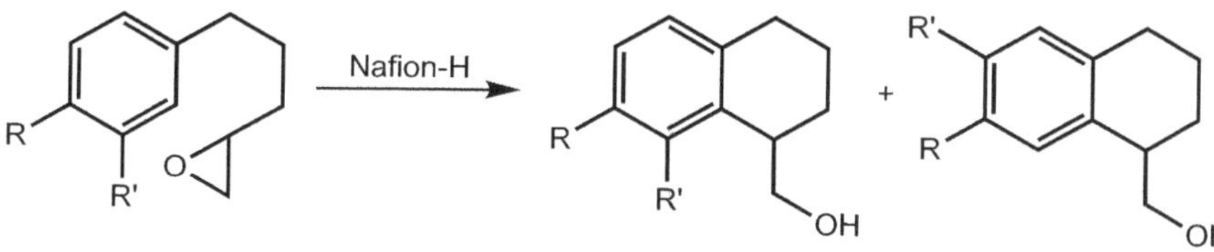

R = H, Me, MeO, F, Cl
R' = H, Me, MeO

R' = H	42–88% yield	
R = H, R' = Me	35%	43%
R = H, R' = MeO	31%	43%

(5.135)

Doyle et al.[344] and Wee and Liu[345] have reported the ring-closing transformation of α-diazoacetamides **108** and **109** to yield 2(3*H*)-indolinones over Nafion-H [Eq. (5.136)]. In the transformation of compounds **109** the electrophilic intramolecular substitution is followed by decarboxylation.[345] Small amounts of 2-azetidinone derivatives (4–10%) formed through a carbene intermediate were also detected. The yield of products from compounds **108** are even higher than observed in the presence of $Rh(OAc)_2$ often applied in the decomposition of diazo compounds.[344]

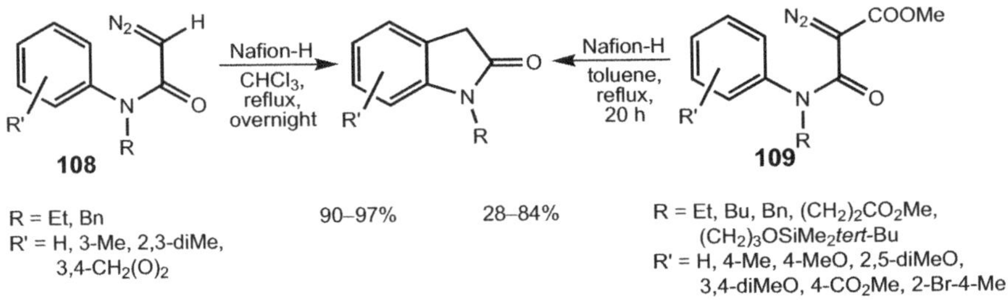

(5.136)

5.3. ACYLATION OF AROMATICS

Friedel–Crafts acylation of aromatics is of considerable practical value owing to the importance of aryl ketones and aldehydes as chemical intermediates.[346] Whereas alkylation of aromatics with alkyl halides requires only catalytic amounts of catalysts, acylation to ketones generally necessitates equimolar or even some excess of the Friedel–Crafts catalysts. Usually one molar equivalent of catalyst combines with an acyl halide, giving a 1:1 addition compound, which then acts as the active acylating agent [Eq. (5.137)].

$$\mathrm{RCOX} + \mathrm{AlX_3} \longrightarrow \mathrm{RCOX} \rightarrow \mathrm{AlX_3} \rightleftharpoons \mathrm{RCO^+}\ \mathrm{AlX_4^-} \qquad (5.137)$$

Evidence supporting the formation of 1:1 addition compounds is further substantiated by the actual isolation of stable acyl cation salts (Chapter 3). Therefore, it is highly desirable to develop methods in which only a catalytic amount of Friedel–Crafts acid catalyst may be used for effective conversion.

Effenberger and Epple[347] showed that alkylbenzenes are effectively acylated when ~1% triflic acid is added to the mixture [Eq. (5.138)] (Table 5.26). It was shown that when other Brønsted or Lewis acids were used, the yield decreased drastically (Table 5.27). Perfluorobutanesulfonic acid was found similarly effective (35–84% yields)[348] [Eq. (5.138)].

Table 5.26. Triflic Acid-Catalyzed Acylation of Aromatics with Acyl Chlorides (RCOCl)[347]

Reactants R	Reactants Aromatics	T (°C)	Time (h)	Product	Yield (%)	*o:p*
C_6H_5	Benzene	80	8.5	Benzophenone	14	
C_6H_5	Chlorobenzene	132	5	2- and 4-Chlorobenzophenone	13	1:3
C_6H_5	Toluene	110	48	2- and 4-Methylbenzophenone	85	1:2
C_6H_5	*para*-Xylene	138	6	2,5-Dimethylbenzophenone	82	
$4\text{-}NO_2C_6H_4$	Benzene	80	4	4-Nitrobenzophenone	82	
Me_3C	Anisole	154	12	*tert*-Bu-4-methoxyphenyl ketone	54	
Me_2CH	Anisole	154	0.2	isoPr-4-methoxyphenyl ketone	46	

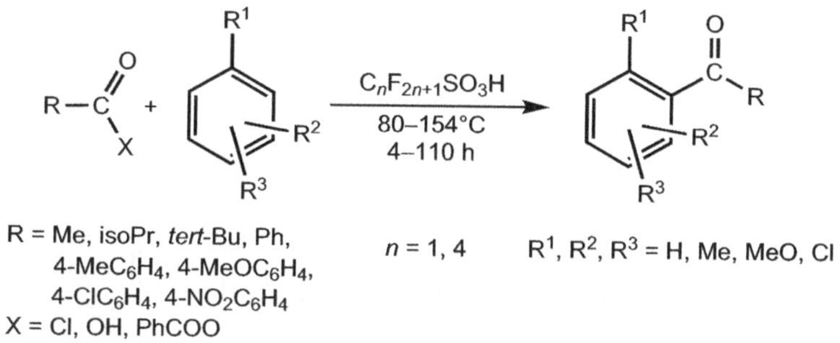

(5.138)

Acyl chlorides were also tested in acylations promoted by $B(OTf)_3$.[231] Acylation of benzene and toluene in competitive reactions (molar ratio = 5:1) with acetyl chloride shows high *para* selectivity (92–95% with 2.5–7% of *meta*, $k_T/k_B = 31\text{–}73$), whereas the *para* isomer is formed only with 72–75% selectivity (8–10% of *meta*, $k_T/k_B = 78$) in benzoylation with benzoyl chloride. Acetylation appears not to be affected by significant isomerization as indicated by isomer distributions and relative reactivity data.

Table 5.27. Catalytic Action of Brønsted or Lewis Acids in the Acylation of *para*-Xylene by Benzoyl Chloride[347]

Catalyst	Amount (%)	Temperature (°C)	Time (min)	Yield (%)
CF_3SO_3H	1	138	6	82
HSO_3F	1	138	6	20
$4\text{-}MeC_6H_4SO_3H$	1	138	6	31
H_2SO_4	1	138	6	28
$HClO_4$	1	138	6	14
CF_3COOH	2.6	138	10	21
$HPOF_2$	3.1	138	10	4
$AlCl_3$	2	138	15	26
$SnCl_4$	2	138	15	30

Triflic acid has been found to be an effective catalyst in the aroylation of fluorobenzene with 3- and 4-trifluoromethyl- and 3,5-bis(trifluoromethyl)benzoyl chloride.[349] Although a prolonged treatment is required (reflux, 144 h), the reactions proceed with high selectivity to give the desired CF_3-sustituted 4-fluorobenzophenones in high yields (79–98%).

Studies on mixed anhydrides of carboxylic acids and triflic acid have shown them to be extremely powerful acylating agents.[350] Similarly, higher perfluoroalkanesulfonic acids also form mixed anhydrides.[351]

Triflic acid has been shown to exhibit high activity in the acylation of aromatics with methyl benzoate to give benzophenone derivatives in good to excellent yields[352] [Eq. (5.139)]. Aromatic carboxamides bearing an additional functional group, when protonated by triflic acid to form activated dicationic intermediates, are capable of reacting with benzene to yield acylated products.[253] Triflic acid also activates β-lactams to react with arenes to form aryl-substituted β-amino ketones[353] [Eq. (5.140)]. Naphthalene, ferrocene, and *N*-substituted pyrroles were also reacted. In all three reactions, diprotonated cationic intermediates were invoked for interpretation of the results.

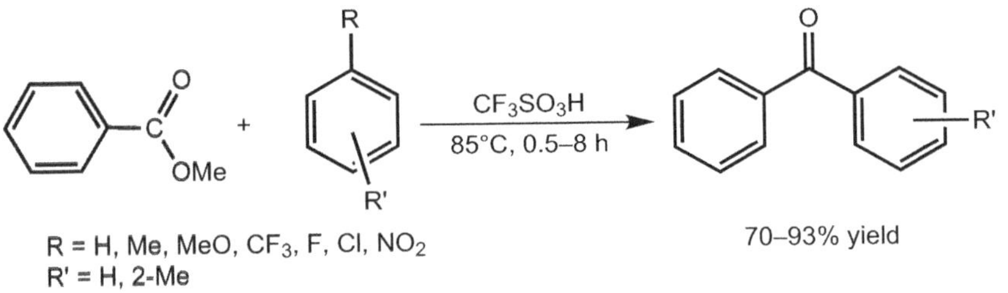

(5.139)

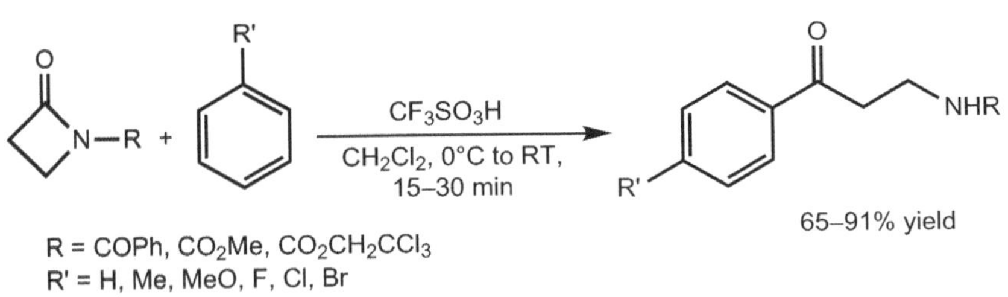

(5.140)

Roberts and Wells[354] were the first to use triflic acid in the acylation of phosphaferrocenes with acetic anhydride and benzoic anhydride to afford ketophosphaferrocenes in good yields (62–78%). Recently, acyl trifluoroacetates in excess triflic acid have been shown to be even more effective reagents[355] [Eq. (5.141)].

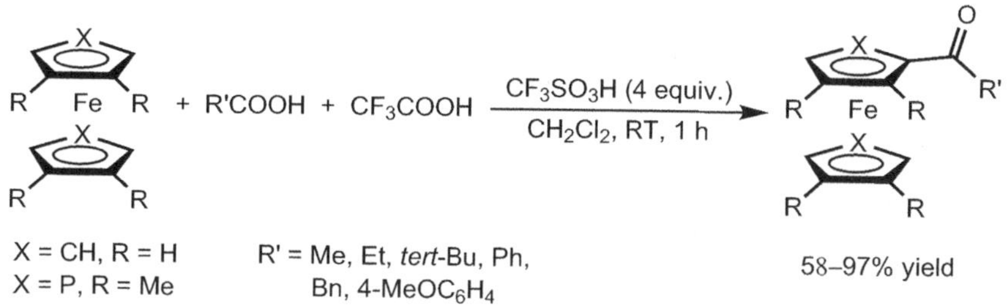

(5.141)

Amorphous and mesostructured ZrO_2 solid catalysts impregnated with various amounts of triflic acid were tested in the acylation of biphenyl[356,357] and toluene[358] (with benzoyl chloride and *para*-toluyl chloride, respectively, nitrobenzene solvent, 170°C and 130°C). All catalysts exhibited lower activity when compared with neat triflic acid. The mesoporous catalysts, however, showed complete selectivity in the formation of *para*-benzoylbiphenyl. A triflic acid–silica catalyst, in turn, prepared using an aminopropyl-modified silica, showed good characteristics in the solvent-less acetylation of anisole and 2-methoxynaphthalene with acetic anhydride.[359,360] The activity of 1,1,2,2-tetrafluoroethanesulfonic acid, either neat or embedded in silica, was found to be similar to that of triflic acid in the acetylation of anisole.[196]

The protonation of a series of α-substituted cinnamic acid in HSO_3F at -78°C has been studied.[361] The protonated (*Z*)-α-phenyl cinnamic acid intermediate **110** undergoes further transformation to yield 2-phenylindenone as a result of intramolecular acylation [Eq. (5.142)].

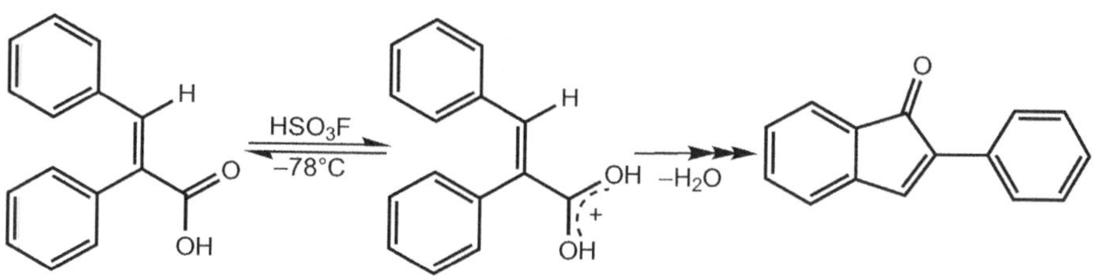

110

(5.142)

Klumpp and co-workers[362] have performed a detailed study of the acylation reactions of benzene with a variety of cinnamic acid derivatives in triflic acid. Cinnamic acids with alkyl or weakly electron-withdrawing groups (F, Br) on the phenyl ring give the corresponding substituted indanones as a result of a two-step

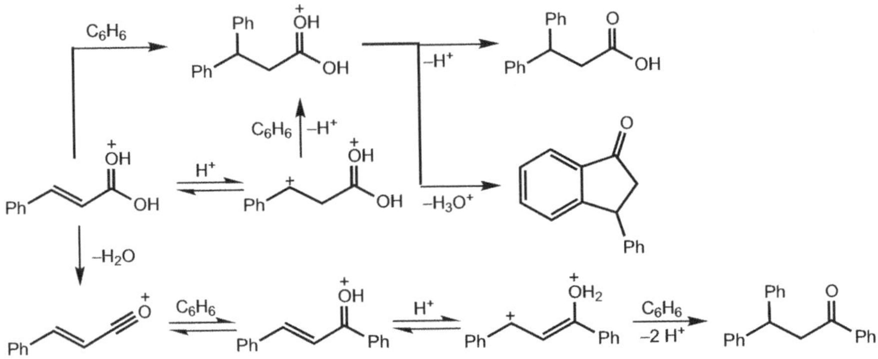

Scheme 5.38

intermolecular and intramolecular acylation (61–98% yield). In addition to indanone derivatives, dihalogenated cinnamic acid derivatives yield chalcones the products of intermolecular acylation of benzene [Eq. (5.143)]. Note that the unsubstituted (i. e., more electron-rich) ring participates in the ring closure. Finally, only chalcones are formed with compounds with strong electron-withdrawing groups. In a large excess of triflic acid (100 equiv.) triarylpropanones are also detected. The suggested mechanism involves diprotonated cationic intermediates with the reactivity of the cationic center greatly enhanced by the adjacent protonated carboxylic or carbonyl group (Scheme 5.38).

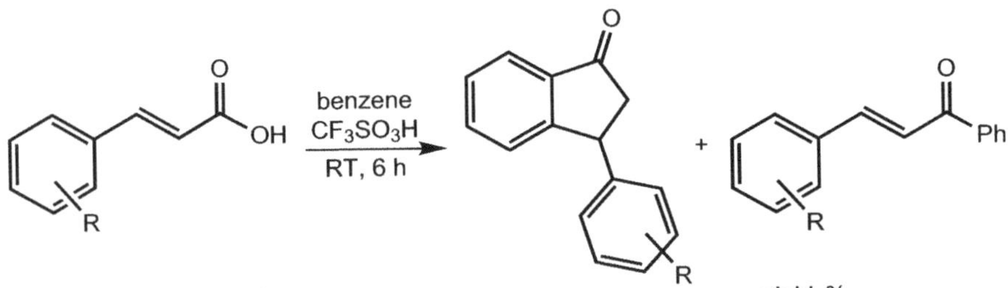

	yield %	
R = 2,6-diF	29	60
R = 2,4-diCl	80	19
R = 3,4-diCl	28	71

(5.143)

The intramolecular acylation of 2-aryloxybenzonitriles allowed the synthesis of molecules with dixanthone skeleton.[363] The procedure, in fact, is an intramolecular Houben–Hoesch reaction[364] to afford the intermediate bisiminium salts [Eq. (5.144)] which, after hydrolysis, gave the final diketo products.

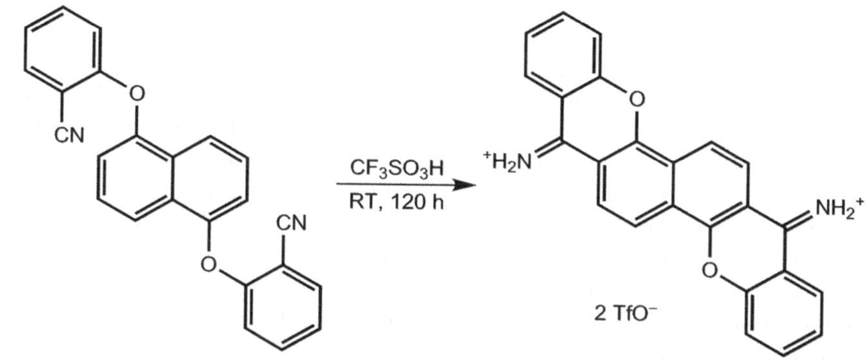

(5.144)

Examples of a two-step process of intermolecular alkylation followed by intramolecular acylation have also been reported. 5-Aryltetrahydrofuran-2-ones undergo ring opening in triflic acid to give the intermediate benzyl cation, which alkylates benzene or chlorobenzene followed by intramolecular acylation to afford tetralone derivatives in low yields[365] [Eq. (5.145)]. The reaction of aromatics with α,β-unsaturated carboxylic acids in triflic acid is a general method to synthesize substituted indanones[366] [Eq. (5.146)]. The alkylation–acylation sequence was deduced from the observations that in certain reactions (benzene with crotonic acid and phenol with styrylacetyic acid), only acylated products are formed. Likewise, 1-tetralones are produced from β,γ-unsaturated carboxylic acids.

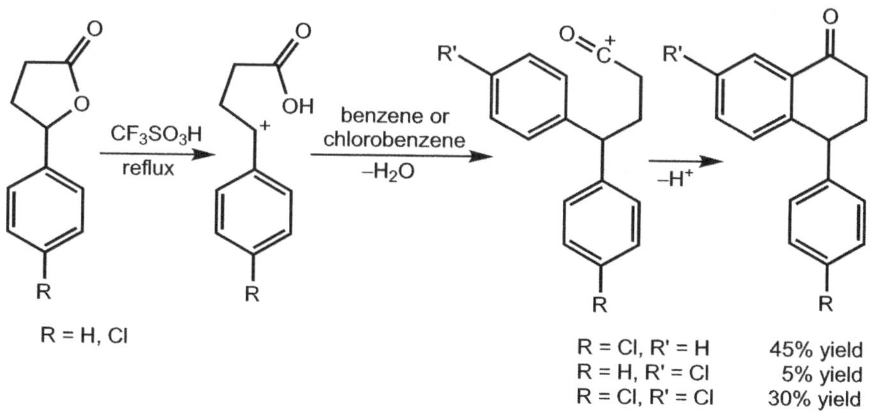

(5.145)

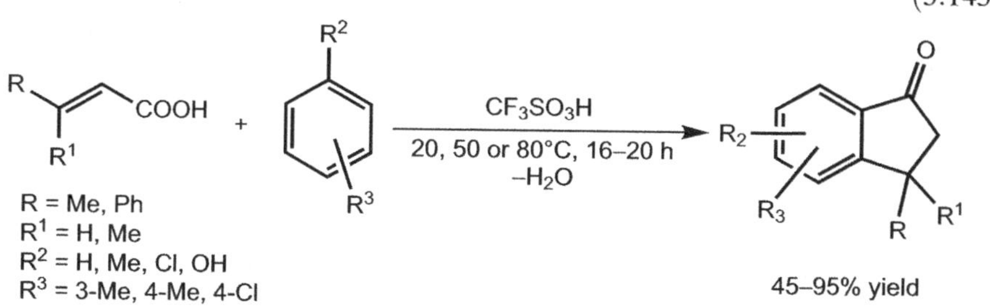

(5.146)

Nafion-H has also been found to be an effective catalyst for heterogeneous acylation of aromatic hydrocarbons with aroyl chlorides and anhydrides.[367]

The reported gas-phase acylations with Nafion-H catalyst were generally carried out at the boiling point of the hydrocarbon to be acylated. The yield of aroylation reaction depends on the relative amount of the catalyst used. Optimum yields were obtained when 10–30% of Nafion-H was employed relative to the aroyl halide. Although this procedure allows very clean reactions with no complex formation and easy work-up procedures, it is presently limited to only aroylation. Attempted acetylation of aromatics with acetyl chloride under similar conditions led to thermal HCl elimination from the latter to form ketene and products thereof. In the reaction of acetyl chloride by itself with Nafion-H, diketene was detected by IR and NMR spectroscopy.[367]

Olah, Prakash, and co-workers[368] have shown that direct aroylation with arenecarboxylic acids including pentafluorobenzoic acid can be performed over Nafion-H. Arenes, with the exception of benzene and toluene, react in the liquid phase at reflux temperature to give benzophenone derivatives in moderate to good yields [Eq. (5.147)]. Intramolecular Friedel–Crafts acylations can be performed under similar conditions by the reaction of arylalkyl-benzoyl chlorides or arylalkyl-benzoic acids in *para*-xylene,[369] or treating benzoic acids bearing an *ortho* substituent with a phenyl group in dichlorobenzene[370] [Eq. (5.148)].

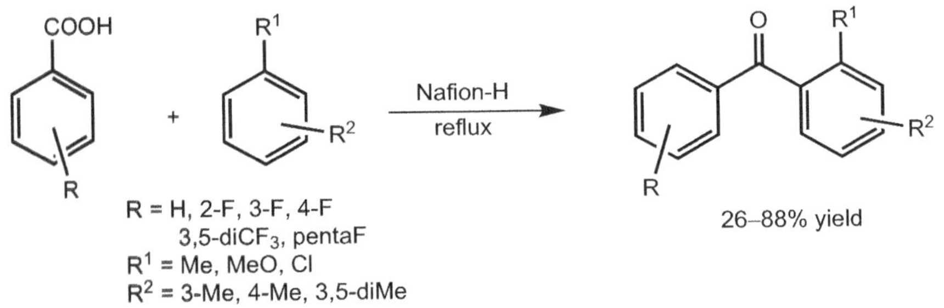

(5.147)

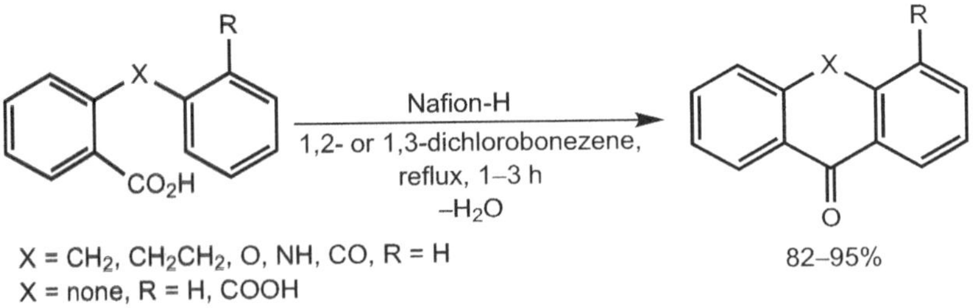

(5.148)

Nafion–silica nanocomposite catalysts have also been tested in the Friedel–Crafts acylation of aromatics with acyl chlorides.[191,194,371] Anisole, toluene, and xylenes, when reacted with phenylacetyl and phenylpropionyl chloride, give the corresponding ketones with high selectivity.[371] Hybrid organic–inorganic silica catalysts modified with superacidic fluoroalkanesulfonic acid groups (catalysts **48** and **49**, Figure 5.15) showed high specific activity exceeding that of Nafion NR50.[192] The rapid formation of the *ortho* isomer was observed in the benzoylation of anisole, followed by a slower isomerization to the *para* isomer.[372] This process, however, was shown to be not an isomerization reaction, but a transacylation occurring between anisole and *ortho*-benzoyl anisole. Furthermore, a catalyst with a more hydrophobic character led to more complete isomerization (*para*/*ortho* = 62.5). Nafion SAC-13, in turn, gave the acylated product in low yield (10%) when *para*-xylene was reacted with heptanoic acid. Among the byproducts, dimeric *para*-xylene was identified.[373]

Acetic anhydride was found to be very effective in the acetylation of anisole over silica with anchored perfluorinated sulfonic acid site **48** (Figure 5.15).[374] Activity of catalysts with high surface concentration of sulfonic acid groups exceeded that of Nafion SAC-13. Rapid deactivation, however, was observed at elevated temperature. A study of Nafion SAC-13 applied in continuous operation[375] (anisole:acetic anhydride molar ratio of 5, 70°C) arrived at the same conclusion: *para*-methoxyacetophenone was formed with high selectivity (>95%), but the initial activity (40–50% conversion) was completely lost in 24 h. Treatment of the spent catalyst with boiling HNO_3 (40% solution), however, successfully restored the activity by removing strongly adsorbed acylated products. Acylation of anisole with octanoic acid was also studied.[376,377] Nafion SAC-25 with mesopores larger than those of SAC-13—and, consequently, more accessible nanoparticles—show the highest specific activities (Table 5.28). Because of competitive adsorption, removal of water formed is essential for optimal catalyst performance.

The reversibility of Friedel–Crafts acylation is only occasionally observed.[378–383] Schlosberg and Woodbury[384] have studied transacylation between tetramethylacetophenones and some arenes in superacidic $HF–SbF_5$ (5:1) and other strong superacid media. In fact, Keumi and co-workers[385] have been able to observe diprotonated acetylpentamethylbenzene intermediate **111** in $HSO_3F–SbF_5–SO_2ClF$ medium at low temperatures, which deacetylates to pentamethylbenzenium ion at more elevated temperatures [Eq. (5.149)].

Table 5.28. Activity of Nafion Catalysts in the Acylation of Anisole with Octanoic Acid[376]

	Reaction Rate		
Catalyst	$[10^{-3}\,L\,(g_{cat}\,h)^{-1}]$	$[10^{-3}\,L\,(g_{Nafion}\,h)^{-1}]$	Selectivity[a] (%)
Nafion beads	28	28	46
Nafion SAC-13	39	280	66
Nafion SAC-25	193	790	64

Reaction conditions: anisole/acid molar ratio = 40, 0.3 g of catalyst, 150°C.
[a]Selectivity of the *para* isomer at 50% conversion.

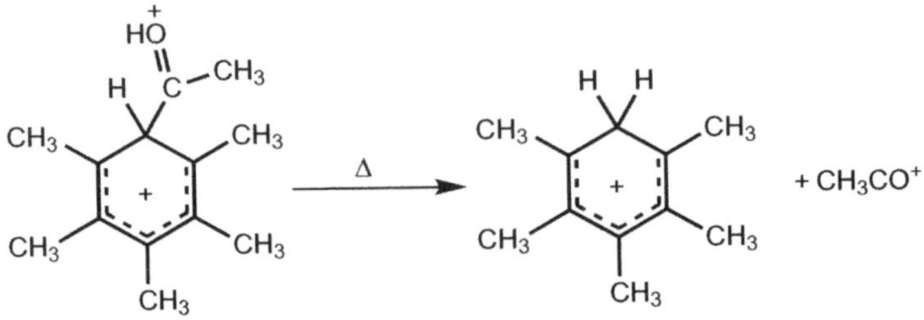

111

(5.149)

Sarca and Laali[386] have developed a convenient process for transacylation of sterically crowded arenes such as acetylmesitylene [Eq. (5.150)] and tetramethyl- and pentamethylacetophenones to activated aromatics using triflic acid in the presence of imidazolium-type ionic liquids under mild conditions. When the reactions are run without an activated arene acceptor, efficient deacylation takes place. Simple 4-methoxyaryl methyl ketones can be transacetylated with toluene and *para*-xylene as acceptors with triflic acid.[387] Nafion-H has been found to be an efficient catalyst for the decarboxylation of aromatic carboxylic acids as well as deacetylation of aromatic ketones.[388]

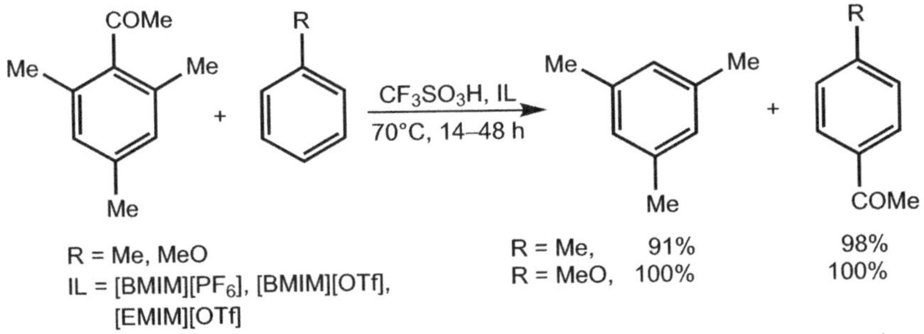

(5.150)

Apparently, only a single example is known for deformylation of aromatics reported by Yamato et al.[389] Treatment of 9-formylanthracene derivatives in the presence of a large excess (200 wt%) of Nafion-H resulted in the formation of the deformylated product (Scheme 5.39). For the di-*tert*-butyl-substituted derivative, selective deformylation and concomitant de-*tert*-butylation could be achieved under appropriately selected reaction conditions.

Akhrem et al.[390] have reported a unique method for the acylation of aromatics. When alkanes and cycloalkanes (propane, butane, cyclopentane, cyclohexane) are treated with CBr_4–$2AlBr_3$ in the presence of carbon monoxide, the intermediate acyl cations react with aromatic silanes to yield acylated products by desilylative acylation [Eq. (5.151)]. The *ipso*-substitution of trimethylsilane takes place regioselectively,

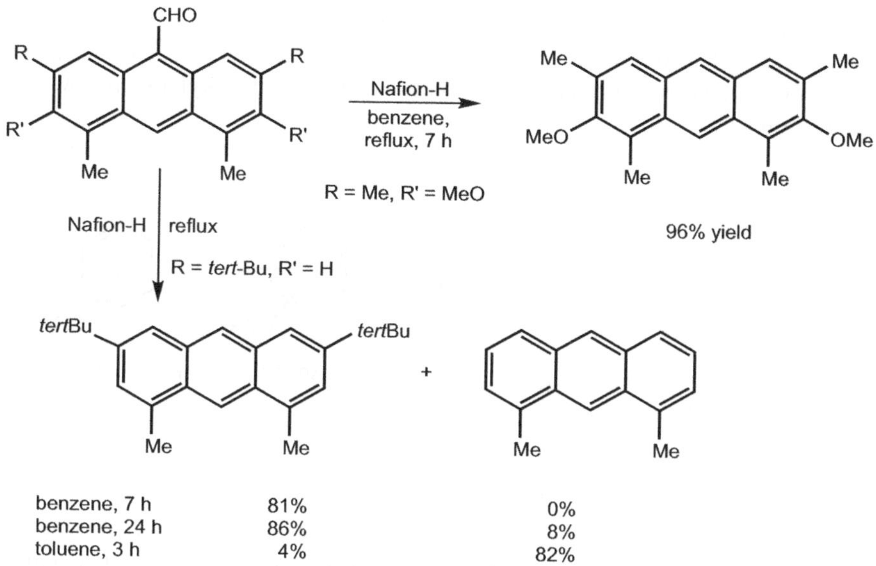

Scheme 5.39

except for *meta*-trimethylsilylanisole, which gives the *para*-acylated product. This was explained by suggesting the initial coordination of the acyl cation to the methoxy group and the resulting steric hindrance for the *ipso* attack. As a result, direct acylation at the *para* position occurs. Consistent with this is the observation that *para*-trimethylsilylanisole was unreactive even in a tenfold excess of the acyl cation. Furthermore, both toluene and anisole could be directly acylated with cyclopentane under identical conditions (92% and 81% yield, respectively).

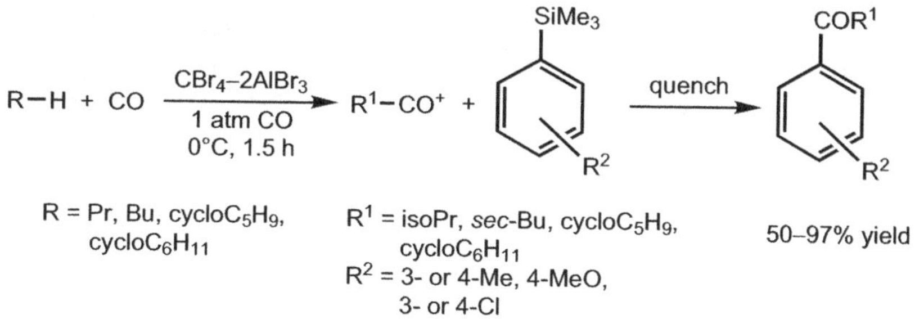

(5.151)

The one-step double functionalization of aromatics—that is, alkylation and acylation—can be accomplished with alkanes or cycloalkanes (in large excess) using the aprotic organic superacids $RCO^+Al_2X_7^-$ (R = alkyl, aryl, X = Cl, Br)[391] [Eq. (5.152)]. However, the method can be used only for benzene and bromobenzene.

Activated arenes (toluene, *meta*-xylene, naphthalene) undergo acylation, whereas strongly deactivated arenes (nitrobenzene, acetophenone) do not react. On the basis of this information, the transformation was suggested to start with alkylation followed by the acylation step.

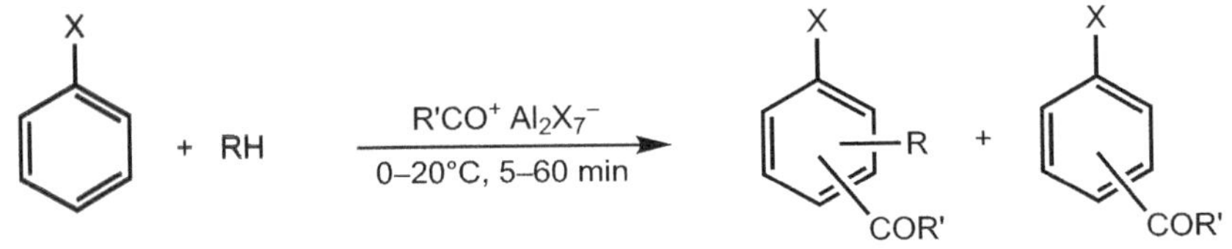

X = H, Br
R = Bu, isoBu, C_5H_{11}, isoC_5H_{11}, isoC_6H_{13}, Me-cycloC_5H_9
R' = Me, Pr, Ph

40–87% traces–41%

(5.152)

Fries rearrangement—that is, the transformation of phenolic esters to isomeric hydroxyphenyl ketones—is related to Friedel–Crafts acylations.[392,393] Olah et al.[394] have found a convenient way to perform the Fries rearrangement of a variety of substituted phenolic esters in the presence of Nafion-H in nitrobenzene as solvent [Eq. (5.153)]. A catalytic amount of Nafion-H is satisfactory, and the catalyst can be recycled. In contrast, Nafion–silica nanocomposites, in general, exhibit low activities in the Fries rearrangement of phenyl acetate to yield isomeric hydroxyacetophenones.[239,395] In a recent study, BF_3–H_2O was found to be highly efficient under mild conditions (80°C, 1 h) to transform phenolic esters of aliphatic and aromatic carboxylic acids to ketones (71–99% yields).[396] In most cases the *para*-hydroxyphenyl isomers are formed with high (up to 94%) selectivity.

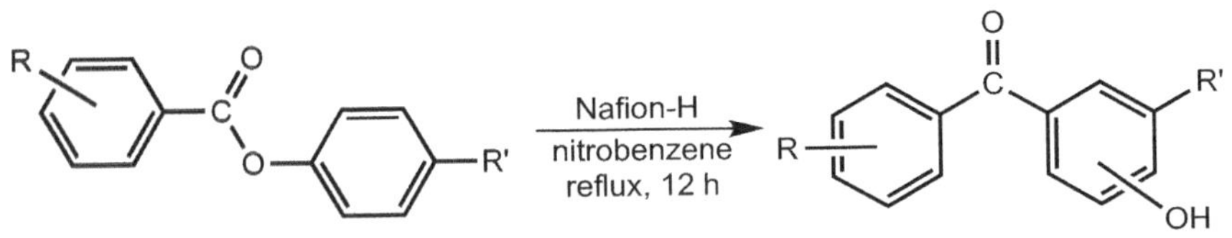

R = H, 4-Me, 3-Cl
R' = H, Me

63–75% yield

(5.153)

5.4. CARBOXYLATION

The Koch–Haaf reaction[397] for the preparation of carboxylic acids from alkenes uses formic acid or carbon monoxide in strongly acidic solutions. The reaction between carbocations and carbon monoxide affording oxo-carbenium ions (acyl cations) is a key step in the Koch–Haaf reaction and the topic has been reviewed by Hogeveen.[398] The original studies used sulfuric acid. Subsequently, the application of liquid superacids in the Koch–Haaf carboxylation met with remarkable success. Triflic acid has been found to be by far superior to 95% H_2SO_4 for carboxylation of alkenes, alcohols, and esters with carbon monoxide at atmospheric pressure. This is attributed to the high acidity of triflic acid and also the higher solubility of CO in this medium as compared with H_2SO_4. Moreover, triflic acid has the advantage that, unlike H_2SO_4, it does not form electrophilic substitution products with aromatics and can be regenerated.[399]

Whereas the C_2–C_4 alcohols are not carboxylated under the usual Koch–Haaf conditions, carboxylation can be achieved in the HF–SbF_5 superacid system under extremely mild conditions.[400] Moreover, Olah and co-workers[401] have shown that even methyl alcohol and dimethyl ether can be carboxylated with the superacidic HF–BF_3 system to form methyl acetate and acetic acid. In the carboxylation of methyl alcohol the quantity of acetic acid increased at the expense of methyl acetate with increase in reaction time and temperature. The quantity of the byproduct dimethyl ether, in turn, decreased. Dimethyl ether gave the desired products in about 90% yield at 250°C (90% conversion, catalyst/substrate ratio = 1:1, 6 h). On the basis of experimental observations, first methyl alcohol is dehydrated to dimethyl ether. Protonated dimethyl ether then reacts with CO to yield methyl acetate [Eq. (5.154)]. The most probable pathway suggested to explain the formation of acetic acid involves the intermediate formation of acetic anhydride through acid-catalyzed ester cleavage without the intervention of CO followed by cleavage with HF [Eq. (5.155)].

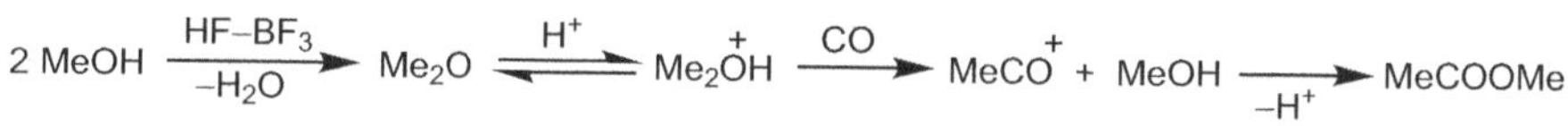

(5.154)

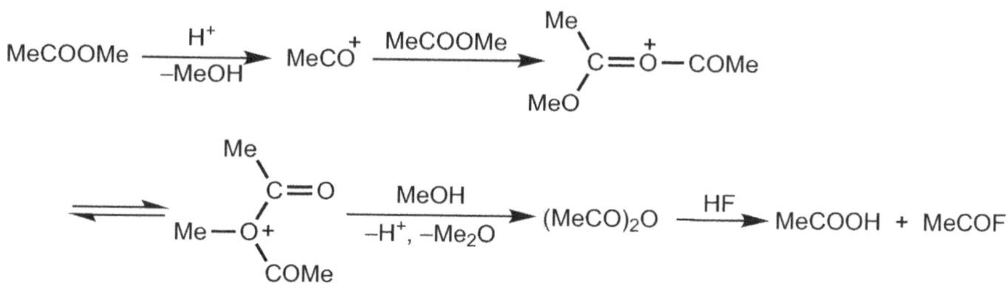

(5.155)

Nafion-H has also been found to be a suitable catalyst to carry out Koch-type carbonylation of a variety primary, secondary, and tertiary alcohols.[402] Under optimal conditions, tertiary carboxylic acids are formed in moderate-to-good yields (Table 5.29). Increased CO pressure was shown to increase acid yields by enhancing the carbonylation of the intermediate tertiary carbocation prior to dimerization and oligomerization. Solvents had a small effect on catalyst performance, which is in contrast to the findings of Lange.[403] Over Nafion NR50, acid yields of 39% and 42%

Table 5.29. Carbonylation of Alcohols over Nafion-H[402]

Alcohol	Main Product	Yield (%)	Yield of Total Acids (%)
tert-Butyl alcohol[a]	2,2-Dimethylpropanoic acid	53	63
1-Pentanol	2,2-Dimethylbutanoic acid	48	65
2-Pentanol	2,2-Dimethylbutanoic acid	40	53
1-Hexanol	2,2-Dimethylpentanoic acid	26	58
	2-Methyl-2-ethylbutanoic acid	23	
1-Octanol	2,2-Dimethylheptanoic acid	27	57
	2-Methyl-2-ethylhexanoic acid	22	
1-Adamantanol[a]	1-Adamantanecarboxylic acid	77	77

Reaction conditions: 2 g of Nafion-H, 20 mmol of alcohol, hexane, 160°C, 9 MPa CO, 22 h.
[a] $CHCl_3$ as solvent.

in heptanoic acid and pivalic acid, respectively, were reported in the carbonylation of 2,6-dimethyl-1-heptanol (150°C, 80 bar CO, 5 h), whereas only 1% of acid was formed in dodecane (99% conversion, 17.5 h).

Olah and Bukala[404] have developed a method for the oxidative carboxylation of methyl halides with CO and copper oxides or Cu and oxygen over SbF_5–graphite [Eq. (5.156)]. Time-dependence studies indicated that the three products—methyl acetate, dimethyl ether, and methyl fluoride—were formed in parallel reactions. The reactivity of methyl halides shows the decreasing order MeBr > MeCl > MeF.

$$\mathrm{MeBr} + \mathrm{CO} \xrightarrow[270^\circ\mathrm{C}]{\mathrm{SbF_5\text{–}graphite}} \mathrm{MeCOOMe} + \mathrm{Me_2O} + \mathrm{MeF} \tag{5.156}$$

	MeCOOMe	Me_2O	MeF
CuO, 150 atm, 26 h	45%	21%	11%
Cu_2O, 140 atm, 24 h	41%	47%	9%
Cu+O_2, 140 atm, 22 h	51%	5%	6%

Using HF–SbF_5, Yoneda et al.[405] have obtained dicarboxylic acids from diols by reaction with CO under mild conditions. Some cyclization products were also obtained. Scheme 5.40 was suggested for the reaction.

The formation of C_6 and C_7 acids along with some ketones was reported in the reaction of isopentane, along with methylcyclopentane and cyclohexane with CO in HF–SbF_5 at ambient temperatures and atmospheric pressure.[406] Yoneda et al.[407] have also found that other alkanes can be carboxylated as well with CO in HF–SbF_5. Tertiary carbenium ions, which are produced by protolysis of C–H bonds of branched alkanes in HF–SbF_5, undergo skeletal isomerization and disproportionation before reacting with CO. Formation of the tertiary carboxylic acids in the

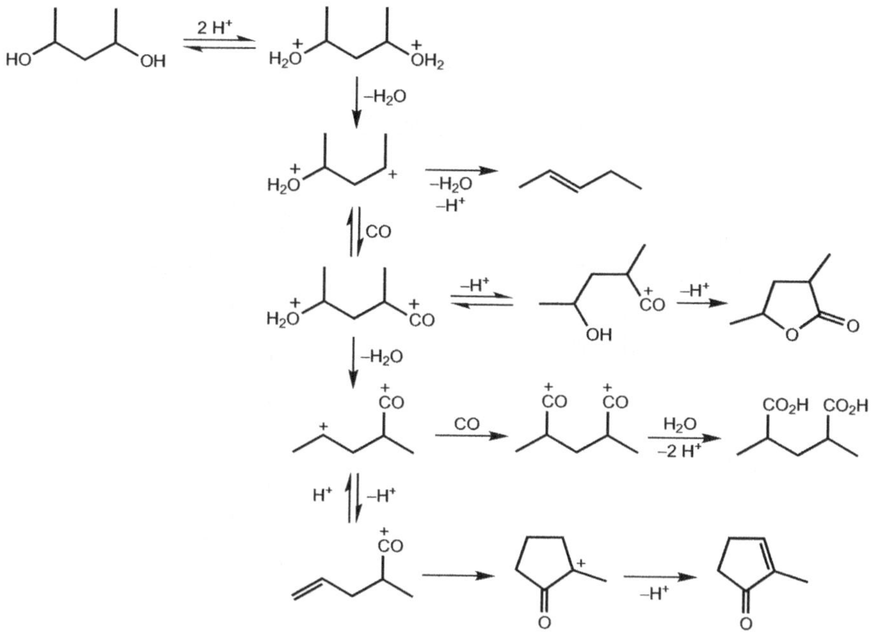

Scheme 5.40

superacid HSOF–SbF_5 was shown to be accelerated by the addition of Cu_2O to the reaction mixture.[408] It was also found that alkyl methyl ketones react with CO in the HF–SbF_5 superacid system to form oxocarboxylic acids after hydrolysis (Scheme 5.41). Alkyl methyl ketones with a short alkyl chain (less than C_4) do not react under these conditions due to the proximity of the positive charge on the protonated ketone and the developing carbenium ion.[409]

Sommer and co-workers[410,411] have made detailed studies of the carbonylation of lower alkanes in superacidic media. Carbonylation of propane was carried out by bubbling mixtures of propane and CO through a HF–SbF_5 solution (molar ratio = 4:1) at −10°C and monitoring the composition of the solution by ^{1}H NMR spectroscopy.[412] Two acylium (acyloxonium) ions, propanoyl cation (ethylcarboxonium ion, **112**) and isobutyryl cation (isopropylcarboxonium ion, **113**), were detected which, upon quenching (ethanol or $NaHCO_3/H_2O$), give the corresponding acids or esters [Eq. (5.157)]. The two products were formed in a ratio of 2:3 using a CO/propane mixture of 3 with a propane conversion of 4%. The gas phase was shown to contain a large excess of methane (H_2/methane/ethane 4:89:7). In a large-scale test the ratio of the two acids were 1:6 (propane conversion = 94%). All experiments with excess CO showed the predominant formation of the ethylcarboxonium ion (**112**); that is, preferential C–C bond cleavage takes place.

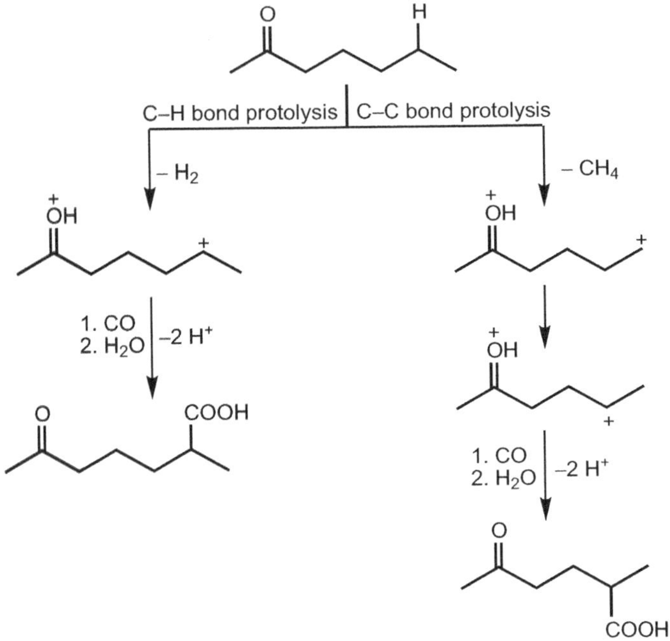

Scheme 5.41

$$\mathrm{CH_3CH_2CH_3} + \mathrm{CO} \xrightarrow[-10^{\circ}\mathrm{C}]{\mathrm{HF{-}SbF_5}} \underset{\mathbf{112}}{\mathrm{CH_3CH_2{-}CO^+}} + \underset{\mathbf{113}}{\mathrm{(CH_3)_2CH{-}CO^+}} \xrightarrow[-\mathrm{H^+}]{\text{quench}} \mathrm{CH_3CH_2COOR} + \mathrm{(CH_3)_2CHCOOR} \quad \mathrm{R = H, Et} \qquad (5.157)$$

Considering all possibilities, the protolytic cleavage of propane can be summarized according to Scheme 5.42. Since the butyryl cation was not detected, path ***a*** (involvement of *n*-propyl cation **34**) can be excluded. Pathway ***b*** (protonation of the secondary C—H bond) is kinetically disfavored compared with protonation of the more electron-rich C—C bond (pathways ***c*** and ***d***). The large amount of methane leaves only path ***c*** as the major activation route of propane leading to the formation of ethylcarboxonium ion **112** (formation of the ethyl cation followed by carbonylation with CO).

It was also found that the ratio of the two esters is highly dependent on the propane/CO ratio. In contrast to the preferential C—C bond cleavage and formation of ethylcarboxonium ion (**112**) observed at high CO/propane ratios, increasing amounts of propane result in increasing selectivities of the formation of the isopropylcarboxonium ion **113** (Figure 5.16). Under such conditions, formation of the isopropyl cation

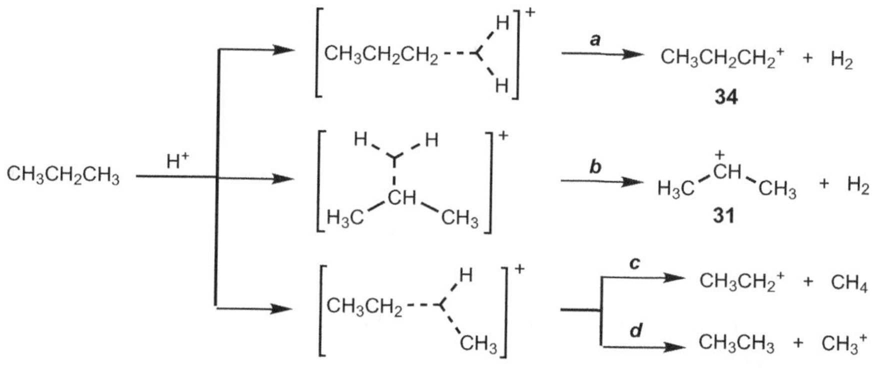

Scheme 5.42

31 via route ***b*** (Scheme 5.42, direct protolysis of the secondary C—H bond) can be neglected in comparison with the activation of propane by the ethyl cation. That is, the isopropyl cation is formed by the hydride transfer between the ethyl cation and propane (Scheme 5.43).

Addition of small amounts of bromine to the reaction mixture before adding the propane/CO mixture resulted in both an increase in conversion and increasing selectivity of the isopropylcarboxonium ion independently of the propane/CO ratio.[413,414] This was accounted for by the involvement of the bromocarbonyl cation $BrCO^+$, which, upon protosolvation, acts as a superelectrophile exhibiting high

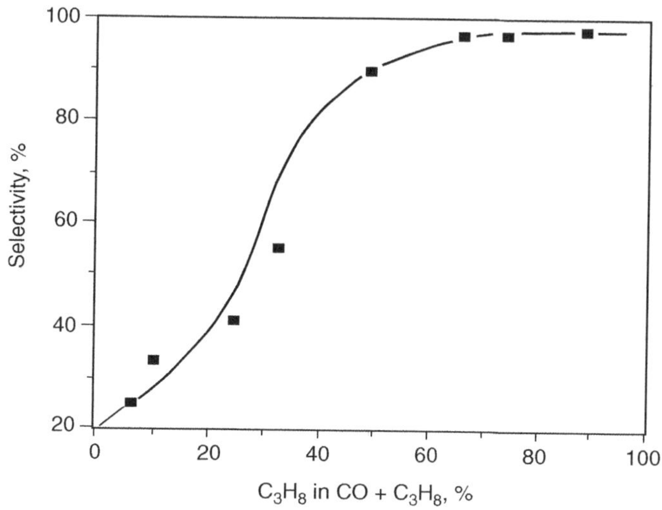

Figure 5.16. Selectivity of the formation of the isopropylcarboxonium (**113**) ion as a function of the propane/CO ratio (selectivity: [**113**]/[**113** + **112**]).[412]

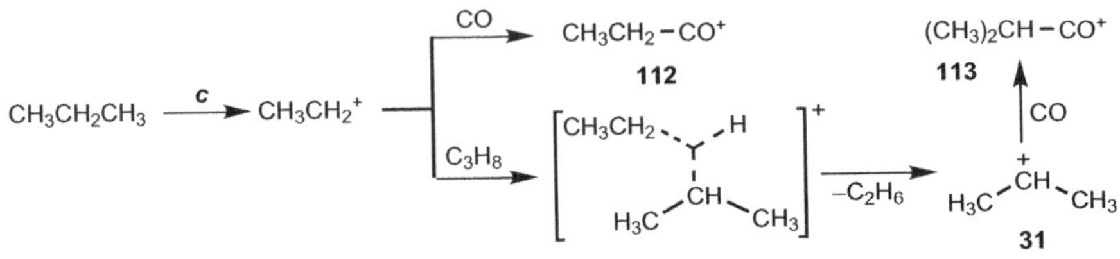

Scheme 5.43

reactivity toward propane. Furthermore, exclusive formation of the isopropylcarboxonium ion (**113**) was observed when carbonylation was performed in the presence of carbon tetrachloride or chloroform.[415] This was explained by the regioselective C–H bond activation by the halomethyl cations CCl_3^+ and $HCCl_2^+$. Carbonylation of isobutane in HF–SbF_5 selectively and almost stoichiometrically occurs through the protolytic cleavage of the tertiary C–H bond to form the *tert*-butyl cation and subsequently pivaloic ester in high yield.[410]

In a subsequent study, Sommer et al.[416] have carried out the carbonylation of methylcyclopentane under two reaction conditions. When CO is bubbled through the HF–SbF_5 solution at −40°C, the methylcyclopentyl cation (**23**) yields quantitatively the corresponding carboxonium ion and, upon quenching with ethanol, gives ethyl 1-methylcyclopentanecarboxylate (80% yield based on SbF_5) (Scheme 5.44). Carbonylation at 0°C, in turn, leads to the formation of ethyl cyclohexanecarboxylate (60% yield). Under such conditions, where carbonylation is a reversible process, the cyclohexyl carbenium ion (**22**), which is in equilibrium with the more stable methylcyclopentyl cation **23**, has a much higher reactivity toward CO. As a result, the reaction mixture becomes enriched in the cyclohexyl carboxonium ion, the quenching of which gives the corresponding ester (Scheme 5.44).

Akhrem and co-workers[417] have successfully applied aprotic organic superacids in the carbonylation of a series of alkanes. Butane was transformed into isomeric carboxylic acids depending on the superelectrophilic reagent and isolated as the

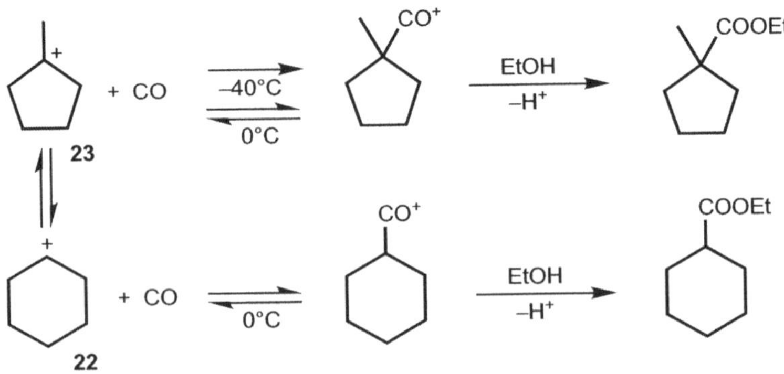

Scheme 5.44

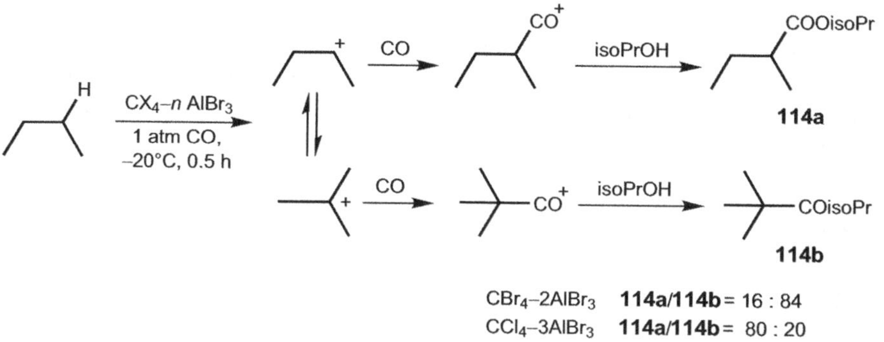

Scheme 5.45

esters **114a** and **114b** (Scheme 5.45), whereas pentane gave 2,2-dimethylbutanoic acid selectively in almost quantitative yield in the presence of CBr_4–$2AlBr_3$.[418]

Isopropyl esters of cyclohexanecarboxylic acids have been synthesized by using CBr_4–$2AlBr_3$ (1 atm CO pressure, −40°C).[419] Isopropyl 1-methylcyclohexanecarboxylate was isolated in the reaction of cycloheptane and methylcyclohexane (82% and 73% yield, respectively), whereas isopropyl 1-ethylcyclohexancarboxylate was formed from cyclooctane and ethylcyclohexane (67% and 69% yield, respectively).

It has been demonstrated by Olah et al.[420] that α,β-unsaturated ketones are *O*-protonated in HF–SbF_5 to form hydroxyallylic cations, which were directly observed by NMR spectroscopy. Jacquesy and Coustard have found indirect evidence for diprotonation of α,β-unsaturated ketones (enones) by trapping the dication with CO.[421] The resulting acylium ion centers are then quenched with methanol or benzene. An interesting synthetic method was therefore developed for carboxylation of bicyclic enones in superacid media at atmospheric pressure [Eqs. (5.158) and (5.159)].

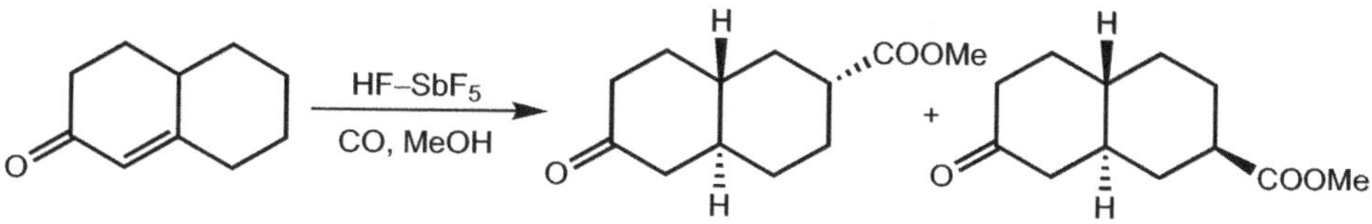

(5.158)

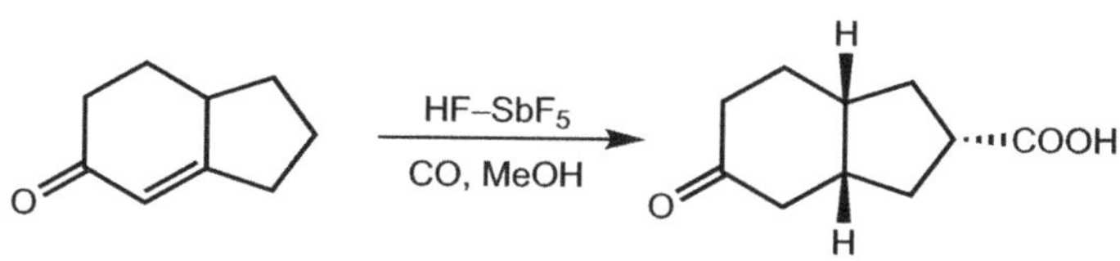

(5.159)

When cyclohexene is mixed with anhydrous triflic acid under a high pressure of carbon monoxide (120 atm) followed by the addition of benzene, cyclohexyl phenyl ketone and the isomeric cyclohexenyl cyclohexyl ketones are obtained with little isomerization of the initially formed cyclohexyl cation **22** to methylcyclopentyl cation **23** (Scheme 5.46).[422]

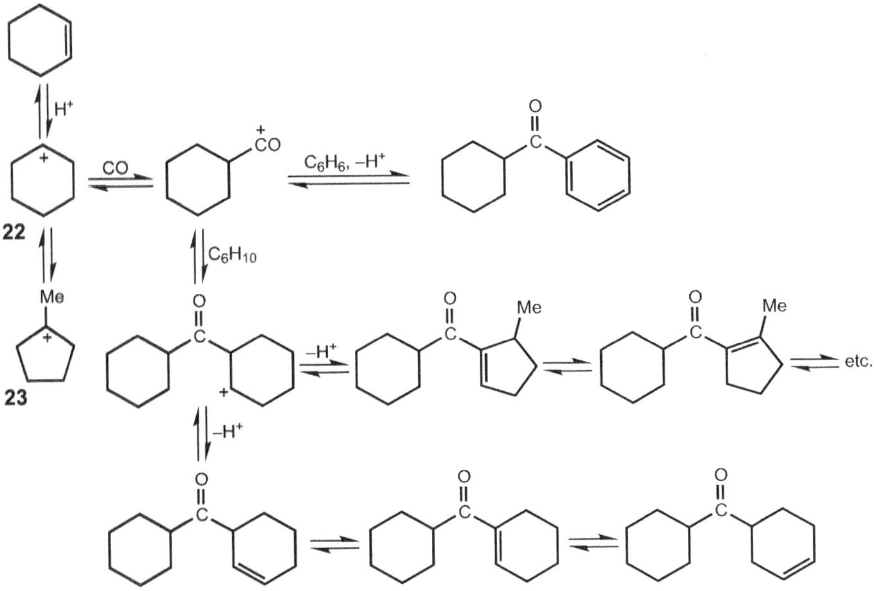

Scheme 5.46

Lee et al.[423] have reported the carbonylation of formaldehyde over various ion-exchange resins. Nafion NR50 showed the highest specific activity (moles of formaldehyde converted per proton site) yielding methyl glycolate with high selectivity [Eq. (5.160)].

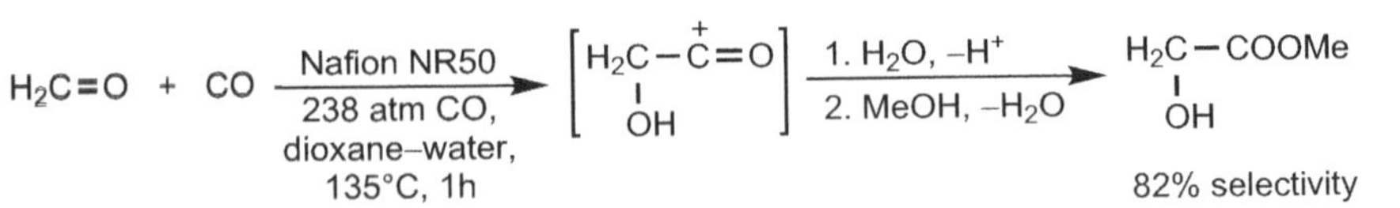

(5.160)

Oxidative carbonylation of alkynyltungsten(II) complexes in excess triflic acid leads to formation of indanone derivatives[424] [Eq. (5.161)]. Elucidation of the reaction mechanism was made by isolation and characterization of acyltungsten(IV) species indicating the involvement of the η^1-vinylidene cation **115**.

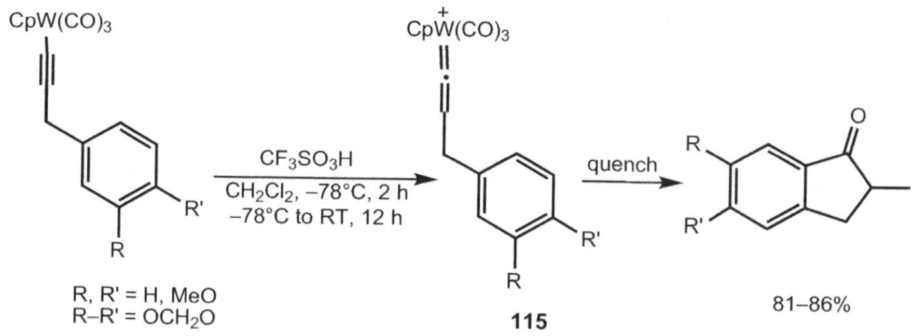

(5.161)

Carboxylation of aromatics with carbon dioxide with Al_2Cl_6/Al has been studied by Olah, Prakash, and co-workers[425] and shown to be a chemoselective process to give aromatic carboxylic acids in good to excellent yields (20–80°C, CO pressure = 57 atm). Two possible mechanistic pathways with the involment of organoaluminium intermediates and complexes of CO_2 with $AlCl_3$ were postulated. On the basis of extensive experimental studies and theoretical calculations, the authors concluded that the most feasible mechanism involves CO_2 activated with superelectrophilic aluminum chloride. Complex **116** reacts with aromatics in a typical electrophilic substitution.

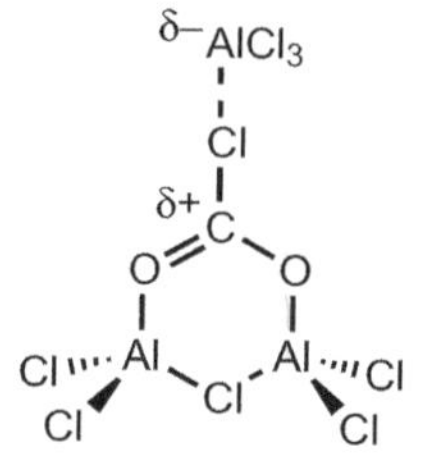

116

5.5. FORMYLATION

Aromatic formylation reactions are known to occur under Gattermann–Koch conditions using mixtures of CO + HCl + $AlCl_3$ and $CuCl_2$.[426–428] The use of superacidic HF–BF_3 and HF–SbF_5 as catalysts for aromatic formylation has been demonstrated.[429–431] Mechanistic studies by Olah et al.[432] have shown that selectivity in formylation reactions strongly depends on the nature of the formylating agent.

Among the most frequently used formylation methods, the Gattermann–Koch reaction shows the highest selectivity reflected both in the observed high $k_{toluene}$:$k_{benzene}$ rate ratios as well as a high degree of *para* substitution (Table 5.30).

Table 5.30. Selectivities in Various Formylation Reactions[432]

Method	Reagent	Catalyst	$k_{toluene}:k_{benzene}$	*para* Product (%)
Gattermann–Koch	CO	$HCl+AlCl_3+CuCl_2$	155–860	88.7–96
Gross	Cl_2CHOCH_3	$AlCl_3$	119	60.4
Gattermann	$Zn(CN)_2$	$AlCl_3$	93–128	57.8–63.9
Friedel–Crafts	HCOF	BF_3	34.6	53
Superacidic[a]	CO	$HF-SbF_5-SO_2ClF$	1.6	52.1
Superacidic[b]	CO	$HF-SbF_5$	~25	89.8

[a] At −95°C.
[b] At 0°C.

Gross formylation with dichloromethyl methyl ether[433] is somewhat less selective, as is the Gattermann synthesis using $Zn(CN)_2$ and $AlCl_3$.[428] Friedel–Crafts-type formylation with formyl fluoride[434] gives a much lower selectivity indicating that the $HCOF–BF_3$ system produces a more reactive electrophile ($HCOF \cdot BF_3$ complex, but not necessarily a free formyl cation, HCO^+).

The lowest selectivity was observed in the case of $HF–SbF_5$-catalyzed formylation with CO in SO_2ClF solution at −95°C, which gave a very low $k_{toluene}:k_{benzene}$ ratio (Table 5.30) and an isomer distribution of 45% *ortho*-, 2.7% *meta*-, and 52.1% *para*-tolualdehydes.[432] Under the superacidic conditions studied, CO is protonated to give rapidly equilibrating (with the solvent acid system) protosolvated formyl cation, an obviously very reactive electrophilic reagent. When the reaction is carried out at 0°C using only excess aromatics as solvent, the selectivity becomes higher and giving an isomer distribution of 7.5% *ortho*-, 2.8% *meta*-, and 89.8% *para*-tolualdehydes.

Subsequently, Olah et al.[435] made a detailed study of the formylation of a variety of arenes (benzene, toluene, ethylbenzene, xylenes, mesitylene) in the superacidic catalyst systems triflic acid–HF–BF_3 and triflic acid–SbF_5. Increasing acidity was found to result in increasing yields of aromatic aldehydes. Good to high yields of aldehydes (59–78%), in general, could be achieved under mild conditions (atmospheric CO pressure, room temperature). Toluene gave isomeric ditolylmethanes through the reaction of intermediate protonated *para*-tolualdehyde with excess toluene. High positional selectivity (91–94% *para* isomer) was observed in both catalyst systems studied. Substrate selectivities (k_T/k_B), in turn, were lower (21 in triflic acid–HF–BF_3 and 15 in triflic acid–SbF_5) as compared to those in Gattermann–Koch reaction. The elusive formyl cation could not be observed (triflic acid–SbF_5–SO_2ClF solution, −80°C) only exchanging ^{13}CO could be detected suggesting the existence of a rapidly equilibrating protosolvated ion.

The formylation of hexadeuteriobenzene, C_6D_6, with HCOF–BF_3 shows a kinetic hydrogen isotope effect of $k_H/k_D = 2.68$, based on comparison of the reactivity of $C_6H_6/CH_3C_6H_5$ and $C_6D_6/CH_3C_6H_5$.[432] This isotope effect is similar to that observed in Friedel–Crafts acetylation and propanoylation reactions, and it indicates that the proton elimination step is at least partially rate-determining. The low substrate selectivity formylation with the CO–HF–SbF_5 system, however, shows no primary isotope effect.

Tanaka et al.[436] have studied the formylation of alkylbenzenes, halobenzenes, indane, and tetralin in HSO_3F–SbF_5 under atmospheric CO pressure and observed both formylation and sulfonation. Time-dependence studies with *meta*-xylene showed that the formylation product is formed first and then transformed into the sulfonated aldehyde with increasing reaction time. However, selectivity could be controlled by changing the acid strength of the system, and formylated products could be obtained in high yields with high *para* selectivity under appropriately selected reaction conditions (high acidity, short reaction time). On the basis of additional studies by comparing formylation and sulfonation, it was concluded[437] that the formyl cation has dual reactivity and can act as an electrophile or Brønsted acid. On the other hand, protonated aromatics can also act as Brønsted acids to produce formyl cations. Under typical electrophilic conditions, where most of the arenes are protonated, formyl cations are produced close to the aromatic ring by the protonated aromatics. Formylation, consequently, has priority over sulfonation.

The HF–SbF_5 system could be used for the synthesis of dialdehydes of bicyclic aromatic compounds.[438,439] With an SbF_5/arene ratio of 2, naphthalene, biphenyl, diphenylmethane, and bibenzyl gave dialdehydes with high positional selectivity (31–98% yield). It was surprising to find in this study, however, that methylnaphthalenes react slowly and exhibit low positional selectivity. Subsequently, the kinetics and regioselectivity of the formylation of 1-methylnaphthalene and *meta*-xylene in the HF–SbF_5 and CF_3SO_3H–SbF_5 acid systems were studied.[440–442] Dependence of the rate on the SbF_5/arene ratio revealed that formylation can be explained by taking into account the protonation equilibrium of *meta*-xylene and the apparent formylation rate decreasing by protonation. The extent of inhibition by protonation is related to the extent of the transformation of the arene to an inactive σ-complex. The apparent formylation rate of arenes, consequently, is not generally proportional to their relative basicities. Furthermore, high *para* selectivity of the formylation of 1-methylnaphthalene was observed at SbF_5/substrate molar ratios around 1, but increasing SbF_5 content resulted in decreasing selectivity. On the basis of this information, an intracomplex reaction mechanism was suggested [Eq. (5.162)] with the arenium ion protonating CO to form the *para*-oriented π-complex **117**. At high SbF_5/substrate molar ratios, protonated CO (formyl cation) is the actual formylating agent and the regioselectivity reflects the ratio of the two mechanisms.

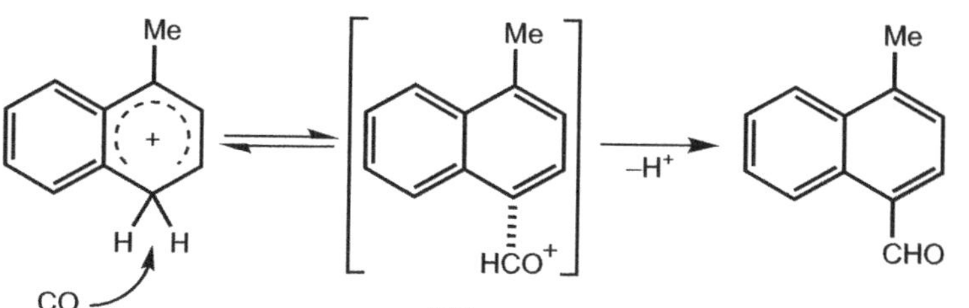

117

(5.162)

Unprecedented high *ortho* selectivities were observed in the monoformylation of bibenzyl in HF–Lewis acid systems (Lewis acid: SbF_5, TaF_5, BF_3, NbF_5) in contrast to biphenyl, diphenylmethane, and 1,3-diphenylpropane.[443] The selectivity increased with decreasing SbF_5/Lewis acid molar ratios, and with the strength of the Lewis acid used. The *ortho* monoformylation was explained to take place with the participation of the sandwich-like complex **118** formed from the monocation, whereas dication **119** gives the *para*-diformyl product.

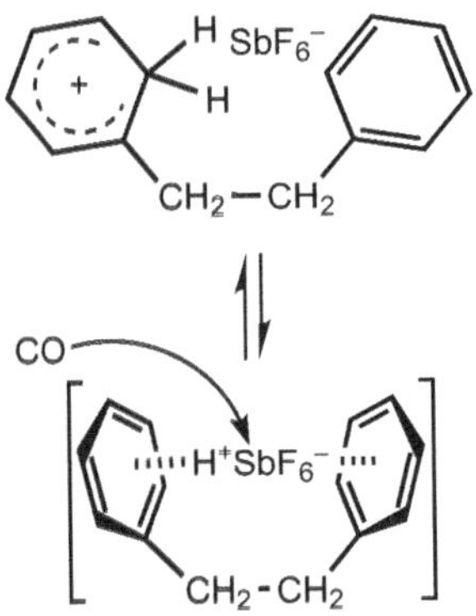

118

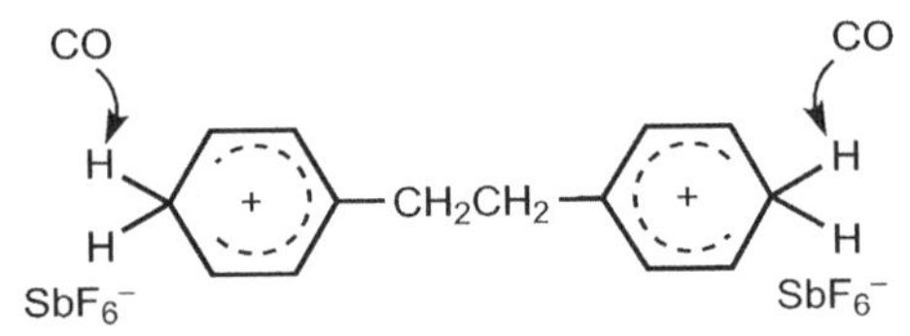

119

For nearly a century, Friedel–Crafts acylations were considered to give nearly exclusive *para* substitution of toluene. The reason accounting for this fact was considered to be steric. The present-day understanding of the mechanism of electrophilic aromatic substitution indicates that this is not necessarily the only reason. *para* Substitution is greatly favored if the transition state of highest energy is an intermediate arenium ion (σ-complex) like, where a *para* methyl group is more stabilizing than an *ortho* (and much more than a *meta*). However, when the highest transition state is becoming increasingly "early" on the reaction path, the ratio of *ortho*:*para* substitution increases. *meta* Substitution always stays relatively low, generally less than 5–6% varying with the reactivity of the reagent within this limit. This substitution pattern is also observed in Friedel–Crafts-type formylation reactions. In these reactions, the involved substituting agents are obviously less space demanding than those of other acylation reactions. Steric effects consequently cannot be a significant factor affecting selectivity, which is primarily reflected in the changing *ortho*:*para* isomer ratio. The methyl group always remains predominantly *ortho*:*para* directing, even in very low substrate selectivity reactions, and the *meta* isomer does not increase above 4%.

Regioselective formylation of toluene, *meta*- and *para*-xylene, and mesitylene has been achieved by carbonylation in triflic acid at CO pressures of 90–125 atm. However, the use of six- to sevenfold excess of acid over arene is required to obtain high yields of the aldehydes.[422] Recently, a *para*-tolualdehyde yield of 99.1% has been reported (triflic acid/toluene molar ratio = 20, CO pressure = 70 atm, room temperature, 30 min).[444]

Formylation of the less reactive phenol and anisole with CO in HF–BF_3 was found to require at least stoichiometric amount of the acid for effective transformation (50 equiv. of HF, 2 equiv. of BF_3, 50 bar CO, 45°C).[445] Conversion increases with increasing reaction time but results in decreasing *para*/*ortho* ratios suggesting a change from kinetic control to thermodynamic control and the reversibility of formylation. Furthermore, the amount of byproducts (mainly diphenylmethane derivatives) originating from reactions between substrates and products also increases. Additional studies in ionic liquids showed that imidazolium cations with increased chain lengths—for example, 1-octyl-3-methylimidazolium salts—are effective in the formylation process. This was attributed to the enhanced solubility of CO in the ionic liquid medium. Tris(dichloromethyl)amine, triformamide, and tris (diformylamino)methane have recently been applied in the formylation of activated aromatic compounds in the presence of triflic acid at low temperature (−10 to −20°C) albeit yields are moderate.[446]

While studying the reaction of adamantane with carbon monoxide under superacidic catalysis, formylation (formation of 1-adamantanecarboxaldehyde) was found by Olah and co-workers[447] to effectively compete with Koch–Haaf carboxylation (formation of 1-adamantanecarboxylic acid, major product formed in 60–75% yield). On the basis of results acquired by the reaction of 1,3,5,7-tetradeuteroadamantane, formylation was interpreted by insertion of the formyl cation into the tertiary C–H σ-bond [Eq. (5.163)].

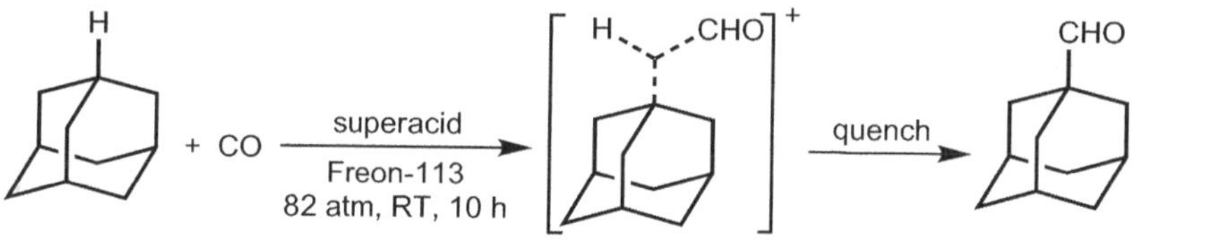

CF_3SO_3H 0.2% yield
CF_3SO_3H–$B(OTf)_3$ 3.4% yield
CF_3SO_3H–SbF_5 8.2% yield

(5.163)

A more effective formylation of adamantane has been developed by Vol'pin, Akhrem, and co-workers[448,449] using the aprotic organic superacids CBr_4–$2AlBr_3$ and CH_2Br_2–$2AlX_3$ (X = Cl, Br). With methylcyclopentane as the hydride source, 1-adamantanecarboxaldehyde was isolated in 100% yield under optimized conditions (substrate:superacid:methylcyclopentane molar ratio = 1:1:2, CH_2Br_2–$2AlCl_3$, atmospheric CO pressure, 20°C, 2 h).

Olah et al.[450] have recently described a new, highly efficient superelectrophilic formylation–rearrangement of isoalkanes. Branched ketones are formed in high yields and with high selectivity with no detectable branched acids (Koch products) in the presence of moderately strong superacids such as HF–BF_3 or triflic acid–BF_3. Carbonylation of isobutane under such conditions gives isopropyl methyl ketone in high yield [Eq. (5.164)] The transformation was interpreted with the involvement of the pentacoordinate carbodication **120a** formed by the attack of reactive protosolvated formyl cation $[HC{=}COH]^{2+}$ on the tertiary C–H bond of isobutane. According to subsequent calculations [MP2(full)/6-31G*],[451] this species is a minimum on the potential energy surface and can be considered as the enolic form of 2-methylpropanal with a proton on one face and a methyl cation on the other face of the C–C double bond (**120c**). This dication can rearrange in a single step to a distonic dication, which leads to the final product isopropyl methyl ketone (see Section 5.17.3).

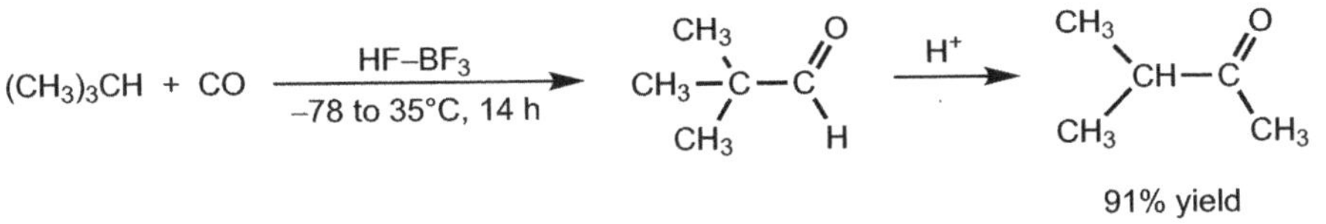

(5.164)

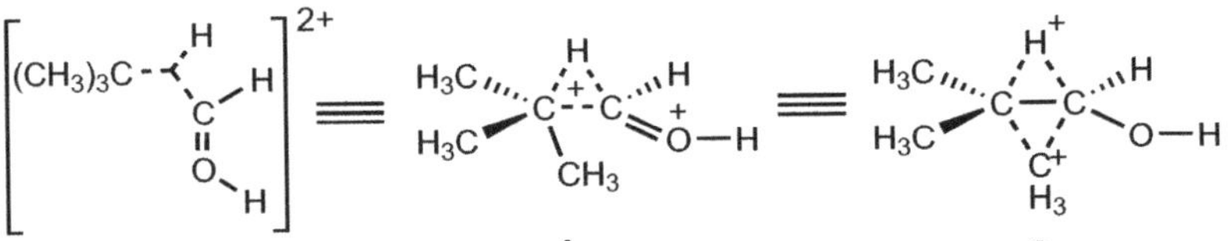

a　　b　　c

120

Attempts have been made[435] to observe the long-lived formyl cation under stable ion conditions using ^{13}C-enriched carbon monoxide. However, even at very low temperatures proton exchange with the superacid solvent is fast on the NMR time scale. In a recent study, however, de Rege, Gladysz, and Horváth[452] have observed protonated carbon monoxide by IR and NMR spectroscopy at high CO pressure in $HF{-}SbF_5$. They were also able to demonstrate[453] the carbonylation of methane with carbon monoxide under similar conditions ($HF{-}SbF_5$ or $HSO_3F{-}SbF_5$, 80°C, pressure), which leads to the exclusive and quantitative formation of the acetylium ion CH_3CO^+. Quenching the reaction mixture with water gives acetic acid in quantitative yield.

5.6. THIO- AND DITHIOCARBOXYLATION

Aromatic carboxylic acid derivatives are generally prepared by Friedel–Crafts methods using phosgene, oxalyl chloride, or carbamoyl chlorides.[397] Carbon disulfide reacts with arenes in the presence of excess $AlCl_3$ catalyst to give dithiocarboxylic acids.[454–456] However, these reactions generally require at least 2 mol excess of the strong Lewis acid catalyst, and significant side reactions occur.

Olah et al.[457] have developed a mild method for the preparation of methyl and ethyl thio(dithio)benzoates. They prepared *S*-methyl (*S*-ethyl) thiocarboxonium and dithiocarboxonium fluoroantimonates **121a** and **121b** by methylating (ethylating) carbonyl sulfide and carbon disulfide, respectively, with methyl(ethyl)fluoride–antimony pentafluoride complexes in SO_2 solution [Eq. (5.165)]. Then these reagents were used under mild conditions for the preparation of methyl and ethyl thio(dithio) benzoates by electrophilic substitution of aromatic hydrocarbons [Eq. (5.166)].

$$[\mathrm{R{-}O{=}S{=}O}]^+ \mathrm{SbF_6}^- + \mathrm{S{=}C{=}X} \xrightarrow[-25^\circ\mathrm{C}]{} \mathrm{R{-}S{=}\overset{+}{C}{=}X}\ \mathrm{SbF_6}^- \qquad (5.165)$$

X = O, S

121 **a** = Me, Et, X = O; **b** = Me, Et, X = S

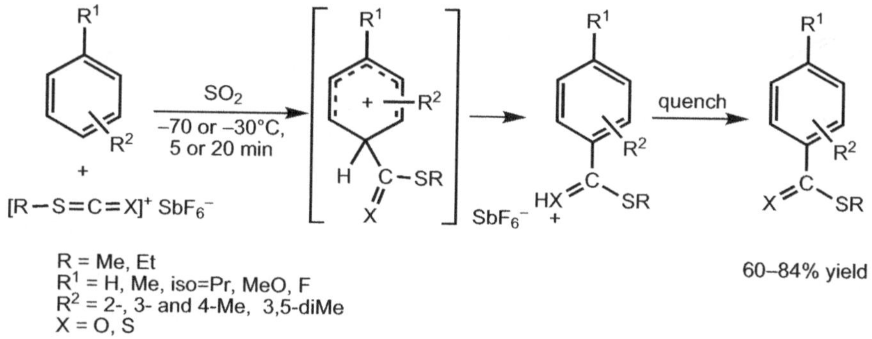

(5.166)

5.7. SULFONATION AND SULFONYLATION

Sulfonation of aromatic compounds is generally carried out with sulfuric acid, halosulfuric acids, or sulfur trioxide as reagent with or without solvent.[458,459] Friedel–Crafts catalysts such as aluminum chloride and boron trifluoride are effective catalysts in certain sulfonations with sulfuric acid and chlorosulfuric acid.

When SO_3 is used in fairly dilute solution, the attacking species is SO_3 itself. In concentrated sulfuric acid, however, the mechanism is more complex. Fuming sulfuric acid (in which the molar fraction of $SO_3 > 0.5$) is actually a mixture of SO_3 and ionized or nonionized monomers, dimers ($H_2S_2O_7$, with), trimers ($H_2S_3O_{10}$), and tetramers ($H_2S_4O_{13}$) of H_2SO_4 (the latter three formed by dehydration). At higher water content, the tetramer and trimer disappear and the amount of dimer decreases. The reactive species in sulfuric acid thus depends on the amount of water in the acid and on the reactivity of the substrate. The reactive species in aqueous sulfuric acid are H_2SO_4 and $H_2S_2O_7$, with the latter being more important at higher acid concentrations. In fuming sulfuric acid, $H_3S_2O_7^+$ and $H_2S_4O_{13}$ are also involved.[460]

Chlorosulfuric acid (HSO_3Cl) reacts with aromatic hydrocarbons to give sulfonic acids, sulfonyl chlorides, and sulfones, with the relative yields depending on the reaction conditions. The reaction with benzene with an equimolar amount of chlorosulfuric acid in sulfur dioxide as solvent at −8°C yields mainly benzenesulfonic acid with only minor amounts of diphenyl sulfone.[461] However, with excess of the acid, arylsulfonyl chloride is also formed.[462] Compared with HSO_3Cl, HSO_3F is a poorer sulfonating agent and tends to give arylsulfonyl fluorides more easily; excess of halosulfuric acids gives halosulfonation.

Studying the formylation of alkylbenzenes in $HSO_3F–SbF_5$, Tanaka et al.[436] have observed both formylation and sulfonation. However, in the presence of HSO_3F, that is at low acidity level, only sulfonyl compounds were obtained [Eq. (5.167)], whereas increasing acidity (with added SbF_5) resulted in the formation of products of formylation.

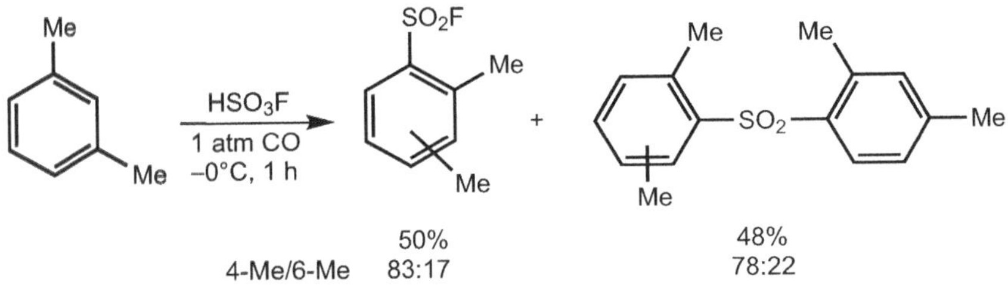

(5.167)

Herlem et al.[463] have observed that asphaltene is dissolved in fluorosulfuric acid and the process is accompanied by strong redox reactions (SO_2 and HF evolution). The products are mainly functionalized by SO_3H groups, but SO_2F groups were also detected by XPS. Indeed, model studies with benzene showed the formation of benzenesulfonic acid, diphenylsulfone, and benzenesulfonyl fluoride. For alkylbenzenes, sulfonation was not accompanied by cracking of the alkyl chain.

Nafion-H has also been used as sulfonation catalyst.[464] When oleum and long-chain alkylbenzenes were separated from each other by a Nafion-H membrane, the membrane transported the sulfonating agent into the alkylbenzene at a rate convenient for dissipating the heat of the reaction. Reported yields of sulfonation products were 34% (4 h), 63% (6 h), and 86% after 22.5 h.

Friedel–Crafts-type intermolecular sulfonylation of aromatics can also be conveniently carried out over Nafion-H by reacting aromatics with arenesulfonic acids at reflux temperature with azeotropic water removal[465] [Eq. (5.168)]. Methanesulfonic acid also reacts with *para*-xylene to yield 2-methanesulfonyl-1,4-dimethylbenzene in much lower yield (30%).

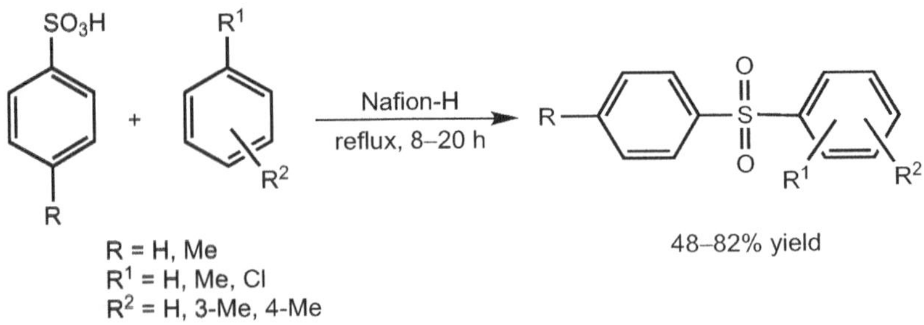

(5.168)

Olah et al.[466] observed immediate formation of protonated benzenesulfinic acid upon addition of SO_2 to benzenium ion formed in HSO_3F–SbF_5–SO_2ClF solution at $-78°C$. Based upon this observation, Laali and Nagvekar[467] developed a method for the synthesis of aromatic sulfoxides [Eq. (5.169)]. Product formation was interpreted in terms of dehydration of protonated benzenesulfinic acid followed by nucleophilic attack by the aromatic to the formed arenesulfinyl cation. Mixed sulfoxides (4-fluorophenyl-4-methylphenyl and 4-fluorophenyl-3-trifluoromethyl sulfoxides) were also prepared by sequential addition of the two aromatics. The direct synthesis of symmetric diaryl sulfoxides in high yields (room temperature, 2–48 h, 50–95%) has been reported through the electrophilic activation of thionyl chloride with triflic acid.[468]

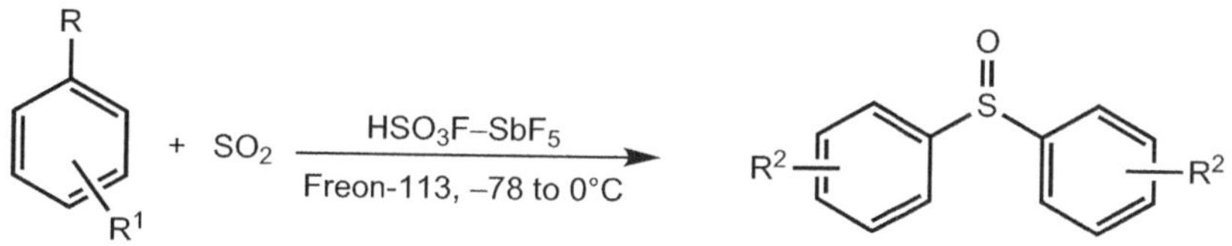

R = Me, Et, F, CF_3,
R^1 = Me, Et, F, Cl

R^2 = 4-Me, 4-Et, 4-F, 3-CF_3, 2,4-diMe, 2,4-diEt, 4-F-2-Me, 4-F-3-Me, 4-Cl-3-Me

32–100% yield

(5.169)

Triflic acid itself is a poor catalyst for the sulfonylation of aromatics with arenesulfonyl chlorides to give diarylsulfones. Addition of $BiCl_3$, however, dramatically increases the activity (10 mol% TfOH, 5 mol% $BiCl_3$, 120°C, 65–97% yields).[469] A similar synergistic effect has been observed with $BiCl_3$ or $SbCl_3$ in the sulfonylation of both activated and deactivated aromatics with methanesulfonyl chloride[470] [Eq. (5.170)]. The triflic acid–$BiCl_3$ system (10 mol% each) proved to be more efficient in most cases; however, only the triflic acid–$SbCl_3$ system could be used for fluorinated arenes because of solubility reason. The actual active species was suggested to be the mixed anhydride $MeSO_2OTf$.

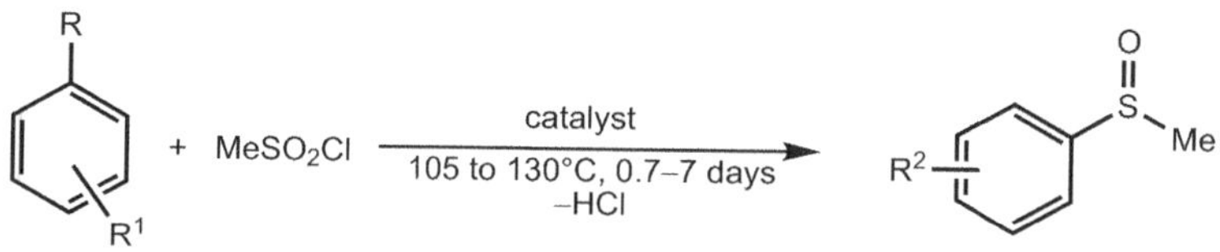

R = H, Me, F, Cl
R^1 = H, 2-, 3-, 4-Me, 3,5-diMe, 2-, 3-, 4-F, 2-Cl

R^2 = 2-, 3-, 4-Me, diMe, 2- or 4-F, 2,4-diF, F+Me, 2- or 4-Cl, 2,3- or 2,4-diCl

$BiCl_3$+CF_3SO_3H 19–94% yield
$SbCl_3$+CF_3SO_3H 8–97% yield

(5.170)

5.8. NITRATION

Convenient nitration of aromatic compounds is carried out using a mixture of nitric acid and sulfuric acid (mixed acid). There are, however, difficulties associated with the use of mixed acid.[459,471–473] In particular, the water formed as the reaction proceeds dilutes the acid and therefore reduces its strength. Also, the strong oxidizing ability of a mixed acid system makes it unsuitable to nitrate many acid-sensitive compounds. The disposal of the spent acid also poses a significant environmental problem. To overcome these difficulties of anhydrous Friedel–Crafts-type nitrations catalyzed by strong acids, nitronium salts (such as $NO_2^+BF_4^-$, $NO_2^+PF_6^-$, $NO_2^+SbF_6^-$, etc.) were developed[474] which are extremely powerful nitrating agents.

Stable nitronium salts, which are readily prepared from nitric acid (or nitrates) with HF and BF_3 (and other Lewis acids such as PF_5, SbF_5, etc.) [Eqs. (5.171) and (5.172)], will nitrate aromatics in organic solvents generally with close to quantitative yield.[472,473] Because HF and PF_5 (or BF_3) can be easily recovered and recycled, the method can be considered as a nitric acid nitration using a superacid catalyst [Eq. (5.173)].

$$HNO_3 + HF + 2\,BF_3 \rightleftharpoons NO_2^+\,BF_4^- + BF_3\text{–}OH_2 \qquad (5.171)$$

$$RONO_2 + HF + 2\,BF_3 \rightleftharpoons NO_2^+\,BF_4^- + BF_3\text{–}ROH \qquad (5.172)$$

$$ArH + NO_2^+\,MF_{n+1}^- \longrightarrow ArNO_2 + HF + MF_n \qquad (5.173)$$

$$MF_{n+1} = BF_4,\ PF_6$$

The powerful nature of nitronium salts as nitrating agents is demonstrated in their ability to affect even trinitration of benzene to trinitrobenzene [Eq. (5.174)].[475] Nitronium salts enable nitration of every conceivable aromatic substrate.

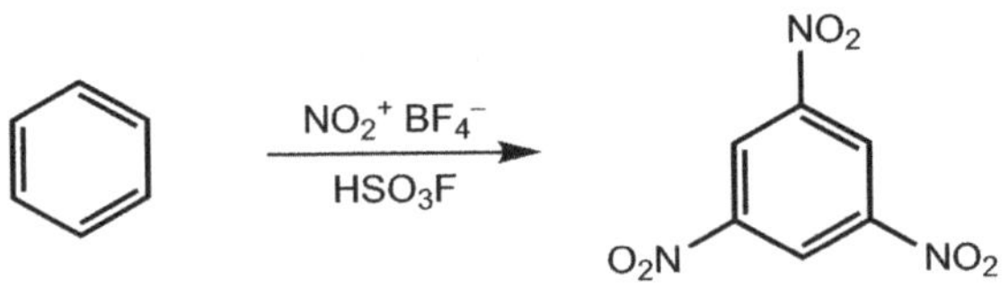

(5.174)

The powerful nitronium salts are also capable of reacting with aliphatics. Electrophilic nitration of alkanes and cycloalkanes has been carried out with $NO_2^+PF_6^-$, $NO_2^+SbF_6^-$, or $NO_2^+BF_4^-$ salts in CH_2Cl_2–tetramethylenesulfolane or HSO_3F solution.[476] Table 5.31 lists some representative reactions of nitronium ion with various alkanes and cycloalkanes. The formation of nitroaliphatics indicates the insertion of nitronium ion into aliphatic σ-bonds involving two-electron, three-center-bonded five-coordinate carbocations as indicated in the nitration of adamantane[477] [Eq. (5.175)].

Table 5.31. Nitration and Nitrolysis of Alkanes and Cycloalkanes with $NO_2^+PF_6^-$[476]

Hydrocarbon	Nitroalkane Products (molar ratio)
Methane	CH_3NO_2
Ethane	$CH_3NO_2 > CH_3CH_2NO_2$ (2.9:1)
Propane	$CH_3NO_2 > CH_3CH_2NO_2 > 2\text{-}NO_2C_3H_7 > 1\text{-}NO_2C_3H_7$ (2.8:1:0.5:0.1)
n-Butane	$CH_3NO_2 > CH_3CH_2NO_2 > 2\text{-}NO_2C_4H_9 \sim 1\text{-}NO_2C_4H_9$ (5: 4:1.5:1)
Isobutane	*tert*-$C_4H_9NO_2 > CH_3NO_2$ (3:1)
Neopentane	$CH_3NO_2 >$ *tert*-$NO_2C_4H_9$ (3.3:1)
Cyclohexane	Nitrocyclohexane
Adamantane	1-Nitroadamantane > 2-Nitroadamantane (17.5:1)

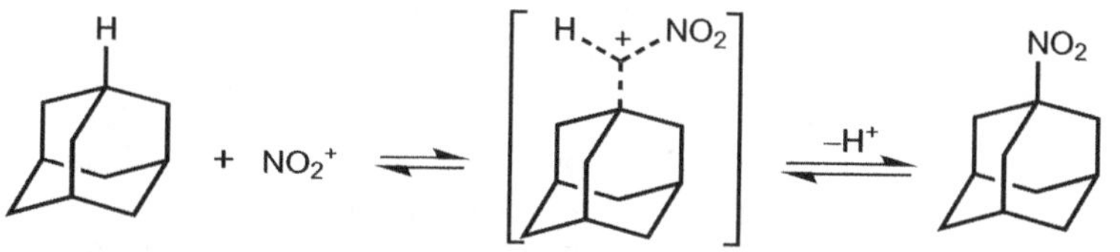

(5.175)

Nitronium ion, which is linear with *sp*-hybridized nitrogen, is not very reactive in aprotic media and not capable of nitrating deactivated aromatics. The reactivity of nitronium salts, however, can be further increased, when nitration is carried out in superacid solution (HF, HSO_3F). The enhanced reactivity is attributed to protosolvation,[3–5]—that is, to the formation of protonitronium dication (**122**, NO_2H^{2+}) [Eq. (5.176)], suggested as early as 1975 by Olah et al.[1] The interaction of the superacid with the nitronium ion weakens the N—O π-bond character, resulting in the bending of the linear ion and rehybridization of the N from *sp* to sp^2. Nitronium salts under superacidic conditions were shown to react even with methane proceeding through the $3c$–$2e$ bound carbocation transition state formed by the insertion of protonitronium dication **122** into the C—H bond[478] [Eq. (5.177)].

122

(5.176)

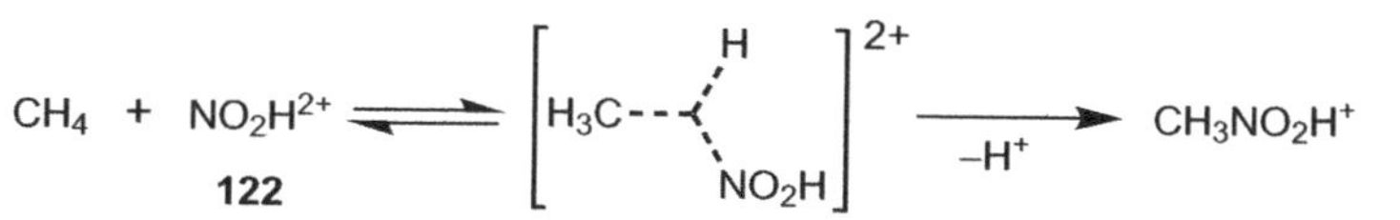

(5.177)

This has been demonstrated in a comparative study[479] using nitronium salts in the nitration of deactivated polyfluoronitrobenzenes performed in dichloromethane and sulfolane, as well as triflic acid. When 2,4-difluoronitrobenzene, 2,3,4-trifluoronitrobenzene, and 1,3,5-trifluorobenzene were reacted with $NO_2^+BF_4^-$ in sulfolane at 70°C, no product was formed, whereas 2,4-difluoronitrobenzene and 1,3,5-trifluorobenzene gave the corresponding nitrated derivative in dichloromethane under reflux (30% and 90% yield, respectively). In contrast, all model compounds underwent nitration in triflic acid at 70°C; moreover, 1,3,5-trifluorobenzene gave the dinitro derivative (yields are 34%, 65%, and 29%). In addition to better solubility and higher dissociation of the nitronium salt, the enhanced reactivity was attributed to protosolvation of the NO_2^+ ion by the superacid.

Olah et al.[480] have found that nitration of aromatics with potassium nitrate or nitric acid catalyzed by boron trifluoride–monohydrate BF_3–H_2O proceeds with good to excellent yield and is capable of nitrating even some deactivated aromatics [Eq. (5.178)]. Later, it was shown that potassium nitrate with the complex boron trifluoride–trifluoroethanol BF_3–$2CF_3CH_2OH$ is equally effective[481] [Eq. (5.178)].

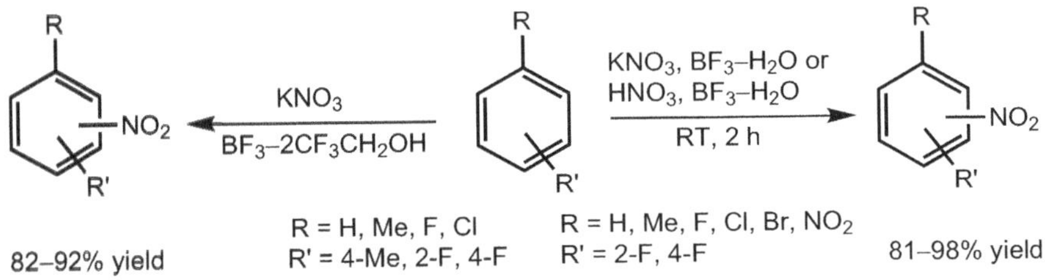

(5.178)

Nitration of strongly deactivated aromatics was carried out with nitric acid mixed with triflatoboric superacid.[482] The method is characterized by high yields (pentafluorobenzene, 99%, 1,2,3,5-tetrafluorobenzene, 89%; 2,3,4-trifluoronitrobenzene, 96%; methyl phenyl sulfone, 78%), usually high regioselectivity, and mild reaction conditions (room temperature), and it also tolerates many functional groups.

The anhydride of nitric acid and triflic acid prepared according to Eq. (5.179) was also shown to be a highly effective nitrating agent in various solvents (nitromethane, nitroethane, $CFCl_3$) to afford nitro products in high yields (67–99%).[483] According to ^{15}N NMR spectroscopy of ^{15}N-enriched nitric acid in triflic acid, trifluoromethanesulfonyl nitrate formed is predominantly covalent in nature. However, triflic acid formed in the system allows protolytic cleavage of the O–NO_2 bond. Competitive nitration of benzene–toluene mixtures may shed light to the nature of the nitrating agent. Preformed, highly reactive nitronium salts, in general, show very low substrate selectivity ($k_T/k_B < 2$), whereas less reactive covalent nitrates exhibit higher k_T/k_B values (nitric acid/acetic acid, 24; trifluoroacetyl nitrate, 28). The value for nitric acid/triflic anhydride (benzene–toluene, 1:1 molar ratio) is 36. Kinetic studies of nitration with nitric acid in neat triflic acid and over triflic acid on silica showed[484,485] that triflic acid is less effective in nonaqueous solution, which is attributed to ion pairing and strong interaction between ionic species.

$$(CF_3SO_2)_2O + HNO_3 \longrightarrow CF_3SO_3NO_2 + CF_3SO_3H \qquad (5.179)$$

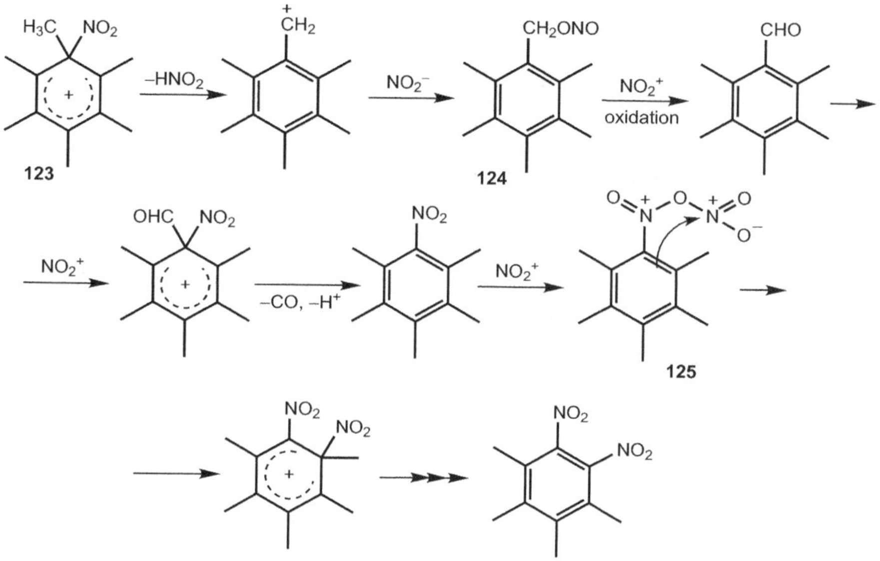

Scheme 5.47

Nitronium tetrafluoroborate used in large excess (>6 equiv.) is able to transform hexamethylbenzene and its derivatives to dinitroprehnitene (1,2,3,4-tetramethyl-5,6-dinitrobenzene) in a highly selective nitration process[486] [Eq. (5.180)]. Scheme 5.47 summarizes the key steps of the mechanistic proposal, including the *ipso*-nitroarenium ion **123**, the formation of benzyl nitrite **124**, and the complexation of the NO_2^+ ion to form the mononitro intermediate (**125**) facilitating the attack to the *ortho* position resulting in the formation of the 1,2-dinitro product.

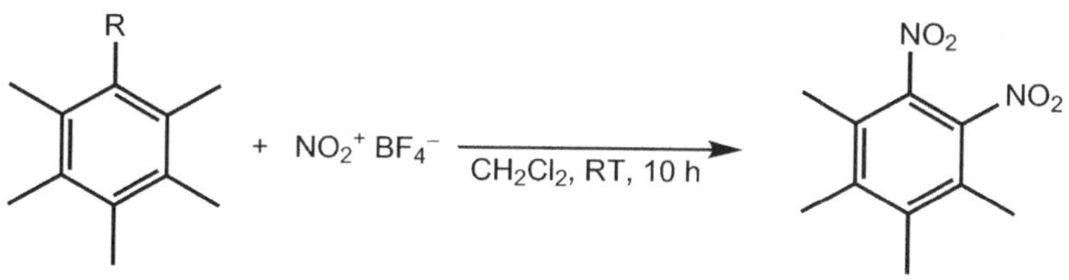

R = Me, Br, NO_2, COOH

90–96% yield

(5.180)

Excellent results have been reported on the nitration with the $NO_2Cl-3MX_n$ ($X_n = AlCl_3$, $SbCl_5$) superelectrophilic aprotic nitrating agent.[487] Strongly deactivated arenes, including benzotrifluoride, polyhalogenated arenes, and aroyl derivatives, give the corresponding nitro derivatives in high, often quantitative yields under mild conditions (CH_2Cl_2, 0–20°C) in short reaction times (15–180 min).

Nitronium ions such as *N*-nitropyridinium salts, which are readily prepared from the corresponding pyridine and nitronium salts,[488,489] act as convenient transfer nitrating agents. Transfer nitrations are applicable to *C*-nitrations [Eq. (5.181)] as well as to a variety of heteroatom nitrations. For example, they allow safe, acid-free preparation of alkyl nitrates and polynitrates from alcohols (polyols) in nearly quantitative yield[490] [Eq. (5.182)].

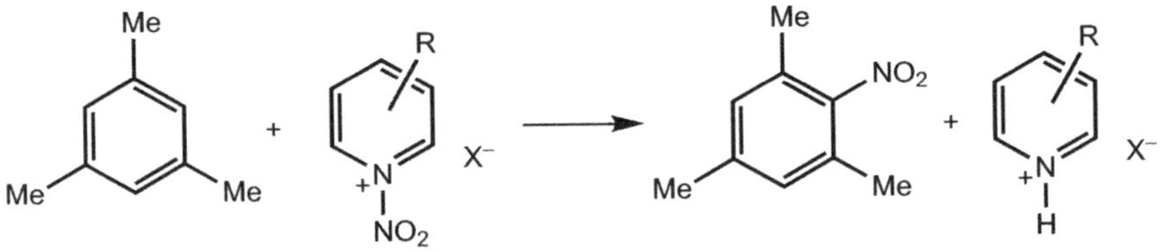

R = 2-Me, 2,6-diMe, 2,4,6-triMe, 4-MeO-2,6-diMe
X = PF_6^-, BF_4^-

(5.181)

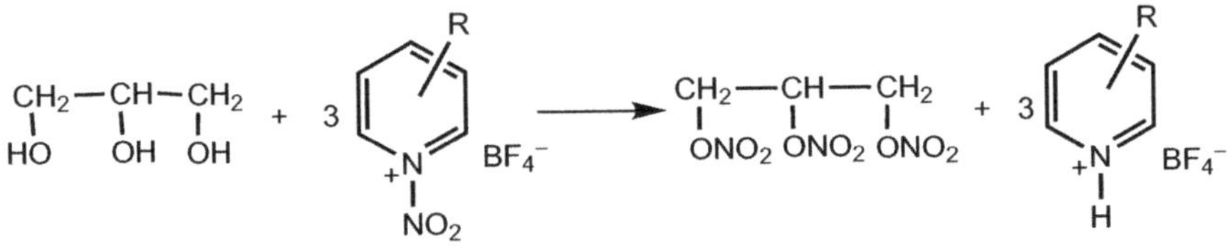

R = H, Me, iso=Pr, *tert*-Bu

(5.182)

Electrophilic nitration of olefins can also be carried out with nitronium salts in pyridinium poly(hydrogen fluoride) (PPHF) solution[491] (which also acts as solvent) to give high yields of nitrofluorinated alkanes. In the presence of added halide ions (iodide, bromide, chloride) the related haloalkanes are formed, and these can be dehydrohalogenated to nitroalkenes[492] [Eq. (5.183)].

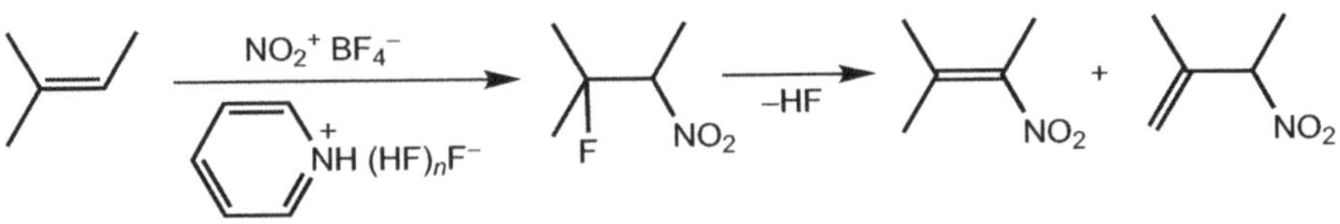

(5.183)

Originally, the nitronium ion was recognized only as a nitrating agent.[493] Subsequently, it was found[494–496] to possess significant ambident reactivity and thus be capable of acting as an oxidizing agent. Dialkyl (diaryl) sulfides [Eq. (5.184)] and selenides [Eq. (5.185)] as well as trialkyl (triaryl) phosphines

[Eq. (5.186)], triaryl arsines, and triaryl stibines react with nitronium salts to give the corresponding oxides.

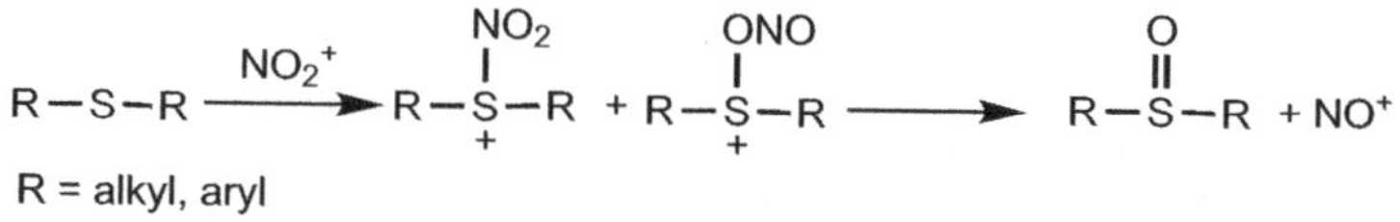

(5.184)

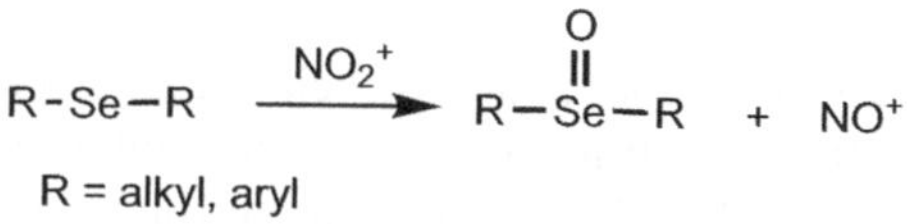

(5.185)

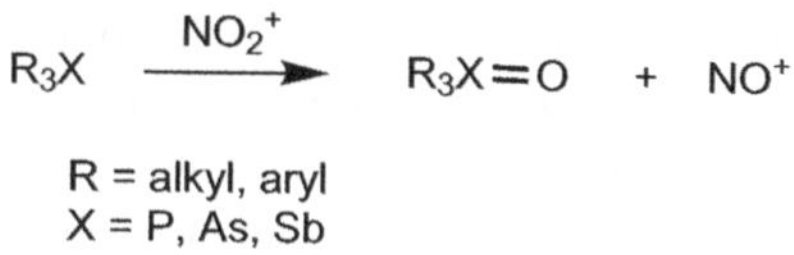

(5.186)

Stable nitronium (NO_2^+) salts, particularly with PF_6^- and BF_4^- counterions, can act as mild, selective oxidative cleavage reagents for a wide variety of functional groups.[497–499] Examples such as the oxidation of methyl ethers [Eq. (5.187)], oximes [Eq. (5.188)], dimethylhydrazones [Eq. (5.189)], and thioacetals [Eq. (5.190)] illustrate the utility of these methods.

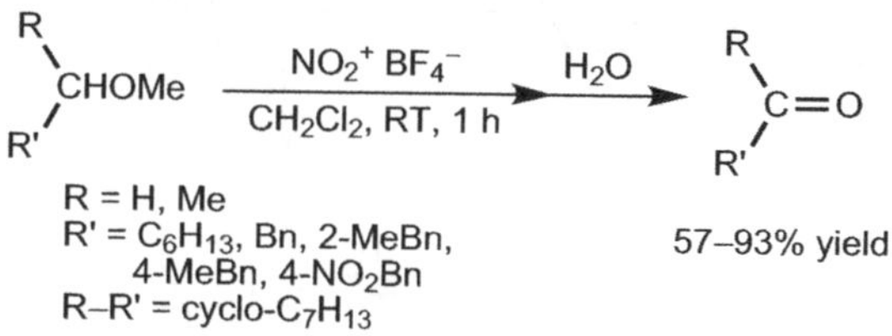

(5.187)

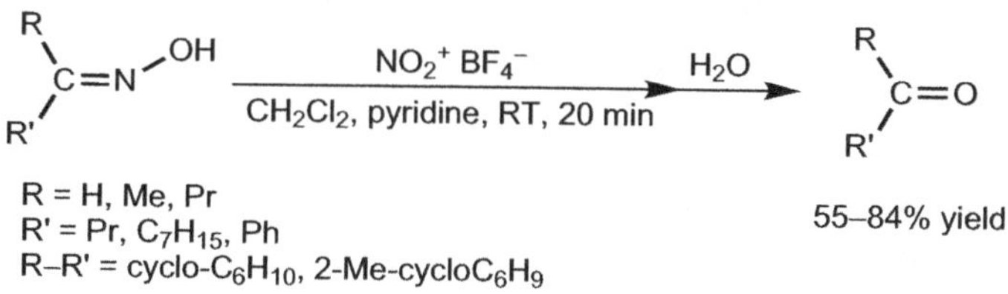

(5.188)

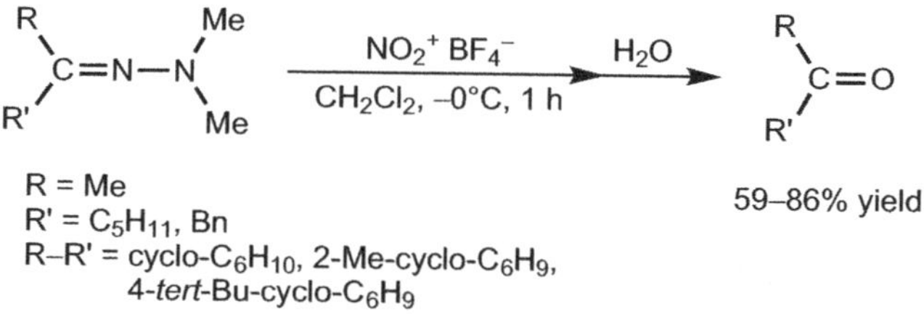

(5.189)

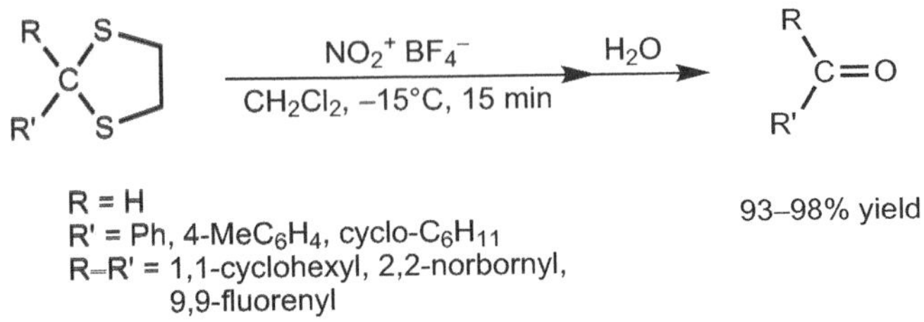

(5.190)

Bicyclo[2.2.1]heptane (norbornane) and bicyclo[2.2.2]octane, when treated with nitronium tetrafluoroborate in nitrile-free nitroethane, unexpectedly gave no nitro products. Instead, only bicyclo[2.2.1]heptane-2-one and bicyclo[2.2.2]octan-1-ol were isolated, respectively.[500] Observation of bicyclo[2.2.1]heptane-2-yl nitrite as an intermediate and additional information led to the suggestion of the mechanism depicted in Scheme 5.48. In the transformation of norbornane the first intermediates are the 2-norbornyl cation **126** formed by hydride abstraction and nonclassical cation **127** formed through insertion of NO_2^+ into the secondary C—H bond. In the case of bicyclo [2.2.2]octane, the oxidation of bridgehead tertiary C—H bond takes place and no further transformation can occur under the reaction conditions. Again these electrophilic oxygenation reactions testify to the ambident character of the nitronium ion.

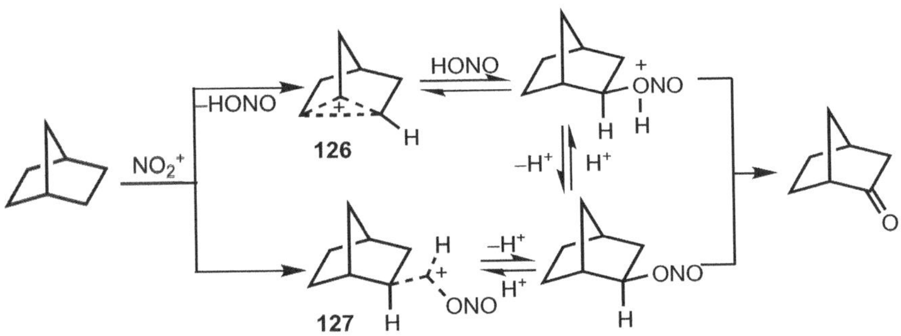

Scheme 5.48

Mononitro compounds could be selectively obtained in the nitration of phenylpropiolic acid derivatives with HNO_3–HSO_3F at low temperature[501] [Eq. (5.191)]. Addition of fluorosulfuric acid to the triple bond, however, may also take place with substituted derivatives.

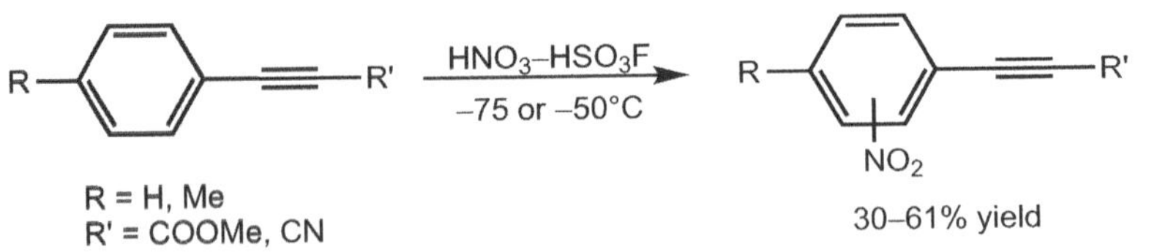

(5.191)

The nitrolysis of the highly energetic *gem*-(bisfluoroamino) derivative **128** has proved to be difficult.[502] The functionalization of the sterically hindered amido groups with HNO_3–triflic acid required long reaction time at elevated temperature [Eq. (5.192)]. Addition of SbF_5 to the nitrating system, in turn, resulted in a faster reaction under milder conditions.

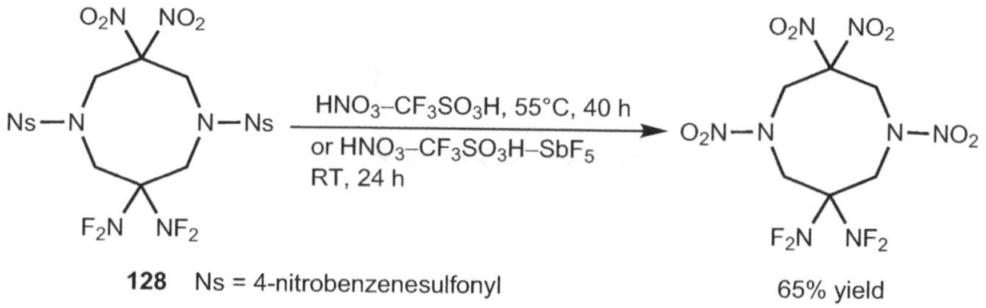

(5.192)

The solid superacidic Nafion-H has also been found to catalyze effectively nitration reactions with various reagents.[503] The nitrating agents employed were *n*-butyl nitrate, acetone cyanohydrin nitrate, and fuming nitric acid. In nitric acid nitrations, sulfuric acid can be substituted by Nafion-H and the water formed is azeotropically removed during the reaction (azeotropic nitration).[503]

Nitrobenzene was also prepared in a liquid flow system using a hollow tube of Nafion-H membrane. Benzene was passed over the outside of tube containing 70% nitric acid. The yield of nitrobenzene was 19%.[504] The nitration of aromatic compounds has also been carried out with Nafion-H in the presence of mercury nitrate as well as Hg^{2+}-impregnated Nafion-H catalyst.[505] These studies indicated that Nafion-H–HNO_3 and Nafion-H–(HNO_3)Hg^{2+} nitrate by different mechanisms. It was proposed tat the mercury-containing catalyst operates in part by mercurating the arene followed by nitrodemercuration of the initial product.

Generally, nitration of aromatic compounds is considered to be an irreversible reaction. However, the reversibility of the reaction has also been demonstrated.[506] 9-Nitroanthracene and pentamethylnitrobenzene transnitrate benzene, toluene, and mesitylene in the presence of HF–SbF_5 as well as Nafion-H [Eq. (5.193)].

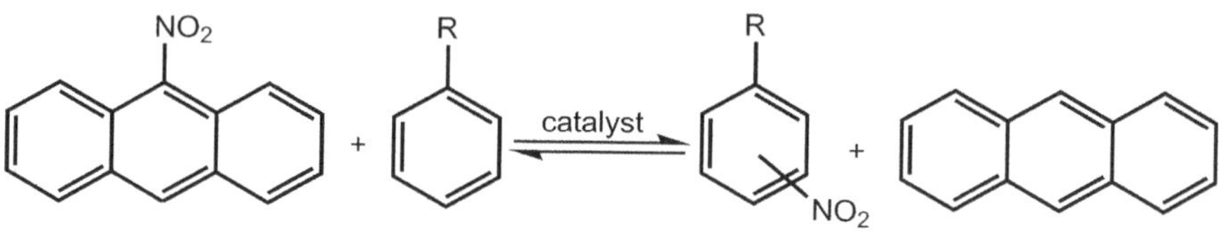

(5.193)

5.9. NITROSONIUM ION (NO^+)-INDUCED REACTIONS

Nitrosonium ion (NO^+) is the electrophilic species formed from nitrous acid media, which is responsible for such reactions as diazotization of amines. Nitrosonium ion has

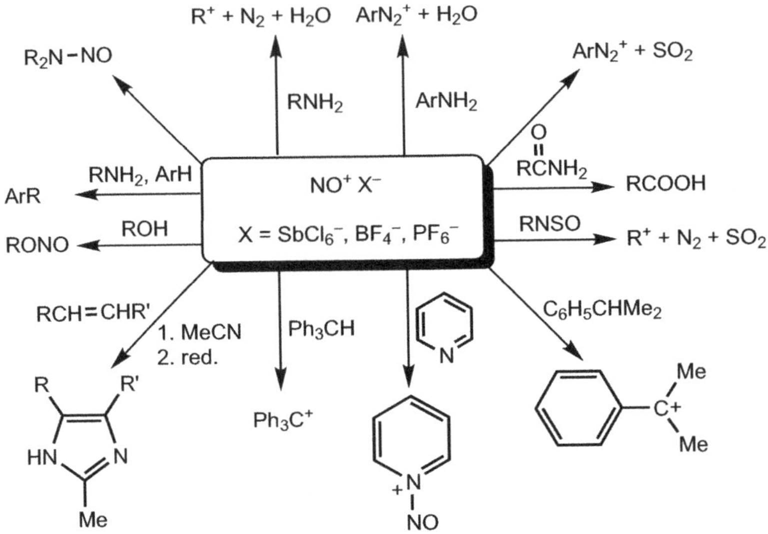

Scheme 5.49

been isolated as salts with a wide variety of counterions such as BF_4^-, PF_6^-, and $SbCl_6^-$. The nitrosonium ion does not react with aromatics except in the case of activated systems (such as *N,N*-dimethylaniline or phenols).[507,508] Frequently, they form colored π-complexes. Nitrosonium ion is a powerful hydride abstracting agent. Cumene reacts with NO^+ to give various condensation products, which involves intermediate formation of cumyl cation.[509] Similarly, the nitrosonium ion is employed in the preparation of a variety of stabilized carbocations.[510] Some of the reactions of nitrosonium ion are depicted in Scheme 5.49.[492]

The hydride abstraction reaction of NO^+ has been employed in a modified Ritter-type reaction[511] (Scheme 5.50, route ***a***) as well as in ionic fluorination[512] of bridgehead hydrocarbons (Scheme 5.50, route ***b***).

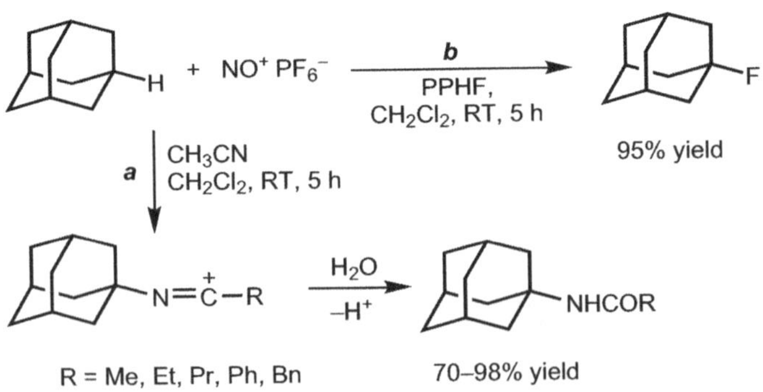

Scheme 5.50

Similarly, NO^+ is also capable of halogen abstraction from alkyl halides.[513,514] In the presence of a suitable oxygen donor such as dimethylsulfoxide, nitrosonium ion can act as a nitrating agent[493,494] [Eq. (5.194)].

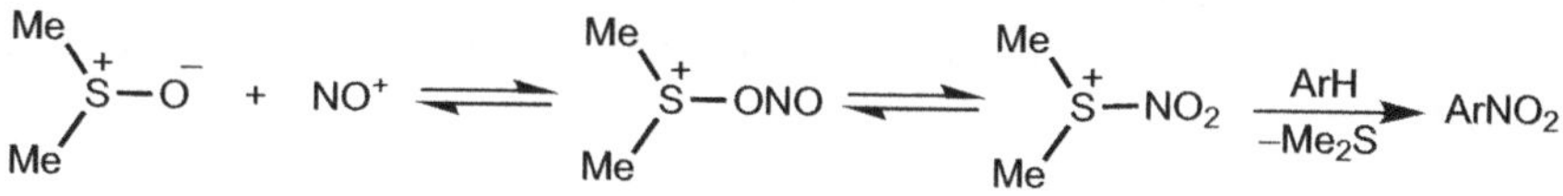

(5.194)

Nitrosonium ion can also act as a mild and selective oxidizing agent. It has been used to cleave oxidatively oximes, hydrazones,[498] and thioketals to their corresponding carbonyl compounds,[499] to cleave benzylic esters[515] [Eq. (5.195)], and to oxidize *O*-tributylstannyl and *O*-trimethylsilyl ethers and benzylic alcohols[516] [Eq. (5.196)].

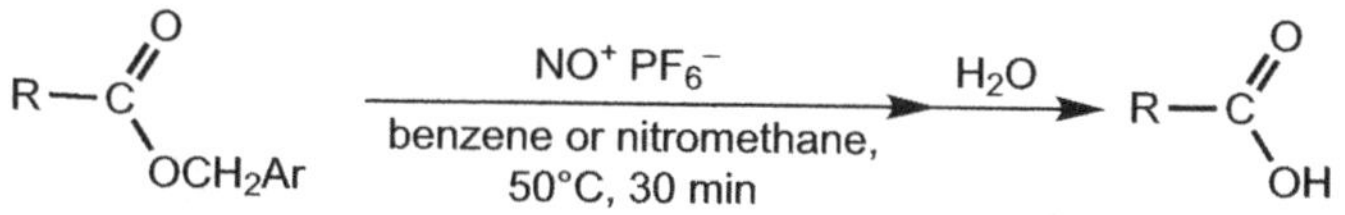

(5.195)

R = $C_{10}H_{21}$, cyclo-C_6H_{11}Ph, 4-ClC_6H_4, 4-$MeOC_6H_4$
Ar = Ph, benzhydryl

82–98% yield

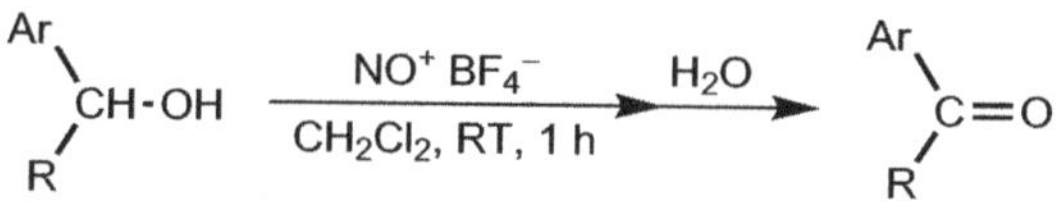

(5.196)

R = H, Me, Ph
Ar = Ph, 4-FC_6H_4, 4-MeC_6H_4

60–78% yield

An additional useful synthetic application is the oxidative nitrolysis with sodium nitrite/pyridinium poly(hydrogen fluoride) (PPHF)[517] of methyl ketones with only the methyl group being enolizable. The transformation, analogous to the haloform reaction, yields carboxylic acids [Eq. (5.197)] and is suggested to proceed through nitrosation of the enolic form of the ketones to form the corresponding oximes via tautomeric rearrangement followed by C–C bond scission.

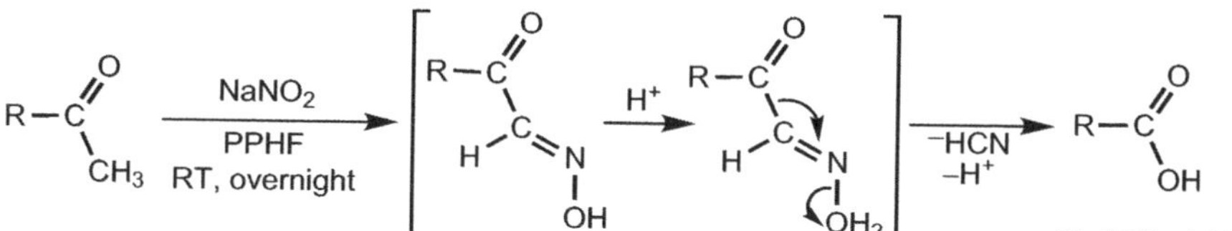

44–85% yield

R = *tert*-Bu, 1-adamantyl, Ph, 2-Me-, 3-MeC_6H_4, 2-MeO-, 3-$MeOC_6H_4$, 4-ClC_6H_4, 4-BrC_6H_4

(5.197)

Nitrosonium ion was found to insert into the Cr–C bond[518] [Eq. (5.198)] and into the C–C bond of the cyclopropane ring[519] [Eq. (5.199)]. The latter reaction opens up a new route to 2-oxazolines. On the basis of experimental observations, particularly the similarity of product distributions in the analogous photoinsertion of NO, the involvement of the intermediate cationic species **129** and **130** was suggested.

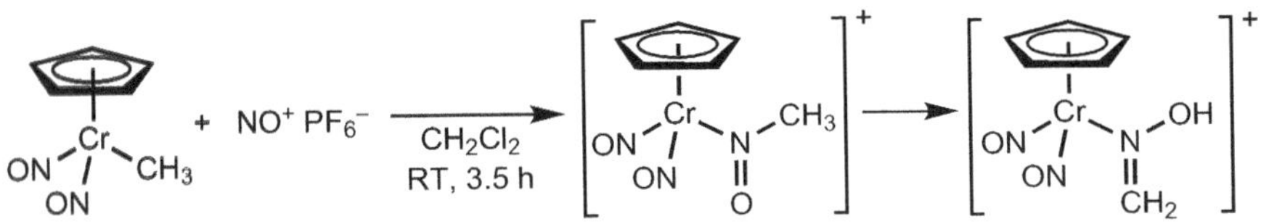

(5.198)

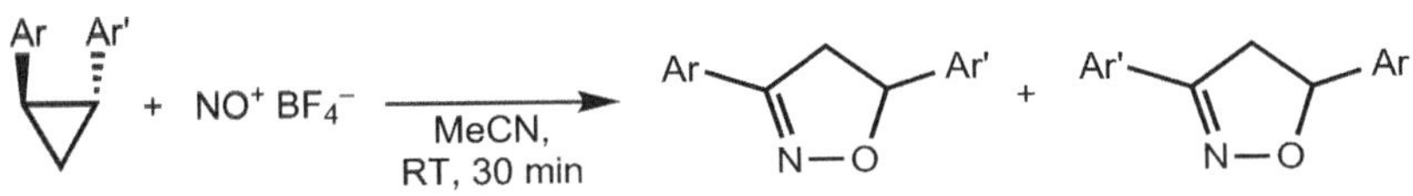

Ar, Ar' = Ph, 4-MeC$_6$H$_4$, 4-MeOC$_6$H$_4$, 4-ClC$_6$H$_4$

26–88% yield
8:2–7:3 isomer ratios

(5.199)

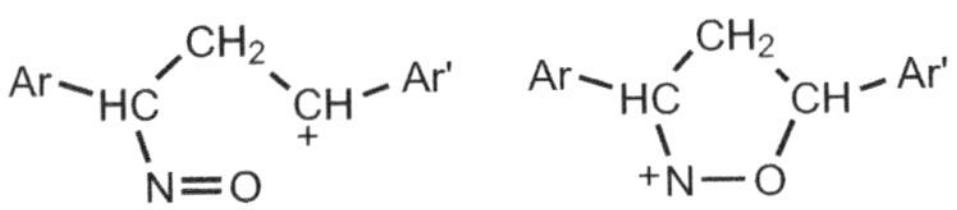

129 **130**

Direct fluorination of diarylacetylenes to the corresponding tetrafluoroethane derivatives has been accomplished with nitrosonium tetrafluoroborate and pyridinium poly(hydrogen fluoride)[520] [Eq. (5.200)].

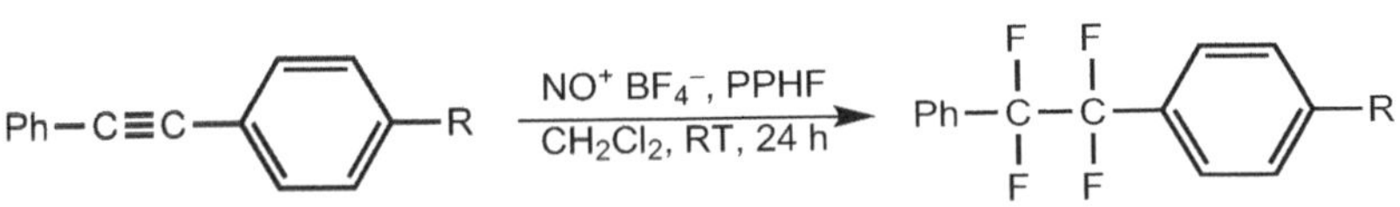

R = H, Me, MeO, F, CF$_3$

38–75% yield

(5.200)

The same reagent combination—that is, NO$^+$BF$_4^-$ (used in an excess of 1.2 equivalent) in conjunction with PPHF—also induces desulfurative fluorination of phenylsulfides[521] (Scheme 5.51). A possible mechanism includes the intermediate cation **131**, which undergoes fluorination via an S$_N$2 or S$_N$1 pathway. Dithiolanes,

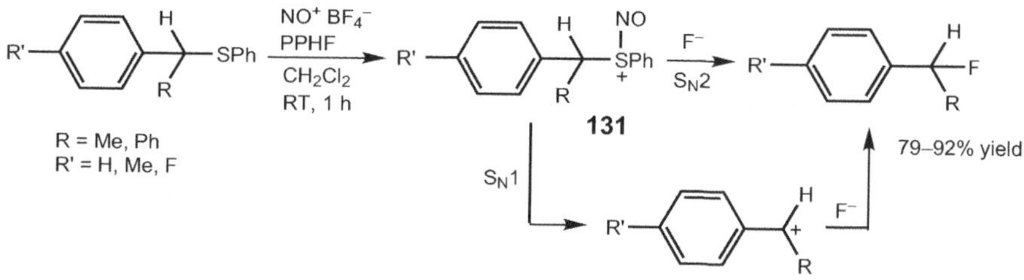

Scheme 5.51

under the same conditions ($NO^+BF_4^-$ in an excess of 2.2 equivalents), afford the corresponding *gem*-difluorides [Eq. (5.201)].

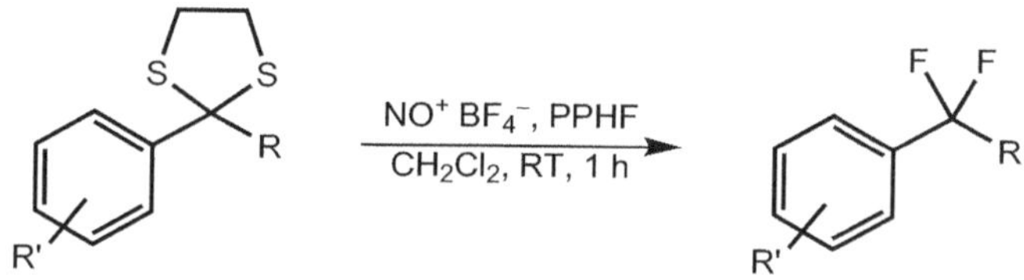

(5.201)

R = Me, Ph, 4-FC_6H_4, 2,4,6-triMeC_6H_2
R' = H, 4-Me, 4-F, 2-MeO, 4-MeO

71–96% yield

Trialkylsilanes and triarylsilanes are fluorinated to the corresponding tertiary organofluorosilanes with $NO^+BF_4^-$ or $NO_2^+BF_4^-$ applied in a slight excess (2–20%)[522] [Eq. (5.202)]. The nitrosonium or nitronium ion induces hydride abstraction followed by fast quenching of the tertiary silicenium ion intermediate by fluoride to give the products in high yields.

$$R_3Si{-}H \xrightarrow[\text{MeCN, 0°C, 2–12 h}]{NO^+BF_4^- \text{ or } NO_2^+BF_4^-} R_3Si{-}F \qquad (5.202)$$

R = Me, Et, isoPr, Bu, *tert*-Bu, Ph, Bn

$NO^+BF_4^-$ 91–98% yield
$NO_2^+BF_4^-$ 88–95% yield

5.10. HALOGENATION

5.10.1. Halogenation of Nonaromatic Compounds

The halogenation of saturated aliphatic hydrocarbons is usually achieved by free radical processes.[523] Ionic halogenation of alkanes has been reported under superacid catalysis. Olah and co-workers[524,525] have carried out chlorination and chlorolysis of

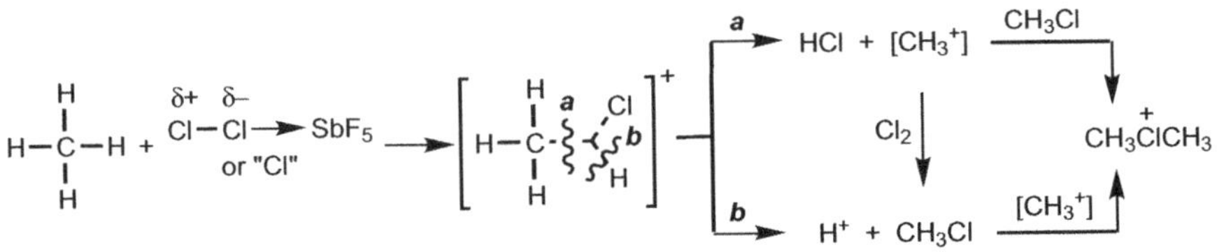

Scheme 5.52

alkanes in the presence of SbF_5, Al_2Cl_6, and $AgSbF_6$ catalysts. As a representative, the reaction of methane with Cl_2–SbF_5 is depicted in Scheme 5.52. The results of $AgSbF_6$-catalyzed chlorination are shown in Table 5.32.

Subsequently, selective ionic chlorination of methane to methyl chloride was achieved in the gas phase over solid superacid catalysts.[526] For example, chlorination of methane in excess chlorine over Nafion-H and SbF_5–graphite gave methyl chloride with 88% and 98% selectivity, respectively (185°C and 180°C, 18% and 7% conversion).[527] Similarly, electrophilic bromination of alkanes has also been carried out[528] (Scheme 5.53).

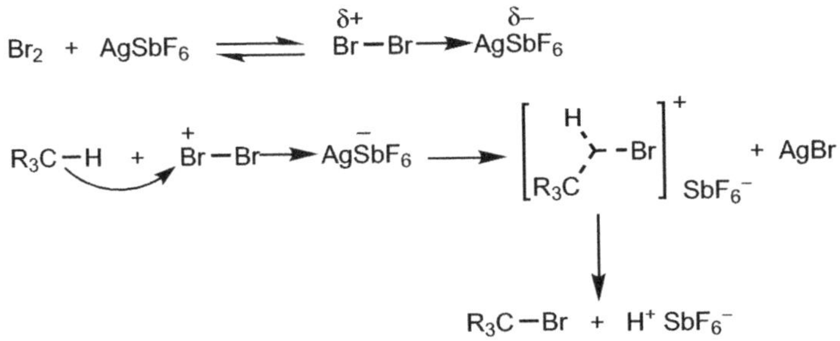

Scheme 5.53

Even electrophilic fluorination of alkanes is possible. F_2 and fluoroxytrifluoromethane have been used to fluorinate tertiary centers in steroids and adamantanes by Barton and co-workers.[529] The strong influence of electron-withdrawing substituents on the reaction rate, as well as reaction selectivity, in the presence of radical inhibitors seems to suggest the electrophilic nature of the reaction involving polarized, but not cationic, fluorine species. Claims for the latter have been refuted.[530] Gal and Rozen[531,532] have carried out direct electrophilic fluorination of hydrocarbons in the presence of chloroform. Fluorine appears to be strongly polarized in chloroform (hydrogen bonding with acidic proton of chloroform). However, positively charged fluorine species (i.e., fluoronium ions) are rare, and only a few examples are known in solution chemistry (see Section 4.2.4.2).

Olah, Prakash, and co-workers[533] have reported the electrophilic fluorination of methane with N_2F^+ and NF_4^+ salts serving as "F^+" equivalents in pyridinium

Table 5.32. $AgSbF_6$-Induced Chlorination of Alkanes with Chlorine in the Dark[525]

Alkanes	Temperature (°C), Time (min)	Reaction Products[a] (%)
Isobutane	−15, 10	*tert*-Butyl chloride (7.1)
	0, 10	*tert*-Butyl chloride (5.3)
Isopentane	−15, 10	*tert*-Pentyl chloride (0.7)
	0, 10	*tert*-Pentyl chloride (1.4)
	10, 10	*tert*-Pentyl chloride (5.0)
Cyclopropane	−15, 10	*n*-Propyl chloride (40.0), Isopropyl chloride (31.5), 1,3-Dichloroporpane (1.2)
	−10, 10	*n*-Propyl chloride (39.2), Isopropyl chloride (36.0), 1,3-Dichloroporpane (1.4)
Cyclopentane	−15, 10	Cyclopentyl chloride (0.5), 1,2-Dichlorocyclopentane (1.8)
	−15, 20	Cyclopentyl chloride (0.3), 1,2-Dichlorocyclopentane (1.9)
	−15, 60	Cyclopentyl chloride (0.5), 1,2-Dichlorocyclopentane (7.1)
	25, 180	Cyclopentyl chloride (0.3), 1,2-Dichlorocyclopentane (3.0)
Cyclohexane	−15, 10	Cyclohexyl chloride (5.1)
	−15, 60	Cyclohexyl chloride (3.3)
Norbornane	25, 15	*exo*-2-Chloronorbornane (66.6), *endo*-2-Chloronorbornane (15.8), 7-Chloronorbornane (7.6)
	25, 30	*exo*-2-Chloronorbornane (69.5), *endo*-2-Chloronorbornane (18.3), 7-Chloronorbornane (6.0)
Adamantane	−15, 5	1-Chloroadamantane (62.4), 2-Chloroadamantane (1.2), 1-Hydroxyadamantane (36.3)

Reaction conditions: Cl_2:$AgSbF_6$:alkane molar ratio = 5:1:10, CH_2Cl_2 as solvent. Analysis by gas chromatography.

[a]Yields of the reaction products based on the amount of $AgSbF_6$.

poly(hydrogen fluoride) (PPHF) [Eq. (5.203)]. Exclusive formation of methyl fluoride was observed with $NF_4^+SbF_6^-$ in a large excess of methane (16 equivalents), whereas the other two reagents ($NF_4^+AsF_6^-$ and $N_2F^+AsF_6^+$) gave methyl fluoride with high selectivity in a methane excess of 4. Calculations with respect to a model of the $CH_4 + F^+$ reaction (QCISD/6-31G*//QCISD/6-31G*+ZPE level) showed that structure **132** (a complex between CH_3^+ and HF) is a global minimum on the potential energy surface. It is 27.7 kcal mol^{-1} more stable than structure **133** (a loose complex between CH_2F^+ and H_2). Formation of cation **132** from CH_4 and F^+ is exothermic by 293.1 kcal mol^{-1}.

$$CH_4 + \text{"}F^{+}\text{"} \xrightarrow[\text{RT}]{\text{HF or PPHF}} CH_3F + CH_2F_2 + CHF_3 \qquad (5.203)$$

	CH_3F	CH_2F_2	CHF_3
$NF_4^+\ SbF_6^-$, 24 h	100%		
$NF_4^+\ AsF_6^-$, 4 h	92%	6%	2%
$N_2F^+\ AsF_6^-$, 4 h	89%	8%	3%

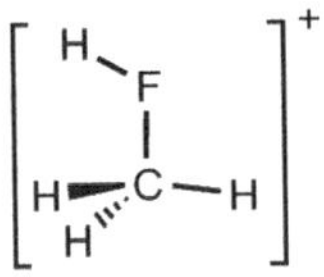

132

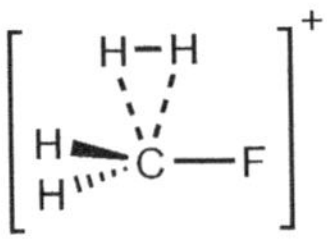

133

Nonactivated C–H bonds in imines can be selectively monofluorinated in HF–SbF_5 in the presence of CCl_4 to yield the corresponding fluoroketones when the reaction mixture is quenched with HF–pyridine[534] (Scheme 5.54). The transformation is initiated by hydride abstraction with CCl_3^+ from the most reactive carbon farthest from the functional group and involves dicationic intermediates **134** and **135**.

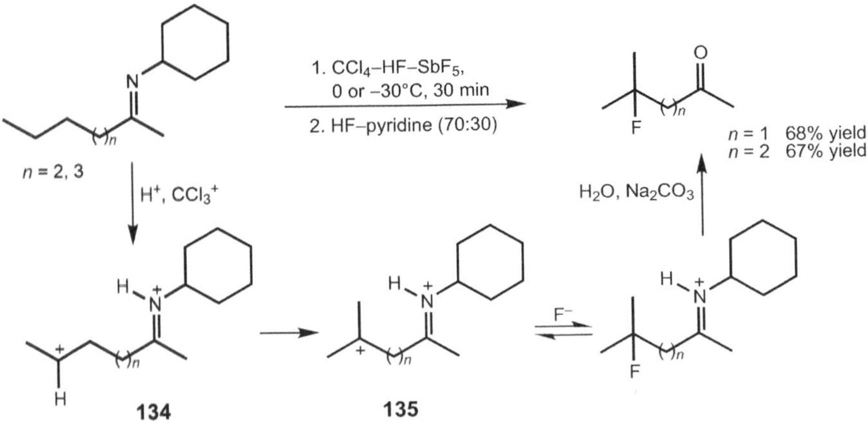

Scheme 5.54

Selective monohalogenation (chlorination and bromination) can be accomplished with excess methylene halides in the presence of SbF_5.[535] Adamantane, for example, gives 1-chloroadamantane in 80% yield at ambient temperature in 24 h.

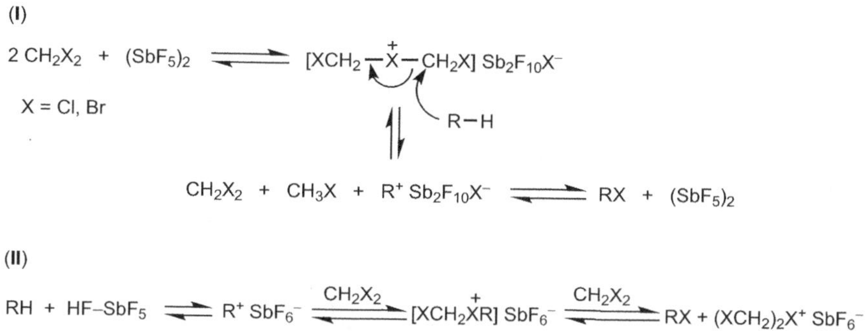

Scheme 5.55

Propane and cyclopentane give isopropyl chloride and cyclopentyl chloride, respectively, whereas isobutane is transformed to *tert*-butyl chloride under the same reaction conditions (yields are 69%, 74%, and 76%, respectively). Neopentane undergoes isomerization to yield 2-chloro-2-butane (88%). When saturated, hydrocarbons were allowed to react with methylene bromide and SbF_5 bromoalkanes were obtained in comparable yields (64–75%). Formation of the halogenated product can be best explained by the mechanistic pathway (**I**) depicted in Scheme 5.55. Since SbF_5 always contains some HF, mechanism (**II**) may also contribute to product formation (Scheme 5.55).

The aprotic organic superacid systems polyhalomethane–$n$$AlX_3$ ($X = Cl$, Br) developed by Vol'pin, Akhrem, and co-workers[536–538] can promote ionic halogenation of saturated hydrocarbons. The bromination of ethane catalyzed by CBr_4–$2AlBr_3$ leads to the formation of 1,2-dibromoethane with high selectivity (55–65°C, CH_2Br_2 solution or solvent-free conditions). Propane yields isopropyl bromide selectively, whereas butane gives isomeric monobromides at −20°C. Cycloalkanes (cyclopentane, cyclohexane, methylcyclopentane) can be monobrominated effectively and selectively at −20°C. Saturated hydrocarbons (propane, cyclopentane, cyclohexane, norbornane, adamantane) also undergo monoiodination with I_2 in the presence of CCl_4–$2AlI_3$ (CH_2Br_2 solution, −20°C, 1.5–2 h, 50–79% yield).[539] The related superacid system AcBr–$2AlX_3$ ($X = Cl$, Br) is able to effectively promote selective monobromination of C_4–C_7 alkanes and cycloalkanes with molecular bromine between −20°C and 20°C.[540] Surprisingly, $AlBr_3$ alone with Br_2 also exhibits activity in bromination.

1-Bromopropargylic imides [Eq. (5.204)] and amines can be transformed to the corresponding fluoromethyl derivatives in HF–SbF_5.[541] The reaction is fast but in some cases affords a mixture of CF_2Br- and CF_3-substituted products. Subsequent treatment with HF–pyridine, however, allows complete bromine–fluorine exchange to get the trifluoromethyl derivatives in high yields (60–96%). A new synthesis of trifluoroalkanes from 1,1-dichloro-1-alkenes under similar conditions has recently been reported.[542] HF in high excess in combination with $SbCl_5$ transforms 1,3-bis(trichloromethyl)benzene to 1,3-bis(trifluoromethyl)benzene in high yield (80%) with high selectivity (99%) (50°C, 1 h).[543] With increasing temperature, redistribution of halogen atoms in the partially fluorinated products becomes significant to form the starting material and the hexafluorinated product. 1-Trichloromethyl-3-trifluoromethylbenzene, the most desired product, can be produced with the highest selectivity in the presence of HF alone. Complete halogen exchange in chloromethoxybenzene is also achieved with HF and a catalytic amount of $SbCl_5$.[544]

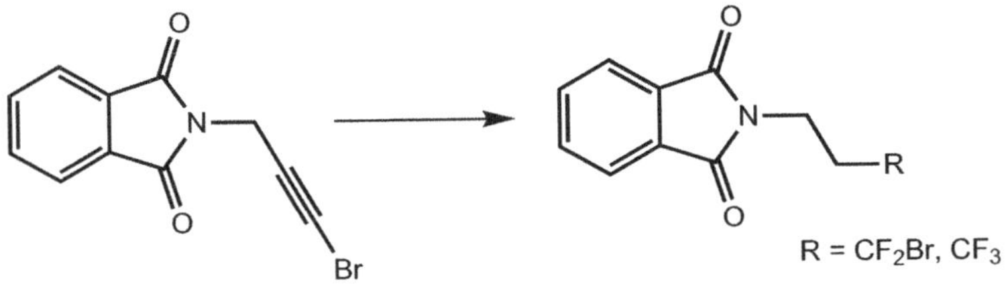

$HF–SbF_5$, −60°C, 10 min — CF_2Br/CF_3 ratio = 11, 77% yield

$HF–SbF_5$, −40°C, 10 min; HF–pyridine, 12 h — $R = CF_3$ 88% yield

(5.204)

The $HF–SbF_5$ superacid finds numerous additional useful applications in other fluorination reactions. Allylic amines are hydrofluorinated in $HF–SbF_5$ (7:1) to give β-fluoro-substituted products (−20°C, 24–85% yields).[545] $HF–SbF_5$ in combination with *N*-bromosuccinimide transforms allylic amines and haloalkyl amines into *gem*-difluoro derivatives (0°C or 20°C, 43–70% yield).[546]

In a search for new fluorinated compounds with biological activity, Jacquesy and co-workers have reported in a series of papers the fluorination of various *Cinchona* and *Vinca* alkaloids and derivatives under superacid conditions. When quinidinone was treated in $HF–SbF_5$ at −78°C, the single monofluoro compound **137** was isolated in 90% yield[547] [Eq. (5.205)]. The reaction is interpreted by invoking the formation of the intermediate cyclic carboxonium ion **136**, which undergoes a concerted rearrangement (ring enlargement) and fluorination to give the final product [Eq. (5.205)].

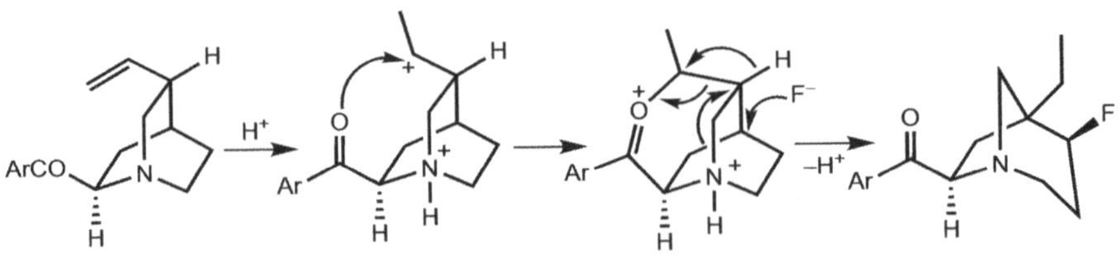

Ar = 6'-MeO-quinolin-4'-yl **136** **137**

(5.205)

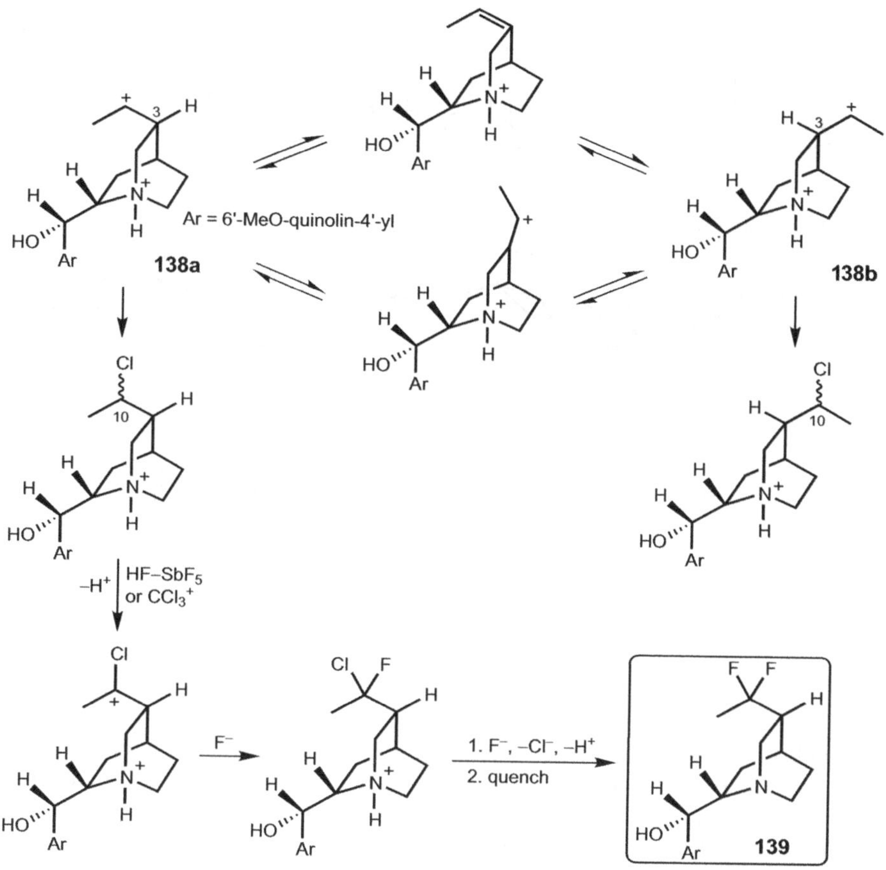

Scheme 5.56

Quinine acetate, 9-epiquinine, and 9-epiquinine acetate give complex mixtures upon treatment in HF–SbF_5 at $-30°C$.[548] In the presence of a chloride source (2 equivalents of CCl_4 or reacting the dihydrochlorides), however, equal amounts of two 10,10-difluoro derivatives, epimeric at C(3), could be isolated in 60% yield. The suggested mechanism, shown in Scheme 5.56 for 9-epiquinine, implies that dication **138a** formed by protonation undergoes isomerization to give dication **138b** epimeric at C(3). The ions are trapped by the complex SbF_5Cl^- to form epimeric intermediates chlorinated at C(10). This is followed by hydride abstraction, reaction with F^-, and finally halogen exchange to give the 10,10-difluorinated products (**139**; only one of the epimers is shown).

Likewise, dihydrochlorides of quinidine and epiquinidine acetate give epimeric 10,10-difluoro derivatives in HF–SbF_5 in the presence of CCl_4 (39% and 30%,

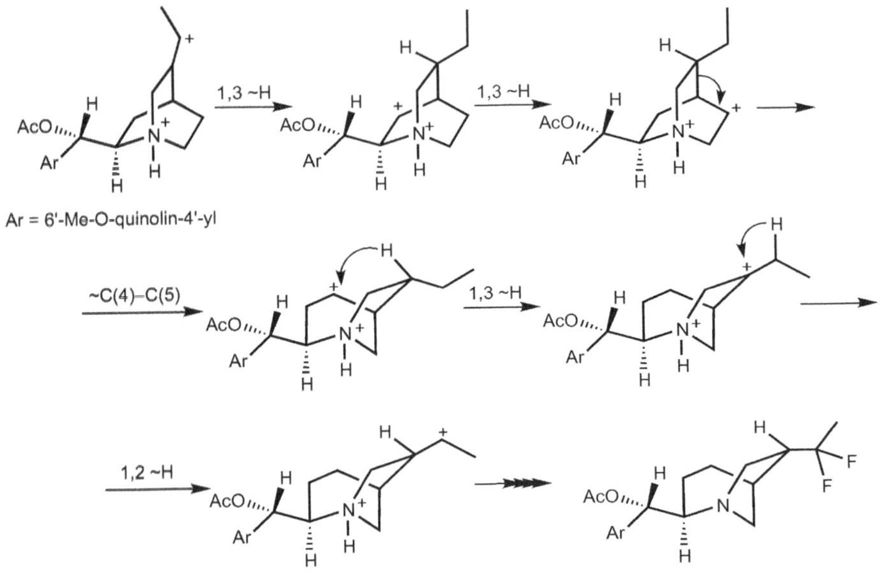

Scheme 5.57

respectively).[549] The pathway given in Scheme 5.56 can also explain the formation of these products. In addition, however, rearranged difluoro derivatives were also obtained in 46% and 45% yield, respectively. This novel rearrangement shown for quinidine acetate (Scheme 5.57) involves 1,2- and 1,3-hydride shifts and a carbon shift.

Two dimeric *Vinca* alkaloids, vinorelbine and anhydrovinblastine, have also been investigated using the $CHCl_3$–HF–SbF_5 system.[550–552] In both cases, the corresponding products difluorinated in the ethyl side chain (**140**) were isolated in modest yields [Eq. (5.206)]. The mechanistic pathway suggested that the transformation includes steps already depicted in Schemes 5.57 and 5.58.

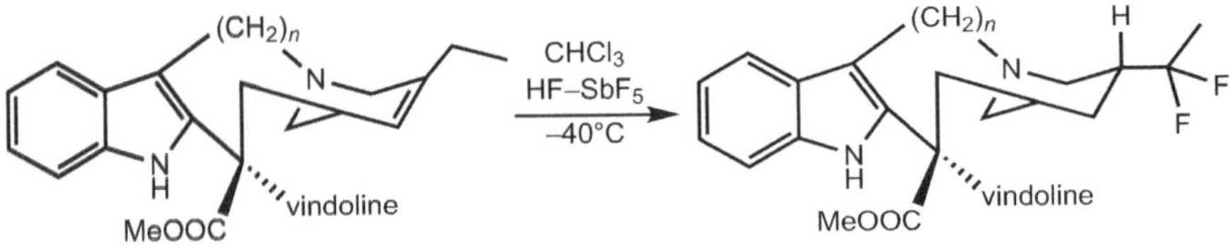

$n = 1$ vinorelbine
$n = 2$ anhydrovinblastine

140
$n = 1$ 35% yield
$n = 2$ 40% yield

(5.206)

Stable dialkyl ether poly(hydrogen fluoride) complexes have been shown to be convenient and effective fluorinating agents.[553] Various open-chain and cyclic alkenes undergo hydrofluorination with dimethyl ether–5 HF (DMEPHF) at room temperature to furnish the corresponding fluoro derivatives in high yields (73–94%) with excellent selectivities. The fluorination of secondary and tertiary alcohols exhibit similar features. Bromofluorination of alkenes can also carried out with DMEPHF in combination with *N*-bromosuccinimide. The homologous diethyl ether and dipropyl ether complexes are also suitable for fluorinations.

5.10.2. Halogenation of Aromatic Compounds

The halogenation of a wide variety of aromatic compounds proceeds readily in the presence of ferric chloride, aluminum chloride, and related Friedel–Crafts catalysts. Halogenating agents generally used are elemental chlorine, bromine, or iodine and interhalogen compounds (such as iodine monochloride, bromine monochloride, etc.). These reactions were reviewed[554] and are outside the scope of the present discussion.

Generally, electrophilic halogenation of phenols leads to the corresponding *ortho*- and *para*-substituted products. The synthesis of *meta*-substituted products is considered difficult.[474,555]

The protonation of phenols and alkyl phenyl ethers in superacids has been extensively investigated.[420] Both oxygen protonation as well as ring protonation occurs (Scheme 5.58). In general, oxygen protonation is observed for unsubstituted phenols and unsubstituted alkyl phenyl ethers, usually at low temperatures. Substitution of electron-releasing groups in the aromatic ring causes ring carbon protonation to predominate.

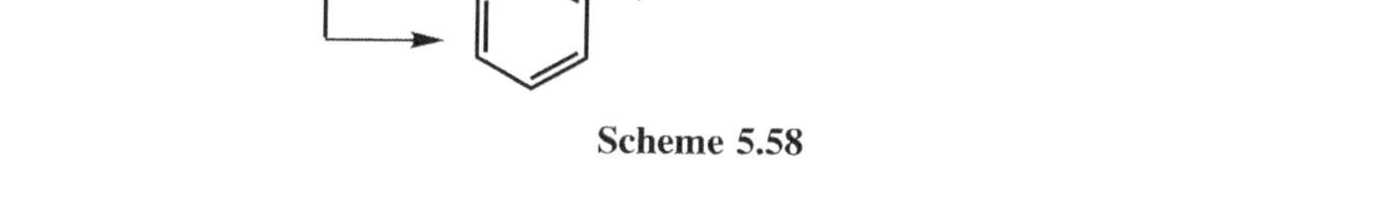

Scheme 5.58

Jacquesy et al.[556] have succeeded in preparing *meta*-bromophenols from phenols using the Br_2–HF–SbF_5 system. The *ortho*-protonated phenol or alkyl phenyl ether, which is in equilibrium with the neutral precursor, reacts with the reactive Br^+ in the HF–SbF_5 medium, leading to only *meta*-brominated phenols (Scheme 5.59). The ring protonated phenols are unreactive toward electrophilic bromine in the superacid

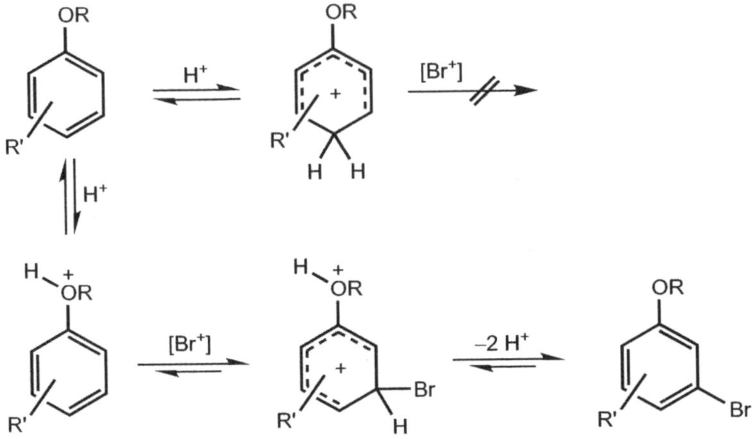

Scheme 5.59

medium. Subsequently, the mechanism of isomerization of *ortho*- and *para*-bromophenols to *meta*-bromophenols using HF–SbF_5 and CF_3SO_3H superacid systems were also studied.[557,558] The Br_2–HF–SbF_5 system was also applied in the bromination of 2,3-dihydro-1*H*-indoles and 1,2,3,4-tetrahydroquinoline to yield isomeric monoboromo derivatives with the bromine always occupying *meta* positions from the N atom (*ortho* or *para* to the side chain).[559]

Later, Sommer and co-workers[560] used NaBr in HF–SbF_5 to brominate anisole and isomeric methylanisoles. High *meta* selectivity was observed only for *para*-methylanisole. A comparative kinetic study showed that bromine is more effective (faster reaction, higher yields) than the NaBr–HF–SbF_5 system, but the product distribution is practically the same. Upon dissolution of NaBr in HF–SbF_5, a strong red color developed and SbF_3 was also detected. On the basis of these observations, the involvement of Br_2^+ formed via oxidation with SbF_5 was suggested.

Jacquesy and co-workers[561] have reacted aniline with Br_2 in the presence HF–SbF_5 and isolated monobrominated anilines (*ortho*/*meta*/*para* = 3:24:50). Isomerization was not observed under the reaction conditions (−40°C), but the products equilibrate at room temperature through 1,2-Br shift. Dibromination yields 2,5-dibromoaniline (10%) and 3,4-dibromoaniline (55%), with the second bromination being controlled by the first bromine introduced.

para-Alkylphenols (*para*-cresol, indan-5-ol, 6-hydroxytetralin) or methyl ethers can be transformed to 4-chloro-2,5-cyclohexadienones via *ipso*-chlorination with the polychloromethane–SbF_5 system (−55 to 0°C, 1.5–90 min, 43–92% yield).[562] Chloromethyl cations $(ClCH_2)_2Cl^+$, $CHCl_2^+$, and CCl_3^+, formed in situ from methylene chloride, chloroform, and carbon tetrachloride, respectively, induce electrophilic chlorination to yield ionic species **141** characterized by NMR spectroscopy [Eq. (5.207)].

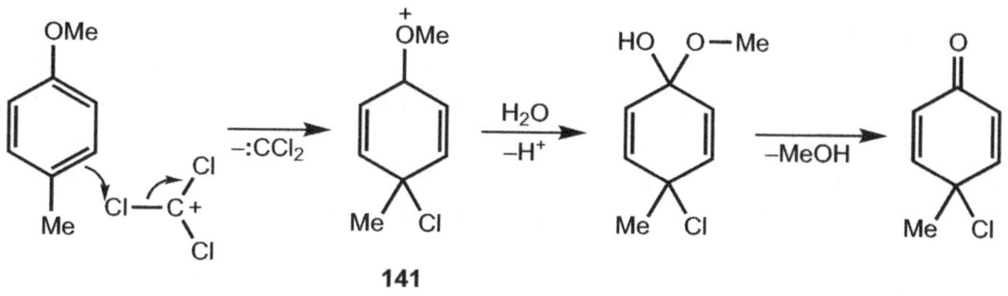

(5.207)

In 1993, Olah, Prakash, and co-workers[563] reported the ability of *N*-iodosuccinimide (NIS) to efficiently iodinate deactivated aromatics in the presence of triflic acid. Deactivated arenes, such as nitrobenzene, halogenated nitrobenzenes, trifluoromethylbenzene, polyhalogenated benzenes, acetophenone, and benzophenone, react readily with NIS in triflic acid (two- or fivefold excess) to give the corresponding iodoarenes in good yields (53–91%) under mild conditions (room temperature, 2 h). The ^{13}C NMR spectrum of NIS–triflic acid showed carbonyl resonances at δ^{13}C 189 and 192 deshielded by 11 ppm from the corresponding peak in NIS. This indicates the presence of both protonated NIS and succinimide [Eq. (5.208)]. The de facto iodinating agent is suggested to be protosolvated superelectrophilic iodine(I) triflate **142**.

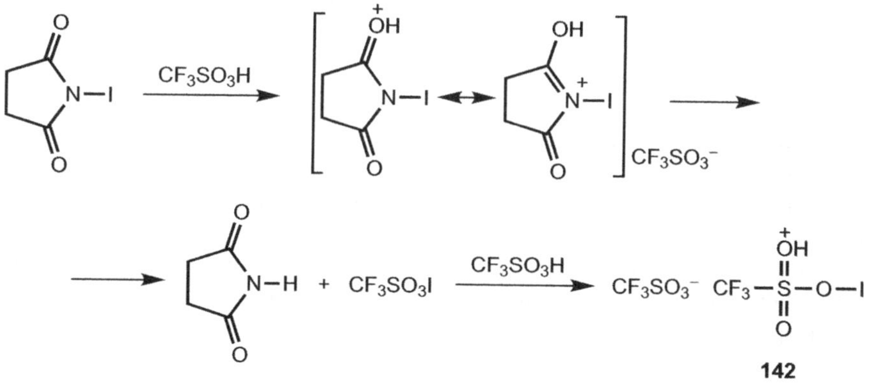

(5.208)

Later, the reagent combination *N*-halosuccinimides with BF_3–H_2O were found to exhibit similar efficiency in chlorination, bromination, and iodination of aromatics[564] [Eq. (5.209)]. Even pentafluorobenzene could readily be halogenated in high yields (83–96%). Most reactions are performed under mild conditions; however, chlorination and bromination of nitro compounds required elevated temperature (100–105°C). DFT calculations (B3LYP/6-311++G**//B3LYP/6-31G* level) for *N*-chlorosuccinimide suggest that multiple protonation can take place at increasing acidities, which may lead to ring opening to reduce internal Coulombic repulsion. The most stable structure with intact ring was found to be dication **143**, which is slightly less stable than the ring-opened dication **144**. Increasing degree of protonation of the reagent leads to higher degree of charge–charge repulsion, that is, increasing destabilization. Consequently, ground-state destabilization is the driving force to transfer the X^+ moiety to the aromatic nucleophile.

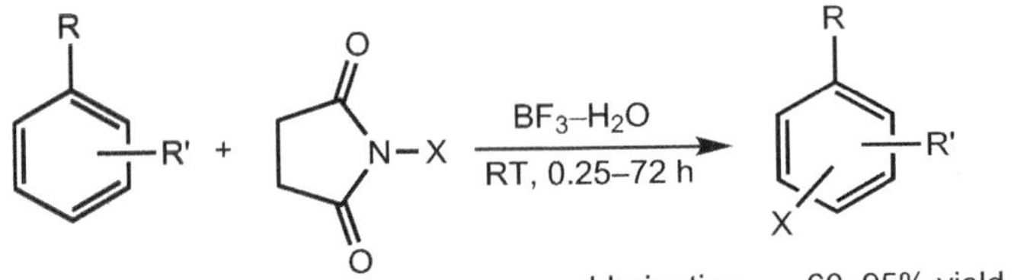

(5.209)

R = F, Cl, Br, I, NO_2
R' = 2-F, 4-F, 3-Br
X = Cl, Br, I

chlorination	60–95% yield
bromination	77–96% yield
iodination	74–98% yield

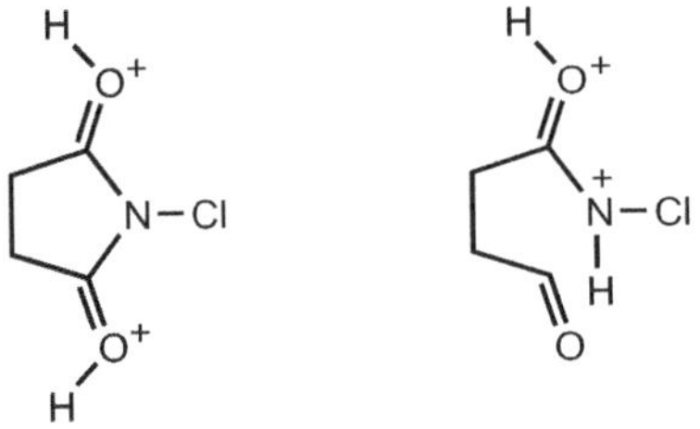

143 **144**

Barluenga et al.[565] have reported the selective monoiodination of arenes with bis (pyridine)iodonium(I) tetrafluoroborate [$I(py_2)BF_4$] in excess superacids (2 equiv.) [Eq. (5.210)]. Comparable results were found for activated compounds with both HBF_4 and triflic acid, whereas triflic acid was more effective in the iodination of deactivated aromatics. For example, nitrobenzene and methyl benzoate are unreactive in HBF_4 but give the corresponding iodo derivatives in triflic acid (83% and 84% yields, respectively, in 14 h). Iodination of phenol required low temperature (–60°C).

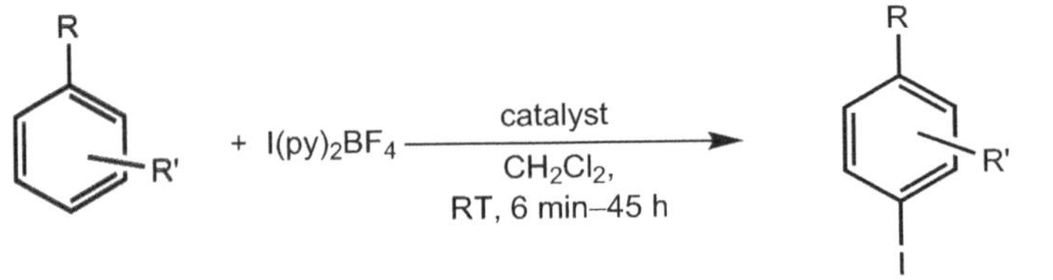

(5.210)

R = H, Me, *tert*-Bu, OH, MeO,
Cl, Br, NH_2, NMe_2, NO_2, CHO, COOH
R' = H, 3-Me, 2-NO_2, 4-Br
2,3-diMe, 3,5-diMe, 3,5-diMeO,

HBF_4	28–99% yields
CF_3SO_3H	40–99% yields

5.11. AMINATION

Electrophilic amination of aromatics is not a widely used reaction. However, attempts have been made with reagents such as hydroxyl ammonium salts,[566–571] hydroxylamine-*O*-sulfonic acid,[572,573] hydrazoic acid,[574–576] and organic azides,[577–580] mostly under Lewis acid-catalyzed conditions (and also with thermal initiation[581–587] or photolysis[588,589]). Kovacic and co-workers[590,591] have found that haloamines can be used as reagents for aromatic amination giving preferential *meta* substitution in the presence of large excess of $AlCl_3$ catalyst. They have also investigated[592] the aromatic amination reaction with hydrazoic acid catalyzed by $AlCl_3$ or H_2SO_4, but the reaction conditions necessitate a 2:1 ratio of catalyst to azide at reflux temperature, giving only modest yields.

Olah and co-workers[593] have carried out a comprehensive investigation of aminodiazonium ions under superacid conditions using ^{13}C and ^{15}N NMR spectroscopic methods (see Section 4.2.5.2). They also studied the electrophilic amination ability of these aminodiazonium ions. It has been found[594,595] that NaN_3 (or trimethylsilyl azide) reacts with $AlCl_3$ and dry HCl in situ to form aminodiazonium tetrachloroaluminate **145** [Eq. (5.211)], which reacts with a variety of arenes to give the corresponding aromatic amines [Eq. (5.212)]. Best results for aromatic aminations were obtained when an excess of the aromatic substrate itself was used as the reaction medium. The results are summarized in Table 5.33.

$$NaN_3 + AlCl_3 \xrightarrow[-NaCl]{} AlCl_2N_3 \xrightarrow{HCl} \underset{\mathbf{145}}{H_2N{-}\overset{+}{N}{\equiv}N\ AlCl_4^-} \qquad (5.211)$$

$$\mathbf{135} + ArH \longrightarrow ArNH_2{\cdot}HCl + N_2\ AlCl_3 \qquad (5.212)$$

Table 5.33. Yield and Isomer Distribution of Aromatic Amines Obtained by Amination of Aromatic Substrates with Aminodiazonium Tetrachloroaluminate[593]

Aromatics	Yield (%)	Isomer Distribution (%) *ortho*/*meta*/*para*
Benzene	63.3	
Toluene	72.6	47.3:13.8:38.9
Mesitylene	77.8	
1,2,3,4-Tetramethylbenzene	39.0	
Chlorobenzene	25.1	28.5:15.3:56.2
para-Xylene	69.8	
Anisole	48.7	74.4:4:21.6
Nitrobenzene	1.5	

Table 5.34. Yield and Isomer Distribution of Aromatic Phenylamines Obtained with Phenylazide–Triflic Acid

	Time (h)	Temperature (°C)	Yield (%)	Isomer Distribution (%), *ortho/meta/para*
Benzene[a]	1	55	82	
Toluene[a]	1	60	94	36:1:63
ortho-Xylene[a]	1	50	90	37.5:0:62.5
Chlorobenzene[a]	2	65	82	20:3:77
Bromobenzene[a]	2	60	88	14.6:0.8:84.6
Naphthalene[b]	2	50	84[c]	
Anisole[b]	2	42	75	35:0:65
Adamantane[b]	2	50	[d]	
Benzonitrile[b]	2	42	Trace	
Nitrobenzene[b]	2	42	[d]	

[a]Reactant was used as solvent.
[b]In CH_2Cl_2 as solvent.
[c]1-Phenylaminonaphthaline was obtained in >95% yield.
[d]No reaction.

Subsequently they showed that trimethylsilylazide is an even more effective reagent when applied in the presence of triflic acid[596] to afford the corresponding aminated products in high yields (73–95%, 40–70°C, 50–90 min). The suggested reaction pathway involves the formation of the aminodiazonium ion intermediate upon protonation with triflic acid, which then reacts with the arenes. This method has been successfully applied in the surface functionalization of polystyrene nanospheres (PS) with a loading of 0.19 mmol of amine group g^{-1} [Eq. (5.213)].[597]

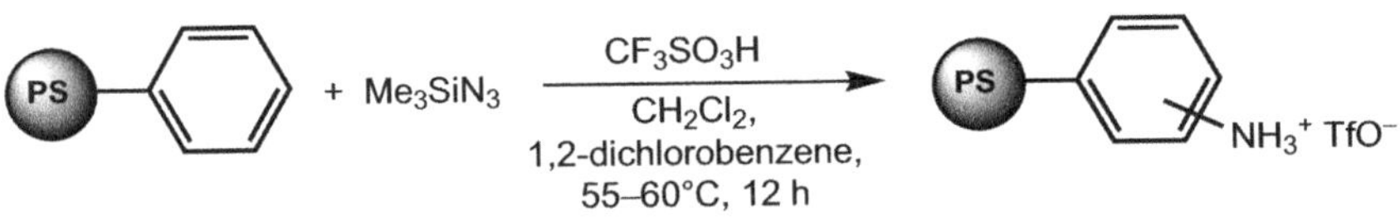

(5.213)

Triflic acid effectively promotes the phenylamination of aromatics with phenylazide in a fast, convenient, high-yield process[598] (Table 5.34). The high *ortho*/*para* selectivity with only a small amount of *meta* product and high substrate selectivity ($k_T/k_B = 11$) indicate the involvement of a substantially electron-deficient species, the phenylaminodiazonium ion intermediate with the possible protosolvation by triflic acid.

5.12. OXYFUNCTIONALIZATION

Converting alkanes and aromatics in a controlled way into their oxygenated compounds is of substantial interest. The discovery and development of superacidic systems and weakly nucleophilic solvent such as HSO_3F–SbF_5–SO_2, HSO_3F–SbF_5–SO_2ClF, and HF–SbF_5–SO_2ClF have enabled the preparation and study of a variety of carbocations (see Chapter 3). In connection with these studies, it was also found that electrophilic oxygenation of alkanes with ozone (O_3) and hydrogen peroxide (H_2O_2) takes place readily in the presence of superacids under typical electrophilic conditions.[599] The reactions giving oxyfunctionalized products of alkanes can be explained as proceeding via initial electrophilic attack by protonated ozone HO_3^+ or the hydrogen peroxonium ion $H_3O_2^+$, respectively, on the σ-bond of alkanes through pentacoordinated carbonium ions.

In recent years, oxyfunctionalization of various natural products (steroids, alkaloids) under superacidic conditions have also been explored. In addition, Nafion resins in combination with various oxidizing agents have also been used in the oxygenations.

5.12.1. Oxygenation with Hydrogen Peroxide

5.12.1.1. Oxygenation of Alkanes Hydrogen peroxide (H_2O_2) in superacidic media is protonated to hydrogen peroxonium ion ($H_3O_2^+$) [Eq. (5.214)]. Christe et al.[600] have reported characterization and even isolation of several peroxonium salts. The ^{17}O NMR spectrum of $H_3O_2^+$ has also been obtained.[601]

$$\text{HOOH} \xrightleftharpoons{H^+} \text{H}\overset{\text{H}}{\underset{+}{\text{O}}}\text{OH} \equiv [\overset{+}{\text{O}}\text{H}\ \text{H}_2\text{O}] \qquad (5.214)$$

The hydrogen peroxonium ion may be considered as an incipient OH^+ ion capable of electrophilic hydroxylation of single (σ) bonds in alkanes and, thus, be able to effect reactions similar to such previously described electrophilic reactions as protolysis, alkylation, chlorination (chlorolysis), and nitration (nitrolysis).

The reaction of branched-chain alkanes with hydrogen peroxide in Magic Acid–SO_2ClF solution has been carried out with various ratios of alkane and hydrogen peroxide and at different temperatures.[602] Some of the results are summarized in Table 5.35.

Because neither hydrogen peroxide nor Magic Acid–SO_2ClF alone led to any reaction under the conditions employed, the reaction must be considered to proceed via electrophilic hydroxylation. Protonated hydrogen peroxide inserts into the C–H bond of the alkane. A typical reaction path is as depicted in Scheme 5.60 for isobutane.

The reaction proceeds via a pentacoordinate hydroxycarbonium ion transition state, which cleaves to either *tert*-butyl alcohol or the *tert*-butyl cation. Since 1 mol of isobutane requires 2 mol of hydrogen peroxide to complete the reaction, one can conclude that the intermediate alcohol or carbocation reacts with excess hydrogen peroxide, giving *tert*-butyl hydroperoxide. The superacid-induced rearrangement and cleavage of the hydroperoxide results in very rapid formation of the dimethylmethylcarboxonium ion, which, upon hydrolysis, gives acetone and methyl alcohol.

When the reaction is carried out at room temperature, by means of passing isobutane into a solution of Magic Acid and excess hydrogen peroxide, the formation

Table 5.35. Products of the Reaction of Branched-Chain Alkanes with H_2O_2 in $HSO_3F–SbF_5–SO_2ClF$ Solution[602]

Alkane	Alkane mmol	H_2O_2 mmol	Temperature[a] (°C)	Major Products
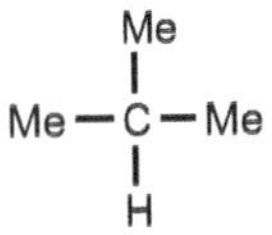	2	2	−78 to −20	$(Me_2C{=}OMe)^+$
	2	4	−78 to −20	$(Me_2C{=}OMe)^+$
	2	6	−78 to −20	$(Me_2C{=}OMe)^+$
	2	6	+20	$(Me_2C{=}OMe)^+$ (trace), DAP[b] (25%), MeOH (50%), $MeCO_2Me$(25%)
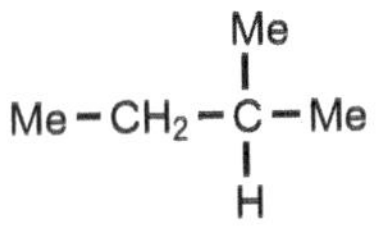	2	3	−78 to −20	$(Me_2C{=}OEt)^+$, $EtMe_2C^+$
	2	6	−78 to −20	$(Me_2C{=}OEt)^+$
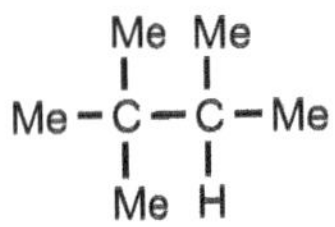	2	4	−78	$(Me_2C{=}OMe)^+$ (50%), $(Me_2C{=}OH)^+$ (50%)
	2	6	−40	$(Me_2C{=}OMe)^+$ (50%), DAP[b] (50%)

[a] ^1H NMR probe temperature.
[b] DAP: dimeric acetone peroxide.

of methyl alcohol, methyl acetate, and some dimethylmethylcarboxonium ion together with dimeric acetone peroxide was observed. These results clearly show that the products observed can be rationalized as arising from hydrolysis of the carboxonium ion and from Baeyer–Villiger oxidation of acetone.

The mechanism was substantiated by independent treatment of alkane hydroperoxides with Magic Acid.[603] Similarly, Baeyer–Villiger oxidation of several ketones in the presence of H_2O_2 and superacids gave similar product compositions.

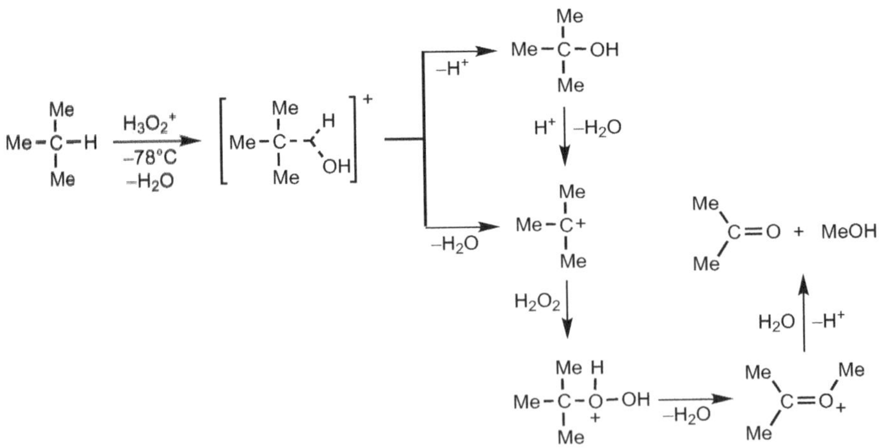

Scheme 5.60

Under the same reaction conditions as employed for branched-chain alkanes, straight-chain alkanes such as ethane, propane, butane, and even methane gave related products.[602]

Methane, when reacted with hydrogen peroxide–Magic Acid above 0°C, gave mainly methyl alcohol. A similar result was obtained with hydrogen peroxide–HSO_3F at 60°C. Ethane with hydrogen peroxide–Magic Acid at −40°C gave ethyl alcohol. The reaction of propane with hydrogen peroxide takes place more easily than that of methane or ethane and yields isopropyl alcohol as the initial oxidation product. On raising the temperature, isopropyl alcohol gave acetone, which underwent further oxidation with hydrogen peroxide, giving dimeric acetone peroxide, methyl acetate, methyl alcohol, and acetic acid.

The activation barriers for the hydroxylation of simple alkanes (methane, ethane, propane, butane, and isobutane) with the hydroperoxonium ion were calculated to be 5.26, 0.16, −4.64, −4.74, and −4.98 kcal mol^{-1}, respectively (MP4//MP2/6-31G** level).[604] The same order of reactivity has been found at the B3LYP/6-31G(d,p) level (the corresponding activation barrier values for the hydroxylation of methane and ethane are 2.58 and 1.40 kcal mol^{-1}, respectively).[605] Both studies showed that the high reactivity of the $H_3O_2^+$ ion originates from its relatively low LUMO energy (−9.18 eV). This value is approximately the same or even lower than the HOMO energy of alkanes (−10.8 to −8.0−eV). This facilitates a nucleophilic electron transfer of σ electrons to the σ^* orbital of the peroxo bond. Reaction path calculations showed that the electrophilic oxygen atom of the hydroperoxonium ion attacks the hydrogen atom of the C−H bond.[605] Consequently, dehydration takes place without the intervention of the pentacoordinate carbocation, and the carbenium ion formed is then hydrated to form protonated alcohol.

5.12.1.2. *Oxygenation of Aromatics.* Although there have been reports of the direct, one-step hydroxylation of aromatic compounds with peracids in the presence of acid catalysts, monohydroxylated products (i.e., phenols) have generally been obtained in only low yields.[606–615] Although moderate-to-good yields of phenols, based on the amount of hydrogen peroxide used, were reported for the $AlCl_3$-catalyzed reaction of simple aromatics with hydrogen peroxide, a 10-fold excess of aromatics was used over hydrogen peroxide.[615] The conversion of the aromatics thus was low, probably due to the fact that introduction of an OH group into the aromatic ring markedly increases its reactivity and thus tends to promote further side reactions.[616]

It is well-recognized that phenols are completely protonated in superacidic solutions.[420] This raised the possibility that protonated phenols, once formed in these media, might resist further electrophilic attack. Electrophilic hydroxylations of aromatics with hydrogen peroxide (98%) in superacidic media has been achieved by Olah and Ohnishi[617] in Magic Acid, which allows clean, high-yield preparation of monohydroxylated products. Benzene, alkylbenzenes, and halobenzenes are efficiently hydroxylated at low temperatures. The obtained yields and isomer distributions are shown in Table 5.36. Subsequently, Olah et al.[618] found that benzene and

Table 5.36. Yields and Isomer Distributions of the Hydroxylation of Aromatics[617]

Aromatic Substrate	Isomer Distribution[a] (%)			Yield[b] (%)
Benzene				67
Fluorobenzene	24 (2)	3 (3)	73 (4)	82
Chlorobenzene	28 (2)	7 (3)	65 (4)	53
Toluene	71 (2)	6 (3)	23 (4)	67
Ethylbenzene	68 (2)	6 (3)	26 (4)	70
sec-Butylbenzene	49 (2)	11 (3)	40 (4)	55
Isobutylbenzene	65 (2)	7 (3)	28 (4)	83
n-Amylbenzene	64 (2)	7 (3)	29 (4)	67
ortho-Xylene	12 (2,6)	59 (2,3)	29 (3,4)	63
meta-Xylene	16 (2,6)	2 (2,5)	82 (2,4) 1 (2,3)	73
para-Xylene	64 (2,5)	36 (2,4)		65
1,2,3-Trimethylbenzene	3 (2,3,6)	91 (2,3,4)	6 (3,4,5)	43
1,2,4-Trimethylbenzene	9 (2,4,6)	30 (2,3,6)	61 (2,3,5 + 3,4,6)	57
1,3,5-Trimethylbenzene	100 (2,4,6)			57

Reactions were performed in HSO_3F-SO_2ClF solution at −78°C.
[a]Based on chromatographic analysis of quenched phenolic products. Parentheses show position of substituent(s).
[b]Based on aromatics used.

alkylbenzenes are smoothly hydroxylated using 30% H_2O_2 in $HF-BF_3$ acid system at low temperatures. The above method is particularly attractive because the acid system is recoverable and recyclable.

All these reactions involve hydroxylation of aromatics by hydrogen peroxonium ion [Eq. (5.215)].

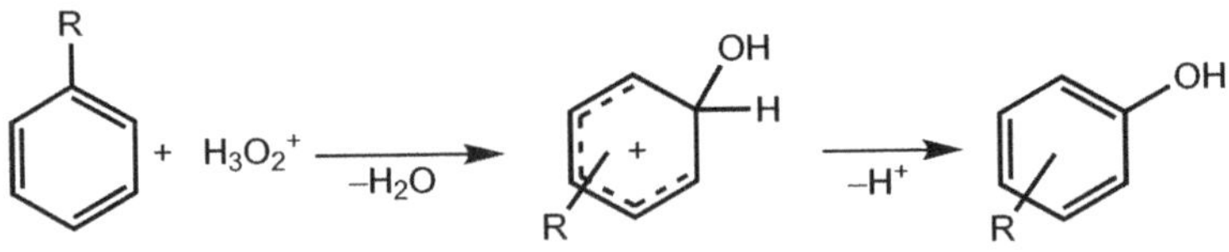

(5.215)

Olah et al.[619] have also hydroxylated naphthalene to the corresponding naphthols. Depending upon the acid strength of the medium employed, preferentially α- or β-naphthol is formed. In hydroxylation with 90% H_2O_2 in hydrogen fluoride or HF–solvent systems the actual electrophilic hydroxylating agent is the hydroperoxonium ion ($H_3O_2^+$), resulting in the formation of α-naphthol by the usual aromatic electrophilic substitution mechanism. In superacids ($HF-BF_3$, $HF-SbF_5$, $HF-TaF_5$, HSO_3F, HSO_3F-SbF_5), in contrast, β-naphthol is formed through the attack of hydrogen peroxide to protonated naphthalene according to Eq. (5.216).

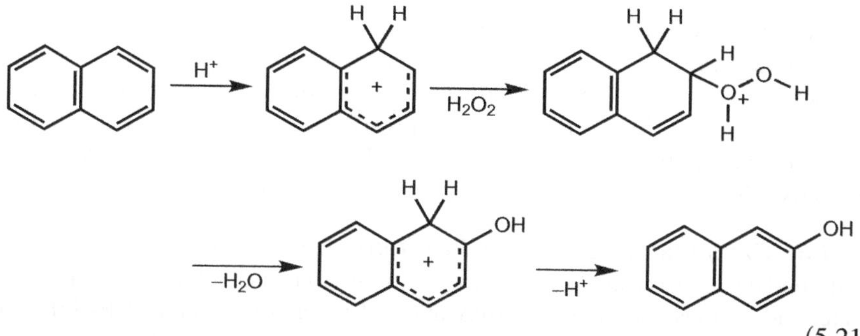

(5.216)

Jacquesy and co-workers[620] have developed a new route to resorcinols by reacting alkyl-substituted phenols or their ethers with H_2O_2 in HF–SbF_5. The products observed are formed by the reaction of $H_3O_2^+$ either on neutral substrate or on the corresponding *O*-protonated ions. When the *C*-protonated form is highly stabilized by alkyl substituents, no hydroxylation occurs. Ring-substituted higher phenol ethers (R = Et, *n*-Pr) isomerize[621] and are then hydroxylated to dialkyl resorcinol.[622] Even hydroxylation of aromatic aldehydes and ketones in HF–SbF_5 medium has been achieved with H_2O_2.[623] No Baeyer–Villiger oxidation products were detected in these reactions.

The H_2O_2–HF–SbF_5 system has been applied by Jacquesy and co-workers in the hydroxylation of a variety of functionalized arenes.[624] Hydroxylation of phenyl esters has been shown to afford the *meta* and *para* isomers as main products[625] [Eq. (5.217)]. Substantial amounts of the deacylated derivatives were obtained in the reaction of phenyl formate and diphenyl carbonate. In the hydroxylation of 2-chlorophenyl and 4-chlorophenyl acetate, regioselectivity is controlled by the chlorine substituent with the hydroxyl entering into the *meta* position to the ester group.[626] A similar effect was observed in the hydroxylation of anilines and anilides.

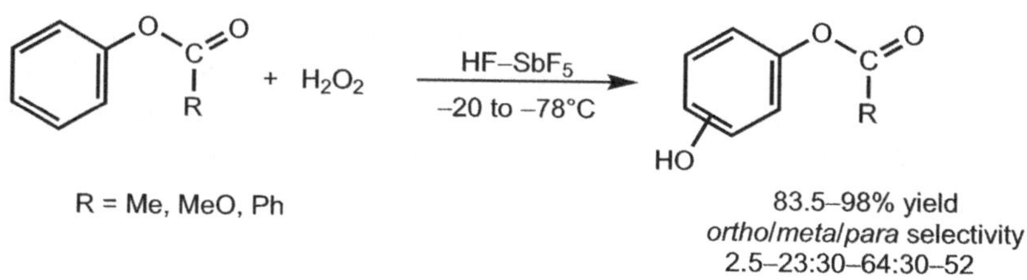

(5.217)

The procedure has also been applied for the hydroxylation of aromatic amines. Aniline and its *N*-alkyl-substituted derivatives show similar behavior under similar conditions to afford the *meta*-substituted aminophenols as the major hydroxylated product.[627] Product formation was interpreted by the attack of protonated hydrogen peroxide on the anilinium ion protected by *N*-protonation from oxidation or degradation. Indoles, indolines, and tetrahydroquinoline have also been successfully hydroxylated with H_2O_2 in HF–SbF_5 with the hydroxyl group *meta* to the nitrogen function.[559,628] Hydroxylation of tryptophane and tryptamine derivatives affords pretonine and serotonine derivatives in 42% and 38% yields, respectively.[629]

It is noteworthy to point out that polyhydroxylation in most cases is practically suppressed under the reaction conditions because phenols are totally *O*-protonated in the superacid media and thus deactivated against further electrophilic attack or secondary oxidation. However, it has been shown by Jacquesy et al.[630] that hydroxylation of α- and β-naphthols can be achieved in H_2O_2–HF–SbF_5 under certain conditions. The hydroxylation occurs selectively on the nonphenolic ring.

5.12.1.3. *Oxygenation of Natural Products* Jacquesy and co-workers[631] have also studied the hydroxylation of a series of natural products, including estrone and its acetate, and various alkaloids. Hydroxylation of vincadifformine in HF–SbF_5 with H_2O_2 (4 equiv., 0°C, 30 min) yields a mixture of two monohydroxylated derivatives in 60% yields, whereas products with exclusive substitution at C(11) (**146**) are obtained in the reaction of 2,16-dihydrovincadifformine and its 14β-hydroxy analog.[632] This corresponds to the same regioselectivity observed for indoles and indolines. A mixture of all four possible hydroxylated derivatives (**147**) were obtained in a combined yield of 69% in the oxidation of yohimbine with Na_2SO_5 (HF–SbF_5, −35°C, 1 h).[633]

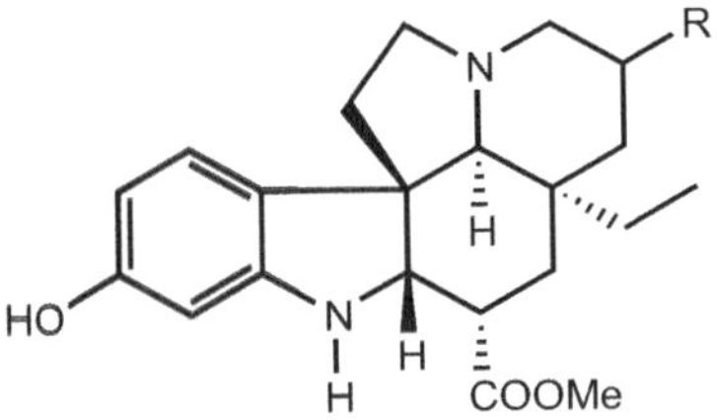

146a R = H 40% yield
146b R = OH 37% yield

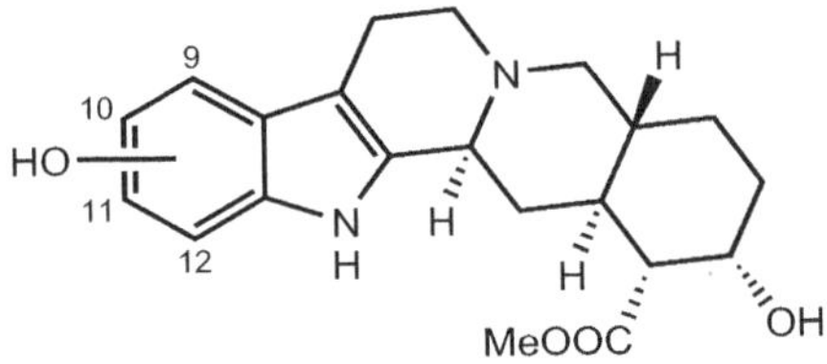

147
9-OH 11%, 10-OH 32%
11-OH 23%, 12-OH 3%

Quinine acetate, 9-epiquinine, and 9-epiquinine acetate give mixtures of 10-keto (**148**) and 10-fluoro-3-hydroxy (**149**) derivatives upon treatment with H_2O_2 in HF–SbF_5 (Scheme 5.61).[634] Each product is epimeric at C(3). The initial isomerization (double bond migration) to form the exocyclic alkene **150** followed by the attack of electrophile "OH^+" equivalent to yield isomeric protonated epoxides **151** are the key steps in product formation [Eq. (5.218)]. Isomerization of epoxides affords the ketones and *trans* ring opening by F^- gives the fluorohydrins.

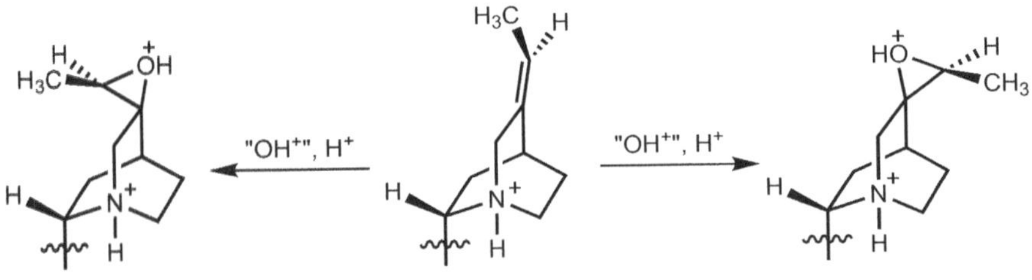

151a **150** **151b**

(5.218)

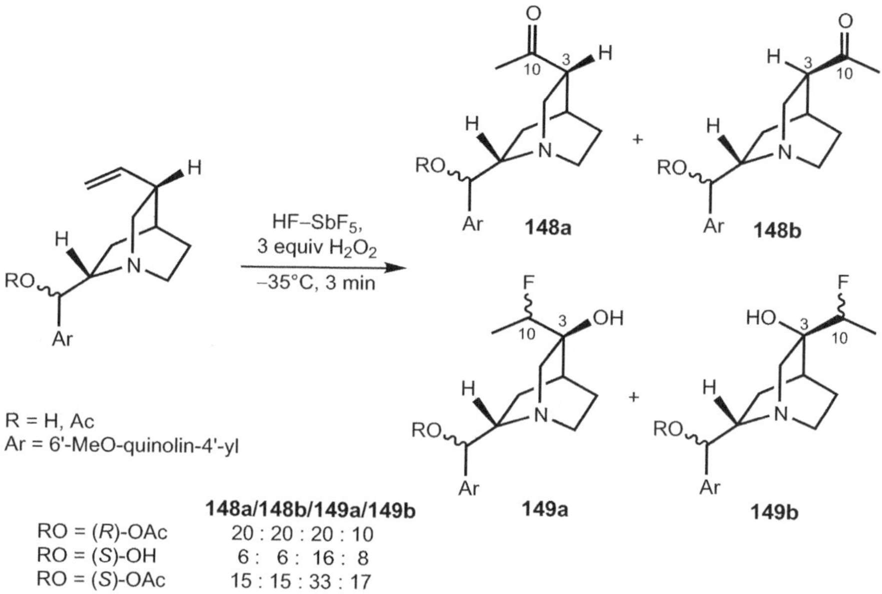

Scheme 5.61

5.12.2. Oxygenation with Ozone

Ozone can be depicted as the resonance hybrid of canonical structures **152a–152d**,[635] and one might expect ozone to react as a 1,3-dipole, an electrophile, or a nucleophile. The electrophilic nature of ozone has been recognized in its reactions toward alkenes, alkynes, arenes, amines, sulfides, phosphines, and so on.[636–640] Reactions of ozone as a nucleophile, however, are less well documented.[641]

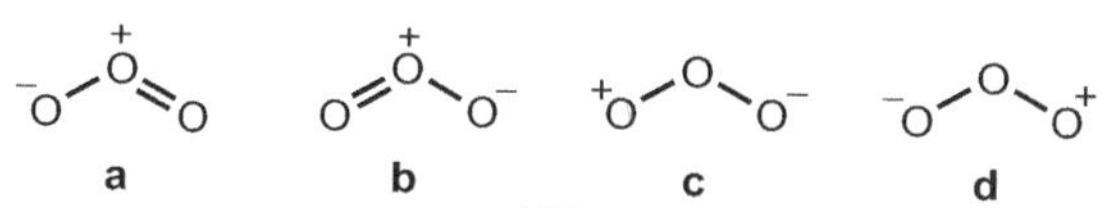

152

It was shown in the reaction of ozone with carbenium ions[642] that the initial attack as expected is alkylation of ozone giving rise to intermediate trioxide, which then undergoes carbon to oxygen alkyl group migration with simultaneous cleavage of a molecule of oxygen, similar to the acid-catalyzed rearrangement of hydroperoxides to carboxonium ions (Hock reaction) [Eq. (5.219)].

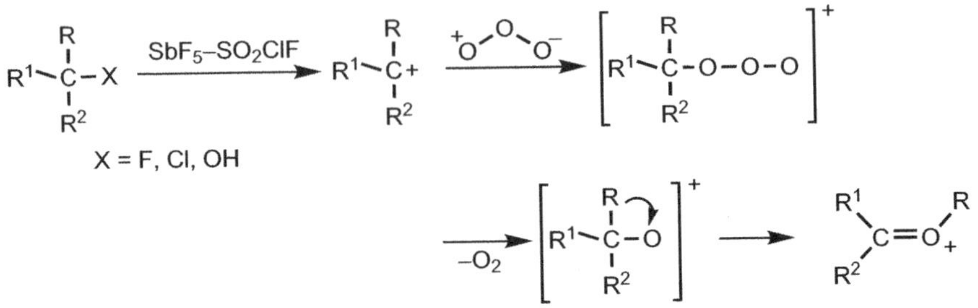

(5.219)

When a stream of oxygen containing 15% ozone was passed through a solution of isobutane in HSO_3F–SbF_5–SO_2ClF solution held at −78°C, the colorless solution immediately turned brown in color. ^1H and ^{13}C NMR spectra of the resultant solution were consistent with the formation of the dimethylmethylcarboxonium ion in 45% yield together with trace amounts of acetylium ion (CH_3CO^+). Further oxidation products (i.e., acetylium ion and CO_2) were reported to be observed in a number of reactions studied. Such secondary oxidation products, however, are not induced by ozone. Similar treatment of isopentane, 2,3-dimethylbutane, and 2,2,3-trimethylbutane resulted in formation of related carboxonium ions as the major products (Table 5.37).

For the reaction of ozone with alkanes under superacidic conditions, two mechanistic pathways could be considered. The first possible pathway is the formation of an

Table 5.37. Products of the Reaction of Branched Alkanes with Ozone in Magic Acid–SO_2ClF at −78°C[599,642]

Alkane	Products
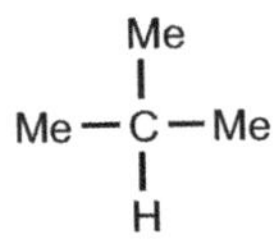	$(Me_2C{=}OMe)^+$
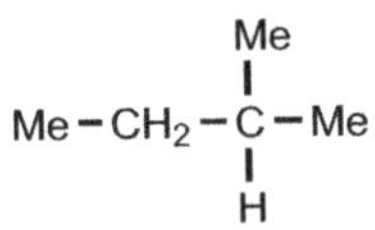	$(Me_2C{=}OEt)^+$
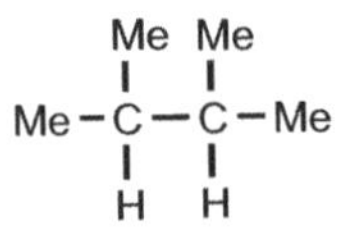	$(Me_2C{=}OH)^+$ (40%) $(Me_2C{=}OCHMe_2)^+$ (60%)
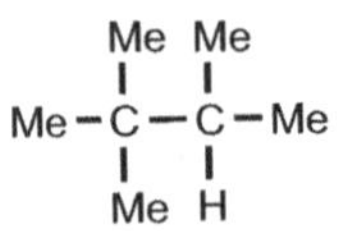	$(Me_2C{=}OMe)^+$ (50%) $(Me_2C{=}OH)^+$ (50%)

alkylcarbenium ion via protolysis of the alkane prior to quenching of the ion by ozone, as shown in Eq. (5.220). Alkylcarbenium ions may also be generated via initial oxidation of the alkane to an alcohol followed by protonation and ionization [Eq. (5.221)]. There have already been a number of reports of ozone reacting with alkanes to give alcohols and ketones.[643–645] In both cases, intermediate alkylcarbenium ions would then undergo nucleophilic reaction with ozone as described earlier [Eq. (5.219)].

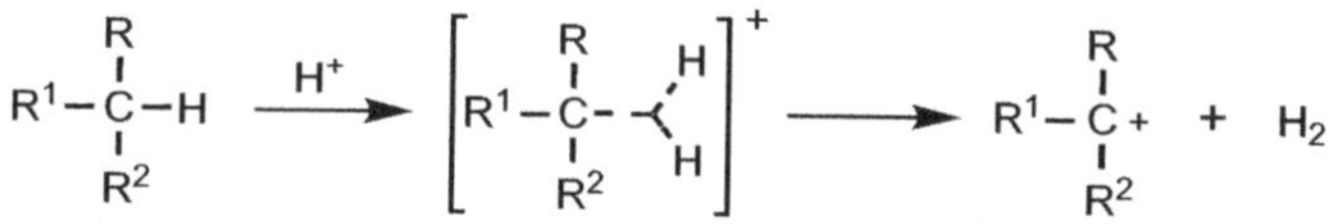

(5.220)

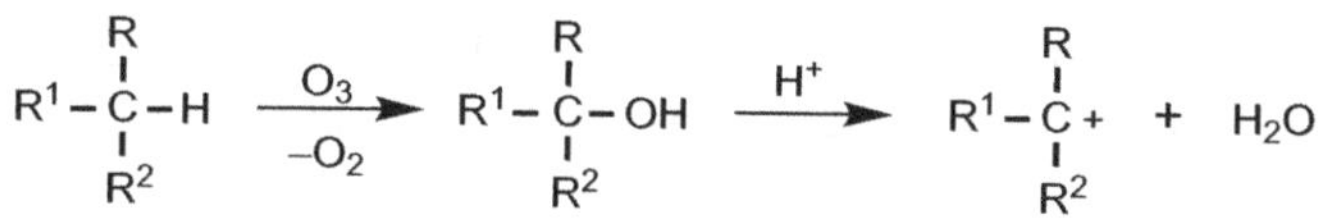

(5.221)

The products obtained from isobutane and isoalkanes (Table 5.37) are in accord with the above-discussed mechanism. However, the relative rate of formation of the dimethylmethylcarboxonium ion from isobutane is considerably faster than that of the *tert*-butyl cation from isobutane in the absence of ozone under the same conditions.[642] Indeed, a solution of isobutane in excess Magic Acid–SO_2ClF solution showed only trace amounts of the *tert*-butyl cation after standing for 5 h at −78°C. Passage of a stream of oxygen gas through the solution for 10 times longer a period than in the ozonization experiment showed no effect. It was only when ozone was introduced into the system, that rapid reaction took place.

On the other hand, *tert*-butyl alcohol itself in Magic Acid–SO_2ClF solution gave the *tert*-butyl cation readily and quantitatively, even at −78°C. In the presence of ozone, however, under the same conditions it gave dimethylmethylcarboxonium ion.

Although isobutane does not give any oxidation products in the absence of Magic Acid under the same low-temperature ozonization conditions, it was not possible for the authors to determine[642] whether formation of intermediate oxidation products, such as alcohols, plays any role in the ozonization of alkanes in Magic Acid. There is no experimental evidence for reactions proceeding via the intermediacy of carbenium ions; whether the initial oxidation step of alkanes to alcohols is important. This oxidation, indeed, was found to be extremely slow in the acidic media studied.

The most probable reaction path postulated for these reactions is electrophilic attack by protonated ozone [Eq. (5.222)] on alkanes, resulting in oxygen insertion into the involved σ-bond and cleavage of H_2O_2 from pentacoordinated trioxide insertion transition state giving a highly reactive oxonium ion intermediate, which immediately rearranges to the corresponding carboxonium ion, which can be hydrolyzed to ketone and alcoholic products [Eq. (5.223)].

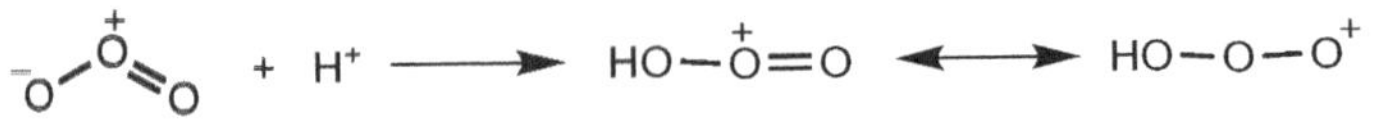

(5.222)

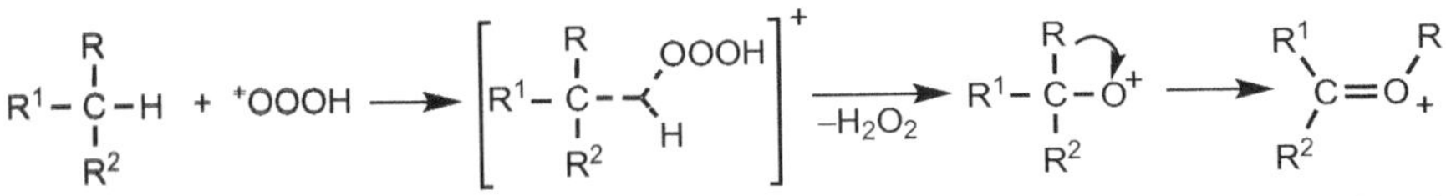

(5.223)

Since ozone is a strong 1,3-dipole,[635] or at least has a strong polarizability (even if a singlet biradical structure is also feasible), it is expected to be readily protonated in superacids, in manner analogous to its alkylation by alkylcarbenium ions. Protonated ozone HO_3^+, once formed, should have a much higher affinity (i.e., be a more powerful electrophile) for σ-donor single bonds in alkanes than neutral ozone.

Attempts to directly observe protonated ozone by ^1H NMR spectroscopy were inconclusive[642] because of probable fast hydrogen exchange with the acid system (may be through diprotonated ozone $H_2O_3^+$) and also the difficulty in differentiating between shifts of HO_3^+ and H_3O^+.

Straight-chain alkanes also efficiently react with ozone in Magic Acid at −78°C in SO_2ClF solution. Ethane gave protonated acetaldehyde as the major reaction product together with some acetylium ion (Scheme 5.62). Reaction of methane, however, is rather complex and involves oxidative oligocondensation to *tert*-butyl cation, which reacts with ozone to give methylated acetone (Scheme 5.63).

Similar reactions have been investigated with a wide variety of alkanes.[599,642] Cycloalkanes in particular give cyclic carboxonium ions along with protonated ketones. The reaction of cyclopentane is shown in Scheme 5.64.

Even electrophilic oxygenation of functionalized compounds has been achieved.[599] Alcohols are oxidized to the corresponding keto-alcohols [Eq. (5.224)].

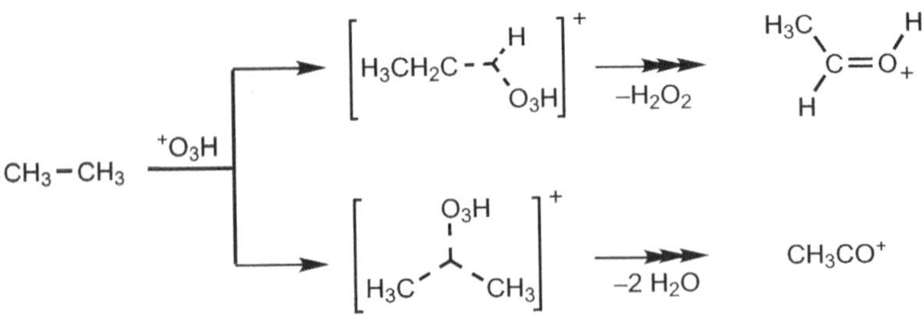

Scheme 5.62

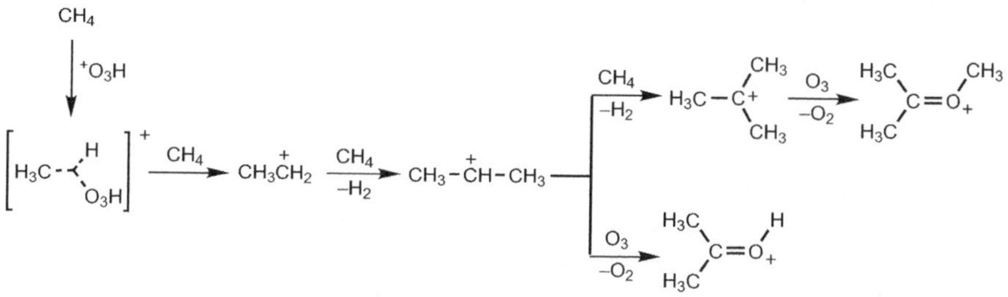

Scheme 5.63

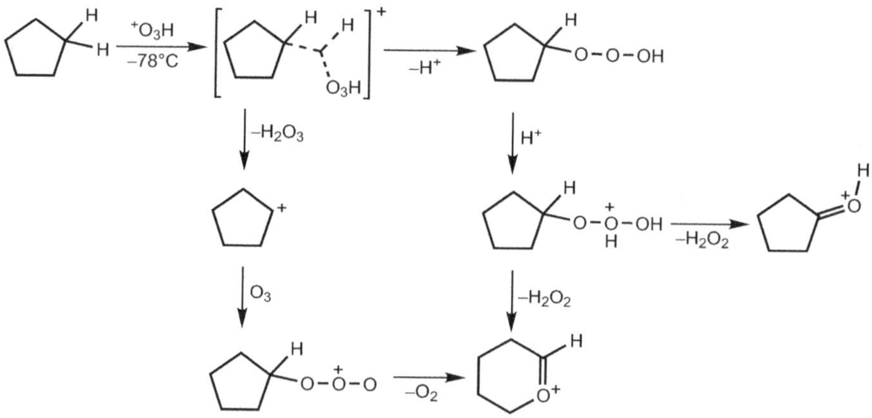

Scheme 5.64

Ketones and aldehydes are oxidized to their corresponding dicarbonyl compounds [Eq. (5.225)].

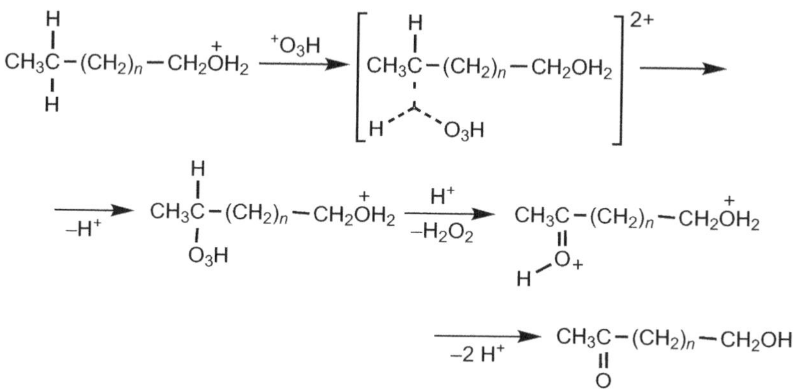

(5.224)

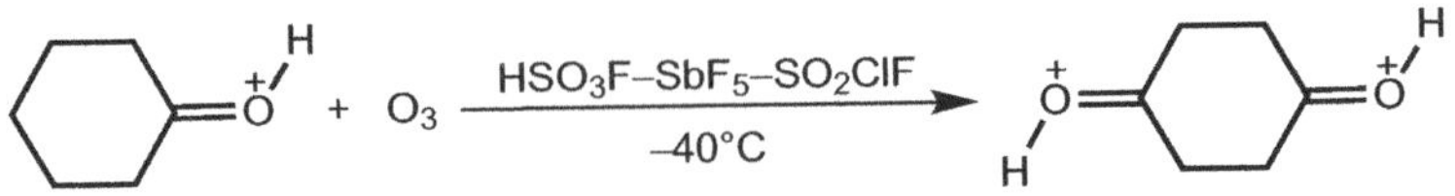

(5.225)

In the case of carbonyl compounds, the C–H bond located farther than γ-position seems to react with ozone in the presence of Magic Acid. It appears that the strong electron-withdrawing effect of protonated carbonyl group is sufficient to inhibit reaction of these C–H bonds (in α-, β-, and γ-positions). Jacquesy and co-workers[646,647] have shown that protonated ozone in HF–SbF_5 reacts with 3-keto-steroids bearing various substituents such as carbonyl, hydroxy, and acetoxy groups at the 17 positions. In situ diprotonation of the substrate in the superacid medium directs the electrophilic attack of ozone to the B- or C-ring methylenes. The position of oxidation depends both on the steric hindrance of the corresponding axial C–H bond and the remoteness of the positive charges initially present (diprotonated species) in the molecule. For example, 3,17-diketosteroids (R=O) only lead to the formation of the corresponding 3,6,17- and 3,7,17-triketones (**153a** and **153b**, respectively, R=O) (Scheme 5.65), whereas oxidation at the 11 or 12 position is also observed when an OH or OAc group is present at the 17β position (**153c**, R=βOH and **153d**, R=βOAc, respectively) (Scheme 5.65). In all cases, the 6-keto to 7-keto ratio appears to be higher for substrates having a *cis* A/B ring junction as compared with those with *trans* one. Low reactivity of the 11 position, especially for substrates with *cis* A/B ring junction, was observed. Oxidation of the tertiary carbon atoms, however, was not observed in any of these systems.

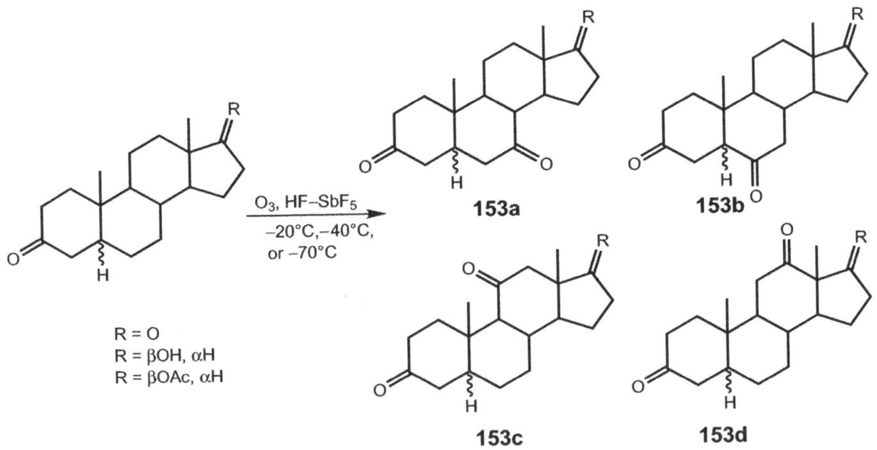

Scheme 5.65

5.12.3. Oxygenation Induced by Nafion Resins

Frusteri and co-workers[648–651] have evaluated the activity of Nafion-based membranes in the oxidation of light alkanes. The membranes were prepared by depositing a carbon–Teflon paste on carbon paper followed by activation and subsequent impregnation by a Nafion-H solution in isopropyl alcohol. They were evaluated in a three-way catalytic reactor using Fe^{2+}–H_2O_2 as the oxidant. Of various acidic catalysts studied, Nafion-H proved to be the best catalyst to afford the selective oxidation of methane (to methanol), ethane (ethanol and acetaldehyde), and propane (1- and 2-propanol, propanal, acetone) under mild conditions (80–110°C, 1.4 bar). The high efficiency of Nafion-H was attributed to its high chemical stability toward H_2O_2.

Nafion-H is an effective catalyst in the Baeyer–Villiger oxidation of cyclic and polycyclic ketones to form lactones in high yields (86–100%) with H_2O_2 or *meta*-chlorobenzoic acid in refluxing dichloromethane.[652] Nonsymmetric ketones, such as 2-methylcyclopentanone, give isomeric mixtures depending on the migratory aptitude of the group adjacent to the carbonyl function [Eq. (5.226)].

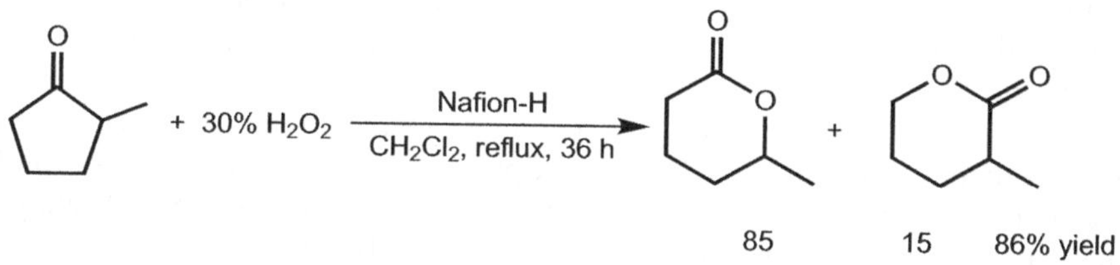

(5.226)

When Nafion SAC-13 and SAC-40 were used in the Baeyer–Villiger oxidation of cyclopentanone, the peroxidic byproduct dicyclopentylidenediperoxide **154** was obtained with unusually high selectivity (24% yield)[653] [Eq. (5.227)]. This was interpreted in terms of the high acid strength of Nafion leading to the preferable formation of $H_3O_2^+$ species able to form the peroxidic byproduct. Nafion NR50 has been found to be an efficient and recyclable catalyst in the oxidation of various organic compounds with 30% H_2O_2 under mild conditions.[654] High yields were achieved in the oxidation of secondary alcohols to esters or lactones, aldehydes in methanol to methyl esters, tertiary amines to *N*-oxides, and sulfides to sulfoxides.

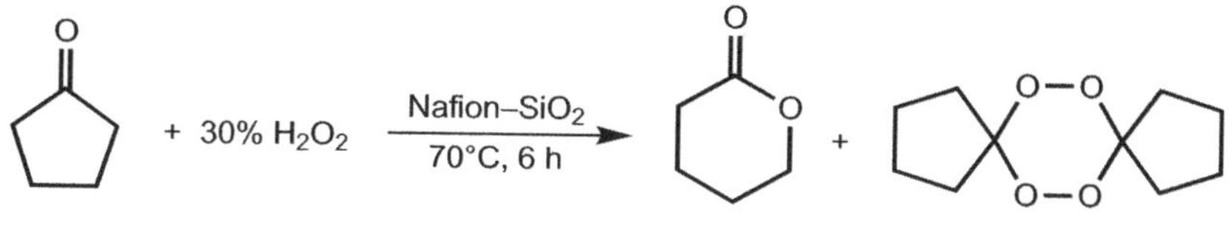

154
24% yield

(5.227)

A clean, solvent-free method has been developed for the bis-hydroxylation of alkenes by the use of Nafion-based acidic catalysts and 30% H_2O_2.[655] Nafion NR50 and SAC-13 exhibited high activity in the oxidation of isomeric C_6 alkenes, cyclohexene [Eq. (5.228)], 1,4-cyclohexadiene, and allylic alcohols in the temperature range of 70–90°C to afford the corresponding 1,2-diols in moderate to excellent yields (40–100%).

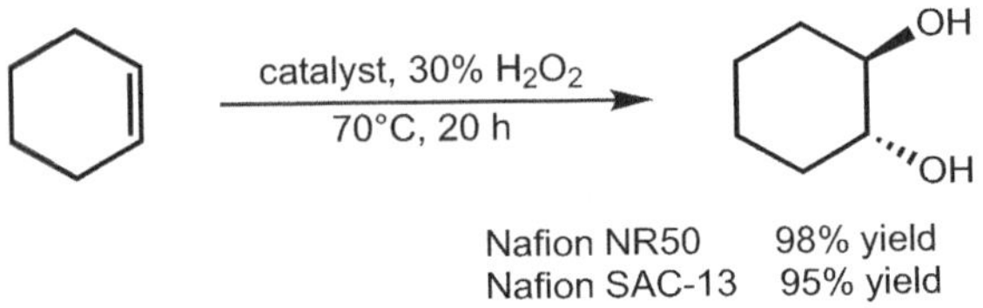

(5.228)

Hoelderich and co-workers[656,657] have applied various acidic resins loaded with Pt and Pd to perform direct hydroxylation of benzene to produce phenol. Among various Nafion SAC catalysts (loadings between 8 and 80%), SAC-13 gave the best results. The highest phenol yield of 7.6% was found using a semicontinuous process over a pre-reduced catalyst (0.5 wt% Pd + 0.5 wt% Pt/SAC-13, 30–40°C, 1.3 L h^{-1} H_2 and O_2 flow, and 1.4 L h^{-1} N_2 flow, 1400 kPa, water–methanol solvent).

Ohsaka and co-workers[658] have reported a new method to prepare peroxyacetic acid by oxidizing acetic acid with H_2O_2 in the presence of Nafion catalysts. Nafion-H proved to be superior to Nafion SAC-13 to give a conversion of 16% (5% of catalyst, initial reactant concentrations: $[CH_3COOH] = 1.65$ M, $[H_2O_2] = 2.85$ M, 17 h).

5.12.4. Oxygenation by Other Methods

The CCl_4–HF–SbF_5 system developed by Jouannetaud and co-workers and used in the selective fluorination of imines (see Section 5.10.1) can be applied in the oxygenation of ketones and carboxamides as well. The hydroxylation of ketones is selective [Eq. (5.229)], provided that a five- or six-membered cyclic carboxonium ion preventing fluorination is involved.[534,659] Fluorination, however, may be a side reaction with product distributions depending on quenching conditions (aqueous Na_2CO_3 or HF–pyridine). Similar features are characteristic of the transformation of carboxamides.[659]

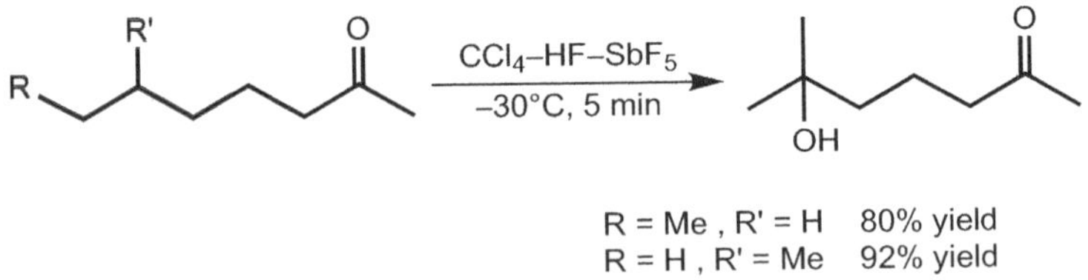

(5.229)

Karpov and co-workers developed two methods to transform perfluorinated cyclic compounds to the corresponding oxo-derivatives. When perfluoroindane (**155**) is treated in the presence of SiO_2 and SbF_5, perfluoroindan-1-one is formed in high yield[660] [Eq. (5.230)]. According to the suggested mechanism, SiO_2 serves as the oxygen source by reacting with the carbocation intermediate formed by F^- loss. The selectivity of oxygenation is highly sensitive to the reaction conditions since the product is prone to undergo complex transformations. Perfluorinated benzocyclobutene, indane and their perfluoroalkyl-substituted derivatives (**156**) [Eq. (5.231)], and perfluoroalkylbenzenes are transformed in $CF_3COOH–SbF_5$ to give carbonyl derivatives under mild conditions in high yields.[661] The key step in product formation, in this case, is the acyloxylation of the carbocation intermediate by CF_3COOH.

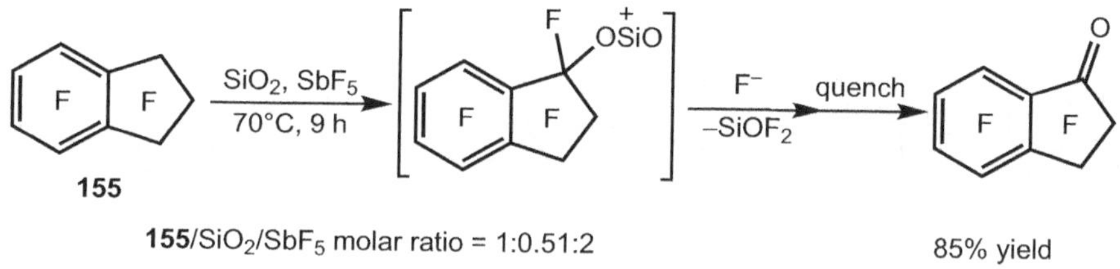

(5.230)

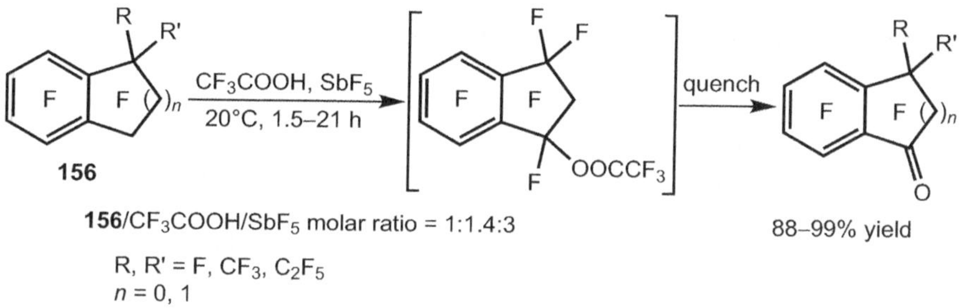

(5.231)

Bis(trimethylsilyl) peroxide[662] and sodium perborate[663] are also efficient electrophilic oxygenating agents in the presence of triflic acid. Both reagents are capable of the selective monohydroxylation of aromatics to the corresponding phenols in high yields [Eq. (5.232)]. Isomer distributions are in accordance wit the electronic nature of the reactions. Moderate substrate selectivity ($k_T/k_B = 20$) was found with bis(trimethylsilyl)peroxide in the competitive hydroxylation of benzene and toluene at −40°C. The byproduct (trimethylsilyloxy)arenes support a mechanistic pathway with the involvement of (trimethylsilyloxy)arenium ions. Bis(trimethylsilyl) peroxide was also used in the hydroxylation of polystyrene nanospheres promoted by triflic acid.[597]

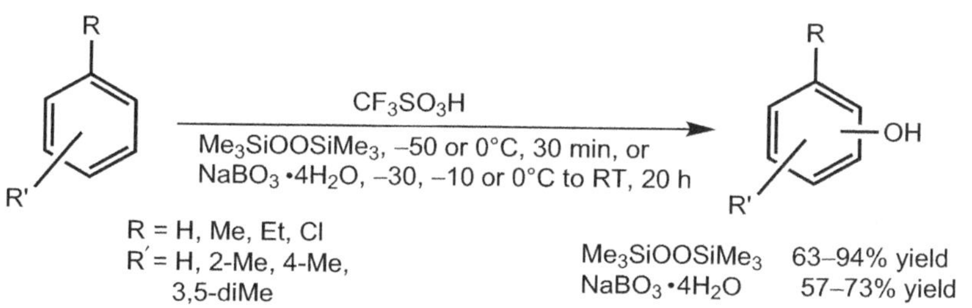

(5.232)

Adamantane can be transformed to 4-oxahomoadamantane in good yields by bis (trimethylsilyl)peroxide[664] and sodium percarbonate[665] in the presence of triflic acid. Product formation is explained by C–C σ-bond insertion [Eq. (5.233)]. Diamantane is transformed into isomeric oxahomodiamantanes (C–C insertion) and bridgehead diamantanols (C–H insertion).[664]

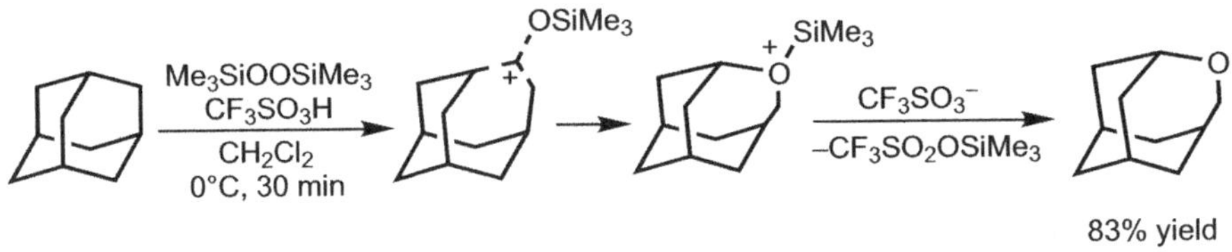

(5.233)

5.13. SUPERACIDS IN PROTECTION GROUP CHEMISTRY

Olah et al.[666] have developed a highly efficient thioacetalization method using boron trifluoride monohydrate BF_3–H_2O. A variety of carbonyl compounds, aromatic and aliphatic aldehydes, and ketones were readily transformed to the corresponding 1,3-dithiolanes in high yields (90–99%) under mild conditions (CH_2Cl_2 solution, 0°C, 15–30 min). A new method for the synthesis of stannyl ethers for further use in glycosylation has been reported by Yamago, Yoshida, and co-workers[667] [Eq. (5.234)]. The efficiency of the method applicable for primary and secondary alcohols was tested by benzoylation.

$$\mathrm{R{-}OH} + \mathrm{Bu_3SnR'} \xrightarrow[\mathrm{CH_2Cl_2,\ RT,\ 1\text{–}25\ h}]{\mathrm{CF_3SO_3H}} \mathrm{R{-}OSnBu_3} \xrightarrow{\mathrm{BzCl}} \underset{75\text{–}98\%\ \text{yield}}{\mathrm{R{-}OBz}} \qquad (5.234)$$

R = Me, isoPr
R' = H, allyl

A newly developed stable dimethyl ether–HF complex (1:15),[668] which can be stored at room temperature, is highly effective for the deprotection and cleavage of peptides from Merrifield resins (isolated peptide yields = 88–94%). It can easily be handled, and it is a very useful and convenient hydrogen fluoride equivalent.

Amorphous and mesostructured ZrO_2 solid catalysts loaded with various amounts of triflic acid (5–30%) were tested in the acetalization of ethyl acetoacetate to form fructone[357] [Eq. (5.235)]. Whereas the amorphous samples [ZrO_2(a)–TfOH] were nonselective and induced the formation of a byproduct, the mesoporous catalysts [ZrO_2(m)–TfOH] showed complete selectivity to give the desired product in high yields.

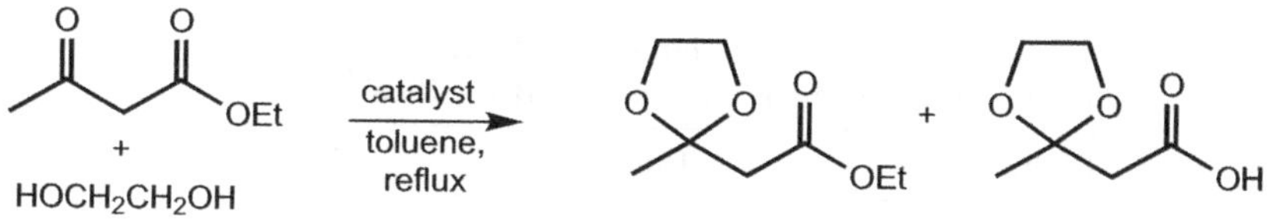

	Conversion (%)	Selectivity (%)	
ZrO_2(a)–TfOH	83–95	64–97	3–36
ZrO_2(m)–TfOH	82–93	100	

(5.235)

Fluoroboric acid supported on silica (HBF_4–silica) has recently been found to be a highly efficient catalyst in the protection of various functional groups. Structurally diverse alcohols, phenols, thiophenols, and anilines can be acylated under solvent-free conditions at room temperature.[669] Even acid-sensitive tertiary alcohols (1-alkylcyclohexanols) and sterically hindered compounds, such as *endo*-borneol, give the acylated products in high yields. A triflic acid–silica catalyst also shows high activity in the *O*-acetylation with Ac_2O of alcohols and phenols.[359]

The transformation of carbonyl compounds to *O*,*S*-acetals and *S*,*S*-acetals can be carried out without solvent under mild conditions (1 mol% of catalyst, room temperature, 5–40 min) to afford the products in high yields (70–97%).[670] *O*,*O*- and *O*,*S*-acetals are also readily transformed to *S*,*S*-acetals [Eq. (5.236)].

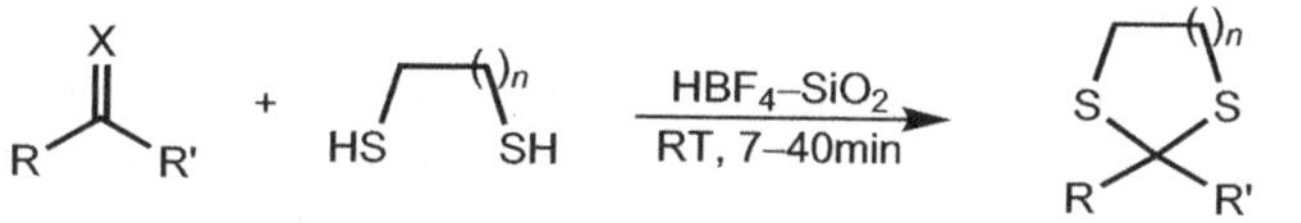

R = C_9H_{19}, Ph, 4-MeC_6H_4, 4-HOC_6H_4,
4-ClC_6H_4, 4-$MeOC_6H_4$, 4-$NO_2C_6H_4$,
4-$BnOC_6H_4$, $PhCH_2CH_2$, PhCH=CH
R' = H, Me
X = $(OMe)_2$, $(OEt)_2$, $O(CH_2)_nO$, $O(CH_2)_nS$
n = 1, 2

89–98% yields (5.236)

HBF_4–silica also exhibits high activity in the protection of aldehydes as 1,1-diacetates (acylals)[671] [Eq. (5.237), Table 5.38]. Both aliphatic and aromatic aldehydes react readily to form diacetates in high yields. The catalyst can be reused with marginal decrease in activity.

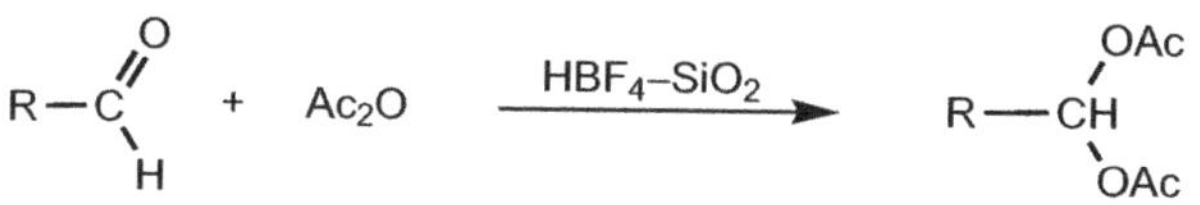

(5.237)

Nafion-H has been shown to be effective in a variety of protection–deprotection reactions including *O*-trialkylsilylation of alcohols, phenols, and carboxylic acids, as well as the preparation and methanolysis of tetrahydropyranyl (THP) ethers.[672] However, when compared, for example, with HBF_4–silica or Nafion nanocomposites,

Table 5.38. Transformation of Aldehydes to Diacetates in the Presence of HBF_4–Silica[671]

R Group	Time (min)	Yield (%)
Propyl	1	98
Pentyl	1	96
Heptyl	1	97
Phenyl	2	99
4-F-phenyl	2	98
4-Me-phenyl	2	97
4-MeO-phenyl	2	97
4-CN-phenyl	2	96
1-Naphthyl	2	97
2-Naphthyl	2	96
2-Furyl	5	97
2-Thienyl	4	98
$CH_3CH{=}CH$	2	96
PhCH=CH	7	98

Reaction conditions: aldehyde:Ac_2O = 1, 0.1 mol% of catalyst, room temperature.

Nafion-H exhibits lower activity. In the acetylation of alcohols,[673] in the transformation of aldehydes and ketones with trimethyl orthoformate to the corresponding dimethylacetals,[674] and in the formation of acylals,[675] longer reaction times (several hours) are required to achieve high yields at room temperature. Furthermore, the formation of ethylenedithioacetals in benzene[674] and the direct transformation of acetals to thioacetals with ethane-1,2-thiol in dichloromethane[676] can only be performed at reflux temperature.

The catalytic performance of Nafion SAC-13 in the formation of 1,1-diacetates,[677] in turn, is very similar to that of HBF_4–silica. In the acetalization of carbonyl compounds with ethane-1,2-diol and propane-1,3-diol, products are isolated in good to excellent yields. The formation of THP ethers of alcohols is fast and protected alcohols are isolated in high yields [Eq. (5.238)]. Nafion SAC-13 can also be used in the removal of the THP ether group[677] although the transformation requires somewhat longer reaction times (30 min–6 h, 81–97% yield). Furthermore, the catalyst could be recycled in all three processes with practically no loss of activity.

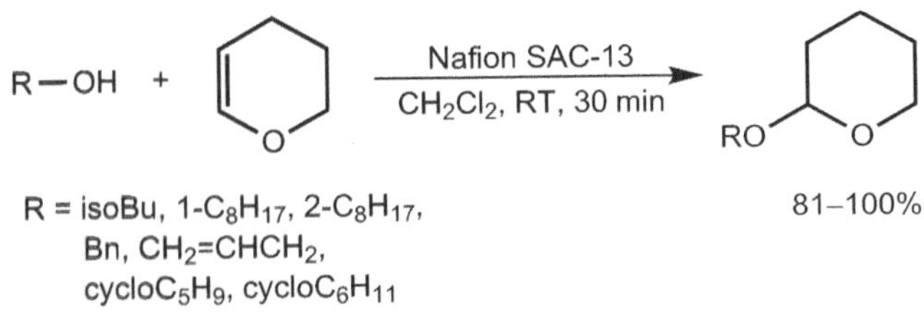

(5.238)

Nafion SAC-13 was also demonstrated to be a superior catalyst in the formation of mixed acetals in the reaction of alcohols with dialkoxymethanes[678] [Eq. (5.239)]. The product acetals are isolated in good to excellent yields in an almost instantaneous transformation and the catalyst can be reused with minor loss of activity.

$$R{-}OH \; + \; (R'O)_2CH_2 \xrightarrow[CH_2Cl_2,\ \text{reflux, 5 min}]{\text{Nafion SAC-13}} RO{-}CH_2{-}OR' \qquad (5.239)$$

R = 1-C_8H_{17}, 2-C_8H_{17}, Bn, $PhCH_2CH_2$, cycloC_6H_{11}, cholesteryl
R' = isoPr, Bu

83–100%

Again, the activity of Nafion-H in the above protection processes (transformation of alcohols to THP ethers[672] and methoxymethyl ethers[679]) is lower, and consequently, longer reaction times and elevated temperatures are required to achieve yields comparable to those over Nafion nanocomposites. Obviously, this is due to the low specific surface area and low accessibility of the active sites of Nafion-H as compared with Nafion SAC-13.

A simple process for the trimethylsilylation of alcohols, phenols, and carboxylic acids in refluxing CH_2Cl_2 over Nafion-H has been reported by Olah et al.[672] [Eq. (5.240)]. The transformation of phenols and carboxylic acids requires higher temperature (CCl_4 reflux).

$$R{-}OH \; + \; CH_2{=}CHCH_2SiMe_3 \xrightarrow[\substack{CH_2Cl_2 \text{ or } CCl_4,\\ \text{reflux, 1–3.5 h}\\ -CH_3CH{=}CH_2}]{\text{Nafion-H}} R{-}O{-}SiMe_3 \qquad (5.240)$$

R = 1-C_7H_{15}, 2-C_9H_{19}, Bn, cycloC_6H_{11}, cyclo$C_6H_{11}CH_2$, Ph, 2,6-diMeC_6H_3, $C_5H_{11}CO$, C_6H_5CO

89–98%

Selective deprotection of trialkylsilyl ethers can also be accomplished by Nafion-H. Trimethylsilyl ethers are cleaved to the corresponding alcohols under mild conditions[680] [Eq. (5.241)]. Nafion-H with NaI (1 equiv.) in methanol was shown to readily cleave *tert*-butyldimethylsilyl ethers (room temperature, 4–25 h, 65–99% yields).[681]

$$RCH_2{-}O{-}SiMe_3 \xrightarrow[\substack{\text{hexane,}\\ \text{RT, 15–45 min}}]{\text{Nafion-H}} R{-}OH \qquad (5.241)$$

R = C_6H_{13}, C_7H_{15}, cyclo-C_6H_{11}, cycloC_7H_{13}, substituted Ph, PhCH=CH, PhCHMe, 1-indanyl, 2-thienyl

90–98%

Nafion-H has also been applied in the transformation of protected organic compounds. Acetals and ketals can be reductively cleaved very efficiently to the corresponding ethers with triethylsilane (CH_2Cl_2 reflux, 1–4 h, 82–99% yields).[682] The combination of acetal synthesis with the use of trimethyl orthoformate,[674] and this cleavage method allows the one-pot direct conversion of carbonyl compounds to their alkyl ethers. Recently, Nafion-H has been applied in the conversion of primary and secondary trimethylsilyl ethers to the corresponding symmetric ethers[683] (hexane, RT, 70–90% yield).

Hyroxamic acids protected in the form of dioxazoles can be readily recovered by treatment with Nafion-H in isopropyl alcohol[684] [Eq. (5.242)]. The method is applicable to primary, secondary, tertiary, and aromatic hydroxamic acids and the byproduct isopropyl esters are formed only in negligible amounts (0–4%).

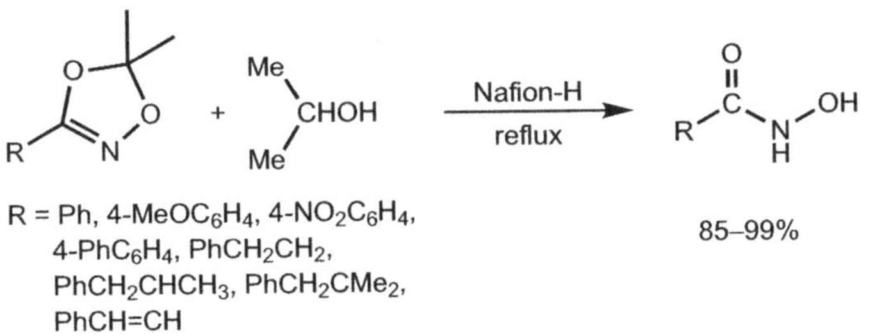

(5.242)

Protection groups related to carbohydrate chemistry are discussed in Section 5.16.

5.14. SUPERACIDS IN HETEROCYCLIC CHEMISTRY

5.14.1. Synthesis of Heterocycles

5.14.1.1. Preparation of Oxacycloalkanes. Unsaturated alcohols can be cyclized under superacid conditions to yield oxolane derivatives. Laali et al.[685] have studied the protonation of homoallylic adamantylideneadamantyl alcohols. The pseudo-axial alcohol **157** was protonated in $HSO_3F–SO_2ClF$ to give the intermediate protonated cyclic ether observed by ^1H and ^{13}C NMR spectroscopy, which, upon quenching, furnished the corresponding ether [Eq. (5.243)].

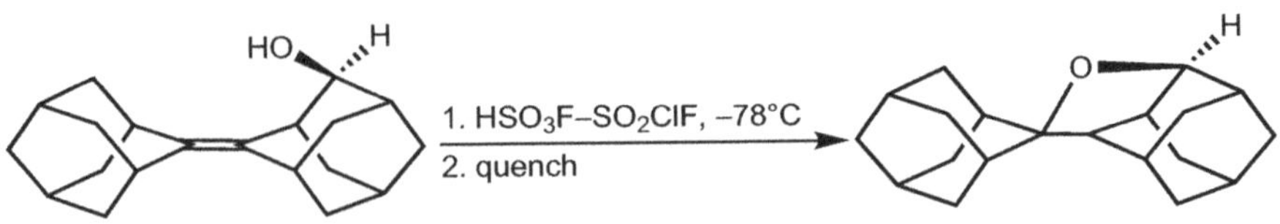

157

(5.243)

Derivatives with five- and six-membered oxygen-containing rings have been isolated when vindoline and deacetylvindoline were treated under superacid conditions.[686] Formation of the unexpected oxolane derivatives (R = Ac: 4%, R = H: 18%) was accounted for by rearrangement through ethyl and hydride shifts (Scheme 5.66).

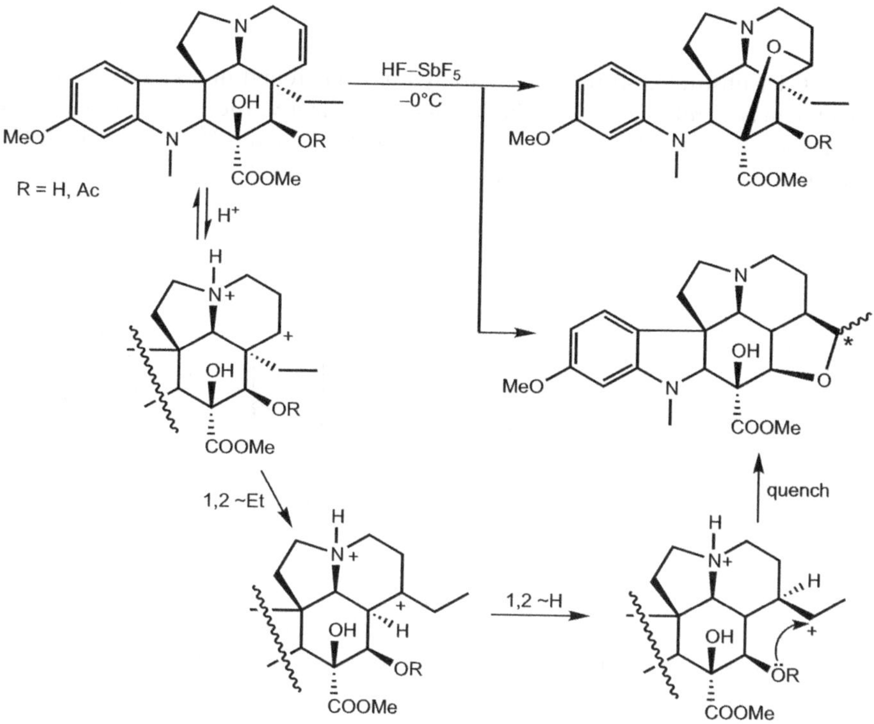

Scheme 5.66

The ring-closure (cyclodehydration) of dihydroxy compounds affords five-, six-, seven-, and eight-membered cyclic ethers. The reaction can efficiently and conveniently be carried out with Nafion-H.[687,688] Even diphenols (2,2′-dihydroxybiphenyls) undergo dehydration to afford oxolane derivatives[276] [Eq. (5.244)]. The low yields of the *tert*-butyl-substituted compounds result from trans-*tert*-butylation (see Section 5.2.6). Cyclodehydration of stereoisomeric 2,5-hexanediols was shown to be stereospecific[688]: The racemic mixture yields *cis*-2,5-dimethyloxolane, whereas the *meso* diol gives the *trans* compound. This indicates that the transformation is an intramolecular S_N2 process without the involvement of carbocationic intermediates.

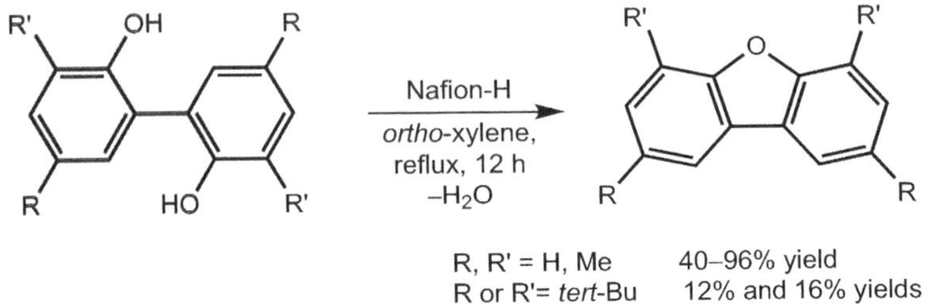

(5.244)

cis-2,6-Disubstituted 4-methylenetetrahydropyrans found in a variety of natural products can be synthesized by the highly diastereoselective (de > 98%) ring closure of enol ethers developed by Schrock, Hoveyda, and co-workers[689] [Eq. (5.245)]. The use of triflic acid in very low amounts (0.01–0.1 mol%) is necessary to avoid polymerization.

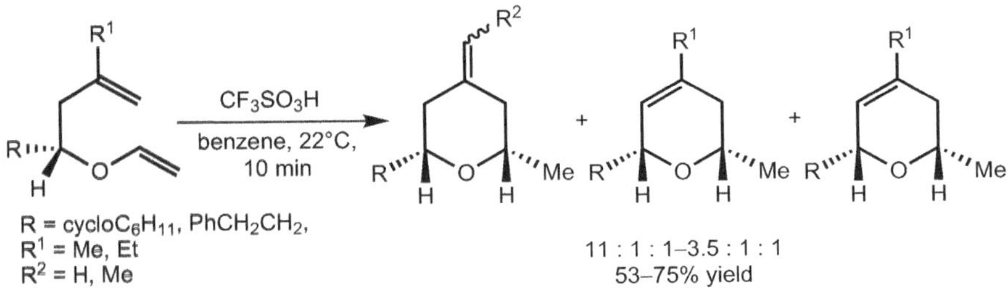

(5.245)

Triflic acid supported on titania has proved to be an effective catalyst to transform hydroxyphenyl-1,3-propanediones to chromone derivatives in high yield and with high selectivity[690] [Eq. (5.246)].

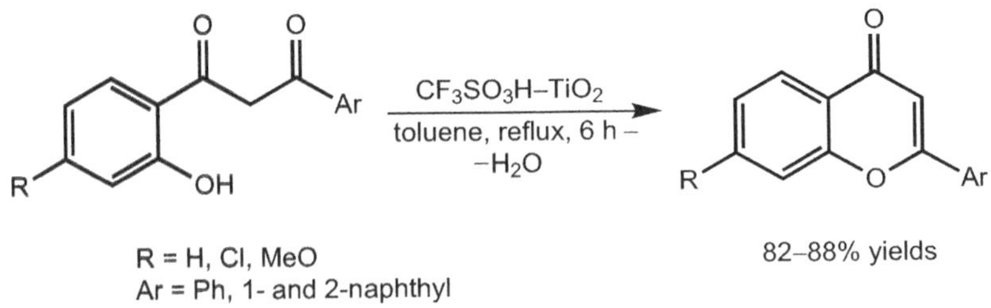

(5.246)

Nafion–silica nanocomposites exhibit high selectivity in the synthesis of substituted 7-hydroxychromanones **158** in high yields[691] [Eq. (5.247)]. Conversions increase with increasing Nafion loading and, consequently, Nafion SAC-80 (Nafion–silica with a Nafion content of 80 wt%) affords the highest yields. The catalysts could be recycled after treatment with nitric acid or hydrogen peroxide.

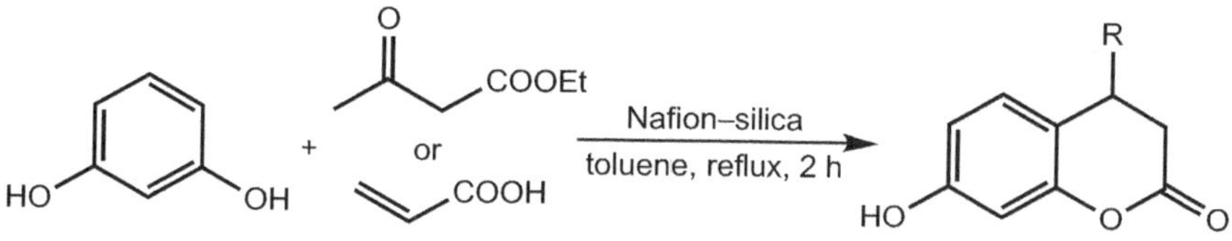

R = H **158a** 78–88% yields
R = Me **158b** 66–96% yields

(5.247)

A low amount (1 mol%) of triflic acid is sufficient to carry out the hetero Diels–Alder reaction of aromatic aldehydes with simple dienes to furnish 3,6-dihyro-2*H*-pyran derivatives in moderate to good yields[692] [Eq. (5.248)]. The strongly deactivated *para*-methoxybenzaldehyde and pentanal gave the products in very low yields.

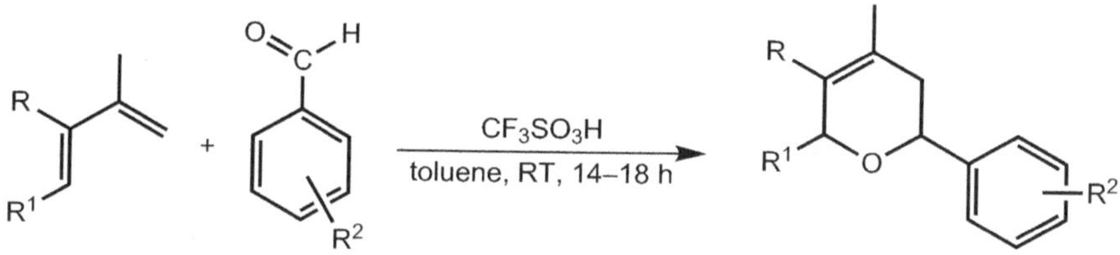

R, R^1 = H, Me
R^2 = H, 4-Me, 4-Cl, 4-NO_2, 3-MeO

40–85% yields

(5.248)

3-Alkenamides react with trioxane in triflic acid to yield 3,6-dihydro-2*H*-pyran derivatives[693] [Eq. (5.249)]. An oxo–ene reaction to form a homoallyl alcohol followed by a second oxo–ene reaction and subsequent dehydration were suggested to explain the product formation. Interestingly, lactam formation was not observed.

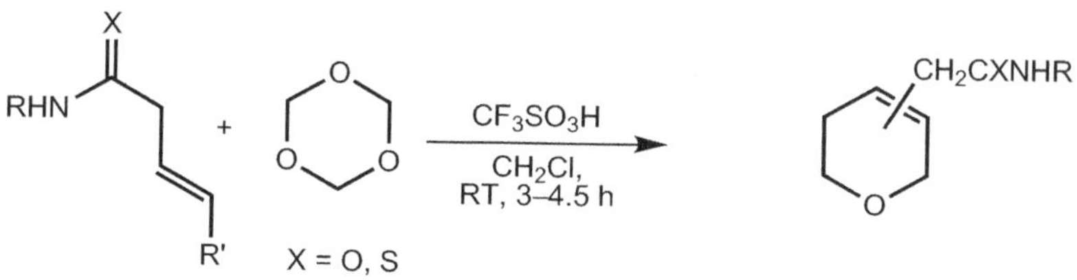

(5.249)

X = O, S
R = H, Bu
R' = H, Me

38–58% yield

Triflic acid has been successfully used in the stereocontrolled synthesis of substituted tetrahydropyrans. 2,4,6-Trisubstituted tetrahydropyrans have been synthesized by an intramolecular Prins reaction–pinacol sequence[694] [Eq. (5.250)].

The 2,6-*cis* isomers are formed exclusively and in all cases the equatorial C(4) epimer **159a** is the major product.

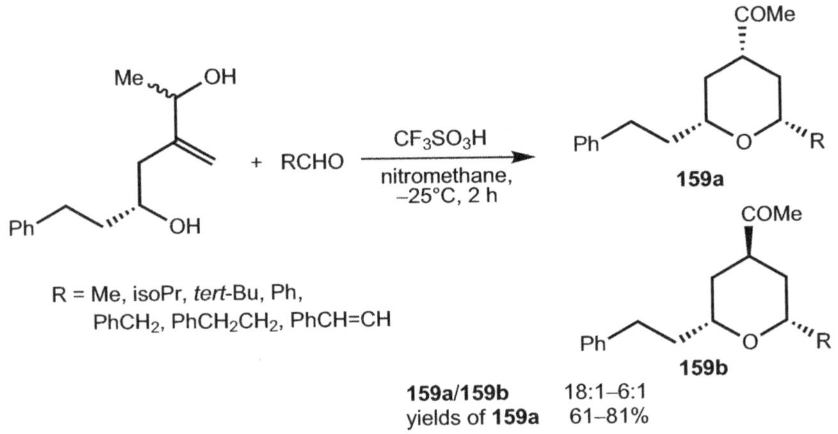

(5.250)

Nafion–silica nanocomposites (5%, 13%, 20%, 40%, and 80% loading) have been applied in the synthesis of α-tocopherol (**160**) in various solvents[695,696] [Eq. (5.251)]. Under optimized conditions, Nafion SAC-40 proved to be the best catalyst (91% yield of α-tocopherol using 0.6 wt% of catalyst). Catalyst recycling was possible after reactivation with oxidizing agents (nitric acid or hydrogen peroxide).

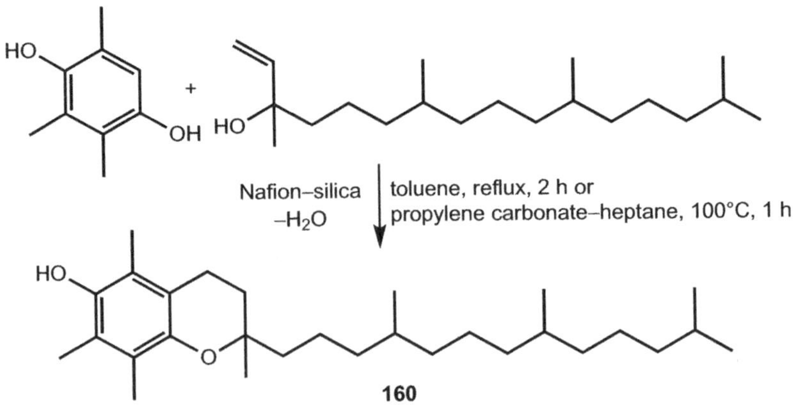

(5.251)

Bright, Coxon, and Steel[697] reported the cyclization of a methallyl carbinol in HSO_3F via the intermediacy of a protonated oxetane involving a unique oxygen atom migration [Eq. (5.252)].

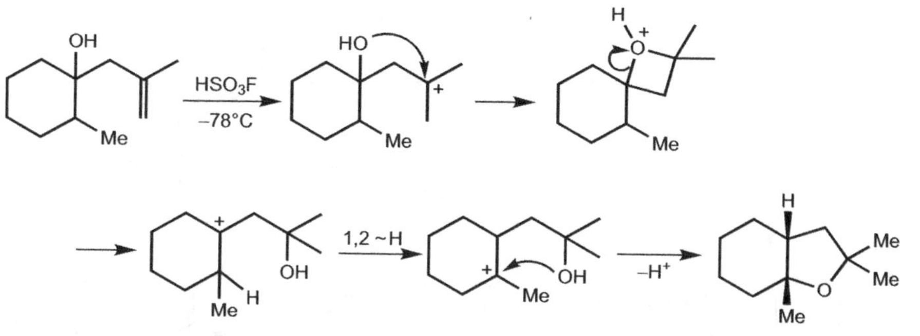

(5.252)

Additional examples for the formation of oxacycloalkanes under superacidic conditions are to be found in Section 5.17.1.

5.14.1.2. *Synthesis of Nitrogen Heterocycles.* Pyrrolidines and piperidines are synthesized by intramolecular hydroaminaton of *N*-protected alkenylamines.[698] Both triflic acid and conc. sulfuric acid are active; however, triflic acid gives better yields at much shorter reaction times [Eq. (5.253)]. 4-(*para*-Nitrophenyl)but-3-enylamine was unreactive, whereas the *para*-methoxy-substituted compound decomposed. The transformation of the tosylate of 1-isobutyl-pent-4-enylamine gave the 2,5-disubstituted pyrrolidine with modest selectivity (*trans*/*cis* = 68:32) The possible reaction pathway includes an intramolecular proton transfer from the protonated protected amino group to the double bond followed by trapping of the formed carbocation by the sulfonamide function.

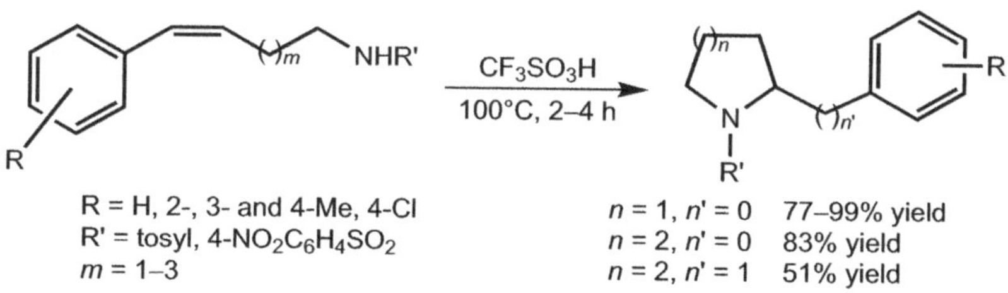

(5.253)

The tandem cyclopropyl ring-opening/nitrilium ring-closing reaction in the presence of triflic acid results in the stereoselective formation of cyclic imines **161** (Scheme 5.67).[699]

Cyclic imines are obtained in the related transformation of methylenecyclpropanes with nitriles mediated by triflic acid[700] (Scheme 5.68). The reaction pathway suggested to interpret product formation is similar to that in Scheme 5.67. The reaction of intermediate **162** with water may give Ritter products (carboxamides) isolated in some cases.

Similar carbocationic intermediates may be involved in the transformation of substituted ethyl pentadienoates under Ritter conditions in excess triflic acid (3 equiv.) to furnish the bicyclic aza compounds **163** [Eq. (5.254)].[701]

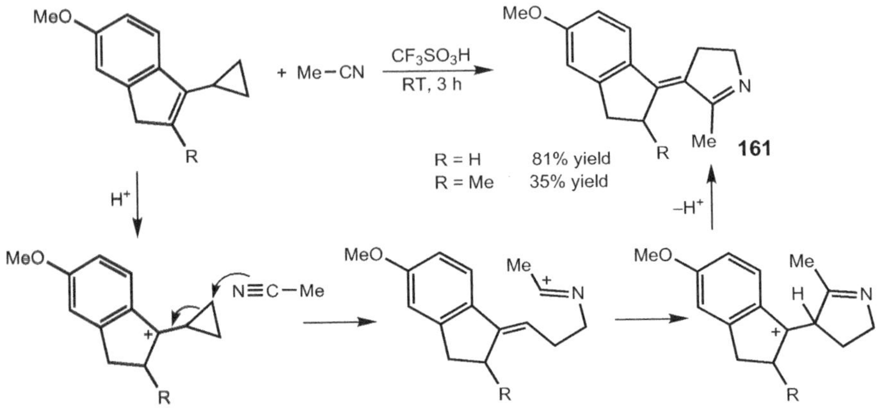

Scheme 5.67

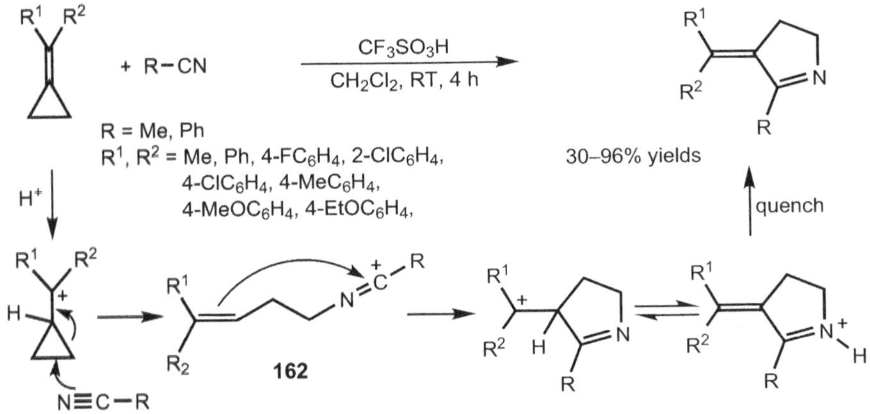

Scheme 5.68

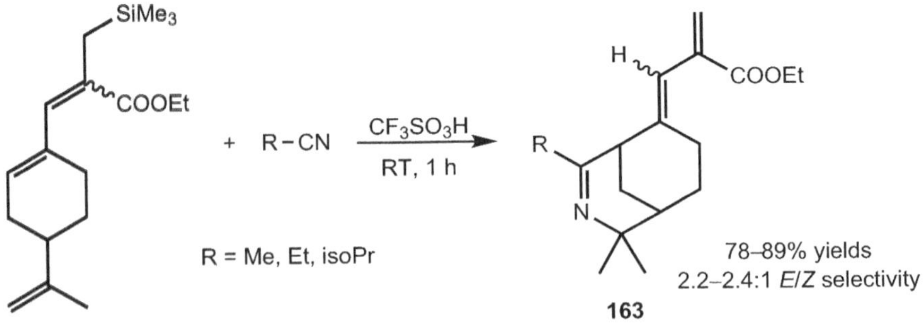

(5.254)

Dihydroquinoline derivatives are synthesized in intramolecular hetero Diels–Alder reaction from α-(alkynylsiloxy)aldimines in triflic acid[702] [Eq. (5.255)].

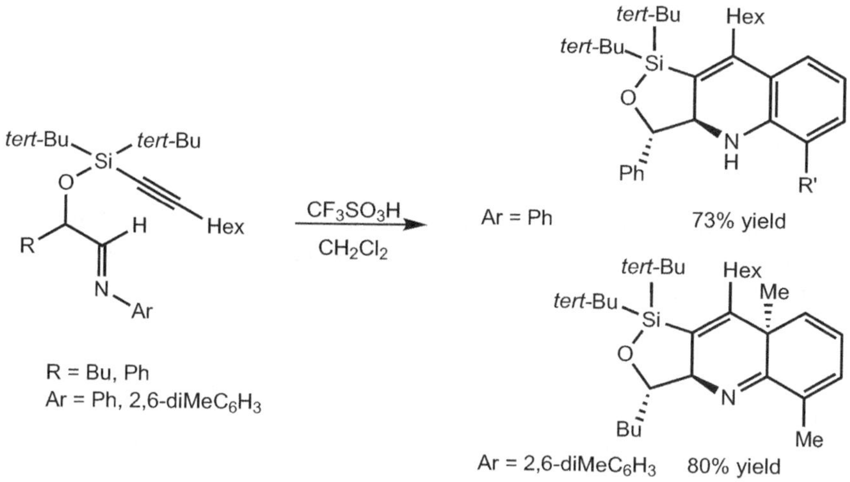

(5.255)

Efficient synthetic methods have been developed for the synthesis of 4-functionalized quinolines mediated by triflic acid.[703] Electron-rich, highly reactive ethynyl ketene-*S,S*-acetals react readily with arylamines and aldehydes in an aza-Diels–Alder reaction to afford the desired products [Eq. (5.256)]. Arylimines and ethynyl ketene-*S,S*-acetals react similarly (60–70% yields).

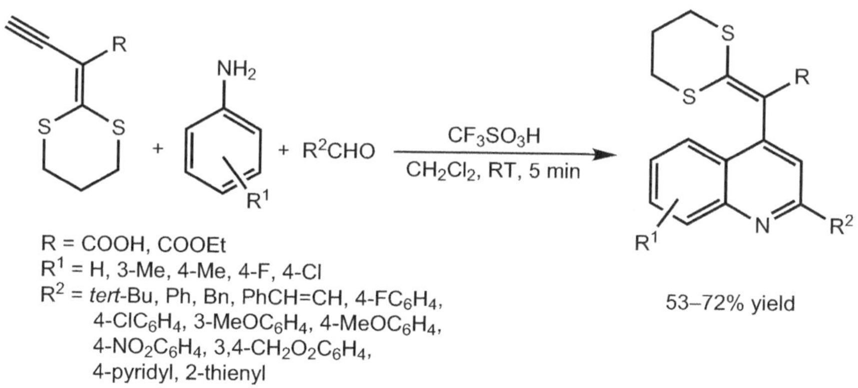

(5.256)

Knorr cyclization of a range of *N*-susbtituted butyramides in triflic acid yields 4-methyl-1*H*-quinolin-2-one derivatives[704] [Eq. (5.257)]. Suggested intermediates of the transformation directly observed by low-temperature multinuclear NMR spectroscopy ($HSO_3F-SbF_5-SO_2ClF$, $-40°C$) are distonic superelectrophiles formed by diprotonation of the two carbonyl oxygen atoms.

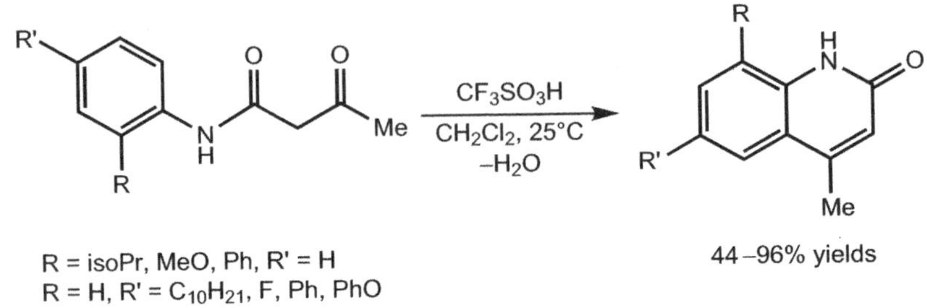

(5.257)

When aliphatic azido compounds with a suitably placed alkene or alcohol function are treated with triflic acid, nitrogen heterocycles are formed. This intramolecular version of the Schmidt reaction affords the synthesis of 2-substituted pyrrolines from 4-substituted-3-butenyl azides,[705] whereas fused or bridged aza compounds are obtained from exomethylene cycloalkanes[706] [Eq. (5.258)] or cycloalkene derivatives[707] [Eq. (5.259)]. The precondition for the reactions is the formation of stable (tertiary, allylic, benzylic, propargylic) carbocations, which are captured by the azido group followed by rearrangement to yield the final product after quenching [Eq. (5.259)]. Azidoalkyl-substituted ketals or enol ethers of cycloalkanones also react under similar conditions to form bicyclic lactams.[708]

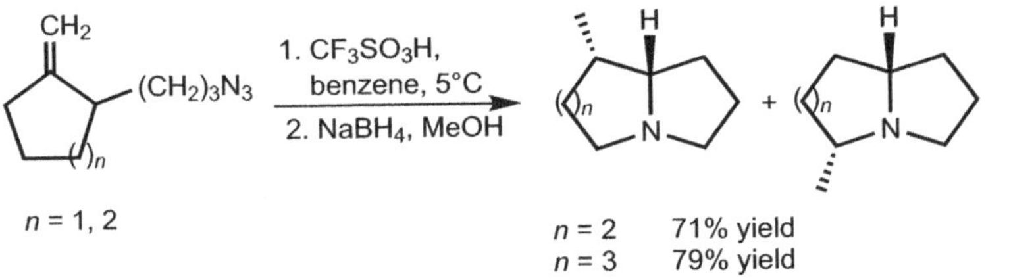

(5.258)

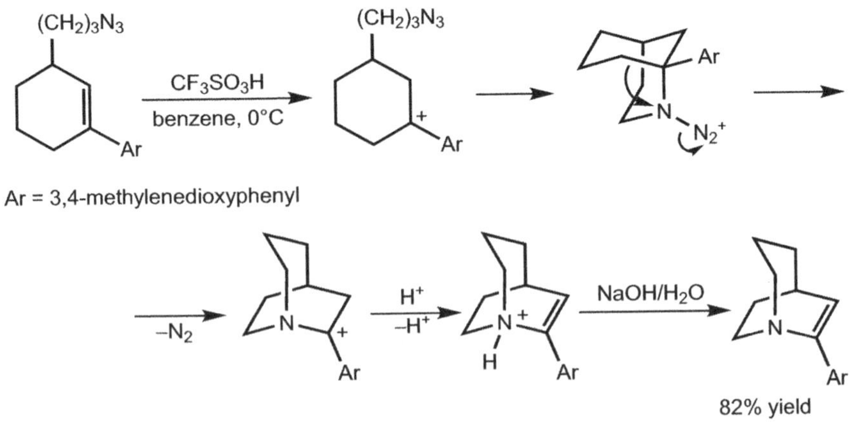

(5.259)

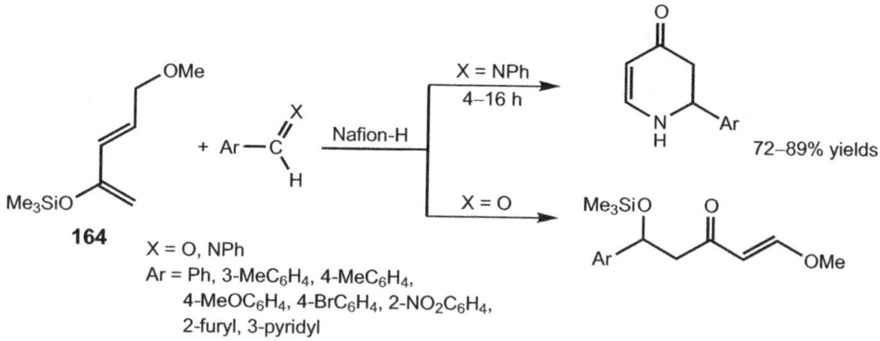

Scheme 5.69

Vankar and co-workers[709] have shown that Nafion-H can catalyze the hetero Diels–Alder reaction between the Danisefsky diene **164** and aromatic imines to form 2,3-dihydro-γ-pyridones (Scheme 5.69). The reaction with aromatic aldehydes, however, yields only the Mukaiyama aldol condensation products.

In analogy with the ring closure of 2,2′-dihydroxybiphenyls [Eq. (5.244)], Nafion-H catalyzes the formation of carbazole derivatives from 2,2′-diaminobiphenyls.[710] The synthesis, however, requires higher temperature and 4-*tert*-butyl-*ortho*-xylene was found to be the best solvent [Eq. (5.260)].

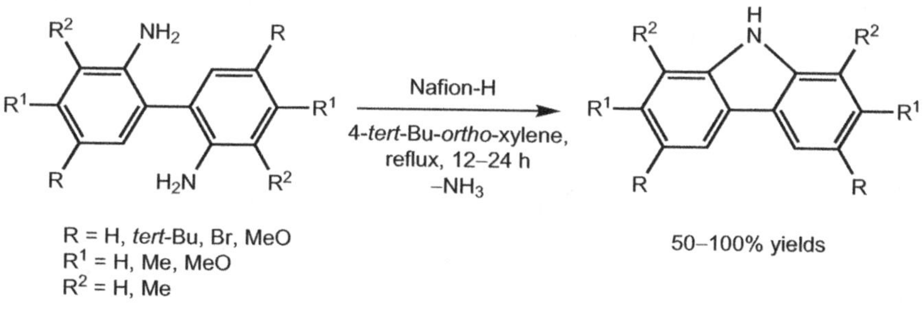

(5.260)

5.14.1.3. Heterocycles with Two or Three Heteroatoms. As already discussed (see Sections 5.10.1 and 5.12.1.3), Jacquesy and co-workers[548,549,634,711] have studied the transformations of various alkaloids in superacids. They have found that quinine and quinidine, but not epiquinine and epiquinidine, afford cyclization products resulting from the reaction between the 9-OH group and the rearranged quinuclidine skeleton ($HF–SbF_5$ or $H_2O_2–HF–SbF_5$). The transformation illustrated by the reaction of quinine in Scheme 5.70 involves polyprotonated quinine, which undergoes a series of rearrangements before the final ring-closing step to yield the oxazapolycyclic product.

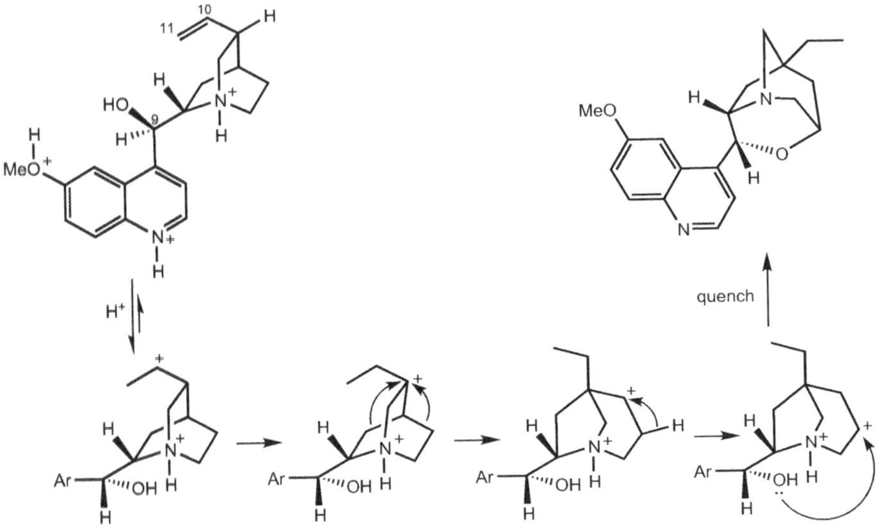

Scheme 5.70

Austin and Ridd[712] have described an interesting ring-closing transformation of 1-ethyl-2-nitrobenzene to 3-methylbenzo[*c*]isoxazole in triflic acid. On the basis of deuterium labeling and the large substituent effect observed (the reaction of 1-methyl-2-nitrobenzene is extremely slow), the mechanism shown in Eq. (5.261) including a hydride transfer from the benzyl carbon to one of the oxygens of the protonated nitro group as the rate-determining step has been proposed.

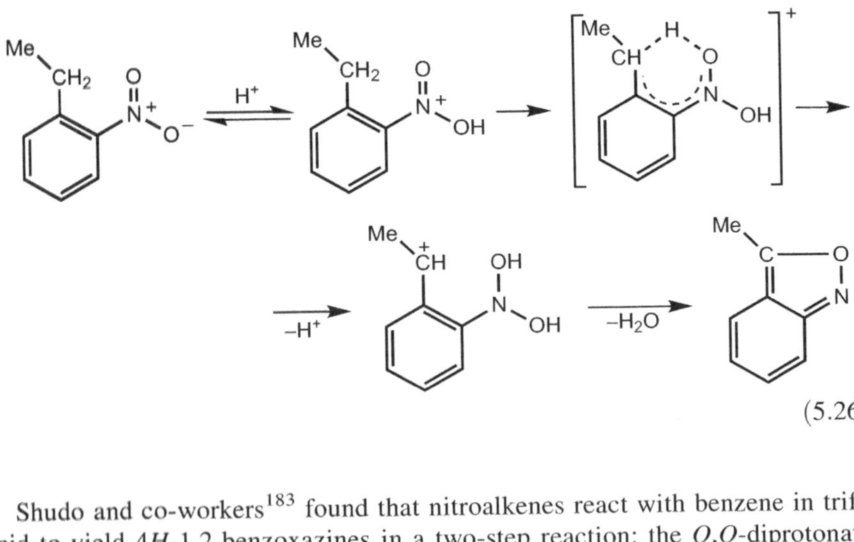

(5.261)

Shudo and co-workers[183] found that nitroalkenes react with benzene in triflic acid to yield 4*H*-1,2-benzoxazines in a two-step reaction: the *O*,*O*-diprotonated nitroalkene alkylates benzene (see Section 5.2.1) and then the diprotonated oxime intermediate undergoes ring closure to afford 4*H*-1,2-benzoxazines. A detailed study carried out recently by Ohwada and co-workers[713] has shown that triflic acid in large excess (10 equiv.) can induce a similar intramolecular cyclization of methyl 3-aryl-2-nitropropionates to form 4*H*-1,2-benzoxazines [Eq. (5.262)]. Compounds with electron-donating substituents (*para*-methyl- and *para*-methoxyphenyl derivatives) and 3-naphthyl- and 3-indolyl-2-nitropropionates are either unreactive or give products in very low yields. The use of triflic acid–TFA mixtures showed that the transformation is highly sensitive to acid strength: An acidity of $H_0 = -9.1$ (TfOH/TFA = 9:1) is required, which was interpreted in terms of activation by protonation of the methyl ester group.

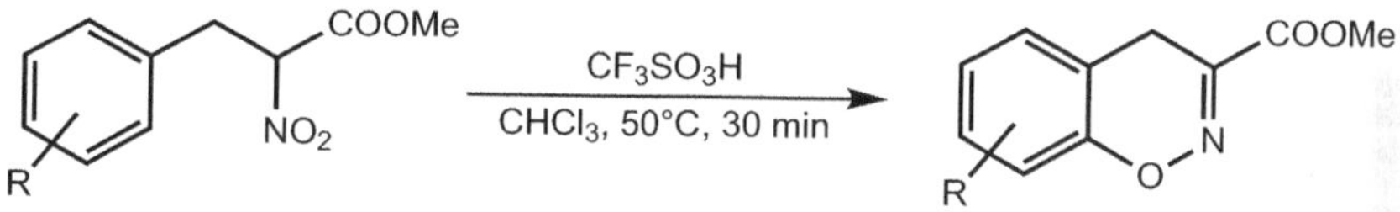

R = H, 3-Cl, 3-NO$_2$, 4-F, 4-Cl, 4-Br, 4-CF$_3$, 4-NO$_2$, 4-NC

74-93% yield

(5.262)

meta-Substituted derivatives give isomeric products, whereas *ortho*-halogen-substituted compounds yield products formed via halogen shift [Eq. (5.263)]. It was also found that the ring-closing is characteristic of only 3-aryl-2-nitropropionates (formation of six-membered ring), whereas analogous compounds with longer alkyl chain (methyl phenylnitro-acetate, -butanoate, and -pentanoate) decomposed under identical reaction conditions.

X, X' = F: 10 eqiv. of acid, 30 min
X = H, X' = Cl: 50 eqiv. of acid, 1.5 h
X, X' = Cl or Br: 50 eqiv. of acid, 2.5 h

	yield, %
X = F, X' = TfO	46
X or X' = Cl	27 and 41
X, X' = Cl	50
X, X' = Br	68

(5.263)

Deuterium labeling studies indicated that the electrocyclization mechanism necessitating the removal of a benzylic hydrogen, suggested earlier by Shudo and co-workers,[183] is not operative here. Instead, experimental observations indicate the

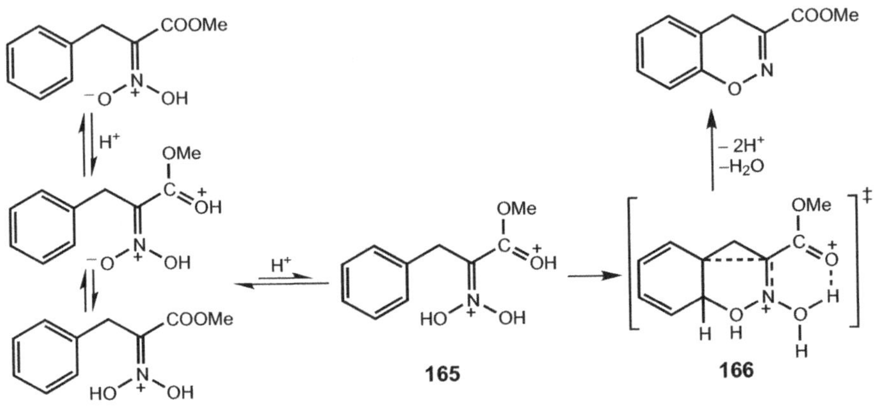

Scheme 5.71

involvement of the superelectrophilic diprotonated intermediate **165** stabilized by an intramolecular hydrogen bond (Scheme 5.71). Indeed, theoretical calculations showed that the activation energy of transition state **166** is significantly reduced compared to those of monocations (28.7 kcal mol^{-1} versus 46.2 and 47.4 kcal mol^{-1}).

The transformation of optically active epoxides with acetonitrile into optically active oxazolines (**167**, **168**) can be induced by various superacids[714] [Eq. (5.264)]. The reaction proceeds with inversion of the asymmetric center with high stereospecificity with anhydrous HF and $AlCl_3$, whereas partial racemization is observed in triflic acid (Table 5.39).

Table 5.39. Synthesis of Optically Active Oxazolines[714]

Epoxide			Temperature	Time	Yield		
R	R′	Acid	(°C)	(h)	(%)	**167/168**	%ee[a]
C_6H_{13}	H	HF	0	2.5	68	73:37	91
		CF_3SO_3H	RT	4.5	47	74:26	53
		$AlCl_3$	0	1	45	52:48	91
$C_6H_{13}OCH_2$	H	HF	0	1.5	91	27:73	89
n-C_4H_9	CH_3	CF_3SO_3H	0	1.5	45	100:0	36
		$AlCl_3$	0	2	66	100:0	88
C_6F_5	H	HF	0	1	74	100:0	35
		CF_3SO_3H	0	1	67	100:0	23
		$AlCl_3$	0	1.5	18	100:0	91

[a]Enantiomeric excess.

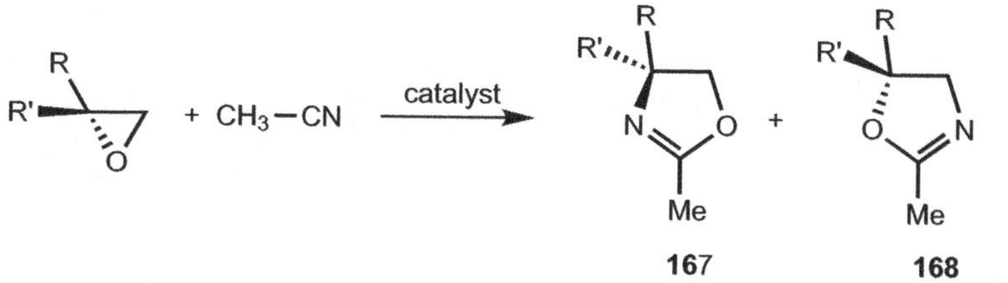

(5.264)

A one-step synthesis of tricyclic diazadihydroacenaphthylenes with an isoxazoline ring has been developed from 1-benzylamino-1-methylthio-2-nitroethene derivatives induced by a large excess of triflic acid[715] [Eq. (5.265)]. Dications **169**, similar to those detected by Coustard,[197] were observed by NMR spectroscopy. Quenching with water gives a reactive intermediate nitrile oxide, which undergoes an intramolecular cyclization to furnish the final products in fair yields.

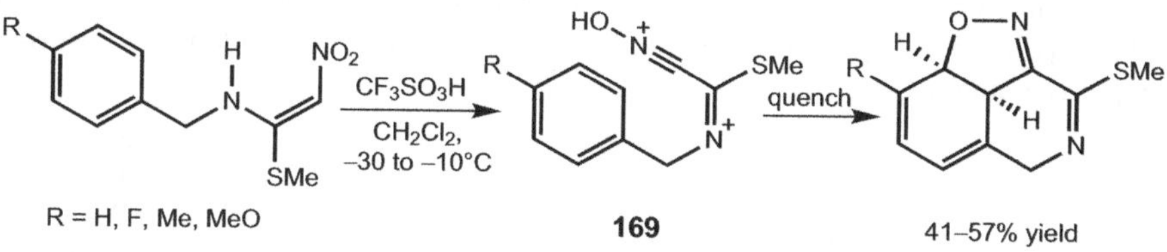

(5.265)

An unusual oxetane cyclization of compound **170** with the participation of the neighboring carbonyl group to yield bicyclic acetal **171** can be induced by triflic acid[716] [Eq. (5.266)].

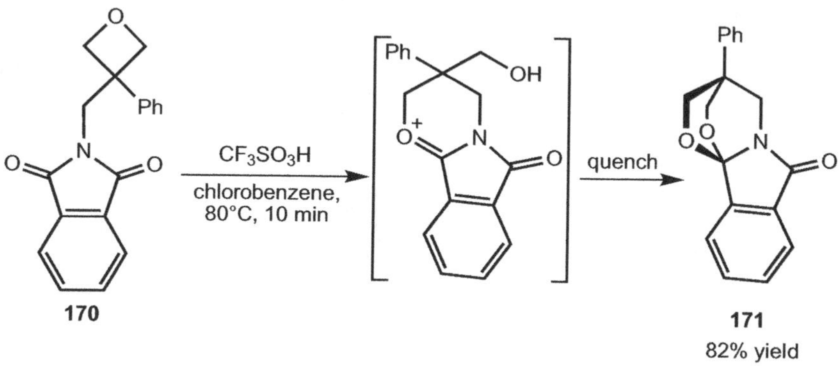

(5.266)

1,3-Dioxanes can be efficiently synthesized from styrenes using formalin as the formaldehyde source (Prins reaction) with triflic acid as catalyst[717] [Eq. (5.267)].

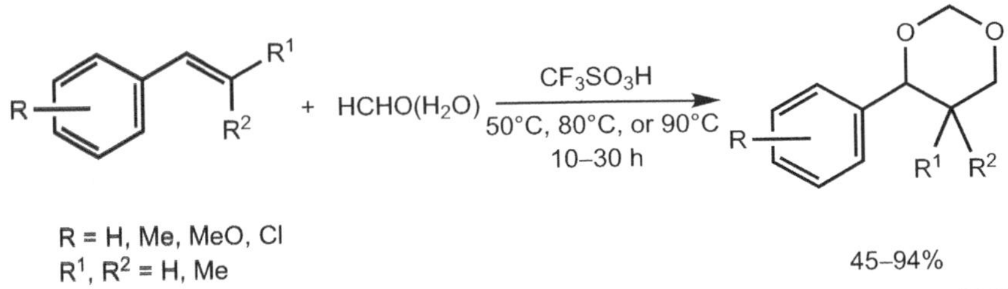

(5.267)

Superacids have been found to be efficient catalysts for the three-component condensation of aldehydes with 1,3-dicarbonyl compounds and urea (thiourea) to synthesize dihydropyrimidin-2(1*H*)-one derivatives (Biginelli reaction). Compared to the traditional process, the use of superacids results in high yields in shorter reaction times. Fluoroboric acid[718] [Eq. (5.268)] and Nafion NR50 (acetonitrile, reflux, 74–96%)[719] have been shown to give excellent results even with aliphatic aldehydes. The use of cyclic 1,3-diketones allows the synthesis of octahydroquinazolinone derivatives in good yields[720] [Eq. (5.269)]. Unexpectedly, 3- and 4-pyridincarboxaldehydes afforded hexahydroxanthenes.

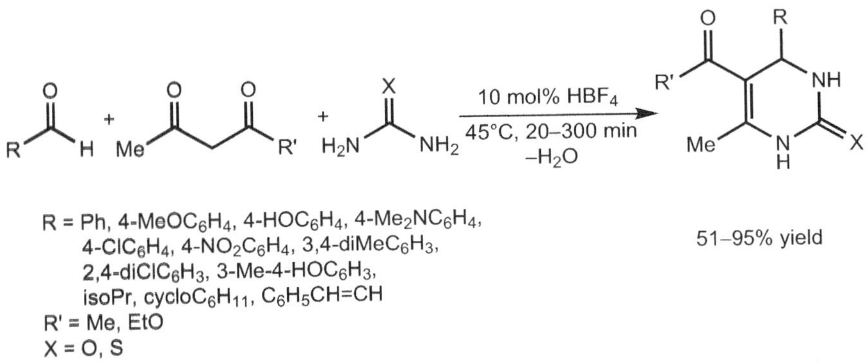

(5.268)

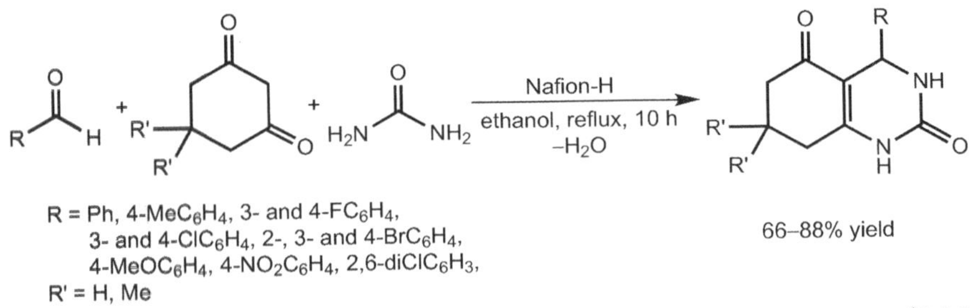

(5.269)

Nafion-H is highly efficient to induce the one-pot, three-component synthesis of 2,3-disubstiuted 4-(3*H*)-quinazolines under solvent-free conditions with microwave irradiation[721] [Eq. (5.270)]. Reactions are complete in a few minutes to afford the products in high yields.

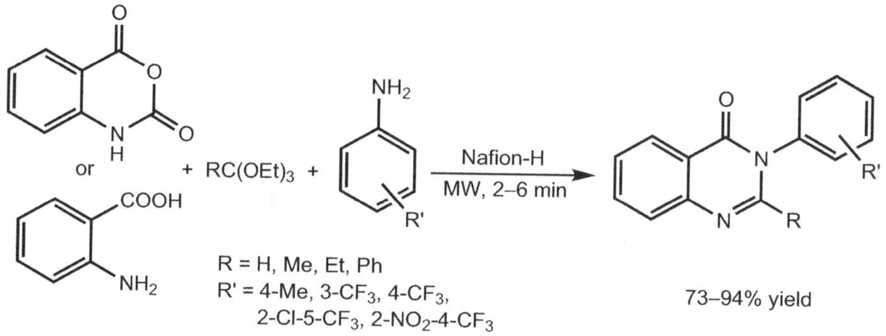

(5.270)

The condensation–cyclization of fluoromethyl ketones with bifunctional aniline derivatives or anthranilic acid gives a range of fluorinated heterocycles (benzimidazolines, benzothiazolines, benzoxazolines and dihydrobenzoxazinones) over Nafion-H and Nafion SAC-13 both exhibiting high catalytic activity.[722] Products are formed under mild conditions in high yields with high selectivity and purity. Illustrative is the transformation shown in Eq. (5.271).

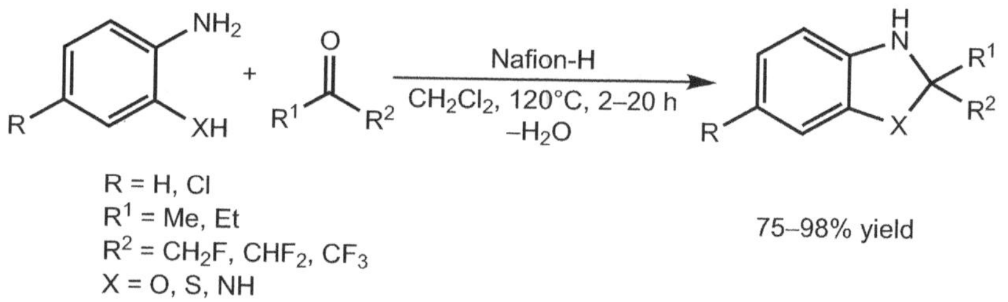

(5.271)

Fluoroboric acid supported on silica is also highly efficient for the synthesis of 1,5-benzodiazepines under solvent-free conditions[723] [Eq. (5.272)].

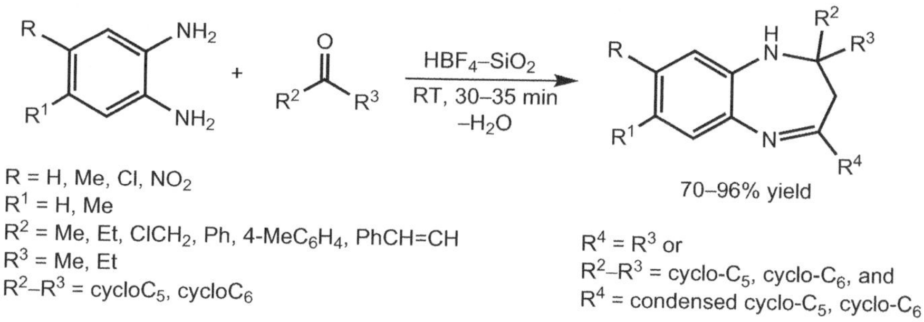

(5.272)

5.14.2. Ring Opening of Oxygen Heterocycles

Olah et al.[724] have shown that Nafion-H induces the ring opening of oxiranes under mild conditions to afford various products. Substituted oxiranes undergo hydrolysis or alcoholysis to yield 1,2-diols or 1,2-diol monoethers, when treated with Nafion-H under mild conditions in the presence of water or alcohols, respectively. Cycloalkene oxides give the corresponding *trans* products stereoselectively [Eq. (5.273)].

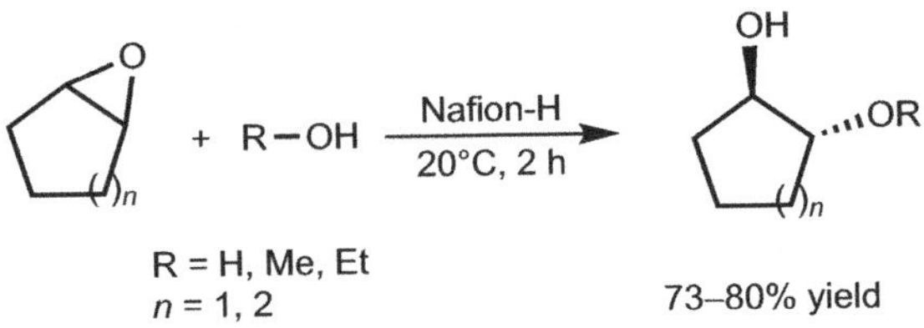

(5.273)

Ring opening of epoxidized fatty esters with alcohols has been studied to explore the synthesis of vicinal hydroxy ethers, which are potential lubricants[725] [Eq. (5.274)]. In a comparative study of ring-opening with methanol, Nafion SAC-13 showed superior activity (TOF = 1 min^{-1} versus 0.04 min^{-1} after 0.5 h at 60°C) when compared to Amberlist 15 (TOF = 0.04 min^{-1}) both exhibiting equally high selectivity (>98%).

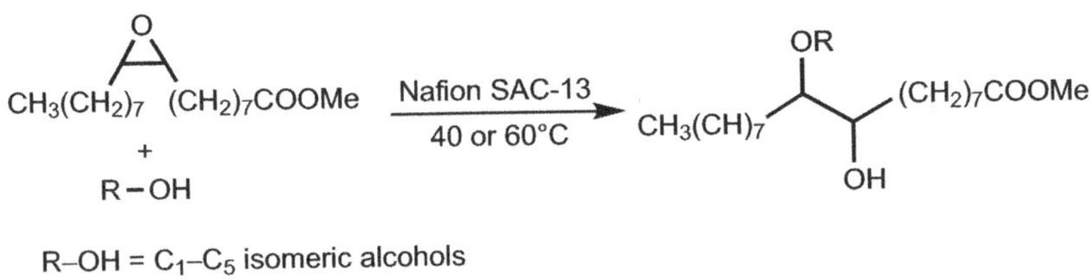

(5.274)

Isomerization of substituted styrene oxides allows the synthesis of aldehydes in high yields[726] [Eq. (5.275)]. Cycloalkene oxides do not react under these conditions, whereas 2,2,3-trimethyloxirane gives isopropyl methyl ketone (85% yield). Isomerization of oxiranes to carbonyl compounds is mechanistically similar to the pinacol rearrangement involving either the formation of an intermediate carbocation or a concerted mechanism may also be operative. Glycidic esters are transformed to α-hydroxy-β,γ-unsaturated esters in the presence of Nafion-H[727] [Eq. (5.276)].

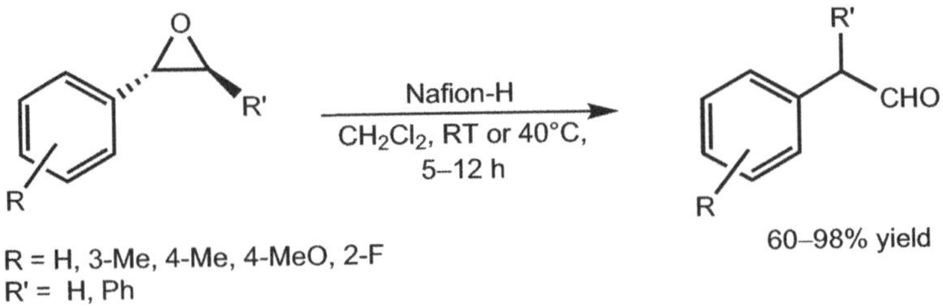

(5.275)

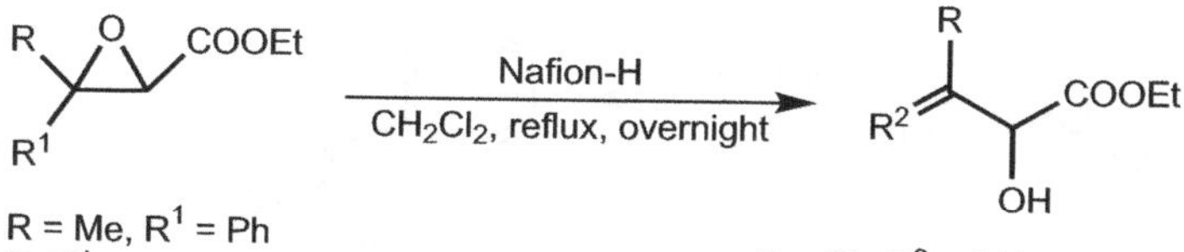

R = Me, R^1 = Ph
R–R^1 = cycloC_6H_{10}, 2-Me-cycloC_6H_9, 4-Me-cycloC_6H_9, cycloC_7H_{12}, 1-decalyl

R = Ph, R^2 = CH_2
R–R^2 = cycloalkenyl

69–99% yield

(5.276)

Methyloxirane showed a complex pattern of transformations over various solid acids in a pulse microreactor study.[728] Nafion-H was found to exhibit the highest activity (30% conversion at 90°C) to produce propanal (about 22% selectivity) and cyclic dimers (substituted 1,3-dioxacycloalkanes, 60% selectivity).

Ring opening of chiral epoxy alcohols with long-chain *n*-alcohols was induced by triflic acid (65°C, 4–24 h) to afford mono-*O*-alkylated glycerols (**172**).[729] The products are formed in good yields (36–83%) with high regiospecificity [ratio of C(3)/C(2) attack > 10, ee's > 95%]. 3,3-*O*-polymethylene glycerols (**173**) were synthesized in a similar manner using α,ω-diols (neat, 82°C, 8–10 h or $CHCl_3$, reflux, 48 h, 36–68% yield).

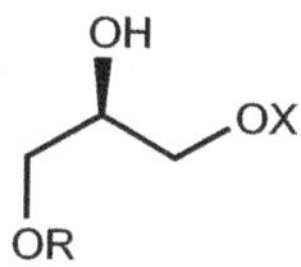

172

R = $C_{16}H_{33}$, $C_{18}H_{37}$, $C_{20}H_{41}$
X = 3-nirobenzenesulfonyl
n = 16, 20

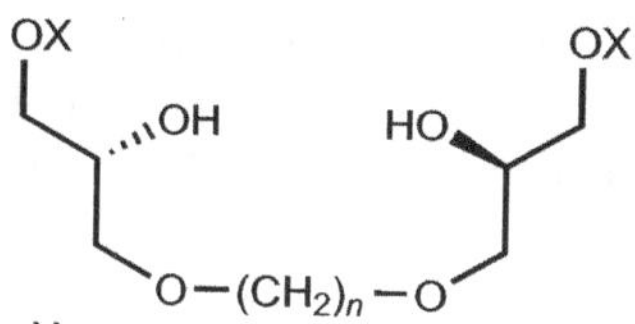

173

The ring opening of oxanorbornadiene **174** in triflic acid results in the formation of isomeric phenols[730] [Eq. (5.277)]. Products are formed directly via the rearrangement of the ring *O*-protonated oxonium ion intermediate **175** or Wheland intermediates may also be involved. The related transformation of the diepoxytetralin **176**, which includes a Wagner–Meerwein phenyl migration to a carbocation generated by oxirane ring opening, was performed with Nafion-H[731] [Eq. (5.278)].

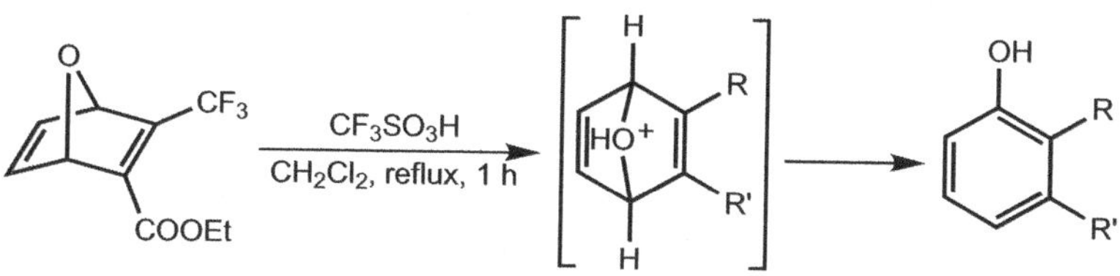

174 **175**

R = COOEt, R' = CF_3 31% yield
R = CF_3, R' = COOEt 48% yield

(5.277)

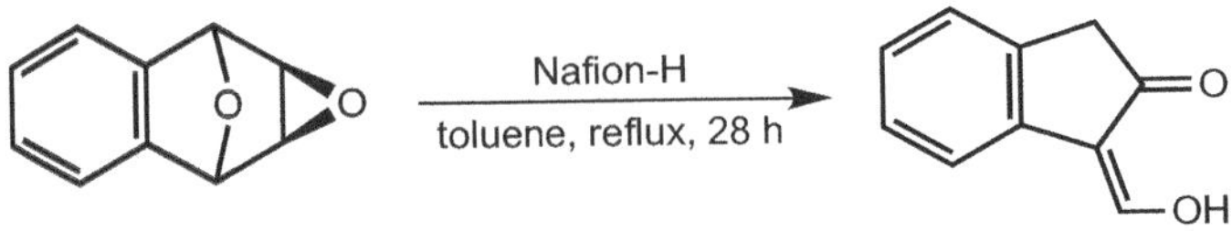

176 62% yield (5.278)

5.15. DEHYDRATION

1,2-Diols are known to undergo the pinacol–pinacolone rearrangement under acidic conditions[732] including Nafion-H[733] to yield carbonyl compounds with the concomitant elimination of water. Olah, Klumpp, and co-workers[734,735] have, however, observed that aryl pinacols and epoxides are cleanly and selectively afford condensed aromatics in superacidic triflic acid [Eq. (5.279)]. Although a mixture of isomers were obtained from 1,2-bis(4-fluorophenyl)-1,2-diphenyl-1,2-ethanediol, the transformation of other 1,2-bishalophenyl derivatives and 1,2-bis(2,4-difluorophenyl) diol was regioselective. Acidity-dependence studies with benzpinacol, tetraphenyloxirane, and triphenylacetophenone, the corresponding pinacol rearrangement product ketone, showed that they are all converted to 9,10-diphenylphenanthrene in acid systems with acid strengths greater than $H_0 = -12$—that is, with superacids. This led to the suggestion of a reaction mechanism with the involvement of superelectrophilic protosolvated reactive intermediate **177** [Eq. (5.280)]. Similar observations were made with superacidic BF_3–$2CF_3CH_2OH$ in benzene solution.[481]

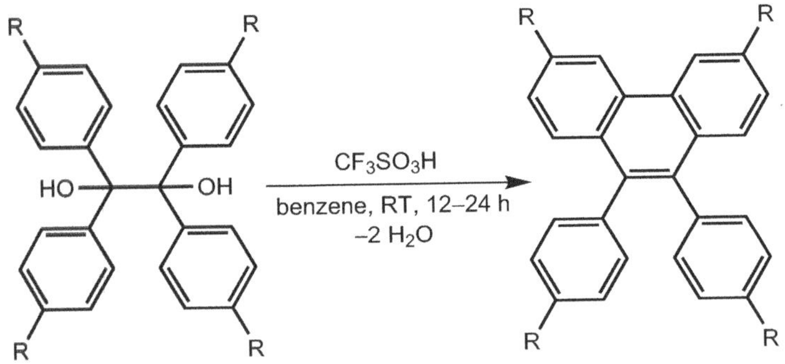

R = H, Me, F, Cl 90–100% yield (5.279)

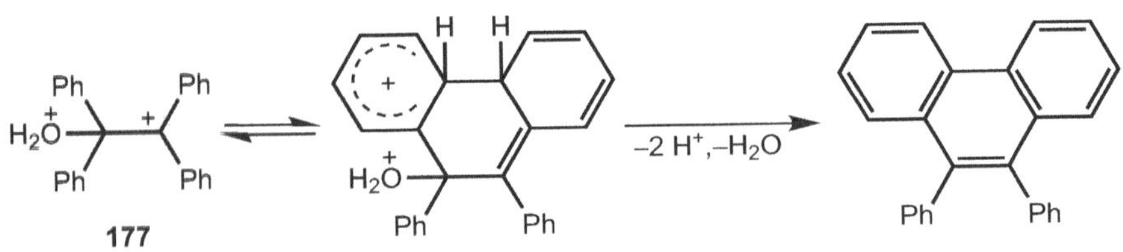

177

(5.280)

Takeuchi and co-workers[736] have reported that ionization of 3,4-dimethyl-4-homoadamantanol in Magic Acid results in ionization and rearrangement to yield the 3-ethyl-5-methyl-1-adamantyl cation **178** observed by ^{13}C NMR spectroscopy at −30°C, which, after quenching with methanol, gives ether **179** [Eq. (5.281)]. A series of known rearrangement steps and intermediates including protoadamantyl cations can account for the observation.

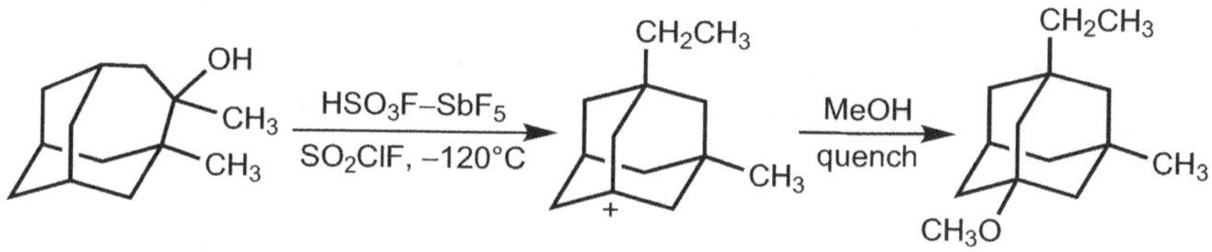

178

179
71% yield

(5.281)

Dehydration of alcohol to alkenes usually does not require superacidic conditions. Surprisingly, however, perfluroalkylcyclopentanols could not be dehydrated with *para*-toluenesulfonic acid, but alkenes could be isolated with the use of the stronger acid Nafion-H albeit in low yields[737] [Eq. (5.282)].

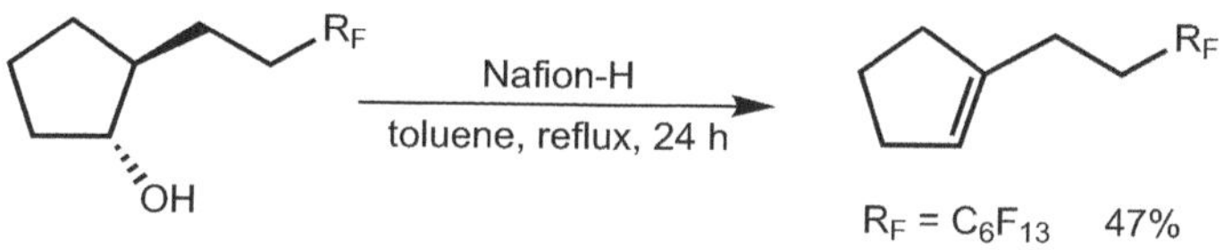

R_F = C_6F_{13} 47%
C_8F_{17} 20%

(5.282)

Alcohols can undergo acid-catalyzed dehydration to give either the corresponding alkenes or the corresponding ethers. The product distribution of the dehydration of alcohols over Nafion-H catalyst shows temperature dependence[187] (Table 5.40). Alcohols are thus efficiently dehydrated in the gas phase over Nafion-H under relatively mild conditions with no evidence for any side reactions such as dehydrogenation or decomposition. At higher temperature, olefin formation predominates.

Table 5.40. Dehydration of Alcohols over Nafion-H[187]

Alcohol	Temperature (°C)	Contact Time (s)	Dehydration (%)	Product	
				Alkene (%)	Ether (%)
isoPrOH	100	10	9		100
	130	9	28	45	55
	160	8	>97	100	
n-PrOH	130	4.5	8	47	53
	160	8	96	100	
tert-BuOH	120	5	100	100	

Nafion SAC-13 exhibited the highest activity and selectivity in the dehydration of 1- and 2-hexanol and cyclohexanol to alkenes in the gas phase over various solids in the temperature range of 200–300°C.[738] Furthermore, it did not show the dramatic decrease in activity when compared to zeolites; rather, it exhibited an initial activity increase. This was attributed to solvation of the polymeric matrix by the water vapor formed in the process. This served to swell the resin, rendering previously inaccessible sites available to the reactants.

Low temperature and high alcohol partial pressure were found to favor ether formation in the dehydration of methanol and isobutyl alcohol over Nafion-H.[739] The mixed methyl isobutyl ether as the dominant product, the absence of methyl *tert*-butyl ether, and large amounts of 2,5-dimethylhexene indicate that free carbenium ions are not involved in the transformations. In a comparative study, Nafion NR50 showed the highest selectivity to form dipentyl ether from 1-pentanol (150–190°C), but its activity was lower than that of Amberlyst resins calculated on a weight basis.[740,741] Nafion-H has been found to be an excellent catalyst to transform primary alcohols to the corresponding symmetric ethers (reflux temperature, 7–12 h, 92–98% yield),[742] whereas mixed diphenylmethyl ethers of both primary and secondary alcohols were isolated in good to high yields under mild conditions[743] [Eq. (5.283)].

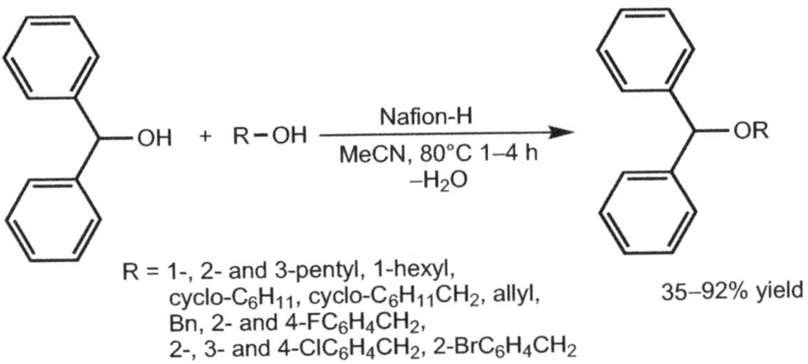

(5.283)

5.16. SUPERACIDS IN CARBOHYDRATE CHEMISTRY

O-Glycosylation—that is, the condensation of a sugar derivative bearing an appropriate functional group (donor) with an appropriately protected other sugar derivative with a free hydroxyl group (acceptor)—is a crucial methodology in carbohydrate and natural product chemistry.[744]

Glycosyl fluorides, widely used reagents in carbohydrate and natural product chemistry,[745,746] can be used to carry out stereoselective synthesis of glycosides with a catalytic amount (5 mol%) of triflic acid. The appropriately protected β-D-glycosyl fluorides of glucose and galactose as donor molecules, when applied in dichloro-

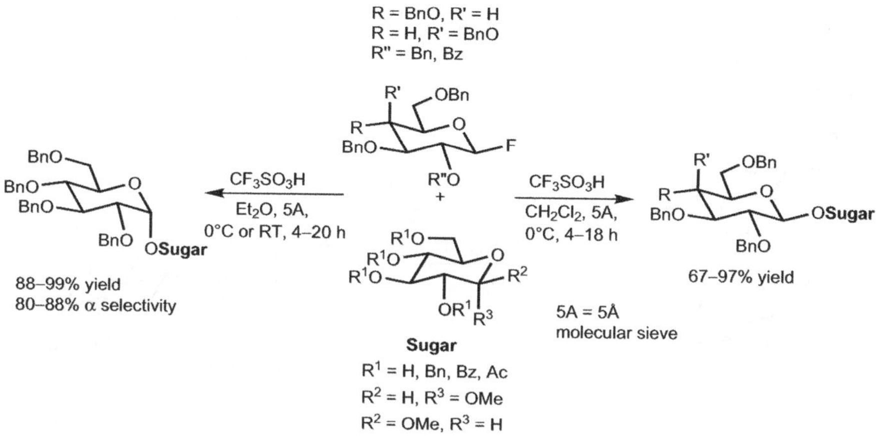

Scheme 5.72

methane as solvent, afford the corresponding β-D-glycosides[747,748] (Scheme 5.72). In contrast, the same combination of reagents in diethyl ether solvent leads to the formation of the α-linked disaccharides[749] (Scheme 5.72). One-pot sequential stereoselective glycosylation based on this method has been applied for the convergent total synthesis of oligosaccharides.[750]

Mukaiyama and co-workers[751] have tested a variety of donor compounds in glycosylation of various acceptor molecules in triflic acid. Glucosyl α-thioformimidates are highly reactive donors and the stereochemistry can be controlled by reaction conditions: In methyl *tert*-butyl ether at 0°C, α-glycosylation takes place with selectivities up to 90%, whereas selective β-glycosylation was observed in propionitrile (−78°C, up to 98% selectivities). In a comparative study with various donors in triflic acid, 6-nitro-2-benzothiazolyl α-glucoside showed the highest β-selectivity[752] [Eq. (5.284)].

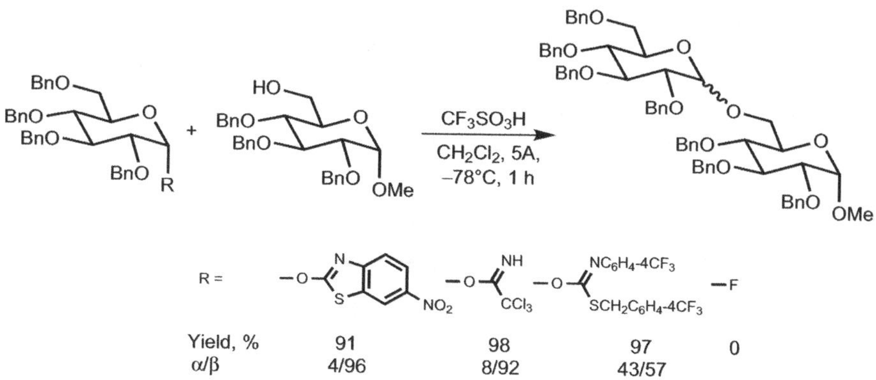

(5.284)

Wong and co-workers[753] and Hashimoto and co-workers[754] have demonstrated that triflic acid is an ideal promoter in chemo- and stereoselective glycosylation with phosphites [Eq. (5.285)].

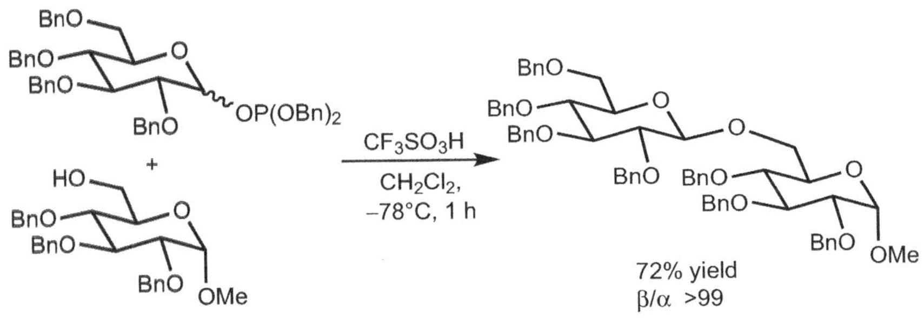

(5.285)

Franck and Marzabadi[755] have developed the heterocyclic donor **180** and used it in coupling with a variety of alcohol acceptors to obtain β-glycosides in good yields with excellent stereoselectivities using equimolar amount of triflic acid [Eq. (5.286)].

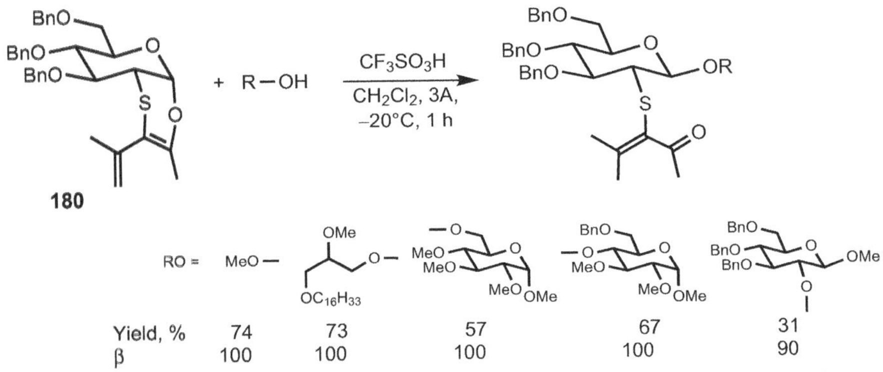

(5.286)

Catalytic amount (5 mol%) of triflic acid has been shown to be effective in the intramolecular condensation–dehydration-type glycosylation to form anhydroketo-pyranoses without the necessity to introduce any leaving groups[756] [Eq. (5.287)]. Interestingly, the twist-boat conformers were also detected in the transformation of the 1-phenyl- and 1-benzyl-susbtituted compounds.

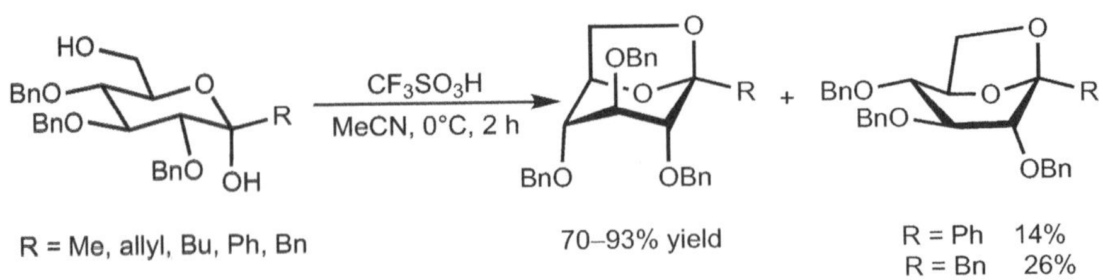

(5.287)

In another application of catalytic glycosylation with triflic acid, two disaccharides were synthesized with donors bearing trichloroacetimidate function and then the disaccharides thus prepared were condensed in a similar manner to a tetrasaccharide[757] (1,2-dichloroethane, 55°C, 1.5 h, 50–85% yields). The same protocol was used in the synthesis of a decasaccharide from mono-, tetra-, and pentasaccharide building blocks.[758] The trichloroacetimidate procedure has been successfully used in the glycosylation of β-cyclodextrins promoted by triflic acid.[759,760]

Glycosylations can also be effectively induced with Nafion-H.[761–763] Protected mannopyranosyl 1-α-sulfoxide reacts with alcohol acceptors to yield α-glycosides with high stereoselectivity[762] [Eq. (5.288)]. In fact, the sulfoxide method using arylsulfoxide derivatives as the donor and an appropriate scavenger in triflic acid has been successfully applied in the synthesis of di- and oligosaccharides[764] and in the coupling of unreactive nucleophiles as glycosyl acceptors.[765]

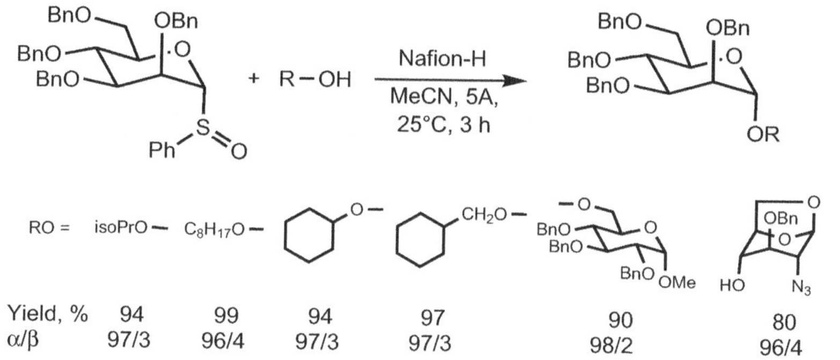

(5.288)

Neighboring group participation by substituents such as a PhS goup at an adjacent carbon atom has long been considered to control stereochemistry in glycosylation. Sulfonium ions, for example, thiiranium ion **181**, have often been postulated to be involved; moreover, such ions have been observed by low-temperature NMR spectroscopy (see Section 4.2.2.4). The participation of sulfonium ions as reactive intermediates in glycosylation, however, has recently been disputed. Woerpel and co-workers[766] have analyzed the stereochemistry of various glycosylation reactions and computed possible reaction intermediates. They concluded that the bridged form (**181**) equilibrates with the open carboxonium ion form (**182**), and the latter should be the prescursor to the products if electron donation from oxygen is more stabilizing than anchimeric assistance from sulfur.

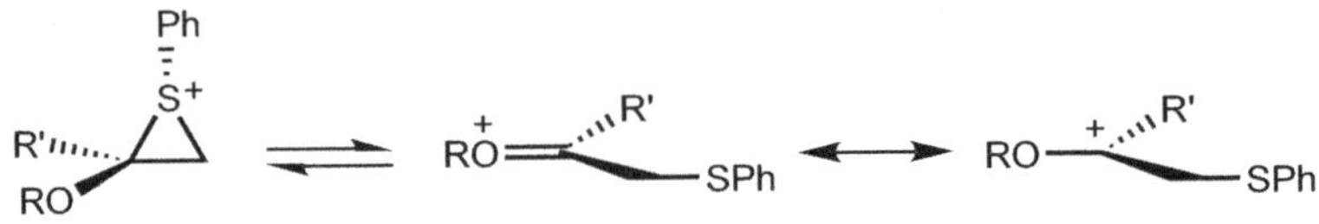

181 **182**

C-2-Methylene α-glycosides **183** and **184** are readily synthesized by reacting *C*-2-acetoxymethyl glycals with alcohols in the presence of Nafion-H[763] [Eq. (5.289)]. Unexpectedly, *para*-cresol gave the corresponding β-glycoside, whereas β-naphthol afforded a *C*-glycoside. Facile chemoselective deprotection at room temperature of terminal isopropylidene acetals (2–4 h, 68–96% yields) and trityl ethers (7–14 h, 75–92% yields) with Nafion-H in methanol has also been reported.[767] Nafion SAC-13 showed high stereoselectivity in the condensation of glucosyl imidate with protected glucose at low temperature (66% yield, $\beta/\alpha = 13.3$, $-20°C$, 4 h).[768]

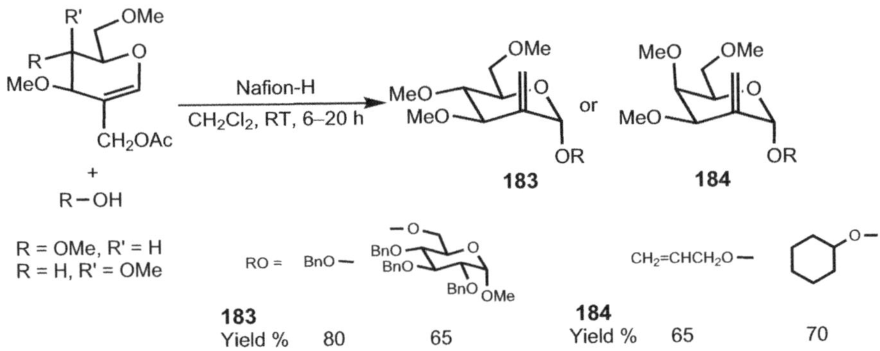

(5.289)

Triflic acid has been applied in the transformation of monosaccharides into spiroketals (dianhydrides).[769,770] 3,4,6-Tri-*O*-protected 1,2-*O*-isopropylidene-β-D-fructofuranoses undergo a tandem acetal cleavage–intermolecular glycosylation–intramolecular spiroketalization process in the presence of triflic acid to give binary mixtures of α,α and α,β diastereomeric products [Eq. (5.290)]. The isomer ratio strongly depends on the nature of the protecting groups. Ether groups prefer the nonsymmetric α,β distereomer, whereas the benzoyl group favors the symmetric α,α diastereomer. Similar observations were made in the transformation of D-fructopyranose derivatives. The method has recently been used in the synthesis of D-fructose spiroketals attached to xylylene tethers.[771] Anhydrous hydrogen fluoride and pyridinium poly(hydrogen fluoride) (PPHF) were also used in the synthesis of oligosaccharides with spiroketal moieties.[772]

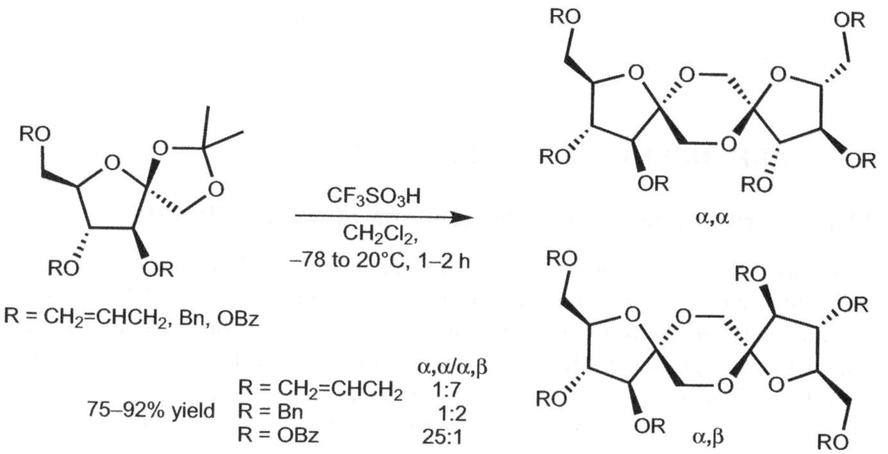

(5.290)

Fused or spiro sugar oxazolines have been synthesized by treating D-fructopyranose or D-fructofuranose 1,2-*O*-acetonides with nitriles in triflic acid[773] (Scheme 5.73). First the activation of the anomeric center takes place with simultaneous isopropylidene cleavage to form the oxocarbenium ion intermediate, which is attacked by the nitrile. The resulting nitrilium ion is then trapped by the hydroxyl group in an intramolecular Ritter-like reaction to yield the final product.

Anhydrous hydrogen fluoride has been successfully applied in the structural analysis of polysaccharides to cleave glycosidic linkages since the 1980s. Recently, triflic acid has been found to be a more potent and more selective agent.[774] In structural studies, for example, polysaccharides isolated from bacteria were stable toward

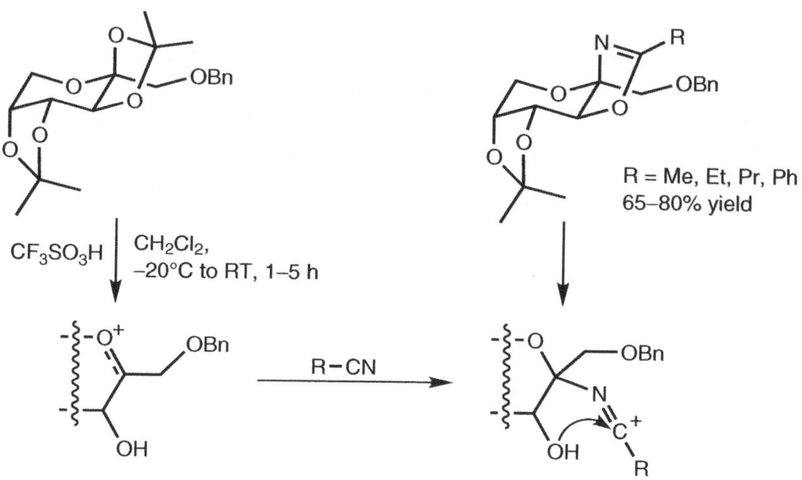

Scheme 5.73

solvolysis in anhydrous HF. Triflic acid, in turn, selectively cleaved glycosidic linkages without affecting amide-like substituents.[775,776]

5.17. REARRANGEMENTS AND CYCLIZATIONS

5.17.1. Rearrangements and Cyclizations of Natural Products

Acid-catalyzed rearrangements of natural products, particularly those of terpenes, which often involve cyclization, have been studied extensively.[777–784] This research has unraveled numerous unique rearrangements that are both synthetically and mechanistically significant. Moreover, the recognition that cationic cyclizations play a key role in biogenesis[777,785–789] of isoprenoids (terpenes) has provided incentive to mimic in vitro[790–792] many of these rearrangements by generating the appropriate carbocations.

The literature abounds with examples of rearrangement of a wide variety of terpenoids with varying acid catalysts.[777–784] One of the most significant areas is the synthesis of triterpenoids by van Tamelen,[790] Johnson,[791] and others. However, most of these rearrangements have been studied under mild conditions, in nucleophilic media with relatively weak acids. Under these conditions of low acidity, only a very minute amount of the reactants exists as carbocation at any time because the electron-deficient intermediates react immediately with available nucleophiles, both external and internal via a number of competitive, irreversible reactions. In other words, only short-lived, transient intermediates are involved. Consequently, quite often, complex mixtures of products are formed in the rearrangements during mild acid catalysis, because the products obtained are generally comparable in energy content. In addition, products with more than two carbocycles could not be obtained in high yields.

The use of carbocation-stabilizing superacids such as HSO_3F, HSO_3F-SbF_5, CF_3SO_3H, $HF-SbF_5$, $HF-BF_3$, and so on, to alter the normal course of acid-catalyzed rearrangements has offered unique possibilities, since the stable carbocation would have time to explore many internal escape paths (through a probably shallow potential energy surface) but would not convert to neutral products. In recent decades, considerable progress has been made in this area, especially by the pioneering work of Jacquesy[784] and the wide-ranging studies by Vlad[793] and Barkhash and Polovinka.[794]

As early as 1893, it was discovered[795] that camphor **185** rearranges to 3,4-dimethylacetophenone **186** in concentrated H_2SO_4 and the mechanism of the reaction has been elucidated by Roberts[796] and Rodig[797] [Eq. (5.291)].

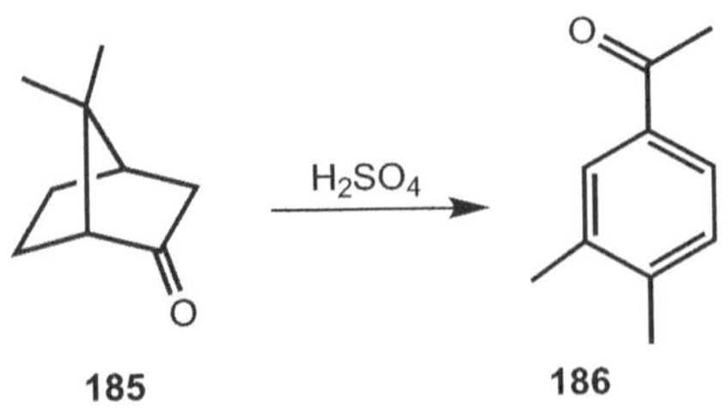

(5.291)

185 **186**

However, Jacquesy and co-workers[784,798] have shown that in superacidic HF–SbF_5 medium, the reaction takes a different course. Camphor gives a mixture of three ketones [Eq. (5.292)].

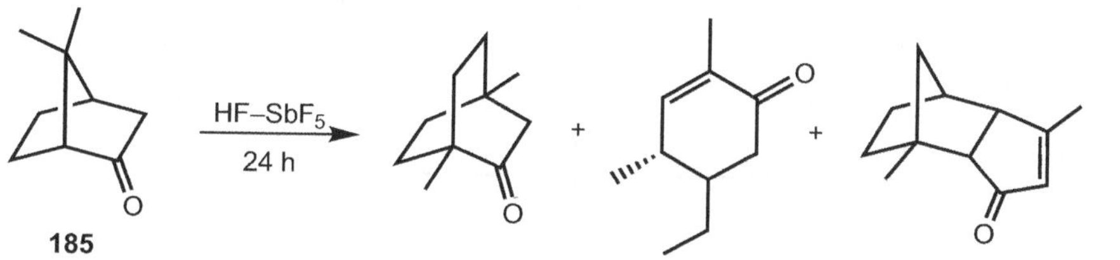

(5.292)

Sorensen and co-workers[799–801] have studied the fate of observable camphene hydrocation **190** prepared from isoborneol **187**, camphene **188**, or tricyclene **189** in HSO_3F acid medium (Scheme 5.74). The intermediate cycloalkenyl cation **191** can also be prepared by protonation of α-terpineol (**192**), sabine (**193**), and β-pinene (**194**) (Scheme 5.75).

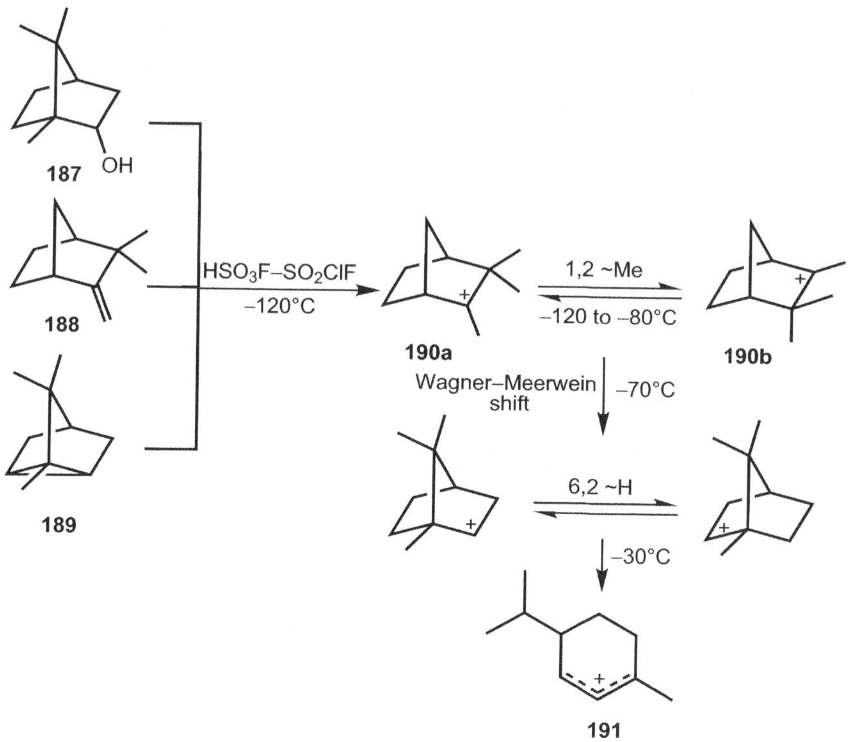

Scheme 5.74

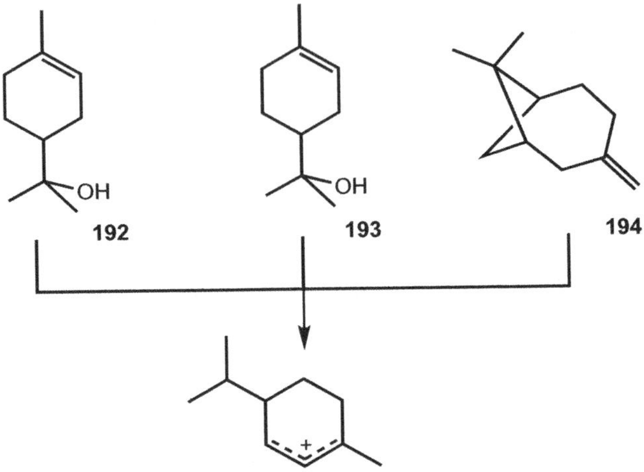

Scheme 5.75

An early example is the fascinating rearrangement[802] of acyclic monoterpenes geraniol **195** and nerol **196** to a stable observable oxonium ion in $HSO_3F-SO_2-CS_2$ at $-78°C$, which represents yet another dramatic example of altered course of terpene rearrangement in superacid medium. Careful quenching of this ion led to the isolation of 3*β*,6*α*,6a*β*-trimethyl-*cis*-perhydrocyclopenta[*b*]furan **197** in excellent yield (Scheme 5.76).

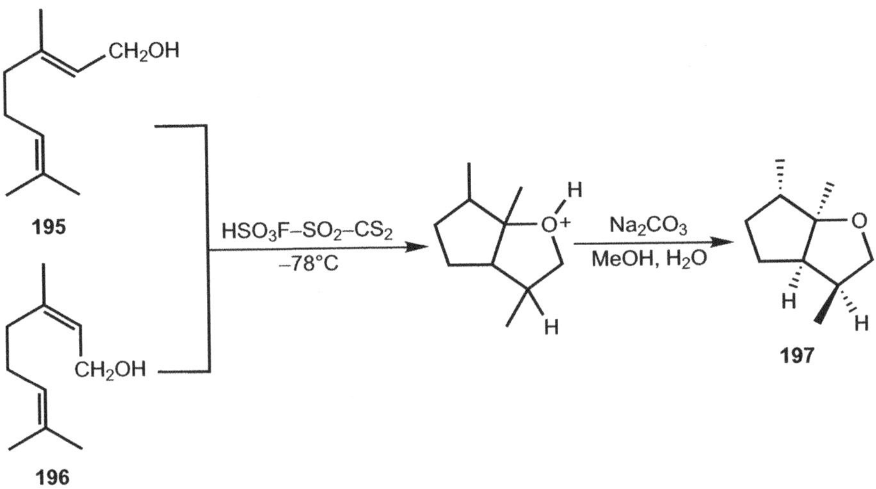

Scheme 5.76

In a subsequent study, however, Whittaker and co-workers[803] showed that the product formed from geraniol, in fact, is the isomeric bicyclic ether **198**. It was also shown that cyclization of citronellal in $HSO_3F{-}SO_2$ at $-78°C$ affords pulegol and neoisopulegol; that is, the transformation parallels closely the reaction in normal acids.[804]

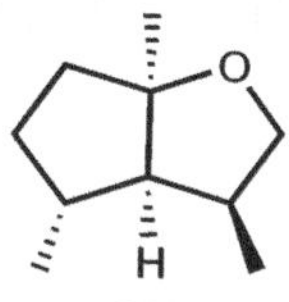

198

Whittaker and co-workers[805,806] have also prepared a number of terpenoid bicyclic ethers, such as isomeric 1,6-dihydrocarveols **199** ($HSO_3F{-}SO_2$, 78°C) from unsaturated alcohols or diols.[805,806] *para*-Menth-1-en-9-ol and a related diol afforded the seven-membered ring systems **200**. The oxolane moiety in compound **201** was generated from the corresponding unsaturated alcohol precursor in 2 equivalents of triflic acid.[807]

199

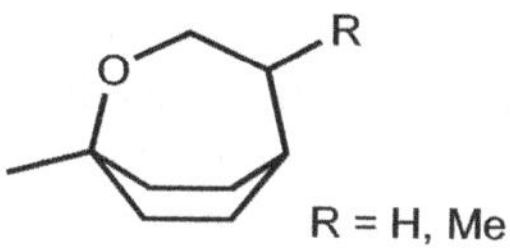

200

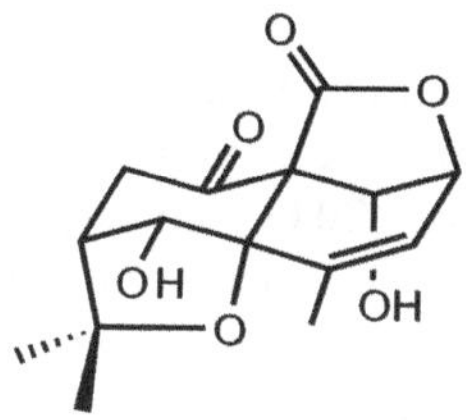

201

Fluorosulfuric acid is a widely used, highly effective reagent[808] in electrophilic polyene cyclizations to synthesize fully cyclized compounds in a selective and stereospecific way. Furthermore, an internal nucleophile, most often oxygen, allows the construction of polycyclic skeletons with a heteroatom (*vide supra*). Snowden and co-workers[809] transformed trienone **202** (a mixture of 4 diastereomers) into three isomeric irone derivatives [Eq. (5.293)]. Other acids gave inferior results.

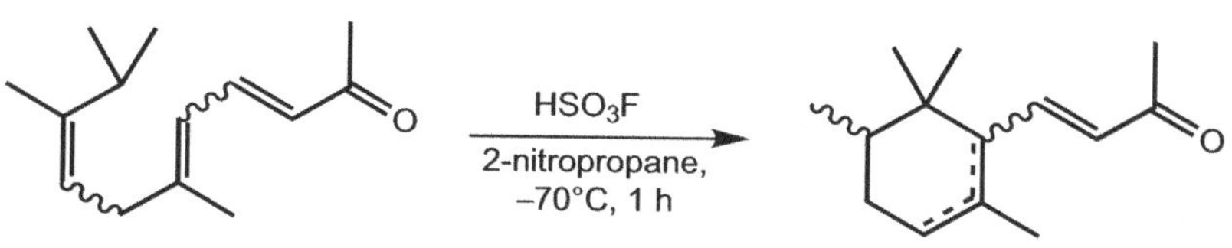

202

70% yield

(5.293)

Structurally related dienols and acyclic trienols, when reacted in fluorosulfuric acid, give tricyclic ether derivatives in kinetically controlled cyclization.[810,811] The stereospecific product formation is rationalized by synchronous internal *anti*-addition via chair-like conformations of the protonated cyclohexene ring, resulting in ring closure with equatorial C–C bond formation and concomitant internal nucleophilic termination by *anti*-addition of the OH group [Eq. (5.294)]. *Z*/*E* isomerization may be competitive with cyclization.

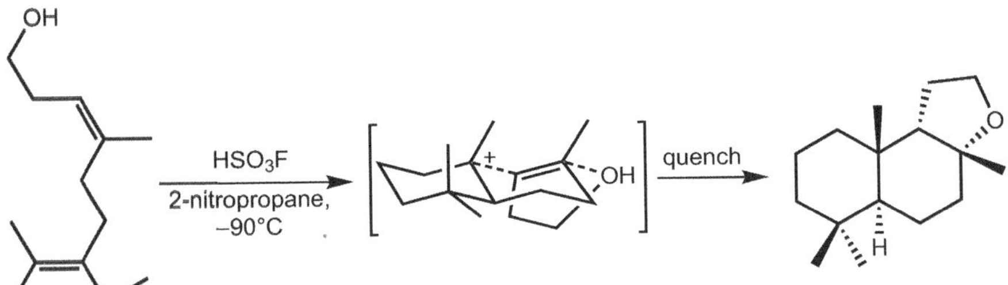

80% yield

(5.294)

Chlorosulfuric acid used less frequently has been shown to be equally effective in related transformations. Altarejos, Barrero, and co-workers[812,813] carried out superacid-mediated cyclizations including the transformation of trienol **203** to racemic ambrox (**204a**) and *epi*-ambrox (**204b**) constituents of ambergris fragrances [Eq. (5.295)]. Isomeric β-monocyclofarnesol and -nerolidol and their acetates gave a single unique rearranged ocathydromethanonaphthalenol product in chlorosulfuric acid under similar conditions (2-nitropropane, –78°C, 10 min)[814] [Eq. (5.296)].

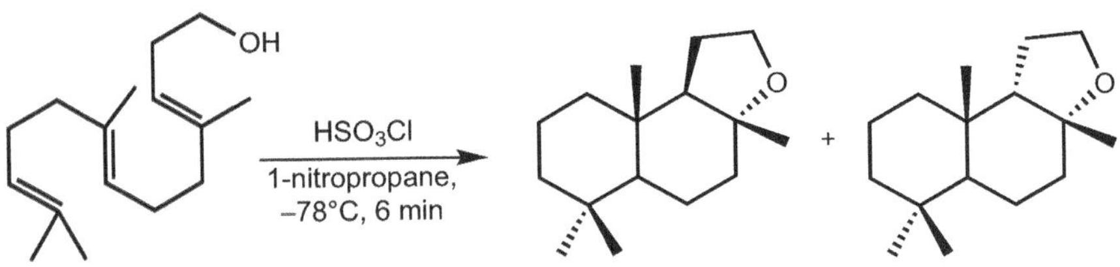

203 **a** **204** **b**

41% 34%

(5.295)

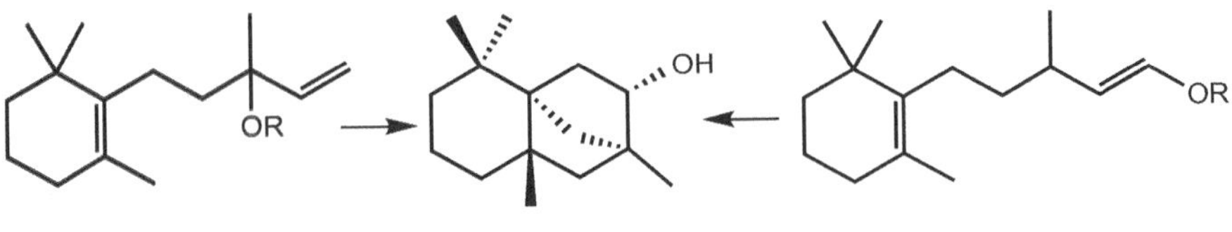

R = H, Ac 28–60% yield

(5.296)

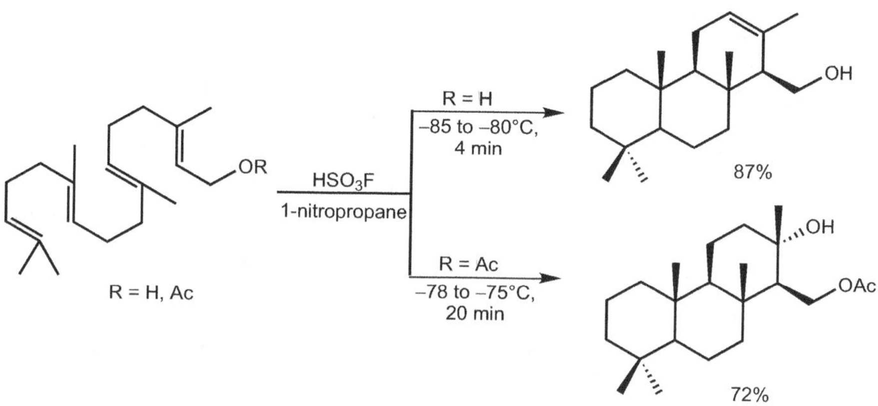

Scheme 5.77

Vlad and co-workers have made extensive studies of the superacidic cyclization of terpenoids. Early results have been summarized in a review.[793] Subsequently, they reported the superacidic low-temperature cyclization of terpenols and terpenol acetates to form homoallylic alcohols or hydroxy acetates of cyclized terpenols, respectively[815] (Schemes 5.77 and 5.78). The transformations are highly efficient, chemo- and structurally selective, and stereospecific. The configuration of the hydroxymethyl (acetoxymethyl) group is determined by the configuration of the allylic double bond of the precursor (Schemes 5.77 and 5.78). Terpenoid acids and esters,[816,817] as well as phenylsulfones,[818,819] also undergo stereoselective cyclization in the presence of HSO_3F. In bicyclic stereoisomeric compounds, the internal (13*Z*) C=C double bonds were shown to affect the selectivity of cyclization.[817] In the transformation of isomeric methyl (6*Z*)-geranylfarnesoates in HSO_3F the (6*Z*) double bond plays a key role to form steroisomeric tricyclic products.[820]

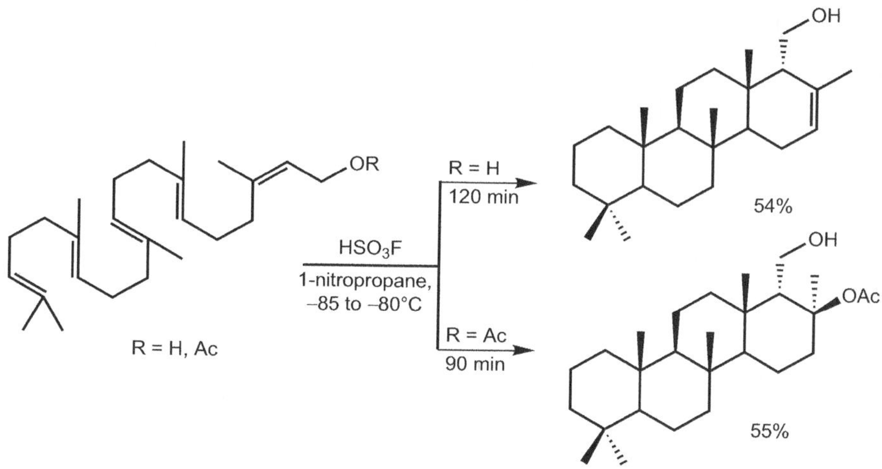

Scheme 5.78

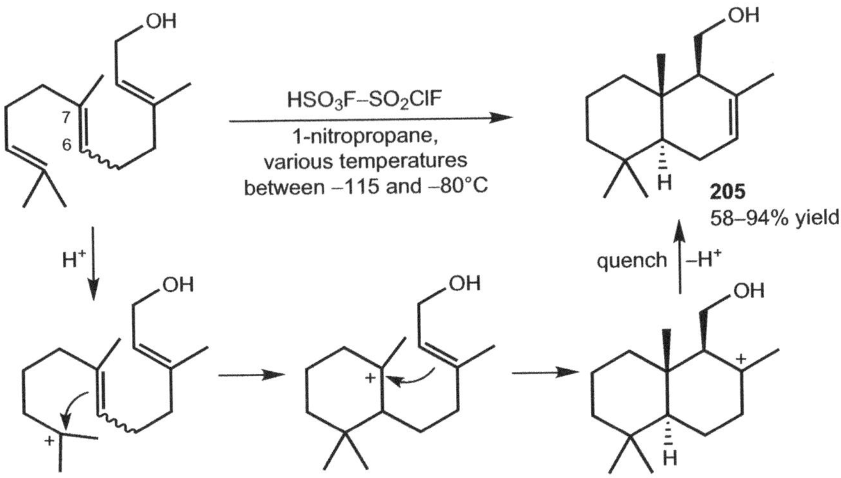

Scheme 5.79

Vlad and co-workers[821] have also found that the cyclization of stereoisomeric farnesols proceeds regioselectively and stereospecifically in HSO_3F-SO_2ClF to yield drimenol **205** (Scheme 5.79) and *epi*-drimenol, respectively. The configuration of the C(6)–C(7) double bond does not affect the stereochemistry. When **205** or *trans,trans*-farnesol **206** was dissolved in HSO_3F-SO_2ClF at −115°C and the solution was warmed to −70°C, the tricyclic oxolane derivative **207** was isolated (Scheme 5.80). Recently, they have successfully carried out the cyclization of several aliphatic sesquiterpene derivatives (*E,E*-farnesol and its acetate, *E,E*-farnesyl phenylsulfone, and the methyl ester of *E,E*-farnesylic acid) with HSO_3F in ionic liquids ([BMIM][BF$_4$], [BMIM][PF$_6$]).[822]

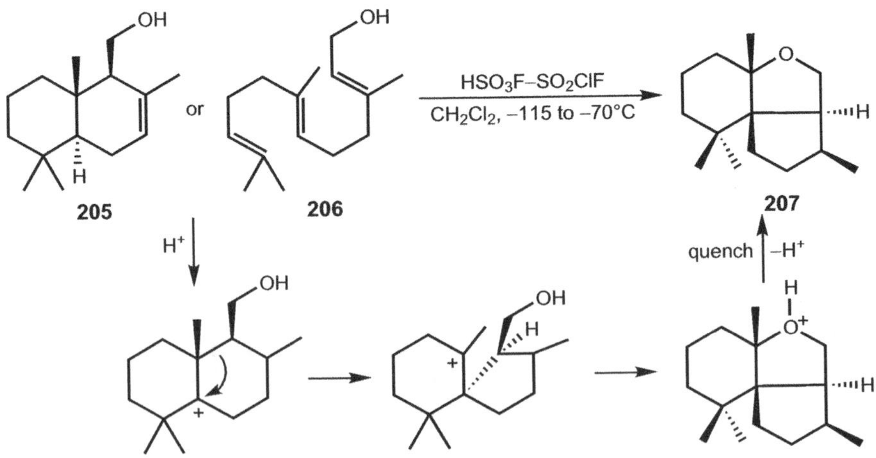

Scheme 5.80

They have also reported the transformation of all-*trans*-ω-acetoxyfarnesol benzyl ether to afford a mixture of two stereoisomers in a clean reaction[823] [Eq. (5.297)]. In sharp contrast, the hydroxy compound gave a complex mixture of products. The selective reaction of the acetoxy derivative was interpreted as resulting from a rare example of initial protonation of the internal C(6)–C(7) double bond.

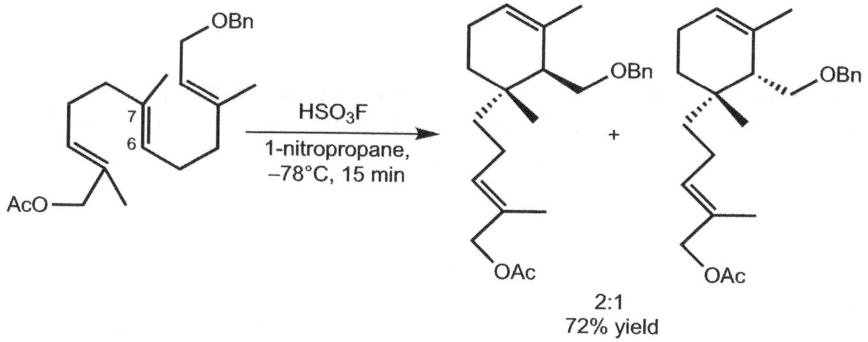

(5.297)

Barkhash and Polovinka[794] have reported in a series of papers the results of their extensive studies of the transformations of a large number of terpenoids induced by homogeneous and heterogeneous acids including fluorosulfuric acid. 2,3-Epoxygeraniol **208** was shown previously by Whittaker and co-workers[824] to afford epimeric oxabicyclooctanes in $HSO_3F–SO_2$ at 70°C [Eq. (5.298)]. Repeating the transformation at lower temperature (–100°C) and quenching the reaction mixture by methanol/ether, Barkhash and co-workers isolated the tetrahydropyranyl derivative **209** (Scheme 5.81).[825] Quenching at –80°C gave, in addition to compound **209**, one of the oxabicyclooctane isomers and isomeric dehydrooxepanes.

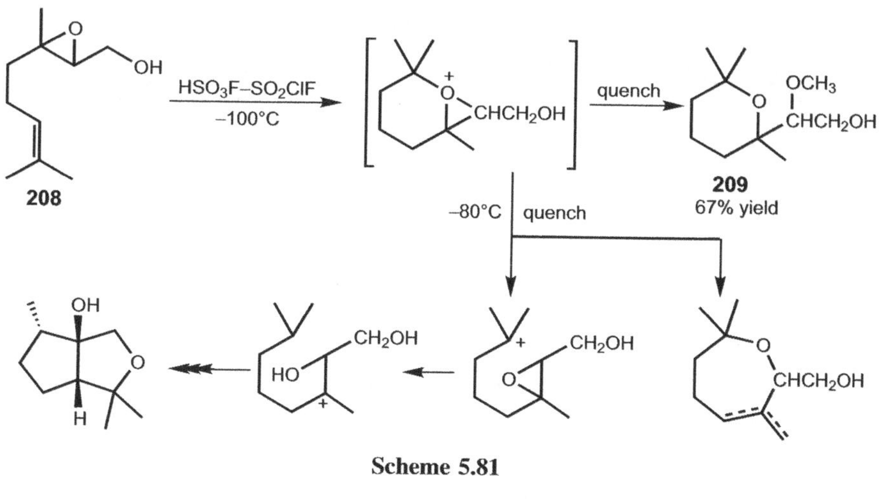

Scheme 5.81

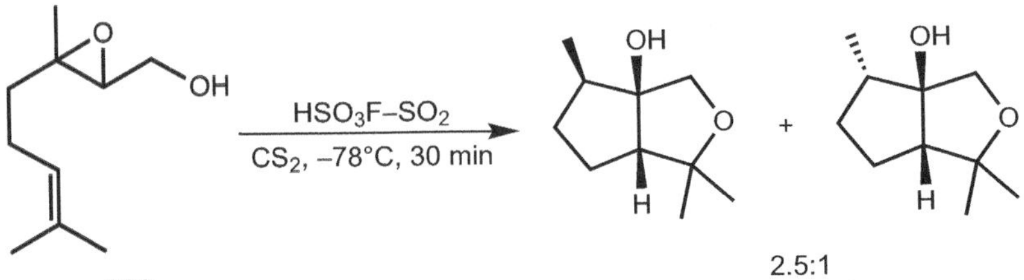

208

(5.298)

Cyclic epoxide derivatives, such as 2,3-epoxy-*cis*-pinane **210** [Eq. (5.299)][826] and isomeric caryophyllene diepoxides **211** [Eq. (5.300)],[827] undergo varied transformations including ring contraction under similar conditions.

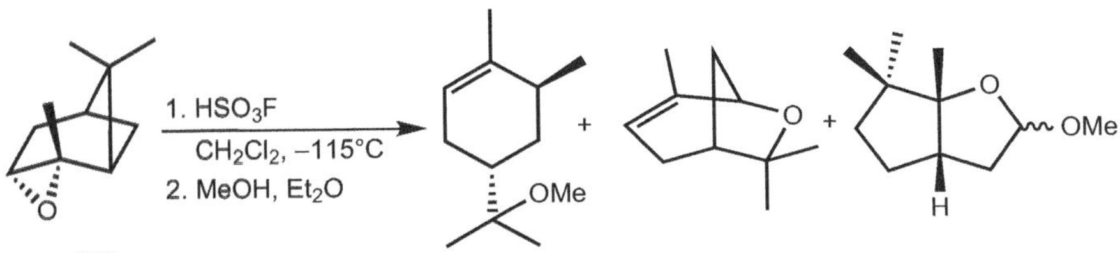

210

(5.299)

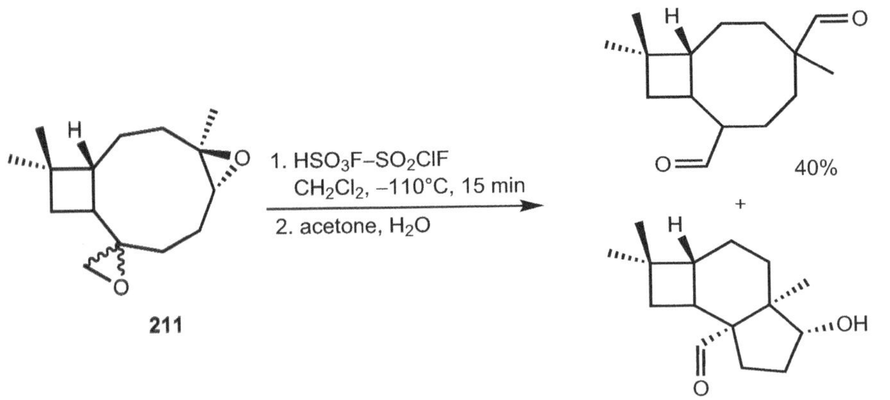

211

(5.300)

Alkenes of the aromadendrane series rearrange into tricyclic ring systems in fluorosulfuric acid[828] [Eq. (5.301)]. The transformation of related compounds under similar conditions and calculations suggest the involvement of cation **212** as the key intermediate.

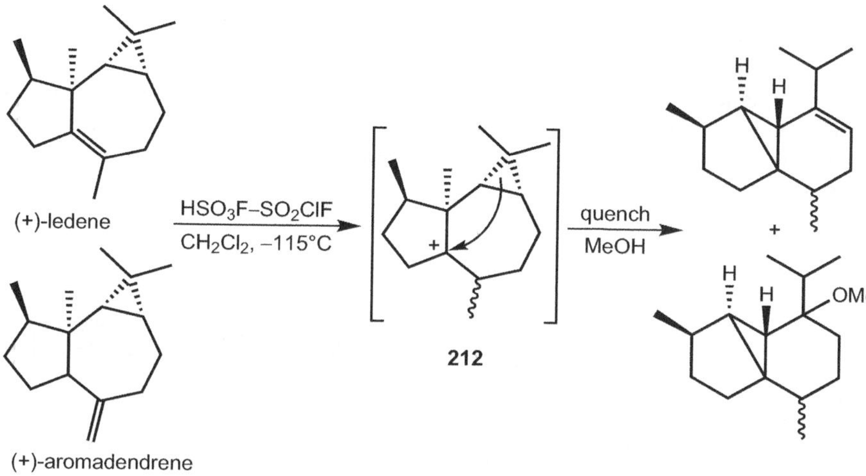

(5.301)

Additional examples are the clean ring opening transformation of *cis*- and *trans*-3-methyl-*cis*-verbanone[829] [Eq. (5.302)]. Product structures indicate that there is no epimerization at C(3) (in the starting materials) and C(6) (in the products) despite the strongly acidic conditions. The ring opening and rearrangement shown in Eq. (5.303) are key steps in the synthesis of a marine *nor*-sesquiterpene.[830]

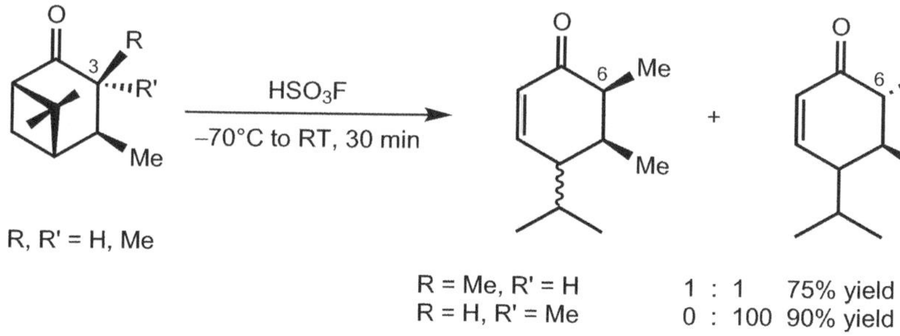

(5.302)

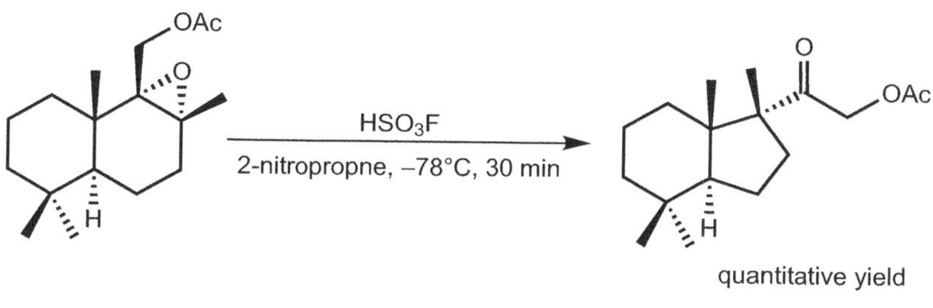

(5.303)

Collado and co-workers made detailed studies of the chemistry of the sesquiterpenoid caryophyllene and its hydroxylated products including rearrangements induced by superacids.[831] They have recently reported[832] novel rearrangements of the sesquiterpenoid panasinsane derivatives **213** to provide three products and interpreted the transformations by the involvement of the common carbocationic intermediate **214** [Eq. (5.304)].

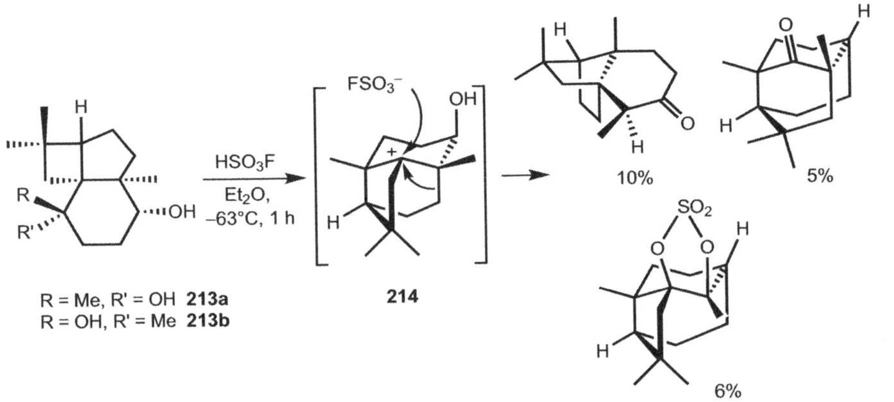

(5.304)

Triflic acid has been used to induce cyclization of polyisoprene (xylene, 75°C, 5–30 min) and the various ring moieties (**215–217**) were identified by ^{1}H NMR.[833]

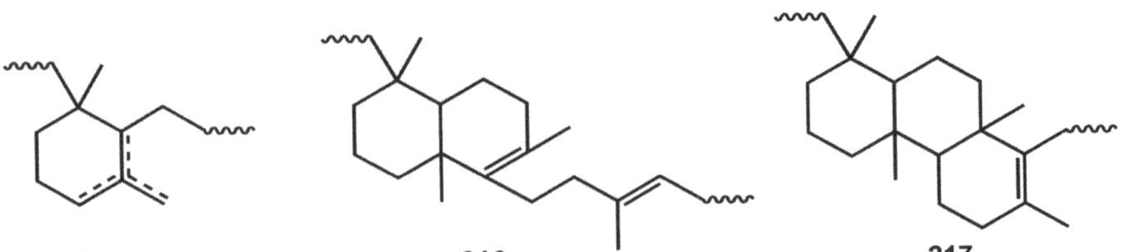

215 **216** **217**

Farnum, Mehta, and co-workers[834,835] have shown that longifolene **218** and isolongifolene **219** in HSO_3F media give a mixture of cyclohexenyl cations at different temperatures. Quenching these cations provides some unusual sesquiterpene like C_{15}-hexahydronaphthalenes **220–225** (Scheme 5.82).

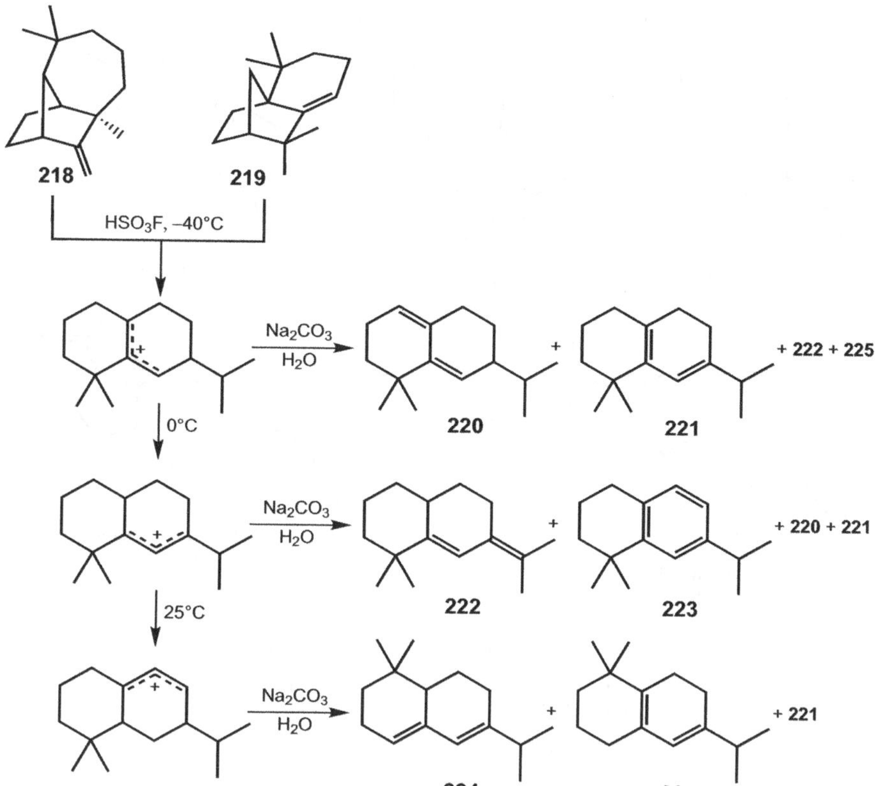

Scheme 5.82

The rearrangement of some resin acids **226**, **227**, and **228** in superacidic HSO_3F and HSO_3Cl media has also been studied.[836,837] Jacquesy and et al.[838,839] have developed a novel isomerization of pregnan-3,20-diones **229** to mixture of isomers that also contains 13α isomers. The reaction is proposed to occur through the cleavage of C(13)–C(17) bond (Scheme 5.83).

Similarly, $HF-SbF_5$-induced isomerization of androsta-4,6-diene-3,17-dione **230** has been studied in detail,[840] which led to a new entry into the 9-methylsterane series **231**. Also, methods have been developed for the synthesis of isosterane derivatives[841] and other methyl-substituted estrane dione derivatives of unnatural configurations.[842] $HF-SbF_5$ superacid medium is also capable of demethylating aromatic ethers. This reaction has been successfully employed in the synthesis of 11-deoxyanthracyclines **232** [Eq. (5.305)].[843]

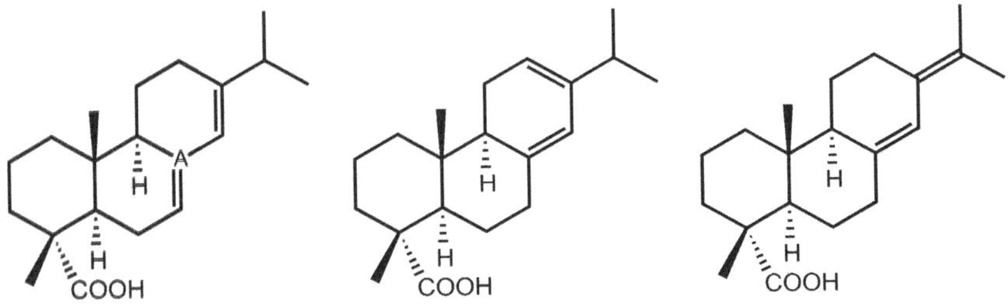

226 **227** **228**

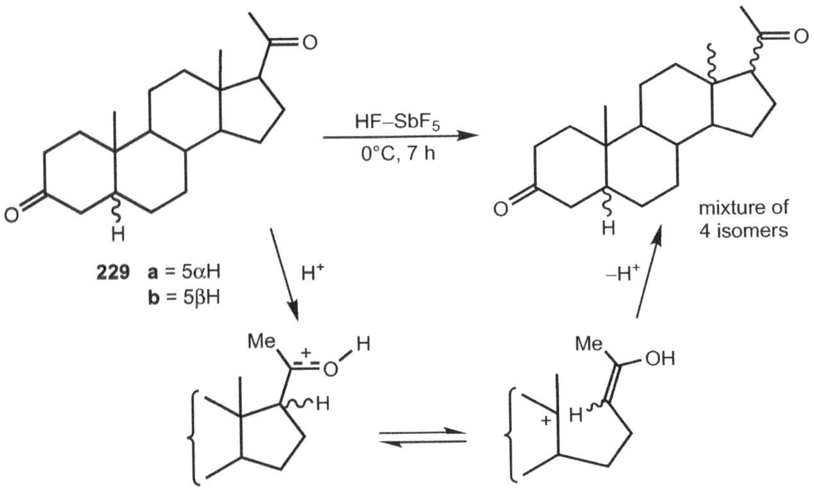

Scheme 5.83

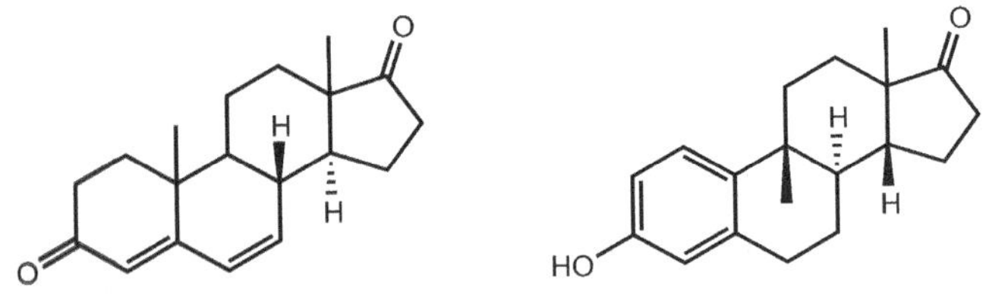

230 **231**

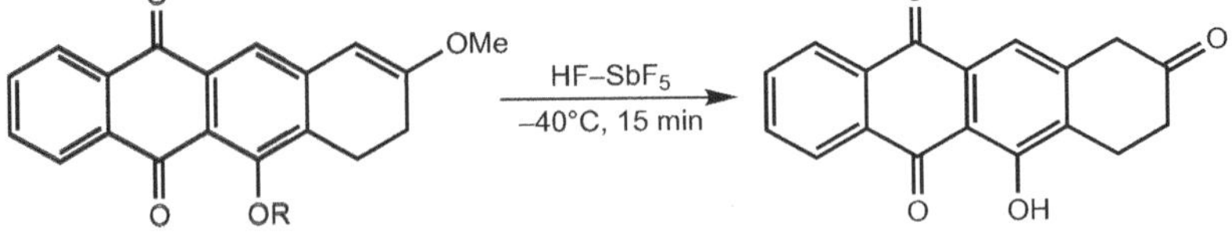

(5.305)

R = H, Et **232**

In a series of papers, Jacquesy and Jouannetaud have demonstrated the efficacy of aromatic cyclization reactions in HF–SbF_5 medium.[784] 1,3-Bis(methoxyphenyl) propanes and other substituted phenylpropanes **233** give substituted indenes (**234**) and tricyclic spiro enones (**235**) in the superacid medium[844] (Scheme 5.84).

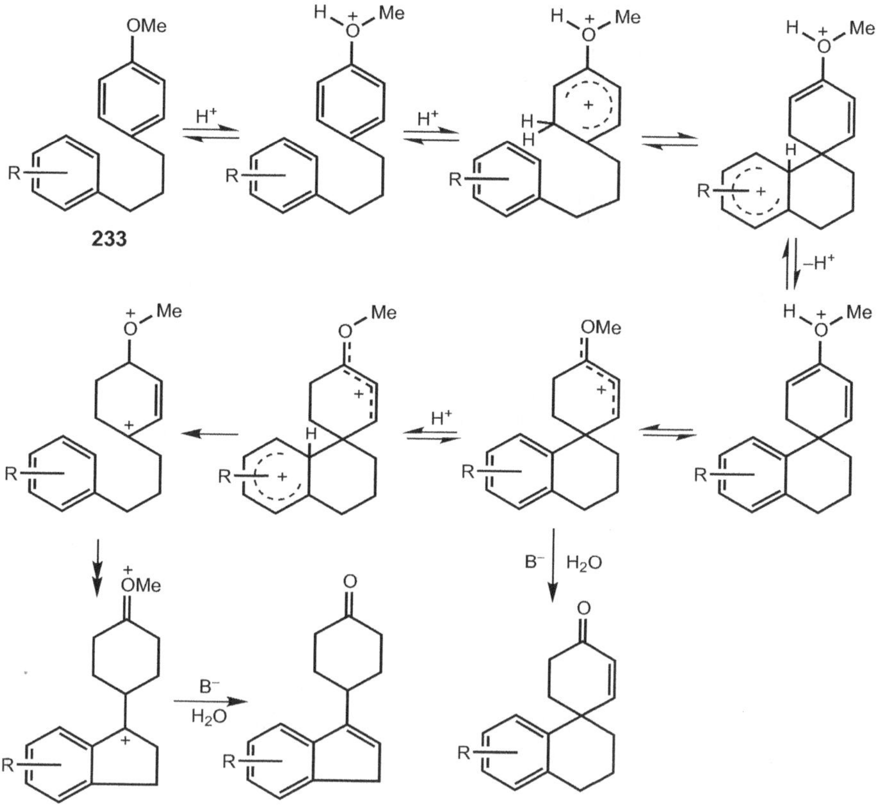

233

234

235

Scheme 5.84

1,2-Bis(methoxyphenyl)ethanes subjected to a similar treatment (HF–SbF_5, 0°C, 5–270 min) gave tetrahydrophenanthrenone derivatives[845] [Eq. (5.306)].

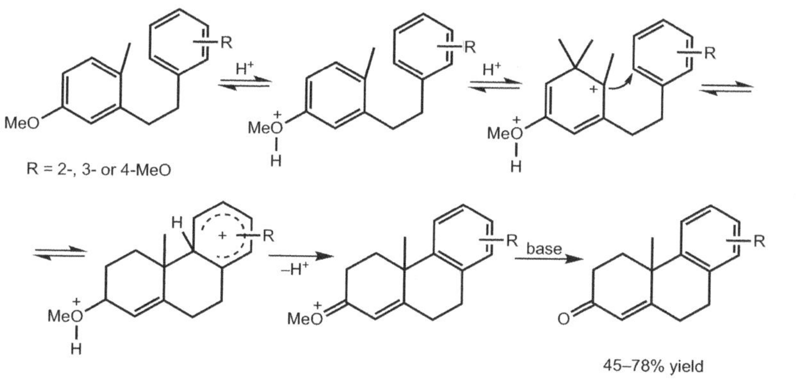

(5.306)

This cyclization method has been applied in the synthesis of the C-ring aromatic analog, 4-androstene-3,17-dione derivative **236**. Two cyclizations and the subsequent hydroxylation take place under superacidic conditions[846] [Eq. (5.307)].

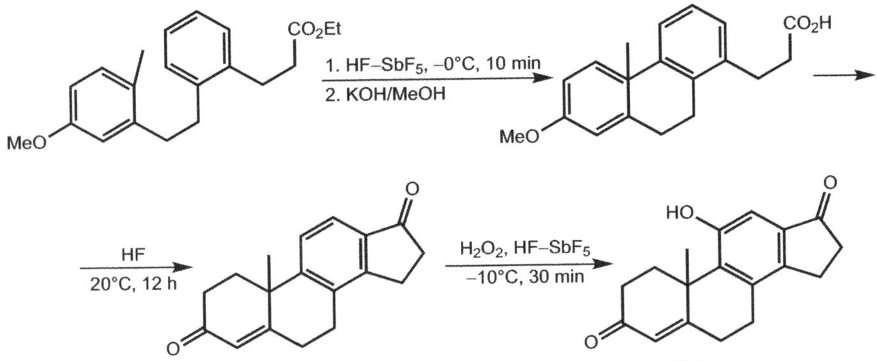

236

(5.307)

Monocyclic phenols and their methyl ethers react with benzene in HF–SbF_5 medium to provide 4,4-disubstituted cyclohexenones **237** [Eq. (5.308)].[301] *para*-Methylanisole gives three products: two cyclohexenone derivatives [see Eq. (5.112)] and an interesting tricyclic ketone **238**.

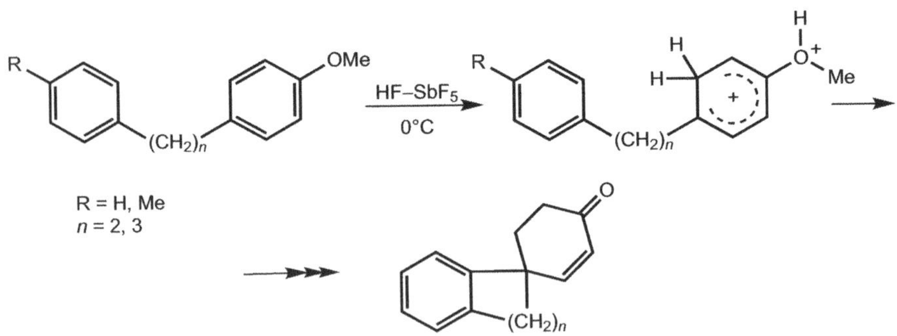

237 20–75%

(5.308)

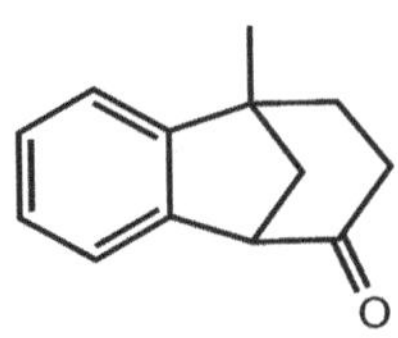

238

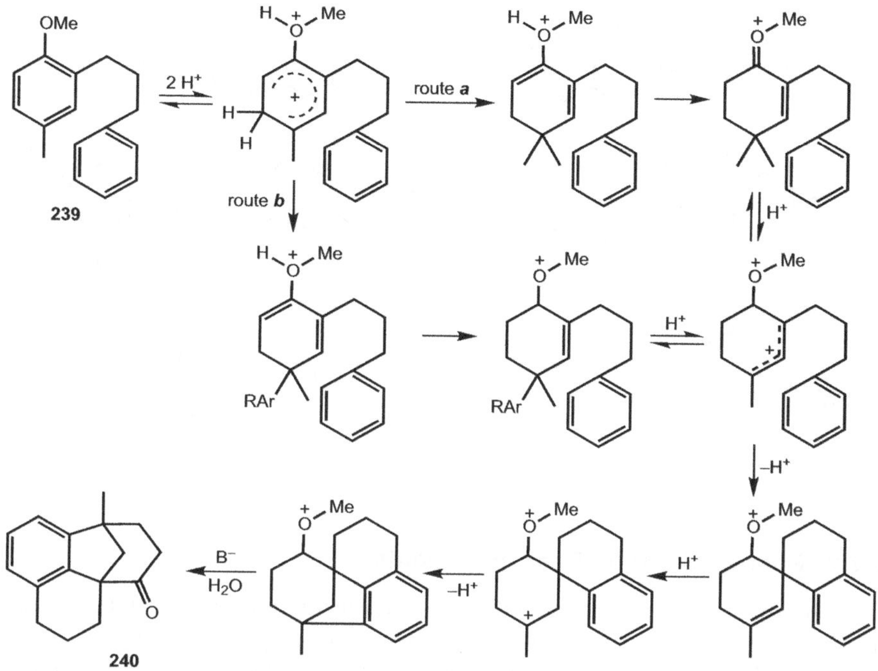

Scheme 5.85

By utilizing the above methods, a convenient route to the tetracyclic ketone **240**, a derivative of the benzobicyclo[3.2.1]octane skeleton, has been developed[847] using 2′-methoxy-5′-methyl-1,3-diphenylpropane **239** (Scheme 5.85).

Zhao and co-workers[848] synthesized macroreticular *para*-(ω-sulfonic-perfluoroalkylated) polystyrene (FPS) and used it in the cyclization of pseudoionone into α-ionone **241a** an important fragrance material. The ring closure induced by the catalyst led to complete conversion and the product was formed in low yield but selectively (β-ionone was not observed) [Eq. (5.309)]. Amberlyst proved to be less active and less selective. A new catalyst loaded with perfluoroalkanesulfonic as well as phenylsulfonic acid groups (FPSS) exhibited improved performance[849] [Eq. (5.309)].

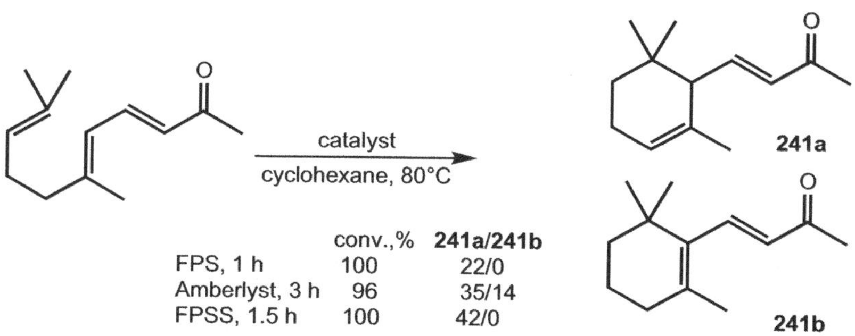

	conv.,%	**241a/241b**
FPS, 1 h	100	22/0
Amberlyst, 3 h	96	35/14
FPSS, 1.5 h	100	42/0

(5.309)

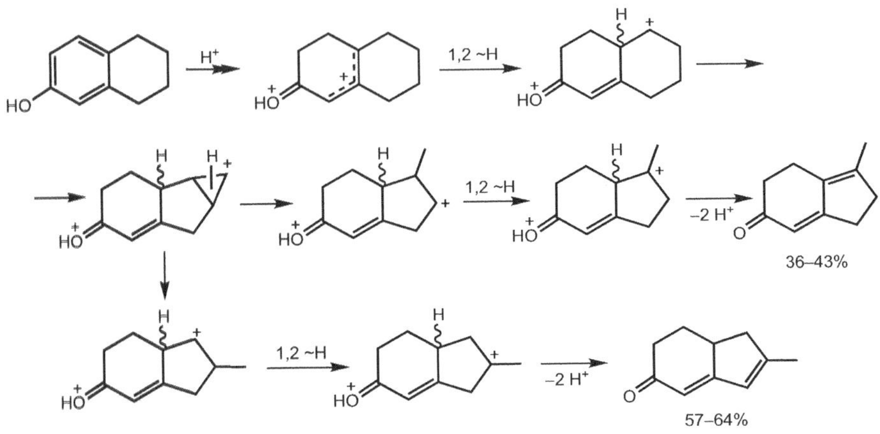

Scheme 5.86

5.17.2. Phenol–Dienone Rearrangements

Conversion of phenol to a dienone and vice versa is an important transformation and is widely used in organic natural product synthesis. Jacquesy and co-workers[850] have demonstrated that this rearrangement occurs very efficiently in superacid solutions. Treatment of estrone derivatives **242** in $HF-SbF_5$ followed by aqueous bicarbonate work up led to estra-4,9-diene-3,7-dione **243** [Eq. (5.310)]. The reaction also occurred in HSO_3F-SbF_5 medium. The intermediate tricationic species and their isomers of the above rearrangement process have been characterized by ^1H NMR spectroscopy.[851,852] Coustard and Jacquesy[853] have also investigated as models, the rearrangement of simple bicyclic phenols and phenolic ethers in $HF-SbF_5$ or HSO_3F-SbF_5 medium [Eq. (5.311) and Scheme 5.86].

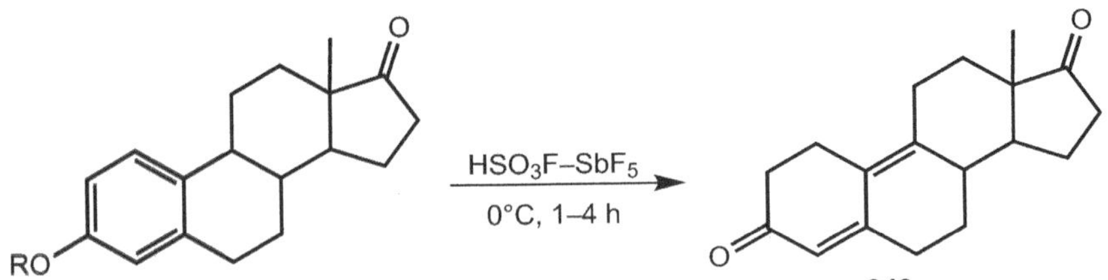

242
R = H, Me, Ac

243
70–80% yield

(5.310)

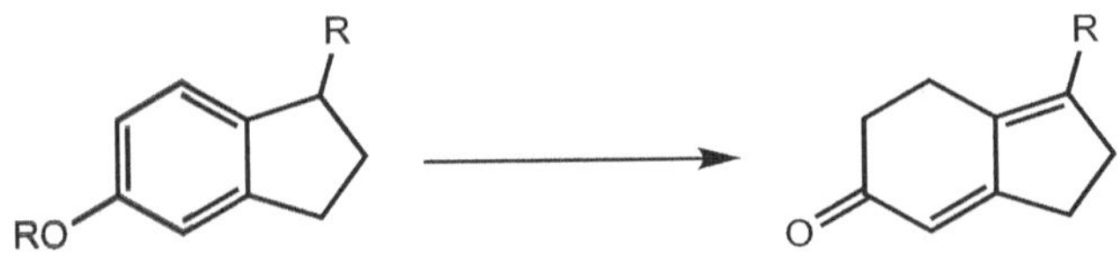

(5.311)

R = H, Me

87–98%

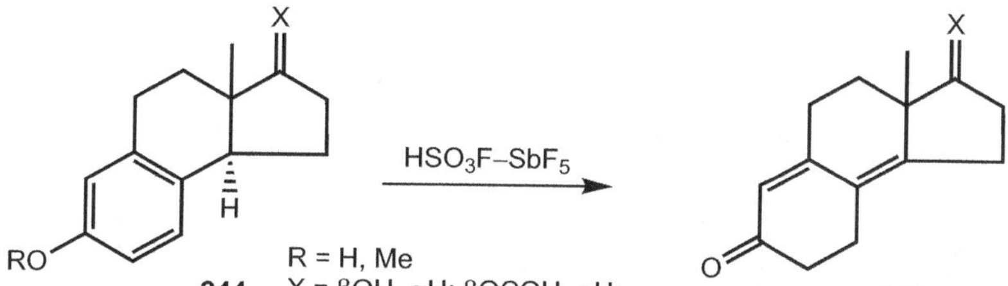

244 R = H, Me
X = βOH, αH; βOCOH, αH; βOAc, αH; βCOMe, αH

245

(5.312)

Similarly, many A-norsteroids **244** have been subjected to phenol–dienone rearrangement in HF–SbF_5 medium[854] [Eq. (5.312)]. ^{1}H NMR spectroscopic studies at low temperature confirms the formation of *O*-protonated intermediates (Scheme 5.87), which subsequently rearrange to diprotonated precursors of the dienones **245**.

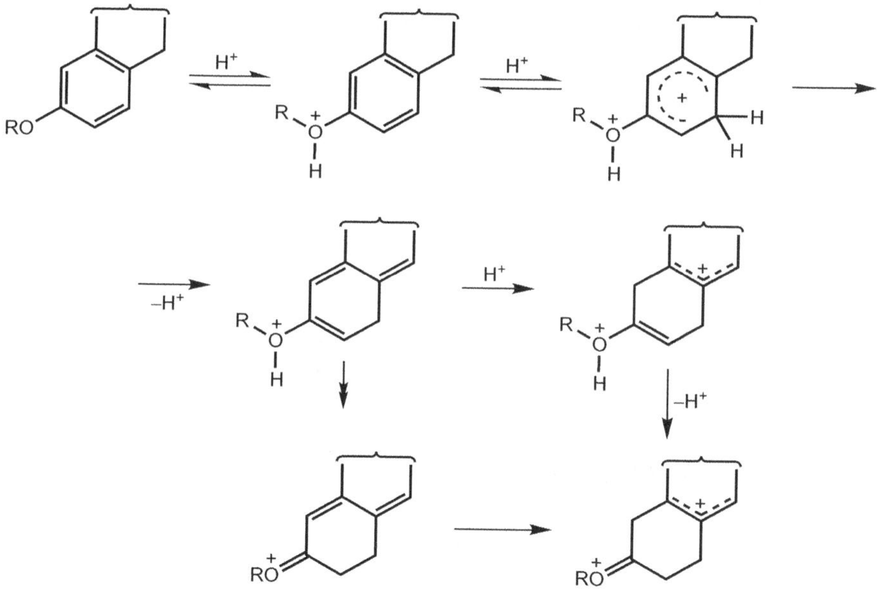

Scheme 5.87

Androsta-1,4,6-triene-3,17-dione **246**, when treated with ordinary acid catalysts such as acetic anhydride–*para*-toluenesulfonic acid, gives the *meta* phenolic product **247** (Scheme 5.88). However, under HF–SbF_5 catalysis at −50°C the phenolic product **248** is obtained in 75% yield[855] (Scheme 5.88). The mechanism of the above discussed phenol–dienone and dienone–phenol rearrangement has been investigated in detail.[856,857]

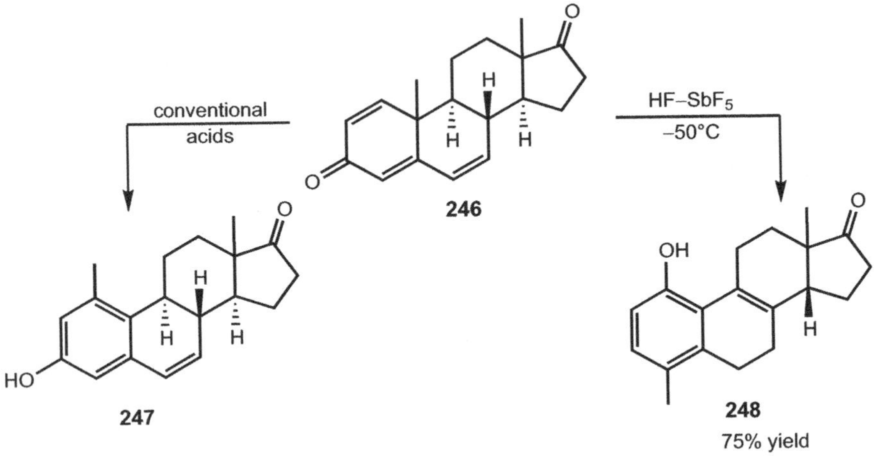

Scheme 5.88

A related transformation is the rearrangement–aromatization of ketoisophorone to trimethylhydroquinone diacetate, an intermediate in the industrial synthesis of (all-*rac*)-α-tocopherol. Of heterogeneous catalysts, Nafion–silica exhibited the best catalyst performance[858] [Eq. (5.313)], but activities decreased with repeated use because of the leaching of Nafion resin.

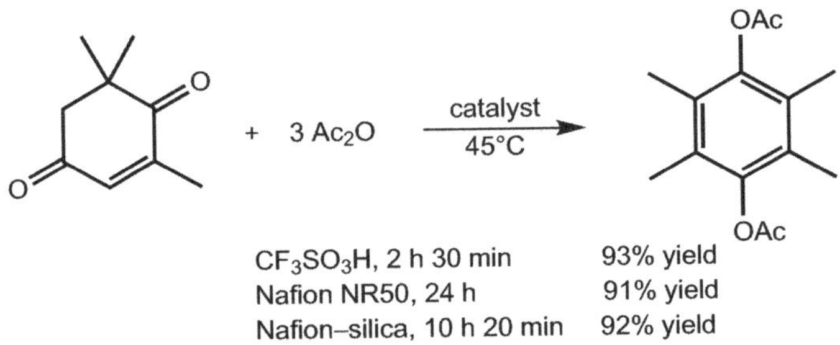

(5.313)

5.17.3. Other Rearrangements and Cyclizations

Sheldon and co-workers[859] have performed the cyclization of *N*-formyl enamide **249** in triflic acid or over Nafion-H to form the corresponding *N*-formyloctahydroisoquinolines starting compounds in the synthesis of *N*-formylmorphinanes [Eq. (5.314)].

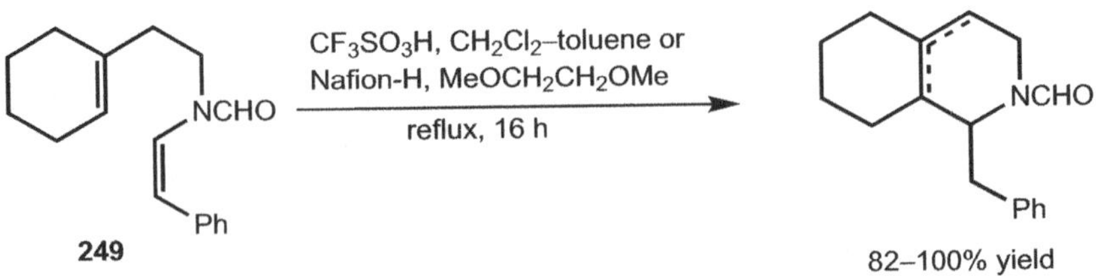

(5.314)

A novel one-pot tandem oxidation–cyclization–oxidation process was successfully applied in the transformation of unsaturated alcohols **250** [Eq. (5.315)].[860] The intermediate aldehyde formed by oxidation with pyridinium chlorochromate (PCC) undergoes a carbonyl–ene cyclization followed by an additional oxidation to form 3-substituted piperidinones.

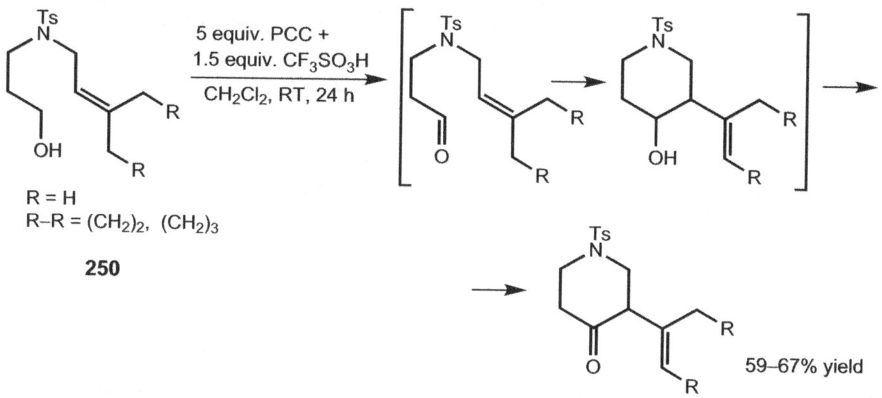

(5.315)

The allyl-transfer reaction based on 2-oxonia Cope rearrangement allows highly stereocontrolled chirality transfer. Triflic acid has been shown to induce the rearrangement of the **251** allyl sterols into 22-homoallylic sterols with high stereoselectivity without side reactions[861] [Eq. (5.316)]. The protocol, however, is not effective for *syn* substrates (for example, **251**, R = H, R′ = COOEt).

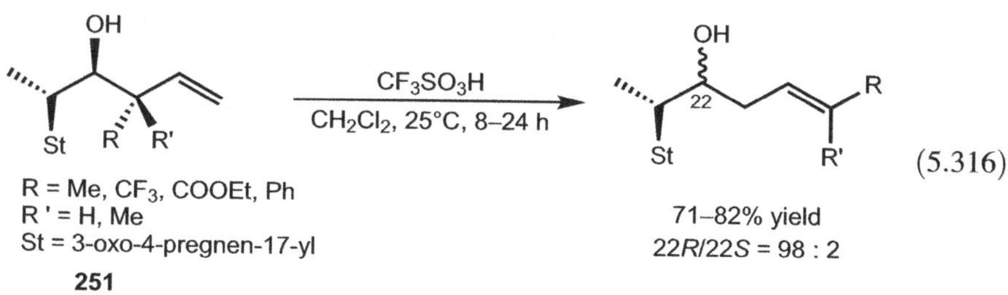

(5.316)

Isomerization of pivalaldehyde to yield isopropyl methyl ketone has been shown by Olah et al.[451] to strongly depend on the acidity. Triflic acid, anhydrous HF, and $BF_3-2CF_3CH_2OH$, but not trifluoroacetic acid, induce quantitative isomerization; that is, acids with $H_0 \leq -11$ are required for complete isomerization. These observations and DFT calculations (B3LYP/6-31G* level) suggest that protonation (protosolvation) of protonated pivalaldehyde forms the reactive superelectrophilic species **252** [Eq. (5.317)]. Gitonic dication **252**, however, is not a minimum on the potential energy surface and is transformed to the most stable distonic dication **253** (the global minimum) in a barrierless process. Formation of dication **252** increases the electrophilic character of the protonated carbonyl oxygen leading to methyl shift, which alleviates the electron demand in the adjacent position. Gitonic dication **252** was estimated to be 48 kcal mol^{-1} less stable than distonic dication **253**.

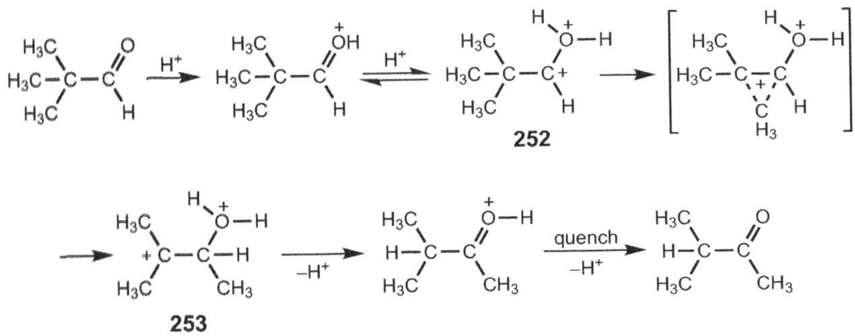

(5.317)

By studying stable ion chemistry of polycyclic aromatic systems, Laali et al.[862] observed the ring closure of dicyanometacyclophanediene **254** with the involvement of diprotonated intermediate **255** [Eq. (5.318)]. When product **256** was treated again in superacids under different conditions, rearrangement took place to yield 1-cyanopyrene through mono- and diprotonated intermediates [Eq. (5.319)].

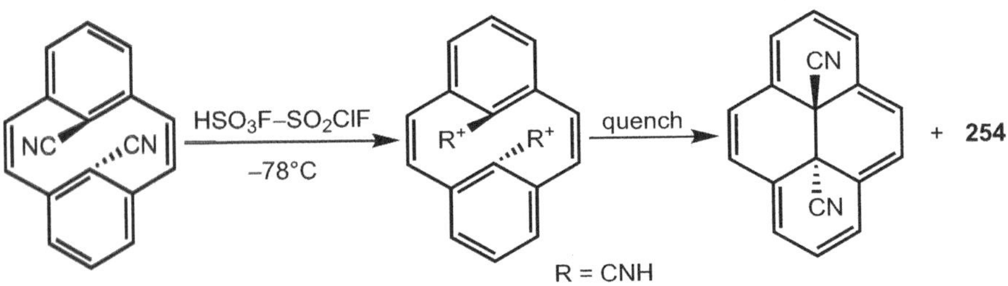

254 **255** **256**

(5.318)

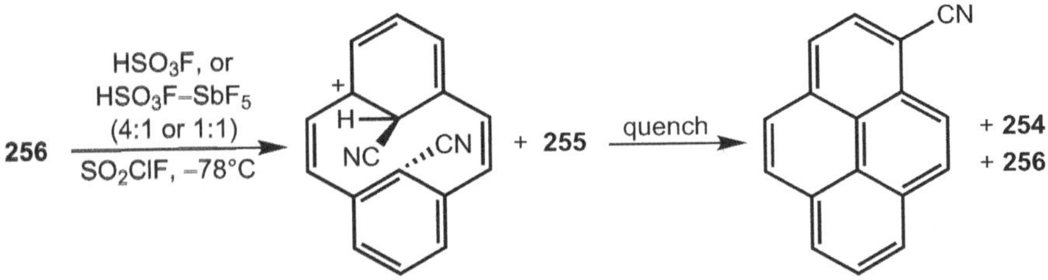

major product

(5.319)

Karpov and co-workers[290,660,863,864] have performed extensive studies with respect to the transformation of perfluorinated benzocyclobutene, indane, tetralin, their perfluoroalkyl-substituted, and mono- and dioxo derivatives in $HF-SbF_5$ or SbF_5 at elevated temperatures. Complex product mixtures are usually obtained resulting from ring cleavage, ring expansion and ring contraction. Product distributions strongly depend on temperature and the quantity of SbF_5 used. For example, compounds **257**, **258**, and **259** could be isolated in modest yields (10–58%), when perfluoro-3-ethylindan-1-one was treated in excess SbF_5 followed by quenching with hydrochloric acid.[865] Cations **260** and **261** were detected by ^{19}F NMR spectroscopy.

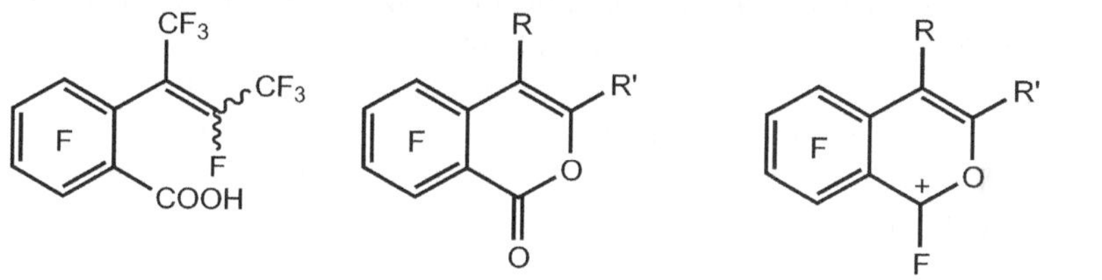

257

258 R = R' = CF_3
259 R = CF_3CF_2, R' = F

260 R = R' = CF_3
261 R = CF_3CF_2, R' = F

In the presence of SiO_2 serving as the oxygen source, complex mixtures of rearranged oxygenated compounds are formed. As the transformation of perfluoro-1-ethylindane (**262**) shows, however, products may be isolated in high yields under appropriate conditions[865] [Eq. (5.320)].

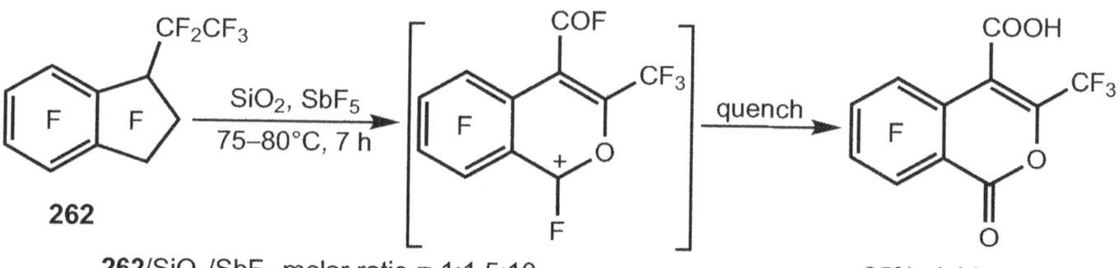

262

262/SiO_2/SbF_5 molar ratio = 1:1.5:10

85% yield

(5.320)

5.18. IONIC HYDROGENATION

In catalytic hydrogenations, usually hydrogen is activated (generally by noble metal catalysts). In contrast, in ionic hydrogenations an acid catalyst activates the hydrocarbon-forming carbocationic sites, which then are quenched by hydrogen to reduced products. A significant number of ionic hydrogenations were studied and a review of earlier work is available.[866] Hydride transfer in superacidic media is a very versatile reaction, especially in isomerization and cracking of hydrocarbons (Sections 5.1.3 and 5.1.4). The reaction has been extensively employed in liquefaction of coal using tetralin, methylcycolpentane, and so on, as hydride donors in the presence of strong Lewis acids.[867] In fact, Olah and co-workers[130–132] have found that coal can be depolymerized in the presence of HF–BF_3 under hydrogen pressure at modest temperatures (100–170°C). Addition of isopentane to this reaction substantially improved depolymerization (*vide supra*). The related hydrogenation of benzene has been studied using HF–TaF_5 and HF–SbF_5 as catalyst.[868] Another early example is the use of BF_3–H_2O in combination with triethylsilane to reduce anthracene, naphthancene, and hydroxy- and methoxynaphthalenes to the corresponding di- or tetrahydroderivatives.[869] 2-Acetonaphthalene and adamantanone were transformed to the corresponding parent hydrocarbons, whereas benzene, substituted benzene derivatives, naphthalene, and phenantrene could not be reduced.

The hydride transfer reaction catalyzed by strong acids has also been successfully adapted in the natural product chemistry. Jacquesy et al.[870] have found that protonated dienones and enones of steroid nucleus can be conveniently reduced in HF–SbF_5 medium under hydrogen pressure. The reaction has also been carried out with added hydrocarbons (methylcyclopentane, cyclohexane) as hydride donors.[871] The proposed mechanism is depicted in Scheme 5.89.

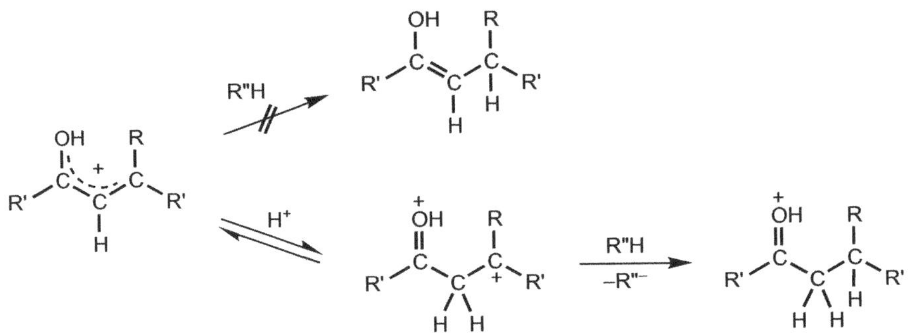

Scheme 5.89

As model studies, several bicyclic compounds have also been reduced using cyclohexane in HF–SbF_5[872] [Eq. (5.321)]. 2-Hydroxytetralin undergoes phenol–dienone rearrangement before reduction. Estrone and an unsaturated estrane have been reduced in HF–SbF_5 medium under hydrogen pressure yielding an anthrasteroid, whose structure and absolute configuration have been determined.[873]

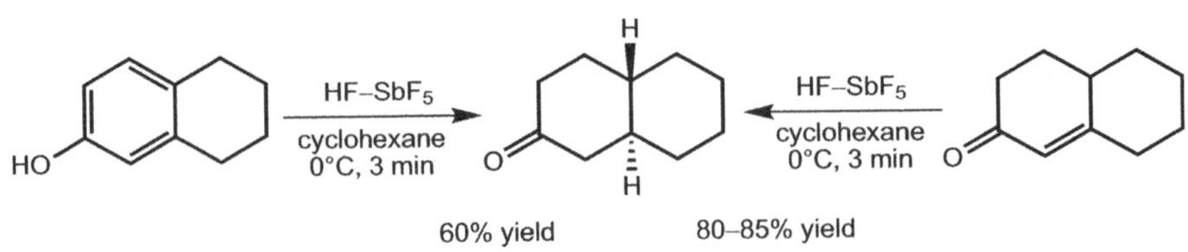

(5.321)

In a series of papers, Koltunov, Repinskaya, and co-workers[874–877] have reported the ionic hydrogenation of isomeric naphthols and dihydroxynaphthalenes with alkanes in the presence of aluminum halides. 1-Naphthol and substituted derivatives undergo regioselective reduction under mild conditions in excess alkane and aluminum halides with the involvement of various reactive intermediates to yield α-tetralone derivatives[874] [Eq. (5.322)]. Byproducts are 3-, 6-, and 7-alkyl-substituted derivatives. Mechanistic studies[875] with cyclohexane-d_{12} showed that deuterium incorporation takes place exclusively at C(4), indicating the involvement of superelectrophilic dication **263**. 2-Naphthol is much less reactive and complete conversion cannot be achieved.[876]

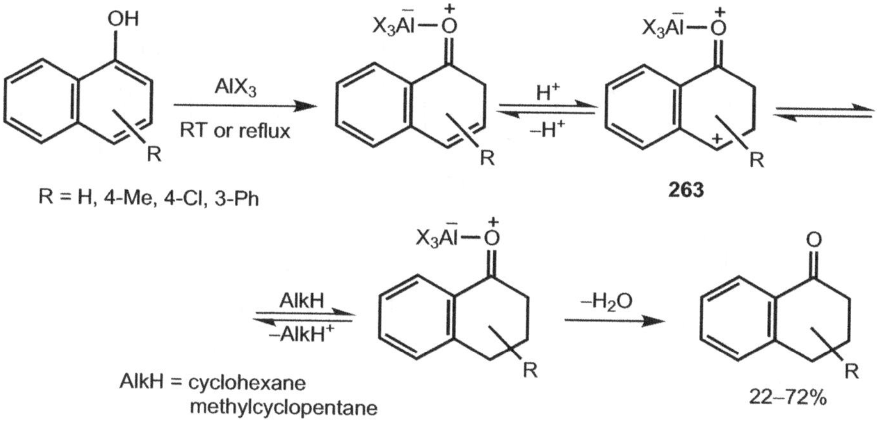

(5.322)

When isomeric 1,5-, 1,6-, and 1,7-dihydroxynaphthalenes are reacted under similar conditions,[877] quantitative formation of the corresponding hydroxy-substituted α-tetralones were observed. Results were interpreted by invoking the superelectrophilic tricationic complex intermediate **264**. In a similar manner, the 1-hydroxy-substituted ring is reduced regioselectively when 5-amino-1-naphthol is reacted with cyclohexane in the presence of superacidic $AlCl_3$, $AlBr_3$, triflic acid or triflic acid–SbF_5.[293] In the Brønsted superacids, *N,C*-diprotonated dication **265** is suggested as the superelectrophilic intermediate.

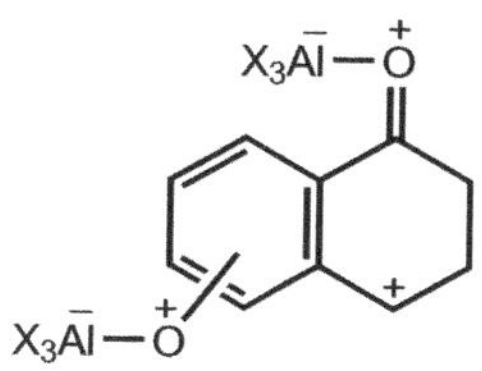

264

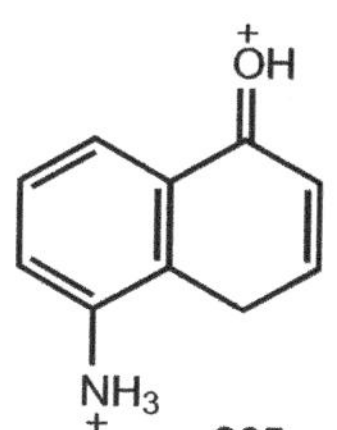

265

Detailed studies have been reported of the ionic hydrogenation of isomeric hydroxyquinolines and hydroxyisoquinolines. 5-, 7-, and 8-hydroxyquinolines are selectively reduced with cyclohexane in the presence of aluminum chloride to yield 5,6,7,8-tetrahydroquinoline.[295] 5-Hydroxyisoquinoline shows a similar transformation in triflic acid–SbF_5.[294] As shown by the example of 5-hydroxyquinoline, the probable mechanism (Scheme 5.90) involves the stepwise selective ionic hydrogenation of superelectrophilic dications **266**, **267**, and **268**. 5- and 8-hydroxyquinolines exhibited the highest reactivity, whereas 6-hydroxyquinoline proved to be inert toward cyclohexane. This is in agreement with calculations showing 6-hydroxyquinoline to be the weakest nucleophile on the basis of LUMO energies (B3LYP/6-31G*) and atomic charges (NBO analysis) on the reaction center.

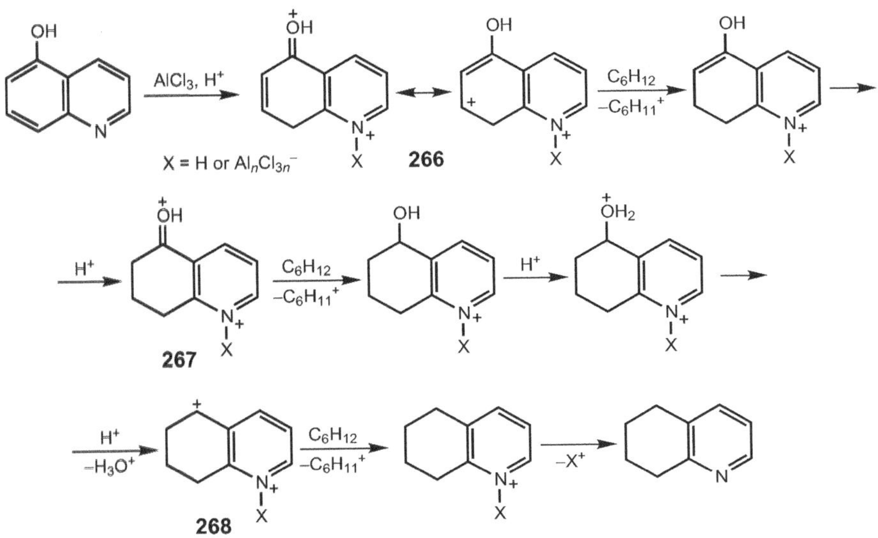

Scheme 5.90

Isomeric 2- and 4-hydroxyquinolines undergo selective ionic hydrogenation with cyclohexane in triflic acid–SbF_5 to produce the corresponding 5,6,7,8-tetrahydroquinolinones (83% and 92% yields), whereas 3-hydroxyquinoline gives 5,6,7,8-tetrahydro-3-quinolinol in 72% yield.[878] NMR spectroscopy showed that the three isomeric hydroxyquinolines undergo monoprotonation in triflic acid, whereas 3-hydroxyquinoline gives the *N*,*C*-diprotonated species **269** in the more acidic triflic acid–SbF_5 system.

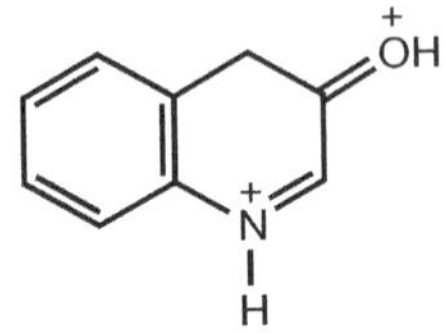

269

When 1-, 3-, and 8-hydroxyisoquinolines are treated with $AlCl_3$[296,870] isomeric tetrahydroisoquinolinones **270** and **271** are formed, whereas selective formation of the corresponding 5,6,7,8-tetrahydroisoquinolinone **270** is found in the presence of triflic acid–SbF_5 [Eq. (5.323)]. Quinoline also undergoes hydrogenation upon treatment in CF_3SO_3H–SbF_5 system (room temperature, 24 h) or in HBr–$AlCl_3$–CH_2B_2 (70°C, 150 h) to yield 5,6,7,8-tetrahydroisoquinoline as the sole product.[297] Isoquinoline exhibits higher reactivity in CF_3SO_3H–SbF_5 (94% yield in 2 h). The observations, again, were interpreted by invoking superelectrophilic dicationic intermediates.

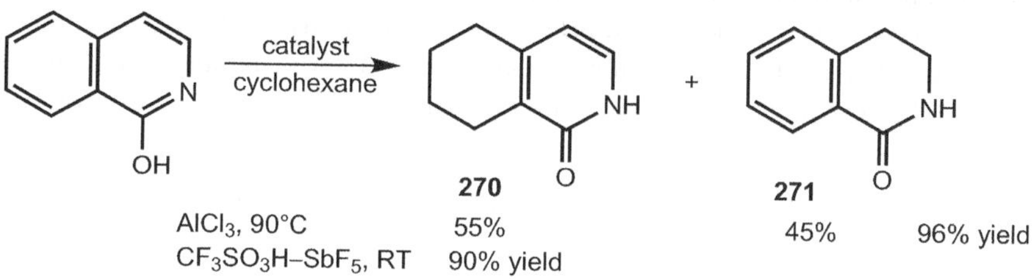

(5.323)

During the reductive isomerization of 7β-methyl-14-isoestr-4-ene-3,17-dione **272** in HF–SbF_5/methylcyclopentane at 0°C, it was found[879] that a 1,3-hydride shift occurs followed by kinetically controlled hydride transfer (Scheme 5.91). The mechanism of the reaction was confirmed by employing the deuteriated donor cyclohexane-d_{12} as well as a specifically deuterium-labeled starting steroid.

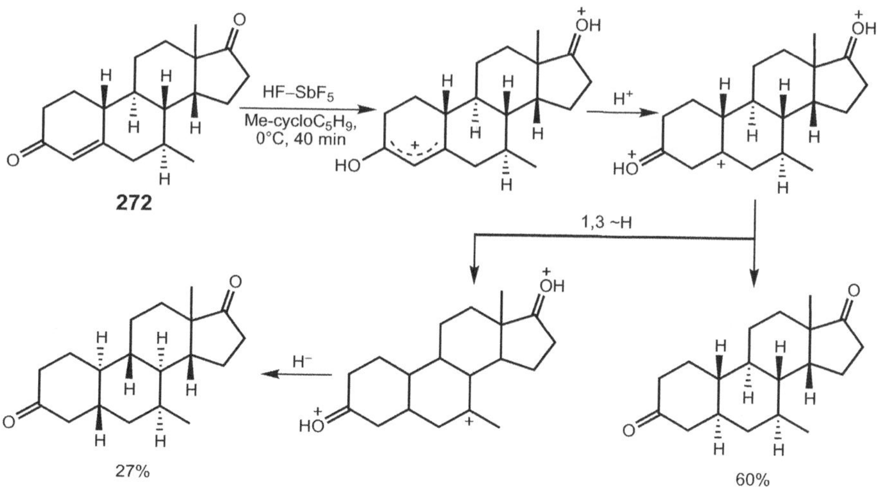

Scheme 5.91

The ionic transfer hydrogenation in superacidic media has also been studied in the case of bicyclo[4.n.0]enones under both kinetic[880] and thermodynamic control.[881]

Kinetically controlled hydride transfer between the hydrocarbon hydride source and the substrate often involves skeletal rearrangements. This has also been demonstrated in the reduction of androsta-4,6-diene-3,17-dione **230**.[882]

Ionic hydrogenation of α,β-unsaturated ketones has been studied by Koltunov et al.[883] A variety of α,β-unsaturated ketones reacts with cyclohexane in the presence of excess aluminum chloride or bromide to yield the corresponding saturated ketones in high yields [Eq. (5.324)]. Similar observations were made using triflic acid–SbF_5. The only exception is 4-phenyl-2-butanone, which produces a mixture of products including alkylated derivatives. The reactive intermediates are likely to be *C*-protonated complexes (when aluminum halides are used) or the *O*,*C*-protonated analogs (in triflic acid–SbF_5). A few α,β-unsaturated carboxamides could also be reduced to the corresponding saturated amides with cyclohexane in excess $AlCl_3$.[178]

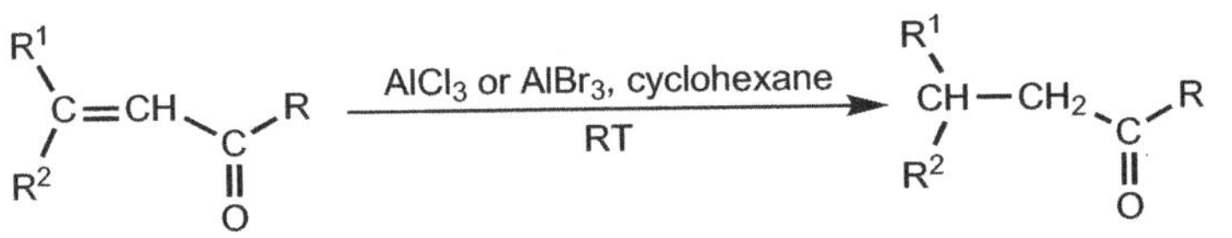

(5.324)

R = Me, Ph; R^1 = H, Me
R^2 = Me, Pr, Ph, 2,4-diClC_6H_3, 4-$MeOC_6H_4$, 2-HOC_6H_4, 4-HOC_6H_4, 4-$Me_2NC_6H_4$

Jacquesy and co-workers[884] have reported the stereoselective ionic hydrogenation of *Vinca* alkaloids to produce the corresponding 4′*R* reduced analogs (Scheme 5.92). Mechanistic studies with cyclohexane-d_{12} showed that deuterium incorporation takes place exclusively at C(20′).

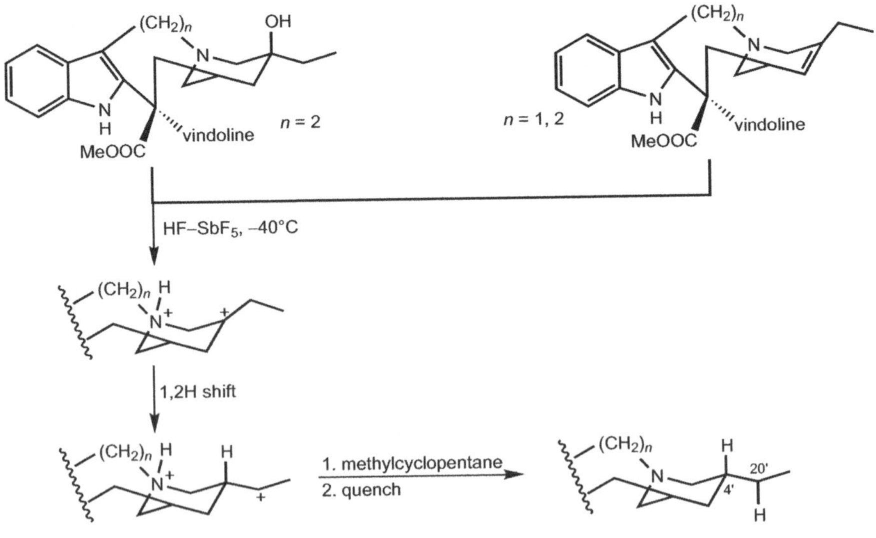

Scheme 5.92

Ionic hydrogenation can be used to reduce alcohols and it is a convenient way to transform carbonyl compounds to the corresponding methylene derivatives. Olah et al.[885] have reported that 2-aryladamantanes and 3-aryldiamantanes were isolated in near-quantitative yields by reducing the corresponding tertiary alcohols with $NaBH_4$ or formic acid in superacidic triflic acid. The $NaBH_4$–triflic acid system was also used to carry out a single-step reductive isomerization of unsaturated polycyclics to $C_{4n+6}H_{4n+12}$ diamondoid cage hydrocarbons.[886] Tertiary and benzylic alcohols are also reduced with trialkylboron–triflic acid[887] [Eq. (5.325)]. Trimethylboron is nonselective producing about 1:1 mixtures of the methylated and reduced products. Triisopropylboron–triflic acid was also reported to selectively convert a variety of hydroxy-substituted carbonyl compounds (carboxylic acids, aldehydes, ketones) to the corresponding carbonyl compounds.[888] $NaBH_4$–triflic acid is also effective in methanation; that is, it reduces CO, CO_2, CS, CS_2, methanol, and formic acid to methane under mild conditions.[889]

$$\text{R–OH} + \text{R}'_3\text{B} \xrightarrow[\text{Freon-113, }-30^\circ\text{C to RT, 5–6 h}]{\text{CF}_3\text{SO}_3\text{H}} \underset{\text{90–100\% yield}}{\text{R–H}} \qquad (5.325)$$

R = *tert*-Bu, Bn, Ph_2CH, Ph_3C, 1-adamantyl
R' = Et, isoPr

Various reagents such as Et_3SiH–triflic acid,[890] Et_3SiH with the complex $BF_3-2CF_3CH_2OH$,[481] and cyclohexane in the presence of triflic acid–SbF_5 (Scheme 5.93)[298] are capable of transforming the carbonyl group to methylene group. The mechanism depicted for the reduction of maleimide is interpreted by the potential involvement of the superelectrophilic dicationic intermediates. Phthalimide is reduced similarly both with triflic acid–SbF_5 and in the presence of an excess of $AlCl_3$ at elevated temperature. Aromatics (acenaphthylene, anthracene, benz[*a*]anthracene, dibenz[*a*,*h*]anthracene) can also be reduced with Et_3SiH–BF_3–$2CF_3CH_2OH$.[481]

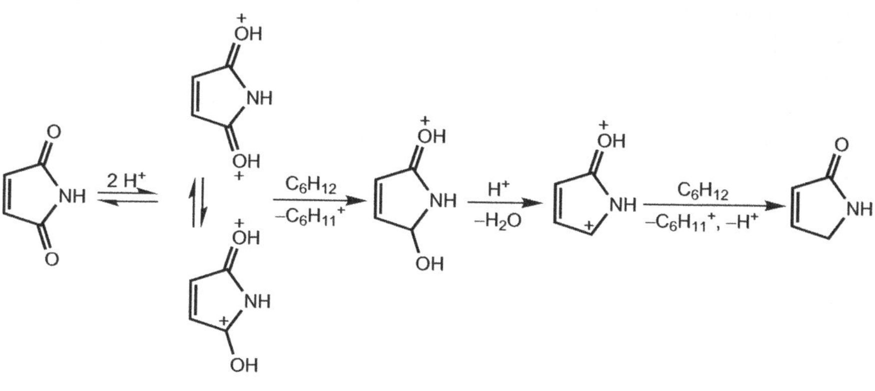

Scheme 5.93

5.19. ESTERIFICATION AND ESTER CLEAVAGE

Esterifications do not require superacid catalysis. However, various Nafion preparations have been tested in ester formation. In fact, Olah et al.[891] showed in 1978 that Nafion-H offers a convenient and improved method for direct esterification with unchanged catalytic activity for prolonged periods of operation. Recently, esterification of cyclic alkenes and dienes with saturated and unsaturated carboxylic acids have been reported.[892] Nafion nanocomposites exhibited activities about two orders of magnitude higher than Amberlyst 15 and gave the product esters with higher selectivities. Silica with anchored perfluorinated sulfonic acid sites (catalyst **49**, Figure 5.15) showed high activity for the esterification of long-chain fatty acids with ethanol and long-chain alcohols.[374] Nafion preparations have also been tested in transesterification of triacetin (1,2,3-triacetoxypropane) in connection with biodiesel production.[893,894] In a comparative study, Nafion SAC-13, and zeolite Hβ exhibited the same activity in gas-phase esterification of acetic acid with methanol when calculated on a weight basis.[895] On a rate-per-active sites basis, however, all catalysts including sulfated and tungstated zirconia showed similar activities. On the basis of rate data acquired on Nafion SAC-13, they concluded that the reaction occurs between acetic acid adsorbed on active sites and alcohol from the bulk phase.[896] Nafion SAC-13 has recently been found by Meunier and Ni to be a promising candidate to esterify free fatty acids found in vegetable oils, thereby obtaining fatty-acid free oil.[897] SAC-13 does not show mass transport limitations and can be used in batch and fixed-bed reactors, and its mild poisoning by water is reversible.

Carr and Whittaker[898] studied lactone formation of 1-hydroxycycloalkanecarboxylic acids in superacidic media. When 1-hydroxycyclohexanecarboxylic acid was treated in HSO_3F, the protonated acid observed by ^1H and ^{13}C NMR spectroscopy at −78°C slowly transformed to 1,4-lactone **275** [Eq. (5.326)]. When the 2,2,6,6-tetratedutero derivative was treated under identical conditions, lactone formation was not detected. This was interpreted as indicating that the intermediate dication **273** is transformed to dication **274** through 1,2-hydride shifts and the presence of deuterium hinders the hydride shifts. The surprising feature of the transformation is that the cyclohexyl cation does not undergo ring contraction even though cyclization requires the flipping of the chair form of cyclohexane to the boat conformation having a high energy barrier. Ring contraction of 1-hydroxycycloheptanecarboxylic acid, in turn, did take place to yield 1-methylcyclohexanecarboxylic acid 1,4-lactone as the major product. Triflic acid has been reported to be highly active and selective in the transformation of unsaturated carboxylic acids to lactones.[899]

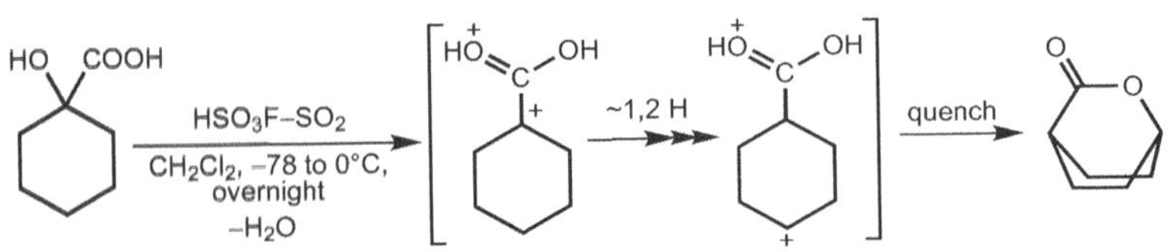

273 274 275

(5.326)

Triflic acid has been used in the selective esterification (*O*-acylation) of a series of aminoalcohols in the presence of a crown ether (DB24C8) to prepare rotaxanes.[900] The test reaction of diethanolamine with bulky anhydride, crown ether, and triflic acid (molar ratio = 1:2:2:1.5) gave the rotaxane **276** in high yield in a clean reaction [Eq. (5.327)]. *N*-Arylmethylaminoalcohols were similarly transformed (85–92% yields).

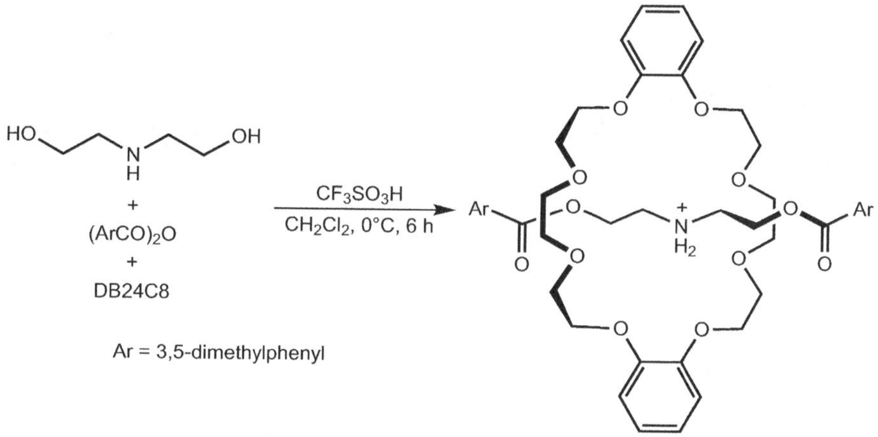

276 84% yield

(5.327)

Olah, Prakash, and co-workers[901] have studied ester cleavage in superacidic media. Protonated methyl acetate was found to undergo slow acyl–oxygen cleavage in $HSO_3F-SbF_5-SO_2$ solution even at −78°C to give acetyl cation and methyloxonium ion. Diprotonation of methyl acetate to form the distonic dication **277** was found to be a thermodynamically favorable process by *ab initio* calculations [MP4(SDTQ)/6-31G*//MP2/6-31G* level of theory], which allowed the authors to suggest a new mechanism for the acid-catalyzed ester cleavage in superacidic media [Eq. (5.328)].

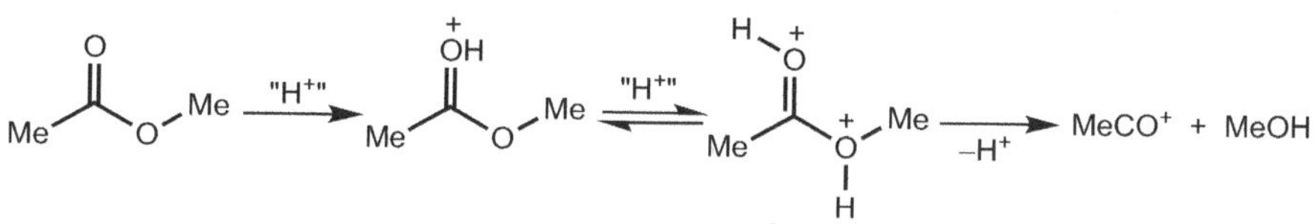

277

(5.328)

5.20. ADDITIONS

5.20.1. Cycloadditions

Olah et al.[902] demonstrated in 1979 that Nafion-H is able to catalyze Diels–Alder reaction of anthracene with a number of dienophile (maleic anhydride [Eq. (5.329)], *para*-benzoquinone, dimethyl maleate, dimethyl fumarate) in chloroform or benzene as solvent to afford the corresponding adducts in excellent yields (87–95%). In addition, reactive dienes such as isoprene, 2,3-dimethylbutadiene, and 1,3-cyclohexadiene reacted with benzoquinone, naphthoquinone, and acrolein, respectively, at room temperature without polymerization (80–93% yields). Nafion-H proved to be an excellent catalyst in the Diels–Alder reaction of olefinic acetals to provide cycloadducts in good to excellent yields and with high *endo* selectivities[903] [Eq. (5.330)].

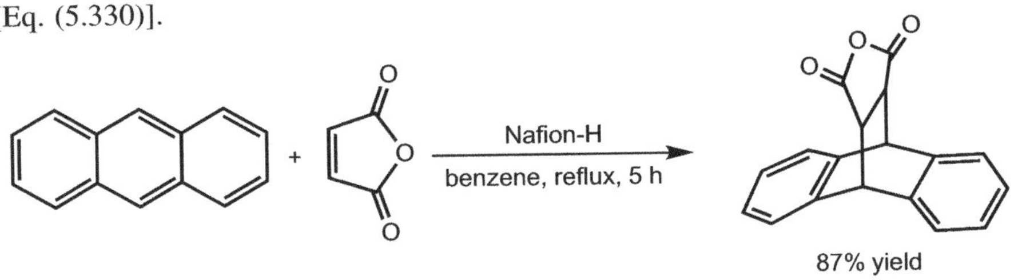

(5.329)

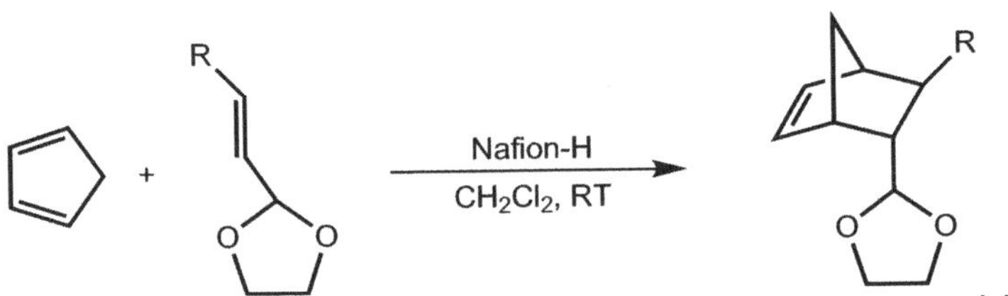

	yield, %	*endo*/*exo*
R = Me, 10 h	81	85/15
R = Ph, 12 h	84	90/10

(5.330)

The aza-Diels–Alder reaction in Eq. (5.331) catalyzed by triflic acid has been carried out to synthesize adduct **278a**, an intermediate in the total synthesis of microfungal alkaloid (±)-lapatin.[904] Whereas the reaction of the substituted azadienes led to the exclusive formation of *exo* compounds, the unsubstituted parent compound gave a 1:1 mixture of the isomers.

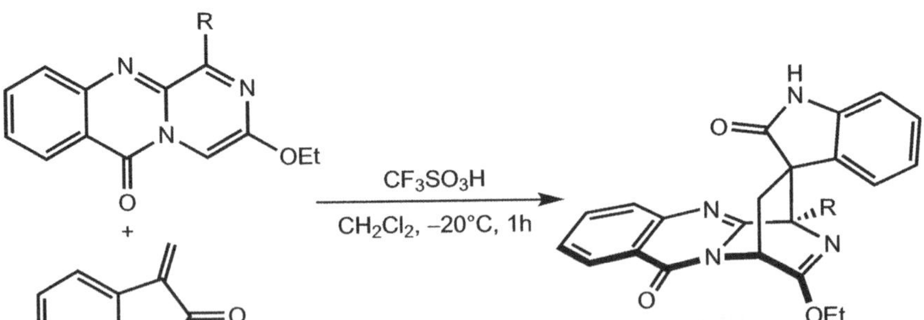

278

	yield, %	*exo*/*endo*
R = H, **278a**	55	50/50
Me	68	100/0
isoPr	75	100/0

(5.331)

Gorman and Gassman[905] have shown that undecatetraenes undergo cyclization (intramolecular Diels–Alder reaction) in the presence of triflic acid to provide bicyclo[4.3.0]nonyl [Eq. (5.332)], bicyclo[4.4.0]decyl, and bicyclo[5.4.0]undecyl [Eq. (5.333)] ring systems, depending on the methyl-substitution pattern. On the basis of a comparative study with varied tetraenes, they concluded that product formation, at least in some cases, could be best interpreted by a stepwise process.[905,906]

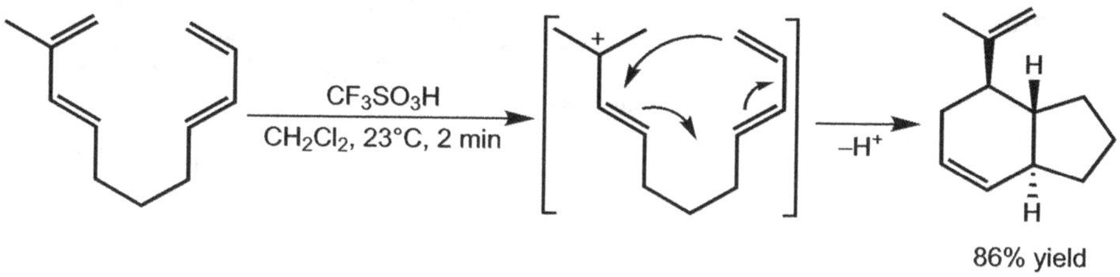

(5.332)

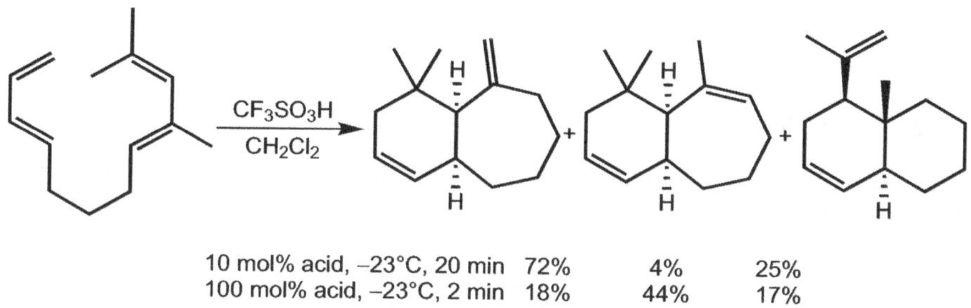

(5.333)

Recently, Shin and co-workers[907] have studied the cyclization of the trienol **279** and isolated two compounds (**280** and **281**) (Scheme 5.94). Temperature-dependence studies and the transformation of related compounds resulted in the conclusion that both a concerted process (route ***a***) and a stepwise mechanism are operative (route ***b***).

The formation of polysubstituted cycloheptadienones depicted in Eq. (5.334) constitutes the first example of an acid-induced ketene–diene cycloaddition.[908]

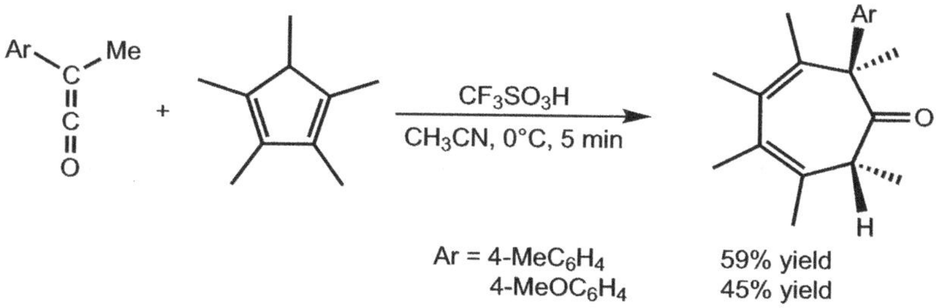

(5.334)

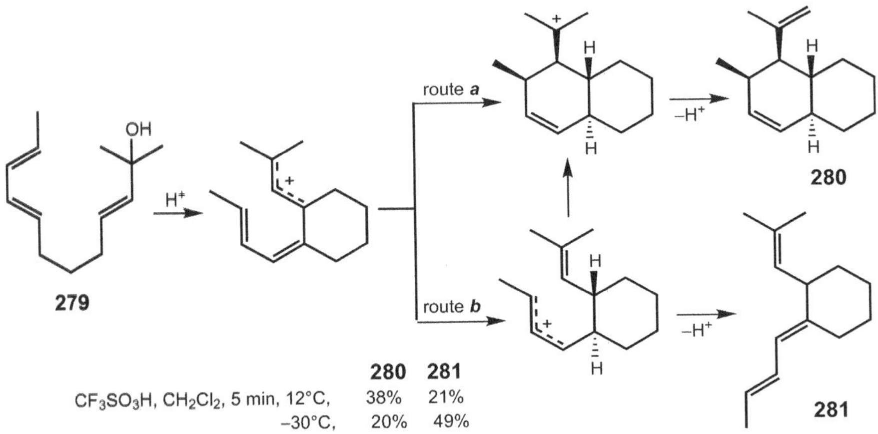

Scheme 5.94

5.20.2. Other Additions

Kotsuki et al.[909] have developed a method to effect the Michael addition of β-ketoesters with ethyl acrylate in the presence of triflic acid under solvent-free conditions [Eq. (5.335)]. Nonactivated cyclohexanones as Michael donors and α,β-unsaturated ketones as acceptors are also reactive. The use of menthyl acrylates did not result in any significant asymmetric induction.

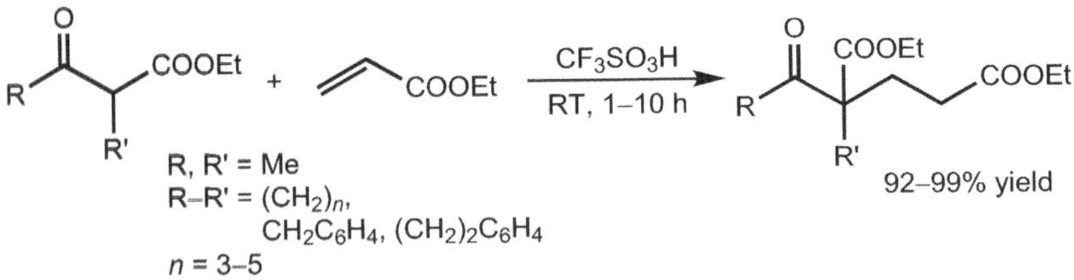

(5.335)

The addition of phenols [Eq. (5.336)], carboxylic acids, and sulfonamides [Eq. (5.337)] to alkenes can be induced by triflic acid.[910,911] Whereas the yields are low with the use of 10–15 mol% of acid, triflic acid in catalytic amounts (1–5 mol%) allow the isolation of addition products in good to high yields.

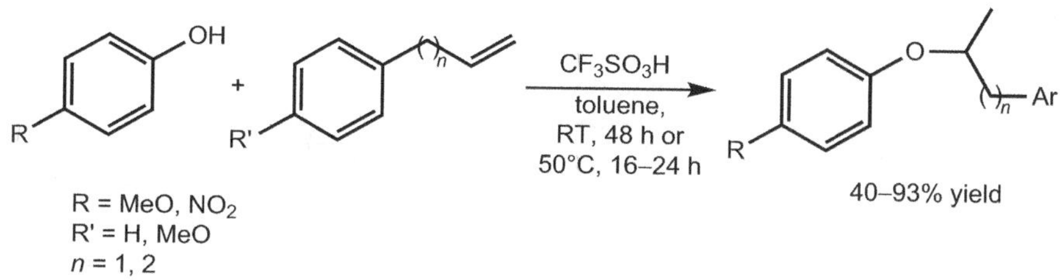

(5.336)

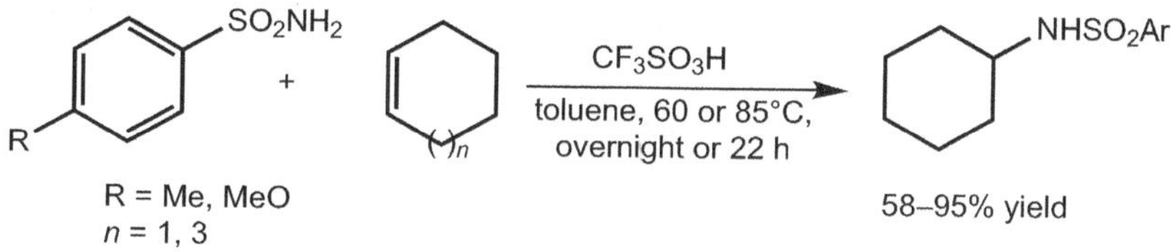

(5.337)

Amino-sulfonation of alkenes has been performed in a three-component reaction with SO_3–dimethylformamide complex (SO_3·DMF) and acetonitrile followed by hydrolysis.[912] Whereas amino-sulfonation occurs without the use of triflic acid, the acid accelerates the reaction considerably and prevents the formation of byproducts. The X-ray structure of intermediate **282** provided evidence that the addition is completely regio- and stereoselective [Eq. (5.338)].

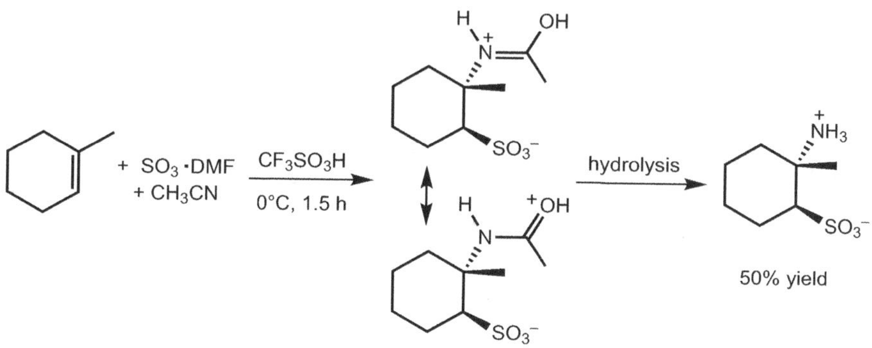

282

(5.338)

The superacid $CF_3SO_3H_2^+$–$B(OSO_2CF_3)_4^-$ was shown to catalyze the addition of allylsilanes to aldehydes and cyclohexanone to form homoallylic alcohols[913] (CH_2Cl_2 solvent, room temperature, 60–95% yields). 2-Alkoxyallylboronates add

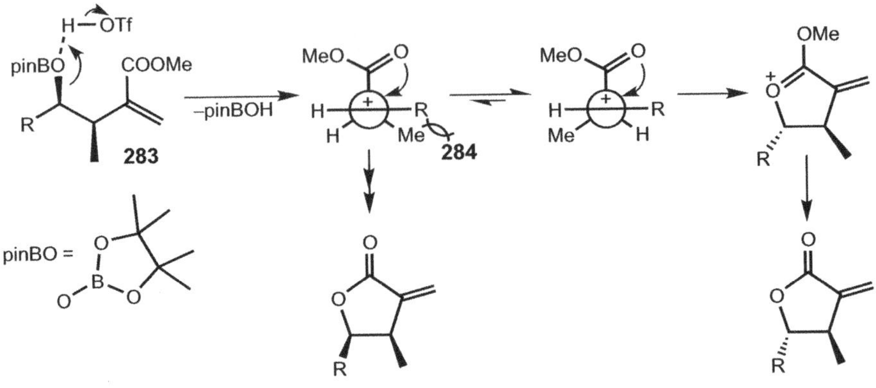

Scheme 5.95

to aldehydes in the presence of triflic acid to afford β,γ-disubstituted δ-lactones with an α-exo-methylene group[914] [Eq. (5.339)]. Aromatic, straight-chain and cyclic aliphatic aldehydes are all suitable for allylboronation. While all reactions with *Z*-alkenes gave *trans*-lactones, the reversal of stereochemistry was observed for some of the *E*-alkenes. The suggested mechanism includes the borate intermediate **283**, which is transformed to carbocation **284** (Scheme 5.95). When **284** is trapped by the ester before bond rotation, the expected *cis* stereoisomer is formed. In turn, bond rotation to form the favorable conformer followed by trapping by the ester group results in the *trans* isomer. ^{18}O labeling studies indicated the loss of aldehyde oxygen. Additional results with respect to the chemo- and stereoselectivity of allylboronation promoted by triflic acid have been reported.[915,916]

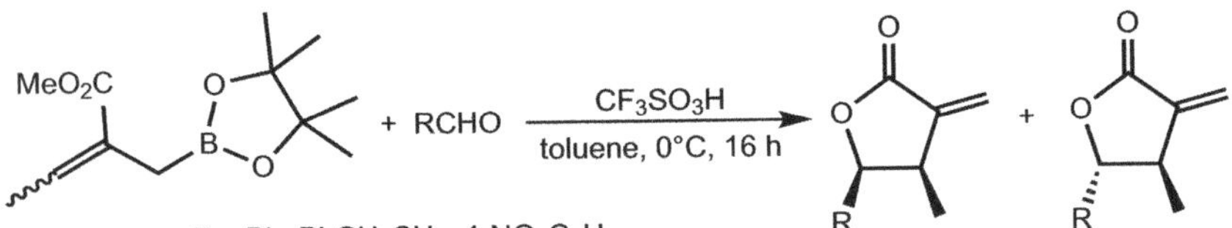

R = Ph, $PhCH_2CH_2$, 4-$NO_2C_6H_4$, 2-Me and 4-MeC_6H_4, 4-BrC_6H_4, 4-$MeOC_6H_4$, *E*-$CH_2CH{=}CH$, cycloC_6H_{11}

	% yield	*cis*/*trans*
E-alkene	11–79	> 19:1 or < 1:19
Z-alkene	26–91	< 1:19

(5.339)

A highly selective synthesis of homoallylic alcohols has been reported by Tietze et al.,[917] who reacted methyl ketones, the chiral norpseudoephedrine derivative **285**, and an allylsilane in the presence of a catalytic amount (0.2 mol%) of triflic acid [Eq. (5.340)]. The transformation was interpreted as an S_N2 attack of the allylsilane to the protonated mixed acetal **286**. The obtained ethers were then cleaved to the final product, homoallylic alcohols.

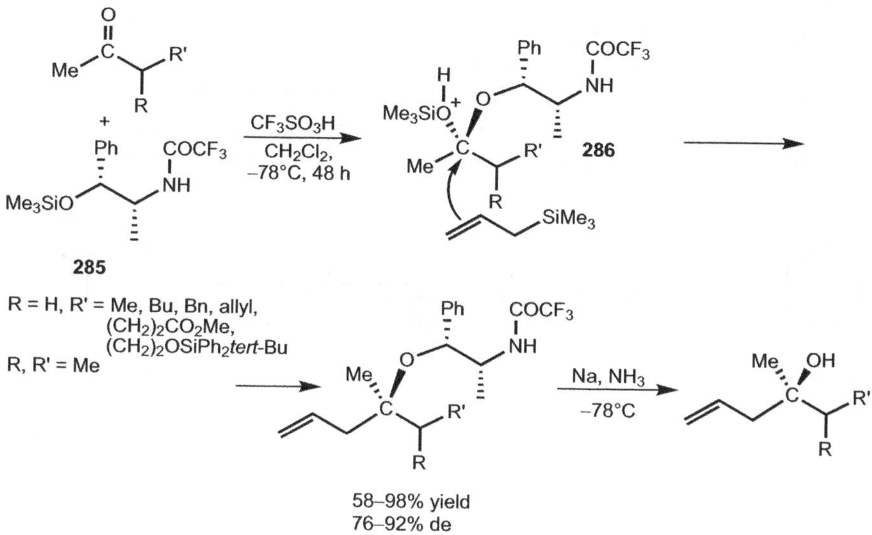

(5.340)

Fluorosulfuric acid and triflic acid add stereoselectively to acetylenic acids, esters, and ketones to form vinyl sulfonates and vinyl triflates, respectively.[918] Acids and esters yield the corresponding *E* derivatives with high selectivity [Eq. (5.341)]. In contrast, exclusive formation of the *Z* compounds was observed in the transformation of keto derivatives ($R = 4\text{-}FC_6H_4$, $R^1 = Me$). The stereochemistry is accounted for by postulating the attack of the sulfonate anion from the sterically less hindered side of diprotonated intermediate **92** to furnish the *E* isomers. Accordingly, the stereoselectivity is always better for the triflate addition. Isomerization to the thermodynamically more stable *Z* isomers, however, was observed in prolonged reactions at increased temperature. Furthermore, compounds that are unreactive under the above conditions (methyl oct-2-ynoate, hex-3-yn-2-one, 4-phenylbut-3-yn-2-one) react in stronger acid ($CF_3SO_3H-SbF_5$, H_0 -20) and give *Z* compounds. Acidity, consequently, also plays a key role in the isomerization process. Isomerization appears to be slow for acids and esters but fast for keto derivatives.

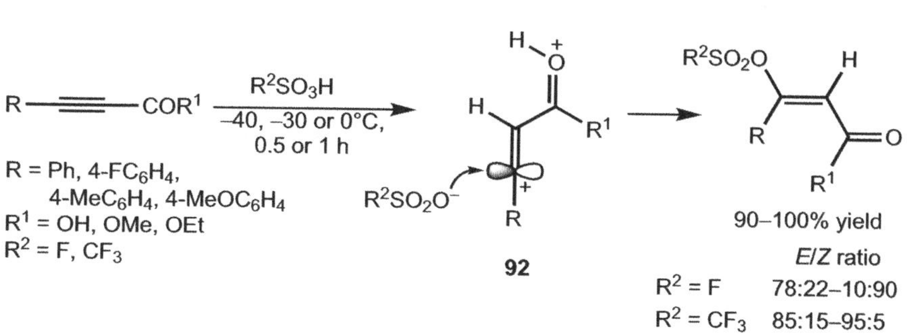

(5.341)

The efficient azidobromination of cycloalkenes [Eq. (5.342)] and open-chain alkenes to give β-bromoalkyl azides with *N*-bromosuccinimide and azidotrimethylsilane is catalyzed by Nafion-H.[919] Terminal alkenes and alkenes with bulky substituents do not react, whereas 2,3-trimethylbutene-2 reacts without catalysis. The stereochemistry of the process suggests the involvement of a bromonium ion intermediate.

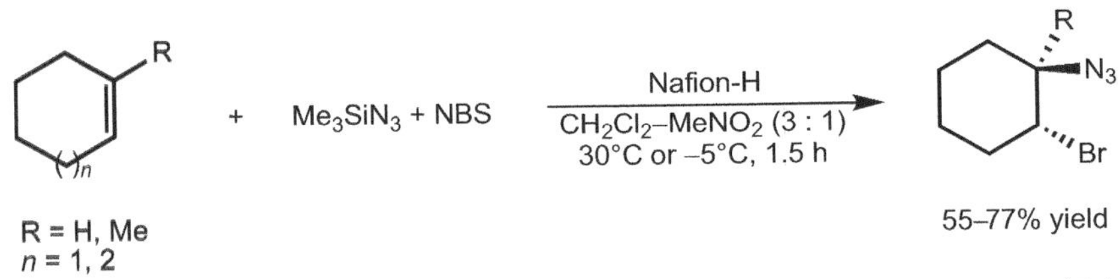

(5.342)

5.21. RITTER REACTIONS

In addition to Ritter-type transformations already discussed in previous sections, there are additional examples of the Ritter reaction—that is, the transformation of alkenes or alcohols with nitriles to give carboxamides carried out under superacidic conditions.

Ritter reaction of the triene **287** in triflic acid, performed to accomplish the synthesis of a marine sesquiterpene, gave the product acetamide derivative via a predominant *trans* antiparallel addition of H^+ and acetonitrile to the endocyclic double bond[920] [Eq. (5.343)].

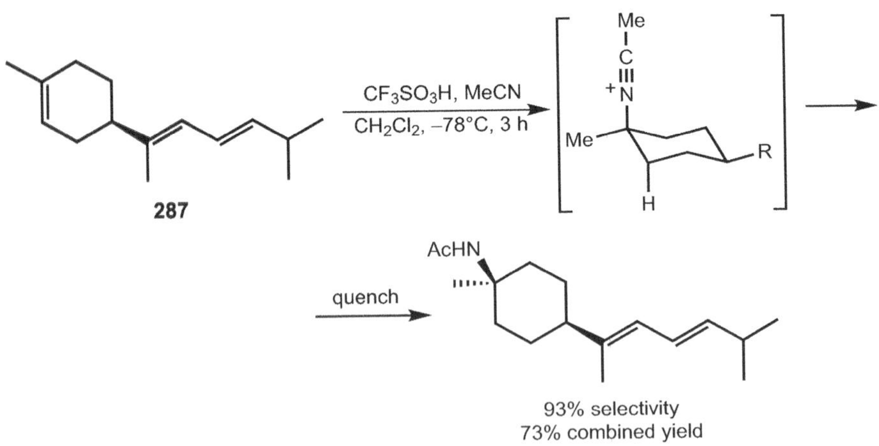

(5.343)

2-(Arylmethylene)cyclopropylmethanols have been reported to react with acetonitrile in triflic acid to give ring-enlarged *N*-(arylmethylidenecyclobutyl)acetamides[921]

[Eq. (5.344)]. The suggested mechanistic pathway is a Ritter-type reaction with the involvement of the bicyclobutonium ion **288**.

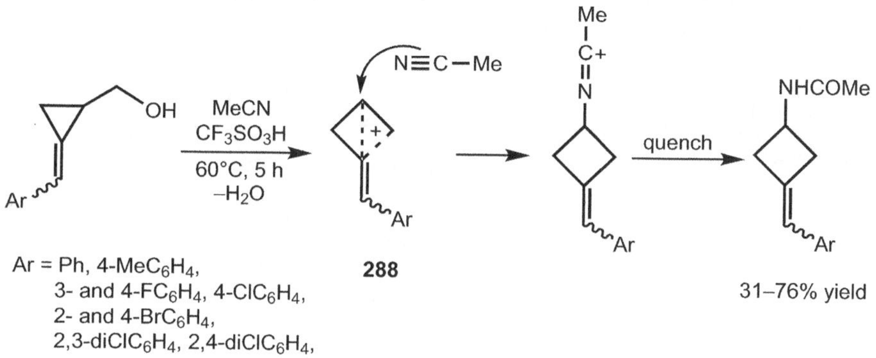

(5.344)

Triflic acid has been applied in a three-component condensation of phenols or 2-naphthol, aromatic aldehydes, and alkyl nitriles to form amidoalkylphenols under mild conditions in good to high yields[922] [Eq. (5.345)]. The reaction involves a Ritter-type step, wherein the intermediate condensation product reacts with the nitrile component.

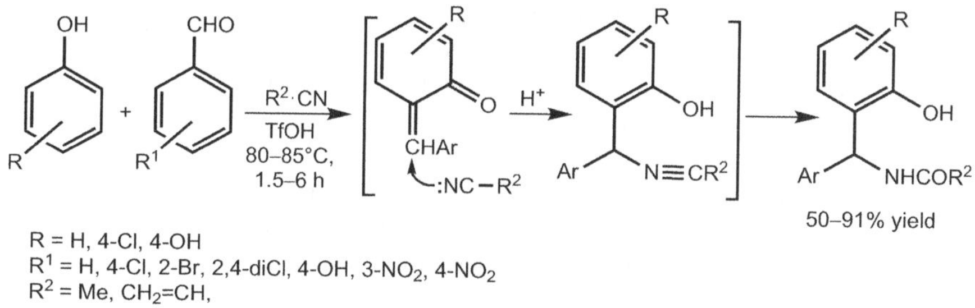

(5.345)

Nafion-H has also been shown to be active in the Ritter reaction. Treatment of alcohols, such as benzyl alcohols, 1-adamantanol [Eq. (5.346)], and 2-norborneols with acetonitrile or benzonitrile in the presence of Nafion-H under forcing conditions (140–145°C, 18–48 h), affords the corresponding carboxamides in moderate to excellent yields (40–99%).[652] Nafion-H can be reused after a simple activation (washing with deionized water and acetone followed by drying at 105°C).

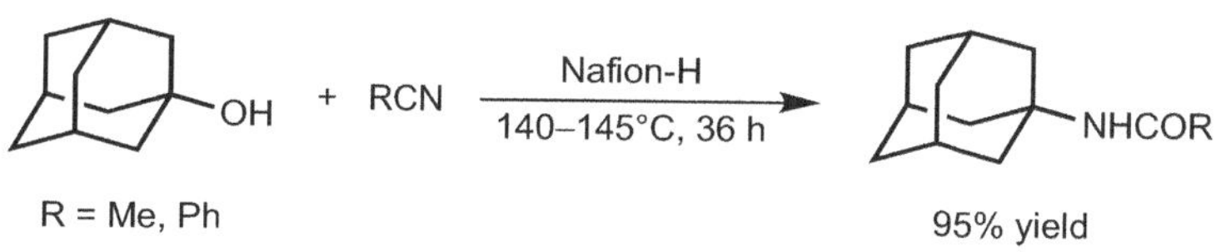

(5.346)

Nafion beads, Nafion–SiO_2, and Aciplex–SiO_2 were tested by Okuhara and coworkers[923,924] in the Ritter reaction between 1-adamantanol and acrylonitrile [Eq. (5.347)]. Nafion SAC-13 exhibited the highest specific activity calculated by taking into account the number of acidic sites.

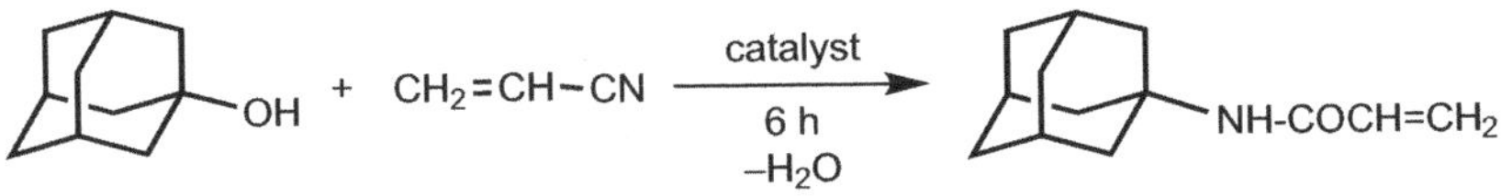

	100°C	60°C
Nafion NR-50	98%	17%
SAC-13	97%	29%
Aciplex-SiO_2		27%

(5.347)

5.22. POLYMERIZATION

The key initiation step in cationic polymerization of alkenes is the formation of a carbocationic intermediate, which can then react with excess monomer to start propagation. The kinetics and mechanisms of cationic polymerization and polycondensation have been studied extensively.[925–928] Kennedy and Maréchal[926] have pointed out that only cations of moderate reactivity are useful initiators, since stable ions such as arenium ions were found to be unreactive for olefin polymerization. On the other hand, energetic alkyl cations such as $CH_3CH_2^+$ were too reactive and gave side products.

It has been shown by Olah et al.[928,929] that cationic polymerization of alkenes can be initiated by stable alkyl or acyl cations as well as nitronium ion salts.

Trivalent carbenium ions play a key role, not only in the acid-catalyzed polymerization of alkenes [Eq. (5.348)] but also in the polycondensation of arenes (π-bonded monomers) as well as in the cationic polymerization of ethers, sulfides, and nitrogen compounds (nonbonded electron-pair donor monomers). On the other hand, pentacoordinated carbonium ions play the key role in the electrophilic reactions of σ-bonds (single bonds), including the oligocondensation of alkanes and alkenes (Section 5.1.5).

$$RF + BF_3 \longrightarrow R^+ BF_4^- + C_2H_5CH{=}CH_2 \longrightarrow C_2H_5\overset{+}{C}HCH_2R\ BF_4^- \longrightarrow$$

$$\xrightarrow{n\,(C_2H_5CH=CH_2)} C_2H_5\underset{}{\overset{CH_2R}{\overset{|}{C}}}H_2\text{—}[CH_2\overset{C_2H_5}{\overset{|}{C}}H\text{—}]_{n-1}\text{—}CH_2\overset{+}{C}HC_2H_5\ BF_4^- \longrightarrow$$

$$\longrightarrow C_2H_5\overset{CH_2R}{\overset{|}{C}}H_2\text{—}[CH_2\overset{C_2H_5}{\overset{|}{C}}H\text{—}]_{n-1}\text{—}CH{=}C_2H_5 + HF\ \ BF_3$$

(5.348)

In general, the cationic polymerization of olefins should be considered as a typical example of general carbocationic reactivity in electrophilic reactions, and all other suggested mechanisms can be looked upon as only differing in the nature of the initial electrophiles, always leading to the key trivalent alkyl cation, which then initiates the polymerization reaction [Eq. (5.349)].

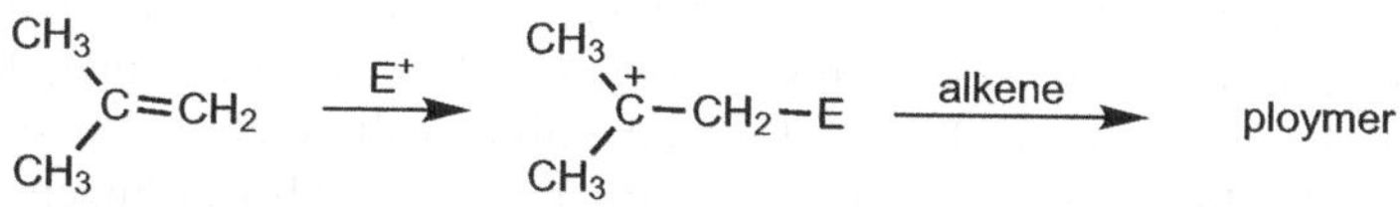

(5.349)

The solid superacid Nafion-H is also a good polymerization catalyst. Isobutylene has been polymerized with Nafion-H. At 145°C, only oligomers (dimers and tetramers) were obtained; decreasing the temperature increased the molecular weight of the oligomers.[930] The oligomerization reaction has also been studied for higher alkenes. For example, in the liquid phase, the addition of 1% by weight Nafion-H to 1-decene at 150°C gave, after 5 h, a 65% conversion to oligomeric products, of which 55% consisted of the trimer.[931] Under similar conditions, 5-decene was converted to a 4:1 mixture of dimers and trimers, which was hydrogenated to a lubricating-type oil.[932]

Nafion-H has also been used as a catalyst for the oligomerization of styrene. The reaction was studied by Higashimura and co-workers.[933,934] Hydroxy-terminated poly (alkylene)oxides were prepared by condensation–polymerization of 1,8-octanediol and 1,10-decanediol in the presence of Nafion-H.[935] It showed higher activity than sulfuric acid; consequently, polymerization could be carried out at lower temperature.

In the patent literature, there are several reports of the cationic polymerization of tetrahydrofuran (THF) with Nafion-H. In most cases, small amounts of acetic anhydride were added so the initial polymer had a terminal acetate group that could be hydrolyzed to the free hydroxyl. THF has also been homopolymerized[936–938] and copolymerized with ethylene oxide and propylene oxide in the presence of Nafion-H.[939–941]

Triflic acid has become a widely applied catalyst in various polymerization processes, and a few selected characteristic examples are discussed here. Additional basic information, examples of practical significance, and recent trends of cationic polymerization can be found in books and monographs.[942–945]

A polyaldolic condensation of acetone can be induced by triflic acid to give a solid resin with a polyenic structure resembling poly(methylacetylene) having some functional groups[946] (**289**).

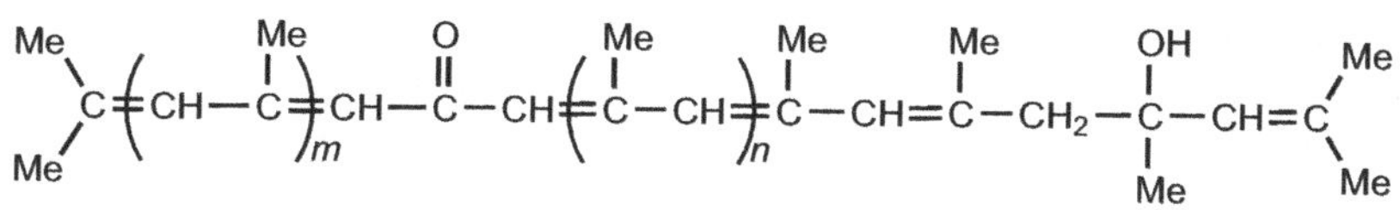

289

Zolotukhin and co-workers[947] have performed experimental and theoretical studies with respect to the use of superelectrophiles generated by triflic acid in the polycondensation of carbonyl compounds such as isatins,[948] fluorinated ketones,[949,950] and acenaphthenequinone[951] with aromatics (biphenyl, diphenyl ether, 4,4′-diphenoxyacetophenone, etc.) (Scheme 5.96). It has been concluded in recent studies (PBE0/aug-cc-PVTZ//PBE0/6-31+G* level of theory)[255,952] that *C,O* and *O,O* diprotonated species derived from molecules with electron-withdrawing or moderately electron-donating substituents (for example, hexafluoroacetone) do not participate in polycondensation. This is due to the high positive Gibbs energy for their formation. Acidity dependence is accounted for by a multistep mechanism involving only monoprotonated species. Monoprotonated species of molecules with strongly electron-donating substituents (for example, acetophenone), in turn, are highly stable, and their transformation is thermodynamically unfavorable.

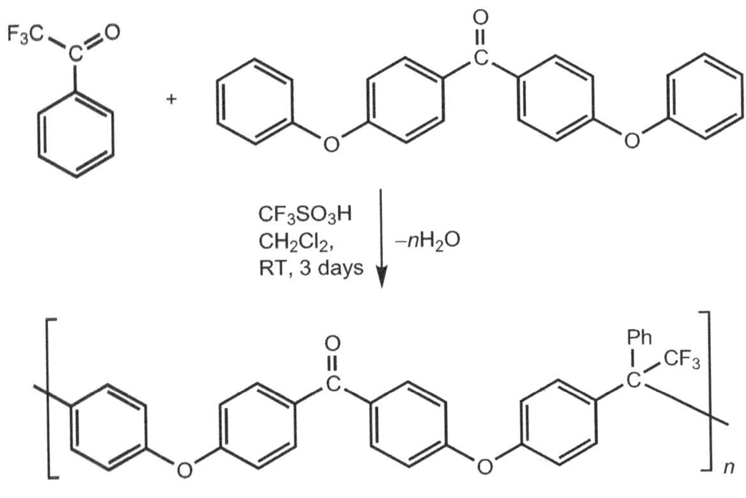

Scheme 5.96

Ring-opening polymerizations can also be induced by triflic acid. It is often used for the initiation of the ring-opening polymerization of oxacycloalkanes,[953,954] dioxacycloalkanes,[955,956] and cyclosiloxanes.[957–961] There has been a growing interest in poly(hydroxyalkanoates) and, in particular, polylactides as biodegradable and biocompatible polymers. A combination of triflic acid and a protic solvent (alcohols or water) has been recently reported to initiate the cationic polymerization of lactide at room temperature.[962] Polymerization was shown to occur selectively via acyl–oxygen cleavage [Eq. (5.350)]. Polymer characteristics show that triflic acid preferentially activates the monomer compared with the polymer chain. Polymerization of L-lactide resulted in the formation of a perfectly isotactic polymer, indicating that epimerization did not occur. In a similar manner, ring-opening polymerization of (*R*)-β-hydroxybutyrolactone catalyzed by triflic acid in toluene afforded chiral poly(β-hydroxybutyrate). The polymerization process proceeds with the retention of configuration.[963] The cyclic thiocarbonate **290** undergoes a ring-opening polymerization into a poly(monothio)ester with the adamantane moieties in the backbone initiated by triflic acid[964] [Eq. (5.351)].

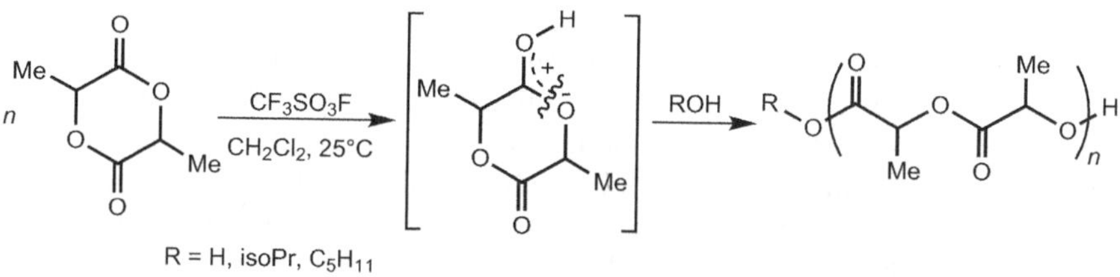

(5.350)

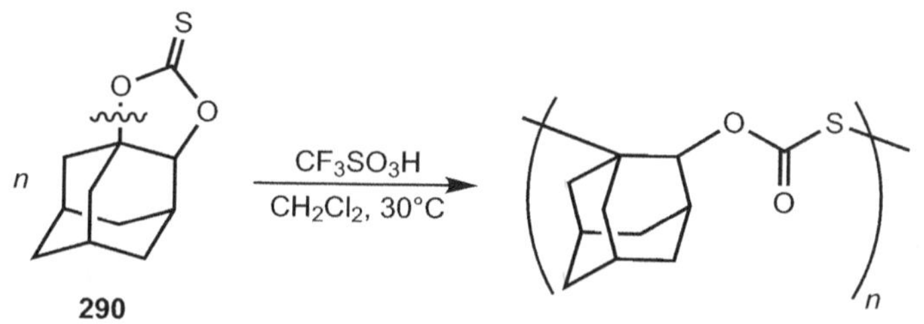

290

(5.351)

Polymerization of tetrahydrofuran to produce linear polyethers or cyclic oligomers is initiated by the protonation of the ring oxygen[965] (Scheme 5.97). The formed ion then reacts with the monomer to give tertiary oxonium ion intermediates. Macrocyclic oligomer formation proceeds by intramolecular attacks involving backbiting (route ***a***) or tailbiting mechanism (route ***b***). The ability of triflic acid to form tertiary oxonium

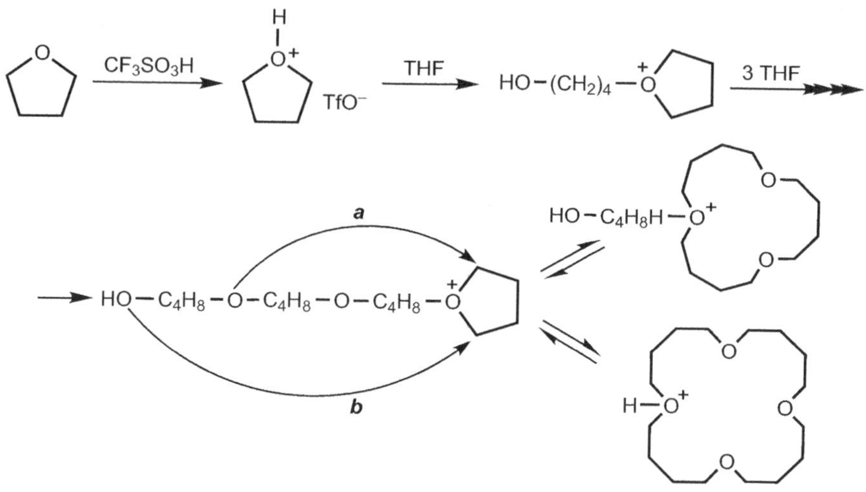

Scheme 5.97

ion intermediates, participating in chain growth instead of ester formation, was shown to be the highest among various initiators studied ($CF_3SO_3H > HSO_3F > MeOTF >>$ ScOF). The highest concentration of crown ethers ranging from 15-crown-3 to 100-crown-20 [$(C_4H_8)_n$, $n = 3$–20] was generated in nitromethane.

Kanoh et al.[966] have reported unusual monomer isomerization–polymerization processes of oxetanes with varied substituents. Cyclic imide-substituted monomers, such as **170**, undergo isomerization catalyzed by triflic acid as shown previously in Eq. (5.266) and then polymerize. Polyacetals **291** are formed below room temperature (single ring opening), whereas at elevated temperature (120–130°C) polyethers **292** are isolated (double ring-opening). Oxetanes with ester substituents exhibit similar dual characteristics.[967] Monomer **170**, however, showed an exceptional behavior yielding the oligoindene derivative **293** with a carbon backbone.[968]

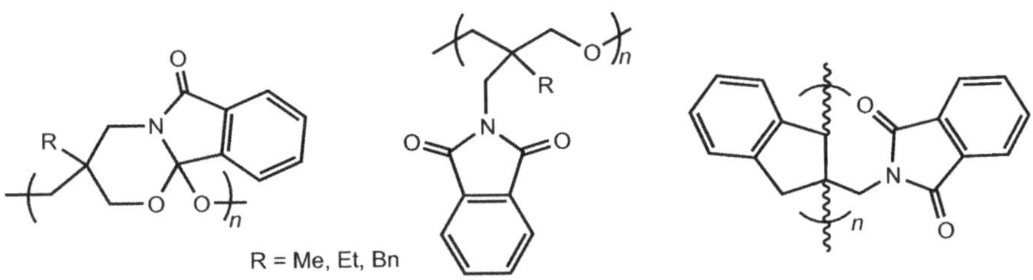

291 **292** **293**

Hyperbranched carbohydrate polymers have been synthesized through ring-opening multibranching polymerization of 1,4-anhydroerythritol and 1,4-anhydro-L-threitol using triflic acid and fluorosulfuric acid as cationic initiators.[969] HSO_3F resulted in a polymer of lower molecular weight. The proposed mechanism shown for 1,4-anhydroerythritol includes ring opening with the protonated monomer initiator, and branching is induced by proton transfer reaction (Scheme 5.98).

Photoinitiated cationic polymerization is in widespread use in UV curing and photoresist technology. Crivello and Lam[970] were the first to report that diaryliodonium salts undergo photodecomposition to generate Brønsted superacids (Scheme 5.99), which are capable of initiating the polymerization of suitable monomers. Since then, triarylsulfonium and triarylphosphonium and other salts with weakly coordinating anions have also been applied.[971,972]

A rare example of cationic polymerization of emulsified epoxy resins has been reported by Walker et al.[973] Polymerization of water emulsion of epoxy resins with a variety of superacids (triflic acid, $HClO_4$, HBF_4, HPF_6) results in polyols with two glycidyl units (**294**) in contrast to commercial epoxy resins with one unit separating the aromatic moieties. The level of residual glycidyl ether and Bisphenol-A units is also much lower than in conventional epoxy resins.

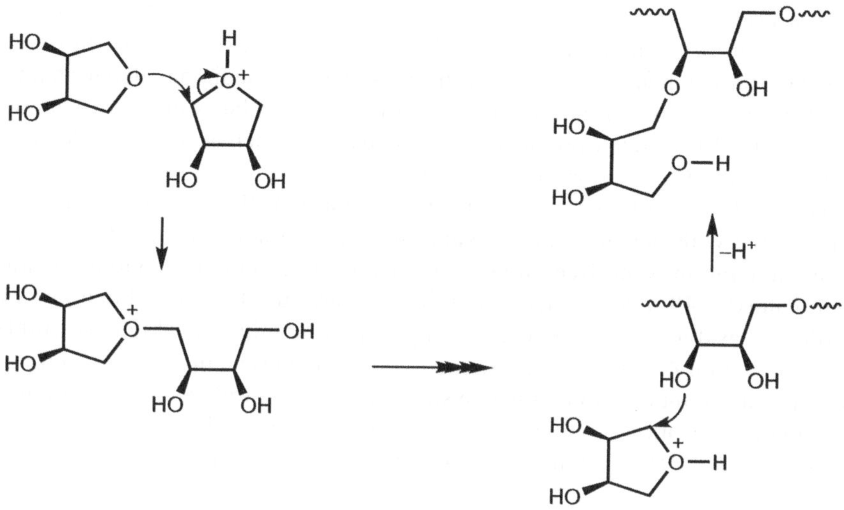

Scheme 5.98

(I)

$$Ar-\overset{+}{I}-Ar'\ X^- \xrightarrow{h\nu} ArI^{+\bullet}X^- + Ar'^{\bullet}$$

X = BF_4, SbF_6, AsF_6, PF_6, R_FSO_3, etc.

(II)

$$ArI^{+\bullet}X^- + R-H \longrightarrow H^+X^- + PhI + R^{\bullet}$$

RH = solvent or monomer

Scheme 5.99

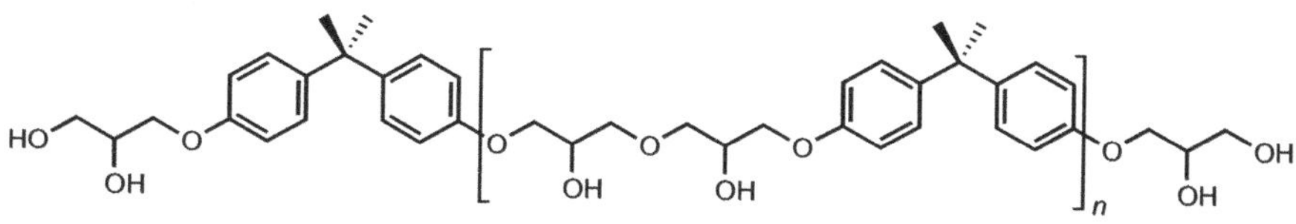

294

Lacaze and co-workers[974] have prepared poly(*para*-phenylene) films by electropolymerization of benzene in strong acidic media including triflic acid and fluorosulfuric acid. The films grown in superacids are characterized by sharp redox peaks in contrast to films prepared in other acids, which exhibit wide, ill-defined peaks. Poly(*para*-phenylene) prepared in the superacid media has linear chains, a high degree of polymerization, and a narrow molecular weight distribution. Extremely well-organized polymers were prepared by performing the electropolymerization in SO_2 or in organic solvents.[975]

Jones and co-workers[976] have recently reported the use of catalyst **49** (Figure 5.15) with perfluorinated alkanesulfonic acid sites anchored to SBA-15 as a methylaluminoxane-free supported cocatalyst for ethylene polymerization. When catalyst **49** and trimethylaluminum were used in combination with $Cp^*{}_2ZrMe_2$ as the metallocene precatalyst, productivities as high as 1000 kg polyethylene mol Zr^{-1} h^{-1} were obtained without experiencing reactor fouling.

In a recent communication, a microsystem allowing controlled polymerization and block copolymerization of vinyl ethers with triflic acid as the initiator at −25°C has been described.[977] The system allows a high level of control on molecular weight distribution.

5.23. MISCELLANEOUS REACTIONS

Isomerization of 3,3′- and 4,4′-dimethylbiphenyl in triflic acid results in the formation of only 3 of the possible isomers: 3,3′-dimethylbiphenyl (~40%), 3,4′-dimethylbiphenyl (~55%), and 4,4′-dimethylbiphenyl (~5%).[978] This product composition reflects the relative stability of the intermediate carbocations. Isomerization was suggested to occur by *ipso* protonation to the methyl group followed by a slow methyl migration as the rate-determining step [Eq. (5.352)]. The intermediate formed by *ipso* protonation to the phenyl group was calculated to have higher energy by 5.4 kcal mol^{-1} (AM1 method).

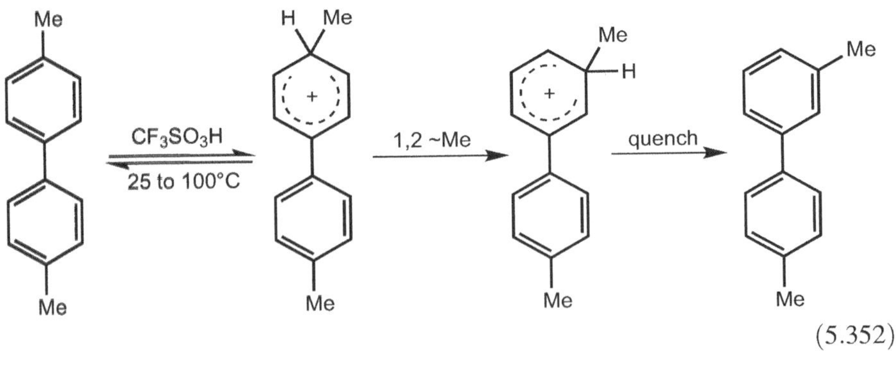

(5.352)

Triflic acid has proved to be an efficient catalyst to promote direct *C*-alkylation of 1,3-dicarbonyl compounds with benzyl alcohols[979] [Eq. (5.353)]. Alkylation of hydroquinone by *tert*-butyl alcohol in a solventless reaction yields 2,5-di-*tert*-butylhydroquinone (93% selectivity at 90% conversion, 150°C), whereas 2-*tert*-butylhydroquinone is the main product in 1,4-dioxane (79.5% selectivity at 39% conversion).[980]

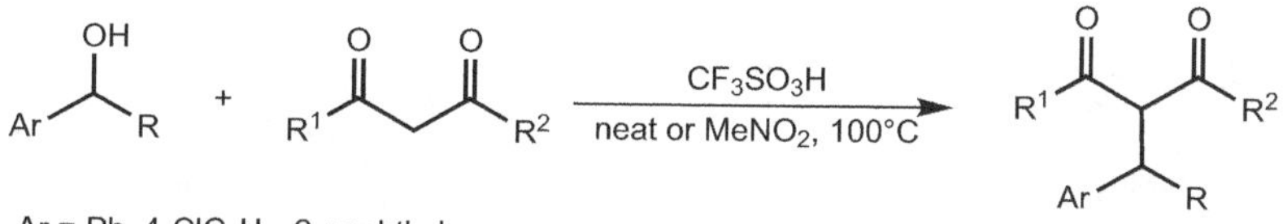

(5.353)

Ar = Ph, 4-ClC_6H_4, 2-naphthyl
2-, 3- and 4-BrC_6H_4,
R = Me, Et, Pr, Bu, $CH{=}CH_2$
R^1 = Me, Ph
R^2 = Me, Ph, OEt

53–90% yield

Aubé and co-workers[981,982] have found that enolizable ketones react with benzyl azide in triflic acid to yield *N*-(phenylamino)-methylated products [Eq. (5.354)]. The transformation is an aza-Mannich reaction interpreted with the involvement of the Mannich reagent *N*-phenyl iminium ion **295** formed *in situ* in a Schmidt rearrangement. Cyclic tertiary alcohols react with alkyl azides in triflic acid to yield *N*-alkylamines (**296**, **297**)[983] [Eq. (5.355)]. The Schmidt rearrangement was used to transform Merrifield resin into amino-polystyrene resin by reacting the azido derivative in excess triflic acid (CH_2Cl_2, 0°C).[984]

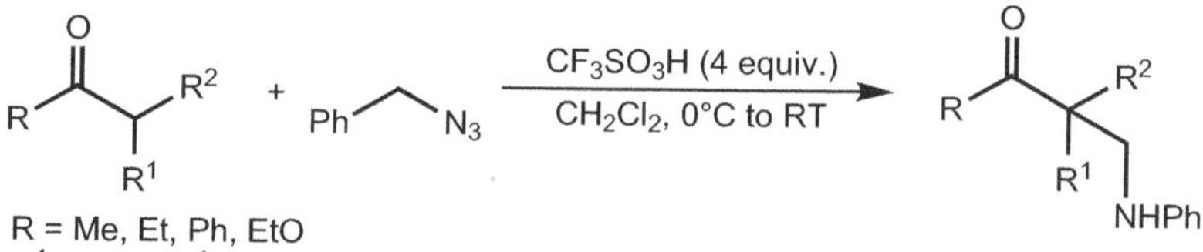

(5.354)

R = Me, Et, Ph, EtO
R^1 = H, R–R^1 = C_4H_8, C_5H_{10}
R^2 = H, Me, Pr, CO_2Et

50–98% yield

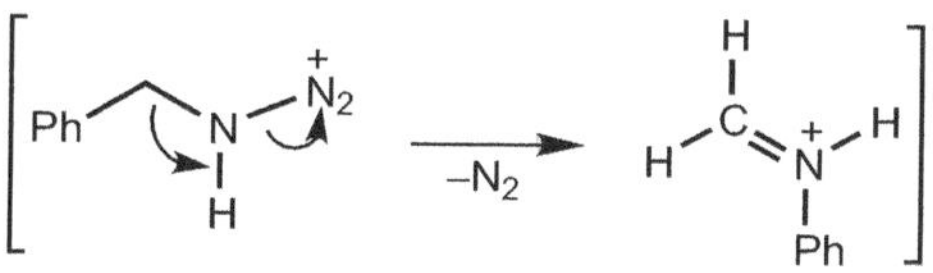

295

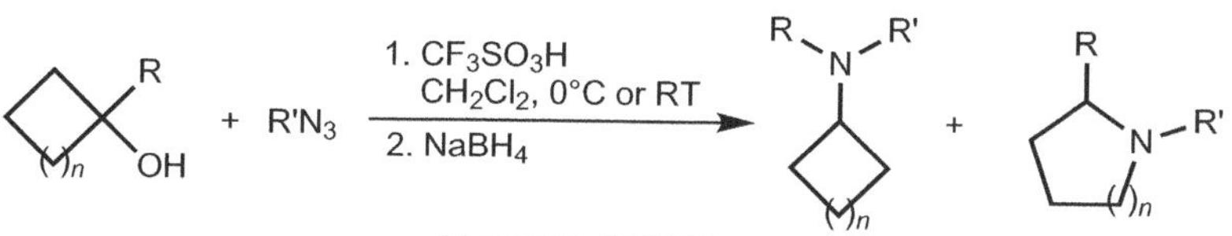

296 **297** (5.355)

	Yield (%)	**296/297**
R = Ph, R' = Bu, *n* = 1	97	16 : 84
n = 2	66	52 : 48
n = 3	82	71 : 29

R = Me, *n* = 2, R' = Bu **297** 95% yield
R' = Bn **297** 88% yield

N-Alkylation of methyl *N*-(trimethylsilyl)pyroglutamate with benzhydryl chlorides or trimethylsilyl benzhydryl ethers can be carried out in almost quantitative yields with the use of catalytic amounts of triflic acid (0.3–3 mol%)[985] [Eq. (5.356)].

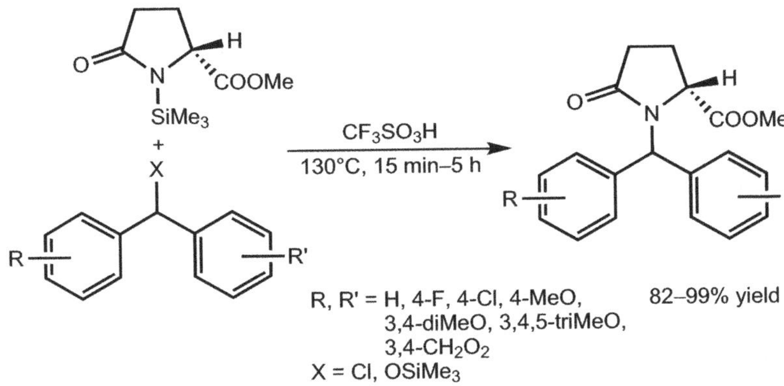

(5.356)

The reaction of chiral aldimines with 2-silyloxybutadienes in the presence of triflic acid affords novel Mannich-type products with high diastereoselectivity[986] [Eq. (5.357)].

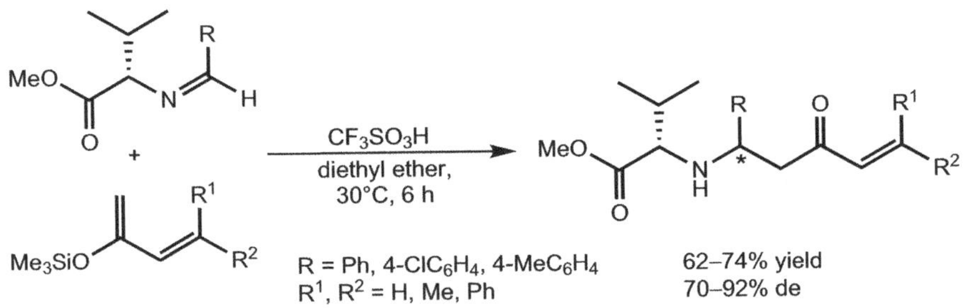

(5.357)

α-Hydroxy ketones have been shown by Olah and Wu[987] to undergo various transformations in triflic acid depending on their substitution pattern. Compounds without β-hydrogen give fragmentative products (Scheme 5.100, route ***a***). Since formation of the intermediate carbocation is not favored because of the lack of stabilizing substituents, a concerted mechanism was suggested. A similar mechanism accounts for the exclusive formation of α,β-unsaturated ketones under similar conditions (Scheme 5.100, route ***b***). The involvement of the stable benzylic cation, in turn, allows the formation of the cyclized product as shown in route ***c***.

In the 1980s, Rudenko et al.[988,989] made pioneering studies on the electrochemical oxidation of various aromatic compounds in HSO_3F and $HSO_3F{-}SbF_5$. Recently,

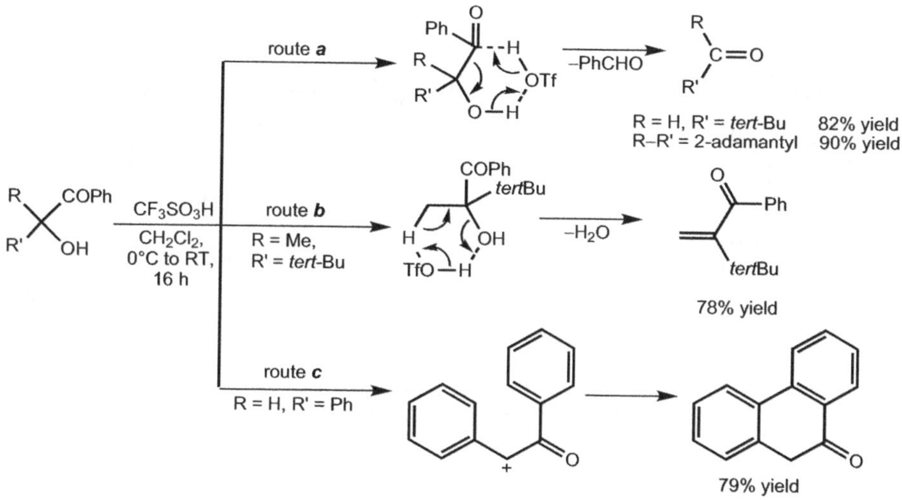

Scheme 5.100

oxidations using the HSO_3F-PbO_2 system have been reported in a long series of papers. The studies, among others, included the oxidation of substituted nitrobenzenes and anilines,[990] benzoic acids,[991] phenols and derivatives,[992] and so on, in the temperature range $-75°C$ to $0°C$. Product formation is interpreted by two one-electron oxidation steps followed by a proton loss. The formed intermediate reacts with the fluorosulfate anion and the work-up procedure leads to the final chloro-substituted product as shown for 2-methylbenzonitrile[993,994] [Eq. (5.358)]. In most cases, a range of products including various dimers are formed usually in low yields, but selective transformations may also be observed. Oxidation of 2,4,6-trisubstituted benzonitriles, for example, affords carboxamides in high yields formed in a Ritter-type transformation[993,994] [Eq. (5.359)]. Benzonitriles could be transformed into 2,4,6-triaryltriazines (**298**) in HSO_3F alone at higher temperatures (20°C, 3 h, 49–65% yield).[993] The elementary steps, the nature of the reaction intermediates, and general trends of chemical and electrochemical oxidation of aromatics in HSO_3F have been discussed in a review paper.[995]

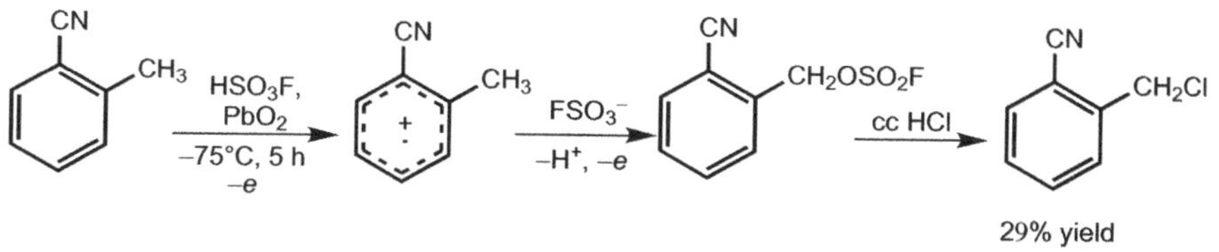

(5.358)

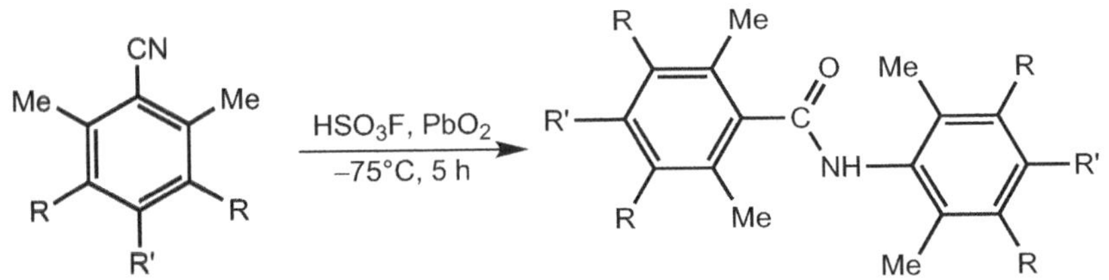

R = H, Me
R' = H, Me, *tert*-Bu

75–94% yield

(5.359)

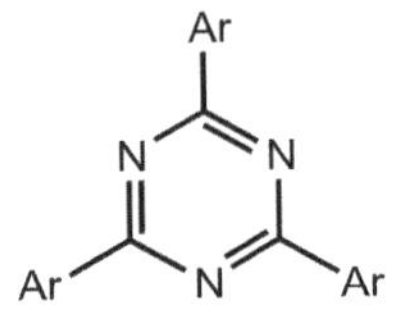

Ar = 4-MeC_6H_4,
4-*tert*-BuC_6H_4,
3,5-$diMeC_6H_3$

298

Harmer and co-workers[996,997] tested various Nafion-based samples in the dimerization of α-methylstyrene to form isomeric pentenes and a cyclic dimer, which are of industrial interest: 13% and 40% Nafion–silica nanocomposites exhibited near-complete conversion and gave the cyclic dimer **300** with high selectivity [Eq. (5.360)], whereas compound **299a** was the main product over Nafion NR50. In sharp contrast, isomeric pentenes **299** could be obtained in a continuous process (86% selectivity at conversion >95%). Similar findings were reported with Nafion immobilized in MCM-41 mesoporous silica.[227] In kinetic studies, 13% Nafion–silica[996] and the catalyst with anchored perfluorinated sulfonic acid site (catalyst **48**, Figure 5.15)[192] showed the highest activity. Dimerization of isobutylene has been studied in a forced-flow catalytic membrane reactor using Nafion-based catalytic membranes made with various binders.[998] The best results were found for Nafion SAC-13 exhibiting high selectivity (80%) at low conversion (45%). The fast removal of product isooctene inhibits secondary processes, whereas byproducts are purged from the active sites, thereby preventing catalyst deactivation.

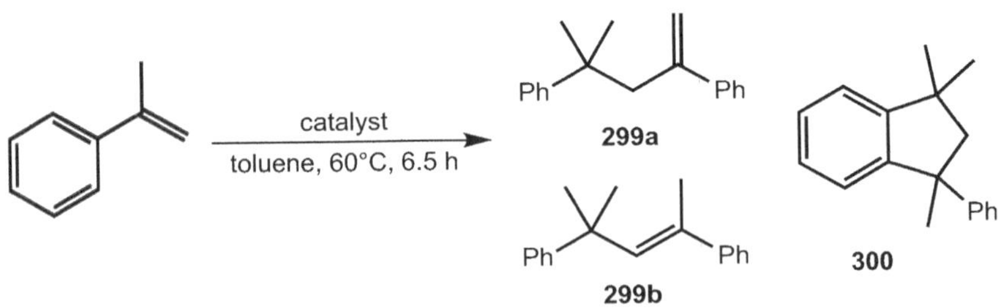

		Selectivity, %
13% Nafion-SiO_2	**300**	91
40% Nafion-SiO_2	**300**	99
Nafion NR50	**299a/299b/300**	47:15:24

(5.360)

Olah and Ip[999] showed that Nafion-H catalyzes condensation of acetone to give mesitylene[999] [Eq. (5.361)]. In contrast to phosphoric acid and sulfuric acid, the reaction carried out in a flow system gives the product with high selectivity. Likewise, substituted aryl methyl ketones can be transformed to 1,3,5-triarylbenzenes[1000] [Eq. (5.361)]. *ortho*-Substituted acetophenones, however, do not react.

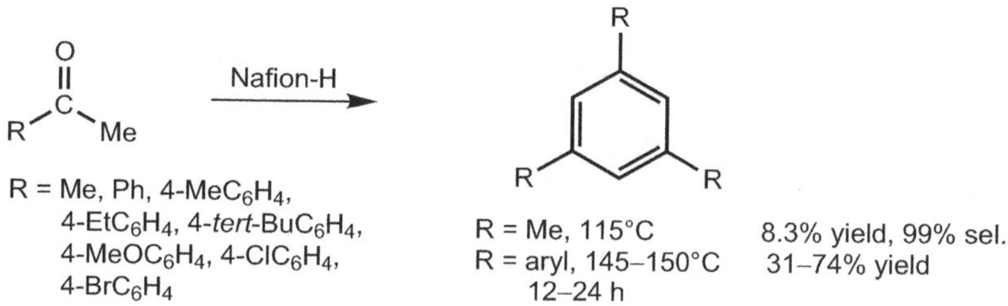

(5.361)

The Peterson silyl-Wittig methylenation of carbonyl compounds has been significantly improved by performing the elimination of the trimethylsilanol form of the intermediate β-hydroxysilanes with Nafion-H under mild conditions[1001] [Eq. (5.362)].

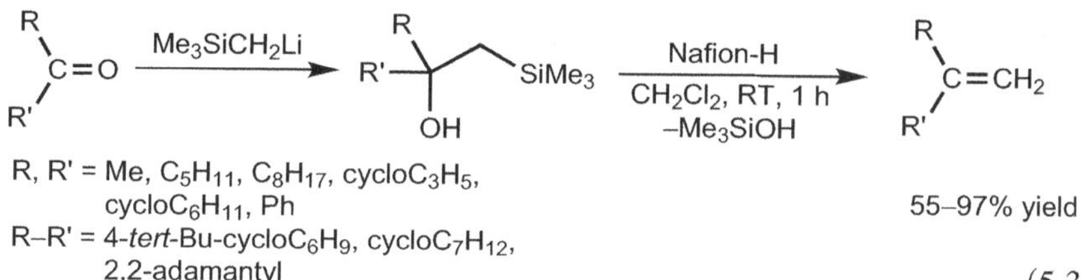

(5.362)

β-Acetamido ketones have been prepared in a multicomponent reaction from aromatic aldehydes, enolizable ketones (acetophenone and propiophenone), and acetyl chloride in acetonitrile over Nafion-H[1002] [Eq. (5.363)]. High yields are achieved under mild conditions and the catalyst proved to be recyclable.

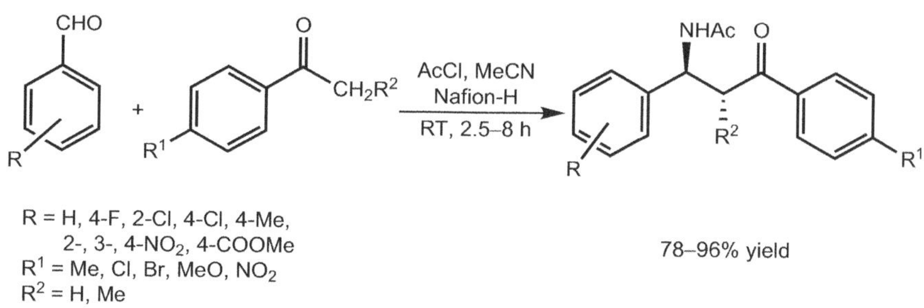

(5.363)

REFERENCES

1. G. A. Olah, A. Germain, H. C. Lin, and D. A. Forsyth, *J. Am. Chem. Soc.* **97**, 2928 (1975).
2. G. A. Olah, G. K. S. Prakash, and K. Lammertsma, *Res. Chem. Intermed.* **12**, 141 (1989).
3. G. A. Olah, *Angew. Chem. Int. Ed. Engl.* **32**, 767 (1993).
4. G. A. Olah and D. A. Klumpp, *Acc. Chem. Res.* **37**, 211 (2004).
5. G. A. Olah and D. A. Klumpp, *Superelectrophiles and Their Chemistry*, Wiley-Interscience, Hoboken, NJ, 2008.
6. D. A. Klumpp, *Chem. Eur. J.* **14**, 2004 (2008).
7. H. Pines, *The Chemistry of Catalytic Hydrocarbon Conversions*, Academic Press, New York, 1981.
8. G. A. Olah and Á. Molnár, *Hydrocarbon Chemistry*, 2nd ed. Wiley-Interscience, Hoboken, 2003.
9. K. Tanabe and W. F. Hölderich, *Appl. Catal., A* **181**, 399 (1999).
10. F. Asinger, *Paraffins, Chemistry and Technology*, Pergamon Press, New York, 1965.
11. A. Corma, *Chem. Rev.* **95**, 559 (1995).
12. X. Song and A. Sayari, *Catal. Rev.—Sci. Eng.* **38**, 329 (1996).
13. T. Yamaguchi, *Appl. Catal.* **61**, 1 (1990).
14. J. A. Martens and P. A. Jacobs, in *Handbook of Heterogeneous Catalysis*, Vol. 3, G. Ertl, H. Knözinger, and J. Weitkamp eds., VCH-Wiley, Weinheim, 1997, Chapter 5.4.2, p. 1137.
15. *Acidity in Aluminas, Amorphous and Crystalline Silico-Aluminas,* J.J. Fripiat and J.A. Dumesic Eds., *Topics Catal.* **4**, 1–171 (1997).
16. R. J. Gorte, *Catal. Lett.* **62**, 1 (1999).
17. H. S. Bloch, H. Pines, and L. Schmerling, *J. Am. Chem. Soc.* **68**, 153 (1946).
18. G. A. Olah and J. Lukas, *J. Am. Chem. Soc.* **89**, 2227, 4739 (1967).
19. A. F. Bickel, C. J. Gaasbeek, H. Hogeveen, J. M. Oelderik, and J. C. Platteeuw, *Chem. Commun.* 634 (1967).
20. H. Hogeveen and A. F. Bickel, *Chem. Commun.* 635 (1967).
21. G. K. S. Prakash, A. Husain, and G. A. Olah, *Angew. Chem.* **95**, 51 (1983).
22. G. A. Olah and J. Lukas, *J. Am. Chem. Soc.* **90**, 933 (1968).
23. G. A. Olah and R. H. Schlosberg, *J. Am. Chem. Soc.* **90**, 2726 (1968).
24. H. Hogeveen and C. J. Gaasbeek, *Recl. Trav. Chim. Pays-Bas* **87**, 319 (1968).
25. G. A. Olah, G. Klopman, and R. H. Schlosberg, *J. Am. Chem. Soc.* **91**, 3261 (1969).
26. G. A. Olah, J. Shen, and R. H. Schlosberg, *J. Am. Chem. Soc.* **92**, 3831 (1970).
27. T. Oka, *Phys. Rev. Lett.* **43**, 531 (1980).
28. H. Hogeveen and A. F. Bickel, *Recl. Trav. Chim. Pays-Bas* **86**, 1313 (1967).
29. D. M. Brouwer and H. Hogeveen, *Prog. Phys. Org. Chem.* **9**, 184 (1972).
30. J. Sommer, J. Bukala, M. Hachoumy, and R. Jost, *J. Am. Chem. Soc.* **119**, 3274 (1997).
31. For a review, see J. Sommer, and R. Jost, *Pure Appl. Chem.* **72**, 2309 (2000).
32. H. H. Schobert, *Energy and Society an Introduction*, Taylor and Francis, New York, 2002.

33. J. C. Mackie, *Catal. Rev.—Sci. Eng.* **33**, 169 (1991).
34. E. E. Wolf, *Methane Conversion by Oxidative Processes: Fundamental and Engineering Aspects*, Van Nostrand Reinhold, New York, 1992.
35. A. E. Shilov, *Activation of Saturated Hydrocarbons by Transition Metal Complexes*, Reidel, Dordrecht, 1984, Chapter 5, p. 142.
36. G. A. Olah, *Acc. Chem. Res.* **20**, 422 (1987).
37. A. Shen, *Acc. Chem. Res.* **31**, 550 (1998).
38. B. L. Conley, W. J. Tenn III, K. J. H. Young, S. K. Ganesh, S. K. Meier, V. R. Ziatdinov, O. Mironov, J. Oxgaard, J. Gonzales, W. A. Goddard III, and R. A. Periana, *J. Mol. Catal. A: Chem.* **251**, 8 (2006).
39. G. A. Olah, *Angew. Chem. Int. Ed. Engl.* **12**, 173 (1973).
40. H. Hogeveen and A. F. Bickel, *Recl. Trav. Chim. Pays-Bays* **88**, 371 (1969).
41. V. L. Tal'roze and A. K. Lyubimova, *Dokl. Akad. Nauk SSSR* **86**, 909 (1952).
42. V. L. Tal'roze and A. K. Ljubimova, *J. Mass. Spectrom.* **33**, 502 (1998).
43. P. R. Schreiner, S. -J. Kim, H. F. Schaefer III, and P.v.R. Schleyer *J. Chem. Phys.* **99**, 3716 (1993).
44. H. Müller, W. Kutzelnigg, J. Noga, and W. Klopper, *J. Chem. Phys.* **105**, 1863 (1997).
45. D. Marx and A. Savin, *Angew. Chem. Int. Ed. Engl.* **36**, 2077 (1997).
46. P. R. Schreiner, *Angew. Chem. Int. Ed.* **39**, 3239 (2000).
47. D. Marx and M. Parrinello, *Nature* **375**, 216 (1995).
48. E. T. White, J. Tang, and T. Oka, *Science* **284**, 135 (1999).
49. P. Ahlberg, A. Karlsson, A. Goeppert, S. O. Nilsson Lill, P. Dinér, and J. Sommer, *Chem. Eur. J.* **7**, 1936 (2001).
50. P. M. Esteves, A. Ramírez-Solís, and C. J. A. Mota, *J. Phys. Chem. B* **105**, 4331 (2001).
51. S. Walspurger, A. Goeppert, M. Haouas, and J. Sommer, *New J. Chem.* **28**, 266 (2004).
52. A. Goeppert, P. Dinér, P. Ahlberg, and J. Sommer, *Chem. Eur. J.* **8**, 3277 (2002).
53. J. Sommer and J. Bukala, *Acc. Chem. Res.* **26**, 370 (1993).
54. J. Sommer, J. Bukala, S. Rouba, R. Graff, and P. Ahlberg, *J. Am. Chem. Soc.* **114**, 5884 (1992).
55. J. Sommer, R. Jost, and M. Hachoumy, *Catal. Today* **38**, 309 (1997).
56. J. -C. Culmann and J. Sommer, *J. Am. Chem. Soc.* **112**, 4057 (1990).
57. A. Goeppert, A. Sassi, J. Sommer, P. M. Esteves, C. J. Mota, A. Karlson, and P. Ahlberg, *J. Am. Chem. Soc.* **121**, 10628 (1999).
58. A. G. Stepanov, H. Ernst, and D. Freude, *Catal. Lett.* **54**, 1 (1998).
59. I. I. Ivanova, E. B. Pomakhina, A. I. Rebrov, and E. G. Derouane, *Topics Catal.* **6**, 49 (1998).
60. M. Haouas, S. Walspurger, F. Taulelle, and J. Sommer, *J. Am. Chem. Soc.* **126**, 599 (2004).
61. A. Goeppert, and J. Sommer, *New J. Chem.* **26**, 1335 (2002).
62. J. Lukas, P. A. Kramer, and A. P. Kouwenhoven, *Recl. Trav. Chim. Pays-Bas* **92**, 44 (1973).
63. G. A. Olah, Y. Halpern, J. Shen, and Y. K. Mo, *J. Am. Chem. Soc.* **93**, 1251 (1971).
64. J. W. Larsen, *J. Am. Chem. Soc.* **99**, 4379 (1977).

65. J. Sommer, J. Bukala, and M. Hachoumy, *Res. Chem. Intermed.* **22**, 753 (1996).
66. J. W. Otvos, D. P. Stevenson, C. D. Wagner, and O. Beeck, *J. Am. Chem. Soc.* **73**, 5741 (1951).
67. C. D. Nenitzescu, and A. Dragăn, *Ber. Dtsch. Chem. Ges.* **66**, 1892 (1933).
68. J. Sommer, D. Habermacher, R. Jost, A. Sassi, A. G. Stepanov, M. V. Luzgin, D. Freude, H. Ernst, and J. Martens, *J. Catal.* **181**, 265 (1999).
69. M. V. Luzgin, A. G. Stepanov, A. Sassi, and J. Sommer, *Chem. Eur. J.* **6**, 2368 (2000).
70. A. G. Stepanov, M. V. Luzgin, A. V. Krasnoslobodtsev, V. P. Shmachkova, and N. S. Kotsarenko, *Angew. Chem. Int. Ed.* **39**, 3658 (2000).
71. W. Hua, A. Goeppert, J. Sommer, and *Catal Appl. A* **219**, 201 (2001).
72. Y. D. Xia, W. M. Hua, Y. Tang, and Z. Gao, *Chem. Commun.* 1899 (1999).
73. B. Schoofs, J. A. Martens, P. A. Jacobs, and R. Schoonheydt, *J. Catal.* **183**, 355 (1999).
74. J. F. Haw, J. B. Nicholas, T. Xu, L. W. Beck, and D. B. Ferguson, *Acc. Chem. Res.* **29**, 259 (1996).
75. R. A. van Santen and G. J. Kramer, *Chem. Rev.* **95**, 637 (1995).
76. G. J. Kramer, R. A. van Santen, C. A. Emels, and A.K. Nowak, *Nature* **363**, 529 (1993).
77. J. M. Vollmer and T. N. Truong, *J. Phys. Chem. B* **104**, 6308 (2000).
78. J. Bertram, J. P. Coleman, M. Fleischmann, and D. Pletcher, *J. Chem. Soc., Perkin Trans. 2*, 374 (1973).
79. P. -L. Fabre, J. Devynck, and B. Trémillon, *Chem. Rev.* **82**, 591 (1985).
80. J. Devynck, P. -L. Fabre, A. Ben Hadid, and B. Tremillon, *J. Chem. Res. (S)*, 2469 (1979).
81. J. Devynck, A. Ben-Hadid, P. -L. Fabre, and B. Tremillon, and B. Carré, *J. Fluorine Chem.* **35**, 214 (1987).
82. B. Carré, P.-L. Fabre, and J. Devynck, *Bull. Soc. Chim. Fr.* 255 (1987).
83. H. Pines and N. E. Hoffman, in *Friedel–Crafts and Related Reactions*, Vol. 2, G A. Olah, ed., Wiley-Interscience, New York, 1964, Chapter 28, p. 1211.
84. C. D. Nenitzescu, in *Carbonium Ions*, Vol. 2, G.A. Olah, and P.v.R. Schleyer, eds., Wiley-Interscience, New York, 1970, Chapter 13, p. 490.
85. F. Asinger, *Paraffins*, Pergamon Press, New York, 1968, p. 695.
86. G. A. Olah and G. K. S. Prakash, in *The Chemistry of Alkanes and Cycloalkanes*, S. Patai and Z. Rappoport, eds., Wiley, Chichester, 1992, Chapter 13, p. 615.
87. S. T. Sie, in *Handbook of Heterogeneous Catalysis*, Vol. 4, G. Ertl, H. Knözinger, and J. Weitkamp eds., VCH-Wiley, Weinheim, **1997**, Chapter 3.12, p. 1998.
88. Y. Ono, *Catal. Today* **81**, 3 (2003).
89. V. M. Akhmedov and S. H. Al-Khowaiter, *Catal. Rev.—Sci. Eng.* **49**, 33 (2007).
90. R. Bonifay, B. Torck, and M. Hellin, *Bull. Soc. Chim. Fr I.* 808, (1977).
91. R. Bonifay, B. Torck, and M. Hellin, *Bull. Soc. Chim. Fr. I* **1057**, (1977).
92. R. Bonifay, B. Torck, and M. Hellin, *Bull. Soc. Chim. Fr. I* 36 (1978).
93. D. Brunel, J. Itier, A. Commeyras, P. T. Luu, and D. Mathieu, *Bull. Soc. Chim. Fr. II* **249**, 257 (1979).
94. A. Germain, P. Ortega, and A. Commeyras, *Nouv. J. Chim.* **3**, 415 (1979).
95. H. Choukroun, A. Germain, D. Brunel, and A. Commeyras, *Nouv. J. Chim.* **5**, 39 (1981);**7**, 83 (1983).

96. G. A. Olah, U.S. Patent, 4,472,268 (1984).

97. G. A. Olah, U.S. Patent, 3,766,286 (1973).

98. G. A. Olah, O. Farooq, A. Husain, N. Ding, N. J. Trivedi, and J. A. Olah, *Catal. Lett.* **10**, 239 (1991).

99. D. Fărcaşiu and P. Lukinskas, *J. Chem. Soc., Perkin Trans. 2* 1609 (1999).

100. D. Fărcaşiu and P. Lukinskas, *J. Chem. Soc., Perkin Trans. 2* 2715 (1999).

101. D. Fărcaşiu and P. Lukinskas, *J. Chem. Soc., Perkin Trans. 2* 2295 (2000).

102. D. M. Brouwer and J. M. Oelderik, *Recl. Trav. Chim. Pays-Bas* **87**, 721 (1968).

103. D. Fărcaşiu, *Catal. Lett.* **71**, 95 (2001).

104. F. Le Normand, F. Fajula, F. Gault, and J. Sommer, *Nouv. J. Chim.* **6**, 411 (1982).

105. F. Le Normand, F. Fajula, and J. Sommer, *Nouv. J. Chim.* **6**, 291 (1982).

106. F. Le Normand, F. Fajula, F. Gault, and J. Sommer, *Nouv. J. Chim.* **6**, 417 (1982).

107. R. Jost, K. Laali, and J. Sommer, *Nouv. J. Chim.* **7**, 79 (1983).

108. N. Yoneda, T. Fukuhara, T. Abe, and A. Suzuki, *Chem. Lett.* 1485 (1981).

109. J. J. L. Heinerman and J. Gaaf, *J. Mol. Catal.* **11**, 215 (1981).

110. K. Laali, M. Muller, and J. Sommer, *J. Chem. Soc., Chem. Commun.* 1088 (1980).

111. H. Hattori, O. Takahashi, M. Takagi, and K. Tanabe, *J. Catal.* **68**, 132 (1981).

112. G.A. Olah, U.S. Patent, 4,116,880 (1978).

113. O. Takahashi, T. Yamauchi, T. Sakuhara, H. Hattori, and K. Tanabe, *Bull. Chem. Soc. Jpn.* **53**, 1807 (1980).

114. P.v.R. Schleyer, *J. Am. Chem. Soc.* **79**, 3292 (1957).

115. R. C. Fort, Jr. *Adamantane, the Chemistry of Diamond Molecules*, Marcel Dekker, New York, 1976, Chapter 2, p. 35.

116. G. A. Olah and J. A. Olah, *Synthesis*, 488 (1973).

117. G. A. Olah and O. Farooq, *J. Org. Chem.* **51**, 5410 (1986).

118. O. Farooq, S. M. F. Farnia, M. Stephenson, and G. A. Olah, *J. Org. Chem.* **53**, 2840 (1988).

119. M. Vol'pin, I. Akhrem, and A. Orlinkov, *New J. Chem.* **13**, 771 (1989).

120. I. S. Akhrem and A. V. Orlinkov, *Russ. Chem. Bull. Engl. Transl.* **47**, 740 (1998); *Chem. Rev.* **107**, 2037 (2007).

121. A. V. Orlinkov, I. S. Akhrem, S. V. Vitt, L. V. Afanas'eva, L. I. Zakharkin, S. T. Ovseenko, and M. E. Vol'pin, *Russ. Chem. Bull. Engl. Transl.* **39**, 290 (1990).

122. L. A. Paquette and D. W. Balogh, *J. Am. Chem. Soc.* **104**, 774 (1982).

123. L. Ernst, H. Hopf, and R. Savinsky, *Liebigs Ann. Recueil*, 1915 (1997).

124. J. Scherzer, *Catal Rev.—Sci. Eng.* **31**, 215 (1989); *Octane-Enhancing Zeolitic FCC Catalysts*, Marcel Dekker, New York, 1990.

125. B. W. Wojciechowski, *Catal. Rev.—Sci. Eng.* **40**, 209 (1998).

126. Y. V. Kissin, *Catal. Rev.—Sci. Eng.* **43**, 85 (2001).

127. H. Hogeveen, C. J. Gaasbeek, and A. F. Bickel, *Recl. Trav. Chim. Pays-Bas* **88**, 703 (1969).

128. G. A. Olah, *Carbocations and Electrophilic Reactions*, Verlag Chemie, Wiley, Weinheim, New York, 1974.

129. D. M. Brouwer and H. Hogeveen, *Prog. Phys. Org. Chem.* **9**, 179 (1972).

130. G. A. Olah, M. R. Bruce, E. H. Edelson, and A. Husain, *Fuel* **63**, 1130 (1984).
131. G. A. Olah and A. Husain, *Fuel* **63**, 1427 (1984).
132. G. A. Olah, M. R. Bruce, E. H. Edelson, and A. Husain, *Fuel* **63**, 1432 (1984).
133. O. P. Strausz, T. W. Mojelsky, J. D. Payzant, G. A. Olah, and G. K. S. Prakash, *Energy Fuels* **13**, 558 (1999).
134. K. Shimizu, H. Karamatsu, Y. Iwami, A. Inaba, A. Suganuma, and I. Saito, *Fuel Process. Technol.* **45**, 85 (1995).
135. K. Shimizu, H. Karamatsu, A. Inaba, A. Suganuma, and I. Saito, *Fuel* **74**, 853 (1995).
136. K. Shimizu, K. Miki, and I. Saitou, *Fuel* **76**, 23 (1997).
137. K. Shimizu, Y. Iwami, A. Suganuma, and I. Saito, *Fuel* **76**, 939 (1997).
138. K. Shimizu, I. Saito, H. Kawashima, and S. Sasaki, *Energy Fuels* **13**, 197 (1999).
139. K. Shimizu and H. Kawashima, *Energy Fuels* **13**, 1223 (1999).
140. K. Shimizu and I. Saito, *Energy Fuels* **12**, 115 (1998).
141. D. Fraenkel, V. R. Pradhan, J. W. Tierney, and I. Wender, *Fuel* **70**, 64 (1991).
142. G. A. Olah and J. A. Olah, *J. Am. Chem. Soc.* **93**, 1256 (1971).
143. G. A. Olah, J. R. DeMember, and J. Schlosberg, *J. Am. Chem. Soc.* **91**, 2112 (1969).
144. G. A. Olah, J. R. DeMember, J. Schlosberg, and Y. Halpern, *J. Am. Chem. Soc.* **94**, 156 (1972).
145. G. A. Olah, Y. K. Mo, and J. A. Olah, *J. Am. Chem. Soc.* **95**, 4939 (1973).
146. G. A. Olah, J. R. DeMember, and J. Shen, *J. Am. Chem. Soc.* **95**, 4952 (1973).
147. G. A. Olah and A. M. White, *J. Am. Chem. Soc.* **91**, 5801 (1969).
148. M. Siskin, *J. Am. Chem. Soc.* **98**, 5413 (1976).
149. J. Sommer, M. Muller, and K. Laali, *Nouv. J. Chim.* **6**, 3 (1982).
150. G. A. Olah, J. D. Feldberg, and K. Lammertsma, *J. Am. Chem. Soc.* **105**, 6529 (1983).
151. G. A. Olah, O. Farooq, V. V. Krishnamurthy, G. K. S. Prakash, and K. Laali, *J. Am. Chem. Soc.* **107**, 7541 (1985).
152. M. Boronat, P. Viruela, and A. Corma, *J. Phys. Chem. B.* **103**, 7809 (1999).
153. G. A. Olah, G. K. S. Prakash, B. Török, and M. Török, *Catal. Lett.* **40**, 137 (1996).
154. L. F. Albright, M. A. Spalding, C. G. Kopser, and R. E. Eckert, *Ind. Eng. Chem. Res.* **27**, 386 (1988).
155. A. Corma and A. Martínez, *Catal Rev. —Sci. Eng.* **35**, 483 (1993).
156. S. I. Hommeltoft, O. Ekelund, and J. Zavilla, *Ind. Eng. Chem. Res.* **36**, 3491 (1997).
157. G. A. Olah, P. Batamack, D. Deffieux, B. Török, Q. Wang, Á. Molnár, and G.K. S. Prakash, *Appl. Catal., A* **146**, 107 (1996).
158. G. A. Olah, E. Marinez, B. Török, and G. K. Prakash, S. *Catal. Lett.* **61**, 105 (1999).
159. G. A. Olah, U.S. Patent, 5,073,674 (1991).
160. G. A. Olah, T. Mathew, A. Goeppert, B. Török, I. Bucsi, X. -Y. Li, Q. Wang, E. R. Marinez, P. Batamack, R. Aniszfeld, and G. K. S. Prakash, *J. Am. Chem. Soc.* **127**, 5964 (2005).
161. A. de Angelis, C. Flego, P. Ingallina, L. Montanari, M. G. Clerici, C. Carati, and C. Perego, *Catal. Today* **65**, 363 (2001).
162. T. Rørvik, I. M. Dahl, H. B. Mostad, and O. H. Ellestad, *Catal. Lett.* **33**, 127 (1995).
163. P. Botella, A. Corma, and J. M. López-Nieto, *J. Catal.* **185**, 371 (1999).

164. I. Pálinkó, B. Török, G. K. S. Prakash, and G. A. Olah, *Appl. Catal. A* **174**, 147 (1998); *J. Mol. Struct.* **482–483**, 29 (1999).

165. C. Lyon, B. Subramaniam, and C. Pereira, in *Catalyst Deactivation 2001*, *Studies in Surface Science and Catalysis*, Vol. 139, J. J. Spivey, G. W. Roberts, and B. H. Davis, eds., Elsevier, Amsterdam, 2001, p. 221.

166. C. L. Lyon, V. R. Sarsani, and B. Subramaniam, *Ind. Eng. Chem. Res.* **43**, 4809 (2004).

167. V. R. Sarsani, Y. Wang, and B. Subramaniam, *Ind. Eng. Chem. Res.* **44**, 6491 (2005).

168. D. M. Ginosar, D. N. Thompson, K. Coates, and D. J. Zalewski, *Ind. Eng. Chem. Res.* **41**, 2864 (2002).

169. P. Kumar, W. Vermeiren, J. -P. Dath, and W. F. Hoelderich, *Energy Fuels* **20**, 481 (2006).

170. G. A. Olah, Y. Halpern, J. Shen, and Y. K. Mo, *J. Am. Chem. Soc.* **95**, 4960 (1973).

171. D. T. Roberts, Jr. and L. E. Calihan, *J. Macromol. Sci. Chem.* **A7**, 1629 (1973).

172. D. T. Roberts, Jr. and L. E. Calihan, *J. Macromol. Sci. Chem.* **A7**, 1641 (1973).

173. M. Herlem and A. Jobert-Perol, *J. Electroanal. Chem.* **315**, 87 (1991).

174. I. S. Akhrem, A. V. Orlinkov, E. I. Myshov, and M. E. Vol'pin, *Tetrahedron Lett.* **22**, 3891 (1981).

175. I. S. Akhrem, I. M. Churilova, and S. V. Vitt, *Russ. Chem. Bull., Int. Ed.* **50**, 81 (2001).

176. G. A. Olah, P. S. Iyer, and G. K. S. Prakash, *Synthesis*, 513 (1986).

177. G. A. Olah and Á. Molnár, *Hydrocarbon Chemistry*, 2nd ed., Wiley-Interscience, Hoboken, NJ, 2003, Chapter 5, pp. 257, 263 and references cited therein.

178. K. Yu. Koltunov, S. Walspurger, and J. Sommer, *Eur. J. Org. Chem.* 4039 (2004).

179. K. Yu. Koltunov, S. Walspurger, and J. Sommer, *Tetrahedron Lett.* **45**, 3547 (2004).

180. Y. Zhang, A. McElrea, G. V. Sanchez, Jr. D. Do, A. Gomez, S. L. Aguirre, R. Rendy, and D. A. Klumpp, *J. Org. Chem.* **68**, 5119 (2003).

181. T. Ohwada, N. Yamagata, and K. Shudo, *J. Am. Chem. Soc.* **113**, 1364 (1991).

182. K. Okabe, T. Ohwada, T. Ohta, and K. Shudo, *J. Org. Chem.* **54**, 733 (1989).

183. T. Ohwada, K. Okabe, T. Ohta, and K. Shudo, *Tetrahedron* **46**, 7539 (1990).

184. Y. Zhang, S. L. Aguirre, and D. A. Klumpp, *Tetrahedron Lett.* **43**, 6837 (2002).

185. Y. Zhang and D. A. Klumpp, *Tetrahedron Lett.* **43**, 6841 (2002).

186. D. A. Klumpp, P. S. Beauchamp, G. V. Sanchez, Jr. S. Aguirre, and S. de Leon, *Tetrahedron Lett.* **42**, 5821 (2001).

187. G. A. Olah, J. Kaspi, and J. Bukala, *J. Org. Chem.* **42**, 4187 (1977).

188. J. M. Lalancette, L. Roy, and J. Lafontaine, *Can. J. Chem.* **54**, 2505 (1976).

189. W. M. Kutz, J. E. Nickels, J. J. McGovern, and B. B. Corson, *J. Org. Chem.* **16**, 699 (1951).

190. G. C. Bailey and J. A. Reid, U.S. Patent 2,519,099.

191. M. A. Harmer, W. E. Farneth, and Q. Sun, *J. Am. Chem. Soc.* **118**, 7708 (1996).

192. M. A. Harmer, Q. Sun, M. J. Michalczyk, and Z. Yang, *Chem. Commun.* 1803 (1997).

193. M. A. Harmer and Q. Sun, *Adv. Mater.* **10**, 1255 (1998).

194. M. A. Harmer, Q. Sun, A. J. Vega, W. E. Farneth, A. Heidekum, and W. F. Hoelderich, *Green Chem.* **2**, 7 (2000).

195. H. Wang, B. Q. Xu, M. H. Han, C. Xu, and E. Z. Min, *Chinese Chem. Lett.* **13**, 1121 (2002).

196. M. A. Harmer, C. Junk, V. Rostovtsev, L. G. Carcani, J. Vickery, and Z. Schnepp, *Green Chem.* **9**, 30 (2007).
197. J. -M. Coustard, *Tetrahedron* **52**, 9509 (1996).
198. B. C. Gagea, A. N. Parvulescu, V. I. Parvulescu, A. Auroux, P. Grange, and G. Poncelet, *Catal. Lett.* **91**, 141 (2003).
199. V. D. Sarca and K. K. Laali, *Green Chem.* **8**, 615 (2006).
200. F. Mühlthau, D. Stadler, A. Goeppert, G. A. Olah, G. K. S. Prakash, and T. Bach, *J. Am. Chem. Soc.* **128**, 9668 (2006).
201. J. Le Bras and F. Muzart, *Tetrahedron* **63**, 7942 (2007).
202. S. Das, R. Fröhlich, and A. Pramanik, *J. Chem. Res. (S)*, 5 (2007).
203. H. Kotsuki, T. Ohishi, M. Inoue, and T. Kojima, *Synthesis*, 603 (1999).
204. Á. Molnár, I. Ledneczky, I. Bucsi, and M. Bartók, *Catal. Lett.* **89**, 1 (2003).
205. P. J. Linares-Palomino, G. K. S. Prakash, and G. A. Olah, *Helv. Chim. Acta* **88**, 1221 (2005).
206. G. K. S. Prakash, P. J. Linares-Palomino, K. Glinton, S. Chacko, G. Rasul, T. Mathew, and G. A. Olah, *Synlett* 1158 (2007).
207. J. Kaspi, D. D. Montgomery, and G. A. Olah, *J. Org. Chem.* **43**, 3147 (1978).
208. J. Kaspi and G. A. Olah, *J. Org. Chem.* **43**, 3142 (1978).
209. S. V. Kannan and C. N. Pilali, *Ind. J. Chem.* **8**, 1144 (1970).
210. Y. Ogata, K. Sakanishi, and K. Hosai, *Kogyo Kagaku Zasshi* **72**, 1102 (1969); *Chem. Abstr.* **71**, 80842 (1969).
211. H. L. Schlichting, A. D. Barbopoluos, and W. H. Pahl, U.S. Patent, 3,426,358 (1969).
212. J. A. Sharp and R. E. Dean, British Patent, 1,125,087 (1968); *Chem. Abstr.* **70**, 37452 (1969).
213. T. Kotanigawa, *Bull. Chem. Soc. Jpn.* **47**, 950, 2466 (1975).
214. T. Wada, *Japan Kokai*, 75, 76, 033 (1975); *Chem Abstr.* **83**, 178557 (1975).
215. T. Kotanigawa, H. Yamamoto, H. Shimokawa, and Y. Yoshiba, *Bull. Chem. Soc. Jpn.* **44**, 1961 (1971).
216. Y. Iohikawa, Y. Yamanaka, and O. Kobayashi, *Japan Kokai*, 75 99 129 (1973); *Chem. Abstr.* **80**, 95477 (1974); *Japan Kokai*, 73 96 529 (1973); *Chem. Abstr.* **80**, 95481 (1974); *Japan Kokai*, **73 99 128** (1973); *Chem. Abstr.* **80**, 95486 (1974).
217. W. Funakashi, T. Urasaki, I. Oda, and T. Shima, *Japan Kokai*, 73 97 825 (1973); *Chem. Abstr.* **80**, 59675 (1974); *Japan Kokai*, 74 07 235 (1974); *Chem. Abstr.* **80**, 120534 (1974); *Japan Kokai*, 74 13 128 (1974); *Chem. Abstr.* **80**, 120532 (1974); *Japan Kokai*, 74 14 432 (1974); *Chem. Abstr.* **80**, 120527 (1974); *Japan Kokai*, 74 14 834 (1974); *Chem. Abstr.* **81**, 3586 (1974).
218. G. A. Olah and D. Meidar, unpublished results.
219. T. Yamato, C. Hideshima, G. K. S. Prakash, and G. A. Olah, *J. Org. Chem.* **56**, 2089 (1991).
220. Q. Sun, M. A. Harmer, and W. E. Farneth, *Ind. Eng. Chem. Res.* **36**, 5541 (1997).
221. P. Beltrame and G. Zuretti, *Appl. Catal, A* **283**, 33 (2005).
222. B. Rác, G. Mulas, A. Csongrádi, K. Lóki, and Á. Molnár, *Appl. Catal., A* **282**, 255 (2005).
223. P. Beltrame and G. Zuretti, *Appl. Catal. A* **248**, 75 (2003); **268**, 169, (2004).

224. R. Amandi, J. R. Hyde, S. K. Ross, T. J. Lotz, and M. Poliakoff, *Green Chem.* **7**, 288 (2005).

225. J. R. Hyde, R. A. Bourne, I. Noda, P. Stephenson, and M. Poliakoff, *Anal. Chem.* **76**, 6197 (2004).

226. R. Amandi, P. Licence, S. K. Ross, O. Aaltonen, and M. Poliakoff, *Org. Proc. Res. Dev.* **9**, 451 (2005).

227. M. Fujiwara, K. Shiokawa, and Y. Zhu, *J. Mol. Catal. A: Chem.* **264**, 153 (2007).

228. J. -M. Lalancette, M. -J. Faurrier-Breault, and R. Thiffault, *Can. J. Chem.* **52**, 589 (1974).

229. S. Debarge, B. Violeau, N. Bendaoud, M. -P. Jouannetaud, and J. -C. Jacquesy, *Tetrahedron Lett.* **44**, 1747 (2003).

230. S. Debarge, K. Kassou, H. Carreyre, B. Violeau, M. -P. Jouannetaud, and J. -C. Jacquesy, *Tetrahedron Lett.* **45**, 21 (2004).

231. G. A. Olah, O. Farooq, S. M. F. Farnia, and J. A. Olah, *J. Am. Chem. Soc.* **110**, 2560 (1988).

232. G. A. Olah, O. Farooq, S. M. F. Farnia, and A. Wu, *J. Org. Chem.* **55**, 1516 (1990).

233. G. A. Olah, C. S. Lee, G. K. S. Prakash, R. M. Moriarty, and M. S. C. Rao, *J. Am. Chem. Soc.* **115**, 10728 (1993).

234. K. K. Laali, V. D. Sarca, T. Okazaki, A. Brock, and P. Der, *Org. Biomol. Chem.* **3**, 1034 (2005).

235. G. A. Olah and D. Meidar, *Nouv. J. Chim.* **3**, 269 (1979).

236. G. E. Langlois, E.U.S. Patent, 2,713,600 (1955).

237. K. G. Yang, R. M. Hua, H. Wang, and B. Q. Xu, *Chinese Chem. Lett.* **16**, 527 (2005).

238. G. A. Olah, B. Török, T. Shamma, M. Török, and G. K. S. Prakash, *Catal. Lett.* **42**, 5 (1996).

239. B. Török, I. Kiricsi, Á. Molnár, and G.A. Olah, *J. Catal.* **193**, 132 (2000).

240. T. Beregszászi, B. Török, Á. Molnár, G. A. Olah, and G. K. S. Prakash, *Catal. Lett.* **48**, 83 (1997).

241. J. E. Hofmann and A. Schriesheim, *Friedel—Crafts Related Reactions*, Vol. 2, G.A. Olah, ed., Wiley-Interscience, New York, 1964, Chapter 19, p. 597.

242. G. A. Olah, G. Rasul, C. York, and G. K. S. Prakash, *J. Am. Chem. Soc.* **117**, 11211 (1995).

243. S. Saito, T. Ohwada, and K. Shudo, *J. Am. Chem. Soc.* **117**, 11081 (1995).

244. S. Saito, T. Ohwada, and K. Shudo, *J. Org. Chem.* **61**, 8089 (1996).

245. S. Fukuzawa, T. Tsuchimoto, and T. Hiyama, *J. Org. Chem.* **62**, 151 (1997).

246. D. A. Klumpp, G. V. Sanchez, Jr., S. L. Aguirre, Y. Zhang, and S. de Leon, *J. Org. Chem.* **67**, 5028 (2002).

247. D. A. Klumpp, P. J. Kindelin, and A. Li, *Tetrahedron Lett.* **46**, 2931 (2005).

248. D. A. Klumpp and S. Lau, *J. Org. Chem.* **64**, 7309 (1999).

249. D. A. Klumpp, Y. Zhang, P. J. Kindelin, and S. Lau, *Tetrahedron* **62**, 5915 (2006).

250. K. Yu. Koltunov, S. Walspurger, and J. Sommer, *Catal. Lett.* **98**, 89 (2004).

251. D. A. Klumpp, A. Jones, S. Lau, S. de Leon, and M. Garza, *Synthesis*, 1117, (2000).

252. D. A. Klumpp, M. Garza, A. Jones, and S. Mendoza, *J. Org. Chem.* **64**, 6702 (1999).

253. D. A. Klumpp, R. Rendy, Y. Zhang, A. Gomez, and A. McElrea, *Org. Lett.* **6**, 1789 (2004).

254. D. A. Klumpp, M. Garza, G. V. Sanchez, Jr. S. Lau, and S. de Leon, *J. Org. Chem.* **65**, 8997 (2000).
255. A. L. Lira, M. G. Zolotukhin, L. Fomina, and S. Fomine, *J. Phys. Chem. A* **111**, 13606 (2007).
256. G. K. S. Prakash, P. Yan, B. Török, and G. A. Olah, *Synlett*, 527 (2003).
257. T. Yamazaki, S. Saito, T. Ohwada, and K. Shudo, *Tetrahedron Lett.* **36**, 5749 (1995).
258. D. A. Klumpp, K. Y. Yeung, G. K. S. Prakash, and G. A. Olah, *J. Org. Chem.* **63**, 4481 (1998).
259. D. A. Klumpp, K. Y. Yeung, G. K. S. Prakash, and G. A. Olah, *Synlett*, 918, (1998).
260. D. A. Klumpp, S. Fredrick, S. Lau, K. K. Jin, R. Bau, G. K. S. Prakash, and G. A. Olah, *J. Org. Chem.* **64**, 5152 (1999).
261. D. A. Klumpp, S. Lau, M. Garza, B. Schick, and K. Kantardjieff, *J. Org. Chem.* **64**, 7635 (1999).
262. D. Plażuk and J. Zakrzewski, *J. Org. Chem.* **67**, 8672 (2002).
263. D. Plażuk, A. Rybarczyk-Pirek, and J. Zakrzewski, *J. Organomet. Chem.* **689**, 1165 (2004).
264. Z. Hou and T. Okuhara, *Catal Appl. A* **216**, 147 (2001).
265. G. A. Olah, D. Meidar, R. Malhotra, J. A. Olah, and S. C. Narang, *J. Catal.* **61**, 96 (1980).
266. G. A. Olah, in *Friedel–Crafts and Related Reactions*, Vol. 1, G. A. Olah, ed., Wiley-Interscience, New York, 1963, Chapter 2, pp. 49, 110.
267. P. H. Gore, in *Friedel–Crafts and Related Reactions*, Vol. 3, G. A. Olah, ed., Wiley-Interscience, New York, 1964, Chapter 31, pp. 34, 35.
268. G. A. Olah and J. Kaspi, *Nouv. J. Chim* **2**, 581 (1978).
269. P. Beltrame, P. L. Beltrame, P. Carniti, and M. Magnoni, *Gazz. Chim. Ital.* **108**, 651 (1978).
270. P. Beltrame, P. L. Beltrame, P. Carniti, and G. Nespoli, *Ind. Eng. Chem., Prod. Res. Dev.* **19**, 205 (1980).
271. G. A. Olah and J. Kaspi, *Nouv. J. Chim.* **2**, 585 (1978).
272. G. A. Olah, M. W. Meyer, and N. A. Overchuk, *J. Org. Chem.* **29**, 2313 (1964).
273. D. A. McCaulay, in *Friedel–Crafts and Related Reactions*, Vol. 2, G.A. Olah, ed., Wiley-Interscience, New York, 1964, Chapter 24, p. 1049.
274. H. Matsumoto, J. Take, and Y. Yoneda, *J. Catal.* **11**, 211 (1968); **19**, 113 (1970).
275. G. A. Olah, G. K. S. Prakash, P. S. Iyer, M. Tashiro, and T. Yamato, *J. Org. Chem.* **52**, 1881 (1987).
276. T. Yamato, C. Hideshima, G. K. S. Prakash, and G. A. Olah, *J. Org. Chem.* **56**, 3192 (1991).
277. A. Miyazawa, T. Yamato, and M. Tashiro, *J. Org. Chem.* **56**, 1334 (1991).
278. A. Miyazawa, A. Tsuge, T. Yamato, and M. Tashiro, *J. Org. Chem.* **56**, 4312 (1991).
279. T. Yamato, A. Miyazawa, and M. Tashiro, *J. Chem. Soc., Perkin Trans. 1* 3127 (1993).
280. T. Yamato, M. Komine, and K. Matsuo, *J. Chem. Res. (S).* 82 (1997).
281. T. Yamato, C. Hideshima, A. Miyazawa, M. Tashiro, G. K. S. Prakash, and G. A. Olah, *Catal. Lett.* **6**, 345 (1990).
282. S. G. Rha and S. -K. Chang, *J. Org. Chem.* **63**, 2357 (1998).

283. M. C. Al-Kinany and S. H. Al-Khowaiter, in *Reaction Kinetics and the Development of Catalytic Processes, Studies in Surface Science and Catalysis*, Vol. 122, G. F. Froment and K. C. Waugh, eds., Elsevier, Amsterdam, 1999, p. 375.

284. S. M. Al-Zahrani, M. C. Al-Kinany, K. I. Al-Humaizi, and S. H. Al-Khowaiter, *Chem. Eng. Proces.* **41**, 321 (2002); *Int. J. Chem. Kinet.* **35**, 555 (2003).

285. M. C. Al-Kinany B. Y. Jibril, S. H. Al-Khowaiter, M. A. Al-Dosari, H. A. Al-Megren, S. M. Al-Zahrani, and K. I. Al-Humaizi, *Chem. Eng. Proces.* **44**, 841 (2005).

286. G. A. Olah, unpublished results.

287. G. A. Olah and J. R. DeMember, *J. Am. Chem. Soc.* **91**, 2113 (1969); **92**, 718, 2562 (1970).

288. G. A. Olah, J. R. DeMember, Y. K. Mo, J. J. Svoboda, P. Schilling, and J. A. Olah, *J. Am. Chem. Soc.* **96**, 884 (1974).

289. A. Orlinkov, I. Akhrem, S. Vitt, and M. Vol'pin, *Tetrahedron Lett.* **37**, 3363 (1996).

290. V. M. Karpov, T. V. Mezhenkova, V. E. Platonov, and V. R. Sinyakov, *J. Fluorine Chem.* **107**, 53 (2001).

291. I. B. Repinskaya, K. Yu. Koltunov, M. M. Shakirov, L. N. Shchegoleva, and V. A. Koptyug, *Russ. J. Org. Chem. (Engl. Transl.)* **29**, 803 (1993).

292. J. C. Jacquesy, J. M. Coustard, and M. P. Jouannetaud, *Topics Catal.* **6**, 1 (1998) and references cited therein.

293. K. Yu. Koltunov, G. K. S. Prakash, G. Rasul, and G. A. Olah, *Tetrahedron* **58**, 5423 (2002).

294. K. Yu. Koltunov and I. B. Repinskaya, *Russ. J. Org. Chem. (Engl. Transl.)* **38**, 437 (2002).

295. K. Yu. Koltunov, G. K. S. Prakash, G. Rasul, and G. A. Olah, *J. Org. Chem.* **67**, 4330 (2002).

296. K. Yu. Koltunov, G. K. S. Prakash, G. Rasul, and G. A. Olah, *J. Org. Chem.* **67**, 8943 (2002).

297. K. Yu. Koltunov, G. K. S. Prakash, G. Rasul, and G. A. Olah, *J. Org. Chem.* **72**, 7394 (2007).

298. K. Yu. Koltunov, G. K. S. Prakash, G. Rasul, and G. A. Olah, *Eur. J. Org. Chem.* 4861 (2006).

299. D. A. Klumpp, R. Rendy, and A. McElrea, *Tetrahedron Lett.* **45**, 7959 (2004).

300. I. B. Repinskaya, D. D. Barkhutova, Z. S. Makarova, A. V. Alekseeva, and V. A. Koptyug, *Zh. Org. Khim.* **21**, 836 (1985).

301. J. C. Jacquesy and M. P. Jouannetaud, *Bull. Soc. Chim. Fr. II*, **267**, 295 (1980).

302. G. A. Olah and Y. K. Mo, *J. Am. Chem. Soc.* **94**, 5341 (1972).

303. A. P. Rudenko, P. Yu. Savechenkov, and A. V. Vasil'ev, *Russ. J. Org. Chem. (Engl. Transl.)* **40**, 1377 (2004).

304. P. Yu. Savechenkov, A. P. Rudenko, A. V. Vasil'ev, and G. K. Fukin, *Russ. J. Org. Chem. (Engl. Transl.)* **41**, 1316 (2005).

305. S. A. Aristov, A. V. Vasil'ev, G. K. Fukin, and A. P. Rudenko, *Russ. J. Org. Chem. (Engl. Transl.)* **43**, 691 (2007).

306. S. Walspurger, A. V. Vasil'ev, J. Sommer, P. Pale P. Yu. Savechenkov, and A. P. Rudenko, *Russ. J. Org. Chem. (Engl. Transl.)* **41**, 1485 (2005).

307. P. Yu. Savechenkov, A. P. Rudenko, and A. V. Vasil'ev, *Russ. J. Org. Chem. (Engl. Transl.)* **40**, 1065 (2004).
308. D. A. Klumpp, R. Rendy, Y. Zhang, A. McElrea, A. Gomez, and H. Dang, *J. Org. Chem.* **69**, 8108 (2004).
309. L. R. C. Barclay, in *Friedel–Crafts and Related Reactions*, Vol. 2, G. A. Olah, ed., Wiley-Interscience, New York, 1964, Chapter 22, p. 785.
310. Á. Molnár, B. Török, I. Bucsi, and A. Földvári, *Topics Catal.* **6**, 9 (1998).
311. I. Ledneczki, P. Forgo, and Á. Molnár, *Catal. Lett.* **119**, 296 (2007).
312. M. Abid, L. Teixeira, and B. Török, *Tetrahedron Lett.* **48**, 4047 (2007).
313. Reference 309, p. 859.
314. S. Saito, Y. Sato, T. Ohwada, and K. Shudo, *J. Am. Chem. Soc.* **116**, 2312 (1994).
315. D. P. Schumacher, B. L. Murphy, J. E. Clark, P. Tahbaz, and T. A. Mann, *J. Org. Chem.* **54**, 2242 (1989).
316. T. Suzuki, T. Ohwada, and K. Shudo, *J. Am. Chem. Soc.* **119**, 6774 (1997).
317. T. Ohwada, T. Suzuki, and K. Shudo, *J. Am. Chem. Soc.* **120**, 4629 (1998).
318. N. Yoshida and T. Ohwada, *Synthesis*, 1487 (2001).
319. C. J. Barrow, S. T. Bright, J. M. Coxon, and P. J. Steel, *J. Org. Chem.* **54**, 2542 (1989).
320. S. T. Bright, J. M. Coxon, and P. J. Steel, *J. Org. Chem.* **55**, 1338 (1990).
321. A. V. Vasilyev, S. Walspurger, P. Pale, and J. Sommer, *Tetrahedron Lett.* **45**, 3379 (2004).
322. A. V. Vasil'ev, S. Walspurger, P. Pale, J. Sommer, M. Haouas, and A. P. Rudenko, *Russ. J. Org. Chem.* **40**, 1769 (2004).
323. A. V. Vasilyev, S. Walspurger, M. Haouas, J. Sommer, P. Pale, and A. P. Rudenko, *Org. Biomol. Chem.* **2**, 3483 (2004).
324. S. Walspurger, A. V. Vasilyev, J. Sommer, and P. Pale, *Tetrahedron* **61**, 3559 (2005).
325. V. A. Bushmelev, A. M. Ganaev, S. A. Osadchii, M. M. Shakirov, and V. G. Shubin, *Russ. J. Org. Chem. Engl. Transl.* **39**, 1301 (2003).
326. J. Barluenga, M. Tomás, A. Suárez-Sobrino, and E. Rubio, *J. Org. Chem.* **58**, 700 (1993).
327. J. -M. Coustard, Y. Soro, S. Siaka, F. Bamba, and A. Cousson, *Tetrahedron* **62**, 3320 (2006).
328. J. -M. Coustard, *Eur. J. Org.* Chem. 1525 (2001).
329. A. Yokoyama, T. Ohwada, and K. Shudo, *J. Org. Chem.* **64**, 611 (1999).
330. S. Nakamura, M. Tanaka, T. Taniguchi, M. Uchiyama, and T. Ohwada, *Org. Lett.* **5**, 2087 (2003).
331. L. F. Tietze, R. Schimpf, and J. Wichmann, *Chem. Ber.* **125**, 2571 (1992).
332. E. Reimann, W. Erdle, E. Hargasser, and H. Lotter, *Monatshefte Chem.* **133**, 1017 (2002).
333. B. E. Maryanoff, H. -C. Zhang, J. H. Cohen, I. J. Turchi, and C. A. Maryanoff, *Chem. Rev.* **104**, 1431 (2004).
334. C. M. Marson, J. H. Pink, D. Hall, M. B. Hursthouse, A. Malik, and C. Smith, *J. Org. Chem.* **68**, 792 (2003).
335. T. Okano, T. Sakaida, and S. Eguchi, *Heterocycles* **44**, 227 (1997).
336. S. E. de Laszlo, B. L. Bush, J. J. Doyle, W. J. Greenlee, D. G. Hangauer, T. A. Halgren, R. J. Lynch, T. W. Schorn, and P. K. S. Siegl, *J. Med. Chem.* **35**, 833 (1992).

337. Y. Zhang, P. J. Kindelin, D. J. DeSchepper, C. Zheng, and D. A. Klumpp, *Synthesis*, 1775 (2006).

338. Y. Zhang, D. J. DeSchepper, T. M. Gilbert, K. K. S. Sai, and D. A. Klumpp, *Chem. Commun.* 4032 (2007).

339. D. A. Klumpp, Y. Zhang, M. J. O'Connor, P. M. Esteves, and L. S. de Almeida, *Org. Lett.* **9**, 3085 (2007).

340. M. M. Abelman, J. K. Curtis, and D. R. James, *Tetrahedron Lett.* **44**, 6527 (2003).

341. F. -J. Zhang, C. Cortez, and R. C. Harvey, *J. Org. Chem.* **65**, 3952 (2000).

342. S. K. Taylor, M. G. Dickinson, S. A. May, D. A. Pickering, and P. C. Sadek, *Synthesis* 1133 (1998).

343. P. C. Sadek, S. K. Taylor, and M. G. Dick, *Anal. Lett.* **20**, 169 (1987).

344. M. P. Doyle, M. S. Shanklin, H. Q. Pho, and S. N. Mahapatro, *J. Org. Chem.* **53**, 1017 (1988).

345. A. G. Wee and B. Liu, *Heterocycles* **36**, 445 (1993); *Tetrahedron* **50**, 609 (1994).

346. G. A. Olah and Á. Molnár, *Hydrocarbon Chemistry*, 2nd ed., Wiley-Interscience, Hoboken, NJ, 2003, Chapter 8, p. 407, and references cited therein.

347. F. Effenberger and G. Epple, *Angew. Chem. Int. Ed.* **11**, 300 (1972).

348. F. Effenberger, E. Sohn, and G. Epple, *Chem. Ber.* **116**, 1195 (1983).

349. O. Akiko, M. Katsuya, O. Hideaki, and Y. Noriyuki, *Synth. Commun.* **37**, 2701 (2007).

350. R. D. Howells and J. D. McCown, *Chem. Rev.* **77**, 69 (1977).

351. T. R. Forbus, Jr. and J. C. Martin, *J. Org. Chem.* **44**, 313 (1979).

352. J. P. Hwang, G. K. S. Prakash, and G. A. Olah, *Tetrahedron* **56**, 7199 (2000).

353. K. W. Anderson and J. J. Tepe, *Tetrahedron* **58**, 8475 (2002).

354. R. M. G. Roberts and A. S. Wells, *Inorg. Chim. Acta* **112**, 167 (1986).

355. D. Plażuk and J. Zakrzewski, *Synth. Commun.* **34**, 99 (2004).

356. M. Chidambaram, C. Venkatesan, P. R. Rajamohanan, and A. P. Singh, *Appl. Catal., A* **244**, 27 (2003).

357. M. Chidambaram, D. Curulla-Ferre, A. P. Singh, and B. G. Anderson, *J. Catal.* **220**, 442 (2003).

358. S. M. Landge, M. Chidambaram, and A. P. Singh, *J. Mol. Catal. A: Chem.* **213**, 257 (2004).

359. A. N. Pârvulescu, B. C. Gagea, G. Poncelet, and V. I. Pârvulescu, *Appl. Catal., A* **301**, 133 (2006).

360. A. N. Pârvulescu, B. C. Gagea, V. I. Pârvulescu, D. De Vos, and P. A. Jacobs, *Appl. Catal., A* **306**, 159 (2006).

361. I. Pálinkó, A. Burrichter, G. Rasul, B. Török, G. K. S. Prakash, and G. A. Olah, *J. Chem. Soc., Perkin Trans. 2*, 379 (1998).

362. R. Rendy, Y. Zhang, A. McElrea, A. Gomez, and D. A. Klumpp, *J. Org. Chem.* **69**, 2340 (2004).

363. H. M. Colquhoun, D. F. Lewis, and D. J. Williams, *Org. Lett.* **3**, 2337 (2001).

364. W. Ruske, in *Friedel–Crafts and Related Reactions*, Vol. 3, G. A. Olah, ed., Interscience, New York, 1964, Chapter 32, p. 383.

365. S. -K. Kwon and Y. -N. Park, *Arch. Pharm. Res.* **23**, 329 (2000).

366. G. K. S. Prakash, P. Yan, B. Török, and G. A. Olah, *Catal. Lett.* **87**, 109 (2003).
367. G. A. Olah, R. Malhotra, S. C. Narang, and J. A. Olah, *Synthesis*, 672 (1978).
368. G. K. S. Prakash, T. Mathew, M. Mandal, M. Farnia, and G. A. Olah, *ARKIVOC*, Part 8, 103 (2004).
369. T. Yamato, C. Hideshima, G. K. S. Prakash, and G. A. Olah, *J. Org. Chem.* **56**, 3955 (1991).
370. G. A. Olah, T. Mathew, M. Farnia, and G. K. S. Prakash, *Synlett*, 1067 (1999).
371. A. Heidekum, M. A. Harmer, and W. F. Hoelderich, *J. Catal.* **188**, 230 (1999).
372. D. J. Macquarrie, S. J. Tavener, and M. A. Harmer, *Chem. Commun.* 2363 (2005).
373. M. Kawamura, D. -M. Cui, and S. Shimada, *Tetrahedron* **62**, 9201 (2006).
374. M. Alvaro, A. Corma, D. Das, V. Fornés, and H. García, *J. Catal.* **231**, 48 (2005).
375. V. R. Sarsani, C. J. Lyon, K. W. Hutchenson, M. A. Harmer, and B. Subramaniam, *J. Catal.* **245**, 184 (2007).
376. A. E. W. Beers, I. Hoek, T. A. Nijhuis, R. S. Downing, F. Kapteijn, and J. A. Moulijn, *Topics Catal.* **13**, 275 (2000).
377. A. E. W. Beers, T. A. Nijhuis, F. Kapteijn, and J. A. Moulijn, *Microporous Mesoporous Mater.* **48**, 279 (2001).
378. G. Baddeley and A. G. Pendleton, *J. Chem. Soc.* 807 (1952).
379. W. M. Schubert and H. K. Latourette, *J. Am. Chem. Soc.* **74**, 1829 (1952).
380. I. Agranat, Y. -S. Shih, and Y. Bentor, *J. Am. Chem. Soc.* **96**, 1259 (1974).
381. I. Agranat, Y. Bentor, and Y. -S. Shih, *J. Am. Chem. Soc.* **99**, 7068 (1977).
382. A. D. Andreou, R. V. Bulbulian, and P. H. Gore, *J. Chem. Res.* 225 (1980).
383. A. D. Andreou, R. V. Bulbulian, P. H. Gore, D. F. C. Morris, and E. L. Short, *J. Chem. Soc., Perkin Trans. 2*, 830 (1981).
384. R. H. Schlosberg and R. V. Woodbury, *J. Org. Chem.* **37**, 2627 (1972).
385. T. Keumi, T. Morita, T. Shimada, N. Teshima, H. Kitajima, and G. K. S. Prakash, *J. Chem. Soc., Perkin Trans. 2*, 847 (1986).
386. V. D. Sarca and K. K. Laali, *Green Chem.* **6**, 245 (2004).
387. M. Koželj, and A. Petrič, *Synlett*, 1699 (2007).
388. G. A. Olah, K. Laali, and A. K. Mehrotra, *J. Org. Chem.* **48**, 3360 (1983).
389. T. Yamato, N. Sakaue, N. Shinoda, and K. Matsuo, *J. Chem. Soc., Perkin Trans. 1*, 1193 (1997).
390. I. S. Akhrem, I. M. Churilova, A. V. Orlinkov, L. V. Afanas'eva, S. V. Vitt, and P. V. Petrovskii, *Russ. Chem. Bull. (Engl. Transl.)* **47**, 918 (1998).
391. I. S. Akhrem, A. V. Orlinkov, S. V. Vitt, L. V. Afanas'eva, and M. E. Vol'pin, *Russ. Chem. Bull. (Engl. Transl.)* **42**, 1196 (1993).
392. A. H. Blatt, *Org. React.* **1**, 342 (1960).
393. A. Gerecs, in *Friedel–Crafts and Related Reactions*, G. A. Olah, ed., Wiley-Interscience, New York, 1964, Vol. 3, Chapter 33, p. 499.
394. G. A. Olah, M. Arvanaghi, and V. V. Krishnamurthy, *J. Org. Chem.* **48**, 3359 (1983).
395. A. Heidekum, M. A. Harmer, and W. F. Hoelderich, *J. Catal.* **176**, 260 (1998).
396. G. K. S. Prakash, C. Panja, T. Mathew, and G. A. Olah, *Catal. Lett.* **114**, 24 (2007).

397. G. A. Olah and J. A. Olah, in *Friedel–Crafts and Related Reactions*, Vol. 3, G. A. Olah, ed., Wiley-Interscience, New York, 1964, Chapter 39, 1257.
398. H. Hogeveen, *Adv. Phys. Org. Chem.* **10**, 29 (1973).
399. B. L. Booth and T. A. El-Fekky, *J. Chem. Soc., Perkin Trans. 1*, 2441 (1979).
400. Y. Takahashi, N. Tomita, N. Yoneda, and A. Suzuki, *Chem. Lett.* 997 (1975).
401. A. Bagno, J. Bukala, and G. A. Olah, *J. Org. Chem.* **55**, 4284 (1990).
402. N. Tsumori, Q. Xu, Y. Souma, and H. Mori, *J. Mol. Catal. A: Chem.* **179**, 271 (2002).
403. J.-P. Lange, U.S. Patent, 6,294,691 (2001).
404. G. A. Olah and J. Bukala, *J. Org. Chem.* **55**, 4293 (1990).
405. N. Yoneda, Y. Takahashi, Y. Sakai, and A. Susuki, *Chem. Lett.* 1151 (1978).
406. R. Paatz and G. Weisberger, *Chem. Ber.* **100**, 984 (1967).
407. N. Yoneda, T. Fukuhara, Y. Takahashi, and A. Suzuki, *Chem. Lett.* 17 (1983).
408. Y. Souma, N. Tsumori, H. Willner, Q. Xu, H. Mori, and Y. Morisaki, *J. Mol. Catal. A: Chem.* **189**, 67 (2002).
409. N. Yoneda, H. Sato, T. Fukuhara, Y. Takahashi, and A. Suzuki, *Chem. Lett.* 19 (1983).
410. J. Sommer, in *Stable Carbocation Chemistry*, G.K.S. Prakash and P.v.R. Schleyer, eds., Wiley, New York, 1997, Chapter 15, p. 497.
411. S. Delavarenne, M. Simon, M. Fauconet, and J. Sommer, European Patents 270,398 and 272,945 (1987).
412. S. Delavarenne, M. Simon, M. Fauconet, and J. Sommer, *J. Chem. Soc., Chem. Commun.* 1049 (1989).
413. S. Delavarenne, M. Simon, M. Fauconet, and J. Sommer, *J. Am. Chem. Soc.* **111**, 383 (1989).
414. J. Bukala, J. -C. Culmann, and J. Sommer, *J. Chem. Soc., Chem. Commun.* 481 (1992).
415. J. -C. Culmann, M. Simon, and J. Sommer, *J. Chem. Soc., Chem. Commun.* 1098 (1990).
416. J. Sommer, M. Hachoumy, J. -C. Culmann, and J. Bukala, *New J. Chem.* **21**, 939 (1997).
417. I. S. Akhrem, *Topics Catal.* **6**, 27 (1998); *Russ. Chem. Bull., Int. Ed.* **52**, 2606 (2003).
418. I. Akhrem, A. Orlinkov, L. Afanas'eva, P. Petrovskii, and S. Vitt, *Tetrahedron Lett.* **40**, 5897 (1999).
419. I. Akhrem, L. Afanas'eva, P. Petrovskii, S. Vitt, and A. Orlinkov, *Tetrahedron Lett.* **41**, 9903 (2000); *Russ. Chem. Bull. Int. Ed.* **50**, 2394 (2001).
420. G. A. Olah, A. M. White, and D. M. O'Brien, *Chem. Rev.* **70**, 591 (1970).
421. J. M. Coustard and J. C. Jacquesy, *J. Chem. Res. (S)* 280 (1977).
422. B. L. Booth, T. A. El-Fekky, and G. F. M. Noori, *J. Chem. Soc., Perkin Trans. 1*, 181 (1980).
423. S. Y. Lee, J. C. Kim, J. S. Lee, and Y. G. Kim, *Ind. Eng. Chem. Res.* **32**, 253 (1993).
424. K. -W. Liang, M. Chandrasekharam, C. -L. Li, and R. -S. Liu, *Organometallics* **17**, 2683 (1998).
425. G. A. Olah, B. Török, J. P. Joschek, I. Bucsi, P. M. Esteves, G. Rasul, and G. K. S. Prakash, *J. Am. Chem. Soc.* **124**, 11379 (2002).
426. L. Gattermann and J. A. Koch, *Ber. Dtsch. Chem. Ges.* **30**, 1622 (1897).
427. N. N. Crounse, *Org. React.* **5**, 290 (1949).

428. G. A. Olah and S. J. Kuhn, in *Friedel–Crafts and Related Reactions*, Vol. 3, G.A. Olah, ed., Interscience, New York, 1964, Chapter 38, p. 1153.
429. W. F. Gresham and G. E. Tabet, U.S. Patent, 2,485,237 (1946).
430. J. M. Oelderick and A. Kwantes, British Patent, 1,128,966 (1968); *Chem. Abstr.* **70**, 37472 (1969).
431. S. Fujiyama, M. Takagawa, and S. Kajiyama, Ger. Offen. **2**, 422 197 (1974); *Chem. Abstr.* **83**, 9506 (1975).
432. G. A. Olah, F. Pelizza, S. Kobayashi, and J. A. Olah, *J. Am. Chem. Soc.* **98**, 296 (1976).
433. A. Rieche, H. Gross, and E. Höft, *Chem. Ber.* **93**, 88 (1960).
434. G. A. Olah and S. J. Kuhn, *J. Am. Chem. Soc.* **82**, 2380 (1960).
435. G. A. Olah, K. Laali, and O. Farooq, *J. Org. Chem.* **50**, 1483 (1985).
436. M. Tanaka, J. Iyoda, and Y. Souma, *J. Org. Chem.* **57**, 2677 (1992).
437. M. Tanaka, M. Fujiwara, and H. Ando, *J. Org. Chem.* **60**, 3846 (1995).
438. M. Tanaka and Y. Souma, *J. Chem. Soc. Commun Chem.* 1551 (1991).
439. M. Tanaka, M. Fujiwara, H. Ando, and Y. Souma, *J. Org. Chem.* **58**, 3213 (1993).
440. M. Tanaka, M. Fujiwara, H. Ando, and Y. Souma, *Chem. Commun.* 159 (1996).
441. M. Tanaka, M. Fujiwara, and H. Ando, *J. Org. Chem.* **60**, 2106 (1995).
442. M. Tanaka, M. Fujiwara, Q. Xu, Y. Souma, H. Ando, and K. K. Laali, *J. Am. Chem. Soc.* **119**, 5100 (1997).
443. M. Tanaka, M. Fujiwara, Q. Xu, H. Ando, and T. J. Raeker, *J. Org. Chem.* **63**, 4408 (1998).
444. D. S. Sood, S. C. Sherman, A. V. Iretskii, J. C. Kenvin, D. A. Schiraldi, and M. G. White, *J. Catal.* **199**, 149 (2001).
445. J. Willemse, B. C. B. Bezuidenhoudt, and C. W. Holzapfel, *Synthesis*, 2543 (2006).
446. W. Kantlehner, S. Leonhardt, G. Ziegler, O. Scherr, R. Kress, A. Goeppert, and J. Sommer, *Z. Naturforsch.* **62b**, 995 (2007).
447. O. Farooq, M. Marcelli, G. K. S. Prakash, and G. A. Olah, *J. Am. Chem. Soc.* **110**, 864 (1988).
448. I. Akhrem, I. Churilova, S. Bernadyuk, and M. Vol'pin, *Tetrahedron Lett.* **37**, 5775 (1996).
449. I. S. Akhrem, I. M. Churilova, S. V. Vitt, A. V. Orlinkov, and M. E. Vol'pin, *Russ. Chem. Bull. (Engl. Transl.)* **46**, 491 (1997).
450. G. A. Olah, G. K. S. Prakash, T. Mathew, and E. R. Marinez, *Angew Chem. Int. Ed. Engl.* **39**, 2547 (2000).
451. G. A. Olah, T. Mathew, E. R. Marinez, P. M. Esteves, M. Etzkorn, G. Rasul, and G. K. S. Prakash, *J. Am. Chem. Soc.* **123**, 11556 (2001).
452. P. J. F. de Rege, J. A. Gladysz, and I. T. Horváth, *Science* **276**, 776 (1997).
453. P. J. F. de Rege, J. A. Gladysz, and I. T. Horváth, *Adv. Synth. Catal.* **344**, 1059 (2002).
454. H. Jörg, *Chem. Ber.* **60**, 1466 (1972).
455. A. Treibs and R. Friess, *Justus Liebigs Ann. Chem.* **737**, 173 (1970).
456. P. D. George, *J. Org. Chem.* **26**, 4235 (1961).
457. G. A. Olah, M. R. Bruce, and F. L. Clouet, *J. Org. Chem.* **46**, 438 (1981).
458. H. Cerfontain, *Mechanistic Aspects in Aromatic Sulfonation and Desulfonation*, Interscience, New York, 1968.

459. G. A. Olah and Á. Molnár, *Hydrocarbon Chemistry*, 2nd ed., Wiley-Interscience, Hoboken, NJ, 2003, Chapter 10, and references cited therein.
460. A. Koeberg-Telder and H. Cerfontain, *Recl. Trav. Chim. Pays-Bas* **90**, 193 (1971).
461. L. Leiserson, R. W. Bost, and R. LeBaron, *Ind. Eng. Chem.* **40**, 508 (1948).
462. L. I. Levina, S. N. Patrakova, and D. A. Patrushev, *Zh. Obshch. Khim.* **28**, 2427 (1958).
463. M. Herlem, C. Mathieu, and D. Herlem, *Fuel* **83**, 1665 (2004).
464. R. J. Vaughan, U.S. Patent, 4,308,215 (1981).
465. G. A. Olah, T. Mathew, and G. K. S. Prakash, *Chem. Commun.* 1696 (2001).
466. G. A. Olah, R. H. Schlosberg, D. P. Kelly, and Gh. D. Mateescu, *J. Am. Chem. Soc.* **92**, 2546 (1970).
467. K. K. Laali and D. S. Nagvekar, *J. Org. Chem.* **56**, 1867 (1991).
468. G. A. Olah, E. R. Marinez, and G. K. S. Prakash, *Synlett*, 1397 (1999).
469. S. Répichet, C. Le Roux, and J. Dubac, *Tetrahedron Lett.* **40**, 9233 (1999).
470. M. Peyronneau, M. -T. Boisdon, N. Roques, S. Mazières, and C. Le Roux, *Eur. J. Org. Chem.* 4636 (2004).
471. K. Schofield, *Aromatic Nitration*. Cambridge University Press, Cambridge, 1980.
472. G. A. Olah, S. C. Narang, J. A. Olah, and K. Lammertsma, *Proc. Natl. Acad. Sci. USA* **79**, 4487 (1982).
473. G. A. Olah, R. Malhotra, and S. C. Narang, *Nitration: Methods and Mechanisms*, VCH, New York, 1989.
474. G. A. Olah, S. J. Kuhn, and A. Mlinkó, *J Chem. Soc.* 4257 (1956).
475. G. A. Olah and H. C. Lin, *Synthesis*, 444 (1974).
476. G. A. Olah and H. C. Lin, *J. Am. Chem. Soc.* **93**, 1259 (1971).
477. G. A. Olah, P. Ramaiah, C. B. Rao, G. Sandford, R. Golam, N. J. Trivedi, and J. A. Olah, *J. Am. Chem. Soc.* **115**, 7246 (1993).
478. G. A. Olah, P. Ramaiah, and G. K. S. Prakash, *Proc. Natl. Acad. Sci. USA* **94**, 11783 (1997).
479. G. A. Olah, K. K. Laali, and G. Sandford, *Proc. Natl. Acad. Sci. USA* **89**, 6670 (1992).
480. G. A. Olah, Q. Wang, X. -Y. Li, and I. Bucsi, *Synthesis*, 1085 (1992).
481. G. K. S. Prakash, T. Mathew, E. R. Marinez, P. M. Esteves, G. Rasul, and G. A. Olah, *J. Org. Chem.*, **71**, 3952 (2006).
482. G. A. Olah, A. Orlinkov, A. B. Oxyzoglou, and G. K. S. Prakash, *J. Org. Chem.* **60**, 7348 (1995).
483. G. A. Olah, V. P. Reddy, and G. K. S. Prakash, *Synthesis* 1087 (1992).
484. N. C. Marziano, C. Tortato, and M. Sampoli, *J. Chem. Soc., Perkin Trans. 2*, 423 (1991).
485. N. C. Marziano, L. Ronchin, C. Tortato, A. Zingales, and A. A. Sheikh-Osman, *J. Mol. Catal. A: Chem.* **174**, 265 (2001).
486. G. K. S. Prakash, Q. Wang, X. -Y. Li, and G. A. Olah, *Helv. Chim. Acta* **73**, 1167 (1990).
487. G. A. Olah, A. V. Orlinkov, P. Ramaiah, A. B. Oxyzoglou, and G. K. S. Prakash, *Russ. Chem. Bull. (Engl. Transl.)* **47**, 924 (1998).
488. G. A. Olah, J. Olah, and N. A. Overchuk, *J. Org. Chem.* **30**, 3373 (1965).
489. C. A. Cupas and R. L. Pearson, *J. Am. Chem. Soc.* **90**, 4742 (1968).
490. G. A. Olah, S. C. Narang, R. L. Pearson, and C. A. Cupas, *Synthesis*, 452 (1978).

491. G. A. Olah and M. Nojima, *Synthesis*, 785 (1973).
492. G. A. Olah, *Acc. Chem. Res.* **13**, 330 (1980).
493. G. A. Olah, H. C. Lin, J. A. Olah, and S. C. Narang, *Proc. Natl. Acad. Sci. USA* **75**, 1045 (1978).
494. G. A. Olah, B. G. B. Gupta, and S. C. Narang, *J. Am. Chem. Soc.* **101**, 5317 (1979).
495. G. A. Olah, B. G. B. Gupta, A. Garcia-Luna, and S. C. Narang, *J. Org. Chem.* **48**, 1760 (1983).
496. G. A. Olah and B. G. B. Gupta, *J. Org. Chem.* **48**, 3585 (1983).
497. T. -L. Ho and G. A. Olah, *J. Org. Chem.* **42**, 3097 (1977).
498. G. A. Olah and T. -L. Ho, *Synthesis*, 610 (1976).
499. G. A. Olah, S. C. Narang, G. F. Salem, and B. G. B. Gupta, *Synthesis*, 273 (1979).
500. G. A. Olah and P. Ramaiah, *J. Org. Chem.* **58**, 4639 (1993).
501. P. Yu. Savechenkov, A. P. Rudenko, and A. V. Vasil'ev, *Russ. J. Org. Chem. (Engl. Transl.)* **40**, 1633 (2004).
502. T. Axenrod, X. -P. Guan, J. Sun, L. Qi, R. D. Chapman, and R. D. Gilardi, *Tetrahedron Lett.* **42**, 2621 (2001).
503. G. A. Olah, R. Malhotra, and S. C. Narang, *J. Org. Chem.* **43**, 4628 (1978).
504. R.J. Vaughan, Australian Patent 506423 *Chem. Abstr.* **93**, 46166 (1980).
505. G. A. Olah, V. V. Krishnamurthy, and S. C. Narang, *J. Org. Chem.* **47**, 596 (1982).
506. G. A. Olah, S. C. Narang, R. Malhotra, and J. A. Olah, *J. Am. Chem. Soc.* **101**, 1805 (1979).
507. C. K. Ingold, *Structure Mechanism in Organic Chemistry*, Cornell University Press, Ithaca, NY, 1953.
508. Z. J. Allan, J. Podstata, D. Šnobl, and J. Jarkovský, *Tetrahedron Lett.* 3565 (1965).
509. G. A. Olah and N. Friedman, *J. Am. Chem. Soc.* **88**, 5330 (1966).
510. G. A. Olah, G. Salem, J. S. Staral, and T. -L. Ho, *J. Org. Chem.* **43**, 173 (1978).
511. G. A. Olah and B. G. B. Gupta, *J. Org. Chem.* **45**, 3532 (1980).
512. G. A. Olah, J. G. Shih, B. P. Singh, and B. G. B. Gupta, *J. Org. Chem.* **48**, 3356 (1983).
513. G. A. Olah, B. G. B. Gupta, and S. C. Narang, *Synthesis*, 274 (1979).
514. G. A. Olah, J. G. Shih, B. P. Singh, and B. G. B. Gupta, *Synthesis*, 713 (1983).
515. T. -L. Ho and G. A. Olah, *Synthesis,* 418, (1977).
516. G. A. Olah and T. -L. Ho, *Synthesis,* 609, (1976).
517. G. A. Olah, M. T. Ramos, Q. Wang, and G. K. S. Prakash, *Synlett* 41, (1991).
518. P. Legzdins, B. Wassink, F. W. B. Einstein, and A. C. Willis, *J. Am. Chem. Soc.* **108**, 317 (1986).
519. K. Mizuno, N. Ichinose, T. Tamai, and Y. Otsuji, *J. Org. Chem.* **57**, 4669 (1992).
520. C. York, G. K. S. Prakash, and G. A. Olah, *J. Org. Chem.* **59**, 6493 (1994).
521. C. York, G. K. S. Prakash, and G. A. Olah, *Tetrahedron* **52**, 9 (1996).
522. G. K. S. Prakash, Q. Wang, X. Li, and G. A. Olah, *New J. Chem.* **14**, 791 (1990).
523. M. L. Poutsma, in *Free Radicals*, Vol. 2, J. K. Kochi, ed., Wiley-Interscience, New York, 1973, Chapter 15, p. 159.
524. G. A. Olah and Y. K. Mo, *J. Am. Chem. Soc.* **94**, 6864 (1972).
525. G. A. Olah, R. Renner, P. Schilling, and Y. K. Mo, *J. Am. Chem. Soc.* **95**, 7686 (1973).

526. G. A. Olah, U.S. Patent, 4,523,040 (1985).
527. G. A. Olah, B. Gupta, M. Farina, J. D. Felberg, W. M. Ip, A. Husain, R. Karpeles, K. Lammertsma, A. K. Melhotra, and N. J. Trivedi, *J. Am. Chem. Soc.* **107**, 7097 (1985).
528. G. A. Olah and P. Schilling, *J. Am. Chem. Soc.* **95**, 7680 (1973).
529. D. Alker, D. H. R. Barton, R. H. Hesse, J. Lister-James, R. E. Markwell, M. M. Pechet, S. Rozen, T. Takeshita, and H. T. Toh, *New J. Chem.* **4**, 239 (1980).
530. K. O. Christe, *J. Fluorine Chem.* **22**, 519 (1983); **25**, 269, (1984).
531. C. Gal and S. Rozen, *Tetrahedron Lett.* **25**, 449 (1984).
532. S. Rozen and C. Gal, *J. Org. Chem.* **52**, 2769 (1987).
533. G. A. Olah, N. Hartz, G. Rasul, Q. Wang, G. K. S. Prakash, J. Casanova, and K. O. Christe, *J. Am. Chem. Soc.* **116**, 5671 (1994).
534. S. Thibaudeau, A. Martin-Mingot, M. -P. Jouannetaud, and J. C. Jacquesy, *Tetrahedron* **58**, 6643 (2002).
535. G. A. Olah, A. Wu, and O. Farooq, *J. Org. Chem.* **54**, 1463 (1989).
536. I. Akhrem, A. Orlinkov, and M. Vol'pin, *J. Chem. Soc., Chem. Commun.* 671 (1993).
537. I. S. Akhrem, A. V. Orlinkov, L. V. Afanas'eva, E. I. Mysov, and M. E. Vol'pin, *Tetrahedron Lett.* **36**, 9365 (1995).
538. I. S. Akhrem, A. V. Orlinkov, L. V. Afanas'eva, and M. E. Vol'pin, *Russ. Chem. Bull. (Engl. Transl.)* **45**, 1148 (1996).
539. I. Akhrem, A. Orlinkov, S. Vitt, and A. Chistyakov, *Tetrahedron Lett.* **43**, 1333 (2002).
540. I. S. Akhrem, A. V. Orlinkov, S. Vitt, L. V. Afanas'eva, and M. E. Vol'pin, *Russ. Chem. Bull. (Engl. Transl.)* **38**, 1864 (1989).
541. A. -C. Cantet, H. Carreyre, J. -P. Gesson, B. Renoux, and M. -P. Jouannetaud, *Tetrahedron Lett.* **47**, 5547 (2006).
542. A. -C. Cantet, J. -P. Gesson, B. Renoux, and M. -P. Jouannetaud, *Tetrahedron Lett.* **48**, 5255 (2007).
543. J. Salomé, C. Bachmann, K. De Oliveira-Vigier, S. Burnet, and J. Lopez, *J. Mol. Catal. A: Chem.* **279**, 119 (2008).
544. J. Salomé, C. Mauger, S. Burnet, and V. Schanen, *J. Fluorine Chem.* **125**, 1947 (2004).
545. S. Thibaudeau, A. Martin-Mingot, M. -P. Jouannetaud, O. Karam, and F. Zunino, *Chem. Commun.* 3198 (2007).
546. A. Moine, S. Thibaudeau, A. Martin, M. -P. Jouannetaud, and J. -C. Jacquesy, *Tetrahedron Lett.* **43**, 4119 (2002).
547. V. Chagnault, S. Thibaudeau, M. -P. Jouannetaud, J. -C. Jacquesy, A. Cousson, and C. Bachmann, *J. Fluorine Chem.* **128**, 55 (2007).
548. S. Debarge, S. Thibaudeau, B. Violeau, A. Martin-Mingot, M. -P. Jouannetaud, J. -C. Jacquesy, and A. Cousson, *Tetrahedron* **61**, 2065 (2005).
549. S. Debarge, B. Violeau, M. -P. Jouannetaud, J. -C. Jacquesy, and A. Cousson, *Tetrahedron* **62**, 662 (2006).
550. J. Fahy, A. Duflos, J. -P. Ribet, J. -C. Jacquesy, C. Berrier, M. -P. Jouannetaud, and F. Zunino, *J. Am. Chem. Soc.* **119**, 8576 (1997).
551. J. -C. Jacquesy, *J. Fluorine Chem.* **127**, 1484 (2006).
552. J. -C. Jacquesy, C. Berrier, M. -P. Jouannetaud, F. Zunino, J. Fahy, A. Duflos, and J. -P. Ribet, *J. Fluorine Chem.* **114**, 139 (2002).

553. I. Bucsi, B. Török, A. I. Marco, G. Rasul, G. K. S. Prakash, and G. A. Olah, *J. Am. Chem. Soc.* **124**, 7728 (2002).
554. H. P Braendlin and E. T. McBeein, in *Friedel–Crafts and Related Reactions*, Vol. 3, G. A. Olah, ed., Wiley-Interscience, New York, 1964, Chapter 46, p. 1517.
555. J. M. Brittain, P. B. D. de la Mare, N. S. Isaacs, and P. D. McIntyre, *J. Chem. Soc., Perkin Trans. 2*, 933 (1979).
556. J. -C. Jacquesy, M. -P. Jouannetaud, and S. Makami, *J. Chem. Soc., Chem. Commun., 110 (1980); Nouv. J. Chim.* **4**, 747 (1980).
557. J. M. Brittain, P. B. D. de la Mare, and P. A. Newman, *Tetrahedron Lett.* **21**, 4111 (1980).
558. J. -C. Jacquesy and M. -P. Jouannetaud, *Tetrahedron Lett.* **23**, 1673 (1982).
559. C. Berrier, J. C. Jacquesy, M. P. Jouannetaud, and A. Renoux, *New J. Chem.* **11**, 605 (1987).
560. G. Cherry, J. -C. Culmann, and J. Sommer, *Tetrahedron Lett.* **31**, 2007 (1990).
561. C. Berrier, J. C. Jacquesy, and A. Renoux, *Bull. Soc. Chim. Fr.* **127**, 93 (1990).
562. M. -P. Jouannetaud, J. -C. Jacquesy, O. Karam, J. -M. Coustard, and B. Ferron, *Bull. Soc. Chim. Fr.* **131**, 284 (1994).
563. G. A. Olah, Q. Wang, G. Sandford, and G. K. S. Prakash, *J. Org. Chem.* **58**, 3194 (1993).
564. G. K. S. Prakash, T. Mathew, D. Hoole, P. M. Esteves, Q. Wang, G. Rasul, and G. A. Olah, *J. Am. Chem. Soc.* **126**, 15770 (2004).
565. J. Barluenga, J. M. González, M. G. García-Martín, P. J. Campos, and G. Asensio, *J. Org. Chem.* **58**, 2058 (1993).
566. C. Graebe, *Ber. Dtsch. Chem. Ges.* **34**, 1778 (1901).
567. G. F. Jaubert, *Hebd. Seances Acad. Sci.* **132**, 841 (1901).
568. J. F. de Turski, German Patent, 287 756 (1914), *Chem. Abstr.* **10**, 11280 (1916).
569. J. S. F. Turski, British Patent 626 661 (1949), *Chem. Abstr.* **44**, 13909 (1950).
570. J. S. Turski,U.S. Patent 2,585,355 (1952).
571. A. Bertho, *Ber. Dtsch Chem. Ges.* **59**, 589 (1926).
572. R. N. Keller and P. A. S. Smith, *J. Am. Chem. Soc.* **66**, 1122 (1944).
573. P. A. S. Smith, Ph.D. thesis, University of Michigan *Ann Arbor.* (1944).
574. K. F. Schmidt, *Ber. Dtsch. Chem. Ges.* **57**, 704 (1924); **58**, 2413 (1925).
575. K. F. Schmidt and P. Zutavern, U.S. Patent 1,637,661 (1927).
576. G. M. Hoop and J. M. Tedder, *J. Chem. Soc.* 4685 (1961).
577. W. Borsche and H. Hahn, *Chem. Ber.* **82**, 260 (1949).
578. E. Bamberger *Justus Liebig's Ann. Chem.* **424**, 233 (1921).
579. E. Bamberger and J. Brun, *Helv. Chim. Acta* **6**, 935 (1923).
580. C. L. Arcus and J. V. Evans, *J. Chem. Soc.* 789 (1958).
581. P. A. S. Smith and J. H. Hall, *J. Am. Chem. Soc.* **84**, 480 (1962).
582. G. Smolinsky, *J. Am. Chem. Soc.* **83**, 2489 (1961).
583. T. Curtius, *J. Prakt. Chem.* **125**, 303 (1930).
584. J. F. Heacock and M. T. Edmison, *J. Am. Chem. Soc.* **82**, 3460 (1960).
585. J. F. Tilney-Bassett, *J. Chem. Soc.* 2517 (1962).
586. J. E. Leffler and Y. Tsuno, *J. Org. Chem.* **28**, 902 (1963).

587. Th. Curtius and A. Bertho, *Ber. Dtsch. Chem. Ges.* **59**, 565 (1926).
588. D. H. R. Barton and L. R. Morgan, Jr., *J. Chem. Soc.* 622, (1962).
589. K. Stewart, *Trans. Faraday Soc.* **41**, 663 (1945).
590. P. Kovacic, J. A. Levisky, and C. T. Goralski, *J. Am. Chem. Soc.* **88**, 100 (1966).
591. J. W. Strand and P. Kovacic, *J. Am. Chem. Soc.* **95**, 2977 (1973).
592. P. Kovacic, R. L. Russell, and R. P. Bennett, *J. Am. Chem. Soc.* **86**, 1588 (1964).
593. A. L. Mertens, K. Lammertsma, M. Arvanaghi, and G. A. Olah, *J. Am. Chem. Soc.* **105**, 5657 (1983).
594. E. Wiberg and H. Michaud, *Z. Naturforsch. B* **9**, 495 (1954).
595. E. Wiberg, W. -C. Joo, and K. H. Schmid, *Z. Anorg. Allg. Chem.* **394**, 197 (1972).
596. G. A. Olah and T. D. Ernst, *J. Org. Chem.* **54**, 1203 (1989).
597. G. K. S. Prakash, R. Desousa, and G. A. Olah, *J. Nanosci. Nanotechnol.* **5**, 397 (2005).
598. G. A. Olah, P. Ramaiah, Q. Wang, and G. K. S. Prakash, *J. Org. Chem.* **58**, 6900 (1993).
599. G. A. Olah, D. G. Parker, and N. Yoneda, *Angew. Chem. Int. Ed. Engl.* **17**, 909 (1978).
600. K. O. Christe, W. W. Wilson, and E. C. Curtis, *Inorg. Chem.* **18**, 2578 (1979).
601. G. A. Olah, A. L. Berrier, and G. K. S. Prakash, *J. Am. Chem. Soc.* **104**, 2373 (1982).
602. G. A. Olah, N. Yoneda, and D. G. Parker, *J. Am. Chem. Soc.* **99**, 483 (1977).
603. G. A. Olah, D. G. Parker, N. Yoneda, and F. Pelizza, *J. Am. Chem. Soc.* **98**, 2245 (1976).
604. P. D. Bach and M. -D. Su, *J. Am. Chem. Soc.* **116**, 10103 (1994).
605. H. Shi, Y. Wang, and Z. Zhang, *J. Mol. Catal. A: Chem.* **258**, 35 (2006).
606. D. H. Derbyshire and W. A. Waters, *Nature (London)* **165**, 401 (1950).
607. R. D. Chambers, P. Goggin, and W. K. R. Musgrave, *J. Chem. Soc.* 1804 (1959).
608. J. D. McClure and P. H. Williams, *J. Org. Chem.* **27**, 24, 627 (1962).
609. C. A. Buehler and H. Hart, *J. Am. Chem. Soc.* **85**, 2177 (1963).
610. A. J. Davidson and R. O. C. Norman, *J. Chem. Soc.* 5404 (1964).
611. H. Hart and C. A. Buehler, *J. Org. Chem.* **29**, 2397 (1964).
612. H. Hart, C. A. Buehler, and A. J. Waring, *Adv. Chem. Ser.* **51**, 1 (1965).
613. S. Hashimoto and W. Koike, *Bull. Chem. Soc. Jpn.* **43**, 293 (1970).
614. J. A. Vesely and L. Schmerling, *J. Org. Chem.* **35**, 4028 (1970).
615. M. E. Kurz and G. J. Johnson, *J. Org. Chem.* **36**, 3184 (1971).
616. R. O. C. Norman and R. Taylor, *Electrophilic Substitution in Benzenoid Compounds*, Elsevier, Amsterdam, 1965.
617. G. A. Olah and R. Ohnishi, *J. Org. Chem.* **43**, 865 (1978).
618. G. A. Olah A. P. Fung, and T. Keumi, *J. Org. Chem.* **46**, 4305 (1981).
619. G. A. Olah, T. Keumi, J. C. Lecoq, A. P. Fung, and J. A. Olah, *J. Org. Chem.* **56**, 6148 (1991).
620. J. -P. Gesson, J. -C. Jacquesy, and M. -P. Jouannetaud, *J. Chem. Soc., Chem. Commun.* 1128 (1980).
621. J. -P. Gesson, L. D. Giusto, and J. -C. Jacquesy, *Tetrahedron* **34**, 1715 (1978).
622. J. -P. Gesson, J. -C. Jacquesy, and M. -P. Jouannetaud, *Nouv. J. Chim.* **6**, 477 (1982).
623. J. -P. Gesson, J. -C. Jacquesy, M. -P. Jouannetaud, and G. Morellet, *Tetrahedron Lett.* **24**, 3095 (1983).

624. J. -C. Jacquesy, C. Berrier, J. -P. Gesson, and M. -P. Jouannetaud, *Bull. Soc. Chim. Fr.* **131**, 658 (1994).
625. C. Berrier, J. -C. Jacquesy, M. -P. Jouannetaud, and G. Morellet, *Bull. Soc. Chim. Fr.* 158 (1987).
626. C. Berrier, H. Carreyre, J. -C. Jacquesy, and M. -P. Jouannetaud, *New J. Chem.* **14**, 283 (1990).
627. J. -C. Jacquesy, M. -P. Jouannetaud, G. Morellet, and Y. Vidal, *Tetrahedron Lett.* **25**, 1479 (1984).
628. C. Berrier, J. -C. Jacquesy, M. -P. Jouannetaud, and A. Renoux, *Tetrahedron Lett.* **27**, 4565 (1986); *J. Fluorine Chem.* **35**, 40 (1987).
629. C. Berrier, J. C. Jacquesy, M. P. Jouannetaud, and A. Renoux, *New J. Chem.* **11**, 611 (1987).
630. J. -C. Jacquesy, M. -P. Jouannetaud, and G. Morellet, *Tetrahedron Lett.* **24**, 3099 (1983).
631. C. Berrier, J. C. Jacquesy, and M. P. Jouannetaud, *Tetrahedron* **40**, 5135 (1984).
632. C. Berrier, J. C. Jacquesy, M. P. Jouannetaud, and Y. Vidal, *Tetrahedron* **46**, 827 (1990).
633. A. Duflos, F. Redoules, J. Fahy, J. -C. Jacquesy, and M. -P. Jouannetaud, *J. Nat. Prod.* **64**, 193 (2001).
634. V. Chagnault, M. -P. Jouannetaud, and J. -C. Jacquesy, *Tetrahedron Lett.* **47**, 5723 (2006).
635. R. Trambarulo, S. N. Ghosh, C. A. Jr.Barrus, and W. Gordy, *J. Chem. Phys.* **21**, 851 (1953).
636. P. D. Bartlett, M. Stiles, *J. Am. Chem. Soc.* **77**, 2806 (1955).
637. J. P. Wibaut, F. L. J. Sixma, and L. W. F. Kampschmidt, *Recl. Trav. Chim. Pays-Bas* **69**, 1355 (1950).
638. J. P. Wibaut and F. L. J. Sixma, *Recl. Trav. Chim. Pays-Bas* **71**, 761 (1951).
639. P. S. Bailey, *Chem. Rev.* **58**, 925 (1958).
640. L. Horner, H. Schaefer, and W. Ludwig, *Chem. Ber.* **91**, 75 (1958).
641. P. S. Bailey, J. W. Ward, R. E. Hornish, and F. E. Potts, *Adv. Chem. Ser.* **112**, 1 (1972).
642. G. A. Olah, N. Yoneda, and D. G. Parker, *J. Am. Chem. Soc.* **98**, 5261 (1976).
643. T. M. Hellman and G. A. Hamilton, *J. Am. Chem. Soc.* **96**, 1530 (1974).
644. M. C. Whiting, A. J. N. Bolt, and J. H. Parish, *Adv. Chem. Ser.* **77**, 4 (1968).
645. D. O. Williams and R. J. Cvetanovic, *J. Am. Chem. Soc.* **92**, 2949 (1970).
646. J. -C. Jacquesy, R. Jacquesy, L. Lamande, C. Narbonne, J. -F. Patoiseau, and Y. Vidal, *Nouv. J. Chim.* **6**, 589 (1982).
647. J. -C. Jacquesy and J. -F. Patoiseau, *Tetrahedron Lett.* 1499 (1977).
648. F. Frusteri, F. Arena, C. Espro, N. Mondello, and A. Parmaliana,in *Natural Gas Conversion V, Studies in Surface Science and Catalysis* Vol. 119, A. Parmaliana, D. Sanfilippo, F. Frusteri, A. Vaccari, and F. Arena, eds., Elsevier, Amsterdam, 1998, p. 447.
649. F. Frusteri, F. Arena, S. Bellitto, and A. Parmaliana, *Appl. Catal., A* **180**, 325 (1999).
650. C. Espro, F. Frusteri, F. Arena, and A. Parmaliana, *J. Mol. Catal. A: Chem.* **159**, 359 (2000).
651. F. Frusteri, C. Espro, F. Arena, E. Passalacqua, A. Patti, and A. Parmaliana, *Catal. Today* **61**, 37 (2000).

652. G. A. Olah, T. Yamato, P. S. Iyer, N. J. Trivedi, B. J. Singh, and G. K. S. Prakash, *Mater. Chem. Phys.* **17**, 21 (1987).

653. J. Fischer and W. F. Hölderich, *Appl. Catal., A* **180**, 435 (1999).

654. S. L. Jain and B. Sain, *Appl. Catal., A* **301**, 259 (2006).

655. Y. Usui, K. Sato, and M. Tanaka, *Angew. Chem. Int. Ed.* **42**, 5623 (2003).

656. M. C. Laufer and W. F. Hoelderich, *Chem. Commun.* 1684 (2002).

657. M. C. Laufer, J. P. M. Niederer, and W. F. Hoelderich, *Adv. Synth. Catal.* **344**, 1084 (2002).

658. M. S. Saha, Y. Nishiki, T. Furuta, A. Denggerile, and T. Ohsaka, *Tetrahedron Lett.* **44**, 5535 (2003).

659. A. Martin, M. -P. Jouannetaud, and J. -C. Jacquesy, *Tetrahedron Lett.* **37**, 2967 (1996).

660. Ya. V. Zonov, V. M. Karpov, and V. E. Platonov, *J. Fluorine Chem.* **126**, 437 (2005).

661. Ya. V. Zonov, V. M. Karpov, and V. E. Platonov, *J. Fluorine Chem.* **128**, 1058 (2007).

662. G. A. Olah and T. D. Ernst, *J. Org. Chem.* **54**, 1204 (1989).

663. G. K. S. Prakash, N. Krass, Q. Wang, and G. A. Olah, *Synlett*, 39 (1991).

664. G. A. Olah, T. D. Ernst, C. B. Rao, and G. K. S. Prakash, *New J. Chem.* **13**, 791 (1989).

665. G. A. Olah, Q. Wang, N. Krass, and G. K. S. Prakash, *Rev. Roum. Chim.* **36**, 567 (1991).

666. G. A. Olah, Q. Wang, and G. K. S. Prakash, *Catal. Lett.* **13**, 55 (1992).

667. S. Yamago, T. Yamada, R. Nishimura, H. Ito, Y. Mino, and J. Yoshida, *Chem. Lett.* 152 (2002).

668. B. Török, I. Bucsi, G. K. S. Prakash, and G. A. Olah, *Chem. Commun.* 2882 (2002).

669. A. K. Chakraborti, and R. Gulhane, *Tetrahedron Lett.* **44**, 3521 (2003).

670. V. T. Kamble, B. P. Bandgar, D. B. Muley, and N. S. Joshi, *J. Mol. Catal. A: Chem.* **268**, 70 (2007).

671. V. T. Kamble, B. P. Bandgar, N. S. Joshi, and V. S. Jamode, *Synlett*, 2719 (2006).

672. G. A. Olah, A. Husain, and B. P. Shing, *Synthesis*, 892 (1983).

673. R. Kumareswaran, K. Pachamuthu, and Y. D. Vankar, *Synlett*, 1652 (2000).

674. G. A. Olah, S. C. Narang, D. Meidar, and G. F. Salem, *Synthesis*, 282 (1981).

675. G. A. Olah and A. K. Mehrotra, *Synthesis*, 962 (1982).

676. G. K. S. Prakash, P. Ramaiah, and R. Krishnamurti, *Catal. Lett.* **9**, 59 (1991).

677. I. Ledneczki, M. Darányi, F. Fülöp, and Á. Molnár, *Catal. Today* **100**, 437 (2005).

678. I. Ledneczki and Á. Molnár, *Synth. Commun.* **34**, 3683 (2004).

679. G. A. Olah, A. Husain, B. G. B. Gupta, and S. C. Narang, *Synthesis*, 471 (1981).

680. M. A. Zolfigol, I. Mohammadpoor-Baltork, D. Habibi, B. F. Mirjalili, and A. Bamoniri, *Phosphorus, Sulfur Silicon*, **179**, 2189 (2004).

681. S. Rani, J. L. Babu, and Y. D. Vankar, *Synth. Commun.* **33**, 4043 (2003).

682. G. A. Olah, T. Yamato, P. S. Iyer, and G. K. S. Prakash, *J. Org. Chem.* **51**, 2826 (1986).

683. M. A. Zolfigol, I. Mohammadpoor-Baltork, D. Habibi, B. F. Mirjalili, and A. Bamoniri, *Tetrahedron Lett.* **44**, 8165 (2003).

684. M. Couturier, J. L. Tucker, C. Proulx, C. Boucher, P. Dubé, B. M. Andresen, and A. Ghosh, *J. Org. Chem.* **67**, 4833 (2002).

685. K. K. Laali, T. Okazaki, K. Takeuchi, K. Ogawa, and A. J. Bennet, *J. Chem. Soc., Perkin Trans. 2*, 1105 (2002).

686. C. Lafitte, M. -P. Jouannetaud, J. -C. Jacquesy, and A. Duflos, *Tetrahedron* **55**, 1989 (1999).
687. G. A. Olah, A. P. Fung, and R. Malhotra, *Synthesis*, 474 (1981).
688. I. Bucsi Á. Molnár, M. Bartók and G. A. Olah, *Tetrahedron* **51**, 3319 (1995).
689. A. Puglisi, A. -L. Lee, R. R. Schrock, and A. H. Hoveyda, *Org. Lett.* **8**, 1871 (2006).
690. D. O. Bennardi, G. P. Romanelli, J. C. Autino, and L. R. Pizzio, *Appl. Catal., A* **324**, 62 (2007).
691. M. C. Laufer, H. Hausmann, and W. F. Hölderich, *J. Catal.* **218**, 315 (2003).
692. V. K. Aggarwal, G. P. Vennall, P. N. Davey, and C. Newman, *Tetrahedron Lett.* **38**, 2569 (1997).
693. C. M. Marson and A. Fallah, *Tetrahedron Lett.* **38**, 9057 (1997).
694. M. J. Cloninger and L. E. Overman, *J. Am. Chem. Soc.* **121**, 1092 (1999).
695. H. Wang and B. -Q. Xu, *Appl. Catal., A* **275**, 247 (2004).
696. M. C. Laufer, W. Bonrath, and W. F. Hoelderich, *Catal. Lett.* **100**, 101 (2005).
697. S. T. Bright, J. M. Coxon, and P. J. Steel, *Tetrahedron Lett.* **30**, 5651 (1989).
698. B. Schlummer and J. F. Hartwig, *Org. Lett.* **4**, 1471 (2002).
699. V. Zanirato, G. P. Pollini, C. De Risi, F. Valente, A. Melloni, S. Fusi, J. Barbetti, and M. Olivucci, *Tetrahedron* **63**, 4975 (2007).
700. J. -W. Huang and M. Shi, *Synlett* 2343 (2004).
701. C. Kuroda, N. Mitsumata, and C. Y. Tang, *Bull. Chem. Soc. Jpn.* **69**, 1409 (1996).
702. T. Shimizu, K. Tanino, and I. Kuwajima, *Tetrahedron Lett.* **41**, 5715 (2000).
703. Y. -L. Zhao, W. Zhang, S. Wang, and Q. Liu, *J. Org. Chem.* **72**, 4985 (2007).
704. K. K. S. Sai, T. M. Gilbert, and D. A. Klumpp, *J. Org. Chem.* **72**, 9761 (2007).
705. P. Molina, J. Alcántara, and C. López-Leonardo, *Synlett* 363 (1995).
706. W. H. Pearson, R. Walavalkar, J. M. Schkeryantz, W. -K. Fang, and J. D. Blickensdorf, *J. Am. Chem. Soc.* **115**, 10183 (1993).
707. W. H. Pearson and J. M. Schkeryantz, *Tetrahedron Lett.* **33**, 5291 (1992).
708. C. J. Mossman and J. Aubé, *Tetrahedron* **52**, 3403 (1996).
709. R. Kumareswaran, B. G. Reddy, and Y. D. Vankar, *Tetrahedron Lett.* **42**, 7493 (2001).
710. T. Yamato, C. Hideshima, K. Suehiro, M. Tashiro, G. K. S. Prakash, and G. A. Olah, *J. Org. Chem.* **56**, 6248 (1991).
711. S. Thibaudeau, B. Violeau, A. Martin-Mingot, M. -P. Jouannetaud, and J. -C. Jacquesy, *Tetrahedron Lett.* **43**, 8773 (2002).
712. R. P. Austin and J. H. Ridd, *J. Chem. Soc., Perkin Trans.* 2, 1229 (1993).
713. S. Nakamura, H. Sugimoto, and T. Ohwada, *J. Am. Chem. Soc.* **129**, 1724 (2007).
714. J. Umezawa, O. Takahashi, K. Furuhashi, and H. Nohira, *Tetrahedron: Asymmetry* **5**, 491 (1994).
715. Y. Soro, F. Bamba, S. Siaka, and J. -M. Coustard, *Tetrahedron Lett.* **47**, 3315 (2006).
716. S. Kanoh, T. Nishimura, M. Naka, and M. Motoi, *Tetrahedron* **58**, 7065 (2002).
717. Y. Du and F. Tian, *Catal. Commun.* **8**, 2012 (2007).
718. W. Chen, S. Qin, and J. Jin, *Catal. Commun.* **8**, 123 (2007).
719. J. K. Joseph, S. L. Jain, and B. Sain, *J. Mol. Catal. A: Chem.* **247**, 99 (2006).

720. H. Lin, Q. Zhao, B. Xu, and X. Wang, *J. Mol. Catal. A: Chem.* **268**, 221 (2007).
721. B. V. Lingaiah, G. G. Ezekiel, T. Yakaiah, G. V. Reddy, and P. S. Rao, *Synlett* 2507 (2006).
722. G. K. S. Prakash, H. Vaghoo, C. Panja, Á. Molnár, M. Thomas, and G. A. Olah, *Synthesis*, 897 (2008).
723. B. P. Bandgar, A. V. Patil, and O. S. Chavan, *J. Mol. Catal. A: Chem.* **256**, 99 (2006).
724. G. A. Olah, A. P. Fung, and D. Meidar, *Synthesis* 280 (1981).
725. L. A. Rios, P. P. Weckes, H. Schuster, and W. F. Hoelderich, *Appl. Catal., A* **284**, 155 (2005).
726. G. K. S. Prakash, T. Mathew, S. Krishnaraj, E. R. Marinez, and G. A. Olah, *Appl. Catal., A* **181**, 283 (1999).
727. M. Hachoumy, T. Mathew, E. C. Tongco, Y. D. Vankar, G. K. S. Prakash, and G. A. Olah, *Synlett* 363 (1999).
728. A. Fási, I. Pálinkó, Á. Gömöry, and I. Kiricsi, *J. Mol. Catal. A: Chem.* **208**, 307 (2004).
729. D. T. Thompson, C. B. Svendsen, C. Di Meglio, and V. C. Anderson, *J. Org. Chem.* **59**, 2945 (1994).
730. M. G. Barlow, N. N. E. Suliman, and A. E. Tipping, *J. Fluorine Chem.* **70**, 59 (1995).
731. L. G. French and T. P. Charlton, *Heterocycles* **35**, 305 (1993).
732. M. Bartók and Á. Molnár, *The Chemistry of Functional Groups. Supplement E: The Chemistry of Ethers, Crown Ethers, Hydroxyl Groups and Their Sulphur Analogues*, S. Patai, ed., Wiley, Chichester, 1980, Chapter 16, p. 721.
733. I. Bucsi, Á. Molnár, M. Bartók, and G. A. Olah, *Tetrahedron* **50**, 8195 (1994).
734. G. A. Olah, D. A. Klumpp, G. Neyer, and Q. Wang, *Synthesis* 321 (1996).
735. D. A. Klumpp, D. N. Baek, G. K. S. Prakash, and G. A. Olah, *J. Org. Chem.* **62**, 6666 (1997).
736. T. Okazaki, T. Kitagawa, and K. Takeuchi, *Chem. Lett.* 981 (1995).
737. J. Kvíčala, T. Bříza, P. Paleta, and J. Čermák, *Collect. Czech. Chem. Commun.* **67**, 1345 (2002).
738. C. Park and M. A. Keane, *J. Mol. Catal. A: Chem.* **166**, 303 (2001).
739. J. G. Nunan, K. Klier, and R. G. Herman, *J. Catal.* **139**, 406 (1993).
740. J. Tejero, F. Cunill, M. Iborra, J. F. Izquierdo, and C. Fité, *J. Mol. Catal. A: Chem.* **182–183**, 541 (2002).
741. R. Bringué, M. Iborra, J. Tejero, J. F. Izquierdo, F. Cunill, C. Fité, and V. J. Cruz, *J. Catal.* **244**, 33 (2006).
742. G. A. Olah, T. Shamma, and G. K. S. Prakash, *Catal. Lett.* **46**, 1 (1997).
743. M. A. Stanescu and R. S. Varma, *Tetrahedron Lett.* **43**, 7307 (2007).
744. K. Toshima and K. Tatsuta, *Chem. Rev.* **93**, 1503 (1993).
745. M. Shimizu, H. Togo, and M. Yokoyama, *Synthesis* 799 (1998).
746. K. Toshima, *Carbohyd. Res.* **327**, 15 (2000).
747. T. Mukaiyama and H. Jona, K. Takeuchi, *Chem. Lett.* 696 (2000).
748. T. Mukaiyama, K. Takeuchi, H. Jona, H. Maeshima, and T. Saitoh, *Helv. Chim. Acta* **83**, 1901 (2000).
749. H. Jona, K. Takeuchi, and T. Mukaiyama, *Chem. Lett.* 1278 (2000).

750. T. Hashihayata, K. Ikegai, K. Takeuchi, H. Jona, and T. Mukaiyama, *Bull. Chem. Soc. Jpn.* **76**, 1829 (2003).
751. H. Chiba, S. Funasaka, and T. Mukaiyama, *Bull. Chem. Soc. Jpn.* **76**, 1629 (2003).
752. T. Hashihayata, H. Mandai, and T. Mukaiyama, *Bull. Chem. Soc. Jpn.* **77**, 169 (2004).
753. H. Kondo, S. Aoki, Y. Ichikawa, R. L. Halcomb, H. Ritzen, and C. -H. Wong, *J. Org. Chem.* **59**, 864 (1994).
754. H. Sakamoto, S. Nakamura, T. Tsuda, and S. Hashimoto, *Tetrahedron Lett.* **41**, 7691 (2000).
755. R. W. Franck and C. M. Marzabadi, *J. Org. Chem.* **63**, 2197 (1998).
756. T. Yamanoi, K. Matsumura, S. Matsuda, and Y. Oda, *Synlett* 2973 (2005).
757. F. Bélot, D. Rabuka, M. Fukuda, and O. Hindsgaul, *Tetrahedron Lett.* **43**, 7743 (2002).
758. F. Bélot, C. Guerreiro, F. Baleux, and L. A. Mulard, *Chem. Eur. J.* **11**, 1625 (2005).
759. P. Fügedi, P. Nánási, and J. Szejtli, *Carbohyd. Res.* **175**, 173 (1988).
760. N. Nakanishi, Y. Nagafuchi, and K. Koizumi, *J. Carbohyd. Chem.* **13**, 981 (1994).
761. K. Toshima, K. Kasumi, and S. Matsumura, *Synlett* 643 (1998).
762. H. Nagai, K. Kawahara, S. Matsumura, and K. Toshima, *Tetrahedron Lett.* **42**, 4159 (2001).
763. A. Gupta and Y. D. Vankar, *Tetrahedron* **56**, 8525 (2000).
764. S. Raghavan and D. Kahne, *J. Am. Chem. Soc.* **115**, 1580 (1993) and references cited therein.
765. I. Alonso, N. Khiar, and M. Martín-Lomas, *Tetrahedron Lett.* **37**, 1477 (1996).
766. M. G. Beaver, S. B. Billings, and K. A. Woerpel, *Eur. J. Org. Chem.* 771 (2008). *J. Am. Chem. Soc.* **130**, 2082 (2008).
767. G. K. Rawal, S. Rani, A. Kumar, and Y. D. Vankar, *Tetrahedron Lett.* **47**, 9117 (2006).
768. M. Oikawa, T. Tanaka, N. Fukuda, and S. Kusumoto, *Tetrahedron Lett.* **45**, 4039 (2004).
769. T. Ziegler, M. Vollmer, S. Oberhoffner, and E. Eckhardt, *Liebigs Ann. Chem.* 255 (1993).
770. J. M. Benito, E. Rubio, M. Gómez-García, C. O. Mellet, and J. M. G. Fernández, *Tetrahedron* **60**, 5899 (2004).
771. E. M. Rubio, M. I. García-Moreno, P. Balbuena, F. J. Lahoz, E. Álvarez, C. O. Mellett, and J. M. G. Fernández, *J. Org. Chem.* **71**, 2257 (2006).
772. J. Defaye and J. M. G. Fernández, *Carbohyd. Res.* **237**, 223 (1992); **251**, 1, 17 (1994).
773. J. L. J. Blanco, E. M. Rubio, C. O. Mellett, and J. M. G. Fernández, *Synlett* 2230 (2004).
774. Y. A. Knirel and A. V. Perepelov, *Aust. J. Chem.* **55**, 69 (2002) and references cited therein.
775. A. V. Perepelov, S. N. Senchenkova, A. S. Shashkov, N. A. Komandrova, S. V. Tomshich, L. S. Shevchenko, Y. A. Knirel, and N. K. Kochetkov, *J. Chem. Soc., Perkin Trans. 1*, 363 (2000).
776. A. V. Perepelov, B. Babicka, S. N. Senchenkova, A. S. Shashkov, H. Moll, A. Rozalski, U. Zähringer, and Y. A. Knirel, *Carbohyd. Res.* **331**, 195 (2001).
777. J. L. Simonsen and L. N. Owen, *The Terpenes* Vol. 1, (1947); and Vol. 2, (1949); J. L. Simonsen and R. Barton,Vol. 3, (1949); J. L. Simonsen and W. C. J. Ross,Vols. 4 and 5 Cambridge University Press, Cambridge (1957).
778. J. King and P. de Mayo, in *Molecular Rearrangements*, Vol. 2, P. de Mayo, ed., Interscience, New York, 1964, Chapter 13, p. 771.

779. G. Ourisson, *Proc. Chem. Soc.* 274 (1964).
780. T. S. Santhanakrishnan, *J. Sci. Ind. Res.* **24**, 631 (1965).
781. W. G. Dauben, *J. Agr. Food Chem.* **22**, 156 (1974).
782. T. S. Sorensen, *Acc. Chem. Res.* **9**, 257 (1976).
783. J. K. Sutherland, *Chem. Soc. Rev.* **9**, 265 (1980).
784. J.-C. Jacquesy, in *Stable Carbocation Chemistry*, G. K. S. Prakash and P. v. R. Schleyer, eds., Wiley, New York, 1997, Chapter 17, p. 549.
785. L. Ruzicka, *Experientia* **9**, 357 (1953).
786. L. Ruzicka, *Proc. Chem. Soc.* 341 (1959).
787. J. B. Hendrickson, *Tetrahedron* **7**, 82 (1959).
788. J. H. Richards and J. B. Hendrickson, *The Biosynthesis of Steroids, Terpenes and Acetogenins*, Benjamin, New York, (1964), 173–288.
789. W. Parker, J. S. Roberts, and R. Ramage, *Quart. Rev. London* **21**, 331 (1967).
790. E. E. van Tamelen, *Fortschr. Chem. Org. Naturstoffe* **19**, 245 (1961); *Acc. Chem. Res.* **1**, 111 (1968); **8**, 152 (1975).
791. W. S. Johnson, *Acc. Chem. Res.* **1**, 1 (1968); *Angew Chem. Int. Ed. Engl.* **15**, 9 (1976).
792. S. Dev, *Acc. Chem. Res.* **14**, 82 (1981).
793. P. F. Vlad, *Pure Appl. Chem.* **65**, 1329 (1993).
794. V. A. Barkhash and M. P. Polovinka, *Russ. Chem. Rev. Engl. Transl.* **68**, 393 (1999).
795. H. E. Armstrong and F. S. Kipping, *J. Chem. Soc.* **63**, 75 (1893).
796. R. P. Lutz and J. D. Roberts, *J. Am. Chem. Soc.* **84**, 3715 (1962).
797. O. R. Rodig and R. J. Sysko, *J. Am. Chem. Soc.* **94**, 6475 (1972).
798. J. -C. Jacquesy, R. Jacquesy, and J. -F. Patoiseau, *Tetrahedron* **32**, 1699 (1976).
799. E. Huang, K. Ranganayakulu, and T. S. Sorensen, *J. Am. Chem. Soc.* **94**, 1780 (1972).
800. R. Haseltine, E. Huang, K. Ranganayakulu, and T. S. Sorensen, *Can. J. Chem.* **53**, 1056 (1975).
801. R. P. Haseltine and T. S. Sorensen, *Can. J. Chem.* **53**, 1067 (1975).
802. D. V. Banthorpe, P. A. Boullier, and W. D. Fordham, *J. Chem. Soc., Perkin Trans. 1*, 1637 (1974).
803. G. Carr, C. Dean, and D. Whittaker, *J. Chem. Soc., Perkin Trans. 2*, 71 (1989).
804. C. Dean and D. Whittaker, *J. Chem. Soc., Perkin Trans. 2*, 1275 (1990).
805. G. Carr, C. Dean, and D. Whittaker, *J. Chem. Soc., Perkin Trans. 2*, 351 (1988).
806. G. Carr and D. Whittaker, *J. Chem. Soc., Perkin Trans. 2*, 359 (1989).
807. J. D. White, H. Shin, T. -S. Kim, and N. S. Cutshall, *J. Am. Chem. Soc.* **119**, 2404 (1997).
808. G. A. Olah, G. K. S. Prakash, Q. Wang, and X.-Y. Li, in *Encyclopedia of Reagents for Organic Synthesis*, Vol. 4, L. A. Paquette, ed., Wiley, Chichester, (1995), p. 2567.
809. K. H. Schulte-Elte, H. Pamingle, A. P. Uijttewaal, and R. L. Snowden, *Helv. Chim. Acta* **75**, 759 (1992).
810. R. L. Snowden, J. -C. Eichenberger, S. M. Linder, P. Sonnay, C. Vial, and K. H. Schulte-Elte, *J. Org. Chem.* **57**, 955 (1992).
811. R. L. Snowden, J. -C. Eichenberger, W. Giersch, W. Thommen, and K. H. Schulte-Elte, *Helv. Chim. Acta* **76**, 1608 (1993).

812. A. F. Barrero, J. Altarejos, E. J. Alvarez-Manzaneda, J. M. Ramos, and S. Salido, *J. Org. Chem.* **61**, 2215 (1996).

813. P. J. Linares-Palomino, S. Salido, J. Altarejos, and A. Sánchez, *Tetrahedron Lett.* **44**, 6651 (2003).

814. H. Tanimoto, H. Kiyota, T. Oritani, and K. Matsumoto, *Synlett* 121 (1997).

815. P. F. Vlad, N. D. Ungur, N. V. Hung, and V. B. Perutsky, *Russ. Chem. Bull. Engl. Transl.* **44**, 2390 (1995).

816. P. F. Vlad, N. D. Ungur, and N. V. Tuen, *Russ. Chem. Bull. Engl. Transl.* **44**, 2404 (1995).

817. N. Ungur, V. Kulciki, M. Gavagnin, F. Castelluccio, P. F. Vlad, and G. Cimino, *Tetrahedron* **58**, 10159 (2002).

818. V. Kulciki, N. Ungur, and P. F. Vlad, *Tetrahedron* **54**, 11925 (1998).

819. V. Kulcitki, M. Grin'ko, A. Barba, N. Ungur, and P. F. Vlad, *Chem. Nat. Comp.* **43**, 268 (2007).

820. M. Grinco, V. Kulciki, N. Ungur, W. Jankowski, T. Chojnacki, and P. F. Vlad, *Helv. Chim. Acta* **90**, 1223 (2007).

821. M. P. Polovinka, D. V. Korchagina, Y. V. Gatilov, I. Yu. Bagrianskaya, V. A. Barkhash, V. V. Shcherbukhin, N. S. Zefirov, V. B. Perutskii, N. D. Ungur, and P. F. Vlad, *J. Org. Chem.* **59**, 1509 (1994).

822. M. Grin'ko, V. Kul'chitskii, N. Ungur, and P. F. Vlad, *Chem. Nat. Comp.* **42**, 439 (2006).

823. V. Kulciki, N. Ungur, P. F. Vlad, M. Gavagnin, F. Castelluccio, and G. Cimino, *Synthesis* 407 (2000).

824. M. A. Baig, D. V. Banthorpe, G. Carr, and D. Whittaker, *J. Chem. Soc., Perkin Trans. 2*, 1981 (1989).

825. T. M. Khomenko, D. V. Korchagina, and V. A. Barkhash, *Russ. J. Org. Chem. Engl. Transl.* **37**, 793 (2001).

826. M. P. Polovinka, D. V. Korchagina, Yu. V. Gatilov, O. G. Vyglazov, and V. A. Barkhash, *Russ. J. Org. Chem. Engl. Transl.* **35**, 1292 (1999).

827. O. V. Solomatina, D. V. Korchagina, Yu. V. Gatilov, M. P. Polovinka, and V. A. Barkhash, *Russ. J. Org. Chem. Engl. Transl.* **40**, 1441 (2004).

828. M. P. Polovinka, A. A. Shal'ko, D. V. Korchagina, Yu. V. Gatilov, V. V. Shcherbukhin, G. A. Zenkovets, and V. A. Barkhash, *Russ. J. Org. Chem. Engl. Transl.* **36**, 40 (2000).

829. J. M. Coxon, G. J. Hydes, and P. J. Steel, *Tetrahedron* **41**, 5213 (1985).

830. V. Kulciki, N. Ungur, M. Gavagnin, M. Carbone, and G. Cimino, *Eur. J. Org. Chem.* 1816 (2005).

831. I. G. Collado, J. R. Hanson, and A. J. Macías-Sánchez, *Nat. Prod. Rep.* **15**, 187 (1998).

832. C. F. D. Amigo, I. G. Collado, J. R. Hanson, R. Hernández-Galán, P. B. Hitchcock, A. J. Macías-Sánchez, and D. J. Mobbs, *J. Org. Chem.* **66**, 4327 (2001).

833. C. -K. Chan and C. -Y. Lin, *J. Appl. Polymer Sci.* **101**, 3666 (2006).

834. D. G. Farnum and G. Mehta, *Chem. Commun.* 1643, (1968).

835. D. G. Farnum, R. A. Mader, and G. Mehta, *J. Am. Chem. Soc.* **95**, 8692 (1973).

836. G. Mehta, N. Pattnaik, and S. K. Kapoor, *Tetrahedron Lett.* 4947 (1972).

837. G. Mehta and S. K. Kapoor, *Tetrahedron Lett.* 2385 (1973).

838. J. -C. Jacquesy, R. Jacquesy, S. Moreau, and J. -F. Patoiseau, *Chem. Commun.* 785 (1973).

839. J. -C. Jacquesy, R. Jacquesy, and J. -F. Patoiseau, *Bull. Soc. Chim. Fr. II* 1959 (1974).
840. R. Jacquesy and H. L. Ung, *Tetrahedron* **32**, 1375 (1976).
841. J. -C. Jacquesy, R. Jacquesy, and C. Narbonne, *Bull. Soc. Chim. Fr. II* 1240 (1976).
842. J. -C. Jacquesy and C. Narbonne, *Bull. Soc. Chim. Fr. II* 163 (1978).
843. J. P. Gesson, J. C. Jacquesy, and M. Mondon, *Tetrahedron Lett.* 3351 (1980).
844. J. -C. Jacquesy and M. -P. Jouannetaud, *Bull. Soc. Chim. Fr. II* 202 (1978).
845. C. Berrier, J. C. Jacquesy, J. P. Gesson, and A. Renoux, *Tetrahedron* **40**, 1983 (1984).
846. C. Berrier, J. P. Gesson, J. C. Jacquesy, and A. Renoux, *Tetrahedron* **40**, 4973 (1984).
847. J. P. Gesson, J. -C. Jacquesy, and M. -P. Jouannetaud, *Bull. Soc. Chim. Fr. II* 304 (1980).
848. Z. H. Lin, C. J. Guan, X. L. Feng, and C. X. Zhao, *J. Mol. Catal. A: Chem.* **247**, 19 (2006).
849. Z. Lin, H. Ni, H. Du, and C. Zhao, *Catal. Commun.* **8**, 31 (2007).
850. J. P. Gesson, J. C. Jacquesy, and R. Jacquesy, *Tetrahedron Lett.* 4733 (1971).
851. J. P. Gesson, J. -C. Jacquesy, and R. Jacquesy, *Bull. Soc. Chim. Fr.* 1433 (1973).
852. J. M. Coustard, J. P. Gesson, and J. C. Jacquesy, *Tetrahedron Lett.* 4929 (1972).
853. J. -M. Coustard and J. -C. Jacquesy, *Tetrahedron Lett.* 1341 (1972); *Bull. Soc. Chim. Fr.* 2098 (1973).
854. J. P. Gesson and J. -C. Jacquesy, *Tetrahedron* **29**, 3631 (1973).
855. J. -C. Jacquesy, R. Jacquesy, and U. H. Ly, *Tetrahedron Lett.* 2199 (1974).
856. J. P. Gesson, J. -C. Jacquesy, R. Jacquesy, and G. Joly, *Bull. Soc. Chim. Fr.* 1179 (1975).
857. R. Jacquesy and H. L. Ung, *Tetrahedron* **33**, 2543 (1977).
858. M. Schneider, K. Zimmermann, F. Aquino, and W Bonrath, *Appl. Catal., A* **220**, 51 (2001) and references cited therein.
859. G. J. Meuzelaar, E. Neeleman, L. Maat, and R. A. Sheldon, *Eur. J. Org. Chem.* 2101 (1998).
860. P. S. Bahia and J. S. Snaith, *J. Org. Chem.* **69**, 3226 (2004).
861. T. -P. Loh, Q. -Y. Hu, and L. -T. Ma, *Org. Lett.* **4**, 2389 (2002).
862. K. K. Laali, T. Okazaki, R. H. Mitchell, K. Ayub, R. Zhang, and S. G. Robinson, *J. Org. Chem.* **73**, 457 (2008).
863. V. M. Karpov, T. V. Mezhenkova, V. R. Sinyakov, and V. E. Platonov, *J. Fluorine Chem.* **117**, 73 (2002) and references cited therein.
864. Ya. V. Zonov, V. M. Karpov, V. E. Platonov, T. V. Rybalova, and Y. V. Gatilov, *J. Fluorine Chem.* **127**, 1574 (2006).
865. Ya. V. Zonov, V. M. Karpov, and V. E. Platonov, *J. Fluorine Chem.* **128**, 1065 (2007).
866. D. N. Kursanov, Z. N. Parnes, M. I. Kalinkin, and N. M. Loim, *Ionic Hydrogenation and Related Reactions*. Harwood, Chur, Switzerland, (1985).
867. D. D. Whitehurst, T. O. Mitchell, and M. Fracasiu, *Coal Liquefaction*, Academic Press, New York, (1980).
868. J. Wristers, *J. Am. Chem. Soc.* **97**, 4312 (1975).
869. J. W. Larsen and L. W. Chang, *J. Org. Chem.* **44**, 1168 (1979).
870. J. -C. Jacquesy, R. Jacquesy, and G. Joly, *Tetrahedron Lett.* 4433 (1974).
871. J. -C. Jacquesy, R. Jacquesy, and G. Joly, *Bull. Soc. Chim. Fr.* **2281**, 2289 (1975).
872. J. -M. Coustard, M. -H. Douteau, J. -C. Jacquesy, and R. Jacquesy, *Tetrahedron Lett.* 2029, (1975).
873. J. C. Jacquesy, R. Jacquesy, and G. Joly, *Tetrahedron* **31**, 2237 (1975).

874. K. Yu. Koltunov E. N. Subbotina, and I. B. Repinskaya, *Russ. J. Org. Chem. Engl. Transl.* **33**, 689 (1997).

875. K. Yu. Koltunov, M. M. Shakirov, and I. B. Repinskaya, *Russ. J. Org. Chem. Engl. Transl.* **34**, 595 (1998).

876. K. Yu. Koltunov, L. A. Ostashevskaya, and I. B. Repinskaya, *Russ. J. Org. Chem. Engl. Transl.* **34**, 1796 (1998).

877. L. A. Ostashevskaya, K. Yu. Koltunov, and I. B. Repinskaya, *Russ. J. Org. Chem. Engl. Transl.* **36**, 1474 (2000).

878. K. Yu. Koltunov, G. K. S. Prakash, G. Rasul, and G. A. Olah, *Heterocycles* **62**, 757 (2004).

879. R. Jacquesy and C. Narbonne, *J. Chem. Soc., Chem. Commun.* **765**, (1979).

880. J. M. Coustard, M. H. Douteau, R. Jacquesy, P. Longevialle, and D. Zimmermann, *J. Chem. Res. (S)*, 16 (1978).

881. J. M. Coustard, M. H. Douteau, and R. Jacquesy, *J. Chem. Res. (S)*, 18 (1978).

882. R. Jacquesy, C. Narbonne, and H. L. Ung, *J. Chem. Res. (S)*, 288 (1979).

883. K. Yu. Koltunov, I. B. Repinskaya, and G. I. Borodkin, *Russ. J. Org. Chem. Engl. Transl.* **37**, 1534 (2001).

884. C. Lafitte, M. -P. Jouannetaud, J. -C. Jacquesy, J. Fahy, and A. Duflos, *Tetrahedron Lett.* **39**, 8281 (1998).

885. G. A. Olah, A. Wu, and O. Farooq, *J. Org. Chem.* **53**, 5143 (1988).

886. G. A. Olah, A. Wu, O. Farooq, and G. K. S. Prakash, *J. Org. Chem.* **54**, 1450 (1989).

887. G. A. Olah, A. Wu, and O. Farooq, *J. Org. Chem.* **56**, 2759 (1991).

888. G. A. Olah and A. Wu, *Synthesis* 407 (1991).

889. G. A. Olah and A. Wu, *Synlett* 599 (1990).

890. G. A. Olah, M. Arvanaghi, and L. Ohannessian, *Synthesis* 770 (1986).

891. G. A. Olah, T. Keumi, and D. Meidar, *Synthesis* 929 (1978).

892. A. Heidekum, M. A. Harmer, and W. F. Hoelderich, *J. Catal.* **181**, 217 (1999).

893. D. E. López, J. G. Goodwin, Jr., D. A. Bruce, and E. Lotero, *Appl. Catal. A* **295**, 97(2005).

894. D. E. López, J. G. Goodwin, Jr., and D. A. Bruce, *J. Catal.* **245**, 381 (2007).

895. K. Suwannakarn, E. Lotero, and J. G. Goodwin, Jr., *Catal. Lett.* **114**, 122 (2007).

896. K. Suwannakarn, E. Lotero, and J. G. Goodwin, Jr., *Ind. Eng. Chem. Res.* **46**, 7050 (2007).

897. J. Ni and F. C. Meunier, *Appl. Catal., A* **333**, 122 (2007).

898. G. Carr and Whittaker, *J. Chem. Soc., Perkin Trans. 2*, 1877 (1987).

899. L. Coulombel and E. Dunach, *Synth. Commun.* **35**, 153 (2005).

900. N. Kihara, J. -I. Shin, Y. Ohga, and T. Takata, *Chem. Lett.* 592 (2001).

901. G. A. Olah, N. Hartz, G. Rasul, A. Burrichter, and G. K. S. Prakash, *J. Am. Chem. Soc.* **117**, 6421 (1995).

902. G. A. Olah, D. Meidar, and A. P. Fung, *Synthesis* 270 (1979).

903. R. Kumareswaran, P. S. Vankar, M. V. R. Reddy, S. V. Pitre, R. Roy, and Y. D. Vankar, *Tetrahedron* **55**, 1099 (1999).

904. D. Leca, F. Gaggini, J. Cassayre, O. Loiseleur, S. N. Pieniazek, J. A. R. Luft, and K. Houk, *J. Org. Chem.* **72**, 4284 (2007).

905. D. B. Gorman and P. G. Gassman, *J. Org. Chem.* **60**, 977 (1995).
906. P. G. Gassman and D. B. Gorman, *J. Am. Chem. Soc.* **112**, 8624 (1990).
907. Y. -J. Ko, S. -B. Shim, and J. -H. Shin, *Tetrahedron Lett.* **48**, 863 (2007).
908. H. v. Seggern and M. Schmittel, *Chem. Ber.* **127**, 1269 (1994).
909. H. Kotsuki, K. Arimura, T. Ohishi, and R. Maruzasa, *J. Org. Chem.* **64**, 3770 (1999).
910. Z. Li, J. Zhang, C. Brouwer, C. -G. Yang, N. W. Reich, and C. He, *Org. Lett.* **8**, 4175 (2006).
911. D. C. Rosenfeld, S. Shekhar, A. Takemiya, M. Utsunomiya, and J. F. Hartwig, *Org. Lett.* **8**, 4179 (2006).
912. F. M. Cordero, M. Cacciarini, F. Machetti, and F. De Sarlo, *Eur. J. Org. Chem.* 1407 (2002).
913. A. P. Davis and M. Jaspars, *J. Chem. Soc., Perkin Trans. 1*, 2111 (1992).
914. T. G. Elford, Y. Arimura, S. H. Yu, and D. G. Hall, *J. Org. Chem.* **72**, 1276 (2007).
915. L. Carosi, H. Lachance, and D. G. Hall, *Tetrahedron Lett.* **46**, 8981 (2005).
916. S. H. Yu, M. J. Ferguson, R. McDonald, and D. G. Hall, *J. Am. Chem. Soc.* **127**, 12808 (2005).
917. L. F. Tietze, K. Schiemann, C. Wegner, and C. Wulff, *Chem. Eur. J.* **4**, 1862 (1998).
918. A. V. Vasilyev, S. Walspurger, S. Chassaing, P. Pale, and J. Sommer, *Eur. J. Org. Chem.* 5740 (2007).
919. G. A. Olah, Q. Wang, X. -Y. Li, and G. K. S. Prakash, *Synlett* 487 (1990).
920. Y. Ichikawa, *J. Chem. Soc., Perkin Trans. 1*, 2135 (1992).
921. M. Shi, and G. -Q. Tian, *Tetrahedron Lett.* **47**, 8059 (2006).
922. B. Das, K. Laxminarayana, P. Thirupathi, and B. Ramarao, *Synlett* 3103 (2007).
923. H. Matsuda and T. Okuhara, *Catal. Lett.* **56**, 241 (1998).
924. K. Okuyama, X. Chen, K. Takata, D. Odawara, T. Suzuki, S. Nakata, and T. Okuhara, *Appl. Catal., A* **190**, 253 (2000).
925. J. P. Kennedy, in *Polymer Chemistry of Synthetic Elastomers*, J. P. Kennedy and E. G. M. Törnquist, ed., Wiley-Interscience, New York, Part 1, Chapter 5, (1968), p. 291.
926. J. P. Kennedy and E. Maréchal, *Carbocationic Polymerization*, Wiley-Interscience, New York, 1982.
927. G. Sauvet and P. Sigwalt, *Comprehensive Polymer Science*, G. Allen and J. C. Bevington, in eds., Pergamon Press, Oxford, Chapter 39, (1989), p. 579.
928. G. A. Olah, H. W. Quinn, and S. J. Kuhn, *J. Am. Chem. Soc.* **82**, 426 (1960).
929. G. A. Olah, *Macromol. Chem.* **175**, 1039 (1974).
930. W. R. Cares,U.S. Patent, 4,065, 512 (1977).
931. Sanyo Chem. Industries, Ltd, Japanese Patent, 8 226 626 (1982); *Chem. Abstr.* **97**, 72986b (1982).
932. E. T. Marquis, L. W. Watts, Jr., W. H. Brader, Jr., and J. W. Darden, Belgian Patent 891, 533; *Chem. Abstr.* **91**, 185195m (1982).
933. H. Hasegawa and T. Higashimura, *Polym. J.* **11**, 737 (1979).
934. T. Higashimura and H. Nishii, *J. Polym. Sci., Polym. Chem.* **15**, 329 (1977).
935. P. A. Gunatillake, G. F. Meijs, R. C. Chatelier, D. M. McIntosh, and E. Rizzardo, *Polym. Intern.* **27**, 275 (1992).

936. G. E. Heinsohn, G. Pruckmayr, I. M. Robinson, I. Maxwell, and W. W. Gilbert, German Patent, 2,709,280 (1977); *Chem. Abstr.* **87**, P202353 (1971).

937. G. E. Heinsohn, I. M. Robinson, G. Pruckmayr, and W. W. Gilbert, U.S. Patent 4,163,115 (1979).

938. G. Pruckmayr and R. H. Weir, U.S. Patent 4,120,903 (1978).

939. E. I. DuPont de Nemours, Belgian Patent 868,381 (1978); *Chem. Abstr.* **90**, P204877j (1979).

940. G. Pruckmayr, U.S. Patent 4,139,567 and 4,153,786 (1979); U.K. Patent 2,025,955 (1980); *Chem. Abstr.* **93**, 47469v (1980).

941. E. I. DuPont de Nemours, Fr. Demande 2,429,234 (1980); *Chem. Abstr.* **92**, P199059 (1980).

942. *Comprehensive Polymer Science*, G. Allen and J. C. Bevington, eds., Pergamon Press, Oxford, 1989, Chapters 39–53.

943. V. A. Babkin, G. E. Zaikov, and K. S. Minsker, *Quantum Chemical Aspects of Cationic Polymerization of Olefins*, Nova Science Publishers, New York, 1997.

944. *Cationic Polymerization, Fundamentals and Applications, ACS Symposium Series* Vol. 665, R. Faust and T.D. Shaffer, eds., American Chemical Society, Washington, DC, 1997.

945. *Living and Controlled Polymerization: Synthesis Characterization and Properties of the Respective Polymers and Copolymers*, J. Jagur-Grodzinski Nova Science Publishers, New York, (2006).

946. F. Cataldo, *Angew. Macromol. Chem.* **236**, 1 (1996).

947. E. R. Peña, M. Zolotukhin, and S. Fomine, *Macromolecules* **37**, 6227 (2004). *Polymer* **46**, 7494 (2005).

948. H. M. Colquhoun, M. G. Zolotukhin, L. M. Khalilov and U. M. Dzhemilev, *Macromolecules* **34**, 1122 (2001).

949. M. Zolotukhin, S. Fomine, R. Salcedo, and L. Khalilov, *Chem. Commun.* 1030 (2004).

950. A. M. Diaz, M. G. Zolotukhin, S. Fomine, R. Salcedo, O. Manero, G. Cedillo, V. M. Velasco, M. T. Guzman, D. Fritsch, and A. F. Khalizov, *Macromol. Rapid. Commun.* **28**, 183 (2007).

951. M. G. Zolotukhin, S. Fomine, L. M. Lazo, R. Salcedo, L. E. Sansores, G. G. Cedillo, H. M. Colquhoun, J. M. Fernandez-G., and A. F. Khalizov, *Macromolecules* **38**, 6005 (2005).

952. A. L. Lira, M. G. Zolotukhin, L. Fomina, and S. Fomine, *Macromol. Theory Simul.* **16**, 227 (2007).

953. S. Kobayashi, N. Tsuchida, K. Morikawa, and T. Saegusa, *Macromolecules* **8**, 942 (1975).

954. Y. -L. Liu, G. -H. Hsiue, and Y. -S. Chiu, *J. Polym. Sci. A, Polym. Chem.* **32**, 2543 (1994).

955. C. -Y. Pan, Y. Liu, and W. Liu, *J. Polym. Sci. A, Polym. Chem.* **36**, 2899 (1998).

956. Y. Xu and C. Pan, *J. Polym. Sci. A, Polym. Chem.* **38**, 1232 (2000).

957. P. Nicol, M. Masure, and P. Sigwalt, *Macromol. Chem. Phys.* **195**, 2327 (1994).

958. A. -F. Mingotaud, F. Dargelas, and F. Cansell, *Macromol. Symp.* **153**, 77 (2000).

959. J. K. Paulasaari and W. P. Weber, *Macromolecules* **32**, 5217 (1999).

960. C. J. Teng, W. P. Weber, and G. Cai, *Macromolecules* **36**, 5126 (2003); *Polymer* **44**, 4149 (2003).

961. T. M. Gädda, A. K. Nelson, and W. P. Weber, *J. Polym. Sci. A, Polym. Chem.* **42**, 5235 (2004).

962. D. Bourissou, B. Martin-Vaca, A. Dumitrescu, M. Graullier, and F. Lacombe, *Macromolecules* **38**, 9993 (2005).

963. F. A. Jaipuri, B. D. Bower, and N. L. Pohl, *Tetrahedron: Asymmetry* **14**, 3249 (2003).

964. H. Kameshima, N. Nemoto, F. Sanda, and T. Endo, *Macromolecules* **35**, 5769 (2002).

965. M. J. Poss, M. A. Nordhaus, and K. W. Stagliano, *Synthesis* 1193 (1999) and references cited therein.

966. S. Kanoh, T. Nishimura, T. Tsuchida, H. Senda, M. Motoi, M. Takani, and N. Matsuura, *Macromol. Chem. Phys.* **202**, 2489 (2001).

967. S. Kanoh, M. Naka, T. Yokozuka, S. Itoh, T. Nishimura, M. Honda, M. Motoi, and N. Matsuura, *Macromol. Chem. Phys.* **203**, 511 (2002).

968. S. Kanoh, T. Nishimura, H. Ogawa, M. Motoi, and M. Takani, *Macromolecules* **32**, 1301 (1999).

969. T. Imai, T. Satoh, H. Kaga, N. Kaneko, and T. Kakuchi, *Macromolecules* **37**, 3113 (2004).

970. J. V. Crivello and J. H. W. Lam, *Macromolecules* **10**, 1307 (1977).

971. J. V. Crivello and K. Dietliker, in *Photoinitiators for Free Radical Cationic and Anionic Photopolymerisation*, 2nd ed., G. Bradley, ed., Wiley, Chichester, Chapter 3, (1998).

972. *Photoinitiated Polymerization ACS Symposium Series*, Vol. 847, K. D. Belfield and J. V. Crivello, eds., American Chemical Society, Washington, DC, 2003, Chapters 15–28.

973. F. H. Walker, J. B. Dickenson, C. R. Hegedus, F. R. Pepe, and R. Keller, *J. Coat. Technol.***74**, (928), 33 (2002).

974. L. M. Goldenberg, S. Aeiyach, and P. C. Lacaze, *J. Electroanal. Chem.* **327**, 173 (1992), **335,** 151 (1992); *Synth. Metals* **51**, 343, 349 (1992); **55–57**, 1360 (1993).

975. P. C. Lacaze, S. Hara, P. Soubiran, and S. Aeiyach, *Synth. Metals* **75**, 111 (1995).

976. J. C. Hicks, B. A. Mullis, and C. W. Jones, *J. Am. Chem. Soc.* **129**, 8426 (2007).

977. T. Iwasaki, A. Nagaki, and J. Yoshida, *Chem. Commun.* 1263 (2007).

978. S. C. Sherman, A. V. Iretskii, M. G. White, C. Gumienny, L. M. Tolbert, and D. A. Schiraldi, *J. Org. Chem.* **67**, 2034 (2002).

979. R. Sanz, D. Miguel, A. Martínez, J. M. Álvarez-Gutiérrez, and F. Rodríguez, *Org. Lett.* **9**, 2027 (2007).

980. B. C. Gagea, A. N. Pârvulescu, G. Poncelet, and V. I. Pârvulescu, *Catal. Lett.* **105**, 219 (2005).

981. K. Schildknegt and K. A. Agrios, and J. Aubé, *Tetrahedron Lett.* **39**, 7687 (1998).

982. P. Desai, K. Schildknegt, K. A. Agrios, C. Mossman, G. L. Milligan, and J. Aubé, *J. Am. Chem. Soc.* **122**, 7226 (2000).

983. W. H. Pearson and W. -K. Fang, *J. Org. Chem.* **60**, 4960 (1995).

984. S. Arseniyadis, A. Wagner, and C. Mioskowski, *Tetrahedron Lett.* **43**, 9717 (2002).

985. B. Rigo, P. Gautret, A. Legrand, J. -P. Hénichart, and D. Couturier, *Synlett* 998 (1997).

986. K. Ishimaru and T. Kojima, *Tetrahedron Lett.* **44**, 5441 (2003).

987. G. A. Olah and A. Wu, *J. Org. Chem.* **56**, 2531 (1991).

988. A. P. Rudenko, M. Ja. Sarubin, and F. Pragst, *Z. Chem.* **22**, 386 (1982); *J Prakt. Chem.* **325**, 612 (1983).

989. A. P. Rudenko, M. Ja. Sarubin, S. F. Awerjanow, and F. Pragst, *J. Prakt. Chem.* **327**, 191 (1985) and references cited therein.
990. A. P. Rudenko, N. S. Korovina, and S. F. Aver'yanov, *Russ. J. Org. Chem. Engl. Transl.* **31**, 1077 (1995).
991. A. P. Rudenko and N. S. Korovina, *Russ. J. Org. Chem. Engl. Transl.* **31**, 1084 (1995).
992. S. F. Aver'yanov and A. P. Rudenko, *Russ. J. Org. Chem. Engl. Transl.* **31**, 1089 (1995).
993. A. P. Rudenko, Yu. N. Salfetnikova, and A. V. Vasil'ev, *Russ. J. Org. Chem. Engl. Transl.* **32**, 1447 (1996).
994. S. I. Kol'tsov and Yu. N. Salfetnikova, *Russ. J. Gen. Chem. Engl. Transl.* **73**, 389 (2003).
995. A. P. Rudenko and F. Pragst, *Russ. J. Org. Chem. Engl. Transl.* **34**, 1589 (1998).
996. Q. Sun, W. E. Farneth, and M. A. Harmer, *J. Catal.* **164**, 62 (1996).
997. A. Heidekum, M. Harmer, and W. F. Hoelderich, *Catal. Lett.* **47**, 243 (1997).
998. D. Fritsch, I. Randjelovic, and F. Keil, *Catal. Today* **98**, 295 (2004).
999. G. A. Olah and W. M. Ip, *New J. Chem.* **12**, 299 (1988).
1000. T. Yamato, C. Hideshima, M. Tashiro, G. K. S. Prakash, and G. A. Olah, *Catal. Lett.* **6**, 341 (1990).
1001. G. A. Olah, V. P. Reddy, and G. K. S. Prakash, *Synthesis* 29 (1991).
1002. T. Yakaiah, B. P. V. Lingaiah, G. V. Reddy, B. Narsaiah, and P. S. Rao, *ARKIVOC*, Part 13, 227 (2007).

OUTLOOK

In the first edition of this book we wrote the following:

"Acids have played a fundamental role in the entire history of chemistry. Mineral acids such as sulfuric and perchloric acid were until recently considered as the strongest acids. In the last two decades, emergence of acid systems up to 10^{15} times stronger than 100% sulfuric acid has opened up a whole new area of chemistry. These superacids, reviewed in the present monograph in a systematic way, range from liquid systems such as antimony pentafluoride–hydrogen fluoride, antimony pentafluoride–trifluoromethanesulfonic acid, and related acids containing arsenic, tantalum, and niobium pentafluorides and the like, to solid superacids such as polymeric perfluorinated resinsulfonic acids (Nafion-H), immobilized or intercalated antimony pentafluoride and related Lewis acid-based systems. The exceedingly high acidity of these systems covering a broad range allows many weak bases to undergo acid-catalyzed reactions, which otherwise would not take place using conventional acid catalysts. Many hydrocarbon conversions, including those of methane and alkanes, are some of the examples of applications catalyzed by superacids that in the future are expected to gain increasing importance. Usually, acid-catalyzed reactions such as isomerization of alkanes, carried out with mineral acids or conventional Friedel–Crafts systems, necessitate relatively high temperatures. However, superacids promote such reactions at much lower temperatures, thus allowing favorable isomer distribution of branched hydrocarbons (increasing the octane number significantly).

A whole array of potential weakly basic reagents, such as CO_2, CO, O_3, O_2 (singlet), N_2O, Cl_2, Br_2, and the like, are starting to show electrophilic reactivity under superacid-catalyzed conditions. These reactions could lead to novel and practical applications, and some of them are already emerging. Ionic reactions generally show much higher selectivity than their free radical counterparts. The ability of superacid catalysts to cleave carbon–carbon bonds with ease has resulted in novel hydrocarbon cracking applications, particularly in systems where metal-based catalysts are susceptible to deactivation (poisoning) by impurities such as sulfur and nitrogen compounds. This is of particular interest in processing heavy oils, tar sands, and oil shale, as well as in coal conversion–liquefaction processes.

Superacid Chemistry, Second Edition, George A. Olah, G. K. Surya Prakash, Árpád Molnár, and Jean Sommer

Superacid-catalyzed reactions and ionic reagents (obtained under superacidic, stable ion conditions) are gaining increasing significance in synthetic and natural product chemistry, and this trend is expected to continue and expand. Some special applications, such as use of intercalated antimony or arsenic pentafluoride/graphite (or polyacetylene) as highly conducting (even superconducting) materials, has attracted and will continue to attract considerable attention.

Future development of new and improved superacidic systems, particularly allowing long onstream time in catalytic applications without deactivation and ease of regeneration, is of particular interest. Applications of superacids are foreseen to expand in catalysis and in synthetic chemistry, as well as in preparation and study of reactive ionic intermediates."

It is rewarding to see that many of our expectations have materialized during the last two decades. The use of superacids has become common practice in synthetic organic chemistry as demonstrated by the numerous new areas of applications, including particularly their use in the chemistry of hydrocarbons, protecting groups, heterocycles, and carbohydrates. Furthermore, superacids have also found wide application in the generation of varied inorganic cations and complexes. Additional important developments include that of new weakly coordinating anions and methods to manufacture and use Nafion–silica and other nanocomposites, which show high stability and specific activities in catalytic applications.

Extending the concept of superacids to varied superelectrophiles has emerged as a productive new field in recent years (G. A. Olah and D. A. Klumpp, *Superelectrophiles and Their Chemistry*, Wiley-Interscience, Hoboken, NJ, 2008). Highly reactive and activated protosolvated or multiply charged superelectrophilic intermediates are involved in varied chemical reactions, many of them of substantial practical significance.

INDEX

Superacid Chemistry, Second Edition, George A. Olah, G. K. Surya Prakash, Árpád Molnár, and Jean Sommer

Printed and bound by CPI Group (UK) Ltd, Croydon, CR0 4YY
18/09/2026
14965015-0004